AF509218

Proceedings of the XXI
International Symposium

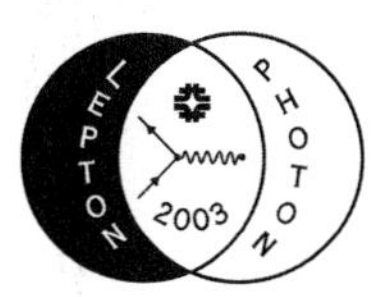

LEPTON AND PHOTON INTERACTIONS AT HIGH ENERGIES

Proceedings of the XXI
International Symposium

LEPTON AND PHOTON INTERACTIONS AT HIGH ENERGIES

Fermi National Accelerator Laboratory, USA 11 – 16 August 2003

EDITORS

Harry W K Cheung
Fermi National Accelerator Laboratory, USA

Tracey S Pratt
University of Liverpool, UK

NEW JERSEY · LONDON · SINGAPORE · SHANGHAI · HONG KONG · TAIPEI · BANGALORE

Published by

World Scientific Publishing Co. Pte. Ltd.

5 Toh Tuck Link, Singapore 596224

USA office: Suite 202, 1060 Main Street, River Edge, NJ 07661

UK office: 57 Shelton Street, Covent Garden, London WC2H 9HE

British Library Cataloguing-in-Publication Data
A catalogue record for this book is available from the British Library.

LEPTON AND PHOTON INTERACTIONS AT HIGH ENERGIES (Lepton–Photon 2003)
Proceedings of the XXI International Symposium

ISBN 981-238-716-1

Printed in Singapore.

Preface

The XXI International Symposium on Lepton and Photon Interactions at High Energies marked the return of the Symposium to Fermilab at the start of the Run II collider era of the Tevatron. As of 11-16th August 2003, the Tevatron has delivered an integrated luminosity surpassing that of the previous collider run. Whereas the IX Symposium, held at Fermilab in 1979, marked the start of construction of what was then called the "Energy Saver/Doubler", this one saw the newest accelerator at Fermilab – the Main Injector – in full operation for both protons and antiprotons. With the Main Injector the instantaneous luminosity of the Tevatron collider is a factor ~ 50 higher than the original design luminosity presented in 1979 for the "Energy Saver/Doubler". Soon to join the presently running MiniBooNE experiment will be a new neutrino program at Fermilab powered by the intense proton beams provided by the Main Injector. This summer saw the completion of the MINOS far detector and the NUMI neutrino beamline will soon be completed.

At the IX Symposium the first evidence for gluons was shown and QCD emerged as the theory that could describe what was known about hadrons. It was hoped that QCD would soon be near to gaining the stature of QED. At this Symposium a major breakthrough was presented. Lattice QCD has reached a new level of precision, where errors on calculated quantities can at last be reliably quantified at the several % level. Soon, instead of testing Lattice QCD we will be using it to help us determine more precisely the parameters of the Standard Model and to look for New Physics. Other highlights for this Symposium included the vast amount of new data from the B-factories with the tantalizing possibility of non-Standard Model CP-violation in ϕK_S^0, and the discovery of a new, as yet unexplained, particle, the $X(3872)$. The remarkable progress in neutrino physics was again evident with the last pieces of the solar neutrino problem falling into place and settling on the LMA-1 solution. However we were reminded how much we still need to understand by the talks on Dark Matter and Dark Energy. The talks also showed the synergy between the different areas of particle physics and between particle physics, astroparticle physics and cosmology, especially in the search for New Physics at the TeV energy scale.

While still following the traditional program of review talks, for this edition of the Symposium a real effort was made to increase active participation throughout the Symposium. This was partly achieved by holding breakout discussion sessions between the formal plenary talks. There, the speakers from several sessions were available for an extended "Question and Answer" session. These new sessions were enormously successful, and the only regret maybe is that we do not have a written record of these discussions. However this probably allowed for a more informal and relaxed atmosphere. Additional participation was also generated by another new introduction to the Symposium, a "Physics Posters" session. There were 66 physics posters presented with obvious enthusiasm by young physicists. These poster presentations were split over two days and included results from 20 different experiments as well as posters on theory, analytical techniques, and detector development. A "Laboratory Posters" day was again held with the participation of 15 laboratories, based in eight different countries. In addition, one evening was devoted to a special session on the Computational Data GRID, comprising an interesting program with both talks and posters.

Another aim for this Symposium was to increase the participation of younger scientists, who are the future of the field. As well as having some excellent young speakers and holding the new "Physics Posters" session, younger scientists were recruited to help throughout the Symposium. Included in this were the local organizing committee, the Scientific Secretaries, and one of the editors can still be considered young!

A major effort at this Symposium was public outreach. Members of the press held special discussion sessions with scientists on various physics topics; there was also a special press conference with the Laboratory Directors. The outreach effort included interaction with the public during several hours in the Field Museum in Chicago, which was covered by the press. There were also excellent "Plain English" articles written on seven different particle physics and cosmology topics made available via the Symposium website. At a public lecture on the last evening of the Symposium, Fermilab director, Mike Witherell, gave an inspiring talk on "New Questions about Matter, Space and Time" to a packed auditorium of the local public. Capping off these outreach efforts was the launch of Interactions.org, a multinational website designed as a central resource for communicators of particle physics.

Following the excellent efforts of LP2001 on keeping a good record of the Symposium, we have tried to provide as good a record as possible via the website and through a DVDROM included with the hardcopy of the proceedings. These will provide a lasting record of the live webcast streaming video for each talk, the talk slides, the writeups of each talk, the public lecture, and the physics and laboratory posters.

The synergy between different areas of particle physics and cosmology that was so evident at LP2003 can be viewed as a broad attack to dispel the veil at the TeV energy scale. This effort will continue to pull at this veil and foreshadows the bursting forth of New Physics that will overwhelm, delight, excite, and inspire us. We hope that LP2005 will bring us a glimpse of this future era.

Harry W. K. Cheung
and
Tracey S. Pratt
Proceedings Editors

Acknowledgments

It is a pleasure to acknowledge the many people who made Lepton-Photon 2003 such a success. Keith Ellis and Joel Butler deserve special mention. They worked with enthusiasm to get Lepton-Photon 2003 hosted at Fermilab and then continued as key members of the Local Organizing Committee. Keith, with the Scientific Program Committee which he chaired, arranged a varied and interesting program. Joel, with his Special Events Committee, organized a full program of extra events, including a special session on the Computational Data Grid.

We thank the excellent Local Organizing Committee who worked very hard on all aspects of the Symposium, including the poster sessions, website, public outreach and contributed paper submission. Thanks to the International Advisory Committee for contributing useful suggestions on the program. The Scientific Secretaries are gratefully acknowledged for their many activities, including classifying all the contributed papers, providing technical help to speakers, and for their support during the sessions.

We thank the many Symposium speakers for their hard work and effort in producing excellent and engrossing reviews, and especially those speakers (almost everyone) who wrote up their talks for these Proceedings. We also thank the session chairs, many of whom also served as moderators in the experimental "breakout sessions". Special thanks go to Harry Cheung and Tracey Pratt for editing these Proceedings and getting them out promptly.

We thank Fred Ullrich and his staff and helpers from the Fermilab Visual Media Services for taking care of the streaming video, both the live webcast and archived version available from the conference website, the talk and poster presentation details, and for capturing the Symposium with a set of splendid photographs. Our thanks go to Judy Jackson, head of the Fermilab Office of Public Affairs, for organizing an extensive program for the press and laboratory public affairs personnel. We are pleased to acknowledge the financial support of Cisco Systems, Hamamatsu and Sun Microsystems. We thank the Illinois State Board of Higher Education /Illinois Consortium for Accelerator Research for their generous support of the poster session.

We received excellent support from Fermilab Director Michael Witherell and from staff in every Fermilab Division and Section. We particularly acknowledge the outstanding work of the Fermilab Conference Office. Cynthia Sazama, head of the Conference Office and a veteran conference organizer, coordinated a vast number of details, with assistance from Suzanne Weber, Patti Poole, and Harry Ferguson. Elaine Phillips (Particle Physics Division Administrative Support) was also particularly helpful.

Finally, for support that made this Symposium possible, our thanks go to the International Union of Pure and Applied Physics, the U. S. Department of Energy, the National Science Foundation, the Universities Research Association, OIST-Okinawa, NASA and of course Fermilab. The Fermi National Accelerator Laboratory is operated by the Universities Research Association, Inc., under contract DE-AC02-76CHO3000 with the U. S. Department of Energy.

C. Newman-Holmes
Chair, Lepton-Photon 2003 Organizing Committee

Editors' Acknowledgments

We would like to thank Cathy Newman-Holmes, Chair of LP2003, for giving us the opportunity and full responsibility for producing these proceedings. We made a special effort to try to publish these proceedings in the shortest possible time and we would like to thank all authors for delivering their contributions in a timely manner, while maintaining high standards in content and format. Regrettably a few contributions were not received. We also thank all authors for letting us try to improve the style, format and the consistency of the whole volume – especially the U. K. authors for letting us change from British to U. S. spelling in their writeups!

We wish to thank the following people for helping out with the production of these proceedings. Jim Shultz and Fred Ullrich for help producing the DVDROM; Cynthia Sazama and Suzanne Weber for help with logistics; the Scientific Secretaries for help with the Discussion sections for each paper; Pushpa Bhat for help getting laboratory poster files; Fred Ullrich for help selecting photos for the cover and endpapers; Keith Ellis, Andreas Kronfeld and Chris Quigg for help trying to get writeup contributions; and the production staff at World Scientific – Karen Yeo, Jimmy Low and Rajesh Babu for producing the cover and for help with various production issues. TSP gratefully acknowledges support from the U. K. Particle Physics and Astronomy Research Council in the form of a PPARC Postdoctoral Fellowship.

Harry W. K. Cheung
and
Tracey S. Pratt

List of Committees

International Advisory Committee

A. Bettini	*Gran Sasso*	W. J. Marciano	*BNL*
H. S. Chen	*IHEP, Beijing*	H. Murayama	*Univ. of California, Berkeley*
M. V. Danilov	*ITEP, Moscow*	K. Peach	*RAL*
M. Davier	*LAL-Orsay*	H. R. Quinn	*SLAC*
J. M. Dorfan	*SLAC*	A. F. S. Santoro	*State Univ. of Rio de Janerio*
E. Fernández	*Barcelona*	A. N. Skrinsky	*Budker INP*
R. M. Godbole	*IISc, Bangalore*	H. Sugawara	*KEK*
W. Haxton	*Univ. of Washington*	M. Tigner	*Cornell Univ.*
E. Iarocci	*INFN*	Y. Totsuka	*Tokyo*
R. K. Keeler	*Victoria*	A. Wagner	*DESY*
L. Maiani	*CERN*	M. S. Witherell	*FNAL*

Local Organizing Committee

Cathy Newman-Holmes	*FNAL, Chair*	Judy Jackson	*FNAL*
Mike Albrow	*FNAL*	Boris Kayser	*FNAL*
Jeffrey A. Appel	*FNAL*	Steve Kent	*Univ. of Chicago, FNAL*
Pushpa Bhat	*FNAL*	Ben Kilminster	*Univ. of Rochester*
Jerry Blazey	*Northern Illinois Univ.*	Patti Poole	*FNAL*
Amber Boehnlein	*FNAL*	Erik Ramberg	*FNAL*
Chuck Brown	*FNAL*	Jose Repond	*ANL*
Liz Buckley-Geer	*FNAL*	Cynthia Sazama	*FNAL*
Joel N. Butler	*FNAL*	Heidi Schellman	*Northwestern Univ.*
Harry W. K. Cheung	*FNAL*	Suzanne Weber	*FNAL*
Ray Culbertson	*FNAL*	Herman White	*FNAL*
Keith Ellis	*FNAL*	John Womersley	*FNAL*
Bill Foster	*FNAL*	G. P. Yeh	*FNAL*

Symposium Session Chairs

Halina Abramowicz	*Tel Aviv*	David Miller	*UC, London*
Marina Artuso	*Syracuse*	Cathy Newman-Holmes	*FNAL*
Ikaros Bigi	*Notre Dame*	Steve Olsen	*Hawaii*
Mikhail Danilov	*ITEP*	Roberto Peccei	*UCLA*
Dong-Sheng Du	*Beijing, IHEP*	Hirotaka Sugawara	*KEK*
Enrique Fernandez	*Barcelona*	Robert Tschirhart	*Fermilab*
Rohini Godbole	*Bangalore*	Albrecht Wagner	*DESY*
Richard Keeler	*Victoria*	Stanley Wojcicki	*Stanford*
William Louis	*LANL*		

Scientific Secretaries

Oleksiy Atramentov	*Iowa State Univ.*	Minsuk Kim	*KCHEP Korea*
Sergei Avvakumov	*Stanford Univ.*	C-J Stephen Lin	*FNAL*
Ken Bloom	*Univ, of Michigan*	Mansoora Shamim	*Kansas State Univ.*
Marc Buehler	*UIC*	Petra Merkel	*FNAL*
Brendan Casey	*Brown Univ.*	Holger Meyer	*FNAL*
Katherine Copic	*Univ. of Michigan*	Martijn Mulders	*FNAL*
Jay Dittmann	*FNAL*	Jason Nielsen	*LBNL*
Bogdan Dobrescu	*FNAL*	Tracey Pratt	*Univ. of Liverpool*
Michael Eads	*Northern Illinois Univ.*	Xiangrong Xi	*FNAL*
Harald Fox	*Northwestern Univ.*	Michiel Sanders	*Univ. of Manchester*
Cristina Galea	*NIKHEF*	Makoto Tomoto	*FNAL*
Adam Gibson	*UC, Berkeley LBNL*	Brigitte Vachon	*FNAL*
Andre de Gouvea	*FNAL*	Michael Weber	*FNAL*
Ulrich Haisch	*FNAL*	Martin Wegner	*RWTH Aachen*
Kazu Hanagaki	*FNAL*	Jodi Witlin	*Boston Univ.*
Hyejoo Kang	*Stanford Univ.*	Lin Zhang	*Texas Tech. Univ.*

Breakout Session Moderators

Halina Abramowicz	*Tel Aviv*	William Louis	*LANL*
Fred Gilman	*CMU*	Kirill Melnikov	*Hawaii*
Rohini Godbole	*Bangalore*	Roberto Peccei	*UCLA*
Robert Kephart	*FNAL*		

Secretariat

Harvey Bruch	Jackie Coleman	Lory Curry	Elizabeth Duty
Jody Federwitz	Harry Ferguson III	Nicole Gee	Terry Grozis
Maxine Hronek	Luz Jaquez	Etta Johnson	Christine Johnson
Odarka Jurkiw	Cindy Kennedy	Barbara Kristen	Cathryn Laue
Elizabeth May	Phyllis Metelmann	Nancy Michael	Dawn Passarella
Marilyn Paul	Barbara Perington	Elaine Phillips	Carol Picciolo
Terry Read	Monica Sasse	Cynthia Sazama (Chair)	Marsha Schmidt
Sue Schultz	Nancy Sells	Leticia Shaddix	Joanne Stern
Mary Sulek	Jacquelyn Sullivan	Julie Toole	Priscilla Trevino
Olivia Vizcarra	Suzanne Weber	Sonya Wright	

Visual Media Services

Streaming Video and Audio Visual Team

Jim Shultz   Al Johnson   Video Impression of Aurora, Illinois   Cindy Arnold   Fred Ullrich

Poster Sessions Team

Reidar Hahn   Deborah Guzman   Cindy Arnold   Diana Canzone   Karen Seifrid   Fred Ullrich

Photo Credits

Reidar Hahn and Deborah Guzman of the Fermilab Visual Media Services took all the photos contained in these proceedings (except for the ones mentioned below). Photos from the Symposium Banquet were provided by Doris Kim and the photos for the first "Behind the Scenes" page were supplied by Leticia Shaddix. The photo for the cover was captured by Fred Ullrich, and Reidar Hahn took the photos displayed on the front and back endpapers. We thank each of them for the use of these excellent photographs. All the Visual Media Services 2003 Lepton-Photon Symposium photos can be obtained from the Fermilab Visual Media Services online photo database: search for captions with "Lepton Photon" at `http://vmsfmp2.fnal.gov/v1/VMSSearch_Online.htm`.

Descriptions of Cover and Endpaper Photos

Cover: Broken Symmetry

The tri-span sculpture, called "Broken Symmetry", rises 50 feet above the road at Fermilab's Pine Street entrance. It exemplifies the notions of symmetry and symmetry-breaking that are so important in the physics of elementary particles. Although asymmetric from most angles, when viewed from directly above or below the apex, the structure appears perfectly symmetric. This 21 ton "gateway" was constructed in 1978 from the steel armor plates of the U. S. Navy aircraft carrier, Princeton, which was decommissioned in May, 1956. The steel from this and other surplus decommissioned U. S. Navy battleships is used at Fermilab for radiation shielding.

Front Endpaper: Aerial View of the Fermilab Accelerator Complex

The large ring on the left in the aerial view of the Fermilab accelerator complex indicates the location of the Main Injector, Fermilab's newest accelerator. The larger four mile ring on the right shows the location of the Tevatron. Robert Rathbun Wilson Hall, the "Highrise", is clearly visible in the center. This is named after Fermilab's founding director.

Back Endpaper: View over Prairie, of Wilson Hall and Lederman Science Center

A look over an open tallgrass prairie which is part of Fermilab's Prairie Reconstruction Project. This is an effort to recreate some of the awesome beauty of the old prairies, as well as the biodiversity of native grassland ecosystems. Wilson Hall can be seen in the background on the right, and on the left is the Lederman Science Center, named after Nobel Laureate and former Fermilab Director, Leon M. Lederman. This visitor center is very popular with the public and surrounding schools for gaining hands-on experience to explore how Fermilab physicists understand nature's secrets.

Contents

ELECTROWEAK PHYSICS AND BEYOND

QCD PHENOMENOLOGY AND THEORY

FLAVOR PHYSICS: RARE AND BEYOND STANDARD MODEL DECAYS

FUTURE FACILITIES

FUTURE DIRECTIONS

WELCOME

WELCOMING REMARKS TO LEPTON PHOTON 2003
THE 21ˢᵗ INTERNATIONAL SYMPOSIUM ON LEPTON AND PHOTON
INTERACTIONS AT HIGH ENERGIES

S. PETER ROSEN

Office of Science, U. S. Department of Energy
E-mail: Peter.Rosen@science.doe.gov

Ladies and Gentlemen, on behalf of the United States Department of Energy, Secretary Spencer Abraham and Director of the Office of Science, Raymond L. Orbach, it is my pleasure to welcome you to Fermilab and to Lepton-Photon 2003. Minasama, Fermilab narabini Lepton-Photon 2003 he Youkoso irassyaimashita. Meine Damen und Herren, willkommen zum Nationalen Fermi-Beschleuniger-Forschungszentrum. Mesdames et Messieurs, bienvenue au Laboratoire National d'Accélératures de Fermi. Signore e Signori, benvenuti al Laboratorio Nazionale Fermilab. Дамы и Господа, добро пожаловать в Fermilab. Damas y Caballeros, bienvenidos al Laboratorio Nacional Fermilab. Please forgive me if I have omitted your own language, but you are equally welcome.

I like to think of Fermilab as the home of the third family of quarks and leptons: the b quark was discovered here, and the first observations of the top quark and tau neutrino were made here. It is also the home of the Tevatron, presently the world's highest energy hadron collider and it will remain so until the LHC comes on-line in 2007. We all look forward to the first results from Run II to be presented at this meeting, and to the exciting physics it will produce in the next five years. The discoveries made here and the lessons learned about extracting physics from hadronic collisions at the current energy frontier will lay the groundwork for exciting and successful hadronic collision programs at the future energy frontier at the LHC. They may also provide hints about the physics we may encounter at a future linear electron-positron collider.

While allowing for the precautions which the United States Government believes to be necessary for security reasons, the spirit of Fermilab, like the spirits of its sister laboratories at Brookhaven and Stanford, and at high energy physics laboratories in Asia and Europe, is that of an international research center. You can see outside the main entrance to Wilson Hall the flags of all the nations from which physicists come to perform their experiments. About one half of the 1100 or so users at Fermilab are from foreign countries, and the same is true of the users at RHIC and at SLAC. We welcome this participation by our colleagues from abroad, just as we appreciate the welcome given to U. S. scientists at facilities in other parts of the world.

This international, or rather global nature of our field has grown strong roots in the world high energy physics community and it serves as a model for other scientific fields, especially those that require large, global-scale facilities. The high energy community has been a pioneer in the development of major facilities, for example super-scale detectors like CMS and ATLAS. The next steps that I believe it needs to pioneer are planning on a global scale, and the formation of a laboratory for the next frontier facility, which has to be global from its inception, through construction and into its operating phase.

Turning to the physics at this meeting, we look forward to the results to be extracted from the wealth of data accumulated by the B-factories at KEK and SLAC, something like 100 inverse femtobarns apiece. Will they reveal new sources of CP-violation beyond the Standard Model, or give hints of other new physics through precision measurements of exclusive decay modes? Neutrino physics has been extremely exciting in the past five years: what new insights will we learn about the number of neutrinos, the masses and mixing angles? What hints will we hear from the Tevatron about Supersymmetry and extra dimensions? Will new light shine on Dark Matter and Dark Energy? Obviously there is an abundance of exciting questions which will make for an exciting conference, and so with no more ado, I bid you welcome and good *physicking*!

ELECTROWEAK PHYSICS AND BEYOND

TOP QUARK MEASUREMENTS AT THE FERMILAB TEVATRON

P. AZZI

I.N.F.N.-Sezione di Padova, Via Marzolo 8, 35100 Padova, Italy
E-mail: azzi@pd.infn.it

The top quark, discovered at the Tevatron in 1995, is a very interesting particle. Precise measurement of the top properties using large data samples will allow stringent tests of the Standard Model and offer a unique window on new physics. This report contains a review of the status of the current knowledge of the top quark as provided by the Run I results of the CDF and D0 experiments. A first look at various preliminary measurements obtained with data collected during Run II will also be presented.

1. Introduction

The top quark is one of the building blocks of the Standard Model of Electroweak interactions as the partner of the bottom quark in the SU(2) isospin doublet of the third family of quarks. After having eluded experimentalists for many years, the CDF experiment found evidence for its existence in 1994[1] followed soon by its discovery by the CDF[2] and D0[3] experiments in 1995.

While the data set collected at the Tevatron $p\bar{p}$ collider during Run I ($\mathcal{L}_{int} = 110\,\mathrm{pb}^{-1}$) has been sufficient for the discovery of the top quark, it is still too small to allow stringent tests of the Standard Model in top enriched samples. The ensemble of available top measurements up-to-date is thus consistent with expectations, however Run I left us a few interesting discrepancies that will need to be addressed in Run II. The large Run II dataset will allow one to achieve a greater precision and probe possible deviations from the Standard Model in a significant way.

2. Top Production and Decay

At the Tevatron the top quark is produced mainly in pairs through the process $q\bar{q}, gg \to t\bar{t}$ with a cross section of about $6.7\,\mathrm{pb}$[4] for $m_t = 175\,\mathrm{GeV}/c^2$ and $\sqrt{s} = 1.96\,\mathrm{TeV}$. Within the Standard Model framework the top decays almost 100% of the time via $t \to Wb$. Therefore it is customary to classify the $t\bar{t}$ final states based on the W decays modes: dileptons ($\ell = e, \mu$), lepton+jets ($\ell = e, \mu$), all-jets and inclusive τ (hadronically decaying) final-state events.

Even if its production cross section is much smaller compared to other Standard Model background processes, top events have very distinctive signatures that guide the overall analysis strategy: given the large top mass the decay products (leptons, jets) have large p_T's and the event topology is central and spherical, and there are always two b jets in the final state. Excellent lepton identification, energy resolution and b identification capabilities are essential for a successful Run II top physics program. The full kinematical and heavy flavor characterization of top enriched data sets in terms of all the known Standard Model processes is important not only for precision measurements but also to test for any new phenomena.

3. The Run II Tevatron Collider and Detector Upgrades

For Run II the Fermilab accelerator complex underwent a major upgrade. As a result the Tevatron operates at a higher center-of-mass energy $\sqrt{s} = 1.96\,\mathrm{TeV}$, with a bunch crossing interval of $396\,\mathrm{ns}$ with a goal of $\mathcal{L}_{inst} = 33 \times 10^{31}\,\mathrm{cm}^{-2}\,\mathrm{s}^{-1}$. By August 2003, the total integrated luminosity delivered by the Tevatron amounted to about $\mathcal{L}_{int} = 300\,\mathrm{pb}^{-1}$ with a typical $\mathcal{L}_{inst} = 3-4.5 \times 10^{31}\,\mathrm{cm}^{-2}\,\mathrm{s}^{-1}$.

The CDF and D0 detectors underwent extensive upgrades for Run II motivated by the need to improve overall acceptance, secondary vertex capabilities and to cope with the upgraded Tevatron performance. CDF has retained its central calorimeter and part of the muon detectors, while it has replaced the central drift chamber (COT) and the silicon tracking system (L00, SVXII, ISL). New plug calorimeters and additional muon coverage allow CDF to extend lepton identification in the forward region. The most important upgrade of the D0 detector is the new tracking system which consists of a fiber tracker plus a silicon tracker immersed in a new $2\,\mathrm{T}$ superconducting solenoid. D0 has also improved the muon coverage and added new preshower detectors. Both CDF and D0 have new DAQ and trigger systems to cope with the shorter interbunch time.

3.1. *Silicon Detectors Upgrades and High p_T b-tagging*

The feasibility and the techniques of b hadron identification at hadron machines have been firmly established in Run I. The use of b-tagging has been crucial for top discovery and it is an essential piece of the Run II top and exotic physics program. Both CDF and D0 silicon detectors upgrades followed the same guidelines: provide good coverage of the long (~ 30 cm) Tevatron luminous region, extend acceptance in the forward region, and with 3D reconstruction capabilities plus excellent impact parameter resolution achieve large signal efficiency and good background rejection.

The "silicon vertex" methods for identifying a b jet in a top event exploit the b hadron's long lifetime, large boost and significant charged track multiplicity of the decay. However, alternative methods ("soft lepton") that exploit the softer p_T spectrum and low isolation properties of b semileptonic decay modes are also employed successfully by CDF and D0.

4. Top Cross Section Measurement

Precise measurements of the $t\bar{t}$ production cross section and the branching ratios in all the decay channels provide a stringent test for the presence of new physics phenomena. Top-color and SUSY models predict not only alternative top production processes but extra decay modes that can alter the branching ratios of the various channels. The Run I top cross section measurements are summarized in Table 1. The goal for Run II ($\mathcal{L}_{int} = 2\,\text{fb}^{-1}$) is to achieve a relative uncertainty of about 10% or less on $\sigma_{t\bar{t}}$, this will be possible not only through the increased detector acceptance and efficiencies but also because the main data driven systematic uncertainties (jet energy scale, ISR/FSR, ϵ_{btag}) will scale with the size of the control sample used for their determination.

4.1. *Run II Cross Section Measurement*

The first Run II measurements of $\sigma_{t\bar{t}}$ have been focused on the channels with the highest signal-to-background ratio, namely the dilepton and lepton

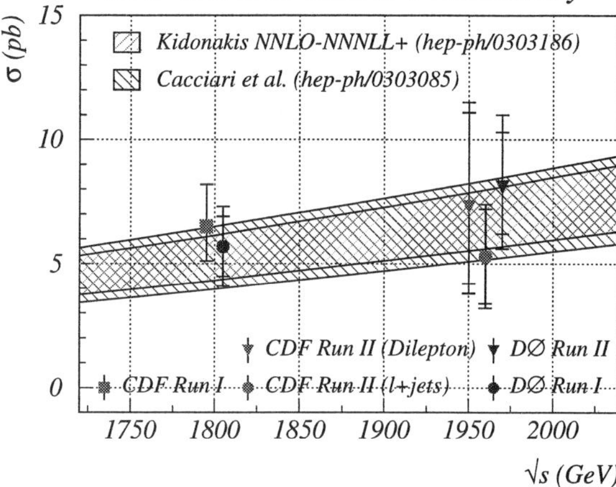

Figure 1. Summary of $\sigma_{t\bar{t}}$ versus center-of-mass energy.

plus jets ($\ell=e, \mu$). The results presented have been summarized in Table 2 and they are shown in Fig. 1 as a function of the center-of-mass energy.

4.2. *Dilepton Channel*

This final state is characterized by the presence of two high p_T leptons (a reconstructed e, μ, τ or an isolated track), large missing transverse energy from the missing neutrinos and two or more central jets. The main sources of background for this channel come from other Standard Model processes with similar signatures (Drell-Yan $\gamma^*/Z^0 \rightarrow e^+e^-, \mu^+\mu^-$, $Z^0 \rightarrow \tau\tau$, $W^+W^-/W^\pm Z^0$) and processes with one real lepton and another object that fakes the second lepton. The dilepton event selection starts with two oppositely charged high p_T leptons, asking that one or both are well isolated from nearby track activity. Different techniques are employed in order to reduce the contribution from Z^0 events without reducing the signal acceptance. $\not{E}_T$ is required to be large given the two neutrinos from W decay and the presence of two central jets accounts for the two b's from the top decay. Other kinematical and topological cuts (H_T, total energy of all the object in the event, $\Delta\phi(\not{E}_T, object)$) are finally employed in order to reduce the remaining backgrounds. The comparison of the remaining events after selection with the total Standard Model expectation (background plus signal) is shown in Fig. 2 and Fig. 3.

Table 1. Summary table of Run I $\sigma_{t\bar{t}}$ measurements, $\mathcal{L}_{int} = 110\,\mathrm{pb}^{-1}$.

$\sigma_{t\bar{t}}$(pb) Channel	CDF measurement	D0 measurement
Dilepton	$8.4^{+4.5}_{-3.5}$	6.4 ± 3.4
Lepton+jets	$5.7^{+1.9}_{-1.5}$	5.2 ± 1.8
All jets	$7.6^{+3.5}_{-2.7}$	7.1 ± 3.2
Combined	$6.5^{+1.7}_{-1.4}$	5.9 ± 1.7

Table 2. Summary table of Run II $\sigma_{t\bar{t}}$ measurements. CDF and D0 preliminary results.

Decay Channel	Method	$\sigma_{t\bar{t}}$(pb)	$\mathcal{L}_{int}$ (pb^{-1})	Experiment
Dilepton	$\ell\ell$	$8.7^{+6.4}_{-4.7}(stat) +^{+2.7}_{-2.0}(syst) \pm 0.9(lum)$	$90-107$	D0
Dilepton	$\ell\ell$	$7.6^{+3.8}_{-3.1}(stat)^{+1.5}_{-1.9}(syst)$	126	CDF
Dilepton	ℓ+track	$7.3 \pm 3.4(stat) \pm 1.7(syst)$	126	CDF
ℓ+jets	CSIP	$7.4^{+4.4}_{-3.6}(stat)^{+2.1}_{-1.6}(syst) \pm 0.7(lum)$	45	D0
ℓ+jets	SVT	$10.8^{+4.9}_{-4.0}(stat)^{+2.1}_{-2.0}(syst) \pm 1.1(lum)$	45	D0
ℓ+jets	topo	$4.6^{+3.1}_{-2.7}(stat)^{+2.1}_{-2.0}(syst) \pm 0.5(lum)$	92	D0
ℓ+jets	SMT	$11.4^{+4.1}_{-3.5}(stat)^{+2.0}_{-1.8}(syst) \pm 1.1(lum)$	92	D0
ℓ+jets	combined	$8.0^{+2.4}_{-2.1}(stat)^{+1.7}_{-1.5}(syst) \pm 0.8(lum)$	92	D0
ℓ+jets	SVX	$5.3 \pm 1.9(stat) \pm 0.9(syst)$	57	CDF
ℓ+jets	H_T	$5.1 \pm 1.8(stat) \pm 2.1(syst)$	126	CDF
Dilepton,ℓ+jets	combined	$8.1^{+2.2}_{-2.0}(stat)^{+1.6}_{-1.4}(syst) \pm 0.8(lum)$	$90-107$	D0

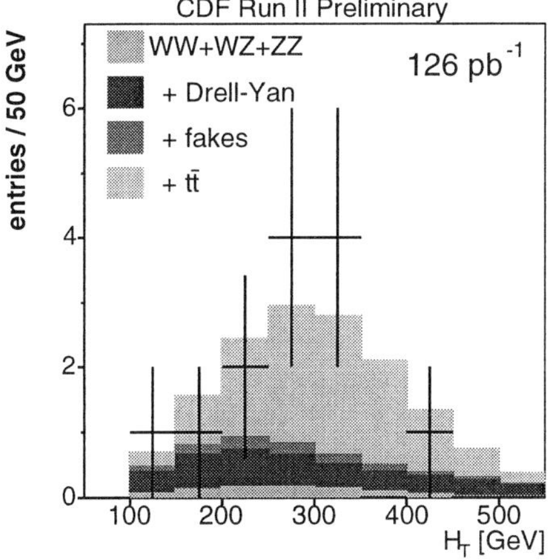

Figure 2. Run II data H_T distribution of dilepton events compared to SM expectation (CDF).

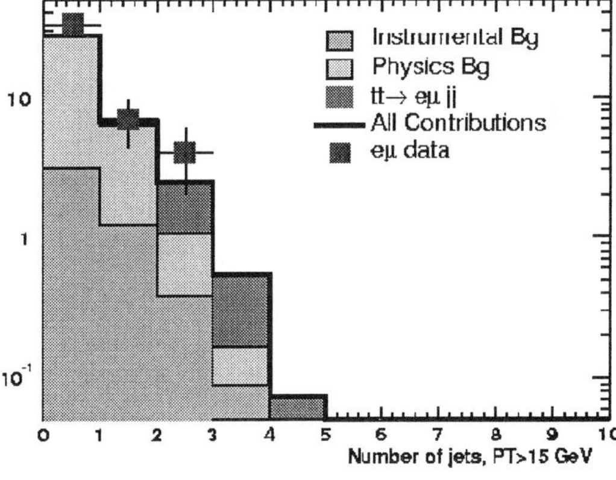

Figure 3. Run II data jet multiplicity distribution for dilepton events compared to SM expectation (D0).

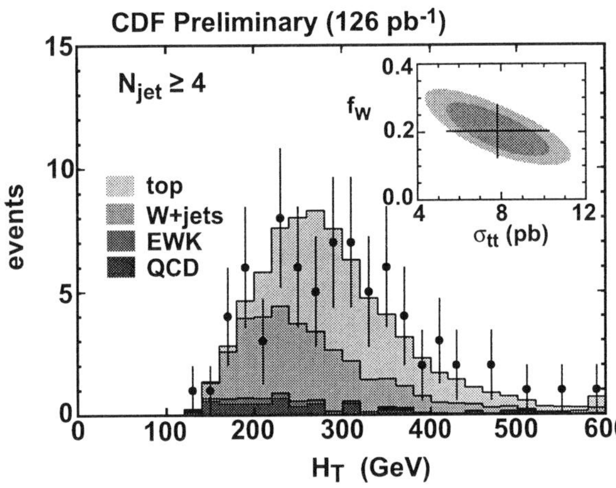

Figure 4. Run II data H_T distribution for lepton plus jet events with at least four jets compared to SM expectation (CDF).

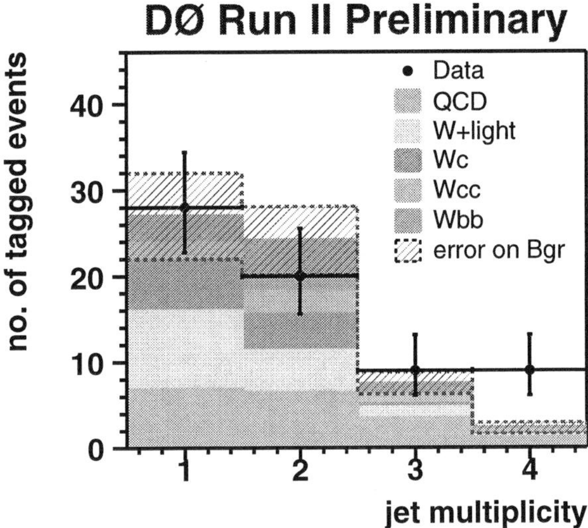

Figure 5. Run II data jet multiplicity distribution in the lepton plus jets channel with a b-tagged jet compared with SM expectations (D0).

4.3. *Lepton Plus Jets Channel*

The lepton plus jets signature is characterized by the presence of one high p_T lepton and large $\not{E}_T$ due to the leptonic W decay plus three or more jets from the hadronically decaying W and the b jets. This final state suffers from large $W + jets$ background. However, kinematical and topological properties of the $t\bar{t}$ signal or its heavy flavor content (or both) provide a good separation from the background processes. In the kinematical approach, after the basic event selection, variables such as H_T, the scalar sum of all the objects' transverse energies in the event, or the aplanarity $\mathcal{A}$, a measure of the event shape, are found to be the most discriminant, see for example Fig. 4. A complementary approach is to exploit the heavy flavor content of signal events and the large b-tagging efficiency compared to the low fake rate. There are several b identification algorithms available at the moment, some employ the silicon vertex detector information, while another category focuses on the peculiar properties of leptons from b semileptonic decays. In Fig. 5 the jet multiplicity in $W+jets$ events is shown after the requirement of an identified secondary vertex: the points are the data compared to the background Standard Model expectation. The excess due to the $t\bar{t}$ signal is visible in the three or more jet bins.

4.4. *All Hadronic Channel*

The all-jets final state where both W decay hadronically is a very challenging signature of six central and energetic jets, swamped by a QCD multijet background of several orders of magnitude bigger than the $t\bar{t}$ signal. However, in Run I both experiments succeeded in isolating the signal and measuring cross section and mass in this channel as well.[5,6] The D0 experiment has taken a first look at this channel in Run II repeating the Run I neural network analysis. A small excess of 78 events with a background of 68 ± 1.6 is found in $\mathcal{L}_{int} = 80.7\,\mathrm{pb}^{-1}$ of data, consistent with Standard Model expectations. The measurement of the cross section in this channel is in progress.

4.5. *Test for New Physics in $t\bar{t}$ Production*

Both CDF[7] and D0[8] have searched for $t\bar{t}$ resonances using the Run I data sample. Models with a dynamically broken EW symmetry (technicolor) predict a top-quark condensate, X, that decays to a $t\bar{t}$ pair. By searching for narrow $t\bar{t}$ resonances this limit becomes model independent, see Fig. 6. However, 95% CL limits have been placed on a leptophobic $Z' \to t\bar{t}$ with a large cross section for $m(Z') < 560\,\mathrm{GeV/c^2}$ by both CDF and D0, see Fig. 7.

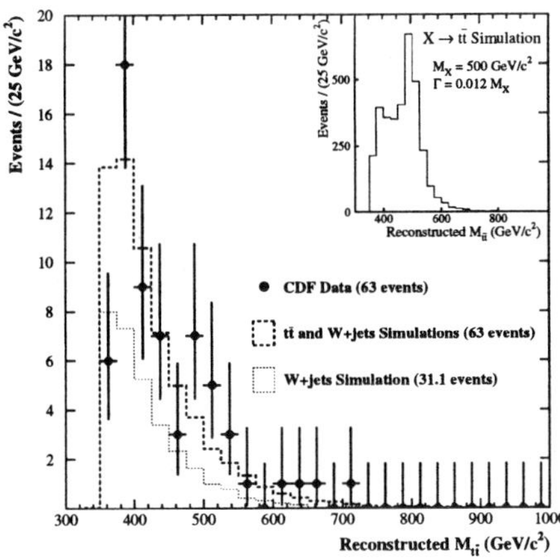

Figure 6. Run I data $M(t\bar{t})$ distribution (CDF).

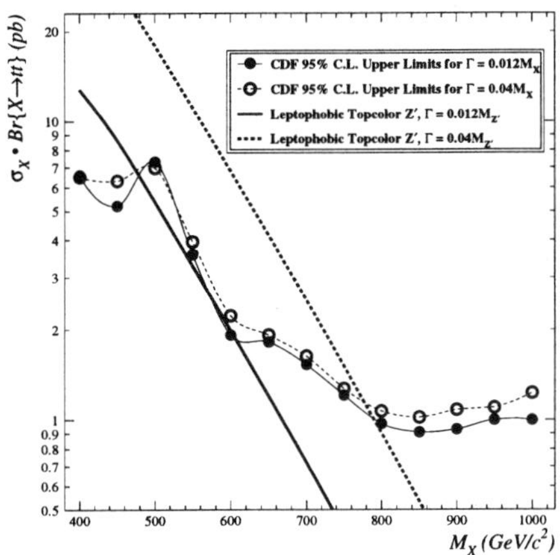

Figure 7. Run I limit on a leptophobic $Z' \to t\bar{t}$ (CDF).

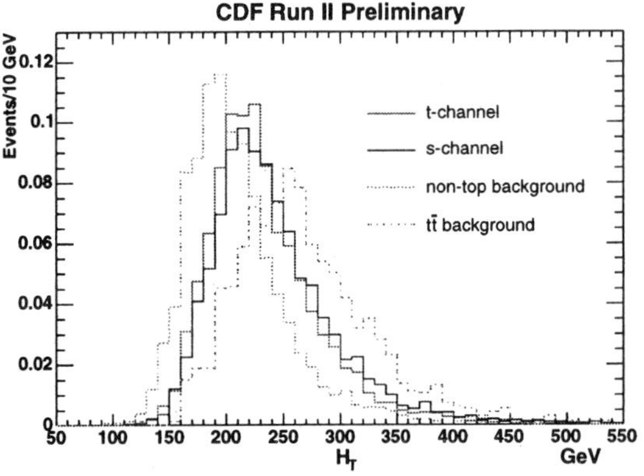

Figure 8. H_T distribution for MC single top events compared to $t\bar{t}$ and $W + jets$ (CDF).

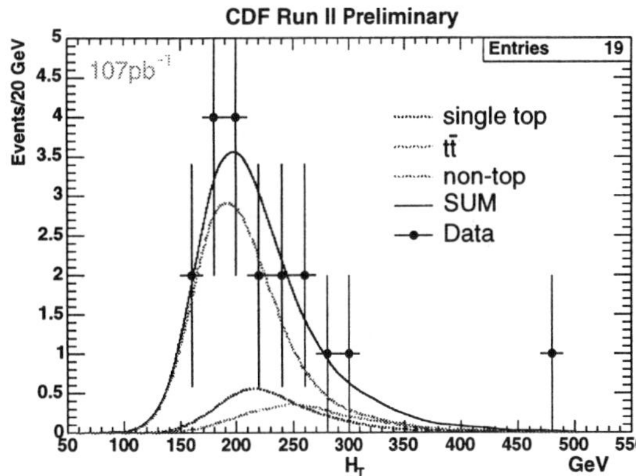

Figure 9. Run II data (points) H_T distribution compared to the expected SM background plus $t\bar{t}$ and single top production (CDF).

5. Single Top Physics

In addition to pair production, single top quarks can be produced by weak interaction by a virtual W or through Wg fusion, with a total cross section of about $\sigma_{tX} = 2.9\,\mathrm{pb}$.[9] Single top production is interesting in its own right: a precise measurement of the cross section would provide a direct determination of $|V_{tb}|$ with a 14% uncertainty expected for $\mathcal{L}_{int} = 2\,\mathrm{fb}^{-1}$ of data. Moreover single top events have the same final-state experimental signature as the Standard Model Higgs associated production process ($HW \to b\bar{b}\ell\nu_\ell$). The extraction of a single top signal is more challenging than the pair produced case since there are fewer objects in the final state and the overall event properties are less distinct from the $W + jets$ background, see Fig. 8. In Run I searches for single top production (in the s- and t-channels separately and combined) were performed by both CDF[10] and D0.[11] The same search method has been applied by CDF on a Run II dataset of about $\mathcal{L}_{int} = 107\,\mathrm{pb}^{-1}$ and the preliminary result is still consistent with the Run I cross section limit, $\sigma_{tX}^{RunII}(comb) < 17.5\,\mathrm{pb} \,@\, 95\%\mathrm{CL}$. The H_T distribution for the candidate events compared to Standard Model expectation for signal and background is shown in Fig. 9.

6. *W* Helicity in Top Decays

Since the top is the only quark that decays free without hadronizing, the decay products carry its

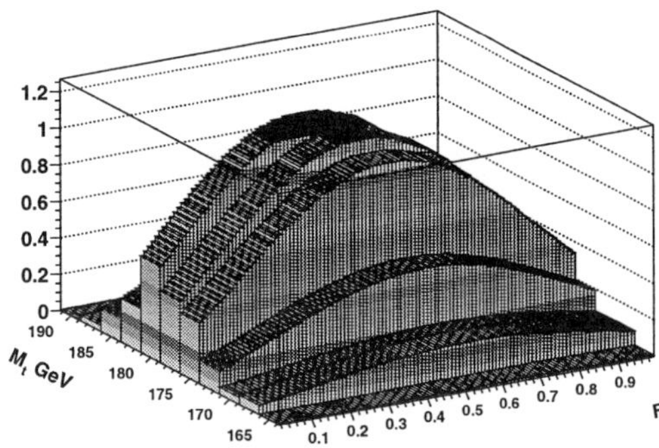

Figure 10. Two dimensional (m_t, f_0) probability distribution of Run I lepton plus jets data (D0).

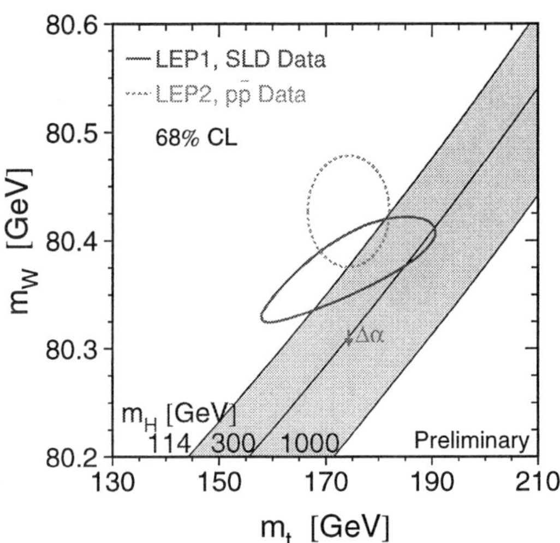

Figure 11. The closed curves representing experimental measurements of m_t and m_W constrain the SM Higgs mass. The shaded band shows the allowed combinations of m_t and m_W for various m_H.

polarization information. In the Standard Model the top quark decays only to longitudinal or left-handed W's, where the ratio is predicted to be about $f_0 = \frac{W_{long}}{W_{left}} = 70\%$ in the case of a $m_t = 175\,\text{GeV}/c^2$. The helicity information is reflected in several kinematical properties of the decay products (W lepton p_T, $M(\ell b)$) that are traditionally used to extract an experimental measurement of f_0. However, D0 has performed a new measurement using the Run I datasets: $f_0 = 0.56 \pm 0.31(stat) \pm 0.04(syst)$ under the assumption of $m_t = 175\,\text{GeV}/c^2$, see in Fig. 10 the two dimensional (m_t, f_0) probability distribution. The likelihood method used here makes better use of the event information thus greatly improving the statistical uncertainty. This method is used also in the measurement of the top mass and will be discussed more in Sec. 7.

7. Top Mass

The top quark mass is a fundamental Standard Model parameter that needs to be measured with the greatest possible precision. It is needed to determine the strength of the ttH coupling and it has a substantial effect on radiative corrections. In fact, an uncertainty of $2\,\text{GeV}/c^2$ on the top mass would constrain the Higgs mass to 35%, see Fig. 11. It is not an easy task to achieve such a small uncertainty, but several experimental handles are available in Run II. On one hand the increased detector acceptance and large data sample will allow one to select purer samples less sensitive to systematic uncertainties: for instance requiring events with well mea-

sured jets (lowers the energy scale uncertainty) and two b-tagged jets (lowers the overall background). On the other hand, since most of the systematics are data driven, their uncertainty will scale approximately with $1/\sqrt{N}$, N being the number of events of the control samples themselves.

7.1. *Run I Measurement Summary*

Using the Run I dataset the CDF and D0 experiment have measured directly the top mass in channels (lepton+jets, dilepton and all hadronic) employing different methods and techniques. The results from the two experiments and their combination are summarized in Table 3. However, the single most precise measurement on the Run I data comes from the latest measurement from the D0 experiment in the lepton plus jets channel[12] of $m_t = 180.1 \pm 5.4\,\text{GeV}/c^2$. The likelihood method employed for this measurement was originally proposed for the mass reconstruction in dilepton events[13−15] where the system is underconstrained for a simple kinematic fit. However, the technique is very useful also in the case of lepton plus jets events, since a better use of the event information effectively increases the statistical power of the data sample itself. Each event has an associated probability to be signal or background defined

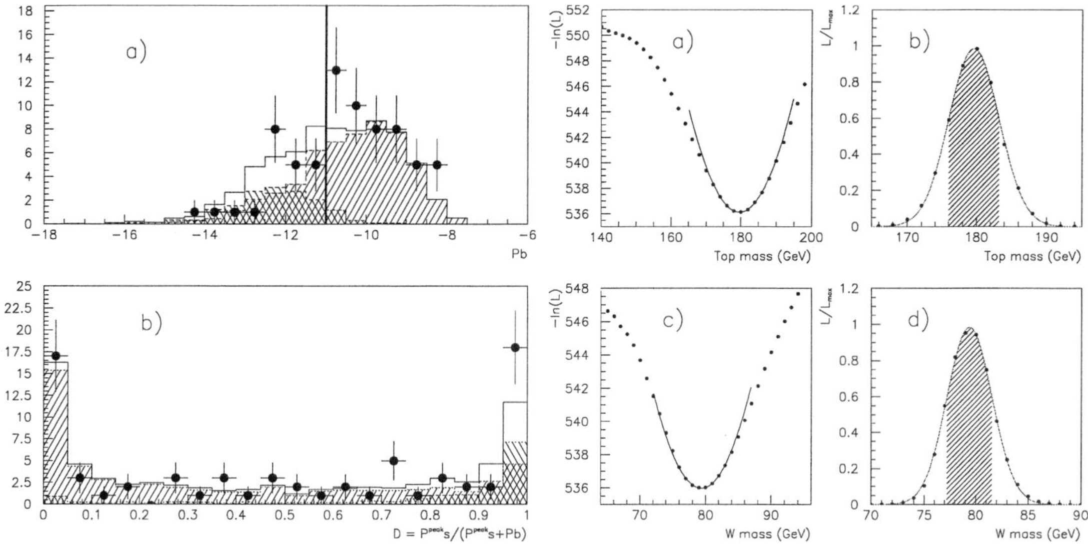

Figure 12. New D0 Run I top mass measurement. Left: a) background probability distribution; b) ratio $P_{top}/(P_{bkg} + P_{top})$. Right: a) and b) fitted top mass value and its uncertainty; c) and d) fitted W mass when the top mass is fixed to its fitted value.

in terms of the matrix element information, and this probability is convolved with a transfer function that relates the object at the parton level to the object after reconstruction. The only background considered is $W + 4\,jets$, which makes up 80% of the total, and cleanup cuts are added to the data to further reduce the absolute background contribution. The remaining 22 events in $\mathcal{L} = 125\,\mathrm{pb}^{-1}$ of data are then used in a global likelihood fit to extract the top mass and the W mass (or the f_0 at the same time), as shown in Fig. 12. The large reduction in statistical uncertainty, corresponding to an effective increase of the data set by a factor 2.4, is mainly due to the more complete use of the event information.

7.2. *Issues for Precision Top Mass in Run II*

A precise measurement of the top mass combines cutting edge theoretical knowledge with state-of-the-art detector calibration. The highest contribution to the systematic uncertainty still comes from the jet energy scale. With the statistics available now the best calibration sample consists of events where a jet is recoiling against a well measured photon, with larger statistics the sample where a jet recoils against a reconstructed Z can be used. Finally the hadronic W in lepton plus jets events can provide an in situ calibration for the light quark jets, while the $Z \to b\bar{b}$ sig-

nal would be used for the heavy flavor ones. A large amount of data will allow one to not only reduce the systematics above but also to pick the best measured event categories with smaller backgrounds and that are less sensitive to systematics uncertainties. However, to achieve the ultimate precision excellent Monte Carlo generators implementing the latest theory knowledge and understanding of all the various effects (ISR, FSR, PDF's) plus an accurate detector simulation are essential.

While work is in progress on all these fronts, preliminary measurements of the top mass, still dominated by large systematic uncertainties, have been performed using the Run II data sample. In the lepton plus four jets channel with at least one secondary vertex b-tagged jet a value of $m_t = 177.5^{+12.7}_{-9.4}(stat) \pm 7.1(syst)\,\mathrm{GeV}/c^2$ is found using 22 candidate events shown in Fig. 13. In the dilepton channel a preliminary measurement of $m_t = 175.0^{+17.4}_{-16.9}(stat) \pm 7.9(syst)\,\mathrm{GeV}/c^2$, is obtained using 6 candidate events, shown in Fig. 14.

8. Conclusions

The top is still a very young particle and our current knowledge about its properties comes from the Run I Tevatron measurements. This accelerator and its two experiments, CDF and D0, are the place for top physics still for years to come. In Run II a data

Table 3. Summary table of Run I m_t measurements.

m_t (GeV/c^2) Channel	CDF measurement	D0 measurement
Dilepton	167.4 ± 11.4	168.4 ± 12.8
Lepton+jets	175.9 ± 7.1	173.3 ± 7.8
All jets	186.0 ± 11.5	–
Combined	176.0 ± 6.5	172.1 ± 7.1
CDF+D0 Combined	174.3 ± 5.1	
Lepton+jets (New)	–	180.1 ± 5.4

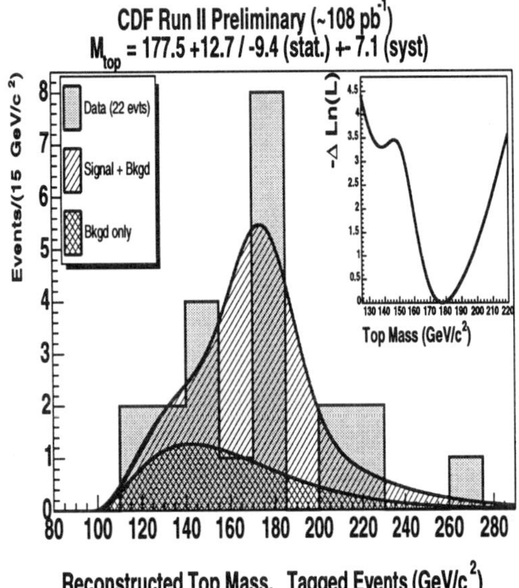

Figure 13. Reconstructed m_t distribution in the lepton+jets channel with a b-tagged jet in Run II data (CDF).

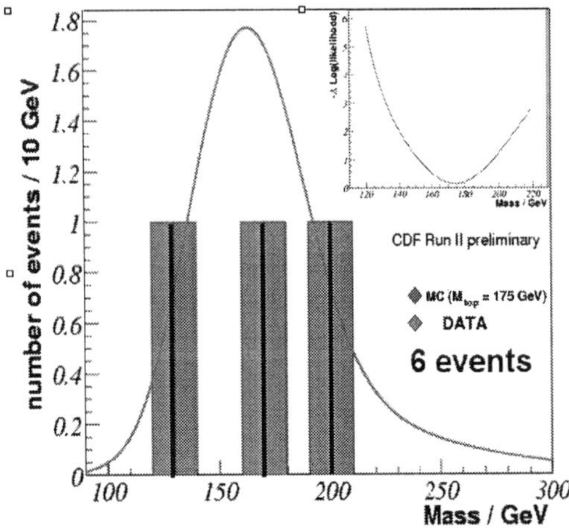

Figure 14. Reconstructed m_t in the dilepton channel in Run II data (CDF).

sample about 50 times the Run I statistics will be collected. It will be possible to achieve better precision in the measurements and perform significant tests of the Standard Model expectations. Maybe there will be surprises ahead... The first preliminary round of Run II measurements of the production cross section and mass covers already a variety of channels and very soon the uncertainties will start to drop. The two experiments are exploiting all their new upgraded detector features and a very exciting top physics program lies ahead.

Acknowledgments

I'd like to thank Keith Ellis and the LP2003 conference organizers for this great opportunity. I'd like also to thank my friends and colleagues Marumi Kado, Ela Barberis, Aurelio Juste, Jaco Konigsberg and Simona Rolli for their help in the preparation of this report.

References

1. F.Abe *et al.*, *Phys. Rev. Lett.* **73**, 225 (1994).
2. F.Abe *et al.*, *Phys. Rev. Lett* **74**, 2626 (1995).
3. S.Abachi *et al.*, *Phys. Rev. Lett.* **74**, 2632 (1995).
4. N.Kidonakis and R.Vogt, *hep-ph/0308222*.
5. F.Abe *et al.*, *Phys. Rev. Lett.* **79**, 1992 (1997).
6. D0 Collaboration, *Phys. Rev. D* **60**, 012001 (1999).
7. T.Affolder *et al.*, *Phys. Rev. Lett.* **85**, 2062 (2000).
8. V.M.Abazov *et al.*, *hep-ex/0307079*2003.
9. B.W.Harris *et al.*, *Phys. Rev. D* **66**, 054024 (2002).
10. D.Acosta *et al.*, *Phys. Rev. D* **65**, 091102 (2002).
11. V.Abazov *et al.*, *Phys. Lett.* B **517**, 282 (2001).
12. J. Estrada *for D0 Coll.*, *hep-ex/03/020312*002.
13. R.H.Dalitz and G.R.Goldstein, *Proc. R. Soc. Lond.* **A445**, 2803 (1999).
14. K. Kondo *et al.*, *J.Phys.Soc.Jap* **62**, 1177 (1993).
15. B. Abbott *et al.*, *Phys. Rev D* **60**, 052001 (1999).

DISCUSSION

S. Heinemeyer (LMU Munich): What is the new combined value for m_t from CDF Run I plus Run II plus new D0 Run I result?

Patrizia Azzi: The combined number is not available yet.

R. Keeler (U. Victoria): Was the choice of the "future position" of the result from the collider on the m_t versus m_W plane motivated by any experimental knowledge?

Patrizia Azzi: No. It was for illustration only.

ELECTROWEAK MEASUREMENTS FROM RUN II AT THE TEVATRON

T. R. WYATT

Department of Physics and Astronomy, University of Manchester, Manchester M13 9PL, UK
Email: twyatt@fnal.gov

The CDF and DØ detectors were fully commissioned for physics running in Run II at the Tevatron $p\bar{p}$ collider in early 2002. Since then both experiments have collected data samples corresponding to an integrated luminosity of around $\int L = 200$ pb^{-1} at a $p\bar{p}$ centre-of-mass energy of $\sqrt{s} = 1.96$ TeV. Datasets corresponding $\int L = 120$ pb^{-1} have been analyzed for physics so far. Recent electroweak measurements from Run II are reviewed. Cross section times branching ratio measurements ($\sigma \cdot$ Br) are presented for the intermediate vector bosons (IVB's) in their leptonic decay modes: $W \to \ell\nu$ and $Z \to \ell^+\ell^-$. For the first time, a combination of the $\sigma \cdot$ Br results from the CDF and DØ experiments is made; this includes using a consistent choice of the total inelastic $p\bar{p}$ cross section for the luminosity determinations of the two experiments. Quantities derived from these $\sigma \cdot$ Br values are also updated. These include: R_ℓ the ratio of the $\sigma \cdot$ Br values for W and Z; Br($W \to \ell\nu$), the leptonic branching ratio of the W; and Γ_W, the total decay width of the W. Other measurements using events containing W and Z leptonic decays are presented, including studies that probe the QCD phenomenology of W/Z production and searches for events containing two intermediate vector bosons.

1. Experimental Measurements of $\sigma \cdot$ Br for $Z \to \ell^+\ell^-$ and $W \to \ell\nu$

1.1. *Introduction*

Figure 1 shows the mechanism for IVB production in $p\bar{p}$ collisions.

The experimental signature for $Z \to \ell^+\ell^-$ is illustrated in Fig. 2. We observe a pair of oppositely charged leptons that have high p_T with respect to the beam direction and are isolated with respect to other energetic particles in the event. The presence of two high p_T leptons in the event leads to a high degree of redundancy in the trigger and offline selection. This leads to low backgrounds and excellent control of systematic uncertainties.

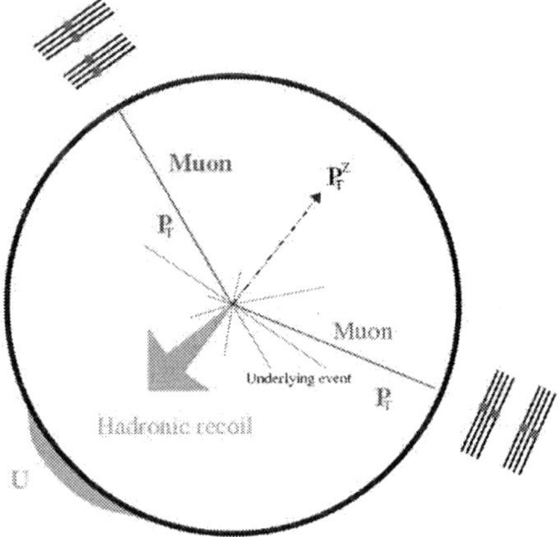

Figure 2. An illustration of the experimental signature for $Z \to \ell^+\ell^-$ in $p\bar{p}$ collisions.

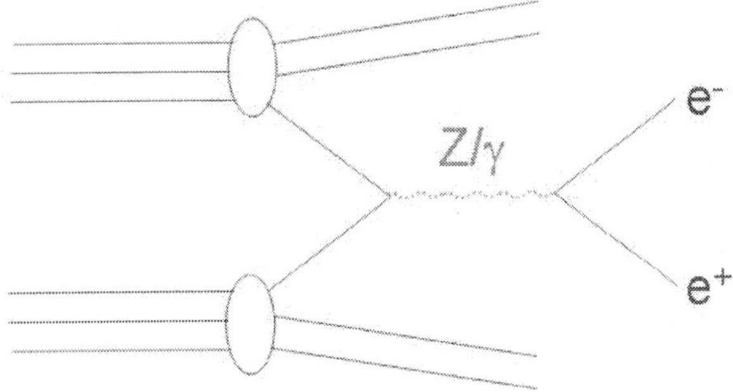

Figure 1. The mechanism for IVB production in $p\bar{p}$ collisions.

The experimental signature for $W \to \ell\nu$ is illustrated in Fig. 3. We observe a single high p_T isolated charged lepton plus missing transverse momentum, E_T^{miss}, carried away by the unobserved neutrino. The presence of only one high p_T lepton in $W \to \ell\nu$ events leads to less redundancy in the trigger and offline selection than for $Z \to \ell^+\ell^-$. In addition, the measurement of E_T^{miss} requires us to understand the measurement of the p_T of the hadrons recoiling against the W. These issues make it more difficult to control backgrounds and systematic uncertainties in the analysis of W's than is the case for Z's. Of course, from the point of view of electroweak physics, measurements at the Tevatron on W's are much more interesting than those on Z's, since the properties of the Z have been so well understood at LEP and SLC. However, the samples of Z events are extremely useful as a means to measure experimental efficiencies and control phenomenological systematic uncer-

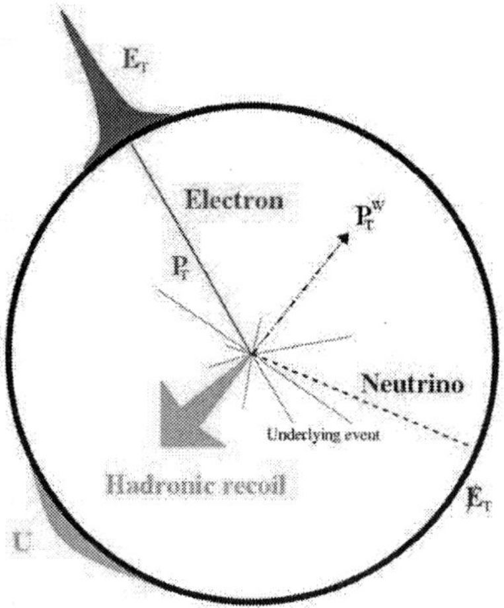

Figure 3. An illustration of the experimental signature for $W \to \ell\nu$ in $p\bar{p}$ collisions.

tainties. Most experimental systematics currently quoted are limited by the size of the data samples currently available and will decrease as larger data samples are collected and analyzed.

In measuring the rate of production of W and Z events at the Tevatron the dominant systematic error arises from the determination of the $p\bar{p}$ luminosity. For CDF this uncertainty is quoted at 6% and for DØ it is quoted as 10% for the preliminary results presented so far. These respective uncertainties are, of course, completely correlated amongst all of the measurements made by an individual experiment. A large part of the above uncertainties is correlated between the two experiments. The treatment of the luminosity scale and uncertainties for the purposes of combining the CDF and DØ results will be discussed in Sec. 2.1.

The other significant source of systematic uncertainty that introduces correlations among the measurements arises from uncertainties in the parton distribution functions (PDF's). Experimentally the observed charged leptons from IVB decay are required to lie within a given range of pseudorapidity (η). The probability for the leptons to lie within this acceptance depends on the degree to which the IVB's are boosted along the beam direction, and this in turn depends on the PDF's. Accurate knowledge of the PDF's is therefore essential in order to determine the experimental acceptance. Of equal importance, a quantitative assessment of the uncertainties in the PDF's is essential in order to evaluate the result-

ing uncertainty in the experimental acceptance. The uncertainties in the $\sigma \cdot \mathrm{Br}$ measurements of both experiments have been evaluated using the PDF error sets provided as part of the CTEQ6 PDF's.[1]

When determining $\sigma \cdot \mathrm{Br}$ for $Z \to \ell^+\ell^-$ it must be borne in mind that the physically observed process is $p\bar{p} \to \ell^+\ell^- X$; the $\ell^+\ell^-$ system may couple to a Z or a γ. In order to determine the (unphysical) quantity $\sigma_Z \cdot \mathrm{Br}(Z \to \ell^+\ell^-)$ the number of observed $\ell^+\ell^-$ events must therefore be "corrected" by the factor $\sigma_Z/\sigma_{Z\gamma}$, where $\sigma_{Z\gamma}$ is the full Standard Model cross section including Z, γ and Z-γ interference, and σ_Z is the cross section calculated using Z exchange only.

1.2. DØ: $Z \to \mu^+\mu^-$

The DØ experiment has updated its measurement of $\sigma_Z \cdot \mathrm{Br}(Z \to \mu^+\mu^-)$ for this conference using a dataset corresponding to $\int L = 117\,\mathrm{pb}^{-1}$. The event selection cuts require two oppositely charged central tracks with $p_\mathrm{T} > 15$ GeV. In order to maintain a high selection efficiency the tracks are required to satisfy only loose requirements on muon identification and only one of the muons is required to be isolated. Events are selected over a wide angular range $|\eta| < 1.8$ and a loose cut on the invariant mass of the $\mu^+\mu^-$ system, $\mathrm{M}_{\mu\mu} > 30$ GeV, is made. Cosmic ray muons are rejected by cuts on scintillator timing and the distance of closest approach of the muons to the beam crossing point. The invariant mass of the 6126 DØ $\mu^+\mu^-$ candidates is shown in Fig. 4. The dominant backgrounds arise from QCD $(0.6 \pm 0.3)\%$ and $Z \to \tau^+\tau^-$ $(0.5 \pm 0.1)\%$.

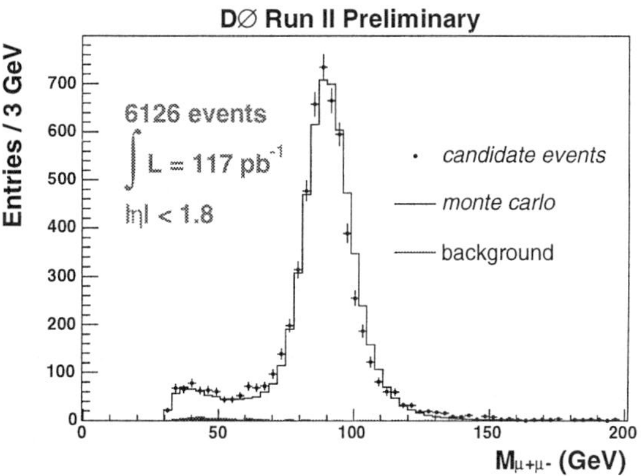

Figure 4. The invariant mass of DØ $\mu^+\mu^-$ candidates.

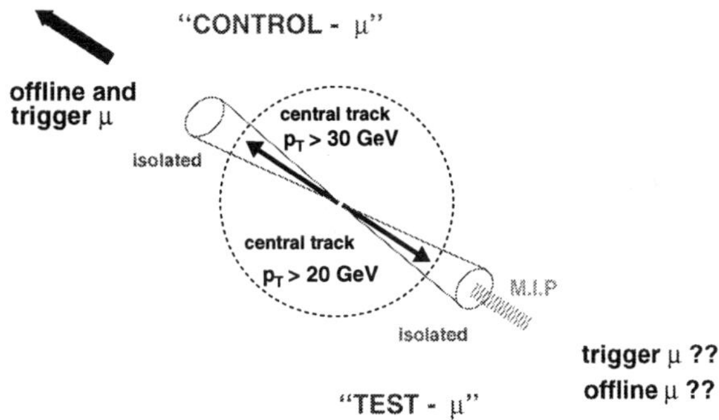

Figure 5. Illustration of the methods used to evaluate experimental efficiencies using the $Z \to \ell^+\ell^-$ data. In this particular figure, the muon trigger efficiency for the "test" muon is evaluated. This measurement uses a pure sample of $Z \to \mu^+\mu^-$ events that is triggered and selected without any bias as to whether or not the "test" muon is detected by the muon system.

An illustration of the methods used to evaluate experimental efficiencies using the $Z \to \ell^+\ell^-$ data is given in Fig. 5. The basic idea is that a very pure sample of $Z \to \mu^+\mu^-$ events can be selected by making rather tight cuts on one "control" muon and only very loose cuts on the second "test" muon. In Fig. 5, the muon trigger efficiency for the "test" muon is evaluated using a sample of $Z \to \mu^+\mu^-$ events that is triggered and selected without any bias as to whether or not the "test" muon is detected by the muon system. In making such measurements it is important to demonstrate that a very pure sample of $Z \to \mu^+\mu^-$ events can be selected even though only loose cuts are made on the "test" muon. This is illustrated by Fig. 6, which shows a comparison of the shapes of the $\mu^+\mu^-$ invariant mass distributions for the two relevant sub-samples of DØ $Z \to \mu^+\mu^-$ events: the points with error bars show the events in which the test muon did not fire the Level-1 trigger; the line histogram shows the events in which the test muon did fire the Level-1 trigger. The shapes of the two distributions are very similar, both being dominated by $Z \to \mu^+\mu^-$. The resulting efficiency per muon of the DØ Level-1 muon trigger as a function of η is shown in Fig. 7. The efficiencies measured in the data for the trigger, tracking and muon identification are used as inputs to a Monte Carlo simulation that is used to evaluate the overall event acceptance $\times$ efficiency. The total acceptance $\times$ efficiency for a $Z \to \mu^+\mu^-$ event to be triggered and selected is 19%.

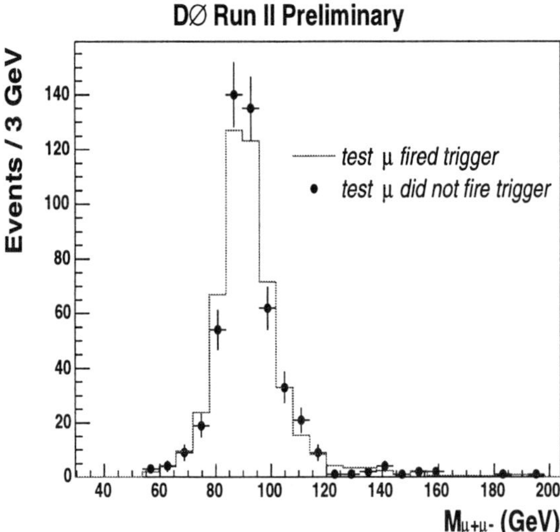

Figure 6. Comparison of the shapes of the invariant mass distributions for two samples of DØ $Z \to \mu^+\mu^-$ events: *points with error bars:* test muon did not fire the Level-1 trigger; *line histogram:* test muon did fire the Level-1 trigger.

The dominant experimental systematic uncertainties on $\sigma_Z \cdot \mathrm{Br}(Z \to \mu^+\mu^-)$ arise from the limited size of the Z data sample currently available to make such efficiency measurements ($\pm$ 3.3%) and from PDF's ($\pm$ 1.6%). The preliminary result is:

$$\sigma_Z \cdot \mathrm{Br}(Z \to \mu^+\mu^-) =$$
$$261.8 \pm 5.0(\text{stat.}) \pm 8.9(\text{syst.}) \pm 26.2(\text{lum.}) \text{ pb.}$$

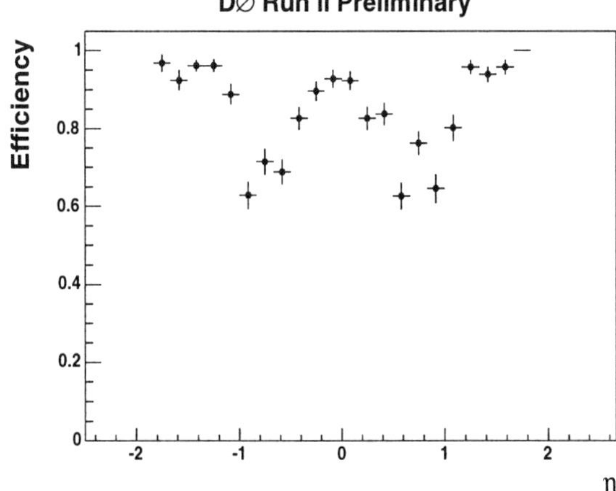

Figure 7. Efficiency per muon of the DØ Level-1 muon trigger as a function of η.

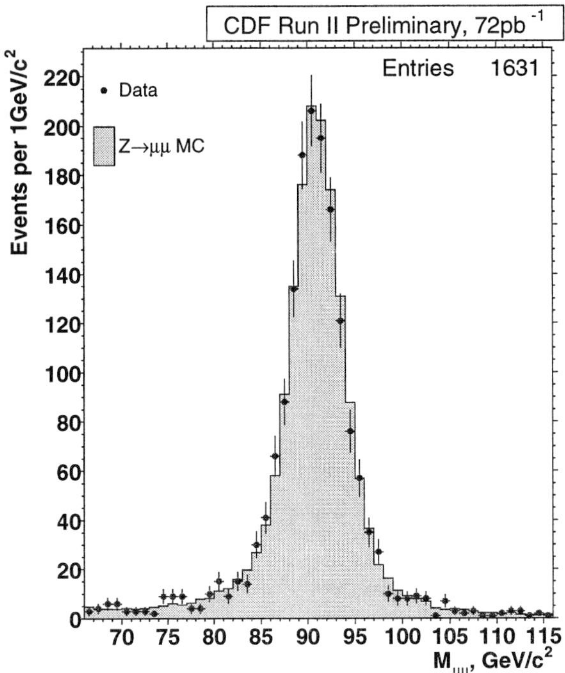

Figure 8. The invariant mass of CDF $\mu^+\mu^-$ candidates.

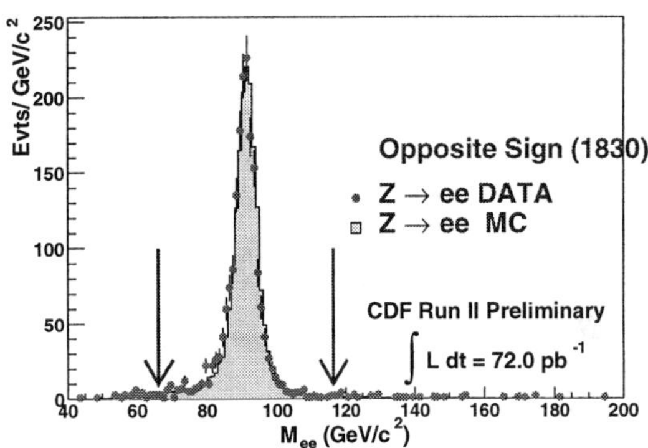

Figure 9. The invariant mass of CDF e^+e^- candidates.

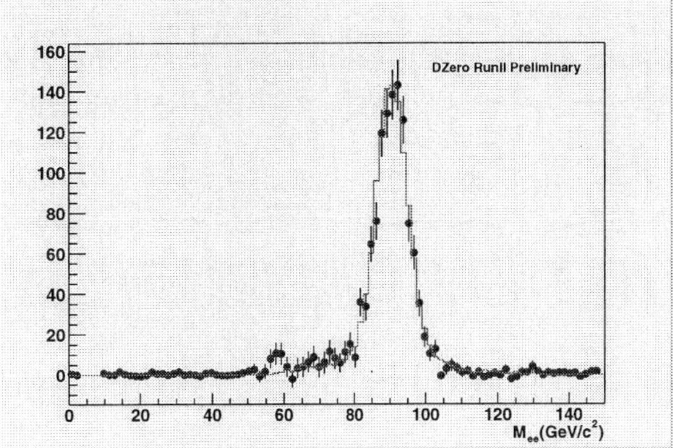

Figure 10. The invariant mass of DØ e^+e^- data *(points with error bars)* compared to Monte Carlo *(line histogram)*.

1.3. *CDF:* $Z \to \mu^+\mu^-$

All of the CDF measurements of $\sigma \cdot \mathrm{Br}$ for Z and W at this conference correspond to $\int L = 72$ pb^{-1}. The event selection cuts for $Z \to \mu^+\mu^-$ require two oppositely charged central tracks that are identified as muons and have $p_\mathrm{T} > 20$ GeV. Both of the muons are required to be isolated. Events are selected over a fairly restricted angular range: at least one muon is required to satisfy $|\eta| < 0.6$ and both muons are required to satisfy $|\eta| < 1.0$. A cut on the invariant mass of the $\mu^+\mu^-$ system around the Z mass is made: $66 < \mathrm{M}_{\mu\mu} < 116$ GeV. The total acceptance $\times$ efficiency for a $Z \to \mu^+\mu^-$ event to be triggered and selected is 9% and the candidate event sample comprises 1631 events. The invariant mass of the CDF $\mu^+\mu^-$ candidates is shown in Fig. 8. The dominant backgrounds arise from cosmic ray muons $(0.9 \pm 0.9)\%$. The largest experimental systematic uncertainty arises from PDF's ($\pm$ 3%); this is larger than for the corresponding DØ analysis due to the more restricted angular acceptance of the CDF event selection. The preliminary result is:

$$\sigma_\mathrm{Z} \cdot \mathrm{Br}(Z \to \mu^+\mu^-) =$$
$$246 \pm 6(\text{stat.}) \pm 12(\text{syst.}) \pm 15(\text{lum.}) \text{ pb.}$$

1.4. *CDF and DØ:* $Z \to e^+e^-$

CDF and DØ employ very similar cuts to select candidate $Z \to e^+e^-$ events: two isolated electron candidates are required with $\mathrm{E_T} > 25$ GeV and $|\eta| < 1.1$. The invariant mass of the 1830 CDF e^+e^- candidates is shown in Fig. 9. The CDF result is:

$$\sigma_\mathrm{Z} \cdot \mathrm{Br}(Z \to e^+e^-) =$$
$$267.0 \pm 6.3(\text{stat.}) \pm 15.2(\text{syst.}) \pm 16.0(\text{lum.}) \text{ pb.}$$

The invariant mass of the 1631 DØ e^+e^- candidates from $\int L = 42$ pb^{-1} is shown in Fig. 10. The DØ result is:

$$\sigma_\mathrm{Z} \cdot \mathrm{Br}(Z \to e^+e^-) =$$
$$275 \pm 9(\text{stat.}) \pm 9(\text{syst.}) \pm 28(\text{lum.}) \text{ pb.}$$

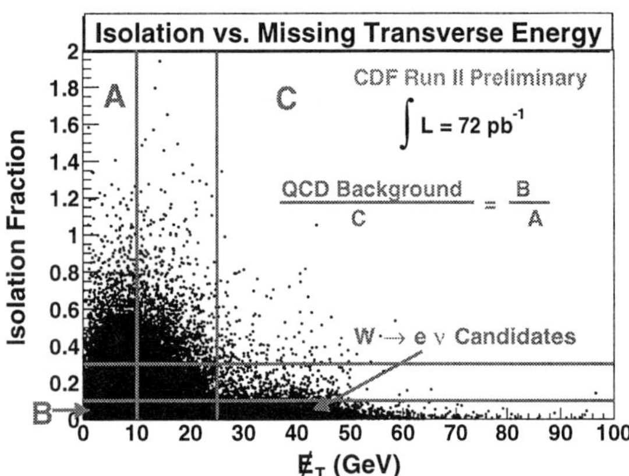

Figure 11. CDF electron data: the degree to which the electron is isolated vs. E_T^{miss}.

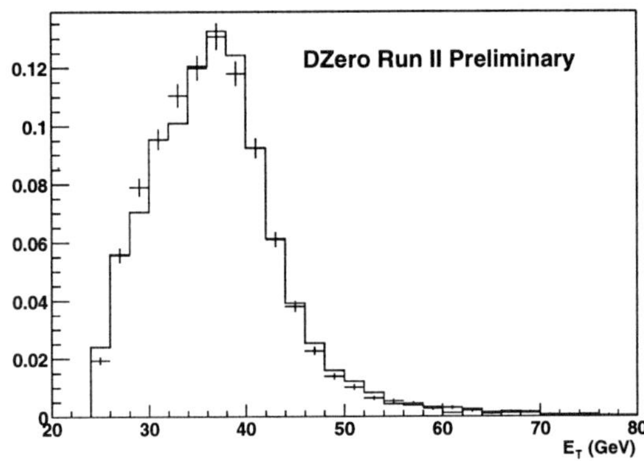

Figure 12. The electron E_T distribution in DØ $W \to e\nu$ data *(points with error bars)* compared to Monte Carlo *(line histogram)*.

1.5. CDF and DØ: $W \to e\nu$

In selecting candidate $W \to e\nu$ events both CDF and DØ require an isolated electron candidate with $E_T > 25$ GeV and $E_T^{miss} > 25$ GeV. The background to $W \to \ell\nu$ is dominated by QCD events in which a jet fakes the isolated lepton signal. Monte Carlos cannot be trusted to provide an adequate description of the background processes and so the level of background is estimated using the data. A method used to estimate the background is illustrated by Fig. 11, which shows for the CDF electron data the degree to which the electron is isolated vs. E_T^{miss}. The candidate $W \to e\nu$ events occupy the lower right area of the plot, which corresponds to isolated electrons and high E_T^{miss}. The rest of the plot is dominated by background. The probability for the electron candidate in a background event to appear to be isolated is estimated by taking the ratio of the numbers of events in regions A and B in Fig. 11, which are both at low E_T^{miss}. The number of background events in the signal region is estimated by applying this factor to the number of non-isolated events with high E_T^{miss} (region C). The accuracy of this method is limited by kinematic correlations between isolation and E_T^{miss} for the background events. CDF quotes an estimated background of $(3.5 \pm 1.7)\%$; the 50% uncertainty is evaluated by making large variations in the boundaries of the regions A, B and C and seeing by how much the estimated background changes. With 38628 candidate $W \to e\nu$ events the CDF result is:

$$\sigma_W \cdot \mathrm{Br}(W \to e\nu) =$$
$$2.64 \pm 0.01(\mathrm{stat.}) \pm 0.09(\mathrm{syst.}) \pm 0.16(\mathrm{lum.}) \text{ nb}.$$

Figure 12 shows the electron E_T distribution in DØ $W \to e\nu$ data *(points with error bars)* compared to Monte Carlo *(line histogram)*. 27370 events are selected from $\int L = 42$ pb^{-1} and the result is:

$$\sigma_W \cdot \mathrm{Br}(W \to e\nu) =$$
$$2.88 \pm 0.02(\mathrm{stat.}) \pm 0.13(\mathrm{syst.}) \pm 0.29(\mathrm{lum.}) \text{ nb}.$$

1.6. CDF and DØ: $W \to \mu\nu$

In selecting candidate $W \to \mu\nu$ events both CDF and DØ require an isolated muon candidate with $p_T > 20$ GeV and $E_T^{miss} > 20$ GeV. At present, the background to $W \to \mu\nu$ for both experiments has a large contribution from $Z \to \mu^+\mu^-$ events in which one of the muons is not reconstructed, as well as from QCD events in which a muon in a jet fakes the isolated muon signal.

The "transverse mass", M_T, is given by:

$$M_T = \sqrt{2p_T E_T^{miss}(1 - \cos\Delta\phi)},$$

where $\Delta\phi$ is the difference in azimuthal angle between the p_T of the charged lepton candidate and the missing transverse momentum vector. M_T corresponds to the invariant mass of the muon-neutrino system, taking only their momentum components in the plane perpendicular to the beam direction into account. Figure 13 shows the M_T of CDF $W \to \mu\nu$ candidates. The total background is estimated to be $(10.8 \pm 1.1)\%$, the number of candidate events is 21599 and the result is:

$$\sigma_W \cdot \mathrm{Br}(W \to \mu\nu) =$$
$$2.64 \pm 0.02(\mathrm{stat.}) \pm 0.12(\mathrm{syst.}) \pm 0.16(\mathrm{lum.}) \text{ nb}.$$

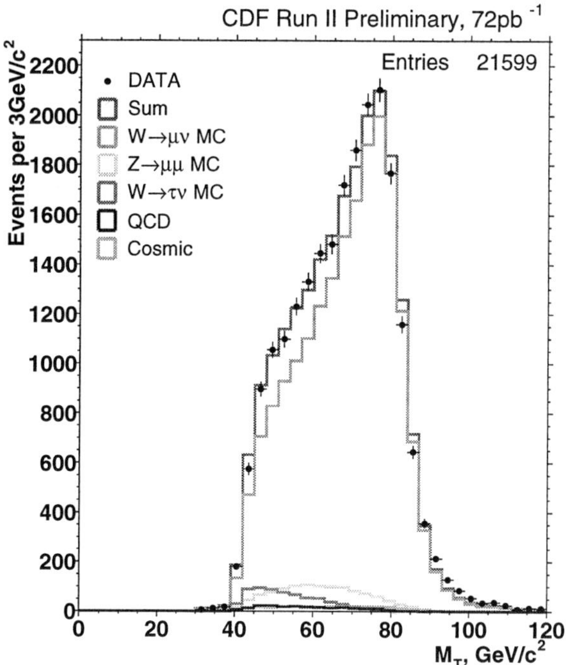

Figure 13. The M$_T$ of CDF $W \to \mu\nu$ candidates.

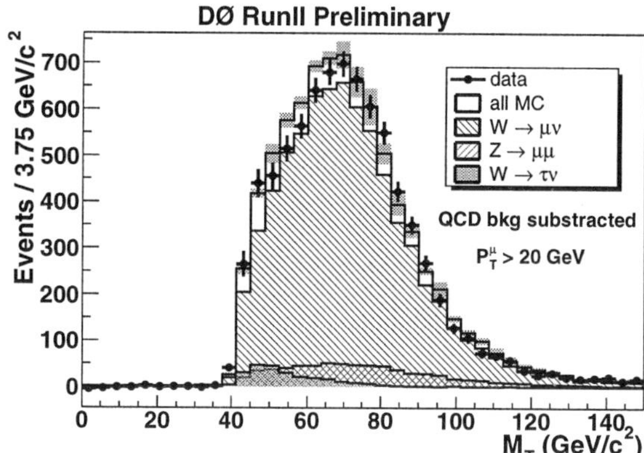

Figure 14. The M$_T$ of DØ $W \to \mu\nu$ candidates.

Figure 14 shows the M$_T$ of DØ $W \to \mu\nu$ candidates. The total background is estimated to be $(11.4 \pm 1.8)\%$, the number of candidate events is 7352 from $\int L = 17$ pb^{-1} and the result is:

$$\sigma_{\mathrm{W}} \cdot \mathrm{Br}(W \to \mu\nu) =$$
$$3.23 \pm 0.13(\mathrm{stat.}) \pm 0.10(\mathrm{syst.}) \pm 0.32(\mathrm{lum.})\ \mathrm{nb}.$$

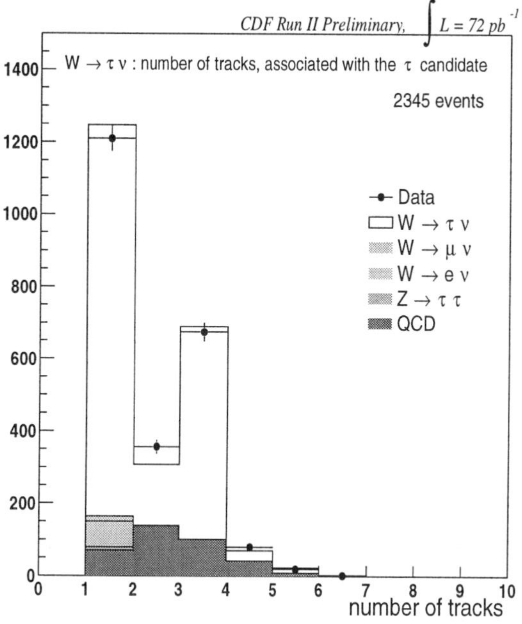

Figure 15. The number of charged tracks associated with CDF $W \to \tau\nu$ candidates.

1.7. *CDF: $W \to \tau\nu$*

CDF selects candidate hadronic tau decays by looking for a narrow jet that is contained within a cone of half-width 10° and is isolated within a wider cone of half-width 30°. Candidate $W \to \tau\nu$ events are selected by requiring $p_T > 25$ GeV for the hadronic tau candidate and $E_T^{\mathrm{miss}} > 25$ GeV. The estimated contributions to the selected sample of 2345 events are illustrated in Fig. 15, which shows the number of charged tracks associated with the tau candidates. The result is:

$$\sigma_{\mathrm{W}} \cdot \mathrm{Br}(W \to \tau\nu) =$$
$$2.67 \pm 0.07(\mathrm{stat.}) \pm 0.21(\mathrm{syst.}) \pm 0.16(\mathrm{lum.})\ \mathrm{nb}.$$

2. Combination of $\sigma \cdot$ Br Results from CDF and DØ

2.1. *Luminosity Determination*

CDF and DØ determine the delivered luminosity by measuring the total rate of inelastic $p\bar{p}$ collisions. The luminosity determination therefore requires knowledge of the total inelastic cross section, $\sigma_{\mathrm{inelastic}}$. This cross section has been measured during Tevatron Run I at $\sqrt{s} = 1.8$ TeV by two experiments: CDF and E811. These two measurements disagree at the level of three standard deviations. There is some ambiguity as to how to perform an average of these two inconsistent values – different methods lead to results in the range $59.1 < \sigma_{\text{inelastic}} < 60.7$ mb (2.7% difference) when extrapolated to $\sqrt{s} = 1.96$ TeV.

For the $\sigma \cdot$ Br results reported in the preceding sections CDF uses $\sigma_{\text{inelastic}} = 60.7$ mb and DØ uses $\sigma_{\text{inelastic}} = 57.6$ mb (which corresponds to a 5.3% difference). For the combinations presented below I have chosen[a] to:

- scale the reported $\sigma \cdot$ Br values to correspond to a consistent value of $\sigma_{\text{inelastic}}$.

  – I have chosen: $\sigma_{\text{inelastic}} = 60.7$ mb, the value used by CDF.

  – This choice corresponds to multiplying the DØ $\sigma \cdot$ Br values reported in the preceding sections by a factor 1.053.

- quote an additional 2.7% systematic error to cover the ambiguity in the choice of $\sigma_{\text{inelastic}}$. This leads to a total error of $(4.0 \oplus 2.7 = 4.8)\%$ assumed for $\sigma_{\text{inelastic}}$, which is 100% correlated between CDF and DØ.

2.2. *Combined* $\sigma \cdot$ **Br** *Results*

The $\sigma \cdot$ Br values given in Sec. 1 have been combined. The luminosity scale and uncertainty are treated as described in Sec. 2.1. At the present level of accuracy the only other source of systematic uncertainty that introduces significant correlations among the measurements arises from the PDF's.

Figure 16 shows the resulting combined CDF and DØ measurement of $\sigma_Z \cdot \text{Br}(Z \to \ell^+\ell^-)$:

$$\sigma_Z \cdot \text{Br}(Z \to \ell^+\ell^-) = 258 \pm 10(\text{expt.}) \pm 16(\text{lum.}) \text{ pb.}$$

Also shown are the individual CDF and DØ measurements of $\sigma_Z \cdot \text{Br}(Z \to \mu^+\mu^-)$ and $\sigma_Z \cdot \text{Br}(Z \to e^+e^-)$. These values are compared to the Standard Model NNLO expectation:[2]

$$\sigma_Z \cdot \text{Br}(Z \to \ell^+\ell^-) = 252 \pm 9 \text{ pb.}$$

---

[a] These issues have been discussed within the Tevatron Electroweak Working Group (TeVEWWG), but no official policy has yet been agreed by CDF and DØ. The results given below labelled as "my combination" should be taken as the responsibility of this review speaker and not officially sanctioned CDF/DØ results.

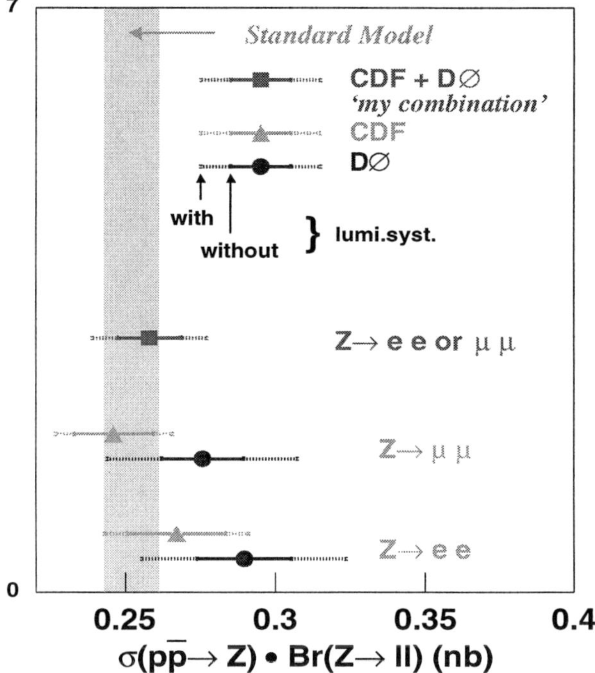

Figure 16. Combined CDF and DØ measurement of $\sigma_Z \cdot \text{Br}(Z \to \ell^+\ell^-)$ compared with the Standard Model expectation. Also shown are the individual measurements of $\sigma_Z \cdot \text{Br}(Z \to \mu^+\mu^-)$ and $\sigma_Z \cdot \text{Br}(Z \to e^+e^-)$.

Figure 17 shows the combined CDF and DØ measurement of $\sigma_W \cdot \text{Br}(W \to \ell\nu)$:

$$\sigma_W \cdot \text{Br}(W \to \ell\nu) = 2.69 \pm 0.09(\text{expt.}) \pm 0.17(\text{lum.}) \text{ nb.}$$

Also shown are the individual CDF and DØ measurements of $\sigma_W \cdot \text{Br}(W \to \mu\nu)$ and $\sigma_W \cdot \text{Br}(W \to e\nu)$. These values are compared to the Standard Model NNLO expectation:[2]

$$\sigma_W \cdot \text{Br}(W \to \ell\nu) = 2.72 \pm 0.10 \text{ pb.}$$

In the Tevatron combined $\sigma \cdot$ Br measurements for both Z and W, the experimental (non lumi.) error is dominated by uncertainties of a statistical nature.

2.3. *Quantities Derived from the* *Combined* $\sigma \cdot$ **Br** *Results*

A number of interesting quantities may be derived from the above $\sigma \cdot$ Br results. It is useful to define the ratio of the $\sigma \cdot$ Br values for W and Z:

$$R_\ell = \frac{\sigma_W \cdot \text{Br}(W \to \ell\nu)}{\sigma_Z \cdot \text{Br}(Z \to \ell^+\ell^-)}.$$

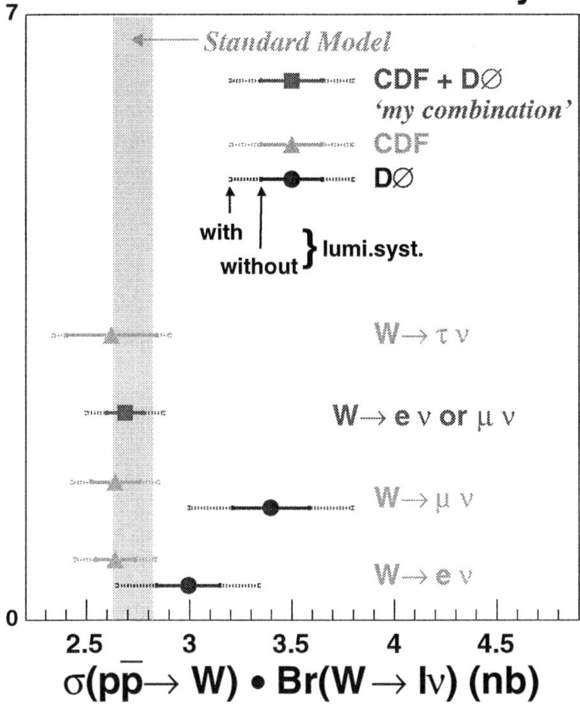

Figure 17. Combined CDF and DØ measurement of $\sigma_W \cdot Br(W \to \ell\nu)$ compared with the Standard Model expectation. Also shown are the individual measurements of $\sigma_W \cdot Br(W \to \mu\nu)$ and $\sigma_W \cdot Br(W \to e\nu)$.

In taking this ratio the luminosity uncertainties cancel and other important systematic uncertainties partially cancel, for example, those arising from PDF's and the efficiencies to trigger and select high p_T, isolated leptons. The Tevatron Electroweak Working Group (TeVEWWG) has evaluated the correlated systematic uncertainties and has previously averaged the CDF electron and muon and the DØ electron results.[4] For this conference the value of R_ℓ has been updated[b] to include a value of R_ℓ extracted[c] from the DØ muon results: $R_\mu = 12.32 \pm 0.73$. Figure 18 shows the updated combined CDF and DØ measurement of R_ℓ from Run II:

$$R_\ell = 10.61 \pm 0.30,$$

which when combined with the values from Run I yields the value:

$$R_\ell = 10.59 \pm 0.20.$$

As can be seen in the figure, these results are in agreement the Standard Model expectation.

The value of R_ℓ can be used to make an indirect determination of the leptonic branching ratio of the W, $Br(W \to \ell\nu)$. This follows from the definition of R_ℓ given above, with the ratio of the W and Z production cross sections input from a NNLO calculation and the value of $Br(Z \to \ell^+\ell^-)$ as measured at LEP. Figure 19 shows the values of $Br(W \to e\nu)$, $Br(W \to \mu\nu)$ and $Br(W \to \ell\nu)$ extracted from the Tevatron-combined values of R given in Fig. 18. These results are compared with the measurements made at LEP and with the Standard Model expectation.

The W leptonic branching ratio may be expressed as:

$$Br(W \to \ell\nu) = \Gamma(W \to \ell\nu)/\Gamma_W.$$

Since the W leptonic partial width, $\Gamma(W \to \ell\nu)$, can be predicted very accurately within the SM, $Br(W \to \ell\nu)$ may thus be interpreted as an indirect measurement of the W total width, Γ_W. The Tevatron combined Run I plus Run II indirect measurement using this technique is[d]:

$$\Gamma_W = 2.135 \pm 0.053 \text{ GeV}.$$

This may be compared with the direct measurement of Γ_W from the W lineshape, combining LEP plus the Tevatron Run I, of:

$$\Gamma_W = 2.139 \pm 0.069 \text{ GeV}.$$

2.4. *Future Prospects*

There is a promising future for further improvements in the accuracy of such "ratio" measurements at the Tevatron. Although at present the $Br(W \to \ell\nu)$ measurements from LEP are the most accurate available, these measurements are limited in precision by the statistical uncertainties from samples of only $O(10^3)$ leptonic W decays per channel and per experiment. With a dataset of $O(1 \text{ fb}^{-1})$ within the next couple of years at the Tevatron we expect to select

---

[b]The updated combination follows exactly the method used previously by TeVEWWG. However, since there was insufficient time for this new combination to pass through the official approval procedures of the collaborations it should be regarded as the responsibility of the speaker.
[c]"my combination"

[d]"my combination"

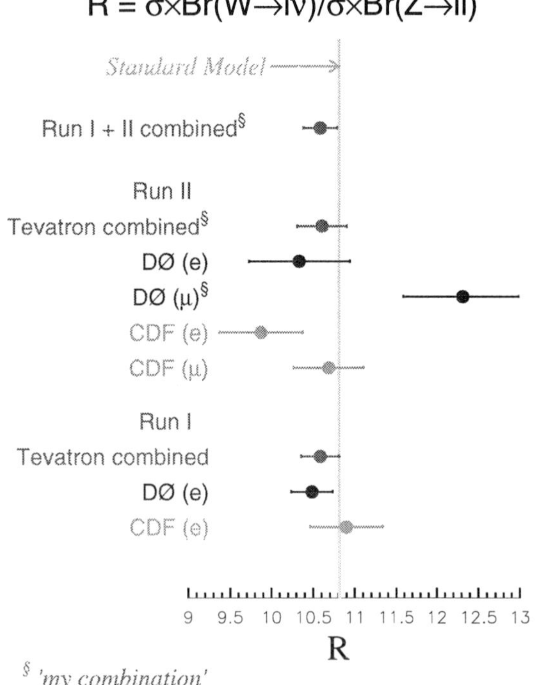

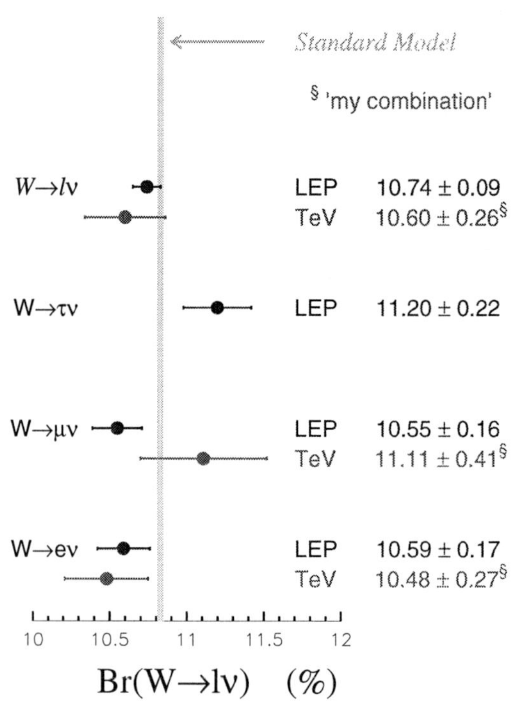

Figure 18. Combined CDF and DØ measurement of R_ℓ from Run II and Run I compared with the Standard Model expectation. Also shown are the individual measurements of R_ℓ in the electron and muon channels.

Figure 19. Measurements of $Br(W \to \ell\nu)$ from LEP and the Tevatron compared with the Standard Model expectation.

$O(10^6)$ $W \to \ell\nu$ events per channel and per experiment. It will, of course, be a considerable challenge to beat systematic uncertainties down to the few per mille level to keep pace with the statistical errors. However, the samples of $O(10^5)$ $Z \to \ell^+\ell^-$ events that are expected per channel and per experiment will play a large part in achieving the precise systematic understanding of detector performance and phenomenology that will be necessary. In addition, considerable skill will be needed to design and implement experimental triggers and event selections with sufficient redundancy to achieve the necessary precision.

In parallel to the expected increase in experimental precision, much theoretical progress is currently being made in understanding, for example, NNLO cross section calculations and PDF's, and in quantifying PDF uncertainties. This should allow the production cross sections of W and Z to be predicted at the level of about 1% and their ratio at the level

of few per mille.[5] This offers the prospect that the experimental luminosity for the rest of the Tevatron physics programme could be determined with a better precision than can ever be expected from the luminosity determinations based on the total rate of inelastic collisions.

3. Other Measurements with Events Containing W and Z Bosons

With sizeable samples of W and Z events now becoming available, many other interesting measurements are starting to be made. Studies sensitive to possible new physics include measurements of the high mass tail of the $\ell^+\ell^-$ invariant mass distribution (see Figs. 20 and 21) and the forward-backward charge asymmetry of $\ell^+\ell^-$ events (see Fig. 22).

Understanding the QCD phenomenology of IVB production is an important part of the physics programme with W and Z events. This is of interest both as a topic in its own right and also because

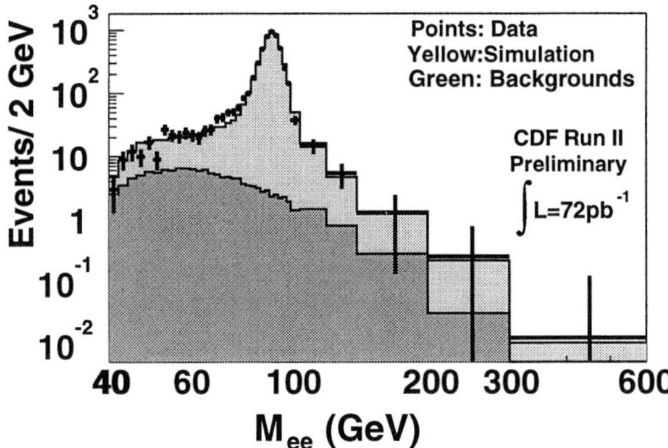

Figure 20. The invariant mass of CDF e^+e^- candidates showing the high mass region.

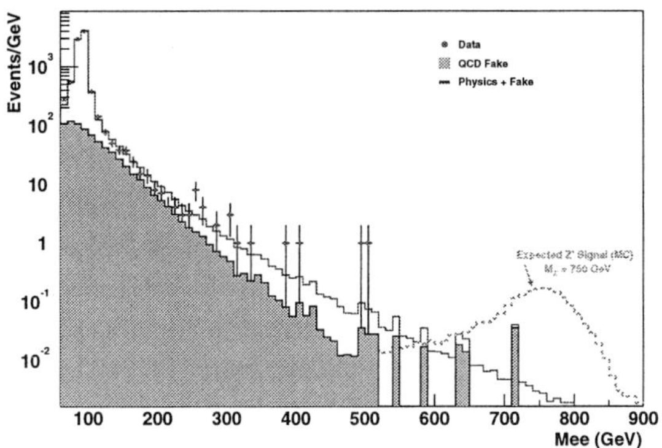

Figure 21. The invariant mass of DØ e^+e^- candidates showing the high mass region.

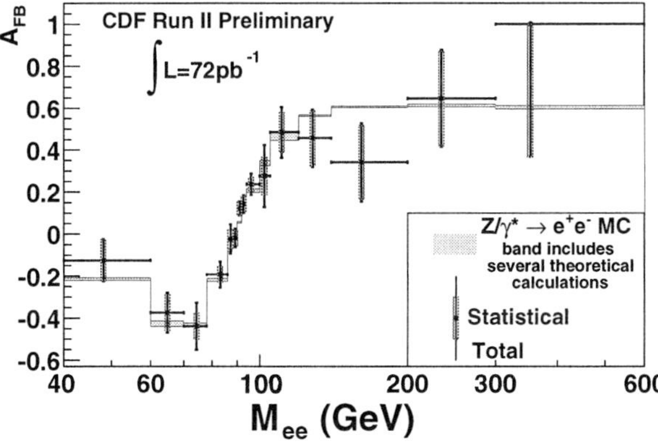

Figure 22. The A_{fb} of CDF e^+e^- candidates as a function of the e^+e^- invariant mass.

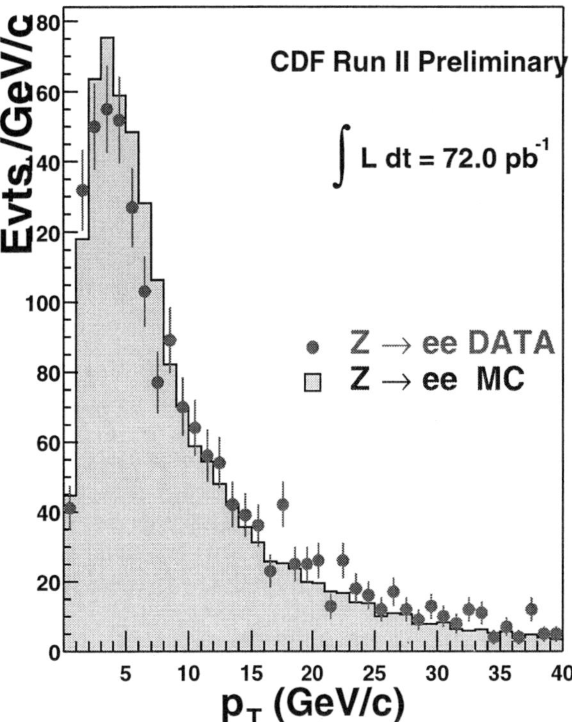

Figure 23. The p_T of CDF e^+e^- candidates.

a precise understanding of this phenomenology and the degree to which QCD Monte Carlos describe the data will be necessary to control systematic uncertainties in much of the physics programme at CDF and DØ (e.g. measuring the W and top masses). As an example of first steps in this direction, Figs. 23 and 24 show the p_T of e^+e^- candidates in CDF and $\mu^+\mu^-$ candidates in DØ, respectively, compared to Monte Carlo simulations. Other measurements that probe PDF's and QCD phenomenology that can be expected in the future include: the rapidity distribution of Z's, the p_T distribution of W's, and the charge asymmetry in $W \to \ell\nu$ events.

4. Selection of Events Containing Two Electroweak Bosons

The selection of events containing two electroweak IVB's are of interest, for example, because they allow measurements of the self-coupling of the IVB's and they allow searches to be made for new particles or interactions. The SM cross sections for events con-

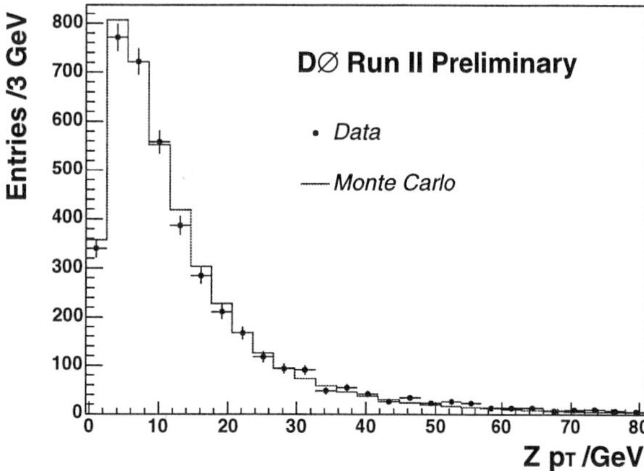

Figure 24. The p_T of DØ $\mu^+\mu^-$ candidates.

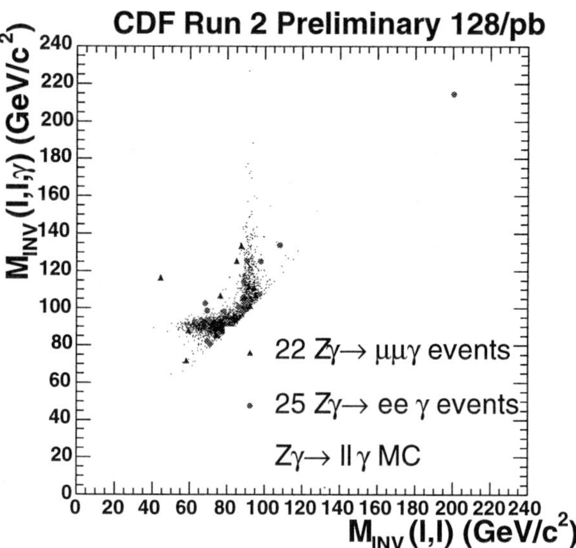

Figure 25. Invariant mass of $\ell^+\ell^-\gamma$ system vs. invariant mass of $\ell^+\ell^-$ system for CDF $Z\gamma$ candidates.

taining two electroweak IVB's are very small and so only small samples of such events are expected with the present datasets. However, CDF has updated its results in this area with samples that correspond to $\int L$ of around 126 pb^{-1}, and this promises to be an area of considerable activity in the future.

4.1. $Z\gamma$ and $W\gamma$

CDF selects $Z\gamma$ and $W\gamma$ events by requiring, in addition to the standard Z and W event selections, the presence of a central photon candidate with $E_T > 7$ GeV that is spatially separated from the charged lepton(s) in the event according to:

$$\Delta R = \sqrt{\Delta\phi^2 + \Delta\eta^2} > 0.7,$$

where $\Delta\phi$ and $\Delta\eta$ are the separation in azimuthal angle and pseudo-rapidity between the photon and the nearest charged lepton. Figure 25 shows the invariant mass of the $\ell^+\ell^-\gamma$ system vs. the invariant mass of the $\ell^+\ell^-$ system for CDF $Z\gamma$ candidates. The concentration of 3-body masses at M_Z seen in the Monte Carlo events is due to final-state radiation from one of the charged leptons. The concentration of 2-body masses at M_Z is due to initial-state radiation. 47 events are observed as compared with 43 events expected. CDF quotes a $\sigma \cdot$ Br value for $Z \to \ell^+\ell^-$ containing a photon satisfying the above kinematic cuts on E_T and ΔR:

$$\sigma \cdot \mathrm{Br} = 5.8 \pm 1.0(\mathrm{stat.}) \pm 0.4(\mathrm{syst.}) \pm 0.4(\mathrm{lum.}) \text{ pb.}$$

Figure 26 shows the M_T of the $\ell\nu\gamma$ system vs. the M_T of the $\ell\nu$ system for CDF $W\gamma$ candidates. 133

events are observed as compared with 141 events expected. CDF quotes a $\sigma \cdot$ Br value for $W \to \ell\nu$ containing a photon satisfying the above kinematic cuts on E_T and ΔR:

$$\sigma \cdot \mathrm{Br} = 17.2 \pm 2.2(\mathrm{stat.}) \pm 2.0(\mathrm{syst.}) \pm 1.2(\mathrm{lum.}) \text{ pb.}$$

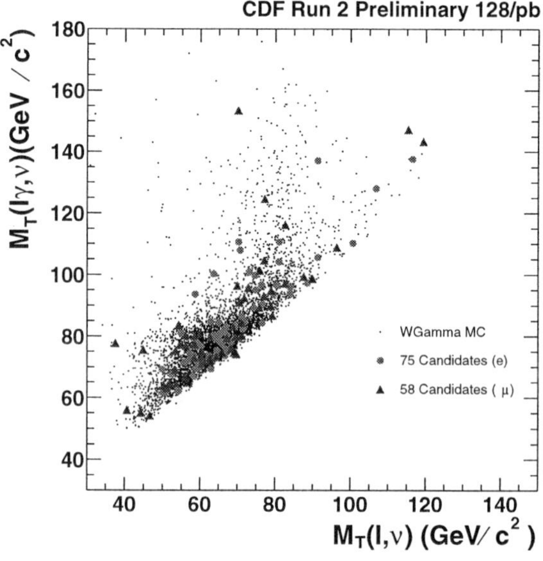

Figure 26. M_T of $\ell\nu\gamma$ vs. M_T of $\ell\nu$ for CDF $W\gamma$ candidates.

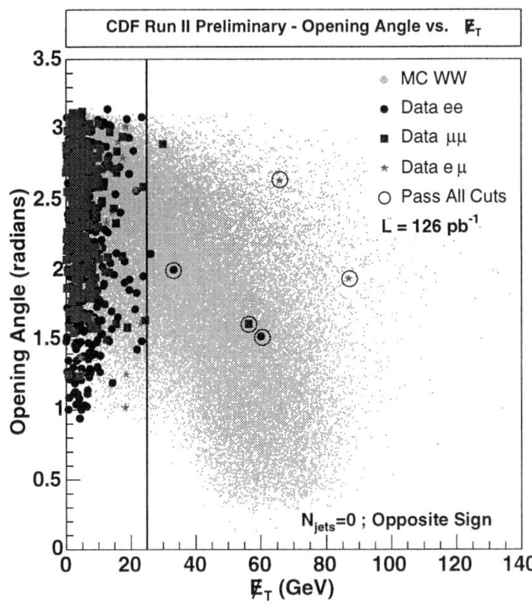

Figure 27. CDF WW selection: the opening angle between the two leptons vs. E_T^{miss}. The 5 selected signal events are indicated as points within a circle.

4.2. *WW Search*

CDF has performed a search for WW events in the channels e^+e^-, $\mu^+\mu^-$ and $e^\pm\mu^\mp$. A pair of oppositely charged, isolated, high p_T lepton candidates is required in events with high E_T^{miss}. In the e^+e^- and $\mu^+\mu^-$ channels, events are removed if the lepton pair mass is consistent with M_Z. Events containing hadronic jets are removed. Figure 27 shows the opening angle between the two leptons vs. E_T^{miss}. 5 signal events are selected, to be compared with 9.2 events expected, of which 2.3 are background and 6.9 are signal. Clearly, it is too early to claim observation of WW in Run II.

4.3. *Summary*

Analyses of events containing W and Z are now becoming available from CDF and DØ with datasets from Run II corresponding to $\int L$ of around $120\,\text{pb}^{-1}$ – well in excess of the $\int L$ collected in Run I. So far, these analyses have concentrated on selection of the event samples and measuring the relevant cross section times branching ratios. The detailed understanding of the performance of the detectors, trig-

gers, event reconstruction algorithms, calibrations, and Monte Carlos required by such measurements benefits the entire physics programmes of the two experiments. Detailed measurements of the properties of the selected events have started and should lead to interesting results in the near future.

It is in the interests of the two Tevatron experiments and Fermilab that prompt and authoritative combinations of the latest results from CDF and DØ are provided. I hope that I am the last review speaker at a major international conference who has to present combinations of Tevatron results with the proviso "my combination". Clearly the individual experiments have complete responsibility for deciding which of their results are released into the public domain. In addition, the *procedures* for combining results from the two experiments have to be subjected to the full scrutiny of internal collaboration review. However, once this has occurred I hope that in future any *particular combination of the latest set of results* can be performed by the TeVEWWG and presented in public without the need for an additional protracted period of internal review by the individual collaborations.

Acknowledgments

I am grateful to the CDF and DØ collaborations for providing me with their preliminary results and descriptions of the associated analyses. I thank the other members of the Tevatron Electroweak Working Group for useful discussions and for establishing the procedures used to combine the values of R_ℓ, $\text{Br}(W \to \ell\nu)$ and Γ_W: Sarah Eno, Harald Fox, Martin Grünewald, Eva Halkiadakis, Eric James, Ashutosh Kotwal, Giulia Manca, Sean Mattingly, Pasha Murat, Emily Nurse, Michael Schmitt, Georg Steinbrück, Paul Telford, Alexei Varganov, Marco Verzocchi, and Junjie Zhu. In particular, I thank Martin Grünewald for his invaluable help in making the combinations given in Sec. 2; they should perhaps more properly have been labelled "our combination". I am grateful to Martin Grünewald and Emily Nurse for their useful comments on a draft version of this note.

I gratefully acknowledge support from the UK Particle Physics and Astronomy Research Council in the form of a PPARC Senior Research Fellowship.

References

1. J. Pumplin *et al.*, hep-ph/0201195.
2. The Standard Model NNLO expected $\sigma \cdot$ Br values are calculated using: R. Hamberg *et al.*, *Nucl. Phys.* B **359**, 1991 (343). The central value uses the MRST2002 NNLO PDF's.[3] The 3.5% uncertainty is assessed using the CTEQ6 error PDF's. The LEP value $\text{Br}(Z \to \ell^+\ell^-) = 0.03366 \pm 0.00002$ and the SM value $\text{Br}(W \to \ell\nu) = 0.1082 \pm 0.0002$ are used.
3. A. D. Martin *et al.*, hep-ph/0211080.
4. The Tevatron Electroweak Working Group, http://tevewwg.fnal.gov/wz/eps2003/
5. See, for example, the talk by Robert Thorne in these proceedings.

DISCUSSION

Rohini Godbole (Indian Institute of Science, Bangalore): What is the expected error on the M_W measurement now that the upgraded detectors are in operation, and how does it compare with the 30 MeV given in the studies made before Run II started?

Terry Wyatt: The description of the data by the Monte Carlos is adequate for the purposes of estimating event selection efficiencies and backgrounds in the $\sigma \cdot \mathrm{Br}$ measurements presented here. However, as I mentioned in my talk already, at the level of detail needed for M_W measurements there are considerable disagreements (see, for example, Figs. 12 and 13). A great deal of careful work to understand detector performance and phenomenology will be needed before M_W measurements with the current data will be feasible. At present there is not much one can say beyond the estimates made before Run II of around 30 MeV.

Thomas Gehrmann (Zurich Univ.): You briefly mentioned the option of using W and Z cross sections to determine the luminosity of the Tevatron. What are the actual prospects for this to become the default luminosity calibration?

Terry Wyatt: From the experimental point of view I am very confident that we shall understand efficiencies and backgrounds for $Z \to \ell^+\ell^-$ and $W \to \ell\nu$ selections at the level of better than 1%. As I also mentioned, there is currently a lot of theoretical activity that should allow the production cross sections of W and Z to be predicted at the level of about 1%.[5] Therefore, there is every likelihood that the *overall* luminosity scale for the general physics programme will be set by measuring the numbers of observed $Z \to \ell^+\ell^-$ and $W \to \ell\nu$ events. Of course, we still need as precise as possible measurements of the total rate of inelastic collisions and knowledge of $\sigma_{\mathrm{inelastic}}$. Measurements of the total inelastic rate will certainly be needed for measurement of *relative* luminosities, minute by minute and run by run. They are also essential for the kind of detailed book-keeping and stability checks needed for precision measurements. In addition, we should like to test the predictions for the Z and W production cross sections by having an alternative absolute luminosity measurement (albeit with an accuracy likely to be limited at the level of about 5%).

Un-Ki Yang (Univ. of Chicago): The DØ measurement of $R_\mu = \dfrac{\sigma_W \cdot \mathrm{Br}(W \to \mu\nu)}{\sigma_Z \cdot \mathrm{Br}(Z \to \mu^+\mu^-)}$ seems a little high, because of a high $\sigma_W \cdot \mathrm{Br}(W \to \mu\nu)$. In CDF we learned that there are many high p_T muons from low p_T kaon decays inside the tracking volume. I wonder how this background is removed.

Terry Wyatt: Firstly, the value of R_μ from DØ is not inconsistent with the other measured values of R_ℓ. A Kolmogorov-Smirnov test gives a 20% probability for consistency of the set of R_ℓ measurements. Secondly, the level of background from pion and kaon decay in flight is very small in DØ due to the small radius of the tracking detectors. We'd expect the methods used to evaluate QCD backgrounds to the $W \to \mu\nu$ sample to take any residual background into account.

THE TOP PRIORITY: PRECISION ELECTROWEAK PHYSICS FROM LOW TO HIGH ENERGY

P. GAMBINO

Theory Division - CERN, CH 1211, Geneve 23, Switzerland
E-mail: paolo.gambino@cern.ch

Overall, the Standard Model describes electroweak precision data rather well. There are however a few areas of tension (charged current universality, NuTeV, $(g-2)_\mu$, b quark asymmetries), which I review critically, emphasizing recent theoretical and experimental progress. I also summarize what precision data tell us about the Higgs boson and new physics scenarios. In this context, the role of a precise measurement of the top mass is crucial.

Precision electroweak physics lies at the intersection of many specialized fields and involves experiments performed at hugely different energies. In testing the consistency between all data within the Standard Model (SM) framework, we hope to uncover signs of physics beyond the SM. However, as we will see, the main problem is the precision of the theoretical predictions with which we confront the experimental data. Almost invariably, long-distance hadronic interactions enter the game, so we often take great pains to try to make sense of extremely precise experiments.

In the following I will try to summarize the main recent progress in the field, concentrating on the unsettled questions. For some of the topics I will not have space to cover, see the References.[1,2]

1. Parity Violation in Møller Scattering

Let us start from low-energy experiments. The E158 experiment at SLAC[3] has measured for the first time parity violation (PV) in polarized Møller (e^-e^-) scattering. The PV asymmetry $A_{LR} = (\sigma_L - \sigma_R)/(\sigma_L + \sigma_R)$ is extremely small in the SM, $\approx 10^{-7}$, due to an extra suppression factor $1/4 - \sin^2\theta_W$. It can be measured at SLAC thanks to the huge luminosity and the high polarization of the beam. A_{LR} is very sensitive to $\sin^2\theta_W$ and the goal of E158 is to measure it with 8% precision, equivalent to an error of 0.001 on $\sin^2\theta_W$. Such a precision is not competitive with LEP and SLD determinations, but one should keep in mind that a low-energy measurement would test completely different radiative corrections, and would be sensitive to new physics complementary or orthogonal to collider experiments.

E158 is currently performing a last and third run and expect to be able to reach the aimed precision. The preliminary result of Run I (at $Q^2 = 0.027$ GeV^2),

$$A_{LR} = [151.9 \pm 29.0(stat) \pm 32.5(syst)] \times 10^{-9},$$

translates into $\sin^2\hat{\theta}_W^{\overline{MS}}(M_Z) = 0.2296 \pm 0.0038$, in good agreement with the global average, $\sin^2\hat{\theta}_W^{\overline{MS}}(M_Z) = 0.2312 \pm 0.0003$. Radiative corrections[4] reduce A_{LR} by about 40%. A large theoretical uncertainty comes from the $\gamma - Z$ hadronic vacuum polarization, which cannot be computed perturbatively. The current estimate, inducing $\approx 5\%$ error on A_{LR}, can and should be updated in view of E158's final result, expected next year.

2. Universality of Charged Currents

This is a very old subject.[5] Universality in the leptonic sector is verified at the 0.2% level.[6] Charged currents in the quark sector, on the other hand, involve also the CKM matrix elements. One can however test accurately the unitarity relation

$$|V_{ud}|^2 + |V_{us}|^2 + |V_{ub}|^2 = 1. \qquad (1)$$

Since the last term on the lhs is $O(10^{-5})$, the test concerns the consistency of Cabibbo angle measurements from V_{ud} and V_{us}.

The most precise method to measure V_{ud} is to use Superallowed Fermi Transitions, i.e. $0^- \to 0^-$ nuclear β decays. There are several experiments in good agreement, yielding $\delta V_{ud} \sim 0.0005$. Neutron β decay is also becoming competitive: the present $\delta V_{ud} \sim 0.0013$ will be improved at PERKEO.[7] A promising mode is pion decay, currently at $\delta V_{ud} \sim 0.005$, which is theoretically cleaner and will soon be improved by PIBETA.[8] The consistent picture that emerges from these experiments can be expressed, using Eq. (1), as

$$|V_{us}|(\text{unitarity}) = 0.2269 \pm 0.0021. \qquad (2)$$

The most precise direct measurement of $|V_{us}|$ is given on the other hand by $K \to \pi\ell\nu$ decays ($K_{\ell3}$). Here the experimental situation is not as consistent as for V_{ud}: the recent E865 result for K^+ decays disagrees with a series of old experiments by more than 2σ. While the E865 result agrees well with the unitarity prediction, Eq. (2), the older results and a recent preliminary K^0 measurement by KLOE all yield a smaller Cabibbo angle. Upcoming analyses from KLOE, NA48, and KTeV should tell us whether grossly underestimated isospin breaking corrections are the cause of this situation, or there is an experimental problem. Averaging the old published data only, one obtains

$$|V_{us}|_{K_{\ell3}} = 0.2201 \pm 0.0024,$$

but the result changes little if one includes also E865 and KLOE results. Alternative promising strategies to extract $|V_{us}|$ are provided by τ and hyperon decays. In particular, measurements of the τ spectral functions at the B factories will make the first method competitive with $K_{\ell3}$, while the use of hyperon decays requires a careful assessment of SU(3) breaking effects, which could be helped by lattice simulations.

In summary, a puzzling violation of unitarity persists at the level of $\sim 2\sigma$, despite new data. Fortunately, upcoming experimental results are likely to shed light on this problem. For a more detailed discussion, see elsewhere.[9]

3. The NuTeV Electroweak Result

NuTeV measures ratios of Neutral (NC) to Charged Current (CC) cross sections in νN DIS.[10] Ideally, in the parton model with only one generation of quarks and an isoscalar target

$$R_\nu \equiv \frac{\sigma(\nu N \to \nu X)}{\sigma(\nu N \to \mu X)} = g_L^2 + r g_R^2,$$

$$R_{\bar\nu} \equiv \frac{\sigma(\bar\nu N \to \bar\nu X)}{\sigma(\bar\nu N \to \bar\mu X)} = g_L^2 + \frac{1}{r} g_R^2,$$

where $r \equiv \frac{\sigma(\bar\nu N \to \bar\mu X)}{\sigma(\nu N \to \mu X)}$ and $g_{L,R}^2$ are average effective left- and right-handed ν-quark couplings. The actual experimental ratios $R_{\nu,\bar\nu}^{exp}$ differ from $R_{\nu,\bar\nu}$ because of ν_e contamination, experimental cuts, NC/CC misidentification, the presence of second generation quarks, the non-isoscalarity of steel target, QCD and electroweak corrections, etc.. In the NuTeV analysis, a Monte Carlo including most of these effects relates $R_{\nu,\bar\nu}^{exp}$ to $R_{\nu,\bar\nu}$. It is useful to note that most uncertainties and $O(\alpha_s)$ effects drop out in the Paschos-Wolfenstein (PW) ratio[11]

$$R_{PW} \equiv \frac{R_\nu - r R_{\bar\nu}}{1 - r} = \frac{\sigma(\nu N \to \nu X) - \sigma(\bar\nu N \to \bar\nu X)}{\sigma(\nu N \to \ell X) - \sigma(\bar\nu N \to \bar\ell X)}$$

which equals $g_L^2 - g_R^2 = \frac{1}{2} - \sin^2\theta_W$ and therefore could provide a clean measurement of $\sin^2\theta_W$, if experimentally accessible. NuTeV do not measure R_{PW} directly, but, using the fact that $R_{\bar\nu}$ is almost insensitive to $\sin^2\theta_W$, they extract from it the main hadronic uncertainty, an effective charm mass. The weak mixing angle is then obtained from R_ν. In practice, NuTeV fit for m_c^{eff} and $\sin^2\theta_W$. This procedure certainly approximates a measurement of R_{PW}, but it is not clear to what extent exactly.

The NuTeV result provides a test of the on-shell $s_W^2 \equiv 1 - M_W^2/M_Z^2$ definition of $\sin^2\theta_W$:

$$s_W^2(\text{NuTeV}) = 0.2276 \pm 0.0013 \pm 0.0006 \pm 0.0006, \quad (3)$$

where the three errors are statistical, systematic and theoretical respectively. Because of accidental cancellations, the choice of the on-shell scheme implies very small top and Higgs mass dependences in Eq. (3). The above value must be compared to the result of the global fit, $s_W^2 = 0.2229 \pm 0.0004$, which is 2.8σ away.

NuTeV works at Leading Order (LO) in QCD in the context of a *cross section model* which effectively introduces some Next-to-Leading-Order (NLO) improvement. They use LO PDF's self-consistently fitted in the experiment, with little external input. There are a number of theoretical systematics which could have been underestimated in Eq. (3), and considerable work has been devoted to study the most obvious among them.

i) *Uncertainties in the parton distribution functions* (PDF's): neglecting for the moment asymmetric sea contributions (see later) they are small in R_{PW} with the cuts used.[12,13]

ii) *NLO QCD corrections:*[13−15] vanish in R_{PW}, and effects introduced by asymmetric cuts and differences in the $\nu, \bar\nu$ energy spectra seem small. Again, this refers to the ideal observable R_{PW}.

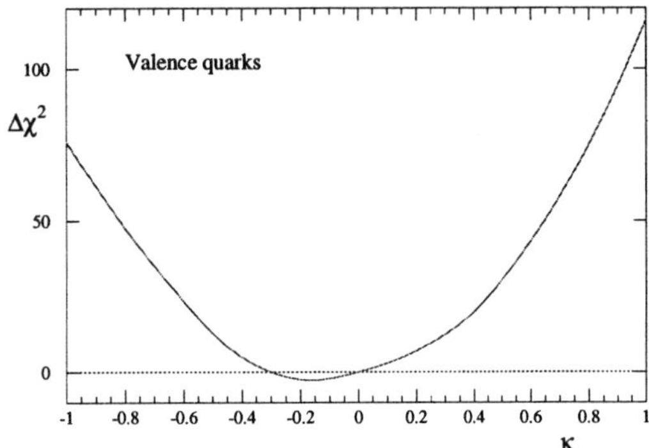

Figure 1. MRST fit of isospin violation in valence PDF's.

Only a complete NLO analysis can ensure that the same conclusions apply to the NuTeV fit. For instance, the phenomenological cross section model used by NuTeV may distort in an important way cancellations among QCD corrections.[15] Estimating the actual effect on s_W^2 would require refitting the PDF's at NLO. In summary, the analysis needs to be consistently upgraded to NLO, and the NuTeV collaboration is investigating this possibility.

iii) Electroweak corrections (mainly photonic): the NuTeV analysis is largely based on very old code, which needs to be checked against recently developed tools.[16]

3.1. *Asymmetric Sea*

I have so far used the assumptions, generally made in the extraction of PDF's from the data, of isospin symmetry and of a symmetric strange and charm sea ($s = \bar{s}$, $c = \bar{c}$). If we drop these assumptions, the PW relation is explicitly violated by new terms[12]

$$R_{PW} = \frac{1}{2} - s_W^2 + \frac{\tilde{g}^2}{Q^-}\,(u^- - d^- + c^- - s^-), \quad (4)$$

where q^- is the asymmetry in the momentum carried by the quark species q in an isoscalar target, $q^- = \int_0^1 x\,[q(x) - \bar{q}(x)]\,dx$, $\tilde{g}^2 \approx 0.23$ a coupling factor, and $Q^- = (u^- + d^-)/2 \approx 0.18$. The non-isoscalarity of the target gives a contribution to $u^- - d^-$ that is obviously taken into account by NuTeV, although the uncertainty on this correction seems to have been somewhat underestimated.[17] There are, however, less standard and potentially more dangerous contribu-

tions: there is no reason in QCD to expect $s^- = 0$, and for an isoscalar target $u^- - d^-$ is of the order of the isospin violation. Eq. (4) tells us that even quite small values of these two asymmetries could change significantly the value of s_W^2 measured by NuTeV.[a]

A violation of isospin of the form $u_p(x) \neq d_n(x)$ would induce a u^- different from d^- even in an isoscalar target and affect the PW relation according to Eq. (4). A rough estimate for its size is $(m_u - m_d)/\Lambda_{QCD} \approx 1\%$. This could explain a fraction of the anomaly – about a third, according to Eq. (4). Isospin violation is very weakly constrained by experiment, as demonstrated by a new MRST analysis.[18] MRST have performed a global fit to the PDF's deforming the valence distributions by a contribution proportional to a function, $f(x)$, with zero first moment: $u_n^-(x) = d_p^-(x) + \kappa f(x)$ and $d_n^-(x) = u_p^-(x) - \kappa f(x)$. The fit to κ, shown in Fig. 1, gives a mild indication for a negative κ, but with very large uncertainty (MRST use $\Delta\chi^2 = 50$ to define a 90% CL). The central value $\kappa \approx -0.2$ corresponds to a reduction of the NuTeV anomaly by about a third, and has the expected order of magnitude. Amusingly, the MRST central value leads to a shift in s_W^2 very close to that of a recent analysis in the context of nucleon models.[19] Using similar models, NuTeV claim a much smaller isospin breaking shift.[20] In any case, it is clear that model calculations,[21] though sometimes useful to understand the size of an effect, cannot be relied upon for a precision measurement. We are therefore left with a substantial uncertainty unaccounted for in Eq. (3).

What do we know about the strange quark asymmetry? An asymmetry s^- of the sign needed to explain NuTeV can be induced non-perturbatively (*intrinsic strange*) by fluctuations of the kind $p \leftrightarrow \Lambda K^+$.[22] Unfortunately, the strange quark sea is mainly constrained by (mostly old) νN DIS data, which are usually not included in standard PDF's fits. In fact, MRST and CTEQ use an *ansatz* $s = \bar{s} = (\bar{u} + \bar{d})/4$. Barone *et al.* (BPZ)[23] reanalyzed, a few years ago, a host of νN DIS together with ℓN and Drell-Yan data at NLO. Allowing for a strange asymmetry improved the BPZ best fit drastically and could explain a large fraction of the NuTeV discrepancy. The result, $s^- \approx 0.0018 \pm 0.0005$, was compat-

---

[a] These effects are somewhat diluted in the actual NuTeV analysis compared to the direct use of Eq. (4),[20] precisely because NuTeV differs from a measurement of R_{PW}.

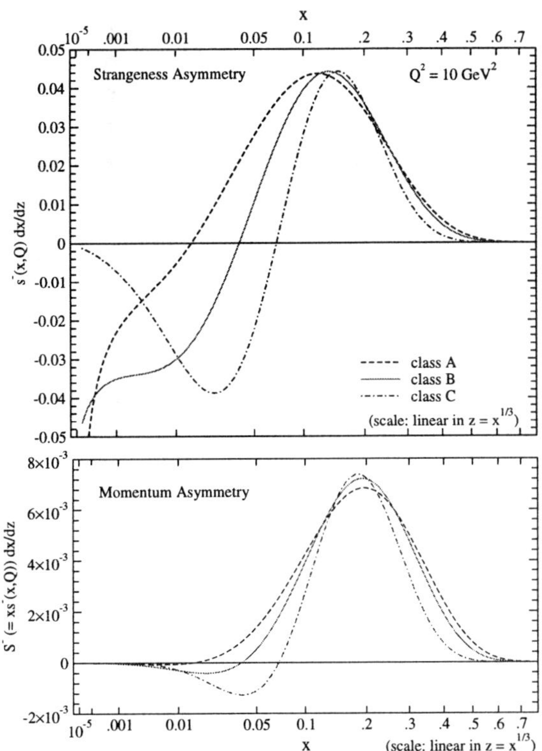

Figure 2. CTEQ fit of the strangeness asymmetry using different low-x behaviors.

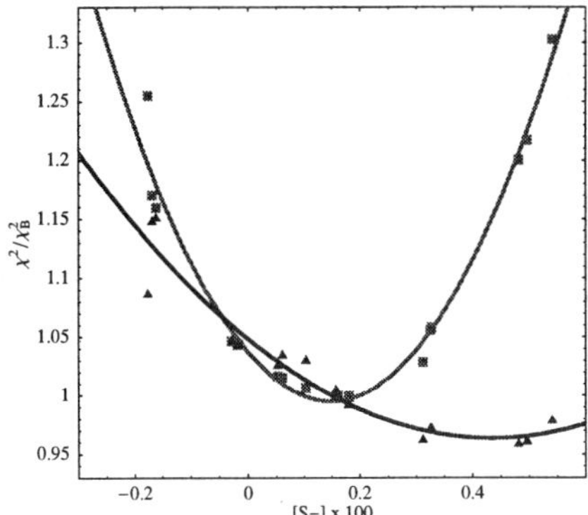

Figure 3. CTEQ fit of the strange momentum asymmetry s^-.

ible with theory estimates[22] and was driven by cross section measurements by CDHSW (νN) and BCDMS (μp). The BPZ analysis was recently updated with the inclusion of CCFR cross sections, leading to a quite different result, $s^- \approx 0.0002 \pm 0.0004$.

The inclusive analysis pioneered by BPZ should however be supplemented by data on dimuon events (tagged charm production), a rather sensitive probe of the strange sea. The most precise dimuon data come from the CCFR/NuTeV Collaboration,[24] which has analyzed them at LO with the specific aim of constraining the strange asymmetry. Their result, $s^- = -0.0027 \pm 0.0013$, would $increase$ the anomaly to 3.7σ,[20] but it suffers from various shortcomings, detailed in the note added to S. Davidson $et\ al.$[12] and elsewhere.[25] The main problem is in the parameterization, which does not satisfy the condition

$$\int_0^1 dx[s(x) - \bar{s}(x)] = 0 \qquad (5)$$

that ensures zero strangeness quantum number for the nucleon. As the dimuon data are concentrated at $x < 0.3$, the evidence for a small negative strange asymmetry at low x would imply, if the condition

given by Eq. (5) is imposed, a $positive$ asymmetry at large x, and hence a $positive$ momentum asymmetry. This is illustrated in Fig. 2,[25] which shows strange asymmetries with the above qualitative features but different shapes. The NuTeV analysis of dimuon data is not reliable.

A dedicated global fit that employed both inclusive and dimuon data in a consistent way was therefore necessary. The CTEQ Collaboration has presented at this conference the preliminary results of one such analysis.[25] The inclusion of the CCFR-NuTeV dimuon data in the CTEQ global fit is presently done using NuTeV software developed at LO in QCD. Dimuon data are therefore included at LO, which should not influence the main qualitative conclusions. CTEQ explored the full range of parameterizations of $s(x) - \bar{s}(x)$ that satisfy Eq. (5), studying for instance different low-x behavior, as shown in Fig. 2. They perform a new global fit to all PDF's using all available inclusive and dimuon data, although they do not reanalyze old νN data in detail, as was done by BPZ. The preliminary result of the s^- fit is shown in Fig. 3 for the best performing (class B) parameterization. While inclusive data alone show only a mild preference for a positive s^-, the dimuon data have real discriminating power. The central value of the global class B fit is $s^- \approx 0.002$, and corresponds to the indicated line in Fig. 2. In general, all acceptable fits have central values $0.001 < s^- < 0.003$. Negative s^- are disfavored, but $s^- = 0$ cannot be excluded. CTEQ estimate that the likely impact on the NuTeV s^2_W extraction would be a reduction of s^2_W by 0.0012 to 0.0037. Note that if a strange asymmetry shifted s^2_W by 0.002 ± 0.002, the NuTeV result would be at 1σ from the SM. Although a more detailed study is under way with the active participation of the NuTeV Collaboration, two firm conclusions are that: i) the strange asymmetry is a strong candidate to explain part or most of the NuTeV anomaly; and ii) one cannot avoid the related, substantial uncertainty.

Given the present understanding of hadron structure, R_{PW} does not seem to be a good place for high precision electroweak physics. In fact, the relevant momentum asymmetries in the quark sea induce an error in the extraction of s_W^2 of the same order as the experimental error. Improved analyses of dimuon data would certainly constrain s^- better, and data from CHORUS might also be useful – if not for measuring s_W^2, at least for constraining the sea asymmetries.[26] Useful input could also come from associated charm-W production at the Tevatron and RHIC. In the long term, a precise $s(x), \bar{s}(x)$ determination will be possible at a neutrino factory.[27]

I should also mention that several attempts at explaining the NuTeV anomaly with nuclear effects like nuclear shadowing have been made,[28] but no convincing case has so far been presented.

3.2. *New Physics vs NuTeV*

A new physics explanation of the NuTeV anomaly requires a $\sim$ 1-2% effect, and naturally calls for tree level physics. It is very difficult to build realistic models that satisfy all present experimental constraints and explain a large fraction of the anomaly.[12]

In particular, Supersymmetry, with or without R-parity, cannot help, because it is strongly constrained by other precision measurements (often at the 10^{-3} level) and by direct searches.[12,29] The same is generally true of models inducing only oblique corrections or only anomalous Z couplings.[12] Realistic and well-motivated examples of the latter are models with ν_R mixing.[12,31] Models with ν_R mixing *and* oblique corrections have been considered by W. Loinaz *et al.*[30] and found to fit well all data including NuTeV.[b] However finding sensible new physics that provides oblique corrections in the preferred range is far from obvious.

____

[b]Can the necessary oblique corrections be provided by a heavy SM Higgs boson? No, the only way to obtain an acceptable fit with a preference for both ν mixing and a heavy Higgs is to exclude M_W from the data.[30] However, solving the NuTeV anomaly at the expense of the very precise measurement of M_W is hardly an improvement.

On the other hand, the required new physics can be parameterized by a contact interaction of the form $[\bar{L}_2\gamma_\mu L_2][\bar{Q}_1\gamma_\mu Q_1]$. This operator might be induced by different kinds of short-distance physics. Leptoquarks generally also induce another operator which over-contributes to $\pi \to \mu\bar{\nu}_\mu$, or have the wrong sign, but SU(2) triplet leptoquarks with non-degenerate masses could fit NuTeV, albeit not very naturally. Another possible new physics contribution inducing the above contact interactions is an unmixed Z' boson. It could be either light ($2 \lesssim M_{Z'} \lesssim 10$ GeV) and super-weakly coupled, or heavy ($M_{Z'} \gtrsim 600$ GeV). The Z' must have very small mixing with the Z^0 because of the bounds on oblique parameters and on the anomalous Z couplings[12,2] (see E. Ma and D. P. Roy[46] for an explicit $L_\mu - L_\tau$ model and R. S. Chivukula and E. H. Simmons[47] for technicolor models).

4. The Ups and Downs of $(g - 2)_\mu$

The anomalous magnetic moment of the muon is an excellent place to look for new physics: it probes unexplored loop effects proportional to m_μ^2/Λ^2, where Λ is the mass scale characteristic of the new physics. Given the present experimental resolution, in order for us to observe large deviations from the SM, the new physics we need must have a chiral enhancement, of the kind naturally emerging in Supersymmetric models with large $\tan\beta$. Conversely, no deviation from the SM would impose severe constraints on these models. This is at the origin of the great attention this observable has recently received.

The last few years have seen a dramatic progress in the measurement of a_μ, driven by the $g - 2$ experiment at Brookhaven. The present world average

$$a_\mu(\text{w.a.}) = 11659203(8) \times 10^{-10}$$

is dominated by their latest μ^+ result,[32] released in 2002. The results of the 2001 Run, performed with μ^-, should reduce the error by $\sim 30\%$ and are expected soon.

Figure 4 summarizes the evolution of the measurement and of the theoretical estimates of a_μ. As you will see in a moment, the theoretical prediction of this quantity depends heavily on other experimental results, so the ups and downs are mostly due to the evolution of data and to the corrections of some unfortunate mistakes.

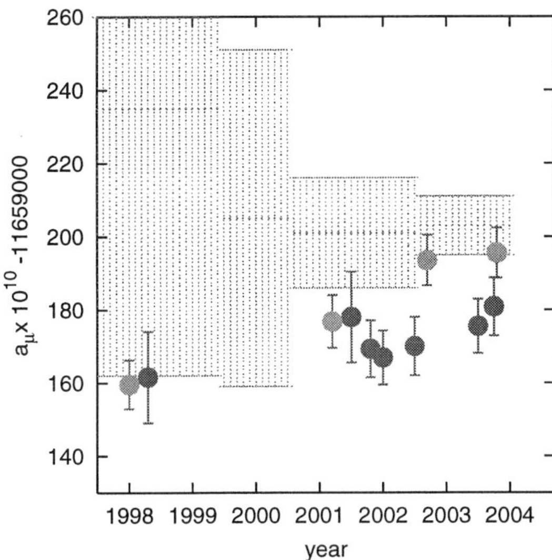

Figure 4. Historical evolution of the measurement and prediction of a_μ. The lighter (green) dots represent estimates based also on τ decay data. The compilation of theoretical estimates is not complete.

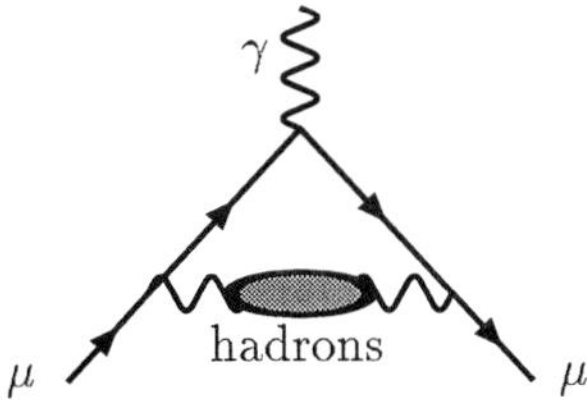

Figure 5. Lowest order vacuum polarization insertion.

While most of us have computed the lowest-order QED contribution to a_μ at graduate school, a calculation of a_μ at the current level of precision is a very involved and sophisticated enterprise (there are excellent reviews,[33] with references to the original literature). Here I will concentrate only on the general aspects and on recent developments. The various contributions to a_μ, listed with their estimated errors, are:

$$
\begin{aligned}
a_\mu = {}& 11\,658\,470.35(28) \times 10^{-10} \, (\text{QED}) \\
& + 694(7) \times 10^{-10} \, (\text{had, Leading Order}) \\
& - 10.0(6) \times 10^{-10} \, (\text{had, Higher Order}) \quad (6) \\
& + 8(4) \times 10^{-10} \, (\text{had, Light by Light}) \\
& + 15.4(2) \times 10^{-10} \, (\text{EW})
\end{aligned}
$$

The main component comes from QED without hadronic loops. The four-loop contribution[34] is not so small, $\sim 40 \times 10^{-10}$, and has never been checked. But these heroic calculations at least can be done. Not so for the hadronic contributions: hadronic loops

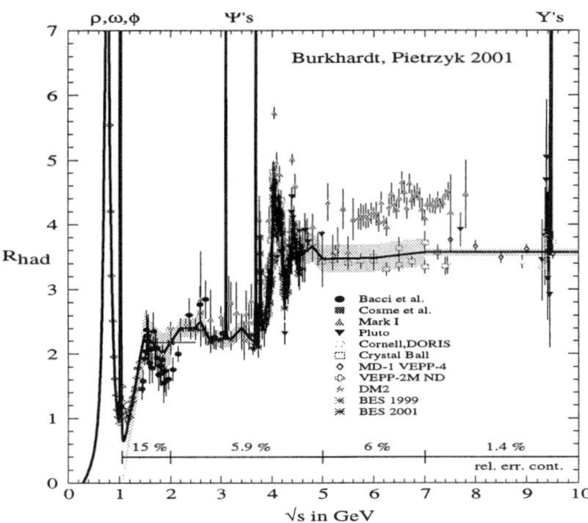

Figure 6. The spectral function.

enter the second order diagram of Fig. 5 and are characterized by the scale $\Lambda_{QCD} \approx 300$ MeV. They provide the largest uncertainty to the determination of a_μ. As the energy scale is too low to employ perturbative methods, the usual route is to use a dispersive integral of the vector spectral function,

$$
a_\mu^{\text{LO,had}} = \frac{1}{4\pi^3} \int_{4m_\pi^2}^{\infty} R_{\text{had}}(s) K(s) ds \qquad (7)
$$

where the spectral function $R_{had}(s)$ is measured from the total hadronic cross section in $e^+ e^-$ collisions. A number of experiments have contributed to its measurement, most recently CMD-2, SND, and BES, leading to the situation summarized in Fig. 6. Different strategies are also available to combine the

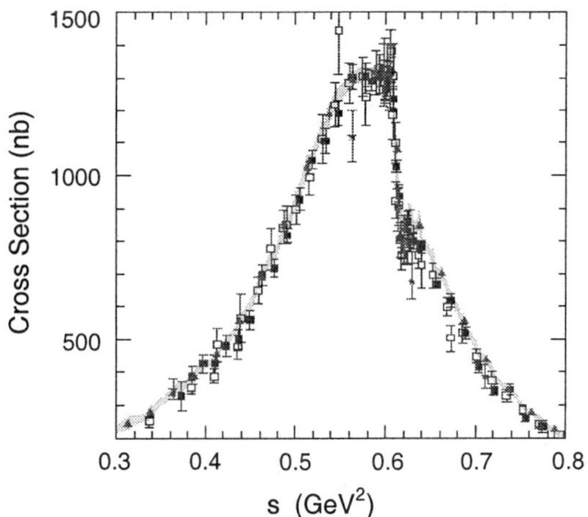

Figure 7. The pion form factor.

data and their errors – see the References[35−38] for

the most recent and complete analyses. Because of the weight function $K(s)$, the integral given by Eq. (7) is dominated by the low energy region, and in particular by the ρ resonance in the $\pi\pi$ channel. Indeed, the pion form factor (see Fig. 7) alone contributes more than 70% of $a_\mu^{\mathrm{LO,had}}$. The recent CMD-2 reanalysis[41] of their very precise $\pi\pi$ data, with a revised treatment of QED corrections, is therefore of the utmost importance. It is included in the following updated estimates:[36−38]

$$a_\mu^{\mathrm{LO,had}}(\mathrm{HMNT}) = (691.8 \pm 5.8_{exp} \pm 2.0_{r.c.}) \times 10^{-10}$$
$$a_\mu^{\mathrm{LO,had}}(\mathrm{DEHZ}) = (696.3 \pm 6.2_{exp} \pm 3.6_{r.c.}) \times 10^{-10}$$
$$a_\mu^{\mathrm{LO,had}}(\mathrm{GJ}) = (694.8 \pm 8.6) \times 10^{-10} \qquad (8)$$

where the r.c. error is mostly due to uncertainty in correcting old data for missing radiative corrections. Adding all other SM contributions, this translates into a 1.9-2.5σ discrepancy between SM prediction and experiment.

A second way of measuring the spectral function in the crucial region below 1.8 GeV consists of relating the τ hadronic decays to the e^+e^- hadronic cross section using CVC and isospin symmetry, as schematically illustrated in Fig. 8. This method has

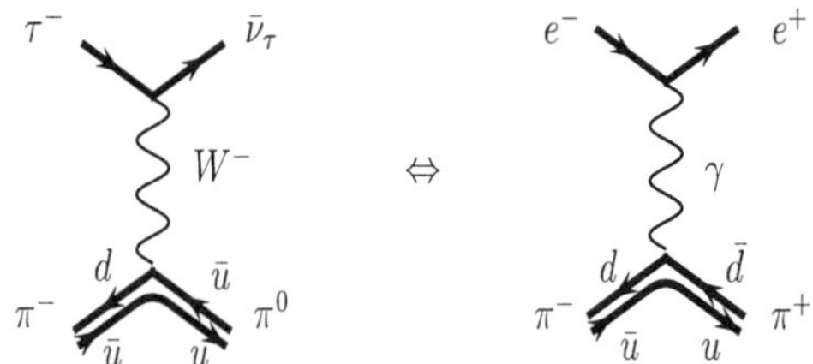

Figure 8. The diagrams relating τ hadronic decays to the e^+e^- hadronic cross section.

been explored by M. Davier *et al.*[37] with data from Aleph, CLEO, and Opal. A series of corrections have been implemented,[39] leading to

$$a_\mu^{\mathrm{LO,had}}(\mathrm{DEHZ},\tau) = (709.0 \pm 5.1_{exp} \pm 1.2_{r.c.} \pm 2.8_{\mathrm{SU}(2)})$$

where the last uncertainty refers to the isospin corrections. This determination is competitive with e^+e^- and leads to a prediction of a_μ in much better agreement with experiment (0.7σ). Figure 9 from M. Davier *et al.*[37] shows a comparison of the spectral function extracted from e^+e^- and τ data: although the CMD2 revision has much improved the situation below 850 MeV, there is still a discrepancy between 0.85 and 1 GeV. The problem could be in the data

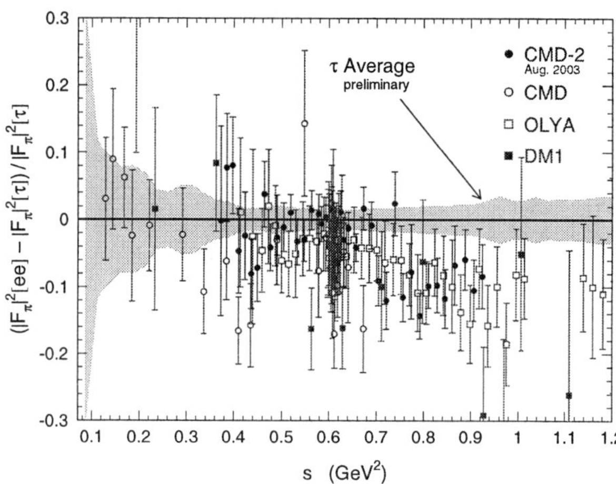

Figure 9. Comparison of the spectral function extracted from τ decay and e^+e^- data (relative difference).

or in the theoretical framework. While a recent paper advocates the second possibility,[38,40] hinting at underestimated isospin breaking, an important check of the CMD2 e^+e^- results has come from the first results of a third method to measure the spectral function, the radiative return.

The idea behind radiative return is that a photon radiated off the initial e^+ or e^- (ISR) reduces the effective energy of the collision, see Fig. 10. Provided

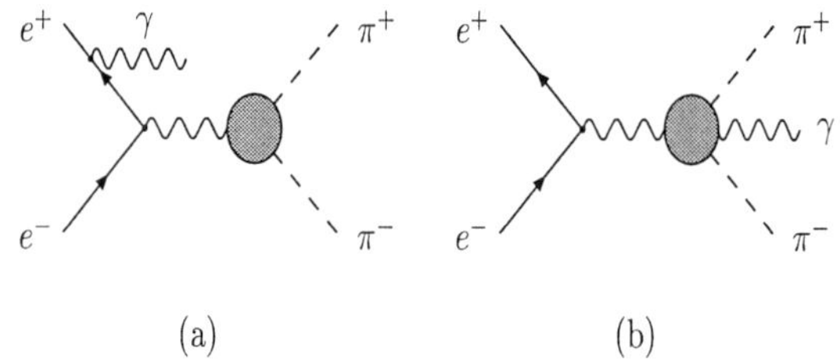

Figure 10. Examples of ISR and FSR.

the photon momentum is measured, a fixed energy collider can investigate a whole q^2 range, with obvious advantages over the energy scan experiments. The large luminosities at DAΦNE and at the B-factories compensate the radiative suppression. The potential pollution from FSR at low-energy (see Fig. 10(b)) is circumvented by kinematic cuts. Radiative corrections[43] play a crucial role here, as they do anyway in the energy scan case. KLOE has announced[42] the first preliminary results of radiative return: the contribution of the two pions chan-

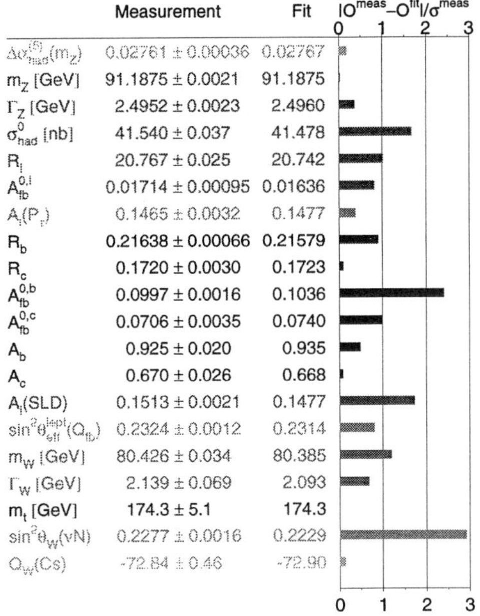

Figure 11. Pulls in the summer 2003 fit by the LEP Electroweak Working Group.

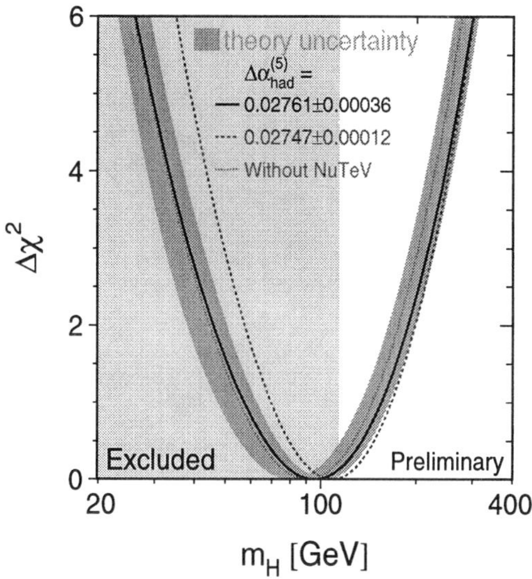

Figure 12. The parabolic darker (blue) band summarizes our indirect information on M_H, while the lighter shaded (yellow) area is excluded by LEP.

nel to $a_\mu^{\mathrm{LO,hadr}}$ in the range $0.37 < q^2 < 0.93$ GeV^2 is, in units 10^{-10}, $378.4 \pm 0.8_{stat} \pm 4.5_{syst} \pm 2.6_{th} \pm 3.8_{FSR}$, to be compared with the *new* CMD2 result $378.6 \pm 2.6_{stat} \pm 2.2_{syst\&th}$. KLOE agrees well with CMD2. The systematic error will soon be reduced by $\sim$25%. Radiative return analyses are also expected from Babar, Belle and CLEO-c.

5. The SM Fit and the Higgs Boson Mass

The latest compilation of electroweak data of the LEP Electroweak Working Group[1] is shown in Fig. 11, where the data are compared with the results of a global fit. The main changes with respect to last year are: a revised (lower) M_W value from Aleph, that draws the world average down by 0.5σ, to $M_W(w.a.) = 80.426 \pm 0.034$ GeV and improves the consistency of the global fit; small shifts in the heavy flavor observables; and a new value of atomic PV, due to revised (and hopefully converging) theoretical calculations. The value for M_t, 174.3 ± 5.1 GeV, is the old one, and does not include the new D0 analysis.[44] Also the value of $\alpha(M_Z)$, from the conservative estimate,[45] has not yet been updated to reflect the new CMD2 data, a rather small effect anyway (the new value is $\Delta\alpha_{had} = 0.02768 \pm 0.00036$).

Indeed, the spectral function discussed in the previous section enters also the determination of $\alpha(M_Z)$, but higher energy data have more weight. Considerable progress has been achieved in the last few years, and this uncertainty is no more a bottleneck for the present bounds on M_H.

The χ^2/d.o.f. of the global fit is 25.4/15, corresponding to 4.5% probability. The NuTeV result shares the responsibility for the degradation of the fit with another deviant measurement, that of the bottom quark Forward-Backward asymmetry, A_{FB}^b, at LEP. The best fit[1] points to a fairly light Higgs boson, with mass $M_H = 96$ GeV, while the 95% CL upper bound on M_H, including an estimate of theoretical uncertainty, is about 220 GeV. As the uncertainty used for NuTeV is the one given by the experiment, let us consider the fit performed excluding this result. The information on the Higgs mass is almost insensitive to the NuTeV result ($M_H^{fit} = 91$ GeV, $M_H < 202$ GeV at 95%), but of course the quality of the fit improves significantly, with χ^2/d.o.f.=16.8/14, corresponding to 26.5%. One would conclude that the SM fit is quite satisfactory. The direct and indirect information on the Higgs mass are summarized in Fig. 12, where the lighter shaded (yellow) area, $M_H < 114.4$ GeV, is excluded by LEP.

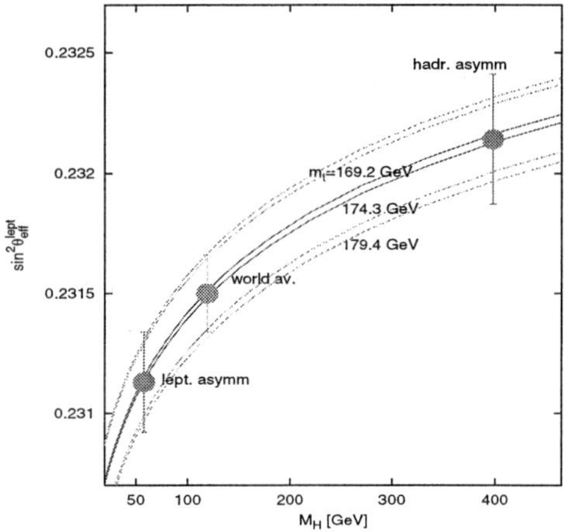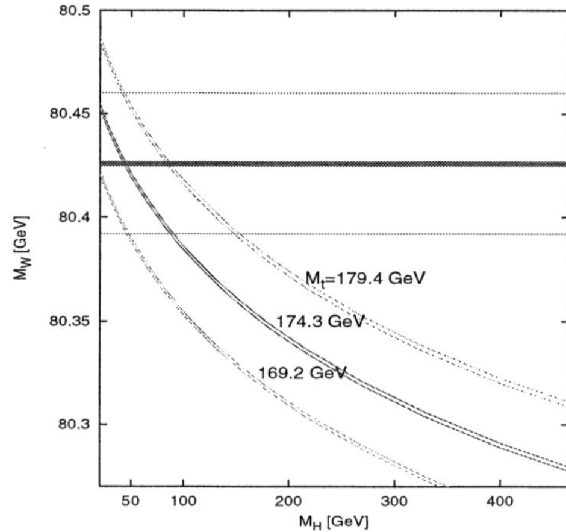

Figure 13. Higgs mass dependence of $\sin^2\theta_{\text{eff}}^{lept}$ – extracted from leptonic and hadronic asymmetries – and of M_W, for three M_t values, compared with the experimental values.

We have noted that, excluding NuTeV, the data are rather consistent. But the table in Fig. 11 contains an arbitrary set of observables. For example, it does not include a_μ, or $B \to X_s\gamma$, which are important and precise data. The overall conclusion would not change, in my view. However, if we are interested in extracting information on the Higgs mass, we should concentrate only on the subset of observables that are really sensitive to M_H and, because of a strong correlation, to the top mass, M_t. Using only M_W, M_t, Γ_ℓ, the Z-pole asymmetries, and R_b, one obtains $M_H^{fit} = 98$ GeV, $M_H < 210$ GeV at 95% CL, and $\chi^2/\text{dof}=11/4$, corresponding to 2.6% probability. In other words, the restricted fit gives the same constraints on M_H of the global fit. However, it is now obvious that the SM fit to the Higgs mass is *not* really satisfactory.

The root of the problem is an old 3σ discrepancy between the Left-Right asymmetry, A_{LR}, measured by SLD and the Forward-Backward b quark asymmetry, A_{FB}^b, measured by the LEP experiments. In the SM these asymmetries measure the *same* quantity, $\sin^2\theta_{\text{eff}}^{lept}$, related to the lepton couplings to an on-shell Z^0. It now happens that all leptonic asymmetries, measured both at LEP and SLD, are mutually consistent and prefer a very *light Higgs* mass – see Fig. 13. In this sense, they are also consistent with M_W measured at LEP and Tevatron. Only the asymmetries into hadronic final states prefer a *heavy Higgs* (see Fig. 13).

Since the hadronic asymmetries are dominated by A_{FB}^b, and the third generation is naturally sin-

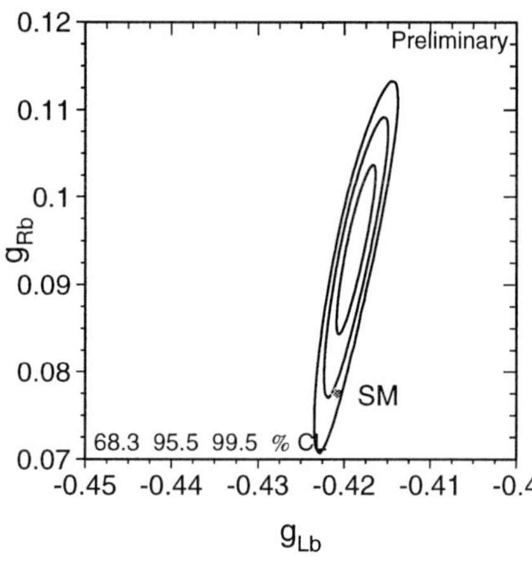

Figure 14. Fit to the left- and right-handed b quark neutral current couplings.

gled out in many extensions of the SM, could this be a signal of new physics in the b couplings? After all, QCD and experimental systematics in A_{FB}^b have been carefully considered.[1] New physics in the b couplings seems unlikely for several reasons: (i) fixing $\sin^2\theta_{\text{eff}}^{lept}$ at the value measured by the leptonic asymmetries, A_{FB}^b corresponds to a measurement of a combination of b couplings, $\mathcal{A}_b(A_{FB}^b) = 0.886 \pm 0.017$; the same combination is also tested by A_{LR}^{FB} at SLD, yielding $\mathcal{A}_b(A_{LR}^{FB}) = 0.922 \pm 0.020$. One should compare these two values to the very precise SM prediction, $\mathcal{A}_b^{SM} = 0.935 \pm 0.002$: the SLD result is compatible with the SM and at 1.4σ from the value extracted from A_{FB}^b; (ii) the value of $\mathcal{A}_b$ extracted from A_{FB}^b would require a $\sim 25\%$ correction to the b vertex, i.e. tree level physics; and (iii)

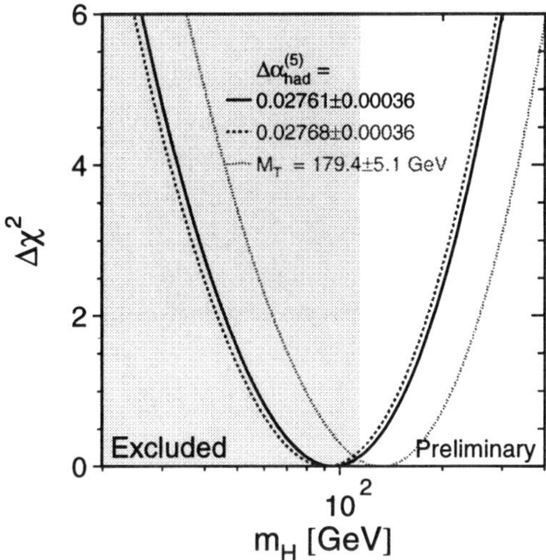

Figure 15. Effect of a 1σ change in M_t on the Higgs mass constraints.

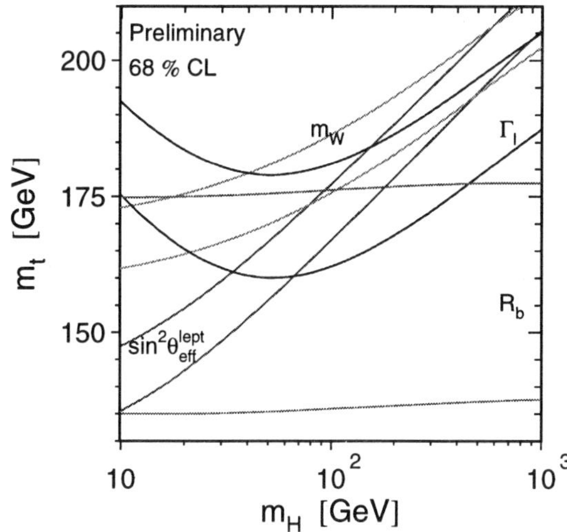

Figure 16. Constraints in the M_t, M_H plane from different observables.

R_b agrees well with the SM and tests an orthogonal combination of b couplings; it follows that new physics should predominantly affect the right-handed b coupling, $|\delta g_R^b| \gg |\delta g_L^b|$, see Fig. 14. All this places strong restrictions on the extensions of the SM that can explain A_{FB}^b. Exotic scenarios that shift only the b_R coupling include mirror vector-like fermions mixing with the b quark,[48] and LR models that single out the third generation,[49] but even these *ad hoc* models have problems in passing all experimental tests.

We have seen that their preference for a heavy Higgs really singles out the hadronic asymmetries. This brings us to what can be called the *Chanowitz argument*:[50,51] there are two possibilities, both involving new physics:

(a) A_{FB}^b points to new physics; or

(b) A_{FB}^b is a fluctuation or is due to unknown systematics.

In the second case it is interesting to see what happens if one excludes the hadronic asymmetries from the above restricted Higgs mass fit. Not surprisingly, a consistent picture emerges: a very light Higgs with $M_H^{fit} = 42$ GeV fits perfectly all data and one obtains an upper bound $M_H < 120$ GeV at 95% CL. This would suggest new physics because the direct lower bound on the Higgs boson in the SM is $M_H > 114$ GeV.[50,51]

Although it may be the ringing bell for something more spectacular, the inconsistency with the

direct lower bound is statistically rather weak at the moment. It also marginally depends on the value of the hadronic contributions to $\alpha(M_Z)$ used in the fit, although we are already employing the most unfavorable estimate. Similarly, current estimates of the theoretical error agree that it cannot shift up $M_H^{95\%}$ more than ~ 20 GeV.[52] The inconsistency would be alleviated if the top mass turned out to be heavier than the present central value, a possibility suggested by the latest D0 analysis of Run-I data (yielding $M_t = 180.1 \pm 5.4$ GeV[44]) and soon to be tested at the Tevatron. Figure 15 illustrates this point by showing the result of a global fit with $M_t = 179.4 \pm 5.1$ GeV.

We have seen that excluding A_{FB}^b (and NuTeV) from the fit the quality of the fit improves considerably, but M_H^{fit} becomes very small. Finding new physics that simulates a very light Higgs is much easier than fixing the two anomalies. An example are oblique corrections: in general it just requires $S < 0$ $(T > 0)$ or $\epsilon_{2,3} < 0$.[50,51] A non-degenerate unmixed fourth generation with a heavy neutrino with $m_N \approx 50$ GeV would easily work. More interestingly, the MSSM offers rapid decoupling (small corrections), M_W always higher than in the SM, and $\sin^2\theta_{eff}^{lept}$ lower than in the SM. A plausible MSSM scenario involves light sneutrinos and sleptons, heavy squarks, and $\tan\beta \gtrsim 5$.[51]

As illustrated in Fig. 15, the Higgs indirect determination depends strongly on the top mass: a shift of $+5$ GeV in M_t would imply $M_H < 280$ GeV instead of 200 GeV. A factor 2 improvement in the determination of $\alpha(M_Z)$ would lower the 95% CL upper bound on M_H by only about 5 GeV. A factor 2 improvement in the measurement of M_t would lower the 95% CL upper bound on M_H by about 35 GeV. Figure 16 is also instructive: all the main precision observables define almost parallel bands in the M_t, M_H plane. The only important piece of information that can, in the near future, significantly improve the Higgs mass constraints is the top mass. A better M_t measurement would also help clarify the fate of the *Chanowitz argument*.

In the future, interesting new data will come from the Tevatron (M_t and M_W), from E158 and QWeak, and later from the LHC and possibly from a Linear Collider. Running the latter on the Z^0 peak (the *Giga-Z* option) would reach a new frontier in precision physics. We will be able to exploit this precision only with a major effort on the theoretical side. After many years of studies and despite some progress,[53] automatic two-loop calculations in the electroweak sector are nowhere in sight: the complete two-loop calculation of the relation between M_W, M_Z and G_μ has just been completed,[54] but the analogous calculation for $\sin^2\theta_{lept}^{\text{eff}}$ is not yet available.

6. Conclusions

The SM works fine, but there are several areas of tension in the data. None of them gives a convincing indication of new physics. Though each of them could, depending on the evolution of data and theory.

For what concerns the tests of charged current universality, an odd discrepancy persists between the measurements of the Cabibbo angle from $K_{\ell 3}$ and nuclear β decays. The situation, possibly due to underestimated theoretical uncertainties, should soon be clarified by a number of upcoming measurements.

A new global analysis of PDF's favors a positive strange quark asymmetry in the nucleon, that would reduce the NuTeV anomaly. This effect and isospin violation in the PDF's add a substantial uncertainty to the NuTeV result. Given our present understanding of the nucleon structure, the Paschos-Wolfenstein relation is probably not a good place for electroweak precision physics: NuTeV may end up teaching us more about hadronic structure than short-distance physics.

Revised CMD-2 data have reduced to $\approx 2\sigma$ the discrepancy between the experimental result for $(g-2)_\mu$ and the SM prediction based on e^+e^- data. KLOE has given the first results with the method of radiative return, confirming within errors CMD-2. On the other hand, the spectral function extracted from τ decays still deviates significantly from e^+e^- data in a small $\sqrt{s}$ window, a rather odd result that needs to be confirmed and understood, probably in terms of isospin breaking.

Although the SM fit shows a clear preference for a light Higgs boson, what we know of the Higgs mass and of the kind of new physics we might expect depends heavily on conflicting experimental data. Removing the most deviant result from the SM fit leads to a mild inconsistency with the direct lower bound on M_H. The *top priority* here is a precise measurement of the top mass, and we all expect interesting results from the Tevatron soon.

Acknowledgments

I am grateful to A. Ferroglia, S. Forte, M. Grünewald, G. Isidori, S. Kretzer, K. McFarland, B. Pietrzyk, A. Polosa, G. Rodrigo, A. Strumia, T. Teubner, Wu-Ki Tung and P. Wells for useful discussions and communications. This work is supported by a Marie Curie Fellowship, contract No. HPMF-CT-2000-01048.

References

1. P. Wells, talk at HEP 2003, Aachen, July 17th-23rd 2003, see http://eps2003.physik.rwth-aachen.de/.
2. J. Erler and P. Langacker, in K. Hagiwara *et al.* [PDG], *Phys. Rev.* D **66**, 010001 (2002); P. Langacker, hep-ph/0308145.
3. E158 Coll., see http://www.slac.stanford.edu/exp/e158/talks.html.
4. A. Czarnecki and W. Marciano, *Phys. Rev.* D **53**, 1066 (1996); A. Denner and S. Pozzorini, *Eur. Phys. J.* C **7**, 185 (1999), [hep-ph/9807446]; F. J. Petriello, *Phys. Rev.* D **67**, 033006 (2003), [hep-ph/0210259]; A. Ferroglia, G. Ossola and A. Sirlin, hep-ph/0307200.
5. A. Sirlin, hep-ph/0309187, and refs. therein.
6. W. Marciano, *Phys. Rev.* D **60**, 093006 (1999).
7. H. Abele, hep-ex/0308062.
8. See http://pibeta.web.psi.ch.
9. K. Schubert, these proceedings; Battaglia *et al.*, hep-ph/0304132; G. Isidori, hep-ph/0311044.

10. G. P. Zeller *et al.* [NuTeV Coll.], *Phys. Rev. Lett.* **88**, 091802 (2002), [hep-ex/0110059].
11. E. A. Paschos and L. Wolfenstein, *Phys. Rev.* D **7**, 91 (1973).
12. S. Davidson *et al.*, *JHEP* **0202**, 037 (2002), [hep-ph/0112302].
13. S. Kretzer and M. H. Reno, hep-ph/0307023.
14. K. S. McFarland and S. O. Moch, hep-ph/0306052.
15. B.A. Dobrescu and R.K. Ellis, hep-ph/0310154.
16. D. Yu. Bardin and V. A. Dokuchaeva, JINR-E2-86-260 (1986); K. P. O. Diener, S. Dittmaier, W. Hollik, hep-ph/0310364.
17. S. A. Kulagin, *Phys. Rev.* D **67**, 091301 (2003), [hep-ph/0301045].
18. R. Thorne, these proceedings, hep-ph/0309343; A. D. Martin, R. G. Roberts, W. J. Stirling and R. S. Thorne, hep-ph/0308087.
19. J. T. Londergan and A. W. Thomas, *Phys. Rev.* D **67**, 111901 (2003), [hep-ph/0303155].
20. G. P. Zeller *et al.* [NuTeV Coll.], *Phys. Rev.* D **65**, 111103 (2002), [hep-ex/0203004].
21. E. Sather, *Phys. Lett.* B **274**, 433 (1992); F. G. Cao and A. I. Signal, *Phys. Rev.* C **62**, 015203 (2000); E. N. Rodionov, A. W. Thomas and J. T. Londergan, *Mod. Phys. Lett.* A **9**, 1799 (1994).
22. S.J. Brodsky and B.Q. Ma, *Phys. Lett.* B **381**, 317 (1996); A. I. Signal and A. W. Thomas, *Phys. Lett.* B **191**, 205 (1987).
23. V. Barone, C. Pascaud and F. Zomer, *Eur. Phys. J.* C **12**, 243 (2000), [hep-ph/9907512].
24. M. Goncharov *et al.* [NuTeV Coll.], *Phys. Rev.* D **64**, 112006 (2001), [hep-ex/0102049].
25. S. Kretzer *et al.*, BNL-NT-03/16, RBRC-328; F. Olness *et al.*, MSUHEP-030703; contributed papers to LP2003 n.292-293.
26. R. Oldeman, PhD Thesis, 2000, University of Amsterdam.
27. M. L. Mangano *et al.*, hep-ph/0105155.
28. G. A. Miller and A. W. Thomas, hep-ex/0204007; S. Kovalenko, I. Schmidt and J. J. Yang, *Phys. Lett.* B **546**, 68 (2002), [hep-ph/0207158]; S. Kumano, hep-ph/0209200.
29. A. Kurylov, M. J. Ramsey-Musolf and S. Su, *Nucl. Phys.* B **667**,321 (2003), [hep-ph/0301208].
30. W. Loinaz, *et al.*, hep-ph/0210193.
31. K. S. Babu and J. C. Pati, hep-ph/0203029.
32. G. W. Bennett *et al.* [Muon g-2 Coll.], *Phys. Rev. Lett.* **89**, 101804 (2002), [Erratum-ibid. **89**, 129903 (2002), [hep-ex/0208001].
33. M. Knecht, hep-ph/0307239; W. J. Marciano, *J. Phys.* G **29**, 23 (2003); A. Czarnecki, eConf **C0209101** (2002) WE09; K. Melnikov, *Int. J. Mod. Phys.* A **16**, 4591 (2001), [hep-ph/0105267]; A. Czarnecki and W. J. Marciano, *Phys. Rev.* D **64**, 013014 (2001), [hep-ph/0102122]; V. W. Hughes and T. Kinoshita, *Rev. Mod. Phys.* **71**, S133 (1999).
34. T. Kinoshita, B. Nizic and Y. Okamoto, *Phys. Rev.* D **41**, 593 (1990).
35. K. Hagiwara, *et al.*, *Phys. Lett.* B **557**, 69 (2003), [hep-ph/0209187].
36. T. Teubner, talk at HEP 2003, Aachen, July 17th-23rd 2003, see http://eps2003.physik.rwth-aachen.de/.
37. M. Davier *et al.*, hep-ph/0308213 and *Eur. Phys. J.* C **27**, 497 (2003), [hep-ph/0208177].
38. S. Ghozzi and F. Jegerlehner, hep-ph/0310181; see also F. Jegerlehner, *J. Phys.* G **29**, 101 (2003), [hep-ph/0104304] and [hep-ph/0310234].
39. V. Cirigliano, G. Ecker and H. Neufeld, *Phys. Lett.* B **513**, 361 (2001), [hep-ph/0104267]; and *JHEP* **0208**, 002 (2002), [hep-ph/0207310]; J. Erler, hep-ph/0211345.
40. See also B. A. Li and J. X. Wang, *Phys. Lett.* B **543**, 48 (2002), [hep-ph/0207185].
41. R.R. Akhmetshin *et al.*, hep-ex/0308008.
42. A. Aloisio *et al.* [KLOE Coll.], hep-ex/0307051; B. Valeriani, talk at Sighad03, Pisa, October 2003.
43. H. Czyz, *et al.*, hep-ph/0308312 and *Eur. Phys. J.* C **27**, 563 (2003), [hep-ph/0212225].
44. P. Azzi, these proceedings.
45. H. Burkhardt and B. Pietrzyk, *Phys. Lett.* B **513**, 46 (2001); update summer 2003 in T. Teubner.[36]
46. E. Ma and D. P. Roy, *Phys. Rev.* D **65**, 075021 (2002).
47. R. S. Chivukula and E. H. Simmons, *Phys. Rev.* D **66**, 015006 (2002), [hep-ph/0205064].
48. D.Choudhury, T.Tait and C.Wagner, *Phys. Rev.* D **65**, 053002 (2002), [hep-ph/0109097].
49. X. G. He and G. Valencia, *Phys. Rev.* D **66**, 013004 (2002), *Err-ibid.* D **66**, 079901 (2002); *Phys. Rev.* D **68**, 033011 (2003), [hep-ph/0304215].
50. M. S. Chanowitz, *Phys. Rev.* D **66**, 073002 (2002), [hep-ph/0207123]; *Phys. Rev. Lett.* **87**, 231802 (2001), [hep-ph/0104024].
51. G. Altarelli *et al. JHEP* **0106**, 018 (2001), [hep-ph/0106029].
52. Freitas *et al.*, hep-ph/0202131; P. Gambino, hep-ph/9812332.
53. A. Andonov *et al.*, hep-ph/0209297.
54. M. Awramik and M. Czakon, *Phys. Lett.* B **568**, 48 (2003), [hep-ph/0305248]; and *Phys. Rev. Lett.* **89**, 241801 (2002), [hep-ph/0208113]; A. Onishchenko and OI. Veretin, *Phys. Lett.* B **551**, 111 (2003), [hep-ph/0209010]; A. Freitas *et al.*, *Phys. Lett.* B **495**, 338 (2000), *E: ibidem* **570**, 260 (2003).

Robert Bernstein (Fermilab): I'm going to address the NuTeV discussion. First of all the NuTeV dimuon analysis directly measures the momentum asymmetry in the strange sea in the right kinematic range and with a statistically independent data sample and in that sense, I think, it is superior to extrapolated global fits. The second point on that is, the only asymmetry evidence in the global fits comes from CDHS data which is a predecessor of CCFR at much lower energy and about a factor 12 less data in the relevant region. The data themselves are inconsistent with QCD and so I find it very odd, people insisting on using that for a delicate measurement like the strange sea asymmetry. Third, the updated Barone, Pascaud and Zomer fits you didn't show have no asymmetry in them. The fourth thing is that the asymmetry you ascribed to us in the strange sea is barely a 2σ effect, which we've stated to you many times we think is consistent with zero, and you characterized it as a large negative asymmetry. The other thing is, please stop referring to our fits as leading order because they are structure function fits to a leading order parameterization but in any case we are doing a fit with the Fermilab theory group to Next-to-Leading-Order. So given all this, here is the question for you. I wonder why you made this totally cavalier dismissal as this not being a good place [for precision electroweak physics - Eds.]?

Paolo Gambino: [provided for the Proceedings as this discussion was cut short – Eds.] I certainly agree with you that dimuons provide a superior method to study the strange sea and its asymmetry. On the other hand, let me stress that the CDHS data are irrelevant to my conclusions. I also agree that the updated analysis by BPZ is compatible with zero asymmetry, but it does not include dimuons and can be compared only to the inclusive CTEQ fit: I wonder why they do not agree. Concerning the published NuTeV analysis of dimuons, I have already explained why I do not think it is reliable. Finally, during my talk I have listed a number of theoretical uncertainties which I believe affect the NuTeV result at the level of the experimental error or more. In this unfortunate situation, where the $\sim 3\sigma$ discrepancy can easily be explained by Standard Physics, it is difficult to talk about *precision* tests.

HIGGS AND SUPERSYMMETRY

M. SCHMITT

Northwestern University, 2145 Sheridan Road, Evanston, IL 60208, USA
E-mail: schmittm@lotus.phys.northwestern.edu

NO CONTRIBUTION RECEIVED

SEARCHES FOR NEW PHENOMENA AT COLLIDERS

E. PEREZ

CE-Saclay, DSM/DAPNIA/Spp, F-91191 Gif-sur-Yvette, France
E-mail: eperez@hep.saclay.cea.fr

An overview of recent experimental results on searches for new phenomena at the LEP, HERA and Tevatron high energy colliders is presented, including in particular new results obtained from the analysis of the Run II data at the Tevatron. No significant evidence for physics beyond the Standard Model has been found and limits at the 95% confidence level have been set on the mass and couplings of several new particles. The complementarity between the different experiments is discussed, as well as future prospects for ongoing and future experiments.

1. Introduction

Although remarkably confirmed by low and high energy experiments over the last 30 years, the Standard Model (SM) of strong, weak and electromagnetic interactions remains unsatisfactory and incomplete. Due to the huge hierarchy between the electroweak and the Planck scales, breaking the electroweak symmetry spontaneously via a fundamental scalar field imposes an extreme fine-tuning of the mass of the latter field. One of the most popular solutions to that problem is brought by Supersymmetry (SUSY), which allows some cancellation between the loop corrections responsible for that fine-tuning.[a] Alternative solutions may be obtained by breaking the electroweak symmetry in a dynamical way, or by extending the space-time by additional space dimensions.

In addition, there are many questions which are not answered by the SM or its simplest supersymmetric extensions. For example, the SM does not explain the quantization of the electromagnetic charge, or the observed "replication" of three fermion families. It does not explain either the origin of fermion masses or the observed hierarchy between them. Many models of "new physics" have been proposed to address these issues.

Experimentally, new physics might manifest itself in rare meson decays, in precision measurements, in the search for Lepton Flavor Violating processes such as the $\mu \to e\gamma$ conversion, in the search for Cold Dark Matter, etc.. High energy colliders also provide a very high sensitivity to new phenomena, allowing new massive particles to be directly looked for, or the effect of new interactions interfering with known SM processes to be studied.

The LEP (Large Electron Positron) collider ran until end 2000 at a centre-of-mass energy $\sqrt{s}$ up to 209 GeV, and delivered about 1 fb^{-1} to the ALEPH, DELPHI, L3 and OPAL experiments. Extensive searches for new phenomena have been carried out and the final analyses are being completed. Stringent constraints have been set especially on new particles coupling preferably to the electroweak bosons.

The Tevatron proton-antiproton collider delivers data to the D0 and CDF experiments. During the first phase ("Run I") which ended in 1996, both experiments collected about 120 pb^{-1} of luminosity. The Tevatron has restarted in May 2001 with a larger centre-of-mass energy of 2 TeV. The increased energy, the improved detector capabilities, and the much larger expected luminosities of several fb^{-1} provide a large discovery potential for new phenomena. About 300 pb^{-1} has been delivered so far, with more than 200 pb^{-1} collected by the experiments with fully operational detectors. The Run II results presented in the following were obtained with a luminosity of about $100 - 130$ pb^{-1}.

At HERA (Hadron Electron Ring Accelerator), electrons (or positrons) collide with protons at $\sqrt{s} \simeq$ 320 GeV. During the first phase of data taking which ended in mid-2000, the two colliding experiments H1 and ZEUS collected about 120 pb^{-1} of luminosity. The restart of HERA in the fall of 2001 has been more difficult than expected, but the machine should now be able to deliver again high luminosities. Besides its lower energy and the limited possibility to produce new particles by pair, HERA appears very well suited to look for particles coupling to an electron and a first generation quark, or to search for new phenomena for which the SM backgrounds at the Tevatron are not easy to handle.

---

[a] An overview of experimental results on searches for Supersymmetry is given elsewhere in these proceedings.[1]

2. Any Hints for New Physics?

The most convincing evidence so far for physics beyond the Standard Model is the observation that neutrinos oscillate.[2] However, the SM can easily accomodate neutrino masses and the current observations do not provide a strong guidance to which kind of new physics should extend the SM.

On the other hand, some precision measurements deviate slightly from the corresponding SM expectations[b] and some "interesting" events have been observed at high energy colliders. In the next section I shortly review some of these observations which, if confirmed, might provide hints for new physics.

2.1. *Precision Measurements*

In atomic physics, the interference of photon and Z exchange between the valence electron and the nucleus leads to transitions which would be forbidden by QED alone. Such transitions allow one to measure a (small) parity violating asymmetry giving access to the so-called weak charge, which might be affected by the exchange of new particles, such as new Z bosons or leptoquark bosons. The weak charge has been measured to 0.6% in Cesium atoms.[4] Until spring 2003, this measurement showed a discrepancy of more than 2σ with the SM prediction. The latter has been revisited recently[5] and the new calculation agrees very well with the experimental value.

The precise measurement of the anomalous magnetic moment of the muon is known to be very sensitive to new physics effects, such as loops induced by sleptons and charginos in supersymmetric theories. The experimental world average, largely dominated by the measurement done by the $(g - 2)_\mu$ Collaboration at BNL, reaches the impressive precision of 0.7 ppm. On the other hand, the theoretical expectation still suffers from uncertainties related to the evaluation of the hadronic vacuum polarization. This can be estimated by using either hadronic cross section measurements done in low energy e^+e^- collisions, or the spectral functions of hadronic tau decays. In the latter case, the resulting SM prediction for $(g - 2)_\mu$ agrees well with the measurement, while some discrepancy exists when using low energy e^+e^- data.[6] The CMD-2 Collaboration revisited its

---

[b] A detailed review of precision measurements including topics not discussed here can be found elsewhere.[3]

$\pi\pi$ cross sections recently.[7] Taking into account these new measurements, the discrepancy remains of the order of 2σ.

At the ICHEP'02 conference, both the BaBar and BELLE Collaborations reported a discrepancy between the measurements of $\sin 2\beta$ in the $B \to J/\psi K_s$ and $B \to \Phi K_s$ modes, with the combined effect being at the level of 2.7σ. Updated results were presented at this symposium[8] and the BELLE experiment confirms this deviation. This observation has already triggered quite some speculations, although it does not seem easy for some "new physics" to account for it, without contradicting measurements on the $b \to s\gamma$ decay or on the B_d mixing parameter.

2.2. *Tevatron Events with Lepton(s) and Photon(s)*

In 1995, the CDF experiment observed a spectacular event with two electrons, two photons, and a large amount of missing transverse energy in the final state.[9] Due to the very low expectation for such events within the SM, this triggered a lot of activity and in particular revived somehow the supersymmetric models where SUSY breaking is mediated by gauge interactions. Both the D0 and CDF experiments looked for such events in a sample of about 100 pb^{-1} of Run II data, and no new candidate has been observed. One can note that with their improved hermiticity, the Run II detectors are very well suited to study such final states.

The CDF data from Run I also showed a slight excess of events with a lepton and a photon, both at high transverse momentum P_T, together with some large missing transverse energy.[10] This excess is not confirmed in the data from Run II, and the current CDF measurement of $W\gamma$ production shows a very good agreement with the SM expectation.

2.3. *The "Superjets" Observed at CDF Run I*

In the data from Run I, the CDF experiment reported an excess of events with 2 or 3 jets and a W boson, where one of the jets was emerging from a secondary vertex and also contained a soft lepton (such jets, doubly identified as b candidates, were called "superjets").[11] The experiment observed 13 such events for a SM expectation of 4.4 ± 0.6. These events were seen to have atypical kinematic proper-

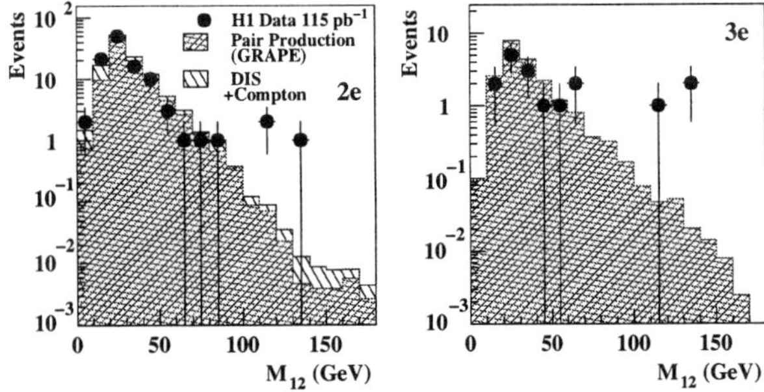

Figure 1. Distribution of the invariant mass of the two highest P_T electrons in events with (left) exactly two and (right) exactly three measured electrons. The H1 data (symbols) are compared with the prediction from the SM, dominated by two photon processes.

ties and many checks were performed to try to understand them, which did not show any experimental bias. So far, there is no statement from Run II data on such events. Work is going on in both D0 and CDF experiments to study the correlations between various b-tagging algorithms.

2.4. HERA Multilepton Events

The H1 Collaboration recently published the first measurement of multilepton production cross section in ep collisions.[12] At HERA multilepton events are mainly produced via $\gamma\gamma$ collisions, with the incoming electron being often scattered at very small angles and thus undetected. Figure 1 shows the distribution of the invariant mass M_{12} of the two highest P_T electrons, for final states with (left) exactly two and (right) exactly three measured electrons.[c] The data are well described by the SM expectation besides at largest masses. For $M_{12} > 100\,\mathrm{GeV}$, 6 events are observed for a SM expectation of 0.53 ± 0.06. On the other hand, a preliminary analysis of the ZEUS data[13] does not show any significant excess.

2.5. HERA Events with a Lepton and Large Missing Transverse Energy

In 1994, the H1 Collaboration reported the observation of an $e^+ p \rightarrow \mu^+ X$ event with high transverse momenta. Since then, other similar events with a high P_T isolated lepton, a large amount of missing transverse energy, and a high P_T hadronic final state have been observed in the H1 data[14] and the study

---

[c] "Electron" actually stands here for e^- and e^+.

of such final states has been of highest interest in both the H1 and ZEUS Collaborations. The main SM contribution to such final states comes from W production, for which the cross section at HERA is about 1 pb. Although the events observed in H1 look compatible with the W hypothesis, Fig. 2 shows that an excess of events is observed in the data for high values of the transverse momentum P_T^X of the hadronic final state. For $P_T^X > 25\,\mathrm{GeV}$, 10 events are observed where the isolated lepton is a muon or an electron, for a SM expectation of 2.92 ± 0.49. For $P_T^X > 40\,\mathrm{GeV}$ 6 events are still observed while the SM expectation is 1.08 ± 0.22. The ZEUS experiment does not observe a significant excess of events with the lepton being an electron or a muon, but reports the observation of 2 events at $P_T^X > 25\,\mathrm{GeV}$ with a τ lepton decaying hadronically.[15] That is in slight excess of the SM expectation of 0.12 ± 0.02 event.

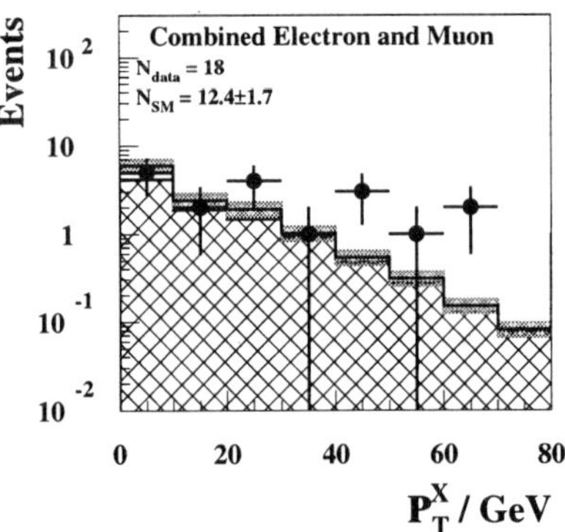

Figure 2. Distribution of the transverse momentum of the hadronic final state for events with a high P_T isolated e or μ and missing transverse energy. The H1 data (symbols) are compared with the SM expectation, dominated by W production (hatched histogram).

2.6. Comments on the Above Observations

The excesses reported in the three previous paragraphs are clearly statistically limited. We are looking forward to seeing new results on events with "superjets" in the Tevatron experiments. H1 should be able to clarify the two excesses observed with the much larger luminosities expected within the next years. Meanwhile, assuming that the "anomalies" seen at H1 are a sign for some new physics, one may wonder whether something should also have been seen by the LEP and Tevatron experiments. New physics in ep collisions might proceed via a lepton-quark interaction, or via $e\gamma$ or γp collisions.

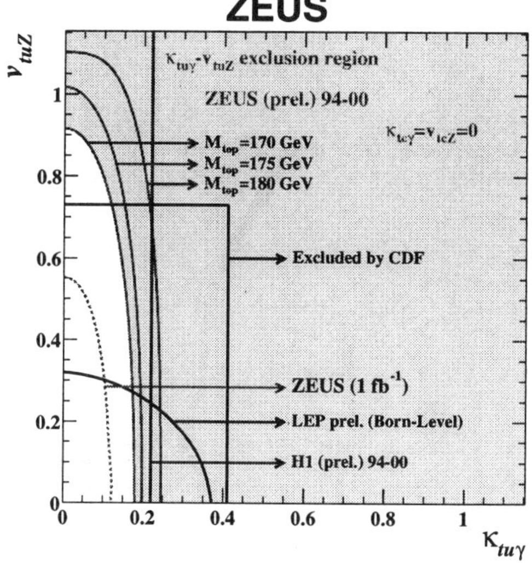

Figure 3. Existing constraints on the anomalous FCNC magnetic (vectorial) coupling of a top quark to a u quark and a photon (a Z) derived from the HERA, Tevatron and LEP experiments. The expected sensitivity of HERA experiments with 1 fb^{-1} is also shown as the dashed curve.

The first case would have very model dependent consequences at other colliders. The "partonic" $e\gamma$ (γq) luminosities are much larger at LEP II (Tevatron) than at HERA. Thus it is likely that any new physics proceeding via $e\gamma$ (γq) collisions would have a much larger cross section at LEP II (Tevatron). On the other hand the corresponding SM backgrounds should also be much larger. This may make the signal very difficult to single out from the background.

An example of that is provided by the "anomalous single top" interpretation of the events reported in Sec. 2.5. Anomalous couplings between a light quark (u or c), a gauge boson and a top quark are present in several extensions of the SM. These might allow for single top production at LEP and at HERA (which, otherwise, has a tiny cross section in the SM); for an enhanced single top production rate at the Tevatron; for rare decays $t \to q\gamma(Z)$. Amongst the events reported in Sec. 2.5, 5 candidates fulfil dedicated criteria designed to select single top events, while the SM expectation is 1.31 ± 0.22. No significant excess is observed in the hadronic channel. Using the convention given in,[16] an anomalous coupling $\kappa_{tu\gamma}$ of $0.20^{+0.05}_{-0.06}$ would be needed to account for these observations.[17] As shown in Fig. 3 this range of coupling values is not yet ruled out by other experiments.[18−20]

At the Tevatron, the anomalous single top production cross section induced by a $\kappa_{tu\gamma}$ coupling could be quite large, e.g. about 2 pb for $\kappa_{tu\gamma} = 0.2$. However this rate is similar to that of SM single top production, the observation of which is challenging due to the huge W+jets background.[d] As a result, the future sensitivity on $\kappa_{tu\gamma}$ of the Tevatron experiments with a luminosity of 2 fb^{-1} will remain driven by the study of decays $t \to q\gamma$. It should reach about 0.1, similar to the expected sensitivity of HERA experiments with a luminosity of 1 fb^{-1}.

3. Searches for New Resonances

3.1. *Excited Fermions*

The observed replication of three fermion families motivates the possibility of a new scale of matter yet unobserved. Deep-inelastic scattering (DIS) experiments have been known for a long time to be very well suited to study the structure of hadronic matter.[e] At HERA, where the exchanged boson probes the proton with a resolution corresponding to scales up to 10^5 GeV^2, a finite quark radius would reduce the DIS cross section compared to its SM values, this effect being more prominent when the square of the four-momentum carried by the gauge boson (Q^2) is large. The study of the full statistics of high Q^2 DIS data allowed both the H1 and ZEUS experiments[23] to rule out quark radii smaller than about 10^{-18} m, assuming the electron to be point-like.

Besides these indirect effects, an unambiguous signature for a new scale of matter would be the direct observation of excited states of fermions (f^*), via their decay into a gauge boson. Effective models describing the interactions of excited fermions with standard matter have been proposed.[24−26] In the most commonly used model,[24,25] the interaction of an f^* with a gauge boson is described by a magnetic coupling (i.e. a dimension-five operator) proportional to $1/\Lambda$ where Λ is a new scale. Proportionality constants f, f' and f_s result in different couplings to $U(1)$, $SU(2)$ and $SU(3)$ gauge bosons. Existing constraints on excited electrons are shown in Fig. 4, under the assumption that $f = f'$. Searches for pair produced e^* at LEP[27] allowed masses below

---

[d]At Run I, only an upper bound of about 15 pb could be set on the cross section of that process.[21]

[e]More generally two fermion production processes are sensitive probes to fermion compositeness.[22]

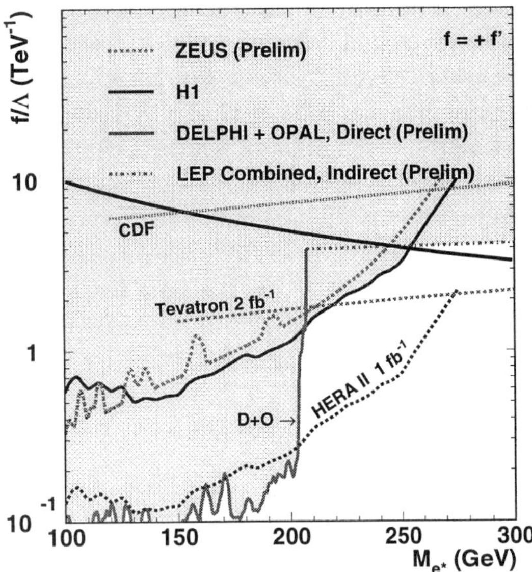

Figure 4. Existing constraints on excited electron masses and couplings, assuming that $f = f'$. The decreasing curve shows the hyperbola $f/\Lambda = 1/M_{e^*}$. The uppermost dotted line shows the translation of the CDF Run II constraint coming from single e^* search. Also shown as dotted curves are the future sensitivities of HERA and Tevatron experiments.

about 103 GeV to be ruled out, independently of the value of the coupling f/Λ. In contrast, searches for single e^* production at LEP[28] and at HERA[29] set mass bounds which depend on f/Λ. The best sensitivity at highest masses is provided by looking for indirect effects in $e^+e^- \to \gamma\gamma$ which might be induced by e^* t-channel exchange.[30]

Recently, the CDF experiment looked for the single production of an excited electron focusing on the decay $e^* \to e\gamma$. In the formalism of Baur *et al.*,[26] which differs from that used by others[24,25] mainly by a different normalization, e^* masses M_{e^*} up to 863 GeV can be ruled out assuming $\Lambda = M_{e^*}$. My interpretation of this constraint in the model of[24,25] is shown in Fig. 4 as the upper dashed-dotted curve. Also shown in Fig. 4 are the future sensitivities of HERA and Tevatron. Although e^* have been severely constrained at LEP, a much larger discovery potential remains at the Tevatron for other generations of excited leptons.

3.2. *Dijet Resonances*

The Tevatron is very well suited to look for excited quarks and other new particles coupling strongly to quarks or gluons, which might manifest themselves as

dijet resonances. With a resolution of about 10% of the dijet mass, CDF looked for narrow resonances in the dijet spectrum measured in the data from Run II. No signal was observed and mass bounds were derived for several new particles, as shown in Fig. 5. In particular, the data allow to exclude axigluons and (color universal) colorons[31] with masses below 1130 GeV, under the assumption that the coupling of these new colored bosons to a $q\bar{q}$ pair is the standard strong coupling. This is the first direct mass bound above 1 TeV obtained so far.

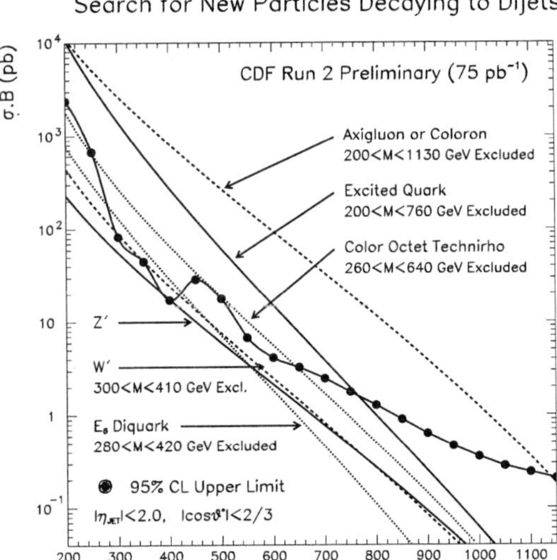

Figure 5. Mass dependent upper bound on the production cross section of a resonance times the branching ratio of its decay into two jets. The experimental constraint (symbols) is compared to the prediction of several models allowing lower mass bounds to be set on the new resonances they predict.

3.3. *Leptoquarks*

An intriguing characteristic of the Standard Model is the observed symmetry between the lepton and the quark sectors, which is manifest in the representation of the fermion fields under the SM gauge groups, and in their replication over three family generations. This could be a possible indication of a new symmetry between the lepton and quark sectors, leading to "lepto-quark" interactions. Leptoquarks (LQs) are new scalar or vector color-triplet bosons, carrying a fractional electromagnetic charge and both a baryon and a lepton number. Several types of LQs might exist, differing in their quantum numbers. A classification of LQs has been proposed by Buchmüller, Rückl and Wyler (BRW)[32] under the assumptions that LQs have pure chiral couplings to SM fermions, and that a given LQ couples only to fermions of a given family. The interaction of the LQ with a lepton-quark pair is of Yukawa or vector nature and is parameterized by a coupling λ. Assuming that LQs belonging to the same isospin multiplet are mass degenerate, the LQ decays only into a quark and a neutrino or a charged lepton, and the branching ratio β of the latter decay mode is 1 or 1/2.

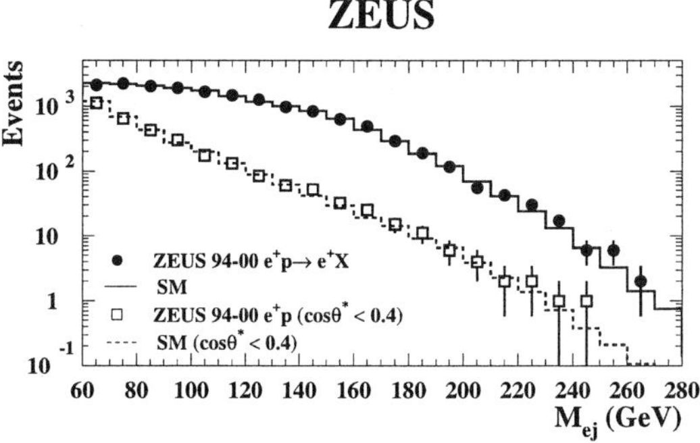

Figure 6. Distribution of the electron-jet invariant mass for high Q^2 Neutral Current DIS candidate events. The ZEUS data (symbols) are compared with the SM prediction (histograms), before and after applying an angular cut designed to maximize the sensitivity to a scalar LQ signal.

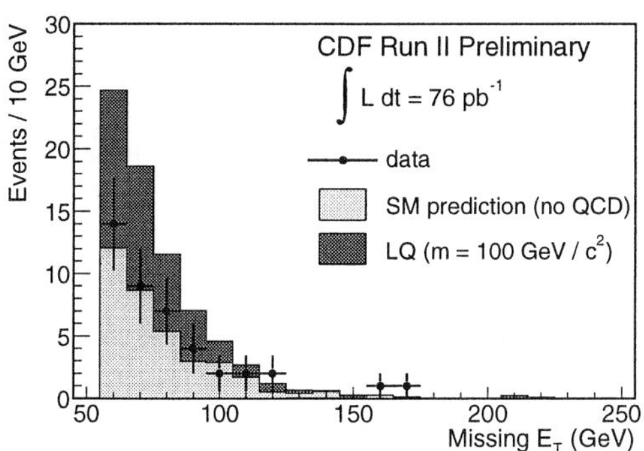

Figure 7. Distribution of the missing transverse energy measured by the CDF experiment for events with at least two hard jets. The additional expected contribution of pair produced scalar leptoquarks with mass 100 GeV is also shown as the dark histogram.

In electron-proton collisions, first generation LQs might be singly produced via the fusion of the incoming lepton with a quark coming from the proton. Hence they might be observed as a resonant peak in the distribution of the lepton-jet mass spec-

trum of Neutral or Charged Current DIS events. No such signal has been observed by the H1 and ZEUS experiments. As an example, Fig. 6 shows the invariant mass distribution of Neutral Current DIS events measured in the ZEUS data,[33] which agrees well with the SM expectation up to the highest masses.

At the Tevatron, leptoquarks are mainly pair produced via their coupling to gluons. First generation LQs have been looked for in Run II data in the $eejj$, $e\nu jj$ and $\nu\nu jj$ final states, with the letter j denoting generically a jet. In the first two cases, a large amount of SM background comes from QCD events where a jet "fakes" an electron and has to be determined from the data. Searching for the signal in the last two channels requires a good understanding of the measurement of missing transverse energy. Figure 7 shows that indeed the distribution of missing transverse energy is well controlled.

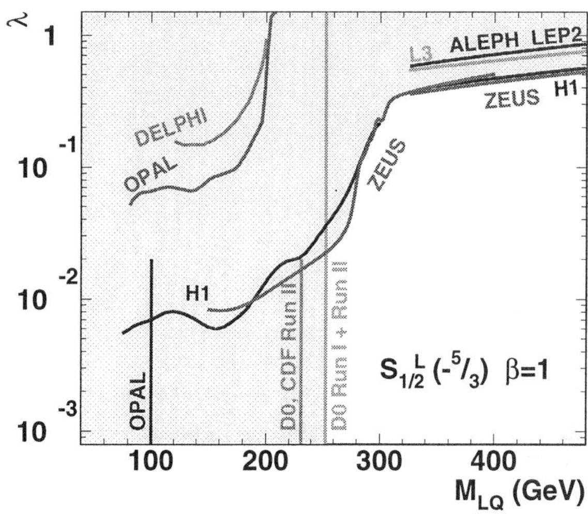

Figure 8. Example mass-dependent upper bounds on the Yukawa coupling λ of a first generation leptoquark to the electron-quark pair. These are shown for a scalar LQ which decays exclusively into an electron and a quark.

Existing constraints on a scalar leptoquark which decays solely into an electron and a quark are summarized in Fig. 8. For an electromagnetic strength of the coupling λ $(\lambda^2/4\pi = \alpha_{em})$, HERA experiments[33,34] rule out LQ masses below $\sim 290\,\mathrm{GeV}$. Constraints on high mass LQs were also obtained by the LEP experiments, where the $t-$channel LQ exchange might affect the observed cross section for the process $e^+e^- \rightarrow q\bar{q}$. In contrast, constraints derived from the search for pair produced LQs at the Tevatron do not depend on the coupling

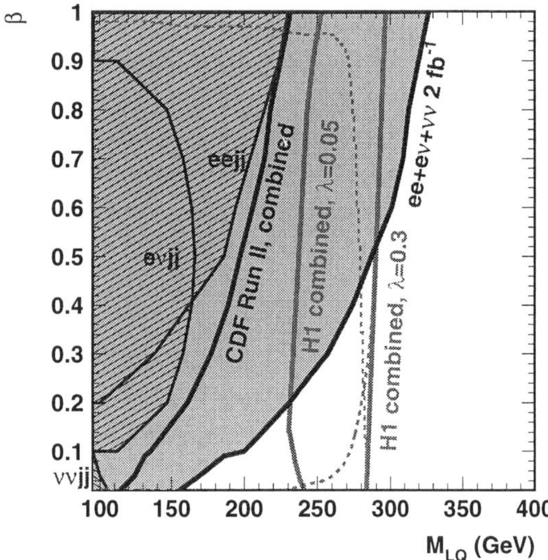

Figure 9. Example constraints on first generation scalar leptoquarks decaying exclusively into eq and νq. For $\lambda = 0.3$, the H1 constraints obtained from the ej and νj analyses are indicated by the dashed curves, and the combined bound is shown for two values of the coupling. Also shown is the future sensitivity of the Tevatron for a luminosity of 2 fb^{-1}.

λ. By combining data from Run I[35] and Run II, D0 rules out LQ masses below 253 GeV, which is the most stringent bound existing so far on scalar LQs decaying only into an electron and a quark.

Relaxing some of the BRW assumptions, more general LQ models might be constructed where the branching ratio β for the LQ to decay into an electron and a quark is a free parameter of the model.[f] Example constraints on such LQ models are shown in Fig. 9 assuming that the LQ decays exclusively into eq and νq. Leptoquarks decaying with a large branching ratio into νq are not easily probed at the Tevatron due to the large background. HERA experiments can provide a better sensitivity in such cases if the Yukawa coupling λ is reasonably large.

Leptoquarks coupling to second or third generation fermions and with masses above 100 GeV can be directly searched for only at the Tevatron. Pair produced LQs leading to final states with two muons and two jets have been searched for in the Run II data by the D0 experiment. The resulting mass bound of 186 GeV is close to that obtained with Run I data using a much more involved analysis.[36]

---

[f] An example of such LQs is provided by supersymmetric partners of quarks possessing R-parity violating couplings to a lepton-quark pair.

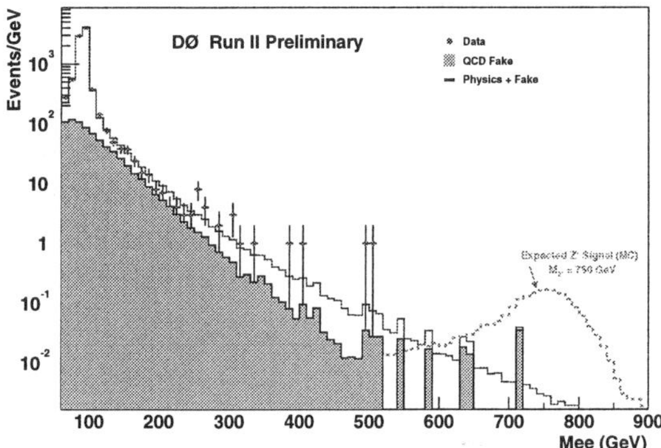

Figure 10. Distribution of the dielectron invariant mass observed by the D0 experiment. Above the Z peak, a large contribution to the SM expectation (white histogram) comes from QCD processes where a jet "fakes" an electron (shaded histogram).

3.4. Searches for New Gauge Bosons

New resonances coupling to leptons are predicted in several extensions of the Standard Model. Extending the gauge group of the SM often leads to additional Z' bosons, with most popular models being left-right symmetric or based on the E_6 Grand Unification group.

Both the D0 and CDF experiments looked for such dilepton resonances[g] in the data from Run II. As an example, Fig. 10 shows the dielectron invariant mass distribution measured by the D0 experiment. The measurement agrees well with the SM prediction up to the highest masses. A similar agreement is observed in the dimuon channel. Combining the dielectron and dimuon channels bounds on the Z' mass have been obtained in several models, some of which are shown in Fig. 11.

They range between 545 and 730 GeV, with the most stringent bound obtained in the "Sequential Standard Model" where the new Z' boson couples to fermions in the same way as the standard Z. In that case however, indirect bounds coming from the measurements of cross sections and forward-backward asymmetries at LEP provide stronger constraints. In contrast, in the E_6 inspired types of models, the direct bounds obtained at the Tevatron are the most stringent and less dependent on the model parameters. While the ultimate sensitivity of the Tevatron experiments should approach 1 TeV, the LHC will be able to probe Z' masses up to $4 - 5$ TeV.

---

[g] As shown in Fig. 5, the search for a new boson in the dijet channel has a limited sensitivity.

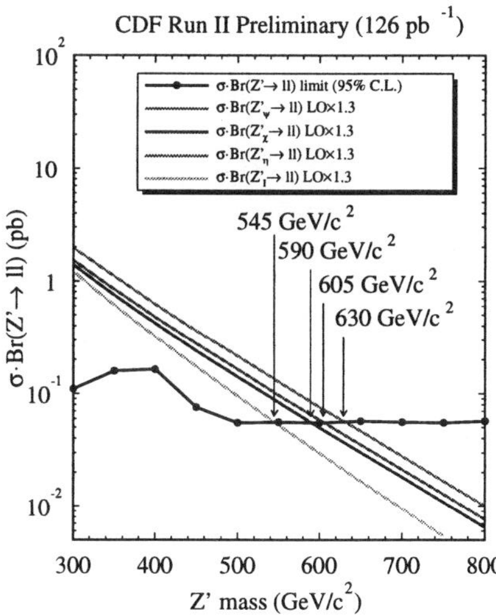

Figure 11. Upper bounds on the production cross section of a new Z' boson times its branching ratio to decay into a dilepton pair. Both the ee and $\mu\mu$ channels are combined. Resulting bounds in several E_6-inspired Z' models are also shown.

3.5. *Doubly-charged Higgses*

In addition to dilepton resonances possessing no net lepton number, new particles coupling to two leptons of the same charge might also be considered. Such particles are predicted in models with an exotic extended Higgs sector, such as left-right symmetric models where the symmetry $SU(2)_L \times SU(2)_R$ is broken by a triplet of $Y = 2$ scalar fields, containing a doubly-charged Higgs field.[37] Such models might naturally provide (small) Majorana mass terms for the neutrinos. The pair production of doubly-charged Higgses has been looked for by DELPHI and OPAL,[38] considering all possible dilepton decay modes for the $H^{\pm\pm}$. This yields mass limits of about 100 GeV. Doubly-charged Higgses might also be singly produced at LEP and HERA[39] via the process $e^\pm\gamma \to e^- H^{\pm\pm}$. In that case, the production cross section depends, in addition to the $H^{\pm\pm}$ mass, on its unknown coupling h_{ee} to a dielectron pair.[h] A dedicated analysis of the H1 events described in Sec. 2.4 does not support the $H^{\pm\pm}$ hypothesis for these events.[40] This is confirmed by the search for singly produced charged Higgses carried out by OPAL,[41] which sets strong bounds on the single $H^{\pm\pm}$ production cross section in $e\gamma$ collisions.

---

[h]Note that this Yukawa coupling is not expected to be vanishingly small since the Higgs triplet is not involved in the generation of fermion Dirac mass terms.

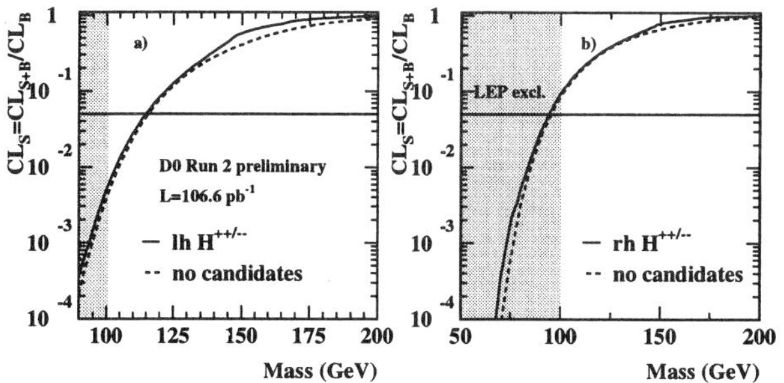

Figure 12. Confidence level of the doubly-charged Higgs hypothesis as a function of the $H^{\pm\pm}$ mass, as obtained from the D0 search for pair-produced $H^{\pm\pm}$ decaying into muons.

Pair production of doubly charged Higgses has also been looked for at the Tevatron experiments using the data from Run II. CDF searched for a signal in the $e^-e^-e^+e^+$ final state. Although no mass bound could be set with the current statistics, the future sensitivity should reach about 180 GeV. In contrast, doubly-charged Higgses coupling to muons are already constrained by the Tevatron. As shown in Fig 12, masses below 116 (95) GeV can be ruled out for $H^{\pm\pm}$ coupling to left- (right-) handed muons.

3.6. *Resonant Kaluza-Klein Gravitons*

The phenomenological and experimental interest in higher-dimensional physics has been considerably renewed recently, after it was realized that compactified extra dimensions could yield observable effects at the current or foreseen experiments. Models with extra space dimensions try to address the problem of the huge hierarchy between the Planck and the electroweak scales. The experimental constraints on models with "large" extra dimensions, resulting in a "strong" gravity at the TeV scale, will be reviewed in Sec. 4. This paragraph considers the case of "small" extra dimensions, where the gravity is "localized" on a brane, apart from another brane where the SM fields are confined. The propagation of gravity in the extra dimension is exponentially damped due to a fine-tuned space-time metric as proposed by Randall and Sundrum,[42] resulting in the weakness of gravity when observed from the SM brane.

Since it propagates in the extra dimension, the graviton observed in $4d$ manifests itself as a "tower" of Kaluza-Klein (KK) modes, with the zeroth mode corresponding to the massless graviton. In models of localized gravity the first graviton excitation $G^{(1)}$ is

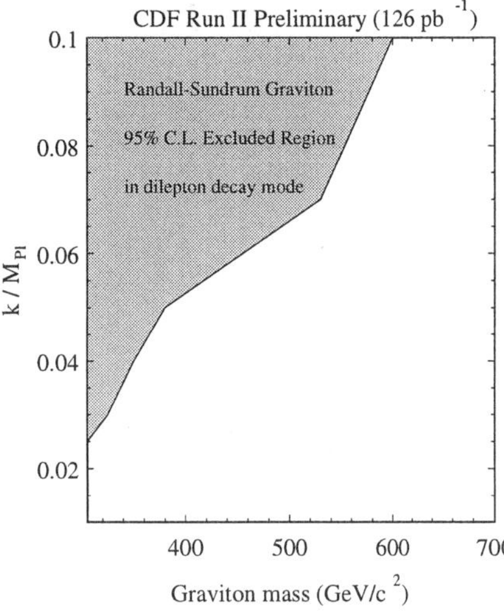

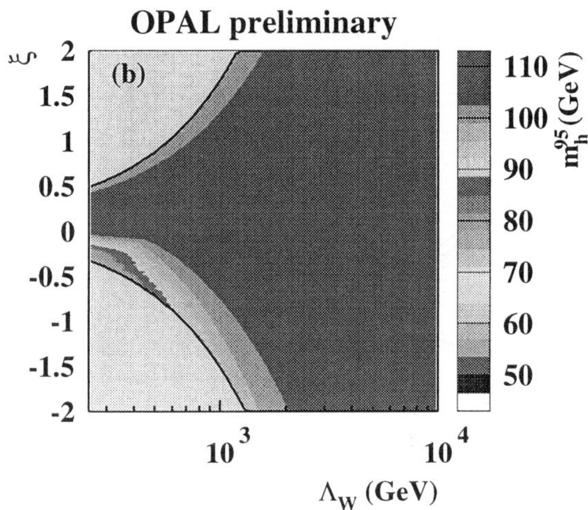

Figure 13. Constraints set on Kaluza-Klein graviton mass and coupling obtained by combining the ee and $\mu\mu$ channels.

Figure 14. Lower limits on the mass of the Higgs-like scalar field as a function of the mixing parameter ξ between the Higgs and the radion and of the scale parameter Λ_W corresponding to the vacuum expectation value of the radion.

expected to be heavy, with a mass around the TeV scale. Its coupling to SM particles is determined by a model parameter (k/M_{Pl}) and is expected to be sizeable. Hence, resonant production of $G^{(1)}$ might be observed at the Tevatron or future colliders. Using the data from Run II, the CDF experiment looked for such a signal when the $G^{(1)}$ decays hadronically or leptonically into ee or $\mu\mu$, with the latter modes providing the largest sensitivity. Resulting bounds are shown in Fig. 13. These are the first direct constraints on Randall-Sundrum models.

3.7. Searches for Radions

In models with extra dimensions the excitations of the metric tensor along the new coordinates yield new scalar fields. In the simplest version of the Randall-Sundrum model with five dimensions there is only one such scalar field. For each mode k its kth Kaluza-Klein excitation is "eaten" to provide the additional degrees of freedom to the massive $G^{(k)}$, such that only the zeroth mode, the so-called "radion", remains.[43,44] The radion might mix with the SM Higgs boson.[44] Since the radion couples strongly to gluons the lightest scalar field resulting from this mixing may have dominant decay modes which differ from those of the SM Higgs. The OPAL experiment reinterpreted its flavor-independent Higgs searches to set first constraints on such models.[45] An example of those is shown in Fig. 14.

4. Effects of Additional Large Space Dimensions

The phenomenology of models with large extra dimensions[46,47,43] is widely different from that of Randall-Sundrum models. Here, the weakness of gravity in the $4d$ world can be accounted for by the "dilution" of gravity flux lines in large extra dimensions. The fundamental scale characterizing the gravity in the full space, M_D, might be of the order of the TeV for a compactification radius of the extra dimensions up to about 0.1 mm. Such large radii are actually not ruled out by experiments measuring gravity at submillimeter distances.

4.1. Direct Searches for KK Gravitons

In these models the Kaluza-Klein excitations of gravitons form a quasi continuum in mass. Hence, although the coupling of $G^{(k)}$ to SM fields is tiny, the production cross section of Kaluza-Klein gravitons might be large due to the huge multiplicity of accessible states. Once produced a $G^{(k)}$ would escape detection because of its very small coupling such that the typical signature of graviton production is the presence of missing energy. Most sensitive direct bounds come from the LEP experiments which searched for $e^+e^- \to \gamma G^{(k)}$. The L3 experiment[48] rules out M_D values lower than 1.5 TeV (0.9 TeV) in the case of $n = 2$ (4) extra dimensions. At hadronic colliders, final states with a jet and missing energy are the most sensitive channels to directly search for KK gravitons. Data from the Run I of the Tevatron[49] allowed to set a lower bound of about 1 TeV on M_D, while the LHC experiments with 100 fb^{-1} should reach a sensitivity of 7 to 8 TeV.[47]

4.2. *Indirect Effects of KK Graviton Exchange*

Another way to search for large extra dimensions is to look for the effect of graviton exchange and its interference with SM processes. Calculations need to be performed by introducing an effective coupling λ/M_s^4, where the parameter λ is expected to be close to one and the effective scale M_s should not differ too much from the fundamental scale M_D.[i] Several formalisms have been proposed for this effective coupling.[47−51] Bounds given below use the convention of Giudice *et al.*[47] Combining all LEP data on Bhabha scattering and $\gamma\gamma$ production, a lower bound on M_s of 1.35 TeV can be set.[52] The study of NC DIS at HERA rules out M_s values below 0.82 TeV.[23] The D0 experiment searched for indirect effects induced by KK graviton exchange in the ee, $\gamma\gamma$ and, for the first time, in the $\mu\mu$ final states. Figure 15 shows the dimuon invariant mass measured by D0 together with the effect of additional contribution of graviton exchange for several values of M_s. Combining the ee and $\gamma\gamma$ analyses from Run I[53] and Run II, the constraint $M_s > 1.38$ TeV is obtained, which is the most stringent bound so far.

4.3. *Searches for Branons*

The assumption that the brane on which the SM fields are confined is allowed to "vibrate" in the extra dimensions might modify the phenomenology of extra dimensional models.[54] A small brane tension would result in a large suppression of the production rate of $G^{(k)}$ for high modes (large $|k|$), such that the "standard" signal of direct graviton production might not be observable. The counterpart is that the scalar fields associated to the brane vibration, called "branons", might in turn be produced with a sizeable rate. The experimental signature would remain the same as for graviton production, i.e. missing energy.

---

[i]Indirect effects induced by graviton exchanges thus allow to probe the scale M_s but do not provide a direct probe of M_D.

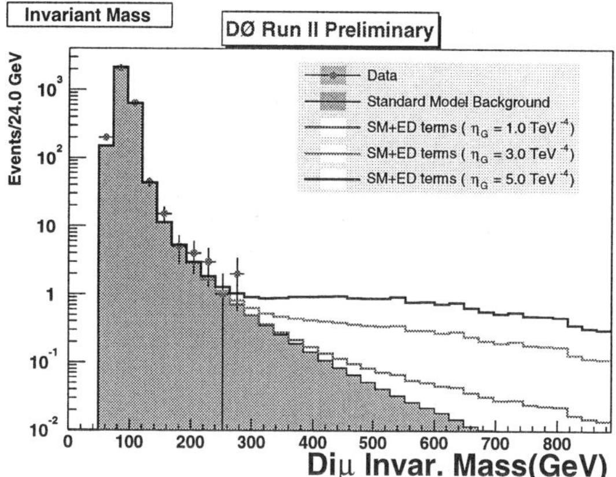

Figure 15. Distribution of the dimuon invariant mass measured by the D0 experiment. The data (symbols) are compared to the SM expectation (shaded histogram). Several examples of the deviation from the SM prediction induced by Kaluza-Klein graviton exchange are also shown.

A re-interpretation of the $\gamma + E_{missing}$ analysis has been performed by L3[55] to set some bounds on the branon mass depending on the brane tension. These are shown in Fig. 16. For very elastic branes, $f \to 0$, branons with a mass below 103 GeV are ruled out.

5. Model-independent Searches

The searches for new phenomena reported in the previous sections are driven by specific models: within a given model, the analysis is optimized in order to maximize the sensitivity to the "known" signal. Many models are investigated by experiments, and models themselves actually offer quite some freedom via their free parameters. However, additional information can be brought by a complementary approach consisting in a broad-range search for deviations from the SM prediction. Such an approach should consider phase space regions where the SM prediction is sufficiently reliable, e.g. those where the particles in the final state have a large transverse momentum.

This approach has been pionneered by the D0 Collaboration using the full statistics of Run I data.[56] It has been recently applied to the full sample of data collected by H1.[57] The strategy begins with definition criteria for "objects", as electrons, muons, photons, jets, W/Z, or neutrinos. Channels are then defined as final states with at least two objects, and the data are compared to the SM expectation in each channel. For example, Fig. 17 shows that an overall good agreement is observed in all channels considered in

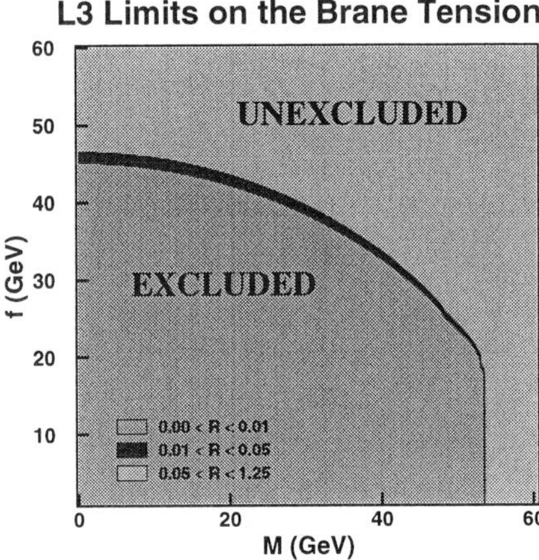

Figure 16. Constraints on the branon mass as a function of the brane tension, obtained from the analysis of L3 single photon data.

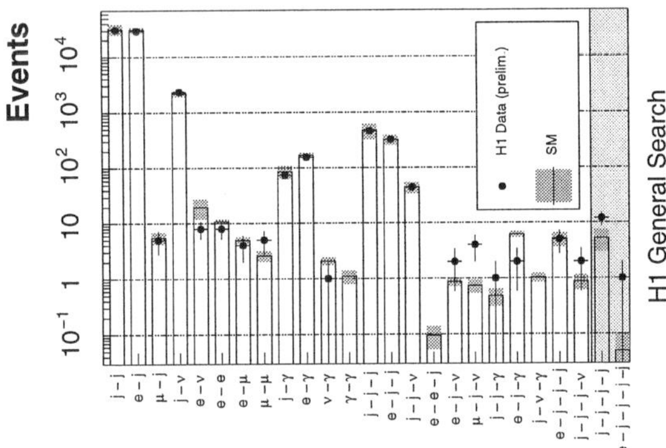

Figure 17. The number of observed and expected events for all event classes considered in the H1 general search for new phenomena.

the H1 analysis. In each channel, the part of the phase space where the data shows the largest deviation with the SM prediction is then identified, and the significance of this deviation is estimated using Monte Carlo experiments. In the H1 analysis, this procedure allows indeed to retrieve the "anomalies" reported in Secs. 2.4 and 2.5. No other significant deviation is observed by this analysis.

Such an approach is likely to be very helpful at the LHC experiments, where new physics might manifest itself in various final states. However such analyses require a very good understanding of the detectors and of the backgrounds, which will probably be achieved after some time of data taking.

6. Conclusions

Physics beyond the Standard Model is believed to possibly manifest itself at the TeV scale and many searches for new phenomena are carried out at the current experiments, some of which were presented in this non-exhaustive review. Although no convincing evidence for new physics has been observed so far, some "anomalies" remain to be clarified in the near future at the Tevatron and HERA colliders. Recent results coming from the Run II data at the Tevatron set the most stringent constraints on many models and the much larger luminosities expected within the next years will allow the TeV scale to be thoroughly probed.

Acknowledgments

I am very grateful to W. Addams, L. Bellagamba, M. Besançon, F. Boudjema, G. Broojmans, N. Castro, D. Dannheim, C. Diaconu, E. Gallo, F. Gianotti, G. Landsberg, N.S. Lockyer, C. Mariotti, H.U. Martyn, S. Mele, L. Poggioli, S. Rolli, A. Schöning, T. Sloan, I. Trigger, S. Worm, and A.F. Zarnecki for useful discussions or providing figures. I also thank the organizers for this very interesting symposium and the Scientific Secretaries for their help in preparing this talk.

References

1. M. Schmitt, these proceedings.
2. For a review, see A. Bellerive, these proceedings.
3. P. Gambino, these proceedings.
4. C.S. Wood *et al.*, *Science* **275**, 1759 (1997).
5. M. Kuchiev and V. Flambaum, hep-ph/0305053.
6. M. Davier *et al.*, *Eur. Phys. J.* C **27**, 497 (2003).
7. CMD-2 Collab., R.R. Akhmetshin *et al.*, hep-ex/0308008.
8. T. Browder, these proceedings.
9. CDF Collab., F. Abe *et al.*, *Phys. Rev.* D **59**, 092002 (1999).
10. CDF Collab., D. Acosta *et al.*, *Phys. Rev.* D **66**, 012004 (2002).
11. CDF Collab., D. Acosta *et al.*, *Phys. Rev.* D **65**, 052007 (2002); CDF Collab., G. Apollinari *et al.*, *Phys. Rev.* D **65**, 032004 (2002).
12. H1 Collab., A. Aktas *et al.*, DESY-03-082, to be published in *Eur. Phys. J.*
13. ZEUS Collab., "Study of multi-lepton production in $e^{\pm}p$ interactions at HERA", Contributed Paper. #910 to the ICHEP'02 Conference, Amsterdam (The Netherlands), July 2002.

14. H1 Collab., V. Andreev *et al.*, *Phys. Lett.* B **561**, 241 (2003).

15. ZEUS Collab., Contributed Paper. #909 to the ICHEP'02 Conference, Amsterdam (The Netherlands), July 2002.

16. T. Han, J. Hewett, *Phys. Rev.* D **60**, 074015 (1999).

17. H1 Collab., A. Aktas *et al.*, DESY-03-132, submitted to *Eur. Phys. J.*

18. ZEUS Collab., S. Chekanov *et al.*, *Phys. Lett.* B **559**, 153 (2003).

19. CDF Collab., *Phys. Rev. Lett.* **80**, 2525 (1998).

20. LEP Exotica Working Group, Note 2001-01.

21. CDF Collab., D. Acosta *et al.*, *Phys. Rev.* D **65**, 091120 (2002); D0 Collab., V.M. Abazov *et al.*, *Phys. Lett.* B **517**, 282 (2001).

22. E. Eichten, K. Lane and M. Peskin, *Phys. Rev. Lett.* **50**, 811 (1983).

23. H1 Collab., C. Adloff *et al.*, *Phys. Lett.* B **568**, 35 (2003); ZEUS Collab., "Search for large extra dimensions, finite quark radius and contact interactions in *ep* collisions at HERA", Contributed paper #602 to the EPS'01 Conference, Budapest (Hungary), July 2001.

24. K. Hagiwara, D. Zeppenfeld and S. Komamiya, *Zeit. Phys.* C **29**, 115 (1985).

25. F. Boudjema, A. Djouadi and J.L. Kneur, *Zeit. Phys.* C **57**, 425 (1993).

26. U. Baur, M. Spira and P.M. Zerwas, *Phys. Rev.* D **42**, 815 (1990).

27. DELPHI Collab., "Search for excited leptons with the DELPHI detector at LEP", Contributed Paper #291 to the ICHEP'02 Conference, Amsterdam (The Netherlands), July 2002; OPAL Collab., G. Abbiendi *et al.*, *Phys. Lett.* B **544**, 57 (2002).

28. LEP Exotica Working Group, Note 2001-02.

29. H1 Collab., C. Adloff *et al.*, *Phys. Lett.* B **548**, 35 (2002); ZEUS Collab., Contributed Paper #607 to the EPS'01 Conference, Budapest (Hungary), July 2001.

30. LEP II Diphoton WG, LEP2FF/02-02 (July 2002).

31. J. Bagger, C. Schmidt and S. King, *Phys. Rev.* D **37**, 1188 (1988); P.H. Frampton and S.L. Glashow, *Phys. Lett.* B **190**, 157 (1987); E.H. Simmons, *Phys. Rev.* D **55**, 1678 (1997).

32. W. Buchmüller, R. Rückl and D. Wyler, *Phys. Lett.* B **191**, 442 (1987); *Err. ibid* B **448**, 320 (1999).

33. ZEUS Collab., S. Chekanov *et al.*, *Phys. Rev.* D **68**, 052004 (2003).

34. H1 Collab., "A search for leptoquarks at HERA", Contributed Paper #185 to the LP'03 Symposium, Fermilab, August 2003; H1 Collab., C. Adloff *et al.*, *Phys. Lett.* B **523**, 234 (2001).

35. D0 Collab., B. Abbott *et al.*, *Phys. Rev. Lett.* **80**, 2051 (1998).

36. D0 Collab., *Phys. Rev. Lett.* **84**, 2088 (2000).

37. G.S. and R.N. Mohapatra, *Phys. Rev.* D **12**, 1502 (1975); R.N. Mohapatra and R.E. Marshak, *Phys. Rev. Lett.* **44**, 1316 (1980).

38. OPAL Collab., "Search for Doubly Charged Higgs Bosons with the OPAL detector at LEP", Contributed Paper #215 to the LP'03 Symposium, Fermilab, August 2003; DELPHI Collab., J. Abdallah *et al.*, *Phys. Lett.* B **552**, 127 (2003).

39. S. Godfrey, P. Kalyniak, and N. Romanenko, *Phys. Rev.* D **65**, 033009 (2002).

40. H1 Collab., "Search for doubly charged Higgs production at HERA", Contributed Paper #184 to the LP'03 Symposium, Fermilab, August 2003.

41. OPAL Collab., "Search for the Single Production of Doubly-Charged Higgs Bosons and Constraints on its Couplings from Bhabha Scattering", Contributed Paper #221 to the LP'03 Symposium.

42. L. Randall and R. Sundrum, *Phys. Rev. Lett.* **83**, 3370 (1999); idem, *Phys. Rev. Lett.* **83**, 4690 (1999).

43. T. Han, J.D. Lykken, R. Zhang, *Phys. Rev.* D **59**, 105006 (1999).

44. G.F. Giudice, R. Rattazzi and J.D. Wells, *Nucl. Phys.* B **595**, 250 (2001).

45. OPAL Collab., "Searches for Radions at LEP II", Contributed Paper #238 to the LP'03 Symposium, Fermilab, August 2003.

46. N. Arkani-Hamed, S. Dimopoulos and G. Dvali, *Phys. Lett.* B **429**, 263 (1998); I. Antoniadis, N. Arkani-Hamed, S. Dimopoulos and G. Dvali, *Phys. Lett.* B **436**, 257 (1998).

47. G.F. Giudice, R. Rattazzi and J.D. Wells, *Nucl. Phys.* B **544**, 3 (1999).

48. L3 Collab., "Single and Multi-Photon Events with Missing Energy in e^+e^- Collisions at $\sqrt{s} = 189 - 208$ GeV", L3 Note 2811, July 2003.

49. D0 Collab., V.M. Abazov *et al.*, *Phys. Rev. Lett.* **90**, 251802 (2003); CDF Collab., D. Acosta *et al.*, FERMILAB-PUB-03/285-E., submitted to *Phys. Rev. Lett.*, Sep. 2003.

50. J.L. Hewett, *Phys. Rev. Lett.* **82**, 4765 (1999).

51. K. Cheung and G. Landsberg, *Phys. Rev.* D **62**, 076003 (2000).

52. LEP Electroweak Working Group ($f\bar{f}$ subgroup), C. Geweniger *et al.*, LEP2FF/02-03 note.

53. D0 Collab., *Phys. Rev. Lett.* **86**, 1156 (2001).

54. A. Dobado and A.L. Maroto, *Nucl. Phys.* B **592**, 203 (2001); H. Murayama and J.D. Wells, *Phys. Rev.* D **65**, 056011 (2002).

55. L3 Collab., "Searches for Branons at LEP", Contributed Paper #829 to the EPS'03 Conference, Aachen (Germany), July 2003.

56. D0 Collab., *Phys. Rev.* D **64**, 012004 (2001).

57. H1 Collab., "General Search for New Phenomena in *ep* scattering at HERA", Contributed Paper #195 to the LP'03 Symposium, Fermilab, August 2003.

DISCUSSION

Peter Rosen (Department of Energy): Looking at like-sign dileptons, can you get better limits from nuclear double-beta decay?

Emmanuelle Perez: I was not aware of the answer at the time of the talk. This is what I found afterwards, going through the relevant literature.

With the standard gauge structure of the SM, the contribution of doubly-charged Higgses to the process $dd \to uue^-e^-$, via the trilinear interaction of H^{--} with the singly charged field belonging to the $SU(2)$ Higgs doublet, was first investigated by R.N. Mohapatra and J.D. Vergados in *Phys. Rev. Lett.* **47**, 1713 (1981). It was shown however that the corresponding amplitude is largely suppressed, being proportional to the square of the d quark mass (J. Schechter and J.W.F. Valle, *Phys. Rev.* D **25**, 2591 (1982); L. Wolfenstein, *Phys. Rev.* D **26**, 2507 (1982)). Schechter and Valle also showed that the contribution induced by $W^-W^- \to H^{--} \to e^-e^-$ is negligible as well, with a suppression proportional to the Majorana neutrino mass.

In left-right symmetric models the dominant contribution is induced by the H^{--} which belongs to the $SU(2)_R$ triplet, via its coupling to the right-handed W. However, as shown by M. Hirsch *et al.* in *Phys. Lett.* B **374**, 7 (1996), this contribution turns out to be numerically small unless the doubly-charged Higgs boson is very light.

THEORETICAL PREDICTIONS FOR COLLIDER SEARCHES

G. F. GIUDICE

Theoretical Physics Division, CERN, CH-1211 Geneva 23, Switzerland
E-mail: gian.giudice@cern.ch

I review recent developments in extensions of the Standard Model that address the question of electroweak symmetry breaking and discuss how these theories can be tested at future colliders.

In their search for understanding the fundamental laws of elementary particles, high energy physicists are eager to prove that one of the most successful and elegant scientific theories ever formulated – the Standard Model (SM) – actually fails at distances smaller than about a hundred zeptometers (10^{-19} m). A generation of high energy colliders, culminating with the LHC under construction at CERN, has been designed to achieve this goal. Here I will briefly review the present status of the theoretical speculations for new physics within reach of the LHC. In the spirit of this conference, and because of limited space, I will consider only topics in which, in my opinion, there have been recent developments, leaving aside other possibilities which, although interesting, have known only limited progress. The list of references is also largely incomplete.

1. "Big" and "Little" Hierarchy Problems

The hierarchy problem[1] is not necessarily the most puzzling open question of the SM, but it is certainly the most relevant to collider experiments, since its resolution lies – most likely – in the TeV energy range. Its formulation is well known. Treating the SM as an effective theory valid up to a scale Λ_{SM}, and cutting off momenta in loop integrals at the same scale, we find that the dominant radiative correction to the Higgs mass is given by

$$\delta m_H^2 = \frac{3G_F}{4\sqrt{2}\pi^2}\left(2m_W^2 + m_Z^2 + m_H^2 - 4m_t^2\right)\Lambda_{\mathrm{SM}}^2$$

$$= -\left(200\,\mathrm{GeV}\,\frac{\Lambda_{\mathrm{SM}}}{0.7\,\mathrm{TeV}}\right)^2. \tag{1}$$

The request of no fine-tuning between the tree-level and one-loop contributions to m_H implies $\Lambda_{\mathrm{SM}} \lesssim$ TeV. In other words, the SM cannot be valid beyond the TeV, and new physics should appear to modify the ultraviolet behavior. I will refer to this result as the "big" hierarchy, since other fundamental scales, such as the Planck mass M_{Pl}, are known to be much larger than Λ_{SM}.

Table 1. 90% CL limits on the scale Λ_{LH} (in TeV) of dimension-six operators $\mathcal{O}$ in the effective Lagrangian $\mathcal{L} = \pm\Lambda_{\mathrm{LH}}^{-2}\mathcal{O}$, in both cases of constructive and destructive interference with the SM contribution ($\pm$). The limits on the operators relevant to LEP1 are derived under the assumption of a light Higgs.

		$-$	$+$
LEP1[3]	$H^\dagger\tau^a H W_{\mu\nu}^a B^{\mu\nu}$	10	9.7
	$|H^\dagger D_\mu H|^2$	5.6	4.6
	$iH^\dagger D_\mu H\bar{L}\gamma^\mu L$	9.2	7.3
LEP2[4]	$\bar{e}\gamma_\mu e\bar{\ell}\gamma^\mu\ell$	6.1	4.5
	$\bar{e}\gamma_\mu\gamma_5 e\bar{b}\gamma^\mu\gamma_5 b$	4.3	3.2
MFV[2]	$\frac{1}{2}(\bar{q}_L\lambda_u\lambda_u^\dagger\gamma_\mu q)^2$	6.4	5.0
	$H^\dagger\bar{d}_R\lambda_d\lambda_u\lambda_u^\dagger\sigma_{\mu\nu}q_L F^{\mu\nu}$	9.3	12.4

As a word of caution, I should recall that, if I apply the same naturalness argument to the quartic divergences, then the present upper bound on the cosmological constant implies that the corresponding cut-off has to be smaller than 10^{-3} eV. Although we cannot fully rule out the possibility that the ultraviolet behavior of gravity is prematurely modified at about 10^{-3} eV, this embarrassing result undoubtedly casts a grievous shadow over the hierarchy problem.

Let me come back to the SM as an effective theory. Unknown new physics at the cut-off is parametrized in terms of non-renormalizable operators. To be most conservative, I will add only operators that preserve all local and global SM symmetries and satisfy the criterion of minimal flavor violation.[2] As shown in Table 1, typical limits on Λ_{LH}, defined as the effective scale of the new dimension-six operators ($\mathcal{L} = \pm\Lambda_{\mathrm{LH}}^{-2}\mathcal{O}$), are $\Lambda_{\mathrm{LH}} > 5$–$10$ TeV.

I will refer to the "little" hierarchy (LH) problem as the tension between the constraints on Λ_{LH} – which imply that new-physics virtual effects could only emerge at energies larger than 5–10 TeV – and the no-fine-tuning condition – which requires the presence of new dynamics at the scale Λ_{SM}, below the TeV. To some readers this could seem like a marginal problem, but I believe it provides a very useful guideline in the search for the correct theory beyond the SM. It is a typical LEP heritage, where any new-physics theory has to be confronted with very precise data.

The "little" hierarchy between Λ_{SM} and Λ_{LH} is already giving us some lessons on the hypothetical theory beyond the SM. First of all, new physics at Λ_{SM} is most likely weakly-interacting, or else effective operators would be generated with $\Lambda_{\mathrm{LH}} \simeq \Lambda_{\mathrm{SM}}$. Strong dynamics, if any, can only appear at scales larger than Λ_{LH}. Moreover, even for weak dynamics, physics at Λ_{SM} should not induce (sizable) tree-level contributions to effective operators. Recently there has been growing interest in constructing new theories that describe the physics between Λ_{SM} and Λ_{LH}, as I will report in Secs. 4, 5, and 6.

2. Supersymmetry

Supersymmetry[5] provides an elegant solution to the "big" hierarchy problem, by eliminating quadratic divergences with a symmetry principle. In softly-broken supersymmetry, the superpartner masses (generated by gauge-invariant terms) effectively play the rôle of Λ_{SM}. Because of the absence of quadratic divergences, the validity of the theory can be extended to energy scales as large as M_{Pl}, without encountering naturalness problems. This extension has several welcome consequences. It offers the possibility to link the SM to speculative ideas about quantum gravity at M_{Pl}. It provides a viable setting for GUT's (with successful unification of gauge coupling constants), a framework for neutrino masses, and for suppressing proton decay. Finally, it allows an implementation of cosmological mechanisms, such as inflation or baryogenesis, which benefit from extrapolations of particle physics models to high energies.

Incidentally, the need for an extrapolation of the SM up to super-heavy scales has been challenged by recent research aiming at constructing new models whose validity cut-off is not very far from the electroweak scale. In these scenarios, at first sight, the advantages of the energy "desert" are lost. Nevertheless, theoreticians have suggested new ways of recovering some of the positive aspects. Just as an example, let me consider gauge coupling unification. In theories with extra dimensions, Kaluza–Klein excitations of SM particles can accelerate the running, possibly leading to a precocious gauge unification,[6] although predictivity is lost in the power running. Alternatively, one can consider an N-fold replication of the SM gauge group, where the unification scale becomes $10^{13/N}$ TeV.[7] Yet, one can abandon the usual tree-level expression for the weak mixing angle in GUT's: $\sin^2 \theta_W = \mathrm{Tr}\ I_3^2 / \mathrm{Tr}\ Q^2 = 3/8$ (where the trace is over any GUT irreducible representation). Taking an electroweak group $SU_3 \times SU_2 \times U_1$ broken to $SU_2 \times U_1$, in the limit $\tilde{g}_2, \tilde{g}_1 \gg \tilde{g}_3$ (where $\tilde{g}_i$ are the coupling constants of the high energy group), one finds $\sin^2 \theta_W = 1/4$.[8] This is much closer to the experimental value $\sin^2 \theta_W = 0.231$ than 3/8, and therefore little running is needed.

Aside from this connection to super-high energies, supersymmetry has several interesting and successful features in the low-energy domain. *(i)* Gauge-coupling unification leads to a prediction of a SM parameter: $\alpha_s(M_Z) = 0.124$ (for typical supersymmetric thresholds at 1 TeV) and 0.130 (for thresholds at 250 GeV). Given the intrinsic uncertainty of GUT thresholds, this is consistent with the PDG value $\alpha_s(M_Z) = 0.117 \pm 0.002$.[9] *(ii)* Electroweak symmetry is triggered radiatively by the top Yukawa interaction, with the broken group dynamically chosen (color SU_3 is preserved since squarks do not develop negative square masses, while weak SU_2 is broken). *(iii)* The Higgs boson is predicted to be lighter than about 130 GeV, compatibly with the indications from electroweak precision data. *(iv)* The "little" hierarchy is satisfied, if one assumes R-parity conservation. In this case, supersymmetric virtual effects are loop-suppressed and $\Lambda_{\mathrm{LH}} \sim 4\pi\Lambda_{\mathrm{SM}}$. *(v)* There is a natural candidate for cold dark matter, if R-parity is conserved.

These positive aspects of low-energy supersymmetry are well known, but let me spend a few words on the negative aspects. The first obvious drawback of low-energy supersymmetry is that no superpartner has been observed. This is in contradiction with the requirement of less than 10% fine-tuning, which predicted a discovery at LEP2.[11] In conventional models of low-energy supersymmetry, the most strin-

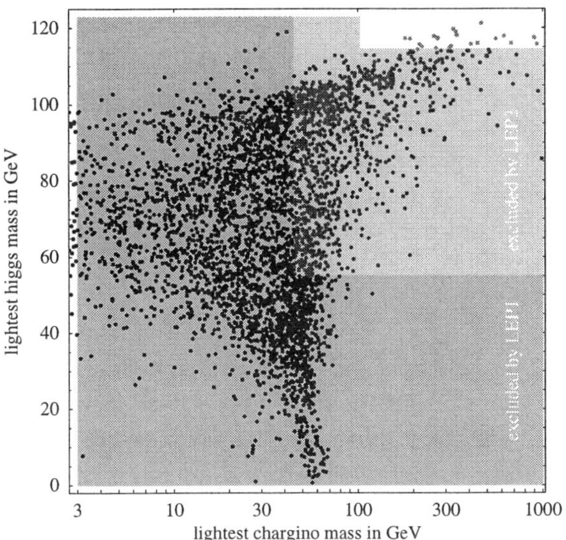

Figure 1. Scatter plot of the chargino and Higgs mass obtained by sampling the parameter space of minimal supergravity. The regions excluded by LEP1 and LEP2 searches are shown. This figure was produced by A. Strumia, updating earlier results.[10]

gent direct bounds come from chargino ($m_{\chi^+} > 103.5$ GeV),[12] charged slepton ($m_{\tilde{e}_R} > 99.4$ GeV),[13] and Higgs searches. The SM Higgs mass bound of $114.4\,\text{GeV}$[14] applies only in the limit of a heavy pseudoscalar, while the general bound is 91.0 GeV[15] (the final combined LEP analysis on the supersymmetric Higgs has not appeared yet). These limits imply a fine-tuning of the order of the per cent. This is illustrated in Fig. 1, which shows the values of Higgs and chargino masses obtained by a random sample of supersymmetric models with universal boundary conditions at the GUT scale and correct electroweak breaking.[10] Less than 1% of the points have simultaneously $m_{\chi^+} > 103$ GeV and $m_H > 114$ GeV (for pseudoscalar Higgs masses close to m_H, the bound becomes weaker but the allowed region of Fig. 1 remains essentially the same).

This is a very disappointing result of supersymmetry, leading to three possible explanations. *(i)* Supersymmetry is not realized at the weak scale. *(ii)* Low-energy supersymmetry is correct, but parameters are fine-tuned at the level of the per cent (and LHC will discover the superpartners). In this case, you may be relieved to know that there is one example in nature of an accidental correlation

of this order of magnitude. The apparent sizes of the Sun and the Moon in the sky are roughly equal. If you applied the naturalness criterion to conclude that their actual sizes are roughly equal too, you would be quite wrong. Their radius and distance to the Earth are "fine-tuned" to give, on average, $(\theta_{\text{Sun}} - \theta_{\text{Moon}})/\theta_{\text{Sun}} \sim 3\%$, where $\theta_{\text{Sun,Moon}}$ are the solar and lunar angular sizes. *(ii)* Low-energy supersymmetry exists, but its realization is different from what we had imagined. Let me expand on this.

The origin of the fine-tuning is linked to the success of supersymmetry to trigger the electroweak breaking. Indeed, from the Higgs mechanism we know that $M_Z^2 = -g^2\mu_H^2/\lambda$, where μ_H^2 and λ are the coefficients of the quadratic and quartic terms in the Higgs potential. In supersymmetry, $\lambda \sim g^2$, and therefore $M_Z^2 \sim -\mu_H^2$. The effective μ_H^2 has a tree-level direct contribution from the soft terms and a one-loop negative contribution roughly proportional to the stop mass (which, in turn, is usually determined by the contribution proportional to the gluino mass). In the conventional supergravity approach, the one-loop contribution has a large logarithm and it overcomes the tree-level effect. Therefore $|\mu_H|$ is of the order of the supersymmetric masses, which are expected to lie close to M_Z. One may think that the situation improves in gauge-mediated models,[16] since the logarithm can be made small with sufficiently light messengers. However, this is not the case, because gauge mediation gives a boundary condition for the stop mass at the messenger scale, which is significantly larger than the tree-level contribution to μ_H^2. Actually, in gauge mediation, the fine-tuning is usually more acute, because the right-handed selectron is a factor of about 2 lighter than the Higgs mass parameter, and because of the limit on the visible decay of the lightest neutralino.[10] The situation could improve in models where the stop and gluino masses are not large (with respect to the other soft masses) or in models with a low supersymmetry-breaking scale,[17] or in models where μ_H^2 is truly a one-loop factor smaller than all supersymmetric masses (see for instance ref.[18]).

The second drawback of low-energy supersymmetry is the lack of predictivity in the supersymmetry-breaking sector, which translates into the "supersymmetric flavor problem" (i.e. the large flavor violations present with generic soft terms). We are now aware of various schemes, alternative to the conventional supergravity scenario, which are more predictive and address the flavor problem: gauge mediation,[16] anomaly mediation,[19] gaugino mediation.[20] However, at present, we cannot say that one scheme is preferable to the others. For instance, in operator-based schemes (like supergravity) the Higgs mixing mass μ can be properly accounted for,[21] while in other schemes where the soft terms are calculable, the Higgs mass parameters μ and $B\mu$ typically appear at an undesired order in perturbation theory. Luckily, the different schemes usually have quite distinct features both in the mass spectrum and in the nature of the lightest supersymmetric particles, leading to distinct signals at colliders. Therefore, in the case of a discovery, the question of the correct scheme of supersymmetry-breaking terms can be settled experimentally.

Recently many new models have been proposed, where supersymmetry at the weak scale is implemented in unconventional ways. These proposals make use of new ingredients, coming from extra dimensions. Before discussing them, let me first introduce extra dimensions.

3. Extra Dimensions

The hierarchy problem can motivate the existence of experimentally-accessible extra dimensions.[22,23] The scenarios are by now well known: our three-dimensional space is embedded in a larger D-dimensional space-time. While SM fields are confined on the three-dimensional brane, gravity is described by the geometry of the full space. The crucial assumption is that the fundamental Planck scale of the D-dimensional theory M_D is of the order of the weak scale, avoiding any hierarchy. Any distance scale shorter than TeV^{-1} (associated with Newton's constant, with GUT's, right-handed neutrinos, etc.) has to be explained by some geometrical property.

For flat and compact extra dimensions, the observed weakness of gravity is translated into a large value of the compactification radius R, since the ordinary Planck mass is given by[22] $M_{\text{Pl}} \sim R^{\delta/2} M_D^{1+\delta/2}$, where $D = 4 + \delta$. The interpretation is that we observe a weak gravitational force not because gravity is intrinsically weak ($M_D \sim \text{TeV}$), but because of the small overlap between our world and the graviton wave function, which is spread in a very large volume.

For spaces with non-factorizable metrics, the large hierarchy M_{Pl}/M_W can be reproduced with much smaller values of R, exploiting the strong functional dependence in the warp factor. In this case, the interpretation is that, because of the non-trivial geometry, the zero-mode graviton wave function is peaked on a brane far from ours and only its exponential tail overlaps with our world, leading to the observed weakness of gravity. The hierarchy is the result of the familiar gravitational redshift in general relativity. A photon that climbs out of a gravitational potential sees its measured energy reduced. It is known that for time-independent metrics with $g_{0j} = 0$, the product $E\sqrt{|g_{00}|}$ is conserved, rather than the energy E. Therefore, a photon emitted at a distance r from a spherically symmetric gravitational source is observed at infinity to be redshifted by an amount $(E_{obs} - E_{em})/E_{em} = \sqrt{|g_{00}|} - 1 \simeq -G_N M/r$, since the Schwarzschild metrics generated by a mass M has $g_{00} = 1 - 2G_N M/r$. Similarly, in the Randall Sundrum set-up, masses in the hidden (Planck) brane are blueshifted by a warp factor, when measured on the observable (TeV) brane, as a result of the non-trivial gravitational field.

Although the scenarios with a quantum-gravity scale at the TeV have various theoretical and cosmological difficulties, they represent a wonderful possibility for new-physics searches at future colliders. At the LHC, graviton emission is observed in jet plus missing-energy events[24,25] (for flat dimensions) or as resonances in Drell–Yan[26] (for warped dimensions). Tree-level graviton exchange leads to an effective dimension-8 operator $T_{\mu\nu}T^{\mu\nu}$, which predicts correlated signals in diphoton and dilepton final states.[24,27] Graviton loops give rise to dimension-6 operators, with a flavor-universal axial-vector contact interaction playing a special rôle.[28] Because of the lower dimensionality of the induced operators, loop effects are generally more important than tree-level effects, unless the short-distance behavior of gravity is modified at an energy scale significantly smaller than M_D. The radion, a scalar field contained in the extra-dimensional graviton, can mix with the Higgs boson, giving rise to interesting variations in the Higgs searches.[29]

The theoretical description of the experimental signals described above relies upon linearized gravity, whose validity ends as the relevant energy of the process ($\sqrt{s}$) approaches the fundamental gravity scale

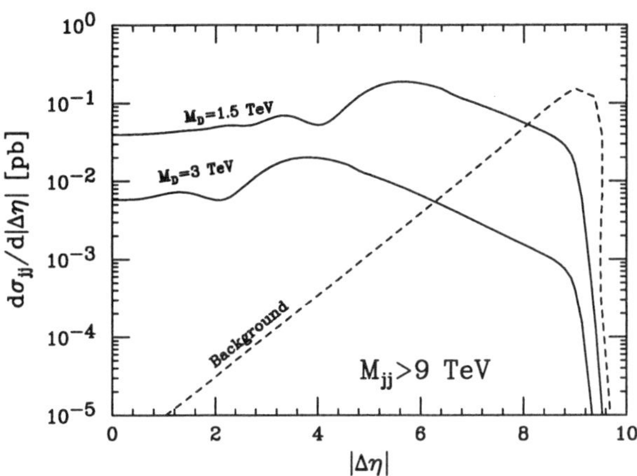

Figure 2. The di-jet differential cross section $d\sigma_{jj}/d|\Delta\eta|$ versus the jet rapidity separation $\Delta\eta$ at the LHC from transplanckian gravitational scattering for $\delta = 6$, $M_D = 1.5$ TeV and 3 TeV. Cuts are chosen such that the di-jet invariant mass $M_{jj} > 9$ TeV, and both jets have $|\eta| < 5$ and $p_T > 100$ GeV. The dashed line is the expected background from QCD.

M_D, and we enter the region where experiments are directly sensitive to the underlying quantum-gravity theory. However, it is interesting that there exists another kinematical regime where the theoretical description is tractable. This is the transplanckian region $\sqrt{s} \gg M_D$, where a semi-classical description is appropriate. This can be understood by noticing that, in the transplanckian region, the classical length associated with a gravitational scattering – the Schwarzschild radius $R_S \sim (G_D\sqrt{s})^{1/(\delta+1)}$, where $G_D \sim M_D^{-(\delta+2)}$ is the D-dimensional Newton constant – becomes larger than the Planck length M_D^{-1} associated to quantum-gravity effects.

When the collision impact parameter b is larger than R_S, the non-linearity of the gravitational field is negligible, and the gravitational scattering, although non-perturbative, can be theoretically calculated. This leads to a prediction at hadron colliders for two-jet final states at large invariant mass and large rapidity separation with characteristic distributions[30] (an example is shown in Fig. 2). The peculiarity in the rapidity distribution comes from a diffractive pattern associated with a new scale in the gravitational potential $b_c \sim (G_D s)^{1/\delta}$, which exists in any dimensions different from 4 (i.e. $\delta \neq 0$) – the familiar Coulomb-like potential is actually very special. These features can be used to confirm the gravitational nature of possible discoveries of new physics.

As the impact parameter b approaches the

Schwarzschild radius R_S, the gravitational field becomes strong, and analytic calculations cannot be done. However, there are solid indications[31] that, when $b < R_S$, black holes are formed.[32] This opens the exciting possibility that (unstable) black holes can be observed at future colliders, with spectacular experimental signals.

I should remark that given the present bounds on M_D from direct searches, and the "limited" energy available at the LHC, the window for observing transplanckian phenomena is rather narrow and, more importantly, the separation between classical and quantum scale may not be sufficient to avoid pollution from quantum-gravity effects. Although it is of course important to pursue these searches at the LHC, transplanckian phenomena can offer a good motivation to consider future hadron colliders, such as the VLHC, operating at $\sqrt{s} \sim 100$–200 TeV.

As we have considered extra-dimensional gravity, it is possible that also SM particles live in (part of) the extra-dimensional compactified space.[33] Collider experiments can look for their Kaluza–Klein (KK) excitations. If only gauge bosons live in five dimensions, while fermions are conventional, a combination of present direct and indirect searches yield a 95% CL lower limit on the compactification scale M_C (the mass of the first excited mode) of 6.8 TeV.[34] Searches at the LHC have a sensitivity reach of up to 13–15 TeV. Much weaker bounds ($M_C > 0.3$ TeV) apply to the case of universal extra dimensions,[35] where all SM fields live in five dimensions. This is because, after compactification, the momentum conservation along the fifth dimension corresponds to a conserved KK number. This plays a rôle analogous to R-parity conservation in supersymmetry. Indeed KK particles can only be pair-produced and cannot contribute to SM processes by virtual tree-level exchange. Moreover, the lightest KK mode, corresponding to the hypercharge gauge boson, is stable and is a possible candidate for cold dark matter.[36]

4. Supersymmetry Breaking and Extra Dimensions

At the end of Sec. 2, I anticipated that extra dimensions bring new elements in the construction of low-energy supersymmetric models. Let me briefly discuss these new ingredients, which are related to the issues of supersymmetry and gauge-symmetry breaking, orbifold projection, and the AdS/CFT correspondence.

Consider a field Φ, which lives in a five-dimensional space with the fifth dimension y compactified on a circle of radius R. Because of periodicity, the field can be expanded in discrete Fourier modes in y, which correspond to the familiar KK tower of four-dimensional fields with masses $m_n = n/R$, for integers n. Let us now impose the nontrivial boundary condition that the field Φ picks up a phase as it goes around the circle:

$$\Phi(x, y + 2\pi R) = e^{2\pi i Q_\Phi}\Phi(x, y). \qquad (2)$$

The KK expansion consistent with condition (2) is

$$\Phi(x, y) = e^{iQ_\Phi y/R} \sum_{n=-\infty}^{+\infty} e^{iny/R}\Phi_n(x). \qquad (3)$$

The mass of the n-th mode (i.e. its momentum along the fifth coordinate) is shifted by the boundary condition to the value $m_n = (n+Q_\Phi)/R$. Since the mass spectrum of different fields can be shifted in different ways, symmetries of the five-dimensional theory can be hidden if we look at its four-dimensional truncated version, obtained by keeping only light modes. If the phase Q_Φ corresponds to an R-charge (i.e. if particles inside the same supermultiplet have different boundary conditions), then supersymmetry is broken in the four-dimensional effective low-energy theory. This is the Scherk–Schwarz symmetry-breaking mechanism through boundary conditions in the extra dimensions.[37] Notice that the mechanism is intrinsically non-local, since it involves the global structure of the compactified space. Therefore, if we look at small distances (smaller than R), any symmetry-breaking effect should disappear. This is a very useful property. For instance, in order to stabilize the weak scale it is essential to maintain the cancellation of quadratic divergences in the ultraviolet, and therefore we want to recover supersymmetry at short distances.

The compactification of the fifth dimension into a circle was obtained by identifying the point y with $y + 2\pi R$. We can further restrict the space by identifying y with $-y$: the circle collapses into a segment. The KK decomposition (take Eq. (3) with $Q_\Phi = 0$ for simplicity) can be written as $\Phi = \Phi^{(+)} + i\Phi^{(-)}$, where

$$\Phi^{(+)}(x, y) = \sum_{n=0}^{\infty} [\Phi_n(x) + \Phi_{-n}(x)] \cos(ny/R)$$

$$\Phi^{(-)}(x, y) = \sum_{n=1}^{\infty} [\Phi_n(x) - \Phi_{-n}(x)] \sin(ny/R). \qquad (4)$$

The fields $\Phi^{(+)}$ and $\Phi^{(-)}$ are even and odd, respectively, under the symmetry $y \to -y$. However, only the even tower has a massless zero mode ($n = 0$). This projection into an orbifold (essentially a manifold with boundaries) is very useful in model building. In particular,[38] starting from vector-like fermion representations in extra dimensions, one can retain only some chiral components in the truncated four-dimensional theory, where all massive modes have been integrated out, and construct realistic models for weak interactions. Combinations of Scherk–Schwarz boundary conditions and orbifold projections give rise to many interesting possibilities for reducing the symmetry and the particle content of extra-dimensional theories.

The next ingredient I want to discuss is the AdS/CFT correspondence,[39] which is the conjecture that properties of conformal field theories in D dimensions are related to those of $(D+1)$-dimensional theories with gravity in anti-de Sitter space. It has also been suggested that the non-compact version of the RS model (RS2)[40] is dual to a four-dimensional strongly-coupled conformal theory with gravity.[41] One can take advantage of these speculations to extract some physical information on the structure of the RS1 set-up with two branes in a slice of AdS$_5$, discussed in Sec. 3.[42]

The line element of the RS set-up is given by

$$ds^2 = \exp\left(-\frac{2y}{L}\right) dx^\mu dx_\mu + dy^2, \qquad (5)$$

where y is the extra coordinate and L is the AdS radius. From Eq. (5) we observe that a scale transformation $x \to \xi x$ can be reabsorbed by a shift of the coordinate y. This is suggestive of the holographic interpretation of the fifth coordinate as the renormalization scale of the four-dimensional theory. The hidden (Planck) brane corresponds to the ultraviolet cut-off of the conformal theory and to the addition of gravity. Indeed, gravity decouples from the four-dimensional theory as the Planck brane is moved to infinity. The visible (TeV) brane is the infrared cut-off, where the observable fields live. It corresponds to a spontaneous breaking of the conformal symmetry. Local gauge symmetries in the bulk of the five-dimensional theory are matched to global symmetries of the conformal theory. This connection between higher-dimensional and quasi-conformal theories could be very useful for gaining new insights on some properties of the strongly-coupled regime, which can be accessed by perturbative calculations in the dual theory.

The ingredients presented above have been used by many authors to construct new implementations of supersymmetry at the weak scale. Although I find many of these attempts quite interesting, I do not intend to give a review of these proposals here. Instead I will take only two examples, to show the use of the higher-dimensional ingredients.

The first example[43] is based on the extension of the supersymmetric model into five dimensions, one of which is compactified on a circle with two orbifold projections, $S^1/(Z_2 \times Z_2)$. In five dimensions, there are two supersymmetries. Each boundary of the orbifold breaks one supersymmetry and therefore the effective low-energy theory is non-supersymmetric. However, supersymmetry is recovered at distances smaller than the compactification radius R, leaving only a *finite* contribution to the Higgs mass, and predicting[43] $m_H = 127 \pm 10$ GeV. It is remarkable that one is able to compute a SM parameter (most new-physics theories add new parameters instead of predicting the values of SM quantities!). Unfortunately a quadratic sensitivity to the ultraviolet cut-off is actually introduced by a one-loop contribution to the Fayet–Iliopoulos term.[44] Also the contribution to the ρ parameter is too large, unless we tune the coefficient of an ultraviolet-sensitive operator. Finally unitarity is violated at a scale of 1.7 TeV, where an ultraviolet completion is needed. Variations of this model have been proposed,[45] in which some of these problems can be alleviated, in particular by raising the cut-off into the 5–10 TeV region.

In this model we can clearly appreciate the attractive features of Scherk–Schwarz supersymmetry breaking and orbifolding. Note also that, from the point of view of collider searches, these models are completely distinct from the conventional low-energy supersymmetric ones. Indeed, here there is only one Higgs doublet and two superpartners for each SM particle (the remnant of supersymmetry in five dimensions). The lightest supersymmetric particle is a stable (or metastable, if small R-parity violations are included) stop with a mass of about 210 GeV.

The second example[46] I want to present exploits the possibility of coexistence of supersymmetric and non-supersymmetric sectors in extra dimensions.[47] Let us consider a supersymmetric RS1 set-up where the Higgs sector lies on the TeV brane and the other SM multiplets live in the bulk. A source of supersymmetry breaking is localized on the Planck brane. The bulk fields are directly coupled to the supersymmetry-breaking sector and superpartners (squarks, sleptons, and gauginos) acquire masses of the order of M_{Pl}: supersymmetry is badly broken. However the Higgs sector can only feel supersymmetry-breaking effects through loops involving brane-to-brane mediation of bulk particles. Since this corresponds to a non-local interaction, the contribution is finite. Moreover, because of the gravitational redshift discussed in Sec. 3, the Higgs sector feels that the supersymmetry-breaking source is a warp factor smaller than M_{Pl}. Using the AdS/CFT correspondence, we can interpret the theory in four dimensions. The SM fields are elementary, while the Higgs sector corresponds to bound states of a quasi-conformal field theory. The electroweak scale is generated at the one-loop level, and therefore $\langle H \rangle \sim (1/4\pi)L^{-1}$, where L (assumed to be TeV^{-1}) is the size of the bound states. The loop factor is very welcome to explain the little hierarchy discussed in Sec. 1. The collider phenomenology is again quite distinct from usual low-energy supersymmetry. Only the Higgs has supersymmetric partners (higgsinos), while a tower of CFT bound states will appear at energies of the order of L^{-1}. This interpretation allows a connection of this set-up with models of composite Higgs bosons.[48] Unfortunately, precision electroweak data set a lower bound on L^{-1} of the order of 9 TeV, implying a certain degree of fine-tuning in the generation of the weak scale. I hope this example has clarified how AdS/CFT can be used in model building to extract information in theories with strong interaction and composite Higgs bosons.

In conclusion, extra dimensions have brought in the game new ingredients for low-energy supersymmetric models. Supersymmetry could reveal itself at colliders in a variety of ways: not only with missing energy, as in supergravity models, but also accompanied by leptons, or photons, or stable charged particles (as in gauge mediation[16]) or with nearly-degenerate $\tilde{W}^{\pm,0}$ (as in anomaly mediation[49]), or with new characteristic signals (stable stop, partially supersymmetric spectrum, etc.).

5. Electroweak Breaking and Extra Dimensions

The ingredients that I have discussed in the previous section can also be used to break gauge symmetries, and it is therefore a logical possibility to use extra dimensions to break the electroweak group.

The hierarchy problem is solved by a symmetry that forbids the Higgs mass term. Supersymmetry (in conjunction with chiral symmetry) is an example. Another example could be gauge symmetry $A_\mu \to A_\mu + \partial_\mu \Lambda$ that forbids a mass term $m^2 A_\mu^2$, but the Higgs boson is not a gauge particle. However, this is true in four dimensions, but not necessarily in more dimensions. A gauge boson in $D = 4 + \delta$ dimensions can be decomposed as $A_M = (A_\mu, A_j)$, $j = 1, \ldots, \delta$, where A_j are scalar fields under the Lorentz group. The extra-dimensional components could be interpreted as Higgs bosons.[50,51] This provides a Higgs–gauge unification completely analogous to the old version of the KK unification of electromagnetism and gravity, in which photon and graviton are different components of the same field – the five-dimensional metrics.

Since, unlike the gauge boson, the Higgs does not transform under the gauge group as the adjoint representation, one has to rely on the orbifolding discussed in Sec. 4 to project out unwanted states. More difficult is to generate the Higgs Yukawa and quartic couplings without reintroducing quadratic divergences. This problem is typically harder than the similiar one encountered in Little Higgs theories (discussed in Sec. 6), since the structure of the Higgs interactions is tightly constrained by the embedding in of extra dimensions. Nevertheless, there has been progress towards the construction of realistic models.[52]

Following a different approach, one can try to break the electroweak group by appropriate boundary conditions of the gauge bosons in the extra dimensions, and eliminate the Higgs altogether. Of course, in this case, one expects KK excitations for the W and Z bosons with masses too low to satisfy present collider bounds.[53] By warping the extra dimension, one can lift the masses of the excited KK modes with respect to the zero modes (to be identified with ordinary gauge bosons) and make these particles phenomenologically acceptable,[54,55] opening a path for the construction of realistic models for elec-troweak breaking with no Higgs (neither fundamental nor composite). Notice that these proposals differ substantially from the case of a Higgsless SM where the gauge symmetry is non-linearly realized. In the latter case unitarity in scattering processes involving longitudinally-polarized W bosons is violated at an energy scale $G_F^{-1/2}$, below the TeV. On the contrary, violation of unitarity in higher-dimensional theories is postponed to energy scales in the 10 TeV region.[56] This is sufficient to allow these theories to pass the test of the little hierarchy (see Sec. 1). Effectively, the tower of gauge boson KK modes (partly) plays the rôle of the Higgs boson in modifying the cross section high-energy behavior.

At present the main obstacle for Higgsless extra-dimensional theories seems to come from electroweak precision data. Contributions to the ρ parameter can be made small by imposing a gauge custodial symmetry in the bulk (which has the holographic interpretation of a global custodial symmetry in the four-dimensional theory, see Sec. 4). However, contributions to the oblique parameter S[57] (or ϵ_3[58]) appear to be positive and unacceptably large.[59]

6. Little Higgs

We have seen that supersymmetry and gauge symmetry are two examples of principles that can be used to forbid a mass term for the Higgs boson and solve the hierarchy model. A third example is a non-linearly realized global symmetry, where the Higgs has to be identified with a Goldstone boson.

Consider a scalar field Φ, transforming under an abelian global symmetry as $\Phi \to e^{ia}\Phi$, and assume that the symmetry is spontaneously broken by $\langle\Phi\rangle = f$. I can parametrize the complex field Φ in terms of two real fields ρ and θ as $\Phi = (\rho + f)e^{i\theta/f}/\sqrt{2}$. The symmetry transformation properties are $\rho \to \rho$ and $\theta \to \theta + a$. If I identify θ (the Goldstone boson) with the Higgs, the symmetry forbids a mass term $m^2\theta^2$.

The non-trivial part of the problem is to generate the gauge, Yukawa, and self-interaction of the Higgs (which are non-derivative couplings and therefore forbidden by the symmetry) without reintroducing quadratic divergences. Some (not entirely successful) attempts to construct realistic models for the Higgs as a pseudo-Goldstone boson were made in the past.[60]

Recently, a much less ambitious program was

proposed, which is known under the name of "Little Higgs".[61] The proposal is to solve only the little hierarchy problem; in other words, one searches for a description valid only up to 10 TeV or so. Recall, from Sec. 1, that the big hierarchy problem is formulated in terms of one-loop corrections to the Higgs mass:

$$\delta m_H^2 \sim \frac{G_F}{\pi^2} m_{\rm SM}^2 \Lambda_{\rm SM}^2. \tag{6}$$

If we require this correction not to exceed the masses of the SM particles $m_{\rm SM}$, we obtain

$$\Lambda_{\rm SM} < \frac{\pi}{\sqrt{G_F}} \sim \text{TeV}. \tag{7}$$

Suppose that at the scale $\Lambda_{\rm SM}$ ($< \text{TeV}$) new physics cancels the one-loop power divergences. Then we are left with one-loop logarithmic divergences and two-loop power divergences of the form

$$\delta m_H^2 \sim \frac{G_F^2}{\pi^4} m_{\rm SM}^4 \Lambda^2. \tag{8}$$

The same naturalness argument now implies

$$\Lambda < \frac{\pi^2}{G_F m_{\rm SM}} \sim 10 \text{ TeV}, \tag{9}$$

which is an energy scale of the order of $\Lambda_{\rm LH}$ (defined in Sec. 1). Therefore, cancelling only one-loop divergences is just sufficient to postpone the big hierarchy problem beyond the scale of the little hierarchy problem, thus being consistent with the phenomenological constraints.

In order to achieve the cancellation, at least at one-loop order, we can use the mechanism of "collective breaking". There are two (or more) approximate global symmetries under which the Higgs is a Goldstone boson. Therefore, each symmetry protects the Higgs mass and one requires that no single term in the Lagrangian simultaneously break all these symmetries. More than one term is necessary to generate the Higgs mass, and therefore the corresponding Feynman diagram has more than one loop.

The collective breaking can be achieved with gauge-group replication and, in this sense, the Little Higgs can be interpreted as a model with discrete extra dimensions through deconstruction.[62] The model-building recipe is the following. One starts from a non-linear sigma model of the Goldstone bosons in the coset G/H. The group G has a weakly-gauged subgroup $G_1 \times \ldots \times G_n$, where $n \geq 2$. Each of the gauge groups G_j preserves a different non-linear global symmetry under which the Higgs transforms

as a Goldstone boson. The SM group is a subgroup of $G_1 \times \ldots \times G_n$ which breaks all the global symmetries. Divergent contributions to the Higgs mass necessarily involve the gauge couplings from all the G_j groups and are therefore generated only at the n-th loop:

$$\delta m_H^2 \sim \frac{g_1^2}{(4\pi)^2} \cdots \frac{g_n^2}{(4\pi)^2} \Lambda^2. \tag{10}$$

Instead of using gauge-group replication, one can obtain collective breaking by replicating the field content. Consider the example[63] of an $SU(N)$ gauge group with two scalar fields $\Phi_{1,2}$ in the fundamental representation. Take a scalar potential of the form $V(\Phi_1^\dagger \Phi_1, \Phi_2^\dagger \Phi_2)$ such that both $\Phi_{1,2}$ acquire vacuum expectation values, spontaneously breaking the $SU(N)$ gauge symmetry. In the limit in which you turn off the gauge coupling to Φ_1, the theory has a local $SU(N)$ under which only Φ_2 transforms, and a global $SU(N)$ under which only Φ_1 transforms. Both symmetries are spontaneously broken and the spectrum contains a Goldstone boson. In the limit in which you turn off the gauge coupling to Φ_2, the situation is analogous after replacing the indices 1 and 2, and therefore we still find a massless Goldstone boson. This means that power-divergent contributions can only appear in diagrams with gauge couplings to *both* Φ_1 and Φ_2, i.e. at two loops:

$$\delta m_H^2 \sim \frac{g^4}{(4\pi)^4} \Lambda^2. \tag{11}$$

Along these lines, for both gauge-group and field replications, many models have been constructed. They are quite interesting although, admittedly, rather elaborate.[64,63,65] Effectively, the common operative mechanism that leads to the one-loop cancellation of power divergences is the introduction, at the scale $f \sim \Lambda_{\rm SM}$, of new degrees of freedom. One-loop diagrams from each SM particle are compensated by one-loop diagrams involving a new particle, much in the same fashion as supersymmetry cancels quadratic divergences. The peculiarity, however, is that in Little-Higgs theories the cancellation occurs between particles with the same spin. For instance the top-quark contribution to the Higgs mass is cancelled by a loop of a new vector-like charge-2/3 fermion with a dimension-five effective coupling with the Higgs. The W and Z contribution is cancelled by new electroweak triplet and singlet gauge bosons with negative couplings to the Higgs. The content of scalar particles is more model-dependent and, in some cases, new electroweak triplet scalar fields are necessary for the cancellation. All these new degrees of freedom represent the signature of Little-Higgs theories, which can be searched for in future colliders, in particular the LHC.[66]

Tevatron searches for new gauge bosons and LEP precision data (in particular $\Delta\rho$ contributions from the new top-like fermion, from mixing with the new gauge bosons and, possibly, from the scalar triplet) provide significant constraints on Little-Higgs models. In the minimal model of Arkani-Hamed *et al.*,[64] the symmetry-breaking scale f can be, at best, as low as 5 TeV.[67] This leads to a lower bound on the mass of the new top-like quark $m_{t'} > 2\sqrt{2}(m_t/v)f = 14$ TeV$(f/5$ TeV$)$. Comparing with the top contribution to the Higgs mass, we find that this implies a fine-tuning of at least one part in a thousand. Variations of the minimal model can significantly reduce the amount of fine-tuning needed.

The validity of the Little-Higgs theories extends up to a region of about 10 TeV, where we expect an embedding into a more fundamental region. The construction of such an ultraviolet completion is still an open and compelling question.[68]

7. Conclusions

During the last few years we have significantly enriched our basket of theories for the electroweak scale, substantially deepened our understanding of them, and constructed many previously-undiscovered variations. Nevertheless, none of the known models is fully satisfying or totally free from fine-tuning. The excessive fine-tuning may well be just a fortuitous accident, or it is a sign that some theoretical ingredient is still missing (or, possibly, that we are on a completely wrong track!)

A great distinction exists between theories with and without a "desert". I define "desert" the case in which some new dynamics modify the ultraviolet behavior of the Higgs mass parameter below the TeV, but then physics can be extrapolated up to a very large energy scale without inconsistencies. Conventional low-energy supersymmetry is the best known example. As previously discussed, the desert scenario has some indisputable advantages.

- It allows a connection between SM physics and quantum gravity or other speculative theories at very short distances, such as GUT's or strings.

- It offers the possibility of predicting some SM parameters (α_S, m_b/m_τ) through GUT relations. In particular, gauge-coupling unification in low-energy supersymmetry is certainly the most remarkable information on new physics we have.

- It naturally embeds the existence of a new mass scale Λ, as suggested by the evidence for neutrino masses and for a new dimension-5 operator $(1/\Lambda)\ell\ell HH$ to be included in the SM Lagrangian.

- It allows a natural suppression of fast proton decay and unwanted flavor violations.

- It provides a set-up for interesting cosmological theories (inflation, baryogenesis, etc.), which require an extrapolation of particle physics to very small distances.

However, for each of these points, alternative (and more or less attractive) explanations have been proposed in scenarios with a low cut-off. In my presentation I did not have the time to enter into the various solutions, but at present it is fair to conclude that the verdict for desert versus non-desert is still pending.

The interest in constructing theories that extend the SM at the TeV, but are valid only up to the 10 TeV region, has both experimental and theoretical roots. The experimental LEP data have convinced us that the scale at which new-physics virtual effects appear (Λ_{LH}) cannot be the same as the scale at which the ultraviolet behavior of the Higgs mass is modified (Λ_{SM}). This has forced us to abandon old versions of electroweak-breaking theories with strong dynamics. From a theoretical perspective, extra dimensions have brought new tools into the game and there has been a great urge to apply them at the weak scale. However, extra-dimensional theories are non-renormalizable, and unitarity is violated at an energy of typically few times the mass of the first KK mode. This implies that the cut-off is not far from the electroweak scale and it opens the urgent question of finding acceptable ultraviolet completions. The tools from extra dimensions have also given new hope for constructing theories of composite Higgs bosons and to better understand their properties.

Of course, in both desert and non-desert scenarios, many fundamental questions are postponed to physics at the cut-off, be it M_{Pl}, 10 TeV, or 1 TeV. Although theoretically this is equally unsatisfactory, the value of the cut-off makes a big difference from a phenomenological point of view. When I discussed the little hierarchy problem, I restricted myself to operators that satisfy the same symmetries as the SM Lagrangian. Had I introduced operators that arbitrarily violate flavor, CP, lepton or baryon number, I would have obtained much more stringent bounds on the corresponding mass scale. These bounds pose severe constraints on the dynamics of an ultraviolet completion at 10 TeV, while they are irrelevant for physics at M_{Pl}.

Finally, I want to point out that the question of desert versus non-desert has important implications for the strategy of future collider planning. In a scenario like conventional low-energy supersymmetry, a multi-TeV linear collider seems the most appropriate next step after the LHC, because precise studies of new-particle masses and properties are going to be of the utmost importance. On the other hand, in non-desert scenarios, the new physics to be discovered at the LHC is just the first shell of a more complicated structure. Moving towards the highest possible energy then becomes a priority, motivating research for hadron colliders in the 100–200 TeV region, like the VLHC.

Acknowledgements

I wish to thank my colleagues at CERN for various discussions and, in particular, R. Rattazzi for useful comments and A. Strumia for producing Fig. 1.

References

1. S. Weinberg, *Phys. Rev.* D **13**, 974 (1976); *Phys. Rev.* D **19**, 1277 (1979); L. Susskind, *Phys. Rev.* D **20**, 2619 (1979); G. 't Hooft, PRINT-80-0083 (UTRECHT), *Lecture given at Cargese Summer Inst., Cargese, France, Aug 26 - Sep 8, 1979.*
2. G. D'Ambrosio, G. F. Giudice, G. Isidori and A. Strumia, *Nucl. Phys.* B **645**, 155 (2002) [arXiv:hep-ph/0207036].
3. R. Barbieri and A. Strumia, arXiv:hep-ph/0007265.
4. LEP Difermion Working Group, LEP2FF/02-03.
5. Y. A. Golfand and E. P. Likhtman, *JETP Lett.* **13**, 323 (1971) [*Pisma Zh. Eksp. Teor. Fiz.* **13**, 452 (1971)]; D. V. Volkov and V. P. Akulov, *Phys. Lett.* B **46**, 109 (1973); J. Wess and B. Zumino, *Nucl. Phys.* B **70**, 39 (1974).
6. K. R. Dienes, E. Dudas and T. Gherghetta, *Nucl. Phys.* B **537**, 47 (1999) [arXiv:hep-ph/9806292].
7. N. Arkani-Hamed, A. G. Cohen and H. Georgi, arXiv:hep-th/0108089.
8. S. Dimopoulos and D. E. Kaplan, *Phys. Lett.* B **531**, 127 (2002) [arXiv:hep-ph/0201148].
9. K. Hagiwara *et al.* [Particle Data Group Collaboration], *Phys. Rev.* D **66**, 010001 (2002).
10. L. Giusti, A. Romanino and A. Strumia, *Nucl. Phys.* B **550**, 3 (1999) [arXiv:hep-ph/9811386]; A. Strumia, arXiv:hep-ph/9904247.
11. R. Barbieri and G. F. Giudice, *Nucl. Phys.* B **306**, 63 (1988); G. W. Anderson and D. J. Castano, *Phys. Rev.* D **52**, 1693 (1995) [arXiv:hep-ph/9412322]; S. Dimopoulos and G. F. Giudice, *Phys. Lett.* B **357**, 573 (1995) [arXiv:hep-ph/9507282].
12. LEP SUSY Working Group, LEPSUSYWG/01-03.1.
13. LEP SUSY Working Group, LEPSUSYWG/02-01.1.
14. LEP Higgs Working Group, CERN-EP/2003-011.
15. LEP Higgs Working Group, LHWG Note 2001-04.
16. M. Dine and A. E. Nelson, *Phys. Rev.* D **48**, 1277 (1993) [arXiv:hep-ph/9303230]; M. Dine, A. E. Nelson and Y. Shirman, *Phys. Rev.* D **51**, 1362 (1995) [arXiv:hep-ph/9408384]; M. Dine, A. E. Nelson, Y. Nir and Y. Shirman, *Phys. Rev.* D **53**, 2658 (1996) [arXiv:hep-ph/9507378]; G. F. Giudice and R. Rattazzi, *Phys. Rept.* **322**, 419 (1999) [arXiv:hep-ph/9801271].
17. J. A. Casas, J. R. Espinosa and I. Hidalgo, arXiv:hep-ph/0310137.
18. R. Barbieri and A. Strumia, *Phys. Lett.* B **490**, 247 (2000) [arXiv:hep-ph/0005203].
19. L. Randall and R. Sundrum, *Nucl. Phys.* B **557**, 79 (1999) [arXiv:hep-th/9810155]; G. F. Giudice, M. A. Luty, H. Murayama and R. Rattazzi, *JHEP* **9812**, 027 (1998) [arXiv:hep-ph/9810442].
20. D. E. Kaplan, G. D. Kribs and M. Schmaltz, *Phys. Rev.* D **62**, 035010 (2000) [arXiv:hep-ph/9911293]; Z. Chacko, M. A. Luty, A. E. Nelson and E. Ponton, *JHEP* **0001**, 003 (2000) [arXiv:hep-ph/9911323].
21. G. F. Giudice and A. Masiero, *Phys. Lett.* B **206**, 480 (1988).
22. N. Arkani-Hamed, S. Dimopoulos and G. R. Dvali, *Phys. Lett.* B **429**, 263 (1998) [arXiv:hep-ph/9803315].
23. L. Randall and R. Sundrum, *Phys. Rev. Lett.* **83**, 3370 (1999) [arXiv:hep-ph/9905221].
24. G. F. Giudice, R. Rattazzi and J. D. Wells, *Nucl. Phys.* B **544**, 3 (1999) [arXiv:hep-ph/9811291].
25. E. A. Mirabelli, M. Perelstein and M. E. Peskin, *Phys. Rev. Lett.* **82**, 2236 (1999) [arXiv:hep-ph/9811337].
26. H. Davoudiasl, J. L. Hewett and T. G. Rizzo, *Phys. Rev. Lett.* **84**, 2080 (2000) [arXiv:hep-ph/9909255].
27. T. Han, J. D. Lykken and R. J. Zhang, *Phys. Rev.* D **59**, 105006 (1999) [arXiv:hep-ph/9811350];

J. L. Hewett, *Phys. Rev. Lett.* **82**, 4765 (1999) [arXiv:hep-ph/9811356].

28. G. F. Giudice and A. Strumia, *Nucl. Phys.* B **663**, 377 (2003) [arXiv:hep-ph/0301232].

29. G. F. Giudice, R. Rattazzi and J. D. Wells, *Nucl. Phys.* B **595**, 250 (2001) [arXiv:hep-ph/0002178].

30. G. F. Giudice, R. Rattazzi and J. D. Wells, *Nucl. Phys.* B **630**, 293 (2002) [arXiv:hep-ph/0112161].

31. R. Penrose, unpublished; D. M. Eardley and S. B. Giddings, *Phys. Rev.* D **66**, 044011 (2002) [arXiv:gr-qc/0201034]; E. Kohlprath and G. Veneziano, *JHEP* **0206**, 057 (2002) [arXiv:gr-qc/0203093].

32. S. B. Giddings and S. Thomas, *Phys. Rev.* D **65**, 056010 (2002) [arXiv:hep-ph/0106219]; S. Dimopoulos and G. Landsberg, *Phys. Rev. Lett.* **87**, 161602 (2001) [arXiv:hep-ph/0106295].

33. I. Antoniadis, *Phys. Lett.* B **246**, 377 (1990).

34. K. Cheung and G. Landsberg, *Phys. Rev.* D **65**, 076003 (2002) [arXiv:hep-ph/0110346].

35. T. Appelquist, H. C. Cheng and B. A. Dobrescu, *Phys. Rev.* D **64**, 035002 (2001) [arXiv:hep-ph/0012100].

36. G. Servant and T. M. Tait, *Nucl. Phys.* B **650**, 391 (2003) [arXiv:hep-ph/0206071].

37. J. Scherk and J. H. Schwarz, *Phys. Lett.* B **82**, 60 (1979); *Nucl. Phys.* B **153**, 61 (1979).

38. L. J. Dixon, J. A. Harvey, C. Vafa and E. Witten, *Nucl. Phys.* B **261**, 678 (1985) and B **274**, 285 (1986).

39. J. M. Maldacena, *Adv. Theor. Math. Phys.* **2**, 231 (1998) [arXiv:hep-th/9711200]; S. S. Gubser, I. R. Klebanov and A. M. Polyakov, *Phys. Lett.* B **428**, 105 (1998) [arXiv:hep-th/9802109]; E. Witten, *Adv. Theor. Math. Phys.* **2**, 253 (1998) [arXiv:hep-th/9802150].

40. L. Randall and R. Sundrum, *Phys. Rev. Lett.* **83**, 4690 (1999) [arXiv:hep-th/9906064].

41. J. M. Maldacena, unpublished; E. Witten, talk at ITP Santa Barbara conference *New Dimensions in Field Theory and String Theory*; S. S. Gubser, *Phys. Rev.* D **63**, 084017 (2001) [arXiv:hep-th/9912001]; H. Verlinde, *Nucl. Phys.* B **580**, 264 (2000) [arXiv:hep-th/9906182].

42. N. Arkani-Hamed, M. Porrati and L. Randall, *JHEP* **0108**, 017 (2001) [arXiv:hep-th/0012148]; R. Rattazzi and A. Zaffaroni, *JHEP* **0104**, 021 (2001) [arXiv:hep-th/0012248]; M. Pérez-Victoria, *JHEP* **0105**, 064 (2001) [arXiv:hep-th/0105048].

43. R. Barbieri, L. J. Hall and Y. Nomura, *Phys. Rev.* D **63**, 105007 (2001) [arXiv:hep-ph/0011311].

44. D. M. Ghilencea, S. Groot Nibbelink and H. P. Nilles, *Nucl. Phys.* B **619**, 385 (2001) [arXiv:hep-th/0108184].

45. R. Barbieri, L. J. Hall, G. Marandella, Y. Nomura, T. Okui, S. J. Oliver and M. Papucci, *Nucl. Phys.* B **663**, 141 (2003) [arXiv:hep-ph/0208153]; R. Barbieri, G. Marandella and M. Papucci, *Nucl. Phys.* B **668**, 273 (2003) [arXiv:hep-ph/0305044].

46. T. Gherghetta and A. Pomarol, *Phys. Rev.* D **67**, 085018 (2003) [arXiv:hep-ph/0302001].

47. M. A. Luty, *Phys. Rev. Lett.* **89**, 141801 (2002) [arXiv:hep-th/0205077].

48. R. Contino, Y. Nomura and A. Pomarol, arXiv:hep-ph/0306259.

49. J. L. Feng, T. Moroi, L. Randall, M. Strassler and S. f. Su, *Phys. Rev. Lett.* **83**, 1731 (1999) [arXiv:hep-ph/9904250]; T. Gherghetta, G. F. Giudice and J. D. Wells, *Nucl. Phys.* B **559**, 27 (1999) [arXiv:hep-ph/9904378].

50. N. S. Manton, *Nucl. Phys.* B **158**, 141 (1979); S. Randjbar-Daemi, A. Salam and J. Strathdee, *Nucl. Phys.* B **214**, 491 (1983).

51. Y. Hosotani, Ann. Phys. **190**, 233 (1989).

52. C. Csaki, C. Grojean and H. Murayama, *Phys. Rev.* D **67**, 085012 (2003) [arXiv:hep-ph/0210133]; G. Burdman and Y. Nomura, *Nucl. Phys.* B **656**, 3 (2003) [arXiv:hep-ph/0210257]; C. A. Scrucca, M. Serone and L. Silvestrini, *Nucl. Phys.* B **669**, 128 (2003) [arXiv:hep-ph/0304220].

53. C. Csaki, C. Grojean, H. Murayama, L. Pilo and J. Terning, arXiv:hep-ph/0305237.

54. K. Agashe, A. Delgado, M. J. May and R. Sundrum, *JHEP* **0308**, 050 (2003) [arXiv:hep-ph/0308036].

55. C. Csaki, C. Grojean, L. Pilo and J. Terning, arXiv:hep-ph/0308038.

56. R. S. Chivukula, D. A. Dicus and H. J. He, *Phys. Lett.* B **525**, 175 (2002) [arXiv:hep-ph/0111016].

57. M. E. Peskin and T. Takeuchi, *Phys. Rev.* D **46**, 381 (1992).

58. G. Altarelli and R. Barbieri, *Phys. Lett.* B **253**, 161 (1991).

59. R. Barbieri, A. Pomarol and R. Rattazzi, arXiv:hep-ph/0310285.

60. H. Georgi and A. Pais, *Phys. Rev.* D **12**, 508 (1975); D. B. Kaplan, H. Georgi and S. Dimopoulos, *Phys. Lett.* B **136**, 187 (1984).

61. N. Arkani-Hamed, A. G. Cohen and H. Georgi, *Phys. Lett.* B **513**, 232 (2001) [arXiv:hep-ph/0105239].

62. N. Arkani-Hamed, A. G. Cohen and H. Georgi, *Phys. Rev. Lett.* **86**, 4757 (2001) [arXiv:hep-th/0104005]; C. T. Hill, S. Pokorski and J. Wang, *Phys. Rev.* D **64**, 105005 (2001) [arXiv:hep-th/0104035].

63. D. E. Kaplan and M. Schmaltz, arXiv:hep-ph/0302049.

64. N. Arkani-Hamed, A. G. Cohen, E. Katz and A. E. Nelson, *JHEP* **0207**, 034 (2002) [arXiv:hep-ph/0206021].

65. N. Arkani-Hamed, A. G. Cohen, E. Katz, A. E. Nelson, T. Gregoire and J. G. Wacker, *JHEP* **0208**, 021 (2002) [arXiv:hep-ph/0206020]; I. Low, W. Skiba and D. Smith, *Phys. Rev.* D **66**, 072001 (2002) [arXiv:hep-ph/0207243]; S. Chang and J. G. Wacker, arXiv:hep-ph/0303001; C. Csaki, J. Hubisz, G. D. Kribs, P. Meade and J. Terning, *Phys. Rev.* D **68**, 035009 (2003) [arXiv:hep-

ph/0303236]; T. Gregoire, D. R. Smith and J. G. Wacker, arXiv:hep-ph/0305275; W. Skiba and J. Terning, *Phys. Rev.* D **68**, 075001 (2003) [arXiv:hep-ph/0305302].

66. G. Burdman, M. Perelstein and A. Pierce, *Phys. Rev. Lett.* **90**, 241802 (2003) [arXiv:hep-ph/0212228]; T. Han, H. E. Logan, B. McElrath and L. T. Wang, *Phys. Rev.* D **67**, 095004 (2003) [arXiv:hep-ph/0301040]; M. Perelstein, M. E. Peskin and A. Pierce, arXiv:hep-ph/0310039.

67. C. Csaki, J. Hubisz, G. D. Kribs, P. Meade and J. Terning, *Phys. Rev.* D **67**, 115002 (2003) [arXiv:hep-ph/0211124]; J. L. Hewett, F. J. Petriello and T. G. Rizzo, arXiv:hep-ph/0211218.

68. A. E. Nelson, arXiv:hep-ph/0304036.

QCD PHENOMENOLOGY AND THEORY

QCD: THEORETICAL DEVELOPMENTS

T. GEHRMANN

Institut für Theoretische Physik, Universität Zürich, Winterthurerstraße 190, CH-8057 Zürich, Switzerland
E-mail: gehrt@physik.unizh.ch

I review recent theoretical advances in quantum chromodynamics. Particular emphasis is put on developments related to the precise prediction and interpretation of experimental data from present and future high energy colliders.

1. Introduction

Quantum Chromodynamics (QCD) is well established as the theory of strong interactions through a large number of experimental verifications. The era of "testing QCD" is clearly finished, and QCD today is becoming precision physics. The next generation of high energy collider experiments are all performed at hadron colliders, where (in contrast to LEP and SLC) QCD is ubiquitous. Any precision measurement (strong coupling constant, quark masses, electroweak parameters, parton distributions) at the Tevatron and the LHC, as well as any prediction of new physics effects and their backgrounds, relies on the understanding of QCD effects on the observable under consideration.

The derivation of precise QCD predictions for collider observables poses several theoretical and computational challenges. The most important challenge is the fact that QCD describes quarks and gluons, while experiments observe hadrons. This mismatch is either accounted for by a description of the parton to hadron transition through fragmentation functions or by defining sufficiently inclusive final-state observables, such as jets. Moreover, the strong coupling constant is considerably larger than the electromagnetic coupling constant at scales typically probed at colliders: $\alpha_s(M_Z) \simeq 15\,\alpha_{\rm em}(M_Z)$, resulting in a slower convergence of the perturbative expansion. As a consequence, a precise description of QCD observables (precise means here that the theoretical uncertainty becomes similar to the achieved or projected experimental errors) is obtained only by including higher order corrections, often requiring beyond the next-to-leading-order. The largeness of the strong coupling constant also implies that multiparticle final states occur rather frequently. Finally, many collider observables involve largely different scales, such as quark masses, transverse momenta and vector boson masses. These give rise to potentially large logarithms, which might spoil the convergence of the perturbative series and need to be resummed to all orders.

In this talk, I shall try to highlight recent theoretical progress towards precision QCD at colliders, focusing on heavy quark production in Sec. 2, on jets and multiparton final states in Sec. 3, on photons in Sec. 4 and electroweak bosons in Sec. 5. Finally, a summary of the current state-of-the-art and as yet open issues is given in Sec. 6.

2. Heavy Quarks

Heavy quark production is one of the main topics investigated at high energy collider experiments. Heavy quarks are of particular interest to probe the flavor sector of the Standard Model, which is less well tested than the gauge sector. Also, many approaches to physics beyond the Standard Model, often related to electroweak symmetry breaking and mass generation, predict new effects to be most pronounced in observables involving heavy quarks.

2.1. *Total Cross Sections*

The total cross sections for the production of heavy quarks can be computed reliably within perturbation theory. The current state-of-the-art is a next-to-leading-order (NLO) calculation,[1] which is further improved by summing large logarithms due to soft gluon emission up to the next-to-leading (NLL)[2] and next-to-next-to-leading-logarithmic (NNLL)[3] level. As can be seen from Fig. 1, these predictions are in good agreement with experimental data on the total $t\bar{t}$ cross section at the Tevatron[4] and the total $b\bar{b}$ cross section at HERA-B[5] (which both actually refer to similar kinematical values of $m_Q/\sqrt{s}$). The theoretical uncertainty on the prediction for HERA-B is larger for two reasons: the larger value of the strong coupling at m_b than at m_t and the dominance

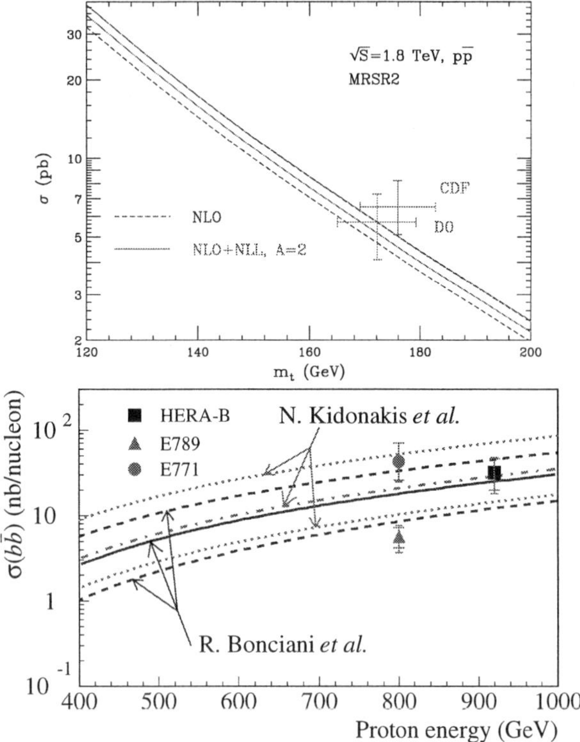

Figure 1. Total cross sections for $t\bar{t}$ at the Tevatron[4] and $b\bar{b}$ at HERA-B.[5]

of gg initial states in pN collisions (HERA-B) compared to $q\bar{q}$ dominance in $p\bar{p}$ collisions (Tevatron). The effects due to soft gluon resummation turn out to be moderate, but do yield a significant decrease in the uncertainty of the theoretical prediction. Further uncertainties on the prediction of the $t\bar{t}$ cross section come from the parton distribution functions.[6]

2.2. *Transverse Momentum Distributions*

Differential distributions of hadrons containing b quarks measured in hadron-hadron, photon-hadron or photon-photon collisions have been in apparent discrepancy with theoretical predictions for quite some time. The spectrum of $B^{\pm}$ hadrons measured at CDF[7] is one of the most recent examples of this discrepancy.

The theoretical prediction for B meson production involves a convolution of the hard matrix element for heavy quark production in parton-parton scattering with initial parton distributions and final-state fragmentation functions describing the non-perturbative transition from a b quark to a B hadron. It is in particular the latter which is suspected to

account for the discrepancy between the theoretical prediction and experimental data, especially since it has been observed[8] that the transverse momentum distribution of b-tagged jets[9] (which has little sensitivity to fragmentation functions) is in much better agreement with theoretical predictions.

The definition of heavy quark fragmentation functions is not free from ambiguities, since some aspects of these functions are actually calculable in perturbation theory.[10] In extracting these fragmentation functions from data on B hadron production in $e^{+}e^{-}$ collisions, several choices are made, related to the order of perturbation theory, the incorporation of mass effects in the matrix elements, the resummation of potentially large perturbative terms, the inclusion of power corrections,[11] the correction of data for parton showers and the parametric form of the ansatz used in the determination. Unfortunately, the sensitivity of the fragmentation function on the assumptions used in the extraction from $e^{+}e^{-}$ spectra is often overlooked when using this fragmentation function to compute heavy hadron spectra at colliders.

Recently, an approach incorporating quark mass effects, perturbatively calculable components of the heavy quark fragmentation function[10] and resummation of large logarithms up to the next-to-leading-logarithmic level has been put forward with the fixed-order next-to-leading-log (FONLL) scheme.[12] This approach requires only a small, genuinely non-perturbative component of the fragmentation function to be fitted to $e^{+}e^{-}$ data. In order to expose the information content actually relevant to heavy hadron spectra at hadron colliders, this fit is done in moment space.

In view of new data from ALEPH,[13] a phenomenological study of B hadron production at colliders based on the FONLL scheme was performed.[14] It was shown that the consistent treatment of the fragmentation function in extraction and prediction reduced the discrepancy between data and theoretical prediction considerably. The theoretical prediction is however still falling somewhat short of the experimental data, which is probably due to currently uncalculated corrections beyond NLO. More recently, the same framework was applied to charmed hadron production at hadron colliders.[15] In this case, one also observes that the experimental data[16] exceed the theoretical prediction, Fig. 2, although the

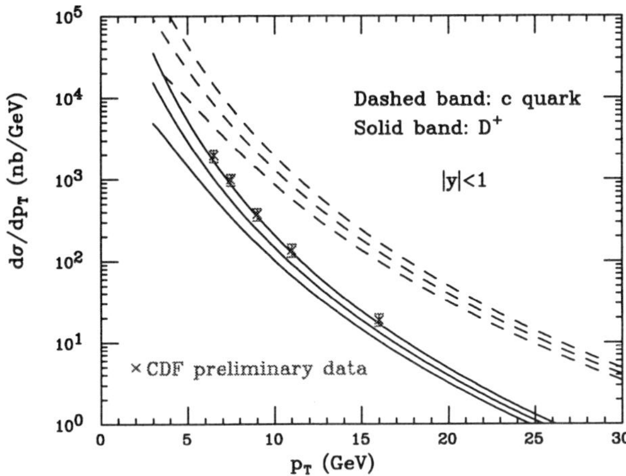

Figure 2. Transverse momentum spectrum of D^+ hadrons at CDF, compared to calculation using FONLL fragmentation functions.[15]

effect is less pronounced than for bottom hadron production. Comparison of massless[17] and massive[15] calculations for hadronic charm production[16] indicates only moderate mass effects, which are smaller than the theoretical uncertainty.

Many collider experiments also report an excess in the b quark production spectra. In interpreting these data, it must always be kept in mind that it is not the b quark but rather the B hadron production which is observed in the experiment. Information on b quark production is only inferred from these data using some model for the heavy quark fragmentation. As discussed above, there are numerous ambiguities, which can yield inconsistent predictions if not implemented consistently. In view of the rather sizable effects due to a consistent treatment of the fragmentation function observed on the B hadron spectra at CDF, it might be that the data sets on b quark spectra have to be reanalyzed incorporating the new experimental and theoretical information on the b quark fragmentation functions in a consistent manner.

3. Jets and Multiparticle Production

Hadronic jets at large transverse momenta are produced very copiously at colliders. Final states with a small number of jets are measured to very high experimental accuracy, such that they can be used for precision measurements of the strong coupling constant and of parton distribution functions. For these mea-

surements, the uncertainty on the theoretical prediction is often the dominant source of error, and one would consequently like to have more accurate theoretical calculations, which implies in general an extension towards higher perturbative orders. Multiparton final states, involving a large number of jets, can on the other hand mimic final-state signatures induced by physics beyond the Standard Model, thus forming an irreducible background to searches. For these, QCD predictions serve as a guidance to devise search strategies, and one demands QCD to yield a description of the full hadronic final state.

3.1. *Leading-order Calculations*

Multiparton final states are described using leading-order QCD predictions, implemented in flexible multiparton matrix element generators. These programs evaluate the scattering amplitudes using efficient representations of helicity amplitudes or fully numerically from the interaction Lagrangian. Examples of these codes are VECBOS,[18] COMPHEP,[19] MADGRAPH,[20] GRACE,[21] HELAC,[22] ALPHGEN,[23] and AMEGIC++.[24] Using these, the computation of $2 \to 8$ reactions is feasible on current computers. These programs are then combined with automatic integration over multiparticle phase space, using for example RAMBO,[25] PHEGAS,[26] or MADEVENT.[27] Most programs can be interfaced to hadronization models using standard interfaces.[28]

Matrix element calculations accurately include large angle single gluon radiation. At small angles from the emitting particles, one does however encounter multiple gluon radiation, which can be accounted for by parton showers. Recently, a generic procedure was devised to combine both descriptions for multiparton final states in a "modified matrix element plus vetoed parton shower".[29]

Including these developments, leading-order QCD provides the basis of Monte Carlo event generators. However, its predictions contain large (and non-quantifiable) errors due to the setting of renormalization and factorization scales. Leading-order QCD is therefore a good tool to estimate relative magnitudes of processes and to design searches. Once precision is required (e.g. to identify a discovery with a particular model), it is not sufficient.

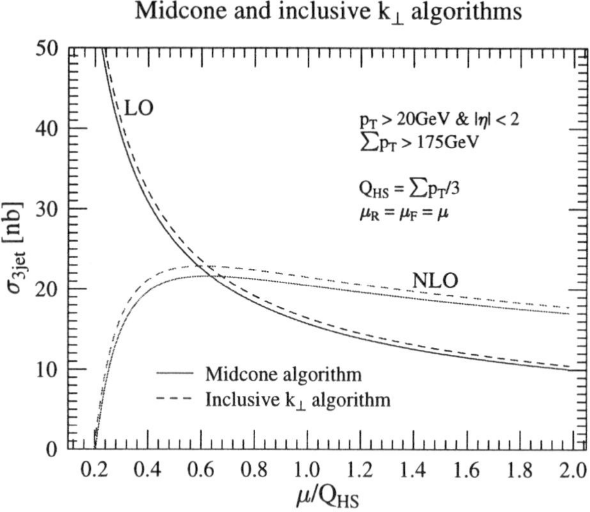

Figure 3. Dependence of the NLO prediction for the $3j$ cross section at the Tevatron on renormalization and factorization scale.[32]

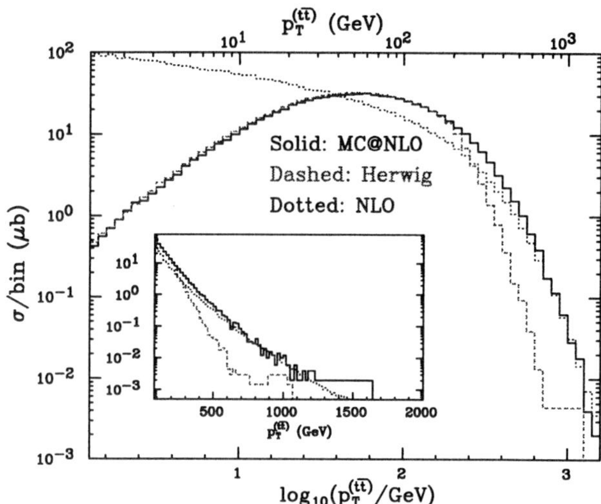

Figure 4. Transverse momentum distribution of top quark pairs at the LHC as predicted by MC@NLO.[42]

3.2. *Next-to-leading-order Calculations*

Including next-to-leading-order QCD corrections improves the theoretical predictions in numerous ways by reducing the renormalization scale uncertainty, and by providing reliable normalizations of cross sections and reliable error estimates. Moreover, NLO is the first order where differences between jet algorithms show up. In contrast to the leading order, there is no generic procedure for doing NLO calculations, such that each new process under consideration implies a completely new calculation.

For hadron colliders, NLO results are available for all relevant $2 \to 2$ reactions; $2 \to 3$ reactions are the current frontier. A number of $2 \to 3$ results (each involving several man-years of work) became available recently: $pp \to V + 2j$,[30] $ep \to (3 + 1)j$,[31] $pp \to 3j$,[32] $pp \to \gamma\gamma + j$,[33] $pp \to t\bar{t}H$,[34] and the vector boson fusion processes $pp \to H + 2j$,[35] $pp \to V + 2j$.[36] Some of the features of NLO calculations, such as the improved scale dependence and the differences between jet algorithms are illustrated in Fig. 3.

To overcome the large amount of work required for each NLO calculation, efforts are under way towards their automatization. The NLO calculation for an n parton reaction contains the one-loop n parton matrix elements, the tree level $n + 1$ parton matrix elements and a procedure to extract the infrared singularities from both and to combine them. While this procedure was automatized for the tree level real radiation matrix elements long ago,[37] there is at present no automatic procedure to compute one-loop integrals. Very recently several algorithms were proposed, including the analytic reduction of hexagon integrals,[38] a subtraction formalism for virtual corrections[39] and the numerical evaluation of hexagon integrals.[40]

Another important development is the combination of NLO calculations with parton showers, as realized in the MC@NLO approach.[41] This approach introduces a modified NLO subtraction method, where both real and virtual contributions become initial conditions for the parton shower. In this, hard radiation is accurately described by the NLO matrix element, while multiple soft radiation is accounted for by the parton shower; a double counting of contributions is avoided. So far, this formalism has been applied to VV, $b\bar{b}$ and $t\bar{t}$ production at hadron colliders. Figure 4 illustrates that MC@NLO[42] smoothly connects the kinematic region dominated by multiple radiation at small transverse momenta to the region controlled by single hard radiation at large transverse momenta.

3.3. *Next-to-next-to-leading-order Calculations*

Despite the evidently good agreement of NLO QCD with experimental data on jet production rates, predictions to this order are insufficient for many appli-

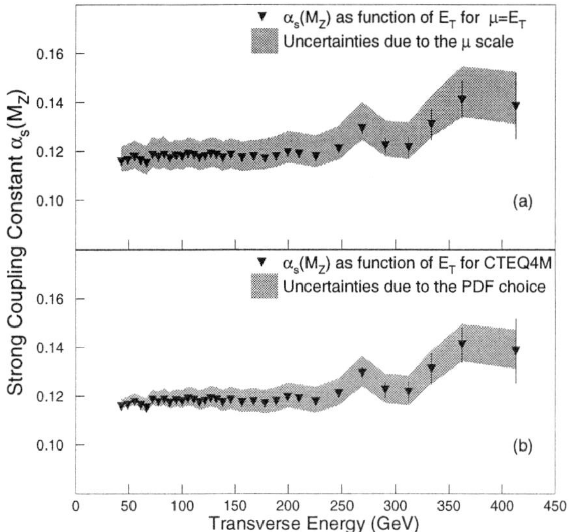

Figure 5. Errors on extraction of α_s from single jet inclusive cross section at CDF.[43]

cations. For example, if one uses data on the single jet inclusive cross section[43] compared to the NLO theoretical prediction[44,45] to determine the strong coupling constant α_s, it turns out that the dominant source of error on this extraction is due to unknown higher order corrections. Given that the theoretical prediction to infinite order in perturbation theory should be independent of the choice of renormalization and factorization scale, this error can be estimated from the variation of the extracted α_s under variation of these scales, as seen in Fig. 5. As a result, CDF find from their Run I data

$$\alpha_s(M_Z) = 0.1178 \pm 0.0001(\text{stat})^{+0.0081}_{-0.0095}(\text{sys})$$
$$^{+0.0071}_{-0.0047}(\text{scale}) \pm 0.0059(\text{pdf}).$$

It can be seen that the statistical error is already negligible; improvements in the systematic error can be anticipated in the near future. To lower the theoretical error, it is mandatory to compute next-to-next-to-leading-order (NNLO) corrections to the single jet inclusive cross section.

A similar picture is true in e^+e^- annihilation into three jets and deep-inelastic $(2+1)$ jet production, where the error on the extraction of α_s from experimentally measured jet shape observables[46] is completely dominated by the theoretical uncertainty inherent in the NLO QCD calculations.

Besides lowering the theoretical error, there is a number of other reasons to go beyond NLO in the description of jet observables. While jets at NLO are modeled theoretically by at most two partons, NNLO allows up to three partons in a single jet, thus improving the matching of experimental and theoretical jet definitions and resolving the internal jet structure. At hadron colliders, NNLO also accounts for double initial state radiation, thus providing a perturbative description for the transverse momentum of the hard final state. Finally, including jet data in a global NNLO fit of parton distribution functions, one anticipates a lower error on the prediction of benchmark processes at colliders.

The calculation of jet observables at NNLO requires a number of different ingredients. To compute the corrections to an n-jet observable, one needs the two-loop n parton matrix elements, the one-loop $n+1$ parton matrix elements and the tree level $n+2$ parton matrix elements. Since the latter two contain infrared singularities due to one or two partons becoming theoretically unresolved (soft or collinear), one needs to find one- and two-particle subtraction terms, which account for these singularities in the matrix elements, and are sufficiently simple to be integrated analytically over the unresolved phase space. One-particle subtraction at tree level is well understood from NLO calculations[37] and general algorithms are available for one-particle subtraction at one loop,[47] in a form that could recently be integrated analytically.[48] Tree level two-particle subtraction terms have been extensively studied in the literature,[49] their integration over the unresolved phase space was up to now made only in one particular infrared subtraction scheme in the calculation of higher order corrections to the photon-plus-one-jet rate in e^+e^- annihilation.[50] The same techniques (and the same scheme) were used very recently in the rederivation of the time-like gluon-to-gluon splitting function from splitting amplitudes.[51] A general two-particle subtraction procedure is still lacking at the moment, although progress on this is anticipated in the near future.

Concerning virtual two-loop corrections to jet-observables related to $2 \to 2$ scattering and $1 \to 3$ decay processes, enormous progress has been made in the past years. Much of this progress is due to several technical developments concerning the evaluation of two-loop multi-leg integrals. Using iterative algorithms,[52] one can reduce the large number of two-loop integrals by means of integration-by-parts[53] and Lorentz invariance[54] identities to a small number of master integrals. The master integrals relevant to two-loop jet physics are two-loop four-point functions with all legs on-shell[55] or one leg off-shell,[56] which were computed using explicit integration or implicitly from their differential equations.[54]

Combing the reduction scheme with the master integrals, it is straightforward to compute the two-loop matrix elements relevant to jet observables using computer algebra.[57] Following this procedure, massless two-loop matrix elements were obtained for Bhabha scattering,[58] parton-parton scattering into two partons,[59] parton-parton scattering into two photons,[60] as well as light-by-light scattering.[61] Two-loop corrections were also computed for the off-shell process $\gamma^* \to q\bar{q}g$,[62] relevant to $e^+e^- \to 3j$. Part of these results were already confirmed[63] using an independent method.[64] Related to $e^+e^- \to 3j$ by analytic continuation[65] are $(2+1)j$ production in ep collisions and $V+j$ production at hadron colliders. A strong check on all these two-loop results is provided by the agreement of the singularity structure with predictions obtained from an infrared factorization formula.[66]

More recently, first results were obtained for master integrals involving massive internal propagators, as appearing in the two-loop QED corrections to the $\gamma^* \to Q\bar{Q}$ vertex,[67] in the two-loop electroweak corrections to the $V \to q\bar{q}$ vertex[68] and in the QED corrections to massive Bhabha scattering.[69]

4. Photons

Photons and gauge bosons provide very prominent final-state signatures at colliders. Their study allows the precise determination of electroweak parameters at hadron colliders, and their final-state signatures are often background to searches, such as photon pair production to the Higgs search in the lower mass range.

4.1. *Isolated Photons*

Photons produced in hadronic collisions arise essentially from two different sources: "direct" or "prompt" photon production via hard partonic processes such as $qg \to q\gamma$ and $q\bar{q} \to g\gamma$ or through the "fragmentation" of a hadronic jet into a single photon carrying a large fraction of the jet energy. The former gives rise to perturbatively calculable short-distance contributions whereas the latter is primarily a long distance process which cannot be calculated perturbatively and is described in terms of the quark-to-photon fragmentation function. In principle, this fragmentation contribution could be suppressed to a certain extent by imposing isolation cuts on the photon. Commonly used isolation cuts are defined by admitting only a maximum amount of hadronic energy in a cone of a given radius around the photon. An alternative procedure is the democratic clustering approach,[70] which applies standard jet clustering algorithms to events with final-state photons, treating the photon like any other hadron in the clustering procedure. Isolated photons are then defined to be photons carrying more than some large, predefined amount of the jet energy.

Both types of isolation criteria are infrared safe, although the matching of experimental and theoretical implementations of these criteria is in general far from trivial. It was pointed out recently[71] that cone-based isolation criteria fail for small cone sizes R (once $\alpha_s \ln R^{-2} \sim 1$), since the isolated photon cross section exceeds the inclusive photon cross section. This problem can only be overcome by a resummation of the large logarithms induced by the cone size parameter. The contribution from photon fragmentation to isolated photon cross sections at hadron colliders is sensitive (for both types of isolation criteria) to the photon fragmentation function at large momentum transfer, which has up to now been measured only at LEP.[72,73] Further information on the photon fragmentation function at large momentum transfer might be gained from yet unanalyzed LEP data or from the study of photon-plus-jet final states in deep-inelastic scattering at HERA,[74] where first data are now becoming available.[75]

4.2. *Photon Pairs*

One of the most promising channels for the discovery of a light Higgs boson ($m_H \lesssim 140$ GeV) at the LHC is based largely on the observation of the rare decay to two photons. To perform an accurate background subtraction for this observable, one requires a precise prediction for QCD reactions yielding di-photon final states. At first sight, the $\mathcal{O}(\alpha_s^0)$ process $q\bar{q} \to \gamma\gamma$ yields the leading contribution. However, due to the large gluon luminosity at the LHC, both $qg \to q\gamma\gamma$ ($\mathcal{O}(\alpha_s^1)$) and $gg \to \gamma\gamma$ ($\mathcal{O}(\alpha_s^2)$) subpro-

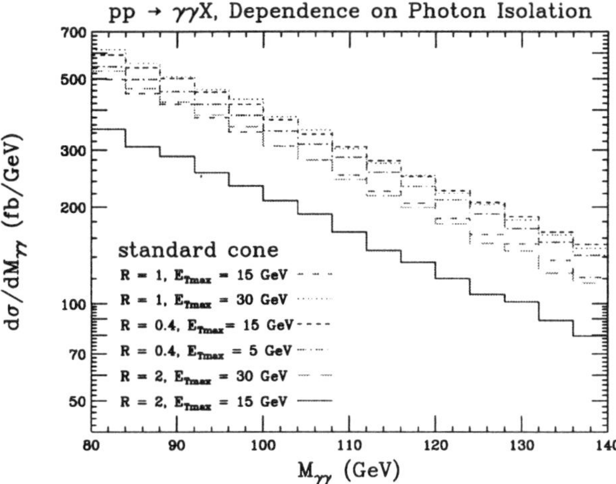

Figure 6. Photon pair production for different isolation criteria.[77]

cesses yield contributions of comparable magnitude. The NLO corrections to the $q\bar{q}$ and qg subprocesses have been known for quite some time, these are implemented in the flexible parton level event generator DIPHOX.[76] Most recently, NLO corrections were also derived for the gg subprocess.[77] Since the lowest order contribution to this process is already mediated by a quark loop, this calculation contains some of the features appearing in jet physics only at NNLO, such as two-loop QCD amplitudes and unresolved limits of one-loop amplitudes (see Sec. 3.3 above). Another important new result are the NLO corrections to two-photon-plus-jet production,[33] forming the background to Higgs boson detection at large transverse momenta.

It must be kept in mind that the di-photon cross sections are highly sensitive on the isolation criteria applied to the photons, Fig. 6, with a substantial contribution arising from photon fragmentation at large momentum transfers.[78] Moreover, it is experimentally difficult to distinguish photons from highly energetic neutral pions which decay into a closely collimated photon pair, mimicking a single photon signature. The pion background in photon pair production has been studied to NLO[79] and implemented in DIPHOX, showing that in particular the $\pi^0\gamma$ channel remains comparable to the $\gamma\gamma$ channel even for tight isolation criteria.

5. Higgs and Gauge Boson Production

The search for the Higgs boson is one of the primary goals of present and future hadron collider experiments, where one expects the main production channel to be gluon fusion, mediated through a top quark loop. To a good approximation,[80] one can use an effective gluon-gluon-Higgs coupling to describe this process in perturbative QCD (provided the leading-order mass dependence is factored out explicitly). In this approximation, the calculation of higher order corrections to inclusive Higgs production becomes very similar to the analogous calculation for gauge boson production.

Inclusive vector boson production has been computed to NNLO[81] already more than ten years ago. Very recently, these results have been verified for the first time in an independent calculation,[82] carried out in the context of the derivation of NNLO corrections to inclusive Higgs boson production.

5.1. *Higgs Boson*

The NNLO corrections to the Higgs production cross section were obtained first in the soft/collinear approximation;[83] shortly thereafter, the full coefficient functions were obtained by expansion around the soft limit,[82] and fully analytically[84] by extending the IBP/LI reduction method and the differential equation technique (see Sec. 3.3) to compute double real emission contributions. These results were confirmed independently[85] using the techniques of the original vector boson calculation. It turned out that inclusion of NNLO corrections yields a sizable enhancement of the Higgs production cross section, Fig. 7, and a reduction of the uncertainty due to the renormalization and factorization scale. Recently, this calculation was further improved by the inclusion of effects due to soft gluon resummation.[86] Further NNLO results on inclusive Higgs boson production involve pseudoscalar Higgs production through gluon fusion,[87] Higgs production in bottom quark fusion[88] and Higgs-strahlung off a vector boson.[89]

Since hadron collider experiments only cover a limited range of the final-state phase space, it is very desirable to have not only predictions for the inclusive Higgs production cross sections, but also for differential distributions in rapidity and transverse momentum. Next-to-leading-order corrections to

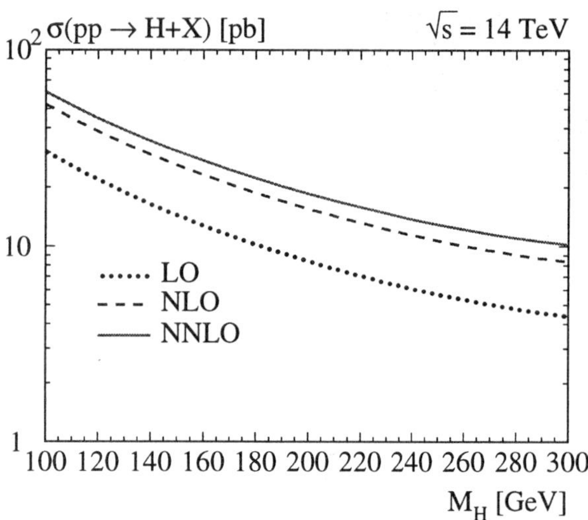

Figure 7. Inclusive Higgs production at the LHC.[82]

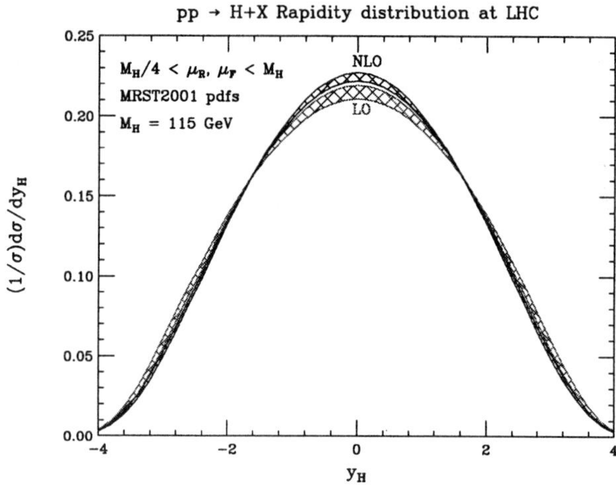

Figure 8. Rapidity distribution of Higgs bosons at the LHC.[90]

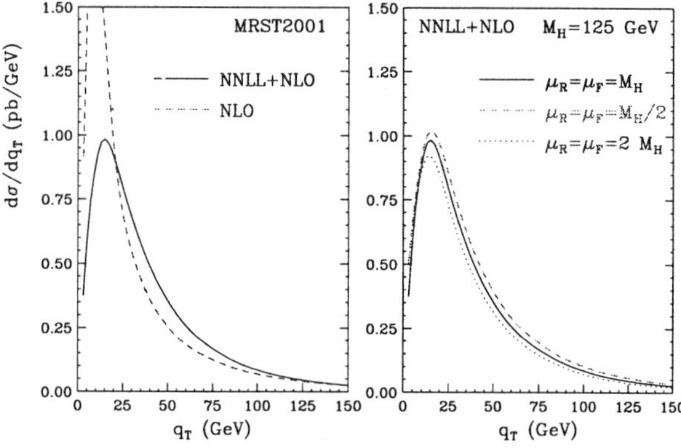

Figure 9. Transverse momentum spectrum of Higgs bosons at the LHC, including soft gluon resummation.[93]

both distributions[90,91,92] became available recently. The rapidity distribution, Fig. 8, is only moderately modified, but extends out significantly beyond the experimental coverage. The calculation of corrections to transverse momentum distributions is reliable only for transverse momenta larger or equal to the Higgs boson mass, while multiple soft gluon radiation plays a crucial role at lower transverse momenta. The resummation of these corrections was performed recently[93] to NNLL accuracy, Fig. 9.

5.2. *Gauge Bosons*

The production cross sections for $W^{\pm}$ and Z^0 bosons at hadron colliders are well understood both experimentally and theoretically. At present, these cross sections are measured to an error of about 10% from Tevatron Run I, largely limited by statistics. A considerable reduction of the experimental error is anticipated from Run II and for the LHC. On the theory side, inclusive vector boson production has been computed to NNLO.[81,82]

Given the good theoretical and experimental understanding of $W^{\pm}$ and Z^0 boson production, it has been suggested to use these for a determination of the LHC luminosity.[94] In practice, it turns out that it is not possible to measure the fully inclusive production cross sections, but only cross sections integrated over a restricted range in rapidity, for which NNLO corrections were also computed very recently.[95] For the $W^{\pm}$ production, which is observed only through the

$l\nu$ decay channel, it is moreover mandatory to compute the spatial distribution of the decay products, which is however only known to NLO.[45]

A crucial ingredient to precision NNLO predictions of these cross sections at the LHC are parton distributions accurate to this order. The determination of parton distributions from a global fit to experimental data is described in great detail in Robert Thorne's talk at this conference.[96] To perform a fit at NNLO, one needs on the one hand the coefficient functions for all contributing observables (deep-inelastic scattering, the Drell-Yan process, jet production and possibly direct photon production) to NNLO. At present, only the Drell-Yan process and deep-inelastic scattering are known to this order. On the other hand, also the partonic splitting kernels (Altarelli-Parisi splitting functions) are required to NNLO accuracy. At the moment, this calculation is ongoing. The method applied in this calculation is the determination of the splitting functions from the forward photon-parton scattering amplitude at three loops, evaluated in moments of the external partonic momentum. Intermediate results involve some fixed moments,[97] as well as all moments for the non-singlet fermionic loops.[98]

6. Conclusions and Outlook

QCD affects all observables studied at present and future hadron colliders. Given the anticipated luminosity of these machines, QCD reactions will become precision physics, very much like electroweak physics was precision physics in the LEP era. The study of many of the standard scattering reactions will allow a precise determination of the strong coupling constant, electroweak parameters, quark masses and parton distribution functions. In turn, this information translates into improved predictions for new physics signals and their backgrounds.

Both the precision determination of Standard Model parameters (and auxiliary quantities) and the design of search strategies for new physics effects require substantial input from theoretical calculations. For the purpose of drafting searches for physics beyond the Standard Model, one often requires predictions for multiparticle final states, which are at present only available at leading order. Such accuracy is in general sufficient for this purpose, given that the available generic programs also provide interfaces to partonic showers and hadronization models, thus predicting fully hadronic events, which can be further processed through detector simulations. Next-to-leading-order calculations will be important to refine searches, and to identify potential new signals, since quantitative predictions start to become reliable only at this order. The current frontier of NLO calculations are $2 \to 3$ reactions, where some of the most prominent observables are known. Extension to $2 \to 4$ reactions will require new theoretical tools, in particular towards generic, process independent algorithms. Another important development in NLO calculations is the interface to partonic showers, which has recently been devised. For the precise extraction of Standard Model parameters from benchmark reactions, NNLO calculations will be mandatory, since these reactions are (already at present colliders) measured to a level of accuracy at which the theoretical error on the NLO calculation becomes the dominant source of uncertainty. First NNLO calculations were already performed for $2 \to 1$ reactions, and now $2 \to 2$ calculations are well under way. For applications at hadron colliders, NNLO calculations only become meaningful if augmented by parton distributions accurate to NNLO, which require the knowledge of the three-loop splitting functions, which are also calculated at present.

Many observables do moreover require the resummation of large logarithms spoiling the convergence of the perturbative series. Fragmentation effects enter many observables with identified particles in the final state. In particular, a consistent treatment of heavy quark fragmentation effects can account for a large part of the observed discrepancy in B hadron spectra, and quark-to-photon as well as quark-to-pion fragmentation yield important contributions to photon pair final states forming an important background to Higgs searches. Much valuable information on these fragmentation functions is contained in data from LEP, and should be extracted (as long as this is still a feasible task) to improve predictions for collider observables.

References

1. P. Nason, S. Dawson and R. K. Ellis, *Nucl. Phys.* B **303**, 607 (1988); W. Beenakker, W. L. van Neerven, R. Meng, G. A. Schuler and J. Smith, *Nucl. Phys.* B **351**, 507 (1991).
2. R. Bonciani, S. Catani, M. L. Mangano and P. Nason, *Nucl. Phys.* B **529**, 424 (1998) [arXiv:hep-ph/9801375]; arXiv:hep-ph/0307035.
3. N. Kidonakis, E. Laenen, S. Moch and R. Vogt, *Phys. Rev.* D **64**, 114001 (2001) [arXiv:hep-ph/0105041].
4. T. Affolder *et al.* [CDF Collaboration], *Phys. Rev.* D **64**, 032002 (2001) [arXiv:hep-ex/0101036]; V. M. Abazov *et al.* [D0 Collaboration], *Phys. Rev.* D **67**, 012004 (2003) [arXiv:hep-ex/0205019].
5. I. Abt *et al.* [HERA-B Collaboration], *Eur. Phys. J.* C **26**, 345 (2003) [arXiv:hep-ex/0205106].
6. M. Cacciari, S. Frixione, M. L. Mangano, P. Nason and G. Ridolfi, arXiv:hep-ph/0303085.
7. D. Acosta *et al.* [CDF Collaboration], *Phys. Rev.* D **65**, 052005 (2002) [arXiv:hep-ph/0111359].
8. B. Abbott *et al.* [D0 Collaboration], *Phys. Rev. Lett.* **85**, 5068 (2000) [arXiv:hep-ex/0008021].
9. S. Frixione and M.L. Mangano, *Nucl. Phys.* B **483**, 321 (1997) [arXiv:hep-ph/9605270].
10. B. Mele and P. Nason, *Nucl. Phys.* B **361**, 626 (1991).

11. M. Cacciari and E. Gardi, *Nucl. Phys.* B **664**, 299 (2003) [arXiv:hep-ph/0301047].

12. M. Cacciari, M. Greco and P. Nason, *JHEP* **9805**, 007 (1998) [arXiv:hep-ph/9803400].

13. A. Heister *et al.* [ALEPH Collaboration], *Phys. Lett.* B **512**, 30 (2001) [arXiv:hep-ex/0106051].

14. M. Cacciari and P. Nason, *Phys. Rev. Lett.* **89**, 122003 (2002) [arXiv:hep-ph/0204025].

15. M. Cacciari and P. Nason, *JHEP* **0309**, 006 (2003) [arXiv:hep-ph/0306212].

16. D. Acosta *et al.* [CDF Collaboration], arXiv:hep-ex/0307080.

17. B. A. Kniehl, G. Kramer and B. Pötter, *Nucl. Phys.* B **597**, 337 (2001) [arXiv:hep-ph/0011155].

18. F. A. Berends, H. Kuijf, B. Tausk and W. T. Giele, *Nucl. Phys.* B **357**, 32 (1991).

19. A. Pukhov *et al.*, arXiv:hep-ph/9908288.

20. T. Stelzer and W. F. Long, *Comput. Phys. Commun.* **81**, 357 (1994) [arXiv:hep-ph/9401258].

21. F. Yuasa *et al.*, *Prog. Theor. Phys. Suppl.* **138**, 18 (2000) [arXiv:hep-ph/0007053].

22. A. Kanaki and C. G. Papadopoulos, *Comput. Phys. Commun.* **132**, 306 (2000) [arXiv:hep-ph/0002082].

23. M. L. Mangano, M. Moretti, F. Piccinini, R. Pittau and A. D. Polosa, *JHEP* **0307**, 001 (2003) [arXiv:hep-ph/0206293].

24. F. Krauss, R. Kuhn and G. Soff, *JHEP* **0202**, 044 (2002) [arXiv:hep-ph/0109036].

25. R. Kleiss, W. J. Stirling and S. D. Ellis, *Comput. Phys. Commun.* **40**, 359 (1986).

26. C. G. Papadopoulos, *Comput. Phys. Commun.* **137**, 247 (2001) [arXiv:hep-ph/0007335].

27. F. Maltoni and T. Stelzer, *JHEP* **0302**, 027 (2003) [arXiv:hep-ph/0208156].

28. W. Giele *et al.*, "The QCD/SM working group: Summary report", Proceedings of Les Houches Workshop on "Physics at TeV Colliders", 2001, arXiv:hep-ph/0204316.

29. S. Catani, F. Krauss, R. Kuhn and B. R. Webber, *JHEP* **0111**, 063 (2001) [arXiv:hep-ph/0109231].

30. J. Campbell and R.K. Ellis, *Phys. Rev.* D **65**, 113007 (2002) [arXiv:hep-ph/0202176].

31. Z. Nagy and Z. Trocsanyi, *Phys. Rev. Lett.* **87**, 082001 (2001) [arXiv:hep-ph/0104315].

32. Z. Nagy, *Phys. Rev. Lett.* **88**, 122003 (2002) [arXiv:hep-ph/0110315]; arXiv:hep-ph/0307268.

33. V. Del Duca, F. Maltoni, Z. Nagy and Z. Trocsanyi, *JHEP* **0304**, 059 (2003) [arXiv:hep-ph/0303012].

34. W. Beenakker, S. Dittmaier, M. Krämer, B. Plümper, M. Spira and P.M. Zerwas, *Phys. Rev. Lett.* **87** 201805 (2001) [arXiv:hep-ph/0107081]; *Nucl. Phys.* B **653** 151 (2003) [arXiv:hep-ph/0211352]; L. Reina and S. Dawson, *Phys. Rev. Lett.* **87**, 201804 (2001) [arXiv:hep-ph/0107101]; S. Dawson, C. Jackson, L. H. Orr, L. Reina and D. Wackeroth, *Phys. Rev.* D **68**, 034022 (2003) [arXiv:hep-ph/0305087].

35. T. Figy, C. Oleari and D. Zeppenfeld, arXiv:hep-ph/0306109.

36. C. Oleari and D. Zeppenfeld, arXiv:hep-ph/0310156.

37. W.T. Giele and E.W.N. Glover, *Phys. Rev.* D **46**, 1980 (1992); S. Catani and M. H. Seymour, *Nucl. Phys.* B **485**, 291 (1997) [Erratum-ibid. B **510**, 503 (1997)] [arXiv:hep-ph/9605323].

38. T. Binoth, J. P. Guillet, G. Heinrich and C. Schubert, *Nucl. Phys.* B **615**, 385 (2001) [arXiv:hep-ph/0106243].

39. Z. Nagy and D. E. Soper, *JHEP* **0309**, 055 (2003) [arXiv:hep-ph/0308127].

40. T. Binoth, G. Heinrich and N. Kauer, *Nucl. Phys.* B **654**, 277 (2003) [arXiv:hep-ph/0210023].

41. S. Frixione and B. R. Webber, *JHEP* **0206**, 029 (2002) [arXiv:hep-ph/0204244]; arXiv:hep-ph/0207182; arXiv:hep-ph/0307146; arXiv:hep-ph/0309186.

42. S. Frixione, P. Nason and B. R. Webber, *JHEP* **0308**, 007 (2003) [arXiv:hep-ph/0305252].

43. T. Affolder *et al.* [CDF Collaboration], *Phys. Rev. Lett.* **88**, 042001 (2002) [arXiv:hep-ex/0108034].

44. Z. Kunszt and D. E. Soper, *Phys. Rev.* D **46**, 192 (1992); S. D. Ellis, Z. Kunszt and D. E. Soper, *Phys. Rev. Lett.* **69**, 3615 (1992) [arXiv:hep-ph/9208249].

45. W.T. Giele, E.W.N. Glover and D.A. Kosower, *Nucl. Phys.* B **403**, 633 (1993) [arXiv:hep-ph/9302225].

46. R. Hirosky, these proceedings.

47. Z. Bern, L.J. Dixon, D.C. Dunbar and D.A. Kosower, *Nucl. Phys.* B **425**, 217 (1994) [arXiv:hep-ph/9403226]; D.A. Kosower, *Nucl. Phys.* B **552**, 319 (1999) [arXiv:hep-ph/9901201]; D.A. Kosower and P. Uwer, *Nucl. Phys.* B **563**, 477 (1999) [arXiv:hep-ph/9903515]; Z. Bern, V. Del Duca and C.R. Schmidt, *Phys. Lett.* B **445**, 168 (1998) [arXiv:hep-ph/9810409]; Z. Bern, V. Del Duca, W.B. Kilgore and C.R. Schmidt, *Phys. Rev.* D **60**, 116001 (1999) [arXiv:hep-ph/9903516]; S. Catani and M. Grazzini, *Nucl. Phys.* B **591**, 435 (2000) [arXiv:hep-ph/0007142].

48. D. A. Kosower, *Phys. Rev. Lett.* **91**, 061602 (2003) [arXiv:hep-ph/0301069]; S. Weinzierl, *JHEP* **0307**, 052 (2003) [arXiv:hep-ph/0306248].

49. J.M. Campbell and E.W.N. Glover, *Nucl. Phys.* B **527** (1998) 264 [arXiv:hep-ph/9710255]; S. Catani and M. Grazzini, *Phys. Lett.* B **446**, 143 (1999) [arXiv:hep-ph/9810389]; *Nucl. Phys.* B **570**, 287 (2000) [arXiv:hep-ph/9908523]; F.A. Berends and W.T. Giele, *Nucl. Phys.* B **313**, 595 (1989); D. A. Kosower, *Phys. Rev.* D **67**, 116003 (2003) [arXiv:hep-ph/0212097]; S. Weinzierl, *JHEP* **0303** (2003) 062 [arXiv:hep-ph/0302180].

50. A. Gehrmann-De Ridder, T. Gehrmann and E.W.N. Glover, *Phys. Lett.* B **414**, 354 (1997) [arXiv:hep-ph/9705305]; A. Gehrmann-De Ridder and E.W.N. Glover, *Nucl. Phys.* B **517**, 269 (1998) [arXiv:hep-ph/9707224].

51. D.A. Kosower and P. Uwer, arXiv:hep-ph/0307031.

52. S. Laporta, *Int. J. Mod. Phys.* A **15**, 5087 (2000)

[arXiv:hep-ph/0102033].

53. F.V. Tkachov, *Phys. Lett.* B **100**, 65 (1981); K.G. Chetyrkin and F.V. Tkachov, *Nucl. Phys.* B **192**, 159 (1981).

54. T. Gehrmann and E. Remiddi, *Nucl. Phys.* B **580**, 485 (2000) [arXiv:hep-ph/9912329].

55. V.A. Smirnov, *Phys. Lett.* B **460**, 397 (1999) [arXiv:hep-ph/9905323]; V.A. Smirnov and O.L. Veretin, *Nucl. Phys.* B **566**, 469 (2000) [arXiv:hep-ph/9907385]; J.B. Tausk, *Phys. Lett.* B **469**, 225 (1999) [arXiv:hep-ph/9909506]; C. Anastasiou, T. Gehrmann, C. Oleari, E. Remiddi and J.B. Tausk, *Nucl. Phys.* B **580**, 577 (2000) [arXiv:hep-ph/0003261]; T. Gehrmann and E. Remiddi, *Nucl. Phys.* B (Proc. Suppl.) **89**, 251 (2000) [arXiv:hep-ph/0005232]; C. Anastasiou, J.B. Tausk and M.E. Tejeda-Yeomans, *Nucl. Phys.* B (Proc. Suppl.) **89**, 262 (2000) [arXiv:hep-ph/0005328].

56. T. Gehrmann and E. Remiddi, *Nucl. Phys.* B **601**, 248 (2001) [arXiv:hep-ph/0008287] and **601**, 287 (2001) [arXiv:hep-ph/0101124].

57. T. Gehrmann, *Nucl. Phys.* Proc. Suppl. **116**, 13 (2003) [arXiv:hep-ph/0210157].

58. Z. Bern, L. Dixon and A. Ghinculov, *Phys. Rev.* D **63**, 053007 (2001) [arXiv:hep-ph/0010075].

59. C. Anastasiou, E.W.N. Glover, C. Oleari and M.E. Tejeda-Yeomans, *Nucl. Phys.* B **601**, 318 (2001) [arXiv:hep-ph/0010212]; **601**, 347 (2001) [arXiv:hep-ph/0011094]; **605**, 486 (2001) [arXiv:hep-ph/0101304]; E.W.N. Glover, C. Oleari and M.E. Tejeda-Yeomans, *Nucl. Phys.* **605**, 467 (2001) [arXiv:hep-ph/0102201]; E.W.N. Glover and M.E. Tejeda-Yeomans, *JHEP* **0306**, 033 (2003) [arXiv:hep-ph/0304169]; Z. Bern, A. De Freitas and L. Dixon, *JHEP* **0203**, 018 (2002) [arXiv:hep-ph/0201161]; Z. Bern, A. De Freitas and L. Dixon, *JHEP* **0306**, 028 (2003) [arXiv:hep-ph/0304168].

60. Z. Bern, A. De Freitas and L. J. Dixon, *JHEP* **0109**, 037 (2001) [arXiv:hep-ph/0109078]; C. Anastasiou, E.W.N. Glover and M.E. Tejeda-Yeomans, *Nucl. Phys.* B **629**, 255 (2002) [arXiv:hep-ph/0201274].

61. Z. Bern, A. De Freitas, L.J. Dixon, A. Ghinculov and H.L. Wong, *JHEP* **0111**, 031 (2001) [arXiv:hep-ph/0109079]; T. Binoth, E.W.N. Glover, P. Marquard and J.J. van der Bij, *JHEP* **0205**, 060 (2002) [arXiv:hep-ph/0202266].

62. L.W. Garland, T. Gehrmann, E.W.N. Glover, A. Koukoutsakis and E. Remiddi, *Nucl. Phys.* B **627**, 107 (2002) [arXiv:hep-ph/0112081] and **642**, 227 (2002) [arXiv:hep-ph/0206067].

63. S. Moch, P. Uwer and S. Weinzierl, *Phys. Rev.* D **66**, 114001 (2002) [arXiv:hep-ph/0207043].

64. S. Moch, P. Uwer and S. Weinzierl, *J. Math. Phys.* **43**, 3363 (2002) [arXiv:hep-ph/0110083]; S. Weinzierl, *Comput. Phys. Commun.* **145**, 357 (2002) [arXiv:math-ph/0201011].

65. T. Gehrmann and E. Remiddi, *Nucl. Phys.* B **640**, 379 (2002) [arXiv:hep-ph/0207020].

66. S. Catani, *Phys. Lett.* B **427**, 161 (1998) [arXiv:hep-ph/9802439]; G. Sterman and M. E. Tejeda-Yeomans, *Phys. Lett.* B **552**, 48 (2003) [arXiv:hep-ph/0210130].

67. R. Bonciani, P. Mastrolia and E. Remiddi, *Nucl. Phys.* B **661**, 289 (2003) [arXiv:hep-ph/0301170]; arXiv:hep-ph/0307295.

68. U. Aglietti and R. Bonciani, *Nucl. Phys.* B **668**, 3 (2003) [arXiv:hep-ph/0304028].

69. R. Bonciani, A. Ferroglia, P. Mastrolia, E. Remiddi, J.J. van der Bij, arXiv:hep-ph/0310333.

70. E.W.N. Glover and A.G. Morgan, *Z. Phys.* C **62**, 311 (1994).

71. S. Catani, M. Fontannaz, J. P. Guillet and E. Pilon, *JHEP* **0205**, 028 (2002) [arXiv:hep-ph/0204023].

72. D. Buskulic *et al.* [ALEPH Collaboration], Z. Phys. C **69**, 365 (1996).

73. K. Ackerstaff *et al.* [OPAL Collaboration], *Eur. Phys. J.* C **2**, 39 (1998) [arXiv:hep-ex/9708020].

74. A. Gehrmann-De Ridder, G. Kramer and H. Spiesberger, *Nucl. Phys.* B **578**, 326 (2000) [arXiv:hep-ph/0003082].

75. R. Lemrani [H1 Collaboration], arXiv:hep-ex/0308066.

76. T. Binoth, J.P. Guillet, E. Pilon and M. Werlen, *Eur. Phys. J.* C **16**, 311 (2000) [arXiv:hep-ph/9911340].

77. Z. Bern, L. Dixon and C. Schmidt, *Phys. Rev.* D **66**, 074018 (2002) [arXiv:hep-ph/0206194].

78. T. Binoth, J. P. Guillet, E. Pilon and M. Werlen, *Phys. Rev.* D **63**, 114016 (2001) [arXiv:hep-ph/0012191].

79. T. Binoth, J. P. Guillet, E. Pilon and M. Werlen, *Eur. Phys. J.* C **24**, 245 (2002) [arXiv:hep-ph/0111043].

80. M. Spira, A. Djouadi, D. Graudenz and P. M. Zerwas, *Nucl. Phys.* B **453**, 17 (1995) [arXiv:hep-ph/9504378]; M. Spira, *Fortsch. Phys.* **46**, 203 (1998) [arXiv:hep-ph/9705337].

81. R. Hamberg, W. L. van Neerven and T. Matsuura, *Nucl. Phys.* B **359**, 343 (1991) [Erratum-ibid. B **644**, 403 (2002)].

82. R.V. Harlander and W.B. Kilgore, *Phys. Rev. Lett.* **88**, 201801 (2002) [arXiv:hep-ph/0201206].

83. R.V. Harlander, *Phys. Lett.* B **492**, 74 (2000) [arXiv:hep-ph/0007289]; S. Catani, D. de Florian and M. Grazzini, *JHEP* **0105**, 025 (2001) [arXiv:hep-ph/0102227]; R.V. Harlander and W.B. Kilgore, *Phys. Rev.* D **64**, 013015 (2001) [arXiv:hep-ph/0102241].

84. C. Anastasiou and K. Melnikov, *Nucl. Phys.* B **646**, 220 (2002) [arXiv:hep-ph/0207004].

85. V. Ravindran, J. Smith and W. L. van Neerven, *Nucl. Phys.* B **665**, 325 (2003) [arXiv:hep-ph/0302135].

86. S. Catani, D. de Florian, M. Grazzini and P. Nason, *JHEP* **0307**, 028 (2003) [arXiv:hep-ph/0306211].

87. R.V. Harlander and W.B. Kilgore, *JHEP* **0210**, 017 (2002) [arXiv:hep-ph/0208096]; C. Anastasiou and

K. Melnikov, arXiv:hep-ph/0208115.

88. R. V. Harlander and W. B. Kilgore, *Phys. Rev.* D **68**, 013001 (2003) [arXiv:hep-ph/0304035].

89. O. Brein, A. Djouadi and R. Harlander, arXiv:hep-ph/0307206.

90. C. Anastasiou, L. Dixon and K. Melnikov, *Nucl. Phys.* Proc. Suppl. **116**, 193 (2003) [arXiv:hep-ph/0211141].

91. D. de Florian, M. Grazzini and Z. Kunszt, *Phys. Rev. Lett.* **82**, 5209 (1999) [arXiv:hep-ph/9902483].

92. V. Ravindran, J. Smith and W. L. Van Neerven, *Nucl. Phys.* B **634**, 247 (2002) [arXiv:hep-ph/0201114]; C. J. Glosser and C. R. Schmidt, *JHEP* **0212**, 016 (2002) [arXiv:hep-ph/0209248].

93. G. Bozzi, S. Catani, D. de Florian and M. Grazzini, *Phys. Lett.* B **564**, 65 (2003) [arXiv:hep-ph/0302104].

94. M. Dittmar, F. Pauss and D. Zürcher, *Phys. Rev.* D **56**, 7284 (1997) [arXiv:hep-ex/9705004]; A.D. Martin, R.G. Roberts, W.J. Stirling and R.S. Thorne, *Eur. Phys. J.* C **14**, 133 (2000) [arXiv:hep-ph/9907231].

95. C. Anastasiou, L. Dixon, K. Melnikov and F. Petriello, arXiv:hep-ph/0306192.

96. R. Thorne, these proceedings, arXiv:hep-ph/0309343.

97. S. A. Larin, P. Nogueira, T. van Ritbergen and J. A. Vermaseren, *Nucl. Phys.* B **492**, 338 (1997) [arXiv:hep-ph/9605317]; A. Retey and J. A. Vermaseren, *Nucl. Phys.* B **604**, 281 (2001) [arXiv:hep-ph/0007294].

98. S. Moch, J. A. Vermaseren and A. Vogt, *Nucl. Phys.* B **646**, 181 (2002) [arXiv:hep-ph/0209100].

DISCUSSION

Sungwon Lee (Texas A&M University): Currently there are large discrepancies between data and NLO QCD predictions for direct photon production. Has there been any theoretical progress on the intrinsic (effective parton) k_t issue for direct photon production?

Thomas Gehrmann: Understanding of k_t effects is one motivation for doing NNLO calculations. At NNLO, for the first time, we start to fully model k_t effects due to hard parton emission in the initial state because we allow for either double emission of one of the incoming legs, or for double uncorrelated emission. So, once you have NNLO calculations for $2 \to 2$ scattering processes available, you will really have a theoretical tool for computing k_t effects from perturbation theory.

Ikaros Bigi (Notre Dame University): You showed this very instructive curve about b quark frag-mentation, where you showed that the Peterson *et al.* prediction is less than optimal. Do you have (or does someone have) a similar curve for charm fragmentation?

Thomas Gehrmann: The recent work by Cacciari and Nason addresses this issue. They refit charm fragmentation functions using a new parameterization in moment space trying to expose the information content of LEP data relevant to proton-antiproton colliders. Comparison with old parameterizations is however not made. The basic message is that when you start fitting fragmentation functions, you have to be extremely careful that you are not introducing artifacts from the choice of parameterization which, although giving you a minimum χ^2 fit, do not really reproduce the data because you are starting with too stiff an initial form.

QCD AT COLLIDERS

R. HIROSKY

University of Virginia, Dept. of Physics, Charlottesville, VA 22904, USA
E-mail: hirosky@virginia.edu

The success of the theory of Quantum Chromodynamics (QCD) in describing processes controlled by the strong interaction is generally seen as a triumph of modern particle physics. This paper reviews recent QCD measurements using hadronic jet final states from the Fermilab Tevatron, DESY's HERA, and CERN's LEP colliders. Recent advancements in the measurements of jet production cross sections, events shapes, and energy flow, along with improved theoretical calculations, allow for new levels of precision in the study of the physics of strong interactions and point to areas in need of further refinement.

1. Introduction

In the framework of quantum chromodynamics (QCD) outgoing hadronic jets are a key signature of strong interactions between constituent partons in inelastic hadron collisions. We have witnessed substantial progress in both the theoretical and experimental understanding of such processes throughout the past decade. QCD predicts the partonic cross sections[1] for hard scattering at large momenta transfers. The determination of the production amplitudes for various final states requires the convolution of accurately determined parton distribution functions (PDF's)[2] with the partonic level cross sections to model the initial state parton momenta within the hadron beams. Pictorially jet production in hadron collisions can be modeled as in Fig. 1. Matrix elements for the hard interaction are available to (N)NLO/(N)NLL for many processes and phenomenological models tuned to data can be used to account for hadronization effects.

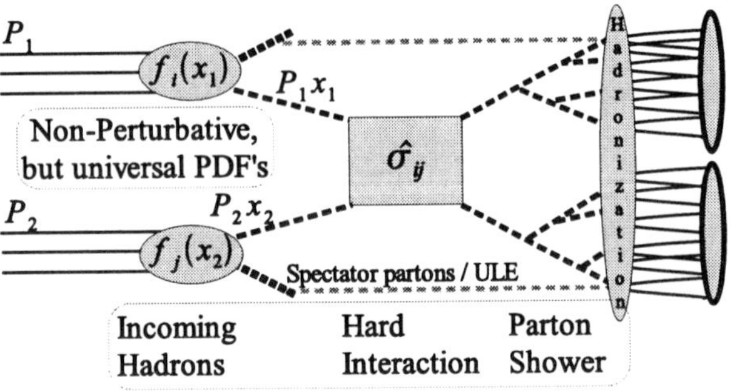

Figure 1. Jet production in hadron collisions. Matrix elements for the hard scatter are available at (N)NLO/(N)NLL for many processes under study. Non-perturbative parton distribution functions are factorized from the hard scattering cross section in calculating production rates at colliders.

This paper reviews the status of recent QCD studies from collider experiments at the Tevatron, HERA, and LEP.

2. Jet Production at the Tevatron

The available center-of-mass (cms) energy of $\sqrt{s} = (1.8)1.96$ TeV for $p\bar{p}$ collisions at the Tevatron allows the experimental probe of distance scales down to $\sim 10^{-17}$ cm in measures of inclusive single jet and di-jet cross sections. This section summarizes the status of QCD jet analyses for the CDF[3] and DØ[4] experiments. Figures 2 and 3 show the kinematic regions accessed in measurements[5] of jet production in Run I at the Tevatron.[a]

The Run I history of QCD studies at the Tevatron includes the observation of a prominent excess of high-p_T central jet production (as compared to NLO predictions and contemporary PDF's) initially observed by the CDF experiment[6] and contrasted by the DØ observation with good agreement in shape and normalization between the data and theory.[7] (Figure 2 shows NLO QCD predictions for the DØ cross sections calculated with JETRAD.[8]) Analyses of experimental uncertainties showed the two data sets to be consistent and analyses of constraints in global fits for PDF's found that uncertainties in the determination of large-x gluon distributions, underlying assumptions in PDF parameterizations, and scale choices, allowed a very wide range of large-x behavior.[9] The dominant effect is due to poor constraints on the gluons at large-x from the data available for global fits. Ensuing discussions of these is-

---

[a] Run I refers to the data taking periods ending in 1996 with cms energy $\sqrt{s} = 1.8$ TeV. Run II refers to the current data run starting in 2001 with cms energy $\sqrt{s} = 1.96$ TeV.

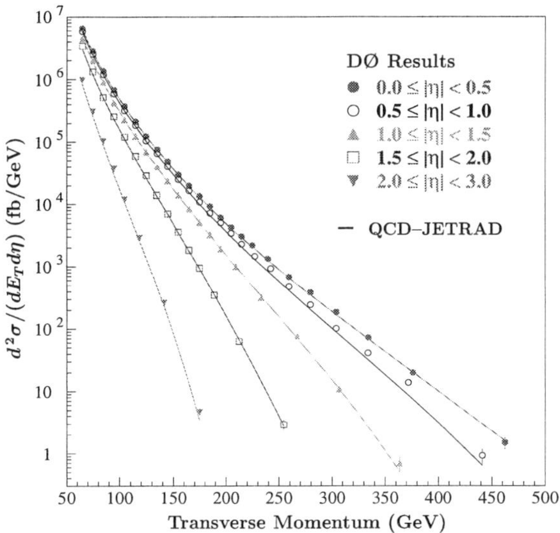

Figure 2. The single jet inclusive cross section measured in five bins of pseudorapidity from the DØ Run I data. Data are shown compared to NLO predictions for the cross sections.

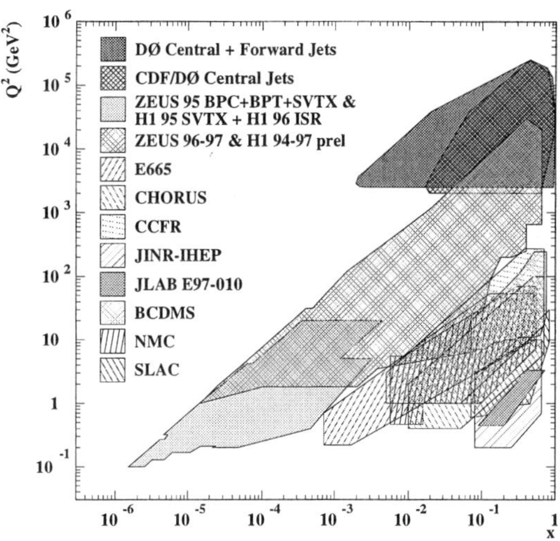

Figure 3. Kinematic reach of high-p_T jet data at the Tevatron compared to other colliders and fixed target experiments in the plane of parton momentum fraction x and square of the momentum transfer Q^2.

sues have directly contributed to an improved state-of-the-art in the reporting and applications of PDF uncertainties.[10]

In Run I both CDF and DØ performed most jet analyses using iterative cone algorithms based on the Snowmass standard.[11] Both experiments defined jets with cone radius $R = \sqrt{\Delta\eta^2 + \Delta\phi^2} = 0.7$. Where the polar (pseudorapidity, η^b) and azimuthal (ϕ) angles are defined by E_T weighted averages of all energy depositions associated with the jet.

Run II at the Tevatron brings numerous detector improvements to both DØ and CDF.[12,13] Additionally, the physics reach is extended with an increased center-of-mass energy of $1.96\,\mathrm{TeV}$ and an expected factor of 50 (minimum) increase in total integrated luminosity. An improved iterative cone algorithm based on 4-momenta clusters (as opposed to the E_T weighting schemes defined at Snowmass) has been proposed[14] to address IR divergences in the Run I schemes and to fully specify the treatment of overlapping cones. DØ has implemented this improved cone algorithm for all Run II cone-jet analyses. In the preliminary CDF analyses jets have been reconstructed using their legacy Run I algorithm.

CDF has presented results of their measure of the inclusive jet cross section in the central region

defined by $0.1 < |\eta^{jet}| < 0.7$ for $177\,\mathrm{pb}^{-1}$ of accumulated Run II data. The effect of the increased Run II cms energy on CDF's central jet cross section is shown in Fig. 4(a). The observed increase in jet production cross section is in agreement with NLO predictions within experimental uncertainties. In Fig. 4(b) the data are compared to the NLO QCD calculation using the full set of CTEQ6.1[15] parameterizations of the parton density functions. The range of NLO predictions defined by the full set of parameterizations is contained in the envelope shown by the solid lines. The preliminary DØ central jet inclusive cross section shown in Fig. 5 is measured in the region $|\eta^{jet}| < 0.5$. Both experiments are in agreement with NLO predictions within experimental uncertainties. Due to the increased cms energy at the Tevatron the reach of this cross section measurement has already been extended by $150\,\mathrm{GeV}$ with respect to Run II (see Fig.4).

Successive recombination or k_T algorithms have been introduced to resolve problems of infrared singularities in parton jet clustering and to provide a theoretically "clean" procedure, free from ad-hoc or non-physical definitions. Both DØ[16] and CDF have repeated their central jet inclusive cross section measurement using the Ellis-Soper k_T algorithm[17] for jet reconstruction. The DØ data for the central ($|\eta^{jet}| < 0.5$) inclusive jet cross section for k_T jets are compared to the NLO QCD prediction and to

---

$^b\eta = -ln(tan(\theta/2))$, where θ is the polar angle relative to the beamline.

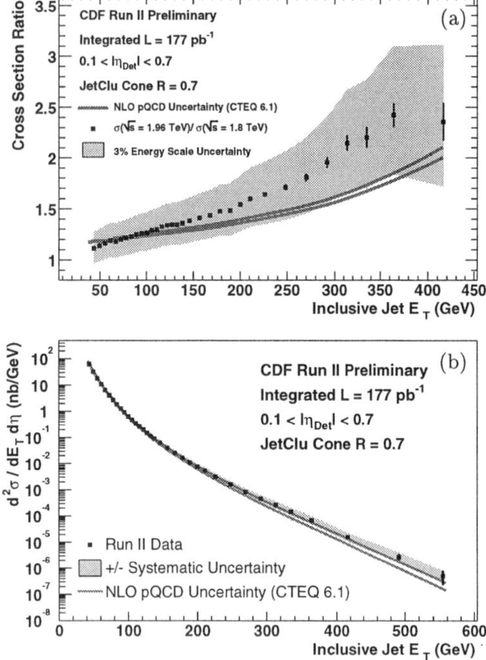

Figure 4. (a) Ratio of CDF's central inclusive jet cross section (Run II/Run I). (b) CDF's central inclusive jet cross section (circles) compared to NLO QCD (lines). The two lines represent limits on the ranges of predictions determined from the set of CTEQ6 parton density functions. Experimental systematic uncertainties are shown by the shaded band.

the cone jet cross section in the same η region in Fig. 6. The D-parameter used to control the size of the k_T jets was set to 1.0 for this analysis. The analysis used $88\,\mathrm{pb}^{-1}$ from the Run I data sample. Both DØ and CDF (not shown) find only marginal agreement with the NLO prediction at the lower p_T end of the spectrum. For the clustering algorithms in the DØ measurement, the NLO QCD predictions for the k_T and cone jet cross sections agree to within 1%. To identify the source of the difference between the data and theory, hadronization effects were evaluated using HERWIG to correct particle level jets back to the parton level. While k_T jets are found to appear more energetic at the hadron level as compared to cone jets, the magnitude of the difference was only found to be about one third of the observed difference in the data.

The di-jet mass cross section may also be used as a test of QCD, to constrain allowable PDF models, and to search for evidence of interactions beyond the Standard Model. The most strict limits on quark compositeness from Run I were obtained using di-

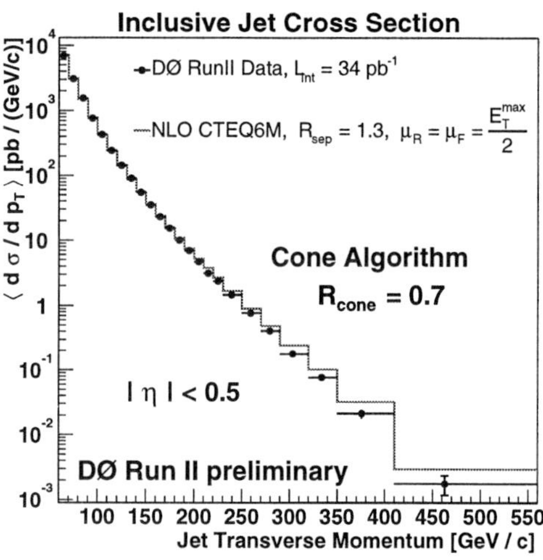

Figure 5. DØ's central inclusive jet cross section $d\sigma/dp_T$ as a function of p_T^{jet} integrated over $|\eta^{jet}| < 0.5$. Also shown is the NLO calculation using the CTEQ6M parton distributions.

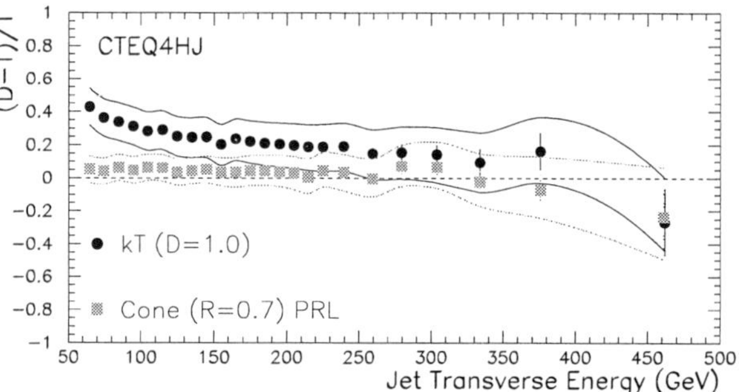

Figure 6. Fractional difference between DØ measurement of the inclusive jet cross section (D) compared to NLO QCD prediction (T) from Run I data. Jets reconstructed with k_T (circles) and cone (squares) algorithms are compared. Experimental systematic uncertainties (excluding luminosity) are shown by solid lines.

jet measurements.[18] In di-jet analyses, reduction in experimental systematics due to detector resolution effects can compensate for reduced statistics due to the two jet requirement. Further, measuring the angular dependence of the jet production can reduce effects of PDF uncertainties when searching for new physics with relatively isotropic production angles. The preliminary DØ results for the di-jet mass cross section for Run II are summarized in Fig. 7. The data are in agreement with NLO QCD predictions within the experimental uncertainties. Preliminary results from CDF are shown in Fig. 8. A comparison with the Run I cross section is included to show the increased reach at Run II.

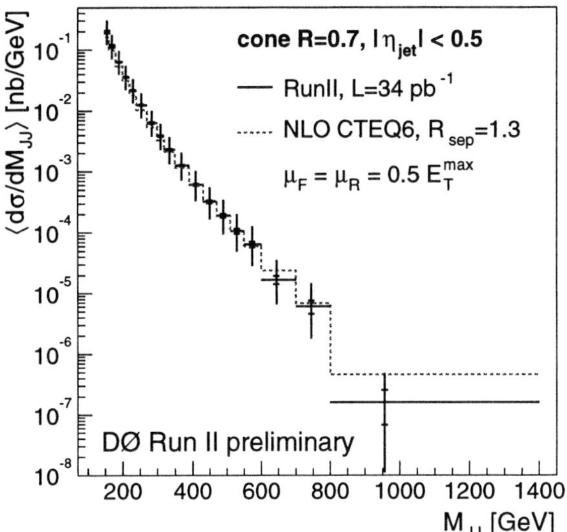

Figure 7. DØ dijet mass cross section. $d^3\sigma/dM\,d\eta_1\,d\eta_2$ for $|\eta_{jet}| < 0.5$. Inner (outer) error bars show statistical (total) experimental uncertainties. The NLO calculation is shown by the dashed line.

3. Jets in DIS and Photoproduction

This section summarizes recent experimental tests of QCD with jet production from the H1[19] and ZEUS[20] experiments at HERA. The HERA collider at DESY produces $e^{(+/-)}p$ collisions with a center-of-mass energy of 318 GeV. Precision measurements of the proton structure in ep deep-inelastic scattering (DIS) provide a strong verification of perturbative QCD, with a hard scale defined by Q^2, the virtuality of the photon exchange. Jet production gives direct access to the underlying parton dynamics and provides an additional scale, the jet transverse energy E_T, to describe the interaction within the framework of perturbative QCD. Figure 9 shows $O(\alpha_s)$ diagrams for jet production in ep DIS. Photoproduction will be discussed in Sec. 3.2.

3.1. Jets in DIS

Previous analyses of inclusive and multi-jet production in neutral current DIS at large virtualities[21] ($Q^2 > 125\,\mathrm{GeV}^2$) have shown excellent agreement with NLO QCD predictions. H1 has extended their measured inclusive jet cross sections to low Q^2 values in the range $5\,\mathrm{GeV}^2 < Q^2 < 100\,\mathrm{GeV}^2$. The low Q^2 analysis[22] shows the onset of discrepancies with NLO

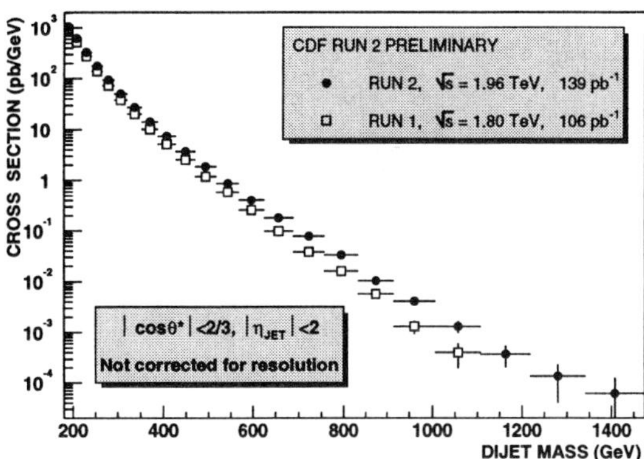

Figure 8. Di-jet mass cross section for CDF. The Run I data are overlaid to show the increased production rates in the Run II experiment.

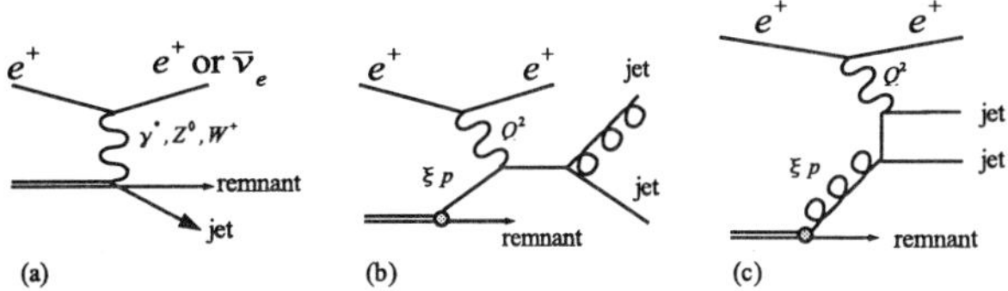

Figure 9. Order α_s diagrams for jet production in deep inelastic scattering. (a) Born, (b) QCD-Compton and (c) Boson-gluon fusion.

QCD predictions as demonstrated in Fig. 10. Jets are reconstructed using the inclusive k_T algorithm[17] and transverse energies are calculated in the Breit frame.[c] NLO calculations of jets observables were calculated using the DISENT program.[23] Corrections to the partonic cross section due to hadronization effects were estimated using a variety of Monte Carlo generators implementing the Lund color string model.[24]

Figure 10 shows the inclusive jet cross section versus jet E_T for three regions of pseudorapidity η_{lab}. Positive or forward η_{lab} is defined to be in the direction of the proton. Good agreement between the data and NLO QCD is observed in the backward region for all E_T's, while discrepancies become apparent in more forward regions, especially for jets with low E_T. The estimated renormalization scale uncertainties do not cover the discrepancies at lowest E_T in this region, where NLO corrections and scale sensitivity are also at their largest.

---

[c]In the Breit frame the photon and parton collide head-on ($2x\overline{p} + \overline{q} = 0$, where x is the Bjorken scaling variable, and $\overline{p}, \overline{q}$ are the proton and γ^* momentum, respectively). In this frame a jet produced at the Born level is defined to have $p_T = 0$.

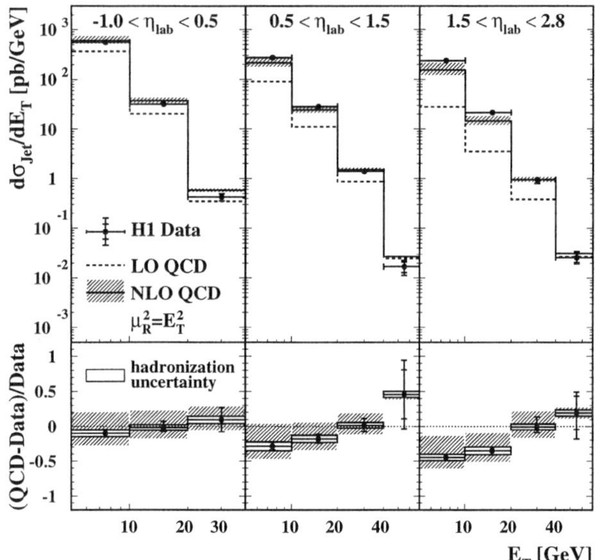

Figure 10. H1 measurements of inclusive jet cross sections $d\sigma/dE_T$ integrated over the region $5 < Q^2 < 100\,\mathrm{GeV}^2$. The data are shown as points. Error bars include statistical and systematic uncertainties. Top frame: Data are compared to DISENT NLO QCD calculations using CTEQ5M PDF's (solid line) and DISENT LO calculations using CTEQ5L (dashed line), with renormalization scale $\mu_R = E_T$ (no hadronization corrections are applied). The effects of varying μ_R by factors of ± 2 are shown in the hatched band. Bottom frame: (QCD-Data)/Data, where "QCD" denotes NLO QCD corrected for hadronization effects. Hadronization uncertainties are denoted by the inner white band.

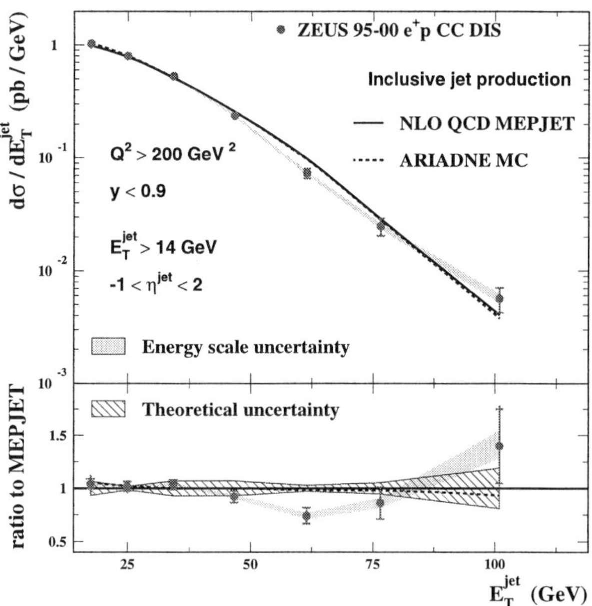

Figure 11. (a) Differential cross section, $d\sigma/dE_T$ for inclusive jet production measured at ZEUS (dots) with $E_T^{jet} > 14\,\mathrm{GeV}$ and $-1 < \eta^{jet} < 2$ with $Q^2 > 200\,\mathrm{GeV}^2$. The solid band shows the uncertainty due to the calorimeter energy scale. Error bars are statistical (inner) and statistical $\oplus$ systematic (outer), exclusive of the scale error. (b) Ratio of the measured cross section to NLO calculation. The hatched band represents the theoretical uncertainty.

The ZEUS experiment has measured inclusive and di-jet production rates[25] and jet substructures in charged current (CC) DIS for $Q^2 > 200\,\mathrm{GeV}^2$. Jet production in CC DIS provides a testing ground for both QCD and the electroweak sector of the Standard Model. Up to leading-order in α_s CC DIS proceeds via the QCD-Compton ($Wq \to q'g$), W-gluon-fusion ($Wg \to q\bar{q}'$), or the pure electroweak process ($Wq \to q'$). The inclusive jet cross section for $E_T^{jet} > 14\,\mathrm{GeV}$ and $-1 < \eta^{jet} < 2$ is shown in Fig. 11. The data sample corresponds to an integrated luminosity of $110.5\,\mathrm{pb}^{-1}$. Jets were reconstructed using the invariant k_T algorithm in the laboratory frame. Both the NLO QCD calculation of MEPJET[26] and the LO ARIADNE MC[27] models give a good description of the jet cross section. The NLO calculations were corrected to the hadron level using ARIADNE and LEPTO[28] models to estimate the uncertainty on the procedure.

It is interesting to contrast the agreement of the HERA cross sections and NLO QCD for low E_T jets defined using the k_T algorithm with the Tevatron k_T jet measurements. Correction for hadronization effects at the Tevatron does not recover good agreement at lower E_T values. This may suggest complications due to more energetic underlying events in the $p\bar{p}$ collisions at the Tevatron.

3.2. *Jets in Photoproduction*

A comparison of measurements of jet production at different center-of-mass energies demonstrates scaling violations due to the evolution of structure functions and the running of the strong coupling constant α_s. In the parton model the scaled jet invariant cross section, $S(x_T) = (E_T^{jet})^4 E^{jet} d^3\sigma/dp_X^{jet} dp_Y^{jet} dp_Z^{jet}$, should be independent of W, the center-of-mass energy, when plotted against the dimensionless variable $x_T = 2E_T^{jet}/W$. Scaling violations have been observed in $p\bar{p}$ collisions at center-of-mass energies of 546(630) and 1800 GeV.[29] Photoproduction of jets in ep collisions can be used to perform similar tests. Photoproduction is differentiated from neutral current DIS, by requiring a small virtuality of the exchanged photon, typically $Q^2 < 1\,\text{GeV}^2$. Hence the photons are quasi-real and γp collisions may be studied. The photon may act as a point-like particle in an interaction with a parton carrying a fraction x_p of the proton momentum (direct process). Alternatively the photon may develop a hadronic structure (resolved process) where a parton carrying a fraction x_γ of the photon momentum interacts with a parton in the proton.

ZEUS has measured the scaled jet invariant cross section[30] in γp collisions for the γp center-of-mass energy range $142 < W_{\gamma p} < 293\,\text{GeV}$ for the pseudorapidity range $-2 < \eta^{jet} < 0$. Event selection requires $Q^2 < 1\,\text{GeV}^2$, yielding a median $Q^2 \sim 10^{-3}\,\text{GeV}^2$. The data are divided into two center-of-mass ranges with $\langle W_{\gamma p} \rangle = 180$ and $255\,\text{GeV}$. Figure 12 shows the ratio of the scaled invariant cross sections as a function of x_T. The clear deviation from unity is in agreement with NLO QCD predictions, which include the running of α_s and evolution of PDF's with the scale. These data constitute the first observation of scaling violations in γp collisions. The cross section $d\sigma/dE_T^{jet}$ as a function of E_T^{jet} was used to determine $\alpha_s(M_Z)$.[30] The value $\alpha_s(M_Z)$ was obtained in each bin of the measured cross section and from χ^2 fits to all the data: $\alpha_s(M_Z) = 0.1224 \pm 0.0001(\text{stat.})^{+0.0022}_{-0.0019}(\text{exp.})^{+0.0054}_{-0.0042}(\text{th.})$. NLO calculations with hadronization corrections are used throughout the analysis. The largest contribution to the experimental uncertainty ($\pm 1.5\%$) arises from the jet energy scale. The largest contribution to the theoretical uncertainty arises from terms beyond NLO $\left(^{+4.2}_{-3.3}\%\right)$. For a full discussion see the References.[31]

H1 has also reported a new measurement of inclusive jet production in photoproduction[32] based on $24.1\,\text{pb}^{-1}$ of e^+p data. Jets were reconstructed using the k_T algorithm for $-1 < \eta^{jet} < 2.5$ in the laboratory frame. Events were selected with $Q^2 < 1\,\text{GeV}^2$ and $E_T^{jet} \geq 21\,\text{GeV}$. The γp cms energy range for these events was $95 \leq W_{\gamma p} \leq 285\,\text{GeV}$. The measurement was extended down to $E_T^{jet} \geq 5\,\text{GeV}$ using a $0.47\,\text{pb}^{-1}$ collected via a dedicated trigger. The kinematic range for these data is $Q^2 \leq 0.01\,\text{GeV}^2$ and $164 \leq W_{\gamma p} \leq 242\,\text{GeV}$. Figure 13 shows a measurement of the inclusive jet cross section over the entire E_T range after combining the two samples. The samples were combined by applying a common $W_{\gamma p}$ cut ($164 \leq W_{\gamma p} \leq 242\,\text{GeV}$) and by correcting

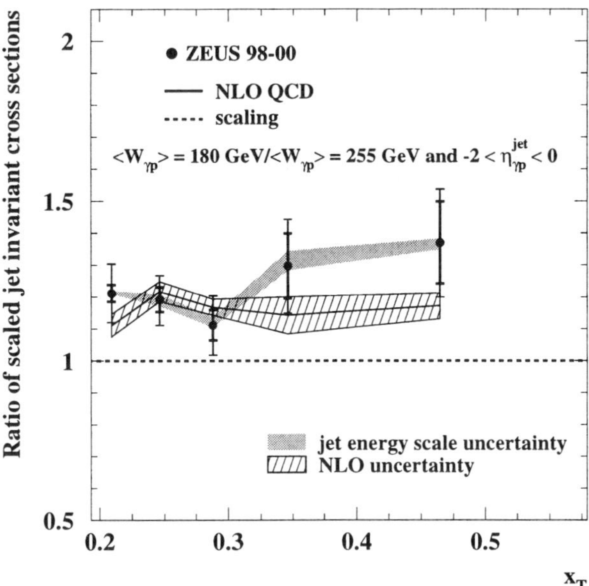

Figure 12. Ratio of scaled jet invariant cross sections versus x_T for two intervals ($W_{\gamma p}$) of γp center-of-mass energies in photoproduction measured at ZEUS.

the low E_T sample by the difference in photon flux $f_{\gamma/e}(y,Q^2)$ over ranges $Q^2 < Q^2_{max}$ for the two samples. The fraction of the electron energy carried by the photon is given by $y = E_\gamma/E_e$. The measured cross section $d\sigma/dE_T^{jet}$ falls by more than 6 orders of magnitude and is well produced by the theoretical prediction. Hadronization corrections are needed in addition to NLO contributions for good agreement with the data at low E_T^{jet}.

The photoproduction results are compared with measurements in $p\bar{p}$ collisions to observe differences arising from the structures of the photon and the proton. The differential cross section was redetermined using a cone algorithm with radius $R = 1$ to match the procedure used for available $p\bar{p}$ data at a comparable cms $\sqrt{s} = 200\,\text{GeV}$.[33] To compare with $p\bar{p}$ measurements at different energies, the cross section was measured with the cone algorithm in the restricted range $1.5 \leq \eta^{jet} \leq 2.5$ and for $E_T^{jet} > 8\,GeV$ and scaled to the invariant cross section at fixed $W_{\gamma p} = 200\,\text{GeV}$ averaged over the cms pseudorapidity range $|\eta^*| \leq 0.5$ using Monte Carlo models to evaluate the correction factors. The invariant scaled cross section is compared to data from UA2[33] and DØ[5,34] in Fig. 14. The $p\bar{p}$ data were transformed into $S(x_T)$ using the central bin values and were scaled by factors of $O(\alpha_{em}/\alpha_s)$ to match the photoproduction

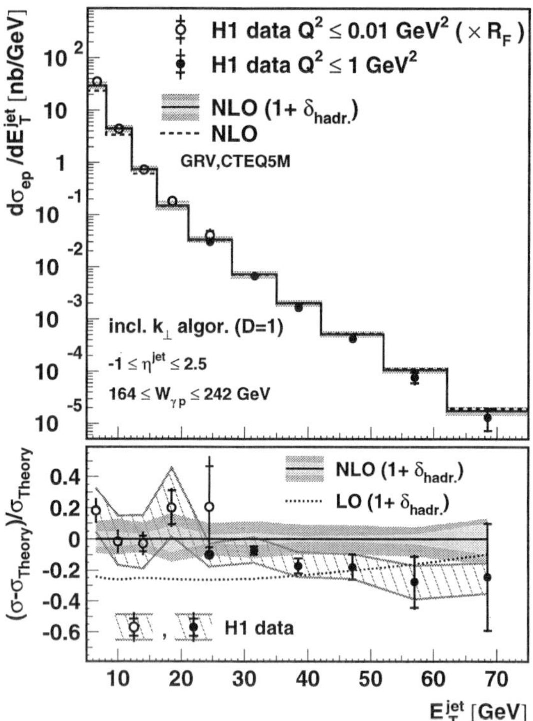

Figure 13. H1 inclusive jet photoproduction. Top: the differential e^+p cross section $d\sigma/dE_t^{jet}$ for inclusive jet production integrated over $-1 \leq \eta^{jet} \leq 2.5$ and $Q^2 \leq 1\,\mathrm{GeV}^2$ (filled circles). Low E_T data measured for $Q^2 \leq 0.01\,\mathrm{GeV}^2$ shown by open circles are corrected for the ratio of photon fluxes in the two Q^2 regions. Bottom: fractional difference between the data (or LO QCD prediction) and the NLO calculation with hadronization corrections. The PDF's used for the photon and proton were GRV and CTEQ5M, respectively.

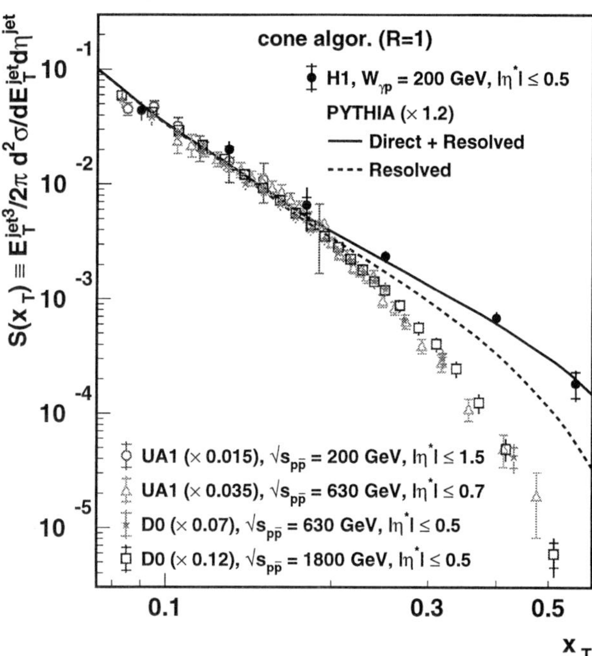

Figure 14. H1 inclusive jet photoproduction. The scaled γp cross section at $W_{\gamma p} = 200\,\mathrm{GeV}$ for inclusive jet production as a function of x_T for $|\eta^*| < 0.5$. Jets are found with a cone algorithm ($R = 1$) and the data are compared with measurements from UA2 and DØ of the inclusive jet cross section at various cms energies. Predictions of PYTHIA for the direct γp and resolved photon contributions are also shown (a normalization factor of 1.2 is applied).

data at $x_T \sim 0.1$. Despite differences in η intervals and analysis procedures, all $p\bar{p}$ data are in approximate agreement. The γp data are compatible with the $p\bar{p}$ data in the region $x_T \lesssim 0.2$, where the resolved photon leads to scaling behavior similar to that for a hadron. At larger x_T the shape of the γp cross section deviates from the $p\bar{p}$ measurements. This is a result of the enhanced quark density of the resolved photon relative to that of the proton at large momentum fractions. Dominating the scaled cross section at largest x_T, the direct photon contribution involves the convolution of only the proton PDF's.

4. $\gamma\gamma$ Collisions at LEP

High energy hadrons at LEP are mainly produced via the process $e^+e^- \to e^+e^-\gamma^*\gamma^* \to e^+e^- + hadrons$, where hadrons of large transverse momentum are produced directly from the virtual photons (QED

process $\gamma^*\gamma^* \to q\bar{q}$) or via QCD processes if the partonic content of the photons is resolved. L3 has measured the differential cross sections $d\sigma/dp_T$ for the production of high-p_T hadrons and jets in two-photon collisions. Distributions for charged pions and jets are shown in Fig. 15. Events were selected based on low photon virtuality $\langle Q^2 \rangle \simeq 0.2\,\mathrm{GeV}$ and an effective mass of the $\gamma\gamma$ system $W_{\gamma\gamma} \geq 5\,\mathrm{GeV}$. The published results for charged pion production[35] show a clear excess of events as compared to NLO QCD for $p_T > 5\,\mathrm{GeV}$. A recent analysis performed for jets confirms the observation of excess production at high p_T. Both analyses were restricted to the region $|\eta| < 1.0$ and jets were reconstructed using the k_T algorithm[17] with a minimum p_T requirement of 3 GeV with a D-cut of 1.0.

In Fig. 15(b) the jet data are compared to NLO analytical QCD predictions.[36] The flux of quasi-real photons was obtained via the Weizsäcker-Williams formula.[37] Both the direct and resolved production processes were included in the NLO calcula-

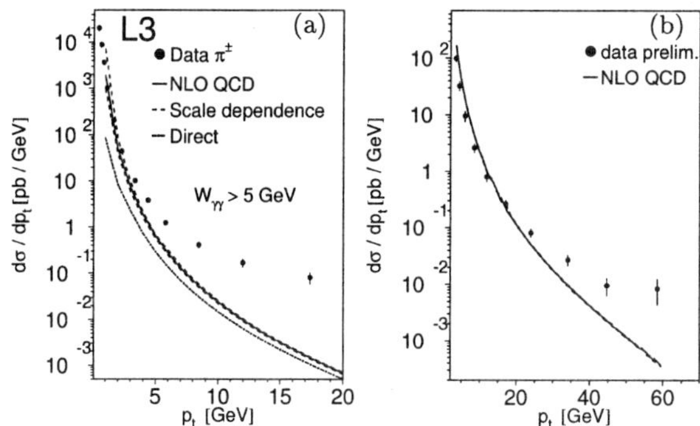

Figure 15. L3 measurements of inclusive differential cross sections $d\sigma/dp_T$ in $\gamma\gamma$ collisions for charged pion production (a) and jet production (b). The data (preliminary) are compared to NLO calculations (solid lines). Scale uncertainties are shown in dashed lines. The dashed-dot line corresponds to the direct process ((a) only).

tion. Parton densities were modeled using GRV-HO.[38] Scale uncertainties calculated by varying the scale up/downward by a factor of two are small. Poor agreement is observed in the high-p_T jet region as in other channels. The cause of this excess has yet to be explained.

The OPAL experiment has measured the production of di-jets at center-of-mass energies $\sqrt{s_{ee}}$ from 189-209 GeV. Jets were reconstructed using the k_T algorithm and the event selection required at least 2 jets with $|\eta^{jet}| < 2$, $E_T^{jet} > 3$ GeV. A cut of $Q^2 < 4.5$ GeV^2 was used to select quasi-real photons. Backgrounds from hadronic decays of the Z^0, $\gamma\gamma \to \tau\tau$, and $\gamma\gamma^*$ collisions were subtracted. For single or double resolved processes the variables $x_\gamma^\pm$ estimate the fraction of the photon's momentum participating in the hard scattering. At LO all of the photon's energy goes into the two jet final state ($x_\gamma^\pm = 1$). Whereas for single and double resolved events one or both x_γ values will be less than 1, with:

$$x_\gamma^\pm = \frac{\sum_{jet_{1,2}} E \pm p_z}{\sum_{hadrons} E \pm p_z}. \qquad (1)$$

The data are compared to LO and NLO QCD predictions[39] in Fig. 16. The cross section is plotted versus the mean jet transverse energy $\langle E_T^{1,2}\rangle$. At large $\langle E_T^{1,2}\rangle$ the cross section is expected to be dominated by direct processes, consequently a softer spectrum is observed in single (double) resolved enhanced subsamples. The NLO calculation using GRV-HO parton densities is in good agreement with the full

data sample and for single resolved enhanced events (x_γ^+ or $x_\gamma^- < 0.75$). Predictions for a double resolved enhanced sample (x_γ^+ and $x_\gamma^- < 0.75$) are below the measurement. Good agreement is found with PYTHIA and SaS 1D[40] parton densities. Observables in the double resolved region may be used to study effects of multiple parton interactions in greater detail.

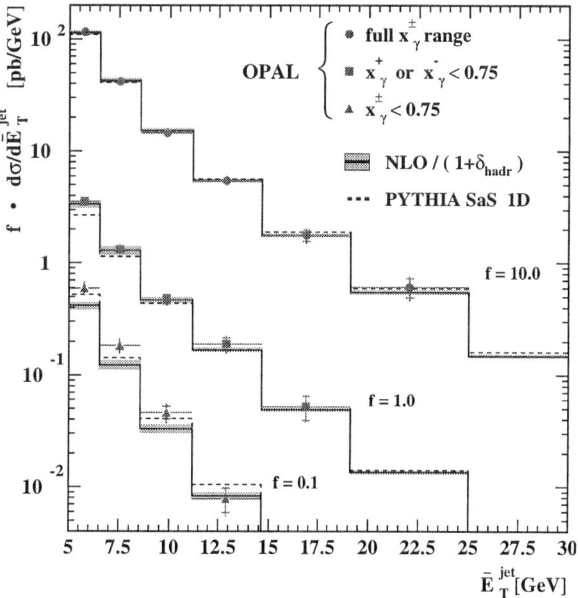

Figure 16. Di-jet cross section versus mean jet transverse energy $\langle E_T^{1,2}\rangle$ for three regions of x_γ. Distributions are shown for the full sample, single resolved enhanced (x_γ^+ or $x_\gamma^- < 0.75$), and double resolved enhanced subsamples (x_γ^+ and $x_\gamma^- < 0.75$). The data are compared to PYTHIA using SaS 1D parton densities and a NLO perturbative prediction using GRV-HO densities.

5. Event Shapes and α_s

Distributions of event shape variables are predicted to various orders in pQCD. Fits to measurements of these distributions may be used to determine the value of the strong coupling constant. Studies at LEP and HERA have used this technique to extract measurements of α_s.

Preliminary results from the LEP QCD Working Group[41] were presented at the conference. The Working Group is charged with combining the various event shape analyses from the four LEP experiments to obtain the best statistical precision for this measurement of α_s. Combining four equally weighted experiments would naïvely reduce statistical uncertainties by a factor of two and combining different events shape measurements would reduce the uncertainty further. However, different event shape variables within a single experiment will not be statistically independent in general, and consistent treatments of systematic uncertainties must be applied to combine results across experiments. The LEP event shape analysis combines a total of 194 measurements spanning 4 experiments, 15 cms energies, and 6 shape variables.

The event shape variables studied include:

Thrust, $T = \left(\dfrac{\sum_i |\vec{p_i} \cdot \vec{n}|}{\sum_i |\vec{p_i}|} \right)$: $\vec{n}$ is chosen to be the thrust axis $\vec{n_T}$ such that the quantity T is maximized.

Heavy Jet Mass, $M_H/\sqrt{s}$: Using a plane through the origin and perpendicular to $\vec{n_T}$ the event is split into two hemispheres and invariant masses are calculated using the particles in each hemisphere. M_H is the larger of the two mass values.

Jet Broadening, B_T and B_W: Measure the broadening of particles in transverse momentum w.r.t. the thrust axis for all particles and those in the hemisphere with the most broadening respectively.

C-parameter, C: Based on eigenvalues of the momentum tensor of the event.[42]

Two-to-3 jet transition parameter, y_3: Based on y_{cut}, the jet resolution parameter in the Durham scheme for combining particles into jet clusters. y_3 is defined as the maximum value of y_{cut} that produces three separate jets.

Each experiment provided results for the observables: thrust, heavy jet mass, C-parameter, wide and total jet broadening. ALEPH and OPAL provided measurements of $-log(y_3)$. Most measurements were performed for cms energies between 91 and 206 GeV. L3 additionally performed measurements between 41 and 85 GeV using radiative events at the Z^0 peak.

Experimental uncertainties are averaged over all experiments for each variable to prevent bias in favor of experiments with the least conservative error estimates. Monte Carlo generators are used to account for non-perturbative effects due to hadronization effects. Hadronization uncertainties are estimated based on the differences among the three generators PYTHIA, HERWIG, and ARIADNE. Theoretical uncertainties are estimated by measuring the

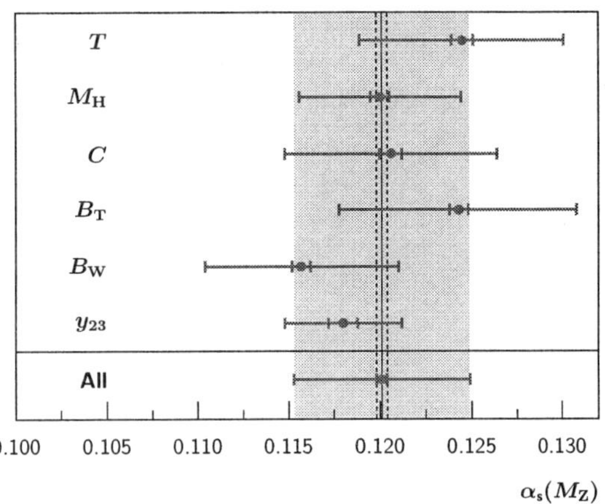

Figure 17. LEP combined $\alpha_s(M_Z^0)$ from event shape analyses (preliminary) broken down as a function of each event shape variable. The inner error bars and dashed band represent statistical errors. The shaded band and outer error bars represent the total uncertainty in the measurement.

scale dependence of the fits. Details may be found in the References.[43] The combination procedure takes correlations between the measurements into account. A 194×194 covariance matrix V is defined relating the uncertainties for all pairs of $\alpha_s(M_Z^0)$ measurements. The covariance matrix is expressed as the sum of four sources of error:

$$V_{ij}^{total} = V_{ij}^{stat} + V_{ij}^{exp} + V_{ij}^{hadr} + V_{ij}^{theo}. \qquad (2)$$

Then the method of least-squares is used to calculate the maximum-likelihood value of $\alpha_s(M_Z^0)$. Figure 17 shows the combined value of $\alpha_s(M_Z^0)$ and individual values broken down as a function of the six event shape variables. The preliminary result for the combined measurements is: $\alpha_s(M_Z^0) = 0.1201 \pm 0.0003(\text{stat.}) \pm 0.0048(\text{syst.})$.

Figure 18 shows a summary of $\alpha_s(M_Z^0)$ results from hadronic processes with the new LEP measurement included. While the most precise measurements of $\alpha_s(M_Z)$ come from $\Gamma(Z \to \tau\tau)$ and τ decays, the hadronic processes offer a strong verification of the Standard Model and in many cases their ultimate precision is limited only by the availability of higher order theoretical corrections.

Analyses have also been performed at LEP[44] and HERA[45,46] to extract $\alpha_s(M_Z^0)$ from event shapes by modeling non-perturbative effects via QCD inspired power corrections. The non-perturbative effects in event shape variables are taken to scale with powers of $1/Q$ and the model of Dokshitzer and Webber[47] is used to parameterize the infrared behavior of the

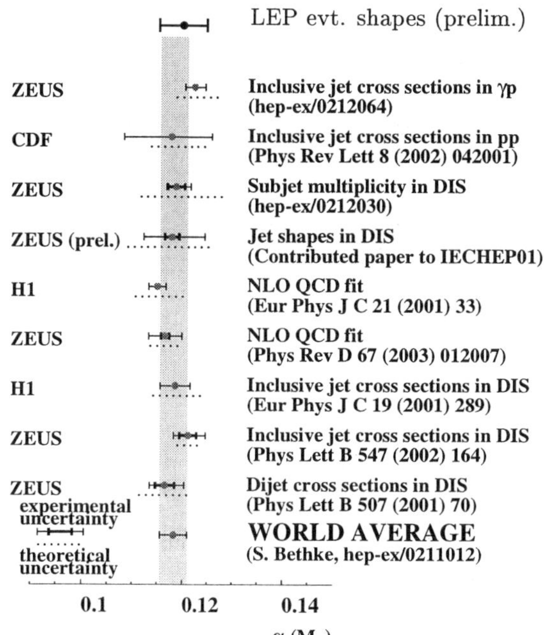

Figure 18. Summary of measurements of $\alpha_s(M_Z^0)$ from hadronic processes. Inner (outer) error bars represent statistical (statistical+systematic) errors. Dotted lines show theoretical uncertainties. All uncertainties are combined in the LEP point.

strong coupling by its mean value below a matching scale μ_I. A recent fit[45] from H1 using power law corrections to event shape distributions is shown in Fig. 19. The event shape variables follow similar definitions to the LEP analyses, for details see the References.[45] This result verifies the universality of power law corrections for event shape distributions.

6. Heavy Flavor Production

Figure 20 shows a preliminary measurement of the inclusive b-jet inclusive cross section measured at DØ using $3.4\,\mathrm{pb}^{-1}$ of integrated luminosity from the Run II data. Events were selected with $|\eta^{jet}| < 0.6$, $E_T^{jet} > 20\,\mathrm{GeV}$ and at least one muon satisfying $|\eta^\mu| < 0.8$ and $E_T^\mu > 6\,\mathrm{GeV}$. Backgrounds to muons from b decays were subtracted by fitting[48] the p_T^{rel} distribution[d] in each bin with signal and background components of the total p_T^{rel} calculated with PYTHIA. The Run II measurement is compared to PYTHIA, and as observed previously,[48] the cross section is ~ 2 times larger than predictions.

CDF has shown a preliminary measure of charm production in the cross sections for the D^0, D^+ and

---

[d] p_T of muon measured relative to jet axis.

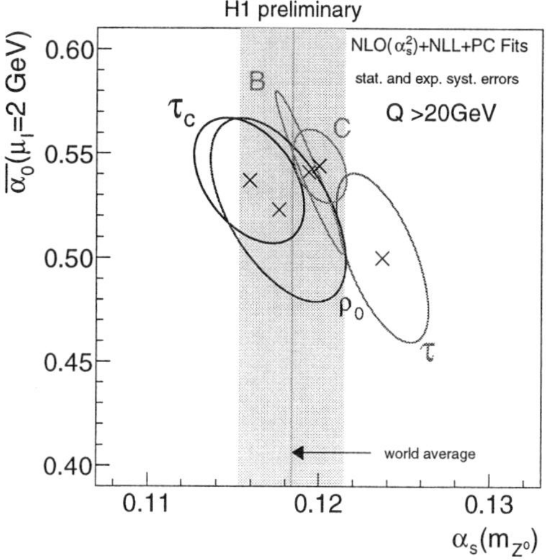

Figure 19. Mean values and 1σ contours of $\alpha_s(M_Z)$ and $\overline{\alpha}_0$ fitted to distributions of event shape variables from H1. $\overline{\alpha}_0$ is the effective value of the strong coupling constant below the cut-off scale μ_I.

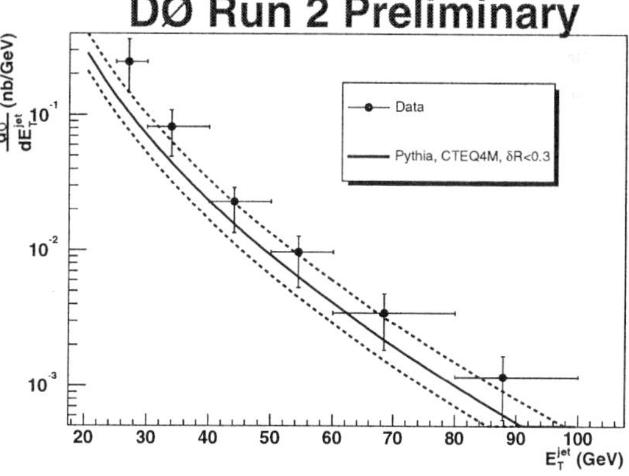

Figure 20. DØ preliminary measurement of b production cross section compared to a PYTHIA+CTEQ4M calculation. Only the LO prediction is shown. Scale uncertainties are represented by the dashed lines.

D^{*+}. A measure of the cross section $d\sigma/dp_T(D^0)$ for D^0 production versus transverse energy is shown in Fig. 21. The data are observed to lie above the NLO calculations[49] by a factor of ~ 1.7.

Both H1 and ZEUS have measured beauty production[50] in photoproduction and DIS events. Preliminary results for beauty in photoproduction for the process $ep \to eb\bar{b}X \to ejj\mu X$ is shown in Fig. 22. All data points lie above the NLO QCD prediction, but are in agreement within experimental errors.

Open charm and beauty production have been studied at LEP in two photon collisions. The two main contributions to the cross sections $\sigma(e^+e^- \to$

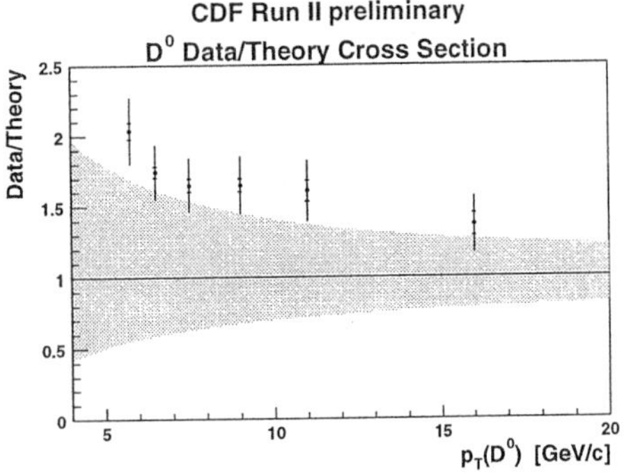

Figure 21. The CDF ratio of cross section for D^0 production to NLO QCD prediction $d\sigma/dp_T(D^0)$.

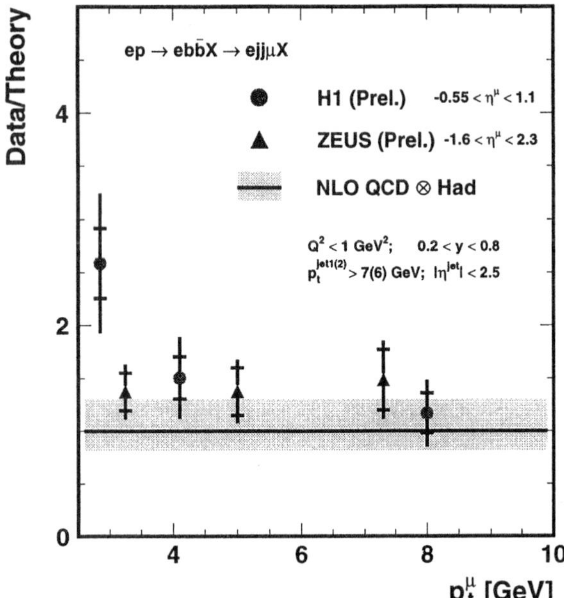

Figure 22. Ratio of the differential di-jet muon beauty cross section $d\sigma/dp_T^\mu$ ($ep \rightarrow eb\bar{b}X \rightarrow ejj\mu X$) from H1 ($-0.55 < \eta^\mu < 1.1$) and ZEUS ($-1.6 < \eta^\mu < 2.3$) to NLO QCD with hadronization corrections. The shaded band shows the uncertainty in the calculation due to scale variations.

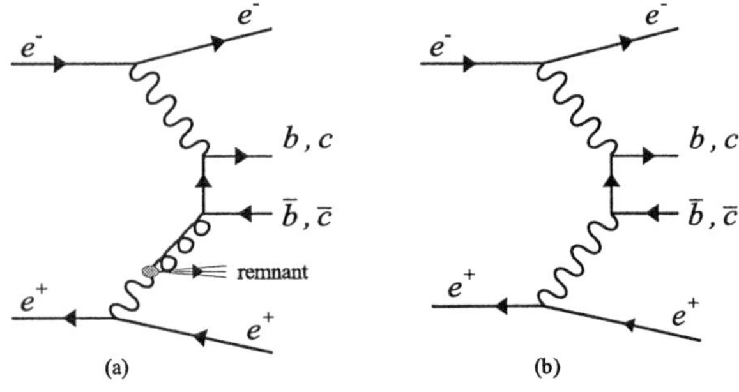

Figure 23. Production of open charm and beauty at LEP2 in two photon collisions proceeds primarily through either the single resolved (a) or direct (b) processes shown above.

$e^+e^- c\bar{c}X$) and $\sigma(e^+e^- \rightarrow e^+e^- b\bar{b}X)$ are the direct and single resolved processes illustrated in Fig. 23. Figure 24 shows the LEP measurements of open charm and beauty compared with the NLO QCD predictions.[51] The charm rate agrees well with the data, but requires the inclusion of the single resolved process for agreement. Beauty production rates are about a factor of two larger than predictions.

7. Summary

We continue to see tremendous progress in the experimental study of QCD at the world's colliders. The theory survives admirably, though increasingly precise data highlight the need for further progress in the theoretical calculations, particularly the need for higher order corrections to confront increasingly precise jet data. It will be interesting to see if other advances, perhaps in resummation techniques, can explain the trends toward higher cross sections for heavy flavor production. On the experimental side, work clearly remains on understanding k_T jets at the hadron colliders.

Acknowledgments

I wish to thank the numerous physics conveners and collaborators from ALPEH, DELPHI, L3, OPAL,

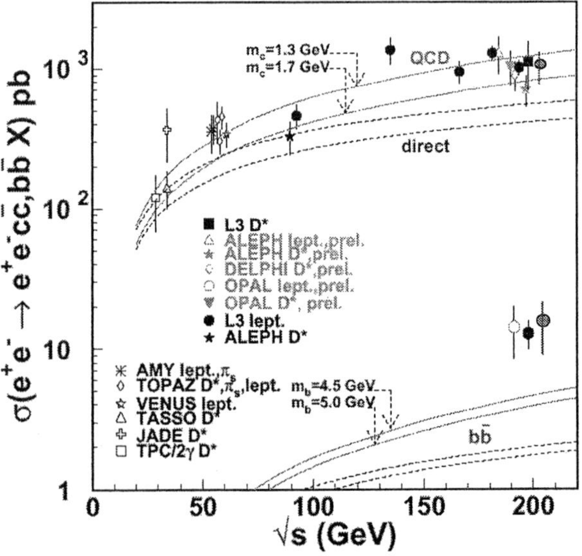

Figure 24. Cross sections for production of open charm and beauty at LEP2

H1, ZEUS, CDF and DØ for their help in preparing this talk.

References

1. F. Aversa et al., *Phys. Rev. Lett* **65**, 401 (1990);
   W. T. Giele, E. W. N. Glover, and D. A. Kosower, *Phys. Rev. Lett* **73**, 2019 (1994);
   D. Ellis, Z. Kunszt, and D. E. Soper, *Phys. Rev. Lett* **64**, 2121 (1990).
2. P. Newman, these proceedings.
3. F. Abe, *Nucl. Instru. Meth* A **271**, 387 (1988).
4. S. Abachi et al., *Nucl. Instru. Meth* A **338**, 185 (1994).
5. B. Abbott et al., *Phys. Rev. Lett.* **86**, 1707 (2001).
6. F. Abe et al., *Phys. Rev. Lett* **77**, 438 (1996).
7. B. Abbott et al., *Phys. Rev. Lett.* **82**, 2451 (1999).
8. W. T. Giele, E. W. N. Glover, and D. A. Kosower, *Nucl. Phys.* B **403**, 633 (1993); *Phys. Rev. Lett* **73**, 2019 (1994).
9. S. Kuhlmann, H. L. Lai, W. K. Tung, *Phys. Lett.* B **409**, 271 (1997).
10. R. Thorne these proceedings.
11. J. E. Huth, et al., in Proceedings of Research Directions for the Decade: Snowmass 1990, July 1990, edited by E. L. Berger (World Scientific, Singapore, 1992) p. 134.
12. P. Petroff, hep-ex/9910028.
13. M. Paulini, FERMILAB-CONF-03/010-E.
14. G. C. Blazey et al., Run II jet physics, hep-ex/0005012, 2000.
15. D. Stump et al., hep-ph/0303013.
16. V. M. Abazov et al., *Phys. Lett.* B **525**, 211 (2002).
17. S. D. Ellis, D. E. Soper, *Phys. Rev.* D **48**, 3160 (1993);
   S. Catani et al., *Nucl. Phys.* B **406**, 187 (1993).
18. B. Abbott et al., *Phys. Rev. Lett.* **82**, 2457 (1999); *Phys. Rev.* D **64**, 032003 (2001).
19. I. Abt et al., *Nucl. Instru. Meth* A **386**, 310 (1997);
   I. Abt et al., *Nucl. Instru. Meth* A **386**, 348 (1997).
20. ZEUS Collaboration, U. Holm (ed.), *The ZEUS Detector* Status Report (unpublished), DESY (1993), http://www-zeus.desy.de/bluebook/bluebook.html.
21. *Eur. Phys. J.* C **19**, 289 (2001);
   Phys. Lett. B **515**, 17 (2001);
   Eur. Phys. J. C **23**, 13 (2002);
   Phys. Lett. B **507**, 70 (2001).
22. H1 Collaboration, DESY-02-079.
23. S. Catani and M. H. Seymore, *Nucl. Phys* B **485**, 45 (1999).
24. B. Andersson et al., *Phys. Rept.* **97**, 31 (1983).
25. DESY-03-055, June 2003.
26. E. Mirkes and D. Zepenfeld, *Phys. Lett.* B **380**, 205 (1996).
27. L. Lönnbald, *Comp. Phys. Comm* **71**, 15 (1992);
   L. Lönnbald, *Z. Phys.* C **65**, 285 (1995).
28. G. Ingleman, A. Edin and J. Rathsman, *Comp. Phys. Comm* **101**, 108 (97).
29. F. Abe et al., *Phys. Rev. Lett* **70**, 1376 (1993);
   B. Abbott et al., *Phys. Rev. Lett* **86**, 2523 (2001).
30. J. Breitweg et al., *Phys. Lett.* B **507**, 70 (2001);
   S. Chekanov et al., *Phys. Lett.* B **547**, 164 (2002).
31. DESY-02-228, December 2002.
32. H1 Collaboration DESY-02-225.
33. C. Albajar et al., *Nucl. Phys.* B **309**, 405 (1988);
   G. Arinson et al.; *Phys. Lett.* B **172**, 461 (1986).
34. B. Abbott et al., *Phys. Rev. Lett.* **82**, 2451 (1999).
35. The L3 Collaboration, *Phys. Lett.* B **554**, 105 (2003).
36. S. Frixione, *Nucl. Physics* B **507**, 295 (1997);
   S. Frixione and G. Ridolfi, *Nucl. Physics* B **507**, 315 (1997).
37. C. F. von Weizsäcker, *Z Phys.* **88**, 612 (1934);
   E. J. Williams, *Phys. Rev.* **45**, 729 (1934).
38. M. Gluck, E. Reya and A. Vogt, *Phys. Rev.* D **45**, 3986 (1992);
   M. Gluck, E. Reya and A. Vogt, *Phys. Rev.* D **46**, 1973 (1992).
39. M. Klasen, T. Kleinwort, G. Kramer *Euro. Phys. J Direct* **C1**, 1 (1998);
   B. Pötter *Euro. Phys. J Direct* C **5**, 1 (1998).
40. G. A. Schuler and T. Sjöstrand *Z. Phys.* C **68**, 607 (1995).
41. The LEP QCD Working Group: S. Banerjee, M. Ford, R. Jones, G. Salam, H. Stenzel and D. Wicke; contact Roger.Jones@cern.ch.
42. G. Parisi, *Phys. Lett.* B **74**, 65 (1978);
   J. F. Donoghue, F. E. Low and S. Y. Pi, *Phys. Rev.* D **20**, 2759 (1979).
43. R. W. L. Jones, M. Ford, G. P. Salam, H. Stenzel, in preparation.
44. DELPHI 2003-019 CONF 639, June 2003.
45. H1 Note H1prelim-03-033.
46. ZEUS Collaboration, DESY-02-198.
47. Y. L. Dokshitzer and B. R. Webber, *Phys. Lett.* B **404**, 321 (1997).
48. B. Abbott et al., *Phys. Rev. Lett.* **85**, 5068 (2000).
49. P. Nason and M. Cacciari, hep-ph/0306212.
50. See list of contributed papers to this conference.
51. Drees, Kramer, Zunft and Zerwas, *Phys. Lett.* B **306**, (1993).

DISCUSSION

Lance Dixon (SLAC): I think you only showed us Tevatron jet data using the Run I algorithm. Are there Run II data using improved jet algorithms, say, infrared-safe algorithms?

Robert Hirosky: Yes, in my discussion of the DØ data, I mentioned that jets are reconstructed using the Run II mid-point algorithm. That is an infrared-safe algorithm.

Doris Kim (University of Illinois): You mentioned that the charm and beauty cross section measurements are larger in the data than in the theory. I was wondering if, in your opinion, this could be corrected by obtaining more accurate parton distributions, or maybe there are some missing charmonium processes, or if this should be corrected by NLO QCD?

Robert Hirosky: I think the recent work in resummation and fragmentation functions that was mentioned in Thomas Gehrmann's talk holds some promise and it will be interesting to first see the effects of these improvements. For example, the b-jet production cross section is typically compared to calculations for b quarks. Thus, fragmentation effects must be well understood. Resummation effects are also important due to the two scales in the problem. In this case we have a jet of a certain p_T and a muon with a second p_T scale defined relative to the jet. So I consider work in these two areas to be well motivated.

HIGH PRECISION LATTICE QCD MEETS EXPERIMENT

P. LEPAGE

Laboratory for Elementary-Particle Physics, Cornell University,Ithaca, NY 14853, USA
E-mail: gpl@mail.lns.cornell.edu

C. DAVIES

Dept. of Physics and Astronomy, University of Glasgow, Glasgow, G12 8QQ, UK
E-mail: c.davies@physics.gla.ac.uk

We review recent results in lattice QCD from numerical simulations that allow for a much more realistic QCD vacuum than has been possible before. Comparison with experiment for a variety of hadronic quantities gives agreement to within statistical and systematic errors of 3%. We discuss the implications of this for future calculations in lattice QCD, particularly those which will provide input for B-factory experiments.

1. Introduction

In this talk we report on recent progress in lattice QCD simulations that finally allows precise calculations of a broad, but restricted, range of important non-perturbative quantities. For most of its 25 year history, high precision calculations in lattice QCD have been stymied by our inability to include realistic effects from light-quark vacuum polarization. Small quark masses, for u and d quarks in particular, were too expensive. Consequently quark vacuum polarization was omitted ("quenched QCD") in most work. When vacuum polarization was included, usually only u and d quarks were kept (no s) and these had masses 10–20× too large. Such approximations led to uncontrolled systematic errors that could be 10–30% or larger in almost all lattice QCD calculations.

During the past three years, several lattice QCD groups have been exploring a new discretization of the quark action that is 50–1000 times faster than previous discretizations, but also highly accurate. These investigations have led to a series of simulations that include u, d, and s quark vacuum polarization, with u and d masses that are 3–5× smaller than before. These masses are still unrealistically large, but they are small enough to allow accurate extrapolations to the physical masses. Consequently lattice QCD errors can be reduced to a few percent, and high precision non-perturbative QCD is now possible for the first time.

High precision non-perturbative QCD is essential to the experimental study of the Standard Model. The CKM parameters ρ and η, for example, are constrained by experimental and theoretical results for $B\text{-}\overline{B}$ mixing, $B \to \pi l \nu$, $K\text{-}\overline{K}$ mixing,.... Each of these quantities has a non-perturbative QCD part and a weak interaction part. Current uncertainties, of order 20%, in the QCD parts dominate the uncertainties in ρ and η. It is critically important that non-perturbative QCD errors be reduced to a few percent, which is of order the experimental errors expected from B-factories, CLEO-c, and the hadron colliders.

High precision non-perturbative quantum field theory may well be important for beyond the Standard Model, as well. Two of the three known interactions (QCD and gravity) are strongly coupled. New strongly-coupled field theories are quite possible, even likely, at LHC energies and/or beyond. Strong coupling is generic at low energies in non-abelian gauge theories, unless the gauge symmetry is spontaneously broken; and, even then, the most likely symmetry breaking mechanisms are dynamical, which again requires strong coupling.

Here we will review the new developments in lattice QCD techniques, and discuss recent calculations, and possibilities (and limitations) for the future.

2. Lattice QCD Calculations

Lattice QCD calculations proceed by the discretization of a 4-d box of space-time into a lattice. The QCD Lagrangian is then discretized onto that lattice. The spacing between the points of the lattice, a, is ≈ 0.1 fm in current calculations and the length of a side of the box is $L \approx 3.0$ fm. Thus our simulations can cover energy scales from ≈ 2 GeV down to ≈ 100 MeV.

The Feynman Path Integral is evaluated numerically in a two-stage process. In the first stage, sets of gluon fields ("configurations") are created which are representative "vacuum snapshots". In the second stage, quarks are allowed to propagate on these background gluon field and hadron correlators are calculated. The dependence of the correlators on lattice time is exponential. From the exponent the masses of hadrons of a particular J^{PC} can be extracted, and from the amplitude, simple matrix elements.

QCD as a theory has a number of unknown parameters, the overall dimensionful scale of QCD ($\equiv$ the bare coupling constant) and the bare quark masses. To make predictions, these parameters must be fixed from experiment. In lattice QCD we do this by using one hadron mass for each parameter. The quantity which is equivalent to the overall scale of QCD on the lattice is the lattice spacing.

Lattice calculations are hard and time consuming. Progress has occurred in the last thirty years through gains in computer power but also, more importantly, through gains in calculational efficiency and physical understanding. One particular area which revolutionized the field from the mid-1980s was the understanding of the origin of discretization errors and their removal by improving the lattice QCD Lagrangian. Discretization errors appear whenever equations are discretized and solved numerically. They manifest themselves as a dependence of the physical result on the unphysical lattice spacing. In lattice QCD, as elsewhere, they are corrected by the adoption of a higher order discretization scheme. The complication in a quantum field theory like QCD is the presence of radiative corrections to the coefficients in the higher order scheme which must be determined.

Determining such radiative corrections is a major challenge for lattice QCD theorists. These involve physics at momentum scales of order π/a, where a is the lattice spacing, and, therefore they can be analyzed using perturbation theory[1] when a is small enough, because of asymptotic freedom. The one- and two-loop perturbative analyses that are required are complicated by the exceedingly unwieldy Feynman rules of lattice QCD. The rule for a single vertex can run to several hundred *pages* of 6 pt text. Computer automation is essential. Progress on these calculations is currently the limiting factor in most high precision work that is relevant to tests of the Standard Model.

Physical understanding of heavy quark physics on the lattice has also made a huge difference to the feasibility of calculating matrix elements relevant to the B-factory program on the lattice. The use of non-relativistic effective theories requires the lattice to handle only scales appropriate to the physics of the non-relativistic bound states and not the (large) scale associated with the b quark mass. B physics is now one of the areas where lattice QCD can make the most impact.

One area which has remained problematic, but which this year's results have addressed successfully, is the handling of light quarks on the lattice. In particular the problem is how to include the dynamical (sea) $u/d/s$ quark pairs that appear as a result of energy fluctuations in the vacuum. We can safely ignore $b/c/t$ quarks in the vacuum because they are so heavy, but we know that light quark pairs have significant effects, for example in screening the running of the coupling constant and in generating Zweig-allowed decay modes for unstable mesons.

Because quarks are fermions, they cannot be simulated directly on the computer, but must be "integrated out" of the Feynman Path Integral. This leaves a QCD Lagrangian in terms of gluon fields which includes $\ln(\det(M))$ where M is an enormous ($10^7 \times 10^7$) sparse matrix. The inclusion of dynamical quarks is then numerically very expensive, particularly as the quark mass is reduced towards the small values which we know the u and d quarks have.

Many calculations even today use the "quenched approximation" in which the light quark pairs are ignored. Results then suffer from a systematic error of $\mathcal{O}(20\%)$. A serious problem with the quenched approximation is the lack of internal consistency which means that the results depend on the hadrons that were used to fix the parameters of QCD. This ambiguity plagues the lattice literature.

Other calculations have included 2 flavors of degenerate dynamical quarks, i.e. u and d, but with masses 10-20$\times$ the physical ones. This approximation is better than the quenched approximation but large uncertainties remain because the s quark is omitted. Results must also be extrapolated to the physical u/d quark mass and chiral perturbation theory is a good tool for this. However, chiral perturbation theory only works well if the u/d quark mass is light enough and, for errors at the few percent level, this means less than $m_s/2$. This has been impossible to achieve in most calculations.

New results this year[2] have included u, d and s quarks in the vacuum, with light enough u/d masses to perform accurate chiral extrapolations. The results use a new discretization of the quark action - the numerically fast improved staggered formalism. This formalism is well-matched to the supercomputing power of a few Tflops that is currently achievable.

2.1. *Improved Staggered Quarks*

The starting point for the staggered quark formalism is the naïve discretization of the Dirac quark action onto a lattice. This action has good features: chiral symmetry and discretization errors that appear only as the square and higher powers of the lattice spacing. The naïve discretization suffers from the notorious doubling problem, however. A single quark species on the lattice gives rise to 16 quark species, or tastes, on a 4-d lattice. The additional tastes appear around the edges of the Brilliouin zone, where $p \approx \pi/a$, as copies of a $p \approx 0$ quark. This would not be a problem if there were no interaction between the different tastes since the quark action would then fall apart into 16 different pieces in an appropriate basis and we could take $\det(M)^{(1/16)}$ in simulations to give the effect of 1 quark flavor.

There is interaction between the different tastes, however. It is mediated by highly virtual gluons, with momenta around π/a. A quark of one taste can absorb or emit such a high momentum gluon and turn into a quark of another taste. The effects of this taste-changing interaction are quite severe for the naïve action, giving rise to large discretization errors (even though formally of $\mathcal{O}(a^2)$) and large perturbative renormalization factors, e.g. for the quark mass, when translating from the lattice scheme to the continuum. The degeneracy in mass of mesons made from quarks of different taste is lost. This is most noticeable for the pions because there is a light Goldstone boson.

Because the taste-changing interaction is a high momentum one it can be understood in lattice perturbation theory. In particular, the effects can be significantly improved by suppressing the coupling of quarks to gluons of momenta π/a in any direction. This is achieved by "smearing" the gluon field in the action in a particular way,[3,4] and can be thought of

as part of the standard Symanzik program for systematically removing discretization errors from lattice actions.

It is simple to "stagger" the naïve action and its improved variant to remove an exact degeneracy of a factor of 4 in tastes which arises from the spin degree of freedom. This results in an action with 4 doublers which can be simulated on the lattice using $\det(M)^{(1/4)}$ per flavor. It is very fast numerically because there is only one spin degree of freedom per site and the eigenvalues of M are well behaved. This is what has allowed the MILC collaboration to generate ensembles of configurations which include $u, d,$ and s quarks in the vacuum with much more realistic masses than before.[5]

Some worries remain about potential non-locality in the action as the result of taking the fourth root. However, this causes no problem in perturbative QCD where a simple power series in x is obtained for an action with $\det(M)^x$. Stringent non-perturbative tests are also then needed. Luckily these tests are possible in this formalism with present day computers because of their speed, and are exactly the calculations required to test (lattice) QCD. The results, shown in the next section, speak for themselves.

3. Recent Results

The MILC collaboration have made sets of ensembles of gluon field configurations which include 2 degenerate light dynamical quarks (u, d) and 1 heavier one (s).[5] Taking the u and d masses as the same makes the lattice calculation much faster and leads to negligible errors in isospin-averaged quantities. The dynamical s quark mass is chosen to be approximately correct based on earlier studies (in fact the subsequent analysis shows that it was slightly high and further ensembles are now being made with a lower value). The dynamical u and d quarks take a range of masses, down as low as a sixth of the (real) m_s. The sets of ensembles divide into two different values of the lattice spacing, 0.13 fm and 0.09 fm, and the spatial lattice volume is $(2.5 \text{ fm})^3$ reasonably large. Analysis of hadronic quantities on these ensembles has been done by the MILC and HPQCD collaborations.[2]

There are 5 bare parameters of QCD relevant to this analysis: $\alpha_s, m_{u/d}, m_s, m_c$ and m_b. The lattice

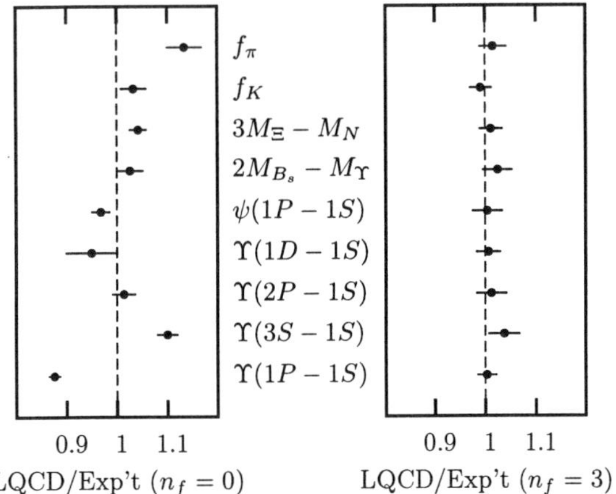

Figure 1. Lattice QCD results divided by experiment for a range of "gold-plated" quantities which cover the full range of hadronic physics.[2] The unquenched calculations on the right show agreement with experiment across the board, whereas the quenched approximation on the left give systematic errors of $\mathcal{O}$(10-20%).

spacing takes the place of α_s in lattice QCD. It is important that these parameters are fixed using the masses of "gold-plated" hadrons, i.e. hadrons which are well below their strong decay thresholds. Such hadrons are well-defined experimentally and theoretically and should be accurately calculable in lattice QCD. Using them to fix parameters will not then introduce unnecessary additional systematic errors into lattice results for other quantities. This has not always been done in past lattice calculations, particularly in the quenched approximation. It becomes an important issue when lattice QCD is to be used as a precision calculational tool. We use the radial excitation energy in the Υ system (i.e. the mass splitting between the Υ' and the Υ) to fix the lattice spacing and m_π, m_K, m_{D_s} and m_Υ to fix the quark masses.

We can then focus on the calculation of other gold-plated masses and decay constants. If QCD is correct and lattice QCD is to work it must reproduce the experimental results for these quantities precisely. Figure 1 shows that this indeed works for the unquenched calculations with u, d and s quarks in the vacuum. A range of gold-plated hadrons are chosen which range from decay constants for light hadrons through heavy-light masses to heavyonium. This tests QCD in different regimes in which the sources of systematic error are very different and stresses the point that QCD predicts a huge range of physics with a small set of parameters.

References[6-9] give more details on the quantities shown in Fig. 1. Here we will discuss some of these. Figure 2 shows the radial and orbital splittings in the $b\bar{b}$ (Υ) system for the quenched approximation ($n_f = 0$) and with the dynamical MILC configurations with 3 flavors of dynamical quarks. Our physical understanding of the Υ system is very good and there are a lot of gold-plated states well below decay thresholds, which makes it a valuable system for lattice QCD tests. We use the standard lattice NRQCD effective theory for the valence b quarks, which takes advantage of the non-relativistic nature of the bound states. The lattice NRQCD action is accurate through v^4 where v is the velocity of the b quark in its bound state. This means that spin-independent splittings, such as radial and orbital excitations, are simulated through next-to-leading-order and should be accurate to $\approx$ 1%. Thus the test of QCD using these splittings is a very accurate one. The fine structure in the spectrum is only correct through leading-order at present and more work must be done to bring this to the same level and allow tests against, for example, the splittings between the different χ_b states.[7]

The Υ system is a good one for looking at the effects of dynamical quarks because we do not expect it to be very sensitive to dynamical quark masses. The momentum transfer inside an Υ is larger than any of the u, d or s masses and so we expect the radial and orbital splittings to simply "count" the number of dynamical quarks once we have reasonably light dynamical quark masses. The lower plot of Fig. 2 shows this to be true - the splittings are independent of the dynamical u/d quark mass in the region we are working in (and therefore for the points plotted in the left-hand figure of Fig. 1 and in Fig. 2).

The π and K decay constants are important light hadron matrix elements, related to the purely leptonic decay rate via a W, and experimentally well-known. These are very sensitive to light quark masses and require a well-controlled extrapolation in the u/d quark mass and interpolation in the s quark mass to get accurate results to compare to experiment. Chiral perturbation theory can be used to perform the u/d quark mass extrapolation provided the masses used on the lattice are small enough for the expansion in powers of quark mass ($\equiv m_\pi^2/(1\mathrm{GeV}^2)$) and its logarithms to work well. In practice this means that second-order chiral perturbation theory

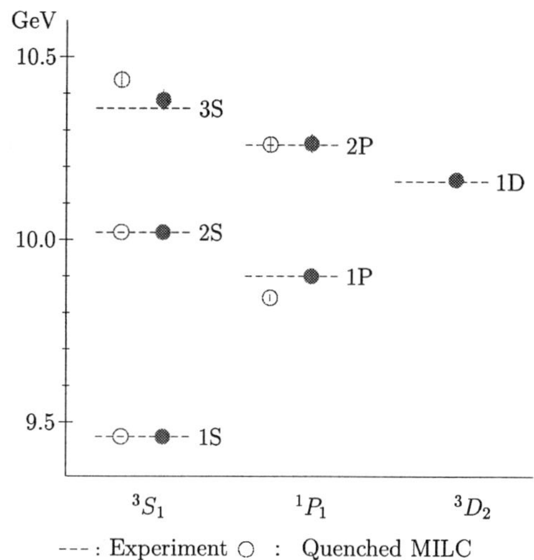

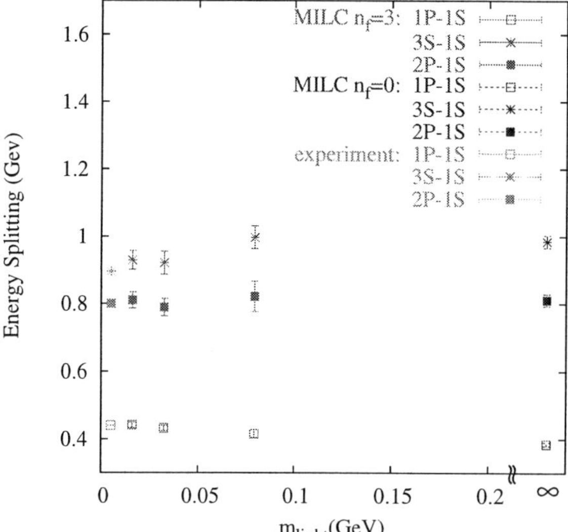

Figure 2. Radial and orbital splittings in the Υ system from lattice QCD in the quenched approximation and including u, d and s dynamical quarks. In this plot the lattice spacing was fixed from the radial excitation energy, i.e. the splitting between the Υ' and the Υ and the b quark mass was tuned to get the Υ mass correct. The bottom plot shows these splittings plotted as a function of the bare dynamical u/d quark mass for several ensembles of MILC configurations. The leftmost lattice points are the ones used in the top plot and in Fig. 1.

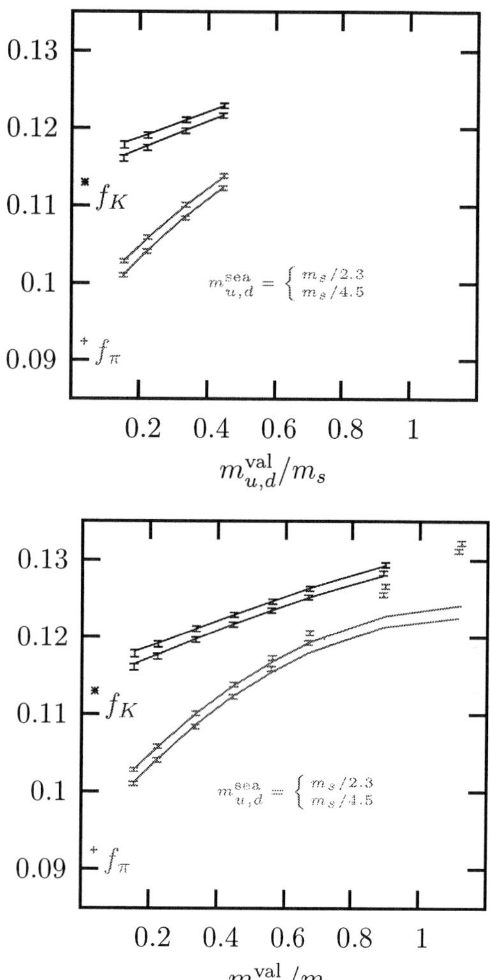

Figure 3. Results for the π and K decay constants as a function of the light quark mass for two dynamical MILC ensembles at a lattice spacing of 0.09 fm. The top plot shows the chiral extrapolation using only results with valence u/d quark masses $< m_s/2$.[2] The chiral extrapolation must subsequently be corrected for the incorrect valence and sea s quark mass to give the results in Fig. 1. The bottom plot shows that this chiral fit from light u/d quark masses does not agree well with the data for $m_{u/d} > m_s/2$.

should work at the 2% level for $m_{u/d} < m_s/2$. Note that the error is set by the largest quark mass used in the chiral fits, *not* the smallest.

Figure 3 shows the results and chiral extrapolation for the decay constants on the ensembles of MILC configurations with $m_{u/d}^{sea} = m_s/2.3$ and $m_s/4.5$ at a lattice spacing of 0.09 fm. The curves in the top plot show the chiral extrapolation using only results with $m_{u/d}^{valence} < m_s/2$. This extrapolation has to be corrected, using the lattice results, to interpolate to the physical s quark mass for both sea and valence s quarks. This then gives the results shown in Fig. 1 which agree with experiment. The lower plot shows what happens when the chiral extrapolation fit obtained in the top plot is evaluated for larger valence $m_{u/d}$. The f_π results start to show clear disagreement for $m_{u/d} > m_s/2$, which makes the problem of performing accurate chiral extrapolations using results with $m_{u/d} > m_s/2$ obvious. Previous lattice calculations have been forced by computing cost to work only in this regime, with the added problem that the sea $m_{u/d}$ is also large.[10]

Another gold-plated hadron mass is that of the nucleon. A full chiral extrapolation of the results for this on the MILC configurations has not yet been done. The upper plot of Fig. 4 shows very encouraging signs that an answer in agreement with experiment will be found.[6] There is a clear sign of dependence on the lattice spacing, however, which will have to be taken into account. Combinations of baryon masses can be made which are relatively insensitive to u/d quark masses and other effects, and it is one of these, $3m_\Xi - m_N$, which is plotted in Fig. 1.

It is important to realize that accurate lattice QCD results are not going to be obtainable in the near future for every hadronic quantity of interest. What these results show is that "gold-plated" quantities should now work. Gold-plated hadrons are those well below decay threshold for strong decays. Unstable hadrons, or even those within 100 MeV or so of Zweig-allowed decay modes, have a strong coupling to their real or virtual decay channel which is not correctly simulated on the lattice. The problem is that, with the lattice volumes being used, the allowed nonzero momenta are typically greater than 400 MeV and this significantly distorts the decay channel contribution. Much larger simulations will be necessary to handle these hadrons.

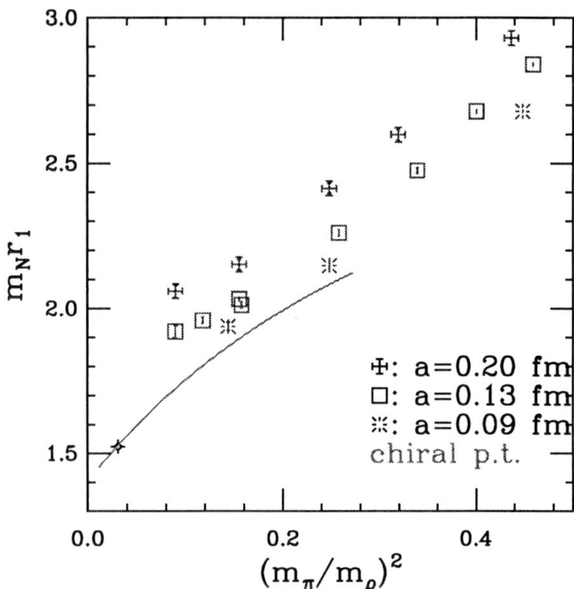

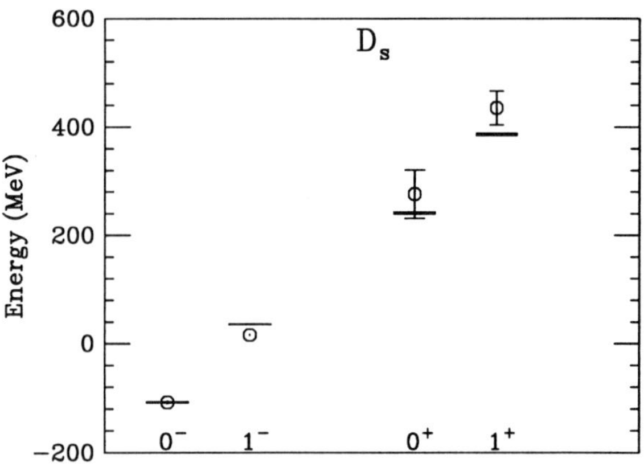

Figure 4. The top plot shows results for the nucleon mass on MILC ensembles for different lattice spacings and dynamical quark masses. The nucleon mass is given in units of r_1, a parameter from the heavy quark potential whose physical value is 0.32 fm. The dynamical quark mass is indicated by the variable m_π^2/m_ρ^2. The curve roughly indicates chiral perturbation theory.[6] The bottom plot shows the spectrum of D_s states obtained from the MILC dynamical configurations with $m_{u/d} = m_s/4$ and lattice spacing 0.13 fm.[11]

Gold-plated hadrons include: π, K, D, D_s, J/ψ, Υ, B, B_s, p, n, Λ, Ω, etc.. The following are *not* gold-plated: ρ, ϕ, D^*, D_{sJ}, Δ, N^*, pentaquarks, glueballs and hybrids in general. Lattice calculations will not get the masses right for non-gold-plated hadrons even when light dynamical quarks are included. This does not preclude lattice calculations giving useful qualitative results and insight but these points should be borne in mind for any quantitative comparison.

Figure 4 also shows the spectrum of D_s states obtained on the dynamical MILC configurations.[11] The valence c quarks are simulated using an effective theory which, in a similar way to the Υ above, should be accurate for spin-independent splittings and not quite so accurate for fine structure in the spectrum. The hyperfine splitting between the D_s and D_s^*, for example, is currently missing a radiative correction to the term in the action proportional to the spin coupling to the chromo-magnetic field. This is being calculated in lattice perturbation theory.[1] Also shown are the scalar and axial vector orbital excitations compared to the recent experimental results for these mesons. The lattice calculation is giving a high result, albeit with large statistical errors at present. However, a high result is consistent with the fact that these mesons are not gold-plated and the lattice calculation does not currently include correctly the coupling to their decay modes.

Decay rates which can be accurately calculated for gold-plated hadrons are those in which there is at most one (gold-plated) hadron in the final state. This therefore includes leptonic and semi-leptonic decays and the mixing of neutral B and K mesons. Luckily there is a gold-plated decay mode available to extract each element (except V_{tb}) of the CKM matrix which mixes quark flavors under the weak interactions in the Standard Model:

$$\begin{pmatrix} \mathbf{V_{ud}} & \mathbf{V_{us}} & \mathbf{V_{ub}} \\ \pi \to l\nu & K \to l\nu & B \to \pi l\nu \\ & K \to \pi l\nu & \\ \mathbf{V_{cd}} & \mathbf{V_{cs}} & \mathbf{V_{cb}} \\ D \to l\nu & D_s \to l\nu & B \to Dl\nu \\ D \to \pi l\nu & D \to Kl\nu & \\ \mathbf{V_{td}} & \mathbf{V_{ts}} & \mathbf{V_{tb}} \\ \langle B_d | \overline{B}_d \rangle & \langle B_s | \overline{B}_s \rangle & \end{pmatrix}.$$

As described earlier, the determination of the CKM elements and tests of the self-consistency of the CKM matrix are the current focus for the search for beyond

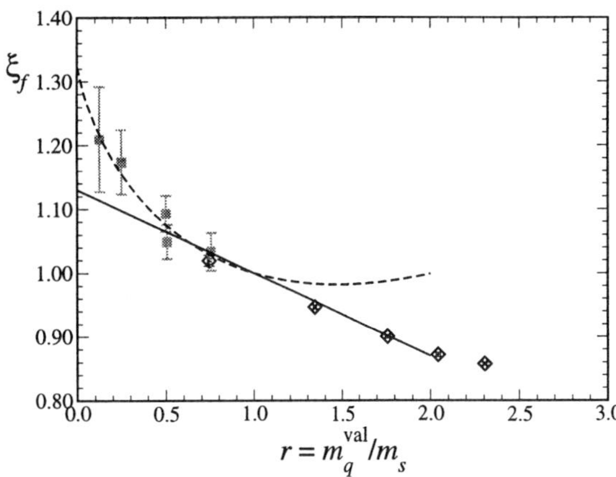

Figure 5. Results for the ratio of f_{B_s}/f_{B_d}, as a function of valence u/d quark mass in units of m_s.[12] The (grey) squares are from the dynamical MILC ensembles including u, d and s dynamical quarks.[13] The (black) diamonds are from the previous best calculation which included 2 flavors of dynamical quarks with masses $> m_s/2$.[10] The straight line is a linear extrapolation for the 2 flavor results, the curve includes the possibility of logarithms from chiral perturbation theory. This ratio, for physical u/d masses, appears in the ratio of oscillation frequencies for B_s and B_d mesons, which it is hoped to measure experimentally.

the Standard Model physics and lattice calculations of these decay rates will be a key factor in the precision with which this can be done.

The first calculations on the dynamical MILC configurations have concentrated on the B and B_s leptonic decay rates,[13,11] because these are the simplest. They are parameterized by the decay constants, f_B and f_{B_s}, and these are an important component of the mixing rate for these mesons, which constrains V_{ts} and V_{td}. Again one issue in extracting reliable lattice results for f_B and f_{B_s} is the chiral extrapolation in the u/d quark mass. Figure 5 shows results on the MILC configurations for the ratio of f_{B_s}/f_{B_d} plotted against the valence u/d quark mass.[12] The data extend into the region $m_{u/d} < m_s/2$ which will allow an accurate chiral extrapolation for the first time. Although the statistical errors are currently rather large, it seems likely that the result for this ratio will be larger than previous estimates based on extrapolations from larger u/d masses, and including only two flavors of dynamical quarks.[10] Further calculations of gold-plated matrix elements are in progress.[14]

4. Conclusions

The possibility of few percent errors in lattice QCD simulations creates a superb opportunity for lattice QCD to have an impact on particle physics. Lattice QCD is essential to the high precision B/D physics programs at BaBar, Belle, CLEO-c, Fermilab, ... *Predicting* BaBar/Belle, and especially CLEO-c results at the few percent level will give much needed credibility to lattice QCD. It is critical in such tests and applications to focus on gold-plated quantities.

High precision non-perturbative QCD is a landmark in the history of quantum field theory, and it is an essential first step in our preparation for strong-coupling beyond the Standard Model.

Acknowledgments

We are grateful to PPARC, NSF, DoE and the EU for funding this work, and to all our collaborators[2] for many useful discussions. A version of this talk was also presented at the Hadron 2003 Conference and is published in the proceedings of that meeting.

References

1. H. Trottier, Proceedings of LAT03, hep-lat/0310044.
2. C. Davies *et al.*, MILC/HPQCD/FNAL/UKQCD collaborations, hep-lat/0304004.
3. G. P. Lepage, *Phys. Rev.* D **59**, 074502 (1999), hep-lat/9809157.
4. K. Orginos, D. Toussaint and R. L. Sugar, *Phys. Rev.* D **60**, 054503 (1999), hep-lat/9903032.
5. C. Bernard *et al.*, *Phys. Rev.* D **64**, 054506 (2001), hep-lat/0104002.
6. S. Gottlieb, Proceedings of LAT03, hep-lat/0310041.
7. A. Gray *et al.*, *Nucl. Phys.* B (Proc. Suppl. **119**), 592 (2003), hep-lat/0209022; C. Davies *et al.*, *ibid* 595, hep-lat/0209122.
8. M. di Pierro *et al.*, Proceedings of LAT03, hep-lat/0310042.
9. C. Aubin *et al.*, Proceedings of LAT03, hep-lat/0309088.
10. S. Aoki *et al.*, hep-ph/0307039.
11. M. di Pierro *et al.*, Proceedings of LAT03, hep-lat/0310045.
12. A. Kronfeld, Proceedings of LAT03, hep-lat/0310063.
13. M. Wingate, Proceedings of LAT03, hep-lat/0309092.
14. J. Shigemitsu *et al.*, Proceedings of LAT03, hep-lat/0309039; M. Okamato *et al.*, Proceedings of LAT03, hep-lat/0309107.

DISCUSSION

Jeff Appel (Fermilab): Congratulations on your progress on masses and engineering numbers needed for CKM measurements. The major issues in QCD itself, however, lie in the area you described as still hard after current successes: e.g. light scalar mesons, glueballs, hybrids, and production processes. When might we expect progress in these areas? Are the new improvements enough or are yet newer techniques needed?

Peter Lepage: It's difficult. You can do some resonance work. There are techniques for low-mass resonances so things like that ρ and the ϕ can probably be nailed – we have the technology. Things like glueballs – which would be really neat – are much, much harder and people have really just scratched the surface in thinking about new ways of doing it. So I don't know what to promise there. The one thing I can say, though, about things like glueballs is that in lattice gauge theory, we really haven't been sure that it was working for anything up until very, very recently. And when you're not sure that it's working for anything, even simple things, it's really, really hard to devote a lot of time to something like the glueball which you know is much much harder, given that you're not even clear that it can get the nucleon mass right yet. And, so at some level, getting a foundation that is really, really solid and well-established – at the couple of percent level where everyone agrees that this is the real thing – is bound to improve the odds for doing something non-trivial like figuring out how to do a glueball decay width. I can't predict what it is, I'm just saying that we're in vastly better shape by virtue of having gotten to first base. We have a much better chance of getting to second base if we've already made it to first base. And what we're doing now is trying to get to first base, and looking for ways in which we can have a big impact. And that's why there's a lot of focus on B physics, because there we really can have a big impact on the scientific program. This is not to say that there's nothing you can get from lattice gauge theory. I mean, probably the most compelling studies so far for things like glueballs come from lattice gauge theory. And people have fooled around with techniques for trying to estimate decay widths and so on – even of glueballs. I don't know if those techniques are reliable or not, but there are some ideas that you can try out. And if we can sort of pin down the rest of the territory then it makes it more likely that we'll be able to figure our way out of that problem. So I'm not being too encouraging, but I think we're in much better shape even for the harder problems.

Enrico Predazzi (INFN, Turin): When you mentioned in the beginning that the old lattice did not really represent QCD, you mentioned that that was, among other things, because of not incorporating quark loops. Now, in the new lattice, you have sea quarks but you have no actual gluons or anything of the kind.

Peter Lepage: No, we have gluons.

Enrico Predazzi (INFN, Turin): How does that come in?

Peter Lepage: Basically, we're evaluating the path integral of QCD numerically so we're integrating over all values of the gluon field. I should be clear for people who don't know lattice QCD. What it is is taking the path integral of the quantum field theory – literally the thing you read in a textbook – and evaluating that integral numerically. It's a numerical approximation to that integral. It's a very hard integral to do, and it's taken us a while to figure out how to do it effectively, but it is just the path integral that's the input. That's what the computer program is munching on. When I say bare quark mass and bare coupling as the inputs, those are literally the things that appear in the Lagrangian. We're working with the real thing here. I showed you upsilon spectra – it's not a quark model with a potential – there are only 5 numbers that went into all the data I showed you, and those are those masses and the bare thing and the rest of it is a numerical path integral.

Alberto Sirlin (New York University): One calculation that would be very interesting for the problem of universality is the calculation of $f^+(0)$ in the $K_{\ell 3}$ decays – the form factor. It seems to fall into the category of your gold-plated...could that be possible? That would be very important.

Peter Lepage: Yes. There's a big experimental community in B and D physics, but we know that there's also some really interesting stuff in kaon physics. So we're aware of it. Whether it gets done this year or next year is another question. It might actually be useful to talk a little bit – maybe you should visit Cornell?

FLAVOR PHYSICS: RARE AND BEYOND STANDARD MODEL DECAYS

RARE KAON DECAY PHYSICS

A. CECCUCCI

CERN, CH-1211 Geneva 23
E-mail: augusto.ceccucci@cern.ch

Recent progress in the field of rare kaon decays will be described. A brief summary of future experimental activities will also be given. Among the highlights, the improved upper limit on $K_L \to \pi^0 e^+ e^-$, and the first observation of the $K_S \to \pi^0 e^+ e^-$ are presented. A precise measurement of $K_S \to \gamma\gamma$ and high statistics studies of the Dalitz kaon decays have recently been reported and will be briefly described.

1. Introduction

1.1. *Why Study Rare Kaon Decays?*

There are four main reasons to study rare kaon decays:

1. study of the explicit violation of the Standard Model (SM) like Lepton Flavor Violation;

2. probe of the flavor sector by means of Flavor Changing Neutral Currents (FCNC);

3. test of fundamental symmetries such as CP and CPT; and

4. study of the strong interaction at low energy in exclusive processes.

1.2. *Organization of the Paper*

This paper reviews the recent experimental progress in the field of rare kaon decays: in addition to K_L and K^+ rare decays that have been the subject of precise studies over many years, K_S rare decays have started to be studied to sensitivities below 10^{-8}. The paper is organized as follows: in Sec. 2 recent progress on $K_{L,S} \to \pi^0 e^+ e^-$ and $K^+ \to \pi^+ \nu \bar{\nu}$ is reported: these decays, together with $K_L \to \pi^0 \nu \bar{\nu}$ form the core of the program to challenge the Cabibbo–Kobayashi–Maskawa (CKM)[1,2] description of CP-violation and quark mixing using kaon decays. Tests of CP- and CPT-symmetry not strictly related to the CKM description are described in Sec. 3: they include the study of $K_{S,L} \to \pi^+ \pi^- e^+ e^-$, $K_S \to 3\pi^0$ and the study of semileptonic K_S decays. Progress on the study of the kaon Dalitz decays is reported in Sec. 4 while new results on non-leptonic neutral kaon decays are reported in Sec. 5. The new initiatives that should lead to significant advances in this area of re-

search by the end of the decade are briefly outlined in Sec. 6.

2. Tests of the Standard Model with Rare Kaon Decays

The $K \to \pi \nu \bar{\nu}$ and the $K_L \to \pi^0 l^+ l^-$ decays allow the study of the coupling of the top quark by means of $s \to d$ virtual transitions which are very suppressed in the Standard Model. In the framework of the CKM description, CP-violation appears naturally from the presence of a complex irreducible phase in the mixing between the first and the third generation of quarks. Without a fourth generation, there is only one irreducible phase and, in the absence of new physics appearing in the virtual loops, the matrix must be unitary. The six unitarity relations can be displayed in the form of triangles on a complex plane. The triangles are all born equal in the sense that they have the same area which is the only *real* measure of CP-violation in the SM.[3] However, some relations look more *triangular* than others because all sides have similar length. For this reason it is common to display the

$$V_{ud}V_{ub}^* + V_{cd}V_{cb}^* + V_{td}V_{tb}^* = 0 \qquad (1)$$

unitarity relation on the $\bar{\rho}$ and $\bar{\eta}$ complex plane following the Wolfenstein parametrization.[4] The kaon decay modes relevant to the study of the unitarity triangle are shown on Fig. 1. The area of the triangle is expressed by the Jarlskog invariant, which for kaons can be written as:

$$\begin{aligned}
J_{CP} &= \Im V_{ud}V_{us}^* V_{td}V_{ts}^* \\
&= \sin\theta_C \cos\theta_C \Im V_{td}V_{ts}^* \\
&= \sin\theta_C \cos\theta_C \Im \lambda_t \;,
\end{aligned} \qquad (2)$$

where θ_C is the Cabibbo angle[1] and λ_t is a shorthand notation for $V_{td}V_{ts}^*$. It is useful to relate $\Im \lambda_t$

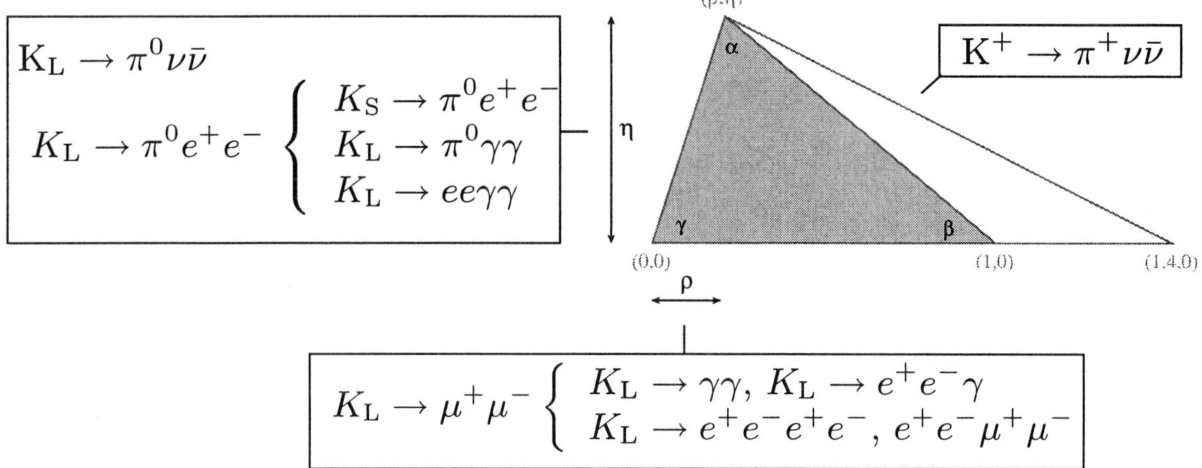

Figure 1. The rare kaon decays related to the study of the unitarity triangle.

and $\Re\lambda_t$ to the $\bar\eta$ and $\bar\rho$ quantities in the Wolfenstein parametrization:

$$\Im\lambda_t = A^2\lambda^5\bar\eta$$
$$\Re\lambda_t = A^2\lambda^5\bar\rho \, . \tag{3}$$

2.1. $K_L \to \pi^0\nu\bar\nu$

Even though no recent results have been reported on the subject at this conference, this article would not be complete without mentioning the status of the $K_L \to \pi^0\nu\bar\nu$ decay which is purely CP-violating[5] without long-distance contaminations. This decay has been dubbed the *holy grail* of kaon physics.[6] Measuring it is equivalent to a measurement of the height of the unitarity triangle and it can be predicted precisely to a few percent uncertainty because the hadronic matrix element can be taken from the well measured $Ke3$ decay. The current upper limit $< 5.9 \times 10^{-7}$ @ 90% CL[7] is more than one order of magnitude less sensitive than the model-independent limit[8] extracted from $K^+ \to \pi^+\nu\bar\nu$ which reads:

$$BR(K_L \to \pi^0\nu\bar\nu) < 1.7 \times 10^{-9} \quad 90\% \text{ CL} \, . \tag{4}$$

Prospects for further improving this channel are discussed in the last section.

2.2. $K_L \to \pi^0 e^+ e^-$

The physics interest is the same as $K_L \to \pi^0\nu\bar\nu$ but the experimental challenges, while very different, are still formidable. In addition, the interpretation of a possible observation is complicated by long-distance effects which require the study of additional reactions as depicted in Fig. 1. There has been a recent re-analysis of the decay by Buchalla, D'Ambrosio and Isidori,[9] motivated in part by a series of recent experimental results on the ancillary modes. The direct CP-violating short-distance contribution can be predicted taking $\Im\lambda_t$ from SM fits to the CKM unitarity triangle. For example taking[10]

$$\Im\lambda_t = 1.36 \times 10^{-4}, \tag{5}$$

they find:

$$BR(K_L \to \pi^0 e^+ e^-)^{Direct}_{CPV} = (3.2 \pm 0.4) \times 10^{-12} \tag{6}$$

which is indeed very small. In addition to the short-distance, there is another CP-violating contribution coming from the CP-even component of the K_L and proportional to $BR(K_S \to \pi^0 e^+ e^-)$. This component can be written as:

$$BR(K_L \to \pi^0 e^+ e^-)^{Indirect}_{CPV} = |\epsilon|^2 \frac{\tau_L}{\tau_S} BR(K_S \to \pi^0 e^+ e^-). \tag{7}$$

There are predictions of $BR(K_S \to \pi^0 e^+ e^-)$ spanning over one order of magnitude[11−13] and a measurement or a significant upper bound is necessary to fix this long-distance-dominated process.

The direct and the indirect CP-violating contributions can interfere, further complicating the extraction of $\Im\lambda_t$. However, if the indirect CP-violating component is large and if the interference term is constructive, the sensitivity to the short-distance physics is enhanced.

Last but not least, a CP-conserving amplitude has also to be taken into account. Information on this component can be extracted from the study of

Table 1. Limits on $K_L \to \pi^0 e^+ e^-$ and $K_L \to \pi^0 \mu^+ \mu^-$ published by the KTeV Collaboration.

Mode	BR @ 90% CL	Ref.
$K_L \to \pi^0 e^+ e^-$	$< 5.1 \times 10^{-10}$	18
$K_L \to \pi^0 \mu^+ \mu^-$	$< 3.8 \times 10^{-10}$	19

$K_L \to \pi^0 \gamma\gamma$. It has been pointed out[14] that reliable estimates of this contribution must be performed in a model-independent way without imposing Vector Meson Dominance (VMD) a priori. This point of view has been followed[9]: determining the three counter-terms from $K_L \to \pi^0 \gamma\gamma$[15] and $K_S \to \gamma\gamma$[16] data, the CP-conserving contribution to $K_L \to \pi^0 e^+ e^-$ is found to be:

$$BR(K_L \to \pi^0 e^+ e^-)_{CPC}^{Conserving} < 3 \times 10^{-12}, \quad (8)$$

much smaller than the total CP-violating term. All the considerations presented in this section apply also to the study of $K_L \to \pi^0 \mu^+ \mu^-$ but the CP-conserving term in this case needs more attention. The best limits on $K_L \to \pi^0 e^+ e^-$ and $K_L \to \pi^0 \mu^+ \mu^-$ are those obtained by the KTeV experiment at Fermilab. KTeV has two purposes: one experiment, E832, is devoted to the study of direct CP-violation in two-pion decays of the neutral kaon[17] (ϵ'/ϵ). The other, E799 II, is devoted to the study of rare kaon decays. E799 II used the extracted 800 GeV Tevatron proton beam and took data in 1997 and 1999 collecting about 7×10^{11} K_L decays in a relatively short running period. The experiment employs state-of-the-art CsI eletromagnetic calorimetry and a transition radiation detector to improve the pion/electron separation. The experimental difficulty resides in the presence of irreducible backgrounds from the radiative decay $K_L \to e^+ e^- \gamma\gamma$ which has a branching ratio of about $\approx 6 \times 10^{-7}$ and, to a lesser extent, $K_L \to \mu^+ \mu^- \gamma\gamma$. Only a very good two-photon mass resolution and kinematic cuts are available to suppress these backgrounds. As a consequence, in order to keep the backgrounds to the ≤ 1 event level, one is forced to reduce the acceptance quite significantly. Future searches will be background dominated and the progress will not be linear with the accumulated kaon flux. The published limits by KTeV are reported in Table 1. KTeV has also analyzed the data collected in 1999 to search

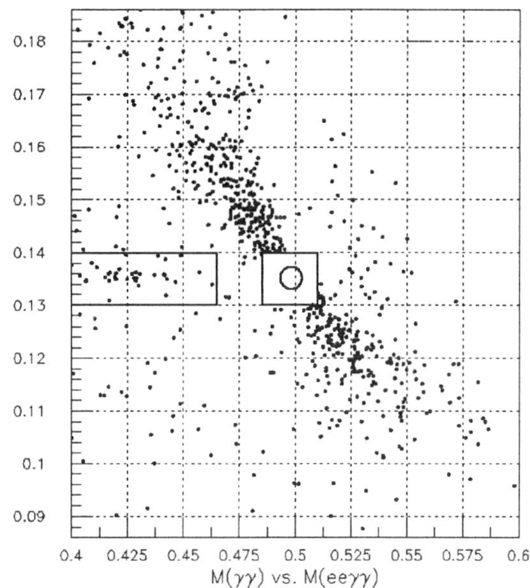

Figure 2. $M(\gamma\gamma)$ vs. $M(e^+ e^- \gamma\gamma)$ scatter plot for the 1999 KTeV data. The signal (circle) and control (square) regions are masked.

for $K_L \to \pi^0 e^+ e^-$. Details on the analysis can be found in a preprint that has just become available.[20] In Fig. 2 the $M(\gamma\gamma)$ vs. $M(e^+ e^- \gamma\gamma)$ scatter plot for the 1999 KTeV data is shown. In this figure the invariant mass of the four particles $M(e^+ e^- \gamma\gamma)$ is plotted under the hypothesis that the two photons come from a π^0 decay. This improves the mass resolution but makes the radiative background appear as a diagonal swath in the scatter plot. Backgrounds not related to $e^+ e^- \gamma\gamma$ are quite small. In particular, one notes that the excellent energy resolution of the detector allows the background $K_L \to \pi^0 e^+ e^- \gamma$ to be suppressed very efficiently. To further suppress the radiative background, one makes use of the different kinematics of the signal and of the background as studied by Greenlee.[21] The two useful variables are the minimum angle between any electron and any photon (θ_{min}), and the minimum angle of any photon with respect to the π^0 direction (θ_{π^0}) which, for a genuine π^0 decay, should be isotropically distributed but not for a radiative decay. The two distributions are shown in Fig. 3.

The values for the cuts on these kinematic variables were chosen to minimize the $BR(K_L \to \pi^0 e^+ e^-)$ upper limit in the absence of signal. After inspection of the signal box, one candidate event was found, which is consistent with the expected

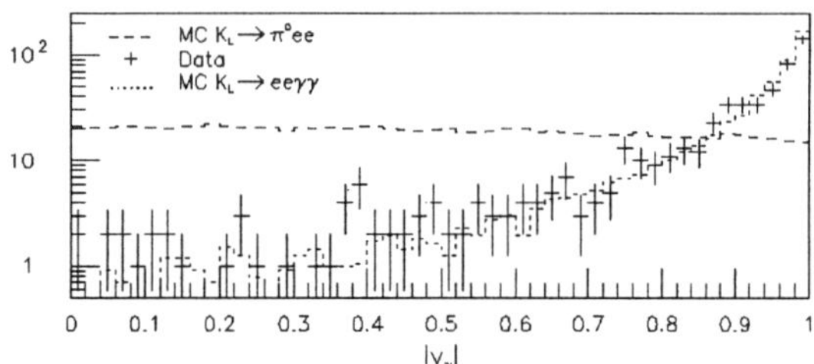

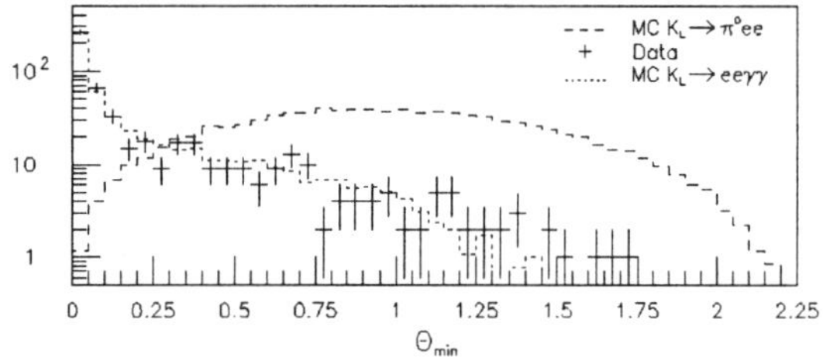

Figure 3. Comparison of $y_\gamma = \cos\theta_{\pi^0}$ (top) and θ_{min} (bottom) for $e^+e^-\gamma\gamma$ and $\pi^0e^+e^-$ final states.

background of 0.99 ± 0.35 events. The $M(\gamma\gamma)$ vs. $M(e^+e^-\gamma\gamma)$ scatter plot, with the unmasked signal region and after all cuts are applied, is shown in Fig. 4.

The KTeV result for the 1999 data is $BR(K_L \to \pi^0e^+e^-) < 3.5 \times 10^{-10}$ @ 90% CL. Combining this result with the previous search leads to the final KTeV result:

$$BR(K_L \to \pi^0e^+e^-) < 2.8 \times 10^{-10} \ @ \ 90\% \ \text{CL}. \quad (9)$$

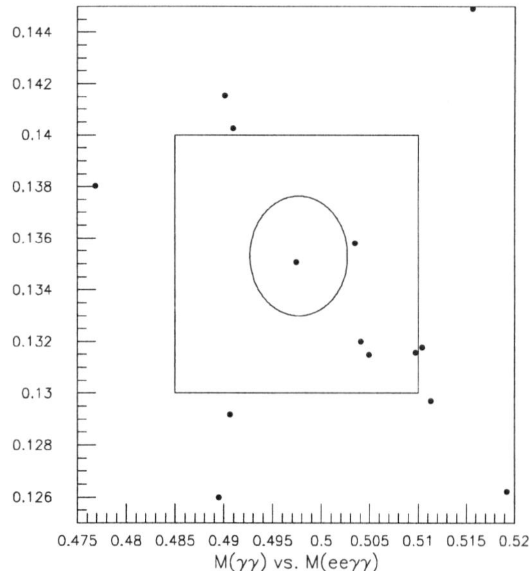

Figure 4. $M(\gamma\gamma)$ vs. $M(e^+e^-\gamma\gamma)$ scatter plot for the $K_L \to \pi^0e^+e^-$ (1999 data).

2.3. Observation of $K_S \to \pi^0e^+e^-$

As described in the previous section, in order to extract the short-distance information from $K_L \to \pi^0l^+l^-$, one has to measure or at least bound to the appropriate level the $K_S \to \pi^0l^+l^-$ reaction. The opportunity to study $K_S \to \pi^0l^+l^-$ took place at CERN after the NA48 experiment completed the measurement of ϵ'/ϵ.[22,23] The NA48 detectors with upgraded readouts and a modified K_S beam line were used to collect data from an intense, short neutral beam. The neutral beam was produced by striking 400 GeV protons from the CERN SPS onto a Be target 40 cm long. The beam emerges from a 6 m long collimator and enters a 90 m long decay tank. The same number of K_L as K_S is produced in the target and the K_L decays have to be taken into account in the background subtraction. The mean energy of the kaons is about 100 GeV. During the year 2002 NA48/1 collected more than 3×10^{10} K_S decays in the fiducial volume. Previously, during the year 2000, NA48/1 collected data for final states consisting of only photons since the charged spectrometer was not available that year. Here we briefly describe the $K_S \to \pi^0e^+e^-$ analysis based on the 2002 data. More details on the analysis can be found elsewhere.[24] Events with e^+e^- invariant mass ($M(ee)$) smaller than 0.165 GeV/c^2 are cut to reject $K_S \to \pi^0\pi^0_D$ decays with a missing photon which could otherwise mimic the signal. In this notation, π^0_D is the shorthand for the $\pi^0 \to e^+e^-\gamma$ (Dalitz) decay. A comparison between the data and the Monte Carlo is shown in Fig. 5 for the events satisfying the relation: 0.09 GeV/$c^2 < M_{ee} < 0.165$ GeV/c^2. These events have the same topology as the signal but fall outside the masked search region and can be used to check the reconstruction procedure. The data is well reproduced by the simulation (the normalization is absolute) and confirms that $M_{ee} > 0.165$ GeV/c^2 to define the signal region is well justified. One can notice that the $\pi^0 \to e^+e^-$ ($BR \simeq 6 \times 10^{-8}$) contribution is quite visible and gives an idea of how small the residual backgrounds due to π^0 Dalitz decays and conversions are. This is due, to a large extent, to the very good energy resolution of the detector. No events reconstructed with two electrons of the same sign were found in the corresponding signal region. This demonstrates that $K_S \to 2\pi^0$ events in which one pion undergoes a Dalitz decay

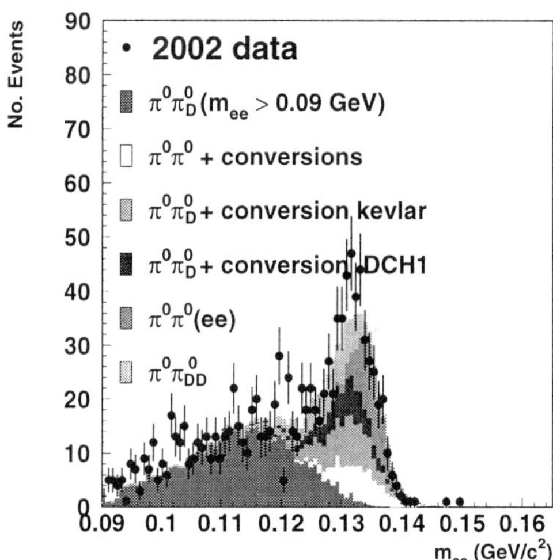

Figure 5. The M_{ee} invariant mass distributions for events that satisfy a $\pm 2.5\sigma$ cut on $M_{ee\gamma\gamma}$ around the kaon mass.

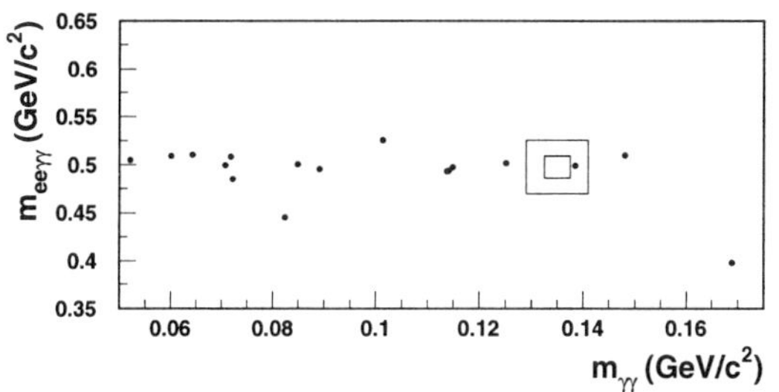

Figure 6. The $M(ee\gamma\gamma)$ vs. $M(\gamma\gamma)$ distribution for the $K_{\rm L}$ data collected in 2001. The sample contains about 10 times the expected number of $e^+e^-\gamma\gamma$ events expected in the 2002 sample and therefore it allows one to determine the radiative background precisely. The signal and background control boxes are indicated.

and a photon converts with subsequent loss of two charged particles of opposite sign are not of concern. The radiative background due to $K_{\rm L,S} \to e^+e^-\gamma\gamma$ was measured analyzing data collected by NA48 in 2001 from a pure $K_{\rm L}$ beam. The data set contains ten times more $K \to e^+e^-\gamma\gamma$ events than the 2002 sample and a precise measurement of the radiative background can therefore be made. The $M(e^+e^-\gamma\gamma)$ vs. $M(\gamma\gamma)$ distribution is shown in Fig. 6. $M(ee\gamma\gamma)$ is plotted by reconstructing the decay vertex with the charged tracks under the constraint that the kaon decay should lie on the straight line joining the target and the reconstructed kaon center-of-gravity. $M(\gamma\gamma)$ is reconstructed assuming the kaon mass. By plotting the data in this way, the two variables are not

Table 2. Sum of the backgrounds to $K_{\rm S} \to \pi^0 e^+e^-$

Source	Control region	Signal region
$K_{\rm S} \to \pi^0_D \pi^0_D$	0.03	< 0.01
$K \to e^+e^-\gamma\gamma$	0.11	0.08
Accidentals	0.19	0.07
Total	0.33	0.15

correlated. Background from the accidental overlap of different kaon decays has also to be taken into account. It typically originates from the time overlap of a $K_{\rm L} \to \pi e\nu$ decay where the pion is misidentified as an electron with a $K_{\rm S} \to \pi^0\pi^0$ decay in which one π^0 is lost outside the acceptance. Given the very good time resolution of the detectors and the wide readout window, this background is measured by counting the events that pass the analysis in the side-bands and extrapolating the result in the signal region. A rectangular signal region (2.5 standard deviations in $M(\gamma\gamma)$ and $M(e^+e^-\gamma\gamma)$ resolution) and a rectangular control region of 6 times 6 standard deviations, both centered around the kaon and π^0 mass, were kept masked until the background studies were completed in order to avoid any human bias. The summary of the expected backgrounds is reported in Table 2. Many other sources of background, for example those originating from neutral cascade decays, were investigated and found to be negligible. The inspection of the control box surrounding the signal region revealed no events, which is consistent with an expected background of 0.33 events. In the signal box seven candidate events were found. The candidates are displayed in Fig. 7. Given the background expectation of 0.15 events, the probability that all seven events are background is negligibly small ($\simeq 10^{-10}$). The events are therefore interpreted as the first observation of the $K_{\rm S} \to \pi^0 e^+e^-$ decay. The flux, measured by counting $\pi^0\pi^0_D$ events which are collected by the same trigger, amounts to 3.51×10^{10} $K_{\rm S}$ decays. After correcting for the acceptance and dividing by the flux, the measured branching ratio for $M(ee) > 0.165$ GeV/c^2 is found to be:

$$BR(K_{\rm S} \to \pi^0 e^+e^-, M(ee) > 0.165 \text{ GeV/c}^2) = (3.0^{+1.5}_{-1.2}(stat) \pm 0.2(syst)) \times 10^{-9}.$$

$$(10)$$

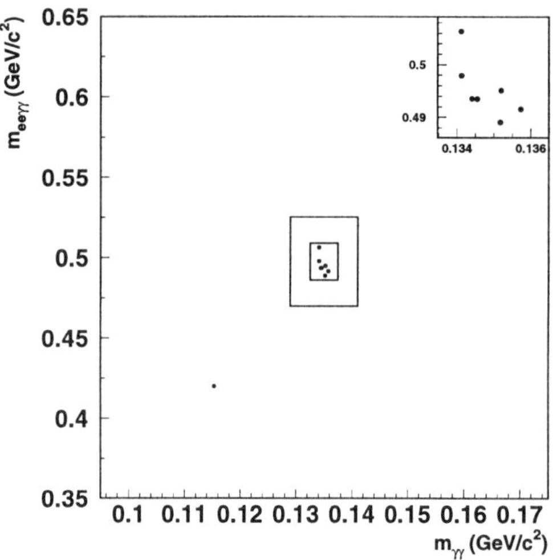

Figure 7. The $M(ee\gamma\gamma)$ vs. $M(\gamma\gamma)$ distribution for the seven candidate events. The same events are shown in the inset in more detail.

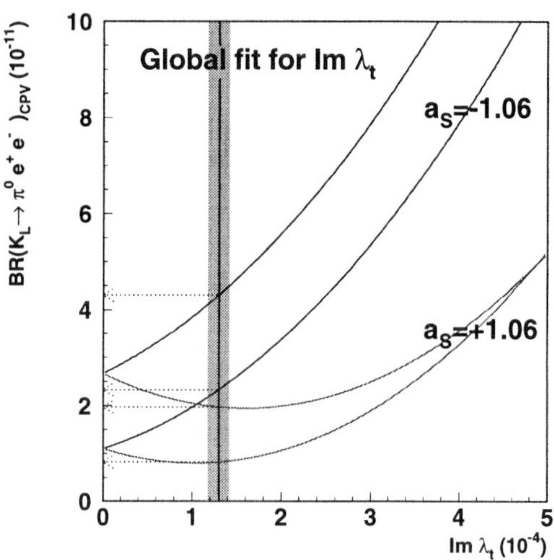

Figure 8. The sensitivity of the total CP-violating $K_L \to \pi^0 e^+ e^-$ branching ratio as a function of $\Im\lambda_t$ for the two signs of the interference term.

Assuming a vector interaction and constant form factor, the result extrapolated to the full phase space becomes:

$$BR(K_S \to \pi^0 e^+ e^-) = (5.8^{+2.8}_{-2.3}(stat) \pm 0.8(syst)) \times 10^{-9}, \quad (11)$$

where the systematic error is totally dominated by the form factor uncertainty. The result is in remarkable agreement with the prediction of L. Sehgal[11] which is 5.5×10^{-9}. In the notation of D'Ambrosio et al.,[25] this measurement can be related to a single parameter:

$$|a_S| = 1.06^{+0.26}_{-0.21} \pm 0.07. \quad (12)$$

Unfortunately, the sign of a_S cannot be extracted from this experiment and the question whether the interference between the short- and long-distance CP-violation is constructive or destructive remains unanswered. Constructive interference is preferred theoretically. [9]

2.4. Outlook for $K_L \to \pi^0 e^+ e^-$ and $K_L \to \pi^0 \mu^+ \mu^-$

The sensitivity of the total CP-violating branching ratio as a function of $\Im\lambda_t$ is shown in Fig. 8 for the two signs of a_S. For constructive interference (i.e. a_S negative in this notation) the sensitivity to $\Im\lambda_t$ is enhanced. The KTeV Collaboration is analyzing the 1999 $K_L \to \pi^0\mu\mu$ data and the sensitivity on this

mode will therefore be improved. The measurement of $K_S \to \pi^0\mu^+\mu^-$ by NA48/1 is in progress. Further searches for $K_L \to \pi^0 e^+ e^-$ and $K_L \to \pi^0\mu^+\mu^-$ will be background dominated. It will be difficult to reduce the background from the radiative decays because KTeV and NA48 already have state-of-the-art calorimeters. The sensitivity to short-distance physics can be further enhanced if the interference between the short-distance and long-distance CP-violation turns out to be constructive. The signal-to-background ratio could be improved by collecting data from the time-dependent $K_L - K_S$ interference region.[26] Further progress relies on the availability of high energy, slowly extracted DC proton beams with intensity in excess of 10^{12} protons/s. *Factory mode* operation of the experiment over a few years has also to be envisaged. Compounding all these factors together one can explore the window of opportunity that spans from the current upper limit to the Standard Model prediction.

2.5. $K^+ \to \pi^+\nu\bar{\nu}$

This decay is a very sensitive probe of the Standard Model because the hadronic matrix element can be extracted from the well measured $K^+ \to \pi^0 e\nu$ decay. There are no long-distance contributions to this mode and the QCD corrections have been calculated to NLO.[27,28] The theoretical error for the extraction

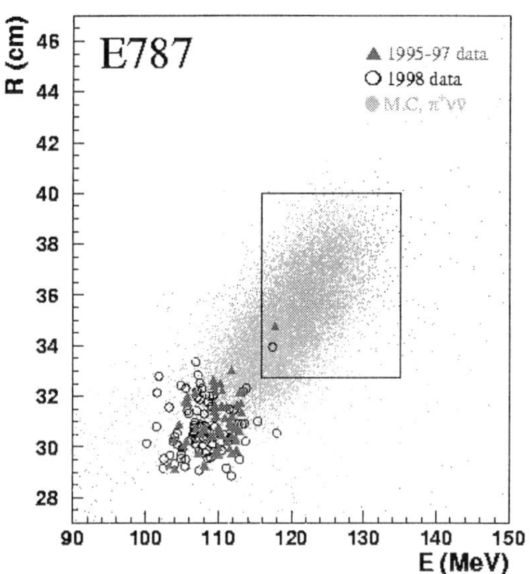

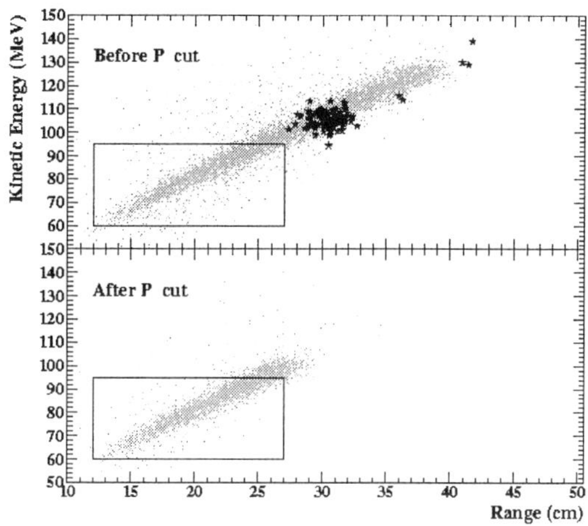

Figure 10. The kinetic energy vs. range scatter plot for the 1997 data in Region II before and after the momentum cut.

Figure 9. The range versus kinetic energy scatter plot. The analysis requires that the pion momentum is contained between 211 and 229 MeV/c. The circles are the 1998 data while the triangles are the 1995–1997 data. The group of events around $E = 108$ MeV corresponds to $K\pi2$ decays. The simulated signal is represented by dots.

of $|\lambda_t|$ is limited to the knowledge of the charm quark contribution which translates into a small residual error on the decay amplitude. Other uncertainties are related to the poor knowledge of the CKM elements and will improve accordingly. One analysis,[29] using ϵ_K and $A_{\mathrm{CP}}(J\psi K_{\mathrm{S}})$ as inputs, predicts in the SM: $BR(K^+ \to \pi^+\nu\bar\nu) = (7.1 \pm 1.0) \times 10^{-10}$. To reach a 10^{-10} Single Event Sensitivity (SES), the analysis is usually restricted to the pion momentum region between the $K^+ \to \pi^+\pi^0$ ($K\pi2$) and $K^+ \to \mu^+\nu$ ($K\mu2$) peaks (Region I). Two events interpreted as a signal were published during the past years by the AGS-E787 experiment.[30] The range versus kinetic energy scatter plot containing the two signal events in the box is shown in Fig. 9. These two events provide a measurement of the branching ratio:

$$BR(K^+ \to \pi^+\nu\bar\nu) = (1.57^{+1.75}_{-0.83}) \times 10^{-10} \qquad (13)$$

which is on the high side of the SM prediction, albeit with very large errors. The analysis of the kinematic region with pion momentum above the $K^+ \to \pi^+\pi^0\pi^0$ endpoint and below the ($K\pi2$) peak (Region II) relies greatly on the photon veto efficiency. There is one published analysis of Region II by AGS-E787,[31] which has been updated for this conference to include the data collected in 1997.[32]

The scatter plot of the kinetic energy versus range before and after the pion momentum cut is shown in Fig. 10 for the 1997 data. The combined 1996 and 1997 analysis yields the preliminary limit:

$$BR(K^+ \to \pi^+\nu\bar\nu) < 22 \times 10^{-10} \ @\ 90\%\ \mathrm{CL}. \qquad (14)$$

Even if the result is still an order of magnitude above the SM prediction, this result is quite important in my opinion because it opens the way to analyzing the decay in regions of larger acceptance, allowing one to test the dynamics of the $K^+ \to \pi^+\nu\bar\nu$ decay (to test, for example, the vector nature of the interaction). Many improvements were made to turn E787 into the successor experiment E949. In particular, the photon veto system was upgraded. E949 took data in 2002 for 12 weeks and results are expected soon. The experiment was originally approved to run for 60 weeks and therefore expects to take more data.

2.6. Outlook

A comparison between the constraint currently provided by K and B mesons on $\bar\eta$ and $\bar\rho$ was prepared by Gino Isidori and is shown in Fig. 11. Obviously, a large window of opportunity exists. Some of the observables to which rare kaon decays have access are well understood theoretically. We expect, within a reasonable time scale of about a decade, that a completely independent and competitive test of the CKM paradigm will be provided by rare kaon decay studies, notably $K \to \pi\nu\bar\nu$.

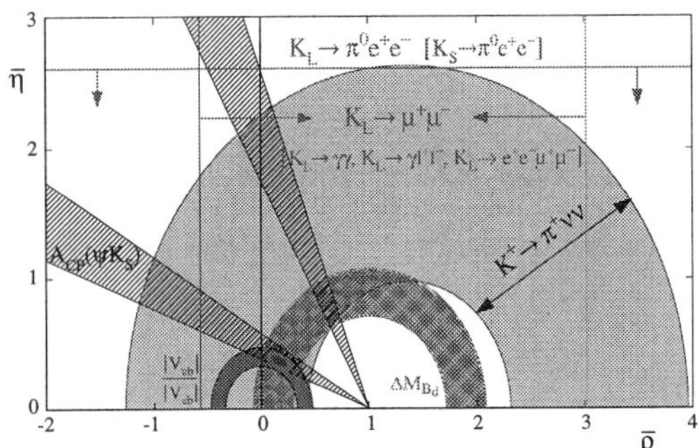

Figure 11. There is a window of opportunity to test the Standard Model with rare kaon decays.

3. Other Tests of CP- and CPT-Violation

3.1. $K_{L,S} \to \pi^+\pi^- e^+ e^-$

Large CP-violating effects are not confined to B decays. It was realized long ago that in the $K \to \pi^+\pi^-\gamma^* \to \pi^+\pi^- e^+ e^-$ decay, large effects due to CP-violation could be observed in the polarization of the photon. The decay has a branching ratio of $\simeq 3 \times 10^{-7}$ and its decay amplitude is dominated by two competing $K_L \to \pi^+\pi^-\gamma^*$ components: one from the CP-violating internal bremsstrahlung and the other from the CP-conserving direct emission associated with a magnetic dipole transition. Sehgal and Wanninger[33] and Heiliger and Sehgal[34] showed that the angular correlation of the $e^+ e^-$ and $\pi^+\pi^-$ planes contains an explicit CP-violating term arising from the interference of the two amplitudes. The CP-asymmetry in the distribution of the angle ϕ between the $e^+ e^-$ and $\pi^+\pi^-$ planes, in the kaon center-of-mass system is:

$$\mathcal{A}_\phi = \frac{\int_0^{\pi/2} \frac{d\Gamma}{d\phi}d\phi - \int_{\pi/2}^{\pi} \frac{d\Gamma}{d\phi}d\phi}{\int_0^{\pi/2} \frac{d\Gamma}{d\phi}d\phi + \int_{\pi/2}^{\pi} \frac{d\Gamma}{d\phi}d\phi}. \qquad (15)$$

This asymmetry, which originates mostly from $K^0\overline{K}^0$ mixing, is predicted to be as large as 14%. The asymmetry was first measured by KTeV.[35] The latest KTeV preliminary result was obtained combining the 1997 and 1999 statistics[36] and it is based on 5056 events. The asymmetry

$$\mathcal{A}_\phi = (13.3 \pm 1.4 \pm 1.0)\% \qquad (16)$$

is shown in Fig. 12 and is in good agreement with the theoretical prediction. The NA48 Collaboration has completed a thorough analysis of asymmetry and branching ratio for both K_L and K_S based on 1162 and 621 events, respectively.[37]

The K_L asymmetry measured by NA48 agrees with the expectation and with KTeV. No asymmetry is observed in K_S decays. This is expected because only one amplitude dominates the $K_S \to \pi^+\pi^- e^+ e^-$ decay. The K_S asymmetry measured by NA48 is shown in Fig. 13.

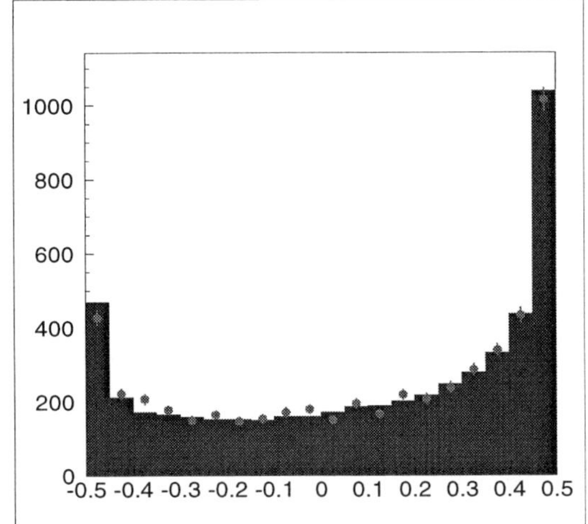

Figure 12. The $K_L \to \pi^+\pi^- e^+ e^-$ CP-violating asymmetry as measured by KTeV (data 1997+1999, preliminary).

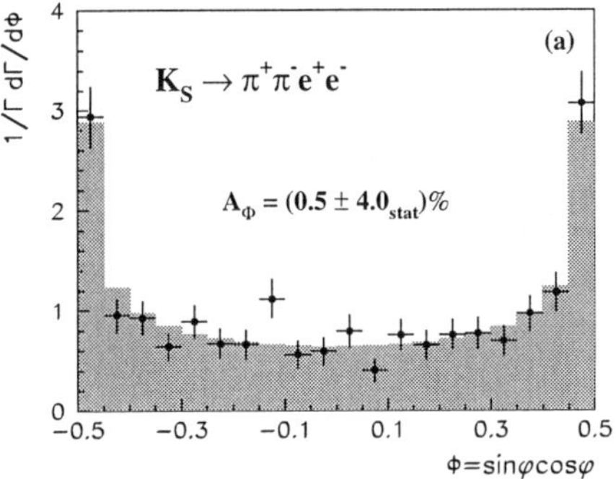

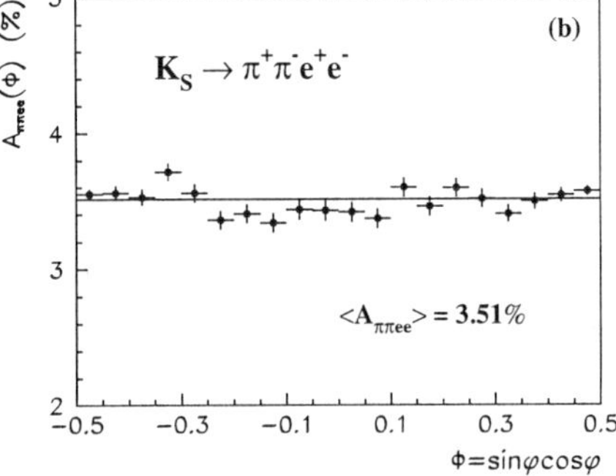

Figure 13. The $K_S \to \pi^+\pi^- e^+ e^-$ angular distribution as measured by NA48. In the bottom plot the acceptance is shown.

Table 3. Overview of current CP-violating results in $K_\mathrm{S} \to 3\pi$ decays.

Experiment	Result
FNAL-E621[40]	$\Im\,\eta_{\pm 0} = (-1.5 \pm 1.7 \pm 2.5) \times 10^{-2}$
CPLEAR[41]	$\Re\,\eta_{\pm 0} = (-2 \pm 7^{+4}_{-1}) \times 10^{-3}$
	$\Im\,\eta_{\pm 0} = (-2 \pm 9^{+2}_{-1}) \times 10^{-3}$
ITEP-761[42]	$\Re\,\eta_{000} = (-8 \pm 18) \times 10^{-2}$
	$\Im\,\eta_{000} = (-5 \pm 27) \times 10^{-2}$
CPLEAR[43]	$\Re\,\eta_{000} = (18 \pm 14 \pm 6) \times 10^{-2}$
	$\Im\,\eta_{000} = (15 \pm 20 \pm 3) \times 10^{-2}$
SND[44]	$BR(K_\mathrm{S} \to 3\pi^0) < 1.4 \times 10^{-5}$ 90% CL

Table 4. Systematic uncertainties.

	$\Re\,\eta_{000}\ (10^{-2})$	$\Im\,\eta_{000}\ (10^{-2})$
Accidentals	± 0.1	± 0.6
Energy scale	± 0.1	± 0.1
Dilution	± 0.3	± 0.4
Acceptance	± 0.3	± 0.8
Binning	± 0.1	± 0.2
Total	± 0.5	± 1.1

3.2. $K_\mathrm{S} \to 3\pi^0$

CP-violation in $K^0 \to 2\pi$ decays is firmly established and the parameters which describe it ($\eta_\pm$, η_{00} and ϵ'/ϵ) are precisely measured.[38] CP-violation in $K_\mathrm{S} \to 3\pi$ is equally allowed in the SM but has been investigated in much less detail owing to the difficulty of the measurements which involve rare kaon decays. A $\pi^+\pi^-\pi^0$ state is mainly CP-odd while a $3\pi^0$ state is purely CP-odd. The equivalent of η_{00} for $K_\mathrm{S} \to 3\pi^0$ decays is $\eta_{000} = A(K_\mathrm{S} \to 3\pi^0)/A(K_\mathrm{L} \to 3\pi^0)$. In the SM $\eta_{000} = \epsilon + i\Im\,a_1/\Re\,a_1$, where a_1 is the isospin 1 amplitude for $K^0 \to \pi^0\pi^0\pi^0$ and $\epsilon = 2/3\eta_\pm + 1/3\eta_{00}$. The current experimental situation is summarized in Table 3. It is worth noticing that the test of CPT-conservation based on the comparison of the K^0 and $\overline{K}^0$ masses is currently limited by the poor knowledge of η_{000}.[39]

During the year 2000 NA48 did not have any drift chambers because they were damaged by the implosion of the carbon fiber beam pipe. So they exploited the excellent energy resolution of the liquid krypton calorimeter (LKr)[45] to collect $3\pi^0$ decays from a short neutral beam to improve the limits on η_{000}. The sensitivity to η_{000} comes from the $K_\mathrm{S} - K_\mathrm{L}$ interference that can be measured by studying the intensity of $3\pi^0$ decays in the short neutral beam as a function of the proper decay time of the kaon. In order to keep the acceptance correction small, the data were normalized using $3\pi^0$ decays collected from the long beam. The long beam is a pure K_L beam for all practical purposes and the $K_\mathrm{S} - K_\mathrm{L}$ interference expected in $3\pi^0$ decays is completely negligible.

A Monte Carlo simulation was used to correct for the residual geometrical difference between the two beams. The decay energy spectra of kaons generated from the near and far beam are not identical. In order to take care of this difference, the analysis is done by fitting the data in 5 GeV wide energy bins according to the function:

$$f(E,t) = \frac{NEAR}{FAR} = A(E)[1 + |\eta_{000}|^2 e^{(\Gamma_L - \Gamma_S)t}$$
$$+ 2D(E)e^{\frac{1}{2}(\Gamma_L - \Gamma_S)t}$$
$$\times (\Re\,\eta_{000}\cos\Delta mt - \Im\,\eta_{000}\sin\Delta mt)\,]$$

where $A(E)$ are normalization constants and $D(E)$ is the so-called $K^0 - \overline{K}^0$ *dilution* describing the excess of K^0 over $\overline{K}^0$ in the incoherent mixture of neutral kaons produced by the 400 GeV protons in the Be target. $D(E)$ is a function of the energy and of the production angle. The dilution values were those measured by the NA31 experiment,[46] slightly adjusted to take into account the different production angle and proton energy.

There are about 5.6×10^6 $3\pi^0$ events from the near beam and in excess of 10^7 from the far beam. The near/far ratio corrected for the beam geometry is shown in Fig. 14 for three energy bins. The position of the collimator and the upstream and downstream boundaries of the fitting region are indicated on the figure. Systematic errors have been evaluated for accidentals, energy scale, K^0 dilution, acceptance, and binning and are reported in Table 4. NA48/1 has made two fits to Eq. (3.2).

1. The Real and the Imaginary part of η_{000} are fitted independently together with the normalization constants. The results are:

$$\Re\,\eta_{000} = -2.6 \pm 1.0(stat) \pm 0.5(syst) \times 10^{-2} \tag{17}$$

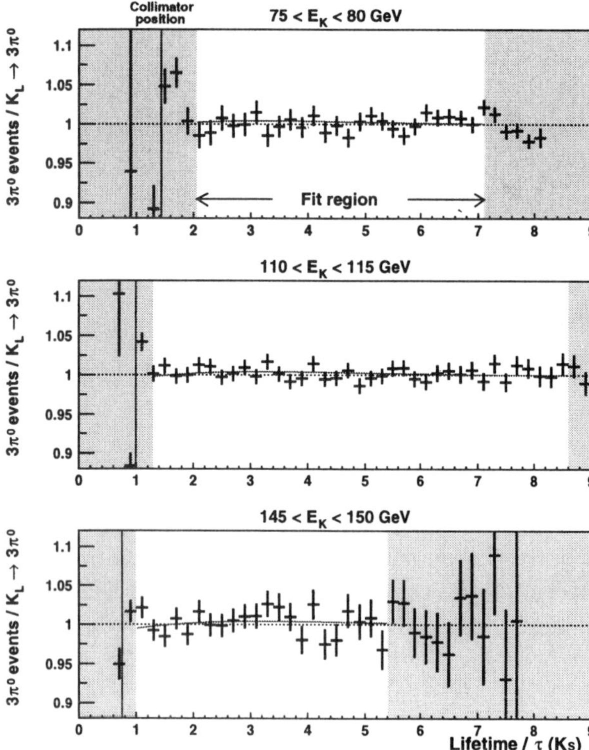

Figure 14. Near over far $3\pi^0$ data after the geometry correction. The position of the end of the collimator and the definition of the upstream and downstream limits of the fitting region are indicated. The upstream cuts were chosen to avoid resolution effects, and the downstream ones to insure good trigger efficiency.

$$\Im\,\eta_{000} = -3.4 \pm 1.0(stat) \pm 1.0(syst) \times 10^{-2}.$$
(18)

The χ^2/ndf is good (415/405) but the two parameters are strongly correlated. The fit is compatible with CP-conservation with a probability of a few percent.

2. To avoid the correlation of the parameters, they imposed CPT-conservation and fixed the Real part of η_{000} to the SM prediction:

$$\Re\,\eta_{000} = \Re\,\epsilon \simeq 1.6 \times 10^{-3}.$$
(19)

Fitting for $\Im\,\eta_{000}$ then yields:

$$\Im\,\eta_{000} = -1.2 \pm 0.7(stat) \pm 1.1(syst) \times 10^{-2}$$
(20)

which is compatible with CP-conservation within errors.

The result of the fits is shown in Fig. 15. The result given by Eq. (20) translates into the limit $BR(K_S \to \pi^0\pi^0\pi^0) < 3.0 \times 10^{-7}$ @ 90%CL which

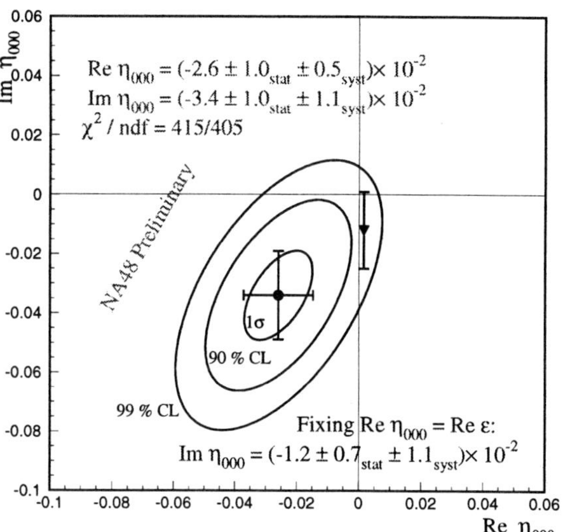

Figure 15. Results of the $3\pi^0$ fits.

improves the current limit by about 50 times and is compatible with the SM prediction of $\sim 1.9 \times 10^{-9}$. Results given by Eqs. (17) and (18) provide input to the Bell–Steinberger unitarity relation,[47] which constrains the CP- and CPT-violating parameter $\Im\,\delta$ under the hypothesis of conservation of probability:

$$(\Re\,\epsilon - i\Im\,\delta) \times (i2\Delta m + (\Gamma_S + \Gamma_L)) = \sum_f A(K_S \to f)^* A(K_L \to f).$$
(21)

Taking into account the NA48/1 result, we obtain:

$$\Im\,\delta = (-1.2 \pm 3.0) \times 10^{-5}$$
(22)

which represents an improvement of about 40% with respect to previous result:[48]

$$\Im\,\delta = (2.4 \pm 5.0) \times 10^{-5}.$$
(23)

Assuming CPT-conservation in the semileptonic K^0 decays, the phase of δ is fixed and the measurement translates into a new limit on the $K^0\overline{K}^0$ mass difference:

$$M(K^0) - M(\overline{K}^0) = (-1.7 \pm 4.2) \times 10^{-19}\ \text{GeV}.$$
(24)

3.3. *Semileptonic K_S Decays*

Semileptonic kaon decays are quite important for the study of fundamental parameters of the SM theory such as V_{us}.[49] Only K_S decays having branching ratios smaller than 10^{-3} can be considered relatively rare. Under the validity of the $\Delta S = \Delta Q$ rule

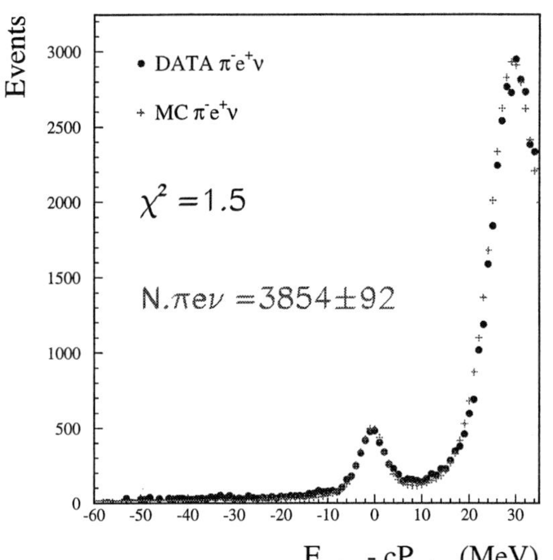

Figure 16. K_S semileptonic decays with positive charge measured by KLOE (2001 statistics, preliminary).

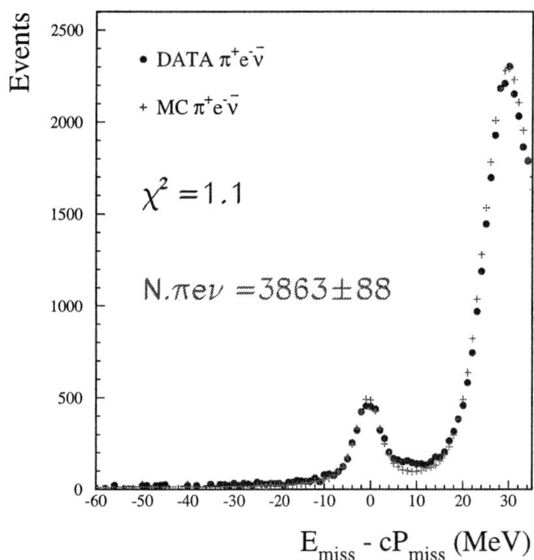

Figure 17. K_S semileptonic decays with negative charge measured by KLOE (2001 statistics, preliminary).

and CPT-conservation, the $K_S \to \pi e \nu$ rate can be derived from that of K_L semileptonic decays using known kaon parameters. The KLOE experiment, installed at the Frascati ϕ factory, can measure the $K_S \to \pi e \nu$ rate in a very straightforward way, using the $\phi \to K_S K_L$ decay. The ϕ decays at rest and therefore the kaons have practically fixed energy, enabling the K_S to be tagged by measuring the time of flight of the accompanying K_L which interacts in the electromagnetic calorimeter. KLOE has already published results on this mode.[50] Here we report the latest preliminary results[51] based on more statistics collected during 2001. A comparison of data and Monte Carlo is shown in Fig. 16 for positive leptons and in Fig. 17 for negative ones. The preliminary results are:

$$BR(K_S \to \pi^+ e^- \bar{\nu}) = (3.44 \pm 0.09 \pm 0.06) \times 10^{-4}$$
$$BR(K_S \to \pi^- e^+ \nu) = (3.31 \pm 0.08 \pm 0.05) \times 10^{-4}$$
$$BR(K_S \to \pi^{\pm} e^{\mp} \nu(\bar{\nu})) = (6.76 \pm 0.12 \pm 0.10) \times 10^{-4}$$

$$(25)$$

On the basis of these results, KLOE has made the first measurement of the CP charge asymmetry for $\mathcal{A}(K_S)$. Deviations of $\mathcal{A}(K_S)$ from the well measured[52] $\mathcal{A}(K_L) = (3.322 \pm 0.055) \times 10^{-3}$ would be a sign of CPT-violation. The preliminary value[51] is $\mathcal{A}(K_S) = (1.9 \pm 1.7 \pm 0.6) \times 10^{-2}$. Conversely, one can assume CPT-conservation and test the $\Delta S = \Delta Q$ rule. The KLOE statistics are improving and are now starting to be competitive with the results obtained by CPLEAR.[53]

4. Kaon Dalitz Decays

4.1. Motivation

Attempts to extract $\Re \lambda_t$ from the decay $K_L \to \mu^+ \mu^-$,[25,54] suffer from the lack of control of the long-distance part of the dispersive amplitude of the decay.[55] The study of the kaon Dalitz decays ($K_L \to \gamma^{(*)} \gamma^*$) allows one in principle to better constrain this long-distance contribution. Significant progress has been achieved on the Dalitz kaon decays as a byproduct of the ϵ'/ϵ experiments. Two models are available in the literature to parametrize the Dalitz form factors. In addition to the BMS model[56] where one parameter (α_K^*) is used to include vector contributions, a parametrization compatible with the chiral expansion to $\mathcal{O}(p^6)$ is available (DIP).[25] The two descriptions are related by the formula

$$\alpha(DIP) = -1 + (3.1 \pm 0.5)\alpha_K^*(BMS), \qquad (26)$$

where the error is due to the different q^2 dependence of the two parametrizations.

4.2. $K_L \to e^+ e^- \gamma$ and $K_L \to e^+ e^- e^+ e^-$

The KTeV experiment has presented a preliminary analysis[57] of the data collected in 1997. The sample includes 93 383 $K_L \to e^+ e^- \gamma$ candidates. Background from $Ke3$ decays where a pion is misidentified as an electron are kept to less than 0.1% by the use of a transition radiation detector. It is very important to understand the tracking for close electron–positron pairs in order to determine the acceptance. The preliminary branching ratio

$$BR(K_{\mathrm{L}} \to e^+e^-\gamma) \times 10^6 = 10.19 \pm 0.04(stat) \pm 0.07(syst) \pm 0.29(norm) \tag{27}$$

is in good agreement with the published NA48 value.[58] The decay rate is dominated by the QED radiative processes and the branching ratio is not very sensitive to the kaon structure. In order to extract the form factor, the q^2 distribution has to be studied. In Fig. 18 the KTeV data are compared with the simulation for different values of α_K^*. The data prefer a mildly negative value.

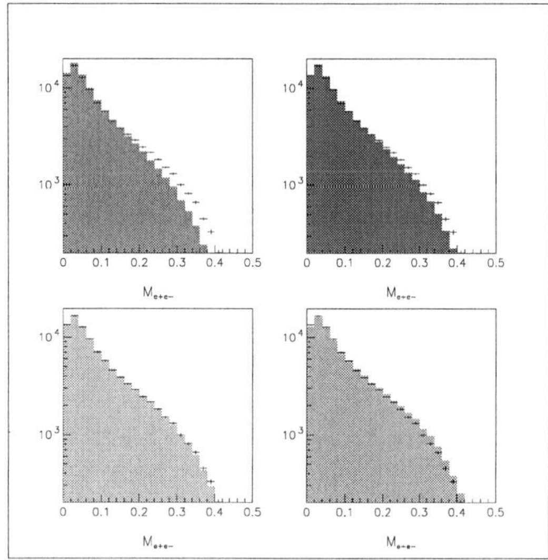

Figure 18. The $M(e^+e^-)$ distribution for $K_{\mathrm{L}} \to e^+e^-\gamma$ candidates as measured by KTeV (1997 statistics, preliminary), compared with several Monte Carlo simulations as a function of different form factors.

Fitting for α_K^* yields: $-0.186 \pm 0.011 \pm 0.09$. This very precise measurement (7% relative error!) is in good agreement with the value measured by KTeV[59] from $K_{\mathrm{L}} \to \mu^+\mu^-\gamma$ and disagrees with the NA48 $e^+e^-\gamma$ measurement, -0.36, at a level of about 2.6 standard deviations.

Extraction of the form factor from the $K_{\mathrm{L}} \to e^+e^-e^+e^-$ decay is complicated by the pairing ambiguity and by the small rate that prevents a precise determination. The branching ratio measurement is already limited by the knowledge of the normalization channel. The preliminary measurement of the branching ratio and of the form factor, based on 1100

(1997 + 1999 statistics) events was reported to be:[60]

$$BR(K_{\mathrm{L}} \to e^+e^-e^+e^-) = (4.16 \pm 0.13 \pm 0.13 \pm 0.17) \times 10^{-8}, \tag{28}$$

$$\alpha_K^* = -0.03 \pm 0.13 \pm 0.04. \tag{29}$$

4.3. $K_{\mathrm{L}} \to e^+e^-\mu^+\mu^-$

This is one of the first results published by KTeV on the full 1997 and 1999 data.[61] It is a very clean measurement with 132 events. Details of the analysis can be found in the publication. Here we recall that the measured branching ratio is

$$BR(K_{\mathrm{L}} \to e^+e^-\mu^+\mu^-) = (2.69 \pm 0.24 \pm 0.12) \times 10^{-9}. \tag{30}$$

A precise measurement of the form factor is prevented by the small statistics:

$$\alpha_K^*(\mu\mu ee) = -0.19 \pm 0.11. \tag{31}$$

Combining K_{L} decays in $\mu\mu\gamma$, $4e$, and $\mu\mu ee$, KTeV has given a first measurement of

$$\alpha(DIP) = -1.53 \pm 0.10. \tag{32}$$

4.4. *Outlook*

The current situation concerning the $K - \gamma^*\gamma^{(*)}$ form factor is summarized in Fig. 19. In the near future new results should become available from the NA48 experiment. They are analyzing $ee\gamma$ and $4e$ collected in 1998–1999 and 2001. Data still look quite scattered. The KTeV experiment has performed two precise measurements which agree between themselves suggesting a consistent description of the form factor. It will be interesting to see if the high statistics measurements from NA48 will finally confirm this picture.

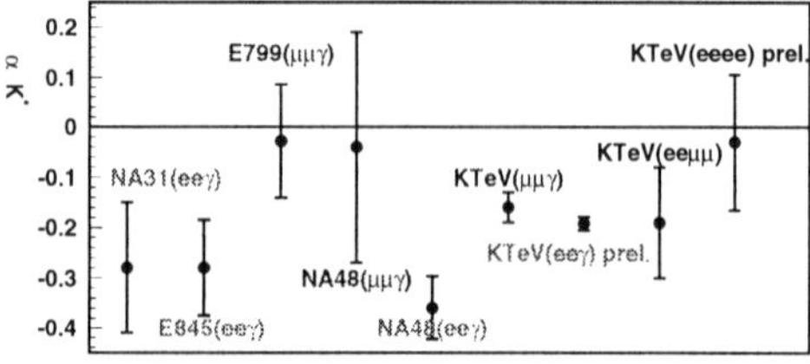

Figure 19. Measurements of the α_K^* form factor. Precise measurements by KTeV on $K_{\mathrm{L}} \to \mu^+\mu^-\gamma$ and from $K_{\mathrm{L}} \to e^+e^-\gamma$ seem to suggest a coherent picture of otherwise scattered data. Measurements from the full statistics of NA48 are eagerly awaited.

5. Nonleptonic Kaon Decays

5.1. $K_{L,S} \to \gamma\gamma$

The study of the decays $K_{S,L} \to \gamma\gamma$ and $K_{S,L} \to \pi^0\gamma\gamma$ is important for understanding the low-energy hadron dynamics of Chiral Perturbation Theory (χPT), since they are sensitive to higher order loop effects.[62] The K_L modes have already been studied precisely. The decay $K_L \to \pi^0\gamma\gamma$ is particularly interesting because it can be used to bound CP-conserving contributions to $K_L \to \pi^0 e^+ e^-$, as explained earlier in this article. Unfortunately the two most recent determinations do not quite agree[63,15] and, given the importance of the measurement, further experimental clarification would be welcome. Two measurements of the $K_L \to \gamma\gamma$ decay normalized to $K_L \to 3\pi^0$ have recently been published. The $K_L \to \gamma\gamma$ reaction is interesting because it provides the normalization to the absorptive amplitude of $K_L \to \mu^+\mu^-$. However, to fully exploit this good precision on $K_L \to \gamma\gamma$, the data should be normalized to $K_L \to \pi^0\pi^0$.

On the K_S side, significant progress has taken place and a precise measurement of $K_S \to \gamma\gamma$ and the first observation of the decay $K_S \to \pi^0\gamma\gamma$ were recently reported. The next sections briefly describe the measurement of $K_L \to \gamma\gamma$ done by KLOE, the precise measurement of $K_S \to \gamma\gamma$ and the first observation of $K_S \to \pi^0\gamma\gamma$ done by NA48/1.

5.2. $K_L \to \gamma\gamma$ and $K_S \to \gamma\gamma$

KLOE at the DAΦNE ϕ factory exploits K_S tagging to study K_L decays without having to subtract the K_S backgrounds. Based on 1.6×10^8 tagged K_L originating from from $10^9\phi$ decays collected during the year 2000 and 2001, KLOE has selected 27 375 $K_L \to \gamma\gamma$ decays. The $\gamma\gamma$ invariant mass $M_{\gamma\gamma}$ is shown in Fig. 20. The background at low $M_{\gamma\gamma}$ is due to $K_L \to 2\pi^0$ and $K_L \to 3\pi^0$ decays. The result, normalized to $K_L \to 3\pi^0$ is:[64]

$$\Gamma(K_L \to \gamma\gamma)/\Gamma(K_L \to \pi^0\pi^0\pi^0) = \\ (2.79 \pm 0.02(stat) \pm 0.02(syst)) \times 10^{-3}. \quad (33)$$

In good agreement with the NA48/1 published result:[16]

$$\Gamma(K_L \to \gamma\gamma)/\Gamma(K_L \to \pi^0\pi^0\pi^0) = \\ (2.81 \pm 0.01(stat) \pm 0.02(syst)) \times 10^{-3}. \quad (34)$$

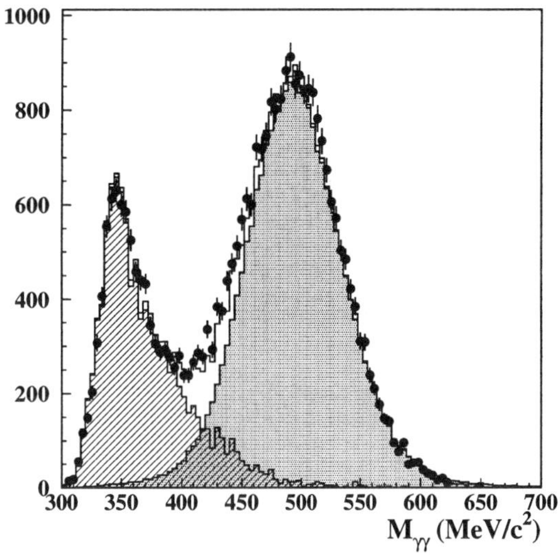

Figure 20. To cope with large backgrounds from $K_L \to 2\pi^0$ and $K_L \to 3\pi^0$ decays, KLOE is obliged to impose that $E_\gamma = M_K/2$ in the K_L rest frame. The energy of the photons can be boosted in the laboratory using the constraint: $\vec{p}(K_L) = \vec{p}(\phi) - \vec{p}(K_S)$.

The decay $K_S \to \gamma\gamma$ is especially interesting because χPT predicts unambiguously that the branching ratio is 2.25×10^{-6}, with an error of less than 10%. NA48/1 has performed a precise measurement of the $BR(K_S \to \gamma\gamma)$ from the data collected from a short neutral beam in the year 2000. The short neutral beam contains a $K_L \to \gamma\gamma$ component that can be precisely subtracted measuring the K_L flux from the number of reconstructed $3\pi^0$ decays and exploiting the precise $K_L \to \gamma\gamma$ to $K_L \to 3\pi^0$ ratio measured by the same experiment employing a long beam. The data is shown in Fig. 21. The NA48/1 $K_S \to \gamma\gamma$ result: [16]

$$BR(K_S \to \gamma\gamma) = \\ (2.78 \pm 0.06(stat) \pm 0.03(syst) \pm 0.02(ext))10^{-6} \quad (35)$$

has an accuracy of about 3%. It differs by 30% from the $\mathcal{O}(p^4)$ prediction of χPT. Progress on this measurement and comparison with the theory is shown in Fig. 22. This measurement was used by Buchalla et al.[9] to extract one of the three counterterms that appear in χPT at $\mathcal{O}(p^6)$.

5.3. $K_S \to \pi^0\gamma\gamma$

The first observation of this decay was reported by NA48/1 from the analysis of the 2000 data.[65] In the decay $K_S \to \pi^0\gamma\gamma$, the photon pair production is

119

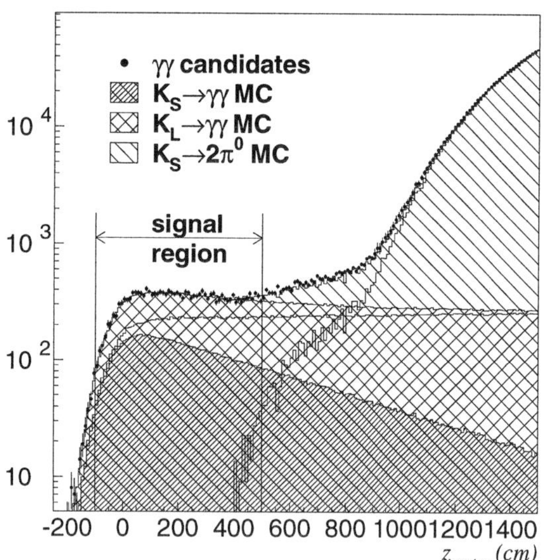

Figure 21. Decay vertex distribution for the $K\gamma\gamma$ candidates measured by NA48/1. Limiting the measurement to a few meter long decay region keeps the background from $K_{\rm S} \to 2\pi^0$ with missing photons in check. The irreducible component due to $K_{\rm L}$ can be precisely subtracted.

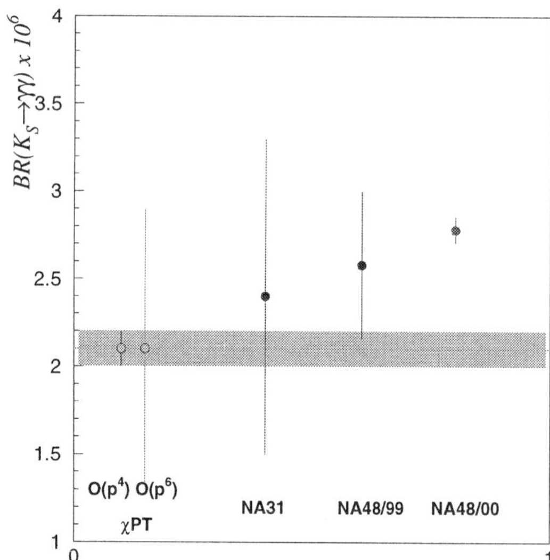

Figure 22. Comparison of the NA48/1 measurement of $BR(K_{\rm S} \to \gamma\gamma)$ with respect to previous measurements and the theoretical predictions.

described by an amplitude dominated by a pseudo-scalar meson pole. In χPT this pole is dominated by π^0 contribution, and the lowest order amplitude is non-vanishing, in contrast to the similar $K_{\rm L} \to \gamma\gamma$ decay. The theoretical prediction for the branching ratio is 3.8×10^{-8} with higher order corrections expected to be small[66] and is quoted in the kinematic region $z = m^2_{\gamma\gamma}/m^2_K > 0.2$ which is free from the overwhelming $K_{\rm S} \to \pi^0\pi^0$ background. A measurement of the branching ratio can provide information about the presence of extra non-pole contributions studied for example by Bijnens et al..[67] In addition, the momentum dependence of the weak vertex which is predicted by χPT can be tested by the measured shape of the z spectrum. The lowest previously published limit on the branching ratio is $BR(K_{\rm S} \to \pi^0\gamma\gamma)_{z>0.2} < 3.3 \times 10^{-7}$ at the 90% confidence level.[68] The main difficulty in analyzing the NA48 data from 2000 was the lack of tracking chambers and the consequent difficulty to veto $\pi^0 \to e^+e^-\gamma$ decays with a missing particle. In Fig. 23 the data is shown with the sum of the background contributions and the excess due to $K_{\rm S} \to \pi^0\gamma\gamma$.

The result, for $z_q = q^2/M^2_k > 0.2$ is:

$$BR(K_{\rm S} \to \pi^0\gamma\gamma) = (4.9 \pm 1.6(stat) \pm 0.9(syst)) \times 10^{-8} \tag{36}$$

in agreement with the theoretical prediction[66] of 3.8×10^{-8}.

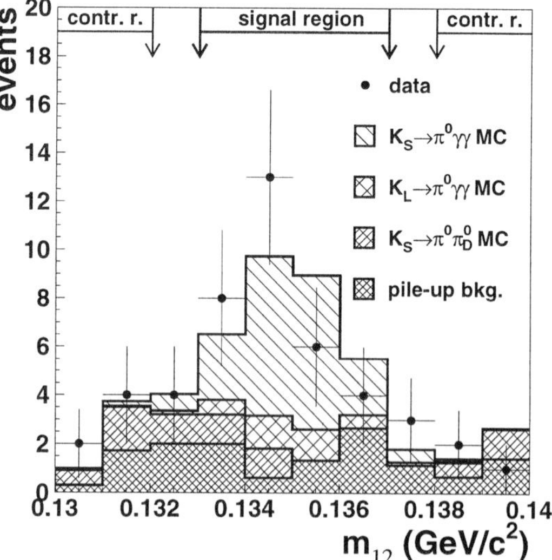

Figure 23. The invariant mass of the two photons forming a π^0. The data is compared with the sum of the predicted background and the $K_{\rm S} \to \pi^0\gamma\gamma$ excess.

6. The Future

New kaon experimental initiatives are planned and significant progress is expected soon. The NA48/2 experiment[69] at CERN is collecting simultaneous K^+ and K^- decays to search for direct CP-violation in the Dalitz plot slope asymmetry. They will also address the study of $\pi\pi$ s-wave scattering by studying a large sample of $Ke4$ decays. In doing so, several rare and medium-rare kaon decays will be thoroughly studied. Fig. 24 shows a sample of approximately 2800 $K^\pm \to \pi^\pm e^+ e^-$ collected by NA48/2 in just one month of data taking. Given that $BR(K^\pm \to \pi^\pm e^+ e^-) \simeq 3 \times 10^{-7}$, the figure clearly demonstrates the capabilities of the experiment.

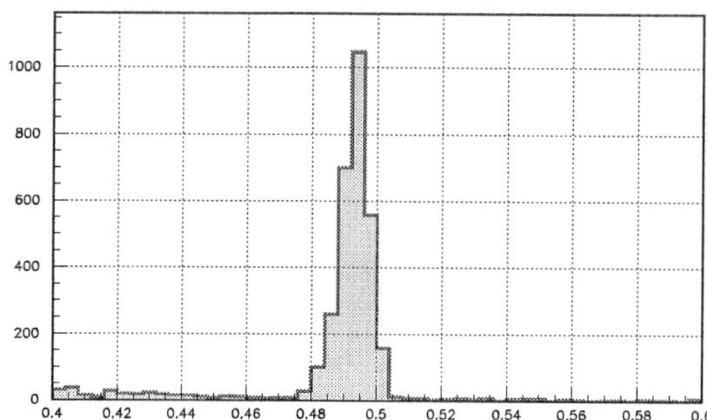

Figure 24. This $K^\pm \to \pi^\pm e^+ e^-$ sample (preliminary) was collected by NA48/2 in about one month. It corresponds to about 1/4 of the world statistics.

Similar sensitivities on charged kaon decays are expected by the OKA experiment at Protvino. CKM at FNAL has proposed[70] to collect 100 $K^+ \to \pi^+ \nu\bar{\nu}$ from decays in flight. The principle of the experiment is to make redundant measurements employing both magnetic spectrometers and tracking RICH's. The experiment is designed to work with a separated kaon beam of 22 GeV/c momentum. The beam design employs superconducting RF cavities of which a full prototype has already been built. Other challenges of CKM are the operation of straw tubes in vacuum and the capability to veto photons with very high efficiency. Both these detector aspects have been successfully prototyped by the Collaboration. The experimental situation on $K_L \to \pi^0 \nu\bar{\nu}$ is poised to improve soon. In 2004 E391a[71] will be the first dedicated experiment to search for $K_L \to \pi^0 \nu\bar{\nu}$. The experiment, which exploits a very well defined neutral pencil beam at the KEK PS and relies on the hermeticity of its veto system, aims to reach a SES of about 3×10^{-10}. It is a pilot experiment for further initiatives at J-PARC. KOPIO[72] at BNL aims to collect about 60 $K_L \to \pi^0 \nu\bar{\nu}$ SM events with a signal-to-background ratio of about 2. The design is based on the principle of measuring as much as possible of the incoming K_L and the outgoing π^0. A micro-bunched proton beam from the AGS coupled with the use of low energy K_L determines the kaon energy by time of flight, allowing one to work in the rest frame of the kaon. Energy, impact position, and direction for each of the two photons from the π^0 decay will be measured.

I wish to conclude this article by stressing the complementarity between the direct searches performed at the energy frontier (at the Tevatron now and later at the LHC) and the precision measurements involving rare virtual processes. I believe that the interest for rare processes will remain or even increase once new phenomenology appears at the colliders. For example, the Minimal Supersymmetric Standard Model allows 43 CP-violating phases.[73] Limiting the parameter space by means of experimental information will remain fundamental. Interesting scenarios on the interplay between new physics and rare kaon decays are described by Grossman in these proceedings.[74]

Acknowledgments

I wish to thank the organizers of Lepton Photon 2003 for the opportunity to present this review at this excellent meeting. I also wish to thank the CERN DTP service for carefully editing the article. It is a pleasure to acknowledge the help of many colleagues who have contributed to this work: most of the credit goes to the members of the collaborations mentioned in this work. In particular, I owe a lot to my NA48 colleagues. Last but not least, a special mention goes to the phenomenologists who have kept my interest in kaon physics alive and kicking.

References

1. N. Cabibbo, *Phys. Rev. Lett.* **10**, 531 (1963).
2. M. Kobayashi and T. Maskawa, *Prog. Theor. Phys.* **49**, 652 (1973).
3. C. Jarlskog, *Z. Phys.* C **29**, 491 (1985).
4. L. Wolfenstein, *Phys. Rev. Lett.* **51**, 1945 (1983).
5. L. S. Littenberg, *Phys. Rev.* D **39**, 3322 (1989).
6. L. Littenberg, arXiv:hep-ex/0010048.

7. A. Alavi-Harati *et al.* [The E799-II/KTeV Collaboration], *Phys. Rev.* D **61**, 072006 (2000) [arXiv:hep-ex/9907014].

8. Y. Grossman and Y. Nir, *Phys. Lett.* B **398**, 163 (1997) [arXiv:hep-ph/9701313].

9. G. Buchalla, G. D'Ambrosio and G. Isidori, arXiv:hep-ph/0308008.

10. M. Battaglia *et al.*, arXiv:hep-ph/0304132.

11. L. M. Sehgal, *Nucl. Phys.* B **19**, 445 (1970).

12. G. Ecker, A. Pich and E. de Rafael, *Nucl. Phys.* B **291**, 692 (1987).

13. C. Bruno and J. Prades, *Z. Phys.* C **57**, 585 (1993) [arXiv:hep-ph/9209231].

14. F. Gabbiani and G. Valencia, *Phys. Rev.* D **66**, 074006 (2002) [arXiv:hep-ph/0207189].

15. A. Lai *et al.* [NA48 Collaboration], *Phys. Lett.* B **536**, 229 (2002) [arXiv:hep-ex/0205010].

16. A. Lai *et al.*, *Phys. Lett.* B **551**, 7 (2003) [arXiv:hep-ex/0210053].

17. A. Alavi-Harati *et al.* [KTeV Collaboration], *Phys. Rev.* D **67**, 012005 (2003) [arXiv:hep-ex/0208007].

18. A. Alavi-Harati *et al.* [KTeV Collaboration], *Phys. Rev. Lett.* **86**, 397 (2001) [arXiv:hep-ex/0009030].

19. A. Alavi-Harati *et al.* [KTEV Collaboration], *Phys. Rev. Lett.* **84**, 5279 (2000) [arXiv:hep-ex/0001006].

20. A. Alavi-Harati *et al.* [KTeV Collaboration], arXiv:hep-ex/0309072.

21. H. B. Greenlee, *Phys. Rev.* D **42**, 3724 (1990).

22. A. Lai *et al.* [NA48 Collaboration], *Eur. Phys. J.* C **22**, 231 (2001) [arXiv:hep-ex/0110019].

23. J. R. Batley *et al.* [NA48 Collaboration], *Phys. Lett.* B **544**, 97 (2002) [arXiv:hep-ex/0208009].

24. J. R. Batley *et al.* [NA48/1 Collaboration], *Phys. Lett.* B **576**, 43 (2003) [arXiv:hep-ex/0309075].

25. G. D'Ambrosio, G. Isidori and J. Portoles, *Phys. Lett.* B **423**, 385 (1998) [arXiv:hep-ph/9708326].

26. A. Belyaev *et al.*, "Kaon physics with a high-intensity proton driver," arXiv:hep-ph/0107046.

27. G. Buchalla and A. J. Buras, *Nucl. Phys.* B **548**, 309 (1999) [arXiv:hep-ph/9901288].

28. M. Misiak and J. Urban, *Phys. Lett.* B **451**, 161 (1999) [arXiv:hep-ph/9901278].

29. S. H. Kettell, L. G. Landsberg and H. H. Nguyen, arXiv:hep-ph/0212321.

30. S. Adler *et al.* [E787 Collaboration], *Phys. Rev. Lett.* **88**, 041803 (2002) [arXiv:hep-ex/0111091].

31. S. Adler *et al.* [E787 Collaboration], *Phys. Lett.* B **537**, 211 (2002) [arXiv:hep-ex/0201037].

32. Bipul Bhuyan, BNL Particle Physics Seminar, July 24, 2003, `http://bnlku28.phy.bnl.gov/bhuyan/talk_bnl/html/talk_bnl.html`

33. L. M. Sehgal and M. Wanninger, *Phys. Rev.* D **46**, 1035 (1992) [Erratum ibid. D **46**, 5209 (1992)].

34. P. Heiliger and L. M. Sehgal, *Phys. Rev.* D **48**, 4146 (1993) [Erratum Phys. Rev. D **60**, 079902 (1999)].

35. A. Alav-Harati *et al.* [KTeV Collaboration], *Phys. Rev. Lett.* **84**, 408 (2000) [arXiv:hep-ex/9908020].

36. A. Golossanov, "CP Violation Measurement using $K_L \to \pi + \pi^- e^+ e^-$ Events Observed by the KTeV (E799 II) Experiment at Fermilab", DPF2002 Meeting, `http://dpf2002.velopers.net/talks_pdf/172talk.pdf`

37. A. Lai *et al.* [NA48 Collaboration], *Eur. Phys. J.* C **30**, 33 (2003).

38. K. Hagiwara *et al.* [Particle Data Group Collaboration], *Phys. Rev.* D **66**, 010001 (2002).

39. I. I. Bigi and A. I. Sanda, *Phys. Lett.* B **466**, 33 (1999).

40. Y. Zou *et al.*, *Phys. Lett.* B **329**, 519 (1994).

41. R. Adler *et al.* [CPLEAR Collaboration], *Phys. Lett.* B **407**, 193 (1997).

42. V. V. Barmin *et al.*, *Phys. Lett.* B **128**, 129 (1983).

43. A. Angelopoulos *et al.* [CPLEAR Collaboration], *Phys. Lett.* B **425**, 391 (1998).

44. M. N. Achasov *et al.*, *Phys. Lett.* B **459**, 674 (1999) [arXiv:hep-ex/9907004].

45. A. Ceccucci, *Nucl. Instrum. Meth.* A **461**, 10 (2001).

46. R. Carosi *et al.* [NA31 Collaboration], *Phys. Lett.* B **237**, 303 (1990).

47. J. S. Bell and J. Steinberger, "Weak Interactions of Kaons," Proc. Oxford International Conference on Elementary Particle Physics (RAL, Chilton, 1966), pp. 195–208 and 221.

48. A. Angelopoulos *et al.* [CPLEAR Collaboration], *Phys. Lett.* B **444**, 52 (1998).

49. Klaus Schubert, these proceedings.

50. A. Aloisio *et al.* [KLOE Collaboration], *Phys. Lett.* B **535**, 37 (2002) [arXiv:hep-ph/0203232].

51. L. Passalacqua, La Thuile 2003, "Recent results from KLOE", `http://www.pi.infn.it/lathuile/2003/talks/Passalacqua.pdf`

52. See for example, I. Mikulec, "CP violation and rare decays in K sector at NA48", `http://www.pi.infn.it/lathuile/2003/talks/Mikulec.pdf`

53. A. Angelopoulos *et al.* [CPLEAR Collaboration], *Phys. Rep.* **374**, 165 (2003).

54. D. Gomez Dumm and A. Pich, *Nucl. Phys. Proc. Suppl.* **74**, 186 (1999) [arXiv:hep-ph/9810523].

55. G. Valencia, *Nucl. Phys.* B **517**, 339 (1998) [arXiv:hep-ph/9711377].

56. L. Bergstrom, E. Masso and P. Singer, *Phys. Lett.* B **131**, 229 (1983).

57. J. LaDue, "Understanding Dalitz Decays of the K_L, in particular the decays of $K_L \to e^+ e^- \gamma$ and $K_L \to e^+ e^- e^+ e^-$", PhD thesis, University of Colorado - Boulder, May 2003.

58. V. Fanti *et al.* [NA48 Collaboration], *Z. Phys.* C **76**, 653 (1997).

59. A. Alavi-Harati *et al.* [KTeV Collaboration], *Phys. Rev. Lett.* **87**, 071801 (2001).

60. T. Barker, "Frontiers of Rarity, Searching for Unusual Kaon Decays at Fermilab", Fermilab seminar, 20 June 2003.

61. A. Alavi-Harati *et al.* [KTeV Collaboration], *Phys. Rev. Lett.* **90**, 141801 (2003) [arXiv:hep-

ex/0212002].

62. J. Kambor and B.R. Holstein, Phys. Rev. D **49**, 2346 (1994).

63. A. Alavi-Harati *et al.* [KTeV Collaboration], *Phys. Rev. Lett.* **83**, 917 (1999) [arXiv:hep-ex/9902029].

64. M. Adinolfi *et al.* [KLOE Collaboration], *Phys. Lett.* B **566**, 61 (2003) [arXiv:hep-ex/0305035].

65. A. Lai *et al.* [NA48 Collaboration], arXiv:hep-ex/0309022.

66. G. Ecker, A. Pich and E. de Rafael, *Phys. Lett.* B **189** 363 (1987).

67. J. Bijnens, E. Pallante and J. Prades, hep-ph/9801326

68. A. Lai *et al.*, *Phys. Lett.* B **556** 105 (2003).

69. R. Batley *et al.*, CERN-SPSC-2000-003.

70. P. S. Cooper [CKM Collaboration], *Nucl. Phys. Proc. Suppl.* **99B**, 121 (2001).

71. T. Inagaki *et al.*, KEK Internal Report 96-13, November 1996.

72. I-H. Chiang *et al.*, "KOPIO–a search for $K_L \rightarrow \pi^0 \nu \bar{\nu}$", in RSVP proposal to NSF (October 1999).

73. J. R. Ellis, *Nucl. Phys. Proc. Suppl.* **99B**, 331 (2001) [arXiv:hep-ph/0011396].

74. Yuval Grossman, these proceedings.

DISCUSSION

Ikaros Bigi (University of Notre Dame):

There is a KEK experiment analyzing the muon transverse polarization in $K\mu3$ decays. What is the status of their analysis and what are their plans?

Augusto Ceccucci: Transverse muon polarization is a sensitive probe of non-standard CP-violation. I have not reviewed the KEK-E246 experiment because it is beyond the scope of a rare kaon decay talk. However, I am not aware of new results from that experiment since the KAON2001 meeting.

Robert Tschirhart (Fermilab): How is a limit placed on $\bar{\eta}$ from $K_{\mathrm{L}} \to \pi^0 e^+ e^-$ limit and $K_{\mathrm{S}} \to \pi^0 e^+ e^-$ measurement? What is assumed about the relative phase between K_{S} and K_{L}?

Augusto Ceccucci: Since the $K_{\mathrm{L}} \to \pi^0 e^+ e^-$ limit is much larger than the indirect CP-violation implied by the K_{S} measurement, the $K_{\mathrm{L}} \to \pi^0 e^+ e^-$ dominates the limit on $\bar{\eta}$. In the figure I presented, which was prepared by Gino Isidori, constructive interference is assumed.

BEYOND THE STANDARD MODEL WITH B AND K PHYSICS

Y. GROSSMAN

Department of Physics, Technion–Israel Institute of Technology, Technion City, 32000 Haifa, Israel *

and

Stanford Linear Accelerator Center, Stanford University, Stanford, CA 94309

and

Santa Cruz Institute for Particle Physics, University of California, Santa Cruz, CA 95064

E-mail: yuval@slac.stanford.edu

In the first part of the talk the flavor physics input to models beyond the Standard Model is described. One specific example of such a new physics model is given: a model with bulk fermions in one non-factorizable extra dimension. In the second part of the talk we discuss several observables that are sensitive to new physics. We explain what type of new physics can produce deviations from the Standard Model predictions in each of these observables.

1. Introduction

The success of the Standard Model (SM) can be seen as a proof that it is an effective low energy description of Nature. Yet, there are many reasons to believe that the SM has to be extended. A partial list includes the hierarchy problem, the strong CP problem, baryogenesis, gauge coupling unification, the flavor puzzle, neutrino masses, and gravity. We are therefore interested in probing the more fundamental theory. One way to go is to search for new particles that can be produced in yet unreached energies. Another way to look for new physics is to search for indirect effects of heavy unknown particles. In this talk we explain how flavor physics is used to probe such indirect signals of physics beyond the SM.

2. New Physics and Flavor

In general, flavor bounds provide strong constraints on new physics models. This fact is called "the new physics flavor problem". The problem is actually the mismatch between the new physics scale that is required in order to solve the hierarchy problem and the one that is needed in order to satisfy the experimental bounds from flavor physics.[1] Here we explain what is the new physics flavor problem, discuss ways to solve it and give one example of a model with interesting, viable, flavor structure.

---

*Permanent address.

2.1. *The New Physics Flavor Problem*

In order to understand what is the new physics flavor problem let us first recall the hierarchy problem.[2] In order to prevent the Higgs mass from getting a large radiative correction, new physics must appear at a scale that is a loop factor above the weak scale

$$\Lambda \lesssim 4\pi m_W \sim 1 \text{ TeV}. \tag{1}$$

Here, and in what follows, Λ represents the new physics scale. Note that such TeV new physics can be directly probed in collider searches.

While the SM scalar sector is unnatural, its flavor sector is impressively successful.[a] This success is linked to the fact that the SM flavor structure is special. First, the charged current interactions are universal. (In the mass basis, this is manifest through the unitarity of the CKM matrix.) Second, Flavor-Changing-Neutral-Currents (FCNCs) are highly suppressed: they are absent at the tree level and at the one loop level they are further suppressed by the GIM mechanism. These special features are important in order to explain the observed pattern of weak decays. Thus, any extension of the SM must conserve these successful features.

Consider a generic new physics model, that is, a model where the only suppression of FCNCs processes is due to the large masses of the particles that

---

[a]The flavor structure of the SM is interesting since the quark masses and mixing angles exhibit hierarchy. These hierarchies are not explained within the SM, and this fact is usually called "the SM flavor puzzle". This puzzle is different from the new physics flavor problem that we are discussing here.

mediate them. Naturally, these masses are of the order of the new physics scale, Λ. Flavor physics, in particular measurements of meson mixing and CP-violation, put severe constraints on Λ.

In order to find these bounds we take an effective field theory approach. At the weak scale we write all the non-renormalizable operators that are consistent with the gauge symmetry of the SM. In particular, flavor-changing four Fermi operators of the form (the Dirac structure is suppressed)

$$\frac{q_1 \bar{q}_2 q_3 \bar{q}_4}{\Lambda^2}, \qquad (2)$$

are allowed. Here q_i can be any quark flavor as long as the electric charges of the four fields in Eq. (2) sum up to zero.[b] The strongest bounds are obtained from meson mixing and CP-violation measurements:

- K physics: $K - \overline{K}$ mixing and CP-violation in K decays imply

$$\frac{\bar{s}d\bar{s}d}{\Lambda^2} \quad \Rightarrow \quad \Lambda \gtrsim 10^4 \text{ TeV.} \qquad (3)$$

- D physics: $D - \overline{D}$ mixing implies

$$\frac{\bar{c}u\bar{c}u}{\Lambda^2} \quad \Rightarrow \quad \Lambda \gtrsim 10^3 \text{ TeV.} \qquad (4)$$

- B physics: $B - \overline{B}$ mixing and CP-violation in B decays imply

$$\frac{\bar{b}d\bar{b}d}{\Lambda^2} \quad \Rightarrow \quad \Lambda \gtrsim 10^3 \text{ TeV.} \qquad (5)$$

Note that the bound from kaon data is the strongest.

There is tension between the new physics scale that is required in order to solve the hierarchy problem, Eq. (1), and the one that is needed in order not to contradict the flavor bounds, Eqs. (3)–(5). The hierarchy problem can be solved with new physics at a scale $\Lambda \sim 1$ TeV. Flavor bounds, on the other hand, require $\Lambda > 10^4$ TeV. This tension implies that any TeV scale new physics cannot have a generic flavor structure. This is the new physics flavor problem.

Flavor physics has been mainly an input to model building, not an output. The flavor predictions of any new physics model are not a consequence of its generic structure but rather of the special structure that is imposed to satisfy the severe existing flavor bounds.

---

[b]We emphasize that there is no exact symmetry that can forbid such operators. This is in contrast to operators that violate baryon or lepton number that can be eliminated by imposing symmetries like $U(1)_{B-L}$ or R-parity.

2.2. Dealing with Flavor

Any viable TeV new physics model has to solve the new physics flavor problem. We now describe several ways to do so that have been used in various models.

(i) Minimal Flavor Violation (MFV) models.[3] In such models the new physics is flavor blind. That is, the only source of flavor violation are the Yukawa couplings. This is not to say that flavor violation arises only from W-exchange diagrams via the CKM matrix elements. Other flavor contributions exist, but they are related to the Yukawa interactions. Examples of such models are gauge mediated Supersymmetry breaking models[4] and models with universal extra dimensions.[5] In general, MFV models predict small effects in flavor physics.

(ii) Models with flavor suppression mainly in the first two generations. The hierarchy problem is connected mainly to the third generation since its couplings to the Higgs field are the largest. Flavor bounds, however, are most severe in processes that involve only the first two generations. Therefore, one way to ameliorate the new physics flavor problem is to keep the effective scale of the new physics in the third generation low, while having the effective new physics of the first two generations at a higher scale. Examples of such models include Supersymmetric models with the first two generations of quarks heavy[6] and Randall-Sundrum models with bulk quarks.[7,8] In general, such models predict large effects in the B and B_s systems, and smaller effects in K and D mixings and decays.

(iii) Flavor suppression mainly in the up sector. Since the flavor bounds are stronger in the down sector, one way to go in order to avoid them is to have new flavor physics mainly in the up sector. Examples of such models are Supersymmetric models with alignment[9] and models with discrete symmetries.[10] In general such models predict large effects in charm physics and small effects in B, B_s and K mixings and decays.

(iv) Generic flavor suppression. In many models some mechanism that suppresses flavor violation for all the quarks is implemented. Examples of such models are Supersymmetric models with spontaneously broken flavor symmetry[11] and models of split fermions in flat extra dimension.[12] In general, such models can be tested with flavor physics.

2.3. An Example: Bulk Quarks in the Randall-Sundrum Model

As discussed above, there are various models that solve the new physics flavor problem in different ways. Here we give one concrete example: the Randall-Sundrum model with bulk quarks[7,8] which belongs to the class of models that treat the third generation differently than the first two. Thus in this model relatively large effects are expected in the B and B_s systems.

The Randall-Sundrum (RS) model solves the hierarchy problem using extra dimensions with non-factorizable geometry. Non-factorizable geometry means that the four-dimensional metric depends on the coordinates of the extra dimensions.[13] In the simplest scenario one considers a single extra dimension, taken to be a S^1/Z_2 orbifold parameterized by a coordinate $y = r_c \phi$, with r_c the radius of the compact dimension, $-\pi \le \phi \le \pi$, and the points (x, ϕ) and $(x, -\phi)$ identified. There are two 3-branes located at the orbifold fixed points: a "visible" brane at $\phi = \pi$ containing the SM Higgs field, and a "hidden" brane at $\phi = 0$. The solution of Einstein's equations for this geometry leads to the non-factorizable metric

$$ds^2 = e^{-2kr_c|\phi|}\, \eta_{\mu\nu}\, dx^\mu dx^\nu - r_c^2\, d\phi^2 \,, \qquad (6)$$

where x^μ are the coordinates on the four-dimensional surfaces of constant ϕ, and the parameter k is of order the fundamental Planck scale M. (This solution can only be trusted if $k < M$, so the bulk curvature is small compared with the fundamental Planck scale.) The two 3-branes carry vacuum energies tuned such that $V_{\rm vis} = -V_{\rm hid} = -24M^3k$, which is required to obtain a solution respecting four-dimensional Poincaré invariance. In between the two branes is a slice of AdS$_5$ space.

With this setup any mass parameter m_0 in the fundamental theory is promoted into an effective mass parameter which depends on the location in the extra dimension, $m = e^{-ky} m_0$. For $y = r_c\pi$ and with $kr_c \approx 12$ this mechanism produces weak scale physical masses at the visible brane from fundamental masses and couplings of order of the Planck scale.

The SM flavor puzzle can be solved by incorporating bulk fermions in the RS model.[14] Then there are several sources for new contributions to FCNC processes. One of these new sources are non-renormalizable operators which appear with scale of

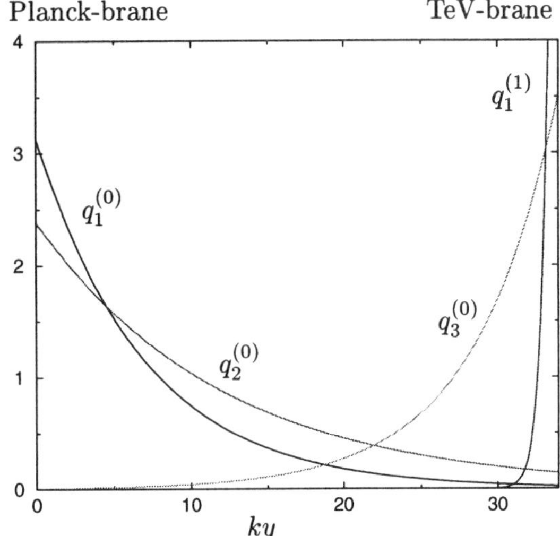

Figure 1. An example of the shape of the fermion field wave functions in the RS model.[8] $q_i^{(0)}$ are the zero mode wave function of the ith generation quark doublet ($i = 1, 2, 3$ where $i = 1$ is the lightest generation). It can be seen that the third generation doublet is localized toward the visible brane while the first two generation doublets are localized toward the hidden brane. This is the reason that the effective scale of the new physics is smaller for the third generation.

order

$$\Lambda \sim M \exp(-ky^f), \qquad (7)$$

where y^f is the "localization" point of the fermion f. In order to reproduce the observed quark masses and mixing angles,[7,8] heavy fermions need to have larger y^f, as can be seen in Fig. 1. Thus, small effects are expected in kaon mixing and decays and large flavor violation effects are expected in b physics.

3. Probing New Physics with Flavor

Any TeV new physics model has to deal with the flavor bounds. Depending on the mechanism that is used to deal with flavor, the prediction of where deviation from the SM can be expected varies. It is important, however, that in many cases large effects are expected. Thus, we hope that we will be able to find such signals.

Generally, it is easier to search for new physics effects where they are relatively large. Namely, in processes that are suppressed in the SM, in particular in:

- meson mixing,

- loop mediated decays, and

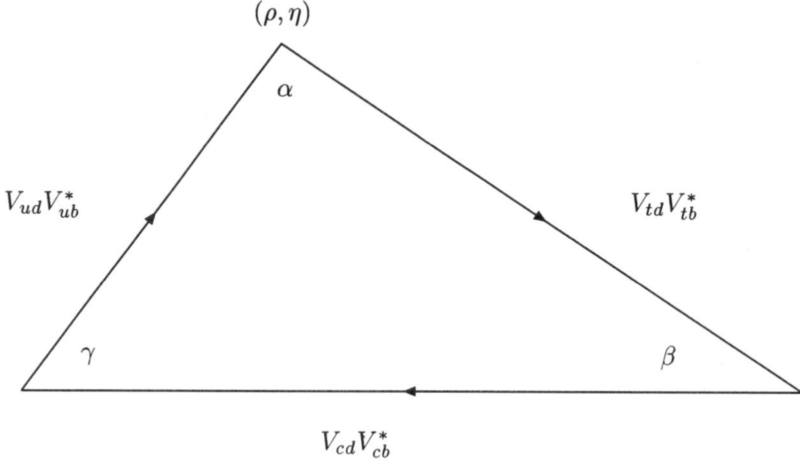

Figure 2. The unitarity triangle.

- CKM suppressed amplitudes.

It is indeed a major part in the B factories' program to study such processes. Below we give several examples for ways to search for new physics.

Before proceeding we emphasize the following point: *at present there is no significant deviation from the SM predictions in the flavor sector.* In the following we give examples of deviations from the SM predictions that are below the 3σ level. In particular, we choose the following possible tests of the SM:

- global fit,

- $a_{\rm CP}(B \to \psi K_S)$ vs $a_{\rm CP}(B \to \phi K_S)$,

- $B \to K\pi$ decays,

- polarization in $B \to VV$ decays, and

- $K \to \pi\nu\bar{\nu}$ vs B and B_s mixing.

There are many more possible tests. Our choice of examples here is partially biased toward cases where the present experimental ranges deviate by more than one standard deviation from the SM predictions. While, as emphasized above, one should not consider these as significant indications for new physics, it should be interesting to follow future improvements in these measurements. Furthermore, it is an instructive exercise to think what one would learn if the central value of these measurements turn out to be correct. As we will see, this would not only indicate new physics, but actually probe the nature of the new physics.

3.1. *Global Fit*

One way to test the SM is to make many measurements that determine the sides and angles of the unitarity triangle (see Fig. 2), namely, to over-constrain it.[15] Another way to put it is that one tries to measure ρ and η in many possible ways. (λ, A, ρ and η are the Wolfenstein parameters.) We emphasize that this is not the only way to look for new physics. It is just one among many possible ways to look for new physics.

The global fit is done using measurements of (or bounds on) $|V_{cb}|$, $|V_{ub}/V_{cb}|$, ε_K, $B - \overline{B}$ mixing, B_s mixing, and $a_{\rm CP}(B \to \psi K_S)$. The fit is very good, as can be seen in Fig. 3. Clearly, there is no indication for new physics from the global fit. There are many more measurements that at present have very little impact on the fit. In the future, such measurements can be included, and then discrepancies may show up.

3.2. *CP-Asymmetries in $b \to s\bar{s}s$ Modes*

The time dependent CP-asymmetry in B decays into a CP eigenstate, f_{CP}, is given by[16]

$$a_{\rm CP}(B \to f_{CP}) \equiv$$

$$\frac{\Gamma(\overline{B}(t) \to f_{CP}) - \Gamma(B(t) \to f_{CP})}{\Gamma(\overline{B}(t) \to f_{CP}) + \Gamma(B(t) \to f_{CP})} =$$

$$-\frac{(1 - |\lambda|^2)\cos(\Delta m_B t) - 2\mathcal{I}m\lambda\sin(\Delta m_B t)}{1 + |\lambda|^2} \equiv$$

$$S\sin(\Delta m_B t) - C\cos(\Delta m_B t). \tag{8}$$

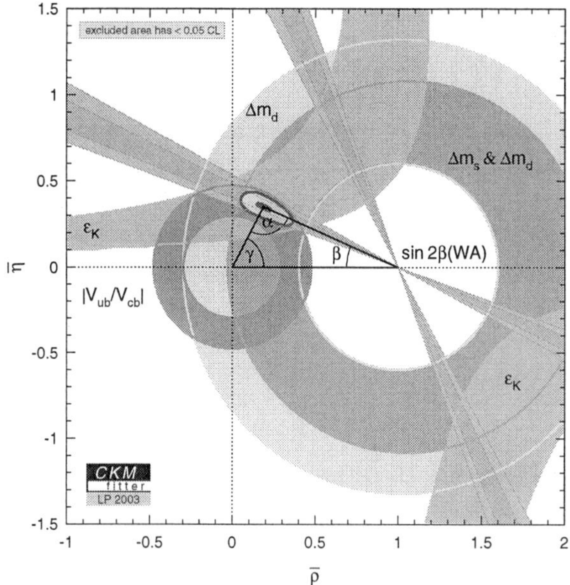

Figure 3. Global fit to the unitarity triangle.[15] The fit is based on the measurements of $|V_{cb}|$, $|V_{ub}/V_{cb}|$, ε_K, $B - \overline{B}$ mixing, and $a_{\rm CP}(B \to \psi K_S)$ and the bound on B_s mixing.

Here $\Delta m_B \equiv m_H - m_L$ and the last line defines S and C. Furthermore,

$$\lambda \equiv \left(\frac{q}{p}\right)\left(\frac{\bar{A}}{A}\right), \tag{9}$$

where $\bar{A} \equiv A(\overline{B} \to f_{CP})$ and $A \equiv A(B \to f_{CP})$. The neutral B meson mass eigenstates are defined in terms of flavor eigenstates as

$$|B_{L,H}\rangle = p|B\rangle \pm q|\overline{B}\rangle. \tag{10}$$

In the $|\lambda| = 1$ limit, which is a very good approximation in many cases, Eq. (8) reduces to the simple form

$$a_{\rm CP}(B \to f_{CP}) = \mathcal{I}m\lambda \sin(\Delta m_B\, t). \tag{11}$$

In that case $\mathcal{I}m\lambda$ is just the sine of the phase between the mixing amplitude and twice the decay amplitude.

In the SM the mixing amplitude is[c]

$$\arg(A_{mix}) = 2\beta. \tag{12}$$

The phase of the decay amplitude depends on the decay mode. $B \to \psi K_S$ is mediated by the tree level quark decay $b \to c\bar{c}s$ which has a real amplitude, namely,

$$\arg(A_{b \to c\bar{c}s}) = 0, \tag{13}$$

___

[c]Here, and in what follows, we use the standard parameterization of the CKM matrix. The results, of course, do not depend on the parameterization we choose.

and therefore $\mathcal{I}m\lambda = \sin 2\beta$. The penguin $b \to s\bar{s}s$ decay amplitude is also real to a good approximation, namely,

$$\arg(A_{b \to s\bar{s}s}) = 0. \tag{14}$$

We learn that also in that case $\mathcal{I}m\lambda = \sin 2\beta$. In particular, the $B \to \phi K_S$, $B \to \eta' K_S$, and $B \to K^+ K^- K_S$ are examples of decays that are dominated by the $b \to s\bar{s}s$ transition. They are of particular interest since their CP-asymmetries have been measured. We conclude that to first approximation the SM predicts

$$S_{\psi K_S} = -S_{K^+ K^- K_S} = S_{\phi K_S} = S_{\eta' K_S}. \tag{15}$$

The theoretical uncertainties in the above predictions are less than $O(1\%)$ for $S_{\psi K_S}$, and of $O(5\%)$ for $S_{\phi K_S}$ and $S_{\eta' K_S}$ and $O(20\%)$ for $S_{K^+ K^- K_S}$.[17] Furthermore, for all these modes the SM predicts $|S| = \sin 2\beta$. Note that in order to violate the predictions of Eq. (15), new physics has to affect the decay amplitudes. New physics in the mixing amplitude shifts all the modes by the same amount, leaving Eq. (15) unaffected.

The data do not show a clear picture yet. Using the most recent results,[18] the world averages of the asymmetries are[d]

$$S_{\psi K_S} = +0.73 \pm 0.05,$$
$$S_{\eta' K_S} = +0.27 \pm 0.21,$$
$$S_{\phi K_S} = -0.15 \pm 0.70,$$
$$-S_{K^+ K^- K_S} = +0.51 \pm 0.26^{+0.18}_{-0.00}. \tag{16}$$

In particular, both $S_{\phi K_S}$ and $S_{\eta' K_S}$ are more then one standard deviation away from $S_{\psi K_S}$. (Since the theoretical errors on $S_{K^+ K^- K_S}$ are large and due to the brief nature of this talk, we do not discuss this mode any further.)

Assuming that these anomalies are confirmed in the future, we ask what can explain them. We have to look for new physics that can generate $S_{\psi K_S} \neq S_{\phi K_S} \neq S_{\eta' K_S}$. Since $B \to \eta' K_S$ and $B \to \phi K_S$ are one loop processes in the SM, we expect new physics to generate large effects in the CP-asymmetries measured in these modes. Moreover, we expect the shift from $\sin 2\beta$ to be different in the two modes since

___

[d]We use the PDG prescription of inflating the errors when combining measurements that are in disagreement.[19] Simply combining the errors there is one change in (16), $S_{\phi K_S} = -0.15 \pm 0.33$.

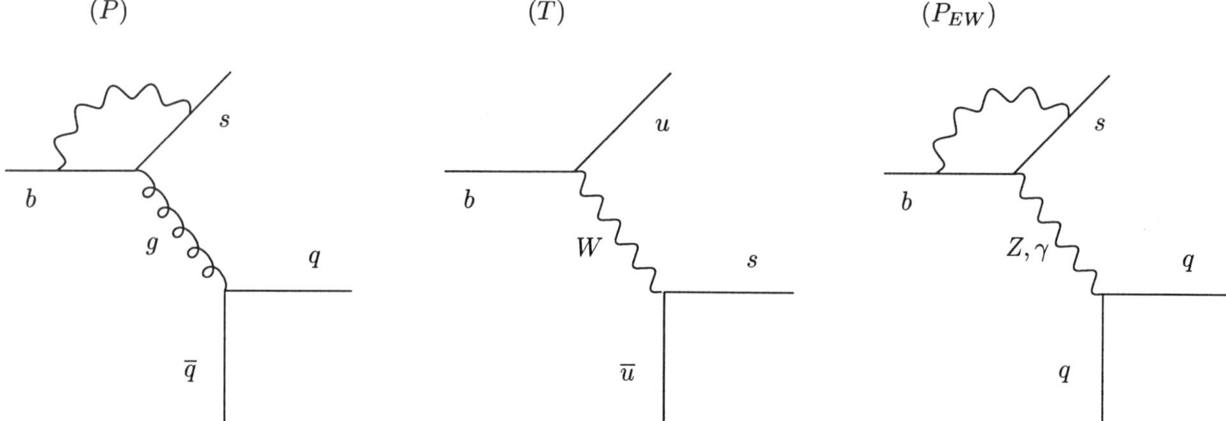

Figure 4. The $B \to K\pi$ amplitudes. The dominant one is the strong penguin amplitude (P), and the sub-dominant ones are the tree amplitude (T) and the electroweak penguin amplitude (P_{EW}).

the ratio of the SM and new physics hadronic matrix elements is in general different. On the contrary, $B \to \psi K_S$ is a CKM favored tree level decay in the SM and thus we do not expect new physics to have significant effects. We conclude that new physics in the $b \to s\bar{s}s$ decay amplitude generally gives $S_{\psi K_S} \neq S_{\phi K_S} \neq S_{\eta' K_S}$.[20]

It is interesting to ask what we would learn if it turns out that $S_{\psi K_S} \neq S_{\phi K_S}$ but $S_{\eta' K_S}$ is consistent with $S_{\psi K_S}$. Such a situation can be the result of new parity conserving penguin diagrams.[21,22] To understand this point note that $B \to \phi K_S$ is parity conserving while $B \to \eta' K_S$ is parity violating. Thus, parity conserving new physics in $b \to s$ penguins only affects $B \to \phi K_S$. While generically new physics models are not parity conserving, there are models that are approximately parity conserving. Supersymmetric $SU(2)_L \times SU(R) \times$ Parity models provide an example of such an approximate parity conserving new physics framework.[21,22]

3.3. $B \to K\pi$

Consider the four $B \to K\pi$ decays and the underlying quark transitions that mediate them:

$$
\begin{aligned}
B^+ &\to K^0\pi^+ & b &\to d\bar{d}s, \\
B^+ &\to K^+\pi^0 & b &\to d\bar{d}s \quad \text{or} \quad b \to u\bar{u}s, \\
B^0 &\to K^+\pi^- & b &\to u\bar{u}s, \\
B^0 &\to K^0\pi^0 & b &\to d\bar{d}s \quad \text{or} \quad b \to u\bar{u}s.
\end{aligned}
\tag{17}
$$

In the SM these modes can be used to measure γ. Moreover, there are many SM relations between these modes that can be used to look for new physics.[23]

There are three main types of diagrams that contribute to these decays. The strong penguin diagram (P), the tree diagram (T) and the EW penguin diagram (P_{EW}); see Fig. 4. It is important to understand the relative magnitudes of these amplitudes. Due to the ratio between the strong and electroweak coupling constants, $P \gg P_{EW}$. The relation between P and T is not as simple. On the one hand, P is a loop amplitude while T is a tree amplitude. On the other hand, the CKM factors in T are $O(\lambda^2) \sim 0.05$ smaller than in P. Thus, it is not clear which amplitude is dominant. Experimentally, it turns out that $P \gg T$. Thus, to first approximation all the four decay rates in Eq. (17) are mediated by the strong penguin amplitude and therefore have the same rate (up to Clebsch-Gordon coefficients). Yet, there are corrections to this expectation due to the sub-leading T and P_{EW} amplitudes.

Due to the hierarchy of amplitudes, there are many approximate relations between the four $B \to K\pi$ decay modes. Let us consider one particular relation, called the Lipkin sum rule.[24] As we explain below the Lipkin sum rule is interesting since the correction to the pure P limit is only second order in the small amplitudes.

The crucial ingredient that is used in order to get useful relations is isospin. Penguin diagrams are pure $\Delta I = 0$ amplitudes, while T and P_{EW} have both $\Delta I = 0$ and $\Delta I = 1$ parts. The Lipkin sum rule, which is based only on isospin, reads[24]

$$
R_L \equiv \frac{2\Gamma(B^+ \to K^+\pi^0) + 2\Gamma(B^0 \to K^0\pi^0)}{\Gamma(B^+ \to K^0\pi^+) + \Gamma(B^0 \to K^+\pi^-)}
$$

$$= 1 + O\left(\frac{P_{EW} + T}{P}\right)^2. \tag{18}$$

Experimentally the ratio was found to be[25]

$$R_L = 1.24 \pm 0.10. \tag{19}$$

Using theoretical estimates[26] that

$$\frac{P_{EW}}{P} \sim \frac{T}{P} \sim 0.1, \tag{20}$$

we expect

$$R_L = 1 + O(10^{-2}). \tag{21}$$

We learn that the observed deviation of R_L from 1 is an $O(2\sigma)$ effect.

What can explain $R_L - 1 \gg 10^{-2}$? First, note that any new $\Delta I = 0$ amplitude cannot significantly modify the Lipkin sum rule since it modifies only P. From the measurement of the four $B \to K\pi$ decay rates we roughly know the value of P. This tells us that new physics cannot modify P in a significant way. What is needed in order to explain $R_L - 1 \gg 10^{-2}$ are new "Trojan penguins", P_{NP}, which are isospin breaking ($\Delta I = 1$) amplitudes. They modify the Lipkin sum rule as follows

$$R_L = 1 + O\left(\frac{P_{NP}}{P}\right)^2. \tag{22}$$

In order to reproduce the observed central value a large effect is needed, $P_{NP} \approx P/2$.[27] In many models there are strong bounds on P_{NP} from $b \to s\ell^+\ell^-$. Leptophobic Z' is an example of a viable model that can accommodate significant Trojan penguins amplitude.[28]

3.4. *Polarization in $B \to VV$ Decays*

Consider B decays into light vectors, in particular,

$$B \to \rho\rho, \qquad B \to \phi K^*, \qquad B \to \rho K^*. \tag{23}$$

Due to the left-handed nature of the weak interaction, in the $m_B \to \infty$ limit we expect[22,29]

$$\frac{R_T}{R_0} = O\left(\frac{1}{m_B^2}\right), \qquad \frac{R_\perp}{R_\parallel} = 1 + O\left(\frac{1}{m_B}\right) \tag{24}$$

where R_0 (R_T, $R_\perp$, $R_\parallel$) is the longitudinal (transverse, perpendicular, parallel) polarization fraction. Recall that $R_T = R_\perp + R_\parallel$ and $R_0 + R_T = 1$.

To understand the above power counting consider for simplicity the pure penguin $\dot{B} \to \phi K^*$ decays. It is convenient to work in the helicity basis

($\mathcal{A}_-$, $\mathcal{A}_+$ and $\mathcal{A}_0$), which is related to the transversity basis via

$$\mathcal{A}_{\parallel,\perp} = \frac{\mathcal{A}_+ \pm \mathcal{A}_-}{\sqrt{2}}, \tag{25}$$

and the longitudinal amplitude is the same in the two bases. We consider the factorizable helicity amplitudes, namely, those contributions which can be written in terms of products of decay constants and form factors. In the SM they are proportional to

$$\mathcal{A}_0 \propto \frac{f_\phi m_B^3}{m_{K^*}}\left[\left(1 + \frac{m_{K^*}}{m_B}\right)A_1 - \left(1 - \frac{m_{K^*}}{m_B}\right)A_2\right] \tag{26}$$

$$\mathcal{A}_\pm \propto f_\phi m_\phi m_B \left[\left(1 + \frac{m_{K^*}}{m_B}\right)A_1 \pm \left(1 - \frac{m_{K^*}}{m_B}\right)V\right],$$

where terms of order $1/m_B^2$ were neglected. The $A_{1,2}$ and V are the $B \to K^*$ form factors, which are all equal in the $m_B \to \infty$ limit.[30] Thus, to leading-order in α_s[31]

$$\frac{A_2}{A_1} \sim \frac{V}{A_1} = 1 + O\left(\frac{1}{m_B}\right). \tag{27}$$

Using Eqs. (26) and (27) we see that the helicity amplitudes exhibit the following hierarchy[22,29]

$$\frac{\mathcal{A}_+}{\mathcal{A}_0} \sim O\left(\frac{1}{m_B}\right), \qquad \frac{\mathcal{A}_-}{\mathcal{A}_0} \sim O\left(\frac{1}{m_B^2}\right). \tag{28}$$

Using Eq. (25) the relations in Eq. (24) immediately follow.

An intuitive understanding of these relations can be obtained by considering the helicities of the $q\bar{q}$ pair that make the vector meson. In the valence quark approximation, when they are both right-handed (left-handed) the vector meson has positive (negative) helicity. When they have opposite helicities the vector meson is longitudinally polarized. In the $m_B \to \infty$ limit the light quarks are ultra relativistic and their helicities are determined by the chiralities of the weak decay operators. Since the weak interaction involves only left-handed b decays, the three outgoing light fermions do not have the same helicities. For example, the leading operator generates decays of the form

$$\bar{b} \to \bar{s}_R s_L \bar{s}_R. \tag{29}$$

(The spectator quark does not have preferred helicity.) Since the ϕ is made from an s quark and an $\bar{s}$ antiquark, in this limit it has longitudinal helicity. For finite m_B each helicity flip reduces the amplitude by a factor of $1/m_B$. To get positive helicities one spin flip, that of the s quark, is required. To get negative helicities, spin flips of the two antiquarks are needed.

The relations in Eq. (24) receive factorizable as well as non-factorizable corrections. Some of these corrections have been calculated, with the result that they do not significantly modify the leading-order results.[29] Still, in order to get a clearer picture, more accurate determinations of the corrections are needed.

Observation of $R_\perp \gg R_\parallel$ would signal the presence of right-handed chirality effective operators in B decays.[21,22] The hierarchy between $\mathcal{A}_+$ and $\mathcal{A}_-$ generated by the opposite chirality operator, $\tilde{Q}_i$, (obtained from Q_i via a parity transformation) is flipped compared to the hierarchy generated by the SM operator. Such right-handed chirality operators lead to an enhancement of R_T and therefore can also upset the first relation in Eq. (24).

The polarization data are as follows.[25] The longitudinal fraction has been measured in several modes

$$R_0(B^0 \to \phi K^{*0}) = 0.58 \pm 0.10,$$
$$R_0(B^+ \to \phi K^{*+}) = 0.46 \pm 0.12,$$
$$R_0(B^+ \to \rho^0 K^{*+}) = 0.96 \pm 0.16,$$
$$R_0(B^+ \to \rho^+ \rho^0) = 0.96 \pm 0.07,$$
$$R_0(B^0 \to \rho^+ \rho^-) = 0.99 \pm 0.08. \tag{30}$$

There is only one measurement of the perpendicular polarization[32]

$$R_\perp(B^0 \to \phi K^{*0}) = 0.41 \pm 0.11. \tag{31}$$

Using $R_0 + R_\perp + R_\parallel = 1$ we extract

$$R_\parallel(B^0 \to \phi K^{*0}) = 0.01 \pm 0.15. \tag{32}$$

We see that in $B \to \rho\rho$ and $B \to K^*\rho$ the SM prediction $R_T/R_0 \ll 1$ is confirmed, although $R_T/R_0 \gg 1/m_B^2$ remains a possibility. Since in these modes R_T is very small, the second SM prediction, $R_\perp \approx R_\parallel$, cannot be tested yet.

The situation is different in $B \to \phi K^*$. First, the data favor $R_T/R_0 = O(1)$, which is not a small number. Second, one also finds that $R_\perp/R_\parallel \gg 1$. Both of these results are in disagreement with the SM predictions in Eq. (24).

It is interesting that the preliminary data indicate that the SM predictions do not hold in $B \to \phi K^*$. This is a pure penguin $b \to s\bar{s}s$ decay. The decays where the SM predictions appear to hold,

$B \to K^*\rho$ and particularly $B \to \rho\rho$, on the other hand, have significant tree contributions. It is thus important to obtain polarization measurements in other modes, especially the pure penguin $b \to s\bar{d}d$ decay $B^+ \to K^{*0}\rho^+$.

With more precise polarization data it may therefore be possible to determine whether or not there are new right-handed currents, and if so whether or not they are only present in $b \to s\bar{s}s$ decays.

3.5. $K \to \pi\nu\bar{\nu}$

The $K \to \pi\nu\bar{\nu}$ decays are very good probes of the unitarity triangle.[33] They are dominated by the $s \to d$ electroweak penguin amplitude with internal top quark which is proportional to $|V_{td}|$. Isospin and perturbative QCD can be used to eliminate almost all the hadronic uncertainties. One more point that makes these modes attractive is that in many cases new physics affects B decays and K decays differently.[34] Then, the apparent determination of the unitarity triangle from these different sources will be different.

Experimentally, there is only a measurement of the decay rate of the charged mode[35]

$$\mathcal{B}(K^+ \to \pi^+\nu\bar{\nu}) = (15.7^{+17.5}_{-8.2}) \times 10^{-11}. \tag{33}$$

The SM prediction is[33]

$$\mathcal{B}^{\mathrm{SM}}(K^+ \to \pi^+\nu\bar{\nu}) = 4.4 \times 10^{-11} \times \left[\eta^2 + (1.4 - \rho)^2\right]. \tag{34}$$

Using the preferred values for ρ and η (see Fig. 2), $\rho \sim 0.15$ and $\eta \sim 0.4$, the central value for the SM prediction is[36]

$$\mathcal{B}^{\mathrm{SM}}(K^+ \to \pi^+\nu\bar{\nu}) \approx 7.7 \times 10^{-11}. \tag{35}$$

We learn that the measurement [Eq. (33)] is in agreement with the SM prediction [Eq. (35)].

It is interesting to ask what one will learn if it turns out that the SM prediction is not confirmed by the data. Let us assume that in the future the measurement of $\mathcal{B}(K^+ \to \pi^+\nu\bar{\nu})$ will converge around its current central value. Inspecting Eq. (34) we learn that in order to get $\mathcal{B}(K^+ \to \pi^+\nu\bar{\nu}) = 15.7 \times 10^{-11}$ we need large η ($\eta \sim 2$) or negative ρ. These possibilities are in conflict with the current global fit of the unitarity triangle; see Fig. 5. Large η is in conflict with the measurement of $|V_{ub}|$. Since $|V_{ub}|$

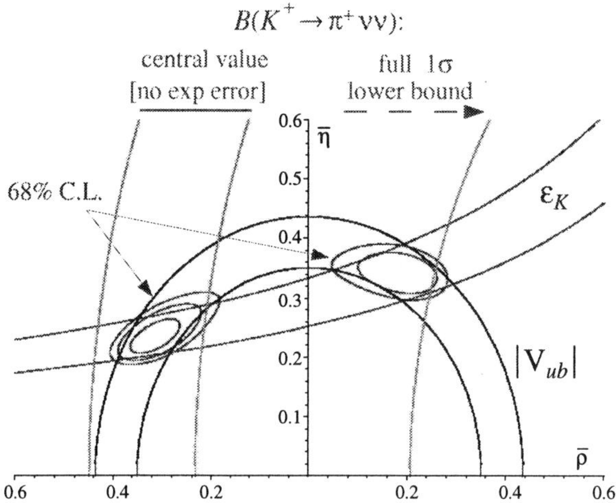

Figure 5. Global fit to the unitarity triangle with the measurement of $\mathcal{B}(K^+ \to \pi^+\nu\bar{\nu})$.[36] It can be seen that the central value of the measurement is inconsistent with the unitarity triangle extracted from the measurement of $B - \bar{B}$ mixing and the bound on $B_s - \bar{B}_s$ mixing.

is extracted from tree level processes, its determination is unlikely to be affected by new physics. On the contrary, $\rho < 0$ is in conflict with the measurement of $B - \bar{B}$ mixing and the bound on $B_s - \bar{B}_s$ mixing. These are loop processes, and can be modified in the presence of new physics. We conclude that new physics in $K^+ \to \pi^+\nu\bar{\nu}$ or $B - \bar{B}$ mixing or $B_s - \bar{B}_s$ mixing can generate such a disagreement.

Higher precision in the measurement of $\mathcal{B}(K^+ \to \pi^+\nu\bar{\nu})$ and a measurement of $\mathcal{B}(K_L \to \pi^0\nu\bar{\nu})$ are important in order to further explore this avenue for searching for new physics.

4. Conclusions

The main goal of high energy physics is to find the theory that extends the SM into shorter distances. Flavor physics is a very good tool for such a mission. Depending on the mechanism for suppressing flavor-changing processes, different patterns of deviation from the SM are expected to be found. In some cases almost no deviations are expected, while in other we expect deviations in specific classes of processes. While there is no signal for such new physics yet, there are intriguing results. More data is needed in order to look further for fundamental physics using low energy flavor-changing processes.

Acknowledgments

I thank Alex Kagan, Yossi Nir, and Martin Schmaltz for helpful comments and discussions. The work of YG is supported in part by a grant from the G.I.F., the German–Israeli Foundation for Scientific Research and Development, by the United States–Israel Binational Science Foundation through grant No. 2000133, by the Israel Science Foundation under grant No. 237/01, by the Department of Energy, contract DE-AC03-76SF00515 and by the Department of Energy under grant No. DE-FG03-92ER40689.

References

1. For a discussion for Supersymmetric models see, for example, Z. Ligeti and Y. Nir, *Nucl. Phys. Proc. Suppl.* **111**, 82 (2002) [hep-ph/0202117]; Y. Grossman, Y. Nir and R. Rattazzi, *Adv. Ser. Direct. High Energy Phys.* **15**, 755 (1998) [hep-ph/9701231].
2. See, for example, M. Schmaltz, *Nucl. Phys. Proc. Suppl.* **117**, 40 (2003) [hep-ph/0210415]; G. F. Giudice, these proceedings.
3. For a review see, for example, A. J. Buras, hep-ph/0307203.
4. For a review see, Y. Shadmi and Y. Shirman, *Rev. Mod. Phys.* **72**, 25 (2000) [hep-th/9907225].
5. Flavor aspects of universal extra dimensions are discussed in A. J. Buras, M. Spranger and A. Weiler, *Nucl. Phys.* B **660**, 225 (2003) [hep-ph/0212143].
6. M. Dine, R. G. Leigh and A. Kagan, *Phys. Rev.* D **48**, 4269 (1993) [hep-ph/9304299]; A. Pomarol and D. Tommasini, *Nucl. Phys.* B **466**, 3 (1996) [hep-ph/9507462]; G. R. Dvali and A. Pomarol, *Phys. Rev. Lett.* **77**, 3728 (1996) [hep-ph/9607383]; A. G. Cohen, D. B. Kaplan and A. E. Nelson, *Phys. Lett.* B **388**, 588 (1996) [hep-ph/9607394].
7. T. Gherghetta and A. Pomarol, *Nucl. Phys.* B **586**, 141 (2000) [hep-ph/0003129]; S. J. Huber and Q. Shafi, *Phys. Lett.* B **498**, 256 (2001) [hep-ph/0010195]; K. Agashe, A. Delgado, M. J. May and R. Sundrum, hep-ph/0308036; G. Burdman, hep-ph/0310144.
8. S. J. Huber, *Nucl. Phys.* B **666**, 269 (2003) [hep-ph/0303183].
9. Y. Nir and N. Seiberg, *Phys. Lett.* B **309**, 337 (1993) [hep-ph/9304307]; M. Leurer, Y. Nir and N. Seiberg, *Nucl. Phys.* B **420**, 468 (1994) [hep-ph/9310320].
10. For one of the early models see S. Pakvasa and H. Sugawara, *Phys. Lett.* B **73**, 61 (1978).
11. M. Leurer, Y. Nir and N. Seiberg, *Nucl. Phys.* B **398**, 319 (1993) [hep-ph/9212278].
12. N. Arkani-Hamed and M. Schmaltz, *Phys. Rev.* D **61**, 033005 (2000) [hep-ph/9903417]; Y. Grossman and G. Perez, *Phys. Rev.* D **67**, 015011 (2003) [hep-ph/0210053].

13. L. Randall and R. Sundrum, *Phys. Rev. Lett.* **83**, 3370 (1999) [hep-ph/9905221].

14. Y. Grossman and M. Neubert, *Phys. Lett.* B **474**, 361 (2000) [hep-ph/9912408].

15. A. Hocker, H. Lacker, S. Laplace and F. Le Diberder, *Eur. Phys. J.* C **21**, 225 (2001) [hep-ph/0104062]. Recent fits can be found in the CKMfitter home page at ckmfitter.in2p3.fr.

16. For a review, notation and formalism, see Y. Nir, *Lectures at XXVII SLAC Summer Institute on Particle Physics*, hep-ph/9911321; G.C. Branco, L. Lavoura and J.P. Silva, "CP violation," *Oxford, UK: Clarendon (1999)*; K. Anikeev *et al.*, hep-ph/0201071.

17. Y. Grossman, G. Isidori and M. P. Worah, *Phys. Rev. D* **58**, 057504 (1998) [hep-ph/9708305]; D. London and A. Soni, *Phys. Lett.* B **407**, 61 (1997) [hep-ph/9704277]; M. Beneke and M. Neubert, *Nucl. Phys.* B **651**, 225 (2003) [hep-ph/0210085]; Y. Grossman, Z. Ligeti, Y. Nir and H. Quinn, *Phys. Rev. D* **68**, 015004 (2003) [hep-ph/0303171]; M. Gronau and J. L. Rosner, *Phys. Lett.* B **564**, 90 (2003) [hep-ph/0304178].

18. T. Browder, these proceedings.

19. K. Hagiwara *et al.*, Particle Data Group, *Phys. Rev. D* **66**, 010001 (2002).

20. See, for example, Y. Grossman and M. P. Worah, *Phys. Lett.* B **395**, 241 (1997) [hep-ph/9612269]; A. Kagan, in proceedings of the 7th International Symposium on Heavy Flavor Physics, Santa Barbara, CA July 1997, hep-ph/9806266; R. Fleischer and T. Mannel, *Phys. Lett.* B **511**, 240 (2001) [hep-ph/0103121]; G. Hiller, *Phys. Rev. D* **66**, 071502 (2002) [hep-ph/0207356]; A. Datta, *Phys. Rev. D* **66**, 071702 (2002) [hep-ph/0208016]; M. Raidal, *Phys. Rev. Lett.* **89**, 231803 (2002) [hep-ph/0208091]; R. Harnik, D. T. Larson, H. Murayama and A. Pierce, hep-ph/0212180; C. W. Chiang and J. L. Rosner, *Phys. Rev. D* **68**, 014007 (2003) [hep-ph/0302094]; G. L. Kane, P. Ko, H. b. Wang, C. Kolda, J. h. Park and L. T. Wang, *Phys. Rev. Lett.* **90**, 141803 (2003) [hep-ph/0304239]; A. K. Giri and R. Mohanta, *Phys. Rev. D* **68**, 014020 (2003) [hep-ph/0306041]; J. F. Cheng, C. S. Huang and X. h. Wu, hep-ph/0306086; R. Arnowitt, B. Dutta and B. Hu, hep-ph/0307152.

21. A. L. Kagan, lecture at SLAC Summer Institute, August 2002, www.slac.stanford.edu/gen/meeting/ssi/2002/kagan1.html#lecture2.

22. A. L. Kagan, talk at first workshop on the discovery potential of an asymmetric B factory at 10^{36} luminosity, May 2003, www.slac.stanford.edu/BFROOT/www/Organization/1036_Study_Group/0303Workshop/index.html

23. For recent reviews, see: R. Fleischer, *Phys. Rept.* **370**, 537 (2002) [hep-ph/0207108]; J. L. Rosner, hep-ph/0304200; M. Gronau, hep-ph/0306308.

24. H. J. Lipkin, hep-ph/9809347; *Phys. Lett.* B **445**, 403 (1999) [hep-ph/9810351]; M. Gronau and J. L. Rosner, *Phys. Rev. D* **59**, 113002 (1999) [hep-ph/9809384].

25. J. Fry, these proceedings.

26. M. Beneke, G. Buchalla, M. Neubert and C. T. Sachrajda, *Nucl. Phys.* B **606**, 245 (2001) [hep-ph/0104110].

27. M. Gronau and J. L. Rosner, hep-ph/0307095; A. J. Buras, R. Fleischer, S. Recksiegel and F. Schwab, hep-ph/0309012.

28. Y. Grossman, M. Neubert and A. L. Kagan, *JHEP* **9910**, 029 (1999) [hep-ph/9909297]; K. Leroux and D. London, *Phys. Lett.* B **526**, 97 (2002) [hep-ph/0111246].

29. A. L. Kagan, in preparation.

30. J. Charles, A. Le Yaouanc, L. Oliver, O. Pene and J. C. Raynal, *Phys. Rev. D* **60**, 014001 (1999) [hep-ph/9812358]; C. W. Bauer, S. Fleming, D. Pirjol and I. W. Stewart, *Phys. Rev. D* **63**, 114020 (2001) [hep-ph/0011336].

31. M. Beneke and T. Feldmann, *Nucl. Phys.* B **592**, 3 (2001) [hep-ph/0008255]; G. Burdman and G. Hiller, *Phys. Rev. D* **63**, 113008 (2001) [hep-ph/0011266].

32. [Belle Collaboration], hep-ex/0307014.

33. G. Buchalla and A. J. Buras, *Phys. Rev. D* **54**, 6782 (1996) [hep-ph/9607447].

34. Y. Grossman and Y. Nir, *Phys. Lett.* B **398**, 163 (1997) [hep-ph/9701313]; Y. Nir and M. P. Worah, *Phys. Lett.* B **423**, 319 (1998) [hep-ph/9711215].

35. S. Adler *et al.* [E787 Collaboration], *Phys. Rev. Lett.* **88**, 041803 (2002) [hep-ex/0111091].

36. G. D'Ambrosio and G. Isidori, *Phys. Lett.* B **530**, 108 (2002) [hep-ph/0112135]; G. Isidori, eConf **C0304052**, WG304 (2003) [hep-ph/0307014].

DISCUSSION

Stephen L. Olsen (Univ. of Hawaii): Don't long-distance effects change the SM predictions for polarization in $B \to VV$?

Yuval Grossman: If m_b tends to infinity, everything is short distance. The full $1/m_b$ corrections have not been calculated yet, but they are naïvely expected to be of the order of Λ_{QCD}/m_b, i.e. about 10%. The more interesting question is why the leading-order predictions in the Standard Model hold in some B decay channels, but not in others. It is hard to believe that long-distance effect can generate such a pattern.

Rajendran Raja (Fermilab): If new physics is found in the K, B, and D sectors, how constraining of the new physics model will it be?

Yuval Grossman: If one measurement is in disagreement with the Standard Model, we won't be able to constrain it to a specific model. However several indications of deviations will be model constraining, because certain models, such as different models of Supersymmetry breaking, predict very specific patterns of deviations from the Standard Model predictions.

RARE HADRONIC B DECAYS

J. R. FRY

Department of Physics, University of Liverpool, PO Box 147 Liverpool, L69 7ZE, England
E-mail: J.R.Fry@liverpool.ac.uk

Recent data from the rare decays of B mesons into hadronic final states is presented from BaBar, Belle, CDF and CLEO. Where possible the data are compared with theoretical calculations, with the twin aims of further testing the Standard Model and searching for evidence of new physics. A brief description is given of some theoretical approaches in order to indicate which decays are the most sensitive for further study.

1. Introduction

Updated branching fractions (BF) and CP-asymmetries (A_{CP}) are presented for rare decays of B_d and B_u mesons into hadronic final states. By rare we typically mean processes having BF of less than 10^{-5}. In the main these rare decay modes are charmless and involve final states in which no charmed quarks are produced. The reason for studying rare decay modes is that the Standard Model is a good approximation to reality at current energies: it gives a very good description of the more common processes in particle interactions, including CP-violation in K^0 and B^0 decays. Thus we need to consider processes where the Standard Model amplitudes are small if we are to be sensitive to new physics. This generally implies decays dominated by (second order) penguin diagrams, or CKM-suppressed decays.

Because of the dependence of the values of BF and A_{CP} on angles of the unitarity triangle in cases where more than one amplitude contributes to the decay process, the study of rare decays gives an alternative route to the measurement of the parameters ρ and η and hence additional constraints on the unitarity triangle. Disagreement between the values of the parameters of the unitarity triangle obtained in this way and those obtained through direct measurement of time-dependent asymmetries could provide an indication for new physics. However, given the difficulties in making theoretical calculations, and the approximations and model-dependent assumptions that are often made, it could also indicate that refinements to our understanding of hadron dynamics are needed. In this situation model-independent calculations are of great value in assessing the difference between experimental measurements and expectation from the Standard Model, even if the constraints they impose are somewhat weaker than those

from QCD-based theories.

1.1. *Direct CP-Violation*

Direct CP-violation is observed when the branching fraction for the decay of a B meson into a particular final state is different from that of its antiparticle into the charge-conjugate final state. It can be measured for both charged and neutral B mesons, although the former is usually easier to do, and gives higher precision, since charged B mesons are self-tagging. It is usual to consider the CP-asymmetry, A_{CP}, which is the difference in branching fractions for charge-conjugate decays divided by the sum, since many acceptance-dependent systematic effects cancel to first order. Direct CP-violation occurs if the decay $B \to f$ (and its charge-conjugate) is mediated by two amplitudes[a] with different strong and weak phases. Writing the decay amplitudes:

$$a_f = a_1 e^{i(\delta_1 + \phi_1)} + a_2 e^{i(\delta_2 + \phi_2)}$$
$$\bar{a}_{\bar{f}} = a_1 e^{i(\delta_1 - \phi_1)} + a_2 e^{i(\delta_2 - \phi_2)}$$

where δ is the (CP-even) strong phase and ϕ the (CP-odd) weak phase, A_{CP} may be written as the difference of the amplitudes-squared divided by the sum:

$$A_{CP} = \frac{|\bar{a}_{\bar{f}}|^2 - |a_f|^2}{|\bar{a}_{\bar{f}}|^2 + |a_f|^2}$$
$$= \frac{2a_1 a_2 sin(\delta_2 - \delta_1) sin(\phi_2 - \phi_1)}{a_1{}^2 + a_2{}^2 + 2a_1 a_2 cos(\delta_2 - \delta_1) cos(\phi_2 - \phi_1)}.$$

If one of the two amplitudes is small compared with the other, then A_{CP} will be small regardless of the values of the weak and strong phases. This is the

---

[a] Using the unitarity relationship $\alpha + \beta + \gamma = \pi$, any number of amplitudes with different strong phases can be written as the sum of two amplitudes with at most two different weak phases.

case for the decay $B \to K\pi$, which is dominated by a penguin diagram, and $B \to \pi^+\pi^0$, which is mediated by tree diagrams. In contrast one would expect a large CP-asymmetry for $B^0 \to \pi^+\pi^-$, unless there is dynamical suppression, since the tree and penguin amplitudes are of comparable size. When a B^0 decays to a self-conjugate final state, like $\pi^+\pi^-$, the value of A_{CP} is simply related to the parameter C describing direct CP-violation in the expression for the time-dependent asymmetry, see Sec. 8.

2. Theoretical Overview

The theoretical problem to be solved is how to calculate the branching fractions and CP-asymmetries for the decay of a B meson to a hadronic final state. For many years the more common two-body and quasi-two-body decays have been understood qualitatively in terms of naïve factorization. Here, the leading quark from the B meson decay is assumed to be in one quark, while the second meson contains the spectator quark. The interaction is calculated using leading-order diagrams only, and the two quarks are assumed to propagate independently of each other. Predictions for BF are made, but without control over, or understanding of, systematics, and all values for A_{CP} are, of course, identically zero. Although useful as a guide to experimental measurements, a major drawback of naïve factorization is its lack of any sound theoretical basis.

2.1. QCD Factorization (QCDF)

Any attempt to calculate BF and A_{CP} from first principles, using QCD, must take into account non-perturbative effects relating mesons to quarks and gluons, higher-order terms resulting from the low energy scale of the interaction, and long-range interactions that are not amenable to a perturbative approach. Such QCD calculations are based on a low-energy effective Hamiltonian written as the sum of generic amplitudes, which are classified as tree-like, penguin-like, electroweak and annihilation. QCD factorization,[1,2] which relies on color transparency and the smallness of the parameter Λ_{QCD} compared with the mass of the B meson, m_B, enables a major simplification of the problem, since the amplitudes of the Hamiltonian factorize to leading order in Λ_{QCD}/m_B and all orders of perturbation theory.

Calculations are done to leading order in Λ_{QCD}/m_B, with non-factorizable corrections calculated to second order in α_S; final-state interactions (FSI) and annihilation contributions are estimated in a model-dependent way. A nice feature of QCDF is that naïve factorization is recovered to leading order in α_S. CP-asymmetries arise naturally from the interference of (leading-order) tree and (second-order) penguin diagrams, with important modifications to the calculated values of branching fractions in some cases.

QCD factorization can be visualized diagrammatically as shown in Fig. 1 where the soft (form-factor and meson-formation amplitudes) and hard-scattering terms factorize. The lower, left-hand diagram represents the two combinations with m and M interchanged. An important feature of QCDF is that interactions between the two-meson systems are dominated by hard gluon exchange, and not soft processes, as shown in the right-hand lower diagram of Fig. 4.

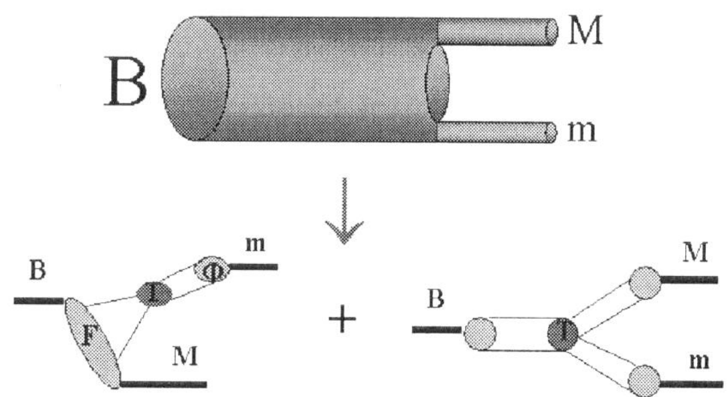

Figure 1. Visualising QCD Factorization in the decay of a B meson to two charmless mesons: M and m. T represents the hard-scattering kernel, F the semi-leptonic form factor and Φ the leading-twist light-cone distribution amplitudes.

One of the important results of QCDF is that the strong phase difference, between tree and penguin diagrams for example, is generally small, and this leads to predictions of small values for A_{CP}. Predictions of BF and A_{CP} are currently made for pseudoscalar-pseudoscalar (PP) and pseudoscalar-vector (PV) mesons. The major thrust of the theory is to calculate the angles α and γ of the unitarity triangle, using a subset of decays where model-dependent effects are well under control. There are significant concerns in applying QCDF to all rare decays, and hence searching for new physics, since it is not clear whether m_B is large enough compared with

Λ_{QCD} for the leading-order expansion to be valid, and whether the model-dependent annihilation terms are small enough to be under control in the calculations.

2.2. *Flavor SU(3) Symmetry*

An alternative approach to the calculation of BF and A_{CP} for rare and unmeasured processes is to use experimental input from selected final states, each having one dominant (hard-scattering) amplitude, to estimate the amplitudes contributing to the rare process. For example the BF for the decay $B \to \pi^+\pi^0$ ($\pi^+\omega$) determines the amplitude for the tree diagram in non-strange PP (PV) final states, while that for $B \to \pi^+K^0$ determines the penguin amplitude for strange PP final states. Using flavor SU(3) symmetry[3,4,5] the tree and penguin amplitudes for strange and non-strange PP and PV final states can then be related to each other, as for example:

$$\left|\frac{p}{p'}\right|^2 = \left|\frac{V_{td}}{V_{ts}}\right|^2 = 0.039; \quad \left|\frac{t}{t'}\right|^2 = \left|\frac{V_{us}}{V_{ud}}\right|^2 \left|\frac{f_K}{f_\pi}\right|^2 = 0.076$$

where p and t represent the penguin and tree amplitudes, respectively, with the primes corresponding to the final state having a strange meson. Electroweak and annihilation contributions, as well as the strong phase difference between dominant hard-scattering diagrams, are then included in such a way as to give the best fit to the more common BF. Such an approach is illustrated in Fig. 2, where the contributions from the soft (non-perturbative) processes are effectively treated as constants subsumed in the measured BF.

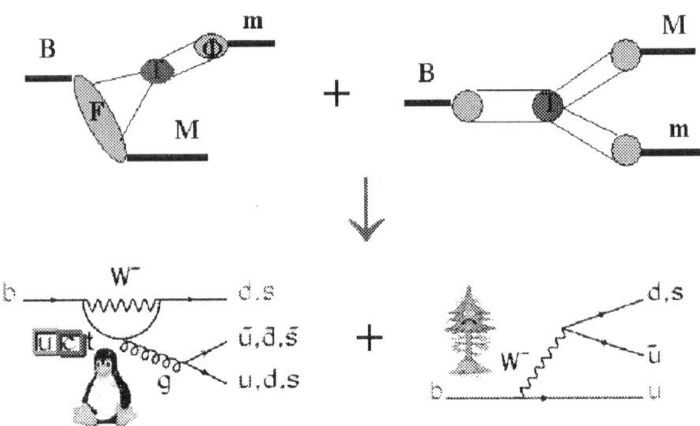

Figure 2. Pictorial representation of phenomenological approaches to B decay such as SU(3) flavor symmetry.

This approach has the enormous merit of classifying and relating measured branching fractions and asymmetries in addition to predicting values for as-yet-unmeasured ones. It gives experimentalists a useful tool in searching for inconsistencies in data and looking for effects that may be due to physics beyond the Standard Model.

2.3. *Model-Independent Calculations*

In the event of any significant disagreement between a measured quantity, confirmed by a second experiment, and a theoretical prediction based on the Standard Model, attention would most likely be focused on the assumptions, approximations and model dependence of the calculations before claiming new physics. It would then be advantageous to consider calculations based only on isospin and SU(3), which are less susceptible to dynamical assumptions. One useful class of such calculations includes the Grossman-Quinn bound[6] and succeeding work,[7] which put limits on possible deviations from simple expectations of the Standard Model calculations of CP-asymmetries.

3. Signal Selection and Background Rejection

When an $\Upsilon(4S)$ is produced in an e^+e^- collision it decays into a pair of B mesons described by a coherent, two-body wave function. At the instant one B decays the other has the opposite flavor. Hence by determining the flavor of one B the other may be tagged, which is essential for the study of neutral final states such as ϕK_s. Tagging and vertex reconstruction are studied using large data samples, where one B meson is a fully reconstructed final state. This minimizes the error from these sources entering the analysis of the small signals from the rare decay processes under study. Two invariant quantities are used to select signal events, m_{ES} the beam energy substituted mass peaking at the B mass, and ΔE the missing energy peaking at zero, defined as:

$$m_{ES} = \sqrt{(E_{beam}^*)^2 - p_B^{*2}}$$
$$\Delta E = E_B^* - E_{beam}^*$$

where p_B and E_B are the momentum and energy of the B meson, E_{beam} the energy of the beam and the asterisk denotes the center-of-mass system. The resolution of m_{ES} is dominated by that of the beam energy and is about 3 MeV for all processes, while the resolution for ΔE depends on the final state but is typically 20 - 30 MeV. Energy resolution is among the many quantities studied by Monte Carlo (MC) simulation, with any small deficiencies in the behaviour of the MC being corrected from the comparison of MC and data for high statistics control samples. Some discrimination against background is given by the dependence of ΔE on the particle types in the final state. Figure 3 shows the experimental data for the control sample $B^0 \to D^-\pi^+$ compared with the two MC distributions $D^-\pi^+$ and D^-K^+, where the misidentification of the π^+ with a K^+ causes a shift in the ΔE distribution.

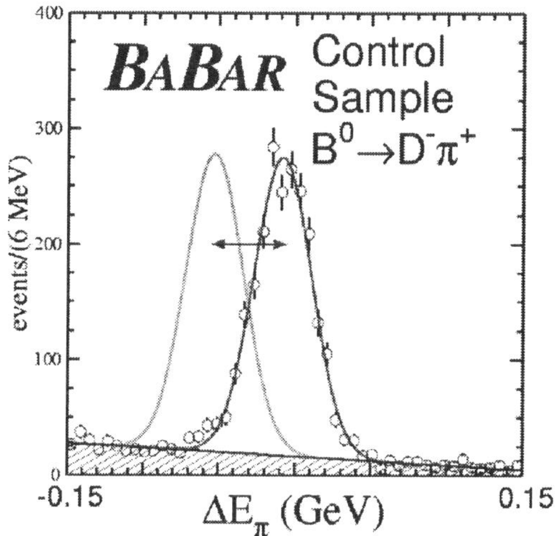

Figure 3. Distribution of ΔE for the control sample $B^0 \to D^-\pi^+$, shown as points with error bars. The curve centred at zero is expected for the correct identification of the bachelor pion, that offset to negative values for the pion misidentified as a kaon.

The dominant source of background in the rare-signal channels arises from the random combinations of particles in continuum events, which happen to satisfy energy and momentum conservation for fake B decays. Since B mesons are produced almost at rest in the center-of-mass they decay rather isotropically, whereas continuum events are produced in narrow, back-to-back jets aligned with the beam axis. Discrimination against background therefore relies on the different angular properties for production and decay of the real and fake B mesons. After preliminary cuts to remove the bulk of the background

with little loss of signal, the angular information for the remaining data sample is combined into a Fisher discriminant, F. The signal and background probability distributions for F, m_{ES} and ΔE are then used in the likelihood fit, together with particle-identification (PID) information, to identify the signal sample. The power of a Čerenkov detector to identify particle types and discriminate signal from background is illustrated in Fig. 4, where a sample of events containing a proton or antiproton is cleanly separated from the rest. BaBar relied on this to set the very small upper limit:[8]

$$BF(B^0 \to \bar{p}p) < 2.7 \times 10^{-7}(90\%\text{C.L.}).$$

The majority of information on baryonic final states has so far been produced by CLEO and Belle,[9] with evidence for the first two-body baryonic B decay presented at EPS[10] by Belle with the measurement:

$$BF(\bar{B}^0 \to \Lambda_c^+ \bar{p}) = (2.19^{+0.56}_{-0.49} \pm 0.32 \pm 0.57) \times 10^{-5}.$$

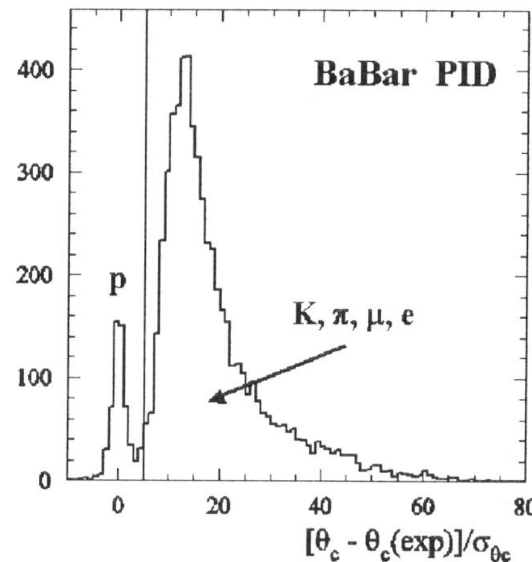

Figure 4. Separation of protons from other particles using angular information from the BaBar DIRC Čerenkov detector. The distribution shows the number of events as a function of the number of standard deviations by which the measured Čerenkov angle differs from that expected for a proton.

4. BF and A_{CP} for $K\pi$, $\pi\pi$ and KK

A summary of branching fractions for $K\pi$, $\pi\pi$ and KK final states[11,9] is given in Table 1. Of particular note is the measurement of $B^0 \to \pi^0\pi^0$. From a sample of 140 fb^{-1} (1.52×10^8 $B\bar{B}$) Belle have a signal of 26 ± 9 events,[12] with a statistical significance of 3.4σ, while BaBar have a signal of $46 \pm 13 \pm 3$ events[13] from 113 fb^{-1} and quote an overall significance of 4.2σ. Also noteworthy is the significant

Table 1. Summary of branching fractions in units of 10^{-6} for $K\pi$, $\pi\pi$ and KK final states.

Mode	BaBar	Belle	CLEO	Average
$K^+\pi^-$	17.9 ± 1.1	18.5 ± 1.2	18.0 ± 2.6	18.2 ± 0.8
$K^0\pi^+$	22.3 ± 2.0	22.0 ± 2.2	18.8 ± 4.3	21.8 ± 1.4
$K^+\pi^0$	12.8 ± 1.6	12.8 ± 1.8	12.9 ± 2.7	12.8 ± 1.1
$K^0\pi^0$	11.4 ± 1.9	12.6 ± 22.8	12.8 ± 4.3	11.9 ± 1.5
$\pi^+\pi^-$	4.7 ± 0.6	4.4 ± 0.7	4.5 ± 1.5	4.6 ± 0.4
$\pi^+\pi^0$	5.5 ± 1.2	5.3 ± 1.4	4.6 ± 19	5.3 ± 0.8
$\pi^0\pi^0$	2.1 ± 0.7	1.7 ± 0.7	< 4.4	1.9 ± 0.5
K^+K^-	< 0.6	< 0.7	< 0.8	< 0.6
$K^+\bar{K}^0$	< 2.5	< 3.4	< 3.3	< 2.5
$K^0\bar{K}^0$	< 1.8	< 3.2	< 3.3	< 1.8

increase in precision of the measurements since the publication of PDG 2002.[14]

Since the decay of a B to $K\pi$ or $\pi\pi$ usually proceeds through both penguin and tree diagrams, there is a significant dependence of the BF for many of the decay modes on the angle γ of the unitarity triangle. Ratios of BF calculated with QCDF (Fig. 14[1]) were in reasonable agreement with the data in 2001 for a value of γ around 75° and remain so despite the increase in precision of the measured quantities. However, the calculated BF of $(0.2 - 0.5) \times 10^{-6}$ is in disagreement with the measured BF of $(1.9 \pm 0.5) \times 10^{-6}$ for the decay $B^0 \to \pi^0\pi^0$. Arising from a color-suppressed diagram, the $\pi^0\pi^0$ BF is not easily amenable to calculation within the framework of QCDF and the authors claim that this result does not discredit the theory. It is worth noting that pQCD predicts an equally small value for the $\pi^0\pi^0$ BF,[15] whereas the other $K\pi$ and $\pi\pi$ branching fractions agree reasonably with the experimental data.

By writing the magnitude of the amplitude as proportional to the square root of the BF, it is apparent that the isospin relationship for $B \to \pi\pi$

$$\sqrt{2}A(\pi^+\pi^0) - A(\pi^+\pi^-) = \sqrt{2}A(\pi^0\pi^0)$$

is satisfied by the experimental data. This is to be expected, since isospin conservation is good to 1-2% in strong interactions, and the expected contribution from electroweak processes is not expected to be greater than about 2%.[16] Using SU(3) arguments in addition to isospin, the following two ratios of BF are expected to be equal[17] so long as the electroweak

Table 2. Average CP-asymmetries (%) for Belle and BaBar data[11] compared with predictions from pQCD[16] and QCDF.[1]

Mode	A_{CP} (Expt)	A_{CP} (pQCD)	A_{CP} (QCDF)
$K^+\pi^-$	-9 ± 3	-13 ↔ -22	+5 ± 10
$K^0\pi^+$	-1 ± 6	-0.6 ↔ -1.5	0 ± 1
$K^+\pi^0$	0 ± 7	-10 ↔ -17	+7 ± 10
$K^0\pi^0$	3 ± 37		-3 ± 4
$\pi^+\pi^-$		16 ↔ 30	-6 ± 13
$\pi^+\pi^0$	-7 ± 14	0	-2 ± 5
$\pi^0\pi^0$			45 ± 60

penguin contribution can be neglected:

$$R_{Neut} = \frac{BF(B^0 \to K^+\pi^-)}{2BF(B^0 \to K^0\pi^0)} = 0.77 \pm 0.10$$

$$R_{Chg} = \frac{2BF(B^+ \to K^+\pi^0)}{2BF(B^+ \to K^0\pi^+)} = 1.17 \pm 0.13.$$

The difference of (0.40 ± 0.16) is not significant, and it will be interesting to see whether the two ratios change with an increase in data.

A comparison of the theoretical predictions for A_{CP} with data is of some interest, since the asymmetries arise as a consequence of the interference of different contributing amplitudes and are zero at leading order. Table 2 shows that most asymmetries are predicted to be small, and are in agreement with data. As yet there is no claimed observation of a non-zero asymmetry, which would indicate direct CP-violation, but the data for $K^+\pi^-$ is tantalizingly

Table 3. Branching fractions (10^{-6}) for data compared with predictions from pQCD.

	Data	pQCD
K^+K^-	< 0.6	0.05
K^+K^0	< 2.5	1.7
K^0K^0	< 1.8	1.8

close with measurements of:

$$A_{CP} = (-8.8 \pm 3.5 \pm 1.8)\% \quad (\text{Belle}[18])$$
$$A_{CP} = (-10.7 \pm 4.1 \pm 1.2)\% \quad (\text{BaBar}[19]).$$

It is of some interest that pQCD and QCDF predict asymmetries with opposite signs, although the theoretical uncertainties may be too large for this ever to become a significant issue.

CDF has evidence for the decays $B_d \to \pi^+\pi^-$ and $K^+\pi^-$, as well as $B_s \to K^+\pi^-$ and K^+K^-.[20] All four decays populate the same mass window (Fig. 5) and are untangled using a combination of dE/dx and the different division of momentum between the particles in the four final states.

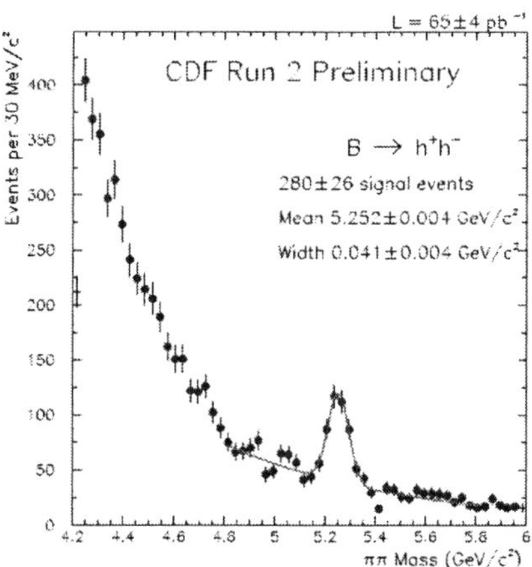

Figure 5. Invariant mass of two charged hadrons, each assumed to be a pion, showing a peak at the B mass in the CDF experiment.[20]

After untangling the samples, the CP-asymmetry for $B_d \to K\pi$ is calculated to be $A_{CP} = (2 \pm 15 \pm 2)\%$, based on 39 ± 14 events, which is in good agreement with results from Belle and BaBar. Rescattering could modify the BF and CP-asymmetries for $K\pi$ and $\pi\pi$ final states, which in turn would complicate the extraction of the unitarity angles γ and α. If it were significant, it might be expected to increase the BF into KK through intermediate $D\bar{D}$ or

Table 4. Branching fractions (10^{-6}) and CP-asymmetries (%) for B decays to pseudoscalar-vector particles for BaBar[21] and Belle.[22]

Mode	BF (BaBar)	BF (Belle)	A_{CP} (BaBar)	A_{CP} (Belle)
$\rho^+\pi^-$	22.6 ± 2.8	29.1 ± 6.4	-11 ± 7	-38 ± 21
ρ^+K^-	7.3 ± 1.8	15.1 ± 4.1	19 ± 18	22 ± 23
$\rho^0\pi^0$	< 2.5	6.0 ± 3.1		
$\rho^+\pi^0$	11.0 ± 2.7		23 ± 17	
$\rho^0\pi^+$	9.3 ± 1.3	8.0 ± 2.3	-17 ± 11	
ρ^0K^+	3.9 ± 3.7	3.9 ± 1.0		
ωK^0	5.3 ± 1.4	4.0 ± 2.0		
ωK^+	5.0 ± 1.1	6.7 ± 1.4	-5 ± 16	6 ± 20
$\omega \pi^0$	< 3	< 1.9		
$\omega \pi^+$	5.4 ± 1.1	5.7 ± 1.5	4 ± 17	48 ± 23

$\pi\pi$ states. Current BF for KK, shown in Table 3, have upper limits consistent with predictions from pQCD, so that there is no evidence for rescattering.

5. BF and A_{CP} for $\rho\pi$, ρK, $\omega\pi$ and ωK

Branching fractions and asymmetries for B decays to $\rho\pi$, ρK, $\omega\pi$ and ωK are shown in Table 4. The recent results for $B^0 \to \rho^0\pi^0$ from Belle and BaBar are consistent, but it is too early to say with confidence what the BF is and whether it is small enough to reduce uncertainties in the extraction of α from the $\rho\pi$ final states. It might be expected from a simple consideration of the contributing diagrams that the BF for ωK would be considerably larger than that for $\omega\pi$; it is not, and we return to this in more detail later.

6. Dalitz Plot Analyses of $K\pi\pi$ and KKK

Understanding the resonance contributions to a three-body final state requires an analysis of the Dalitz plot. With 56.4 fb^{-1} BaBar have made an approximate analysis of the decay $B^+ \to K^+\pi^+\pi^-$ by concentrating on the resonant bands for the $K^*(890)$ and higher-mass K^*, as well as $\rho(770)$, $f_0(980)$ and χ_c. Interference cannot be taken into account in this approach, and to reduce the effects from the domi-

Mode	BaBar	Belle	Average
$K\pi\pi)_{Total}$	59 ± 5	46 ± 5	52.2 ± 3.5
$K\pi\pi)_{NonRes}$	< 17	14 ± 6	
$K^{*0}(890)\pi^+$	15.5 ± 4.4	8.5 ± 1.5	9.0 ± 1.3
$K^{*0}(1400)\pi^+$		40.3 ± 6.5	40.3 ± 6.5
$K^{*0}(1430)\pi^+$		< 10.5	< 10.5
$K^{*0}(1689)\pi^+$		< 21	< 21
$K^+\rho(770)$	3.9 ± 3.7	3.9 ± 1.0	4.1 ± 0.9
$K^+f_0(980)$	9.2 ± 2.9	10.3 ± 2.4	9.9 ± 1.9
$K^+f_2(1270)$		< 6.3	< 6.3
$K^+\chi_c(3400)$	1.46 ± 0.37	1.17 ± 0.42	
$KKK)_{Total}$	29.6 ± 2.6	29.4 ± 2.4	29.5 ± 1.8
$KKK)_{NonRes}$		22.5 ± 4.9	
$K^+\phi(1020)$	10.0 ± 1.0	8.6 ± 1.1	9.0 ± 0.7
$K^+f_2(1525)$		< 12.8	< 12.8
$K^+\chi_c(3400)$		0.85 ± 0.29	

nant D^0, J/Ψ and Ψ' resonances, the overlap region is removed. The branching fractions of the resonant contributions are given in Table 5. Using the larger data sample of 140 fb^{-1}, Belle have made an amplitude analysis of both the $K^+\pi^+\pi^-$ and $K^+K^+K^-$ Dalitz plots, having totals of 2584 and 1400 events, respectively. The background arising from continuum processes is fitted (with high precision) using sideband samples with several times the number of non-signal events than is contained in the Dalitz plot. The background amplitude is parameterized with the following function, which takes into account resonant contributions from $K^*(890)$ and $\rho(770)$ as well as a non-resonant term, which is a function of the invariant mass-squares, s_{ij}, of the two-particle combinations.

$$A_{BG} = \sum_k \alpha_k e^{-\beta s_{ij}} + BW(K^*) + BW(\rho)$$

The signal amplitude is fitted to a sum of resonant (Breit-Wigner) terms and a non-resonant, non-phase-space-like term. For the $K^+\pi^+\pi^-$ analysis the masses and widths are fixed for $K^*(890)$, $K^*(1400)$, $\rho(770)$, $\chi_c(3400)$ and left floating for $f_0(980)$ and a broad distribution described as $X(1350)$; for the $K^+K^+K^-$ analysis the masses and widths are fixed for $\phi(1020)$ and $\chi_c(3400)$ and left floating for the broad distribution described as X(1500). The signal amplitude is described as:

$$A_{Signal} = \sum_R \alpha_R e^{i\delta_R} + \left(\frac{a_1}{s_{12}^{p_1}} e^{\phi_1} + \frac{a_2}{s_{23}^{p_2}} e^{\phi_2} \right)_{NonRes}$$

where the parameters a_R, δ_R, p_1, p_2, ϕ_1 and ϕ_2 are varied to give the best overall fit to the Dalitz plot.

The quality of the fit to the Dalitz plot is appreciated best from the projected fits of invariant mass and helicity for the two-body combinations. Figure 6 shows these for the $K\pi$ and $\pi\pi$ combinations. As expected, the helicity distributions are well described by spin-1 (spin-0) in the vicinity of the $K^*(890)$ ($f_0(980)$) mass regions. The BF for the different contributions to the $K\pi\pi$ and KKK Dalitz plots, together with the total and non-resonant contributions, are given in Table 5.

Despite the excellent fit, the non-phase-space, non-resonant contribution to the Dalitz plot is a cause for concern, since it is not understood. The values of the BF from both BaBar and Belle must therefore be treated with caution at the present time.

7. $B \to VV$ and Longitudinal Polarization

In addition to branching fractions and CP-asymmetry, a measurement of the angular distribution of the two mesons enables further tests of theoretical predictions. Two complementary decompositions can be used: orbital angular momentum states S, P and D, or a longitudinal and two transverse polarization states. In the latter case the longitudinal polarization is a CP-even state while the transverse polarization is mixed, with both CP-odd and CP-even components. On the basis of helicity arguments, assuming short-distance dominance within the framework of perturbative QCD, the longitudinal polarization, f_L, is predicted to be:[25]

$$f_L = 1 - \mathcal{O}(m_V^2/m_B^2)$$

where m_V is the mass of one of the vector mesons. Thus, for decays of a B meson into two mesons of the type ρ, η, ψ or K^* the expectation for f_L is in the region of $95 - 99\%$. There are no simple predictions

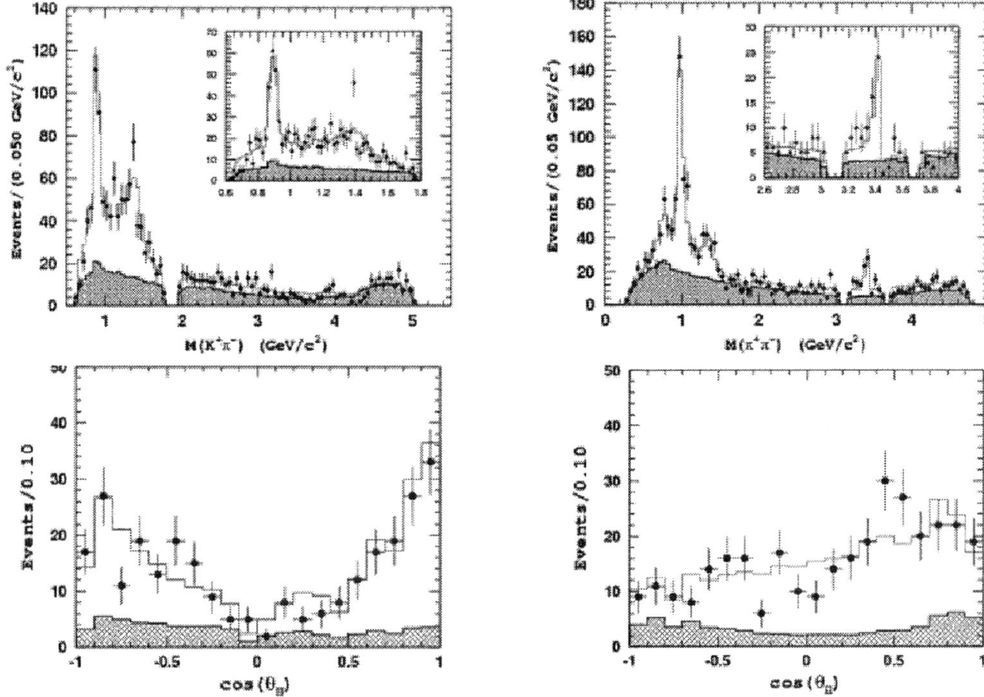

Figure 6. Projections of the $K\pi\pi$ Dalitz plot onto $K\pi$ (left) and $\pi\pi$ (right) showing the mass spectrum (upper) and helicity distribution (lower). The background contribution is shown hatched in all plots. The inset plots show the mass regions around the K^* and χ_c in more detail. The open histogram shows the overall fit to be compared with the data points shown with error bars. The helicity distributions are plotted for a mass interval around the K^* and f_0, respectively, for the $K\pi$ and $\pi\pi$.

for the transverse components, which are expected to be small and will be very difficult to measure with any reasonable precision. The angular distribution used to measure f_L is given by

$$\frac{1}{\Gamma}\frac{d^2\Gamma}{dcos\theta_1 dcos\theta_2} = \frac{1}{4}(1 - f_L)sin^2\theta_1 sin^2\theta_2$$
$$+ f_L cos^2\theta_1 cos^2\theta_2$$

where θ_1 and θ_2 are the helicity angles of the decay products of the two vector mesons. For example, in the decay $B \to \rho^+\rho^-$, then θ_1 would be the angle between the momentum vector of the π^+ and that of the ρ^+, both measured in the rest frame of the ρ^+.

7.1. $B \to \rho\rho$ and ρK^*

Measurements of branching fractions, CP-asymmetries and longitudinal polarizations for the final states $\rho^0\rho^0$, $\rho^+\rho^-$, $\rho^+\rho^0$ and $\rho^0 K^{*+}$ are given in Table 6. A projection of the $\rho^+\rho^-$ onto the π^+ helicity axis is shown in Fig. 7. The structure in the background (dotted curve) under the signal arises from the variation in acceptance.

The observation of $\rho^+\rho^-$ at more than 5σ overall

Table 6. Measurements of branching fraction, CP-asymmetry and longitudinal polarization from BaBar[26] and Belle[27] for $\rho\rho$ and ρK^* final states. All data are from BaBar, alone, except for $\rho^+\rho^0$.

	BF (10^{-6})	A_{CP} (%)	f_L (%)
$\rho^0\rho^0$	< 2.1		
$\rho^+\rho^-$	27 ± 9		99 ± 8
$\rho^+\rho^0$	23 ± 8	-19 ± 23	97 ± 8
Belle	32 ± 10	0 ± 22	95 ± 11
$\rho^0 K^{*+}$	11 ± 4	20 ± 32	96 ± 16

significance, completes the measurement of the $\rho\rho$ final states. All values of longitudinal polarization shown in Table 6 are in agreement with theoretical expectation, and the values of A_{CP} are consistent with zero.

The $\rho\rho$ system is an isospin triplet like the $\pi\pi$ system, and hence the following relationship should hold good:

$$\sqrt{2}A(\rho^+\rho^0) - A(\rho^+\rho^-) = A(\rho^0\rho^0)$$

where A is the amplitude for the decay $B \to \rho\rho$. Using the values from Table 6 gives a value for the LHS of (2.1 ± 1.3), to be compared with < 2.1 for the RHS, where the square root of the BF has been used for each amplitude, and the weighted-average BF for the $\rho^+\rho^0$ was used. Although not inconsistent, the situation is uncomfortable. An optimistic possibility is that the true value for $B(\rho^0\rho^0)$ might turn out to be smaller than presently measured, but it is also possible that there will be adjustments to the BF of either $\rho^+\rho^-$ or $\rho^+\rho^0$, the final states with neutral pions, which are more difficult to distinguish from background.

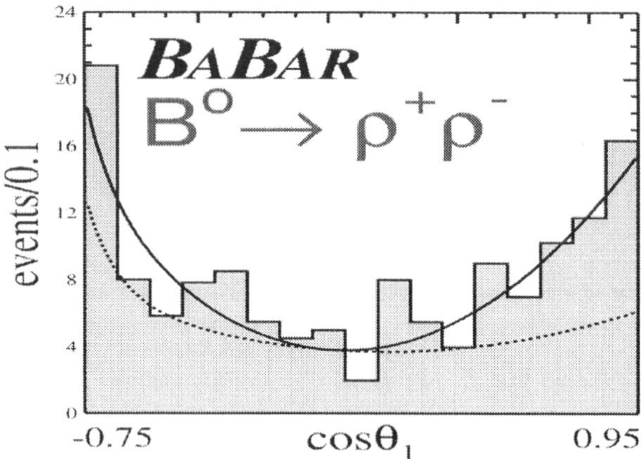

Figure 7. The projected helicity distribution, solid curve, of the π^+ from the fit to the combined angular distribution is compared with the total data sample, histogram. The dotted curve shows the background.

7.2. Bounds on α from $B \to \pi\pi$ and $\rho\rho$

The similarity between the $\pi\pi$ and $\rho\rho$ systems, both consisting of identical bosons of isospin $I = 1$, suggests that one might apply the Grossman-Quinn bound[6] in order to obtain a limit on the difference between the measured unitarity angle, α_{Eff}, and the true angle α. The relationship gives:

$$sin^2(\alpha - \alpha_{eff}) < \frac{BF(B^0 \to \rho^0\rho^0)}{BF(B^0 \to \rho^+\rho^-)} < 0.10$$

leading to a limit on $|\alpha - \alpha_{Eff}|$ of approximately $20°$ at 90% CL,[26] compared with approximately $50°$ for the $\pi\pi$ case. The value is so large for the $\pi\pi$ system as to be of no practical use, whereas the limit on the $\rho\rho$ system – if valid – is of very great interest.

However, the differences between the $\pi\pi$ and $\rho\rho$ systems require some discussion before this bound can be accepted. First, pions are pseudoscalar mesons of definite mass, whereas the ρ is a vector meson with a substantial width. Here, the experimental finding is that the longitudinal polarization is consistent with 100%. This is very important, since it indicates that the $\rho\rho$ system may be described as $\rho_L\rho_L$ - a pure CP-even state, just like the two-pion system. Also, Bose-Einstein statistics would require that a pure $\rho\rho$ system had no contribution from isospin $I = 1$, which is an important condition for the Grossman-Quinn bound to apply. The concern, which is currently receiving a good deal of thought, is to what extent modifications from final-state interactions and the presence of non-resonant background are understood well enough to be taken into account in the analysis.

7.3. Vector and Scalar Couplings

It was mentioned earlier that the branching fractions for ωK and $\omega \pi$ are very similar, with a ratio of 0.9 ± 0.3, whereas on the basis of CKM couplings one would expect the BF for ωK to be much larger than that for $\omega \pi$. Another situation, where simple expectation is confounded, occurs in the comparison of decays to final states $(K^0\pi^+, K^{*0}\pi^+)$ and $(\pi^+\pi^0, \pi^+\rho^0, \rho^+\rho^0)$.

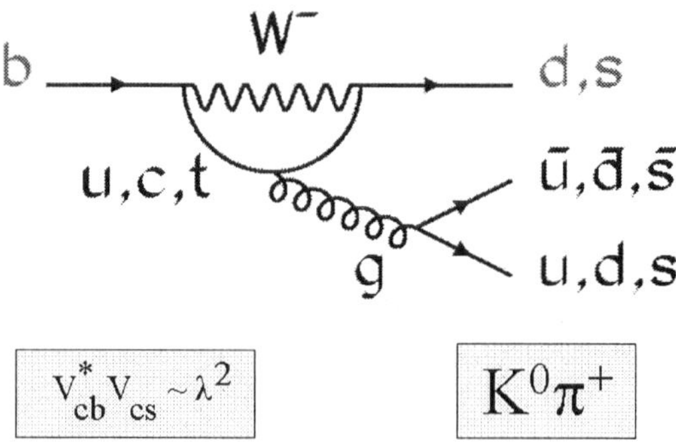

Figure 8. Penguin graph describing $B^+ \to K^0\pi^+$ and $K^{*0}\pi^+$. The dominant contribution is for the c quark coupling to the b and s quarks.

The $K\pi$ decays are dominated by the penguin diagram of Fig. 8, where the neutral kaon includes the leading s quark. Within corrections of 30 - 40%

arising from uncertainties in the form factors, the BF for $K^{*0}\pi^+$ and $K^0\pi^+$ might be expected to be in the ratio of the square of the decay constants, namely 1.85. Experimentally, however, the ratio is very different at 0.65. The decays $\pi^+\pi^0$, $\pi^+\rho^0$ and $\rho^+\rho^0$ are mediated by the tree diagram of Fig. 9, where the leading quark is in the charged meson. Their BFs would be expected to be in the ratio 1: 2.6: 4.4, which agree much better with the experimental ratios of 1: 1.7: 5.0 than those for $K\pi$. It might be thought that the vector boson (W or g) would couple more strongly to a vector meson, and hence an enhancement occur for K^* and ρ production, but the opposite is true for $K^{*0}\pi^+$ and $K^0\pi^+$, while there seems to be little effect for the $\pi^+\pi^0$, $\pi^+\rho^0$ and $\rho^+\rho^0$ decays. Thus, there seems to be no discernible pattern to this behavior.

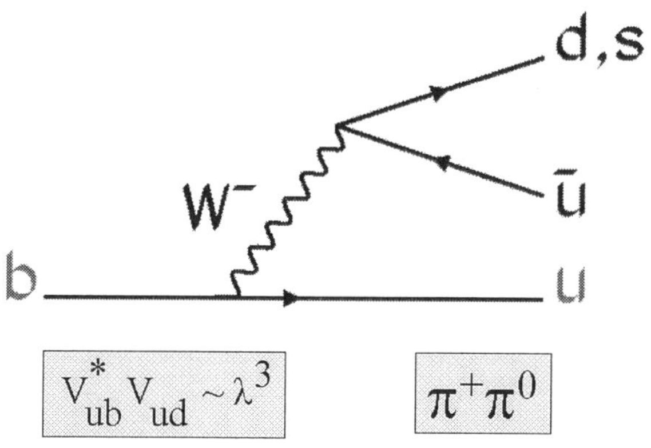

Figure 9. Tree diagram for the decays $B^+ \to \pi^+\pi^0$, $\pi^+\rho^0$ and $\rho^+\rho^0$.

Returning to the consideration of CKM factors, it might be expected that decays mediated by the tree diagram of Fig. 9 would be greatly suppressed relative to those mediated by the penguin diagram of Fig. 8, since the relative CKM factor is $\lambda^2 \approx 0.05$. The experimental ratios R_{Exp}, shown in Table 7, do not indicate such a suppression, and it is clear that an additional enhancement of (approximately) a factor of 10 to the naïve theorectical ratios R_{Th} must occur. This suggests that the ratio of the penguin to tree amplitudes is about 0.3, a value which seems to hold widely in B decays.

In conclusion, a simple pictorial understanding of branching fractions can be misleading, and it is necessary to refer to fundamental theoretical calcu-

Table 7. Ratios of branching fractions from experimental measurements (R_{Exp}) and simple theoretical considerations (R_{Th}) based on CKM factors and decay constants (f_{decay}).

Mode	CKM	$(f_{decay})^2$	R_{Th}	R_{Exp}
$K^0\pi^+$	1	1	1	1
$K^{*0}\pi^+$	1	1.85	1.85	0.65
$\pi^+\pi^0$	λ^2	0.66	0.03	0.27
$\pi^+\rho^0$	λ^2	1.71	0.085	0.46
$\rho^+\rho^0$	λ^2	4.4	0.22	1.35

lations for a quantitative explanation.

7.4. The Decays $B \to \phi K$ and ϕK^*

The final states ϕK (Fig. 10) and ϕK^* are produced through a single penguin diagram, similar to Fig. 8, where the b quark decays to an s quark and the gluon couples to an s-$\bar{s}$ pair, to form a ϕ containing the leading s quark and a K or K^* which includes the spectator quark together with an s quark from the gluon. The dominance of this single diagram means that the branching fractions for all four charged and neutral final states (Table 8) are expected to be equal, within corrections due to different decay constants and form factors, and their CP-asymmetries close to zero. A glance at Table 8 shows that these expectations are met. For completeness we note that CDF see a signal in the decay $B^+ \to \phi K^+$, which they translate into a branching fraction[30] of $(6.9 \pm 2.1 \pm 0.8)\times 10^{-6}$ by normalization with the known BF for $J/\Psi\, K^+$.

However, the measured values of longitudinal polarization for the two vector-vector final states are completely at variance with the expectation of (approximately) 100%. This is not yet understood. It may be an indication for physics beyond the Standard Model in support of the anomalous value for $\sin(2\beta)$ in the channel $B^0 \to \phi K^0$ reported by Belle at this conference,[31] or it could indicate a breakdown of factorization through the theoretical assumption that calculations to leading order in Λ_{QCD}/m_B are sufficient. Whatever the explanation, these decay modes are now of prime interest to experimentalists and theorists alike. Finally, we note that the small value for the upper limit of 4×10^{-7} (at 90% CL) for the branching fraction[32] $B^+ \to \phi\pi^+$ gives no in-

Table 8. Branching fractions, CP-asymmetries and longitudinal polarization for the decay modes $B \to \phi K^0$, ϕK^+, ϕK^{*0} and ϕK^{*+} as measured by BaBar[28] and Belle.[29]

Mode	BF (10^{-6})		A_{CP} (%)		Polarization (%)	
	BaBar	Belle	BaBar	Belle	BaBar	Belle
$\phi\,K^0$	8.4 ± 1.6	9.0 ± 2.2				
$\phi\,K^+$	10.0 ± 1.0	8.6 ± 1.1	4 ± 9	1 ± 13		
$\phi\,K^{*0}$	11.2 ± 1.5	10.0 ± 1.8	4 ± 12	7 ± 16	65 ± 7	43 ± 10
$\phi\,K^{*+}$	12.7 ± 2.4	6.7 ± 2.2	16 ± 17	-13 ± 31	46 ± 12	

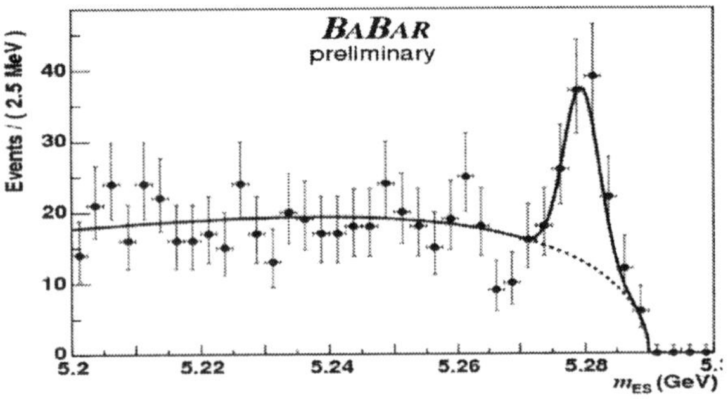

Figure 10. Invariant mass m_{ES} for the final state ϕK_s. The solid curve shows a projection of the likelihood fit for the event sample, with background described by the dotted curve.

Table 9. Branching fractions (10^{-6}) and CP-asymmetries (%) for $B \to \eta K^{(*)}$ and $\eta' K^{(*)}$ decays from BaBar[33] and Belle.[34]

Mode	Belle		BaBar	
	BF	A_{CP}	BF	A_{CP}
$\eta\,K^+$	5.3 ± 1.9		2.8 ± 0.8	-32 ± 22
$\eta\,K^0$			< 4.6	
$\eta'\,K^+$	78 ± 11	-2 ± 7	76.9 ± 5.6	4 ± 5
$\eta'\,K^0$	68 ± 13		60.6 ± 7.2	
$\eta\,K^{*+}$	26.5 ± 8.4		25.7 ± 4.2	15 ± 14
$\eta\,K^{*0}$	16.5 ± 4.8		19.0 ± 2.6	3 ± 11
$\eta'\,K^{*+}$	< 90		< 12	
$\eta'\,K^{*0}$	< 20		< 6.4	

dication of final-state interactions.

7.5. The Decays $B \to \eta K^{(*)}$ and $\eta' K^{(*)}$

The penguin graph of Fig. 11(a), which dominates the B decay to ϕK, might be expected to play a similar role in the decay to ηK final states, in which case branching fractions for ηK, ηK^*, $\eta' K$, and $\eta' K^*$ would all be expected to be equal within 20 - 30%. A glance at Table 9 shows that this expectation is not fulfilled.

Instead, branching fractions for ηK and $\eta' K^*$ are low by comparison with those for ϕK, while those for ηK^* and $\eta' K$ are significantly greater. A possible explanation for this discrepancy was given by Lipkin[35] several years ago. He pointed out that the penguin graph of Fig 11(b) also produces the ηK final states, and that if destructive interference between the amplitudes of Fig. 11(a) and (b) occurred for ηK then it would also occur for $\eta' K^*$, while constructive interference would occur for both $\eta' K$ and ηK^*. Only

recently have experimental measurements and theoretical calculations[2] become precise enough to test whether this explanation is sufficient, and it appears that an additional, flavor-singlet, penguin amplitude is necessary. Theoretical opinion is divided about the relative magnitude of this amplitude,[2,5] but all agree that the CKM-suppressed contribution of Fig. 11 (c) is small, and therefore the charged and neutral final states should have very similar values of branching fractions. That does not appear to be the case and BFs for neutral modes do seem to be smaller than for charged ones, although the disagreement is not yet statistically significant.

Given the dominance of penguin graphs in the decay of B to ηK, one would expect all CP-asymmetries to be very small. Data as yet is sparse, but there is no indication in Table 9 of any non-zero values.

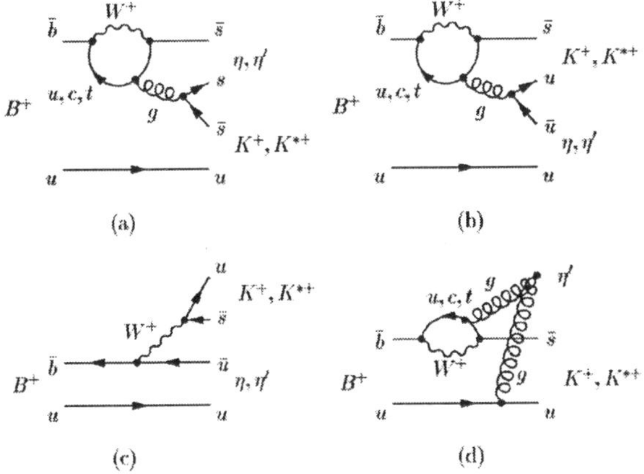

Figure 11. Feynman graphs contributing to the decay of a B meson to $\eta K^{(*)}$ and $\eta' K^{(*)}$ final states: a) Analogue of decay amplitude to ϕK; b) Additional penguin amplitude which cannot occur for ϕK; c) CKM-suppressed tree amplitude; d) Flavor-singlet penguin amplitude. All amplitudes contribute to charged final states, and all but (c) to neutral ones.

7.6. The Decays $B \to \eta^{(\prime)}\pi$ and $\eta^{(\prime)}\rho$

In principle, the Feynman graphs of Fig. 11 are also appropriate to the decays $B \to \eta^{(\prime)}\pi$ and $\eta^{(\prime)}\rho$, once all s quarks are replaced with d quarks – although no-one has yet suggested that a flavor-singlet penguin contribution is necessary. However, the relative magnitude of all penguin graphs is now reduced, and that of the tree amplitude increased, because of the changes in the CKM couplings. This means that in charged modes the effect of interference between penguin graphs (a) and (b) is likely to be reduced, while that between the tree and penguin amplitudes may become significant. These changes could give rise to large differences in branching fractions between the charged and neutral final states, and to measurable CP-asymmetries in charged final states, where both the tree and penguin amplitudes contribute.

Branching fractions and CP-asymmetries are shown in Table 10. BaBar has recently observed decays to the final states $\eta'\pi^+$, $\eta\rho^+$ and $\eta'\rho^+$, with combined statistical and systematic significances of 3.4, 4.8 and 3.8 sigma, respectively. Data on neutral decay modes is sparse, and it is not possible to draw any conclusions about the size of their BFs relative to charged ones. The ratio of $\eta\rho$ to $\eta\pi$ branching fractions is consistent with the expectation from decay constants, and the geometric-mean BF for $\eta\rho$ and $\eta\pi$ is about a factor of three smaller than that for $\eta^{(\prime)}K^{(*)}$. There is an indication of a non-zero

Table 10. Branching fractions (10^{-6}) and CP-asymmetries (%) for $B \to \eta^{(\prime)}\pi$ and $\eta^{(\prime)}\rho$ from BaBar[33] and Belle.[34]

Mode	Belle		BaBar	
	BF	A_{CP}	BF	A_{CP}
$\eta\,\pi^+$	5.4 ± 2.1		4.2 ± 1.0	-51 ± 20
$\eta'\,\pi^+$	< 7		< 4.5	
$\eta\,\rho^+$	< 6.2		10.5 ± 3.4	6 ± 29
$\eta\,\rho^0$	< 5.5			
$\eta'\,\rho^+$			14.0 ± 5.4	
$\eta'\,\rho^0$	< 14			

CP-asymmetry for $\eta\pi^+$, which is not unexpected in light of the discussion above. It is of interest, however, that whereas a large asymmetry is predicted on phenomenological grounds, the same theoretical analysis[5] predicts a small value of A_{CP} for $\eta\rho^+$. These are clearly interesting final states to monitor as statistics increase.

8. The Search for New Physics

Branching fractions for two-body final states are generally in good agreement with theoretical predictions, while measurements of the asymmetry are not yet precise enough to make rigorous tests. We saw earlier that the BF for $B^0 \to \pi^0\pi^0$ is significantly higher than the expectations from QCDF and pQCD, but theorists downplay this disagreement and are at pains to stress the difficulty of such color-suppressed calculations and of estimating theoretical errors. Of more significance is any difference between the measured and calculated ratio $B(B^0 \to \pi^0 K^0)$ / $B(B^+ \to \pi^+ K^0)$. The QCDF calculation is very clean and gives a value of 0.40 ± 0.04,[2] compared with the measured ratio of 0.55 ± 0.08. With a disagreement of less then two standard deviations there is no evidence for new physics.

There are still some worries about the calculations in QCDF of branching fractions for decays to the final states $\phi K^{(*)}$ and $\eta^{(\prime)}K^{(*)}$ where the values are generally lower than experimental measurements and subject to very large errors. One such comparison[36] of experimental and theoretical branching fractions and asymmetries has even suggested that the model-dependent annihilation contribution may be too large for stable solutions to QCDF, and that the basic theoretical assumption of $\Lambda_{QCD}/m_B \ll 1$ is simply not true. Recent modifications to QCDF[5] with a variety of choices of hadronic parameterizations have overcome this objection, but at the expense of weakening the predictive power. It therefore seems unlikely that new physics will be proclaimed on the basis of differences between measurements and QCD predictions, alone.

An alternative approach to the grounds-up QCD calculations is to use minimal theoretical assumptions, such as isospin and flavor-SU(3) symmetry, in an attempt to put limits on possible uncertainties in calculations within the Standard Model. As an example, for all the decay modes $B^0 \to \phi K_s$, $\eta' K_s$ and $K^+ K^- K_s$ the time-dependent asymmetry is expected to have the form:

A(t) = S sin(Δm t) + C cos(Δm t)

where: S = sin(2β) + Δ, C = Δ and Δ = $O(\lambda^2)$.

Given criteria for disagreement with the Standard Model, and hence claiming new physics – for example a five sigma difference between prediction and confirmed experimental measurements – the major question is the size of Δ. The origin of Δ is in the penguin loop of Fig. 12, where the b quark couples to the s quark via the exchange of a virtual W-boson and a u or c quark. The CKM parameters give the factor λ^2, while the interaction amplitudes a^u and a^c, describing b to s quark coupling with the exchange of a u quark and c quark, respectively, are expected to be of similar size. Most theoretical calculations give values of Δ/λ^2 of 0.2 to 1, but an enhancement cannot be ruled out. A method of calculating an upper bound on Δ has been derived within the framework of isospin and SU(3)[7] using ratios of branching fractions. A total of about 20 different branching fractions is used to bound Δ for the three finals states [ϕK_s, $\eta' K_s$ and $K^+ K^- K_s$], with values of Δ ranging from 0.2 to 0.5. Although far too large currently to enable claims of physics beyond the Standard Model for these decays, the values of Δ are not dissimilar from the current experimental precision on S and C, and can be expected to decrease in a similar way as more data is accumulated.

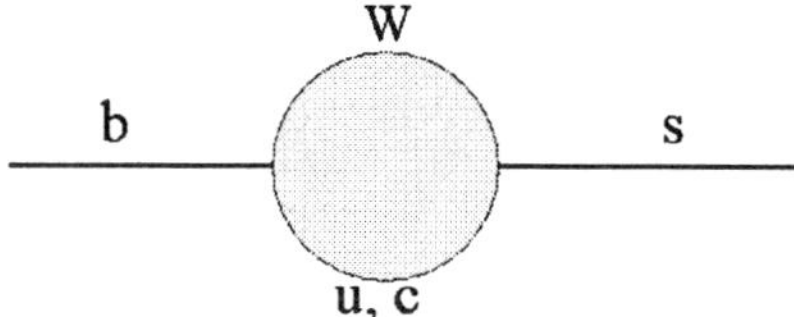

$$\Delta \approx \frac{V_{ub}^* V_{us}}{V_{cb}^* V_{cs}} \frac{a^u}{a^c} = O(\lambda^2)$$

Figure 12. The origin of Δ in the penguin loop coupling the b and s quarks in the flavor-changing neutral decay of the B meson. The amplitudes a^u and a^c describe the dynamics of the quark interactions.

9. Summary and Outlook

The two B factories have made a major step forward in both the precision of measurements and in the wealth of new decay modes that have been analyzed. Real tests have become possible of factorization models and phenomenology. Indeed, theorists have been able to further our understanding of hadron dynamics and to build up a comprehensive picture of branching fractions in B decays, which enables physicists to look for inconsistencies in the experimental measurements of very rare decays. Measurements of CP-asymmetries are starting to become precise, and there is every hope that in the next year or two even more stringent tests of theory will become possible, since the predictions depend on our understanding of both the magnitudes and phases of the interfering amplitudes. In this respect, two final states to watch are $K^+ \pi^-$ and $\eta \pi^+$.

The relationship among the branching fractions of the decay modes $\eta^{(\prime)} K^{(*)}$ is now much better understood than it was two years ago thanks to very precise measurements and considerable theoretical progress both in QCDF and phenomenology. It is still unclear how important the di-gluon, flavor-singlet contribution is by comparison with the other two penguin graphs. Further progress needs a considerable reduction in theoretical uncertainties in the QCDF calculations, as well as an improvement in the precision of measured branching fractions. A major puzzle in the strange final states is the anomalous value of the longitudinal polarization in the ϕK^* final state. A representative value of $(50 \pm 10)\%$ compared with the expectation of almost 100% indicates a worrying lack of theoretical understanding. It would be of real interest to measure the longitudinal polarization in other strange-particle final states to see whether the problem is associated with dominance of penguin graphs, compared with the tree graph for $\rho\rho$ final states.

The measurement of the branching fraction for $B^0 \to \pi^0 \pi^0$ is an experimental triumph. It is surprising that the value is so much higher than previous theoretical estimates, but, as already mentioned, the theoretical calculations of such color-suppressed processes are extremely difficult. One of the unfortunate consequences of this (comparatively) large branching fraction is that its use via the Grossmann-Quinn relationship to limit the theoretical uncertainty on the unitarity angle α is diminished to the point of being useless. By contrast the situation with respect to $\rho\rho$ final states is very encouraging. The time-dependent analyses of $\rho^+\rho^-$ are soon expected to give values of α_{Eff}. Moreover, an improved measurement of the branching fraction for $\rho^0\rho^0$ will enable the difference between the measured and true values of α to be determined more precisely than the current upper limit of about 20°. Together, these measurements will provide an exciting new constraint on the determination of the (ρ, η) apex of the unitarity triangle.

Progress is being made in refining the criteria for claiming new physics on the basis of disagreement with the predictions of the Standard Model. Using measured branching fractions, limits have been calculated on the possible modification to the time-dependent asymmetry parameters for final states ϕK_s, $\eta' K_s$ and $K^+ K^- K_s$ due to the CKM disfavored penguin contribution. As statistics improve so will the precision of the branching fractions used to calculate the limits.

All in all, the expected increase in luminosity of the two B factories promises a continuing, rich harvest of physics.

10. Acknowledgments

I should like to thank Harry Cheung and Tracey Pratt for their help and patience at every stage in the preparation of this paper. My thanks are due to Jim Alexander, Paoti Chang and Jim Smith, authors of the Heavy Flavor Averaging Group, for enabling me to keep abreast of the rapid evolution in physics results before, during and after the Lepton Photon 2003 Symposium, and for making the task of referencing a finite one; any shortcomings are mine alone. My special thanks are due to Jim Smith, who was always at the end of a telephone or computer to help.

References

1. M. Beneke, G. Buchalla, M. Neubert and C. Sachrajda, *Nucl. Phys.* B **606**, 245 (2001).
2. M. Beneke and M. Neubert, *hep-ph/0308039*.
3. C. Chiang and J. Rosner, *Phys. Rev.* D **65**, 074035 (2002).
4. C. Chiang, M. Gronau and J. Rosner, *hep-ph/0306021*.
5. C. Chiang, M. Gronau, Z. Luo, J. Rosner and D. Suprun, *hep-ph/0307395*.
6. Y. Grossman and H. Quinn, *PRD* **58**, 017504 (1988).
7. Y. Grossman, Z. Ligeti, Y. Nir and H. Quinn, *hep-ph/0303171*.
8. BaBar Collaboration: Result contributed to Lepton Photon Symposium 2003.
9. J. Alexander, P. Chang and J. Smith, *Heavy Flavor Averaging Group*, September 2003.
10. H. Huang *et al.*, Belle Collaboration, *Phys. Rev. Lett.* **90**, 121802 (2003).
11. B. Aubert *et al.*, (BaBar Collaboration), *Phys. Rev. Lett.* **89**, 281802 (2001); M. Bona for the BaBar Collaboration, talk at FPCP (2003); B. Aubert *et al.*, (BaBar Collaboration), *hep-ex/0308012* (submitted to *Phys. Rev. Lett.* (2003)); B. Aubert *et al.*, (BaBar Collaboration), *Phys. Rev. Lett.* **91**, 021801 (2003); T. Tomura *et al.*, (Belle Collaboration), *hep-ex/0305036* (2003).
12. Belle Collaboration, presented at Lepton Photon Symposium 2003.
13. B Aubert *et al.*, (BaBar Collaboration), *hep-ex/0308012* (submitted to *Phys. Rev. Lett.* (2003).
14. The Particle Data Group, K. Hagiwara *et al.*, *Phys. Rev.* D **66**, 010001 (2002).
15. Y. Y. Keum and A. L Sanda, *hep-ph/0306004* (2003).
16. M. Gronau, Private communication.
17. M. Gronau and J. Rosner, *hep-ph/0307095* (2003).
18. T. Tomura *et al.*, (Belle Collaboration), *hep-ex/0305036* (2003).
19. B. Aubert *et al.*, (BaBar Collaboration), *Phys. Rev. Lett.* **89**, 281802 (2001).
20. CDF Collaboration, presented at Lepton Photon Symposium 2003.
21. B. Aubert *et al.*, (BaBar Collaboration), *Phys. Rev. Lett.* **87**, 221802 (2001); B. Aubert *et al.*, (BaBar Collaboration), *hep-ex/0306030* (submitted to *Phys.*

Rev. Lett.(2003)); B. Aubert *et al.*, (BaBar Collaboration), *hep-ex/0307087* (2003); B. Aubert *et al.*, (BaBar Collaboration), *hep-ex/0308065* (submitted to *Phys. Rev. Lett.*(2003)); B. Aubert *et al.*, (BaBar Collaboration), *hep-ex/0303040* (2003).

22. H. C. Huang *et al.*, (Belle Collaboration), *hep-ex/0205062* (2002); K. Abe *et al.*, (Belle Collaboration), *BELLE-CONF-0312* EPS (2003); K. Abe *et al.*, (Belle Collaboration), *BELLE-CONF-0318* EPS (2003); A. Gordon *et al.*, (Belle Collaboration), *Phys. Lett.* B **542**, 183 (2002); K. Abe *et al.*, (Belle Collaboration), *BELLE-CONF-0317* EPS (2003); K. Abe *et al.*, (Belle Collaboration), *BELLE-CONF-0318* EPS (2003).

23. B. Aubert *et al.*, (BaBar Collaboration), *Phys. Rev. Lett.* **91**, 051801 (2003).

24. K. Abe *et al.*, (Belle Collaboration), *BELLE-CONF-0338* Lepton Photon Symposium 2003.

25. M. Suzuki, *Phys. Rev.* D **66**, 054018 (2002).

26. B. Aubert *et al.*, (BaBar Collaboration), *hep-ex/0307026* (to be published in *Phys. Rev. Lett.* (2003); B. Aubert *et al.*, (BaBar Collaboration), *hep-ex/0308024* (2003).

27. J. Zhang *et al.*, (Belle Collaboration), *hep-ex/0306007*2003.

28. B. Aubert *et al.*, (BaBar Collaboration), *hep-ex/0307026* (to be published in *Phys. Rev. Lett.* (2003)); B. Aubert *et al.*, (BaBar Collaboration), *hep-ex/0309025* (submitted to *Phys. Rev. Lett.* (2003)).

29. K. F. Chen *et al.*, (Belle Collaboration), *hep-ex/0307014* (submitted to *Phys. Rev. Lett.*(2003)); K. Abe *et al.*, (Belle Collaboration), *BELLE-CONF-0338* Lepton Photon Symposium 2003.

30. CDF Collaboration: presented at Lepton Photon Symposium 2003.

31. T. Browder *et al.*, (Belle Collaboration), *BELLE-CONF-0344* Lepton Photon 2003.

32. B. Aubert *et al.*, (BaBar Collaboration), *hep-ex/0309025* (submitted to *Phys. Rev. Lett.* (2003)).

33. B. Aubert *et al.*, (BaBar Collaboration), *hep-ex/0303046* (to be published in *Phys. Rev. Lett.* (2003)); B. Aubert *et al.*, (BaBar Collaboration), *hep-ex/0308015*2003; B. Aubert *et al.*, (BaBar Collaboration), *hep-ex/0303039* (2003).

34. H. Aihara (for the Belle Collaboration), talk at FPCP 2003.

35. H. Lipkin, *Phys. Lett.* B **254**, 247 (1991).

36. R. Aleksan *et al.*, *hep-ph/0301165* (2003).

DISCUSSION

Gerhard Buchalla (LMU Munich): Concerning the high experimental value of $B(B^0 \to \pi^0\pi^0)$ compared to the theoretical calculations, the prediction is much more uncertain than that of most other decay modes. The reason is that this decay is "color-suppressed" and thus accidentally small and extremely uncertain at LO. It is thus very sensitive to NLO effects and susceptible to other corrections. Very important further information on this issue could also come from comparing the decay $B^+ \to \pi^+\pi^0$ with the semileptonic decay $B \to \pi l\nu$.

John Fry: I accept what you say. However, this should make $B^0 \to \pi^0\pi^0$ an interesting theorectical test bed for NLO effects, in the same way that measurements of A_{CP} are sensitive to them.

George W. S. Hou (National Taiwan University): There is an indication for rescattering, despite the small BF for $B \to K^+K^-$. In $\pi\pi \to \pi\pi$, $K\bar{K}$ rescattering, there are actually two rescattering phases. In a paper with C.K. Chua and K.C. Yang, we find 30% of the parameter space where K^+K^- can be suppressed, while a large $\pi^0\pi^0$ and opposite sign A_{CP} for $K^-\pi^+$ mode can be accounted for. The interest is in fact in CP-asymmetries in the $\pi^+\pi^-$ mode, which can be large.

John Fry: It would be interesting to see quantitative predictions which account for all the data, including the small BF for $B^+ \to \psi\pi^+$.

Jonathan Rosner (University of Chicago): The nearly complete longitudinal polarization in $B \to \rho\rho$ decays is not a surprise since factorization seems to work well in processes dominated by a color-favored tree amplitude. Similarly in $B \to \phi K^*$ the presence of a substantial p-wave component, leading to R (perpendicular) not equal to zero, indicates that the parity conserving component of the $b \to s$ penguin amplitude is more significant than anticipated in factorization, as also seems true in $B \to$ VP decays. The puzzle is $B^+ \to \rho^0 K^{*+}$, which appears to be almost completely longitudinal – though presumably dominated by the same penguin amplitude(s) as $B \to \phi K^*$. It seems hard to describe both $B^+ \to \rho^0 K^{*+}$ and $B \to \phi K^*$ polarizations simultaneously within the Standard Model.

John Fry: Further theoretical guidance will be welcomed, especially if the measurement of additional final states can help to resolve the situation.

RADIATIVE AND ELECTROWEAK RARE B DECAYS

M. NAKAO

KEK, High Energy Accelerator Research Organization, 1–1 Oho, Tsukuba, Ibaraki 305–0801, JAPAN
E-mail: mikihiko.nakao@kek.jp

This report summarizes the latest experimental results on radiative and electroweak rare B meson decays. These rare decay processes proceed through the Flavor-Changing-Neutral-Current processes, and thus are sensitive to the postulated new particles in the theories beyond the Standard Model. Experiments at e^+e^- colliders, Belle, BaBar and CLEO, have been playing the dominant role, while the CDF and D0 experiments have just started to provide new results from Tevatron Run-II. The most significant achievement is the first observation of the decay $B \to K^* \ell^+ \ell^-$, which opens a new window to search for new physics in B meson decays.

1. Introduction

Rare B meson decays that include a photon or a lepton pair in the final state have been the most reliable window—besides the Cabibbo-Kobayashi-Maskawa (CKM) unitarity triangle—to understand the framework of the Standard Model (SM) using the rich sample of B decays, and to search for physics beyond the SM. Belle has just reported that the CP-violating phase in $B \to \phi K_S^0$ may deviate largely from the SM expectation measured using the $B \to J/\psi K_S^0$ and related modes.[1] The former is the $b \to s \bar{s} s$ transition which proceeds presumably through the loop (penguin) diagram for the $b \to s$ Flavor-Changing-Neutral-Current (FCNC) process, while the latter is the $b \to c \bar{c} s$ transition which is dominated by the tree diagram and is unlikely to be interfered with by new physics with a large effect. It is therefore an urgent question whether we can also find a similar deviation from the SM in any other related $b \to s$ transitions using the large samples of $e^+e^- \to \Upsilon(4S) \to B\overline{B}$ data available from two B-factories, Belle and BaBar, in order to investigate the nature of the possible new physics signal.

Radiative B decays with a high energy photon in the final state are a unique probe to explore inside the B meson. In the SM, the high energy photon is radiated through FCNC processes, $b \to s\gamma$ and $b \to d\gamma$. These transitions are forbidden at the tree level and only proceed via penguin loops formed by a virtual top quark and a W boson, or other higher order diagrams. The loop diagram can also be formed by postulated heavy particles if they exist, and is therefore sensitive to physics beyond the SM. The $b \to s\gamma$ decay rate is large enough to have been measured already by CLEO[2] and ALEPH,[3] and then by Belle[4] and BaBar.[5,6] As illustrated in Fig. 1, the $b \to s\gamma$ transition at the quark level can be studied by performing an inclusive measurement for $B \to X_s\gamma$, where X_s is an inclusive state with a strangeness $S = \pm 1$. The photon energy spectrum, which can be characterized by its mean energy and moments, provides a useful constraint to the heavy quark effective theory that essentially helps to reduce the uncertainties in the inclusive semi-leptonic B decay rates and hence the extraction of $|V_{cb}|$ and $|V_{ub}|$. In contrast to the inclusive studies, exclusive decay modes such as $B \to K^*\gamma$ are experimentally much easier to measure and have been extensively explored. However, one has to always consider large model dependent hadronic uncertainties to compare the results with the SM. Such uncertainties largely cancel by searching for CP- and isospin asymmetries. Though the $b \to d\gamma$ transition is suppressed by a CKM factor $|V_{td}/V_{ts}|^2 \sim \mathcal{O}(10^{-2})$ with respect to $b \to s\gamma$, searches are still being pursued for this exclusive decay channel.

Electroweak rare B decays proceeding through a similar FCNC process $b \to s\ell^+\ell^-$ ($\ell = e, \mu$) involves a virtual photon or weak boson, and has sensitivities to new physics that are not covered by $b \to s\gamma$. This process is suppressed with respect to $b \to s\gamma$ by an additional $\alpha_{\rm em}$ factor that has made it inaccessible before the B-factories. Having two leptons

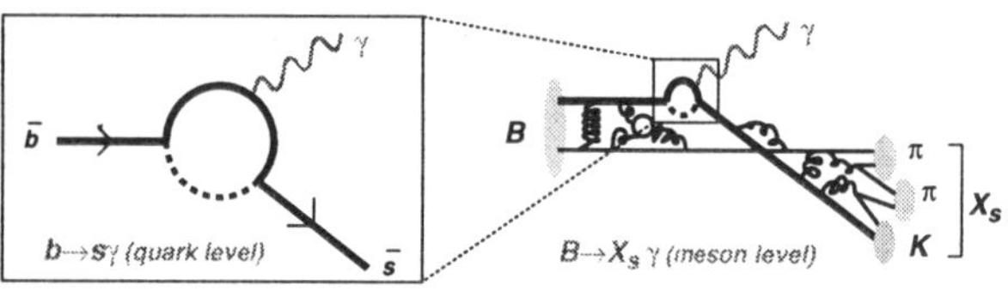

Figure 1. $b \to s\gamma$ and $B \to X_s\gamma$.

"

in the final state, one can measure the dependence on the momentum transfer squared $q^2(= M(\ell^+\ell^-)^2)$ of the virtual γ/Z. Furthermore, measurement of the forward-backward asymmetry of the lepton decay angle will be a unique probe in this electroweak process, with a small theoretical uncertainty even in the exclusive decay $B \to K^*\ell^+\ell^-$. The pure weak process, $b \to s\nu\bar{\nu}$ is experimentally extremely difficult.

Pure leptonic decay $B^0_{d,s} \to \ell^+\ell^-$ is based on the same quark diagram as $b \to (d,s)\ell^+\ell^-$, and hence has a similar sensitivity to new physics. The SM expected branching fractions are beyond the current experimental reach, but new physics may dramatically enhance the decay rate, especially for $B_s \to \mu^+\mu^-$, for which the Tevatron experiments have just restarted to provide new information. The charged counter part, $B^+ \to \tau^+\nu$ and $B^+ \to \ell^+\nu$, are tree level processes. These decays have not been observed yet because of the very small branching fractions due to the GIM suppression mechanism and the experimental difficulty due to the missing neutrino.

In this report, the latest results on radiative (Sec. 2), electroweak (Sec. 3) and pure leptonic (Sec. 4) B decays are reviewed. Belle has analyzed up to 140 fb^{-1} corresponding to 152 million $B\overline{B}$ pairs, while BaBar has analyzed up to 113 fb^{-1} corresponding to 123 million $B\overline{B}$ pairs. The first results from the Tevatron Run-II data from CDF and D0 are also included. Finally Sec. 5 concludes this report.

2. Radiative B Decays

2.1. *Inclusive $B \to X_s\gamma$ Branching Fraction*

Due to the two-body decay nature of the quark level process of $b \to s\gamma$, the photon energy spectrum of $B \to X_s\gamma$ has a peak around half of the b quark mass. This peak is the signature of the fully inclusive $B \to X_s\gamma$ measurement. On top of this signal, there are huge background sources as shown in Fig. 2. The largest contribution is from the continuum process $e^+e^- \to q\bar{q}$ ($q = u,d,s,c$) in which copious $\pi^0 \to \gamma\gamma$ and $\eta \to \gamma\gamma$ are the sources of high energy photons, and the initial-state radiation process $e^+e^- \to q\bar{q}\gamma$. These continuum backgrounds are reliably subtracted by using the off-resonance data sample taken slightly below the $\Upsilon(4S)$ reso-

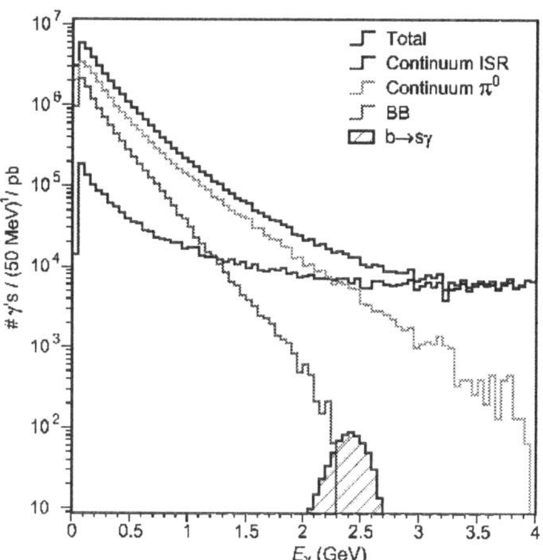

Figure 2. Expected photon energy distribution for $B \to X_s\gamma$ signal and various background sources.

nance. The background from B decay is also significant, especially for lower photon energies. To estimate and subtract the B decay background, one has to largely rely on the Monte Carlo simulation. As the signal rate rapidly decreases and the background rate rapidly increases towards lower photon energies, it is inevitable that one requires a minimum photon energy (E_γ^{min}) and extrapolates the spectrum below E_γ^{min} to obtain the total branching fraction.

An alternative semi-inclusive method is to sum up all the possible fully reconstructed $X_s\gamma$ final states, where X_s is formed from one kaon and up to four pions. In this case, one can require the kinematic constraints on the beam-energy constrained mass $M_{\mathrm{bc}} = \sqrt{E_{\mathrm{beam}}^{*\,2} - p_B^{*\,2}}$ (also denoted as the beam-energy substituted mass M_{ES}) and $\Delta E = E_B^* - E_{\mathrm{beam}}^*$, using the beam energy E_{beam}^* and fully reconstructed momentum p_B^* and energy E_B^* of the B candidate in the center-of-mass (CM) frame. Therefore the large backgrounds can be reduced at the cost of introducing an additional error due to the model dependent hadronization uncertainties.

So far, CLEO[2] and BaBar[5] have performed the fully inclusive measurement, and Belle[4] and BaBar[6] have performed the semi-inclusive measurement. Figure 3 summarizes these results, together with the measurement performed by ALEPH.[3] CLEO has applied the lowest E_γ^{min} of 2.0 GeV and has the smallest error, while BaBar requires $E_\gamma^{\mathrm{min}} = 2.1$ GeV, and Belle requires $M(X_s) < 2.1$ GeV which is roughly equivalent to $E_\gamma^{\mathrm{min}} \sim 2.25$ GeV.

The average of the five measurements, including

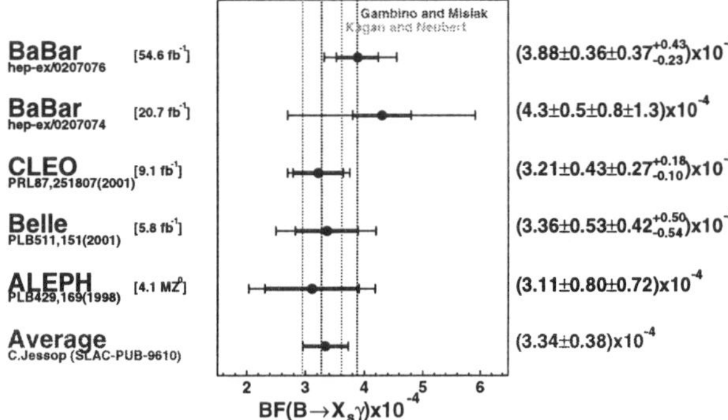

Figure 3. $B \to X_s\gamma$ branching fraction measurements.

the two from BaBar with an overlapping data set, has been calculated taking into account the correlation of the systematic and theory errors, as[7]

$$\mathcal{B}(B \to X_s\gamma) = (3.34 \pm 0.38) \times 10^{-4}. \qquad (1)$$

The latest SM calculation[8] predicts $\mathcal{B}(B \to X_s\gamma) = (3.57 \pm 0.30) \times 10^{-4}$, in very good agreement with the world average. The prediction fully includes up to Next-to-Leading-Order QCD corrections.[9]

This result can be used to constrain new physics hypotheses.[8,10,11] For example, any new physics that has only a constructive interference with the SM amplitude is strongly constrained. The Type-II charged Higgs boson is one such example, and its mass has to be greater than 350 GeV if there is no other destructive amplitude.[8,10] Many SUSY models can, however, have also a destructive amplitude that may cancel the constructive part. The decay amplitude is usually written down using the effective Hamiltonian with Wilson coefficients for the relevant operators. The $B \to X_s\gamma$ result constrains the magnitude of the C_7 Wilson coefficient, that can be a useful measure of a possible deviation from the SM, and also is an input parameter to the constraints provided by other measurements.

In order to further improve the measurement, it is necessary to lower the minimum photon energy. The latest Belle and BaBar data samples, with the largest off-resonance data size, have yet to be analyzed. A new effort to significantly reduce the theory error by including the Next-to-Next-to-Leading-Order QCD correction has also been started.

2.2. Exclusive $B \to K^*\gamma$

The measurement of the $B \to K^*\gamma$ exclusive branching fraction is straightforward, since one can use the

Table 1. $B \to K^*\gamma$ branching fractions

	$B^0 \to K^{*0}\gamma$ [×10⁻⁵]	$B^+ \to K^{*+}\gamma$ [×10⁻⁵]
CLEO	$4.55 \pm 0.70 \pm 0.34$	$3.76 \pm 0.86 \pm 0.28$
BaBar	$4.23 \pm 0.40 \pm 0.22$	$3.83 \pm 0.62 \pm 0.22$
Belle	$4.09 \pm 0.21 \pm 0.19$	$4.40 \pm 0.33 \pm 0.24$

M_{bc}, ΔE and K^* mass constraints. (K^* denotes $K^*(892)$ throughout this report.) The latest Belle measurement (Fig. 4) uses 78 fb⁻¹ data, with a total error of much less than 10% for each of the B^0 and B^+ decays. The results from CLEO,[12] BaBar[13] and Belle[14] are in good agreement and are listed in Table 1. The world averages are calculated as

$$\mathcal{B}(B^0 \to K^{*0}\gamma) = (4.17 \pm 0.23) \times 10^{-5}, \qquad (2)$$

$$\mathcal{B}(B^+ \to K^{*+}\gamma) = (4.18 \pm 0.32) \times 10^{-5}. \qquad (3)$$

The corresponding theoretically predicted branching fraction is about $(7 \pm 2) \times 10^{-5}$, higher than the measurement with a large uncertainty.[15] As the $b \to s\gamma$ transition is well understood by the inclusive measurement, we consider the deviation is due to the ambiguous hadronic form factor, for which the light-cone QCD sum rule result of $F_7^{B \to K^*}(0) = 0.38 \pm 0.05$ is used. However, a recent lattice-QCD calculation[16] is suggesting that the expected form-factor is as small as $F_7^{B \to K^*}(0) = 0.25 \pm 0.04$ and is consistent with the value of $F_7^{B \to K^*}(0) = 0.27 \pm 0.04$ extracted from the measured branching fraction.

A better approach to exploit the $B \to K^*\gamma$ branching fraction measurements is to consider isospin asymmetry.[17] A small difference in the branching fractions between $B^0 \to K^{*0}\gamma$ and $B^+ \to K^{*+}\gamma$ tells us the sign of the combination of the Wilson coefficients, C_6/C_7. Belle has taken into account the correlated systematic errors and performed a measurement as

$$\Delta_{+0} \equiv \frac{(\tau_{B+}/\tau_{B^0})\mathcal{B}(B^0 \to K^{*0}\gamma) - \mathcal{B}(B^+ \to K^{*+}\gamma)}{(\tau_{B+}/\tau_{B^0})\mathcal{B}(B^0 \to K^{*0}\gamma) + \mathcal{B}(B^+ \to K^{*+}\gamma)}$$

$$= (+0.003 \pm 0.045 \pm 0.018), \qquad (4)$$

which is consistent with zero and one cannot tell whether the SM prediction ($\Delta_{+0} > 0$) is correct yet. Here, the lifetime ratio $\tau_{B+}/\tau_{B^0} = 1.083 \pm 0.017$ is used, and the B^0 to B^+ production ratio is assumed to be unity. The latter is measured to be $f_0/f_+ = 1.072 \pm 0.057$ and is a source of an additional systematic error.

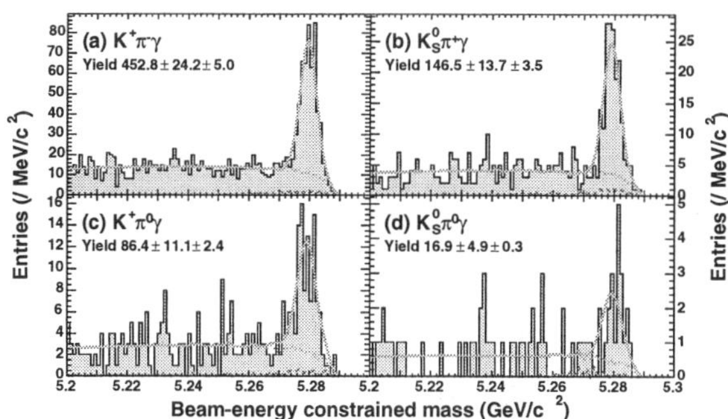

Figure 4. $B \to K^*\gamma$ signal from Belle.

Table 2. $B \to K_2^*(1430)\gamma$ branching fractions.

	$B^0 \to K_2^*(1430)^0\gamma$ $[\times 10^{-6}]$	$B^+ \to K_2^*(1430)^+\gamma$ $[\times 10^{-6}]$
CLEO	$16.6^{+5.9}_{-5.3} \pm 1.3$	
Belle	$13 \pm 5 \pm 1$	—
BaBar	$12.2 \pm 2.5 \pm 1.1$	$14.4 \pm 4.0 \pm 1.3$

2.3. *Other Exclusive Radiative Decays*

The dominant radiative decay channel $B \to K^*\gamma$ covers only 12.5% of the total $B \to X_s\gamma$ branching fraction, and the rest has to be accounted for by decays with higher resonances or multi-body decays. Knowledge of these decay modes will eventually be useful to reduce the systematic error of the inclusive measurement. Some of the decays have a particular property that is useful to search for new physics. As an example, the decay channel $B^0 \to K_1(1270)^0\gamma \to K_S^0\rho^0\gamma$ will be useful to measure the time-dependent CP-asymmetry;[18] while another such measurement is experimentally challenging: $B^0 \to K^{*0}\gamma \to K_S^0\pi^0\gamma$ using the detached $K_S^0 \to \pi^+\pi^-$ decay vertex. Another example is to use the decay $B^+ \to K_1(1400)^+\gamma \to K_S^0\pi^+\pi^0$ for a photon polarization measurement.[19]

The $B \to K_2^*(1430)\gamma$ decay mode is unique since the $K_2^*(1430)$ decays into a $K\pi$ combination, while many other resonances have very small or no decay width to $K\pi$. After measurements by CLEO[12] and Belle,[20] BaBar has also performed a new measurement[21] (Fig. 5). Branching fractions are listed in Table 2. The results are in agreement with the SM predictions,[22] for example, $(17.3 \pm 8.0) \times 10^{-6}$.

Belle has extended the analysis into multi-body decay channels.[20] Using 29 fb^{-1} data, the decay $B^+ \to K^+\pi^+\pi^-\gamma$ is measured to have a branching

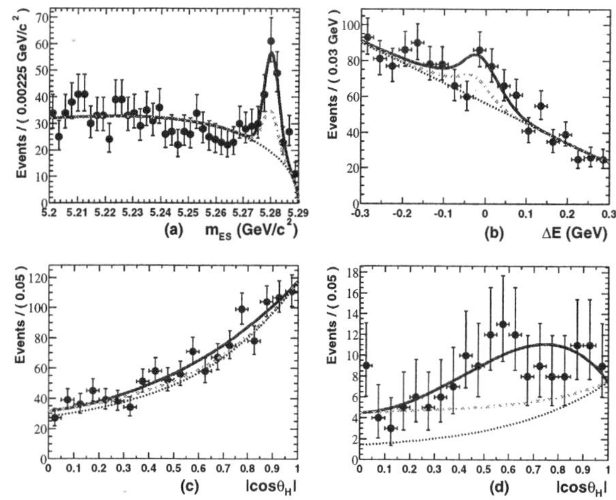

Figure 5. $B^0 \to K_2^*(1400)^0\gamma$ signal by BaBar.

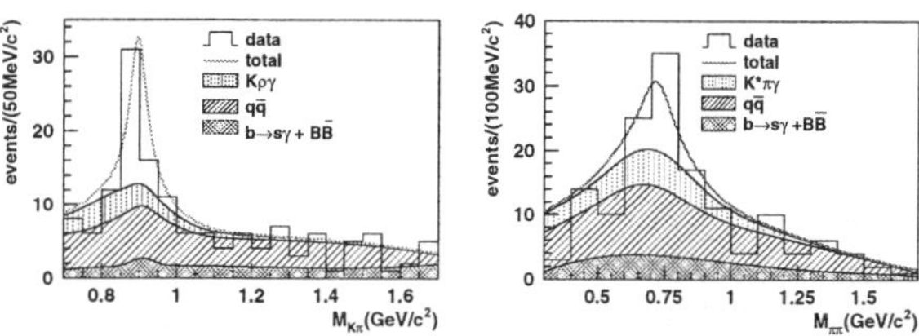

Figure 6. $B^+ \to K^{*0}\pi^+\gamma$ and $B \to K^+\rho^0\gamma$ from Belle.

fraction of $(24 \pm 5 \, ^{+4}_{-2}) \times 10^{-6}$ for $M(K\pi\pi) < 2.4$ GeV. The decay is dominated by $K^{*0}\pi^+\gamma$ and $K^+\rho^0\gamma$ final states that overlap each other as shown in Fig. 6. At this moment, it is not possible to disentangle resonant states that decay into $K^*\pi$ or $K\rho$, such as $K_1(1270)$, $K_1(1400)$, $K^*(1650)$, and so on. A clear $B^+ \to K^+\phi\gamma$ (5.5σ) signal was recently observed by Belle with 90 fb^{-1} data (Fig. 7), together with a 3.3σ evidence for $B^0 \to K_S^0\phi\gamma$. There is no known $K\phi$ resonant state. This is the first example of a $s\bar{s}s\gamma$ final state. Branching fractions are measured to be[23]

$$\mathcal{B}(B^+ \to K^+\phi\gamma) = (3.4 \pm 0.9 \pm 0.4) \times 10^{-6}$$
$$\mathcal{B}(B^0 \to K^0\phi\gamma) = (4.6 \pm 2.4 \pm 0.4) \times 10^{-6} \quad (5)$$
$$< 8.3 \times 10^{-6} \quad (90\% \text{ CL})$$

With more data, one can perform a time-dependent CP-asymmetry measurement with the $K_S^0\phi\gamma$ decay channel.

Radiative decays with baryons in the final state have been searched for by CLEO,[24] in the $B^- \to \Lambda\bar{p}\gamma$ channel for photon energies greater than 2 GeV. The analysis is also sensitive to $B^- \to \Sigma^0\bar{p}\gamma$ with a

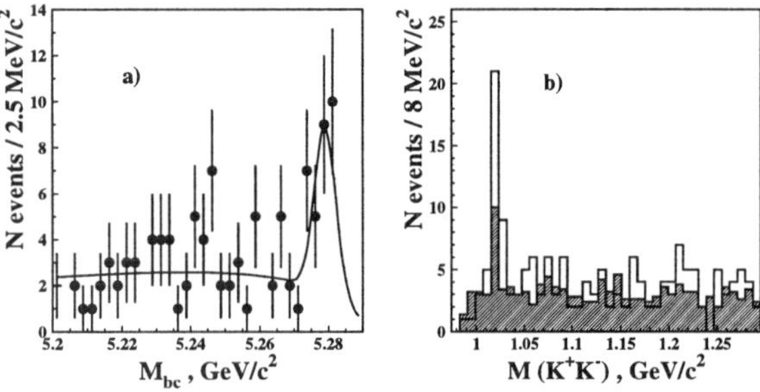

Figure 7. $B^+ \to K\phi\gamma$ from Belle.

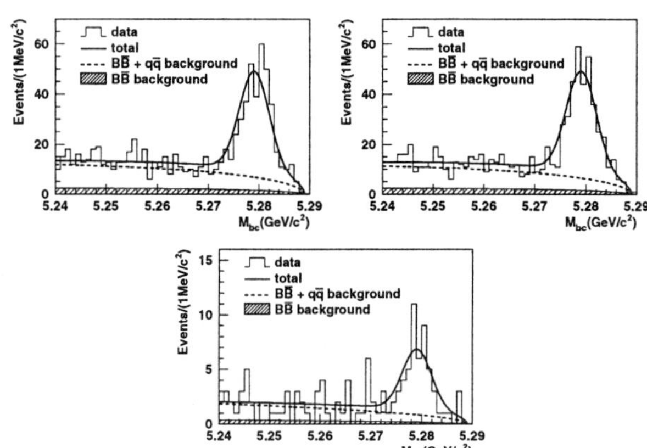

Figure 8. $\overline{B}$-tagged (top-left), B-tagged (top-right) and ambiguous (bottom) $B \to X_s\gamma$ signal from Belle.

slightly shifted ΔE signal window due to the missing soft photon in $\Sigma^0 \to \Lambda\gamma$. Upper limits are given as

$$\mathcal{B}(B^- \to \Lambda\bar{p}\gamma) + 0.3\mathcal{B}(B^- \to \Sigma^0\bar{p}\gamma) < 3.3 \times 10^{-6}$$
$$\mathcal{B}(B^- \to \Sigma^0\bar{p}\gamma) + 0.4\mathcal{B}(B^- \to \Lambda\bar{p}\gamma) < 6.4 \times 10^{-6}. \tag{6}$$

Considering isospin and other resonances such as $N(1232)$, an upper limit on baryonic radiative decay is obtained to be less than 3.8×10^{-5}, or 13% of the total $B \to X_s\gamma$ branching fraction.

In summary, $(35 \pm 6)\%$ of the total $B \to X_s\gamma$ is measured to be one of $B \to K^*\gamma$ (12.5%), $B \to K_2^*(1430)\gamma$ (4% after excluding $K\pi\pi\gamma$), $B \to K^*\pi\gamma$ (9%), $B \to K\rho\gamma$ (9%) or $B \to K\phi\gamma$ (1%). The remaining $(65 \pm 6)\%$ would be accounted for by decays with multi-body final states, baryonic decays, modes with η and η', multi-kaon final states other than $K\phi\gamma$ or in the large X_s mass range.

2.4. Search for Direct CP-asymmetry

Direct CP-asymmetry in $B \to X_s\gamma$ is predicted to be 0.6% in the SM with a small error.[25,26] This is contrary to the other hadronic decay channels with the $b \to s$ transition, for which usually larger SM CP-asymmetries are predicted, however, with large uncertainties. Although such a small SM asymmetry is beyond the sensitivity of the current B-factories, many extensions to the SM predict that it is possible to produce a large CP-asymmetry greater than 10%.[26,27] A large CP-asymmetry will be a clear sign of new physics.

There has been only one measurement by CLEO[28] to search for the direct CP-asymmetry of the radiative decays, which is sensitive also to $B \to X_d\gamma$. The result is expressed as

$$0.965A_{CP}(B \to X_s\gamma) + 0.02A_{CP}(B \to X_d\gamma)$$
$$= (-0.079 \pm 0.108 \pm 0.022) \times (1 \pm 0.030). \tag{7}$$

The SM predicts that $B \to X_d\gamma$ has a much larger A_{CP} with an opposite sign to that of $B \to X_s\gamma$.

A new $A_{CP}(B \to X_s\gamma)$ measurement performed by Belle[29] uses a similar technique to CLEO's, summing up the exclusive modes of one kaon plus up to four pions. In addition, modes with three kaon plus up to one pion are included. Belle's result eliminates $B \to X_d\gamma$ by exploiting particle identification devices for the tagged hadronic recoil system. CLEO requires $E_\gamma^{\min} = 2.2$ GeV while Belle require $M(X_s) < 2.1$ GeV which roughly corresponds to $E_\gamma^{\min} \sim 2.25$ GeV. Events are self-tagged as B candidates (B^0 or B^+) or $\overline{B}$ candidates ($\overline{B}^0$ or B^-), except for ambiguous modes with a K_S^0 and zero net charge. In order to correct the imperfect knowledge of the hadronic final state ingredients, the signal yield for each exclusive mode is used to correct the Monte Carlo multiplicity distribution. The resulting $\overline{B}$-tagged ($342 \pm 23^{+7}_{-14}$ events), B-tagged ($349 \pm 23^{+7}_{-14}$ events) and ambiguous ($47.8 \pm 8.7^{+1.4}_{-1.8}$ events) signals are shown in Fig. 8. Using the wrong-tag fractions of 0.019 ± 0.014 between B- and $\overline{B}$-tagged, 0.240 ± 0.192 from ambiguous to B- or $\overline{B}$-tagged, and 0.0075 ± 0.0079 from B- or $\overline{B}$-tagged to ambiguous samples, the asymmetry is measured to be

$$A_{CP}(B \to X_s\gamma) = 0.004 \pm 0.051 \pm 0.038. \tag{8}$$

The result corresponds to a 90% confidence level limit of $-0.107 < A_{CP}(B \to X_s\gamma) < 0.099$, and therefore already constrains extreme cases of the new physics parameter space.

For exclusive radiative decays, it is straightfor-

Table 3. $B \to K^*\gamma$ direct CP-asymmetry

CLEO (9.1 fb^{-1})	$(8 \pm 13 \pm 3) \times 10^{-2}$
BaBar (20.7 fb^{-1})	$(-4.4 \pm 7.6 \pm 1.2) \times 10^{-2}$
Belle (78 fb^{-1})	$(-0.1 \pm 4.4 \pm 0.8) \times 10^{-2}$

ward to extend the analysis to search for direct CP-asymmetry.[12-14] Particle identification devices of Belle and BaBar resolve the possible ambiguity between $K^{*0} \to K^+\pi^-$ and $\overline{K}^{*0} \to K^-\pi^+$ to an almost negligible level with a reliable estimation of the wrong-tag fraction (0.9% for Belle). The results of the asymmetry measurements are listed in Table 3, whose average is

$$A_{CP}(B \to K^*\gamma) = (-0.5 \pm 3.7) \times 10^{-2}. \quad (9)$$

It is usually considered that the large CP-violation in $B \to K^*\gamma$ is not allowed in the SM and the result may be used to constrain new physics. However, as the strong phase difference involved may not be reliably calculated for exclusive decays, the interpretation may be model dependent.

2.5. Search for $b \to d\gamma$ Final States

There are various interesting aspects in the $b \to d\gamma$ transition. Within the SM, most of the diagrams are a copy of those for $b \to s\gamma$, except for the replacement of the CKM matrix element V_{ts} with V_{td}. A measurement of the $b \to d\gamma$ process will therefore provide the ratio $|V_{td}/V_{ts}|$ without large model dependent uncertainties. This is in contrast with the current best $|V_{td}/V_{ts}|$ limit obtained from Δm_s and Δm_d in B_s and B_d mixing with the help of lattice QCD calculations. Unfortunately, the inclusive $B \to X_d\gamma$ measurement is extremely difficult due to its small rate and the huge $B \to X_s\gamma$ background, and the use of exclusive decay modes such as $B \to \rho\gamma$ involves other model dependences. If the constraints of the SM is relaxed, it is not necessary to retain the CKM structure, and $b \to d\gamma$ becomes a completely new probe to search for new physics effects in the $b \to d$ transition that might be hidden in the B_d mixing and cannot be accessed in the $b \to s$ transition. This mode is also the place where a large direct CP-asymmetry is predicted within and beyond the SM.

The search for the exclusive decay $B \to \rho\gamma$ is as straightforward as the measurement of $B \to K^*\gamma$, except for its small branching fraction, the enormous combinatorial background from copious ρ mesons

Table 4. 90% confidence level upper limits on the $B \to \rho\gamma$ and $\omega\gamma$ branching fractions.

	$\rho^+\gamma$	$\rho^0\gamma$	$\omega\gamma$
CLEO (9.1 fb^{-1})	13×10^{-6}	17×10^{-6}	9.2×10^{-6}
Belle (78 fb^{-1})	2.7×10^{-6}	2.6×10^{-6}	4.4×10^{-6}
BaBar (78 fb^{-1})	2.1×10^{-6}	1.2×10^{-6}	1.0×10^{-6}

and random pions, and the huge $B \to K^*\gamma$ background that overlaps with the $B \to \rho\gamma$ signal window. BaBar has optimized the background suppression algorithm using a neural net technique with input parameters of the event shape, helicity angle and vertex displacement between the signal candidate and the rest of the event, and has optimized the kaon rejection algorithm so that the $B \to K^*\gamma$ background can be suppressed to a negligible level. $B \to \omega\gamma$ is not affected by $B \to K^*\gamma$, but it is still hardly observed. The upper limits obtained by BaBar,[30] Belle[31] and CLEO[12] are summarized in Table 4. The best upper limits by BaBar are still about twice as large as the SM predictions,[15] $(9.0 \pm 3.4) \times 10^{-7}$ for $\rho^+\gamma$, and $(4.9 \pm 1.8) \times 10^{-7}$ for $\rho^0\gamma$ and $\omega\gamma$.

Using the isospin relation $\Gamma(B \to \rho\gamma) \equiv \Gamma(B^+ \to \rho^+\gamma) = 2\Gamma(B^0 \to \rho^0\gamma)$, the combined $B \to \rho\gamma$ upper limit from BaBar becomes $\mathcal{B}(B \to \rho\gamma) < 1.9 \times 10^{-6}$. The ratio of the branching fractions can be expressed as

$$\frac{\mathcal{B}(B \to \rho\gamma)}{\mathcal{B}(B \to K^*\gamma)} = \left|\frac{V_{td}}{V_{ts}}\right|^2 \left(\frac{m_B^2 - m_\rho^2}{m_B^2 - m_{K^*}^2}\right)^3 \zeta^2[1 + \Delta R]$$

$$< 0.047 \ (90\% \ \text{CL}) \quad (10)$$

where $\zeta = 0.76 \pm 0.10$ is the ratio of the form factors obtained from the light-cone QCD sum rule and $\Delta R = 0.0 \pm 0.2$ is to account for $SU(3)$ breaking effects. From this inequality, a bound on V_{td} is given as $|V_{td}/V_{ts}| < 0.34$, which is still a weaker constraint than that given by $\Delta m_s/\Delta m_d$. One can still argue about the validity of the form factor ratio,[32] as a recent lattice QCD calculation[16] gives a value of $\zeta = 0.91 \pm 0.08$ that leads to a different constraint on V_{td} as shown in Fig. 9.

3. Electroweak Rare B Decays

The $b \to s\ell^+\ell^-$ transition has a lepton pair in the final state, which is a clear signature of the decay.

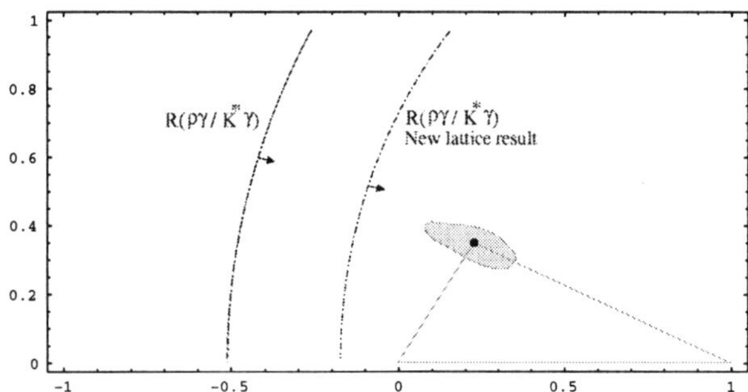

Figure 9. Bound on V_{td} from the ratio of branching fractions of $B \to \rho\gamma$ to $B \to K^*\gamma$.

The decay amplitude is written down using three Wilson coefficients, C_7, C_9 and C_{10}. Although there are three unknown complex coefficients ($|C_7|$ is obtained from $B \to X_s\gamma$), it is possible to disentangle all of them from measurements of the $\hat{s} = q^2/m_b^2$ dependent branching fraction $d\Gamma/d\hat{s}$ and the forward-backward asymmetry $dA_{FB}/d\hat{s}$,

$$\frac{d\Gamma}{d\hat{s}} = \left(\frac{\alpha_{\rm em}}{4\pi}\right)^2 \frac{G_F^2 m_b^5 |V_{ts}^* V_{tb}|^2}{48\pi^3}(1 - \hat{s})^2$$
$$\times \left[(1 + 2\hat{s})\left(|C_9^{\rm eff}|^2 + |C_{10}^{\rm eff}|^2\right) + 4\left(1 + \tfrac{2}{\hat{s}}\right)|C_7^{\rm eff}|^2 \right.$$
$$\left. + 12{\rm Re}\left(C_7^{\rm eff} C_9^{\rm eff}\right)\right] + {\rm corr.}$$

(11)

$$\frac{dA_{FB}}{d\hat{s}} = C_{10}^{\rm eff}(2C_7^{\rm eff} + C_9^{\rm eff}\hat{s})/(d\Gamma/d\hat{s}), \quad (12)$$

where QCD corrections are included in the Wilson coefficients.

There are two amplitudes that interfere with $b \to s\ell^+\ell^-$: one is $b \to s\gamma$ at $q^2 \to 0$ and the other is $b \to (c\bar{c})s$ where $(c\bar{c})$ is a charmonium state such as J/ψ or ψ' that decays into $\ell^+\ell^-$. The latest theory calculation that includes Next-to-Next-to-Leading-Order QCD corrections has been completed for the restricted range of $0.05 < \hat{s} < 0.25$ to avoid these interferences.[33]

Similarly to $b \to s\gamma$, there are a number of extensions to the SM[35] that one may be sensitive to by studying $b \to s\ell^+\ell^-$, and B-factories have just opened the window to search for such effects with a huge sample of B decays that was not available before.

3.1. *Observation of $B \to K^*\ell^+\ell^-$*

The first signal of $B \to K\ell^+\ell^-$ was observed by Belle[36] using 29 fb^{-1} data and confirmed by BaBar[37] with 78 fb^{-1}, while the $B \to K^*\ell^+\ell^-$ signal, which

has a larger expected branching fraction, was not significant with those data samples.

The $B \to K^{(*)}\ell^+\ell^-$ signal is identified with $M_{\rm bc}$, ΔE (and $M(K\pi)$ for $K^*\ell^+\ell^-$). There are five types of background that may contribute. 1) Charmonium decays, $B \to J/\psi K^{(*)}$ and $\psi' K^{(*)}$ have to be removed by the corresponding $M(\ell^+\ell^-)$ veto windows around J/ψ and ψ' masses. Especially for e^+e^- modes, bremsstrahlung has to be taken into account. 2) Hadronic decays, $B \to K^{(*)}\pi^+\pi^-$, are almost completely removed by lepton selection criteria including minimum lepton momentum requirements, but the remaining small contribution has to be evaluated and subtracted from the signal peak. 3) Two leptons from semi-leptonic decays, either in the cascade $b \to c \to s, d$ chain or from two B mesons, combined with a random K^*. This is the dominant combinatorial background that can be reduced for example by using the missing energy of the event. 4) Continuum background, which can be reduced by shape variables. 5) Rare backgrounds, $K^*\gamma$ with a photon conversion to e^+e^-, and $K^{(*)}\pi^0$ with a π^0 decaying into $e^+e^-\gamma$. This background can be removed by requiring a minimum e^+e^- mass as is done by Belle or can be subtracted from the signal as done by BaBar.

Belle has updated the analysis using a 140 fb^{-1} data sample, with a number of improvements in the analysis procedure.[38] The most significant improvement is the lowered minimum lepton momentum of 0.7 (0.4) GeV for muons (electrons) from 1.0 (0.5) GeV to gain 12% (7%) in the total efficiency. In addition, a $K^*\ell^+\ell^-$ combination is removed if there can be an unobserved photon along with one of the leptons that can form $B \to J/\psi K \to \ell^+\ell^-\gamma K$. As a result, the first $B \to K^*\ell^+\ell^-$ signal is observed with a statistical significance of 5.7 from a fit to $M_{\rm bc}$, as shown in Fig. 10, together with the improved $B \to K\ell^+\ell^-$ signal with a significance of 7.4.

BaBar has also updated the analysis using a 113 fb^{-1} data sample,[39] with improvements such as the bremsstrahlung photon recovery to include $K^{(*)}e^+e^-\gamma$ events in the $K^{(*)}e^+e^-$ signal. Evidence for the $B \to K^*\ell^+\ell^-$ signal is also seen with a statistical significance of 3.3 from a simultaneous fit to $M_{\rm bc}$, ΔE and $M(K\pi)$ (Fig. 11 shows their projections). A signal for $B \to K\ell^+\ell^-$ is clearly observed with a significance of ~ 8 (Fig. 12).

The branching fractions obtained are summa-

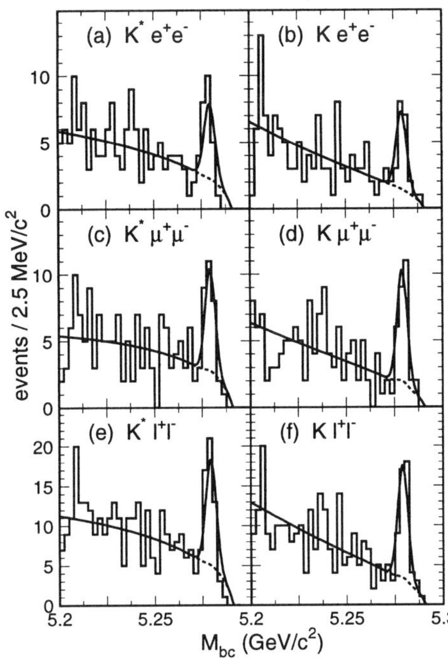

Figure 10. $B \to K^{(*)}\ell^+\ell^-$ signal observed by Belle.

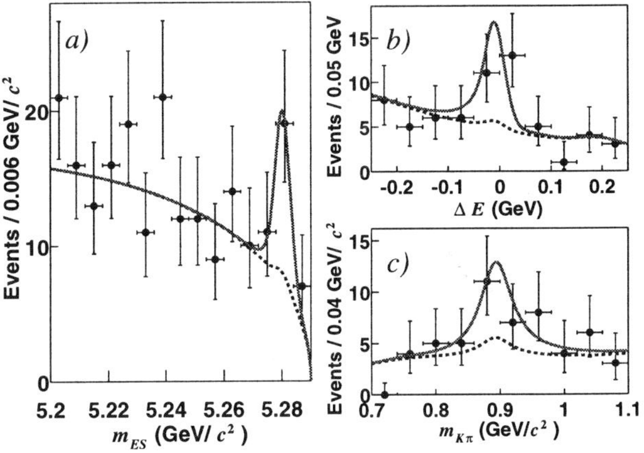

Figure 11. Evidence for $B \to K^*\ell^+\ell^-$ from BaBar.

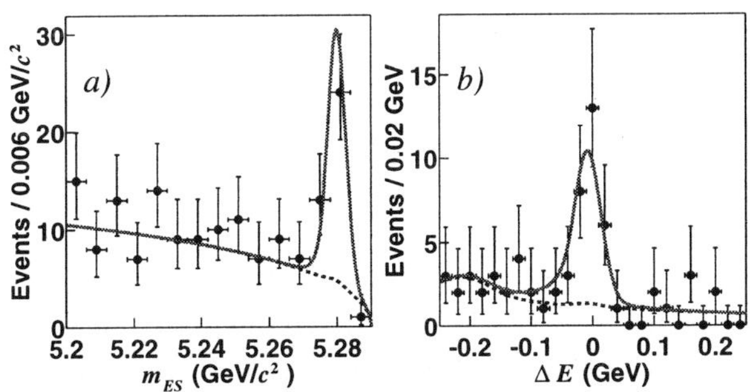

Figure 12. Signal for $B \to K\ell^+\ell^-$ from BaBar.

Table 5. $B \to K^{(*)}\ell^+\ell^-$ branching fractions.

Mode	Belle (140 fb^{-1}) [$\times 10^{-7}$]	BaBar (113 fb^{-1}) [$\times 10^{-7}$]
$B \to K e^+ e^-$	$4.8^{+1.5}_{-1.3} \pm 0.3 \pm 0.1$	$7.9^{+1.9}_{-1.7} \pm 0.7$
$B \to K \mu^+ \mu^-$	$4.8^{+1.3}_{-1.1} \pm 0.3 \pm 0.2$	$4.8^{+2.5}_{-2.0} \pm 0.4$
$B \to K \ell^+ \ell^-$	$4.8^{+1.0}_{-0.9} \pm 0.3 \pm 0.1$	$6.9^{+1.5}_{-1.3} \pm 0.6$
$B \to K^* e^+ e^-$	$14.9^{+5.2+1.1}_{-4.6-1.3} \pm 0.3$	$10.0^{+5.0}_{-4.2} \pm 1.3$
$B \to K^* \mu^+ \mu^-$	$11.7^{+3.6}_{-3.1} \pm 0.8 \pm 0.6$	$12.8^{+7.8}_{-6.2} \pm 1.7$
$B \to K^* \ell^+ \ell^-$	$11.5^{+2.6}_{-2.4} \pm 0.7 \pm 0.4$	$8.9^{+3.4}_{-2.9} \pm 1.1$

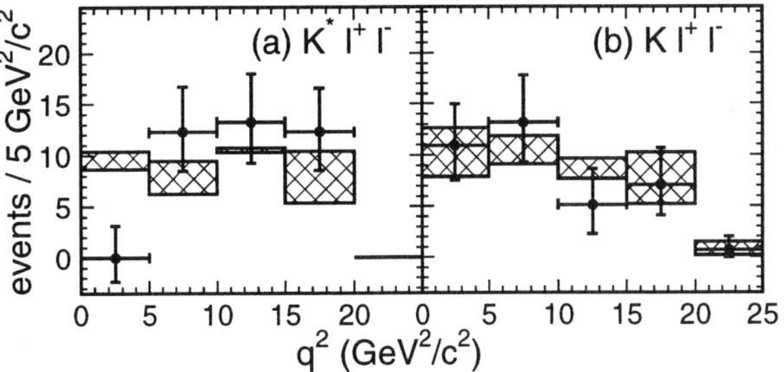

Figure 13. q^2 distributions for $B \to K^{(*)}\ell^+\ell^-$ from Belle.

rized in Table 5. For the combined $B \to K^*\ell^+\ell^-$ results, $\mathcal{B}(B \to K^*\ell^+\ell^-) = \mathcal{B}(B \to K^*\mu^+\mu^-) = 0.75\mathcal{B}(B \to K^*e^+e^-)$ is assumed which compensates the enhancement at the $q^2 = 0$ pole that appears more significantly in $K^*e^+e^-$, using the expected SM ratio.[40] The measured branching fractions are in agreement with the SM, for example[40] $(3.5 \pm 1.2) \times 10^{-7}$ for $B \to K\ell^+\ell^-$ and $(11.9 \pm 3.9) \times 10^{-7}$ for $B \to K^*\ell^+\ell^-$. We note that the experimental errors are already much smaller than the uncertainties in the SM predictions[41] and the variations due to different model-dependent assumptions used to account for the hadronic uncertainties.

It is still too early to fit the q^2 distribution to constrain new physics. First attempts to extract the q^2 distribution using the individual M_{bc} signal yields in q^2 bins has been performed by Belle as shown in Fig. 13.

3.2. Measurement of $B \to X_s\ell^+\ell^-$

The first measurements of the $B \to K^{(*)}\ell^+\ell^-$ branching fractions are consistent with the SM predictions. However since these predictions have uncertainties that are already larger than the measurement errors, the inclusive rate for $B \to X_s\ell^+\ell^-$ be-

Table 6. $B \to X_s \ell^+ \ell^-$ branching fractions.

Mode	Belle (60 fb^{-1}) [×10^{-6}]	BaBar (78 fb^{-1}) [×10^{-6}]
$X_s e^+ e^-$	$5.0 \pm 2.3 ^{+1.3}_{-1.1}$	$6.6 \pm 1.9 ^{+1.9}_{-1.6}$
$X_s \mu^+ \mu^-$	$7.9 \pm 2.1 ^{+2.1}_{-1.5}$	$5.7 \pm 2.8 ^{+1.7}_{-1.4}$
$X_s \ell^+ \ell^-$	$6.1 \pm 1.4 ^{+1.4}_{-1.1}$	$6.3 \pm 1.6 ^{+1.8}_{-1.5}$

comes more important in terms of the search for a deviation from the SM. In contrast to $B \to X_s\gamma$, the lepton pair alone does not provide a sufficient constraint to suppress the largest background from semi-leptonic decays. Therefore, it is only possible to use the semi-inclusive method to sum up the exclusive modes for now.

Belle has successfully measured the inclusive $B \to X_s\ell^+\ell^-$ branching fraction[42] from a 60 fb^{-1} data sample by applying the method to sum up the X_s final state with one kaon (K^+ or K^0_S) and up to four pions, of which one pion is allowed to be π^0. Assuming the K^0_L contribution is the same as K^0_S, this set of final states covers $82 \pm 2\%$ of the signal. In addition, $M(X_s)$ is required to be below 2.1 GeV in order to reduce backgrounds. For leptons, minimum momentum of 0.5 GeV for electrons, 1.0 GeV for muons and $M(\ell^+\ell^-) > 0.2$ GeV are required. Background sources and the suppression techniques are similar to the exclusive decays.

A new result is reported by BaBar with a 78 fb^{-1} data sample, using the same method with slightly different conditions.[43] BaBar includes up to two pions, corresponding to 75% of the signal, and require $M(X_s) < 1.8$ GeV. The minimum muon momentum requirement of 0.8 GeV is lower than Belle's.

The signal of 60 ± 14 events from Belle with a statistical significance of 5.4 is shown in Fig. 14, and 41 ± 10 events from BaBar with a significance of 4.6 is shown in Fig. 15. Corresponding branching fractions are very close to each other as given in Table 6, whose average is

$$\mathcal{B}(B \to X_s\ell^+\ell^-) = (6.2 \pm 1.1 ^{+1.6}_{-1.3}) \times 10^{-6}. \quad (13)$$

The branching fraction results are for the dilepton mass range above $M(\ell^+\ell^-) > 0.2$ GeV and are interpolated in the J/ψ and ψ' regions that are removed from the analysis, assuming no interference with these charmonium states.

The results may be compared with the SM prediction[34] of $(4.2 \pm 0.7) \times 10^{-6}$ integrated over the

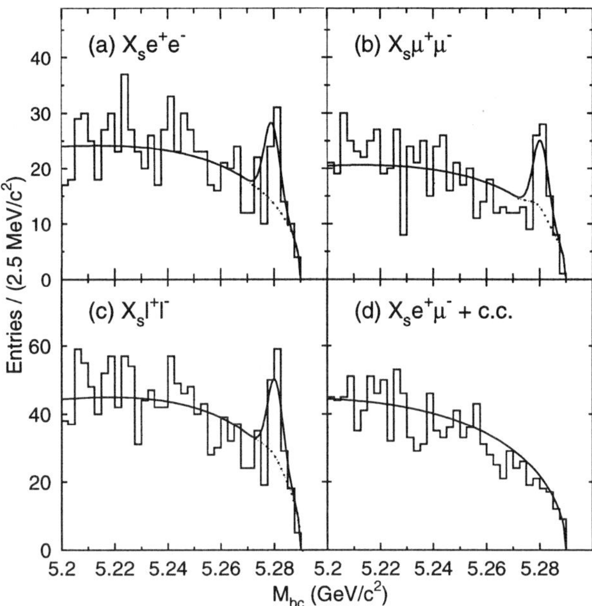

Figure 14. $B \to X_s\ell^+\ell^-$ signal measured from Belle. The $X_s e^+ \mu^-$ sample, which is prohibited in the SM, represents the combinatorial backgrounds.

same dilepton mass range of $M(\ell^+\ell^-) > 0.2$ GeV. With this requirement, the effect of the $q^2 = 0$ pole becomes insignificant, giving almost equal branching fractions for the electron and muon modes. The measured branching fractions are in agreement with the SM, considering the large measurement error. It should be noted that the large systematic error is dominated by the uncertainty in the $M(X_s)$ distribution, in particular the fraction of $B \to K^{(*)}\ell^+\ell^-$, that will be reduced with more statistics. Distributions for $M(X_s)$ and $M(\ell^+\ell^-)$ are shown in Figs. 16 and 17, in which no significant deviation from the SM is observed.

3.3. Search for $B \to K\nu\bar{\nu}$

The $b \to s\nu\bar{\nu}$ channel is sensitive to the weak-boson part of the $b \to s\ell^+\ell^-$ amplitude, and does not involve the $q^2 = 0$ pole and interfering charmonium decays. It is experimentally challenging even for the easiest exclusive $B^+ \to K^+\nu\bar{\nu}$ channel, because there is only one measurable kaon track out of the three-body final state that characterize the signal. In order to identify the signal, the other side B decay has to be tagged, so that there is only one kaon in the rest of the event. The search is attempted by BaBar using two techniques to tag the other B.

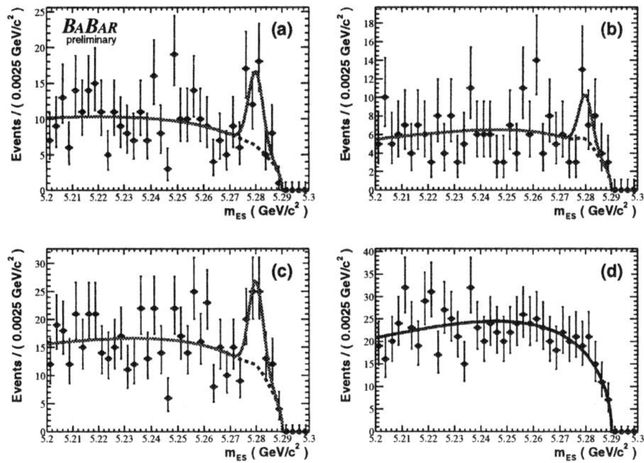

Figure 15. (a) $B \to X_s e^+ e^-$, (b) $B \to X_s \mu^+ \mu^-$, (c) $B \to X_s \ell^+ \ell^-$ signals with the (d) $X_s e^+ \mu^-$ sample from BaBar.

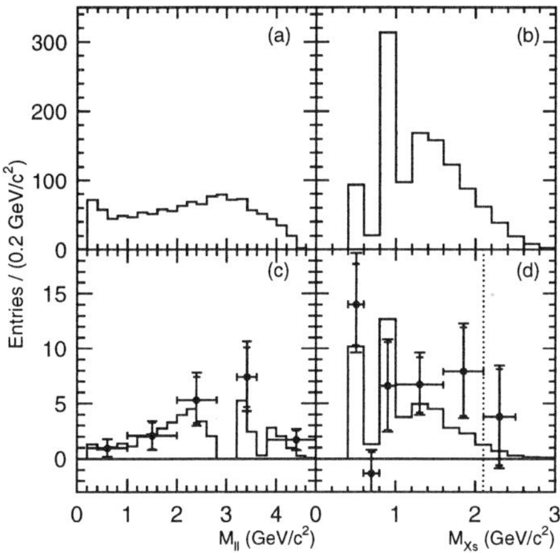

Figure 16. $M(\ell^+\ell^-)$ (left) and $M(X_s)$ (right) distributions for $B \to X_s \ell^+ \ell^-$ from Belle (points with error bars), compared with the SM predictions before (top) and after (bottom) including detector acceptance effects.

One method[44] is to require a D^0 meson and a lepton in the event that come from the $B^- \to D^{(*)0} \ell^- \overline{\nu}$ decay channel to tag the semi-leptonic decay of the other side B. After removing the signal kaon and the tag-side D^0 and lepton, there should be no remaining charged tracks, and the energy in the calorimeter should be at most that from the disregarded soft photon or π^0 from the D^{*0} decay. The signal window is defined in the plane of the remaining energy (less than 0.5 GeV) and the reconstructed D^0 mass (within $\pm 3\sigma$). Two candidates are found using a 51 fb^{-1} data sample (Fig. 18) with a tagging efficiency of 0.5%, where 2.2 background events are expected. This leads to the upper limit

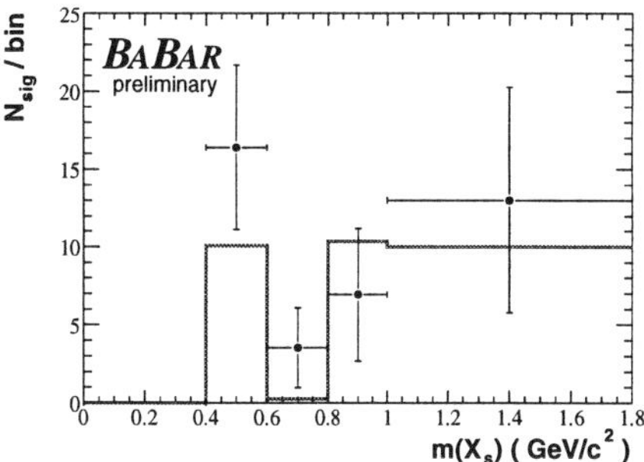

Figure 17. $M(X_s)$ distribution for $B \to X_s \ell^+ \ell^-$ from BaBar.

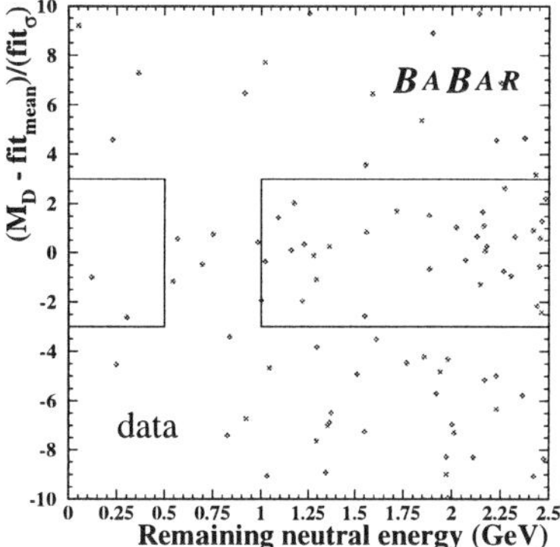

Figure 18. $B \to K \nu \overline{\nu}$ search results from BaBar using the semi-leptonic tag technique.

of $\mathcal{B}(B \to K \nu \overline{\nu}) < 9.4 \times 10^{-5}$ at the 90% confidence level.

The other method[45] is to require a full reconstruction of the hadronic decay $B^- \to D^0 X^-$, where X^- represents a combination of up to three charged pions or kaons and up to two π^0 with a net charge of -1 to tag the hadronic decay of the other side B. In this case, the maximum remaining energy is reduced to 0.3 GeV. The signal is identified with a high energy kaon with more than 1.5 GeV. Three candidates are found using a 80 fb^{-1} data sample with a tagging efficiency of 0.13%, where 2.7 ± 0.8 background events are expected. This leads to the upper limit of $\mathcal{B}(B \to K \nu \overline{\nu}) < 10.5 \times 10^{-5}$ at the 90% confidence level.

Since the two methods use statistically independent sub-samples, the two results can be combined to improve the upper limit as

$$\mathcal{B}(B \to K \nu \overline{\nu}) < 7.0 \times 10^{-5} \ (90\% \ \text{C.L.}), \qquad (14)$$

which is still an order of magnitude higher than the SM prediction[46] of $\mathcal{B}(B \to K \nu \overline{\nu}) = (3.8^{+1.2}_{-0.6}) \times 10^{-6}$.

4. Pure Leptonic B Decays

Leptonic two-body B decays are highly helicity suppressed in the SM due to the large energy release from the B meson decaying into much lighter leptons. The branching fraction for $B^+ \to \ell^+ \nu$ is written down as

$$\mathcal{B}(B^+ \to \ell^+ \nu) = \frac{G_F^2 m_B}{8\pi} m_l^2 \left(1 - \frac{m_l^2}{m_B^2}\right)^2 f_B |V_{ub}|^2 \tau_B \tag{15}$$

which is sensitive to V_{ub} and the B meson decay constant f_B. The experimental sensitivities are still far above the predicted SM branching fractions.

However, if there is a non-SM decay amplitude that is not helicity suppressed, the branching fraction may be accessible by the on-going experiments. The decay modes considered are: $B^+ \to \tau^+ \nu$, $B^+ \to \mu^+ \nu$, $B_d^0 \to \mu^+ \mu^-$, $B_d^0 \to e^+ e^-$ and $B_s^0 \to \mu^+ \mu^-$. The lepton flavor violating $B_d^0 \to e^\pm \mu^\mp$ is also searched for.

4.1. Search for $B \to \tau\nu$ and $B \to \mu\nu$

The decay $B^+ \to \tau^+ \nu$ has been searched for by BaBar using 81 fb^{-1} of data. As there are at least two missing neutrinos, the same two techniques for the $B \to K\nu\bar{\nu}$ search are applied to tag the other side B using semi-leptonic decays and hadronic decays.

In the analysis with the leptonic tag, $\tau^+ \to e^+ \nu_e \bar{\nu}_\tau$ and $\mu^+ \nu_\mu \bar{\nu}_\tau$ are used.[47] A fit to the remaining energy, that can include a soft γ/π^0 in the other side B, shows no significant excess above the expected background (Fig. 19). The upper limit is obtained to be $\mathcal{B}(B^+ \to \tau^+ \nu) < 7.7 \times 10^{-4}$ at the 90% confidence level.

In the analysis with the hadronic tag, hadronic τ decays into $\pi^+ \bar{\nu}_\tau$, $\pi^+ \pi^0 \bar{\nu}_\tau$ and $\pi^+ \pi^- \pi^+ \bar{\nu}_\tau$ are also included.[48] The number of events with the remaining energy less than ~ 100 MeV is counted. In total 35 candidates are found for the expected background of $37.6 \pm 4.7 \pm 1.3$ events. The upper limit is obtained to be $\mathcal{B}(B^+ \to \tau^+ \nu) < 4.9 \times 10^{-4}$ at the 90% confidence level.

These two samples are combined to improve the upper limit,

$$\mathcal{B}(B^+ \to \tau^+ \nu_\tau) < 4.1 \times 10^{-4} \quad (90\% \text{ CL}). \tag{16}$$

This improves the previous upper limit given by L3.[49] The corresponding SM prediction is 7.5×10^{-5} for $\tau_B = 1.674$ ps, $f_B = 198$ MeV and $|V_{ub}| = 0.0036$.

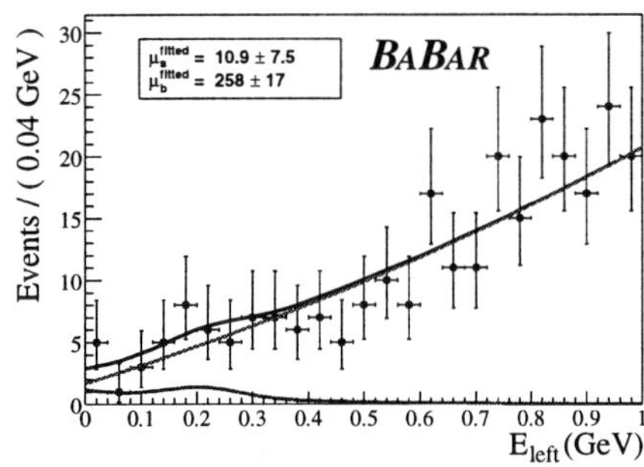

Figure 19. $B^+ \to \tau^+ \nu$ search results from BaBar using the semi-leptonic tag technique.

The decay $B^+ \to \mu^+ \nu$ has been searched for by Belle[50] and BaBar.[51] The analysis technique is to use the "neutrino reconstruction" technique, to determine the neutrino momentum from the missing momentum of the event. The muon momentum is monochromatic, except for the small initial B meson momentum of 340 MeV, while the muon from the dominant background source of semi-leptonic B decays have a smaller momentum. No significant signal excess has been observed; the most stringent upper limit is given by BaBar using 81 fb^{-1},

$$\mathcal{B}(B^+ \to \tau^+ \nu_\tau) < 6.6 \times 10^{-6} \quad (90\% \text{ CL}). \tag{17}$$

The SM predicts an order of magnitude smaller branching fraction of $\sim 4 \times 10^{-7}$.

4.2. Search for $B \to \ell^+ \ell^-$

In the SM, the decay $B_{d,s}^0 \to \ell^+ \ell^-$ occurs through the electroweak penguin transition $b \to (d,s)\ell^+ \ell^-$, and due to the helicity suppression, the expected branching fraction is extremely small:[52] $(2.34 \pm 0.33) \times 10^{-15}$ for $B_d^0 \to e^+ e^-$, $(1.00 \pm 0.14) \times 10^{-10}$ for $B_d^0 \to \mu^+ \mu^-$ and $(3.4 \pm 0.5) \times 10^{-9}$ for $B_s^0 \to \mu^+ \mu^-$. The decay amplitude may be significantly enhanced in some extensions to the SM. For example, these decays are sensitive to the chirality flipping interaction in models with two Higgs doublets, and the branching fractions can be three orders of magnitude larger than the SM at large $\tan\beta$, and may be accessible by the B-factories for B_d^0 decays and by the Tevatron for B_s^0 decays. In this case the $B \to X_s \ell^+ \ell^-$ decay rate may not be affected and can be consistent with the SM. The search can be easily extended to the lepton flavor violating decay $B_d^0 \to e^\pm \mu^\mp$.

Belle has searched for the the decays $B_d^0 \to$

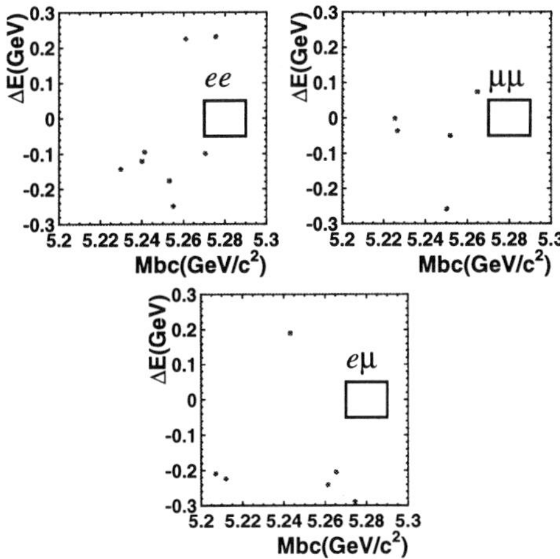

Figure 20. $B_d^0 \to \ell^+\ell^-$ search results from Belle.

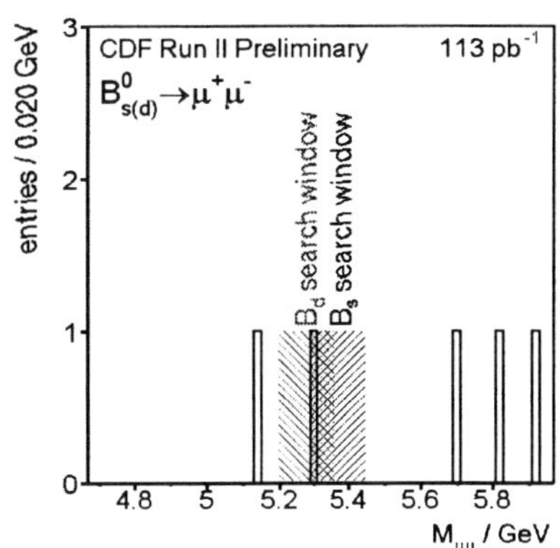

Figure 21. $B_s^0 \to \mu^+\mu^-$ search results from CDF.

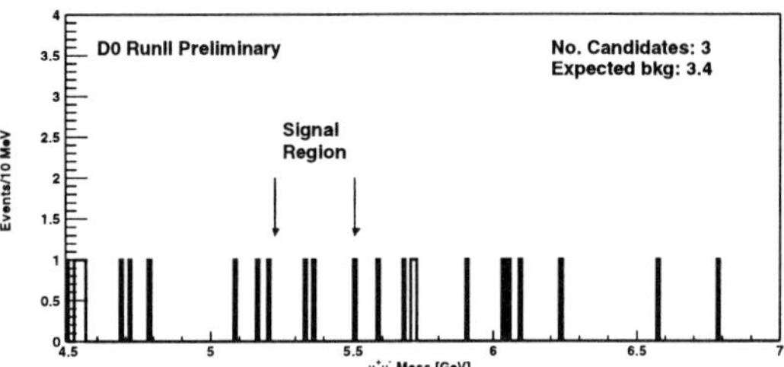

Figure 22. $B_s^0 \to \mu^+\mu^-$ search results from D0.

e^+e^-, $B_d^0 \to \mu^+\mu^-$ and $B_d^0 \to e^\pm\mu^\mp$, using a 78 fb^{-1} data sample.[53] The analysis method is similar to those for the other exclusive decays. The dominant background source is the continuum $e^+e^- \to c\bar{c}$ production in which both charm quarks decay into leptons. Leptons from $e^+e^- \to \tau^+\tau^-$ and two-photon processes can be removed by requiring five or more charged tracks in a event. No event was observed (Fig. 20) for the expected background events of 0.2 to 0.3, and the upper limits set are

$$\mathcal{B}(B_d^0 \to e^+e^-) < 1.9 \times 10^{-7}$$
$$\mathcal{B}(B_d^0 \to \mu^+\mu^-) < 1.6 \times 10^{-7} \qquad (18)$$
$$\mathcal{B}(B_d^0 \to e^\pm\mu^\mp) < 1.7 \times 10^{-7}$$

at the 90% confidence level.

For the B_s^0 decays, 113 pb^{-1} and 100 pb^{-1} of Run-II data from CDF and D0 respectively have been analyzed. Both analyses require three variables to reduce backgrounds and search for the signal in the $\mu^+\mu^-$ mass distribution. CDF uses the proper lifetime $c\tau$, the direction difference in azimuthal angle between the $\mu^+\mu^-$ vertex and momentum directions $\Delta\Phi$, and a measure of isolation of the B_s^0 candidate based on the tracks inside the cone around the B_s^0 direction; D0 also uses a similar set of variables.

CDF and D0 find one and three candidates as shown in Figs. 21 and 22, respectively. The CDF result leads to the upper limit of

$$\mathcal{B}(B_s^0 \to \mu^+\mu^-) < 9.5 \times 10^{-7} \ (90\% \ \text{CL}) \qquad (19)$$

that supersedes the previous CDF Run-I result. D0's limit is $\mathcal{B}(B_s^0 \to \mu^+\mu^-) < 16 \times 10^{-7}$. CDF also reports $\mathcal{B}(B_d^0 \to \mu^+\mu^-) < 2.5 \times 10^{-7}$, which is already competitive with Belle's result.

5. Conclusion

Figure 23 shows the currently measured branching fractions and upper limits for the rare B decays that involve a photon or a lepton pair. Most of the results have been updated rapidly along with the accumulation of the B-factory data: new modes and measurements in the exclusive $b \to s\gamma$ channels, new measurement of the CP-asymmetry in $B \to X_s\gamma$, the first observation of $B \to K^*\ell^+\ell^-$ by Belle and evidence from BaBar, new results on $B \to X_s\ell^+\ell^-$ from BaBar in agreement with Belle's, and new limits on $B \to K\nu\bar{\nu}$ and pure leptonic decays from BaBar, Belle and CDF are included. So far none of these results indicate a deviation from the SM. As $B \to K^*\ell^+\ell^-$ is finally measured, the next target will be the $b \to d\gamma$ transition in the decay $B \to \rho\gamma$.

There are still many programs to be pursued using these already observed rare B decay channels, in addition to the searches for unobserved modes. One

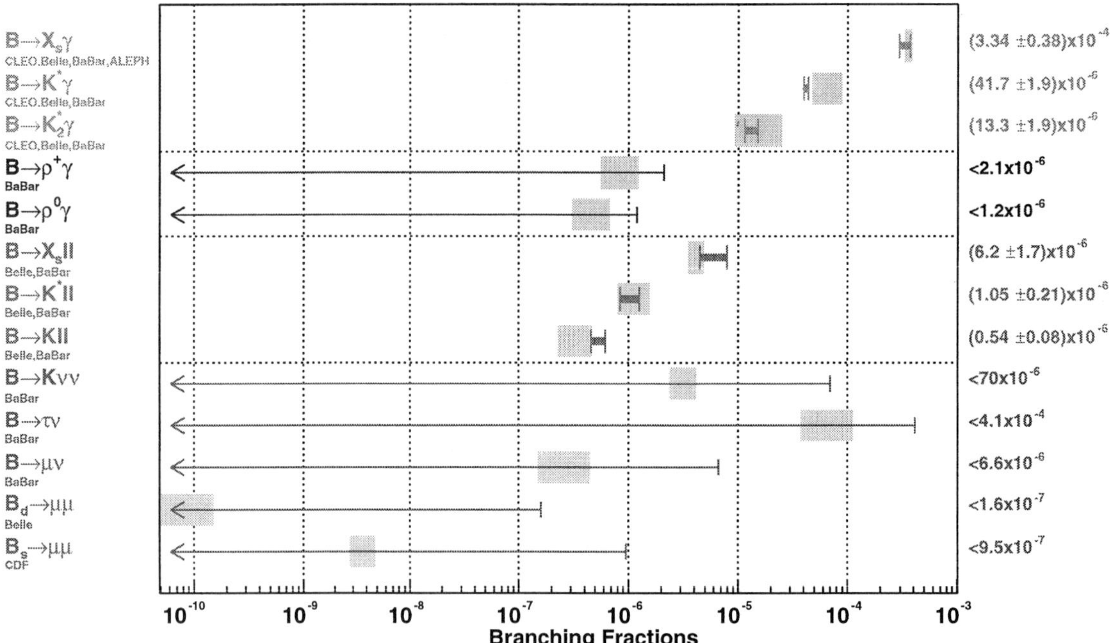

Figure 23. Summary of branching fractions and upper limits compared with the corresponding SM predictions.

example is a measurement for mixing-induced CP-violation in $b \to s\gamma$, for example with $B \to K^{*0}\gamma \to K_S^0\pi^0\gamma$. This channel has been considered to be experimentally challenging due to the displaced K_S^0 decay vertex; however, BaBar has recently demonstrated that it is possible to measure the B decay vertex from K_S^0 in the $B^0 \to K_S^0\pi^0$ channel,[54] and the same technique is applicable to $B \to K^*\gamma$. The other example is the measurement of the forward-backward asymmetry in $B \to K^*\ell^+\ell^-$ or $B \to X_s\ell^+\ell^-$. These examples demand an order of magnitude larger data sample than is available. Fortunately, Belle and BaBar are still collecting more data with improved luminosities expected, and are planning to extend their luminosities by orders of magnitude.

Acknowledgments

I wish to thank all the members of the Belle collaboration, Jeff Richman, Stephane Willocq and Mark Convery for providing the latest BaBar results, Rich Galik for the CLEO results, Majorie Shapiro for the CDF results, Brad Abbott and Vivek Jain for the D0 results, and Paoti Chang, Jim Alexander and Jim Smith of the Heavy Flavor Averaging Group for the averaged numbers from the last minute results. I acknowledge Enrico Lunghi and Mikolaj Misiak

among many theorists for many useful private discussions. I have to note that all the progress would not have been possible at all without excellent accelerator performances by the KEKB and PEP-II accelerator teams. Last but not least, I would like to thank the Lepton-Photon '03 organizers for all their efforts in the excellent conference organization.

References

1. Belle Collaboration, K. Abe *et al.*, arXiv:hep-ex/0308035.
2. CLEO Collaboration, S. Chen *et al.*, *Phys. Rev. Lett.* **87**, 251807 (2001).
3. ALEPH Collaboration, R. Barate *et al.*, *Phys. Lett.* B **429**, 169 (1998).
4. Belle Collaboration, K. Abe *et al.*, *Phys. Lett.* B **511**, 151 (2001).
5. BaBar Collaboration, B. Aubert *et al.*, arXiv:hep-ex/0207076.
6. BaBar Collaboration, B. Aubert *et al.*, arXiv:hep-ex/0207074.
7. C. Jessop, SLAC-PUB-9610 (2002).
8. P. Gambino and M. Misiak, *Nucl. Phys.* B **611**, 338 (2001).
9. K. Chetyrkin, M. Misiak and M. Münz, *Phys. Lett.* B **400**, 206 (1997), Erratum ibid. B **425**, 414 (1998); A. Kagan and M. Neubert, *Eur. Phys. J.* C **7**, 5 (1999).
10. F. Borzumati, C. Greub, *Phys. Rev.* D **58**, 074004 (1998)

11. For example, M. Ciuchini, G. Degrassi, P. Gambino and G. F. Giudice, *Nucl. Phys.* B **534**, 3 (1998); C. Bobeth, M. Misiak and J. Urban, *Nucl. Phys.* B **567**, 153 (2000); M. Carena, D. Garcia, U. Nierste and C. Wagner, *Phys. Lett.* B **499**, 141 (2001).

12. CLEO Collaboration, T. E. Coan *et al.*, *Phys. Rev. Lett.* **84**, 5283 (2000).

13. BaBar Collaboration, B. Aubert *et al.*, *Phys. Rev. Lett.* **88**, 101905 (2002).

14. Belle Collaboration, K. Abe *et al.*, Belle-CONF-0319.

15. S. Bosch and G. Buchalla, *Nucl. Phys.* B **621**, 459 (2002); A. Ali and A. Parkhomenko, *Eur. Phys. J.* C **23**, 89 (2002).

16. D. Becirevic, arXiv:hep-ph/0211340; D. Becirevic, talk given at Ringberg Phenomenology Workshop on Heavy Flavors, Rottach-Egern, Germany, April 27 - May 2, 2003.

17. A. Kagan and M. Neubert, *Phys. Lett.* B **539**, 227 (2002).

18. D. Atwood, M. Gronau and A. Soni, *Phys. Rev. Lett.* **79**, 185 (1997).

19. M. Gronau, Y. Grossman, D. Pirjol and A. Ryd, PRL **88**, 051802 (2002).

20. Belle Collaboration, S. Nishida *et al.*, PRL **89**, 231801 (2002).

21. BaBar Collaboration, B. Aubert *et al.*, arXiv:hep-ex/0308021.

22. S. Veseli and M. G. Olsson, *Phys. Lett.* B **367**, 309 (1996)

23. Belle Collaboration, A. Drutskoy *et al.*, arXiv:hep-ex/0309006.

24. CLEO Collaboration, K. Edwards *et al.*, *Phys. Rev.* D **68**, 011102 (2003)

25. J. Soares, *Nucl. Phys.* B **367**, 575 (1991).

26. A. Kagan and M. Neubert, *Phys. Rev.* D **58**, 094012 (1998).

27. K. Kiers, A. Soni and G. Wu, *Phys. Rev.* D **62**, 116004 (2000); S. Baek and P. Ko, *Phys. Rev. Lett.* **83**, 488 (1998).

28. CLEO Collaboration, T. Coan *et al.*, PRL **86**, 5661 (2001).

29. Belle Collaboration, K. Abe *et al.*, arXiv:hep-ex/0308038.

30. BaBar Collaboration, B. Aubert *et al.*, arXiv:hep-ex/0306038.

31. M. Nakao for the Belle Collaboration, talk given at 2nd Workshop on the CKM Unitarity Triangle, Durham, England, Apr. 2003, arXiv:hep-ex/0307031.

32. T. Hurth and E. Lunghi, arXiv:hep-ex/0307142.

33. C. Bobeth, M. Misiak and J. Urban, *Nucl. Phys.* B **574**, 291 (2000); H. H. Asatrian, H. M. Asatrian, C. Greub and M. Walker, *Phys. Lett.* B **507**, 162 (2001); H. H. Asatrian, H. M. Asatrian, C. Greub and M. Walker, *Phys. Rev.* D **65**, 074004 (2002).

34. A. Ali, E. Lunghi, C. Greub and G. Hiller, PRD **66**, 034002 (2002).

35. For example, E. Lunghi, A. Masiero, I. Scimemi and L. Silverstrini, *Nucl. Phys.* B **568**, 120 (2000); J. L. Hewett and J. D. Wells, *Phys. Rev.* D **55**, 5549 (1997); T. Goto, Y. Okada, Y. Shimizu and M. Tanaka, *Phys. Rev.* D **55**, 4273 (1997); G. Burdman, *Phys. Rev.* D **52**, 6400 (1995); N. G. Deshpande, K. Panose and J. Trampetić, *Phys. Lett.* B **308**, 322 (1993); W. S. Hou, R. S. Willey and A. Soni, *Phys. Rev. Lett.* **58**, 1608 (1987).

36. Belle Collaboration, K. Abe *et al.*, *Phys. Rev. Lett.* **88**, 021801 (2002).

37. BaBar Collaboration, B. Aubert *et al.*, arXiv:hep-ex/0207082.

38. Belle Collaboration, A. Ishikawa *et al.*, arXiv:hep-ex/0308044.

39. BaBar Collaboration, B. Aubert *et al.*, arXiv:hep-ex/0308042.

40. A. Ali, E. Lunghi, C. Greub and G. Hiller, *Phys. Rev.* D **66**, 034002 (2002); E. Lunghi, arXiv:hep-ph/0210379.

41. For example, D. Melikhov, N. Nikitin and S. Simula, *Phys. Lett.* B **410**, 290 (1997); P. Colangelo, F. De Fazio, P. Santorelli and E. Scrimieri, *Phys. Rev.* D **53**, 3672 (1996), Erratum-ibid. D **57**, 3186 (1998); M. Zhong, Y. L. Wu and W. Y. Wang, Int. J. Mod. Phys. A **18**, 1959 (2003); A. Faessler *et al.*, *Eur. Phys. J. direct* C **4**, 18 (2002); T. M. Aliev, C. S. Kim and Y. G. Kim, *Phys. Rev.* D **62**, 014026 (2000); W. Jaus and D. Wyler, *Phys. Rev.* D **41**, 3405 (1990).

42. Belle Collaboration, J. Kaneko *et al.*, *Phys. Rev. Lett.* **90**, 021801 (2003).

43. BaBar Collaboration, B. Aubert *et al.*, arXiv:hep-ex/0308016.

44. BaBar Collaboration, B. Aubert *et al.*, arXiv:hep-ex/0207069.

45. BaBar Collaboration, B. Aubert *et al.*, arXiv:hep-ex/0304020.

46. G. Buchalla, G. Hiller and G. Isidori, *Phys. Rev.* D **63**, 014015 (2001).

47. BaBar Collaboration, B. Aubert *et al.*, arXiv:hep-ex/0303034.

48. BaBar Collaboration, B. Aubert *et al.*, arXiv:hep-ex/0304030.

49. L3 Collaboration, M. Acciarri *et al.*, *Phys. Lett.* B **396**, 327 (1997).

50. Belle Collaboration, K. Abe *et al.*, Belle-CONF-0247.

51. BaBar Collaboration, B. Aubert *et al.*, arXiv:hep-ex/0307047.

52. A. Buras, *Phys. Lett.* B **566**, 115 (2003).

53. Belle Collaboration, M.-C. Chang *et al.*, arXiv:hep-ex/0309069.

54. T. Browder, talk given at XXI International Symposium on Lepton and Photon Interactions at High Energies, Batavia, Illinois, USA, Aug. 2003, in this Proceedings.

DISCUSSION

Marina Artuso (Syracuse University): You speak about the generic needs of a much larger data set size. How about larger continuum samples by changing the ratio 4S:continuum? It would be very beneficial for measurements such as the $b \to s\gamma$ spectrum.

Mikihiko Nakao : We always keep discussing this issue. For the lower photon energy of $b \to s\gamma$, B decay background is the dominant source of uncertainty, for which the continuum sample does not help. Moreover, there are so many other interesting subjects for which the continuum sample is useless.

Ikaros Bigi (University of Notre Dame): I would like to strongly support your call for lowering the cut on the photon energy in $B \to \gamma X_s$. An even more important motivation than determining $\Gamma(B \to \gamma X)$ might be that the photon spectrum below 2 GeV can yield an important calculation of the theoretical understanding of $b \to c$ and $b \to u$ semi-leptonic B decays. Any slice of 100 MeV you could go below 2 GeV is precious.

Mikihiko Nakao : We understand the situation and are working on that. We hope we can go below 2 GeV in the analysis of the latest data sample.

Takuya Morozumi (Hiroshima University): What do you really mean $q^2 = 0$ is removed in the analysis of the dilepton mass distribution for $b \to s e^+ e^-$?

Mikihiko Nakao : We take the lower limit of dilepton mass as $m(\ell^+\ell^-) = \sqrt{q^2} > 0.2\,\mathrm{GeV} \sim 2M_\mu$.

EXPERIMENTAL LIMITS ON NEW PHYSICS FROM CHARM DECAY

B. D. YABSLEY

Virginia Polytechnic Institute and State University, Blacksburg, VA 24061, USA
E-mail: yabsley@bmail.kek.jp
Web: http://belle.kek.jp/~yabsley/

Recent measurements in the charm sector are reviewed, concentrating on results which are sensitive to New Physics effects. The scope of the presentation includes $D^0 - \overline{D}^0$ mixing searches, a CPT / Lorentz invariance study, and a range of searches for rare and forbidden decays. Results from the BaBar, Belle, CDF, CLEO, and FOCUS collaborations are presented, including an important first observation.

1. Introduction

Presentations on "New Physics" can produce a feeling of anti-climax, for surely if there were any New Physics signals to report, the news would have leaked out. One does not expect to hear anything new. This talk does contain at least one first observation, however, and I'll try to maintain some suspense by not mentioning in advance what it is.

1.1. *What I'm Not Talking About*

Today we are at a particular disadvantage because the year's biggest charm news is outside the scope of the talk. BaBar's discovery[1] of a narrow resonance decaying to $D_s \pi^0$ came as a complete surprise—apart from the familiar D_s^*, no mesons decaying to this final state were foreseen—and led to a flurry of speculation that the new $D_{sJ}(2317)$ might be an exotic meson. The honors were shared among the B-factories in an amusing way: CLEO announced the discovery[2] of a second state, the $D_{sJ}(2457)$, decaying to $D_s^* \pi^0$; and Belle, as well as confirming these results,[3] made the first observation of both states in B meson decays.[4] The decay modes and widths of the new D_{sJ} are consistent[a] with these states being the $J^P = 0^+$ ($D_{sJ}(2317)$) and $J^P = 1^+$ ($D_{sJ}(2457)$) members of the $c\bar{s}$ system (with $L = 1$, and "light quark angular momentum" $j_q = \frac{1}{2}$), but their masses are a complete mystery. We thought we understood the $c\bar{q}$ mesons ... but it appears that we don't. This topic will be covered further in Jussara de Miranda's talk on Standard Model charm studies.[5]

The largest discrepancy between theory and experiment in the charm sector is also off-limits, as no-one imagines that new physics is responsible. But we still don't understand why the measured cross-section[6,7] for $e^+ e^- \to \psi \eta_c$ is an order of magnitude larger than the NRQCD prediction; the ingenious suggestion[8] of $e^+ e^- \to \gamma^* \gamma^* \to \psi \psi$ contamination has now been ruled out.[9] The $\psi c\bar{c}$ fraction in $e^+ e^- \to \psi X$ is likewise far "too large", almost saturating ψ production. We had thought that we understood $c\bar{c}$ production at this energy ... but it's pretty clear that we don't. Tomasz Skwarnicki will have something to say about this, and other developments in charmonia, in the next presentation.[10]

1.2. *What I Am Talking About*

After discussing the problems of obtaining clean new physics signatures in the charm sector (Sec. 2), the largest part of the talk treats $D^0 - \overline{D}^0$ mixing searches (Sec. 3); there are important new results on both y_{CP} (Sec 3.3) and $D^0 \to K^+\pi^-$ (Sec. 3.4). The first CPT and Lorentz invariance violation study in charm has recently been published, and we discuss it briefly (Sec. 4). The remainder of the talk is given over to searches for rare and forbidden decays (Sec. 5). There are new results from BaBar, Belle, CDF, CLEO, and FOCUS—including, as I say, a first observation—but let's not get ahead of ourselves.

2. Finding Clear New Physics Signatures

The difficulty with finding a clear signature of New Physics in the charm sector is this: it can be hard to know what the Standard Model (SM) prediction is. One way of thinking about the problem is to consider the masses of the quarks.

---

[a]In the case of the $D_{sJ}(2457)$, the Belle $D_s\gamma$ results[4,3] rule out $J = 0, 2$, and are consistent with $J = 1$. Note that an important "exotic" hypothesis—that the $D_{sJ}(2317)$ is wholly or partly a DK bound state, and the $D_{sJ}(2457)$ likewise a D^*K state—is also consistent with the 0^+ and 1^+ assignments.

- The up and down quark masses are both small compared to the hadronic mass scale: $m_u < m_d \ll \lambda_{QCD}$.[b] Isospin is therefore a rather good symmetry, and (including now the strange quark) $SU(3)$ of flavor, while broken, is useful.

- At the other extreme, the beauty quark has $m_b > \lambda_{QCD}$, and can be considered as a high-energy physics "particle": a "billiard ball with quantum numbers attached". One can think in terms of the Feynman diagrams, in b-sector processes, and not be too seriously misled.

- The charm quark lies between the two extremes: $m_c \gtrsim \lambda_{QCD}$, neither light nor truly heavy.

The awkwardness of the charm mass thus puts limits on both symmetry- and quark-based thinking as guides to charm physics. If light hadron work is like swimming in the ocean, and b-physics is like flying through the air, then in charm studies one is wading knee-deep through the brown muck.

One should really think in terms of hadrons, not quarks, in charm. So-called "long-distance" contributions are important in many processes: quark loops are typically suppressed, so that hadronic processes take a leading role. These are usually difficult to calculate, especially as the charm mass lies in the resonance region. So if some parameter is supposed to be small, but observed to be large, one should be cautious before claiming new physics: perhaps the Standard Model contribution has just been miscalculated. As discussed in Sec. 1.1, there have already been two major surprises in the last two years.

3. $D^0 - \overline{D}^0$ Mixing

3.1. *Mixing in the Standard Model and Beyond*

There are particular pitfalls in the interpretation of charm mixing searches. The SM box diagrams for mixing (e.g. Fig. 1) are doubly Cabibbo-suppressed and suffer from very efficient cancellations (the GIM mechanism): the expected mixing rate due to such processes is negligible. Since most new physics scenarios introduce new particles that couple to the SM fields, they induce new loop diagrams such as Fig. 2,

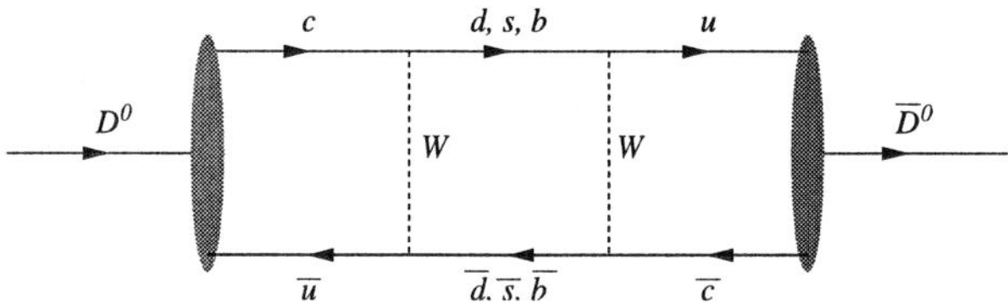

Figure 1. Box diagram for $D^0 - \overline{D}^0$ mixing in the SM.

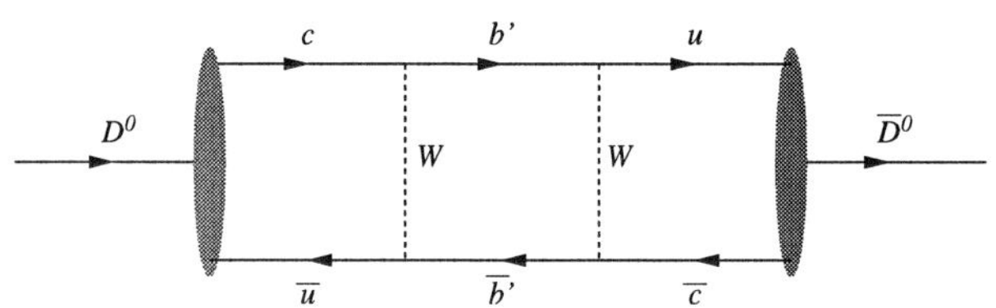

Figure 2. Box diagram for $D^0 - \overline{D}^0$ mixing in a New Physics model with an extra down-type quark.

with no (or a lesser degree of) cancellations. The result is an enhancement of the mass splitting of the $D^0 - \overline{D}^0$ eigenstates, $x \equiv \Delta M/\Gamma$; hence the common statement that "$D^0 - \overline{D}^0$ mixing with measurable x is a signal of New Physics".

As discussed in the previous section, however, hadronic processes cannot be neglected. Final states common to D^0 and $\overline{D}^0$, such as $K\overline{K}$, $\pi\pi$, $K\pi$ and $\overline{K}\pi$, couple the two neutral D's (Fig. 3); such contributions cancel in the $SU(3)_F$ limit, but to the extent that $SU(3)$ of flavor is broken, they induce mixing. One might suppose that only the lifetime splitting parameter $y \equiv \Delta\Gamma/2\Gamma$ would be affected, as the intermediate states are real. The true situation is more complicated. To the extent that quark-hadron duality holds, mixing can be estimated using the Operator Product Expansion: a recent study[11] finds $x \sim y \sim O(10^{-3})$. An alternative approach[12] relying directly on hadronic intermediate states suggests that y may be as large as $O(1\%)$. So as far as x and y are concerned, mixing provides a clean new physics signal only if $x \gg y \sim 10^{-3}$.

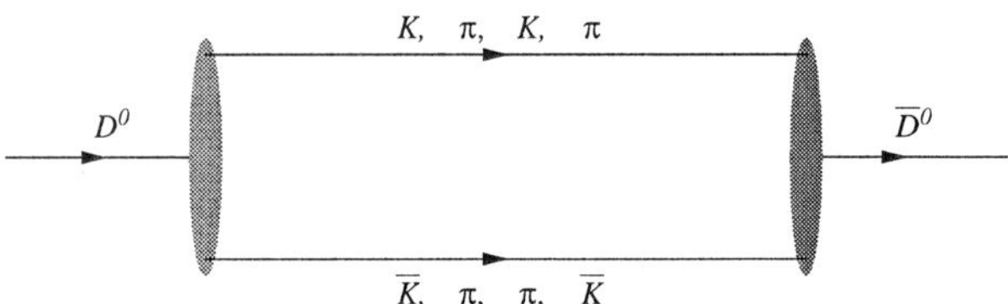

Figure 3. Sample diagram for $D^0 - \overline{D}^0$ mixing due to common hadronic final states.

---

[b]We set aside the related fact that m_u and m_d are elements of the theory, rather than straightforward "observations".

168

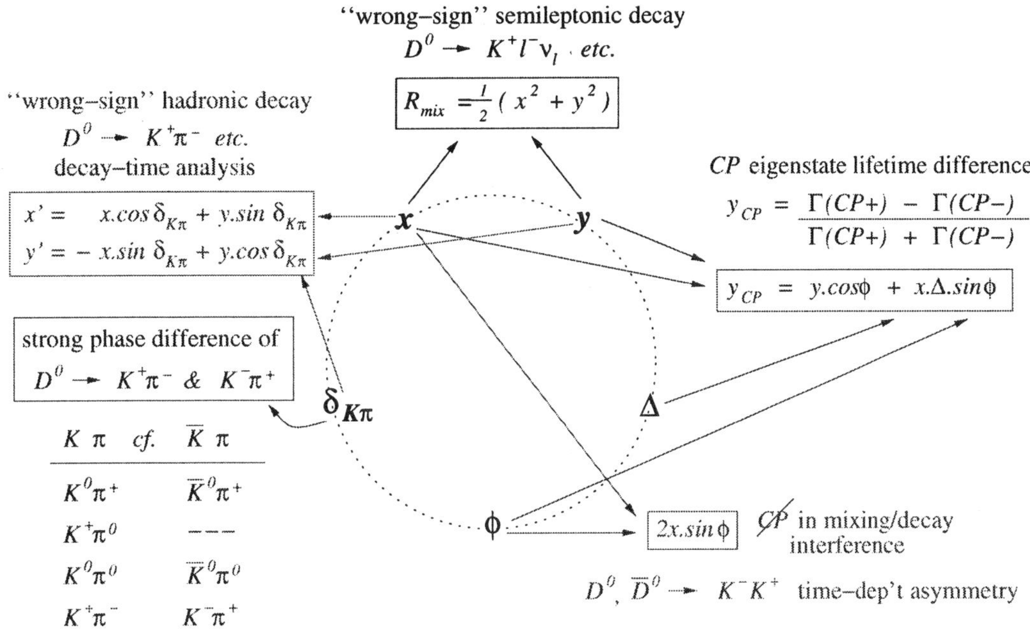

Figure 4. Five parameters describing $D^0 - \overline{D}^0$ mixing, and their relationship with quantities measured in experiment.

3.2. *Mixing Parameters & Measurements*

The full parameter space for mixing is more rich than (x, y): some important parameters are shown in Fig. 4, together with experimentally measurable quantities. The parameters are poorly-known: the strong phase difference $\delta_{K\pi}$ between $D^0 \rightarrow K^+\pi^-$ and $K^-\pi^+$ amplitudes is unconstrained, as is the difference in particle content of the eigenstates $|D_{1,2}\rangle = p|D^0\rangle \pm q|\overline{D}^0\rangle$, $\Delta = (|p|^2 - |q|^2)/(|p|^2 + |q|^2)$. But the news is not all bad. This year has seen the first measurements relevant to the CP violating phase $\phi = \arg\left(q\mathcal{A}(\overline{D}^0 \rightarrow K^-K^+)/p\mathcal{A}(D^0 \rightarrow K^-K^+)\right)$ (Sec. 3.3.6), as well as a major new analysis of $D^0 \rightarrow K^+\pi^-$ (Sec. 3.4).

3.3. *Mixing: Lifetime Difference, y_{CP}, and CP Violation*

The most popular measurement in recent years, however, has been y_{CP}. Defined as the normalized lifetime difference of the D^0-$\overline{D}^0$ CP eigenstates, it is typically measured using the non-eigenstate decay $D^0 \rightarrow K^-\pi^+$ as a convenient reference:

$$y_{CP} \equiv \frac{\Gamma(CP+) - \Gamma(CP-)}{\Gamma(CP+) + \Gamma(CP-)} \approx \frac{\tau(D^0 \rightarrow K^-\pi^+)}{\tau(D^0 \rightarrow K^-K^+)} - 1$$

$$= y\cos\phi + x\Delta\sin\phi,$$

where the last relation holds for small values of the parameters. In the CP-conserving limit $\phi = 0$ and $\Delta = 0$, so $y_{CP} = y$, as one would expect. In this limit y_{CP} is not a new-physics search parameter (since new particles are expected to affect x; Sec. 3.1) but a tool for measuring the level of mixing due to the SM.

3.3.1. y_{CP}: the FOCUS measurement (2000)

Three years ago, the FOCUS collaboration measured y_{CP} using a relatively clean sample of 10,000 $D^0 \rightarrow K^-K^+$ events,[c] and a $K^-\pi^+$ sample ten times that size.[13] Both inclusive and D^*-tagged decays were used, under FOCUS-standard reconstruction, particle identification, and vertex detachment cuts; the result was obtained from a binned maximum-likelihood (ML) fit to the distributions of *reduced proper time* $t' \equiv (l - N\sigma_l)/\beta\gamma c$, where l, σ_l are the D^0 decay length and its error, and N the minimum required detachment of the production and decay vertices.

The result was surprisingly large: $y_{CP} = (3.42 \pm 1.39 \pm 0.74)\%$, over 2σ away from zero. There was considerable excitement at the thought that charm mixing might be within our grasp, partly due to the usual association of mixing with new physics. But as

---

[c]Inclusion of charge-conjugate modes is implied throughout, unless the context makes clear that they are treated separately.

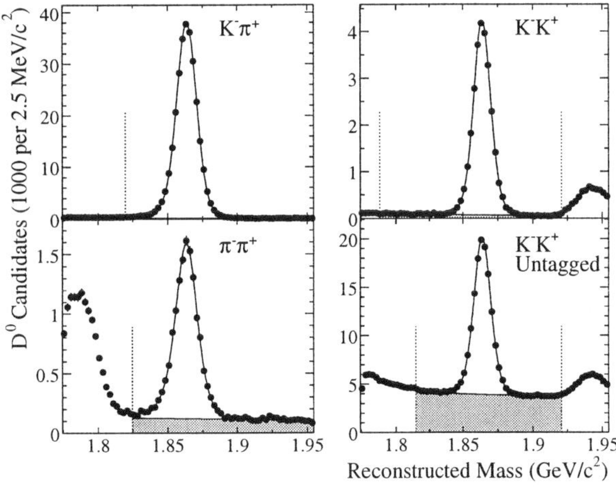

Figure 5. BaBar $y_{\rm CP}$ analysis:[16] invariant mass distribution for data (points), projection of the fit (curve), and fitted background component (shaded) for the four samples.

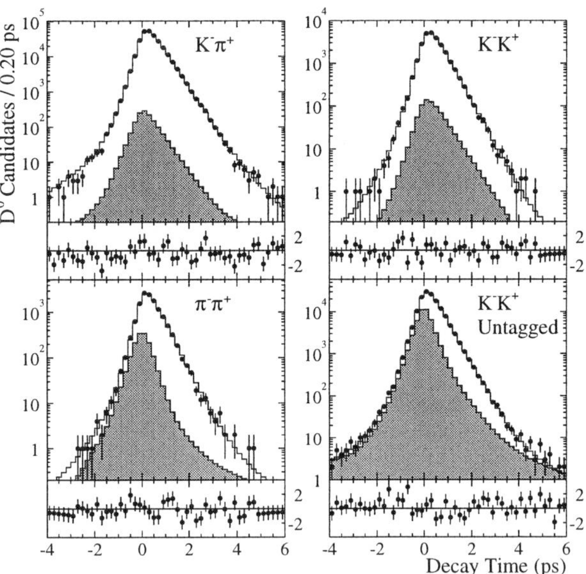

Figure 6. BaBar $y_{\rm CP}$ analysis:[16] proper time distribution for data (points), UML fit projection (open histogram), and fitted background component (shaded) for the four samples. The points beneath each plot show the bin-by-bin differences between the data and the fit, divided by the statistical error.

discussed above, the most natural interpretation of percent-level $y_{\rm CP}$ would be a large lifetime difference parameter y, due to Standard Model effects. The new physics interpretation was possible, but somewhat forced: mixing with $x \gg y$ and large CP violation, $\sin\phi \sim O(1)$, $\Delta \sim O(1)$.

3.3.2. $y_{\rm CP}$: Belle and CLEO (2002)

New measurements have followed in short order, using D^0 produced in $e^+e^- \to c\bar{c}$ interactions at the B-factories. Belle[14] used a sample somewhat larger and cleaner than that of FOCUS, performing an unbinned maximum-likelihood (UML) fit to inclusive $D^0 \to K^-\pi^+$ and K^+K^- decay time distributions, and measuring $y_{\rm CP} = \left(-0.5 \pm 1.0^{+0.7}_{-0.8}\right)\%$. CLEO[15] used a smaller sample (the $9.0\,{\rm fb}^{-1}$ CLEO II.V run), required a D^*-tag, and added $D^0 \to \pi^+\pi^-$ to the usual modes; they measured $y_{\rm CP} = \left(-1.2 \pm 2.5 \pm 1.4\right)\%$. Both results are manifestly consistent with zero, and each other—and the fact that both are negative has led many to discount the FOCUS result. But it's worth noting that the average $y_{\rm CP}$ from the three is positive, and $\sim 1\%$.

3.3.3. $y_{\rm CP}$: BaBar (2003)

This year BaBar has released a comprehensive $y_{\rm CP}$ measurement[16] based on $91\,{\rm fb}^{-1}$ of data including both D^*-tagged $D^0 \to K^-\pi^-$, K^+K^-, $\pi^+\pi^-$ events, and inclusive $D^0 \to K^+K^-$. As usual in e^+e^- analyses, backgrounds are suppressed by a center-of-mass momentum cut and vertex quality cut on the D^0, particle identification (PID) cuts on the daughter tracks, and (for the D^*-tagged samples) cuts on the D^*-decay pion (the "slow pion"). The resulting D^0 samples are shown in Fig. 5. For each sample, the fitted mass distribution is used to determine the event-by-event probability that a D^0 candidate belongs to the signal, as opposed to the background under the peak. This probability is then included in the likelihood function for each event in an UML fit to the proper-time distribution.

These distributions, and the results of the fits for each sample, are shown in Fig. 6. In each fit, the assumed "underlying" distributions of both signal and background are convolved with resolution functions based on a sum of gaussians: most of these terms have widths of the form $S\sigma_t^i$, where σ_t^i is the event-by-event proper time error, and S is a scaling factor, meant to account for deficiencies in modelling of the detector, etc.. This method is common to the earlier Belle[14] and CLEO[15] analyses and reflects a consensus on time-distribution fitting at the B-factories.

A blind analysis was performed to obtain the mixing parameter: the weighted average over the four modes is $y_{\rm CP} = \left(0.8 \pm 0.4^{+0.5}_{-0.4}\right)\%$, the most precise measurement to date.

3.3.4. $y_{\rm CP}$: Belle, D^*-tagged (2003)

A new analysis from Belle, contributed to this symposium,[17] takes a different approach. The idea is to find a robust resolution function which does not rely on the estimated proper-time error: this allows binned ML fits to be used throughout, so that the goodness-of-fit can be explicitly checked. D^*-tagged $D^0 \to K^-\pi^+$ and K^+K^- events from the full Belle dataset of $158\,{\rm fb}^{-1}$ are used, subject to standard reconstruction cuts, and the requirement that the proper time be well-measured.

The D^0 lifetime, $y_{\rm CP}$, and the parameters of the proper-time resolution function are all determined in a single simultaneous binned ML fit to the $K^-\pi^+$ and K^+K^- samples. The form of the resolution function is simple: a sum of five gaussians with a common mean, fixed relative normalizations, and floating widths. The gaussian widths for K^+K^- are constrained to be the same as those for $K^-\pi^+$, up to a single scale factor which is common to all terms. This parameterization has been studied using Monte Carlo (MC) data, and proves to be very stable: all values determined in a full decay-time fit match those fitted to the true resolution function, within their relative errors (Table 1).

Table 1. Belle $y_{\rm CP}$ analysis:[17] comparison of the parameters obtained in MC from a fit to the resolution function, using MC truth information (3rd column) and from the decay time fit, using reconstructed information only (4th column). The fractions of the five gaussian terms, fixed from the resolution function fit, are also shown (2nd column).

par.	fraction (%)	fitted values (fs; except α) resolution fit	lifetime fit
σ_1	26.1	95.1 ± 1.3	94.4 ± 1.7
σ_2	50.4	177.0 ± 2.2	179.0 ± 1.2
σ_3	19.8	328.7 ± 7.4	328.2 ± 2.2
σ_4	3.1	675.7 ± 24.9	664.4 ± 8.5
σ_5	0.6	2199 ± 95	2225 ± 70
X_0 [common shift]		-1.51 ± 0.22	-0.95 ± 0.54
α [$K\pi \to KK$ scale]		1.043 ± 0.004	1.042 ± 0.007

The proper time distribution for data is shown in Fig. 7, together with the result of the binned ML fit; the confidence level is 94%. (All fits in the analysis have an acceptable confidence level.) The fit returns a D^0 lifetime (from the $K^-\pi^+$ sample) of $412.6 \pm 1.1\,{\rm fs}$, consistent with the world average,[18] and a mixing parameter $y_{\rm CP} = (1.15 \pm 0.69 \pm 0.38)\%$; the result is preliminary.

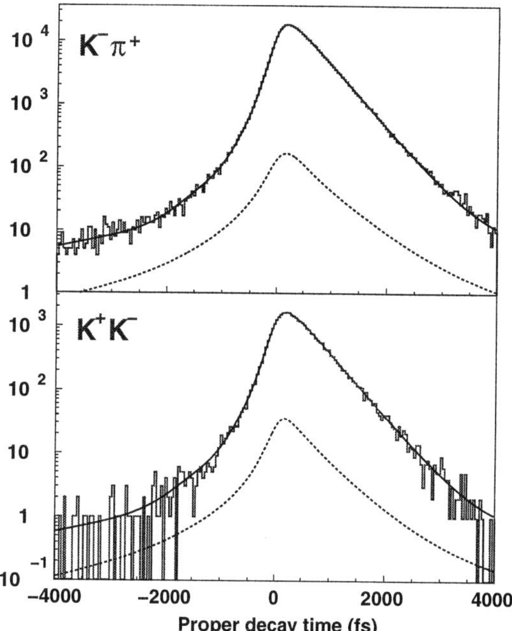

Figure 7. Belle $y_{\rm CP}$ analysis:[17] proper time distribution for data (histogram), binned ML fit (solid curve), and background component (dashed), for the $K^-\pi^+$ (upper) and K^+K^- (lower) samples.

3.3.5. $y_{\rm CP}$: summary of results

For completeness, the $y_{\rm CP}$ results are summarized in Table 2. Note the dominance of the numbers from Belle (the tagged result,[17] as noted, is preliminary) and BaBar. The implications of these results were not addressed at this point in the talk, but in the Discussion at the end. We preserve this order here.

Table 2. Summary of $y_{\rm CP}$ results.

Experiment	$y_{\rm CP}$ (%)
E791[19]	$0.8 \pm 2.9 \pm 1.0$
FOCUS[13]	$3.4 \pm 1.4 \pm 0.7$
Belle, untagged[14]	$-0.5 \pm 1.0 \pm 0.8$
CLEO[15]	$-1.2 \pm 2.5 \pm 1.4$
BaBar[16]	$0.8 \pm 0.4^{+0.5}_{-0.4}$
Belle, tagged[17]	$1.2 \pm 0.7 \pm 0.4$

3.3.6. "$y_{\rm CP}++$": D^0-$\overline{D}^0$ mixing and CPV

In $y_{\rm CP}$ analyses based on D^*-tagged samples, a CP-violation study comes "for free": one can simply compare the lifetime of K^+K^- events with different flavor D-tags, within the same analysis framework used for y_{CP}. Belle[17] defines a parameter

$$A_\Gamma \equiv \frac{\hat{\Gamma}(D \to KK) - \hat{\Gamma}(\overline{D} \to KK)}{\hat{\Gamma}(D \to KK) + \hat{\Gamma}(\overline{D} \to KK)}$$

$$\approx -\Delta . y \cos\phi - x \sin\phi,$$

where the notation $\hat{\Gamma}$ for "effective lifetime" recognises that an exponential is being fitted to distributions which may not be strictly exponential. In the absence of CP violation in mixing, i.e. $\Delta = 0$, the asymmetry parameter $A_\Gamma = -x \sin\phi$, measuring the CP violating phase $\phi = \arg(q\bar{A}/pA)$ due to the interference of decay and mixing.[d] The "ΔY" parameter of BaBar[16] is similar, differing by a factor of $(1 + y_{CP})$. Most systematic errors cancel due to the use of a common final state. The experiments find the following values,

$$\Delta Y = (-0.8 \pm 0.6 \pm 0.2)\% \quad \text{(BaBar)}$$
$$A_\Gamma = (-0.2 \pm 0.6 \pm 0.3)\% \quad \text{(Belle prelim.)},$$

consistent with zero; the measurements are statistically dominated and will continue to improve for the life of the B-factories.

Unlike mixing in general, CP violation associated with mixing is a robust new physics signal:[20] all charm mixing phenomena in the SM are dominated by the first two generations, so CP violation must be small. Even for $x \sim y = O(1\%)$, we expect $A_\Gamma \lesssim 10^{-4}$. So any significant non-zero measurement by this technique would be evidence of new physics contributing to mixing. One can imagine a scenario where *both* Standard Model and new physics processes lead to percent-level values of the mixing parameters (y and x respectively), and the new physics contribution leads to 30%-level CP violation: evidence for both SM mixing (via y_{CP}) and non-SM processes (via A_Γ) would emerge by the end of the B-factory era.

3.4. *Mixing:* $D^0 \to K^+\pi^-$

Another new BaBar analysis,[21] of $D^0 \to K^+\pi^-$ decays, has brought hadronic mixing analyses back into prominence. Pioneered by CLEO,[22] the sophisticated analysis method is sensitive to both mass- (x) and lifetime-splitting (y) of the neutral D eigenstates, and has been considered the technique of

choice for e^+e^- machines. The BaBar analysis has been gestating for some time—there were preliminary presentations (without final fit results) two years ago—and the related analysis at Belle is still underway, with only an intermediate result (the "wrong-sign rate" for $K\pi$) in the public domain.[23]

3.4.1. $D^0 \to K^+\pi^-$: *the analysis method*

"Wrong-sign" hadronic decays such as $D^0 \to K^+\pi^-$ occur via two paths: mixing $D^0 \to \overline{D}^0$ followed by Cabibbo-favoured decay $\overline{D}^0 \to K^+\pi^-$, and directly by doubly-Cabibbo-suppressed (DCS) decay $D^0 \to K^+\pi^-$. The DCS decay thus forms a background to the mixing signal, and the two must be separated by reconstructing the decay time of the D^0 and exploiting the different time distributions: e^{-t} for DCS decay and $t^2 e^{-t}$ for mixing, where the proper time t is in units of the D^0 lifetime. The interference term between the DCS and mixing paths, which goes as te^{-t}, cannot be neglected:[24] in fact it provides most of the mixing sensitivity, since the DCS rate is much larger than the mixing rate and the interference term is thus intermediate in size. A complication of the method is that this term is proportional, not to the lifetime difference parameter y, but to the quantity $y' \equiv y \cos\delta_{K\pi} - x \sin\delta_{K\pi}$ which has been "rotated" by the strong phase difference $\delta_{K\pi}$ between the $D^0 \to K^+\pi^-$ and $K^-\pi^+$ decays.

The subtle difference in time distributions (e^{-t}, te^{-t}, $t^2 e^{-t}$) means that the time structure of background events must be well-understood to avoid faking a mixing signal. This is important as background levels are relatively high. The method used by CLEO,[22] which has been followed by both BaBar[21] and Belle,[23] is to (1) tag the initial D flavor by reconstructing $D^{*+} \to D^0\pi^+$; (2) categorize the backgrounds according to their proper-time distribution; (3) measure their relative levels by fitting the data distribution in (M, Q) where M is the $K\pi$ mass and $Q = M(K^+\pi^-\pi^+) - M(K^+\pi^-) - m_\pi$ is the energy release in D^{*+} decay;[e] and (4) fix the background levels in the fit to proper time. This is difficult, but manageable, and CLEO reported limits at the 95% confidence level of $\frac{1}{2}x'^2 < 0.041\%$ and $-5.8\% < y' < 1.0\%$ based on this method.

---

[d]The analysis is thus a close analogue of the $B^0/\overline{B}^0 \to \psi K^0_S$ analysis in the b-sector, measuring $\sin 2\phi_1$ [$\equiv \sin 2\beta$].

[e]BaBar uses $\delta m \equiv M(K^+\pi^-\pi^+) - M(K^+\pi^-)$, which differs from Q by a constant.

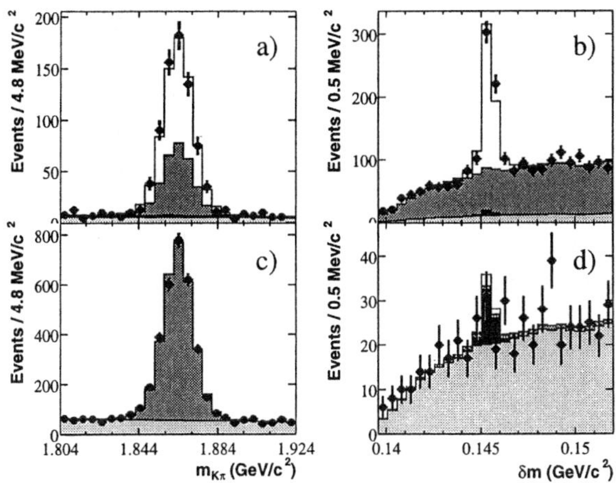

Figure 8. BaBar $D^0 \to K^+\pi^-$ analysis:[21] (a,c) $K^+\pi^-$ mass and (b,d) $D^{*+} - D^0$ mass-difference distributions, for (a,b) signal- and (c,d) background-dominated regions. Shown are the data (points), the projection of the fit (open histogram), and the fitted combinatorial (light), unassociated pion (dark), and double-misidentification background (black).

3.4.2. $D^0 \to K^+\pi^-$: the BaBar analysis

The BaBar analysis follows the CLEO model closely, but with a larger data sample ($57.1\,\mathrm{fb}^{-1}$), lower backgrounds (due to BaBar's superior PID), and a data selection and fitting procedure finalized while remaining "blind" to the mixing results. The $K^+\pi^-$ distributions are shown in Fig. 8: background events are divided into combinatorial, true-D^0-plus-unassociated-pion, and double-misidentification categories. This last type, where $K^-\pi^+$ is misidentified as π^-K^+, is small but significant: see Fig. 8(b,d). Such events are retained to avoid any distortion of the other backgrounds due to targeted rejection cuts.

The fit used to establish the background levels describes the data well.

The mixing parameters are then determined using unbinned, extended maximum-likelihood fits to the $D^0 \to K^+\pi^-$ ("wrong sign") and $D^0 \to K^-\pi^+$ ("right sign") data. The likelihood terms for the signal and various background time distributions are formed from underlying distributions (exponentials or delta functions) convolved with event-dependent resolution functions similar to those of the y_{CP} analysis (Sec. 3.3.3); the event-by-event signal- and background-fractions are determined from the $(m_{K\pi}, \delta m)$ fit. The results are shown in Fig. 9 for regions in $(m_{K\pi}, \delta m)$ dominated by signal (Fig. 9(a)) and background (Fig. 9(b)) events.

Four overall fits allowing DCS decay only; DCS decay and CP violation (treating D^0 and $\overline{D}^0$ separately); DCS decay and mixing, but no CP violation; and all three effects, are performed: the results are shown in Table 3. No evidence for mixing or CP violation is found. A slightly negative (and thus unphysical) value of x'^2 is preferred by the fit: this is taken into account in reporting the results.

Table 3. BaBar $D^0 \to K^+\pi^-$ analysis:[21] parameters returned by the full fit to D^0, $\overline{D}^0$, and combined samples.

Fit case	Para-meter	Fit result ($/10^{-3}$)		
		D^0	$\overline{D}^0$	$D^0 + \overline{D}^0$
Mixing allowed	$R_{WS}^{(\pm)}$	3.9	3.2	3.6
	$x'^{(\pm)2}$	-0.79	-0.17	-0.32
	$y'^{(\pm)}$	17	12	13
No mixing	$R_{WS}^{(\pm)}$	3.9	3.2	3.6

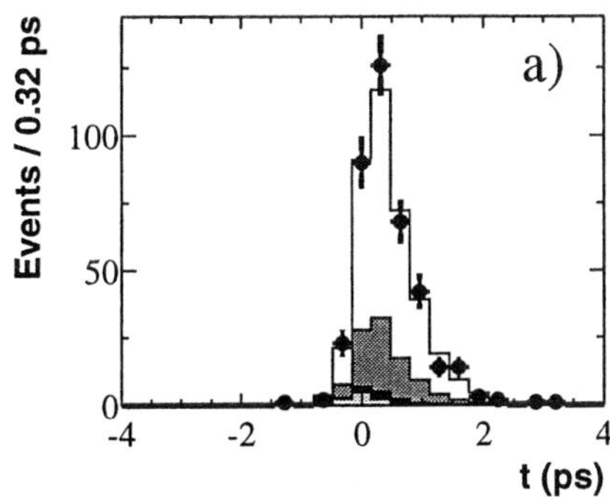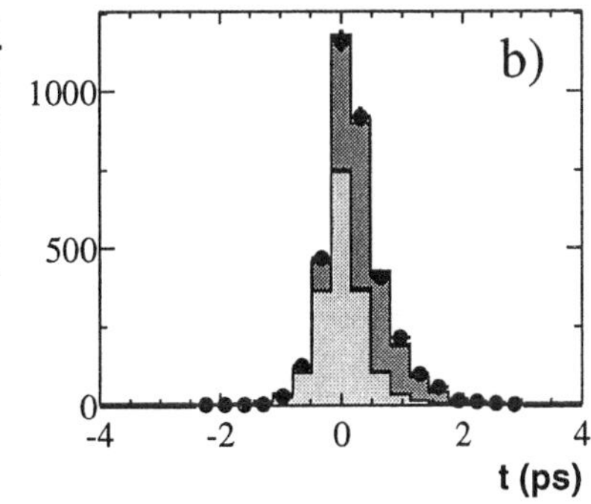

Figure 9. BaBar $D^0 \to K^+\pi^-$ analysis:[21] proper time distributions for (a) signal- and (b) background-dominated regions. Shown are the data (points), the projection of the fit (open histogram), and the fitted combinatorial (light), unassociated pion (dark), and double-misidentification background (black).

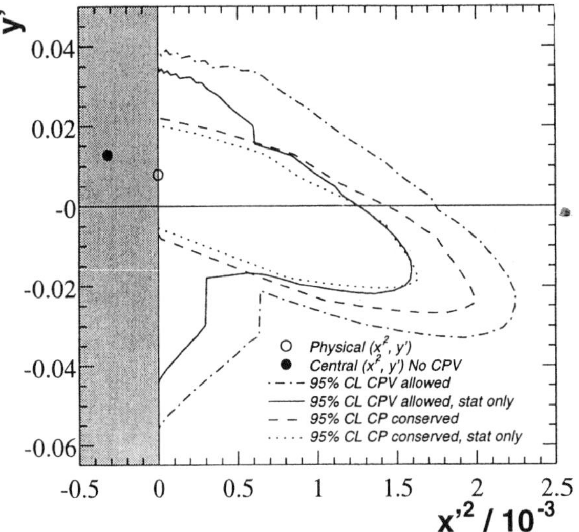

Figure 10. BaBar $D^0 \to K^+\pi^-$ analysis:[21] fit results and confidence intervals for the mixing parameters (x'^2, y').

A rather careful procedure based on toy MC experiments is used to set frequentist confidence intervals in the fitting parameters x'^2 and y': the results are shown in Fig. 10. A remarkable feature of the analysis is that the allowed region in (x'^2, y') is comparable in size to that of CLEO, despite the larger and cleaner dataset. Simulations by both BaBar[25] and Belle show that for a given experiment, when the preferred value has $y' > 0$ (as in the BaBar analysis), the allowed region becomes large compared to that when the data prefers $y' < 0$ (as in CLEO's case).[f] It is thus almost meaningless to "average" the results of experiments whose preferred values fall in different y' regions.

3.5. *Mixing: Issues and Future Measurements*

A combined $D^0 \to K^+\pi^-$ result will require a combined analysis: perhaps there is a need for a joint working group, or at least detailed consultation be-

---

[f] This may also be understood in qualitative terms using the proper time distribution for wrong-sign decays,

$$e^{-t}\left(R_D + \sqrt{R_D}y't + \frac{x'^2 + y'^2}{4}t^2\right),$$

where R_D is the DCS decay rate. For $y' < 0$ there is a partial cancellation between the interference ($\propto te^{-t}$) and mixing ($\propto t^2 e^{-t}$) terms, and the fit becomes sensitive to small changes in both the x'^2 and y' parameters. For $y' > 0$, no such cancellation takes place.

tween experiments, as we enter the CLEO-c era. Comparison of (x', y') and y_{CP} results is more complicated still. The strong phase difference $\delta_{K\pi}$, bequeathed us by the mischievous god of the color force, must first be measured: a significant shift might occur due to final-state interactions (FSI), shown to be significant in $D \to hh$ decays by isospin analyses of CLEO[26] and FOCUS;[27] the latter study provides evidence for inelastic FSI.

The default option is to wait for results from the CLEO-c run at the $\psi(3770)$, as one of the analyses exploiting the coherent $D^0\overline{D}^0$ state promises an error in $\cos\delta_{K\pi}$ of order ± 0.05.[28,29] The other option is a complete measurement of the DCS $D \to K\pi$ decays, of which only $D^0 \to K^+\pi^-$ is currently known. CLEO has recently placed a limit on $D^+ \to K^+\pi^0$;[26] a measurement is presumably within the reach of the other B-factory experiments. Measurement of $D^{+,0} \to K^0\pi^{+,0}$ rates relies on the measurement of $D^{+,0} \to K_L^0\pi^{+,0}$ decays: the asymmetry with the corresponding K_S^0 mode is proportional to the interference between decay amplitudes via K^0 and $\overline{K}^0$. A method for this measurement has been demonstrated by Belle, with a preliminary result for $D^0 \to K_L^0\pi^0$.[30]

A promising new analysis method exploits the $D^0 \to K_S^0\pi^+\pi^-$ final state: Cabibbo-favoured (e.g. $K^{*-}\pi^+$) and doubly-suppressed ($K^{*+}\pi^-$) decays interfere, allowing measurement of the strong phase differences for the various resonant submodes; a study of time-dependence then yields the mixing parameters x and y. This method has the advantages of superior scaling properties (the fit measures x rather than x'^2) and sensitivity to the sign of the mass splitting, in addition to the measurement of phases. It is, however, unproven. It has been championed by CLEO, who have measured the resonant substructure of the decay;[31] a mixing study will presumably require the large samples available at the other B-factories. Those samples may also provide useful sensitivity from semileptonic decays, which have fallen out of favor, although there is an interesting unpublished $D^0 \to K^{(*)+}\ell^-\bar{\nu}_\ell$ analysis by CLEO.[32]

Fits for CP violating effects in mixing are now routine, and will increasingly become the main focus of study. By the next Lepton Photon meeting, the other major development will be CLEO-c analyses exploiting opposite-side tagging, geometric signal-to-background ratios, and a coherent initial state leading to new mixing and CP violation observables.

4. CPT and Lorentz Invariance Violation

Even more general analyses would allow for violation of CPT symmetry, a manifest signal of new physics. While such studies have been carried out for kaons and B mesons, no CPT violation search had been performed in the charm sector until the recent FOCUS analysis.[33] Using D^*-tagged $D^0 \to K^-\pi^+$ decays, they searched for indirect CPT violation parametrized by $\xi \equiv (\Lambda_{11} - \Lambda_{22})/(\lambda_1 - \lambda_2)$, where Λ is the 2×2 effective Hamiltonian governing the time evolution of neutral D mesons, Λ_{ii} are its diagonal elements, and λ_j its eigenvalues. The measured quantity is the time-dependent rate asymmetry

$$A_{\rm CPT}(t) \equiv \frac{\Gamma(\overline{D}^0 \to K^+\pi^-) - \Gamma(D^0 \to K^-\pi^+)}{\Gamma(\overline{D}^0 \to K^+\pi^-) + \Gamma(D^0 \to K^-\pi^+)},$$

which reduces to $(\mathrm{Re}(\xi)y - \mathrm{Im}(\xi)x)\,\Gamma t$ for $xt, yt \ll 1/\Gamma$; x and y are the mixing parameters discussed above. FOCUS finds $(\mathrm{Re}(\xi)y - \mathrm{Im}(\xi)x) = 0.0083 \pm 0.0065 \pm 0.0041$, with a 95% confidence interval $-0.0068 < (\mathrm{Re}(\xi)y - \mathrm{Im}(\xi)x) < 0.0234$, consistent with zero. Their paper cites as an example the case where D^0 and $\overline{D}^0$ mix with parameters $(x, y) = (0.0, 0.01)$: in this scenario, $0.68 < \mathrm{Re}\,\xi < 2.34$.

Within a formalism that allows for violation of Lorentz invariance, ξ may depend on the vector momentum of the studied particles, and on siderial time; the relation is a function of parameters[34] $\Delta a_{0,X,Y,Z}$, where (X, Y, Z) is a non-rotating coordinate system. FOCUS fits for these quantities by measuring the CPT-violating parameter ξ in bins of siderial time: we can summarize the (complicated) result by noting that the various $|\Delta a_\mu| < O\left(10^{-12}\right)$ GeV at 95% confidence for the case $(x, y, \delta_{K\pi}) = (0.01, 0.01, 15°)$. This is to be compared with limits of order 10^{-21} for the K^0-$\overline{K}^0$ system: the difference reflects both the size of the available data samples and the relative strength of mixing in the two systems. The Δa_μ may in principle vary with flavor, so this measurement is important for completeness, even though the sensitivity does not compete with that for kaons.

5. Rare and Forbidden Decays

Moving to less exotic possibilities, flavor-changing neutral currents (FCNC) have not yet been observed in the charm sector. As with mixing, the SM parton-level loop contributions are subject to powerful cancellations, and long-distance contributions dominate.

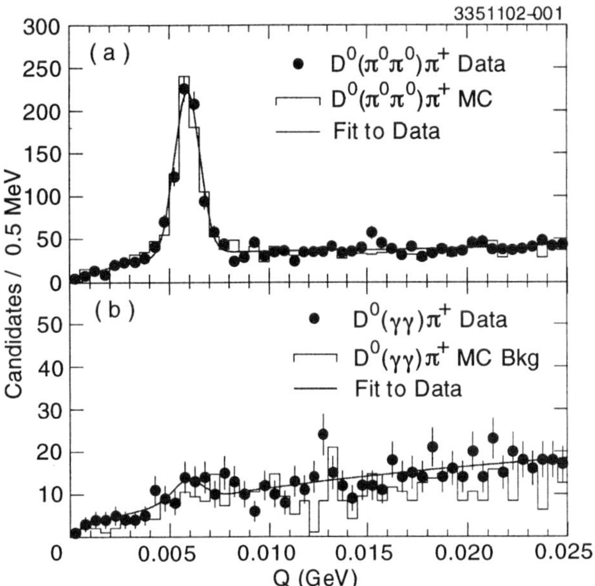

Figure 11. CLEO $D^0 \to \gamma\gamma$ analysis:[36] $D^{*+} - D^0$ mass-difference distributions for (a) $D^0 \to \pi^0\pi^0$ and (b) $D^0 \to \gamma\gamma$.

As a result, it can be difficult to calculate the SM expectation with certainty. The best new physics signals are those decays with extremely small predicted rates, or special features which allow SM and non-SM contributions to be distinguished.

The decay $D^0 \to \gamma\gamma$, with an expected branching fraction of order 10^{-8}, is thus a new physics signal (Sec. 5.1); but $D^0 \to \mu^+\mu^-$ (Sec. 5.2), whose predicted rate is five orders of magnitude smaller, is a more reliable one.[35] In the case of decays $D \to h\ell^+\ell^-$, the most robust new physics signal is not the decay rate, but the dilepton mass spectrum, which exhibits marked differences between SM and new-physics predictions (Sec. 5.3). And in each case, variant modes violating lepton flavor or number conservation can be added "for free" to the analysis, in the spirit of a lamp-post search: there is no uncertainty on Standard Model predictions of zero. We consider each of these cases in turn, before treating decays $D^0 \to V\gamma$ in the final section (Sec. 5.4).

5.1. $D^0 \to \gamma\gamma$ (CLEO)

Decays to photons have very small contributions from parton-level processes, but significant ones from the vector meson dominance (VMD) mechanism: $\mathcal{B}(D^0 \to \gamma\gamma)$ would be $\sim 3 \times 10^{-11}$ if only short-distance mechanisms contributed, but is expected to be of order 10^{-8} due to long-distance SM processes. This mode has not previously been studied. CLEO has recently used $D^{*+} \to D^0 \pi^+$ events in $13.8\,\text{fb}^{-1}$ of data to conduct a $D^0 \to \gamma\gamma$ search, normalizing to the $D^0 \to \pi^0\pi^0$ mode.[36] Under standard cuts, augmented with a $\pi^0 \to \gamma\gamma$ veto on the photons forming the $D^0 \to \gamma\gamma$ candidates, they accumulate fairly clean event samples and then fit the distribution of energy release Q from the D^* decay, to keep differences between the $\pi^0\pi^0$ and $\gamma\gamma$ modes to a minimum. The results are shown in Fig. 11: the agreement between the data and the MC simulation is remarkable.

A limit $\mathcal{B}(D^0 \to \gamma\gamma) < 2.9 \times 10^{-5}$ is found at the 90% confidence level, some three orders of magnitude above the SM prediction. There is thus some room for a new physics signal if future experiments can improve on CLEO's sensitivity.

5.2. $D^0 \to \mu^+\mu^-$ (CDF)

The expected rate for $D^0 \to \mu^+\mu^-$ is much smaller, $O(10^{-13})$, whereas R-parity violating (RPV) Supersymmetry could lead to a branching fraction as high as 3.5×10^{-6}, just smaller than the previous experimental bound.[35] The dimuon decay is thus a straightforward new-physics search mode. Using their upgraded detector and trigger system, which allows them to select a charm decay sample, CDF have conducted a search for this mode in the early Run II data.[37] With a fairly straightforward blind analysis, using D^*-tagged events and $D^0 \to \pi^+\pi^-$ decay as a normalization mode, they observe no $\mu^+\mu^-$ events and set a limit $\mathcal{B}(D^0 \to \mu^+\mu^-) < 2.4 \times 10^{-6}$ at 90% confidence, improving on the previous bound by a factor of two. The $O(1)$ event background estimate relies on interpolation from the sidebands—events with true muon(s) dominate over the misidentification background—and will need to be better understood to significantly improve the limit. Presumably this is achievable, and improvements to this channel will depend on the progress of Run II data-taking.

5.3. $D \to h\ell\ell$ (FOCUS)

The only analysis of a "basket" of rare decay modes in recent times is by FOCUS, who have searched for decays $D^+_{(s)} \to h^\pm \mu\mu$, where $h^\pm = \pi^\pm, K^\pm$.[38] For definiteness we will take $D^+ \to \pi^+\mu^+\mu^-$ as an example. The predicted SM rate for this decay is $O(10^{-6})$, while (e.g.) the allowed R-parity violating

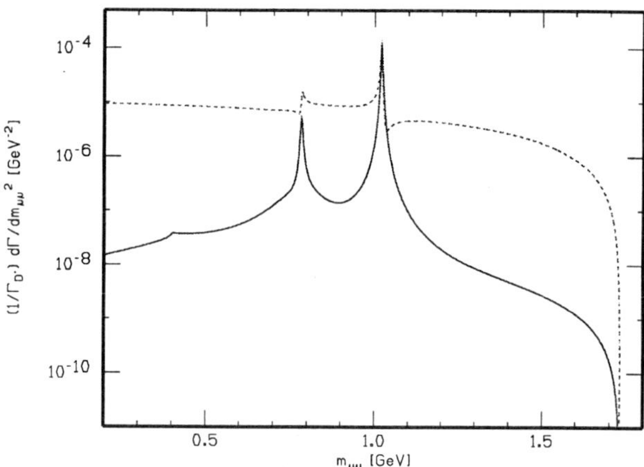

Figure 12. Predicted[35] dimuon mass distributions $M(\mu^+\mu^-)$ for $D^+ \to \pi^+\mu^+\mu^-$. The solid line shows the sum of the short- and long-distance contributions in the SM; the dotted line, the R-parity violating contribution from SUSY at the level allowed prior to the FOCUS measurement—see the text.

contribution,[35] 15×10^{-6}, saturates the previous experimental limit.[39] There would therefore seem to be potential for further restriction of RPV parameters, but not for observation of a new physics signal.

The SM contribution, however, is dominated by the path $D^+ \to \pi^+ V \to \pi^+\mu^+\mu^-$, where the V are vector mesons. The predicted dimuon mass spectrum (Fig. 12) thus shows pronounced peaks at the ρ and ϕ masses, which dominate the SM rate. By contrast the spectrum for RPV is relatively flat, so that a new physics contribution comparable to or even below the SM contribution could be resolved by comparing the $M(\mu^+\mu^-)$ distribution of observed events with the various predictions.

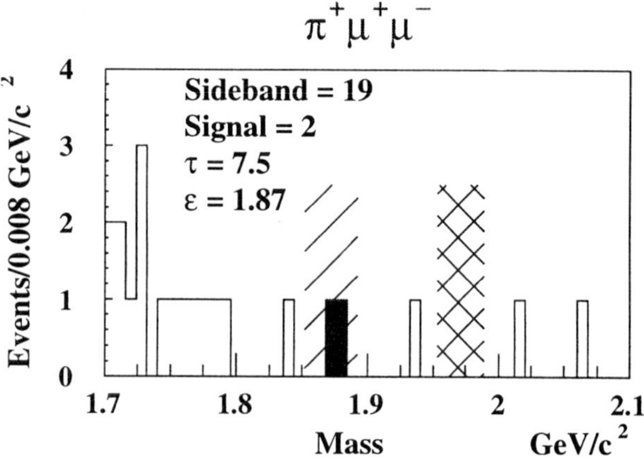

Figure 13. FOCUS $D^+ \to \pi^+\mu^+\mu^-$ analysis:[38] the histogram shows selected events; the filled entry is in the signal region. The signal region (hatched) and the region used for sideband subtraction (cross-hatched) are also shown.

FOCUS is beginning to probe this region. A histogram of selected events for the $D^+ \to \pi^+ \mu^+ \mu^-$ analysis is shown in Fig. 13, and results for all the modes are listed in Table 4. The analysis proceeds using standard FOCUS detached vertex, hadron- and muon-identification cuts; cut values are selected and background estimates calculated using a very careful "dual bootstrap" method, to minimise possible selection biasses. No evidence is seen for any of the decay modes; existing limits are everywhere improved, in some cases by an order of magnitude. For the $D^+ \to \pi^+ \mu^+ \mu^-$ and $D_s^+ \to \pi^+ \mu^+ \mu^-$ modes, the sensitivity is approaching the SM prediction.

Table 4. FOCUS $D \to h\ell\ell$ analysis:[38] measured limit on the branching fraction, SM prediction,[40] previous best limit,[39,41] and expected CLEO-c sensitivity[29] for each mode. (D_s^+ sensitivities are scaled from those of D^+ and are not official CLEO-c numbers.) All entries are ($/10^{-6}$).

Mode	FOCUS	SM	Prev.	CLEO-c
$D^+ \to K^+ \mu^- \mu^+$	9.2	0.007	44	1.5
$D^+ \to K^- \mu^+ \mu^+$	13	-	120	1.5
$D^+ \to \pi^+ \mu^- \mu^+$	8.8	1.0	15	1.5
$D^+ \to \pi^- \mu^+ \mu^+$	4.8	-	17	1.5
$D_s^+ \to K^+ \mu^- \mu^+$	36	0.043	140	15
$D_s^+ \to K^- \mu^+ \mu^+$	13	-	180	15
$D_s^+ \to \pi^+ \mu^- \mu^+$	26	6.1	140	15
$D_s^+ \to \pi^- \mu^+ \mu^+$	29	-	82	15

Further progress is expected at CLEO-c, whose sensitivities[29] are also shown: an improvement by a factor $3 \sim 8$ is foreseen for the D^+ modes, reaching the SM expectation in the case of $D^+ \to \pi^+ \mu^+ \mu^-$ and therefore restricting further the RPV contribution. Any signal in the lepton-number violating modes $D_{(s)}^+ \to h^- \mu^+ \mu^+$, albeit unexpected, would of course be an observation of new physics.

5.4. $D^0 \to \phi\gamma$, $\phi\pi^0$, $\phi\eta$ (Belle)

Finally we turn to radiative decays $D^0 \to V\gamma$, another vector meson dominance process in the Standard Model. The Belle collaboration has conducted a search for $D^0 \to \phi\gamma$, exploiting double kaon identification in $\phi \to K^+ K^-$ to suppress backgrounds.[42] Theoretical estimates for this mode,[43,44] dominated by $D^0 \to VV' \to V\gamma$ (where the $V^{(\prime)}$ are vector mesons), are in the range $(0.04 \sim 3.4) \times 10^{-5}$, well below the previous limit of 1.9×10^{-4} but partially overlapping the sensitivity at the B-factories.

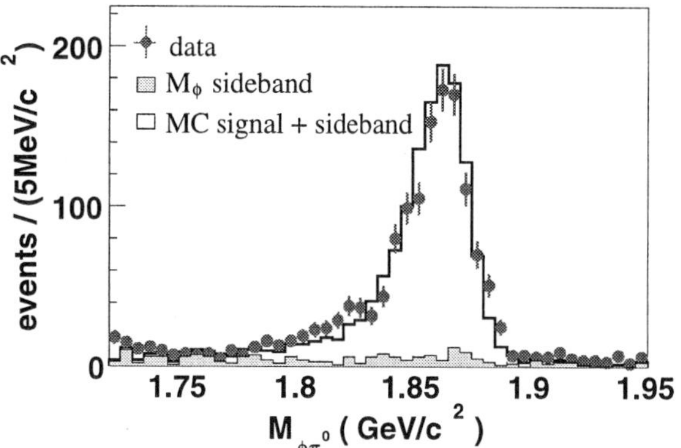

Figure 14. Belle $D^0 \to \phi\gamma$, $\phi\pi^0$, $\phi\eta$ analysis:[42] $\phi\pi^0$ mass distribution for data (points) and MC (histogram); the ϕ-mass sideband is also shown (shaded).

D^*-tagging and cuts on the D^* momentum and γ energy are used to suppress the various combinatorial backgrounds. The dominant remaining background is due to the Cabibbo- and color-suppressed decays $D^0 \to \phi\pi^0$ and $\phi\eta$, which have not previously been observed. With analagous cuts to select $D^0 \to \phi\pi^0$ Belle sees a very clear signal in $M(\phi\pi^0)$ (Fig. 14) and the expected distribution of the helicity angle of the ϕ meson (not shown); the ϕ is polarized in the D^0 decay. A smaller but still clear signal of 31 ± 9.8 events is seen for $D^0 \to \phi\eta$, where a veto is imposed on photons for the $\eta \to \gamma\gamma$ candidate consistent with belonging to a $\pi^0 \to \gamma\gamma$ decay.

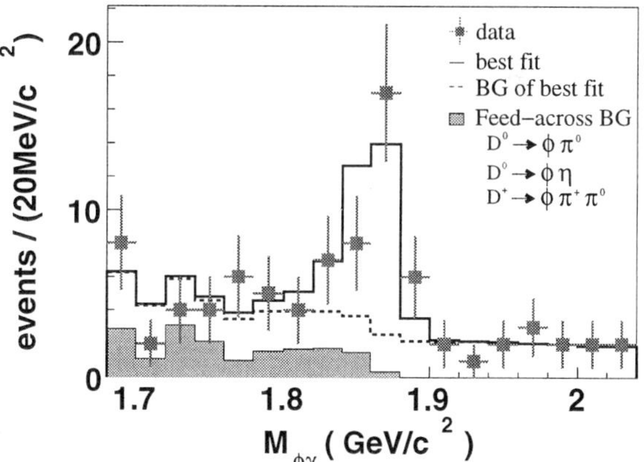

Figure 15. Belle $D^0 \to \phi\gamma$, $\phi\pi^0$, $\phi\eta$ analysis:[42] $\phi\gamma$ mass distribution for data (points), the ML fit (open histogram), the background component of the fit (dashed), and the sum of $D^0 \to \phi\pi^0$, $\phi\eta$ and $D^+ \to \phi\pi^+\pi^0$ backgrounds (shaded).

The contribution of these decays to the $\phi\gamma$ spectrum can then be reliably estimated; it is suppressed by a helicity angle cut $|\cos\theta_{\mathrm{hel}}| < 0.4$, favoring the

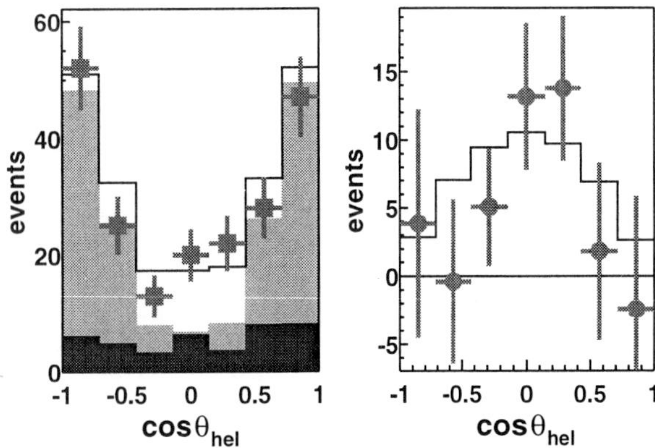

Figure 16. Belle $D^0 \to \phi\gamma$, $\phi\pi^0$, $\phi\eta$ analysis:[42] cross-check of the $\phi\gamma$ result using the $\cos\theta_{\mathrm{hel}}$ distribution. Left plot: data (points), and MC predictions for the total (open histogram), total background (light), and non-$\phi\pi^0$ background (dark). Right plot: background-subtracted distribution for the data (points) and the MC prediction (histogram).

transversely polarized ϕ of $D^0 \to \phi\gamma$ over the longitudinally polarized ϕ of the $\phi\pi^0$ and $\phi\eta$ modes. The resulting $\phi\gamma$ invariant mass spectrum shows a clear $D^0 \to \phi\gamma$ signal of $27.6^{+7.4+0.5}_{-6.5-1.0}$ events (Fig. 15), and is well-described by a linear combinatorial background, the expected $\phi\pi^0$ and $\phi\eta$ contribution, plus the signal. The helicity angle distribution (Fig. 16) is likewise consistent with expectations.

The final results of the analysis are

$$\mathcal{B}(D^0 \to \phi\pi^0) = (8.01 \pm 0.26 \pm 0.46) \times 10^{-4}$$

$$\mathcal{B}(D^0 \to \phi\eta) = (1.48 \pm 0.47 \pm 0.09) \times 10^{-4}$$

$$\mathcal{B}(D^0 \to \phi\gamma) = \left(2.60^{+0.70+0.15}_{-0.61-0.17}\right) \times 10^{-5};$$

the $\phi\gamma$ mode is the first radiative decay, and the first FCNC decay, observed in the D meson system. The measured branching fraction is at the upper end of the VMD predictions—consistent with Standard Model expectations—and is in no sense a new physics measurement, but it provides the first experimental reference point for predictions of other FCNC decays.

6. Summary and Prospect

While all observations are still consistent with Standard Model expectations, there has been significant progress in charm analyses sensitive to new physics effects. The D-mixing measurement using $D^0 \to K^+\pi^-$ is now mature, with a major result released by BaBar, and a Belle analysis in the pipeline; the B-factories will (presumably) exhaust this difficult technique. The sensitivity to y_{CP} continues to improve, and a 1% measurement is within reach of the current facilities during their projected life. Fitting for CP-violating effects has become standard, and as a robust test for new physics—and a technique still dominated by statistical errors—will become increasingly important in the field. And measurements of the new quantities observable at CLEO-c, and the promising but yet-untried $D^0 \to K^0_S\pi^+\pi^-$ mixing analysis, are eagerly awaited.

The first flavor-changing neutral current decay in the charm sector, $D^0 \to \phi\gamma$, has finally been seen, and is consistent with Standard Model expectations due to vector meson dominance. While not the most exciting possible channel—$\gamma\gamma$ or $\mu^+\mu^-$ would have been respectively a shock, and a total revolution in the field—the $\phi\gamma$ observation provides an experimentally-measured point where none previously existed, and will presumably allow more precise SM predictions for other FCNC decays in the future. In the search for other rare and exotic processes, there are continuing improvements in the range of channels studied (with the first $D^0 \to \gamma\gamma$ limit announced, and the first test of CPT/Lorentz invariance conducted) and the reach of existing analyses (with the $D \to h\ell\ell$ results from FOCUS). As with mixing, there will be significant contributions to these searches from CLEO-c in the next two years, and we all—including even the charm coordinators at BaBar, Belle, CDF, and FOCUS—are looking forward to new results in this new era.

Acknowledgments

I would like to thank the Lepton Photon 2003 Symposium organisers for the invitation to present this review; Lin Zhang, Marc Buehler, and Harry Cheung for their technical assistance with the presentation itself and this writeup; and my colleagues in the Belle collaboration, in particular the charm studies group, for providing such a stimulating working environment.

References

1. B. Aubert *et al.* (BaBar Collaboration), *Phys. Rev. Lett.* **90**, 242001 (2003).
2. D. Besson *et al.* (CLEO Collaboration), *Phys. Rev. D* **68**, 032002 (2003).

3. Y. Mikami *et al.* (Belle Collaboration), `arXiv:hep-ex/0307041`, accepted for publication in *Phys. Rev. Lett.*

4. P. Krokovny, A. Bondar *et al.* (Belle Collaboration), `arXiv:hep-ex/0308019`, accepted for publication in *Phys. Rev. Lett.*

5. J. de Miranda, "Lessons from Standard Charm Decays", in these proceeedings.

6. K. Abe *et al.* (Belle Collaboration), *Phys. Rev. Lett.* **89**, 142001 (2002).

7. K. Abe *et al.* (Belle Collaboration), BELLE–CONF–0331.

8. G.T. Bodwin, J. Lee, and E. Braaten, *Phys. Rev.* D **67**, 054023 (2003); *Phys. Rev. Lett.* **90**, 162001 (2003).

9. K. Abe *et al.* (Belle Collaboration), `arXiv:hep-ex/0306015`.

10. T. Skwarnicki, "Heavy quarkonium, Production and Spectroscopy", in these proceeedings; `arXiv:hep-ph/0311243`.

11. I.I. Bigi and N.G. Uraltsev, *Phys. Rev.* B **592**, 92 (2001).

12. A.F. Falk, Y. Grossman, Z. Ligeti, and A.A. Petrov, *Phys. Rev.* D **65**, 054034 (2002).

13. J.M. Link *et al.* (FOCUS Collaboration), *Phys. Lett.* B **485**, 62 (2000).

14. K. Abe *et al.* (Belle Collaboration), *Phys. Rev. Lett.* **88**, 162001 (2002).

15. S.E. Csorna *et al.* (CLEO Collaboration), *Phys. Rev.* D **65**, 092001 (2002).

16. B. Aubert *et al.* (BaBar Collaboration), *Phys. Rev. Lett.* **91**, 121801 (2003).

17. K. Abe *et al.* (Belle Collaboration), BELLE–CONF–0347, `arXiv:hep-ex/0308034`.

18. K. Hagiwara *et al.* (PDG), *Phys. Rev.* D **66**, 1 (2002).

19. E.M. Aitala *et al.* (E791 Collaboration), *Phys. Rev. Lett.* **83**, 32 (1999).

20. I.I. Bigi and A.I. Sanda, *CP violation* (Cambridge University Press, 2000), pp. 252–259.

21. B. Aubert *et al.* (BaBar Collaboration), *Phys. Rev. Lett.* **91**, 171801 (2003).

22. R. Godang *et al.* (CLEO Collaboration), *Phys. Rev. Lett.* **84**, 5038 (2000).

23. K. Abe *et al.* (Belle Collaboration), BELLE–CONF–0254, `arXiv:hep-ex/0208051`.

24. G. Blaylock, A. Seiden, and Y. Nir, *Phys. Lett.* B **355**, 555 (1995).

25. Ulrik Egede, private communication.

26. K. Arms *et al.* (CLEO Collaboration), CLEO–03–10, `arXiv:hep-ex/0309065`, submitted to *Phys. Rev. Lett.*

27. J.M. Link *et al.* (FOCUS Collaboration), *Phys. Lett.* B **555**, 167 (2003).

28. M. Gronau, Y. Grossman, and J.L. Rosner, *Phys. Lett.* B **508**, 37 (2001).

29. R.A. Briere *et al.* (CESR-c and CLEO-c Taskforce), CLNS–01/1742.

30. K. Abe *et al.* (Belle Collaboration), BELLE–CONF–0129, `arXiv:hep-ex/0107078`.

31. H. Muramatsu *et al.* (CLEO Collaboration), *Phys. Rev. Lett.* **89**, 251802 (2002); **90**, 059901(E) (2003); D. Asner *et al.* (CLEO Collaboration), `arXiv:hep-ex/0311033`, submitted to *Phys. Rev. Lett.*

32. S. McGee, to appear in the proceedings of Lake Louise Winter Institute on Fundamental Interactions (LLWI 02), Lake Louise, Alberta, Canada, 17–23 Feb 2002.

33. J.M. Link *et al.* (FOCUS Collaboration), *Phys. Lett.* B **556**, 7 (2003).

34. V.A. Kostelecký, *Phys. Rev.* D **64**, 076001 (2001).

35. G. Burdman, E. Golowich, J. Hewett, and S. Pakvasa, *Phys. Rev.* D **66**, 014009 (2002).

36. T.E. Coan *et al.* (CLEO Collaboration), *Phys. Rev. Lett.* **90**, 101801 (2003).

37. D. Acosta *et al.* (CDF Collaboration), FERMILAB-PUB-03/240-E, `arXiv:hep-ex/0308059`, submitted as a Rapid Communication to *Phys. Rev.* D.

38. J.M. Link *et al.* (FOCUS Collaboration), *Phys. Lett.* B **572**, 21 (2003).

39. E.M. Aitala *et al.* (E791 Collaboration), *Phys. Lett.* B **462**, 401 (1999).

40. S. Fajfer, S. Prelovsek, and P. Singer, *Phys. Rev.* D **64**, 114009 (2001).

41. P.L. Frabetti *et al.* (E687 Collaboration), *Phys. Lett.* B **398**, 239 (1997).

42. K. Abe *et al.* (Belle Collaboration), `arXiv:hep-ex/0308037`, submitted to *Phys. Rev. Lett.*

43. G. Burdman, E. Golowich, J. Hewett, and S. Pakvasa, *Phys. Rev.* D **52**, 6383 (1995).

44. S. Fajfer and P. Singer, *Phys. Rev.* D **56**, 4302 (1997); S. Fajfer, S. Prelovsek, and P. Singer, *Eur. Phys. J.* C **6**, 471 (1999).

DISCUSSION

Hal Evans (Columbia): What is the average y_{CP} measurement? Or is there a reason not to calculate it?

Bruce Yabsley: The average would be dominated by preliminary measurements from BaBar and Belle. Further, results appear to be "clumping" based on measurement technique indicating the possibility of systematic problems. That being said, the speaker's average is $\langle y_{CP} \rangle = (0.9 \pm 0.4)\%$

Further Discussion, Added since LP2003:

It's appropriate to return to this question in more detail for the written version of the talk. I reserved the y_{CP} average for the Discussion—assuming, correctly, that someone would bring it up—partly for lack of time and partly for lack of a clear idea of how to treat it: any average is dominated by a preliminary number from Belle[17] and results from BaBar which, though "final", had not been published at the time of the symposium. Since then, BaBar's paper has appeared in PRL,[16] so my reservations are now diminished. The situation, however, is still unclear.

Table 5. Expanded summary of y_{CP} results.

Technique	Experiment	y_{CP} (%)
Fixed target	E791[19]	$0.8 \pm 2.9 \pm 1.0$
	FOCUS[13]	$3.4 \pm 1.4 \pm 0.7$
	My average:	**2.9 ± 1.4**
e^+e^-, untagged	Belle[14]	$-0.5 \pm 1.0 \pm 0.8$
	CLEO[15]	$-1.2 \pm 2.5 \pm 1.4$
	BaBar[16]	$0.2 \pm 0.5^{+0.5}_{-0.4}$
	My average:	**0.0 ± 0.6**
e^+e^-, D^*-tagged	BaBar,[16] K^+K^-	$1.5 \pm 0.8 \pm 0.5$
	BaBar,[16] $\pi^+\pi^-$	$1.7 \pm 1.2^{+1.2}_{-0.6}$
	Belle,[17] K^+K^-	$1.2 \pm 0.7 \pm 0.4$
	My average:	**1.4 ± 0.6**
Speaker's grand average:		**0.9 ± 0.4**

The results from the various experiments are shown in Table 5, sorted by the type of experiment and the method used: fixed target (E791[19] and FOCUS[13]), e^+e^- with inclusive D^0 samples (Belle,[14] CLEO,[15] and BaBar[16]), and e^+e^- with D^*-tagged samples (BaBar[16] and Belle[17]). The individual results within the BaBar analysis have been listed sep-

arately for this purpose. There is a clear clustering of the results according to technique: fixed target measures high ($\langle y_{CP} \rangle = 2.9\%$), e^+e^- measures null ($\langle y_{CP} \rangle = 0.0\%$), and the D^*-tagged analyses measure an intermediate value ($\langle y_{CP} \rangle = 1.4\%$). These last results are completely dominant, as can be seen in Fig. 17, where the data are shown in the form of the "ideograms" used by the PDG.[18]

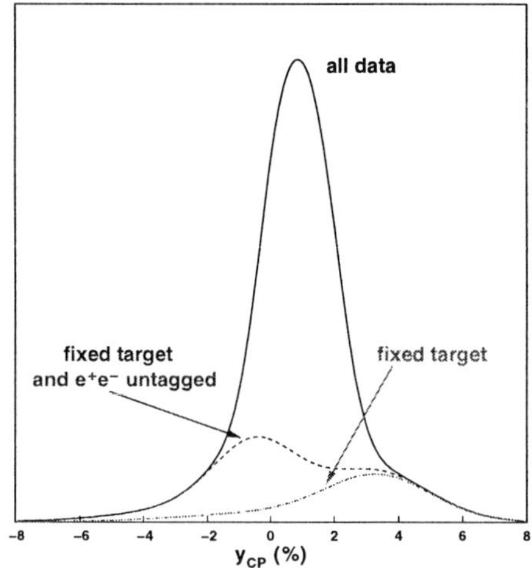

Figure 17. PDG-style "ideogram" of the y_{CP} data in Table 5: each measurement (with mean x_i and total error $\pm \delta x_i$) is represented by a gaussian with central value x_i, error δx_i, and area proportional to $1/\delta x_i$. The sum of all curves (solid), the sum of fixed-target and e^+e^- untagged data (dashed), and the fixed-target data alone (dot-dashed) are shown.

Statistically speaking, the data are consistent, and the average is now 2σ away from zero. But it seems to me a bit worrying that the different techniques don't actively corroborate each other. Put another way, having first become excited about the (positive) FOCUS result, and then rushed to discount it in the light of (negative) e^+e^- results, I think we should be slow to interpret D^*-tagged results which split the difference.

This question will become urgent after the next round of data-taking at the B-factories, if the central value stays at $y_{CP} \gtrsim 1\%$ as the total error shrinks. One convincing cross-check in this situation would be to analyse D^0 decays to CP-odd final states such as $K^0_S(\rho^0, \omega, \phi)$ and even $K^0_S \eta^{(\prime)}$ at the B-factories: for the same y_{CP} value, the shift in the lifetime has the opposite sign. I know of no such plans at this stage.

FLAVOR PHYSICS: CKM AND QCD

RESULTS ON THE CKM ANGLE ϕ_1 (β)

T. E. BROWDER

Department of Physics and Astronomy, University of Hawaii, 2505 Correa Road, Honolulu, HI 96822, USA
E-mail: teb@phys.hawaii.edu

I review results related to the CKM angle $\phi_1(\beta)$. These results include recent measurements of CP-violation from the BaBar and Belle experiments in $b \to c\bar{c}s$, $b \to c\bar{c}d$ and $b \to sq\bar{q}$ processes.

1. Introduction

1.1. *The B Physics Program*

The B physics program addresses several fundamental questions. Is the irreducible phase in the Cabibbo-Kobayashi-Maskawa (CKM) matrix the source of all CP-violating phenomena in the B system?[1] Or is CP-violation, the first manifestation of physics beyond the Standard Model? A related question is whether there are new CP-violating phases from physics beyond the Standard Model.[2]

The unitarity of the CKM matrix implies the existence of three measurable phases. In the convention favored at KEK and Belle, these are denoted

$$\phi_1 \equiv arg\left(-\frac{V_{cd}V_{cb}^*}{V_{td}V_{tb}^*}\right) \tag{1}$$

$$\phi_2 \equiv arg\left(-\frac{V_{ud}V_{ub}^*}{V_{td}V_{tb}^*}\right) \tag{2}$$

$$\phi_3 \equiv arg\left(-\frac{V_{cd}V_{cb}^*}{V_{ud}V_{ub}^*}\right). \tag{3}$$

while at SLAC and at BaBar these angles are usually referred to as β, α and γ, respectively.

As first noted by Bigi, Carter and Sanda,[3] there are large measurable CP-asymmetries in the decays of neutral B mesons to CP-eigenstates. In the decay chain $\Upsilon(4S) \to B^0\bar{B}^0 \to f_{CP}f_{\text{tag}}$, where one of the B mesons decays at time t_{CP} to a final state f_{CP} and the other decays at time t_{tag} to a final state f_{tag} that distinguishes between B^0 and $\bar{B}^0$, the decay rate has a time dependence given by[3]

$$\frac{e^{-\frac{|\Delta t|}{\tau_{B^0}}}}{4\tau_{B^0}}\left\{1 + q \cdot \left[\mathcal{S}\sin(\Delta m_d \Delta t) + \mathcal{A}\cos(\Delta m_d \Delta t)\right]\right\},$$

where τ_{B^0} is the B^0 lifetime, Δm_d is the mass difference between the two B^0 mass eigenstates, $\Delta t = t_{CP} - t_{\text{tag}}$, and the b-flavor charge $q = +1\,(-1)$ when the tagging B meson is a $B^0\,(\bar{B}^0)$. The CP-violation

parameters $\mathcal{S}$ and $\mathcal{A}$ are given by

$$\mathcal{S} \equiv \frac{2\mathcal{I}m(\lambda)}{|\lambda|^2 + 1}, \qquad \mathcal{A} \equiv \frac{|\lambda|^2 - 1}{|\lambda|^2 + 1}, \tag{4}$$

where λ is a complex parameter that depends on both the $B^0\bar{B}^0$ mixing and on the amplitudes for B^0 and $\bar{B}^0$ to decay to f_{CP}. To a good approximation, the SM predicts $\mathcal{S} = -\xi_f \sin 2\phi_1$, where $\xi_f = +1(-1)$ corresponds to CP-even (-odd) final states. Direct CP-violation, $\mathcal{A} = 0$ (or equivalently $|\lambda| = 1$), is expected for both $b \to c\bar{c}s$ and $b \to s\bar{s}s$ transitions.

1.2. *Accelerators and Detectors*

The B-factory accelerators, PEPII[4] and KEKB[5] were commissioned with remarkable speed starting in late 1998. The experiments, BaBar[6] and Belle,[7] started physics data taking in 1999. In the summer of 2001, the two experiments announced the observation of the first statistically significant signals for CP-violation outside of the kaon system.[8,9]

Due to the extraordinary performance of the two accelerators, the most recent results reported in the summer of 2003 at the Lepton-Photon Symposium are based on very large data samples. BaBar has integrated 113 fb^{-1} on the $\Upsilon(4S)$ resonance while Belle has integrated a sample of 140 fb^{-1}. KEK-B also passed a critical milestone for e^+e^- storage rings and achieved a peak luminosity above 1×10^{34} cm$^{-2}s^{-1}$.

1.3. *The Principle of the Measurement*

The measurement of time-dependent CP-asymmetry requires:

- A large sample of $\Upsilon(4S)$ decays into $B^0\bar{B}^0$ pairs. To boost the $\Upsilon(4S)$ decay frame so that the B mesons' flight length can be measured with solid-state vertex detector technology, both the KEKB and PEP-II accelerators use asymmetric energy beams with energies of 8.0 and 3.5 GeV or 9.0 and 3.1 GeV, respectively.

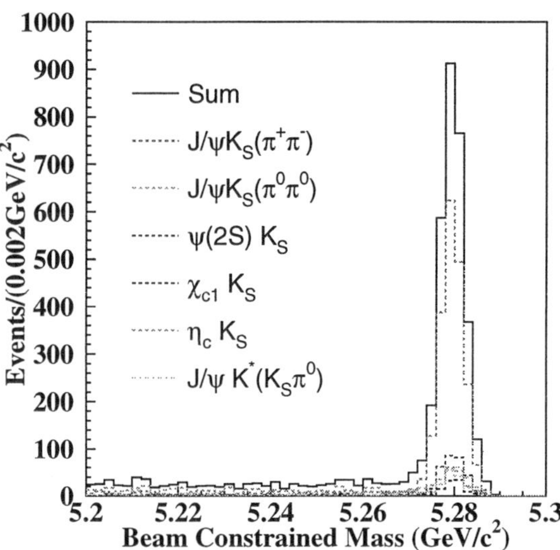

Figure 1. The fully reconstructed CP-eigenstate sample used by Belle. This sample is obtained from a data sample with an integrated luminosity of 140 fb^{-1}.

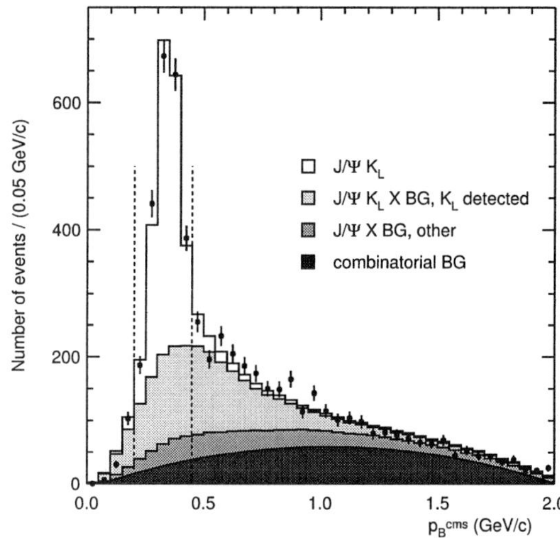

Figure 2. The p_B^* (B momentum in the CM frame) distribution for the $B \to J/\psi K_L$ sample used by Belle. This sample is obtained from a data sample with an integrated luminosity of 140 fb^{-1}. The shaded portions show the contributions of different background components. The vertical dashed lines indicate the signal region.

- Efficient reconstruction of $B \to X_{c\bar{c}}K^0$ decays. This implies accurate measurements of momenta and energies of neutrals using CsI(Tl) crystal calorimeters in addition to good charged particle tracking in small cell drift chambers and efficient identification of leptons and K_S^0 as well as K_L^0 mesons.

- A measurement of Δt. This is related to the measurement of Δz, the spatial distance between the decay vertices and achieved at both experiments by using double-sided silicon strip detectors situated at small radii close to the interaction point.

- A determination of the flavor of the accompanying B ("tagging"); this is based on the identification of electrons, muons and charged kaons and the measurement of their charge.

More detailed descriptions of the detectors[6,7] and the experimental analysis procedure are available elsewhere.[10]

2. Status of CP-Violation in $b \to c\bar{c}s$ Processes

Belle and BaBar reconstruct B^0 decays to the following $b \to c\bar{c}$ CP-eigenstates: $J/\psi K_S$, $\psi(2S)K_S$, $\chi_{c1}K_S$, $\eta_c K_S$ for $\xi_f = -1$ and $J/\psi K_L$ for $\xi_f = +1$.[11] The two classes ($\xi_f = \pm 1$) should have CP-asymmetries that are opposite in sign.

Both experiments also use $B^0 \to J/\psi K^{*0}$ decays where $K^{*0} \to K_S \pi^0$. Here the final state is a mixture of even and odd CP. The CP content can, however, be determined from an angular analysis of other ψK^* decays. The CP-odd fraction is found to be small (i.e. $(19 \pm 4)\%$ $((16 \pm 3.5)\%)$ in the Belle (BaBar) analysis).

The most recent BaBar analysis is based on a data sample with an integrated luminosity of 81 fb^{-1} and was first presented in 2002.[9] There is a corresponding published Belle result also shown in 2002 with 78 fb^{-1}.[8] At this Symposium, Belle provided a new preliminary result for their 140 fb^{-1} sample.[12]

The data sample used for the recent Belle measurement is shown in Fig. 1 and Fig. 2. Table 1 lists the numbers of candidates, N_{ev}, and the estimated signal purity for each f_{CP} mode. It is clear that the CP-eigenstate samples that are used for the CP-violation measurements in $b \to c\bar{c}s$ are large and clean.

In the summer of 2001, the first statistically significant measurements of the CP-violating parameter $\sin 2\phi_1$ were reported by Belle and BaBar. Belle

Table 1. The yields from Belle for reconstructed $B \to f_{CP}$ candidates after flavor tagging and vertex reconstruction, N_{ev}, and the estimated signal purity, p, in the signal region for each f_{CP} mode. J/ψ mesons are reconstructed in $J/\psi \to \mu^+\mu^-$ or e^+e^- decays. Candidate K_S^0 mesons are reconstructed in $K_S^0 \to \pi^+\pi^-$ decays unless otherwise written explicitly.

Mode	ξ_f	N_{ev}	p
$J/\psi K_S^0$	-1	1997	0.976 ± 0.001
$J/\psi K_S^0(\pi^0\pi^0)$	-1	288	0.82 ± 0.02
$\psi(2S)(\ell^+\ell^-)K_S^0$	-1	145	0.93 ± 0.01
$\psi(2S)(J/\psi\pi^+\pi^-)K_S^0$	-1	163	0.88 ± 0.01
$\chi_{c1}(J/\psi\gamma)K_S^0$	-1	101	0.92 ± 0.01
$\eta_c(K_S^0 K^-\pi^+)K_S^0$	-1	123	0.72 ± 0.03
$\eta_c(K^+K^-\pi^0)K_S^0$	-1	74	0.70 ± 0.04
$\eta_c(p\bar{p})K_S^0$	-1	20	0.91 ± 0.02
All with $\xi_f = -1$	-1	2911	0.933 ± 0.002
$J/\psi K^{*0}(K_S^0\pi^0)$	$+1(81\%)$	174	0.93 ± 0.01
$J/\psi K_L^0$	$+1$	2332	0.60 ± 0.03

found

$$\sin 2\phi_1 = 0.99 \pm 0.14 \pm 0.06 \tag{5}$$

while BaBar obtained

$$\sin 2\phi_1 = 0.59 \pm 0.14 \pm 0.05. \tag{6}$$

The results were based on data samples of comparable size (31 million and 32 million $B\bar{B}$ pairs, respectively).

The new Belle data are shown in Fig. 3. This figure shows the Δt distributions where a clear shift between B^0 and $\bar{B}^0$ tags is visible as well as the raw asymmetry plots in two bins of the flavor tagging quality variable r. For low-quality tags ($0 < r < 0.5$), which have a large background dilution, only a modest asymmetry is visble while in the subsample with high quality tags ($0.5 < r < 1.0$), a very clear asymmetry with a sine-like time modulation is present. The final results are extracted from an unbinned maximum-likelihood fit to the Δt distributions that takes into account resolution, mistagging and background dilution. The new Belle result with 140 fb^{-1} (152 million $B\bar{B}$ pairs) is

$$\sin 2\phi_1 = 0.733 \pm 0.057 \pm 0.028. \tag{7}$$

The new Belle result may be compared to the BaBar result with 78 fb^{-1} of

$$\sin 2\phi_1 = 0.741 \pm 0.067 \pm 0.03. \tag{8}$$

Both experiments are now in very good agreement. A new world average can be calculated from these results,

$$\sin 2\phi_1 = 0.736 \pm 0.049. \tag{9}$$

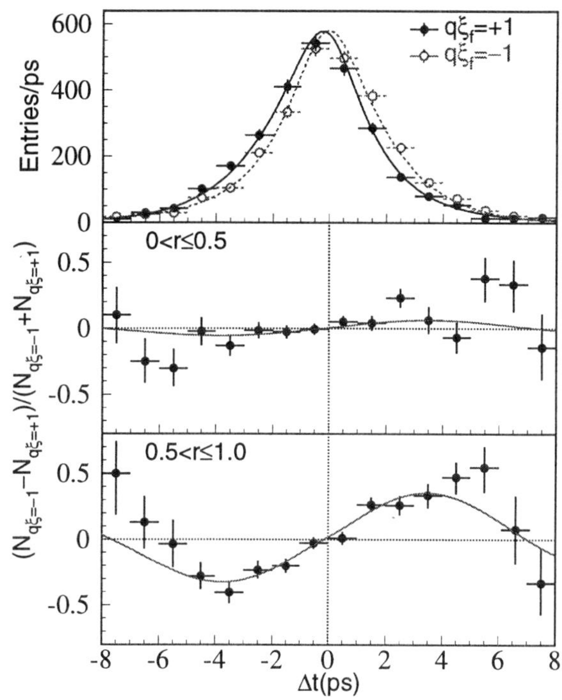

Figure 3. Belle data from 2003: (a) Δt distributions for B^0 and $\bar{B}^0$ tags (b) raw asymmetry for low-quality tags and (c) raw asymmetry for high-quality tags. The smooth curves are projections of the unbinned likelihood fit.

This world average can be interpreted as a constraint on the CKM angle ϕ_1. This constraint can be compared to the indirect determinations on the unitarity triangle.[13] This comparison is shown in Fig. 4 and is consistent with the hypothesis that the Kobayashi-Maskawa phase is the source of CP-violation.

The measurement of $\sin(2\phi_1)$ in $b \to c\bar{c}s$ modes, although still statistically limited, is becoming a precision measurement. The systematics are small and well-understood. Recently, BaBar physicists discovered a new small source of systematic uncertainty due to CP-violation in $b \to c\bar{u}d$ decays on the tagging side.[14]

The presence of an asymmetry with a cosine dependence ($|\lambda| \neq 1$) would indicate direct CP-violation. In order to test for this possibility in $b \to c\bar{c}s$ modes, Belle also performed a fit with $a_{CP} \equiv -\xi_f \mathrm{Im}\lambda/|\lambda|$ and $|\lambda|$ as free parameters, keeping everything else the same. They obtain

$$|\lambda| = 1.007 \pm 0.041(\mathrm{stat}) \tag{10}$$

$$a_{CP} = 0.733 \pm 0.057(\mathrm{stat}),$$

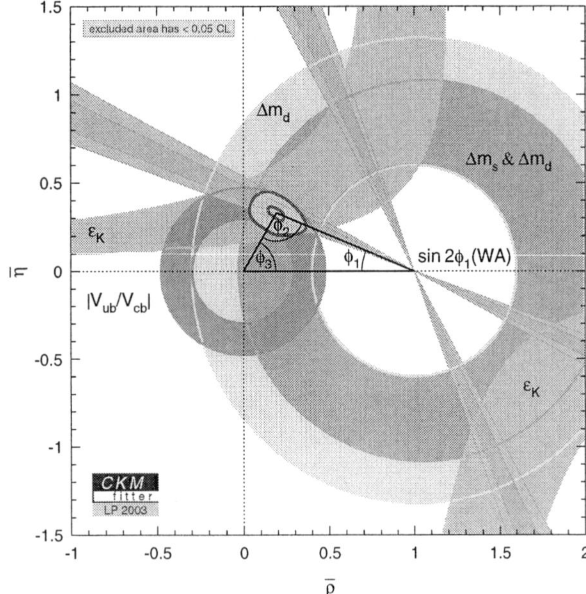

Figure 4. Indirect constraints on the angles of the CKM triangle compared to the most recent direct measurements of ϕ_1 from Belle and BaBar. The theoretical uncertainties in the indirect constraints are conservatively estimated by the CKM fitter group.

for all the $b \to c\bar{c}s$ CP modes combined. This result is consistent with the assumption used in their primary analysis.

3. Studies of CP-Violation in $b \to c\bar{c}d$ Processes

Neutral B decays to CP-eigenstates that proceed by $b \to c\bar{c}d$ processes are expected to have the same CP-violation as $b \to c\bar{c}s$ since both are sensitive to the phase of $B - \bar{B}$ mixing. A small deviation from this expectation is possible because of the contribution of $b \to d$ penguin diagrams (a.k.a. "penguin pollution") in the decay modes that are examined. Penguin pollution may also give rise to direct CP-violation and a CP-violating term with a $\cos(\Delta m_d \Delta t)$ dependence.

The $b \to c\bar{c}d$ decay modes that have been used so far for CP-violation studies are $B \to D^{*+}D^{*-}$, $B \to D^{*+}D^-$, and $B \to J/\psi\pi^0$.[15−18] The effect of penguin pollution might be expected to be the largest in $B \to J/\psi\pi^0$ because the penguin contribution is not color-suppressed in that mode.

For $B \to \psi\pi^0$, with 81 fb^{-1} BaBar has a signal of 40 ± 7 events[15] and finds

$$\sin 2\phi_{1eff}(B \to \psi\pi^0) = 0.05 \pm 0.45 \pm 0.16. \quad (11)$$

The corresponding result from Belle is based on 140 fb^{-1} and uses 89 ± 10 events.[16] They obtain

$$\sin 2\phi_{1eff}(B \to \psi\pi^0) = 0.72^{+0.37}_{-0.42} \pm 0.08. \quad (12)$$

In both cases, the systematic error includes the possibility of CP-violation in a small component of the background that peaks under the signal.

The $b \to c\bar{c}d$ mode $B \to D^{*+}D^{*-}$ has a vector-vector final state and requires special treatment since it includes contributions from both CP-even and odd components. To extract the CP-odd fraction, one fits the angular distribution in the transversity basis. The result from BaBar based on a sample with 156 ± 14 signal events is,

$$R_\perp = 0.063 \pm 0.055 \pm 0.009, \quad (13)$$

where the quantity $R_\perp$ is the fraction of the CP-odd component. The measurement indicates that $B^0 \to D^{*+}D^{*-}$ is mostly CP-even.

The time distributions from BaBar for $B \to D^{*+}D^{*-}$ are shown in Fig. 5. BaBar finds

$$\sin 2\phi_{1eff}(B \to D^{*+}D^{*-}) = -0.05 \pm 0.29 \pm 0.10, \quad (14)$$

which is about 2.5σ from the result in $b \to c\bar{c}s$ modes. This may be a statistical fluctuation or could be an indication that the Standard Model $b \to d$ penguin contribution is large. The fit includes the possibility of direct CP-violation. The parameter λ is found to be $0.75 \pm 0.19 \pm 0.02$, which is consistent with unity, as expected for no direct CP-violation.

Since $B^0 \to D^{*+}D^-$ and its charge conjugate are not CP-eigenstates, a modified treatment is required. There are four rather than two CP-violating observables that are determined from a time-dependent fit to the different D^*D charge states.

BaBar finds,

$$S_{+-} = -0.82 \pm 0.75 \pm 0.14, \quad (15)$$
$$S_{-+} = -0.24 \pm 0.69 \pm 0.12, \quad (16)$$
$$A_{+-} = +0.47 \pm 0.40 \pm 0.12, \quad (17)$$
$$A_{-+} = +0.22 \pm 0.37 \pm 0.10. \quad (18)$$

In the limit of no penguins and assuming factorization in these hadronic decays, $S_{-+} = S_{+-} = -\sin 2\phi_1$ and $A_{+-} = A_{-+} = 0$. The above results for CPV in $B \to D^*D$ decays are consistent with this limit.

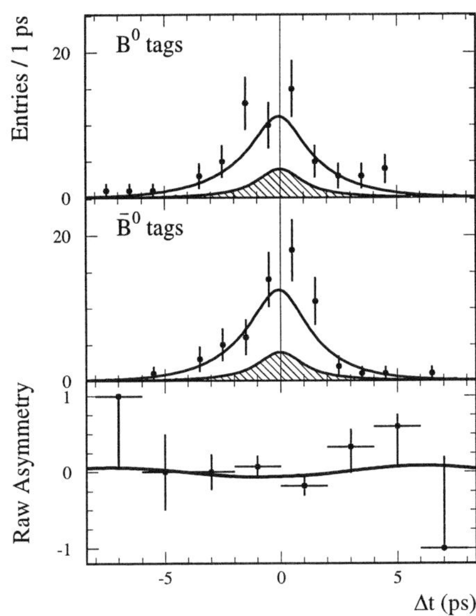

Figure 5. BaBar results on CP-violation in $B \to D^{*+}D^{*-}$. The top two figures show the Δt distributions for B^0 and $\bar{B}^0$ tags. The third plot shows the raw time asymmetry distribution.

Observation of the CP-eigenstate mode $B \to D^+D^-$ was reported by Belle at this conference. With 140 fb^{-1}, the 5σ signal contains 24.3 ± 6.0 events. In the future, this mode can also be used for time-dependent measurements of CPV in $b \to c\bar{c}d$ processes.

The results of CP-violation measurements for $b \to c\bar{c}d$ decays are summarized in Fig. 6. The measurements are not yet precise enough to definitively demonstrate the presence of penguin pollution.

4. Status of CP-Violation in $b \to sq\bar{q}$ Penguin Processes

In addition to the program of measuring the other remaining angles of the unitarity triangle that is discussed in the contribution by Jawahery,[19] there is also the question of whether there are additional CP-violating phases from new interactions or physics beyond the Standard Model. At the moment, such new phases are poorly constrained.

One way to attack this question is to measure the time-dependent CP-asymmetry in penguin-dominated modes such as $B^0 \to \phi K_S^0$, $B^0 \to \eta' K_S^0$ or $B^0 \to K_S^0\pi^0$, where heavy new particles may contribute inside the loop, and compare it to the asymmetry in $B^0 \to J/\psi K_S^0$ and related $b \to c\bar{c}s$ charmonium modes.

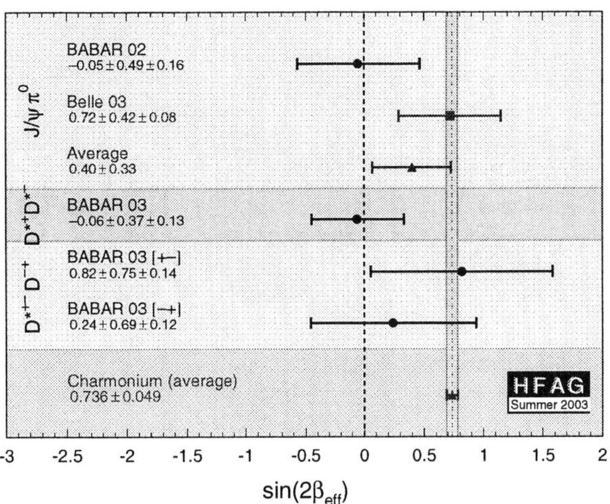

Figure 6. Summary plot of results on CP-violation in $b \to c\bar{c}d$ modes.

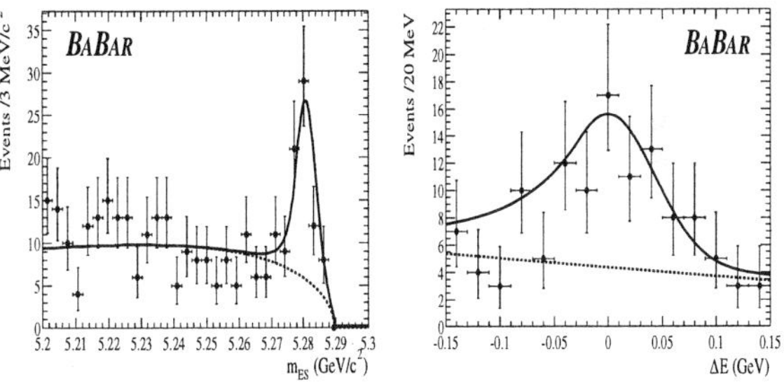

Figure 7. Beam constrained mass and ΔE distributions for $B \to K_S^0\pi^0$ from BaBar.

The mode $B \to K_S\pi^0$ proceeds through a $b \to sd\bar{d}$ transition. The BaBar data on $B \to K_S^0\pi^0$ are shown in Fig. 7. To be useful for time-dependent CPV studies at least one of charged pions from the K_S^0 must be detected in the BaBar silicon vertex detector.[20] There are 123 ± 16 events of this type that are then used to obtain

$$\sin 2\phi_{1eff}(B \to K_S^0\pi^0) = 0.48^{+0.38}_{-0.47} \pm 0.11. \quad (19)$$

The time distributions are shown in Fig. 8. The direct CP-violation parameter is $A = -0.40^{+0.28}_{-0.27} \pm 0.10$.[20] When A is fixed to zero, the value of $S = \sin(2\phi_{1eff})$ shifts slightly to $0.41^{+0.41}_{-0.48} \pm 0.11$. The results for $B \to K_S\pi^0$ are consistent with the value from the $b \to c\bar{c}s$ modes, $\sin 2\phi_1 = 0.736 \pm 0.049$.

The mode $B \to \eta' K_S^0$ is expected to include contributions from $b \to s\bar{u}u$ and $b \to s\bar{d}d$ penguin processes. The beam constrained mass distribution for the $B \to \eta' K_S^0$ sample used by Belle is shown in

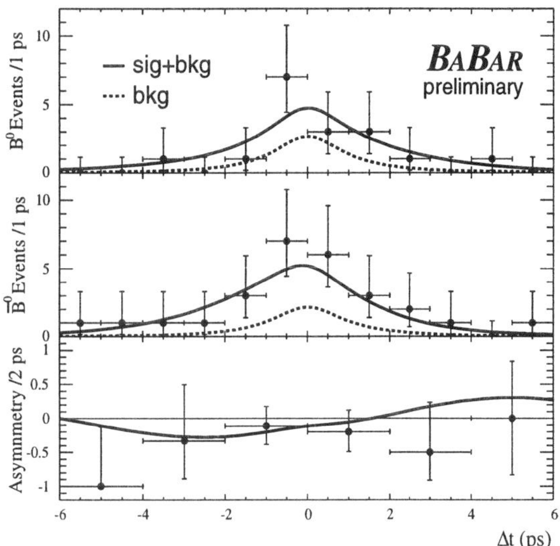

Figure 8. BaBar data on $B \to K_S^0 \pi^0$. The top two figures show the Δt distributions for B^0 and $\bar{B}^0$ tags, separately. The third plot shows the raw time asymmetry distribution.

Fig. 9 and contains 244 ± 21 signal events.[21] Belle finds (Fig. 10),

$$\sin 2\phi_{1eff}(B \to \eta' K_S^0) = 0.43 \pm 0.27 \pm 0.05 \quad (20)$$

The BaBar data is shown in Fig. 11. They obtain,

$$\sin 2\phi_{1eff}(B \to \eta' K_S^0) = 0.02 \pm 0.34 \pm 0.03 \quad (21)$$

The average of these two results for $B \to \eta' K_S^0$ is about 2.2σ from the $b \to c\bar{c}s$ measurement, which is the Standard Model expectation.

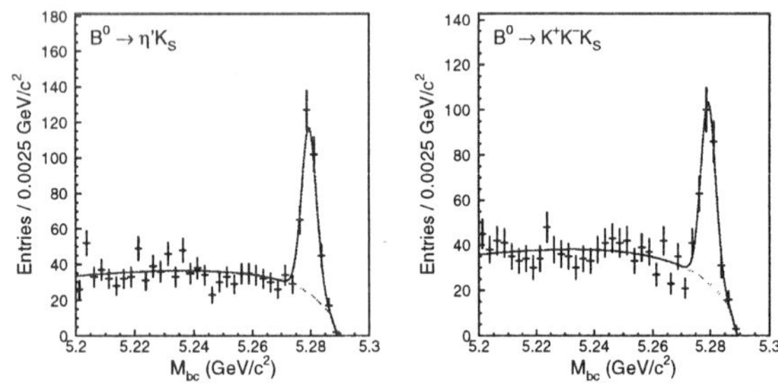

Figure 9. Beam constrained mass distributions for $B \to \eta' K_S^0$ (left) and $B \to K^+ K^- K_S^0$ (right).

The decay mode $B \to K^+ K^- K_S^0$, where $K^+ K^-$ combinations consistent with the ϕ have been removed, is found by Belle to be dominately CP-odd[22] and thus can be treated as a CP-eigenstate and used for studies of time-dependent CP-violation in

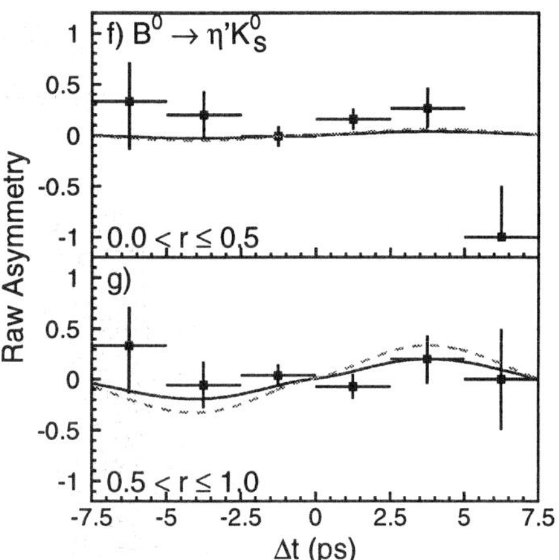

Figure 10. Belle data for the raw asymmetry in $B^0 \to \eta' K_S^0$. The upper plot shows the data for low-quality tags while the lower plot shows the higher quality tags. The dashed curves are the expectations from the Standard Model.

$b \to sq\bar{q}$ processes. The beam constrained mass distribution for the $B \to K^+ K^- K_S^0$ sample used by Belle is shown in Fig. 9. There are 199 ± 18 signal events. Belle obtains,

$$\sin 2\phi_{1eff}(B \to K^+ K^- K_S^0) = 0.51 \pm 0.26 \pm 0.05 ^{+0.18}_{-0.00}, \quad (22)$$

where the third error is due to the uncertainty in the CP content of this final state.[22] The results for $B \to K^+ K^- K_S^0$ are also consistent with $b \to c\bar{c}s$ decays. However, in this decay there is also the possibility of "tree-pollution", the contribution of the $b \to u\bar{u}s$ tree amplitude that may complicate the interpretation of the results.[23]

The $B^0 \to \phi K_S^0$ decay, which is dominated by the $b \to s\bar{s}s$ transition, is an especially unambiguous and sensitive probe of new CP-violating phases from physics beyond the SM.[24] The SM predicts that measurements of CP-violation in this mode should yield $\sin 2\phi_1$ to a very good approximation.[25,23] A significant deviation in the time-dependent CP-asymmetry in this mode from what is observed in $b \to c\bar{c}s$ decays would be evidence for a new CP-violating phase.

The $B \to \phi K_S^0$ sample used by BaBar is shown in Fig. 12. The signal, obtained from a sample with an integrated luminosity of 110 fb^{-1}, contains 70 ± 9 events.[20] The time distributions for the BaBar data are shown in Fig. 13. They obtain

$$\sin 2\phi_{1eff}(B \to \phi K_S^0) = 0.45 \pm 0.43 \pm 0.07. \quad (23)$$

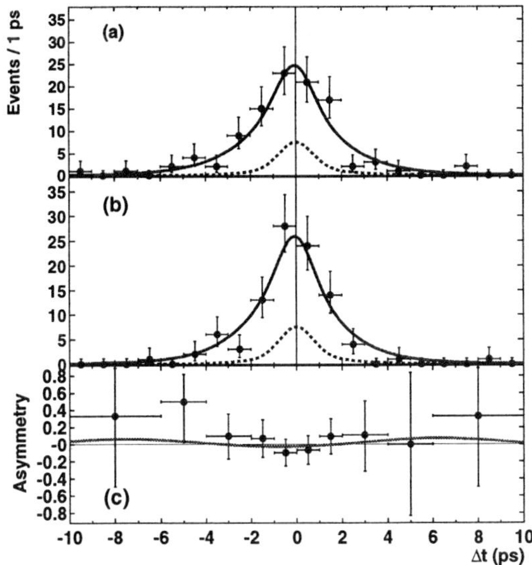

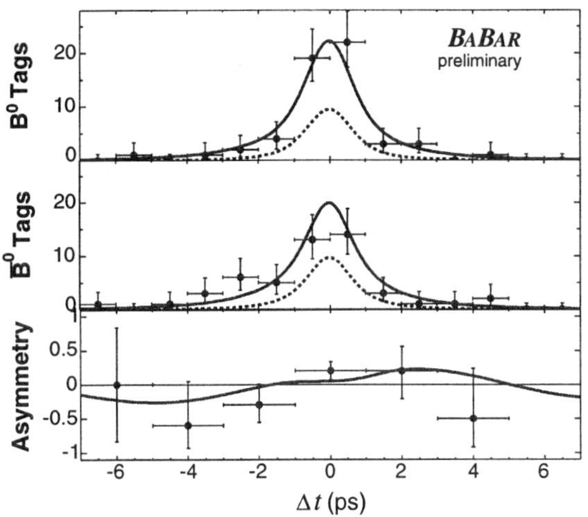

Figure 11. BaBar data on $B \to \eta' K_S^0$. The top two figures show the Δt distributions for B^0 and $\bar{B}^0$ tags, separately. The third plot shows the raw time asymmetry distribution.

Figure 13. BaBar time difference and asymmetry data distributions in $B \to \phi K_S^0$. The top two figures show the Δt distributions for B^0 and $\bar{B}^0$ tags, separately. The third plot shows the raw time asymmetry distribution.

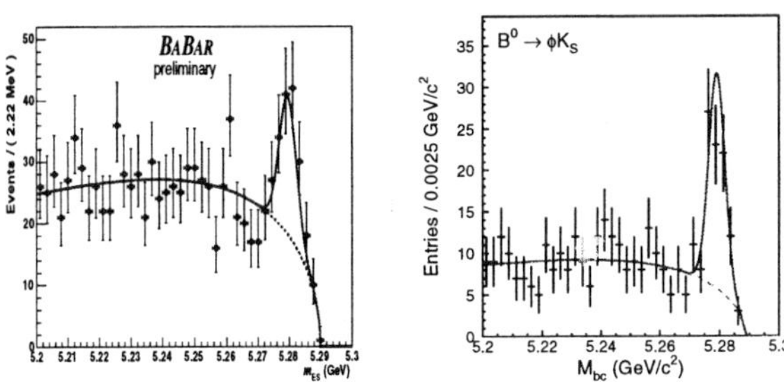

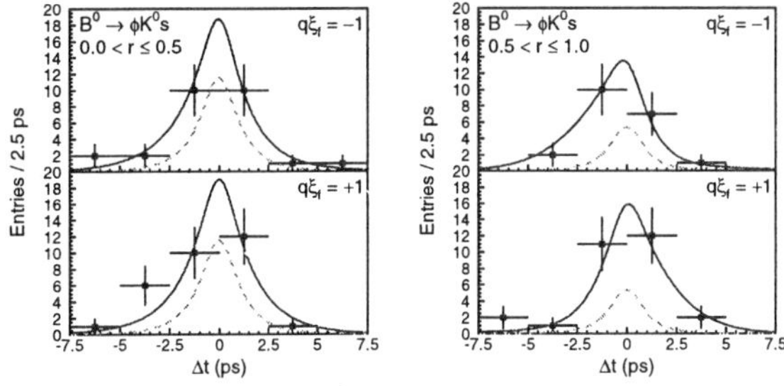

Figure 12. Beam constrained mass distributions for $B \to \phi K_S^0$ from BaBar (left) and Belle(right).

Figure 14. Belle data: (left) Δt distributions for low-quality tags and (right) for high-quality tags. The dashed curves show the background contributions.

This value is consistent with the Standard Model expectation, but is somewhat different from the value obtained with the 81 fb^{-1} sample, which was $\sin 2\phi_{1eff} = -0.18 \pm 0.51 \pm 0.09$. The new result includes more data and a reprocessing of the old data sample. After extensive checks with data and Toy Monte Carlo studies, the large change in the central value is attributed to a 1σ statistical fluctuation.[26]

The $B \to \phi K_S^0$ sample used by Belle is shown in the right panel of Fig. 12. The selection criteria are described in detail elsewhere.[27,28] The signal contains 68 ± 11 events. Figure 15 shows the raw asymmetries from Belle in two regions of the flavor-tagging parameter r. While the numbers of events in the two regions are similar, the effective tagging efficiency is

much larger and the background dilution is smaller in the region $0.5 < r \leq 1.0$. The solid curves show the results of the unbinned maximum-likelihood fit to the Δt distribution.

The observed CP-asymmetry for $B^0 \to \phi K_S^0$ in the region $0.5 < r \leq 1.0$ (Fig. 15 (lower panel)) indicates the difference from the SM expectation (dashed curve). Note that these projections onto the Δt axis do not take into account event-by-event information (such as the signal fraction, the wrong tag fraction and the vertex resolution) that is used in the unbinned maximum likelihood fit.

The contamination of $K^+ K^- K_S^0$ events in the ϕK_S^0 sample $(7.2 \pm 1.7\%)$ is small. Finally, backgrounds from the $B^0 \to f_0(980) K_S^0$ decay, which has the opposite CP-eigenvalue to ϕK_S^0, are found to be

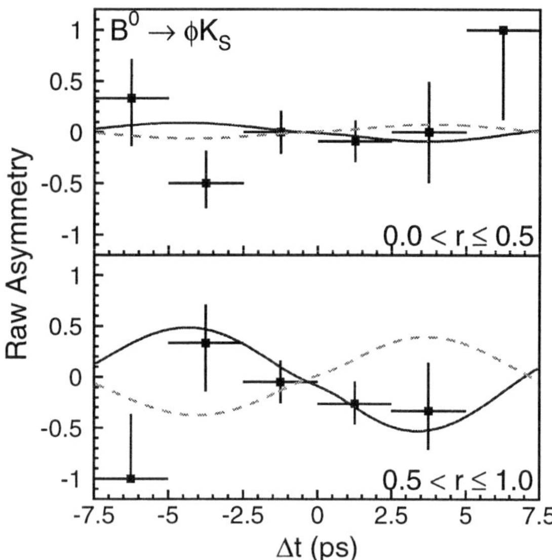

Figure 15. Belle data for the raw asymmetry in $B^0 \to \phi K_S^0$. The upper plot shows the data for low-quality tags while the lower plot shows the higher quality tags. The dashed line is the expectation from the Standard Model.

small ($1.6^{+1.9}_{-1.5}\%$). The influence of these backgrounds is treated as a source of systematic uncertainty.

Belle obtains

$$\sin 2\phi_{1eff}(B \to \phi K_S^0) = -0.96 \pm 0.5^{+0.09}_{-0.11} \quad (24)$$

from their likelihood fit to the ϕK_S^0 data. The likelihood function is parabolic and well-behaved. An evaluation of the significance of the result using the Feldman-Cousins method and allowing for systematic uncertainties shows that this result deviates by 3.5σ from the Standard Model expectation.[28]

The Belle group performed a number of validation checks for their $B \to \phi K_S^0$ CP-violation result. Fits to the same samples with the direct CP-violation parameter $\mathcal{A}$ fixed at zero yield $sin2\phi_{1eff} = -0.99 \pm 0.50(\text{stat})$ for $B^0 \to \phi K_S^0$. As a consistency check for the $\mathcal{S}$ term, the same fit procedure is applied to the charged B meson decays $B^+ \to \phi K^+$. The result is $\mathcal{S} = -0.09 \pm 0.26(\text{stat})$, $\mathcal{A} = +0.18 \pm 0.20(\text{stat})$ for $B^+ \to \phi K^+$ decay. The results for the $\mathcal{S}$ term is consistent with no CP-asymmetry, as expected. The asymmetry distribution is shown in Fig. 16. In addition, the ϕK_S^0 sideband has been examined as shown in Fig. 16. No asymmetry is found in that sample.

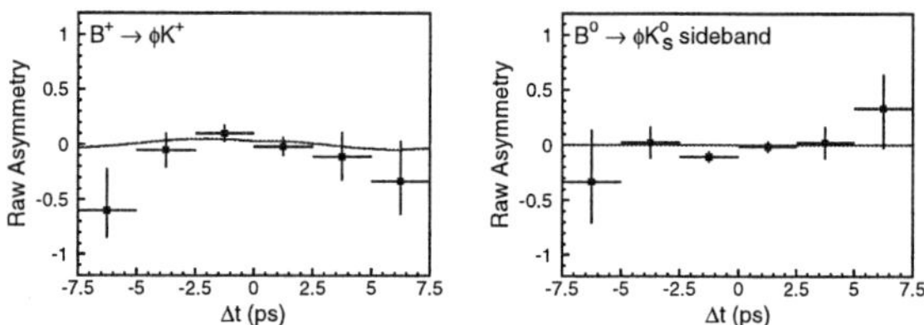

Figure 16. Belle data: consistency checks of the $B \to \phi K_S^0$ analysis. The asymmetries in (a) the $B^\pm \to \phi K^\pm$ sample and (b) the $B \to \phi K_S^0$ sideband sample.

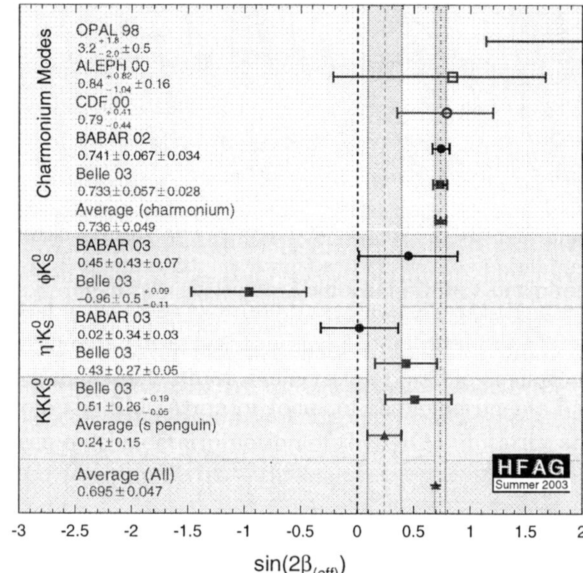

Figure 17. Summary plot of results on $\sin 2\phi_1$ and $\sin 2\phi_{1eff}$ in $b \to c\bar{c}s$ and $b \to s\bar{q}q$ modes.

5. Conclusion

Belle presented a new measurement of time-dependent CP-violation in $b \to c\bar{c}s$ CP-eigenstates. This result and previous results from BaBar are in good agreement with each other and with the hypothesis that the Kobayashi-Maskawa phase is the source of CP-violation.

Studies of CP-violation in $b \to c\bar{c}d$ modes are progressing. In $B \to D^{*+}D^{*-}$ decays, BaBar observes a 2.5σ hint for penguin pollution. More data and measurements are needed to clarify whether penguin pollution is present in this class of decays.

In $B \to \phi K_S^0$ decays there was a surprise. With 140 fb^{-1} Belle observed a 3.5σ deviation from the Standard Model expectation. This could be an indication of new physics from heavy particles in the

$b \to s\bar{s}$ penguin loop. However, BaBar's value moved closer to the Standard Model with the addition of new data and reprocessing. More precise measurements of the other $b \to sq\bar{q}$ modes can further constrain phases from new physics. For example, new physics may contribute differently to pseudoscalar-vector and pseudoscalar-pseudoscalar modes.[2]

The results of CP-violation measurements for $b \to sq\bar{q}$ penguin decays are summarized in Fig. 17. The world average for all $b \to s$ penguin decays (shown by the dotted line) appears to be displaced from the average for $b \to c\bar{c}s$ modes. The high energy physics community will require that this experimental issue be resolved conclusively in the future. This will require large data samples with integrated luminosities of at least 1 ab^{-1} or 1000 fb^{-1}.

Acknowledgments

I thank Harry Cheung for his patience and Andreas Hoecker for his contributions to the figures. I also wish to thank my colleagues at KEK-B, PEP-II, Belle and BaBar for their extraordinary contributions to the work reported here.

References

1. M. Kobayashi and T. Maskawa, *Prog. Theor. Phys.* **49**, 652 (1973).
2. Y. Grossman, contributions to these proceedings.
3. A. B. Carter and A. I. Sanda, *Phys. Rev. D* **23**, 1567 (1981); I. I. Bigi and A. I. Sanda, *Nucl. Phys.* **B193**, 85 (1981).
4. "PEP-II: An asymmetric B Factory", Conceptual Design Report, SLAC-418, LBL-5379 (1993).
5. S. Kurokawa and E. Kikutani *et al.*, *Nucl. Instrum. Methods A* **499**, 1 (2003).
6. Belle Collaboration, A. Palano *et al.*, *Nucl. Instrum. Methods A* **479**, 1 (2002).
7. Belle Collaboration, A. Abashian *et al.*, *Nucl. Instrum. Methods A* **479**, 117 (2002).
8. Belle Collaboration, K. Abe *et al.*, *Phys. Rev. Lett.* **87**, 091802 (2001); *Phys. Rev. D* **66**, 032007 (2002).
9. BaBar Collaboration, B. Aubert *et al.*, *Phys. Rev. Lett.* **87**, 091801 (2001); *Phys. Rev. D* **66**, 032003 (2002); *Phys. Rev. Lett.* **89**, 201802 (2002).
10. T.E. Browder and R. Faccini, Annual Review of Nuclear and Particle Physics, **53**, 353 (2003).
11. The inclusion of the charge conjugate decay mode is implied unless otherwise stated.
12. Belle Collaboration, K. Abe *et al.*, hep-ex/0308036, BELLE-CONF-0353, contributed to the XXI International Symposium on Lepton and Photon Interactions at High Energies, Aug.11-16, 2003, Fermilab, Illinois, U.S.A.
13. A. Hoecker, H. Lacker, S. Laplace and F. LiDiberder, hep-ph/0104062.
14. O. Long, M. Baak, R. N. Cahn and D. Kirkby, hep-ex/0303030 to appear in *Phys. Rev. D*.
15. BaBar Collaboration, B. Aubert *et al.*, hep-ex/0303018, to appear in *Phys. Rev. Lett.*.
16. Belle Collaboration, K. Abe *et al.*, hep-ex/0308053, BELLE-CONF-0342.
17. BaBar Collaboration, B. Aubert *et al.*, hep-ex/0306052, to appear in *Phys. Rev. Lett.*.
18. BaBar Collaboration, B. Aubert *et al.*, hep-ex/0303004, *Phys. Rev. Lett.* **90**, 221801 (2003).
19. H. Jawahery, contributions to these proceedings.
20. BaBar Collaboration, B. Aubert *et al.*, BABAR-PLOT-0053; BABAR-PLOT-0056.
21. The selection requirments for $B \to \eta' K_S^0$ are discussed in Belle Collaboration, K.-F. Chen and K. Hara *et al.*, *Phys. Lett. B* **546**, 196 (2002).
22. Belle Collaboration, A. Garmash *et al.*, hep-ex/0307082 to appear in *Phys. Rev. D*.
23. Y. Grossman, Z. Ligeti, Y. Nir, and H. Quinn, hep-ph/0303171, *Phys. Rev.* **D68**, 015004 (2003).
24. Y. Grossman and M. P. Worah, *Phys. Lett. B* **395**, 241 (1997).
25. D. London and A. Soni, *Phys. Lett. B* **407**, 61 (1997); Y. Grossman, G. Isidori and M. P. Worah, *Phys. Rev. D* **58**, 057504 (1998).
26. Private communication, Livio Lancieri.
27. Belle Collaboration, K.-F. Chen and A. Bozek *et al.*, hep-ex/0307014.
28. Belle Collaboration, K. Abe *et al.*, hep-ex/0308035, to appear in *Phys. Rev. Lett.*.

Stefan Spanier (University of Tennessee):

1) Unfortunately, the plenary session gives the audience only a limited chance to help you to establish the results by asking detailed questions.

2) Knowing the previous value of $S = -0.7 \pm 0.6$ from Belle, the newly added statistics must lead to an unphysical value of $S < -1.4$ leading typically to large correlations in S and C (pathological behavior) in this new sample. How probable is the value in the new sample?

3) How strong is the CP-asymmetry in the background?

Tom Browder:

1) A special breakout session is planned later in the Symposium.

2) For a true value near $S = -1$, the values in the new sample are quite consistent with Toy Monte Carlo studies. There is no statistically pathological behaviour in either old or new data samples. The observed errors are actually slightly larger than expected.

3) The background from $B \to f_0 K_S^0$ and $B \to K^+ K^- K_S^0$ decays is small and the CP-asymmetry from these backgrounds is included in the systematic error.

Alex Kagan (Cincinnati): Can you show the raw BaBar data for $S(\phi K_S^0)$ again?

Tom Browder: Yes. Note that a figure with this data was included in the talk and appears in the Proceedings.

Hitoshi Murayama (Berkeley): On the ϕK_S^0 mode, the change in the BaBar result was attributed to a statistical fluctuation. They have added only 40% more data. How is that possible? Do you have a breakdown of the asymmetry between the previous and new data samples?

Tom Browder: Not only was more data added, but the old BaBar data sample was also reprocessed. After reprocessing, a small number of events changed from B^0 tags to $\bar{B}^0$ tags (or vice versa). This accounts for the shift in the central value.

CKM UNITARITY ANGLES $\alpha(\phi_2)$ AND $\gamma(\phi_3)$

A. JAWAHERY

Department of Physics, University of Maryland, College Park, MD 20742, U.S.A
E-mail: jawahery@physics.umd.edu

I report on the experimental studies of the CKM unitarity angles $\alpha(\phi_2)$ and $\gamma(\phi_3)$ with emphasis on recent measurements by the Belle and BaBar experiments at the B-factory accelerators.

1. Introduction

In the Standard Model, the origin of CP-violation lies in the charged weak interaction sector, which involves transitions amongst quarks of different flavor and charge through the emission of a virtual W. The coupling strengths of these transitions for the 3 generations of quarks form a 3×3 matrix first introduced by M. Kobayashi and T. Maskawa[1] and known as the (CKM) matrix. The CKM matrix is a unitary matrix parameterized by 3 real angles and one complex phase. This single complex phase of the CKM matrix is the source of CP-violation in the Standard Model (SM). A popular parametrization of the CKM matrix, by L. Wolfenstein,[2] is given below, where the 4 parameters are A, λ, ρ and η:

$$V_{\mathrm{CKM}} = \begin{pmatrix} 1 - \frac{\lambda^2}{2} & \lambda & A\lambda^3(\rho - i\eta) \\ -\lambda & 1 - \frac{\lambda^2}{2} & A\lambda^2 \\ A\lambda^3(1 - \rho - i\eta) & -A\lambda^2 & 1 \end{pmatrix}.$$

The parameter η controls the CP-violating effects in this framework. The values of the CKM parameters are not specified by the SM and must be determined from experimental measurements. A review of the measurements of the magnitudes of the CKM matrix elements is given by K. Schubert at this conference, reporting values of $\lambda = 0.2235 \pm 0.0033$ and $A\lambda^2 = 0.0415 \pm 0.0011$. Information on the parameters ρ and η can be obtained from a global fit (hereafter referred to as the "CKM fit") to several measured quantities in the K and B meson systems, including $|V_{ub}|$, $B^0 \leftrightarrow \bar{B}^0$ oscillation frequencies Δm_d and Δm_s, and CP-violation observables in the neutral kaon system (ϵ_K), whose values in the framework of the SM depend on these parameters (Fig. 1).[3] For the (ρ, η) values in the small overlap region in Fig. 1, the SM can accommodate all of the above observables, of which the CPV effect in the kaon system imposes the requirement of a non-zero value for η. An interesting and important conse-

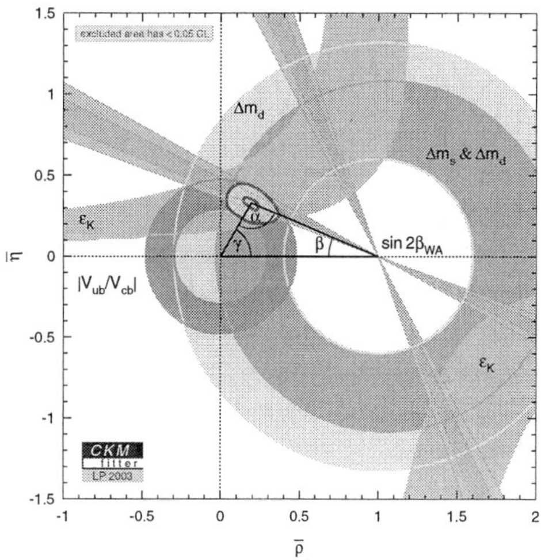

Figure 1. The "CKM fit" to data from the CKMfitter group (http://ckmfitter.in2p3.fr/). See the text for more details.

quence of a non-zero complex parameter in the CKM matrix is CP-violation in other particle systems, notably the decays of charged and neutral B mesons. The primary mission of the B-factory experiments is to search for the breaking of the CP-symmetry in B meson decays and examine the consistency of the measurements with the expected values within the CKM mechanism. A deviation from the SM could be an indication of New Physics effects.

1.1. *Definition of the CKM Unitarity Angles and CPV Observables in B Decays*

The unitarity of the CKM matrix imposes 9 complex relations amongst the matrix elements, the most famous of which is the so-called CKM unitarity triangle, shown below and pictured in Fig. 2:

$$V_{ub}^* V_{ud} + V_{cb}^* V_{cd} + V_{tb}^* V_{td} = 0.$$

The sides of this triangle are determined (or constrained) by the values of the CKM matrix elements.

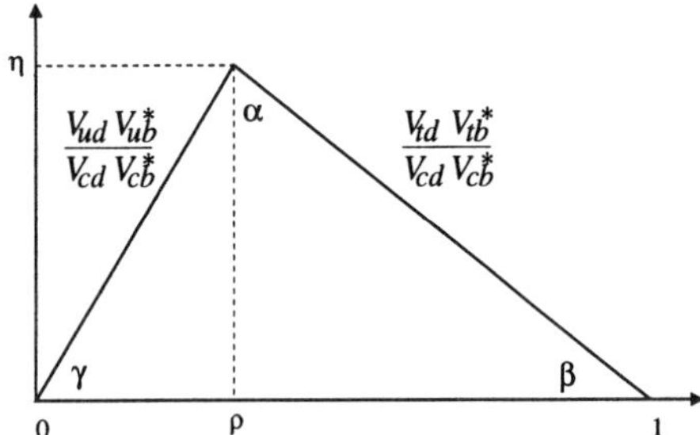

Figure 2. The CKM unitarity triangle.

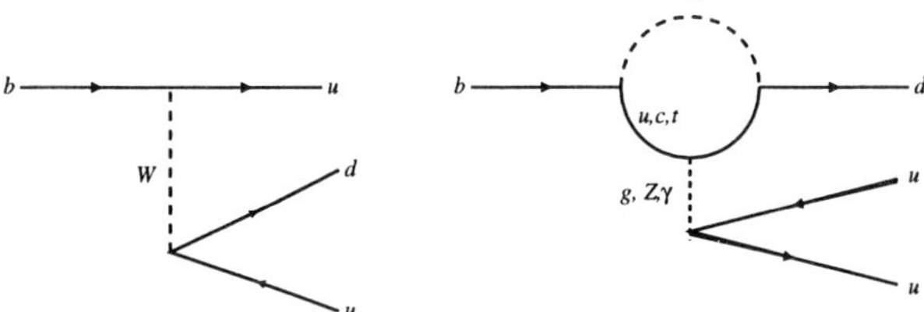

Figure 3. The tree (left) and penguin (right) diagrams contributing to "charmless" B decay ($B \to \pi\pi$, $B \to \rho\rho,...$).

The three angles:

$$\alpha(= \phi_2) = \arg\left[-\frac{V_{td}V_{tb}^*}{V_{ud}V_{ub}^*}\right],$$

$$\beta(= \phi_1) = \arg\left[-\frac{V_{cd}V_{cb}^*}{V_{td}V_{tb}^*}\right], \text{ and}$$

$$\gamma(= \phi_3) = \arg\left[-\frac{V_{ud}V_{ub}^*}{V_{cd}V_{cb}^*}\right]$$

can be extracted from the measurements of CP-violating effects in B meson decays. The experimental goal is to measure the three angles, which along with the knowledge of the sides of the triangle will overconstrain the model and reveal possible deviations from the SM.

The measurements of $\sin2\beta$, the CP-violating asymmetry in B^0 decays to charmonium final states ($b \to c\bar{c}s$), by the BaBar and Belle collaborations established the breaking of CP-symmetry in B decays.[4] The experimental status of $\sin2\beta$ measurements are reviewed at this conference by T. Browder. The knowledge of $\sin 2\beta$ leads to four possible solutions in the (ρ, η) plane as shown in Fig. 1, one of which coincides with the allowed region from the "CKM fit". The fit also yields the allowed ranges $77^o < \alpha(\phi_2) < 122^o$, $37^o < \gamma(\phi_3) < 80^o$ at 95% CL, which set a reference point for comparison with direct measurements.

1.2. Experimental Strategies for Extracting Angles α and γ

Since the angles α and γ essentially measure the phase of the CKM element V_{ub} relative to other matrix elements, any experimental strategy must involve the tiny component of B decays which occur through the $b \to u$ transition. The obvious candidate channels for experimental studies are the so-called "charmless" B decays, such as $B \to \pi\pi$, $B \to \rho\pi$ and $B \to \rho\rho$ which can arise from the tree level transition $b \to u(\bar{u}d)$, carrying the CKM element V_{ub} (Fig. 3 (left)). However, a complication to this approach arises from the presence of loop level (penguin) processes (Fig. 3 (right)), involving different CKM matrix elements, which can also lead to the same final states. The interference of the two diagrams then obscures the connection between the CP observables and the angles α and γ, thus the experimental program must also include a "tree and penguin disentanglement" strategy.

Complementary methods, free of penguin effects, for the determination of γ have also been explored experimentally and will be discussed in Secs. 6 and 7. These approaches involve CPV measurements in decay modes which can occur through both the CKM suppressed $b \to uW^-$ transition and the dominant $b \to cW^-$ diagram, with the interference effects providing the sensitivity to CPV phases.

1.2.1. CPV observables in "charmless" B decays

Two sets of experimental observables are explored for CP-violation studies in charmless B decays. For a given final state f with CP conjugate $\bar{f}$, the mean branching ratio and the time integrated direct CP-asymmetry, defined as:

$$A_{CP}^{direct} = \frac{\Gamma(\bar{B} \to \bar{f}) - \Gamma(B \to f)}{\Gamma(\bar{B} \to \bar{f}) + \Gamma(B \to f)}$$

acquire sensitivity to the angle γ as a result of the interference between the tree and penguin diagrams. This can easily be seen in the dependence of the decay amplitudes for the $B \to f$ and $\bar{B} \to \bar{f}$:

$$A = -(|T|e^{i\gamma} + |P|e^{i\delta}), and$$

$$\bar{A} = -(|T|e^{-i\gamma} + |P|e^{i\delta});$$

resulting in:

$$BR \propto 1 + 2|\frac{P}{T}|cos\delta cos\gamma + |\frac{P}{T}|^2$$

$$A_{cp} = -2|\frac{P}{T}| \sin\delta \sin\gamma$$

where T and P are the amplitudes of the tree and penguin diagrams, respectively, and δ is the relative strong phase of the two amplitudes.

Another set of observables are the CP-violating quantities extracted from the time evolution of neutral B decays to final states that are common to B^0 and $\bar{B}^0$ mesons. Time dependent CP-asymmetry results from the interference of two possible paths for reaching the same final state: $B^0 \to f$ and $B^0 \leftrightarrow \bar{B}^0 \to f$. All CP-violation information is encoded in the parameter:

$$\lambda_f = \eta_{f_{CP}} \frac{p}{q} \frac{\bar{A}}{A},$$

where $A = |\langle f|T|B^0\rangle|$, $\bar{A} = |\langle f|T|\bar{B}^0\rangle|$, $\eta_{f_{CP}}$ is the CP-eigenvalue of the final state and $\frac{q}{p}$ is a measure of CPV in mixing. Observation of either $arg(\lambda_f) \neq 0$ or $|\lambda_f| \neq 1$ would indicate the presence of CPV in the final state f. Time dependent CP-asymmetry follows:

$$A_{cp}(\Delta t) = S_f sin(\Delta m_d \Delta t) - C_f cos(\Delta m_d \Delta t)]$$

where the observables $S_f = \frac{2\Im(\lambda_f)}{1+|\lambda_f|^2}$ and $C_f = \frac{1-|\lambda_f|^2}{1+|\lambda_f|^2}$ measure the so-called indirect CP violation resulting from interference of decay and mixing, and direct CP violation in the decay ($|A| \neq |\bar{A}|$), respectively. In the simplest case where the decay is dominated by a single tree diagram, $\lambda_f = \eta \frac{V_{tb}^* V_{td} V_{ub} V_{ud}^*}{V_{tb} V_{td}^* V_{ub}^* V_{ud}} = e^{i2\alpha}$, $C_f = 0$, thus $S_f = \sin 2\alpha$. However, in general, with the penguin contribution present, one has:

$$\lambda_f = e^{2i\alpha} \frac{1 + |\frac{P}{T}|e^{i\delta}e^{i\gamma}}{1 + |\frac{P}{T}|e^{i\delta}e^{-i\gamma}},$$

resulting in $C_f \neq 1$ and $S_f = \sqrt{1 - C_f^2} \sin 2\alpha_{eff}$ (see the References[14,16] for more details.). The net effect of the penguin presence is to introduce the possibility of direct CP-violation ($C_f \neq 1$) and a possible shift of $\Delta\alpha = \alpha_{eff} - \alpha$, the magnitude of which would depend on the ratio $|P/T|$ and the strong phase δ.

The presence of "non-tree" level diagrams in charmless decays is inferred from the observed pattern of the branching ratios in these modes, as seen in the compilation of the data for the 2-body charmless decays by the Heavy Flavor Averaging Group.[5] At the tree level the expected rate for the CKM suppressed decay $B^0 \to K^-\pi^+$ would be about 5% of the rate for $B^0 \to \pi^+\pi^-$. However, the pattern of the data does not follow this expectation; the measurements (initially by CLEO) give $B(B^0 \to K^-\pi^+) = (18.2 \pm 0.8) \times 10^{-6}$ and $B(B^0 \to \pi^-\pi^+) = (4.6 \pm 0.4) \times 10^{-6}$. This enhancement of

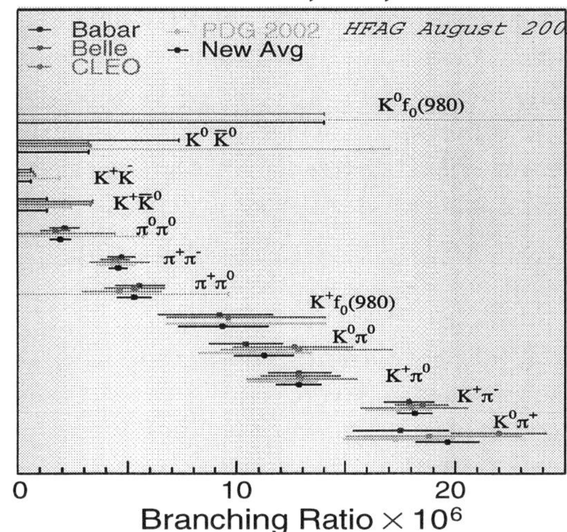

Figure 4. Summary of branching ratios for charmless 2-body B decays by the Heavy Flavor Averaging Group.

the $K\pi$ mode, also observed in the rest of the family of $K\pi$ and $\pi\pi$ modes, is an indication of the presence of another process contributing at a significant level to these decays. The main candidates within the Standard Model are the radiative loop diagrams (penguins), dominated by the QCD penguins. The experimental investigation of charmless decays aiming at the determination of the angles α and γ from these modes is performed with the reality of the penguin contribution in mind. It is clear that no single measurement in these channels is likely to lead to the determination of α and γ. The answer is likely to arise from exploiting the connections (via isospin symmetry, SU(3) of flavor) amongst the rates and CP-asymmetries of many modes, along with the use of factorization and perhaps other more predictive theories such as QCD factorization[6] or perturbative QCD[7] (when independently verified). This requires a comprehensive experimental and theoretical analysis of all pieces of the "charmless" puzzle, which forms the basis of the knowledge of α and γ.

2. Experimental Arrangement at the B-factories

The B-factory experiments, BaBar at the PEP-II machine at SLAC and Belle at the KEKB machine in the KEK laboratory in Japan, have been designed and optimized to study time-dependent CP-violation in B decays at the $\Upsilon(4S)$. The detailed description of the machines and the detectors can be found elsewhere. Here I will briefly summarize the main features of the experimental arrangement, in particular the aspects pertaining to the study of charmless B decays. The B-factories are asymmetric electron-positron colliders operating at a center-of-mass energy near the $\Upsilon(4S)$ (10.58 GeV) resonance. The $\Upsilon(4S)$ is just above the threshold for open Bottom (B) production and decays exclusively to a B^+B^- or $B^0\bar{B}^0$ pair, with nearly equal proportion. Close to the threshold, B's are produced nearly at rest in the center-of-mass frame of the $\Upsilon(4S)$. The $\Upsilon(4S)$ resonance, which produces a cross section peak of about 1 nb^{-1} on a continuum backround of about 3 times larger, has been mined in the past 20 years by the CLEO experiment at the CESR machine at Cornell and the ARGUS experiment at the DORIS-II machine in DESY, for exploring properties of the B mesons. The new machines (PEP-II and KEKB) have earned the title of "Factory" because of their high luminosities, which is required for producing large number of B's for CPV studies. The other new feature is their asymmetric laboratory beam energies: 9 GeV e^- on 3.1 GeV e^+ for PEP-II and 8 GeV e^- on 3.5 GeV e^+ for KEKB, resulting in a relativistic boost to the B mesons, which allows for mapping the time-evolution of "tagged" neutral B decays. "Tagged" here refers to the identification of the initial flavor (b or $\bar{b}$) of the B meson at the $\Delta t = 0$. Since the B^0 meson undergoes flavor oscillation before decaying, information on its initial flavor is extracted from the flavor of the other B meson in the event, taking advantage of the quantum entangled nature of the "$B^0\bar{B}^0$" system in an L=1, CP=-1 state, which by spin statistics forbids the pair to transform into a B^0B^0 or a $\bar{B}^0\bar{B}^0$ state, hence correlating the flavor of the two B mesons until one of them decays. The particle content and the kinematical properties of the decay products of the B meson reveal its (B or $\bar{B}$) flavor. The experiments perform decay time measurements using precision silicon vertex tracking, and particle identification using a variety of techniques, including electromagnetic calorimetry, measurement of dE/dx in the tracking system and Čerenkov detectors. Charmless 2-body B decays pose a particularly difficult experimental problem for hadron identification, as π/K discrimination is crucial to separating the overlapping $B \to \pi\pi$ and $B \to K\pi$ modes, whose decay products populate the momentum range of 1-4 GeV in the laboratory frame.

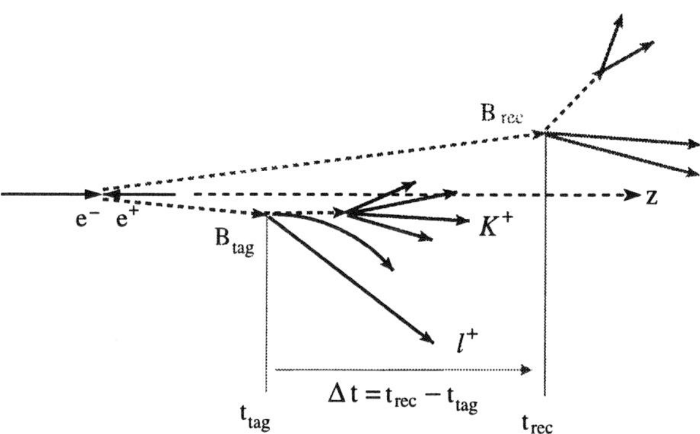

Figure 5. A cartoon of a $B\bar{B}$ event at an asymmetric B-factory. See the text for more detail.

A cartoon of a $B\bar{B}$ event produced in an asymmetric B-factory is shown in Fig. 5, depicting the decay products of a pair of B mesons boosted along the beam direction (the z-axis). In such a scenario one B meson decays into a rare CP-eigenstate such as $B \to \pi^+\pi^-$ (B_{cp}) and the other B in a generic "flavor revealing" mode (B_{tag}). For a fully reconstructed B meson, two kinematical quantities can be constructed from measurements of its decay products and the knowledge of the colliding beam energies, all calculated in the center-of-mass frame: beam energy constrained mass defined as $M_{ES} = \sqrt{E_{beam}^2 - P_B^{*2}}$ and the energy difference $\Delta E = E_{beam} - E_B^*$, where E_{beam}^* is half of the total center-of-mass energy and P_B^* and E_B^* are the reconstructed momentum and energy of the B meson candidate. The experimental resolution of M_{ES} is dominated by the knowledge of the beam energy and is typically around 2 to 3 MeV. The resolution in ΔE is dominated by the E_B^* resolution, thus its value is mode dependent with a range of 15-30 MeV, for modes with only charged tracks in the final state.

Quantities central to a CP analysis of B meson decays are decay vertex separation along the beam axis $\Delta Z = Z_{tag} - Z_{cp}$ and the flavor tagging information from the decay products of the other B in the event. The decay time difference is extracted from the relation $\Delta t \simeq \Delta Z / \gamma\beta$, with a typical resolution of around 180 μm, dominated by the vertex measurement on the tagging side. Flavor tagging is achieved with an effective efficiency $\epsilon(1-2w)^2 \simeq 30\%$, where ϵ

and w are the tagging efficiency and the mis-tagging probability, respectively.

Charmless B decays of interest to CP analysis require additional experimental attention. Typically, the relevant branching ratios are of the order of 10^{-5} or less and their kinematical features are readily mimicked by the copious continuum $e^+e^- \to q\bar{q}$ events. Furthermore, the modes of similar topology, such as $B \to \pi\pi$, $B \to K\pi$ and $B \to KK$ have considerable overlap in the ΔE distribution. A number of experimental handles exploiting the differences in event shape properties of $B\bar{B}$ and continuum $q\bar{q}$ events have been developed to suppress the background effects. The powerful hadron identification capabilities of the detectors are the key to separating modes of similar topologies.

3. CP-violation Studies with 2-body Charmless B Decays

The B-factory experiments have performed comprehensive studies of the branching ratios and CP-asymmetries in 2-body charmless decays $B \to \pi\pi$ and $B \to K\pi$, improving significantly over the previous measurements from the CLEO experiment. Below, I will discuss the latest measurements of the time-dependent CP-asymmetries in $B \to \pi^+\pi^-$ and results on the first observation of the decay $B \to \pi^0\pi^0$, followed by a summary of the measurements of 2-body charmless decays and comments on the implication of these results for the angles α and γ.

3.1. *CP Analysis of the Decay $B \to \pi^+\pi^-$*

Both BaBar[8] and Belle[9] have performed and published CPV analyses of the decay $B \to \pi^+\pi^-$. The published results are based on data samples of 81 fb^{-1} for BaBar and 78 fb^{-1} for Belle, collected up to the summer conferences in 2002. The BaBar collaboration has updated their measurements for this conference by including their Run 3 data and the reprocessed Run 1 and Run 2 data samples, totaling 113 fb^{-1} at the $\Upsilon(4S)$ peak. The measurements from the Belle collaboration were not updated for this conference.

For their 2002 results, each experiment observes about 200 events in this channel, with comparable signal-to-noise ratio of $S/B \simeq 1$. The quality of the signals and the relative contributions of the

backgrounds can be seen in the ΔE distributions in Figs. 6 (Belle) and 7 (BaBar).

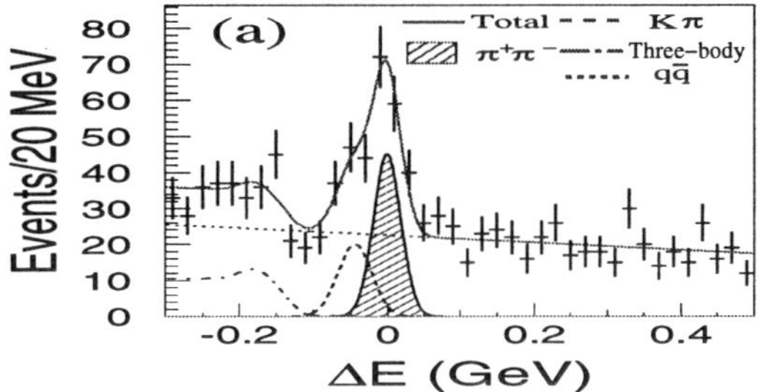

Figure 6. ΔE distribution for $B \to \pi^+\pi^-$ candidates in the Belle data (2002).

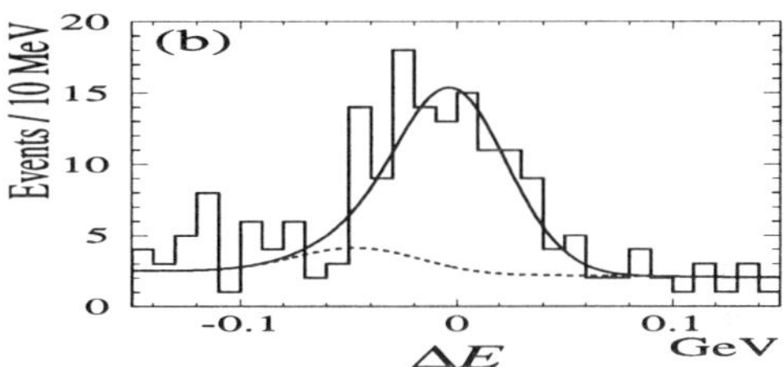

Figure 7. ΔE distribution for $B \to \pi^+\pi^-$ candidates in the BaBar data (2002).

These distributions are obtained after applying cuts on various variables developed for background suppression, including particle identification criteria. The distribution for the BaBar data is a projection plot, where the cut is applied on the signal-to-background likelihood ratio, which is constructed from the distributions of the various quantities used in the analysis. The ΔE distributions show the residual $K\pi$ component, which is present in the distribution of the signal candidates, the size of which depends on the K/π separation of the experiment. In the CP analysis, the Δt and "tagging" information are added to the kinematical and event shape quantities to perform an unbinned maximum-likelihood fit to the event sample in order to extract the values of the CP observables, $S_{\pi\pi}$ and $C_{\pi\pi}$. For details of the methods in each case, see the published papers on the subject.[8,9] The results are summarized in Table 1, including the updated BaBar results with 113 fb^{-1} , which supersedes their previous measurement. The Δt distributions and the CP-asymmetries are shown in Fig. 8 for the BaBar measurement (2003) with their full data sample, including Run 3, and in Fig. 9

197

for the Belle measurement (2002). The BaBar and Belle results have a χ^2 of 6.7 for 2 degrees of freedom, which corresponds to about a 2σ separation. A simple average of the two measurements is given in Table 1. There is no significant evidence for direct CPV ($C_{\pi\pi} \neq 0$) or indirect CPV ($S_{\pi\pi} \neq 0$) in this channel. Further discussion of the implications of the results and possible constraints on the CKM unitarity angle α are presented in Sec. 4.

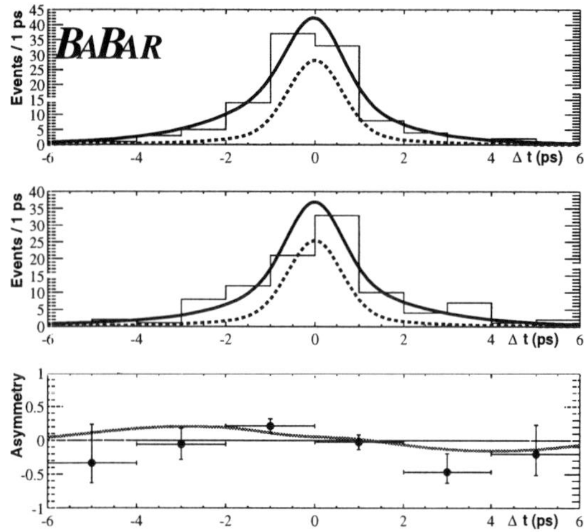

Figure 8. The Δt distributions of $B \to \pi^+\pi^-$ candidates in the BaBar data, including their Run 3 data sample. The measured time dependent asymmetries are shown in the bottom plot.

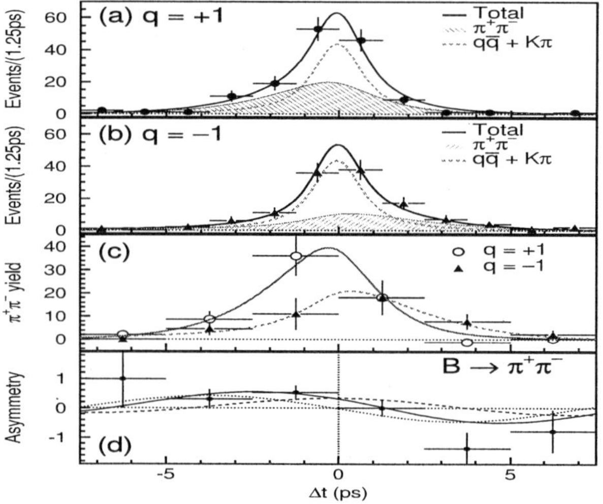

Figure 9. The Δt distributions of $B \to \pi^+\pi^-$ candidates in the Belle data. The measured time dependent CP-asymmetries are shown in the bottom plot.

3.2. Observation of the Decay $B \to \pi^0\pi^0$

Perhaps the most pleasant new input of relevance to the angle α at this conference is the observation of the decay $B \to \pi^0\pi^0$ by the BaBar collaboration,[10] supported by strong evidence from the Belle collaboration.[11] The decay $B \to \pi^0\pi^0$ is one of the components of the isospin analysis, which allows for disentangling the penguin and tree effects in charmless 2-body B decays. Previous results in this mode were limited to upper bounds (at 90% CL) on its branching ratio: $< 3.6 \times 10^{-6}$ for BaBar, $< 6.4 \times 10^{-6}$ for Belle and $< 4.4 \times 10^{-6}$ for CLEO.

The new BaBar result is based on their full data set of 113 fb^{-1}, which represents a 40% enhancement in statistical power of the data. Two major improvements were also introduced in the analysis. A major source of background in this analysis is the decay $B \to \rho^+\pi^0$, whose branching ratio was recently measured by the BaBar collaboration to be: ($B(B \to \rho^+\pi^0) = (11.0 \pm 1.9 \pm 1.9) \times 10^{-6}$), significantly lower than the previous upper bound from the CLEO experiment ($B(B^+ \to \rho^+\pi^0) < 43 \times 10^{-6}$ at 90% CL). This results in a lower level (and lower uncertainty) for the estimated background from this source. The second analysis improvement is the new optimization of the Fischer discriminant (F), which is a linear combination of several signal and continuum background discriminating variables and includes the output of a neural network program on "flavor tagging" analysis. The M_{ES} and (F) distributions of the $\pi^0\pi^0$ candidates are shown in Fig. 10, where a clear peak is evident at the B mass. The small estimated background from $B \to \rho^+\pi^0$ is shown in the dashed curve. The signal is, by choice of binning, distributed uniformly in the F variable (solid histogram), whereas the background $q\bar{q}$ events are represented by the rising dashed histogram. There is a clear excess of events above the expected background in the first bin, which is the most discriminating region of the distribution. From a fit to the data, the BaBar collaboration reports 46^{+14+2}_{-13-3} events in the $B \to \pi^0\pi^0$ decay. The overall significance of the effect, including the systematic uncertainties, is 4.3σ, qualifying it for an "Observation" report. They report a branching ratio of $(2.1 \pm 0.6(stat) \pm 0.3(syst)) \times 10^{-6}$.

The Belle collaboration also presents strong evidence for this mode, using their full data set of 158 fb^{-1} on the $\Upsilon(4S)$ peak. The distributions of

Table 1. Measurements of $S_{\pi\pi}$ and $C_{\pi\pi}$ by the BaBar and Belle experiments.

Experiment	Data sample	$S_{\pi\pi}$	$C_{\pi\pi}$
BaBar(2002)	81 fb^{-1}	$0.02 \pm 0.34 \pm 0.05$	$-0.3 \pm 0.25 \pm 0.04$
Belle(2002)	78 fb^{-1}	$-1.23 \pm 0.41^{+0.08}_{-0.07}$	$-0.77 \pm 0.27 \pm 0.08$
BaBar(2003)	113 fb^{-1}	$-0.40 \pm 0.22 \pm 0.03$	$-0.19 \pm 0.19 \pm 0.05$
Average (BaBar 2003 & Belle 2002)		-0.58 ± 0.20	-0.38 ± 0.16

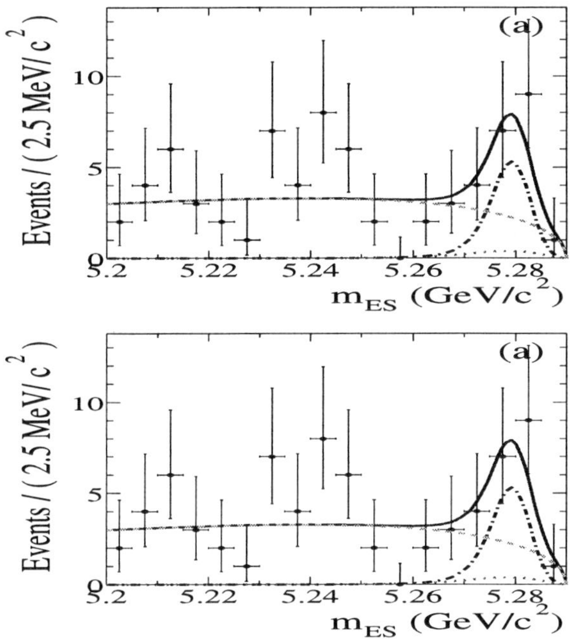

Figure 10. Distribution of M_{ES} and Fischer discriminant (F) for $B \to \pi^0\pi^0$ candidates in the BaBar data.

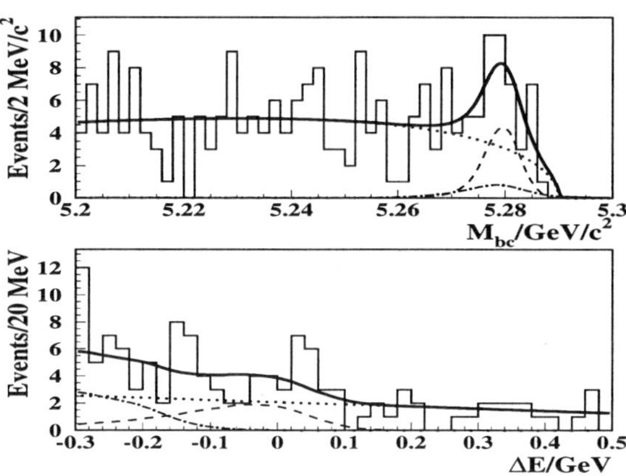

Figure 11. Distribution of M_{ES} and ΔE for $B \to \pi^0\pi^0$ candidates in the Belle data.

their candidate events in M_{ES} and ΔE are shown in Fig. 11, along with the estimated background from $B \to \rho^+\pi^0$ using the new BaBar measurement of this branching ratio, and the estimated background from continuum $q\bar{q}$ events. A fit to the data yields $25.6^{+9.3}_{-8.4}$ events with a significance of 3.4σ. They report a branching ratio of $(1.7 \pm 0.6(stat) \pm 0.3(syst)) \times 10^{-6}$, consistent with the BaBar measurement.

With the observation of the decay $B \to \pi^0\pi^0$, all modes needed for the isospin analysis have now been identified, an important step in the overall program of measuring the angle α. The measurements of the branching ratios are summarized in Table 2, along with the world averages from the Heavy Flavor Averaging Group (HFAG).[5]

3.3. Summary of Measurements of Charmless 2-body Decays

A summary of the measured branching ratios and direct CP-asymmetries in charmless 2-body B decays has been produced by HFAG and is shown in Figs. 4 and 12. Clearly much progress has been made with the data from the B-factories. The new measurements confirm the original finding by the CLEO experiment that the penguin effects play a significant role in rare B decays. Except for the direct CP-asymmetry in the mode $B \to \pi^0\pi^0$, all other components of the isospin analysis for "the tree and penguin disentanglement" in the mode $B \to \pi\pi$ have been identified.

The improvement in the accuracy of time-integrated direct CP-asymmetries is worthy of some attention. In the mode $B^0 \to K^-\pi^+$, the experiments report: $-0.107 \pm 0.041 \pm 0.013$(BaBar) and $-0.086 \pm 0.035 \pm 0.014$ (Belle), each at about 2.5σ from no CPV, with a world average (includ-

Table 2. Summary of the measurements of the branching ratios ($\times 10^{-6}$) of $B \to \pi\pi$ decays.

Mode	CLEO	Belle	BaBar	Average
$\pi^+\pi^-$	$4.5^{+1.4-0.5}_{-1.2-0.4}$	$4.4 \pm 0.6 \pm 0.3$	$4.7 \pm 0.6 \pm 0.2$	4.6 ± 0.4
$\pi^+\pi^0$	$4.6^{+1.8+0.7}_{-1.6-0.6}$	$5.3 \pm 1.3 \pm 0.5$	$5.5^{+1.0}_{-0.9} \pm 0.6$	5.2 ± 0.8
$\pi^0\pi^0$	< 4.4	$1.7 \pm 0.6 \pm 0.3$	$2.1 \pm 0.6 \pm 0.3$	1.97 ± 0.47

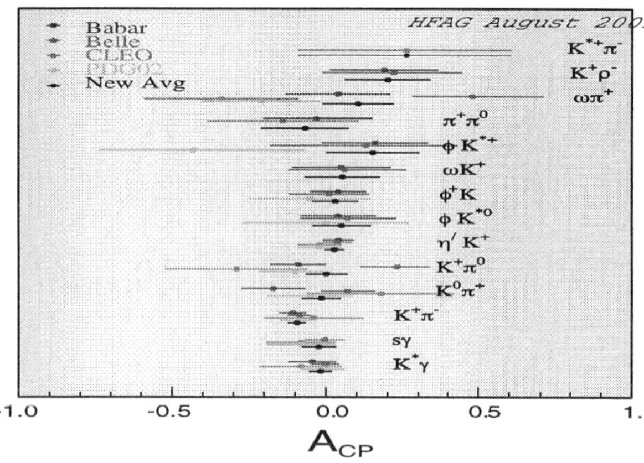

Figure 12. Summary of direct CP-asymmetries for charmless 2-body B decays by the Heavy Flavor Averaging Group (HFAG).

ing the CLEO measurement $-0.04 \pm 0.16 \pm 0.02$) of -0.09 ± 0.03. It is also interesting that at this point there is no evidence for "large" direct CP-violation in any mode.

4. What Have We Learned About α and γ?

To begin with, the consistency of the CPV measurements with the CKM picture in the Standard Model must be examined. I refer the reader for detail discussions of the topic to several recent reviews.[16−18] As discussed in the previous sections, the SM reference points for the purpose of comparison with data are the expected ranges of the angles from the "CKM fit": ($77^o < \alpha(\phi_2) < 122^o$, $37^o < \gamma(\phi_3) < 80^o$ at 95% CL). Following the approach of M. Gronau and J.L. Rosner,[14,16] the CPV measurements are compared with theoretical predictions in the plane of ($S_{\pi\pi}$ and $C_{\pi\pi}$) (Fig. 13). $S_{\pi\pi}$ and $C_{\pi\pi}$ are computed according to the equations in Sec. 1.2, which

require as inputs the ratio of penguin to tree, $|P/T|$, and the relative strong phase of the two amplitudes (δ). In the Gronau and Rosner approach, the value of $|P/T|$ is estimated by employing factorization and SU(3) relations, and exploiting information on 2-body charmless B decays, such as the rates for the penguin dominated decay $B \to K^0\pi^+$, the tree dominated decay $B \to \pi^+\pi^0$ as well as the form factor from the semi-leptonic decay $B \to \pi\ell\nu$. A value of around 0.27 (± 0.03) is favored from these considerations. For the plot presented here a ballpark value of 0.3 is used. No constraints are imposed on δ, allowing it the full range of $-\pi$ to $+\pi$. Figure 13 illustrates the physical boundary imposed by the relation $S^2 + C^2 < 1.0$ (the large circle), the family of $C_{\pi\pi}$ vs $S_{\pi\pi}$ curves (the circles from right to left) for input values of $\alpha = 30$(deg.), 60(deg.), 90(deg.), 105(deg.), 120(deg.) and 140(deg.). The circles corresponding to the two ends of the allowed range of α from the "CKM" fit: 77° and 122° are also labeled. The data points for BaBar (square), Belle (circle) and the "world average"(WA) (diamond) are also shown. Clearly, at the current level of predictive power of the theory and the experimental precision, the SM can accommodate the experimental measurement. Further progress in sharpening this comparison and its implication for the angle α would require independent constraints on the strong phase δ and improved accuracy on the knowledge of the value of $|P/T|$.

4.1. Constraints on α using $S_{\pi\pi}$ and $C_{\pi\pi}$ and Connections amongst 2-body Charmless B Decays

Gronau and London[12] proposed the use of isospin relations amongst rates and CP-asymmetries of $B \to \pi\pi$ decays for extracting the shift $\Delta\alpha = \alpha_{eff} - \alpha$, where α_{eff} is determined from $\sin 2\alpha_{eff} = \frac{S_{\pi\pi}}{\sqrt{1-C_{\pi\pi}^2}}$. The isospin analysis involves construction of sepa-

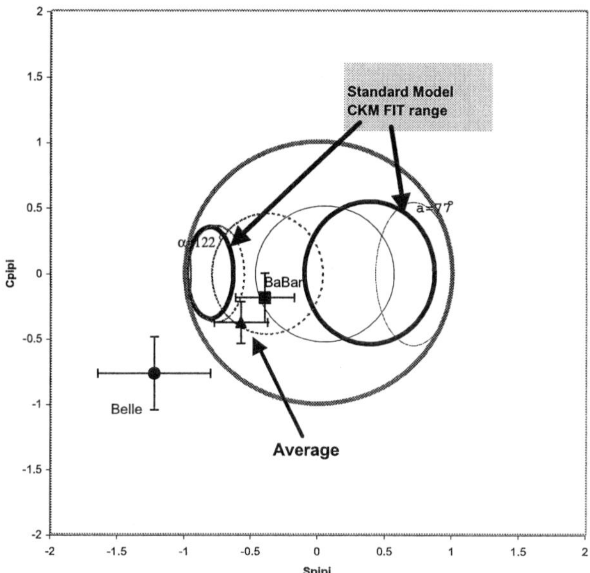

Figure 13. $C_{\pi\pi}$ vs $S_{\pi\pi}$. The physical boundary is confined to inside of the solid enveloping circle. The data points are BaBar (filled square), Belle (filled circle), and average (Diamond). See the text for further description of the graph.

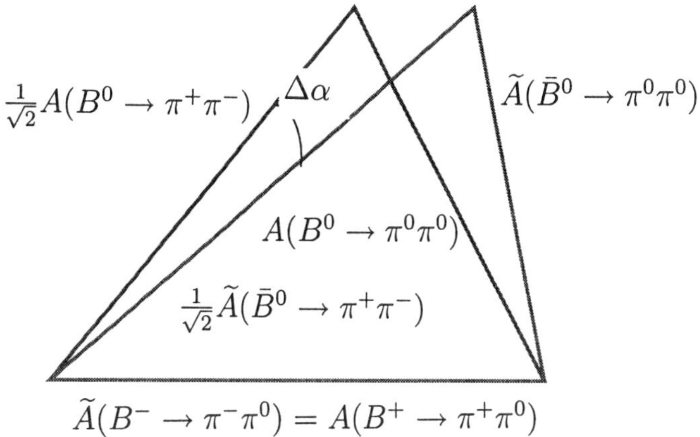

rate triangles from the information on B^0 and $\bar{B}^0$ decays, thus requiring measurement of direct CP-asymmetries in addition to the average branching ratios into $\pi\pi$ final states (Fig. 14). The mismatch between the two triangles yields the value of $|\Delta\alpha|$, with a four-fold ambiguity. Given the current statistical power of the data, we are still some distance from achieving a useful measurement of direct asymmetry in the $\pi^0\pi^0$ mode, thus a full isospin analysis is not yet possible. Quinn and Grossman[19] have shown that the knowledge of the ratio of mean branching fractions for $\bar{B}(B \to \pi^0\pi^0)$ and $\bar{B}(B \to \pi^0\pi^+)$ decays imposes an upper bound on $\Delta\alpha$ according to the following equation:

$$sin^2\Delta\alpha < \frac{\bar{B}(B^0 \to \pi^0\pi^0)}{B(B^\pm \to \pi^\pm\pi^0)}.$$

The current data leads to $|\alpha_{eff} - \alpha| < 48^o$ at 90% CL, which is not a very useful constraint and given the large value of the ratio of branching ratios, no significant improvement is expected in the future. Other relations have also been developed but unfortunately with the current level of accuracy of the measurements none improves significantly over the above limit.[20,21]

A. Hoecker, H. Lacker, M. Pivak and L. Roos[18] report an analysis of the information on charmless 2-body decays, within several theoretical scenarios for connecting the various pieces of the data. These include SU(2) (isospin relations), flavor SU(3), and inputs from QCD factorization[6] on SU(3) breaking and penguin and tree ratios and phases. They employ the tools of the "global CKMfitter" to compute confidence level for the possible values of the angle α. In Fig. 15, is plotted the confidence level vs α from a fit to the data for two scenarios: (top) SU(3) relation among various modes and an estimate of the penguin amplitude from $B \to K^0\pi^+$ and (bottom) using the QCD factorization predictions for P/T and δ. The confidence level plot from the "CKM fit" for the angle α is also shown. In both cases the "global CKM fit" has substantial overlap with fits to the data from the 2-body decays, again indicating that the SM can accommodate the data.

4.1.1. Any constraints on γ?

The decay rates and direct CP-asymmetries in the $B \to K\pi$ channels are sensitive to the angle γ. A number of ratios of branching ratios and pseudo-asymmetries, defined below, have been proposed and examined by several authors for extracting information on γ (Rosner, Gronau, Fleischer, Neubert and Mannel).[22−24] These are:

$$R_0 = \frac{B(B^0 \to K^+\pi^-)+B(\bar{B}^0 \to K^-\pi^+)}{B(B^0 \to K^+\pi^0)+B(\bar{B}^0 \to K^-\pi^0)} \frac{\tau(B^+)}{\tau(B^0)},$$

$$A_0 = \frac{B(B^0 \to K^+\pi^-)-B(\bar{B}^0 \to K^-\pi^+)}{B(B^0 \to K^+\pi^0)+B(\bar{B}^0 \to K^-\pi^0)} \frac{\tau(B^+)}{\tau(B^0)},$$

$$R_c = 2\frac{B(B^+ \to K^+\pi^0)+B(\bar{B}^- \to K^-\pi^0)}{B(B^+ \to K^0\pi^+)+B(\bar{B}^- \to \bar{K}^0\pi^-)},$$

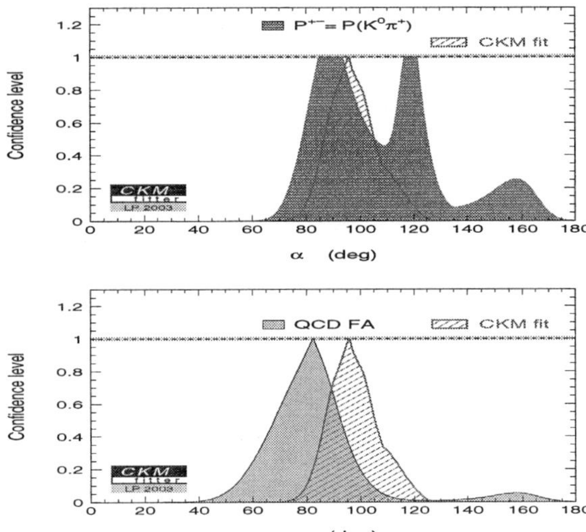

Figure 15. A global fit to the measurements of the $\pi\pi$ system. See the text for detail description of the content of the graphs.

$$A_c = 2\frac{B(B^+\to K^+\pi^0)-B(\bar{B}^-\to K^-\pi^0)}{B(B^+\to K^0\pi^+)+B(\bar{B}^-\to \bar{K}^0\pi^-)},$$

$$R_n = \frac{1}{2}\frac{B(B^0\to K^+\pi^-)+B(\bar{B}^0\to K^-\pi^+)}{B(B^0\to K^0\pi^0)+B(\bar{B}^0\to K^0\pi^0)}, \text{ and}$$

$$A_n = \frac{1}{2}\frac{B(B^0\to K^+\pi^-)-B(\bar{B}^0\to K^-\pi^+)}{B(B^0\to K^0\pi^0)+B(\bar{B}^0\to K^0\pi^0)}.$$

The updated values of these ratios based on the latest world averages (from HFAG) are:

$R_0 = 0.99 \pm 0.09 \quad A_0 = -0.09 \pm 0.03$

$R_c = 1.30 \pm 0.15 \quad A_c = 0.0 \pm 0.06,$

$R_n = 0.8 \pm 0.11 \quad A_n = -0.07 \pm 0.03.$

Comparisons of the data with theoretical relations have been discussed by the authors mentioned above and shown that the data accommodates the range of γ from the "global CKM fit". (For recent reviews see the References.[16,17])

5. Other Channels for Reaching α

Extending the measurement of the angle α to higher multiplicity charmless B decay modes, such as $B \to 3\pi$ and $B \to 4\pi$, is the natural next step. The CPV studies in these channels would require the additional analysis step of isolating the CP-eigenstate components of the multiparticle system. The simplest approach would be to identify quasi-2 body decays into resonances, such as $B \to \rho\pi$, $B \to a_1\pi$, or $B \to \rho\rho$, and perform an angular analysis of the final states to isolate the CP-odd and CP-even components.

The $B \to \rho\pi$ system presents a special case for the neutral B decays: the final states $\rho^+\pi^-$ and $\rho^-\pi^+$, which can be reached by both B^0 and $\bar{B}^0$, have substantial overlap in the Dalitz plot, thus their amplitudes interfere and generate additional dependence on α and the strong phases of the final states. Quinn and Snyder[13] have shown that the interference effect can be exploited to extract the angle α even in the presence of penguins. This would involve an amplitude analysis of the 3π Dalitz distribution, which is a formidable task given the presence of large background from $q\bar{q}$ and generic $B\bar{B}$ events. Studies have shown that the sensitivity of such an analysis begins to become interesting for the B-factory data samples at around $100\,\text{fb}^{-1}$. Studies along these lines are currently underway in both BaBar and Belle with results expected in the next few months.

In the absence of a full amplitude analysis, BaBar has presented a CP analysis of the quasi 2-body decay $B \to \rho^\pm\pi^\mp$[25] and both BaBar and Belle have performed measurements of the branching ratios of the various modes necessary for an isospin analysis of the $\rho\pi$ system, analogous to the $\pi\pi$ system. More details on these measurements are discussed below. Both BaBar and Belle have also presented results on the decay $B \to \rho\rho$, which indicate interesting possibilities for extracting information on the angle α.

5.1. CP Studies of the Decay $B \to \rho\pi$

The system $\rho^\pm\pi^\mp$ is not a CP-eigenstate, but both final states $\rho^+\pi^-$ and $\rho^-\pi^+$ can be reached by B^0 and $\bar{B}^0$, thus it is a candidate for time-dependent CP-asymmetry studies. The time evolution of a neutral B decay in this mode follows:

$$f_{B^0}(\rho^\pm h^\mp)(\Delta t) = (1 \pm A_{cp}(\rho h))e^{-|\Delta|t/\tau}(1 + [(S_{\rho h} \pm \Delta S_{\rho h})sin(\Delta m\Delta t) - (C_{\rho h} \pm \Delta C_{\rho h})cos(\Delta m\Delta t)]),$$

$$f_{\bar{B}^0}(\rho^\pm h^\mp)(\Delta t) = (1 \pm A_{cp}(\rho h))e^{-|\Delta|t/\tau}(1 - [(S_{\rho h} \pm \Delta S_{\rho h})sin(\Delta m\Delta t) - (C_{\rho h} \pm \Delta C_{\rho h})cos(\Delta m\Delta t)]),$$

Where $h = \pi$ or K. Summing over the charge of the ρ meson, results in the time-dependent asymmetry:

$$A_{\rho\pi}(B^0/\bar{B}^0) \simeq S_{\rho\pi}sin(\Delta m\Delta t) - C_{\rho\pi}cos(\Delta m\Delta t).$$

The CP-violating observables are: $S_{\rho\pi}$ for indirect

Table 3. Summary of the measurements of branching ratio ($\times 10^{-6}$) of $B \to \pi\pi$ decays.

$N(\rho\pi)$	$N(\rho K)$	$C_{\rho\pi}$	$S_{\rho\pi}$
804.2 ± 49.2	260.4 ± 31.4	$0.35 \pm 0.13 \pm 0.05$	$-0.13 \pm 0.18 \pm 0.04$

$A_{CP}(\rho\pi)$	$A_{CP}(\rho K)$	ΔC	ΔS
$-0.11 \pm 0.06 \pm 0.03$	$0.18 \pm 0.12 \pm 0.08$	$0.2 \pm 0.13 \pm 0.05$	$0.33 \pm 0.18 \pm 0.03$

CP-violation, and the quantities $C_{\rho\pi}$ and the charge asymmetry $A_{\rho h} = \frac{N(\rho^+ h^-) - N(\rho^- h^+)}{N(\rho^+ h^-) + N(\rho^- h^+)}$ ($A_{\rho\pi}$ and $A_{\rho K}$) are measures of direct CP-violation. The parameters $\Delta C_{\rho\pi}$ and $\Delta S_{\rho\pi}$ are CP conserving quantities and are labeled as dilution factors. The projection plots in Δt and the corresponding time-dependent asymmetry plot are shown in Fig. 16. The results, including the charge asymmetry for the decay $B \to \rho K$, are summarized in Table 3.

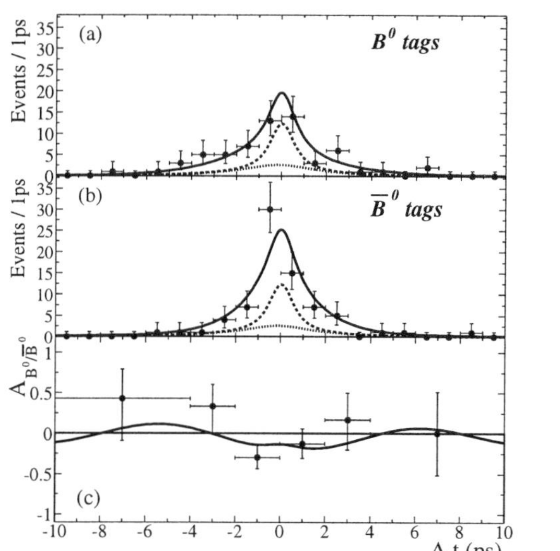

Figure 16. Time-evolution and asymmetry plot for $B \to \rho^\pm \pi^\mp$ candidates in the BaBar data.

There is no significant evidence for CP-violation in this mode, but the values of the direct CPV quantities, $C_{\rho\pi}$ and $A_{CP}(\rho\pi)$, both showing about 2σ deviations from zero, deserve a little more attention. In order to present a statistical statement on direct CPV effects in this channel, these quantities are recast in more familiar forms:

$$A_{+-} = \frac{N(\bar{B}^0 \to \rho^+ \pi^-) - N(B^0 \rho^- \pi^+)}{N(\bar{B}^0 \to \rho^+ \pi^-) + N(B^0 \rho^- \pi^+)}$$
$$= \frac{A_{CP}^{\rho\pi} - C - A_{CP}^{\rho\pi} \cdot \Delta C}{1 - \Delta C - A_{CP}^{\rho\pi} \cdot C},$$

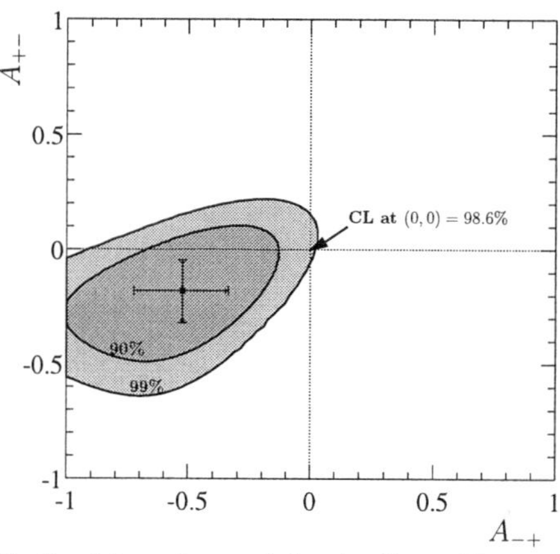

Figure 17. Confidence intervals for the direct CP-asymmetries A_{+-} and A_{-+} using total error bars (statistical and systematic errors in quadrature).

$$A_{-+} = \frac{N(\bar{B}^0 \to \rho^- \pi^+) - N(B^0 \rho^+ \pi^-)}{N(\bar{B}^0 \to \rho^- \pi^+) + N(B^0 \rho^+ \pi^-)}$$
$$= \frac{A_{CP}^{\rho\pi} + C + A_{CP}^{\rho\pi} \cdot \Delta C}{1 + \Delta C + A_{CP}^{\rho\pi} \cdot C}.$$

and plotted in Fig. 17 along with the corresponding confidence level plots. This puts the data point at about 2.5σ away from the origin (zero CPV point).

5.2. Comments on the Decay $B \to \rho\rho$

Both BaBar and Belle have performed extensive studies of the $B \to VV$ decays, where V stands for a vector meson (reviewed in the talk by J. Fry at this conference). An important channel among this class of final states, presenting interesting opportunities for measuring α, is the family of $B \to \rho\rho$ decays, which includes $B^0 \to \rho^+ \rho^-$, $B^+ \to \rho^+ \rho^0$ and $B^0 \to \rho^0 \rho^0$. The $\rho\rho$ system is essentially an analog of the $\pi\pi$ system (with one caveat discussed below), providing a measurement of $\alpha_{eff}^{\rho\rho}$. The decay amplitudes are connected via isospin symmetry and a Quinn-Grossman (type) bound on $|\alpha_{eff}^{\rho\rho} - \alpha|$ can be extracted from information on the mean branching ratios of the decays $B^0 \to \rho^0\rho^0$ and $B^+ \to \rho^+\rho^0$. The main difference from the $\pi\pi$ system is in the CP composition of the final state, which in this case can be a mix of CP-odd and CP-even states. In general, the pair of ρ mesons can be in a state of relative orbital angular momentum L=0 (S-wave), 1 (P-wave) and 2 (D-wave), thus allowing for both a CP-odd state (transversely polarized ρ) and a CP-even state (longitudinally polarized ρ). The angular analysis of the decays $B^+ \to \rho^+\rho^0$ (Belle and BaBar) and $B^+ \to \rho^+\rho^-$ (BaBar) have shown that the decays are dominated by the longitudinal polarization (CP-even component)(f_L).[26−27] This significantly simplifies the time-dependent measurements which are currently underway, with results expected in the near future. Furthermore, applying the Quinn-Grossman relation to the constraints on branching ratios gives:

$$sin^2(\alpha_{eff}^{\rho\rho} - \alpha) \leq \frac{(f_L(\rho^0\rho^0) \times B(\rho^0\rho^0))}{(f_L(\rho^+\rho^0) \times B(\rho^+\rho^0))}$$

resulting in: $|\alpha_{eff}^{\rho\rho} - \alpha| < 19^o$, at 90% CL which is already more restrictive on penguin effects than in the $\pi\pi$ system.

6. Reaching γ via the $B \to DK$ Channel: Interference of $b \to u(W^- \to \bar{c}s)$ and $b \to c(W^- \to \bar{u}s)$

The family of $B \to DK$ decays represents a "penguin free" channel for reaching the angle γ. The responsible diagrams are illustrated in Fig. 18. For details of the methods and formulation of strategies for these measurements see the References.[29] The amplitude relations below show the connection to the angle γ, which enters via V_{ub}:

$$A(B^- \to \bar{D}^0 K^-) = |A_1|e^{i\delta_1}e^{i\gamma}$$
$$A(B^- \to D^0 K^-) = |A_2|e^{i\delta_2},$$

$$A(B^+ \to D^0 K^+) = |A_1|e^{i\delta_1}e^{-i\gamma}$$
$$A(B^+ \to \bar{D}^0 K^+) = |A_2|e^{i\delta_2},$$

where δ_1 and δ_2 are the CP conserving strong phases. For final states where the D meson decays into a CP-eigenstate (such as $\pi^+\pi^-$ or ϕK_s^0), D_{cp}, both diagrams contribute, with the corresponding decay amplitudes:

$$A(B^- \to D_{cp}K^-) = |A_1|e^{i\delta_1}e^{i\gamma} + |A_2|e^{i\delta_2},$$

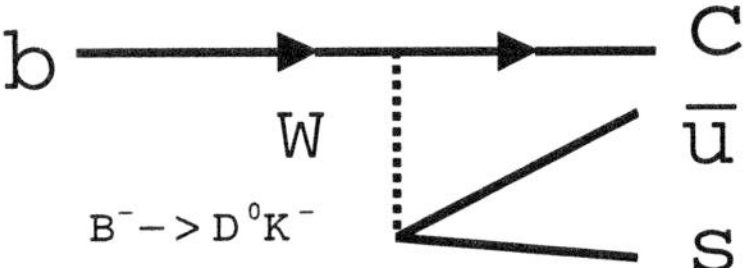

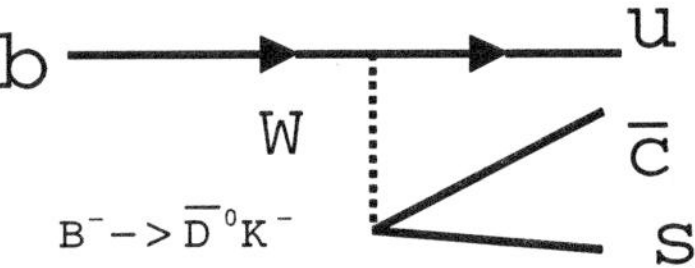

Figure 18. Diagrams of contributions to the decay $B \to D_{cp}K$. See the text for details.

$$A(B^+ \to D_{cp}K^+) = |A_1|e^{i\delta_1}e^{-i\gamma} + |A_2|e^{i\delta_2}.$$

The resulting interference effects preserve the sensitivity to the angle γ in the rates and CPV observables. From the experimental point of view, however, the task is difficult: the expected asymmetries are small (the scale set by the relative sizes of the interfering CKM suppressed and favored diagrams) and the D^0 decays into CP-eigenstates are rare. The original proposal by Wyler et $al.$,[29] involved construction of the two amplitude triangles in the complex plane using the decay rates, and determination of the angle γ from their mismatch. However, it turns out that in practice when hadronic final states are employed for identifying D meson decays, the $b \to u$ channel ($B^+ \to D^0 K^+$) is indistinguishable from the final state $B^+ \to \bar{D}^0 K^+$ ($b \to c$) followed by the Double Cabibbo Suppressed Decay of $\bar{D}^0$. The alternative set of observables, which have been explored in these channels, consists of the rate ratio and the CP-asymmetry defined as :

$$R_{cp} = \frac{B(B^- \to D_{cp}^0 K^-) + B(B^+ \to \bar{D}_{cp}^0 K^+)}{B(B^- \to D^0 K^-) + B(B^+ \to \bar{D}^0 K^+)}$$
$$= 1 + r_{DK}^2 \pm 2r_{DK}\cos\delta_{DK}\cos\gamma,$$

$$A_{cp} = \frac{B(B^- \to D_{cp}^0 K^-) - B(B^+ \to \bar{D}_{cp}^0 K^+)}{B(B^- \to D^0 K^-) + B(B^+ \to \bar{D}^0 K^+)}$$
$$= \frac{\pm 2R_{DK}\sin\delta_{DK}\sin\gamma}{1 + r_{DK}^2 \pm 2r_{DK}\cos\delta_{DK}\cos\gamma},$$

where the unknown parameters are the ratio of the amplitudes of the $b \to u$ and the $b \to c$ transitions (for each decay process), r_{DK}, their relative strong phase δ_{DK} and the angle γ. In principle, all three unknowns can be extracted from the data,

Table 4. Measured Branching ratios for $B \to DK$ decays

Mode	CLEO	Belle	BaBar
$B^- \to D^0 K^-$	$(9.9 \pm 1.3 \pm 0.7)\%$	$(7.7 \pm 0.9 \pm 0.6)\%$	$(8.31 \pm 0.35 \pm 0.2)\%$
$B^- \to D^0 K^{*-}$	$(6.1 \pm 1.6 \pm 1.7) \times 10^{-4}$	$(5.2 \pm 0.5 \pm 0.6) \times 10^{-4}$	$(6.3 \pm 0.7 \pm 0.4) \times 10^{-4}$
$B^- \to D^{*0} K^{*-}$	$(7.7 \pm 2.2 \pm 2.6) \times 10^{-4}$		$(8.3 \pm 1.1 \pm 1.0) \times 10^{-4}$
$B^0 \to \bar{D}^0 K^{*0}$		$(4.8 \pm 1.1 \pm 0.5) \times 10^{-5}$	$(3.0 \pm 1.3 \pm 0.6) \times 10^{-5}$
$B^0 \to \bar{D}^0 K^0$		$(5.0 \pm 1.3 \pm 0.6) \times 10^{-5}$	$(3.4 \pm 1.3 \pm 0.6) \times 10^{-5}$
$B^0 \to \bar{D}^{*0} K^0$		$< 6.6 \times 10^{-5} (90\%\ CL)$	
$B^0 \to \bar{D}^{*0} K^{*0}$		$< 6.9 \times 10^{-5} (90\%\ CL)$	
$B^0 \to D^{*0} K^0$		$< 1.8 \times 10^{-5} (90\%\ CL)$	
$B^0 \to D^{*0} K^{*0}$		$< 4.0 \times 10^{-5} (90\%\ CL)$	

Table 5. Measured CP observables for the $B \to DK$ decays

$B^- \to D^0 K^-$	$R_1(CP = +1)$	$A_1(CP = +1)$	$R_2(CP = -1)$	$A_2(CP = -1)$
Belle	$1.21 \pm 0.25 \pm 0.14$	$0.06 \pm 0.19 \pm 0.04$	$1.41 \pm 0.27 \pm 0.15$	$-0.19 \pm 0.17 \pm 0.05$
BaBar	$1.06 \pm 0.26 \pm 0.17$	$0.17 \pm 0.23 \pm 0.08$		
$B^- \to D^0 K^{*-}$				
Belle		$-0.06 \pm 0.33 \pm 0.07$		$0.19 \pm 0.50 \pm 0.04$

when sufficient precision in the measurements has been achieved. The method has also been extended to B^0 decays.

The experiments have covered many of the possible channels in the $B \to DK$ system,[30-33] as compiled in Table 4 and Table 5, with several incomplete areas waiting for more data. The Belle collaboration has performed an analysis of the decay $B \to D^0 K$, using the 3-body decays of D^0,[34] reporting a 90% CL interval of $61^o < \gamma < 142^o$.

7. Reaching $2\beta + \gamma$ Through the Decay $B \to D^{(*)+}\pi^-$: Interference of the $b \to c$ and $b \to u$ Diagram via $B^0 \to \bar{B}^0$ Mixing

A small interference effect present in the decay $B \to D^{(*)+}\pi$ provides sensitivity to the phase $2\beta + \gamma$.[35] The dominant contributing diagram to this channel is the CKM favored $b \to c(\bar{u}d)$ transition. A small contribution is also expected from the diagram involving $B^0 \leftrightarrow \bar{B}$ oscillation, followed by the CKM suppressed $\bar{b} \to \bar{u}(c\bar{d})$ transition. The interference of the two diagrams generates sensitivity to the phase $2\beta + \gamma$, where the mixing diagram contributes to the phase 2β.

As expected the resulting CP-asymmetry is very small, the scale set by the ratio of the amplitudes: $r_{D\pi} = \frac{A(\bar{B}^0 \to D^- \pi^+)}{A(B^0 \to D^- \pi^+)} \simeq 0.02$. The experimental analysis involves a time-dependent CP analysis of tagged neutral B decays into this final state. The time evolution relations are:

$$P(B^0 \to D^{\mp}\pi^{\pm}, \Delta t) =$$
$$Ne^{-\Gamma|\Delta t|}[1 \pm C_{D\pi}\cos(\Delta m_d \Delta t) + S^{\mp}_{D\pi}\sin(\Delta m_d \Delta t)],$$

$$P(\bar{B}^0 \to D^{\mp}\pi^{\pm}, \Delta t) =$$
$$Ne^{-\Gamma|\Delta t|}[1 \mp C\cos(\Delta m_d \Delta t) - S^{\mp}_{D\pi}\sin(\Delta m_d \Delta t)],$$

where the CP observables are $C_{D\pi} = \frac{1-r^2_{D\pi}}{1+r^2_{D\pi}} \simeq 1$ and

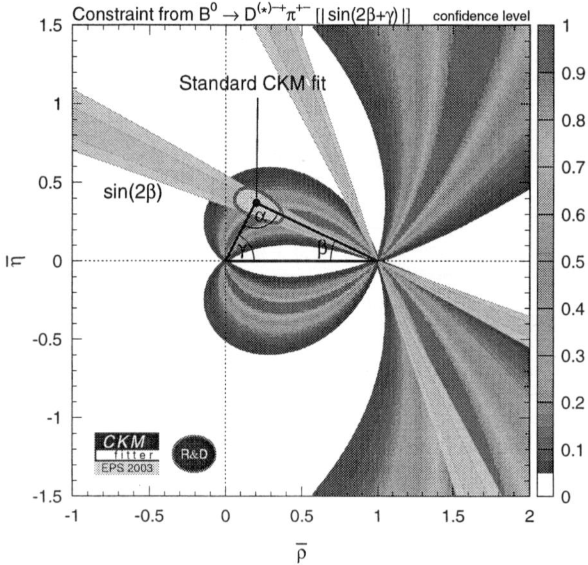

Figure 19. The CKM fit with the constraints from $2\beta + \gamma$ included. (From CKMfitter group (http://ckmfitter.in2p3.fr/.))

$S_{D\pi} = \frac{2r_{D\pi}}{1+r_{D\pi}^2} sin(2\beta + \gamma \pm \delta_{D\pi})$. Besides $2\beta + \gamma$, the other unknowns in this analysis are $r_{D\pi}$ and the relative strong phase $\delta_{D\pi}$. BaBar and Belle presented preliminary results on time-dependent CP-asymmetries in $B \to D^{*+}\pi^-$ and $B \to D^+\pi^-$ at this conference. A summary of the measurements is given below.

BaBar with fully reconstructed decays[36]
$2r_{D^*\pi}sin(2\beta + \gamma)cos\delta_{D^*\pi} = -0.068 \pm 0.038 \pm 0.021$
$2r_{D^*\pi}cos(2\beta + \gamma)sin\delta_{D^*\pi} = -0.031 \pm 0.070 \pm 0.035$
$2r_{D\pi}sin(2\beta + \gamma)cos\delta_{D\pi} = -0.022 \pm 0.038 \pm 0.021$
$2r_{D\pi}cos(2\beta + \gamma)sin\delta_{D\pi} = -0.025 \pm 0.068 \pm 0.035.$

Belle with fully reconstructed decays[37]
$2r_{D^*\pi}sin(2\beta + \gamma + \delta_{D^*\pi})$
$\quad = +0.092 \pm 0.059 \pm 0.016 \pm 0.036(D^*\ell\nu)$
$2r_{D^*\pi}sin(2\beta + \gamma - \delta_{D^*\pi})$
$\quad = +0.033 \pm 0.056 \pm 0.016 \pm 0.036(D^*\ell\nu)$
$2r_{D\pi}sin(2\beta + \gamma + \delta_{D\pi})$
$\quad = +0.094 \pm 0.059 \pm 0.013 \pm 0.036(D^*\ell\nu)$
$2r_{D\pi}sin(2\beta + \gamma - \delta_{D\pi})$
$\quad = +0.022 \pm 0.056 \pm 0.013 \pm 0.036(D^*\ell\nu).$

BaBar with partially reconstructed decays[36]
$2r_{D^*\pi}sin(2\beta + \gamma)cos\delta_{D^*\pi} = -0.064 \pm 0.025 \pm 0.017.$

The BaBar collaboration has presented limits on $sin(2\beta + \gamma)$ using an estimate of $r_{D\pi}$ by employing factorization relations and information from the decay $B \to D_s\pi$, corrected for flavor SU(3) breaking. They perform a χ^2 fit, taking into account estimates of the theoretical and experimental uncertainties to calculate confidence levels as a function of the CPV phase. They set an upper bound of $|sin(2\beta + \gamma)| > 0.76$ at 90% CL Figure 19 shows the implication of the confidence level information on $sin(2\beta + \gamma)$ in the (η, ρ) plane, along with the standard "CKM fit", indicating that the $sin(2\beta + \gamma)$ information also favors the Standard Model (η, ρ) region.

8. Summary and Conclusion

After establishing that CP-symmetry is broken in B meson decays, the B-factory experiments have been hard at work to examine the consistency of the measurements with the Standard Model predictions and search for new physics effects via possible deviations from the SM. A major focus of the experiments has been on the determination of the CKM unitarity angles $\alpha(\phi_2)$ and $\gamma(\phi_3)$, through a variety of channels and the use of theoretical relations to interpret the results. The BaBar and Belle collaborations have performed measurements of the time-dependent CP observables in the channel $B \to \pi^+\pi^-$. The results from the two experiments, which are about 2σ apart, do not yet establish CPV in this channel, but are consistent with the expected values from the SM. Significant progress has also been made in identifying and measuring the components of the 2-body charmless B decays, which together with theoretical models and symmetry relations will eventually lead to the determination of the angles α and γ. The BaBar collaboration announced the observation of the decay $B \to \pi^0\pi^0$, supported by strong evidence from the Belle experiment. New results were also presented on the $B \to \rho\pi$ decays, including measurements of time-dependent CPV observables in the quasi 2-body channel $B \to \rho\pi$ by the BaBar collaboration, with hints (at 2σ level) of possible direct CPV effects. Both experiments expect results with amplitude analyses in this channel in the near future. Measurements of the rates and polarization in the $B \to \rho\rho$ system by Belle and BaBar indicate that this is a potentially powerful channel for the determination of α. The experiments have also presented results, albeit with large errors, on branching ratios and CPV observables in the $B \to DK$ system, which is sensitive to the angle γ, and time-dependent CPV analyses of $B \to D^{*+}\pi^-$ with information on $\sin(2\beta + \gamma)$. At the current level of experimental and theoretical accuracies, all of the measurements in this area can be accommodated in the SM and the information on α and γ are consistent with the CKM picture. However, this is just the begining of the experimental investigation of these channels for CPV information. With the expected significant increase in the data from the B-factories in the next few years, the goal is to extract information on α and γ through as many channels as possible, thus reduce both the experimental and theoretical uncertainties on these quantities.

Acknowledgments

I wish to thank my colleagues on the BaBar and Belle experiments who are responsible for most of the experimental results presented here. I am also grateful to M. Gronau, J. L. Rosner, A. Hoecker, H. Lacker, T. Browder, J. Smith, J. Alexander, P. Chang, B. Cahn, J. Olsen, A. Farbin, E. Rosenberg, D. Roberts and J. Fry for their comments and help in the preparation of the talk and this document. I wish to thank the organizers of the 2003 Lepton-Photon Symposium for their support. This work was partially supported by a grant from the United States Department of Energy.

References

1. M. Kobayashi and T. Maskawa, *Prog. Th. Phys.* **49**, 652 (1973).
2. L. Wolfenstein, *Phys. Rev. Lett.* **51**, 1945 (1983).
3. A. Hoecker, H. Lacker, S. Laplace and F. Le Diberder, *Eur. Phys. Jour.* C **21**, 225 (2001) (hep-ph/0104062).
4. BaBar Collaboration, B. Aubert *et al.*, *Phys. Rev. Lett.* **89**, 201802 (2002); Belle Collaboration, K. Abe *et al.*, *Phys. Rev.* D **66**, 071102 (2002).
5. Heavy Flavor Averaging Group (see the webpage http://www.slac.stanford.edu/xorg/hfag/).
6. M. Beneke, G. Buchalla, M. Neubert and C. T. Sachrajda, *Phys. Rev. Lett.* **83**, 1914 (1999).
7. Y. Keum, H. Li and A. I. Sanda, *Phys. Lett.* B **504**, 6 (2001).
8. B. Aubert *et al.* (BaBar Collaboration), *Phys. Rev. Lett.* **89**, 281802 (2002).
9. K. Abe *et al.* (Belle Collaboration), *Phys. Rev.* D **68**, 012001 (2003).
10. B. Aubert *et al.* (BaBar Collaboration), hep-ex/0308012.
11. S.H. Lee and K. Suzuki *et al.* (Belle Collaboration), hep-ex-0308040.
12. M. Gronau and D. London, *Phys. Rev. Lett.* **65**, 3381 (1990).
13. H. R. Quinn and A.E. Snyder, Phys. Rev. D **48**, 2139 (1993).
14. M. Gronau and J.L.Rosner, *Phys. Rev.* D **65**, 093012 (2002).
15. M. Gronau and J.L.Rosner, *Phys. Rev.* D **65**, 013004 (2001).
16. J.L. Rosner, hep-ph/0305315; M. Gronau, hep-ph/0306308.
17. R. Fleischer, hep-ph/0306270.
18. A. Hoecker, H. Lacker, M. Pivak and L. Roos, LAL 02-103.
19. Y. Grossman and H.R. Quinn, *Phys. Rev.* D **58** 017504 (1998).
20. J. Charles, *Phys. Rev.* D **59** 054007 (1999).
21. M. Gronau, D. London, N. Sinha and R. Sinha, *Phys. Lett.* B **514**, 315 (2000).
22. M. Gronau, J.L. Rosner and D. London, *Phys. Rev. Lett.* **73**, 21 (1994); O. F. Fernandez *et al.*, *Phys. Lett.* B **333**, 500 (1994); R. Fleischer, *Int. Jour. Mod. Phys.* A **12**, 2459 (1997).
23. F. Fleischer and T. Mannel, *Phys. Rev.* D **57**, 2752 (1998).
24. M. Gronau D. London and J.L.Rosner, *Phys. Rev. Lett.* **73**, 21 (1994).
25. B. Aubert *et al.* (BaBar Collaboration), hep-ex/0307087 and hep-ex/0306063.
26. J.Zhang, M.Nakao *et al.* (Belle Collaboration), hep-ex/0306007.
27. B. Aubert *et al.* (BaBar Collaboration), hep-ex/0308026.
28. B. Aubert *et al.* (BaBar Collaboration), hep-ex/0308024.
29. M. Gronau and D. Wyler, *Phys. Lett.* B **265**, 172 (1991); D. Atwood, I. Dunietz and A. Soni, *Phys. Rev. Lett.* **78**, 3257 (1997).
30. S.K.Swain, T.E.Browder *et al.* (Belle Collaboration), *Phys. Rev.* D **68**, 051101 (2003).
31. P.Krokovny *et al.* (Belle Collaboration), *Phys. Rev. Lett.* **90**, 141802 (2003).
32. K. Abe *et al.* (Belle Collaboration), hep-ex/0307074.
33. B. Aubert *et al.* (BaBar Collaboration),hep-ex/0207087 and hep-ex/0307036.
34. K. Abe *et al.* (Belle Collaboration), hep-ex/0308043.
35. R. G. Sachs, Enrico Fermi Institute Report, EFI-85-22 (1985) (unpublished); I. Duniez and R.G. Sachs, *Phys. Rev.* D. **37**, 3186 (1988); *Phys. Lett.* B **427**, 179 (1998).
36. B. Aubert *et al.* (BaBar Collaboration), hep-ex/0308018 and hep-ex/0307036.
37. K. Abe *et al.* (Belle Collaboration), hep-ex/0308048.

Tom Browder (Univ. of Hawaii): This is related to the issue about final-state interactions in $B \to \rho\pi$. Belle has presented evidence for $B \to \rho^0\pi^0$, which is perhaps a hint that there are large final-state phases in that system as well?

Abolhassan Jawahery: That's a good comment. Maybe the theorists want to comment on it. I will leave it at that.

Ikaros Bigi (Univ. of Notre Dame): You talked about the new BaBar result on $B \to \pi\pi$, one of my interests is to show deviations from the Superweak scenario with the new BaBar result. How far away from pure Superweak CP-violation are you?

Abolhassan Jawahery: You mean with pure Superweak we have $S = \sin(2\beta)$, so is $S = \sin(2\beta)$? We cannot say it is not. I think with the numbers we have, we are within 1σ of that.

Klaus Schubert (TU Dresden): This is a comment to Bigi's question. As long as $\cos(2\beta)$ and $\cos(2\alpha_{eff})$ are not measured, direct CP-violation in $\lambda_{(ccK)}/\lambda_{(\pi\pi)}$ can not be established with the present data on $\sin(2\beta)$ and $\sin(2\alpha_{eff})$.

Ikaros Bigi (Univ. of Notre Dame): We can discuss this on Thursday, in the breakout session.

K. T. PITTS

University of Illinois, Department of Physics, 1110 West Green St., Urbana, IL 61801, USA
E-mail: kpitts@uiuc.edu

The Fermilab Tevatron offers unique opportunities to perform measurements of the heavier B hadrons that are not accessible at the $\Upsilon(4S)$ resonance. In this summary, we describe some recent heavy flavor results from the DØ and CDF collaborations and discuss prospects for future measurements.

1. Introduction

In the 1980's and 1990's, the complementary b physics programs of CLEO, ARGUS, the LEP experiments, SLD, DØ and CDF began to make significant contributions to our understanding of the production and decay of B hadrons. With the successful turn-on of the BaBar and Belle experiments in the last few years, many experimental measurements in the b sector have achieved impressive precision.

A number of questions remain, and it will again take an effort of complementary measurements to make further progress on our understanding of the b system. The Fermilab Tevatron, along with the upgraded CDF and DØ detectors, offers a unique opportunity to study heavy flavor production and decay. In many cases, the measurements that can be performed at the Tevatron are complementary to those performed at the e^+e^- B-factories.

In this document, we summarize some of the recent experimental progress in the measurements of B lifetimes, spectroscopy, mixing, and heavy flavor production at the Tevatron. Since the details of many of these analyses are presented in other publications, we will attempt to include some background information in this summary that the reader might not find elsewhere.

2. Overview: B Physics at the Tevatron

In $p\bar{p}$ collisions at $\sqrt{s} = 1.96\,\text{TeV}$, the $b\bar{b}$ cross section is large $\mathcal{O}(50\mu b)$, yet it is only about $1/1000^{\text{th}}$ the total inelastic $p\bar{p}$ cross section. At a typical Tevatron instantaneous luminosity of $\mathcal{L} = 4 \times 10^{31}\,\text{cm}^{-2}\text{s}^{-1}$, we have a $b\bar{b}$ production rate of ~ 2 kHz compared to an inelastic scattering rate of ~ 2 MHz.

The $b\bar{b}$ quarks are produced by the strong interaction, which preserves "bottomness", therefore they

Table 1. B mesons and baryons. This is an incomplete list, as there are excited states of the mesons and baryons (*e.g.* B^{*0}). Also, a large number of b-baryon states (*e.g.* $\Sigma_b^- = |ddb>$) and bottomonium states (*e.g.* η_b, χ_b) are not listed.

Name	$\bar{b}$ hadron	b hadron
charged B meson	$B^+ = |\bar{b}u>$	$B^- = |b\bar{u}>$
neutral B meson	$B^0 = |\bar{b}d>$	$\overline{B^0} = |b\bar{d}>$
B_s (B-sub-s) meson	$B_s^0 = |\bar{b}s>$	$\overline{B_s^0} = |b\bar{s}>$
B_c (B-sub-c) meson	$B_c^+ = |\bar{b}c>$	$B_c^- = |b\bar{c}>$
Λ_b (Lambda-b)	$\overline{\Lambda_b} = |\bar{u}\overline{db}>$	$\Lambda_b = |udb>$
Υ (Upsilon)	$\Upsilon = |\bar{b}b>$	

are always produced in pairs.[a] Unlike e^+e^- collisions on the $\Upsilon(4S)$ resonance, the high energy collisions access all angular momentum states, so the b and $\bar{b}$ are produced incoherently. As a consequence, lifetime, mixing and CP-violation measurements can be performed by reconstructing a single B hadron in an event. The produced b quarks can fragment into all possible species of B hadrons, including B_s, B_c and b-baryons. Table 1 lists the most common species of B hadrons, all of which are accessible at the Tevatron.

The transverse momentum (p_T) spectrum for the produced B hadrons is a steeply falling distribution, which means that most of the B hadrons have very low transverse momentum. For example, a fully reconstructed sample of $B \to J/\psi K$ decays has an average B meson p_T of about $10\,\text{GeV}/c$. As a consequence, the tracks from these decays are typically quite soft, often having $p_T < 1\,\text{GeV}/c$. One of the experimental limitations in reconstructing these modes is the ability to find charged tracks at very low momentum. B hadrons with low transverse momentum

---

[a] It is possible to produce b's singly through weak decays such as $W^- \to \bar{c}b$ and $W^- \to \bar{t}b$. The cross sections for these processes are several orders of magnitude below the cross section for direct $b\bar{b}$ production by the strong interaction.

do not necessarily have low total momentum. Quite frequently, the B mesons have very large longitudinal momentum (the momentum component along the beam axis.) These B hadrons are boosted along the beam axis and are consequently outside the acceptance of the detector.

To reconstruct the B hadrons that do fall within the detector acceptance, the experiments need excellent tracking that extends down to low transverse momentum, excellent vertex detection to identify the long-lived hadrons containing heavy flavor, and high-rate trigger and data acquisition systems to handle the high rates associated with this physics. In the following section, we outline some of the relevant aspects of the Tevatron detectors.

3. The CDF and DØ Detectors

The CDF and DØ detectors are both axially symmetric detectors that cover about 98% of the full 4π solid angle around the proton-antiproton interaction point. For Tevatron Run II, both experiments have axial solenoidal magnetic fields, central tracking, and silicon microvertex detectors. Additional details about the experiments can be found elsewhere.[1,2] The strengths of the detectors are somewhat complementary to one another and are discussed briefly below.

3.1. DØ

The DØ tracking volume features a 2 T solenoid magnet surrounding a scintillating fiber central tracker that covers the region $|\eta| \leq 1.7$, where η is the pseudorapidity, $\eta = -ln(tan(\theta/2))$, and θ the polar angle measured from the beamline. The DØ silicon detector has a barrel geometry interspersed with disk detectors which extends the forward tracking to $|\eta| \leq 3$. In addition, the DØ muon system covers $|\eta| \leq 2$ and the uranium/liquid-argon calorimeter has very good energy resolution for electron, photon and hadronic jet energy measurements.

3.2. CDF

The CDF detector features a 1.4 T solenoid surrounding a silicon microvertex detector and gas-wire drift chamber. The CDF spectrometer has excellent mass resolution. These properties, combined with muon detectors and calorimeters, allow for excellent muon and electron identification, as well as precise tracking and vertex detection for B physics.

3.3. *Triggering*

Both experiments exploit heavy flavor decays which have leptons in the final state. Identification of dimuon events down to very low momentum is possible, allowing for efficient $J/\psi \to \mu^+\mu^-$ triggers. As a consequence, both experiments are able to trigger upon the J/ψ decay, and then fully reconstruct decay modes such as $B_s^0 \to J/\psi\phi$, with $\phi \to K^+K^-$. Triggering on dielectrons to isolate $J/\psi \to e^+e^-$ decays is also possible, although at low momentum the backgrounds become more problematic.

CDF has implemented a $J/\psi \to e^+e^-$ trigger requiring each electron have $p_T > 2\,\text{GeV}/c$. Because the triggering and selection cuts required to reduce background are more stringent than they are for the dimuon mode, the yield for the $J/\psi \to e^+e^-$ mode is about $1/10^{\text{th}}$ the yield for the dimuon mode. However, since selection criteria isolate a dielectron mode is at higher momentum then the dimuon mode, the B purity in the $J/\psi \to e^+e^-$ channel is higher. The analyses shown here use only the dimuon mode, although future analyses will supplement the signal sample with the dielectron mode.

Both experiments also have inclusive lepton triggers designed to accept semileptonic $B \to \ell\nu_\ell X$ decays. DØ has an inclusive muon trigger with excellent acceptance, allowing them to accumulate very large samples of semileptonic decays. The CDF semileptonic triggers require an additional displaced track associated with the lepton, providing cleaner samples with a smaller yields.

New to the CDF detector is the ability to select events based upon track impact parameter. The CDF Silicon Vertex Tracker (SVT) operates as part of the Level 2 trigger system. Tracks identified by the eXtremely Fast Tracker (XFT) are passed to the SVT, which appends silicon hits to the tracks to measure the impact parameter of each track. With the high trigger rate, it is very challenging to extract the data from the silicon detector and perform pattern recognition quickly. The SVT takes on average $25\,\mu s$ per event to extract the data from the silicon, perform silicon clustering and track fitting. As shown in Fig. 1, the impact parameter resolution for tracks with $p_T > 2\,\text{GeV}/c$ is $47\,\mu$m, which is a combination

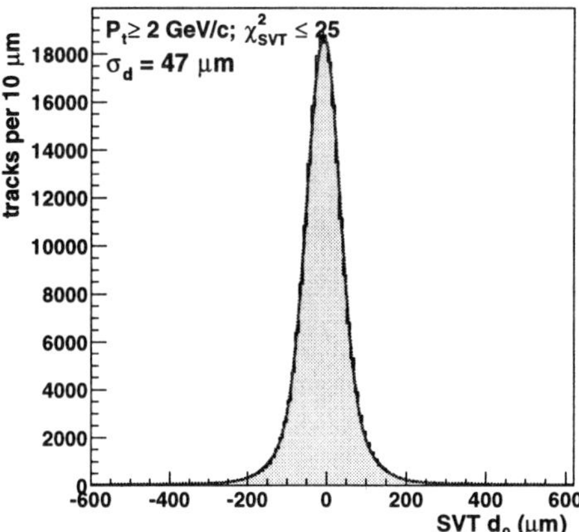

Figure 1. Resolution for the CDF Silicon Vertex Tracker (SVT.) The SVT performs axial tracking in the trigger using the silicon microvertex detector. This plot shows the impact parameter resolution for a sample of tracks. The resolution is $47\,\mu$m. The typical heavy flavor trigger requires two tracks with impact parameter greater than $120\,\mu$m, which has a rejection factor that is 250 times greater than simply demanding charged tracks.

Table 2. Tevatron yields for various fully- and partially-reconstructed heavy flavor modes. The column labeled "trigger" lists the primary signature utilized to select the events. The trigger type d_0 is referring to the impact parameter described in the text. This list is not meant to be exhaustive, only to give the reader a feel for the sample sizes for heavy flavor analyses at the Tevatron. A number of J/ψ, semileptonic and all-hadronic decay modes are not included in this table.

Hadron	Decay mode	Trigger	Yield per pb^{-1}
D^0	$K^-\pi^+$	d_0	6000
D^+	$K^-\pi^+\pi^+$	d_0	5000
B^-	$D^0\pi^-$	d_0	16
B_s^0	$D_s^-\pi^+$	d_0	1
B_s^0	K^+K^-	d_0	1.5
J/ψ	$\mu^+\mu^-$	dimuon	7000
B^+	$J/\psi K^+$	dimuon	11
B_s	$J/\psi\phi$	dimuon	1
Λ_b	$J/\psi\Lambda$	dimuon	0.7
B	$D\ell\nu$	single lepton	400
Λ_b	$\Lambda_c\ell\nu$	single lepton	10

of the primary beam spot size ($30\,\mu m$) and the resolution of the device ($35\,\mu m$.) The CDF SVT has already shown that it will provide a number of new modes in both bottom and charm physics that were previously not accessible. DØ is currently commissioning a displaced track trigger as well.

4. Heavy Flavor Yields

As mentioned in the previous section, the $b\bar{b}$ production cross section is very large at the Tevatron. Although triggering and reconstruction is very challenging, large samples can indeed be acquired. Table 2 shows approximate yields for several different modes at the Tevatron.[b] As of this writing, each experiment has logged approximately $\mathcal{L} = 200\,\mathrm{pb}^{-1}$ of integrated luminosity. The data sample is expected to more than double during the 2004 running period.

5. Prompt Charm Cross Section

Previously published measurements of the b production cross section at the Tevatron have consistently been significantly higher than the Next-to-Leading Order QCD predictions. Although there has been

---

[b]Throughout this note, we will write charge specific decay modes for clarity. All analyses presented here include the charge-conjugate modes.

theoretical activity in this arena and the level of the discrepancy has been reduced, it is not yet clear that the entire scope of the problem is fully understood. Both experiments will again measure the b and $b\bar{b}$ cross sections at higher center-of-mass energy.

To further shed light on this problem, CDF has recently presented a measurement of the charm production cross section.[4] Using the secondary vertex trigger, CDF has been able to reconstruct very large samples of charm decays. Figure 2 shows a fully reconstructed $D^+ \rightarrow K^-\pi^+\pi^+$ signal using $5.8\,\mathrm{pb}^{-1}$ of data from early in the run. In the full data sample available at the time of this Symposium ($\mathcal{L} \sim 200\,\mathrm{pb}^{-1}$), CDF has a D^0 sample exceeding 2 million events. As this sample grows, competitive searches for CP-violation in the charm sector and $D^0/\overline{D^0}$ mixing are anticipated.

Since the events are accepted based upon daughter tracks with large impact parameter, it is clear that the sample of reconstructed charm decays contains charm from bottom ($p\bar{p} \rightarrow b\bar{b}X$, with $b \rightarrow c \rightarrow D$) in addition to prompt charm production ($p\bar{p} \rightarrow c\bar{c}X$, with $c \rightarrow D$.) To extract the charm meson cross section, it is necessary to extract the fraction of D mesons that are coming from prompt charm production. This is done by measuring the impact parameter of the charm meson. If it arises from direct $c\bar{c}$ production, the charm meson will have a small impact parameter pointing back to the point of production, which was the collision vertex. If the charm meson arises from b decay, it will typically not

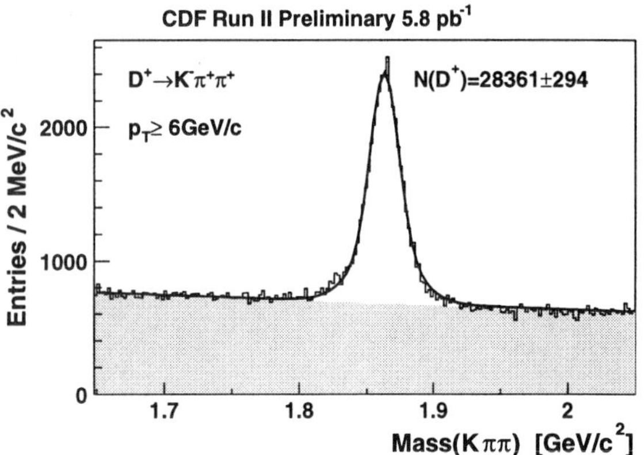

Figure 2. Yield for $D^+ \to K^- \pi^+ \pi^+$ used in the charm cross section analysis. Yields of the size shown in the plot are routinely acquired in a single week of data taking. This sample was acquired using the CDF SVT trigger. About 80% of the signal in this sample arises from direct charm production.

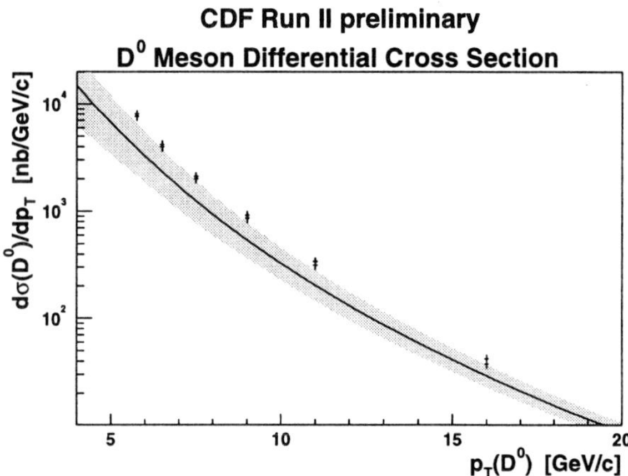

Figure 3. The measured differential cross section for prompt $p\bar{p} \to D^0 X$, with $|y(D^0)| < 1$. The D^0 hadrons arising from B decays have been removed. The NLO calculation is from Cacciari and Nason.[5] The results for the other charm modes are summarized in Table 3.

Table 3. CDF measurement of the direct charm cross section. The results are for D mesons with $|y| < 1$.

Meson	Momentum range	Measured cross section(pb)
D^0	$p_T > 5.5\,\text{GeV}/c$	$13.3 \pm 0.2 \pm 1.5$
D^{*+}	$p_T > 6.0\,\text{GeV}/c$	$5.2 \pm 0.1 \pm 0.8$
D^+	$p_T > 6.0\,\text{GeV}/c$	$4.3 \pm 0.1 \pm 0.7$
D_s^+	$p_T > 8.0\,\text{GeV}/c$	$0.75 \pm 0.05 \pm 0.22$

extrapolate back to the primary vertex.

Using this technique, along with a sample of $K_S^0 \to \pi^+\pi^-$ decays for calibration, it is determined that 80-90% (depending upon the mode) of the charm mesons arise from direct charm production. The shorter charm lifetime is more than compensated by the copious charm production in the high energy collisions.

The full analysis includes measurements of the differential cross sections for prompt D^0, D^+, D^{*+} and D_s^+ meson production. The integrated cross section results of this study are summarized in Table 3. Figure 3 shows the comparison between data and the NLO calculation for the differential D^0 cross section.[5] The trend seen in this figure is the same for the other D species. The prediction seems to follow the measured cross section in shape, but the absolute cross section is low compared to the measured results. This difference in magnitude between the measured and predicted charm meson cross section is similar to the difference in data and theory seen in the B meson cross sections.

As an interesting aside, we can also compare the measured B and D cross sections at the Tevatron.[3] Looking at the charged mesons, the measured cross sections are:

$$\sigma(D^+, p_T \geq 6 \text{ GeV}/c, |y| \leq 1) = 4.3 \pm 0.7 \ \mu\text{b}$$

$$\sigma(B^+, p_T \geq 6 \text{ GeV}/c, |y| \leq 1) = 3.6 \pm 0.6 \ \mu\text{b},$$

where $y = \frac{1}{2}\ln((E+p_z)/(E-p_z))$ is the rapidity. For this momentum range, the D^+ cross section is only

20% larger than the B^+ cross section. At very high transverse momentum (corresponding to high Q^2), we would expect that the mass difference between the bottom and charm quarks to be a small effect, yielding similar production cross sections. However these results show that even at lower p_T, the mass effects are not that significant.

6. B Lifetimes

In the spectator model for meson decays, where the light quark does not participate in the decay, all B lifetimes are equal, since the lifetime is exclusively determined by the lifetime of the b quark. In reality, non-spectator effects such as interference modify this expectation. The Heavy Quark Expansion (HQE) predicts the lifetime hierarchy for the B hadrons as:

$$\tau(B^+) > \tau(B^0) \simeq \tau(B_s) > \tau(\Lambda_b) >> \tau(B_c), \quad (1)$$

where the B_c meson is expected to have the shortest lifetime because both the b and the c quarks are able to decay by the weak interaction.

The lifetimes of the light mesons, B^0 and B^+, are measured with a precision that is better than 1%. This impressive level of precision is dominated by the measurements of the Belle and Babar experiments.[6] The B_s and b-baryon lifetimes have been measured by the LEP, SLD and CDF Run I experiments. One interesting puzzle that persists from those measurements is that the Λ_b lifetime is significantly lower than expectation.[7]

To measure the lifetime of a particle, the experiments utilize their precision silicon tracking to measure the flight distance of the hadron before it decays. At the Tevatron, this is done in the plane transverse to the beamline, and the two-dimensional flight distance is denoted as L_{xy}. Since the particle is moving at a high velocity in the lab frame, the decay time measured in the laboratory is dilated relative to the proper-decay time, which is the decay time of the particle in its rest frame. To extract the proper decay time, we must correct for the time dilation factor:

$$ct_{decay} = \frac{L_{xy}}{(\beta\gamma)_T}, \qquad (2)$$

where t_{decay} is the proper decay time in the rest frame of the particle, c is the speed of light, $\beta\gamma = v/c\sqrt{1 - v^2/c^2}$ is the relativistic correction for the time dilation. We write $(\beta\gamma)_T = p_T/m_B$, with p_T the transverse momentum of the B hadron and m_B the mass of the B hadron. The quantity ct_{decay} is referred to as the proper decay length.

The uncertainty in the measurement of the proper decay length (σ_{ct}) has three terms:

$$\sigma_{ct} = (\frac{m_B}{p_T})\sigma_{L_{xy}} \otimes ct(\frac{\sigma_{p_T}}{p_T}) \otimes (\frac{L_{xy}}{p_T})\sigma_{m_B} \qquad (3)$$

where the $\otimes$ symbol indicates that the terms combine in quadrature. The final term, which is proportional to the uncertainty on the B hadron mass (σ_{m_B}) is negligible in all cases. The first term is the uncertainty on the measured decay length. This depends upon the resolution of the detector as well as the topology and momentum of the decay mode. The middle term, proportional to the uncertainty transverse momentum of the B hadron (σ_{p_T}) is effectively the uncertainty in the time dilation correction. For fully reconstructed modes where all of the B daughter particles are accounted for, this term provides a negligible contribution to the uncertainty on the

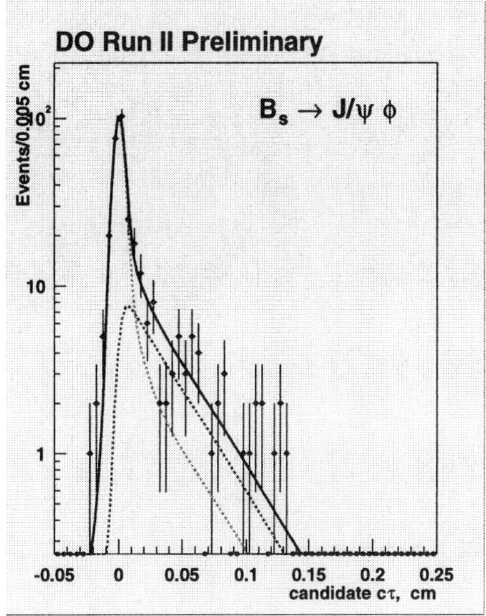

Figure 4. A measurement of the lifetime of the B_s meson from the DØ experiment. This result uses the fully reconstructed $B_s \rightarrow J/\psi\phi$ mode, with $\phi \rightarrow K^+K^-$. The signal yield is 133 ± 17 events in $115\,\mathrm{pb}^{-1}$ of data. The solid curve is the combined signal + background fit to the data, while the dashed curves are the data and background components shown separately. The background dominates at zero decay time, but there is also a background contribution coming from heavy flavor.

proper decay time. In the case of partially reconstructed modes, where some fraction of the B daughters are not reconstructed, the uncertainty on the B hadron momentum becomes a significant contributor to the lifetime uncertainty.

Fully reconstructed J/ψ modes, such as $B^0 \rightarrow J/\psi K^{*0}$, with $K^{*0} \rightarrow K^-\pi^+$ have the advantage of having small uncertainty in the p_T of the B hadron. The drawback, however, is that the signal yields are small due to the small branching ratio into the color-suppressed J/ψ mode. Figure 4 shows a measurement of the B_s lifetime from DØ in the mode $B_s \rightarrow J/\psi\phi$, with $J/\psi \rightarrow \mu^+\mu^-$ and $\phi \rightarrow K^+K^-$. In fitting for the lifetime, it is necessary to account for backgrounds from prompt sources as well as backgrounds from heavy flavor sources. In the case of the fully reconstructed modes, the lifetime fit can additionally utilize reconstructed mass information to properly weight signal versus background events. In the case of J/ψ modes, the dominant backgrounds come from real J/ψ decays from both prompt $c\bar{c}$ production as well as B decays.

213

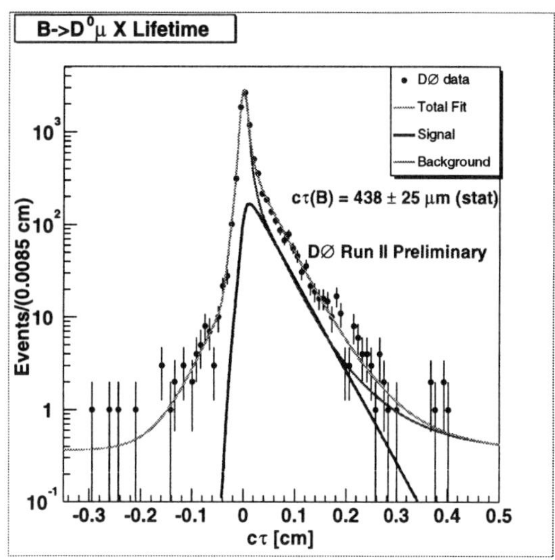

Figure 5. A measurement of the inclusive $B \to \mu D^0 X$ lifetime from the DØ experiment. This inclusive lifetime has contributions from all B modes, although the sample is dominantly B^+ and B^0 decays.

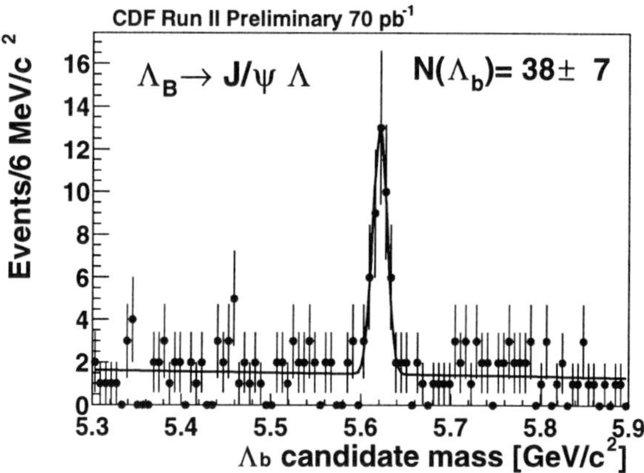

Figure 6. A measurement of the mass of the Λ_b baryon from the CDF experiment. The fully reconstructed $\Lambda_b \to J/\psi \Lambda$, with $J/\psi \to \mu^+\mu^-$ and $\Lambda \to p\pi^-$ is shown here. The mass scale is calibrated with very high precision using J/ψ, $\psi(2S)$ and $\Upsilon \to \mu^+\mu^-$ decays.

The statistical uncertainties on the DØ and CDF lifetime measurements are not yet competitive with the current world average for the B_s and Λ_b lifetimes. With larger data samples over the next 1-2 years, new results from the Tevatron will surpass the current level of precision.

Alternatively, semileptonic decays, such as $B_s^0 \to D_s^- \ell \nu_\ell$, with $D^- \to \phi\pi^-$ provide larger signal yields but suffer from uncertainty in the B hadron p_T due to the unreconstructed neutrino. With large data samples, the semileptonic modes will begin to become systematics limited due to the partial reconstruction, while the statistics limited fully reconstructed modes will continue to provide improved sensitivity.

Figure 5 shows a measurement of the inclusive B lifetime in the μD^0 mode. Since this is a partial reconstruction, backgrounds can be more challenging than they are in the fully reconstructed mode. One technique to suppress backgrounds is to demand the proper charge correlation between the muon and the D^0 meson. The charge of the charm quark is carried through the D^0 decay ($D^0 \to K^-\pi^+$ and $\overline{D^0} \to K^+\pi^-$) so the correlation that the charge of the muon be the same as the charge of the kaon can be enforced to reduce backgrounds. This works even without particle identification, because if the K and π masses are assigned incorrectly they typically do not reconstruct a (narrow) D^0 mass.

One interesting background to the semileptonic analysis is $c\bar{c}$ production where one charm hadron decays semileptonically and the other fragments into a D^0. This background is only a problem if the charm pair is produced at a very small opening angle, which is exactly what occurs when a hard gluon splits into a $c\bar{c}$ pair. This "gluon splitting" contribution produces the "right-sign" charge correlation between the D and muon. Even though the D^0 extrapolates back to the primary vertex in this case, the "fake" μD^0 vertex can look like it arose from a long-lived state. Further studies of $c\bar{c}$ production correlations are needed to fully understand this background source.

7. B Hadron Masses

CDF has performed precision measurements of the masses of B hadrons using fully reconstructed $B \to J/\psi X$ modes. High statistics $J/\psi \to \mu^+\mu^-$ and $\psi(2S) \to J/\psi\pi^+\pi^-$ are used to calibrate tracking momentum scale and material in the tracking volume. The results are tabulated in Table 4. Figure 6 shows the measurement of the Λ_b baryon mass. Even with relatively small statistics (less than 100 events in B_s and Λ_b modes) these new results are the world's best measurements of these masses.

Table 4. CDF results on masses of B hadrons. These results come from fully reconstructed J/ψ modes. The first error is statistical, the second systematic.

B hadron	Decay mode	Measured mass (MeV/c^2)
B^+	$J/\psi K^+$	$5279.32 \pm 0.68 \pm 0.94$
B^0	$J/\psi K^{*0}$	$5280.30 \pm 0.92 \pm 0.96$
B_s^0	$J/\psi\phi$	$5365.50 \pm 1.29 \pm 0.94$
Λ_b	$J/\psi\Lambda$	$5620.4 \pm 1.6 \pm 1.2$

8. Hadronic Branching Ratios

8.1. *Two-body Charmless B Decays*

With the new SVT trigger, CDF has begun to measure B decays with non-leptonic final states. One set of modes of particular interest are the charmless two-body modes. Requiring the final state to consist of two charged hadrons, the following modes can be accessed at the Tevatron:

- $B^0 \to \pi^+\pi^-$, $BR \sim 5 \times 10^{-6}$
- $B^0 \to K^+\pi^-$, $BR \sim 2 \times 10^{-5}$
- $B_s \to K^+K^-$, $BR \sim 1 \times 10^{-5}$
- $B_s \to K^-\pi^+$, $BR \sim 2 \times 10^{-6}$.

The B^0 states are accessible at the e^+e^- facilities, but the B_s modes are exclusive to the Tevatron.

The measurement presented here is the first observation of the decay $B_s \to K^+K^-$. As more data is accumulated, the longer term goal from these modes is to search for direct CP-violation as well as measure the CKM angle $\gamma = \mathrm{Arg}(V_{ub}^*)$.[8]

Figure 7 shows the reconstructed signal where all tracks are assumed to have the mass of the pion. A clear peak is seen, and the width of the peak is significantly larger ($41\,\mathrm{MeV}/c^2$) than the intrinsic resolution of the detector. This additional width is due to the $K^+\pi^-$ and K^+K^- final states from B^0 and B_s decays. To extract the relative contributions, kinematic information and dE/dx particle identification is used.[c] The particle identification is calibrated from a large sample of $D^{*+} \to D^0\pi^+$ decays, with $D^0 \to K^-\pi^+$. The charge of the pion from the D^* uniquely identifies the kaon and pion, providing an excellent calibration sample for the dE/dx system. Although the K-π separation is 1.3σ, this is sufficient to extract the two-body B decay contributions.

---

[c]CDF has a time-of-flight system for particle identification with very good π-K separation for track $p_T < 1.6\,\mathrm{GeV}/c$. The tracks from these two-body decay modes have $p_T > 2\,\mathrm{GeV}/c$ and therefore the time-of-flight system does not provide additional particle identification information for this analysis.

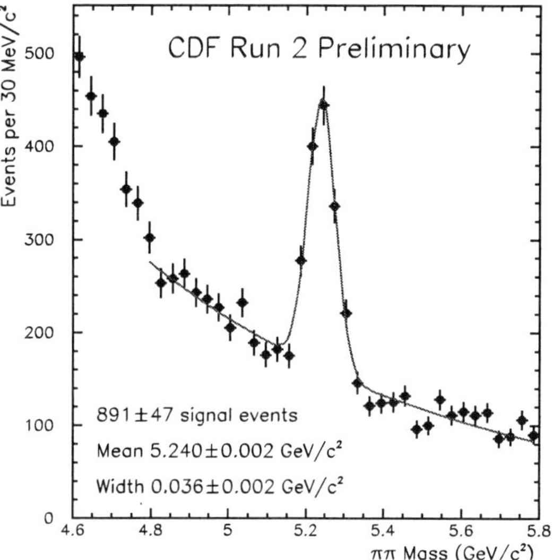

Figure 7. The reconstructed two-body $B \to h^+h^{-\prime}$ ($h,h' = K$ or π) sample from CDF. For this plot, all particles are reconstructed as pions. The peak consists of contributions from the four modes listed in the text. These pieces are extracted using kinematic and particle identification information. The luminosity used for the figure shown here is $\mathcal{L} = 190\,\mathrm{pb}^{-1}$. The results presented in the text are for a subset of this data.

The results from $\mathcal{L} = 65\,\mathrm{pb}^{-1}$ are shown in Table 5. The B^0 modes have been measured by CLEO, Babar and Belle. This is the first observation of the $B_s^0 \to K^+K^-$ decay mode. Turning the yields into a ratio of branching ratios, the result is:

$$\frac{BR(B_s^0 \to K^+K^-)}{BR(B_d^0 \to K^+\pi^-)} = 2.71 \pm 1.15$$

where the error is the combined statistical and systematic uncertainty. To calculate this ratio, information about the relative production rates of B_s^0 and B^0 mesons must be included. The uncertainty includes the uncertainty on the relative production fractions.

8.2. $\Lambda_b \to \Lambda_c\pi^-$ *Branching Ratio*

Using the SVT trigger, CDF has begun to measure b-baryon states. Figure 8 shows a clean signal of the decay $\Lambda_b \to \Lambda_c\pi^-$, with $\Lambda_c \to pK^-\pi^+$. The reconstructed invariant mass plot has a very interesting structure, with almost no background above the peak and a background that rises steeply in going to lower mass. This structure is somewhat unique to baryon modes, which are the most massive weakly

Table 5. CDF results on two-body charmless B decays. The yields reported in this table are extracted by fitting the decay information for the events, including kinematic and particle identification information. Yields shown here are for $\mathcal{L} \sim 65\,\mathrm{pb}^{-1}$.

Mode	Fitted yield (events)
$B^0 \to K^-\pi^+$	$148 \pm 17(stat.) \pm 17(syst.)$
$B^0 \to \pi^+\pi^-$	$39 \pm 14(stat.) \pm 17(syst.)$
$B_s^0 \to K^+K^-$	$90 \pm 17(stat.) \pm 17(syst.)$
$B_s^0 \to K^-\pi^+$	$3 \pm 11(stat.) \pm 17(syst.)$

decaying B hadron states. Because the SVT trigger specifically selects long-lived states, most of the backgrounds are coming from other heavy flavor (b and c) decays. Since there are no weakly decaying B hadrons more massive than the Λ_b, there is very little background above the peak. On the other hand, going to masses below the peak, lighter B mesons begin to contribute. The background in this mode is growing at lower masses because there is more phase space for B^+, B^0, and B_s^0 to contribute.

To extract the number of signal events, $b\bar{b}$ Monte Carlo templates are used to account for the reflections seen in the signal window. The shapes of these templates are fixed by the simulation, but their normalization is allowed to float. The number of fitted signal events in this analysis is $96 \pm 13(stat.)^{+6}_{-7}(syst.)$. The primary result from this analysis is a measurement of the $\Lambda_b \to \Lambda_c\pi^-$ branching ratio relative to the $B^0 \to D^-\pi^+$ mode. We can take that ratio, along with PDG 2002[9] values for measured branching ratios and production fractions, and extract the branching ratio

$$BR(\Lambda_b \to \Lambda_c\pi^-) = (6.5 \pm 1.1 \pm 0.9 \pm 2.3) \times 10^{-3},$$

where the errors listed are statistical, systematic and the final uncertainty is arising from the uncertainty in the $B^0 \to D^-\pi^+$ branching ratio.

8.3. Observation of the X(3872) State

At this Lepton-Photon Symposium, the Belle collaboration announced the observation of a neutral state decaying into $J/\psi\pi^+\pi^-$ with a mass of 3872 MeV.[10] This state may be the 1^3D_2 $c\bar{c}$ bound state, although the observed mass is higher than expected for that state. It has also been hypothesized that this is a loosely bound $D\overline{D}^*$ bound state, since the mass is right at the $D\overline{D}^*$ threshold.

Belle observes this state in B decays. Their ob-

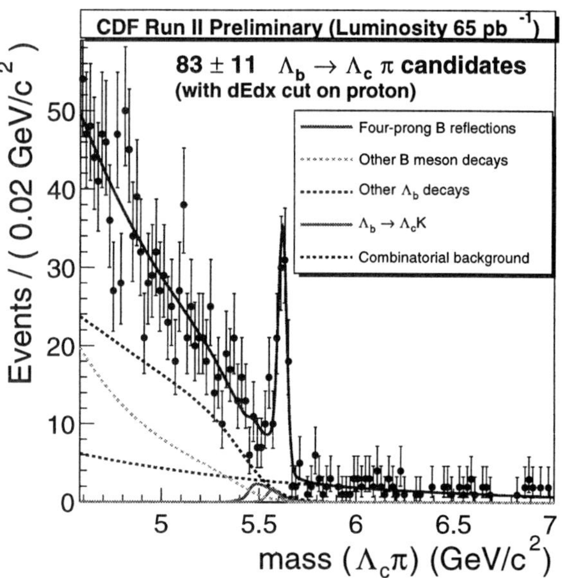

Figure 8. The CDF fully reconstructed $\Lambda_b \to \Lambda_c\pi$, with $\Lambda_c \to pK^-\pi^+$. The features in the sidebands are discussed in the text. For this plot, an additional particle identification cut on the proton was made that was not used in the extraction of the branching ratio.

servation further indicates that the state is narrow, and favors large $\pi^+\pi^-$ mass in the decay.

CDF has searched for this state using a sample of $\mathcal{L} \sim 220\,\mathrm{pb}^{-1}$ and sees a clear signal with a mass of

$$3871.4 \pm 0.7(stat.) \pm 0.4(syst.)\mathrm{MeV}/c^2.$$

Figure 9 shows the CDF $J/\psi\pi^+\pi^-$ mass spectrum. This plot originates from a parent sample of approximately 2.2 million $J/\psi \to \mu^+\mu^-$ decays. The large peak at $3.685\,\mathrm{MeV}/c^2$ is the $\psi(2S)$.[11]

The plot shows events where the dipion mass was required to be greater than $500\,\mathrm{MeV}/c^2$. The significance of the signal without this cut is greater than 10σ. CDF also reports that the state is narrow. The observed width of $4.3\,\mathrm{MeV}/c^2$ is consistent with detector resolution. Further studies are underway to investigate the $\pi^+\pi^-$ mass distribution as well as to determine whether or not the CDF signal is coming from a prompt source or though B decays. Since all angular momentum states are accessible in high energy $p\bar{p}$ collisions, it is possible that this state is directly produced at the Tevatron, while it cannot be directly produced in e^+e^- collisions.

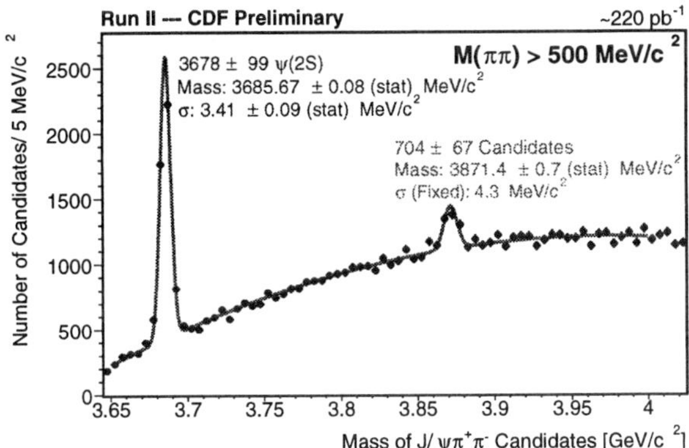

Figure 9. The CDF observation of a neutral state of mass $3871\,\text{MeV}/c^2$ decaying to $J/\psi\pi^+\pi^-$ This is a confirmation of the state first reported by the Belle collaboration at this conference. The distribution shown here includes a requirement that the dipion mass have $M(\pi\pi) > 500\,\text{MeV}/c^2$. The signal is clear and significant without any cut on the dipion mass.

9. Mixing

In the K^0 and B^0 systems, particle-antiparticle mixing has been observed and measured with great precision. This mixing is understood to occur because the weak interaction eigenstates are not the same as the strong interaction (or flavor) eigenstates. The weak eigenstates are then linear combinations of the flavor eigenstates, giving rise to an oscillation frequency that is proportional to the mass difference between the heavy and light states.

Recently, the Babar and Belle experiments have made very precise measurements of $B^0\overline{B^0}$ mixing and have significantly improved the world average. The mixing parameter is typically reported in terms of the heavy/light mass difference. For the B^0 system, the world average is:[7]

$$\Delta m_d = 0.502 \pm 0.006\,\text{ps}^{-1}.$$

From this, we see that a beam of pure B^0 mesons would result in a beam of pure $\overline{B^0}$ mesons in time $\Delta m_d t_{mix} = \pi$, which indicates that $t_{mix} \sim 4.1$ lifetimes, indicating that the oscillation is rather slow.

Mixing proceeds via a second-order weak transition as shown in Fig. 10. The box diagram includes the V_{td} matrix element for $B^0/\overline{B^0}$ mixing, which is replaced by V_{ts} for $B_s^0/\overline{B_s^0}$ mixing. Experimentally, we know that V_{ts} is larger than V_{td}:

$$Re(V_{ts}) \simeq 0.040 > Re(V_{td}) \simeq 0.007,$$

so we expect the B_s^0 system to oscillate with a much

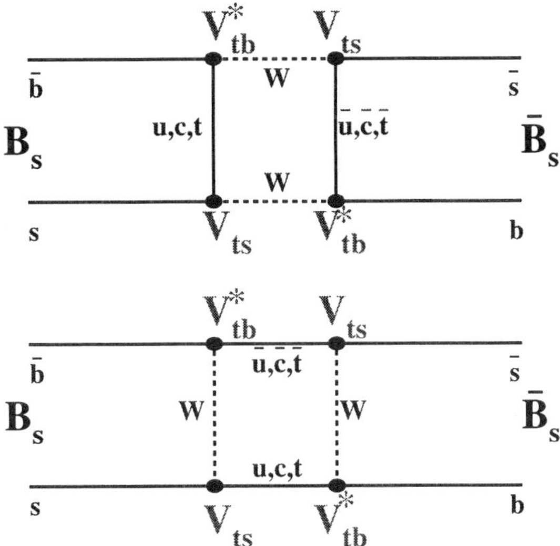

Figure 10. The box diagrams for the $B_s^0 \to \overline{B_s^0}$ transition. All up-type quarks (u,c,t) are included in the box, but since these terms are proportional to the mass of the quark in the box, the top quark dominates. Therefore the V_{ts} element is relevant for B_s mixing and the V_{td} element is relevant for B^0 mixing.

higher frequency than the B^0 system. Indeed this is the case, the B_s^0 system oscillates so quickly that the oscillations have not yet been resolved. The current combined world limit is:[7]

$$\Delta m_s > 14.4\,\text{ps}^{-1}\ @95\%\ \text{CL},$$

which means a beam of B_s^0 mesons would fully become a beam of $\overline{B_s^0}$ mesons in less than $1/7^{\text{th}}$ of one lifetime! For the next several years, the Tevatron will be the exclusive laboratory for B_s meson studies, including the search for B_s mixing.

9.1. Measuring Mixing

To measure B_s mixing, four ingredients are needed.

- **Flavor at the time of production.** It is necessary to know whether the meson was produced as a B_s^0 or a $\overline{B_s^0}$.
- **Flavor at the time of decay.** It is also necessary to know whether the meson was a B_s^0 or $\overline{B_s^0}$ when it decayed. This, combined with the flavor at time of production, tells us whether the B_s had decayed as mixed or unmixed.[d]

---

[d]A meson that "decayed as unmixed" could have mixed and mixed back, undergoing one or more complete cycles. This necessarily comes out of the time-dependent analysis.

- **Proper decay time.** It is necessary to know the proper decay time for the B_s, since we are attempting to measure the probability to mix as a function of decay time. The B_s system mixes too quickly to resolve using time-integrated techniques.
- **Large B_s samples.** We must map out the probability to mix as a function of decay time for at least part of the decay time spectrum. Because each the previous three items have short-comings requiring more statistics, this analysis requires large samples of B_s decays.

In the following subsections, we will discuss each of these pieces necessary to measure B_s mixing.

9.2. *Flavor Tagging*

The first two items in our list of requirements have to do with determining the flavor of the B_s meson at the time of production and at the time of decay, referred to as initial-state and final-state flavor tagging.

For initial-state flavor tagging, we infer the flavor of the B_s meson at the time of production from other information in the event. Here we can take advantage of what we know about $b\bar{b}$ production. By measuring the flavor of the other B hadron in the event, we can infer the flavor of the B_s at the time of production. This technique is imprecise, and also suffers from the fact that quite often ($\sim 75\%$) the other B hadron is outside the acceptance of the detector.

Another technique is to look at fragmentation tracks near the B_s meson. For a $\bar{b}$ quark to become a B_s^0 meson, it must grab an s quark from the vacuum. When the s is popped from the vacuum, an $\bar{s}$ is popped with it, which could potentially turn into a K^+ meson. We can then use the charge of the kaon to infer the flavor of the B hadron. Again, this is an inexact technique, since other fragmentation tracks can confuse this correlation and also the charge information could be lost into neutral particles, like K_S^0. Figure 11 shows an example of same-side tagging. In this case, a fully reconstructed $B^+ \to J/\psi K^+$ sample is used, where the flavor of the B hadron is known. The plot then shows the mass difference between the π-B^+ and the B^+. A clear opposite-sign excess is seen over the entire range of the plot, which is attributed to the fragmentation correlation. In addition, a clear peak is seen near $0.4\,\mathrm{GeV}/c^2$ which is attributed to the B^{**} state.

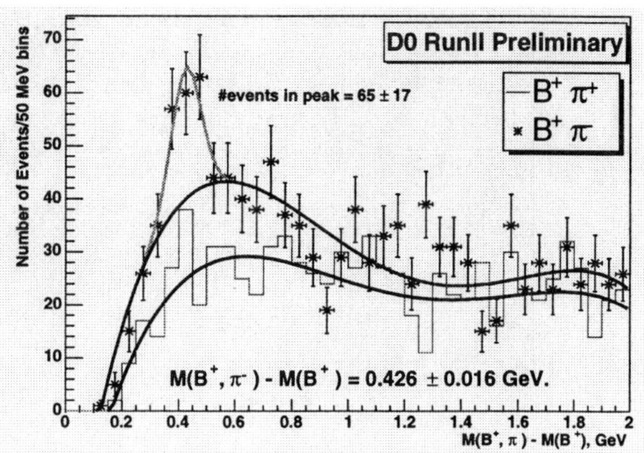

Figure 11. DØ same-side tagging plot.

The efficacy of the initial-state flavor tagging is classified by the tagging power: ϵD^2, where ϵ is the fraction of times the algorithm was able to arrive at a tagging decision and D is the dilution, which is a measure of the probability that the tag is correct. The dilution is written as: $D = (N_R - N_W)/(N_R + N_W)$, where $N_R(N_W)$ are the number of right (wrong) tags. The dilution is related to the mistag fraction, w as $D = 1 - 2w$. Tagging power is proportional to D^2 because incorrectly measuring the sign removes the event from the "correct" charge bin and puts the event into the "wrong" charge bin.

For Babar and Belle, $\epsilon D^2 \simeq 27\%$, whereas for the Tevatron $\epsilon D^2 \simeq 5\%$. The large difference arises because of the nature and cleanliness of the e^+e^- environment. As an example, if an experiment has 1000 signal events with $\epsilon D^2 = 5\%$, then the statistical power of that sample is equivalent to a sample of 50 events where the tag is known absolutely.

For final-state flavor tagging, there are three primary classes of B_s decays:

- $B_s^0 \to D_s^- \pi^+$, with $D_s^- \to \phi\pi^-$,
- $B_s^0 \to D_s^- \ell^+ \nu_\ell$, with $D_s^- \to \phi\pi^-$,
- $B_s^0 \to J/\psi\phi$, with $J/\psi \to \mu^+\mu^-$.

In the first two cases, the flavor of the B_s is immediately evident from the charge of the decay products, which are referred to as a "self-tagging" final states. In the third case, there is no way know whether the meson decayed as a B_s^0 or $\overline{B_s^0}$, so charge-symmetric modes are of no use for the mixing analysis.[e]

---

[e] The charge-neutral B_s modes are important for other analyses, such as the search for a lifetime difference $\Delta\Gamma_s$ in CP-even and CP-odd decays.

9.3. *Proper Decay Time Resolution*

We know the B_s oscillates very quickly, therefore we need proper time resolution that is smaller than the oscillation frequency. As discussed in detail in Sec. 6, the two primary components contributing to the proper time resolution are vertex (L_{xy}) resolution and the time dilation correction ($(\beta\gamma)_T = p_T/m_B$).

For the semileptonic samples, the time dilation correction factor limits the proper time resolution. For fully reconstructed samples (no missing neutrino) the time dilation correction is a negligible effect and only the L_{xy} resolution contributes to the uncertainty on the proper decay time measurement.

If the true value of Δm_s is close to the current limit $\Delta m_s \sim 14-18\,\mathrm{ps}^{-1}$, then both fully reconstructed and semileptonic samples will contribute to the measurement of Δm_s. However, if the true value of Δm_s is $20\,\mathrm{ps}^{-1}$ or higher, then the proper time resolution becomes the limiting factor in resolving the oscillations. At Δm_s values this high, only the fully reconstructed samples are useful, and in fact the vertex resolution becomes the limiting factor.

9.4. *Yields*

As stated previously, the statistics of the semileptonic samples are higher than those of the fully reconstructed modes. For lower values of Δm_s the larger semileptonic event yields somewhat offset the poorer proper time resolution. As a consequence, both CDF and DØ are continuing to acquire semileptonic samples, both for flavor tagging calibration and for the B_s mixing search. A semi-muonic sample enriched in B_s decays is shown in Fig. 12, demonstrating that large semileptonic samples are being acquired.

The fully reconstructed states offer fewer signal events, but the improved proper time resolution compensates for this at values of Δm_s that we are considering. CDF has accumulated a sample of fully reconstructed B_s decays as shown in Figure 13. The plot on the left is the signal, a clear B_s peak can be seen. The broad peak below the B_s is the $B_s \to D_s^* \pi$, where the photon from the D_s^* decay is not reconstructed. The plot on the right shows the expected contributions from a $b\bar{b}$ Monte Carlo. In the data, the signal and sidebands are fit using the shapes from the Monte Carlo letting the normalizations float. The Monte Carlo clearly provides a very good description of the signal and heavy flavor backgrounds. This

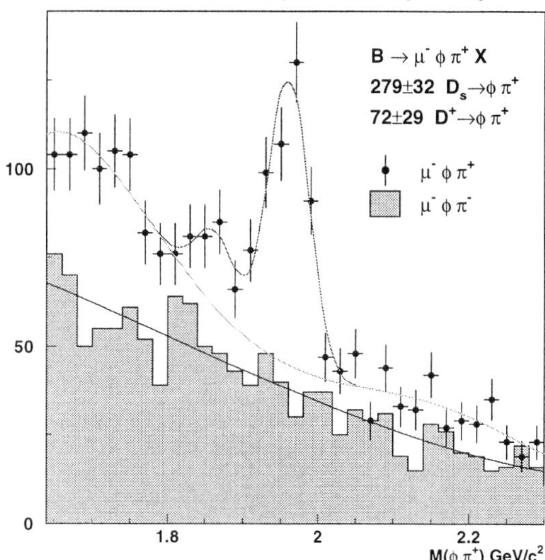

Figure 12. Semileptonic B_s^0 decays can be partially reconstructed in the mode $B_s^0 \to \ell^+ D_s^- X$, with $D_s^- \to \phi\pi^-$ and $\phi \to K^+ K^-$. The distribution shown here is a sample of $\mu\phi\pi$ events from DØ. The two peaks correspond to the D^+ and D_s^+ states, which can both decay to $\phi\pi^+$. This mode will provide a large sample of semileptonic B_s^0 decays to search for $B_s^0/\overline{B_s^0}$ mixing.

is possible because the SVT trigger provides very pure heavy flavor samples. Even though the hadron collider environment is challenging, clean samples of heavy flavor decays can be isolated.

9.5. *B_s Mixing Status and Prospects*

Both experiments have now commissioned the detectors and accumulated the first portion of Tevtron Run 2 data. If we take the current performance of the trigger, reconstruction and flavor tagging, it appears that with a sample of $\mathcal{L} \sim 500\,\mathrm{pb}^{-1}$, the B_s sensitivity will be comparable to the current combined world limit. To observe or exclude a value of $\Delta m_s > 15\,\mathrm{ps}^{-1}$ will require additional data and improvements along with further progress on triggering, reconstruction and flavor tagging.

With modest improvements to the current running configuration, we estimate that it will take 2-3 fb⁻¹ of integrated luminosity to "cover" the region of Δm_s that is currently preferred by indirect fits.[12] It is important to recall that the current combined world limit consists of contributions from 13 different measurements, and is the culmination of the LEP,

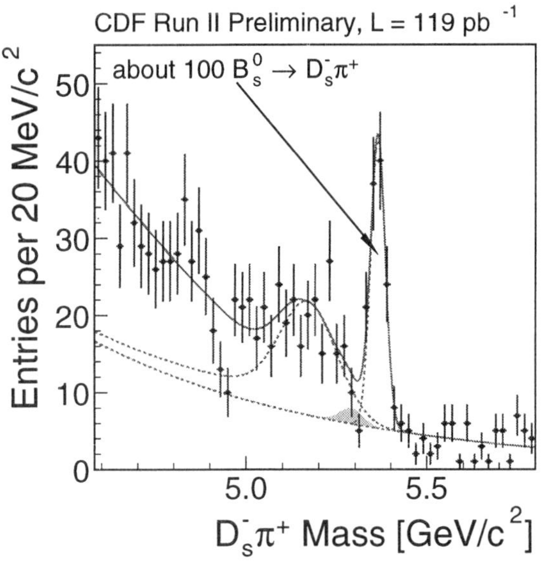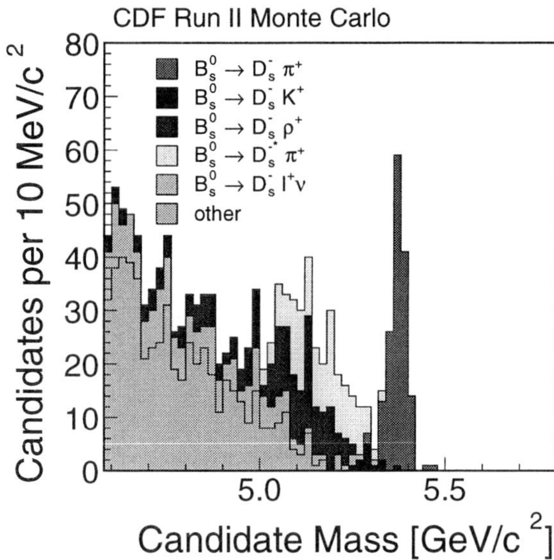

Figure 13. Fully reconstructed B_s^0 decays from CDF using the SVT trigger. The decay chain is $B_s^0 \to D_s^- \pi^+$, with $D_s^- \to \phi\pi^-$ and $\phi \to K^+K^-$. For the mixing analysis, the proper time resolution will be better in this mode, because the error on the time-dilation factor is negligible. The challenge is the limited statistics. CDF is currently trying to reconstruct other modes to supplement this sample.

SLD and CDF Run I programs. Thanks to additional luminosity and upgraded detectors, it will be possible to extend this search for B_s mixing, however this is a very challenging measurement that will take time, effort and a significant data sample.

10. Summary

The results shown in this summary provide a snapshot of the heavy flavor results coming from the Tevatron. The period of commissioning is now complete, and the experiments are slightly more than one year into a multi-year run that will continue to accumulate large data samples. The upgraded detectors are performing well, and many of the upgrades specific to B physics are beginning to pay off. Over the next several years, CDF and DØ will make significant contributions in our understanding of production and decay of heavy flavors.

Acknowledgments

I would like to thank the Lepton-Photon 2003 organizers for the opportunity to speak at this excellent conference. I would also like to thank and acknowledge the collaborators of the Babar, Belle, CDF and DØ experiments. This work is supported by the U.S. Department of Energy Grant DE-FG02-91ER40677.

References

1. DØ Collaboration, FERMILAB-Pub-96/357-E, 1996.
2. CDF Collaboration, FERMILAB-Pub-96/390-E, 1996; Fermilab-Proposal-909, 1998.
3. CDF Collaboration (D. Acosta *et al.*), *Phys. Rev.* **D65**, 052005, (2002).
4. CDF Collaboration (D. Acosta *et al.*), FERMILAB-PUB-03-217-E, hep-ex/0307080, July 2003.
5. M. Cacciari and P. Nason, hep-ph/0306212, *JHEP* 0309:006 (2003), June 2003.
6. BABAR Collaboration (B. Aubert *et al.*), *Phys. Rev. Lett.* **87**, 201803 (2001); BELLE Collaboration (K. Abe *et al.*), *Phys. Rev. Lett.* **88**, 171801 (2002).
7. Heavy Flavor Averaging Group, http://www.slac.stanford.edu/xorg/hfag/index.html.
8. R. Fleischer, *Phys. Lett.* **B459**, 306-320 (1999).
9. Particle Data Group, http://pdg.lbl.gov/.
10. BELLE Collaboration (K. Abe *et al.*), hep-ex/0308029, August 2003; S.-K. Choi *et al.*, hep-ex/0309032, September 2003, submitted to PRL.
11. CDF Collaboration, http://www-cdf.fnal.gov/physics/new/bottom/030224.blessed-x3872/
12. M. Ciuchini, *et al.*, hep-ph/0307195, April 2003.

DISCUSSION

Jonathan L. Rosner (University of Chicago): What are the prospects for seeing fully reconstructed B_c mesons, e.g. in $J/\psi \pi^\pm$?

Kevin Pitts: In Run I with $\mathcal{L} \simeq 100\,\mathrm{pb}^{-1}$, the B_c was observed through the decay $B_c \to J/\psi \ell \nu_\ell$, with $\ell = e, \mu$ and $J/\psi \to \mu^+\mu^-$, providing a tri-lepton final state. In addition, there was a hint of a signal in $B_c \to J/\psi \pi^\pm$. In both of these modes, the b quark decays to charm, providing the J/ψ in the final state. In Run II, with $\mathcal{L} \sim 220\,\mathrm{pb}^{-1}$ already on tape the prospects are quite good both for semileptonic decay and for fully reconstructed decays. In addition, with the enough statistics in the hadronic trigger, it is likely that CDF can reconstruct the B_c in modes where the charm quark decays, such as $B_c^+ \to B_s^0 \pi^+$. This mode might be especially interesting as a new tagging mode for B_s^0 mixing.

Vivek Sharma (University of California at San Diego): What is the timescale for a 10% measurement of Λ_b lifetime? In particular can you use the various fully reconstructed hadronic Λ_b modes given the reflection and impact parameter bias?

Kevin Pitts: Performing a simple average of the CDF and DØ measurements of the Λ_b lifetime in $\Lambda_b \to J/\psi \Lambda$, the result is $\tau_{\Lambda_b} = 1.13 \pm 0.18\,\mathrm{ps}$, which is a 16% measurement on the Λ_b lifetime. These results do not yet use the entire Run II data samples available. Assuming all errors scale like $1/\sqrt{N}$, the combined result from the two experiments using the full $\mathcal{L} \sim 220\,\mathrm{pb}^{-1}$ on tape as of this conference, the combined result would be a 10% measurement on the Λ_b lifetime. With more data coming in, it seems safe to expect that each experiment will have a measurement in the neighborhood of 10% by the summer of 2004.

As for the hadronic modes, significant progress has been made in understanding both the reflections and the impact parameter bias. The understanding of the reflections has been shown in the context of the $\Lambda_b \to \Lambda_c \pi$ branching ratio measurement. Understanding the lifetime bias coming from the SVT trigger is a necessity for B_s mixing, and all lifetime measurements will benefit from ongoing progress on that front.

Vera Luth (SLAC): For charm or B decays you can normalize your measured BR to other measurements. How are you planning to obtain absolute BR and production rates for B_s, Λ_b, etc?

Kevin Pitts: We can still normalize our branching ratios to other modes, but you point out an additional complication coming about due to the lack of precision in our measurements of the relative production fractions. For example, in the CDF measurement of the branching ratio for $\Lambda_b \to \Lambda_c \pi$, the number of signal events in two modes is measured: $N(\Lambda_b \to \Lambda_c \pi)$ and $N(B^0 \to D^-\pi^+)$. The efficiencies and acceptances are calculated, and the PDG values for the $BR(D^- \to K^-\pi^+\pi^+)$ and $BR(\Lambda_c \to pK^-\pi^+)$ are used. Combining this, the measured ratio of corrected yields gives:

$$\frac{f_{baryon} \times BR(\Lambda_b \to \Lambda_c \pi^-)}{f_d \times BR(B^0 \to D^-\pi^+)}$$

where f_{baryon} (f_d) are the fraction of produced Λ_b (B^0) hadrons in $p\bar{p}$ collisions.

The extra piece that comes about in taking this ratio is the ratio of production fractions. These are not known very well, and it is important to improve upon our knowledge of the production fractions. This is easier said than done. One way to improve upon our knowledge of f_s/f_d is to measure $\overline{\chi}$, the time-integrated $B/\overline{B}$ mixing parameter. Other analyses have looked at lepton+D_s, lepton+$D^0/D^\pm$ and lepton+Λ_c correlations to attempt to extract the species fractions. Once statistics have improved, the uncertainty in species fractions will be one of the dominant uncertainties in attempting to extract absolute branching ratios.

CKM MATRIX ELEMENT MAGNITUDES

K. R. SCHUBERT

Technische Universität Dresden, 01062 Dresden, Germany
Email: schubert@physik.tu-dresden.de

The present status of experimental results for the magnitudes of Cabibbo-Kobayashi-Maskawa matrix elements is reviewed and used for a unitarity test. The matrix is found to be unitary within ± 1.8 standard deviations. The matrix violates CP-symmetry and the size of its CP-violation, as derived from only magnitude measurements and unitarity, is in perfect agreement with the observed CP-violations in K and B meson decays.

1. Introduction

The Standard Model Lagrangian for leptons ν_α, ℓ_α and quarks u_α, d_α contains, ignoring right-handed neutrinos, three arbitrary matrices $C_{\alpha\beta}^{(\ell)}$, $C_{\alpha\beta}^{(u)}$, and $C_{\alpha\beta}^{(d)}$ for the coupling of the Higgs doublet to right-handed singlets $\ell_{R\alpha}$, $u_{R\alpha}$, $d_{R\alpha}$ and left-handed doublets $(\nu_{L\beta}, \ell_{L\beta})$, $(u_{L\beta}, d_{L\beta})$. This Yukawa structure does not allow simultaneous diagonalization of the two matrices $C^{(u)}$ and $C^{(d)}$ by "rotations" [a] in three-dimensional family space ($\alpha, \beta = 1, 2, 3$), since the doublet partners $u_{L\alpha}$ and $d_{L\alpha}$ have been coupled by Glashow and, therefore, cannot be rotated separately. Diagonalization of $C^{(u)}$ leads to mass eigenstates u, c, t with Glashow partners d', s', b', and diagonalization of $C^{(d)}$ to d, s, b, u', c', t'. The difference V between the two diagonalizing rotations,

$$
\begin{pmatrix} \begin{pmatrix} u_L \\ d'_L \end{pmatrix} \\ \begin{pmatrix} c_L \\ s'_L \end{pmatrix} \\ \begin{pmatrix} t_L \\ b'_L \end{pmatrix} \end{pmatrix} = V \begin{pmatrix} \begin{pmatrix} u'_L \\ d_L \end{pmatrix} \\ \begin{pmatrix} c'_L \\ s_L \end{pmatrix} \\ \begin{pmatrix} t'_L \\ b_L \end{pmatrix} \end{pmatrix} ,
\tag{1}
$$

is the CKM transformation with $VV^+ = 1$. It was introduced in 1973 by Kobayashi and Maskawa,[1] and in recognition of Cabibbo's[2] early work it has been called V_{CKM} since 1987,[3]

$$
V = V_{\text{CKM}} = \begin{pmatrix} V_{ud} & V_{us} & V_{ub} \\ V_{cd} & V_{cs} & V_{cb} \\ V_{td} & V_{ts} & V_{tb} \end{pmatrix} .
\tag{2}
$$

Since the Standard Model allows arbitrary matrices $C^{(u)}$ and $C^{(d)}$, V is allowed to be complex, leading to CP-violation in the standard weak interaction.

The complex matrix elements V_{ij} are not observables because of unobservable phases in the quark

fields. Transformations $u_\alpha \to u_\alpha \cdot e^{i\phi_\alpha}$ and $d_\beta \to d_\beta \cdot e^{i\phi_\beta}$ lead to $V_{\alpha\beta} \to V_{\alpha\beta} \cdot e^{i(\phi_\alpha - \phi_\beta)}$ with arbitrary ϕ_α and ϕ_β. The observables, i.e. invariants under arbitrary phase transformations, are:

- doublets $V_{ij}V_{ij}^* = |V_{ij}|^2$, content of this talk,

- quartets $V_{ij}V_{kl}V_{il}^*V_{kj}^*$, where modulus and phase are both observable,

- sextets $V_{ij}V_{kl}V_{mn}V_{il}^*V_{kn}^*V_{mj}^*$, and higher n-tets constructed in an analogous way.

The CKM matrix has four observable parameters and an infinite number of choices for these four. One choice is the invariants $|V_{us}|$, $|V_{cb}|$, $|V_{ub}|$, and the phase γ of the quartet $V_{ub}^*V_{ud}V_{cd}^*V_{cb}$. Wolfenstein has named them[4]

$$
|V_{us}| = \lambda , \quad |V_{cb}| = A \cdot \lambda^2 ,
$$

$$
|V_{ub}| \cdot \cos\gamma = A \cdot \lambda^3 \cdot \rho , \quad |V_{ub}| \cdot \sin\gamma = A \cdot \lambda^3 \cdot \eta , \tag{3}
$$

and in addition he has chosen a phase convention where V_{ud}, V_{us}, and V_{cb} are real and positive. In this convention, we have

$$
V \approx \begin{pmatrix} 1 - \lambda^2/2 & \lambda & A\lambda^3(\rho - i\eta) \\ -\lambda & 1 - \lambda^2/2 & A\lambda^2 \\ A\lambda^3(1 - \bar{\rho} - i\bar{\eta}) & -A\lambda^2 & 1 \end{pmatrix}
\tag{4}
$$

with a precision matched to present experiments and

$$
\bar{\rho} = (1 - \lambda^2/2)\rho , \quad \bar{\eta} = (1 - \lambda^2/2)\eta . \tag{5}
$$

Within the Standard Model, the values of the four parameters A, λ, ρ, and η are arbitrary; they have to be determined by experiment. Measurements of the nine magnitudes of V_{ij} test the validity of $VV^+ = 1$ and determine the four parameter values. Discussions of these measurements are the content of this talk.

---

[a] unitary transformations

2. $|V_{\text{ud}}|$

The magnitude of V_{ud} is determined from super-allowed nuclear β^+ decays, from the β^- decay of polarized neutrons, and from the β^+ decay of π^+ mesons. Super-allowed β^+ decays are nuclear $0^+ \to 0^+$, i.e. pure vector, transitions within the same isospin multiplet. A recent review[5] quotes an average of

$$\mathcal{F}t = (3072.2 \pm 0.9 \pm 1.1)\ \text{s} \qquad (6)$$

for nine different decays, where $\mathcal{F}t$ is the product of the Fermi function f, the half-life t, and correction factors for internal bremsstrahlung and isospin symmetry breaking. The first error is experimental, the second is an estimate for the correction-factor error. The magnitude of V_{ud} is obtained from

$$|V_{\text{ud}}|^2 = \frac{2\pi^3\ln 2}{2m_e G_F^2(1+\Delta_{RV})\mathcal{F}t}\ , \qquad (7)$$

where m_e is the electron mass, G_F is Fermi's decay constant taken from muon decay, and $\Delta_{RV} = (2.40\pm 0.08)\%$ is the electroweak correction. This results in

$$|V_{\text{ud}}|_{\text{nuclear}} = 0.9740 \pm 0.0005\ , \qquad (8)$$

where I have combined the experimental error of ± 0.0001, the Δ_{RV} error of ± 0.0004, and the $\mathcal{F}/f$ error of ± 0.0003 quadratically.

Neutron β decays are mixed V and A transitions with two couplings G_V and G_A. Conservation of the vector current leads to $G_V = G_F|V_{\text{ud}}|$ with high precision. Since the axial vector current is only partially conserved, $G_A/G_V = \lambda$ has to be determined experimentally. This is achieved in decays of polarized neutrons, where the angular distribution of decay electrons with respect to the neutron spin direction is given by

$$\frac{dN}{d\cos\theta} = 1 - \frac{v_e}{c}\cdot P\cdot\frac{2\lambda(\lambda+1)}{1+3\lambda^2}\cdot\cos\theta\ , \qquad (9)$$

with the electron velocity v_e and the polarization P. The most recent experiment is PERKEO-II[6] at ILL Grenoble. Cold neutrons of temperature 25 K in a beam with transverse polarization $P = (98.9\pm 0.3)\%$ emit electrons into a 4π detector which separately measures the energy spectra of electrons with $\theta < \pi/2$ and $\theta > \pi/2$. The observed asymmetry results in

$$\lambda = G_A/G_V = -1.2739 \pm 0.0019\ . \qquad (10)$$

Earlier experiments result in higher values of λ; the quoted errors of all experiments[7] give $\chi^2 = 15.5$ for $N(dof) = 4$. Therefore, I prefer to use only the result in Eq. (10), not only because the experiment is the most recent one and quotes the smallest error, but also because it has the highest polarization and the smallest experimental corrections from the raw measured to the final extracted angular asymmetry. Using the mean life[8] of the neutron $\tau_n = (885.7 \pm 0.8)$ s and

$$\frac{1}{\tau_n} = \frac{m_e^5 G_F^2(1+3\lambda^2)|V_{\text{ud}}|^2}{2\pi^3}\cdot f\cdot(1+\delta_R)(1+\Delta_{RV})\ , \qquad (11)$$

where δ_R is a QED correction and Δ_{RV} the same electroweak correction as in Eq. (8), leads to[9]

$$|V_{\text{ud}}|_{\text{neutron}} = 0.9717 \pm 0.0013\ . \qquad (12)$$

Beta decays of π^+ mesons are observed in the recent experiment PIBETA[10] at PSI. The rare decays $\pi^+ \to \pi^0 e^+\nu$ of stopped π^+ are detected by reconstructing $\pi^0 \to \gamma\gamma$ decays in a CsI ball and are normalized by decays $\pi^+ \to e^+\nu$. The preliminary result is a $\pi^0 e^+\nu$ branching fraction of $(1.044 \pm 0.007 \pm 0.009)\times 10^{-8}$, leading to

$$|V_{\text{ud}}|_{\text{pion}} = 0.9765 \pm 0.0056\ . \qquad (13)$$

My combination of the three results in Eqs. (8), (12), and (13) gives

$$|V_{\text{ud}}| = 0.9737 \pm 0.0007\ , \qquad (14)$$

where the error includes a scale factor of 1.3. Prospects for improvement: PIBETA is not expected to reach a competitive precision, the final error may go down to ± 0.0030. The precision of the nuclear result is dominated by radiative corrections. Improvements in the near future are only expected from new neutron experiments. Underway are the experiments PERKEO-III and "NEW PERKEO", advanced plans exist also at LANL, SNS, and Gatchina/PSI.

3. $|V_{\text{us}}|$

Determinations of V_{us} exist from old and new $K_{\ell 3}$ decays, hyperon β decays, and τ decays. $K_{\ell 3}$ decays $K^+ \to \pi^0\ell^+\nu$ and $K^0 \to \pi^-\ell^+\nu$ are $0^- \to 0^-$, i.e. pure vector transitions with $G_V = G_F\cdot|V_{\text{us}}|$. For $P = K^+$ and $P = K^0$ separately, their rates Γ are

Table 1. K lifetimes and $K_{\ell3}$ decays with their branching fractions $\mathcal{B}$ and λ values.

Mode	$\mathcal{B}(\%)$	$10^3\,\lambda_+$	$10^3\,\lambda_0$
K_{e3}^+	4.87 ± 0.06	27.8 ± 1.9	
K_{e3}^0	38.79 ± 0.27	29.1 ± 1.8	
$K_{\mu3}^+$	3.27 ± 0.06	33 ± 10	4 ± 9
$K_{\mu3}^0$	27.18 ± 0.25	33 ± 5	27 ± 6

$$\tau(\mathrm{K}^+) = (1.1384 \pm 0.0024) \times 10^{-8}\,\mathrm{s}$$

$$\tau(\mathrm{K}_L^0) = (5.17 \pm 0.04) \times 10^{-8}\,\mathrm{s}$$

described by two form factors $f_+^{\mathrm{P}}(t)$ and $f_0^{\mathrm{P}}(t)$ in the very good approximation

$$f_{+,0}^{\mathrm{P}}(t) = f_+^{\mathrm{P}}(0)\left(1 + \lambda_{+,0}^{\mathrm{P}} \cdot \frac{t}{m_\pi^2}\right) , \qquad (15)$$

with $t = (p_{\mathrm{P}} - p_\pi)^2$, resulting in

$$\Gamma(\mathrm{P} \to \pi\ell\nu) = \frac{G_{\mathrm{F}}^2 |V_{\mathrm{us}}|^2 m_{\mathrm{P}}^5}{192\pi^3} \cdot C_{\mathrm{P}}^2 \cdot |f_+^{\mathrm{P}}(0)|^2 \cdot I(\lambda_+^{\mathrm{P}}, \lambda_0^{\mathrm{P}}) , \tag{16}$$

where the Clebsch-Gordon coefficient C_{P} is $1/\sqrt{2}$ for both K^+ and K_L^0. Rates (branching fractions/lifetimes) and λ values are determined experimentally, see Table 1. The form-factor values at $t = 0$ have to be obtained theoretically and have essentially remained unchanged[11] since 1984,

$$f_+^{\mathrm{K}^0}(0) = 0.961 , \quad f_+^{\mathrm{K}^+}(0) = 0.982 . \qquad (17)$$

Inclusion of recent radiative corrections by Cirigliano[12] leads to

$$|V_{\mathrm{us}}|_{\mathrm{old}\ K\ell3} = 0.2201 \pm 0.0016 \pm 0.0018 . \qquad (18)$$

This year heralded $K_{\ell3}$ results from two new experiments. BNL-E865[13], whose primary aim is searching for rare decays $\mathrm{K}^+ \to \pi^+\mu^+e^-$, recorded $\mathrm{K}^+ \to \pi^0 e^+\nu$ decays with $\pi^0 \to e^+e^-\gamma$ in a dedicated run of one week in 1998. Normalizing to $\mathrm{K}^+ \to \pi^+\pi^0$ and $\mathrm{K}^+ \to \pi^+\pi^0\pi^0$, they obtain the final branching-fraction result

$$\mathcal{B}(\mathrm{K}^+ \to \pi^0 e^+\nu) = (5.13 \pm 0.02 \pm 0.09 \pm 0.04)\% , \tag{19}$$

which is $2.2\,\sigma$ higher than the old K_{e3}^+ average shown in Table 1. Their λ_+ value is in perfect agreement with that in the Table. Using the correction factors of Cirigliano,[12] the result in Eq. (19) leads to

$$|V_{\mathrm{us}}|_{\mathrm{E865}} = 0.2285 \pm 0.0023 \pm 0.0019 , \qquad (20)$$

which is, using only the statistical errors, $2.2\,\sigma$ higher than the old K_{e3}^+ value and $3.0\,\sigma$ higher than the old $K_{\ell3}^+, K_{\ell3}^0$ average given in Eq. (18). The KLOE experiment at the DAΦNE e^+e^- storage ring in Frascati has presented preliminary $K_{\ell3}^0$ results, with $78\ \mathrm{pb}^{-1}$, using a fraction of their data. At the time of this Symposium, there are no numbers available for $f_+^{\mathrm{K}^0}(0) \cdot |V_{\mathrm{us}}|$, but the results shown graphically[14] agree well with the old result in Eq. (18) and are in poor agreement with BNL-E865.

Hyperon decays have recently been revisited by Cabibbo *et al.*[15] using experimental data from $\mathrm{n} \to \mathrm{pe}\nu$, $\Lambda \to \mathrm{pe}\nu$, $\Sigma^- \to \mathrm{ne}\nu$, $\Xi^- \to \Lambda e\nu$, and $\Xi^0 \to \Sigma^+e\nu$ including experimental values of the form-factor ratios $g_1(0)/f_1(0)$ for each mode separately. Their result is

$$|V_{\mathrm{us}}|_{\mathrm{hyperons}} = 0.2250 \pm 0.0027 , \qquad (21)$$

assuming $f_1(0) = 1.000$ without theoretical uncertainty.

Tau decays are sensitive to $|V_{\mathrm{us}}/V_{\mathrm{ud}}|$ in their branching ratio $\Gamma(\tau^- \to \bar{u}s\nu)/\Gamma(\tau^- \to \bar{u}d\nu)$. Using ALEPH data on $\Gamma(\tau \to \mathrm{K}n\pi\nu)/\Gamma(\tau \to \mathrm{hadrons}\ \nu)$ and $m_{\mathrm{S}}(2\ \mathrm{GeV}) = (105 \pm 20)\ \mathrm{MeV}$, Gamiz *et al.*[16] find

$$|V_{\mathrm{us}}|_\tau = 0.2179 \pm 0.0044 \pm 0.0009 , \qquad (22)$$

where the first error is experimental and the second from theory. Note that the first error dominates. Future measurements with a larger number of tau decays could give smaller errors on $|V_{\mathrm{us}}|$ and, by determining moments of the hadron-mass spectrum, also an independent input value for the s quark mass m_{s}.

Final results from KLOE, also on $K_{\ell3}^+$ rates, are expected soon. Also NA48 and KTeV could analyse these decay modes. For hyperons, more theoretical work on $f_1(0)$ would be welcome. For tau decays, BABAR and BELLE have recorded $10^8\ \tau\tau$ events and should look into their potential to get $\Gamma(\tau^- \to \bar{u}s\nu)$ and hadron-mass moments in this inclusive decay mode. My average from Eqs. (18), (20), and (22) is

$$|V_{\mathrm{us}}| = 0.2210 \pm 0.0023 . \qquad (23)$$

4. $|V_{\mathrm{cd}}|$

There is no new information on this matrix element. Dimuon production by neutrinos and antineutrinos on nuclei has given[17]

$$|V_{cd}| = 0.224 \pm 0.016 . \tag{24}$$

New potentials will be opened by the experiment CLEO-c.

5. $|V_{cs}|$

The magnitude of V_{cs} is obtained from the vector transitions $D^+ \to \overline{K}^0 \ell^+ \nu$ and $D^0 \to K^- \ell^+ \nu$, in complete analogy to the $K_{\ell 3}$ transitions in Eq. (16), and from decays of real W bosons. The first method gives,[18] using $f_+(0) = 0.7 \pm 0.1$,

$$|V_{cs}| = 1.04 \pm 0.16 . \tag{25}$$

The second method is much more precise. Since it requires results from the third quark family, it will be discussed later in Sec. 11.

6. Unitarity Check of the udsc Submatrix

Using the averages for $|V_{ud}|$, $|V_{us}|$, $|V_{cd}|$, and $|V_{cs}|$ in Eqs. (14), (23), (24), and (25), we may check if the mixing matrix of the first two quark families is unitary. The results are

$$|V_{ud}|^2 + |V_{us}|^2 = 0.9969 \pm 0.0017 ,$$
$$|V_{cd}|^2 + |V_{cs}|^2 = 1.13 \pm 0.33 ,$$
$$|V_{ud}|^2 + |V_{cd}|^2 = 0.9983 \pm 0.0073 ,$$
$$|V_{cd}|^2 + |V_{cs}|^2 = 1.13 \pm 0.33 ,$$
$$|V_{ud}V_{cd}| - |V_{us}V_{cs}| = -0.012 \pm 0.039 ,$$
$$|V_{ud}V_{us}| - |V_{cd}V_{cs}| = -0.018 \pm 0.040 . \tag{26}$$

With the exception of only the first line, which fails by $-1.8\,\sigma$, all other checks fulfill unitarity with better than $\pm 0.5\,\sigma$. Within the present experimental precision, we could not predict the existence of a third family from the results discussed so far. A unitarity-constrained fit to the 2×2 matrix gives

$$\lambda_{\text{Wolfenstein}} = 0.2235 \pm 0.0033 , \tag{27}$$

where the error contains a scaling factor of 1.8.

7. $|V_{cb}|$

Sources for the determination of $|V_{cb}|$ are inclusive and exclusive semileptonic B meson decays. The inclusive rate $\Gamma(B \to Xe\nu)$ is proportional to $|V_{cb}|^2$. It is determined from measurements of the B meson

Table 2. Branching fractions of inclusive decays $B \to X\ell\nu$ on the $\Upsilon(4S)$ resonance. These decays are tagged by $B \to X\ell\nu$ decays (ℓ), $B \to Xe\nu$ (e), or full reconstruction (rec) of the second B.

Experiment	Year	Ref.	Tag	Result (%)
ARGUS	1993	[20]	ℓ	$9.75 \pm 0.50 \pm 0.39$
BABAR	2002	[21]	rec	$10.40 \pm 0.50 \pm 0.46$
BELLE	2002	[22]	ℓ	$10.90 \pm 0.12 \pm 0.49$
BABAR	2003	[23]	e	$10.91 \pm 0.18 \pm 0.29$
BELLE	2003	[24]	rec	$11.19 \pm 0.20 \pm 0.31$
CLEO	2003	[25]	ℓ	$10.91 \pm 0.08 \pm 0.30$
Average	07/03	[19]		10.90 ± 0.23

lifetime and the branching fraction $\mathcal{B}(B \to Xe\nu)$, where both are averages from the mixture of B^+ and B^0 in $\Upsilon(4S)$ decays. The lifetimes of B^+ and B^0 are very well known, recent best values of the Heavy Flavor Averaging Group[19] (HFAG) from measurements of ALEPH, BABAR, BELLE, CDF, DELPHI, L3, OPAL, and SLD are

$$\tau(B^0) = (1.534 \pm 0.013) \text{ ps} ,$$
$$\tau(B^+) = (1.653 \pm 0.014) \text{ ps} ; \tag{28}$$

both have reached a precision of $\pm 0.8\%$.

Table 2 lists all results of $\mathcal{B}(B \to X\ell\nu)$ measurements on the $\Upsilon(4S)$ resonance which is known to decay[26] with $(49.0 \pm 1.8)\%$ into $B^0\overline{B}^0$ and with $(51.0 \pm 1.8)\%$ into B^+B^- pairs. Tagging the semileptonic B decay with either fully reconstructed or semileptonically decaying second B mesons in the event has the advantage that the reconstructed flavor of the second B meson allows one to separate primary and secondary leptons in the signal B decay.[20] Figure 1 shows the spectrum of primary ($B \to Xe\nu$) and secondary ($B \to X_cY, X_c \to Ze\nu$) electrons from about 10^6 tagged B decays.[22] Extrapolation of the primary spectrum leads to $\mathcal{B}(B \to X\ell\nu)$. For a determination of $|V_{cb}|$, we have to subtract the small fraction with which the b quark decays into $u\ell\nu$, $\mathcal{B}(B \to X_u\ell\nu) \approx (0.20 \pm 0.05)\%$. The error of this correction is negligible, therefore we have a $\pm 1.1\%$ contribution to the precision of $|V_{cb}|$ from the branching fraction and a $\pm 0.4\%$ contribution from the B meson lifetime. The remaining error is from theory, but here the QCD approximation as an effec-

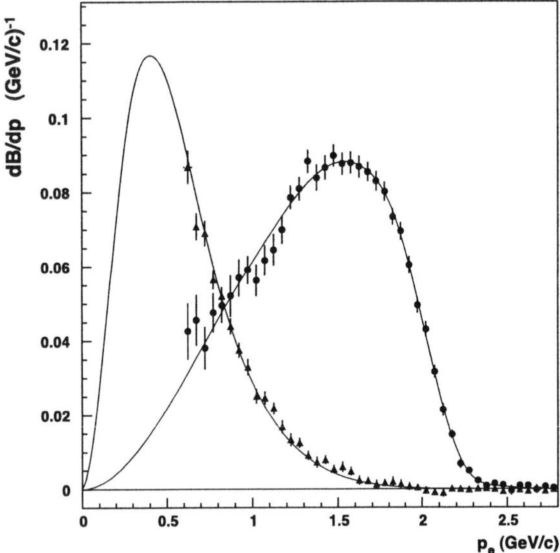

Figure 1. Primary (B $\to$ $Xe\nu$) and secondary (B $\to$ X_cY, $X_c \to Ze\nu$) electrons as measured by BELLE.[22]

tive theory with heavy quarks has brought considerable progress during the last few years.

Twenty years ago, we described the spectrum in Fig. 1 with the ACCMM model[27] which gave the partial rate as $d\Gamma/dp_\ell = |V_{cb}|^2 \cdot f(m_b, m_c, p_F, \alpha_s)$ and allowed $|V_{cb}|$ to be determined with a precision in the order of 10%. The parameter p_F described the "Fermi motion" of the b quark in the B meson. Today, heavy quark effective QCD[28] with its tools Operator Product Expansion (OPE) and Heavy Quark Expansion (HQE) replaces p_F by a more strictly defined matrix element λ_1, uses α_s and quark masses m_b and m_c in a given renormalization scheme, and introduces a few more parameters like the matrix element λ_2 describing QCD magnetism which was absent in ACCMM.

HQE expresses the total rate $\Gamma(\text{B} \to X_c\ell\nu)$ and moments of the partial rate $d^2\Gamma(\text{B} \to X_u\ell\nu)/dm_X^2 dE_\ell$ as functions of the parameters $|V_{cb}|$, m_b, m_c, λ_1, λ_2 and more. Since the dependence on these parameters is different for different moments, a sufficiently large number of observed moments determines all parameters, and a larger number allows to test the consistency of the description. Up to now, measurements have been presented for the "zeroth moments",

$$R(E_0) = \int_{E_0}^{E_{\max}} \frac{d\Gamma}{dE_\ell} dE_\ell , \qquad (29)$$

n-th lepton-energy moments with $n = 1, 2, 3$,

$$M_{En}(E_0) = \frac{1}{R(E_0)} \int_{E_0}^{E_{\max}} E_\ell^n \frac{d\Gamma}{dE_\ell} dE_\ell , \qquad (30)$$

and n-th hadronic-mass moments with $n = 1, 2, 3$,

$$M_{mn}(E_0) = \frac{1}{R(E_0)} \int_{E_0}^{E_{\max}} \int_{m_{\min}^2}^{m_{\max}^2} m_X^n \frac{d^2\Gamma}{dm_X^2 dE_\ell} dE_\ell dm_X^2 . \qquad (31)$$

DELPHI[29] has measured four mass moments ($n = 1, 2, 4, 6$) and three energy moments ($n = 1, 2, 3$) for $E_0 = 0$. One set of parameters fits all observations very well, and in the kinetic renormalization scheme[30] they obtain

$$|V_{cb}| = 0.0429 \times (1 \pm 0.012 \pm 0.019 \pm 0.010) , \quad (32)$$

where the central value is rescaled with $\mathcal{B} = 10.9\%$, the first error originates from the lifetime and branching fraction $\mathcal{B}$, the second from the HQE parameter fit, and the third one is an estimate of the HQE theoretical uncertainty.

Because of the high B meson boost in Z^0 decays, DELPHI is able to determine the six moments in full phase space, i.e. with $E_0 = 0$. This is not possible in the $\Upsilon(4S)$ experiments CLEO and BABAR because of difficulties with lepton identification and background separation at low lepton energies. In 2001, CLEO presented two independent HQE analyses. Using $M_{m2}(1.5\,\text{GeV})$ and the first photon-energy moment in b $\to$ sγ decays which is only sensitive to the HQE parameter m_b, they find[31]

$$|V_{cb}| = 0.0414 \cdot (1 \pm 0.012 \pm 0.022 \pm 0.020) , \quad (33)$$

where the three errors are defined as in Eq. (32) and where I also rescaled with $\mathcal{B} = 10.9\%$. Using $M_{E1}(1.5\,\text{GeV})$ and $R(1.7\,\text{GeV})/R(1.5\,\text{GeV})$, CLEO finds[32]

$$|V_{cb}| = 0.0418 \cdot (1 \pm 0.012 \pm 0.012 \pm 0.022) , \quad (34)$$

with the same comment on errors and the central value. I am not aware of a combined fit with all four CLEO observations, but the extracted HQE parameters m_b and λ_1 in the two fits agree very well.

This summer, BABAR[33] presented E_0-dependent $n = 2$ hadronic-mass moments from a sample of 89 M B$\overline{\text{B}}$ pairs in which one B meson is fully reconstructed in a non-leptonic mode. Requiring that the other B ("signal B") has a lepton (e or μ) with $E_\ell > 0.9$ GeV and good agreement between E_{miss}

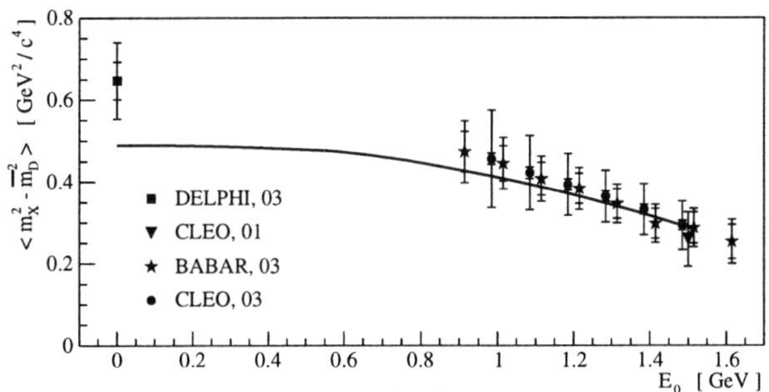

Figure 2. Measurements of hadronic-mass moments M_{m2}, $\overline{m}_D^2 = [(m_D + 3\,m_{D^*})/4]^2$ subtracted, in decays B $\to X_c\ell\nu$ from CLEO,[31,35] DELPHI,[29] and BABAR.[33] The curve is taken from a HQE fit[33,34] to the BABAR points.

and $|\vec{p}_{\text{miss}}|$ as a neutrino signature, the invariant mass of all remaining tracks and neutrals in the event is taken as m_X in a B $\to X\ell\nu$ decay of the signal B. Detailed Monte Carlo studies show that this reconstructed m_X is on average 74% of the true m_X and that the relation between the mean reconstructed and the mean true hadronic mass is very well described by a linear function. Using this Monte Carlo m_X "calibration", BABAR finds the M_{m2} moments as shown in Fig. 2. HQE parameters and $|V_{cb}|$ are fitted in three renormalization schemes. In the 1S scheme, using the HQE of Bauer et al.,[34] the result is

$$|V_{cb}| = 0.0418 \cdot (1 \pm 0.025 \pm 0.017) , \qquad (35)$$

where the first error combines the errors from τ_B, $\mathcal{B}$, and the HQE parameter fit, and the second one estimates the theoretical error of the HQE ansatz for the fit. For this Symposium, CLEO[35] has produced a set of $M_{m2}(E_0)$ values from 10 M B$\overline{\text{B}}$ events. They are in good agreement with the BABAR values, see Fig. 2, but have not yet been used for a new determination of $|V_{cb}|$. To conclude this discussion of inclusive $|V_{cb}|$ determinations, there is surprisingly good agreement between the HQE description of all observed moments. This description results in a ±3.0% precision on $|V_{cb}|$. My average of the results in Eqs. (32), (33), (34), and (35) is

$$|V_{cb}|_{\text{incl}} = 0.0421 \pm 0.0013 . \qquad (36)$$

Determinations of $|V_{cb}|$ with exclusive semileptonic B decays are, and have been for a long time, dominated by analyses of $B^0 \to D^*\ell\nu$ decays, since B $\to$ D$\ell\nu$ is experimentally less background-free and theoretically less certain, knowledge on B $\to$ D$^{**}\ell\nu$

Table 3. Branching fractions $\mathcal{B}$ of the decay $B^0 \to D^{*-}\ell^+\nu$, slopes $\rho_{A_1^2}$ and values at $w = 1$ of the function $A_1(w) \cdot |V_{cb}|$ for this decay from two new experiments presented at this Symposium. $A_1(1) = F(1)$.

	DELPHI[36]	BABAR[37]
$\mathcal{B}$ (%)	$5.90 \pm 0.22 \pm 0.48$	$4.68 \pm 0.03 \pm 0.29$
$\rho_{A_1^2}$	$1.32 \pm 0.15 \pm 0.33$	$1.23 \pm 0.02 \pm 0.28$
$10^3|V_{cb}| \cdot F(1)$	$39.2 \pm 1.8 \pm 2.2$	$34.0 \pm 0.2 \pm 1.3$

and B $\to$ D$^{(*)}\pi\ell\nu$ is very limited, and $B^+ \to D^*\ell\nu$ requires very good photon and π^0 reconstruction.

There are two new analyses presented to this Symposium, from DELPHI and BABAR. DELPHI[36] uses 3.4 M Z^0 decays with 1688 decays of neutral B mesons into $D^{*\pm}\ell\nu$, $\ell = $ e, μ, and BABAR[37] 86 M $\Upsilon(4S)$ with 55700 decays into the same modes. The dominant background in both analyses originates from B $\to$ D$^{**}\ell\nu$ and B $\to$ D$^{(*)}\pi\ell\nu$ events where one or more extra pions, in addition to the well-reconstructed D$^{*\pm}$, are present in a semileptonic decay. DELPHI uses hemisphere and vertex requirements for removing these extra pions, and BABAR uses the kinematic check if $\vec{p}(D^*) + \vec{p}(\ell)$ is compatible with $|\vec{p}(\nu)| = E(B) - E(D^*) - E(\ell)$ and $|\vec{p}(B)|$.

Both analyses determine the partial rate $d\Gamma/dw$, where w is the four-vector product of $p(B)$ and $p(D^*)$. This partial rate is determined by the available phase space and by three form factors $F_i(w)$ which are related to each other and can be parametrized by heavy quark effective QCD (HQET). QCD also predicts $F(1)$, where $F(w)$ is one of the three form factors. Its most precise value is obtained with the help of lattice QCD,[38]

$$F(1) = 0.913\,^{+0.030}_{-0.035} . \qquad (37)$$

Since phase space forces $d\Gamma/dw$ to vanish at $w = 1$, $F(1)$ has to be determined by an extrapolation of the observed $F(w)$, essentially $d\Gamma/dw$ divided by phase space, to $w = 1$. The results of the two new experiments are given in Table 3, where $\rho_{A_1}^2$ is the slope parameter of the form factor $A_1(w)$ in the HQET parametrization of Caprini et al.[39] The slope parameters agree whereas the results for $|V_{cb}|$ and $\mathcal{B}$ disagree by about 2σ. Figure 3 shows the fit results of all D$^*\ell\nu$ analyses selected by HFAG. Earlier work like that of ARGUS[40] is missing because it used a

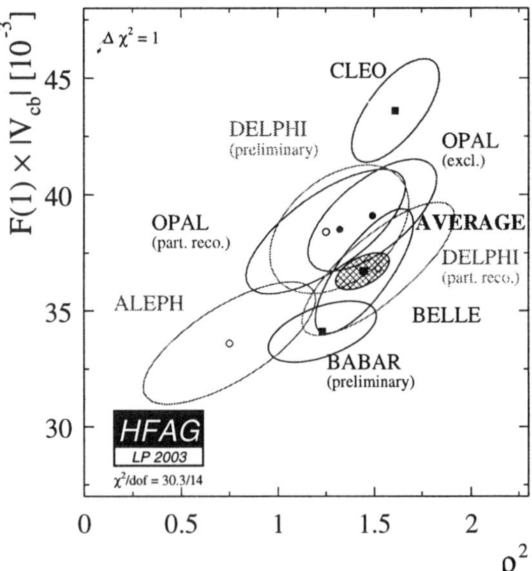

Figure 3. Slopes $\rho^2_{A_1}$ and values at $w = 1$ of the function $A_1(w) \cdot |V_{cb}|$ from eight experiments as compiled by HFAG[19] for this Symposium. The dashed ellipse is the HFAG average with unscaled errors.

different form factor convention. The agreement between the eight results in Fig. 3 is not very good. The average of the $|V_{cb}|$ results is

$$F(1) \cdot |V_{cb}| = 0.0367 \pm 0.0013 , \qquad (38)$$

where I have increased the error by a scale factor of 1.7 since the one-dimensional fit for $F(1) \cdot |V_{cb}|$ gives $\chi^2 = 19.5$ for $N(dof) = 7$. The lattice QCD result in Eq. (37) leads to

$$|V_{cb}|_{\text{excl}} = 0.0402 \pm 0.0020 , \qquad (39)$$

and combined with the compatible inclusive result in Eq. (36) leads to my final best value of

$$|V_{cb}| = A\lambda^2 = 0.0415 \pm 0.0011 . \qquad (40)$$

8. $|V_{ub}|$

Sources for $|V_{ub}|$ are inclusive and exclusive semileptonic B meson decays into charmless final states. Exclusive $|V_{ub}|$ determinations have been published by two experiments, BABAR[41] with $B^0 \to \rho^- e^+ \nu$ and $B^+ \to \rho^0 e^+ \nu$ decays from 55 M $\Upsilon(4S)$ events, and CLEO[42] with decays into $\pi^- \ell^+ \nu$, $\pi^0 \ell^+ \nu$, $\rho^- \ell^+ \nu$, $\rho^0 \ell^+ \nu$, and $\omega \ell^+ \nu$ from 10 M $\Upsilon(4S)$ events. Both

combine modes with the help of isospin and quark-model constraints,

$$\Gamma(\rho^0 \ell \nu) = \Gamma(\omega \ell \nu) = \Gamma(\rho^- \ell \nu)/2 ,$$
$$\Gamma(\pi^0 \ell \nu) = \Gamma(\pi^- \ell \nu)/2 . \qquad (41)$$

Even in the subsamples with high energy leptons, $E_\ell > 2.3$ GeV, there are substantial backgrounds from non-$B\overline{B}$ ("continuum") events, from crossfeed between different signal modes and from "downfeed" where $b \to u\ell\nu$ decays with additional pions are reconstructed in a signal mode. BABAR uses five different form-factor calculations for an extraction of $|V_{ub}|$. They give compatible results, and BABAR[41] quotes an average of

$$|V_{ub}| = (3.64 \pm 0.22 \pm 0.25\,^{+0.39}_{-0.56}) \times 10^{-3} , \qquad (42)$$

where the first error is statistical, the second describes the experimental systematics, and the third one describes the spread between the five different form-factor calculations.

Since the CLEO detector is operated at an energy symmetric e^+e^- storage ring, it has a larger solid angle coverage in the center-of-mass system than BABAR. This larger hermeticity allows better "neutrino reconstruction" and better determination of $q^2 = (p_\nu + p_\ell)^2$. CLEO is therefore able to use events with lower momentum leptons and to divide the data into two samples with $q^2 < 16\,\text{GeV}^2$ where light cone sum rules are expected to predict model-independent form factors, and with $q^2 \geq 16\,\text{GeV}^2$ where lattice QCD calculations predict the form factors well. Both subsamples for both $\pi\ell\nu$ and $\rho\ell\nu$ decays result in four compatible extractions of $|V_{ub}|$; CLEO[42] quotes an average of

$$|V_{ub}| = (3.17 \pm 0.17\,^{+0.16\,+0.53}_{-0.17\,-0.39}) \times 10^{-3} , \qquad (43)$$

with the same error definitions as above in Eq. (42). My average of the two exclusive experiments is

$$|V_{ub}|_{\text{excl}} = (3.40\,^{+0.24}_{-0.33} \pm 0.40) \times 10^{-3} . \qquad (44)$$

Three methods have been tried so far for the inclusive determination of $|V_{ub}|$: (a) the "endpoint" method requiring $B \to Xe\nu$ events with $E_\ell > 2.3$ GeV in order to suppress leptons from $B \to X_c e\nu$ decays; (b) the "low-mass" method allowing a wider range of E_ℓ, e.g. $E_\ell > 1.0$ GeV, and requiring $m_X < 1.5$ GeV in order to suppress hadronic masses from $B \to X_c e\nu$ events; and (c) the "high-q^2" method which combines (b) with an additional requirement of high values of $q^2 = (p_\ell + p_\nu)^2$ since $B \to X_c e\nu$ decays are not only limited to $E_\ell < 2.3\,\text{GeV}$ and $m_X > m_D$ but also to $q^2 < 12\,\text{GeV}^2$.

The first $|V_{ub}|$ measurements[43,44] had been performed with method (a) which has the big disadvantage that extrapolation to $E_\ell = 0$ has large model dependence and little access to stricter QCD estimates. Method (b) is less QCD-dependent, but requires good knowledge of the hadronic mass which can be obtained at the $\Upsilon(4S)$ by tagging the semileptonically decaying B with a fully reconstructed second B. At present, this reduces the number of observed semileptonic decays by a factor of about 10^3. Method (c) is even less QCD-dependent but requires even higher numbers of tagged B decays.

Two new results have been presented to this Symposium. BELLE[45] uses method (a) with 29 M $\Upsilon(4S)$ events and determines the partial rate $\Delta\Gamma(B \to X e\nu, 2.3 < E_\ell < 2.6\,\text{GeV})$. Extrapolation to $E_\ell = 0$ is achieved with the help of the "shape function" concept,

$$\Gamma(B \to X_u e\nu) = \Delta\Gamma(2.3 < E_\ell < 2.6\,\text{GeV})/f_u \, , \quad (45)$$

where the extrapolation factor f_u is obtained using the shape of the partial rate

$$f(E_\gamma) = \mathrm{d}\Gamma(b \to s\gamma)/\mathrm{d}E_\gamma \, , \quad (46)$$

which has recently been measured by CLEO.[46] The shape of $\mathrm{d}\Gamma(B \to X_u \ell\nu)/\mathrm{d}E_\ell$ in the endpoint region is related to $f(E_\gamma)$ with the help of nonlocal operators ("twists") in heavy quark effective QCD.[47,48] This recipe has already been used in the $|V_{ub}|$ endpoint analyses of CLEO[49] and BABAR.[50] In fact, the new BELLE analysis uses CLEO's value and error for f_u in Eq. (45), and BELLE obtains[45]

$$|V_{ub}| = (3.99 \pm 0.17 \pm 0.16 \pm 0.59) \times 10^{-3} \, . \quad (47)$$

The first error is statistical, the second systematic from the experiment, and the third one estimates the uncertainty from using the shape-function concept.

BABAR[51] presents a "low-mass" analysis from 89 M $\Upsilon(4S)$ events resulting in 32000 tagged decays $B \to X e\nu$ with a fully reconstructed B as the tag and with $E_\ell > 1.0\,\text{GeV}$. Reconstruction of the hadronic mass m_X in these decays gives about 600 events with $m_X < 1.6\,\text{GeV}$, 400 of which are estimated to be $B \to X_u \ell\nu$ signal events. Again using

Table 4. Results for $|V_{ub}|$ as compiled by HFAG[19] for this Symposium.

Experiment	Method	$10^3 \cdot |V_{ub}|$
ALEPH		$4.12 \pm 0.67 \pm 0.76$
L3		$5.70 \pm 1.00 \pm 1.40$
DELPHI		$4.07 \pm 0.65 \pm 0.61$
OPAL		$4.00 \pm 0.71 \pm 0.71$
LEP Average		$4.09 \pm 0.37 \pm 0.56$
CLEO	endpoint	$4.08 \pm 0.22 \pm 0.61$
BABAR	endpoint	$4.43 \pm 0.26 \pm 0.67$
CLEO	m_X and q^2	$4.05 \pm 0.61 \pm 0.65$
BELLE	m_X	$5.00 \pm 0.64 \pm 0.53$
BELLE	m_X and q^2	$3.96 \pm 0.47 \pm 0.52$
BABAR	m_X	$4.62 \pm 0.38 \pm 0.49$
BELLE	endpoint	$3.99 \pm 0.25 \pm 0.59$

a shape-function concept, BABAR fits the observed m_X spectrum and extracts

$$|V_{ub}| = (4.62 \pm 0.28 \pm 0.27 \pm 0.40 \pm 0.26) \times 10^{-3} \, . \quad (48)$$

The first two errors are experimental, statistical and systematic, the third estimates the shape-function uncertainty, and the last one estimates the uncertainty in translating $\Gamma(B \to X_u \ell\nu)$ into $|V_{ub}|$.

Table 4 summarizes all inclusive $|V_{ub}|$ results except very early ones. It includes a high-q^2 result from BELLE which will not be discussed here because of its preliminary status. This method, however, seems to be very promising for the future when 3 to 10 times more tagged decays will be available. Finding an objective best value from the entries in Table 4 is difficult, my estimate gives

$$|V_{ub}|_{\text{incl}} = (4.26 \pm 0.13 \pm 0.50) \times 10^{-3} \, , \quad (49)$$

where the first error is meant to have a gaussian shaped likelihood and the second one a rectangular likelihood. Combining this result with the exclusive best estimate in Eq. (44) leads to a best estimate of

$$|V_{ub}| = (3.80\,^{+0.24}_{-0.13} \pm 0.45) \times 10^{-3} \, , \quad (50)$$

as sketched in Fig. 4.

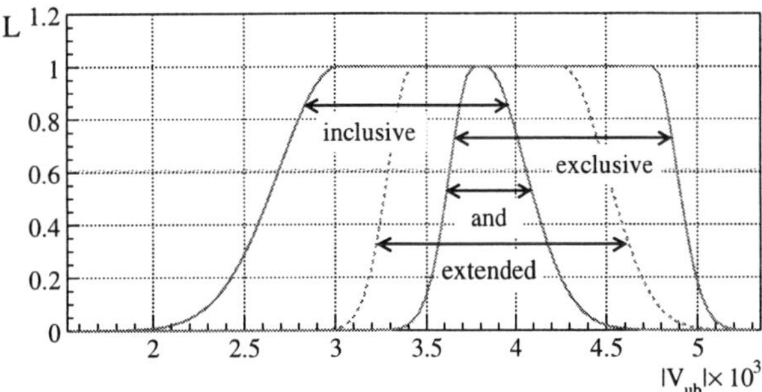

Figure 4. Likelihood functions for $|V_{ub}|$ from exclusive (Eq. (44)) and inclusive (Eq. (49)) measurements. The "and" of both functions is so narrow that an "extended" product, the flat part of which is defined by the centres of $\mathcal{L}_{\mathrm{excl}}$ and $\mathcal{L}_{\mathrm{incl}}$, is used as combined result (Eq. (50)).

9. First Way to $|V_{td}|$

The experimental precision on the mass difference Δm_{d} of the two mass and lifetime eigenstates of the $B_{\mathrm{d}}^0 \overline{B}_{\mathrm{d}}^0$ system has nearly reached 1%. Using all time-dependent measurements of $B_{\mathrm{d}}^0 \overline{B}_{\mathrm{d}}^0$ oscillations, the present best value[7] is

$$\Delta m_{\mathrm{d}} = (0.502 \pm 0.007)\,/\,\mathrm{ps}\,. \tag{51}$$

Using this value, the t quark mass measurement from FNAL[52] and

$$f_{B_{\mathrm{d}}} \sqrt{B_{B_{\mathrm{d}}}} = (233 \pm 13 \pm 12)\,\mathrm{MeV} \tag{52}$$

from lattice QCD,[53] where $f_{B_{\mathrm{d}}}$ is the purely leptonic decay constant of the B^+ meson, and $B_{B_{\mathrm{d}}}$ is the additional non-perturbative "bag factor" in the Standard Model description of $B_{\mathrm{d}}^0 \overline{B}_{\mathrm{d}}^0$ oscillations, gives the result

$$|V_{td}| \cdot |V_{tb}| = (9.2 \pm 1.4 \pm 0.5) \times 10^{-3}\,. \tag{53}$$

The errors are dominated by those in Eq. (52). The first one is the statistical error from the lattice calculation precision with a Gaussian likelihood, and the second one is an estimate of the lattice QCD precision with a rectangular likelihood.

10. $|V_{ts}|$

Information on $|V_{ts}| \cdot |V_{tb}|$ is available from $B_{\mathrm{s}}^0 \overline{B}_{\mathrm{s}}^0$ oscillations and from $B \to X_s \gamma$ decays. The existence of $B_{\mathrm{s}}^0 \overline{B}_{\mathrm{s}}^0$ oscillations is well established. Inclusive time-integrated measurements of e. g. $\Gamma(b\overline{b} \to \ell^{\pm}\ell^{\pm})/\Gamma(b\overline{b} \to \ell^{\pm}\ell^{\mp})$ at LEP lead to

$$f_{\mathrm{s}} \cdot \chi_{\mathrm{s}} = \overline{\chi} - f_{\mathrm{d}} \cdot \chi_{\mathrm{d}} = 0.0509 \pm 0.0060\,, \tag{54}$$

which is more than $8\,\sigma$ above zero; notations and numbers are taken from Schneider.[54] The strength of $B_{\mathrm{s}}^0 \overline{B}_{\mathrm{s}}^0$ oscillations is unknown. Pitts[55] has discussed the status at this Symposium and has presented limits on Δm_{s} as well as improvement prospects. The HFAG 95% limit is now[19]

$$\Delta m_{\mathrm{s}} > 14.4\,\mathrm{ps}^{-1}\,. \tag{55}$$

Using the lattice QCD result[56]

$$\xi = \frac{f_{B_{\mathrm{s}}} \sqrt{B_{B_{\mathrm{s}}}}}{f_{B_{\mathrm{d}}} \sqrt{B_{B_{\mathrm{d}}}}} = 1.24 \pm 0.04 \pm 0.06\,, \tag{56}$$

where $f_{B_{\mathrm{s}}}$ and $B_{B_{\mathrm{s}}}$ have the same meaning for the B_{s} meson as in Eq. (52), this Δm_{s} limit leads to

$$|V_{ts}| \cdot |V_{tb}| > 0.003\,. \tag{57}$$

The branching fraction of inclusive radiative decays $B \to X_s \gamma$ has been determined by CLEO, ALEPH, BABAR, and BELLE. Ali and Misiak[57] use the mean value for an information on $|V_{ts}|$ and derive

$$|V_{ts}| \cdot |V_{tb}| = 0.047 \pm 0.008\,. \tag{58}$$

11. Second Way to $|V_{cs}|$

As already mentioned in Sec. 5, decays of real W bosons at LEP-II[58] allow a better determination of $|V_{cs}|$ than semileptonic decays of D mesons into K mesons. Using α_s corrections, the measured branching ratio $\Gamma(W \to \mathrm{hadrons})/\Gamma(W \to e\nu)$ can be translated into

$$|V_{ud}|^2 + |V_{us}|^2 + |V_{ub}|^2 |V_{cd}|^2 + |V_{cs}|^2 + |V_{cb}|^2$$
$$= 2.039 \pm 0.026\,. \tag{59}$$

Since the five other determinations have a much smaller absolute error than the result on $|V_{cs}|$ in Eq. (25), this LEP-II result can be used to extract

$$|V_{cs}| = 0.995 \pm 0.014\,, \tag{60}$$

which is much more precise than the $K \to D\ell\nu$ result.

12. Unitarity Check of the Full Matrix

Using all final estimates for $|V_{ud}|$, $|V_{us}|$, $|V_{ub}|$, $|V_{cd}|$, $|V_{cs}|$, and $|V_{cb}|$ presented so far, we can check three of the six unitarity constraints which the CKM matrix has to fulfill. I obtain

$$|V_{ud}|^2 + |V_{us}|^2 + |V_{ub}|^2 = 0.9969 \pm 0.0017\,, \tag{61}$$

$$|V_{cd}|^2 + |V_{cs}|^2 + |V_{cb}|^2 = 1.042 \pm 0.026\,, \tag{62}$$

$$|V_{ud}V_{cd}| - |V_{us}V_{cs}| \pm |V_{ub}V_{cb}| = -0.002 \pm 0.016 . \quad (63)$$

The sum of squares in the first row is equal to one with $-1.8\,\sigma$. The sum in the second row, highly correlated with that in the first row because of using Eq. (59), is one within $+1.6\,\sigma$. Eq. (63) estimates the real part of $V_{ud}^{*}V_{cd} + V_{us}^{*}V_{cs} + V_{ub}^{*}V_{cb}$ with any phase γ of the quartet $V_{ub}^{*}V_{cb}V_{ud}V_{cd}^{*}$, see Eq. (3). This upper limit estimate is compatible with zero within $\pm 0.12\,\sigma$.

In conclusion, the observed CKM matrix element magnitudes fulfill unitarity reasonably well. That means, the strengths of all processes which we call "weak" are consistently described by the Standard Model weak interaction in which the CKM matrix is necessarily unitary.

Now assuming unitarity, we can use the relation $|V_{tb}|^2 = 1 - |V_{cb}|^2 - |V_{ub}|^2$ in order to obtain

$$|V_{tb}| = 0.99913 \pm 0.00009 ; \quad (64)$$

i.e. the only unmeasured matrix element has the smallest error. With this result, Eq. (53) translates into

$$|V_{td}| = (9.2 \pm 1.4 \pm 0.5) \times 10^{-3} . \quad (65)$$

And from the relation $|V_{ts}|^2 = |V_{cb}|^2 - |V_{ub}|^2 - |V_{td}|^2$ we then obtain

$$|V_{ts}| = 0.0406 \pm 0.0023 , \quad (66)$$

which is more precise than the result in Eq. (58).

13. $|V_{td}|$ Again

The unitarity result in Eq. (66) and the ratio of $B_d \overline{B}_d$ and $B_s \overline{B}_s$ oscillation strengths,

$$\frac{\Delta m_d}{\Delta m_s} = \frac{m_{B_d}}{m_{Bs}} \cdot \frac{|V_{td}|^2}{|V_{ts}|^2} \cdot \xi^2 , \quad (67)$$

with ξ as defined in Eq. (56), allows a second determination of $|V_{td}|$. The HFAG compilation[19] of $A(\Delta m_s)$ results allows the approximation

$$\mathcal{L}(\Delta m_s) = e^{-(A-1)^2/2\sigma_A^2} \text{ for } \Delta m_s < 20\,\text{ps}$$
$$= 1 \text{ for } \Delta m_s > 20\,\text{ps} , \quad (68)$$

for the likelihood of Δm_s, where A is the amplitude of $B\overline{B}$ oscillations as a function of Δm as taken from HFAG. Combining this likelihood function with Eqs. (66), (67), and with my first $|V_{td}|$ result in Eq. (65) gives $|V_{td}|$ with likelihood contours as shown in Fig. 5 in the $\overline{\rho}, \overline{\eta}$ plane.

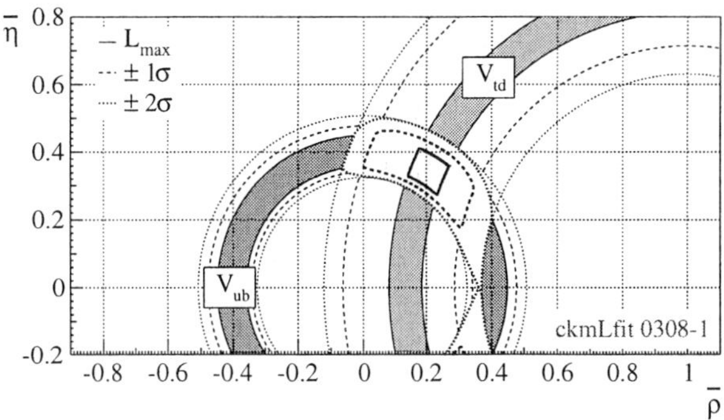

Figure 5. Likelihood contours[60] for $|V_{ub}|$ and $|V_{td}|$ in the $\overline{\rho}$, $\overline{\eta}$ plane. Solid curves are for $\mathcal{L} = \mathcal{L}_{max}$, dashed for $\mathcal{L} = \mathcal{L}_{max} \times e^{-1/2}$ ($\pm 1\,\sigma$), and dotted for $\mathcal{L} = \mathcal{L}_{max} \times e^{-2}$ ($\pm 2\,\sigma$). Also plotted (in white) is the product likelihood for both measurements. It excludes $\overline{\eta} = 0$ with $2.0\,\sigma$.

14. CP-Violation in the CKM Matrix

Figure 5 also shows the likelihood contours of $|V_{ub}|$ as presented at the end of Sec. 8. The two circular bands for $|V_{ub}|$ and $|V_{td}|$ have two intersection regions, one with $\overline{\eta} \approx +0.35$ and one with $\overline{\eta} \approx -0.35$. Both solutions violate $\overline{\eta} = 0$, i.e. CP-symmetry in the Standard Model weak interaction, with 2.0 standard deviations.

Figure 6 superimposes the $\overline{\rho}, \overline{\eta}$ bands of the two well-known CP-violating effects in $K^0 \to \pi\pi$ decays (ϵ_K) and in $B^0 \to (c\overline{c})K$ decays ($\sin 2\beta$). For the $\sin 2\beta$ band, I have used here the new world average

$$\sin 2\beta = 0.737 \pm 0.048 , \quad (69)$$

including the newest result from BELLE as presented by Browder[59] at this Symposium. All four bands intersect perfectly in one solution, i.e. the CP-violation concluded from the magnitudes of V_{ub} and V_{td} describes quantitatively the CP-violation in K and B decays.

15. Final Fit for A, λ, ρ, and η

Also shown in Fig. 6 is a simultaneous fit to the four quantities (ϵ_K, $|V_{ub}|$, $|V_{td}|$ and $\sin 2\beta$) with the help of the program ckmLfit.[60] With the $\sin 2\beta$ value in Eq. (69), the fit results are

$$\overline{\rho} = 0.21 \pm 0.08 \pm 0.05 ,$$
$$\overline{\eta} = 0.35 \pm 0.04 \pm 0.02 . \quad (70)$$

The $\sin 2\beta$ input reduces the error on $|V_{ub}|$ and gives $|V_{ub}|$ a gaussian shaped likelihood since the error on

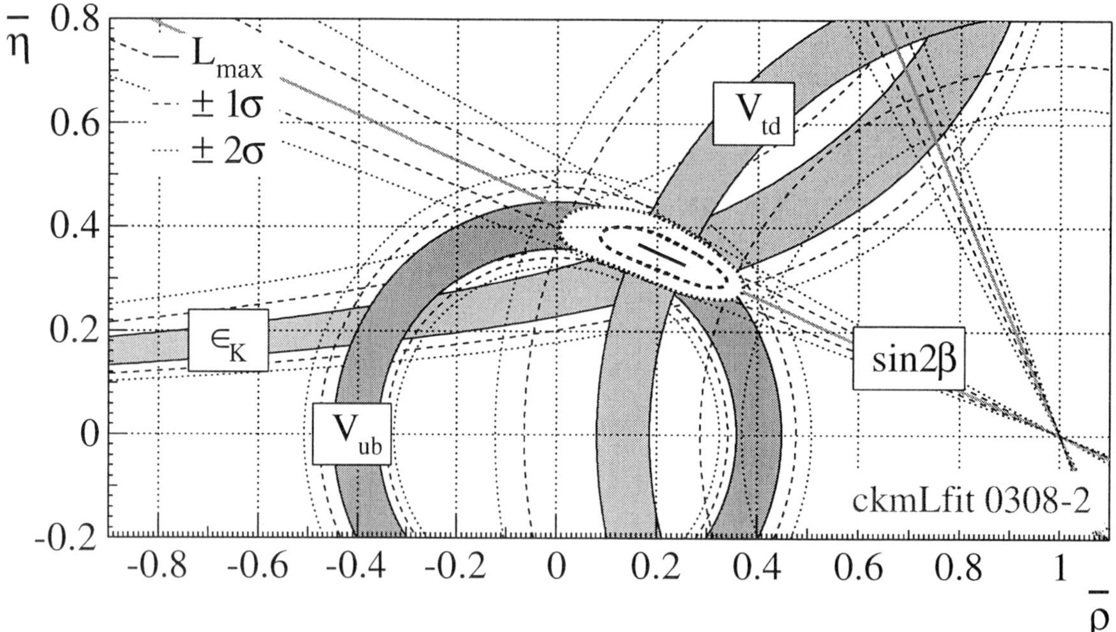

Figure 6. Likelihood contours[60] for ϵ_K and $\sin 2\beta$ superimposed to the contours of Fig. 5. The contours around the white region in the centre show the product of all four likelihoods, leading to the final fit result in Eqs. (70) and (71).

$\sin 2\beta$ is dominated by statistics. Because of the high correlation between the two parameters $\bar{\rho}$ and $\bar{\eta}$, I want to give the final fit results in a different and also extended form:

Taking all magnitude results of this review, combining them with the two explicitly CP-violating results for ϵ_K and $\sin 2\beta$, and constraining them by unitarity, the CKM matrix is given by the four parameters:

$$\lambda = 0.2235 \pm 0.0033\,,$$
$$A\lambda^2 = 0.0415 \pm 0.0011\,,$$
$$A\lambda^3\sqrt{\rho^2 + \eta^2} = (3.85 \pm 0.33)\,10^{-3}\,,$$
$$\mathrm{atan2}(\eta/\rho) = (58 \pm 19)^\circ\,. \qquad (71)$$

The relative errors on these four parameters are $\pm 1.5\%$, $\pm 2.7\%$, $\pm 9\%$, and $\pm 5\%$ of 360°, respectively. At this moment, we know $A\lambda^2$ better than A. For a long time, we were used to the hierarchy of magnitudes 1, λ, λ^2, λ^3 also being a hierarchy of precision. This is no longer so. Therefore, the four parameters in Eq. (71), i.e. the magnitudes $|V_{us}|$, $|V_{cb}|$, $|V_{ub}|$, and the phase of the quartet $V_{ub}^*V_{ud}V_{cd}^*V_{cb}$ may, in future, be four better suited choices than the parameters A, λ, ρ, and η.

Acknowledgements

I would like to thank my BABAR colleagues, especially Th. Brandt, O. Buchmüller, R. Dubitzky, U. Langenegger, and V. Lüth, for many helpful discussions during the preparation of this talk. I also express my thanks to R. Nogowski for his help with the figures and to E. Maly and F. Roschk for reading the final manuscript.

References

1. M. Kobayashi and T. Maskawa, *Progr. Theor. Phys.* **49**, 652 (1973).
2. N. Cabibbo, *Phys. Rev. Lett.* **10**, 531 (1963).
3. K. R. Schubert, *Proc. Int. Europhys. Conf. High En. Phys.*, Uppsala 1987, ed. by O. Botner, p. 791.
4. L. Wolfenstein, *Phys. Rev. Lett.* **51**, 1945 (1983).
5. I. S. Towner and J. C. Hardy, *J. Phys.* G: *Nucl. Part. Phys.* **29**, 197 (2003).
6. H. Abele *et al.* (PERKEO-II), *Phys. Rev. Lett.* **88**, 211801 (2002).
7. K. Hagiwara *et al.* (Particle Data Group),[8] 2003 update, http://pdg.lbl.gov
8. K. Hagiwara *et al.* (Particle Data Group), *Phys. Rev.* D **66**, 010001 (2002).
9. H. Abele, private communcation that the value of the Fermi function f in Ref. 6 has changed.

10. D. Pocanic (PIBETA), hep-ph/0307258 (2003).
11. H. Leutwyler and M. Roos, *Z. Phys.* C **25**, 91 (1984).
12. V. Cirigliano, *Eur. Phys. J.* C **27**, 255 (2003).
13. A. Sher *et al.* (BNL-E865), hep-ex/0305042 (2003).
14. KLOE Collaboration, hep-ex/0307016 v1 (2003).
15. N. Cabibbo *et al.*, hep-ph/0307298 v1 (2003).
16. E. Gamiz *et al.*, *JHEP* **01** 060 (2003) and M. Jamin, hep-ph/0309147 (2003).
17. See F. J. Gilman *et al.*, review 11 in Ref. 8.
18. See F. J. Gilman *et al.*, review 11 in D. E. Groom *et al.* (Particle Data Group), *Eur. Phys. J.* **15**, 1 (2000).
19. J. Alexander *et al.* (HFAG), LP03 listings in http://www.slac.stanford.edu/xorg/hfag
20. H. Albrecht *et al.* (ARGUS), *Phys. Lett.* B **318**, 397 (1993).
21. U. Langenegger (BABAR), hep-ex/0204001 (2002).
22. K. Abe *et al.* (BELLE), *Phys. Lett.* B **547**, 181 (2002).
23. B. Aubert *et al.* (BABAR), *Phys. Rev.* D **67** 031101 (2003).
24. BELLE Collaboration, hep-ex/0306020 (2003).
25. update of B. C. Barish *et al.* (CLEO), *Phys. Rev. Lett.* **76**, 1570 (1996), quoted 2003 by HFAG.[19]
26. S. B. Athar *et al.* (CLEO), *Phys. Rev.* D **66** 052003 (2002).
27. G. Altarelli *et al.*, *Nucl. Phys.* B **208**, 365 (1982).
28. See the textbook "Heavy Quark Physics", A. V. Manohar and M. B. Wise, Cambridge Monogr. Part. Nucl. Phys. (2000).
29. M. Battaglia *et al.*, DELPHI preprint 2003-028 CONF 648 (June 2003).
30. N. Uraltsev, hep-ph/0210413 (2002).
31. D. Cronin-Hennessy *et al.* (CLEO), *Phys. Rev. Lett.* **87**, 251808 (2001).
32. A. Bornheim (CLEO), hep-ex/0307011 (2003).
33. BABAR Collaboration, hep-ex/0307046 (2003).
34. C. W. Bauer *et al.*, *Phys. Rev.* D **67**, 05401 (2003).
35. G. S. Huang *et al.*, CLEO-CONF 03-08 LP-279 (2003).
36. A. Oyanguren *et al.*, DELPHI 2003-011 CONF 631.
37. B. Aubert *et al.* (BABAR), hep-ex/0308027 (2003).
38. S. Hashimoto *et al.*, *Phys. Rev.* D **66** 014503 (2002).
39. I. Caprini *et al.*, *Nucl. Phys.* B **530**, 153 (1998).
40. H. Albrecht *et al.* (ARGUS), *Z. Phys.* C **57**, 533 (1993).
41. B. Aubert *et al.* (BABAR), *Phys. Rev. Lett.* **90**, 181801 (2003).
42. S. B. Athar *et al.* (CLEO) hep-ex/0304019 (2003).
43. H. Albrecht *et al.* (ARGUS), *Phys. Lett.* B **234**, 409 (1990) and B **255**, 297 (1991).
44. R. Fulton *et al.* (CLEO), *Phys. Rev. Lett.* **64**, 2226 (1990).
45. K. Abe *et al.*, BELLE-CONF-0325 (2003).
46. S. Chen *et al.* (CLEO), *Phys. Rev. Lett.* **87**, 251807 (2001).
47. M. Neubert, *Phys. Rev.* D **49**, 3392 (1994) and D **49**, 4623 (1994).
48. Z. Ligeti, hep-ph/0309219 (2003).
49. A. Bornheim *et al.* (CLEO), hep-ex/0202019 (2002).
50. B. Aubert *et al.* (BABAR), hep-ex/0207081 (2002).
51. B. Aubert *et al.* (BABAR), hep-ex/0307062 (2003).
52. T. Affolder *et al.* (CDF), *Phys. Rev.* D **63**, 032003 (2001) and B. Abbott *et al.* (D0), *Phys. Rev.* D **58**, 052001 (1999).
53. Chapter 4 in Proc. CKM Workshop 2002, hep-ph/0304132 v2, and references therein. The rectangular errors have been symmetrized here.
54. O. Schneider, $B^0\overline{B}^0$ mixing review in Ref. 8.
55. K. Pitts, these Proceedings.
56. Chapter 4 in Proc. CKM Workshop 2002, hep-ph/0304132 v2, and references therein.
57. A. Ali and M. Misiak, in Proc. CKM Workshop 2002, hep-ph/0304132.
58. LEP Electroweak Working Group, hep-ex/01120201 v2 (2002).
59. T. Browder (BELLE), these Proceedings.
60. R. Nogowski and K. R. Schubert, Electronic Proceedings of the 2003 CKM Workshop, Durham, http://www.ippp.dur.ac.uk/~ckm/econf/schubert-ckmlfit.pdf

DISCUSSION

Ikaros Bigi (Univ. of Notre Dame): You listed some homework for theorists concerning the calculation of $\Gamma(B \to l\nu X)$. Those items with $O(1/(m_b)^3)$ contributions and $(\alpha_s)^2$ terms have been done already in our scheme in a paper that has been published this year. The limiting factor theoretically is $O(\alpha_s)$ corrections to the nonperturbative contributions.

Alberto Sirlin (NYU): In the case of V_{us}, there is a recent work by Bijnens and Jalavora. They consider two-loop contributions in chiral perturbation theory in the isospin limit. Their result depends on two unknown constants that may, in principle, be determined from the slope and curvature of $K_{\mu 3}$ form factors.

Klaus Schubert: As far as I know this has been included.

Sheldon Stone (Syracuse Univ.): Since Ikaros Bigi wants more homework, let me say that the error due to the "Duality" assumption is still an open question and many of us therefore do not consider it legitimate to average the inclusive and exclusive determinations of V_{cb}, the exclusive having well understood theoretical errors.

Vera Luth (SLAC): Measurements of exclusive charmless semileptonic decays require knowledge of form factors. Experiments will be able to measure rates as a function of q^2. Theoretical predictions over the full range in q^2 would be very welcome, especially for lattice calculations!

QCD AND HEAVY HADRON DECAYS

G. BUCHALLA

Ludwig-Maximilians-Universität München, Sektion Physik, Theresienstraße 37, D-80333 München, Germany
E-mail: Gerhard.Buchalla@physik.uni-muenchen.de

We review recent developments in QCD pertaining to its application to weak decays of heavy hadrons. We concentrate on exclusive rare and nonleptonic B-meson decays, discussing both the theoretical framework and phenomenological issues of current interest.

1. Introduction

Weak decays of heavy hadrons, of B mesons in particular, provide us with essential information on the quark flavor sector. Since the underlying flavor dynamics of the quarks is masked by strong interactions, a sufficiently precise understanding of QCD effects is crucial to extract from weak decays involving hadrons the basic parameters of flavor physics. Much interest in this respect is being devoted to rare B decay modes such as $B \to \pi\pi$, πK, $\pi\rho$, ϕK_S, $K^*\gamma$, $\rho\gamma$ and $K^*l^+l^-$. These decays are a rich source of information on CKM parameters and Flavor-Changing-Neutral-Currents. Many new results are now being obtained from the B meson factories and hadron colliders.[1-6] Both exclusive and inclusive decays can be studied. Roughly speaking, the former are more difficult for theory, the latter for experiment.

In dealing with the presence of strong interactions in these processes the challenge for theory is in general to achieve a systematic separation of long-distance and short-distance contributions in QCD. This separation typically takes the form of representing an amplitude or a cross section as a sum of *products* of long- and short-distance quantities and is commonly referred to as *factorization*. The concept of factorization requires the existence of at least one hard scale, which is large in comparison with the intrinsic scale of QCD. For B decays this scale is given by the b quark mass, $m_b \gg \Lambda_{QCD}$. The asymptotic freedom of QCD allows one to compute the short-distance parts using perturbation theory. Even though the long-distance quantities still need to be dealt with by other means, the procedure usually entails a substantial simplification of the problem.

Various methods, according to the specific nature of the application, have been developed to implement the idea of factorization in the theoretical description of heavy hadron decays. These in-

clude heavy-quark effective theory (HQET), heavy-quark expansion (HQE), factorization in exclusive nonleptonic decays and soft-collinear effective theory (SCET). In particular the latter two topics are more recent developments and are still under active investigation and further study. They play an important role in the exclusive rare B decays listed above. Dynamical calculations based on these tools hold the promise to improve our understanding of QCD in heavy-hadron decays significantly and to facilitate the determination of fundamental weak interaction parameters. A different line of approach is the use of the approximate $SU(2)$ or $SU(3)$ flavor symmetries of QCD in order to isolate the weak couplings in a model-independent way.[7-10] Both strategies, flavor symmetries and dynamical calculations, are complementary to each other and enhance our ability to test quark flavor physics. While the flavor symmetry approach gives constraints free of hadronic input in the symmetry limit, dynamical methods allow us to compute corrections from flavor symmetry breaking.

The following section gives a brief overview of theoretical frameworks for B decays based on the heavy-quark limit. The remainder of this talk then concentrates on the subject of exclusive rare or hadronic decays of B mesons.

2. Tools and Applications

The application of perturbative QCD to hadronic reactions at high energy requires a proper factorization of short-distance and long-distance contributions. One example is given by the operator product expansion (OPE) used to construct effective Hamiltonians for hadronic B decay. This is shown schematically in Fig. 1 for a generic B decay amplitude. The OPE approximates the non-local product of two weak currents, which are connected by W exchange in the full Standard Model, by local 4-quark operators, multi-

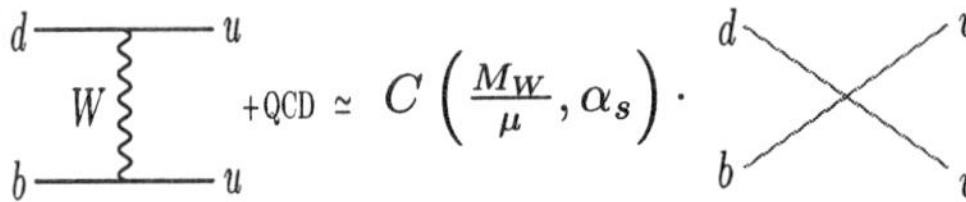

Figure 1. OPE for weak decays.

plied by Wilson coefficients $C(M_W/\mu, \alpha_s)$. In this way the short-distance physics from scales of order M_W (or m_t appearing in penguin loop diagrams) down to a factorization scale $\mu \sim m_b$ is isolated into the coefficient. Determined by high enrgy scales, the coefficient can be computed perturbatively, supplemented by renormalization-group improvement to resum large logarithms $\sim \alpha_s \ln M_W/m_b$. The QCD dynamics from scales below μ is contained within the matrix elements of the local operators. These matrix elements depend on the particular process under consideration, whereas the coefficients are universal. The approximation is valid up to power corrections of order m_b^2/M_W^2.

In the case of B decay amplitudes, the hadronic matrix elements themselves still contain a hard scale $m_b \gg \Lambda_{QCD}$. Contributions of order m_b can be further factorized from the intrinsic long-distance dynamics of QCD. This is implemented by a systematic expansion in Λ_{QCD}/m_b and $\alpha_s(m_b)$ and leads to important simplifications. The detailed formulation of this class of factorization depends on the specific application and can take the form of HQET, HQE, QCD factorization for exclusive hadronic B decays or SCET.

- HQET describes the static approximation for a heavy quark, formulated in a covariant way as an effective field theory.[11,12] It allows for a systematic inclusion of power corrections. Its usefulness is based on two important features. (i) The spin-flavor symmetry of HQET relates form factors in the heavy-quark limit and thus reduces the number of unknown hadronic quantities. (ii) The dependence on the heavy-quark mass is made explicit. Typical applications are (semi)leptonic form factors involving hadrons containing a single heavy quark, such as $B \to D^{(*)}$ form factors in semileptonic $b \to c$ transitions or the decay constant f_B.

- HQE is a theory for inclusive B decays.[13,14] It is based on the optical theorem for inclusive decays and an operator product expansion in Λ_{QCD}/m_b of the transition operator. The heavy-quark expansion justifies the parton model for inclusive decays of heavy hadrons, which it contains as its first approximation. Beyond that it allows us to study non-perturbative power corrections to the partonic picture. The main applications of the HQE method is for processes as $B \to X_{u,c}l\nu$, $B \to X_s\gamma$, $B \to X_s l^+l^-$, and for the lifetimes of b-flavored hadrons.

- QCD factorization refers to a framework for analysing exclusive hadronic B decays with a fast light meson as for instance $B \to D\pi$, $B \to \pi\pi$, $B \to \pi K$ and $B \to V\gamma$. This approach is conceptually similar to the theory of hard exclusive reactions, described for instance by the pion electromagnetic form factor at large momentum transfer.[15,16] The application to B decays requires new elements due to the presence of heavy-light mesons.[17]

- SCET is an effective field theory formulation for transitions of a heavy quark into an energetic light quark.[18] The basic idea is reminiscent of HQET. However, the structure of SCET is more complex because the relevant long-distance physics that needs to be factorized includes both soft and collinear degrees of freedom. Only soft contributions have to be accounted for in HQET. Important applications of SCET are the study of $B \to P, V$ transition form factors at large recoil energy of the light pseudoscalar (P) or vector (V) meson, and formal proofs of QCD factorization in exclusive heavy hadron decays.

There are further methods, which have been useful to obtain information on hadronic quantities relevant to B decays. Of basic importance are computations based on lattice QCD, which can access many quantities needed for B meson phenomenology (see Kronfeld[19] for a recent review). On the other hand, exclusive processes with fast light particles are very difficult to treat within this framework. An important tool to calculate in particular heavy-to-light form factors ($B \to \pi$) at large recoil are QCD sum rules on the light cone.[20,21] We will not discuss those methods here, but refer to the literature for more information.

236

3. Exclusive Hadronic B Decays in QCD

3.1. Factorization

The calculation of B-decay amplitudes, such as $B \to D\pi$, $B \to \pi\pi$ or $B \to \pi K$, starts from an effective Hamiltonian, which has, schematically, the form

$$\mathcal{H}_{eff} = \frac{G_F}{\sqrt{2}} \lambda_{CKM} \, C_i Q_i. \tag{1}$$

Here C_i are the Wilson coefficients at a scale $\mu \sim m_b$, which are known at Next-to-Leading-Order in QCD.[22] Q_i are local, dimension-6 operators and λ_{CKM} represents the appropriate CKM matrix elements. The main theoretical problem is to evaluate the matrix elements of the operators $\langle Q_i \rangle$ between the initial and final hadronic states. A typical matrix element reads $\langle \pi\pi | (\bar{u}b)_{V-A} (\bar{d}u)_{V-A} | B \rangle$.

These matrix elements simplify in the heavy-quark limit, where they can in general be written as the sum of two terms, each of which is factorized into hard scattering functions T^I and T^{II}, respectively, and the nonperturbative, but simpler, form factors F_j and meson light-cone distribution amplitudes Φ_M (Fig. 2).

Important elements of this approach are as follows. i) The expansion in $\Lambda_{QCD}/m_b \ll 1$, consistent power counting, and the identification of the leading power contribution, for which the factorized picture can be expected to hold. ii) Light-cone dynamics, which determines for instance the properties of the fast light mesons. The latter are described by light-cone distribution amplitudes Φ_π of their valence quarks defined as

$$\langle \pi(p) | u(0)\bar{d}(z) | 0 \rangle = \frac{if_\pi}{4} \, \gamma_5 \, \not{p} \int_0^1 dx \, e^{ixpz} \, \Phi_\pi(x) \tag{2}$$

with z on the light cone, $z^2 = 0$. iii) The collinear quark-antiquark pair dominating the interactions of the highly energetic pion decouples from soft gluons (colour transparency). This is the intuitive reason behind factorization. iv) The factorized amplitude consists of hard, short-distance components, and soft, as well as collinear, long-distance contributions. More details on the factorization formalism can be found elsewhere.[17]

An alternative approach to exclusive two-body decays of B mesons, referred to as pQCD, has been proposed.[23] The main hypothesis in this method is that the $B \to \pi$ form factor is not dominated by soft physics, but by hard gluon exchange that can be computed perturbatively. The hypothesis rests on the idea that Sudakov effects will suppress soft endpoint divergences in the convolution integrals. A critical discussion of this framework has been given by Descotes-Genon and Sachrajda.[24]

3.2. CP-Violation in $B \to \pi^+\pi^-$

A framework for systematic computations of heavy-hadron decay amplitudes in a well defined limit clearly has many applications for quark flavor physics with two-body nonleptonic B decays. An important example may serve to illustrate this point. Consider the time-dependent, mixing-induced CP-asymmetry in $B \to \pi^+\pi^-$

$$\mathcal{A}_{CP}(t) = \frac{\Gamma(B(t) \to \pi^+\pi^-) - \Gamma(\bar{B}(t) \to \pi^+\pi^-)}{\Gamma(B(t) \to \pi^+\pi^-) + \Gamma(\bar{B}(t) \to \pi^+\pi^-)}$$
$$= -S \sin(\Delta M_d t) + C \cos(\Delta M_d t). \tag{3}$$

Using the CKM matrix unitarity, the decay amplitude consists of two components with different CKM factors and different hadronic parts, schematically

$$A(B \to \pi^+\pi^-) = \tag{4}$$
$$V_{ub}^* V_{ud}(\text{up} - \text{top}) + V_{cb}^* V_{cd}(\text{charm} - \text{top}).$$

If the penguin contribution $\sim V_{cb}^* V_{cd}$ could be neglected, one would have $C = 0$ and $S = \sin 2\alpha$, hence a direct relation of $\mathcal{A}_{CP}$ to the CKM angle α. In reality the penguin contribution is not negligible compared to the dominant tree contribution $\sim V_{ub}^* V_{ud}$. The ratio of the penguin and tree amplitude, which enters the CP-asymmetry, depends on hadronic physics. This complicates the relation of the observables S and C to the CKM parameters. QCD factorization of B decay matrix elements allows us to compute the required hadronic input and to determine the constraint in the $(\bar{\rho}, \bar{\eta})$ plane implied by measurements of the CP-asymmetry. This is illustrated for S in Fig. 3.

The widths of the bands indicate the theoretical uncertainty.[25] Note that the constraints from S are relatively insensitive to theoretical or experimental uncertainties. The analysis of direct CP-violation measured by C is more complicated due to the importance of strong phases. Recent phenomenological analyses have been performed.[26,27] The current experimental results for S and C are from BaBar[28]

$$S = +0.02 \pm 0.34 \pm 0.05 \tag{5}$$

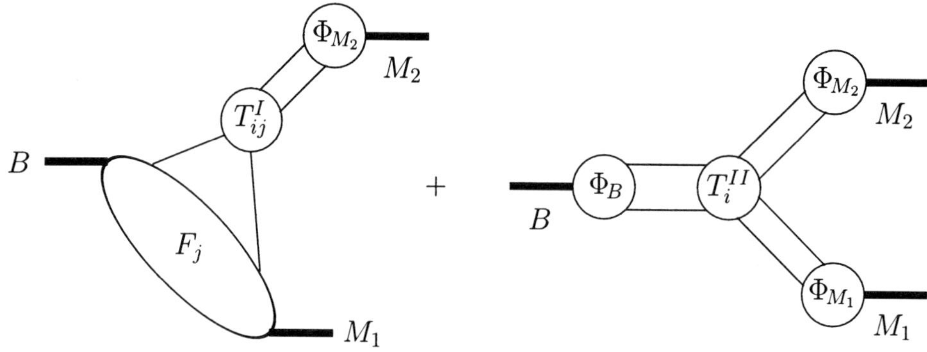

Figure 2. Graphical representation of the factorization formula.

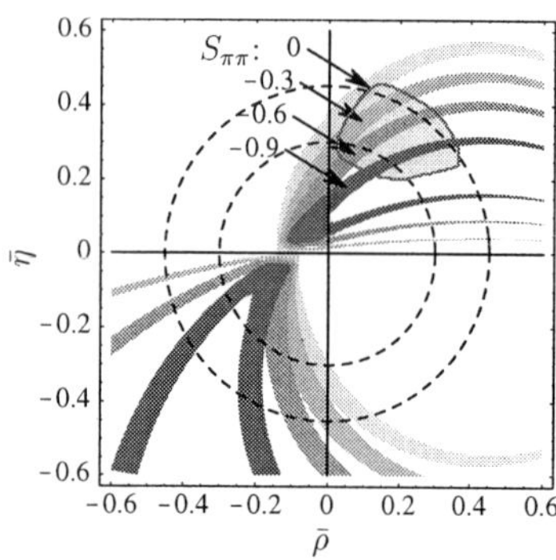

Figure 3. Constraints in the $(\bar{\rho}, \bar{\eta})$ plane from CP-violation observable S in $B \to \pi^+\pi^-$. The constraints from $|V_{ub}/V_{cb}|$ (dashed circles) and from the standard analysis of the unitarity triangle (irregular shaded area) are also shown.

$$C = -0.30 \pm 0.25 \pm 0.04 \qquad (6)$$

and from Belle[29]

$$S = -1.23 \pm 0.41^{+0.08}_{-0.07} \qquad (7)$$

$$C = -0.77 \pm 0.27 \pm 0.08. \qquad (8)$$

A recent preliminary update from BaBar gives[3,30]

$$S = -0.40 \pm 0.22 \pm 0.03 \qquad (9)$$

$$C = -0.19 \pm 0.19 \pm 0.05. \qquad (10)$$

Including the new BaBar results the current world average reads[3]

$$S = -0.58 \pm 0.20 \qquad C = -0.38 \pm 0.16 \qquad (11)$$

which ignores the large χ^2 reflecting the relatively poor agreement between the experiments.

3.3. Current Status

QCD factorization to leading power in Λ/m_b has been demonstrated at $\mathcal{O}(\alpha_s)$ for the important class of decays $B \to \pi\pi, \pi K$. For $B \to D\pi$ (class I), where hard spectator interactions are absent, a proof has been given explicitly at two loops[17] and to all orders in the framework of soft-collinear effective theory (SCET).[31] Complete matrix elements are available at $\mathcal{O}(\alpha_s)$ (NLO) for $B \to \pi\pi, \pi K$, including electroweak penguins.[25] Comprehensive treatments have also been given for $B \to PV$ modes[32] (see also Aleksan et al.[33]) and for B decays into light flavor-singlet mesons.[34] A discussion of two-body B decays into light mesons within SCET has been presented by Chay and Kim.[35]

Power corrections are presently not calculable in general. Their impact has to be estimated and included into the error analysis. Critical issues here are annihilation contributions and certain corrections proportional to $m_\pi^2/((m_u + m_d)m_b)$, which is numerically sizable, even if it is power suppressed. However the large variety of channels available will provide us with important cross-checks, and arguments based on SU(2) or SU(3) flavor symmetries can also be of use in further controlling the uncertainties.

3.4. Phenomenology of $B \to PP, PV$

Two-body B decays into light mesons have been widely discussed in the literature.[36]

In general, a phenomenological analysis of these modes faces the problem of disentangling three very different aspects, which simultaneously affect the observable decay rates and asymmetries. First, there are the CKM couplings that one would like to extract in order to test the Standard Model. Second, it is possible that some observables could be significantly modified by new physics contributions, which would complicate the determination of CKM phases. Third, the short distance physics, CKM quantities and potential new interactions, that one is aiming for, is dressed by the effects of QCD. A priori any discrepancy between data and expectations has to be examined with these points in mind. Fortunately, the large number of different channels with different QCD dynamics and CKM dependence will be very helpful to clarify the phenomenological interpretation. The following examples illustrate how various aspects of the QCD dynamics may be tested independently.

1. Penguin-to-tree ratio. To test predictions of this ratio a useful observable can be built from the mode $B^- \to \pi^- \bar{K}^0$, which is entirely dominated by a penguin contribution, and from the pure tree-type process $B^- \to \pi^- \pi^0$:

$$\left| \frac{penguin}{tree} \right| = \left| \frac{V_{ub}}{V_{cb}} \right| \frac{f_\pi}{f_K} \sqrt{\frac{B(B^- \to \pi^- \bar{K}^0)}{2\, B(B^- \to \pi^- \pi^0)}}. \tag{12}$$

This amplitude ratio is not identical to the P/T ratio required for $B \to \pi^+\pi^-$, but still rather similar to be interesting as a test. Small differences come from $SU(3)$ breaking effects (the dominant ones due to f_π/f_K are already corrected for in Eq. (12)), and weak annihilation corrections in $B \to \pi K$, and from the color-suppressed contribution to $B^- \to \pi^- \pi^0$. Because the $\pi^- \bar{K}^0$ and $\pi^- \pi^0$ channels have only a single amplitude (penguin or tree), no interference is possible and the ratio in Eq. (12) is independent of the CKM phase γ. This is useful for distinguishing QCD effects from CKM issues. A comparison of factorization predictions for the left-hand-side of Eq. (12) with data used to compute the right-hand-side in Eq. (12) is shown in Fig. 4. The agreement is satisfactory within uncertainties.

2. Factorization test for $B^- \to \pi^- \pi^0$. It is of interest to test predictions for the tree amplitude alone using a classical factorization test of the

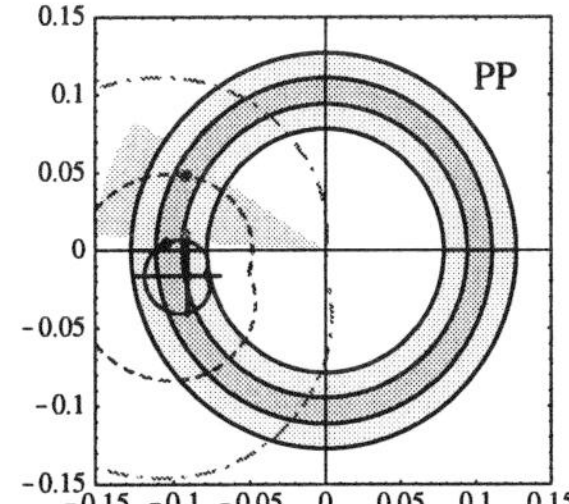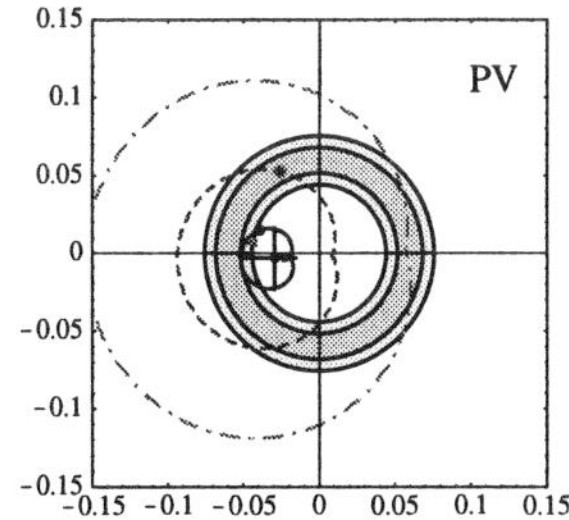

Figure 4. Left panel: Penguin-to-tree ratio extracted from data on $B^- \to \pi^- \bar{K}^0$ and $B^- \to \pi^- \pi^0$ (rings in the center) compared with predictions in QCD factorization (cross). The light (dark) ring is with (without) the uncertainty from $|V_{ub}/V_{cb}|$. The prediction includes a model estimate of power corrections, dominantly from weak annihilation. The solid, dashed, dashed-dotted error contour indicates the uncertainty from assigning 100%, 200%, 300% errors, respectively, to the default annihilation correction. Right panel: The same with the K replaced by K^*. (From Beneke and Neubert.[32])

form

$$B(B^+ \to \pi^+ \pi^0) = 3\pi^2 f_\pi^2 |V_{ud}|^2 \times \tag{13}$$
$$\frac{dB(B_d \to \pi^- l^+ \nu)}{dq^2}\bigg|_{q^2=0} \frac{\tau(B^+)}{\tau(B_d)} |a_1 + a_2|^2$$

where a_1, a_2 are QCD coefficients.[17,25] The advantage of this test is that $B^- \to \pi^- \pi^0$ receives neither penguin nor annihilation contributions. It thus gives information on the other aspects of the QCD dynamics in $B \to \pi\pi$. This test has been discussed recently.[37,32]

3. Direct CP-asymmetries. From the heavy-quark limit one generally expects strong phases to be suppressed, except for a few special cases. This circumstance should suppress direct CP-asymmetries. Of course those also depend sensitively on weak phases and a detailed analysis has to consider individual channels. At present, qualitatively, one may at least say that the non-observation of direct CP-violation in B decays until today, with experimental bounds typically at the 10% level, are not in contradiction with the theoretical expectation.

4. Weak annihilation. Amplitudes from weak annihilation represent power suppressed corrections, which are uncalculable in QCD factorization and so far need to be estimated relying on models. At present there are no indications that annihilation terms would be anomalously large,

but they do contribute to the theoretical uncertainty. Effectively, annihilation corrections may be considered as part of the penguin amplitudes. To some extent, therefore, they are tested with the help of the penguin-to-tree ratio discussed above. Nevertheless, in order to disentangle their impact from other effects it is of great interest to test annihilation separately. This can be done with decay modes that proceed through annihilation or at least have a dominant annihilation component.

An example is the pure annihilation channel $B_d \to D_s^- K^+$. Even though this case is somewhat different from the reactions of primary interest here, because of the charmed meson in the final state, it is still useful to cross-check the typical size of annihilation expected in model calculations. Treating the D meson in the model estimate for annihilation[25] as suggested by Beneke *et al.*,[17] one finds a central (CP-averaged) branching ratio of $B(B_d \to D_s^- K^+) = 1.2 \times 10^{-5}$. Allowing for a 100% uncertainty of the central annihilation estimate, which in the case of the penguin-to-tree ratio shown in Fig. 4 corresponds to the inner (solid) error region around the theoretical value (marked by the cross), gives an upper limit[38] of 5×10^{-5}. This is in agreement with the current experimental result $(3.8 \pm 1.1) \times 10^{-5}$ (see refs. in Beneke[38]).

Additional tests should come from annihilation decays into two light mesons, such as $B \to KK$ modes.[39] These, however, are CKM suppressed and only upper limits are known at present. The $K^+ \bar{K}^0$ and $K^0 \bar{K}^0$ channels have both annihilation and penguin contributions. On the other hand $B \to K^+ K^-$ is a pure weak annihilation process and therefore especially important. Further discussions can be found in the References.[25,32,39]

At present, within current experimental and theoretical uncertainties, there are no clear signals of significant discrepancies between measurements and SM expectations in hadronic B decays, neither with respect to QCD calculations nor suggesting the need for new physics. However, a few experimental results have central values deviating from standard predictions, which attracted some attention in the literature. Even though the discrepancies are not significant at the moment, it will be interesting to follow future developments. We comment on some of those possible hints here, with a view on QCD predictions within the SM.

- As seen in Eq. (11) the measurement of $C = -0.38 \pm 0.16$ suggests the possibility of large direct CP-violation in $B \to \pi^+ \pi^-$ decays. On the other hand, this is largely due to the result from Belle, whereas BaBar gives a smaller effect. In the SM one expects $C \approx 0.1$ with an error of about the same size. It is interesting to note that the perturbative strong interaction phase predicted to lowest order in QCD factorization gives a positive value for C while the measurements seem to prefer negative values. Since the strong phase is a small effect in the heavy-quark limit, uncalculable power corrections could possibly compete with the perturbative contribution. A small negative C is therefore not excluded, but the reliability of a lowest order perturbative calculation of the strong phase would then be in doubt. (A logical possibility for $C < 0$ would be that the positive sign of the strong phase is correct, but the weak phase is negative, which would require new physics in ε_K.) In any case, a clarification of the experimental situation will be important. It may also be noted that the central numbers from Belle, which are large for both S and C, would violate the absolute bound $S^2 + C^2 \leq 1$ when taken at face value.

- Mixing-induced CP-violation S in $B \to \phi K_S$ and $B \to \eta' K_S$, which proceed through the penguin transition $b \to s \bar{s} s$, could be strongly affected by new physics. In the SM one expects $S_{\phi K_S}$ and $S_{\eta' K_S}$ to be close to the benchmark observable $S_{\psi K_S}$ of mixing-induced CP-violation in $B \to \psi K_S$.[40] Hints of deviations in the data from Belle, and to a much lesser extent from BaBar, have motivated several analyses in the literature on this issue.[41−43] Experimentally one finds for the world average[1]

$$S_{\phi K_S} - S_{\psi K_S} = -0.89 \pm 0.33 \qquad (14)$$
$$S_{\eta' K_S} - S_{\psi K_S} = -0.47 \pm 0.22 \qquad (15)$$

where the first result combines the BaBar and Belle values ignoring the rather poor agreement between them. This can be compared with the

SM expectation based on a recent QCD analysis in Beneke and Neubert[32]

$$S_{\phi K_S} - S_{\psi K_S} = 0.025 \pm 0.016 \qquad (16)$$

$$S_{\eta' K_S} - S_{\psi K_S} = 0.011 \pm 0.013. \qquad (17)$$

More information on possible new physics implications can be found in Grossman.[44]

- Current data for the ratio of $B \to \pi^+\pi^-$ and $B \to \pi^+\pi^0$ branching fractions appear to be somewhat low in comparison with theoretical calculations for a CKM phase $\gamma < 90°$ as given by standard fits of the CKM unitarity triangle. This feature is often interpreted[45] as a hint for a larger value of $\gamma > 90°$. Such a value could change a constructive interference of tree and penguin amplitudes in the $\pi^+\pi^-$ mode into a destructive one, and thus reduce the ratio of branching fractions. In Beneke and Neubert[32] a different, QCD related possibility was discussed that could account for the suppression of $B \to \pi^+\pi^-$ relative to $B \to \pi^+\pi^0$, even for $\gamma < 90°$. In this scenario, which can be realized without excessive tuning of input parameters, the factorization coefficient a_2 (color-suppressed tree) is enlarged, while the $B \to \pi$ form factor is somewhat smaller than commonly assumed. This keeps $B \to \pi^+\pi^0$ roughly constant and suppresses $B \to \pi^+\pi^-$, which is independent of a_2. The factorization test mentioned in Point 2 above would be very useful to check such a scenario. This could also help to clarify the situation with $B \to \pi^0\pi^0$, which is very sensitive to a_2 and for which first measurements from BaBar and Belle indicate a substantial branching fraction.[2] Theoretically a_2 is subject to sizable uncertainties, because color suppression strongly reduces the leading order value and makes the prediction sensitive to subleading corrections.

- The ratio (CP-averaged rates are understood)

$$R_{00} = \frac{2\Gamma(\bar{B}^0 \to \pi^0 \bar{K}^0)}{\Gamma(B^- \to \pi^- \bar{K}^0)} \qquad (18)$$

appears to be larger than expected theoretically. This is shown in Fig. 5. The ratio R_{00} is almost insensitive to the CKM angle γ and it is essentially impossible to enhance the prediction in the SM by QCD effects.[32] The discrepancy of about

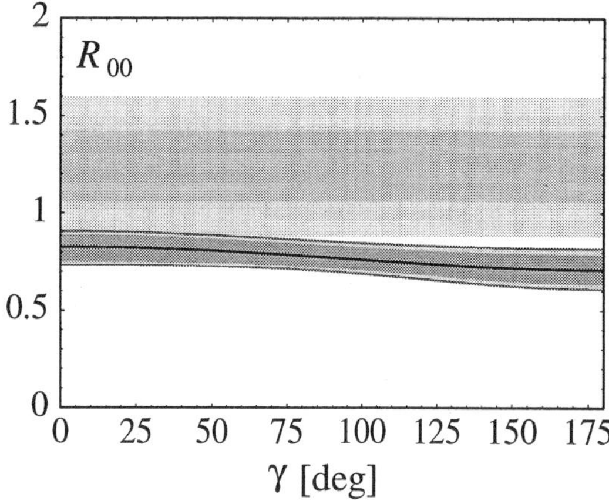

Figure 5. Theoretical prediction for $R_{00} = 2\Gamma(B \to \pi^0 K^0)/\Gamma(B \to \pi^{\pm} K^0)$.[32] The experimental result is indicated by the straight horizontal bands showing the 1σ (dark) and 2σ (light) range.

2σ can also be seen in a different way, using the Lipkin-Gronau-Rosner sum rule, which relates all four πK modes using isospin symmetry.[46] The ratio

$$R_L = \frac{2\Gamma(\bar{B}^0 \to \pi^0 \bar{K}^0) + 2\Gamma(\bar{B}^- \to \pi^0 K^-)}{\Gamma(B^- \to \pi^- \bar{K}^0) + \Gamma(\bar{B}^0 \to \pi^+ \bar{K}^-)} \qquad (19)$$

can be shown to be 1 up to corrections of second order in small quantities. Experimentally it is also about 2σ high.[47] If the discrepancy should become statistically significant, it would be a strong indication of physics beyond the SM.[32,44,47,48]

The status of QCD calculations for $B \to PV$ modes is presented by Beneke and Neubert[32] and a more general discussion of new physics aspects is given by Grossman[44].

4. Rare and Radiative B Decays

4.1. Radiative Decays $B \to V\gamma$

Factorization in the sense of QCD can also be applied to the exclusive radiative decays $B \to V\gamma$ ($V = K^*$, ρ). The factorization formula for the operators in the effective weak Hamiltonian can be written as[49,50]

$$\langle V\gamma(\epsilon)|Q_i|\bar{B}\rangle = \qquad (20)$$

$$\left[F^{B \to V}(0)\, T_i^I + \int_0^1 d\xi\, dv\, T_i^{II}(\xi, v)\, \Phi_B(\xi)\, \Phi_V(v) \right] \cdot \epsilon$$

where ϵ is the photon polarization 4-vector. Here $F^{B \to V}$ is a $B \to V$ transition form factor, and Φ_B, Φ_V are leading-twist light-cone distribution amplitudes (LCDA) of the B meson and the vector meson V, respectively. These quantities describe the long-distance dynamics of the matrix elements, which is factorized from the perturbative, short-distance interactions expressed in the hard-scattering kernels T_i^I and T_i^{II}. The QCD factorization formula Eq. (20) holds up to corrections of relative order Λ_{QCD}/m_b. Annihilation topologies are power-suppressed, but still calculable in some cases. The framework of QCD factorization is necessary to compute exclusive $B \to V\gamma$ decays systematically beyond the leading logarithmic approximation. Results to Next-to-Leading-Order in QCD, based on the heavy quark limit $m_b \gg \Lambda_{QCD}$ have been computed[49,50] (see also Ali and Parkhomenko[51]).

The method defines a systematic, model-independent framework for $B \to V\gamma$. An important conceptual aspect of this analysis is the interpretation of loop contributions with charm and up quarks, which come from leading operators in the effective weak Hamiltonian. These effects are calculable in terms of perturbative hard-scattering functions and universal meson light-cone distribution amplitudes. They are $\mathcal{O}(\alpha_s)$ corrections, but are leading power contributions in the framework of QCD factorization. This picture is in contrast to the common notion that considers charm and up quark loop effects as generic, uncalculable long-distance contributions. Non-factorizable long-distance corrections may still exist, but they are power-suppressed. The improved theoretical understanding of $B \to V\gamma$ decays strengthens the motivation for still more detailed experimental investigations, which will contribute significantly to our knowledge of the flavor sector.

The uncertainty of the branching fractions is currently dominated by the form factors F_{K^*}, F_ρ. A NLO analysis[50] yields (in comparison with the experimental results in brackets) $B(\bar{B} \to \bar{K}^{*0}\gamma)/10^{-5} = 7.1 \pm 2.5$ (4.21 ± 0.29^{52}) and $B(B^- \to \rho^-\gamma)/10^{-6} = 1.6 \pm 0.6$ $(< 2.3^{53})$. Taking the sizable uncertainties into account, the results for $B \to K^*\gamma$ are compatible with the experimental measurements, even though the central theoretical values appear to be somewhat high. $B(B \to \rho\gamma)$ is a sensitive measure of CKM quantities.[50,54,55] This is illustrated in Fig. 6.

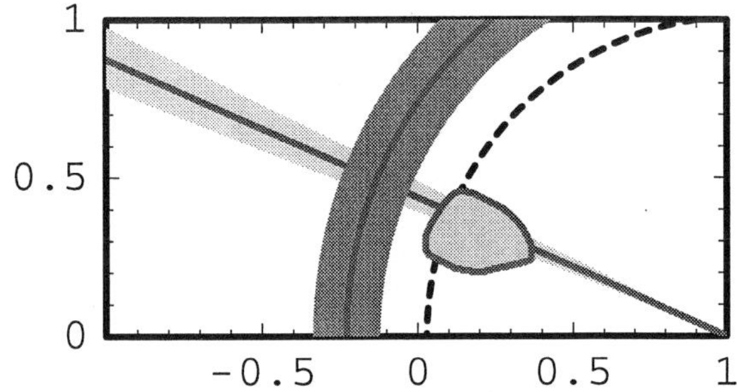

Figure 6. Impact of the current experimental upper limit on $B(B \to \rho\gamma)/B(B \to K^*\gamma)$ in the $(\bar{\rho}, \bar{\eta})$ plane. The area to the left of the dark band is excluded. The width of the dark band reflects the variation of $\xi \equiv F_{K^*}/F_\rho = 1.33 \pm 0.13$ (see Ball and Braun[20]). The case of $\xi = 1$ is illustrated by the dashed curve. The intersection with the light-shaded band from the measurement of $\sin 2\beta$ defines the apex of the unitarity triangle and the length of $R_t = \sqrt{(1 - \bar{\rho})^2 + \bar{\eta}^2} \sim |V_{td}|$, once the upper limit will be turned into a measurement range. The irregular area represents the standard unitarity triangle fit.

4.2. SCET

In the decay processes of B mesons with highly energetic light quarks in the final state, HQET alone is not sufficient to account for the complete long-distance degrees of freedom that need to be represented in an effective theory description. A first step towards implementing the missing ingredients was made by Dugan and Grinstein.[56] In this paper a framework, called large-energy effective theory (LEET), was suggested that describes the interactions of energetic light quarks with soft gluons. To correctly reproduce the infrared structure of QCD, also collinear gluons need to be included, which has been emphasized by a number of authors.[18] These authors[18] constructed an effective theory, the SCET, for soft and collinear gluons, applicable to energetic heavy-to-light transitions. These transitions may be inclusive heavy-to-light processes, such as $b \to u$ decays, but also exclusive $B \to P, V$ form factors at large recoil of the light final state meson. Similarly the SCET is a useful language to investigate factorization properties in hadronic B decays in general terms.

For the construction of the SCET one writes the four-momentum p of an energetic light quark (collinear quark) in light-cone coordinates

$$p^\mu = \frac{1}{\sqrt{2}} \left(p_- n^\mu + p_+ \bar{n}^\mu\right) + p_\perp^\mu \qquad (21)$$

$$p_\pm = \frac{p^0 \pm p^3}{\sqrt{2}} \qquad (22)$$

where n is a light-like four-vector in the direction of the collinear quark and $\bar{n}$ is a similar vector in the opposite direction, that is

$$n^2 = \bar{n}^2 = 0 \qquad n \cdot \bar{n} = 2. \qquad (23)$$

The four-vector $p_\perp$ contains the components of p perpendicular to both n and $\bar{n}$. For p collinear to the light-like direction n the components scale as $p_- \sim M$, $p_\perp \sim M\lambda$, $p_+ \sim M\lambda^2$, where M is the hard scale ($\sim m_b$) and λ is a small parameter, such that $p^2 = 2p_+p_- + p_\perp^2 \sim M^2\lambda^2$. The dependence on the larger components of p, p_- and $p_\perp$ is then removed from the light-quark field $\psi(x)$ in full QCD by writing

$$\psi(x) = \sum_{\tilde{p}} e^{-i\tilde{p}\cdot x}\,\psi_{n,p} \qquad (24)$$

$$\tilde{p} \equiv \frac{1}{\sqrt{2}}p_- n + p_\perp. \qquad (25)$$

This is analogous to the construction of the HQET, where the dependence on the large components v of the heavy-quark velocity is isolated in a similar way. The new fields $\psi_{n,p}$ are then projected onto the spinors

$$\xi_{n,p} = \frac{\slashed{n}\,\slashed{\bar{n}}}{4}\,\psi_{n,p} \qquad \xi_{\bar{n},p} = \frac{\slashed{\bar{n}}\,\slashed{n}}{4}\,\psi_{n,p}. \qquad (26)$$

The field $\xi_{n,p}$ represents the collinear quark in the effective theory. The smaller components $\xi_{\bar{n},p}$ are integrated out in the construction of the effective theory Lagrangian $\mathcal{L}_{SCET}$ from the Lagrangian of full QCD. $\mathcal{L}_{SCET}$ contains collinear quarks $\xi_{n,p}$, the heavy-quark fields from HQET, h_v, and soft and collinear gluons.

A typical application is the analysis of $B \to P$, V form factors at large recoil. Bilinear heavy-to-light currents $\bar{q}\Gamma b$ have to be matched onto operators of the SCET, schematically

$$\bar{q}\Gamma b \to C_i\,\bar{\xi}_{\bar{n},p}\tilde{\Gamma}_i h_v \qquad (27)$$

where the C_i are Wilson coefficient functions. For $B \to P$, V transitions in full QCD there is a total of ten different form factors describing the matrix elements of the possible independent bilinear currents. In SCET the equations of motion

$$\slashed{v}h_v = h_v \qquad \slashed{n}\xi_{n,p} = 0 \qquad (28)$$

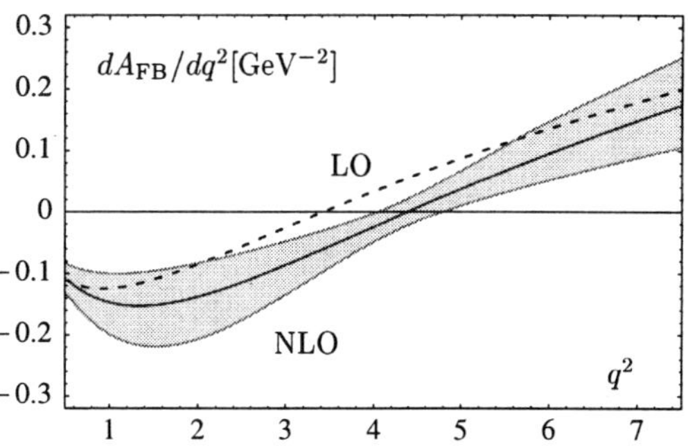

Figure 7. A_{FB} spectrum for $\bar{B} \to K^*l^+l^-$ at Leading- and Next-to-Leading-Order in QCD.[49]

imply constraints, which reduce the number of independent form factors to three, to leading order in the heavy-quark limit. An application to $B \to K^*l^+l^-$ decays will be discussed in the following section. Further developments and applications of the SCET framework to rare, radiative and hadronic B decays can be found in the References.[57−61]

4.3. Forward-Backward Asymmetry Zero in $B \to K^*l^+l^-$

Substantial progress has taken place over the last few years in understanding the QCD dynamics of exclusive B decays. The example of the forward-backward asymmetry in $B \to K^*l^+l^-$ nicely illustrates some aspects of these developments.

The forward-backward asymmetry A_{FB} is the rate difference between forward ($0 < \theta < \pi/2$) and backward ($\pi/2 < \theta < \pi$) going l^+, normalized by the sum, where θ is the angle between the l^+ and B momenta in the center-of-mass frame of the dilepton pair. A_{FB} is usually considered as a function of the dilepton mass q^2. In the Standard Model the spectrum dA_{FB}/dq^2 (Fig. 7) has a characteristic zero at

$$\frac{q_0^2}{m_B^2} = -\alpha_+ \frac{m_b C_7}{m_B C_9^{eff}} \qquad (29)$$

depending on short-distance physics contained in the coefficients C_7 and C_9^{eff}. The factor α_+, on the other hand, is a hadronic quantity containing ratios of form factors.

It was first stressed by Burdman[62] that α_+ is not affected very much by hadronic uncertainties and is very similar in different models for form factors with $\alpha_+ \approx 2$. After relations were found between

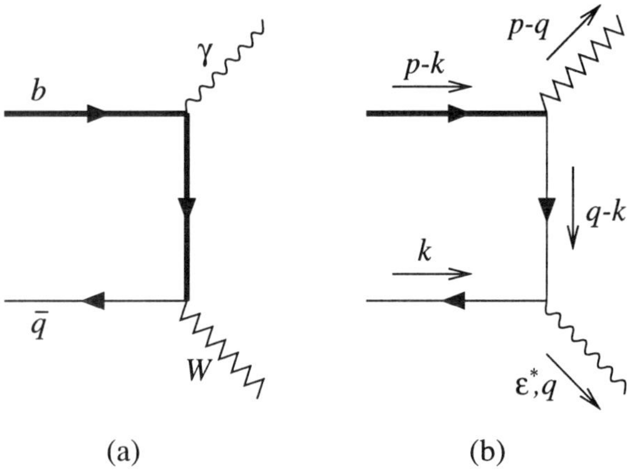

Figure 8. Tree-level diagrams for $B \to l\nu\gamma$. Only diagram (b) contributes at leading power.[67]

different heavy-light form factors ($B \to P$, V) in the heavy-quark limit and at large recoil,[63] it was pointed out by Ali *et al.*[64] that as a consequence $\alpha_+ = 2$ holds exactly in this limit. Subsequently, the results of Charles *et al.*[63] were demonstrated to be valid beyond tree level.[49,18] The use of the A_{FB}-zero as a *clean* test of Standard Model flavor physics was thus put on a firm basis and NLO corrections to Eq. (29) could be computed.[49] More recently the problem of power corrections to heavy-light form factors at large recoil in the heavy-quark limit has also been studied.[57] Besides the value of q_0^2, the sign of the slope of $dA_{FB}(\bar{B})/dq^2$ can also be used as a probe for new physics. For a $\bar{B}$ meson, this slope is predicted to be positive in the Standard Model.[65]

4.4. *Radiative Leptonic Decay $B \to l\nu\gamma$*

The tree-level process $B \to l\nu\gamma$ is not so much of direct interest for flavor physics, but it provides us with an important laboratory for studying QCD dynamics in exclusive B decays that is crucial for many other applications. The leading-power contribution comes from the diagram in Fig. 8(b), which contains a light-quark propagator that is off-shell by an amount $(q - k)^2 \sim q_- k_+$ Here q is the hard, light-like momentum of the photon with components scaling as m_b (this restricts the region of phase-space where the present discussion applies), and k is the soft momentum of the spectator quark. The decay is thus determined by a hard-scattering process, but also depends on the structure of the B meson in a non-trivial way.[66] Recently,[67] it has been proposed,

and shown to one loop in QCD, that the form factors F for this decay factorize as

$$F = \int d\tilde{k}_+ \, \Phi_B(\tilde{k}_+) \, T(\tilde{k}_+) \qquad (30)$$

where T is the hard-scattering kernel and Φ_B the light-cone distribution amplitude of the B meson is defined as

$$\Phi_B(\tilde{k}_+) = \int dz_- \, e^{i\tilde{k}_+ z_-} \, \langle 0|b(0)\bar{u}(z)|B\rangle|_{z_+=z_\perp=0}. \qquad (31)$$

The hard process is characterized by a scale $\mu_F \sim \sqrt{m_b \Lambda}$. At lowest order the form factors are proportional to $\int d\tilde{k}_+ \, \Phi_B(\tilde{k}_+)/\tilde{k}_+ \equiv 1/\lambda_B$, a parameter that enters hard-spectator processes in many other applications. The analysis at NLO requires resummation of large logarithms $\ln(m_b/\tilde{k}_+)$. An extension of the proof of factorization to all orders within the SCET was subsequently given.[68,69]

Progress has also been made recently towards a better understanding of the B meson light cone distribution amplitude itself.[49,70−73]

5. Conclusions

QCD has been very successful as a theory of the strong interaction at high energies, based on expansions in inverse powers of the high-energy-scale and perturbation theory in α_s. This general framework of QCD has recently found new applications in the treatment of exclusive decays of heavy hadrons. It is particularly exciting that these developments come at a time where a large amount of precision data is being collected at the experimental B physics facilities.

Factorization formulas in the heavy-quark limit have been proposed for a large variety of exclusive B decays. They justify in many cases the phenomenological factorization ansatz that has been employed in many applications. In addition they enable consistent and systematic calculations of corrections in powers of α_s. Non-factorizable long-distance effects are not calculable in general but they are suppressed by powers of Λ_{QCD}/m_b. So far, $B \to D^+\pi^-$ decays are probably understood best. Decays with only light hadrons in the final state such as $B \to \pi\pi$, $K^*\gamma$, $\rho\gamma$, or $K^*l^+l^-$ include hard spectator interactions at leading power and are therefore more complicated. An important new tool that has been developed is the soft-collinear effective theory (SCET), which is of use for proofs of factorization and for the theory of heavy-to-light form factors at large recoil. Studies of the process $B \to l\nu\gamma$ have also led to a better understanding of QCD dynamics in exclusive hadronic B decays. These are promising steps towards controling the QCD dynamics in exclusive hadronic or rare B decays in a reliable way. In many cases the required theoretical accuracy is not extremely high, so even moderately precise, but robust predictions will be very helpful. Using all the available tools we can hope to successfully probe CP-violation, weak interaction parameters and new phenomena in the quark-flavor sector.

References

1. T. Browder, these proceedings.
2. J. Fry, these proceedings.
3. H. Jawahery, these proceedings.
4. M. Nakao, these proceedings.
5. K. Pitts, these proceedings.
6. K. Schubert, these proceedings.
7. M. Gronau and D. London, *Phys. Rev. Lett.* **65**, 3381 (1990).
8. M. Gronau, O. F. Hernandez, D. London and J. L. Rosner, *Phys. Rev. D* **52**, 6356 (1995).
9. R. Fleischer, *Phys. Lett. B* **459**, 306 (1999); I. Dunietz, FERMILAB-CONF-93-090-T *Presented at Summer Workshop on B Physics at Hadron Accelerators, Snowmass, CO, 21 Jun - 2 Jul 1993.*
10. C. W. Chiang, M. Gronau, Z. Luo, J. L. Rosner and D. A. Suprun, hep-ph/0307395.
11. N. Isgur and M. B. Wise, *Phys. Lett. B* **232**, 113 (1989); *Phys. Lett. B* **232**, 113 (1989).
12. M. Neubert, *Phys. Rept.* **245**, 259 (1994).
13. I. I. Y. Bigi *et al.*, hep-ph/9401298.
14. I. I. Y. Bigi, M. A. Shifman and N. Uraltsev, *Ann. Rev. Nucl. Part. Sci.* **47**, 591 (1997).
15. A. V. Efremov and A. V. Radyushkin, *Theor. Math. Phys.* **42**, 97 (1980) [*Teor. Mat. Fiz.* **42**, 147 (1980)]; *Phys. Lett. B* **94**, 245 (1980).
16. G. P. Lepage and S. J. Brodsky, *Phys. Rev. D* **22**, 2157 (1980).
17. M. Beneke *et al.*, *Phys. Rev. Lett.* **83**, 1914 (1999); *Nucl. Phys. B* **591**, 313 (2000).
18. C. W. Bauer *et al.*, *Phys. Rev. D* **63**, 114020 (2001); C. W. Bauer, D. Pirjol and I. W. Stewart, *Phys. Rev. D* **65**, 054022 (2002); C. W. Bauer, D. Pirjol and I. W. Stewart, *Phys. Rev. D* **66**, 054005 (2002); C. W. Bauer, D. Pirjol and I. W. Stewart, *Phys. Rev. D* **67**, 071502 (2003).
19. A. S. Kronfeld, hep-lat/0310063.
20. P. Ball, *JHEP* **9809**, 005 (1998); P. Ball and V. M. Braun, *Phys. Rev. D* **58**, 094016 (1998); P. Ball and R. Zwicky, *JHEP* **0110**, 019 (2001); P. Ball, hep-ph/0308249.
21. V. M. Belyaev, A. Khodjamirian and R. Rückl, *Z. Phys. C* **60**, 349 (1993); A. Khodjamirian *et al.*, *Phys. Rev. D* **62**, 114002 (2000).
22. G. Buchalla, A. J. Buras and M. E. Lautenbacher, *Rev. Mod. Phys.* **68**, 1125 (1996).
23. Y. Y. Keum, H. n. Li and A. I. Sanda, *Phys. Lett. B* **504**, 6 (2001); Y. Y. Keum, H. N. Li and A. I. Sanda, *Phys. Rev. D* **63**, 054008 (2001); Y. Y. Keum and A. I. Sanda, *Phys. Rev. D* **67**, 054009 (2003).
24. S. Descotes-Genon and C. T. Sachrajda, *Nucl. Phys. B* **625**, 239 (2002).
25. M. Beneke *et al.*, *Nucl. Phys. B* **606**, 245 (2001).
26. M. Gronau and J. L. Rosner, *Phys. Rev. D* **65**, 093012 (2002).
27. G. Buchalla and A. S. Safir, hep-ph/0310218.
28. B. Aubert *et al.* [BaBar Collaboration], *Phys. Rev. Lett.* **89**, 281802 (2002).
29. K. Abe *et al.* [Belle Collaboration], *Phys. Rev. D* **68**, 012001 (2003).
30. Heavy Flavor Averaging Group, http://www.slac.stanford.edu/xorg/hfag/index.html.
31. C. W. Bauer *et al.*, *Phys. Rev. Lett.* **87**, 201806 (2001).
32. M. Beneke and M. Neubert, hep-ph/0308039.
33. R. Aleksan *et al.*, *Phys. Rev. D* **67**, 094019 (2003).
34. M. Beneke and M. Neubert, *Nucl. Phys. B* **651**, 225 (2003).
35. J. g. Chay and C. Kim, *Phys. Rev. D* **68**, 071502 (2003); hep-ph/0301262; *Phys. Rev. D* **68**, 034013 (2003).
36. J. P. Silva and L. Wolfenstein, *Phys. Rev. D* **49**, 1151 (1994); M. Gronau and J. L. Rosner, *Phys. Rev. Lett.* **76**, 1200 (1996); A. S. Dighe, M. Gronau and J. L. Rosner, *Phys. Rev. D* **54**, 3309 (1996); M. Ciuchini, E. Franco, G. Martinelli and L. Silvestrini, *Nucl. Phys. B* **501**, 271 (1997); A. Ali and C. Greub, *Phys. Rev. D* **57**, 2996 (1998); A. J. Buras and R. Fleischer, *Eur. Phys. J. C* **11**, 93 (1999); Y. F. Zhou *et al.*, *Phys. Rev. D* **63**, 054011 (2001); I. Dunietz, R. Fleischer and U. Nierste, *Phys. Rev. D* **63**, 114015 (2001); M. Ciuchini *et al.*, *Phys. Lett. B* **515**, 33 (2001); R. Fleischer, G. Isidori and J. Matias, *JHEP* **0305**, 053 (2003).
37. Z. Luo and J. L. Rosner, Phys. Rev. D **68**, 074010 (2003) [arXiv:hep-ph/0305262].
38. M. Beneke, p.299, *in*: M. Battaglia *et al.*, proceedings of the Workshop on the Unitarity Triangle (CERN Yellow Report to appear), hep-ph/0304132.
39. M. Gronau and J. L. Rosner, *Phys. Rev. D* **58**, 113005 (1998).
40. Y. Grossman, G. Isidori and M. P. Worah, *Phys. Rev. D* **58**, 057504 (1998).
41. G. Hiller, *Phys. Rev. D* **66**, 071502 (2002);
42. A. Datta, *Phys. Rev. D* **66**, 071702 (2002); M. Ciuchini and L. Silvestrini, *Phys. Rev. Lett.* **89**, 231802 (2002); M. Raidal, *Phys. Rev. Lett.* **89**, 231803

(2002); S. Khalil and E. Kou, *Phys. Rev.* D **67**, 055009 (2003); G. L. Kane *et al.*, *Phys. Rev. Lett.* **90**, 141803 (2003); C. W. Chiang and J. L. Rosner, *Phys. Rev.* D **68**, 014007 (2003).

43. Y. Grossman, Z. Ligeti, Y. Nir and H. Quinn, *Phys. Rev.* D **68**, 015004 (2003).
44. Y. Grossman, these proceedings.
45. W. S. Hou, J. G. Smith and F. Würthwein, hep-ex/9910014.
46. H. J. Lipkin, hep-ph/9809347; H. J. Lipkin, *Phys. Lett.* B **445**, 403 (1999); M. Gronau and J. L. Rosner, *Phys. Rev.* D **59**, 113002 (1999).
47. M. Gronau and J. L. Rosner, *Phys. Lett.* B **572**, 43 (2003).
48. A. J. Buras, R. Fleischer, S. Recksiegel and F. Schwab, hep-ph/0309012.
49. M. Beneke, T. Feldmann and D. Seidel, *Nucl. Phys.* B **612**, 25 (2001); M. Beneke and T. Feldmann, *Nucl. Phys.* B **592**, 3 (2001).
50. S. Bosch and G. Buchalla, *Nucl. Phys.* B **621**, 459 (2002).
51. A. Ali and A. Y. Parkhomenko, *Eur. Phys. J.* C **23**, 89 (2002).
52. T. E. Coan *et al.*, *Phys. Rev. Lett.* **84**, 5283 (2000); B. Aubert *et al.*, *Phys. Rev. Lett.* **88**, 101805 (2002); A. Ishikawa, hep-ex/0205051.
53. B. Aubert *et al.*, hep-ex/0207073.
54. S. W. Bosch, hep-ph/0310317.
55. A. Ali and E. Lunghi, *Eur. Phys. J.* C **26**, 195 (2002).
56. M. J. Dugan and B. Grinstein, *Phys. Lett.* B **255**, 583 (1991).
57. M. Beneke *et al.*, *Nucl. Phys.* B **643**, 431 (2002).
58. M. Beneke and T. Feldmann, *Phys. Lett.* B **553**, 267 (2003).
59. M. Beneke and T. Feldmann, hep-ph/0308303.
60. R. J. Hill and M. Neubert, *Nucl. Phys.* B **657**, 229 (2003); T. Becher, R. J. Hill and M. Neubert, hep-ph/0308122; T. Becher, R. J. Hill, B. O. Lange and M. Neubert, hep-ph/0309227.
61. A. Hardmeier, E. Lunghi, D. Pirjol and D. Wyler, hep-ph/0307171.
62. G. Burdman, *Phys. Rev.* D **57**, 4254 (1998).
63. J. Charles *et al.*, *Phys. Rev.* D **60**, 014001 (1999).
64. A. Ali *et al.*, *Phys. Rev.* D **61**, 074024 (2000).
65. G. Buchalla, G. Hiller and G. Isidori, *Phys. Rev.* D **63**, 014015 (2001).
66. G. P. Korchemsky, D. Pirjol and T. M. Yan, *Phys. Rev.* D **61**, 114510 (2000).
67. S. Descotes-Genon and C. T. Sachrajda, *Nucl. Phys.* B **650**, 356 (2003).
68. E. Lunghi, D. Pirjol and D. Wyler, *Nucl. Phys.* B **649**, 349 (2003).
69. S. W. Bosch, R. J. Hill, B. O. Lange and M. Neubert, *Phys. Rev.* D **67**, 094014 (2003).
70. A. G. Grozin and M. Neubert, *Phys. Rev.* D **55**, 272 (1997).
71. H. Kawamura *et al.*, *Phys. Lett.* B **523**, 111 (2001) [Erratum-ibid. B **536**, 344 (2002)]; H. Kawamura *et al.*, *Mod. Phys. Lett.* A **18**, 799 (2003).
72. B. O. Lange and M. Neubert, *Phys. Rev. Lett.* **91**, 102001 (2003).
73. V. M. Braun, D. Y. Ivanov and G. P. Korchemsky, hep-ph/0309330.

DISCUSSION

Brendan Casey (Brown University): Does the range in predictions for $\bar{B}^0 \to D_s^+ K^-$ of $(1-5) \times 10^{-5}$ correspond to the 1σ contours or to the 5σ contours in the P/T predictions?

Gerhard Buchalla: The default model estimate for the annihilation term gives 1.2×10^{-5} for the branching ratio of $\bar{B}^0 \to D_s^+ K^-$. Allowing for a 100% uncertainty in the default value gives the upper limit of 5×10^{-5}. This corresponds to the inner (solid line) of the three error contours shown in the plot of the P/T prediction (see Fig. 4).

Harry Lipkin (Weizmann Institute): Do you have anything to say about the B decays to the new charmed-strange axial and scalar mesons that have been observed? When I predicted last year a large B decay to the D_s^* axial vector, I was told by HQET experts that this decay would be small.

Gerhard Buchalla: The D_s^* emitted in B decay is a heavy-light meson and therefore represents an extended hadronic object, in contrast to a pion or a similar energetic light meson. The usual factorization formulas do not apply to this situation and it is thus difficult to control QCD uncertainties in the predictions.

Ikaros Bigi (Notre Dame University): When you consider $B \to VV$, like $B \to \rho\rho$, and calculate the polarization of V, there are corrections of order $1/m_b$. These are sensitive to long-distance dynamics, right?

Gerhard Buchalla: That is correct.

HEAVY QUARKONIUM

T. SKWARNICKI

Department of Physics, Syracuse University, NY 13244, USA
E-mail: tomasz@physics.syr.edu

I review heavy quarkonium physics in view of recent experimental results. In particular, I discuss new results on spin singlet states, photon and hadronic transitions, $D-$states and discovery of the yet unexplained narrow $X(3872)$ state.

1. Quarkonia

Quarkonium is a bound state of a quark and its antiquark. Some properties of light and heavy quarkonia are compared to properties of positronium in Table 1. Unlike positronium, light quarkonia are highly relativistic. They also contain mixtures of quarks of different flavor and fall apart easily into other mesons. Charmonium ($c\bar{c}$) was the first heavy quarkonium discovered and is less relativistic; the number of long-lived states below the dissociation energy (*i.e.* the threshold for decay to $D\bar{D}$ meson pairs) equals the number of long-lived positronium states. Bottomonium ($b\bar{b}$) is even more non-relativistic and has a larger number of long-lived states. The toponium system would have been completely non-relativistic. However, weak decays of the top quark will dominate over the strong binding and long-lived states will not be formed. Therefore, charmonium and bottomonium play a special role in probing strong interactions.

The states below open flavor threshold live long enough for electromagnetic transitions between various excitations to occur. The electromagnetic transitions compete with transitions mediated by the emission of soft gluons. The latter materialize as light hadrons. Eventually the heavy quarks must annihilate into two or three hard gluons. Properties of these bound states and their decays are good testing grounds for QCD in both the non-perturbative and perturbative regimes.

The first heavy quarkonium bound state above the $D\bar{D}$ or $B\bar{B}$ threshold acts as a factory of heavy-light mesons. The heavy quarks trapped in these mesons ultimately decay via weak interactions. This is a good place to look for physics beyond the standard model as discussed by Y. Grossman at this Symposium.[1] Many measurements of electroweak parameters are obscured by strong interactions. This provides an important motivation for trying to understand details of strong interaction phenomena. P. Lepage pointed out that at least some of the new interactions to be discovered are likely to be strongly coupled, further motivating detailed studies of QCD.[2]

Heavy quarkonia offer two small parameters: velocities (v) of constituent quarks and the strong coupling constant (α_s) in annihilation and production processes. The expansion of the full theory in these parameters allows for effective theories of strong interactions: in the past - purely phenomenological potential models; more recently - NRQCD.[3] Lattice QCD calculations are also easier for heavy quarkonia than for light hadrons.

2. Hadroproduction

Annihilation of n^3S_{1--} (ψ, Υ) states to $\mu^+\mu^-$ and to a lesser extent e^+e^- makes it possible for experimentalists to fish out these states from large backgrounds in hadroproduction experiments. This is how these states were (co)discovered! There is also some access to the production of the excited states by addition of a transition photon or $\pi^+\pi^-$ pair. So far, except for the initial discovery, the hadroproduction experiments have not played an important role in spectroscopy or decay studies, but have made interesting production measurements. Heavy quarkonium is also a useful probe for determining the structure of the target (e.g. to probe for gluon content or the presence of quark-gluon plasma).

Non-relativistic QCD (NRQCD) is a favorite theoretical approach used to describe the production data.[4] Various diagrams are ordered according to powers of α_s and v. The diagrams can also be classified into color-singlet processes, in which the

Table 1. Some properties of various "onia".

| FORCES | | System | Ground triplet state 1^3S_1 | | | $\left(\frac{v}{c}\right)^2$ | Number of states below dissociation energy | |
binding	decay		Name	Γ (MeV)	Mass (GeV)		n^3S_1	all
POSITRONIUM								
EM	EM	e^+e^-	Ortho-	5×10^{-15}	0.001	~ 0.0	2	8
QUARKONIUM								
Strong	Strong	$u\bar{u}, d\bar{d}$	ρ	150.00	0.8	~ 1.0	0	0
		$s\bar{s}$	ϕ	4.40	1.0	~ 0.8	"1"	"2"
	EM	$c\bar{c}$	ψ	0.09	3.1	~ 0.25	2	8
		$b\bar{b}$	Υ	0.05	9.5	~ 0.08	3	30
	weak	$t\bar{t}$		(3000.0)	(360.)	< 0.01	0	0

heavy $Q\bar{Q}$ pair is produced colorless and therefore can directly form a bound state, and color-octet processes, in which the $Q\bar{Q}$ pair has color which must be emitted away by soft gluon (i.e. long distance) interactions. Initial attempts to describe charmonium production at the Tevatron with only color-singlet processes (CS model)[5] failed spectacularly, revealing the importance of color-octet diagrams. However, the color-evaporation model (CEM),[6] which allows color-octet processes but neglects their ranking in v, also fits the data well. All theoretical approaches involve free parameters, which are fixed by fits to the data, diminishing their effective differences.

Polarization of produced heavy quarkonia should lead to a better discrimination between NRQCD and CEM. NRQCD predicts an increase in polarization with higher p_t, while CEM predicts no polarization at all. The present data do not provide any evidence for increase in polarization with p_t, however the experimental errors are too large to draw any firm conclusions.

There is a large range of kinematical regimes and differential cross sections for charmonium production studies at HERA. Both H1 and ZEUS contributed a number of papers on this topic to this Symposium.[7] NRQCD predicts that matrix elements should be universal. However, it is difficult to reconcile all Tevatron and HERA data with a consistent set of NRQCD matrix elements. It is quite possible that the charm quark is just not heavy enough for the NRQCD approximation to work well. Unfortunately, data on bottomonium production are either non-existent or have large experimental errors.

For a more complete review of this topic see e.g. Krämer.[4]

3. Clean Production Environments

Most of what we know about heavy quarkonium states and their decays comes from experiments at clean production environments, which are time reversals of simple decay modes (see Fig. 1).

Vector states ($J^{PC} = 1^{--}$: n^3S_1, n^3D_1) decay to lepton pairs and thus can be directly formed in e^+e^- collisions. Production rates are large compared to the other e^+e^- processes, thus the backgrounds are small and these states can be studied in both inclusive and exclusive decay modes. Dedicated runs are needed for each vector state. The other states can be reached via photon and hadronic transitions, however, their scope is limited by the transition selection rules and branching ratios.

Spin 0 and 2 states with positive C-parity ($J^{PC} = 0^{-+}, 0^{++}, 2^{++}$: n^1S_0, $n^3P_{0,2}$) decay to two-photons, thus can be formed in two-photon collisions at the e^+e^- colliders. No dedicated runs are needed since the two-photon flux populates a wide range of possible quarkonia masses. However, the flux drops quickly with the energy and production rates for heavy quarkonia are small compared to the dominant e^+e^- processes. To combat these large backgrounds the quarkonia states must be detected in simple exclusive decay modes. So far this formation method has succeeded only for charmonium states.

States of any quantum numbers can decay to $p\bar{p}$ via two- or three-gluon annihilation, however, such couplings are small. Low energy $\bar{p}$ beams annihilating with proton-jet targets were successfully used to form charmonium states. The beam energy must be tuned to form a specific quarkonium state. Since the background cross-sections are very large, the quarko-

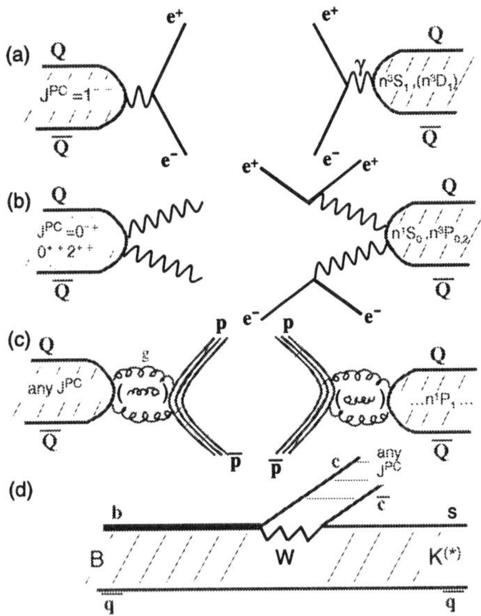

Figure 1. Clean production environments for heavy quarkonia together with corresponding decay modes for most of them.

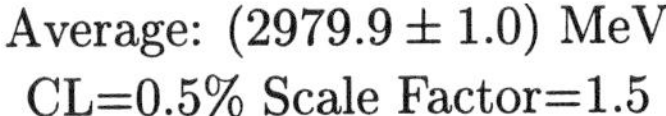
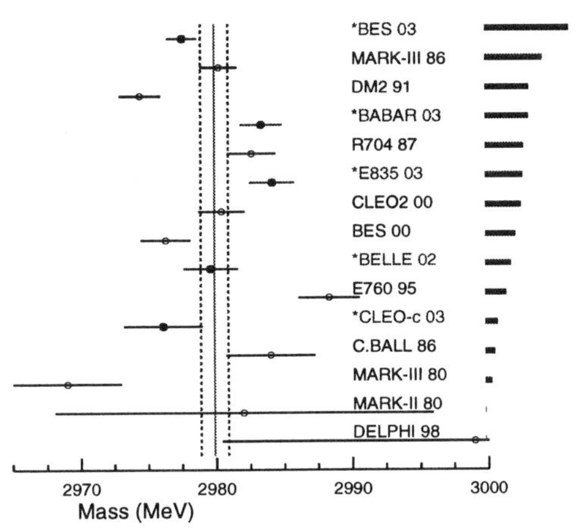

Figure 2. Measurements of the η_c mass. The new measurements (BES 03,[21] BABAR 03,[22] E835 03,[13] BELLE 02[23] and CLEO-c 03[24]) are indicated by a star. See PDG 2002[20] for the references to the older measurements displayed here. The vertical bars indicate the average value (solid) and its error (dashed). The error includes the scale factor. The thick horizontal bars on the right give the relative weight of each experiment into the average value.

nia must be detected in simple exclusive final states to take advantage of the highly constrained kinematics.

Decays of B mesons also offer a clean production environment for charmonium states. Again states of any quantum number can be formed. Since the production rates are rather small, experimentalists must restrict themselves to specific exclusive final states to take advantage of the B mass constraint (and beam energy constraint if produced at the $\Upsilon(4S)$ resonance in e^+e^- collisions).

4. Results from $\bar{p}p$ Experiments

A H_2 gas jet is used as a target for an antiproton beam in the $\bar{p}$ accumulator ring. This technique was pioneered at the CERN-ISR by the R704 experiment (1983-84).[8] It was later used at Fermilab with the E760 detector (1989-91)[9] followed by its major upgrade E835 (1996-97, 2000). Since this was a non-magnetic detector, it was limited to specific final states containing electrons and photons. Charmonium states are observed directly as resonance peaks in the signal event yield measured as function of the center-of-mass energy. Masses and widths of the observed states can be measured with high precision, since the accuracy depends on knowledge of the beam energy rather than resolution of the detector. The Fermilab experiments determined masses and widths

of the χ_{cJ} states with unprecedented accuracy.[9,10] They also studied radiative transitions from these states.[11] The most recent results include detection of the χ_{c0} resonance in decays to $\pi^0\pi^0$, interfering with continuum production of this final state[12] and observation of the singlet $\eta_c(1^1S_0)$ state in annihilation to two photons.[13] The radial excitation of the latter was not found in scans by E760 and E835.[14] Only the region around the mass reported by the Crystal Ball experiment[15] was scanned.

While the singlet $\eta_c(1^1S_0)$ was observed in many different environments, the only hints for the singlet $h_c(1^1P_1)$ state come from the $\bar{p}p$ experiments. R704 provided some inconclusive evidence for this state in 1984.[16] Better evidence for this state, with the mass close to the center-of-gravity mass of the $\chi_c(1^3P_J)$ states, was reported by E760[17] via the decay chain: $h_c \to \pi^0 J/\psi$, $J/\psi \to e^+e^-$. E835 took more scans in this mass range, with larger statistics and an improved detector. Recently, rumors of absence of the h_c state in the E835 data was reported in print by several non-E835 authors.[18] However, the official word from the E835 collaboration[19] is that they are analyzing all available decay channels and are not ready to make any definite statements yet.

5. Charmonium Singlet States

There are many new measurements related to the charmonium singlet states, in addition to the results from the $\bar{p}p$ experiments mentioned in the previous section. For example, there are 5 new measurements of the $\eta_c(1^1S_0)$ mass[13,21–24] (see Fig. 2). BES-II,[21] which reconstructed the η_c in 5 different decay modes in the J/ψ data, achieved the smallest overall error. Consistency of the old and new measurements is questionable. A new preliminary measurement of the η_c width by BABAR[22] constitutes an even bigger experimental puzzle (see Figs. 3-4). The measurement error claimed by BABAR is much smaller than in any other measurement. At the same time the central value, $(33.3 \pm 2.5 \pm 0.5)$ MeV, is the highest ever measured; completely inconsistent with the previous world average value,[20] $(16.0^{+3.6}_{-3.2})$ MeV.

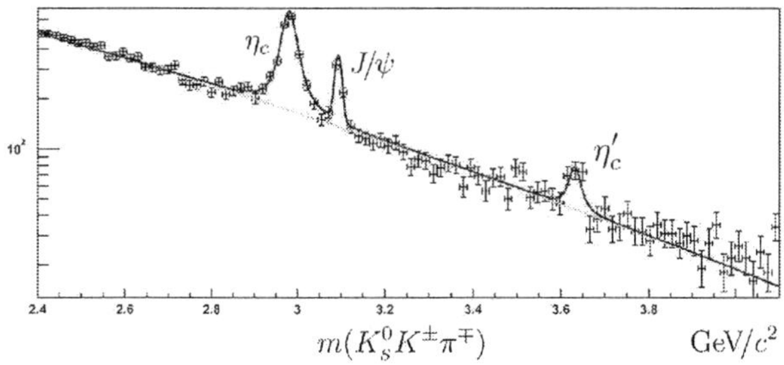

Figure 3. Observation of $\gamma\gamma \to \eta_c(1S)$ and $\gamma\gamma \to \eta_c(2S)$ by the BABAR experiment.[22] A peak due to $e^+e^- \to \gamma J/\psi$ is also visible. These data serve as a precision determination of the $\eta_c(1S)$ width and confirmation of the $\eta_c(2S)$ state at its new mass value.

There have also been dramatic developments concerning the $\eta_c(2^1S_0)$ state. The Crystal Ball experiment at SPEAR claimed observation of this state over 20 years ago.[15] They observed a peak in the inclusive photon energy spectrum from $\psi(2^3S_1)$ decays, which they interpreted as a magnetic dipole photon transition, $\psi(2S) \to \gamma\eta_c(2S)$.[15] The Crystal Ball result corresponded to the hyperfine mass splitting in the radial excitation, which was only slightly smaller than in the ground state (92 MeV vs. 117 MeV). Last year BELLE observed both η_c states in $B \to K\eta_c(nS)$, $\eta_c(nS) \to K_sK^+\pi^-$.[25] The $\eta_c(2S)$ state appeared at much higher mass than in the Crystal Ball measurement, thus reducing the corresponding hyperfine splitting. BELLE also observed both η_c states in the spectrum of the mass recoiling against J/ψ in continuum e^+e^- annihilation to

$J/\psi X$.[26] These results have been updated this year with larger statistics.[27] The $\eta_c(2S)$ mass obtained from these data differs by 1.9 standard deviations from the other BELLE result, but is still significantly higher than the Crystal Ball value.

Average: (25.0 ± 3.3) MeV
CL=0.05% Scale Factor=1.8

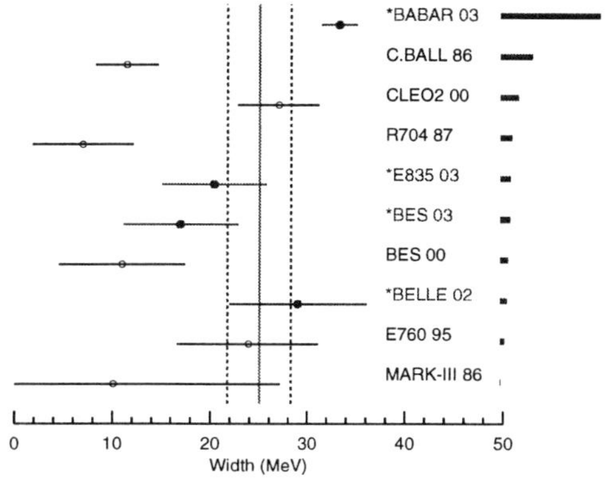

Figure 4. Measurements of the η_c width. See Fig. 2 for references and explanation.

The production rate for charmonium pairs observed by BELLE in the latter analysis is surprisingly high. They also observe a large e^+e^- annihilation rate to J/ψ plus open charm. They claim $(82 \pm 15 \pm 14)\%$ of prompt J/ψ production ($P_{J/\psi} > 2$ GeV) at 10.6 GeV center-of-mass energy is associated with another $c\bar{c}$ pair in the event.[27] This rate is an order of magnitude higher than that predicted theoretically, and constitutes the biggest puzzle in quarkonium production physics.[28]

This year the BABAR and CLEO experiments confirmed the $\eta_c(2S)$ at its heavier mass by formation in the two-photon collision process[22,29] (see Fig. 3). The discrepancy with the Crystal Ball mass measurement is resolved by the CLEO-c experiment, which has been able to remeasure the inclusive photon spectrum from the $\psi(2S)$ for the first time since the Crystal Ball experiment (see the next section). While CLEO-c agrees well with the Crystal Ball on all other E1 and M1 transitions from the $\psi(2S)$, the direct M1 transition to the $\eta_c(2S)$ is not observed at the photon energy claimed by the Crystal Ball. The upper limit on the rate for this transition set by CLEO-c disagrees with the rate measured by the Crystal Ball. Therefore, the peak observed by Crystal Ball could not be due to $\eta_c(2S)$ production.

Measurements of the $\eta_c(2S)$ mass are summarized graphically in Fig. 5. All new mass measurements are consistent with each other. The total width of this state, based on the BELLE[25] and BABAR[22] results, is (19 ± 10) MeV.

The new mass of the $\eta_c(2S)$ state decreases the hyperfine mass splitting for the $2S$-states by a factor of 2 compared to the old value. Since a wrong value prevailed for 20 years, it is interesting to check for an experimental bias on the phenomenological predictions for this splitting. A sample of potential model predictions for the ratio of the hyperfine mass splitting, $R_{HF} \equiv (M_{\psi(2S)} - M_{\eta_c(2S)})/(M_{\psi(1S)} - M_{\eta_c(1S)})$, is compared to the old $(R = 0.79)$ and the new $(R_{HF} = 0.412 \pm 0.028)$ experimental values in Fig. 6. It appears that many old calculations were stretched to accommodate the old result, though there were a few that had the courage to contradict the data. The ratio of hyperfine mass splitting can be related to the ratio of the leptonic widths of the triplet states (Γ_{ee}) using perturbative QCD. The new value agrees well with the prediction by Badalian and Bakker,[31] $R_{HF} = (0.48 \pm 0.07)$, based on this approach. Instead of using the measured values of Γ_{ee}, Recksiegel and Sumino[32] extended the use of perturbative QCD to the extraction of the interquark potential at short distances relevant for the spin-spin forces. Their prediction for the hyperfine mass splitting ratio, $R_{HF} = 0.42$, is in good agreement with the data, but the absolute values for the mass splittings are underestimated. Lattice QCD calculations[33] predicted $R_{HF} = 0.5$, not far from the measured value.

6. Radiative Transitions

Since sizes of heavy quarkonia are small or comparable to the wavelengths of transition photons, selection rules are given by multipole expansion. As the system is non-relativistic, electric dipole transitions ($\Delta L = 1$, $\Delta S = 0$) are much stronger than magnetic dipole transitions ($\Delta L = 0$, $\Delta S = 1$). The latest improvements in measurements of these transitions come mostly from the CLEO experiment, which is equipped with an excellent CsI(Tl) calorimeter. Recently, after increasing statistics of the $\Upsilon(1,2,3S)$ data samples by an order of magnitude, CLEO turned the beam energy down and collected also $\psi(2S)$ data. This was the beginning

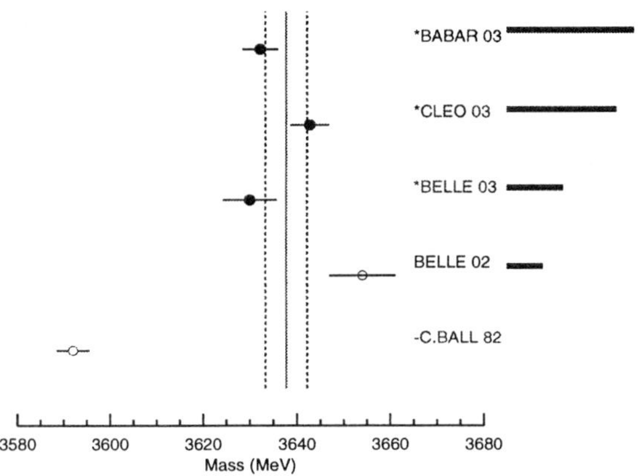

Figure 5. Measurements of the $\eta_c(2S)$ mass (see the text for the references). The thick horizontal bars on the right give the relative weight of each experiment into the average value. The Crystal Ball measurement was excluded from the average.

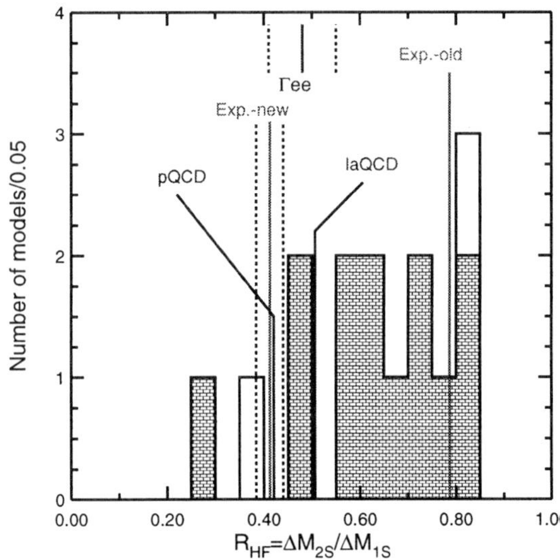

Figure 6. Measurements and predictions for the ratio of the hyperfine mass splittings in the $2S$ and $1S$ states in charmonium. The old and new experimental values are indicated by vertical bars. The dashed lines around the new value show the experimental error. The histogram shows predictions of a sample of potential models.[30] Only models which gave a good fit to the masses of all known charmonium states were included. The shaded part of the histogram represents the models published before the new measurements of the $\eta_c(2S)$ mass. Values predicted by scaling from the measured leptonic widths ratio[31] (Γ_{ee}), by perturbative QCD alone[32] (pQCD) and by lattice QCD[33] (laQCD) are also indicated.

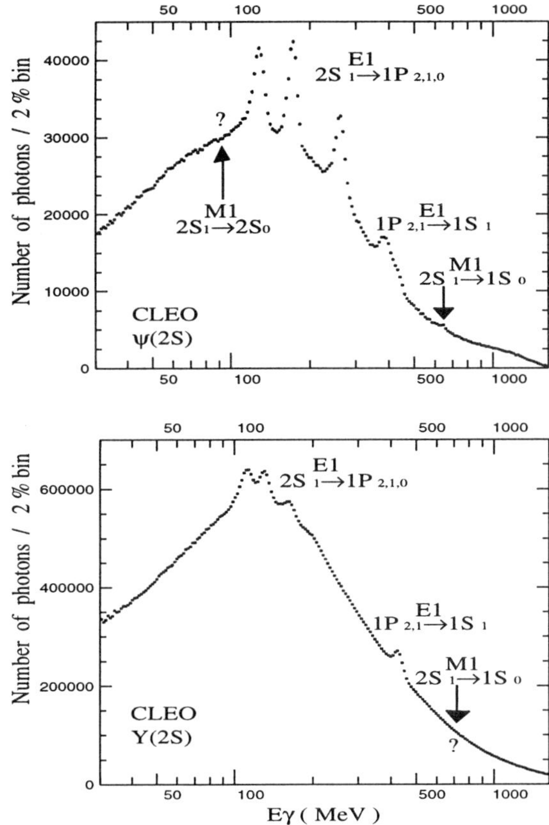

Figure 7. Inclusive photon spectrum from 2^3S_1 decays in the $c\bar{c}$ (top) and $b\bar{b}$ (bottom) systems measured with the CLEO detector. The data correspond to about 1.5M $\psi(2S)$ and 9M $\Upsilon(2S)$ decays.

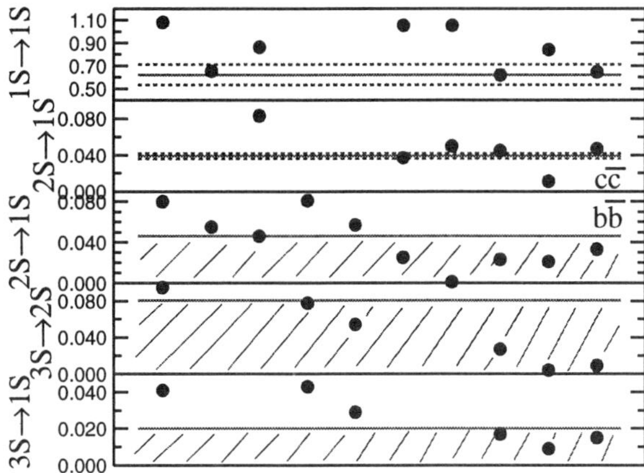

Figure 8. Measured and predicted values of matrix elements for M1 transitions in heavy quarkonia. The measured values are calculated from the world average branching ratios and total widths of the triplet S states. The central value and error bars for the $c\bar{c}$ data are indicated with solid and dashed lines respectively. For the $b\bar{b}$ data, allowed ranges from the preliminary CLEO analysis[45] are shaded (90% CL). The theoretical predictions[35] (points) are ordered according to the publication date.

of the CLEO-c phase. Inclusive photon spectra collected at $\psi(2S)$ and $\Upsilon(2S)$ by CLEO are compared in Fig. 7. The dominant peaks are due to E1 transitions, $2^3S_1 \rightarrow \gamma 1^3P_J$. The peaks in bottomonium system appear smaller because of the increased $\pi^0 \rightarrow \gamma\gamma$ background induced by higher particle multiplicities. The peaks are more crowded together reflecting the smaller fine structure of the 1^3P_J states in the more non-relativistic Υ system. The E1 cascade lines, $1^3P_J \rightarrow \gamma 1^3S_1$ are also visible. The charmonium data also reveal a small peak due to the hindered M1 transition, $2^3S_1 \rightarrow \gamma 1^1S_0$. This is the first confirmation of this transition since the original observation by the Crystal Ball. As discussed in the previous section, the direct M1 transition, $2^3S_1 \rightarrow \gamma 2^1S_0$, is ruled out for the photon energy, $\eta_c(2S)$ width and rate claimed by the Crystal Ball. The preliminary CLEO-c results[45] for the $\psi(2S)$ photon lines are: $\mathcal{B}(\psi(2^3S_1) \rightarrow \gamma\chi_c(1^3P_J)) = \{9.75 \pm 0.14 \pm 1.17, 9.64 \pm 0.11 \pm 0.69, 9.83 \pm 0.13 \pm 0.87\}\%$ for

$J = \{2, 1, 0\}$, $\mathcal{B}(\psi(2^3S_1) \rightarrow \gamma\eta_c(1^1S_0)) = (0.278 \pm 0.033 \pm 0.049)\%$ and $\mathcal{B}(\psi(2^3S_1) \rightarrow \gamma\eta_c(2^1S_0)) < 0.2\%$ for $E_\gamma = (91 \pm 5)$ MeV and $\Gamma(\eta_c(2S)) = 8$ MeV (90% CL).

None of M1 transitions, hindered or direct, are observed for the bottomonium. This is not a big surprise, since the backgrounds are much higher and the expected branching ratios are smaller in the more non-relativistic $b\bar{b}$ system. Searches for the singlet η_b state in two-photon collisions at LEP has also yielded upper limits only,[34] thus no singlet states have been observed in bottomonium so far. The M1 rate measurements can be translated into values of the corresponding matrix elements, which are compared to theoretical predictions in Fig. 8. The matrix elements for the hindered M1 transitions are expected to be very small, since they are generated by relativistic and finite size corrections. Therefore, they are difficult to predict. Only very recent calculations are consistent with all charmonium and bottomonium data. Similar comparisons for the E1 matrix elements is shown in Fig. 9. Non-relativistic calculations (hollow circles) overestimate the E1 rates in charmonium. The predictions with relativistic corrections (filled triangles) can reproduce the data. Relativistic effects in the dominant E1 transitions in bottomonium are much smaller. However, relativistic calculations are needed to reproduce the rare

$3^3S_1 \to \gamma 1^3 P_J$ transition rate. This matrix element is small because of the cancellations in the integral for the E1 operator between the initial and the final state wave functions[39] for which the number of nodes differ by two.

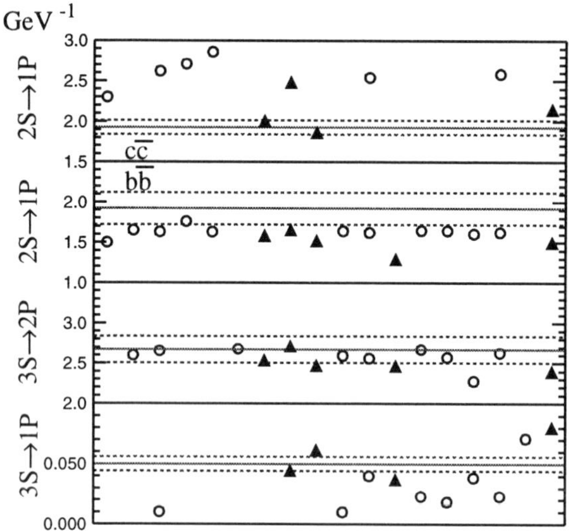

Figure 9. Measured and predicted values of matrix elements for E1 transitions in heavy quarkonia averaged over different spins of the triplet P states. The measured values are calculated from the world average radiative branching ratios and total widths of the triplet S states. The central values and error bars for the measured values are indicated with solid and dashed lines respectively. Circles (triangles) show non-relativistic (relativistic) calculations. The relativistic calculations are averaged over spins with the same weights as the data. The predictions[36] are ordered according to the publication date.

7. Hadronic Transitions

Heavy quarkonia can also change excitation levels by emission of soft gluons, that turn into light hadrons. The multipole expansion approach has proven to be useful also for hadronic transitions[37,38] explaining their gross characteristics. The most prominent are $\pi\pi$ transitions among n^3S_1 states, which can be mediated by two-gluon emission of the E1·E1 type. In fact, these are dominant decays for both $\psi(2S)$ and $\Upsilon(2S)$. The ratio of the measured widths for the these transitions agrees with a suppression by about a factor of 10 predicted by the multipole expansion model[38] due to the smaller size of the $b\bar{b}$ system. The multipole expansion model is also able to explain the $M(\pi\pi)$ distributions for these transitions, which follow the same pattern and peak at high values. Dip-

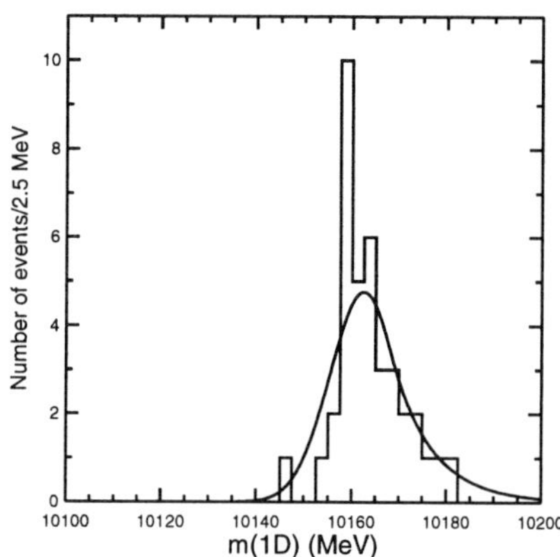

Figure 10. Recoil mass against the lowest energy photons in four-photon cascade between the $\Upsilon(3S)$ and $\Upsilon(1S)$, consistent with $\chi_b(2P_{2,1})$ and $\chi_b(1P_{2,1})$ as intermediate states, as observed by CLEO. The cascades via the $\Upsilon(2S)$ state are suppressed, thus the remaining events are $\Upsilon(1D)$ candidates. The recoil mass is a measure of the $\Upsilon(1D)$ mass. The fit to the data (solid line) assumes production of just one $\Upsilon(1D)$ state, with a natural width much smaller than the experimental resolution. The observed events are most likely dominated by the production of the $\Upsilon(1D_2)$ state. The preliminary CLEO results give $(10161.1 \pm 0.6 \pm 1.6)$ MeV for the mass of this state.

ion transitions from $\Upsilon(3S)$ to $\Upsilon(2S)$ and $\Upsilon(1S)$ are observed too. They have smaller rates either due to the phase space suppressions or cancellations in the dipole matrix element integral for the $3S \to \pi\pi 1S$ transition. The $M(\pi\pi)$ distribution peaks at low and high mass values for the latter transition, which reveals some dynamics beyond the multipole expansion approach. In charmonium, an η transition has been observed between the triplet $2S$ and $1S$ states.[20] This transition is of a magnetic type (E1·M2), thus it has a smaller rate than the $\pi\pi$ transition. A π^0 transition is also observed, but with a tiny rate due to isospin violation. None of these transitions are observed between the Υ states,[45] which is not surprising since there is an additional suppression by at least a factor $1/m_Q^2$ for this type of transition.

The transitions among triplet nS states discussed above were first observed over 20 years ago. A large number of other hadronic transitions are possible,[39] especially for the $b\bar{b}$ system which has a large number of long-lived states. Such transitions had not been observed until recently, when CLEO reported the first observation of $\chi_b(2^3P_{2,1}) \to \omega\Upsilon(1S)$ transitions.[40] The phase-space for these transitions is so small, that this decay is forbidden for the $\chi_b(2^3P_0)$. The measured branching ratios for the

$\chi_b(2^3P_1)$ and $\chi_b(2^3P_2)$ states are of the order of a couple of percent ($(1.6 \pm 0.3 \pm 0.2)\%$ and $(1.1 \pm 0.3 \pm 0.1)\%$ respectively), in spite of the phase space suppression, which reveals that the underlying transition is quite strong. In fact, this transition is of chromo-electric type, E1·E1·E1 (three gluons are needed to generate it). Voloshin[41] pointed out that since the matrix element does not depend on the spin of the $2P_J$ state, transition branching ratios for $J = 1$ and $J = 2$ should be comparable (as the phase space factors approximately cancel the effect due to the smaller total width of the $J = 1$ state). The data are consistent with this prediction.

As hadronization probabilities are difficult to estimate, uncertainties in absolute rate predictions for hadronic transitions are usually very large. When resonances dominate the transition, there are often no theoretical predictions for the rate. For example, there are no rate estimates for the ω transition discussed above. Predictions for other yet unobserved types of transitions vary by orders of magnitude. The predictions based on a model developed by Yan[38] and Kuang[42] tend to be much larger than the predictions based on a different approach introduced by Voloshin, Zakharov, Novikov and Shifman.[43]

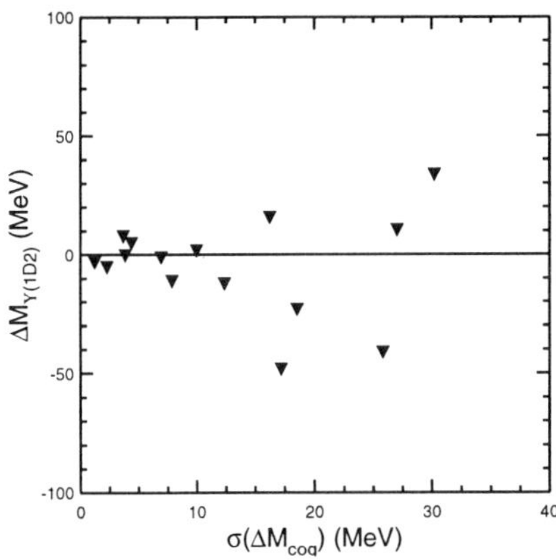

Figure 11. The difference between the predicted and the measured $\Upsilon(1^3D_2)$ mass (vertical axis) versus overall quality of the potential model expressed as the r.m.s. of the differences for the center-of-gravity masses of long-lived $b\bar{b}$ states (horizontal axis). Predictions for the 1S, 2S, 3S, 1P, 2P and 1D masses are included in the r.m.s. calculation for all models displayed here.[46]

One suitable place to test various predictions are dipion transitions from 1^3D_J to 1^3S_1. Last year the $\Upsilon(1^3D_J)$ states were discovered by CLEO in the four-photon cascade:[44] $\Upsilon(3^3S_1) \to \gamma\chi_b(2^3P_{J_2})$,

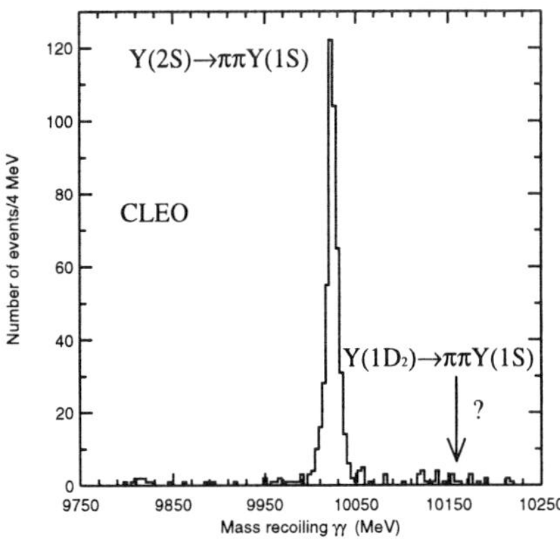

Figure 12. Search for $\Upsilon(1^3D_2) \to \pi^+\pi^-\Upsilon(1S)$ transitions. Recoil mass against two photons in $\gamma\gamma\pi^+\pi^-l^+l^-$ events is plotted. The expected signal position is indicated by the arrow.

$\chi_b(2^3P_{J_2}) \to \gamma\Upsilon(1^3D_{J_D})$, $\Upsilon(1^3D_{J_D}) \to \gamma\chi_b(1^3P_{J_1})$, $\chi_b(1^3P_{J_1}) \to \gamma\Upsilon(1^3S_1)$ followed by $\Upsilon(1^3S_1) \to l^+l^-$. CLEO has analyzed more data since then. The recoil mass distribution against the first two photons in the cascade is consistent with the detector resolution (Fig. 10). Thus the data are dominated by production of just one $\Upsilon(1^3D_{J_D})$ state. Theoretically, production of the $J_D = 2$ state is expected to dominate. This spin assignment is also favored by the experimental data.[44] The measured mass fits the predictions of the potential models well, especially those which provide a good fit to the other narrow $b\bar{b}$ states. This is illustrated in Fig. 11. The lattice QCD calculations also reproduce the observed mass very well.[2] CLEO determines[45] the four-photon cascade branching ratio to be $(2.6 \pm 0.5 \pm 0.5) \times 10^{-5}$. Within the errors this rate is consistent with the predictions by Godfrey and Rosner,[47] 3.8×10^{-5} (2.6×10^{-5} for $J_D = 2$ alone), based on the E1 matrix elements and estimates of hadronic widths of the $\Upsilon(1^3D_{J_D})$ and $\chi_b(2^3P_{J_2})$ states. CLEO also looked for $\Upsilon(1^3D_{J_D}) \to \pi^+\pi^-\Upsilon(1^3S_1)$ replacing the third and fourth photons in the cascade with a $\pi^+\pi^-$ pair. The recoil mass against the remaining two photons in the cascade is plotted in Fig. 12. Since no signal is found, upper limits on the product of branching ratios are set. They are compared in Table 2 to the theoretical predictions calculated by Rosner[48] using predictions for $\Upsilon(1^3D_{J_D}) \to \pi^+\pi^-\Upsilon(1^3S_1)$ width by various authors[42,49,50] together with the E1-photon matrix elements and estimates of hadronic widths of the $\Upsilon(1^3D_{J_D})$ and $\chi_b(2^3P_{J_2})$ states. Large

Table 2. Theoretical predictions and preliminary CLEO upper limits (at 90% CL) on $\mathcal{B}(\Upsilon(3^3S_1) \to \gamma\chi_b(2^3P_{J_2})) \, \mathcal{B}(\chi_b(2^3P_{J_2}) \to \gamma\Upsilon(1^3D_{J_D})) \, \mathcal{B}(\Upsilon(1^3D_{J_D}) \to \pi^+\pi^-\Upsilon(1^3S_1))$ in units of 10^{-4}. The theoretical predictions come from the paper by Rosner.[48] The first row corresponds to the cascade via the $J_D = 2$ state observed in the four-photon cascade by CLEO. The second row corresponds to any $\Upsilon(1^3D_{J_D})$ state with mass in the $10140 - 10180$ MeV interval.

	CLEO	$\Gamma_{\pi\pi}$ model		
		Kuang-Yan[42]	Moxhay[49]	Ko[50]
$\Upsilon(1^3D_2)$	< 1.1	9.2	0.049	0.39
$\Upsilon(1^3D_{J_D})$	< 2.7	17.7	0.094	0.75

$\Upsilon(1^3D_{J_D}) \to \pi^+\pi^-\Upsilon(1^3S_1)$ widths, as predicted by Kuang-Yan,[42] are ruled out. Better experimental sensitivity is needed to test the other models. Voloshin pointed out[51] that $\Upsilon(1^3D_{J_D}) \to \eta\Upsilon(1^3S_1)$ transition may be of comparable strength to the $\pi\pi$ transition between these states. CLEO doesn't find this transition either and sets a preliminary upper limit of $\mathcal{B}(\Upsilon(3^3S_1) \to \gamma\chi_b(2^3P_{J_2})) \, \mathcal{B}(\chi_b(2^3P_{J_2}) \to \gamma\Upsilon(1^3D_{J_D})) \, \mathcal{B}(\Upsilon(1^3D_{J_D}) \to \eta\Upsilon(1^3S_1)) < 2.3\times10^{-4}$ at 90 % CL.

There are also new results on $1^3D_1 \to \pi^+\pi^- 1^3S_1$ transitions in the charmonium system. Here, theoretical situation is complicated by mixing of the $\psi(1^3D_1)$ state with the $\psi(2^3S_1)$. The observed state, $\psi(3770)$, is also above the open flavor threshold, therefore it has a large width for decay to $D\bar{D}$ which suppresses branching ratios for any transitions to the other charmonium states. BES claims observation of such a transition in the data consisting of 5.7×10^4 $\psi(3770)$ decays.[52] Their reconstruction efficiency is 17%. They observe 9 signal candidates with an estimated background of 2.2 ± 0.4 events (see Fig. 13a). The corresponding branching ratio is $(0.59 \pm 0.26 \pm 0.16)\%$. Such a large branching ratio would favor the Kuang-Yan model[53] contrary to the $b\bar{b}$ results discussed above. CLEO-c has also analyzed their first $\psi(3770)$ sample.[45] They have a smaller sample of $\psi(3770)$ decays (4.5×10^4) but larger reconstruction efficiency (37%). They have only 2 events in the signal region, consistent with the estimated background level (see Fig. 13b). They set a preliminary upper limit $\mathcal{B}(\psi(3770) \to \pi^+\pi^- J/\psi) < 0.26\%$ (90% CL). This result is not inconsistent with the BES value because of the large experimental errors in both measurements. However, the CLEO result indicates that it is premature to favor the Kuang-Yan model on the basis of the $\psi(3770)$ data. BES

is already analyzing a larger data sample and CLEO is expected to increase their statistics by an order of magnitude this fall, thus more accurate results are expected next year.

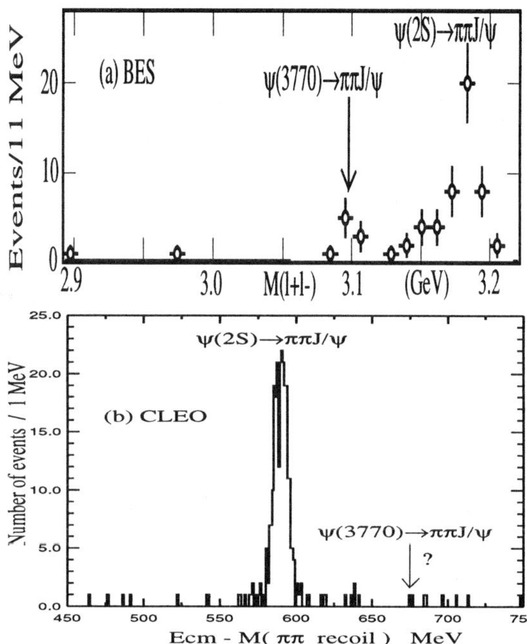

Figure 13. Searches for $\psi(3770) \to \pi^+\pi^- J/\psi$ transitions. The expected signal positions are indicated with arrows. The dominant peak in each distribution is due to $e^+e^- \to \gamma\psi(2S)$, $\psi(2S) \to \pi^+\pi^- J/\psi$. (a) BES results. $M(l^+l^-)$ mass for kinematically constrained $\pi^+\pi^- l^+l^-$ events is plotted. The signal is expected at the J/ψ mass. (b) CLEO results. The difference between the center-of-mass energy and the recoil mass against the $\pi^+\pi^-$ pair is plotted with no kinematic constraints, but after the cut on $M(l^+l^-)$ around the J/ψ mass. The signal is expected at the mass difference between the $\psi(3770)$ and the J/ψ.

8. New Particle Discovered by BELLE

Decays to $\pi^+\pi^- J/\psi$ are also a way to search for other charmonium states, for example $h_c(1^1P_1)$, $\psi(1^3D_2)$, $\psi(1^1D_2)$ etc.. Many of these states cannot be directly formed in e^+e^- annihilation, however they can be produced in $\bar{p}p$ annihilation, B decays or fragmentation processes. BELLE inspected resonance structures in the $\pi^+\pi^- J/\psi$ system produced in the decay $B^\pm \to K^\pm(\pi^+\pi^- J/\psi)$, $J/\psi \to l^+l^-$. The distribution of $\Delta M = M(\pi^+\pi^- J/\psi) - M(J/\psi)$ observed by BELLE[54] in a sample of 3×10^8 B mesons is shown in Fig. 14. The prominent peak is due to the $\psi(2^3S_1)$. There is also a smaller but very significant peak observed at a higher mass. By performing an unbinned maximum likelihood fit to ΔM, the beam-constraint B meson mass and energy difference between the B candidate and the beam energy BELLE finds 35.7 ± 6.8 events in the second peak with a statistical significance of 10.3 standard deviations. The mass determined from the ΔM peak position is $(3872.0 \pm 0.6 \pm 0.5)$ MeV. The observed width of the peak is consistent with the detector resolution, therefore the new state is long-lived ($\Gamma_{tot} < 2.3$ MeV at 90% CL).

The invariant mass of the $\pi^+\pi^-$ system for the signal events is strongly peaked at high mass values. The peaking is stronger than predicted by Yan[38] with the multipole expansion approach for $S \to S$ transitions, and much stronger than predicted for $D \to S$ transitions in the quarkonium system, as illustrated in Fig. 15. The observed dipion mass distribution is suggestive of the isospin violating $X(3872) \to \rho^0 J/\psi$ process instead. Isospin violation can be experimentally verified by measuring $\Gamma(X(3872) \to \pi^0\pi^0 J/\psi)/\Gamma(X(3872) \to \pi^+\pi^- J/\psi)$.[55,56] For isospin conserving $\pi\pi$ transitions, this ratio should be approximately $1/2$, whereas ρ^0 does not decay to $\pi^0\pi^0$.

Soon after BELLE's announcement at the Lepton-Photon Symposium, the $X(3872)$ particle was confirmed by CDF in $\pi^+\pi^- J/\psi$, $J/\psi \to \mu^+\mu^-$.[57] In 220 pb^{-1} of Run-II data they observe 704 ± 67 signal events. The preliminary mass measurement, $3871.4 \pm 0.7 \pm 0.4$ MeV, is consistent with BELLE's result. Their data also show peaking at high dipion mass.

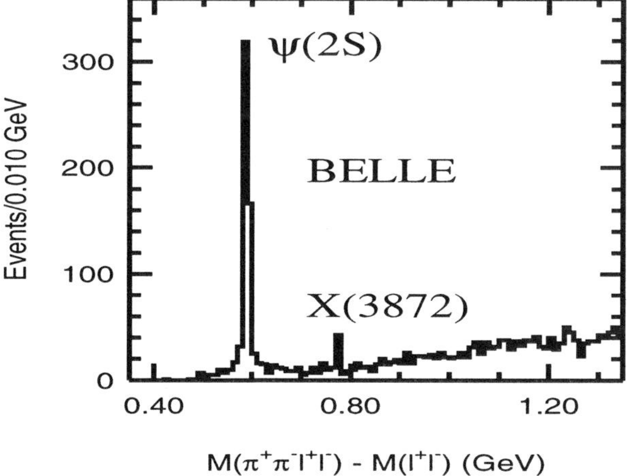

Figure 14. Distribution of $M(\pi^+\pi^- l^+ l^-) - M(l^+ l^-)$ as observed by BELLE for B decay candidates.[54] The first peak corresponds to the $\psi(2S)$ resonances. The second peak represents the $X(3872)$ particle.

Perhaps the most puzzling property of the newly discovered particle is its mass. Within the experimental errors it coincides with the sum of D^0 and D^{*0} masses: $m_X - m_{D^0} - m_{D^{*0}} = 0.2 \pm 0.9$ MeV (using the average of the BELLE and CDF results). One exciting interpretation, which actually predicts this to be the case, is a $D\bar{D}^* (+\bar{D}D^*)$ molecule. Constituent mesons would likely be in a relative S-wave creating a $J^P = 1^+$ state. "Molecular charmonium" was first discussed in the literature in the mid-seventies[58] and was initially introduced to explain the complicated coupling of the e^+e^- resonances above the open flavor threshold to various decay modes involving D and D^* meson pairs. A satisfactory description of the e^+e^- data was later achieved within a simple $c\bar{c}$ bound state model.[59,60] However, since meson molecules were also proposed in the context of light hadron spectroscopy,[61] the concept of molecular charmonium did not go away. Interactions in the $D\bar{D}^*$ system (also $B\bar{B}^*$) were found to be attractive when described by the pion-exchange force.[62,63] No such force would exist in the $D\bar{D}$ (or $B\bar{B}$) system. Quantitative estimates showed that the $D\bar{D}^*$ system could be only loosely bound if at all, with better prospects for the $B\bar{B}^*$ molecule. Tornqvist showed that in the limit of isospin symmetry only an isoscalar molecule would be bound.[62] Isospin is expected to be broken since the binding energy is comparable to the isospin mass splittings (the D^+D^{-*} threshold is 8 MeV above the $D^0\bar{D}^{0*}$ threshold). Close and Page[64] argued that the isospin breaking does not change the conclusion that there is only one molecular system expected near the $D\bar{D}^*$ threshold. The loose binding makes it difficult for the molecule to rearrange the quark content in the meson subcomponents, thus strong decays to a charmonium state plus light hadrons are expected to have small widths. Decays to $\pi^+\pi^- J/\psi$ could proceed via the isospin violating $\rho^0 J/\psi$ channel consistent with the BELLE's dipion mass distribution. The molecular interpretation predicts that decays to $D^0\bar{D}^0\pi^0$ and $D^0\bar{D}^0\gamma$ should occur, since the D^{0*} is almost on shell. The ratio of the widths for these decays should be approximately 3:2 and their sum 60-100 keV.[62,64,65,56] Therefore, the molecular model is consistent with the narrow width of the observed state.

Traditional charmonium states can also be narrow at 3872 MeV if they cannot decay to $D\bar{D}$. Possible candidates, ordered according to increas-

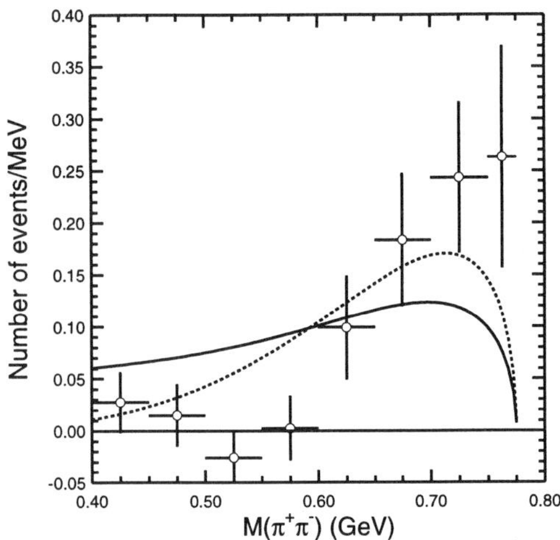

Figure 15. Dipion mass distribution in BELLE's data[66] for $X(3872) \to \pi^+\pi^- J/\psi$. The data are compared to the theoretical shapes predicted by Yan[38] for $D \to S$ (solid line) and $S \to S$ (dashed line) transitions in the quarkonium system. The distributions were normalized to the same number of events. The efficiency dependence on $M(\pi^+\pi^-)$ has been neglected in the comparison.

ing mass predictions are $\psi(1^3D_{2--})$, $\eta_{c2}(1^1D_{2-+})$, $\chi_c(2^3P_{1++})$, $h_c(2^1P_{1+-})$ and $\eta_c(3^1S_{0-+})$.

The lower excitations of the η_c are already broader than the upper limit on the natural width of the new state, thus we can discard the $\eta_c(3^1S_{0-+})$ possibility. The other listed states can satisfy this upper limit. For the remaining options, radiative E1 transitions to the other charmonium states should dominate over $\pi\pi J/\psi$. This is especially true for the singlet states, for which the dipion transition would involve chromomagnetic interactions (spin-flip). Since the $h_c(1^1P_{1+-})$ has not been seen in B decays, the $h_c(2^1P_{1+-})$ hypothesis is unlikely.

BELLE has also just observed $B^\pm \to K^\pm\psi(3770)$ ($\psi(3770) \to D\bar{D}$) for the first time.[67] The measured branching ratio, $(4.8\pm1.1\pm1.2)\times10^{-4}$, is comparable to[20] $\mathcal{B}(B^\pm \to K^\pm\psi(2S)) = (6.8 \pm 0.4) \times 10^{-4}$. Theoretically, $\mathcal{B}(B^\pm \to K^\pm\psi(1^3D_2))$ is expected to be 1.6 times larger than the production of $\psi(3770)$ (assumed to be $\psi(1^3D_1)$).[68] Together, with BELLE's result,[54] $\mathcal{B}(B^\pm \to K^\pm X)\mathcal{B}(X \to \pi^+\pi^- J/\psi) = (0.063 \pm 0.012 \pm 0.007) \times\mathcal{B}(B^\pm \to K^\pm\psi(2S))\mathcal{B}(\psi(2S) \to \pi^+\pi^- J/\psi)$, the $\psi(1^3D_2)$ interpretation requires $\mathcal{B}(\psi(1^3D_2) \to \pi^+\pi^- J/\psi)$ to be about 1 to 3%, much smaller than the naïve expectations[69,56] for the dominant radiative decay of this state, $\mathcal{B}(\psi(1^3D_2) \to \gamma\chi_c(1^3P_1)) > 50\%$. Also, as discussed in the previous section the $\Upsilon(1^3D_2)$

state was observed via its radiative decay, whereas only an upper limit on the decay to $\pi^+\pi^-\Upsilon$ exists. However, BELLE observes no evidence for the photon transitions to $\chi_c(1^3P_1)$ and sets the following 90% CL limit: $\mathcal{B}(X(3872) \to \gamma\chi_c(1^3P_1))/\mathcal{B}(X(3872) \to \pi^+\pi^- J/\psi(1^3S_1)) < 0.89$. Coupled channel effects can be big (proximity of the $D\bar{D}^*$ threshold!) and significantly alter the naïve predictions for the radiative widths.[70] Quantitative estimates are needed.

It will also be useful to resolve the experimental controversy about $\mathcal{B}(\psi(3770) \to \pi^+\pi^- J/\psi)$, since the width for $\psi(1^3D_2) \to \pi^+\pi^- J/\psi$ is related to the width for $\psi(1^3D_1) \to \pi^+\pi^- J/\psi$. The CLEO-c upper limit on the $\psi(3770) \to \pi^+\pi^- J/\psi$ branching ratio (see the previous section) suggests that the latter is not necessarily much larger than the value induced by the $2^3S_1 - 1^3D_1$ mixing, while the BES measurement indicates a rather large rate for direct $\pi\pi$ transitions between the charmonium triplet $1D$ and $1S$ states. The BES result implies[56] $\mathcal{B}(\psi(1^3D_2) \to \pi^+\pi^- J/\psi) \sim (20 - 40)\%$ which would make it easier to accommodate BELLE's results in the 1^3D_2 interpretation of the $X(3872)$. The observed dipion mass distribution does not fit the shape for $1D_2 \to 1S$ transitions predicted by the multipole expansion model (see Fig. 15). However, some dynamical effects beyond this model can alter the dipion distribution. Finally, the $\psi(1^3D_2)$ interpretation is disfavored by the mass predictions from potential model. This is illustrated in Fig. 16. All potential models but one predict the $\psi(1^3D_2)$ mass to be about 70 MeV lower then the $X(3872)$ mass. The model by Fulcher[71] predicts this mass right at the observed value, however it overestimates the $\psi(1^3D_1)$ mass by similar amount. In other words, none of the existing calculations can accommodate the $X(3872)$ and $\psi(3770)$ as spin 2 and 1 members of the 1^3D_J triplet. Coupled channel effects and $1^3D_1 - 2^3S_1$ mixing can increased the mass splitting relatively to the naïve potential model calculations. Quantitative estimates of these effects are needed.

The mass of the $\chi_c(2^3P_{1++})$ state is predicted by the potential models to be higher than the $X(3872)$ mass (see Fig. 16). Thus, significant coupled channel effects would need to be at work for this interpretation as well. Decays of $B^\pm$ to $K^\pm\chi_c(1^3P_1)$ are observed with a rate comparable to $K^\pm\psi(2S)$ and $K^\pm J/\psi(1S)$.[20] Therefore, decays to $K^\pm\chi_c(2^3P_1)$

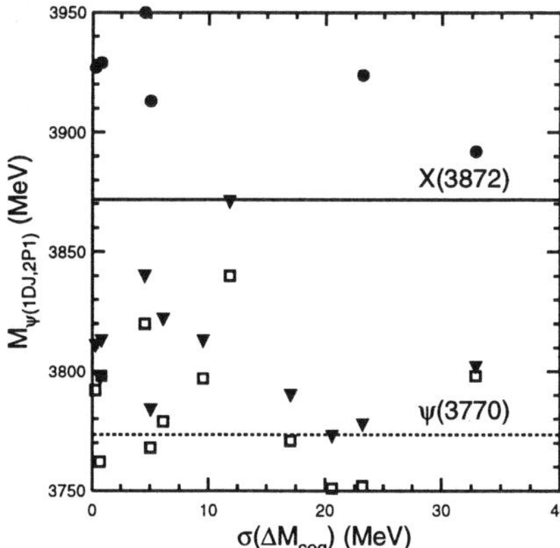

Figure 16. The predicted masses (vertical axis) of the $\psi(1^3D_1)$ (squares), $\psi(1^3D_2)$ (triangles) and $\chi_c(2^3P_1)$ (circles) states versus overall quality of the potential model expressed as RMS of the differences for the center-of-gravity masses of long-lived $c\bar{c}$ states (horizontal axis). The measured masses of $\psi(3770)$ and $X(3872)$ are indicated by horizontal bars. Predictions for the 1S,2S and 1P masses are included in the RMS calculation for all models displayed here.[72]

should also occur. If the $X(3872)$ is the $\chi_c(2^3P_1)$ state then the ratio $\mathcal{B}(X(3872) \to \gamma J/\psi(1^3S_1))/\mathcal{B}(X(3872) \to \pi^+\pi^- J/\psi(1^3S_1))$ should be large. The photon transition here is of an E1 type and is observed with a 21% branching ratio in the $b\bar{b}$ system. Barnes and Godfrey predict 3.5% for the $\chi_c(2^3P_1)$ state.[56] The $\chi_c(2^3P_1) \to \pi^+\pi^- J/\psi(1^3S_1)$ transition is of an isospin violating type and the branching ratio must be small even in the charmonium system. The BELLE experiment does not see evidence for significant $X(3872) \to \gamma J/\psi(1^3S_1)$ branching ratio. Quantitative results should be soon available.[73]

Finally, $X(3872)$ could be a $c\bar{c}g$ hybrid state,[64,56] though the masses of the hybrid states are predicted to be significantly higher than the observed mass.

The molecular, conventional charmonium or hybrid charmonium interpretations of the $X(3872)$ should not be viewed as mutually exclusive options. For example, the production rate for molecular charmonium in B decays is expected to be small. However, the production rate can be enhanced by mixing of the molecular system with conventional charmonium states[74,64,65] (e.g. with $\chi_c(2^3P_1)$ for the 1^{++} molecule). Mixing would influence the pattern of decay branching ratios as well.[74,64,56]

More experimental studies of $X(3872)$ production and decays are needed to clarify nature of this state.

9. Summary and Outlook

Heavy quarkonium physics has been recently experimentally revitalized. Large data samples collected in e^+e^- and $\bar{p}p$ annihilation by BES-II ($c\bar{c}$), CLEO-III ($b\bar{b}$) and E835 ($c\bar{c}$) are still being analyzed. The B-meson gateway to charmonium is now wide open with 3×10^8 B decays recorded by BELLE and BABAR. There has been similar progress in theory by the development of NRQCD and much improved calculations with QCD on the lattice.

More results for charmonium are expected in the future. In addition to more results from the BES-II experiment, the CLEO-c program will contribute greatly. The first CLEO-c results have been presented. Run-II data from CDF are proving to be useful as well. In the farther future, BTeV and LHCb are likely to contribute to quarkonium physics. The BES-III/BEPC-II project was recently approved in China. There is also a proposal for a new dedicated $\bar{p}p$ machine to explore charmonium physics (PANDA at GSI).

Prospects for more $b\bar{b}$ data are not as well defined. In principle, CLEO at CESR could go back to the high energy running and accumulate more statistics for the narrow Υ resonances. BELLE and BABAR could accumulate Υ data with even higher rates if their B-factories are ever utilized to produce these states.

The discovery of the $X(3872)$ particle by BELLE, and its likely compound nature, underlines the importance of the heavy quarkonium systems. Previously charmonium played a crucial role in the solidification of the naïve quark model. It will not be a big surprise if it provides the first convincing proof for the existence of hadronic systems beyond the naïve quark model.

10. Acknowledgments

I would like to thank many colleagues from BABAR, BELLE, BES, CDF, CLEO, E835, H1 and ZEUS for providing me with information contained in this report. I apologize for not being able to discuss all the recent experimental results because of the length limitations. I would also like to thank Steve Godfrey, Jonathan Rosner, Mikhail Voloshin and Tung-Mow Yan for discussions concerning theoretical aspects. Special thanks to Sheldon Stone for help in refining this manuscript.

References

1. Y. Grossman, "Beyond Standard Model sensitivity in K and B Physics", talk presented at Lepton Photon 2003.
2. P. Lepage, "Lattice Gauge Theory", talk presented at Lepton Photon 2003.
3. W. E. Caswell, G. P. Lepage, *Phys. Lett.* **B167**, 437 (1986); B. A. Thacker, G. P. Lepage, *Phys. Rev.* **D43**, 196 (1991); G. T. Bodwin, E. Braaten, G. P. Lepage, *Phys. Rev.* **D51**, 1125 (1995); *erratum* **D55**, 5853 (1997); N. Brambilla, A. Pineda, J. Soto, A. Vairo, *Nucl. Phys.* **B566**, 275 (2000).
4. For a recent review see: M. Krämer, *Prog. Part. Nucl. Phys.* **47**, 141 (2001).
5. E. L. Berger, D. Jones, *Phys. Rev.* **D23**, 1521 (1981); R. Baier, R. Rückl, *Phys. Lett.* **B102**, 364 (1981).
6. H. Fritzsch, *Phys. Lett.* **B67**, 217 (1977); F. Halzen, *Phys. Lett.* **B69**, 105 (1977); M. Glück, J. F. Owens, E. Reya, *Phys. Rev.* **D17**, 2324 (1978); G. A. Schuler, R. Vogt, *Phys. Lett.* **B387**, 181 (1996); J. F. Amundson, O. J. Eboli, E. M. Gregores, F. Halzen, *Phys. Lett.* **B390**, 323 (1997).
7. C. Adloff *et al.* (H1), papers contributed to Lepton-Photon 2003: "Diffractive Photoproduction of $\psi(2S)$ Mesons at HERA", ♯151; "Inelastic Leptoproduction of J/ψ Mesons at HERA", ♯156; "Inelastic Photoproduction of J/ψ Mesons at HERA", ♯157; "Diffractive Photoproduction of J/ψ Mesons with Large Momentum Transfer at HERA", ♯162; "Elastic Photoproduction of J/ψ Mesons", ♯188; S. Chekanov *et al.* (ZEUS), papers contributed to Lepton-Photon 2003: "Measurements of inelastic J/ψ and ψ' photoproduction at HERA", ♯91; "Measurement of proton-dissociative diffractive photoproduction of J/ψ mesons at HERA", ♯106; "Measurement of inelastic J/ψ production in deep inelastic scattering at HERA", ♯109; "Measurement of inelastic J/ψ helicity distribution at HERA", ♯111.
8. C. Baglin *et al.* (R704), *Phys. Lett.* **B172**, 455 (1986).
9. T. A. Armstrong *et al.* (E760), *Nucl. Phys.* **B373**, 35 (1992).
10. S. Bagnasco *et al.* (E835), *Phys. Lett.* **B533**, 237 (2002); N. Pastrone for the E835 at XXXVIIIth Recontres de Moriond, Les Arcs, Savoie, France, March 22-29, 2003, hep-ex/0306032.
11. M. Ambrogiani *et al.* (E835), *Phys. Rev.* **D65**, 052002 (2002); C. Patrignani at International Workshop on Heavy Quarkonium at Fermilab, Batavia, IL, USA, September 20-22, 2003.
12. M. Andreotti *et al.* (E835), *Phys. Rev. Lett.* **91**, 091801 (2003).
13. M. Ambrogiani *et al.* (E835), *Phys. Rev. Lett.* **B566**, 45 (2003).
14. M. Ambrogiani *et al.* (E835), *Phys. Rev.* **D64**, 052003 (2001).
15. C. Edwards *et al.* (Crystal Ball), *Phys. Rev. Lett.* **48**, 70 (1982).
16. C. Baglin *et al.* (R704), *Phys. Lett.* **B171**, 135 (1986).
17. T. A. Armstrong *et al.* (E760), *Phys. Rev. Lett.* **69**, 2337 (1992).
18. See e.g. A. Martin, J.-M. Richard CERN Cour. **43N3**, 17 (2003).
19. Rosanna Cester, private communication.
20. K. Hagiwara *et al.* (PDG), *Phys. Rev.* **D66**, 010001 (2002).
21. J. Z. Bai *et al.* (BES), *Phys. Lett.* **B555**, 174 (2003).
22. C. Wagner for BABAR at XXXVIIIth Recontres de Moriond, Les Arcs, Savoie, France, March 22-29, 2003, hep-ex/0305083; E. Robutti for BABAR at International Europhysics Conference on High Energy Physics EPS, Aachen, Germany, July 17-23, 2003.
23. F. Fang *et al.* (BELLE), *Phys. Rev. Lett.* **90**, 071801 (2003), hep-ex/0208047.
24. The preliminary CLEO-c result for the η_c mass, $(2976.1 \pm 2.4 \pm 3.3)$ MeV, comes from the photon energy measurement in $\psi' \to \gamma\eta_c$ (see the text).
25. S. K. Choi *et al.* (BELLE), *Phys. Rev. Lett.* **89**, 102001 (2002), Erratum-ibid. **89**, 129901 (2002).
26. K. Abe *et al.* (BELLE), *Phys. Rev. Lett.* **89**, 142001 (2002); K. Abe *et al.*, BELLE-PREPRINT-2003-7, KEK-PREPRINT-2003-24, hep-ex/0306015.
27. K. Abe *et al.* (BELLE), paper contributed to Lepton-Photon 2003, "A study of double $c\bar{c}$ production in e^+e^- annihilation at $\sqrt{s} \approx 10.6$ GeV", ♯274; T. Uglov and R. Seuster for BELLE collaboration, talks presented at International Europhysics Conference on High Energy Physics EPS, Aachen, Germany, July 17-23, 2003.
28. E. Braaten, J. Lee, *Phys. Rev.* **D67**, 054007 (2003); K.-Y. Liu, Z.-G. He, K.-T. Chao, *Phys. Lett.* **B557**, 45 (2003); G. T. Bodwin, J. Lee, E. Braaten, *Phys. Rev.* **D67**, 054023 (2003); *Phys. Rev. Lett.* **90**, 162001 (2003); A. B. Kaidalov, *JETP Lett.* **77**, 349 (2003), *Pisma Zh. Eksp. Teor. Fiz.* **77**, 417 (2003), hep-ph/0301246; A. V. Berezhnoy, A. K. Likhoded, hep-ph/0303145; K.-Y. Liu, Z.-G. He, K.-T. Chao, *Phys. Rev.* **D68**, 031501 (2003); K. Hagiwara, E. Kou, C.-F. Qiao, *Phys. Lett.* **B570**, 39 (2003); A. V. Luchinsky, hep-ph/0305253; S. J. Brodsky, A. S. Goldhaber, J. Lee, *Phys. Rev. Lett.* **91**, 112001 (2003); B. L. Ioffe, D. E. Kharzeev, hep-ph/0306062; S. Fleming, A. K. Leibovich, T. Mehen CMU-HEP-03-06, FERMILAB-PUB-03-069-T, hep-ph/0306139; C.-F. Qiao, J.-X. Wang, hep-ph/0308244.
29. J. Ernst *et al.* (CLEO), CLEO-CONF-03-05, hep-ex/0306060, paper contributed to Lepton-Photon 2003, "Observation of η_c' Production in Two-photon Fusion at CLEO", ♯37.
30. References are ordered according to the increasing value of R_{HF} given in square brackets. H. Ito, Prog. of *Theor. Phys.* **84**, 94 (1990) [0.29]; T. A. Lahde, *Nucl. Phys.* **A714**, 183 (2003) [0.37]; S. God-

frey, N.Isgur, *Phys. Rev.* **D32**, 189 (1985) [0.46]; S. N. Jena, *Phys. Lett.* **B123**, 445 (1983) [0.48]; J. Carlson, J. B. Kogut, V.R.Pandharipande, *Phys. Rev.* **D28**, 2807 (1983) [0.55]; T. A. Lahde, C. J. Nyfalt, D. O. Riska *Nucl. Phys.* **A645**, 587 (1999) [0.56]; S. N. Gupta, W. W. Repko, C. J. Suchyta III, *Phys. Rev.* **D39**, 974 (1989) [0.61]; D. Beavis, S.-Y. Chu, B. R. Desai, P. Kaus, *Phys. Rev.* **D20**, 2345 (1979) [0.62]; N. Barik, S. N. Jena, *Phys. Rev.* **D24**, 680 (1981) [0.70]; S. N. Gupta, S. F. Radford, W. W. Repko, *Phys. Rev.* **D26**, 3305 (1982), *Phys. Rev.* **D30**, 2424 (1984) [0.73]; L. P. Fulcher, *Phys. Rev.* **D44**, 2079 (1991) [0.74]; J. S. Kang, *Phys. Rev.* **D20**, 2978 (1979) [0.76]; J. Baacke, Y. Igarashi, G. Kasperidus *Z. Phys.* **C13**, 131 (1982) [0.80]; S. N. Gupta, S. F. Radford, W. W. Repko, *Phys. Rev.* **D34**, 201 (1986) [0.83]; D. Ebert, R. N. Faustov, and V. O. Galkin *Phys. Rev.* **D67**, 014027 (2003) [0.84].

31. A. M. Badalian, B. L. G. Bakker, *Phys. Rev.* **D67**, 071901 (2003).

32. S. Recksiegel, Y. Sumino, TU-691, TUM-HEP-508-03, hep-ph/0305178.

33. M. Okamoto *et al.* (CP-PACS Collaboration) *Phys. Rev.* **D65**, 094508 (2002).

34. A. Heister *et al.* (ALEPH), *Phys. Lett.* **B530**, 56 (2002); M. Chapkine *et al.* (DELPHI), paper contributed to Lepton-Photon 2003, "Search for η_b in two-photon collisions with the DELPHI detector", ♮62.

35. The following sample of potential model predictions for the M1 matrix elements (ordered according to the publication date) is displayed in Fig. 8: V. Zambetakis, N.Byers, *Phys. Rev.* **D28**, 2908 (1983); H. Grotch, D. A. Owen, K. J. Sebastian, *Phys. Rev.*, D30, 1924 (1984) (2 entries: scalar and vector confinement potential); S. Godfrey, N. Isgur, *Phys. Rev.* **D32**, 189 (1985) (2 entries: based on quoted transition moments and wave functions respectively); X. Zhang, K. J. Sebastian, H. Grotch, *Phys. Rev.* **D44**, 1606 (1991) (2 entries: scalar-vector and pure scalar confinement potential); D. Ebert, R. N. Faustov, V. O. Galkin *Phys. Rev.* **D67**, 014027 (2003); T. A. Lahde, *Nucl. Phys.* A714, 183(2003) (2 entries: without and with exchange currents).Values of the M1 matrix elements in the $b\bar{b}$ system are displayed for the photon energies and b quark mass assumed in S. Godfrey, J. L. Rosner, *Phys. Rev.* **D64**, 074011 (2001), Erratum-ibid. **D65**, 039901 (2002).

36. The following sample of potential model predictions for the E1 matrix elements (ordered according to the publication date) is displayed in Fig. 9: D. Pignon, C. A. Piketty, *Phys. Lett.* **74B**, 108 (1978); E. Eichten, K. Gottfried, T. Kinoshita, K. D. Lane, T. M. Yan, *Phys. Rev.* **D21**, 203 (1980); W. Buchmuller, G. Grunberg, S.-H. Tye *Phys. Rev. Lett.* **45**, 103 (1980), *Phys. Rev.* **D24**, 132 (1981); C. Quigg, J. L. Rosner, *Phys. Rev.* **D23**, 2625 (1981)

(2 entries: $c\bar{c}$, $b\bar{b}$ potential respectively); J. Baacke, Y. Igarashi, G. Kasperidus, *Z. Phys.* **C13**, 131 (1982); R. McClary, N. Byers, *Phys. Rev.* **D28**, 1692 (1983); P. Moxhay, J. L. Rosner, *Phys. Rev.* **D28**, 1132 (1983); H. Grotch, D. A. Owen, K. J. Sebastian, *Phys. Rev.* **D30**, 1924 (1984); S. N. Gupta, S. F. Radford, W. W. Repko, *Phys. Rev.* **D26**, 3305 (1982), *Phys. Rev.* **D30**, 2424 (1984); S. N. Gupta, S. F. Radford, W. W. Repko, *Phys. Rev.* **D34**, 201 (1986); M. Bander, D. Silverman, B. Klima, U. Maor, *Phys. Lett.* **B134**, 258 (1984), *Phys. Rev.* **D29**, 2038 (1984), *Phys. Rev.* **D36**, 3401 (1987); W. Kwong, J. L. Rosner, *Phys. Rev.* **D38**, 279 (1988); L. P. Fulcher, *Phys. Rev.* **D37**, 1259 (1988); S. N. Gupta, W. W. Repko, C. J. Suchyta III, *Phys. Rev.* **D39**, 974 (1989); L. P. Fulcher, *Phys. Rev.* **D42**, 2337 (1990); A. K. Grant, J. L. Rosener, E. Rynes, *Phys. Rev.* **D47**, 1981 (1993); T. A. Lahde, *Nucl. Phys.* **A714**, 183(2003).

37. K. Gottfried, *Phys. Rev. Lett.* **40**, 598 (1978); and Yan.[38]

38. T. M. Yan, *Phys. Rev.* **D22**, 1652 (1980).

39. E. Eichten, K. Gottfried, *Phys. Lett.* **B66**, 286 (1977).

40. H. Severini *et al.* (CLEO), CLEO-CONF-03-06, contributed to Lepton Photon 2003, hep-ex/0307034

41. M. B. Voloshin, Mod. *Phys. Lett.* **A18**, 1067 (2003).

42. T. M. Yan, Y. P. Kuang, *Phys. Rev.* **D24**, 2874 (1981).

43. M. B. Voloshin *Nucl. Phys.* **B154**, 365 (1979); M. B. Voloshin, V. Zakharov *Phys. Rev. Lett.* **45**, 688 (1980); V. A. Novikov, M. A. Shifman, *Z. Phys.* **8**, 43 (1981); M. B. Voloshin *Pisma Zh. Eksp. Teor. Fiz.* **37**, 58 (1983) [JETP Lett. 37, 69 (1983)].

44. S. E. Csorna *et al.*, CLEO-CONF-02-06, Contributed to 31st International Conference on High Energy Physics (ICHEP 2002), Amsterdam, The Netherlands, 24-31 Jul 2002, hep-ex/0207060.

45. CLEO collaboration, private communication.

46. The following sample of potential models (ordered according to the increasing value of $\sigma(\Delta M_{cog})$) is displayed in Fig. 11: D. Ebert, R. N. Faustov, V. O. Galkin, *Phys. Rev.* **D62**, 034014 (2000); W. Kwong, J. L. Rosner, *Phys. Rev.* **D38**, 279 (1988); L. P.Fulcher, *Phys. Rev.* **D42**, 2337 (1990); P. Moxhay, J. L. Rosner, *Phys. Rev.* **D28**, 1132 (1983); L. P. Fulcher, *Phys. Rev.* **D44**, 2079 (1991); S. Godfrey, N. Isgur, *Phys. Rev.* **D32**, 189 (1985); E. Eichten, F. Feinberg, *Phys. Rev.* **D23**, 2724 (1981); T. A. Lahde, *Nucl. Phys.* **A714**, 183 (2003); J. R. Hiller, *Phys. Rev.* **D30**, 1520 (1984); W. Buchmuller, *Phys. Lett.* **B112**, 479 (1982); T. A. Lahde, C. J. Nyfalt, D. O. Riska, *Nucl. Phys.* **A645**, 587 (1999); J. S. Kang, *Phys. Rev.* **D20**, 2978 (1979); M. Hirano, *Prog. of Theor. Phys.* **83**, 575 (1990); J. Baacke, Y. Igarashi, G. Kasperidus, *Z. Phys.* **C13**, 131 (1982).

47. S. Godfrey, J. L. Rosner *Phys. Rev.* **D64**, 097501

(2001), Erratum-ibid. **D66**, 059902 (2002).

48. J. L. Rosner, *Phys. Rev.* **D67**, 097504 (2003)

49. P. Moxhay, *Phys. Rev.* **D37**, 2557 (1988).

50. P. Ko, *Phys. Rev.* **D47**, 208 (1993).

51. M. B. Voloshin, *Phys. Lett.* **B562**, 68 (2003).

52. J. Z. Bai *et al.* (BES) hep-ex/0307028.

53. Y. P. Kuang, *Phys. Rev.* **D65**, 094024 (2002); Y. P. Kuang, T.M.Yan, *Phys. Rev.* **D41**, 155 (1990).

54. S. K. Choi *et al.* (BELLE), hep-ex/0309032; K. Abe *et al.* (BELLE), hep-ex/0308029, paper contributed to Lepton-Photon 2003, "Observation of a new narrow charmonium state in exclusive $B^{\pm} \to K^{\pm}\pi^{+}\pi^{-}J/\psi$ decays", ♯307.

55. M. B. Voloshin, hep-ph/0309307.

56. T. Barnes, S. Godfrey, hep-ph/0311162.

57. G. Bauer (CDF), at International Workshop on Heavy Quarkonium at Fermilab, Batavia, IL, USA, September 20-22, 2003.

58. M. Bander, G. L. Shaw, P. Thomas, *Phys. Rev. Lett.* **36**, 695 (1976); M. B. Voloshin, L. B. Okun *JETP Lett.* **23**, 333 (1976), Pisma *Zh. Eksp. Teor. Fiz.* **23**, 369 (1976); A. De Rujula, H. Georgi, S. L. Glashow, *Phys. Rev. Lett.* **38**, 317 (1977).

59. E. Eichten, K. Gottfried, T. Kinoshita, K. D. Lane, Tung-Mow Yan, *Phys. Rev.* **D21**, 203 (1980).

60. A. Bradley, D. Robson, *Phys. Lett.* **B93**, 69 (1980); *Z. Phys.* **C6**, 57 (1980).

61. R. L. Jaffe, *Phys. Rev.* **D15**, 267 (1977); J. Weinstein, N. Isgur, *Phys. Rev. Lett.* **48**, 659 (1982), *Phys. Rev.* **D27**, 588 (1982), *Phys. Rev.* **D41**, 2236 (1990).

62. N. A. Tornqvist, *Phys. Rev. Lett.* **67**, 556 (1991), *Z. Phys.* **C61**, 525 (1994), hep-ph/0308277.

63. A. V. Manohar, M. B. Wise, *Nucl. Phys.* **B339**, 17 (1993); T. E. O. Ericson, G. Karl, *Phys. Lett.* **B309**, 426 (1993).

64. F. E. Close, P. R. Page, hep-ex/0309253.

65. E. Braaten, M. Kusunoki, hep-ph/0311147.

66. The data come from Fig.3a in BELLE's paper.[54] The sideband distribution was subtracted here. The binning and normalization were also changed.

67. K. Abe *et al.* (BELLE), hep-ex/0307061.

68. F. Yuan, C.-F. Qiao , K.-T. Chao, *Phys. Rev.* **D56**, 329 (1997).

69. E. J. Eichten, K. Lane, C. Quigg, *Phys. Rev. Lett.* **89**, 162002 (2002).

70. E. J. Eichten, at International Workshop on Heavy Quarkonium at Fermilab, Batavia, IL, USA, September 20-22, 2003.

71. L. P. Fulcher, *Phys. Rev.* **D44**, 2079 (1991).

72. The following sample of potential models (ordered according to the increasing value of $\sigma(\Delta M_{cog})$) is displayed in Fig. 16: H. Ito, *Prog. of Theor. Phys.* **84**, 94 (1990); D. Ebert, R. N. Faustov, V. O. Galkin, *Phys. Rev.* **D67**, 014027 (2003); E. Eichten, F. Feinberg, *Phys. Rev.* **D23**, 2724 (1981); S. Godfrey, N. Isgur, *Phys. Rev.* **D32**, 189 (1985); T. A. Lahde, *Nucl. Phys.* **A714**, 183 (2003); N. Barik, S. N. Jena, *Phys. Rev.* **D24**, 680 (1981); W. Buchmuller, *Phys. Lett.* **B112**, 479 (1982); L. P. Fulcher, *Phys. Rev.* **D44**, 2079 (1991); J. Baacke, Y. Igarashi, G.Kasperidus, *Z. Phys.* **C13**, 131 (1982); J. R. Hiller, *Phys. Rev.* **D30**, 1520 (1984); M. Hirano, *Prog. of Theor. Phys.* **83**, 575 (1990); T. A. Lahde, C. J. Nyfalt, D. O. Riska, *Nucl. Phys.* **A645**, 587 (1999).

73. BELLE collaboration, private communication.

74. S. Pakvasa, M. Suzuki, hep-ph/0309294.

DISCUSSION

Estia Eichten (FNAL): The effect of decay channels on the mass of the 1^3D_2 $c\bar{c}$ state is not necessarily to reduce the splitting between the 1^3D_2 and 1^3D_1. The effects need to be calculated. Either sign is possible for the effect a priori.

Tomasz Skwarnicki : I fully agree with this comment.

M. Danilov (ITEP, Moscow): I do not think it is possible to say that NRQCD works for charmonium production. The color octet contribution was fixed arbitrary to describe the Tevatron data. There is practically no other process where it is needed. However there are cases with an order of magnitude contradiction to NRQCD.

Tomasz Skwarnicki : Charmonium is likely not heavy enough for the NRQCD approximation to work well.

Kenneth Lane (Boston University): In supposing that the $X(3872)$ is a molecule, you must ask where are all the other possible molecule states - $D\bar{D}$, $D\bar{D}^*$ in $I = 1$ (decaying to $\psi\pi^\pm\pi^0$) etc..

Tomasz Skwarnicki : Using the pion-exchange model, Tornqvist[62] showed that molecular forces are attractive only in an $I = 0$ $D\bar{D}^*$ molecule. Close and Page[64] argued that there should be only one molecular state near the $D\bar{D}^*$ threshold even when isospin violation is taken into account. In the pion-exchange model there is no binding force in the $D\bar{D}$ system, thus no $D\bar{D}$ molecule is expected. If the $X(3872)$ state is a ground state $D\bar{D}^*$ molecule, its binding energy is very small. Thus, any excitations would not be bound.

Michael Albrow (FNAL): If the new BELLE state $X(3872)$ is a $D\bar{D}^*$ bound state, do you expect a similar state in the b-sector?

Tomasz Skwarnicki : Yes absolutely, but maybe it is not produced in e^+e^- annihilations.

LESSONS FROM STANDARD CHARM DECAYS

J. M. DE MIRANDA

Centro Brasileiro de Pesquisas Fisicas, Rua Xavier Sigaud 150, Rio de Janeiro RJ 22290-180, Brazil
E-mail: jussara@cbpf.br

Recent charm results are reviwed with special attention to spectroscopy and hadronic decays. Two new states decaying to $D_s^+\pi^0$ and $D_s^{*+}\pi^0$ were recently discovered by BaBar and CLEO. In the baryon sector, the first observation of a doubly charmed baryon was announced by SELEX. Several amplitude analysis on charm hadronic decays are also discussed.

1. Introduction

Charm physics does not hold the frontier physics place that it once did, nevertheless there is still plenty of room for surprises and much to be understood. There is a consensus that charm is not a "heavy quark" (HQ) in the sense that the "heavy quark" QCD methodology would always work. Rather, charm seems to behave sometimes as a true heavy quark and sometimes not. That is why a large number of theoretical concepts and tools used in high energy physics are relevant for treating charm physics. Due to the richness of the decay pattern and the large clean samples now available charm physics is a very active/attractive research field. B-factories and collider experiments produced some impressive charm data, revealing large potential for high precision and rare processes in charm physics.

Very recently S. Bianco, F. L. Fabbri, D. Benson and I. Bigi wrote a quite complete and self-contained charm physics review.[1] Both theoretical and experimental state-of-the-art aspects of the charm physics are carefully addressed. Here, due to the space restriction, I was forced to impose some stringent selection criteria.

I just mention two results that could not be better discussed. First, I summarize lifetime measurements of all charm hadrons in Fig. 1. The lifetimes are the most inclusive possible decay quantity, they carry information on the underlying dynamics of the weak decays. Although the lifetime hierarchy has been established theoretically by the use of $1/m_Q$ expansions plus QCD corrections techniques, a more fundamental quantitative explanation is still necessary. The ability to make high precision lifetime measurements is essential for the search for mixing and possibly new physics in the charm sector. Traditionally the lifetime measurements have been made

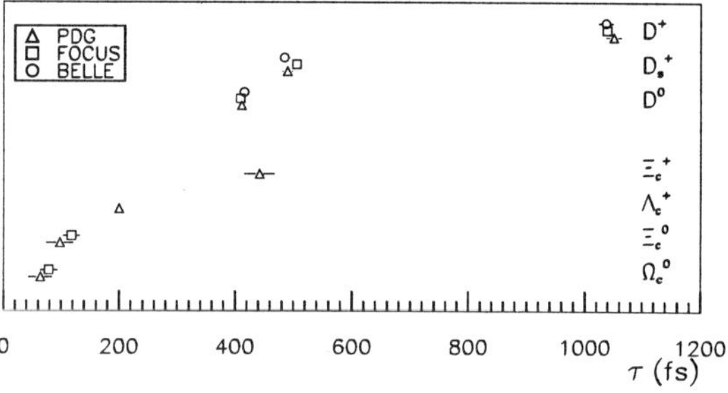

Figure 1. Charm lifetime measurements. Explicitly plotted are FOCUS and Belle measurements which are not yet included in the PDG averages.

by fixed target experiments, favored by the Lorentz boost that enlarges the production-decay displacements, but lately collider experiments have become competitive as is clear from Fig. 1.

The second result that I would like to at least mention is the observation of an unexpected interference phenomena in the semileptonic decay $D^+ \to K^-\pi^+\mu^+\nu$ by the FOCUS experiment.[2] For the past 20 years this decay was believed to occur 100% through the intermediate $D^+ \to \bar{K}^{*0}(892)\mu^+\nu$. During the form factors analysis, FOCUS observed a large discrepancy between data and Monte Carlo. The noticed mismatch was significant only for events in the lower $K\pi$ mass region ($m_{K\pi} < 0.9$ GeV). The simplest explanation for the effect would be the inclusion of a constant scalar amplitude that interferes with the dominant $\bar{K}^{*0}$. Such an amplitude is parameterized as $Ae^{i\delta}$, A and δ being constants with best values measured to be 0.36 and 45° respectively. This solution is plotted in Fig. 2.

The topics I selected to discuss more extensively are: spectroscopy (Sec. 1) and hadronic decays (Sec. 2).

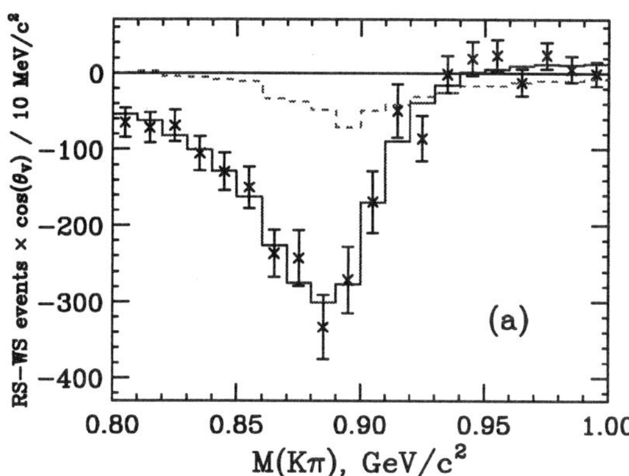

Figure 2. FOCUS $D^+ \to K^- \pi^+ \mu^+ \nu$ asymmetry vs. $K\pi$ invariant mass. The dashed line represents the Monte Carlo simulation with no interfering S-wave. The solid line is the Monte Carlo with constant S-wave amplitude.

Models based on Heavy Quark Symmetry (HQS) were able to predict the first excited states for $[c\bar{u}]$ and $[c\bar{d}]$ and also some of the $[c\bar{s}]$ mesons. Last April, BaBar[3] CLEO[4] and later Belle[5] observed two states decaying predominantly to $D_s^+ \pi^0$ and $D_s^{*+} \pi^0$. Several complementary measurements were performed that favored the interpretation of these being the L=1 $J^P = 0^+, 1^+$ states. This result was a surprise due to the much lower than expected mass observed.

Much more numerous, complex and less studied than the mesons, the baryon spectroscopy sector has also gained some visibility lately with a series of double charm meson states claimed by SELEX.[6,7] Even though some of the proposed states are questionable due to the low statistics, they are worth noticing.

The last selected topic are the hadronic decays, with emphasis on full amplitude analysis. The investigation offers insight into a large number of issues: QCD, on heavy to low quark transitions, lifetime differences, final-state interactions, relative strength on the various decay diagrams and, more recently, it has proved to be a very interesting source of information on light meson spectroscopy. There is no overall interpretation as to why the hadronic charm decay pattern is what it is.

Many results shown here are preliminary results presented in conferences.

2. Spectroscopy

2.1. *Mesons*

Before last April, when the two states $D_{sJ}^*(2317)$ $D_{sJ}^*(2463)$ were observed by BaBar[3] and CLEO,[4] the charm meson spectroscopy was reasonably well understood. The two above signals were first observed in $e^+ e^-$ annihilations in the decays to $D_s^+ \pi^0$ and $D_s^{*+} \pi^0$ respectively. Later Belle[5] also confirmed the results and made the first observation of the channel $D_{sJ}^*(2463) \to D_s \gamma$ and several exclusive $B \to D D_{sJ}^*$.

The charm meson ground states fit in the $SU(4)_{flavor}$ multiplet classification scheme which is not useful for predicting the excited states due to large symmetry breaking ($m_c \ll m_s$). The use of the Heavy Quark Symmetry (HQS) along with QCD based potential models provides a much more useful classification/prediction scheme. In the HQS the charmed meson $[Q\bar{q}]$ is viewed like the hydrogen atom. In the limit $m_c \to \infty$ the spin of the heavy quark (S_Q) decouples from the light quark degrees of freedom, which means that $j_q \equiv S_q + L$ and S_Q are separately conserved. Figure 3 summarizes the $[c\bar{d}]$ and $[c\bar{u}]$ spectroscopy lines. The four $L = 1, n = 1$ particles were observed experimentally. For parity and angular momentum conservations, $j_q = 1/2$ states are forced to decay via S-wave transitions and for that reason they have large widths while the $j_q = 3/2$, that decay via D-wave, are narrow states and much easier to observe. In Fig. 3 many experimental results are plotted, of those I emphasize the measurements made recently by Belle.[8] Belle observed the four excited states by doing the full amplitude analysis of the decays $B^- \to D^{(*)+} \pi^- \pi^-$, which is the same procedure advocated in the next session to study light hadron spectroscopy with Dalitz plot analyses of charmed mesons.

Theory predicts also the existence of two radial excitations ($L = 0, n = 2$), only one of which was seen by Delphi ($D^{*\prime}(2637)$)[9] but not confirmed by a number of other experiments.[10]

The use of the heavy quark symmetry for the $[c\bar{s}]$ excited states means a picture very similar to Fig. 3, only scaled up about 80 MeV. The narrow states have been observed in DK decay modes with mass and width close to the theoretical prediction, see Table 1.

This scenario led to the belief that the states D_{s0}^* and D_{s1}^* ($j_q = 1/2$) expected at masses ~ 2480 and ~ 2570 MeV respectively, would decay through S-wave transitions in the isospin conserving modes DK, and consequently have large widths. Instead, BaBar[3] announced the observation of a very narrow state, $D_{sJ}^*(2317)$ at about 2.32 GeV in the in-

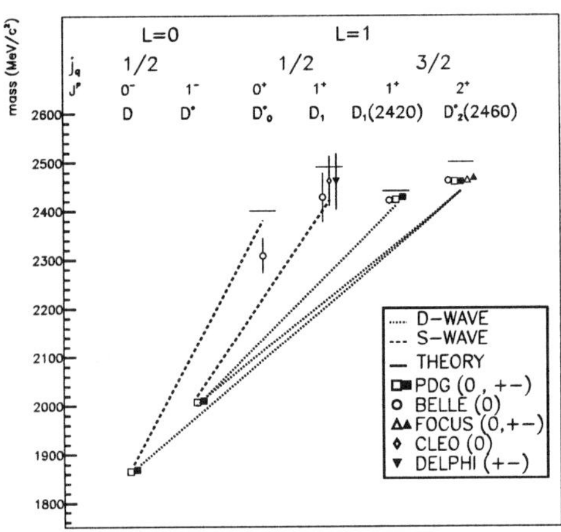

Figure 3. Spectroscopy scheme for $D^{0,\pm}$ mesons.

Table 1. Excited $[c\bar{s}]$ meson $D^*_{SJ}(j_q)$. FOCUS preliminary results.[11]

	Theory	PDG	FOCUS
$D^*_{s1}(2536)(3/2)$			
m (MeV)	~ 2530	2573.3 ± 6	2535.1 ± 0.3
Γ (MeV)	< 1	$< 2.390\%\,CL$	1.6 ± 1
$D^*_{s2}(2573)(3/2)$			
m (MeV)	~ 2590	2572.4 ± 1.5	2567.3 ± 1.4
Γ (MeV)	$10 - 20$	15 ± 5	28 ± 5

clusive invariant mass of $D_s^+\pi^0$. They also noticed an excess of events at the ~ 2.46 GeV region of the $D^*(2112)^+\pi^0$ mass spectrum. This result triggered theoretical[12] and experimental activities to understand, confirm and establish the properties of the new states. CLEO[4] confirmed the $D^*_{sJ}(2317)$ and claimed the existence of a new state $D^*_{sJ}(2463)$ decaying to $D^*(2112)^+\pi^0$. Belle confirmed these prior results and observed the radiative decay mode $D^*_{sJ}(2463) \rightarrow D_s^+\gamma$ and a series of exclusive $B \rightarrow DD^*_{sJ}$, from which they obtained some information on the spin of the new particles.[5]

Using Monte Carlo the three experiments excluded the possibility of any known particle to produce the observed signals. On the other hand the two new states are kinetically very similar, both mass differences $\Delta M_{D^*_{sJ}(2317)} \equiv M_{D_s\pi^0} - M_{D_s}$ and $\Delta M_{D^*_{sJ}(2463)} \equiv M_{D^*_s\pi^0} - M_{D^*_s}$ are of the order of 350 MeV. It is possible that $D^*_{sJ}(2317)$ feeds-up to $D^*_{sJ}(2463)$ by the addition of a random photon consistent with $D^*(2112)^+ \rightarrow D_s\gamma$ or that the

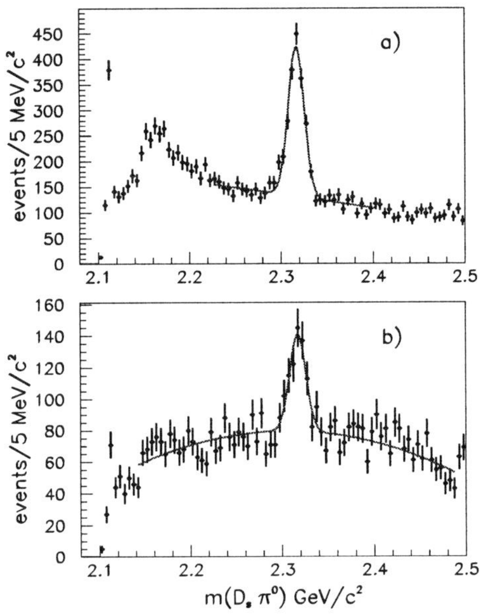

Figure 4. BaBar signal for $D^*_{sJ}(2317) \rightarrow D_s^+\pi^0$ for: a) $D_s^+ \rightarrow K^+K^-\pi^+$ for both $\phi\pi^+$ and $\bar{K}^{*0}K^+$, 1267 ± 63 events in the peak; and b) $D_s^+ \rightarrow K^+K^-\pi^+\pi^0$, 237 ± 33 events.

$D^*_{sJ}(2463)$ feeds-down to $D^*_{sJ}(2317)$ by neglecting the photon in the $D^*(2112) \rightarrow D_s\gamma$ decay. These effects were considered in all analyses. Another possibility also considered is the radiative decay $D^*_{sJ}(2463) \rightarrow D^*_{sJ}(2317)\gamma$.

Figures 4, 5 and 6 show BaBar, CLEO and Belle signals from their respective data set luminosities of 91, 13.5 and 87 fb^{-1}. In Table 2 we summarize their mass measurements. In all cases the width estimated by the experiments is very narrow and compatible with the detector resolution. BaBar and Belle measured the mass of $D^*_{sJ}(2463) \sim 5$ MeV smaller than CLEO.

It is natural to interpret the $D^*_{sJ}(2317)$ and $D^*_{sJ}(2463)$ as the missing J^P 0^+ and 1^+ $[c\bar{s}]$ states. Not having enough phase space to undergo the isospin conserving decay to DK, the decay proceeds by violating isospin, which would explain the narrowness. Several facts support this interpretation: $D^*_{sJ}(2317)$ was not observed in the final states $D_s\pi^+\pi^-, D_s\pi^0\pi^0$ and $D_s\gamma$ which are forbidden for a 0^+ state; $D^*_{sJ}(2463)$ decays to $D_s\gamma$, consequently it is not a 0^+; the helicity distribution for $D^*_{sJ}(2317) \rightarrow D_s\pi^0$ measured by BaBar is uni-

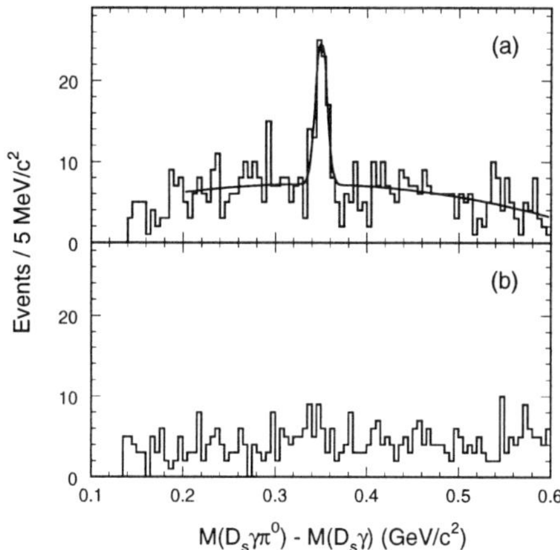

Figure 5. CLEO signal for $D^*_{sJ}(2463)$.

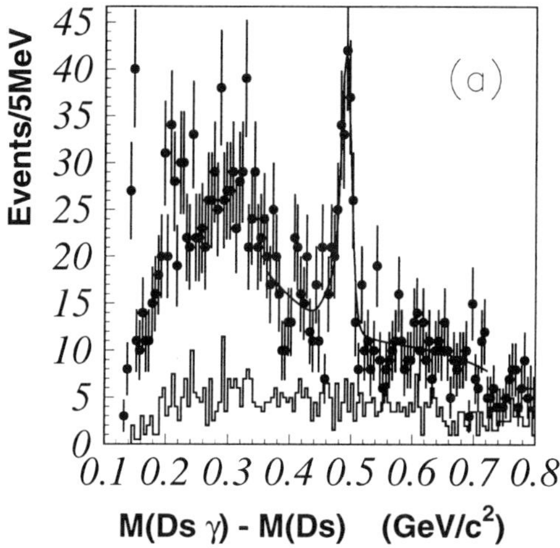

Figure 6. Belle $D^*_{sJ}(2463)$ radiative decay.

form, consistent with 0^+;[13] and Belle finds the $B \to DD^*_{sJ}(2463)$, $D^*_{sJ}(2463) \to D_s\gamma$ exclusive decay helicity angle distribution consistent with $J = 2$.

2.2. *Baryons*

The spectroscopy of the charm baryons is a lot more complex and less studied than the meson sector. SU(4) quark model multiplets are used as a guide to identify the observed states, and none of the J^P values have been directy measured. Of the single charm ground states only the $\Omega^*_{0c}(J^P = 3/2^+)$ remains undetected.

The first observation of the doubly-charmed

Table 2. Mass measurements for $D^*_{sJ}(2317)$ and $D^*_{sJ}(2463)$ states.

	BaBar
$M(D^*_{sJ}(2317))$ MeV	$2316.8 \pm 0.4 \pm 0.3$
$M(D^*_{sJ}(2317)) - M(D_s)$ MeV	$348.4 \pm 0.4 \pm 0.3$
$M(D^*_{sJ}(2463))$ MeV	$2457.0 \pm 1.4 \pm 3$
$M(D^*_{sJ}(2463)) - M(D^*_s)$ MeV	$344.6 \pm 1.2 \pm 3$
	Belle
$M(D^*_{sJ}(2317))$ MeV	$2317.2 \pm 0.5 \pm 0.9$
$M(D^*_{sJ}(2317)) - M(D_s)$ MeV	$348.7 \pm 0.5 \pm 0.7$
$M(D^*_{sJ}(2463))$ MeV	$2456.5 \pm 1.3 \pm 1.1$
$M(D^*_{sJ}(2463)) - M(D^*_s)$ MeV	$344.1 \pm 1.3 \pm 0.9$
	CLEO
$M(D^*_{sJ}(2317))$ MeV	$2318.5 \pm 1.2 \pm 1.1\pm$
$M(D^*_{sJ}(2317)) - M(D_s)$ MeV	$350.0 \pm 1.2 \pm 1.0$
$M(D^*_{sJ}(2463))$ MeV	$2463.6 \pm 1.7 \pm 1.2$
$M(D^*_{sJ}(2463)) - M(D^*_s)$ MeV	$351.2 \pm 1.7 \pm 1.0$

baryon Ξ^+_{cc} was recently published by SELEX.[6] The SELEX experiment uses a 600 GeV charged hyperon beam incident on target foils of Cu or diamond. In the double charm search they look for a secondary vertex of $\Lambda_c K^- \pi^+$ within their sample of 1630 fully reconstructed $\Lambda_c \to pK\pi$ events. Their signal is shown in Fig. 7. It is a 6.3 standard deviation signal of 15.9 events over an estimated background of 6.1 ± 0.5 events. The mass is at 3519 ± 1 MeV identified as $[ccd]^+$.

SELEX have pursued the search for more double charm events by requiring un extra π track on the secondary vertex, and by imposing helicity cuts. With limited statistics SELEX presented preliminary results for 3 more [ccq] candidates.[7] It is suggested that the four candidates are interpreted as the $L = 0$ and $L = 1$ $[ccd]^+$ and $[ccu]^{++}$.

To try and confirm these results the photoproduction experiment FOCUS have made extensive searches in their 19444 ± 262 Λ_c sample. There was no evidence for a doubly charmed baryon.[14]

3. Hadronic Decays

The hadronic decays are responsible for the not so well understood large differences in the lifetimes between the charm hadrons (Fig. 1). The leptonic and semileptonic represent just a small fraction of the total charm decay width. In the semileptonic decays, for example, the hadronic complexity can be isolated in measured form factors. As expected for spectators diagrams, the semileptonic widths for the various charm hadrons are comparable. The hadronic

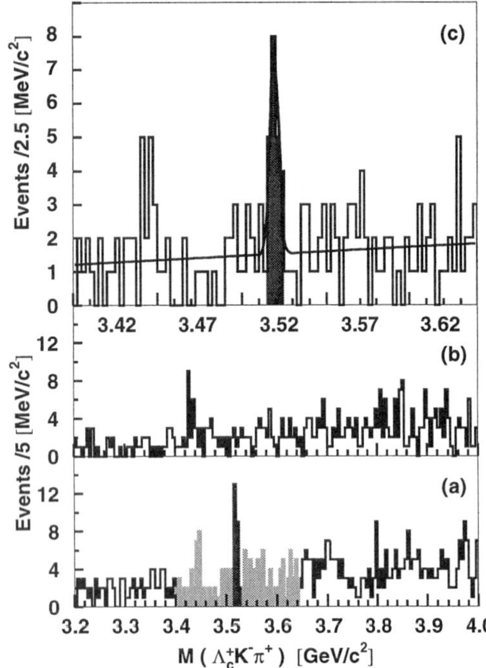

Figure 7. (a) The $\Lambda_c K^- \pi^+$ mass distribution in 5 MeV bins. The shaded region 3.400-3.640 GeV contains the signal peak and is shown in more detail in (c). (b) The wrong-sign combination. (c) The signal (shaded) region (22 events) and sideband mass regions with 162 total events in 2.5 MeV bins.

decays are much more complex and largely influenced by non-perturbative QCD effects. It has been known for a long time that the final-state interactions (FSI) can obscure the interpretation of the results and should be better understood. On the other hand, the hadronic decays are also responsible for the large diversity and rich resonant substructure observed in charm.

The simplest case are the two-body decays. It has been useful for studying final-state interactions and relative decay mechanisms, but gives no information on resonant substructure. For 3 or more bodies in the final state, generally a full amplitude analysis is desired to retrieve complete information on intermediate resonant processes and possible decay mechanisms. This can become too complex to be feasible, but in the 3-body decays involving only scalar/pseudoscalar particles there are only two degrees of freedom.

Next I summarize recent 3-body decay amplitude analyses followed by results on two, four and five bodies in the final state.

3.1. *Three-body Decays*

All results from this session are 3-body full coherent Dalitz plot analyses involving only pseudoscalar particles; $P \to P_1 P_2 P_3$. The Dalitz plot is the scatter plot of $s_{12} \times s_{13}$ [a] and it is proportional to squared decay amplitude, $|\mathcal{H}(s_{12}, s_{13})|^2$. $\mathcal{H}$ is written as a coherent sum of all possible resonant intermediate contributions and a non-resonant term:

$$\mathcal{H}(s_{12}, s_{13}) = \sum a_i \mathcal{A}_i(s_{12}, s_{13}, \vec{\alpha}_i) e^{i\delta_i}$$
$$+ a_{NR} \mathcal{A}_{NR} e^{i\delta_{NR}} \quad (1)$$

where the sum runs over the i possible resonant states; $a_{i,NR}$ are the relative contributions; δ are relative phases that accommodate the final-state interaction effects; and $\mathcal{A}$ is the resonant amplitude that depend on parameters $\vec{\alpha}$.

The parameters $a_{i,NR}$ and δ are extracted from likelihood fits to the data. Usually, the resonances are represented by Breit-Wigners in which case $\vec{\alpha}$ are its mass and width, that may or may not be free parameters of the fits.

There are two limitations on this kind of analysis. First one needs to impose a parametrization for the resonances. Relativistic Breit-Wigner forms are widely used but there are known limitations particularly for broad states near threshold. Second, as the number of states allowed becomes large, the interpretation is not straightforward. The challenge is to define clear quality criteria that, along with physics insights, could guide the analysis.

Nevertheless, the amplitude analysis is a powerful tool to investigate both charm decays and light meson spectroscopy, specially in the case of scalars that seem to be favored in charm decays. It is important to compare and understand the differences between hadron scattering and charm decay environments, for which final-state interactions play an important role.

3.1.1. *E791 - $D^+ \to \pi^- \pi^+ \pi^+$*

In a full Dalitz plot analysis of the decay $D^+ \to \pi^- \pi^+ \pi^+$,[15] E791 found a large contribution of a low mass broad scalar state, that the authors identified with the $\sigma(500)$ meson. Using a standard relativistic

---

[a] $s_{12} = m_{12}^2$ and $s_{13} = m_{13}^2$ are the squared invariant masses of the pairs of particles P_1 and P_2, and P_2 and P_3 respectively.

Breit-Wigner parametrization, they measured $m_0 = 478^{+24}_{-23} \pm 17$ MeV and $\Gamma_0 = 324^{+42}_{-40} \pm 21$ MeV. Using only established resonances, E791 was unable to achieve a satisfactory solution, there was a clear mismatch at low $\pi^+\pi^-$ invariant mass. Likelihood difference tests determined that there is a statistically significant preference seen in the data for a scalar resonance model if compared to a vector, tensor or even just a phaseless structure hypothesis.

The $\sigma(500)$ has a fundamental role in several spontaneous chiral symmetry breaking models and has been actively searched for in scattering experiments but no direct evidence for its existence has been established. The E791 result raised a series of debates on the subject. The strongest criticism, has been the fact that in the Dalitz plot analysis the resonance form is imposed. Moreover, the $\sigma(500)$ is a wide scalar state near threshold, conditions not favored for a Breit-Wigner width parametrization.

To feed in more information on the subject, E791 have applied the recently proposed Amplitude Difference (AD) method[16] to the $D^+ \to \pi^-\pi^+\pi^+$ decay.[17] They perform a direct and model-independent measurement of the phase motion of the amplitude at the low $\pi^+\pi^-$ mass region. Their preliminary results show a strong phase variation compatible with the isoscalar $\sigma(500)$ meson.

The applicability of the AD method requires a clean region of the Dalitz plot where the only contributions are the generic amplitude under study crossing with a well established resonance, represented by a Breit-Wigner. In the $D^+ \to \pi^-\pi^+\pi^+$, the only region that fulfill this condition is the region pictured in Fig. 8 where the low mass region in s_{13} crosses with the $f_2(1270)$ in s_{12}. Under these circumstances, the amplitude is written as:

$$\mathcal{A}(s_{12}, s_{13}) = a_R \mathcal{BW}_{f_2}(s_{12}, m_0, \Gamma_0)\mathcal{M}(s_{12}, s_{13})$$
$$+(a_s/(p^*/\sqrt{s_{13}}))sin\delta(s_{13})e^{i(\delta(s_{13})+\gamma)}, \qquad (2)$$

the first term on the right-hand-side represents the well established $f_2(1270)$ Breit-Wigner resonance of mass m_0, width Γ_0 and angular distribution $\mathcal{M}$, and the second term represents the generic amplitude under study. $(p^*/\sqrt{s_{13}})$ is a kinematic term to make this description compatible with the usual $\pi\pi$ scattering formulation. $\delta(s_{13})$ is the desired phase motion, a_R and a_s are the relative contributions and γ is the final-state interaction phase difference also

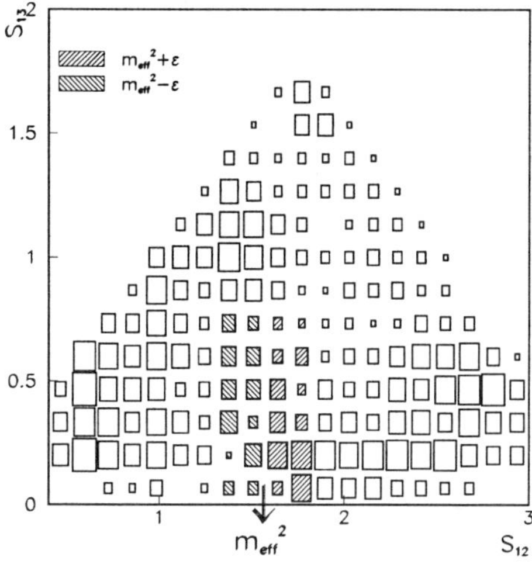

Figure 8. E791 $D^+ \to \pi^-\pi^+\pi^+$ folded Dalitz plot. Dashed region used in the AD method.

present in the full Dalitz plot analysis.

The AD method consists of making the amplitude-squared difference $\Delta|\mathcal{A}|^2 \equiv |\mathcal{A}(m_0^2 + \epsilon, s_{13})|^2 - |\mathcal{A}(m_0^2 - \epsilon, s_{13})|^2$, that takes a very simple functional form s_{13} if the resonance used as the probe is symmetrical with respect to m_0. In the case that the probe is not a narrow scalar resonance, the application of the AD method requires some approximations. Both the angular distribution and the mass dependent width of $f_2(1270)$ affects the symmetry with respect to $m_0^2 = 1.61$ GeV^2. The two approximations used in the E791 application of the method are that $\mathcal{M}(s_{12}, s_{13}) \sim \mathcal{M}(s_{13})$, and instead of the nominal mass they used an effective value $m_{eff}^2 = 1.535$ GeV^2. In this case the integrated amplitude-squared difference is:

$$\Delta \mid \mathcal{A} \mid^2 \sim -\mathcal{C}[sin(2\delta(s_{13}) + \gamma) - sin\gamma]$$
$$\bar{\mathcal{M}}_{f_2(1270)}(s_{13})/(p^*/\sqrt{s_{13}}) \qquad (3)$$

where $\mathcal{C}$ is a constant to be determined experimentally. The difference on the left-hand-side of this equation is taken from the two dashed regions of Fig. 8 where $\epsilon = 0.26$ GeV^2. The amplitude squared difference divided by the angular distribution and phase space factor is shown in Fig. 9a. Notice that the 6th bin has a large error, this is due to a singularity in $\mathcal{M}$. This bin was not further used in the analysis. From the above Eq. (3) it is clear that any variation observed in Fig. 9a reflects a variation in $\delta(s_{13})$.

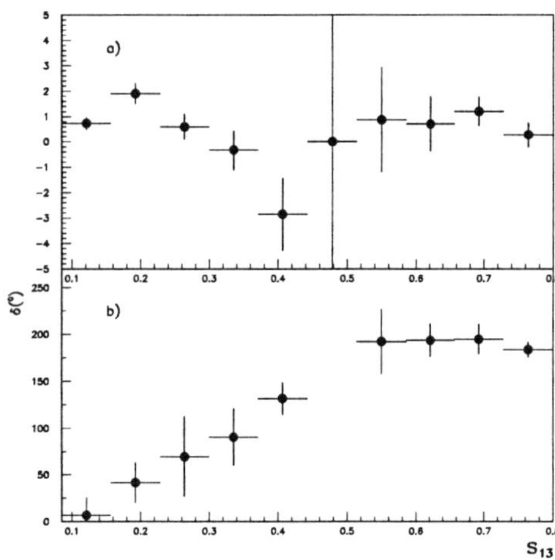

Figure 9. E791 amplitude difference: a) difference of the two shaded areas of Fig. 8 divided by the angular distribution and the phase space factor; and b) phase motion for the low $\pi^+\pi^-$ invariant mass region.

The maximum and minimum values of $\Delta\mathcal{A}/\mathcal{M}$ (in the present case bins 2 and 5 of Fig. 9a) provide two equations that are solved for $\mathcal{C}$ and γ. Eq. (3) is then inverted for $\delta(s_{13})$. There are ambiguities that arise due to the arcsin operations and they are solved by imposing that $\delta(s_{13})$ be a smooth increasing function of s_{13}. Finally, Fig. 9b shows the phase motion of the low $\pi^+\pi^-$ mass region obtained with the AD method. This strong phase motion is compatible with the full Breit-Wigner phase variation confirming the result obtained in the full Dalitz plot analysis. The measured $\gamma = 2.78 \pm 0.33$ rad is also compatible with previous results, $\gamma_{Dalitz} = 2.59 \pm 0.19$ rad.

3.1.2. CLEO - $D^{*+} \to D^0\pi^+; D^0 \to K_s\pi^-\pi^+$

CLEO has reported on the Dalitz plot analysis of the channel $D^{*+} \to D^0\pi^+; D^0 \to K_s\pi^-\pi^+$.[18] Their Dalitz plot are shown in Fig. 10d. The sample is of ~ 5300 events over a small 2% background. The D^*-tag is necessary to identify the favored decays $(\bar{K}^0\pi^-\pi^+)$ and the doubly-Cabibbo-suppressed or mixing contributions $(K^0\pi^+\pi^-)$. This analysis can be very complex because there are many possible resonant states. CLEO has considered up to 20 contributions: a non-resonant; 10 $[\pi^+\pi^-]$; 7 $[K_s\pi^-]$ favored and 2 $[K_s\pi^+]$ suppressed states. Their final result is listed in Table 3 and the model is compared to

Table 3. CLEO - $D^{*+} \to D^0\pi^+; D^0 \to K_s\pi^-\pi^+$. I listed only statistical errors.

Component	Phase	Fit fraction%
$K^*(892)^+\pi^-$	321 ± 10	0.34 ± 0.13
$\bar{K}^0\rho$	$0(fixed)$	26.4 ± 0.9
$\bar{K}^0 w$	114 ± 7	0.72 ± 1.8
$K^*(892)^-\pi^+$	150 ± 2	65.7 ± 1.3
$\bar{K}^0 f_0(980)$	188 ± 4	4.3 ± 0.5
$\bar{K}^0 f_2(1270)$	308 ± 12	0.27 ± 0.15
$\bar{K}^0 f_0(1370)$	85 ± 4	9.9 ± 1.1
$K_0^*(1430)\pi^+$	3 ± 4	7.3 ± 0.7
$K_2^*(1430)\pi^+$	155 ± 7	1.1 ± 0.2
$K^*(1680)\pi^+$	174 ± 6	2.2 ± 0.4
$N.R.$	160 ± 11	0.9 ± 0.4

data in Fig. 10. One can see from Fig. 10b a discrepancy in the low $\pi^+\pi^-$ region. They have opted to not include the $\sigma(500)$ in their final result but when they include it as a regular Breit-Wigner, they observe a sizable fraction of $f_\sigma = 0.57 \pm 0.13$ and measure $m_\sigma = 513 \pm 32$ MeV and $\Gamma_\sigma = 335 \pm 67$ MeV, values compatible with those measured by E791. They do not mention the improvement that the inclusion of the $\sigma(500)$ does to the fit quality.

CLEO sees a small but statistically significant contribution of 5.5 standard deviations from the doubly-Cabibbo-suppressed or D^0-$\bar{D}^0$ mixing $K^*(892)^+\pi^-$ intermediate state. They report for the first time the relative branching ratio, $BR(D^0 \to K^*(892)^+\pi^-)/BR(D^0 \to K^*(892)^-\pi^+) = (0.5 \pm 0.2^{+0.5+0.4}_{-0.1-0.1})\%$. Comparing the phases of the two $K^*(892)\pi$ channels they observe no CP-violating effects.

3.1.3. CLEO - $D^{*+} \to D^0\pi^+; D^0 \to \pi^0\pi^-\pi^+$

CLEO have preliminary results on the Dalitz plot analysis of the decay $D^{*+} \to D^0\pi^+; D^0 \to \pi^0\pi^-\pi^+$.[19] One of the strongest motivations of this analysis is the search for the $\sigma(500)$. In principle, there is no impeachment for the existence of this intermediate channel. The charged intermediate modes, $\rho^+\pi^-$ and $\rho^-\pi^+$, are possible via spectator diagrams and the neutral modes such as $\rho^0\pi^0$, $f_0(980)\pi^0$ or $\sigma(500)\pi^0$ are produced in internal W-emission or W-exchange type diagrams. The D^*-tag is required to distinguish D^0 from $\bar{D}^0$ and to search for the manifestation of CP-violation. Their final signal sample has ~ 1100 events over a $\sim 18\%$ background contamination. The phases and fractions for their preferred model are listed in Table 4. The vec-

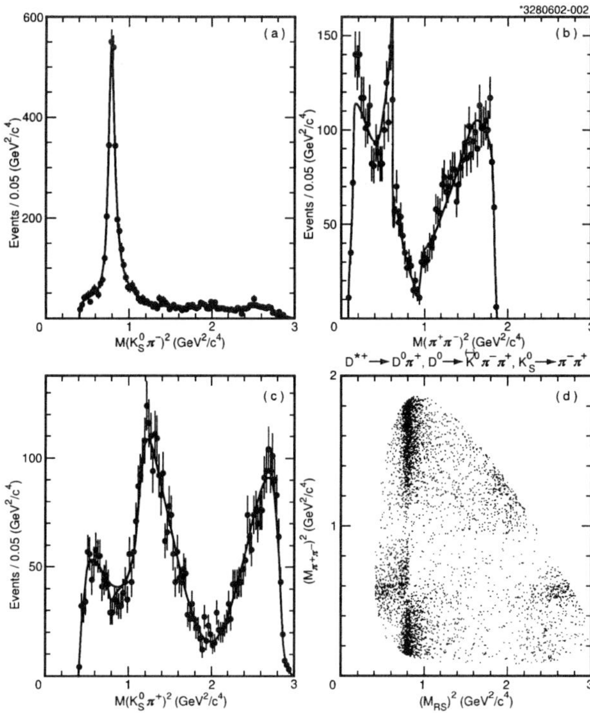

Figure 10. CLEO - $D^0 \to K_s \pi^- \pi^+$ projections compared to the best fit, and the Dalitz plot.

Table 4. CLEO - $D^{*+} \to D^0 \pi^+; D^0 \to \pi^0 \pi^- \pi^+$.

Mode	Phase(^{o})	Fit fraction%
ρ^+	$0(fixed)$	$76.5 \pm 1.8 \pm 4.8$
ρ^0	$10 \pm 3 \pm 3$	$23.9 \pm 1.8 \pm 4.6$
ρ^-	$-4 \pm 3 \pm 4$	$32.3 \pm 2.1 \pm 2.2$
$N.R.$	$77 \pm 8 \pm 11$	$2.7 \pm 0.9 \pm 1.7$

tor $\rho\pi$ contributions dominates. There is a small non-resonant component and no scalar resonant is needed to explain the data. No sign of CP-violation is observed.

3.1.4. BaBar - $D^{*+} \to D^0 \pi^+; D^0 \to K_s H H, (H = K^{\pm}, \pi^{\pm})$

BaBar have nice clean samples of four 3-body decays containing one neutral K_s in the final state, combined with charged kaons or pions.[20] They have preliminary Dalitz plot analysis results for the channels $K^0 K^- \pi^+$, $\bar{K}^0 K^- \pi^+$ and $K^0 K^- K^+$ from 22 fb^{-1} with event samples of 1008, 659 and 1957 events respectively and small background contamination of the order of $\sim 4\%$ or less. They have an amazing $K_s \pi^- \pi^+$ sample of over 15000 events, but no result have so far been released. The difficulty in all these

Table 5. BaBar - $D^{*+} \to D^0 \pi^+; D^0 \to K^0 K^- \pi^+$.

Component	Phase	Fit fraction%
$\bar{K}^0(1430)K^0$	52 ± 27	$4.8 \pm 1.4 \pm 1.6$
$\bar{K}^0(892)K^0$	175 ± 22	$0.8 \pm 0.5 \pm 0.1$
$\bar{K}^{*0}(1680)K^0$	-169 ± 16	$6.9 \pm 1.2 \pm 1.0$
$\bar{K}_2^*(1430)K^0$	51 ± 18	$2.0 \pm 0.6 \pm 0.1$
$K^{*+}(1430)K^-$	-41 ± 25	$13.3 \pm 3.5 \pm 3.9$
$K^{*+}(892)K^-$	$0(fixed)$	$63.6 \pm 5.1 \pm 2.6$
$K^{*+}(1680)K^-$	-178 ± 10	$15.6 \pm 3.0 \pm 1.4$
$K_2^{*+}(1430)K^-$	-52 ± 7	$13.8 \pm 2.6 \pm 7.9$
$a_0^-(980)\pi^+$	-100 ± 13	$2.9 \pm 2.3 \pm 0.7$
$a_0^-(1450)\pi^+$	31 ± 16	$3.1 \pm 1.9 \pm 0.9$
$a_2^-(1310)\pi^+$	-149 ± 27	$0.7 \pm 0.4 \pm 0.1$
NR	-136 ± 23	$2.3 \pm 0.5 \pm 5.6$

analyses is that there are too many possible resonant modes, despite the fact that the Dalitz plots present quite strong contributions of just a few, mostly vector resonances. If many possible states are allowed in the fit, it is quite probable that several different mathematical solutions fit the data with comparable confidence level. The situation gets even more critical because some of the possible resonances are controversial with poorly measured parameters.

In Table 5 we list the $K^0 K^- \pi^+$ proposed solution. There is a very large contribution ($\sim 63\%$) from $K^*(892)^+ K^-$ and eleven more states are allowed. A good $\chi^2/DOF = 46/44$ is obtained, but with a large destructive interference, with the fractions summing to 130$\pm$8%. In this channel the authors do not find a significant contribution of the κ meson.[21] In the case of the channel $K^0 K^+ \pi^-$ listed in Table 6, eleven intermediate states are kept with fractions that sum to 144$\pm$37%. Notice the large NR contribution, unusual for charm decays. The authors point out that if the NR is removed from the fit, it converges, increasing the $K_0^*(1430)^- K^+$ to 26% and decreasing the $a_0^+(980)\pi^-$ to 5%. The two solutions, with and without NR, have approximately the same quality.

Finally BaBar reports on the $3K$ mode where again the preferred solution presents a large destructive interference, see Table 7. Notice that in this solution the contribution of the poorly measured and near-the-threshold contribution of the $a_0(980)$ is dominant, much larger than that of the $\bar{K}^0 \phi$.

3.1.5. E791 - $D^+ \to K^- \pi^+ \pi^+$

The Cabibbo-favored $D^+ \to K^- \pi^+ \pi^+$ was one of the first modes to have a full amplitude analysis done.[21]

271

Table 6. BaBar - $D^{*+} \to D^0 \pi^+$; $D^0 \to \bar{K}^0 K^+ \pi^-$.

Component	Phase	Fit fraction%
$K^{*0}(1430)\bar{K}^0$	-38 ± 22	$26.0 \pm 16.1 \pm 3.3$
$K^{*0}(892)\bar{K}^0$	-126 ± 19	$2.8 \pm 1.4 \pm 0.5$
$K^{*0}(1680)\bar{K}^0$	161 ± 9	$15.2 \pm 11.9 \pm 0.5$
$K_2^{*0}(1430)\bar{K}^0$	53 ± 38	$1.7 \pm 2.5 \pm 0.2$
$K^{*-}(1430)K^+$	-142 ± 115	$2.4 \pm 8.2 \pm 1.0$
$K^{*-}(892)K^+$	$0(fixed)$	$35.6 \pm 7.7 \pm 2.4$
$K^{*-}(1680)K^+$	124 ± 27	$5.1 \pm 5.7 \pm 1.1$
$K_2^{*-}(1430)K^+$	-26 ± 38	$1.0 \pm 1.0 \pm 0.2$
$a_0^+(980)\pi^-$	-160 ± 42	$15.1 \pm 12.5 \pm 0.6$
$a_0^+(1450)\pi^-$	148 ± 25	$2.2 \pm 2.7 \pm 1.2$
NR	-172 ± 13	$36.6 \pm 25.8 \pm 2.7$

Table 7. BaBar - $D^{*+} \to D^0 \pi^+$; $D^0 \to \bar{K}^0 K^+ K^-$.

Component	Phase	Fit fraction%
$\bar{K}^0 \phi$	$0(fixed)$	$45.4 \pm 1.6 \pm 1.0$
$\bar{K}^0 a_0^0(980)$	109 ± 5	$60.9 \pm 7.5 \pm 13.3$
$\bar{K}^0 f_0(980)$	-161 ± 14	$12.2 \pm 3.1 \pm 8.6$
$a_0^+(980)K^-$	-53 ± 4	$34.3 \pm 3.2 \pm 6.8$
$a_0^-(980)K^+$	-13 ± 15	$3.2 \pm 1.9 \pm 0.5$
NR	40 ± 44	$0.4 \pm 0.3 \pm 0.8$

Table 8. Results without the κ and with the κ.

Mode	Fraction (%)	Phase
	No κ	
NR	90.9 ± 2.6	$0°$ (fixed)
$\kappa \pi^+$	$-$	$-$
$\bar{K}^*(892)\pi^+$	13.8 ± 0.5	$(54 \pm 2)°$
$\bar{K}_0^*(1430)\pi^+$	30.6 ± 1.6	$(54 \pm 2)°$
$\bar{K}_2^*(1430)\pi^+$	0.4 ± 0.1	$(33 \pm 8)°$
$\bar{K}^*(1680)\pi^+$	3.2 ± 0.3	$(66 \pm 3)°$
	With κ	
NR	$13.0 \pm 5.8 \pm 4.4$	$(-11 \pm 14 \pm 8)°$
$\kappa \pi^+$	$47.8 \pm 12.1 \pm 5.3$	$(187 \pm 8 \pm 18)°$
$\bar{K}^*(892)\pi^+$	$12.3 \pm 1.0 \pm 0.9$	$0°$ (fixed)
$\bar{K}_0^*(1430)\pi^+$	$12.5 \pm 1.4 \pm 0.5$	$(48 \pm 7 \pm 10)°$
$\bar{K}_2^*(1430)\pi^+$	$0.5 \pm 0.1 \pm 0.2$	$(-54 \pm 8 \pm 7)°$
$\bar{K}^*(1680)\pi^+$	$2.5 \pm 0.7 \pm 0.3$	$(28 \pm 13 \pm 15)°$

E691 and E687[22] with large samples and using only well established resonances observed a large interfering pattern, large NR contributions in bad quality fits. E791, with a large sample of over 15000 events and a small background, 6%, first tried some more sophisticated models for the NR contribution, but were unable to fit the data reasonably. They could only reach a good confidence level when they allowed for the possibility of a new resonant state. Doing extensive fit quality and consistency tests they concluded that the new contribution is a scalar, κ, with mass $797 \pm 19 \pm 43$ MeV and width $410 \pm 43 \pm 87$ MeV. Their result with and without the κ are compared in Table 8. The two solutions are very different. With the inclusion of the new state the NR contribution drops from 90% to 13% following the general trend that charm decays tend to be quasi-two-body. The large destructive interfering pattern is no longer present and the $\chi^2/63$ went from 2.7 to 0.75.

Favoring the new E791 solution is the fit quality and the simplicity in a high statistics sample. The problem with this solution is that the κ is not a well established resonance, it was not observed in $K\pi$ scattering and moreover, it is a wide scalar close to the threshold.

E791 have presented preliminary studies comparing the phase of the S-wave component of the $D^+ \to K^- \pi^+ \pi^+$ decay amplitude with the phase observed in $K\pi$ scattering experiments.[23] In the scattering experiment LASS,[24] a slow phase motion is observed for the scalar partial wave at the low $K\pi$ invariant mass region, not compatible with the variation characteristics of a resonance. The S-wave amplitude used by LASS with a phase space adapted for the D decay is given by:

$$\mathcal{A}_S = \frac{m_{12}}{p_{12}^*} \sin \delta_B \, e^{i\delta_B}$$
$$+ e^{2i\delta_B} \frac{(m_0^2/p_{12}^{*0})\Gamma_0}{m_0^2 - m_{12}^2 - i m_0 \Gamma(m_{12})}$$
$$+ (2 \leftrightarrow 3) \tag{4}$$

with $\Gamma(m) = (m_0/m)(p^*/p^{*0})\Gamma_0$ and the NR term has the effective range form:

$$\cot \delta_B = \frac{1}{a\, p^*} + \frac{1}{2} b\, p^*. \tag{5}$$

The above S-wave parametrization was used to fit E791 data and the measured parameters are: $a = 4.58 \pm 0.33$ c/GeV, $b = -2.94 \pm 0.43$ c/GeV, to be compared with the values $a = 1.95 \pm 0.09$ c/GeV and $b = 1.76 \pm 0.36$ c/GeV from LASS. Figure 11 compares the S-wave phase for the κ Breit-Wigner solution with the scattering parametrization, with a and b fixed at the values measured by LASS and measured by E791. From the plot we see that the scalar sector at low $K\pi$ mass in $D^+ \to K^- \pi^+ \pi^+$ prefers a rapidly varying phase, consistent with the previous indication of the κ state.

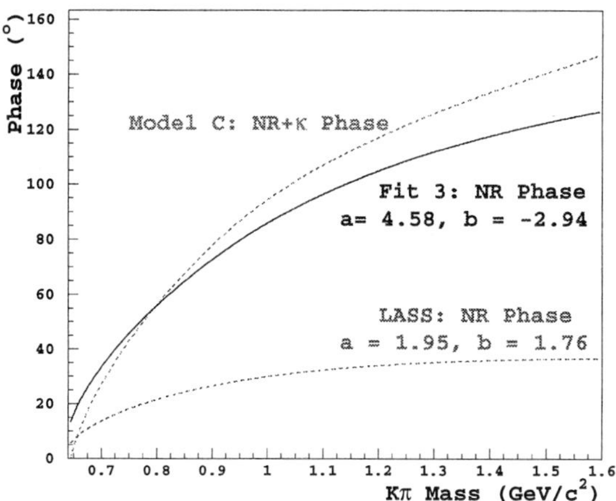

Figure 11. E791 $D^+ \to K^-\pi^+\pi^+$ phase of the scalar amplitude as a function of $K\pi$ invariant mass. Model C: from the Dalitz plot fit of E791 data using a flat NR plus a κ resonance parametrized as a Breit-Wigner. Fit 3: NR alone using the LASS parametrization allowing a and b parameters to float. LASS: for comparison the scalar amplitude as measured by LASS for $K\pi$ scattering.

Table 9. CLEO preliminary relative branching ratios.

D^+	CLEO (%)	PDG (%)
$\dfrac{\Gamma(\pi^+\pi^0)}{\Gamma(K^-\pi^+\pi^+)}$	$1.44 \pm 0.19 \pm 0.10$	$2.08 \pm 0.6 \pm 0.5$
$\dfrac{\Gamma(K^+K_s)}{\Gamma(\pi^+K_s)}$	$18.92 \pm 1.55 \pm 0.73$	28.3 ± 3.5
$\dfrac{\Gamma(K^+\pi^0)}{\Gamma(K^-\pi^+\pi^+)}$	$0.29 \pm 0.18 \pm 0.09$	$--$

3.2. Decays with 2 Bodies or More than 3 Bodies

FOCUS,[25] CLEO[26] and CDF[18] have new numbers on some singly- and doubly-Cabibbo-suppressed decays.

With an integrated luminosity of 13.7 fb^{-1}, CLEO have new numbers for $D^+ \to \pi^+\pi^0$, $D^+ \to K^+K_s$ and $D^+ \to K^+\pi^0$. Figure 12 shows the signals and in Table 9 the relative branching ratios are compared to their previous results. With the PDG value for the $D^+ \to K^-\pi^+\pi^+$ branching ratio, they set the limit $< 4.2 \times 10^{-4}$ at 90% CL for the $D^+ \to K^+\pi^0$ branching ratio.

CDF has preliminary results on $D^0 \to \pi^+\pi^-$ and $D^0 \to K^+K^-$ from 65 pb^{-1} integrated luminosity. They use the D^*-tag to clean their signals shown in Fig. 13. In Table 10, CDF and FOCUS measurements for the relative branching ratios for these channels are compared to previous measurements. The normalizing signals, $D^0 \to K^-\pi^+$, are of 94560 ± 340 and 105030 ± 372 events respectively

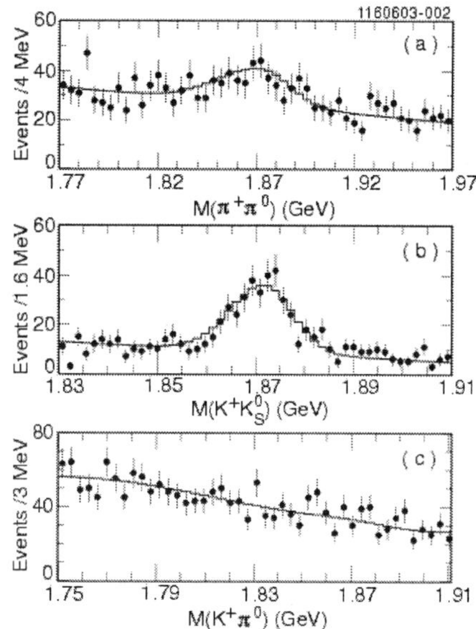

Figure 12. CLEO $D^+ \to \pi^+\pi^0$, K^+K_s, $K^+\pi^0$.

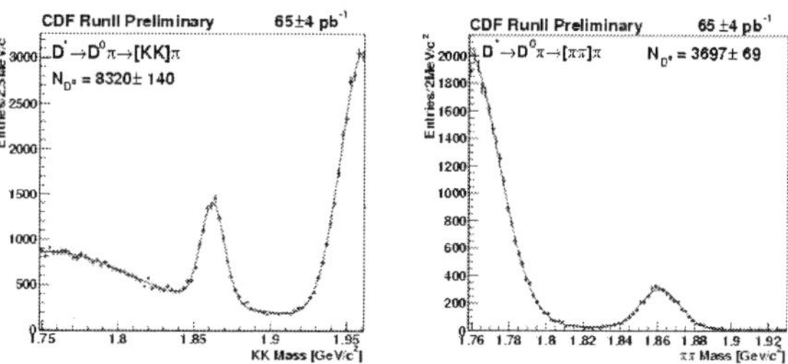

Figure 13. CDF preliminary $D^0 \to \pi^+\pi^-$ and $D^0 \to K^+K^-$.

for CDF and FOCUS. In the $\pi^+\pi^-$ mode, the small errors in the CDF results are a demonstration of the large potential capability they have to do charm physics.

To get insight into the importance of the final-state interactions on charm decays, the $D \to \pi\pi$ and $D \to KK$ decay amplitudes are expressed in terms of isospin amplitudes. The $\pi\pi$ system may have final isospin 0 or 2 while the KK can be in 0 and 1 states. Combining several measurements CLEO and

Table 10. CDF (preliminary) and FOCUS new measurements for D^0 decays.

	$\dfrac{\Gamma(K^+K^-)}{\Gamma(K^-\pi^+)}$ (%)	$\dfrac{\Gamma(\pi^+\pi^-)}{\Gamma(K^-\pi^+)}$ (%)
CDF	$9.38 \pm 0.18 \pm 0.10$	$3.686 \pm 0.076 \pm 0.036$
FOCUS	$9.93 \pm 0.14 \pm 0.14$	$3.53 \pm 0.12 \pm 0.06$
PDG	10.84 ± 0.26	3.76 ± 0.17

273

Table 11. For the isospin amplitude the experiments use new results from Table 10 combined with previous measurements on the channels. CLEO old results are listed for comparison.[10]

$D \to \pi\pi$	$|A_2|/|A_0|$	$(\delta_2 - \delta_0)(^o)$
FOCUS	0.65 ± 0.14	83.6 ± 10.0
CLEO new	0.421 ± 0.040	87.6 ± 4.6
CLEO old	$0.72 \pm 0.13 \pm 0.11$	$82.0 \pm 7.5 \pm 5.2$
$D \to KK$	$|A_1|/|A_0|$	$(\delta_1 - \delta_0)(^o)$
FOCUS	0.56 ± 0.04	37.1 ± 7.5
CLEO old	$0.61^{+0.11}_{0.10}$	$28.4^{12.1}_{9.7}$

FOCUS estimate the ratios of amplitudes and phases differences listed on Table 11. The large phase differences denote the significance of FSI.

The full amplitude analysis of 4-body decay modes is very complex and not much information on the resonant substructure of such decays is available. If the dominance of quasi-two body decays, observed in the 3-body final state, is confirmed for the case of multi-body decays, then one can take a more complete comparison with theoretical models, developed mainly to describe the two-body and quasi-two-body decay modes.

FOCUS has reported on new signals for $D^0 \to K^- K^+ K^- \pi^+$; $K^+ K^- \pi^+ \pi^-$ and $\pi^+ \pi^- \pi^+ \pi^-$.[28] For these they measure the branching ratio relative to the dominant $D^0 \to K^- \pi^+ \pi^- \pi^+$ listed in Table 12 along with previous measurements listed in the PDG. The full coherent amplitude analysis for the channel $D^0 \to K^- K^+ K^- \pi^+$ is now released.[29] The formalism used is a straightforward extension of that described above for the 3-body decays, a coherent sum of Breit-Wigner resonances modulated by angular distribution functions and a constant non-resonant contribution. There are five degrees of freedom, and in principle, many intermediate states possibilities. This decay is Cabibbo-favored but phase space and $s\bar{s}$ creation suppressed. The analysis is based on a sample of 139 signal and 65 estimated background events. While a large number of intermediate states could lead to this final state, phase space restricts the possible contributions. The only resonant mode with nominal mass within the phase space is the $\phi K^- \pi^+$. Wide resonances without a particular angular distribution such as $\bar{K}^0(1430)K^+K^-$ or $\kappa K^+ K^-$ are hard to distinguish from the NR contribution. On the other hand modes with $\bar{K}^0(892)$ with nominal mass just above the kinematic limit have a characteristic angular distribution as a signature and can be identified. The final solution for the best fit is

Table 12. D^0 branching ratios relative to $D^0 \to K^+ \pi^+ \pi^- \pi^+$.

mode	FOCUS	PDG
$[K^- K^+ \pi^- \pi^+]$	2.97 ± 0.10	3.34 ± 0.28
$[K^- K^+ K^- \pi^+]$	$0.257 \pm 0.034 \pm 0.023$	0.32 ± 0.09
$[\pi^- \pi^+ \pi^- \pi^+]$	8.66 ± 0.12	9.8 ± 0.6

Table 13. FOCUS best solution for the $D^0 \to K^- K^+ K^- \pi^+$ amplitude analysis.

Component	Phase $(^o)$	Fit fraction%
$\phi \bar{K}^{*0}(892)$	$0(fixed)$	$48 \pm 6 \pm 1$
$\phi K^- \pi^+$	$194 \pm 24 \pm 8$	$18 \pm 6 \pm 4$
$\bar{K}^{*0}(892) K^- K^+$	$225 \pm 15 \pm 4$	$20 \pm 7 \pm 2$
$N.R.$	$278 \pm 16 \pm 42$	$15 \pm 6 \pm 2$

listed in Table 13. Notice that the two vector mode $\bar{K}^{*0}(892)\phi$ dominates the decay. Similar analysis for the remaining signals are expected soon.

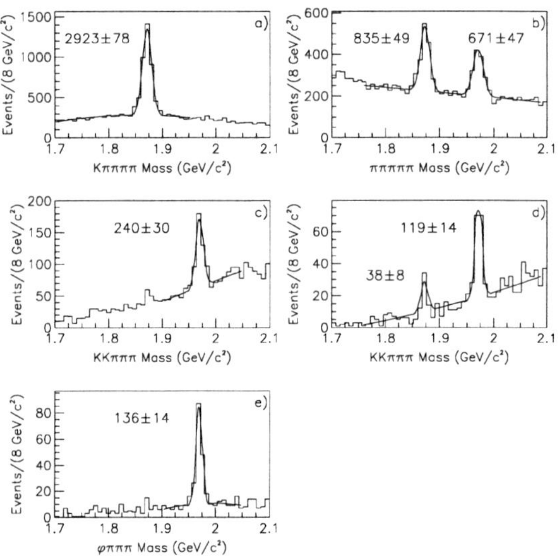

Figure 14. FOCUS five-body signals. Sub-resonant analyses were done for signals in a) and c). c) $K^- K^+ \pi^+ \pi^+ \pi^-$ optimization for D_s^+. d) Optimization for the first observation of the channel $D^+ \to K^- K^+ \pi^+ \pi^+ \pi^-$.

FOCUS have also recently published the 5-body decays of D^+ and D_s^+ into kaons and pions.[30] The inclusive branching ratios of five modes including the first evidence for the decay mode $D^+ \to K^- K^+ \pi^+ \pi^+ \pi^-$ are listed in Table 14 and the signals are in Fig. 14. They study the resonant substructure of the decays $D^+ \to K^- \pi^+ \pi^+ \pi^+ \pi^-$ and $D_s^+ \to K^- K^+ \pi^+ \pi^+ \pi^-$. The full coherent amplitude analysis was not attempted, rather they used a simplified approach where the final state is assumed

Table 14. FOCUS relative branching ratio for 5-body D^+ and D_s^+ decay modes.

Mode	FOCUS
$\dfrac{\Gamma(D^+\to K^-\pi^+\pi^+\pi^+\pi^-)}{\Gamma(D^+\to K^-\pi^+\pi^+)}$	$0.058\pm0.002\pm0.006$
$\dfrac{\Gamma(D^+\to\pi^+\pi^+\pi^+\pi^-\pi^-)}{\Gamma(D^+\to K^-\pi^+\pi^+\pi^+\pi^-)}$	$0.290\pm0.017\pm0.011$
$\dfrac{\Gamma(D_s^+\to\pi^+\pi^+\pi^+\pi^-\pi^-)}{\Gamma(D_s^+\to K^-K^+\pi^+)}$	$0.145\pm0.011\pm0.010$
$\dfrac{\Gamma(D_s^+\to K^+K^-\pi^+\pi^+\pi^-)}{\Gamma(D_s^+\to K^-K^+\pi^+)}$	$0.150\pm0.019\pm0.025$
$\dfrac{\Gamma(D_s^+\to\phi\pi^+\pi^+\pi^-)}{\Gamma(D_s^+\to\phi\pi^+)}$	$0.249\pm0.024\pm0.021$
$\dfrac{\Gamma(D^+\to K^+K^-\pi^+\pi^+\pi^-)}{\Gamma(D^+\to K^-\pi^+\pi^+\pi^+\pi^-)}$	$0.040\pm0.009\pm0.019$

to be an incoherent superposition of sub-resonant decays involving vector resonances. In both cases the non-resonant component is small.

4. Conclusions

The small fraction of recent results discussed here is illustrative of the picture drawn in the introduction: there is a lot to be understood and the conditions are favorable for the development of the field. Several experiments, not specially designed to do charm physics, have now some impressive quality charm data.

References

1. S. Bianco, F. L. Fabbri, D. Benson and I. Bigi, `hep-ex/0309021`.
2. FOCUS Collab., *Phys. Lett.* B **535**, 43-51 (2002), `hep-ex/0203031`.
3. Babar Collab., *Phys. Rev. Lett.* **90**, 242001 (2003), `hep-ex/0304021` and `hep-ex/0309028`.
4. CLEO Collab., `hep-ex/0305100`.
5. Belle Collab., `hep-ex/0307052`, `hep-ex/0307041`.
6. SELEX Collab., *Phys. Rev. Lett.* **89**, 112001 (2003).
7. J. S. Russ for SELEX Collab., at the Fermilab Wine&Cheese seminar (6/13/03).
8. Belle Collab., `hep-ex/0307021`.
9. Delphi Collab., *Phys. Lett.* B **426**, (1998).
10. Particle Data Group, *Phys. Rev.* D **66**, 010001 (2002).
11. R.Kutschke for FOCUS Collab., Frontiers in Contemporary Physics II, Nashville March (2001).
12. Barnes, Close and Lipkin, `hep-ph/0305025`; Bardeen, Eichten and Hill, `hep-ph/0305049`; Nowak, Rho and Zahed, `hep-ph/0307102`; Colangelo and De Fazio, `hep-ph/0305140`; Van Beveran and Rupp, `hep-ph/0308166`; Dai, Huang and Zhu, `hep-ph/0306274`; Browder, Pakvasa and Petrov, `hep-ph/0307054`.
13. A. Palano for BaBar Collab., talk at Physics in Collision, Germany (2003), `hep-ex/0309028`.
14. P. Sheldon for FOCUS Collab., talk at International Workshp on Frontier Science 2002, Frascati, Italy.
15. E791 Collab., *Phys. Rev. Lett.* **86**, 770 (2001).
16. Bediaga and Miranda, *Phys. Lett.* B **550**, 135 (2002), `hep-ph/0211078`.
17. Bediaga for E791 Collab., talk at Scalar Meson Workshop, Utica, May (2003), `hep-ex/0307008`.
18. CLEO Collab., *Phys. Rev. Lett.* **89**, 251802 (2002).
19. V.V. Frolov for CLEO Collab., `hep-ex/0306048`.
20. B. Aubert for Babar Collab., ICHEP2002 `hep-ex/0207089`.
21. E791 Collab., *Phys. Rev. Lett.* **89**, 121801 (2002).
22. E691 Collab., *Phys. Rev.* D **48**, 56 (1995); E687 Collab., *Phys. Lett.* B **331**, 217 (1994).
23. Gobel for E791 Collab., talk at Scalar Meson Workshop, Utica, May (2003), `hep-ex/0307003`.
24. LASS Collab., *Nucl. Phys.* B **296**, 493 (1988).
25. FOCUS Collab., *Phys. Lett.* B **555**, 167 (2003), `hep-ex/0212058`.
26. CLEO Collab., CLEO CONF 03-02, EPS-371.
27. I. Furic for CDF Collab., XVII Rencontres de Physic de la Valle d'Aoeste, March 2003, Italy.
28. K. Stenson, talk "Recent Results in Charm Decays", APS (2003).
29. FOCUS Collab., `hep-ex/0308054`.
30. FOCUS Collab., *Phys. Lett.* B **561**, 225 (2003), `hep-ex/02011056`.

DISCUSSION

Questioner: I want to comment on your $\sigma(500)$. What you have done is really what is traditionally done in partial wave analyses. One does the analysis in partial waves and after obtaining the result, one fits it with a Breit-Wigner or a modified Breit-Wigner. So it is a very conventional sort of thing.

Jussara de Miranda: Not in a Dalitz plot.

Questioner: You can take the events out of a Dalitz plot and do it if you like.

Jussara de Miranda: We have tried this, it is much harder, you have to have one parameter for each slice. I've never seen this in a Dalitz plot. There are people in our collaboration trying to take this out. I understand that is what is done in scattering experiments and traditional partial wave analyses.

Questioner: In a Dalitz plot you cannot take the whole thing as there is crossing and this and that, but really this is very equivalent to what you have done.

Jussara de Miranda: Yes, the idea is the same, the way to go is the difference.

HADRON STRUCTURE

DEEP INELASTIC LEPTON-NUCLEON SCATTERING AT HERA

P. NEWMAN

School of Physics and Astronomy, University of Birmingham, B15 2TT, UK
E-mail: prn@hep.ph.bham.ac.uk

Data from the HERA collider experiments, H1 and ZEUS, have been fundamental to the rapid recent development of our understanding of the partonic composition of the proton and of QCD. This report focuses on inclusive measurements of neutral and charged current cross sections at HERA, using the full available data taken to date. The present precision on the proton parton densities and the further requirements for future measurements at the Tevatron and LHC are explored. Emphasis is also placed on the region of very low Bjorken-x and Q^2. In this region, the 'confinement' transition takes place from partons to hadrons as the relevant degrees of freedom and novel or exotic QCD effects associated with large parton densities are most likely to be observed. Finally, prospects for the second phase of HERA running are discussed.

1. The HERA Collider Experiments

The presence of a point-like probe together with only one initial state hadron makes deep inelastic lepton nucleon scattering (DIS) the ideal environment in which to study the quantum chromodynamics (QCD) of hadronic interactions and to constrain the parton densities of the proton. In the years 1992-2000, the HERA collider experiments, H1 and ZEUS, collected ep data at electron beam energies of 27.5 GeV and proton energies of 820 GeV and 920 GeV, corresponding to ep center-of-mass energies in excess of 300 GeV. The data were split between around 100 pb^{-1} of e^+p collisions and 15 pb^{-1} of e^-p collisions. With the extensions in accessible kinematic phase space afforded by the large center-of-mass energy and the precise electron and hadron reconstruction in the experiments over a wide rapidity range, these data have been used to gain new insights into many aspects of ep collisions.

This article focuses mainly on measurements of inclusive ep cross sections in both neutral current (NC, $ep \rightarrow eX$) and charged current (CC, $ep \rightarrow \nu X$) reactions throughout the available phase space. In most cases, the full available data from the first phase of HERA running are used. At relatively large momentum transfers, the inclusive cross sections yield important information on the parton densities of the proton. At small momentum transfers, they can be used to search for novel effects in quantum chromodynamics (QCD) associated with the large parton densities observed at the previously unexplored low momentum fractions of the struck quark.

Less inclusive measurements concentrating on other aspects of ep scattering are described elsewhere in these proceedings. The precision QCD tests and studies of the QCD evolution of parton cascades that have been possible through jet measurements are covered by Hirosky.[1] Measurements of the structure of diffractive exchanges and insights into the formation of rapidity gaps in hadronic interactions are discussed by Yamazaki.[2] Searches for new physics at HERA, at the highest $\sqrt{s}$ ever accessed in a collider with an initial state lepton, are covered by Perez.[3]

2. Neutral and Charged Current DIS at Large Q^2

The kinematics of inclusive DIS are usually described by the variables Q^2, the modulus of the squared four-momentum transfer carried by the exchanged electroweak gauge boson, and x, the fraction of the proton's longitudinal momentum carried by the quark that couples to the exchanged boson. Figure 1 illustrates the kinematic regions in which inclusive measurements have been made thus far at HERA.

The NC process takes place via the exchange of virtual photon and Z^0 propagators. The cross section can be expressed in the form

$$\frac{d\sigma^{NC}}{dx dQ^2} = 2\pi\alpha_{em}^2 \cdot \left(\frac{1}{Q^2}\right)^2 \cdot \frac{Y_+}{x} \cdot \tilde{\sigma}_{NC} , \qquad (1)$$

where the term α_{em}^2 expresses the dominance of photon exchange over most of the phase space, $1/Q^4$ is the photon propagator term and the reduced cross section $\tilde{\sigma}_{NC}$ contains helicity factors, weak terms due to Z^0 exchange and structure functions related to the parton densities of the proton. The variables $Y_\pm = 1 \pm (1 - y)^2$, dependent on the inelasticity, y, express the helicity dependence of the electroweak

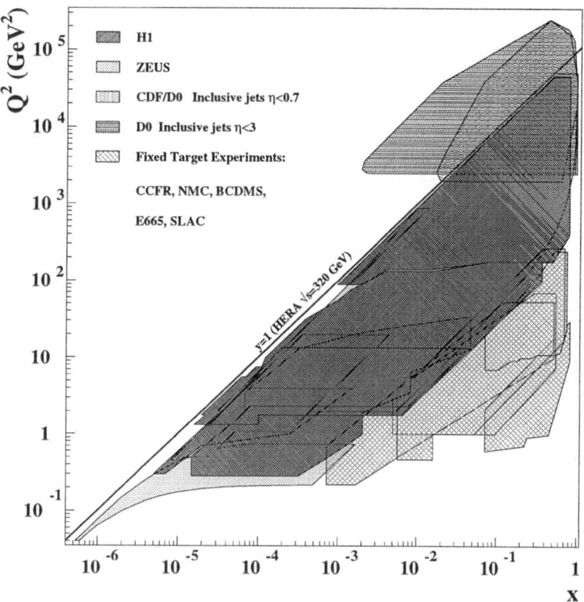

Figure 1. Kinematic plane in x and Q^2 covered by inclusive HERA and fixed-target DIS measurements. The line at inelasticity $y = 1$ represents the kinematic limit. The overlap region with recent jet measurements from the Tevatron[4] is also illustrated.

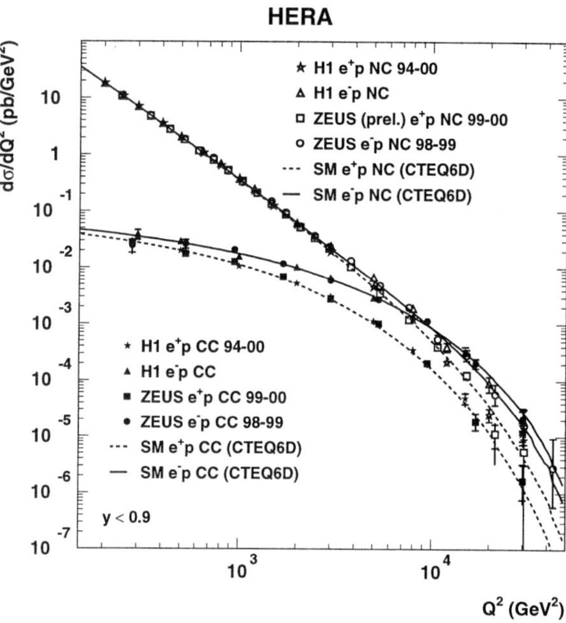

Figure 2. Single differential cross sections for charged and neutral current interactions in $e^\pm p$ collisions, as measured by H1 and ZEUS. The data are compared with the predictions of a recent global QCD fit.[11]

interaction (see also Eq. (3)).

The CC process is purely due to weak interactions. The cross section can be expressed as

$$\frac{\mathrm{d}\sigma^{\mathrm{CC}}}{\mathrm{d}x\mathrm{d}Q^2} = \frac{G_F^2 M_W^4}{2\pi} \cdot \left(\frac{1}{Q^2 + M_W^2}\right)^2 \cdot \frac{1}{x} \cdot \tilde{\sigma}_{\mathrm{CC}} \ , \quad (2)$$

where the coupling and propagator terms are specific to W boson exchange and the reduced cross section term $\tilde{\sigma}_{\mathrm{CC}}$ contains the helicity factors and structure functions.

Figure 2 shows the single differential cross sections measured by H1 and ZEUS for charged and neutral current $e^\pm p$ scattering with $Q^2 > 200$ GeV^2.[5–10] For $Q^2 \ll M_W^2$, the NC cross section dominates heavily due to the differences between the propagator terms in Eqs. (1) and (2). For $Q^2 \gtrsim M_W^2$, the cross sections for NC and CC processes become comparable, providing an illustration of electroweak unification with space-like gauge bosons. The remaining differences between the NC and CC cross sections in this large Q^2 region and the differences between the e^+p and e^-p cross sections can be understood from the structure of the reduced cross sections $\tilde{\sigma}_{\mathrm{NC}}$ and $\tilde{\sigma}_{\mathrm{CC}}$ (see Secs. 3.2 and 3.3).

At the largest Q^2, the data are sensitive to possible new physics beyond the Standard Model, for example due to quark compositeness. The data have been analyzed in the framework of possible contact interactions,[12] by comparing the measured cross sections with predictions based on parton densities constrained mainly by lower Q^2 and non-HERA data. An example of such a study is shown for NC e^+p interactions in Fig. 3. There is good agreement between the data and the predictions based on the CTEQ5D[13] parton densities up to the highest values of $Q^2 \sim 30\,000$ GeV^2. As a result, quark or electron substructure is excluded down to scales of $\sim 1.0 \times 10^{-18}$ m.

3. Parton Densities

3.1. *Current Precision on Parton Densities*

Measuring and understanding the parton densities of the proton over as wide a kinematic range as possible is a central aim of HERA analysis. This is important in its own right as a means of improving our understanding of QCD. It is also crucial for precision measurements and the understanding of backgrounds to searches for new physics at the Tevatron and LHC.

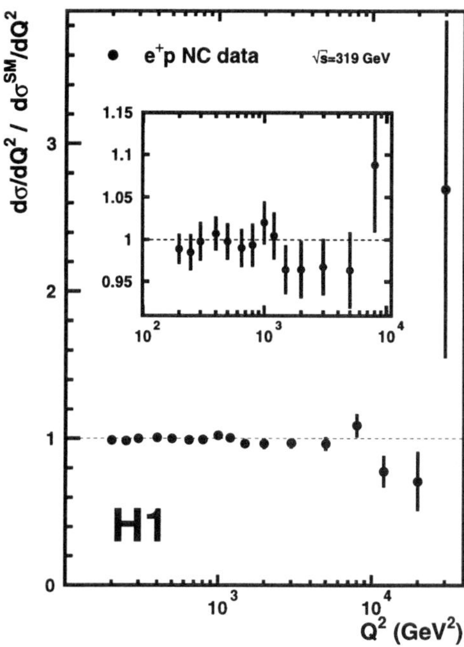

Figure 3. The ratio of H1 data to theoretical prediction for the single differential NC cross section $d\sigma^{\mathrm{NC}}/dQ^2$ in e^+p scattering. The theoretical predictions use the CTEQ5D[13] parton densities, which were obtained using only a small fraction of the HERA data.

As can be seen from consideration of the parton kinematics of pp scattering,[14] accurate knowledge of the quark and gluon densities over a wide range of x is necessary to make precise Standard Model calculations for processes such as weak gauge boson, Higgs, or top quark production over the necessary rapidity ranges at these machines.

Figure 4[11] shows estimates of the precision with which the parton densities are currently known. Over the range $10^{-4} \lesssim x \lesssim 10^{-1}$ the combined experimental and theoretical uncertainties on the u and d quark densities are at the level of a few percent,[a] largely thanks to the low x data provided by HERA. The gluon density is somewhat more poorly constrained over this region. At large x values, the uncertainties on all parton densities rapidly increase. This region is crucial for the understanding of backgrounds to the production of any new particles near to threshold in pp scattering. Due to kinematic con-

straints (Fig. 1), this high x region can only be accessed at HERA at the highest Q^2, where the cross section becomes small (Eq. (1)). Large data samples are therefore needed to improve on the precision obtained from fixed-target or other data in this region. In the following sections, the HERA data used to reach the levels of precision shown in Fig. 4 are presented and discussed in particular in terms of how improvements might be made at large x.

3.2. *Neutral Current Cross Sections*

The reduced neutral current cross section for $e^{\pm}p$ scattering, corrected for QED-radiative effects, can be expressed as

$$\tilde{\sigma}_{\mathrm{NC}}^{\pm} = F_2 \mp \frac{Y_-}{Y_+}\, xF_3 \; - \; \frac{y^2}{Y_+}\, F_L \, , \qquad (3)$$

where F_2, xF_3 and F_L are the generalized unpolarized proton structure functions. Extractions of F_2 and xF_3 are described in this section. HERA F_L data are discussed in Sec. 4.3.

The F_2 term is strongly dominant in most of the measured phase space at HERA. After corrections for Z^0 exchange and interference between the photon and Z^0 contributions, the pure electromagnetic structure function F_2^{em} can be extracted. In the quark-parton model, this structure function can be decomposed as

$$F_2^{\mathrm{em}}(x, Q^2) = x \sum_q e_q^2 \, (q + \bar{q}) \, , \qquad (4)$$

where the sum runs over quark species q of electric charge e_q. F_2^{em} thus provides a squared-charge weighted sum of all quark and antiquark densities. Since $e_u^2 = 4e_d^2$, it yields a particularly strong constraint on the u and $\bar{u}$ contributions.

Figure 5 shows a summary of the F_2^{em} data obtained from e^+p scattering at HERA.[6,10] The structure function is measured over a huge kinematic range and the data are very well described over most of the range by QCD fits (see Sec. (3.4)).[b] The precision reaches $2 - 3\%$ in the bulk of the phase space. However, in the region of the highest x, the precision of the HERA data remains far from that of fixed-target experiments such as BCDMS[16] and NMC,[17]

---

[a]This level of precision is only obtained after assumptions are made on the flavor decomposition at low x, which have yet to be tested. For example, all QCD fits assume that $\bar{u} - \bar{d} \to 0$ as $x \to 0$ and assumptions are necessary on the relative contributions from s and c quarks.

[b]A more complete discussion of the various QCD fits performed to DIS and other data and the estimates of the corresponding uncertainties can be found elsewhere in these proceedings.[15]

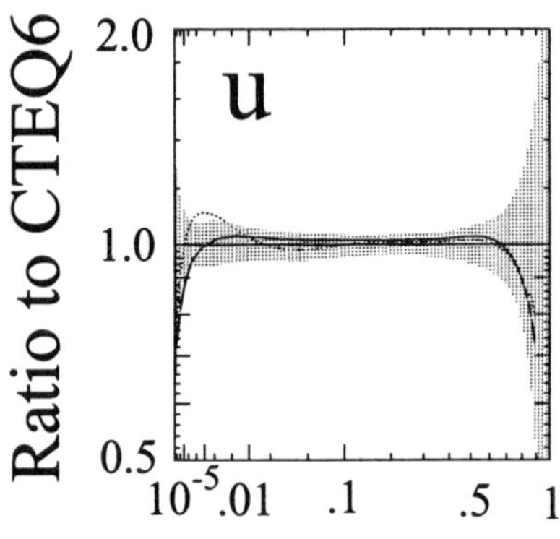
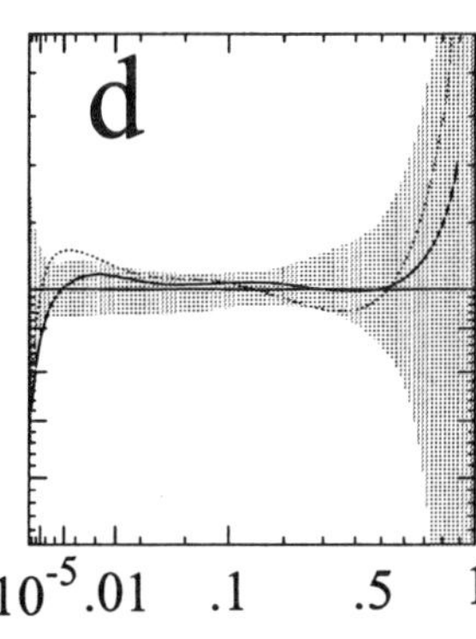
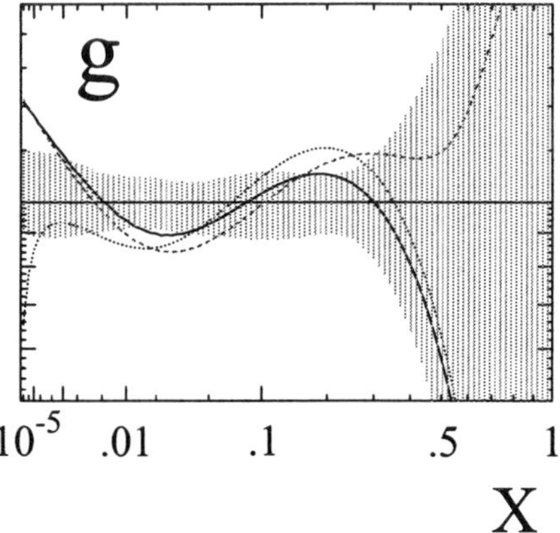

Figure 4. Estimates of the uncertainties on the proton parton densities at $Q^2 = 10$ GeV^2.[11] The bands show the fractional uncertainty as a function of x.

where the smaller center-of-mass energy allows measurements at high x, but lower Q^2. At the highest x, these fixed-target data are not well described by the QCD fits. It is highly desirable to obtain HERA measurements at high x and intermediate Q^2, where the potential problems of the fixed-target data (e.g. higher twist contributions and uncertainties in nuclear corrections) should be absent. This could be achieved with a reasonably large cross section by running HERA with a reduced proton beam energy, as is planned as part of the second phase of HERA running.

The $F_2^{\rm em}$ data also provide the best available constraints on the gluon density via the deviations from Bjorken scaling, caused by gluon radiation. In leading order of QCD, the gluon density can be obtained approximately from $\frac{\partial F_2^{\rm em}(x/2,Q^2)}{\partial \ln Q^2} \sim \alpha_s x g(x)$,[18] such that the strong positive scaling violations at low x in Fig. 5 are indicative of a large and growing gluon density as x becomes small.

The structure function xF_3 arises due to Z^0 exchange. In the HERA phase space, the interference contribution $xF_3^{\gamma Z}$ between the photon and Z^0 exchanges dominates, such that

$$xF_3 = -a_e \, \frac{\kappa Q^2}{Q^2 + M_Z^2} \, xF_3^{\gamma Z} + \Delta(Z^2) \, , \qquad (5)$$

where in the quark-parton model,

$$xF_3^{\gamma Z} = 2x \sum_q e_q a_q (q - \bar{q}) \, . \qquad (6)$$

Here, a_e and a_q are the axial couplings of the Z^0 to electrons and quarks, respectively and $\kappa^{-1} =$

$4\frac{M_W^2}{M_Z^2}(1 - \frac{M_W^2}{M_Z^2})$ in the on-mass-shell scheme. Since xF_3 measures the difference between the quark and antiquark densities, it is uniquely and model independently sensitive to the non-singlet valence quark densities.

Since it contributes with opposite signs to e^+p and e^-p scattering (Eq. (3)), xF_3 can be extracted from the measured differences between the NC e^+p and e^-p cross sections at large Q^2, shown in Fig. 2, according to

$$xF_3 = \frac{Y_+}{2Y_-} \left(\tilde{\sigma}_{\rm NC}^- - \tilde{\sigma}_{\rm NC}^+ \right) \, . \qquad (7)$$

Figure 6 shows the current status of xF_3 data from HERA.[6,8] The data are well described by the predictions of QCD fits in which the valence quark densities are principally constrained by the NC and CC HERA data and measurements from elsewhere, rather than by the differences between the $e^\pm p$ NC cross sections. These measurements thus provide a test of the procedures used and valence densities obtained in QCD fits. The large increases in luminosity expected at HERA-II are required to further exploit this observable.

3.3. Charged Current Cross Sections

In contrast to the NC measurements, where many millions of events are available for analysis, the CC samples collected so far at HERA consist of only around 1500 events (e^+p) and 700 events (e^-p) per experiment. The statistical uncertainties are correspondingly larger than in the NC case, amount-

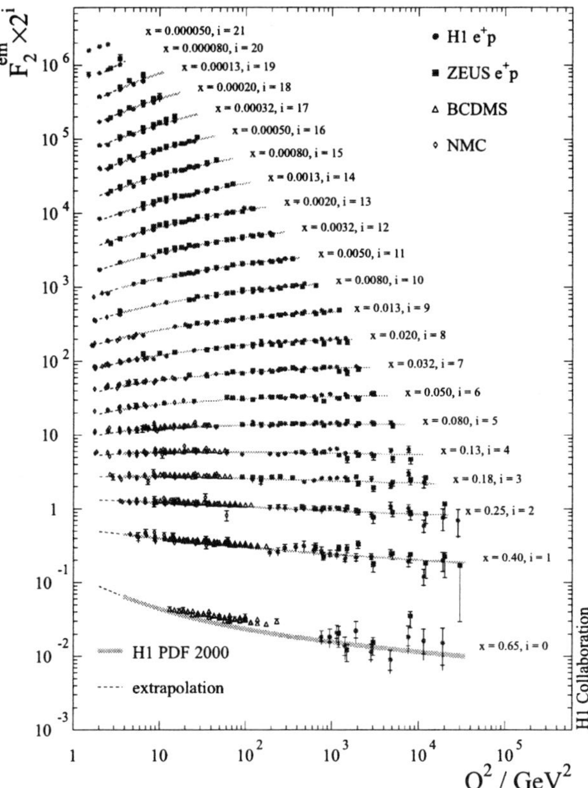

Figure 5. Summary of measurements of the structure function F_2^{em} by H1 and ZEUS. The data are compared with the results of a QCD fit to H1 NC and CC data only.[6]

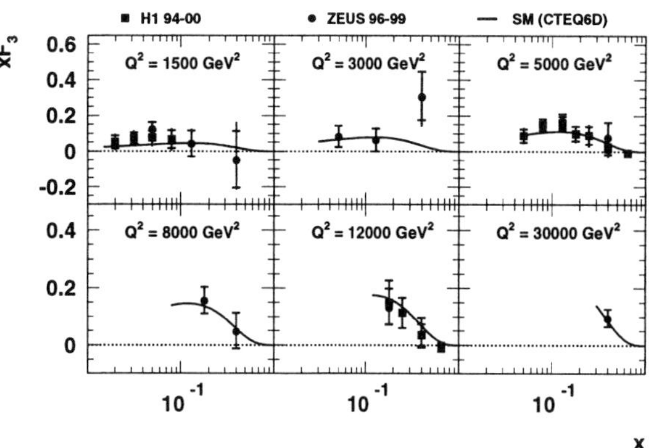

Figure 6. Data on xF_3 from HERA, based on around 100 pb^{-1} of e^+p data and 15 pb^{-1} of e^-p data, compared with the predictions of a recent global fit.[11]

ing to typically 6% for double differential measurements. Charged current cross section measurements do, however, provide important complementary information for the extraction of parton densities, since they are sensitive to particular quark flavors with the correct charges to couple to the exchanged W boson.

As can be seen from Fig. 2, the e^-p CC cross section is significantly larger than the e^+p cross section throughout the measured kinematic range. This is mainly because the e^-p cross section is dominated by the reaction $e^-u \to \nu_e d$, whereas the dominant e^+p process is $e^+d \to \bar{\nu}_e u$, the u density being the larger in the high x region where measurements can be made. In the quark parton model, the charged current reduced cross sections take the form

$$\tilde{\sigma}_{CC}^- = x(u + c) + (1 - y)^2\, x(\bar{d} + \bar{s}) \qquad (8)$$

$$\tilde{\sigma}_{CC}^+ = x(\bar{u} + \bar{c}) + (1 - y)^2\, x(d + s)\,, \qquad (9)$$

where the helicity factor $(1 - y)^2$ implies a further kinematic suppression to the term involving the d density in the e^+p case.

The e^-p CC cross section provides a complementary constraint to F_2^{em} on the u density at high x.[5,7] Despite the unfavorable helicity and smaller cross section, the e^-p CC data yield the most direct available constraint on the d density at large x from HERA. This is illustrated in Fig. 7, which shows the most recent measurements from H1[6] and ZEUS.[9] The data are compared with the predictions of a global QCD fit,[11] in which the u and d densities at large x are constrained mainly by precise fixed-target muon scattering data from protons and deuterons. The calculation from the fit is broken down into the contributions from scattering from d-type and $\bar{u}$-type quarks.

The CC measurements should improve considerably with HERA-II data. However, even with 1 fb^{-1}, the uncertainties will remain large in the important region $x \gtrsim 0.5$. An alternative method of constraining the d density at large x would be to run HERA with deuterons[19] and, using isospin symmetry, to unfold the d/u ratio. Running with deuterons at 920 GeV would also naturally reduce the beam energy per nucleon, such that the benefits of the larger cross section at large x and intermediate Q^2 could be exploited.

The ZEUS collaboration[9] have recently used their charged current data to extract a flavor singlet CC structure function,

$$F_2^{CC} = \frac{2}{Y_+}\left(\tilde{\sigma}_{CC}^+ + \tilde{\sigma}_{CC}^-\right) + \Delta(xF_3^{CC}, F_L^{CC})\,. \quad (10)$$

This structure function is shown in Fig. 8, where the effects of the small xF_3^{CC} and F_L^{CC} correction

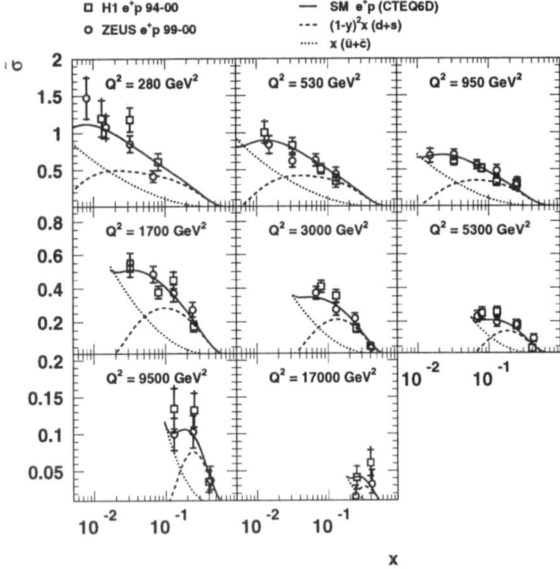

Figure 7. Double differential measurements of the CC e^+p cross section from HERA. The data are compared with the predictions of a recent global fit,[11] which are broken down into contributions from d-type and $\bar{u}$-type quarks.

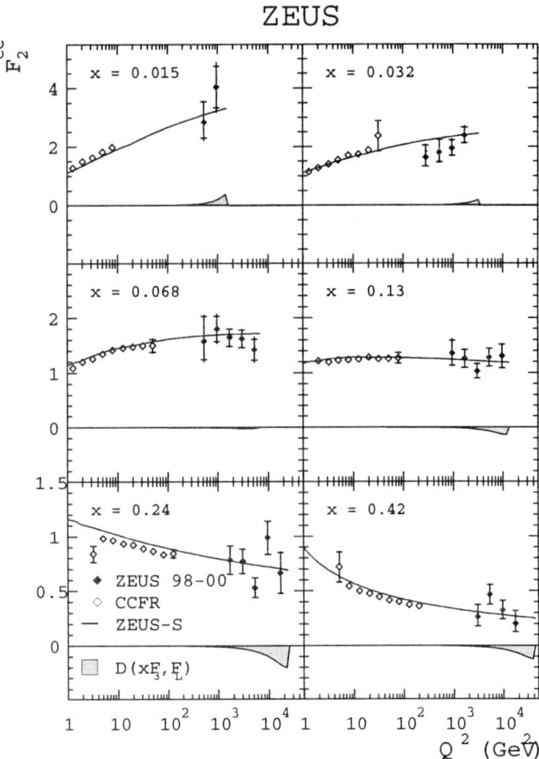

Figure 8. The flavor singlet CC structure function F_2^{CC} as obtained by ZEUS and CCFR. The data are compared with the predictions of a QCD fit to DIS data,[20] which does not include the CC data shown. The corrections necessary for the xF_3^{CC} and F_L^{CC} terms are also illustrated.

terms are also illustrated. The F_2^{CC} results are compared with precise fixed-target neutrino data from CCFR.[21] Viewing the fixed-target and HERA measurements together, the data span more than four orders of magnitude in Q^2 and the influence of gluon radiation on the CC process is clearly visible from the scaling violations. The ZEUS results are well described by the predictions of a QCD fit to various DIS data.[20]

3.4. *Parton Density Extractions*

The H1[6,22] and ZEUS[20] collaborations, along with various other groups,[11,23,24] have performed QCD fits to extract parton densities using various combinations of HERA and other data. The fits are based on the evolution of the parton densities with Q^2 using the DGLAP equations[25] in Next-to-Leading Order (NLO).[26]

With the latest HERA NC and CC data, it is now possible to extract the full set of flavor-separated parton densities from HERA data alone, provided assumptions are made on the validity of DGLAP evolution throughout the fitted phase space and the quark flavor decomposition at low x. Figure 9 shows the results for the valence densities, the sum of all sea quarks and the gluon density, from H1 NC and CC data only,[6] ZEUS NC data together with other fixed-target DIS experiments,[20] and CTEQ,[11] who perform a global fit to many DIS and other data sets. The agreement between the different extractions is reasonable, though there are differences between the H1 and ZEUS valence densities that go beyond the quoted error bands. This is perhaps not surprising, given the very different sources of information that are used to constrain the valence densities in the two fits. H1 use the limited sensitivity to W and Z exchange effects in the HERA data to separate the valence and sea densities, whereas ZEUS rely mainly on xF_3 data from fixed-target νFe and $\bar{\nu} Fe$ scattering.[27] The shapes of the gluon densities are also rather different. This arises from several sources, including different parameterizations of the parton densities at the starting scale for QCD evolution and

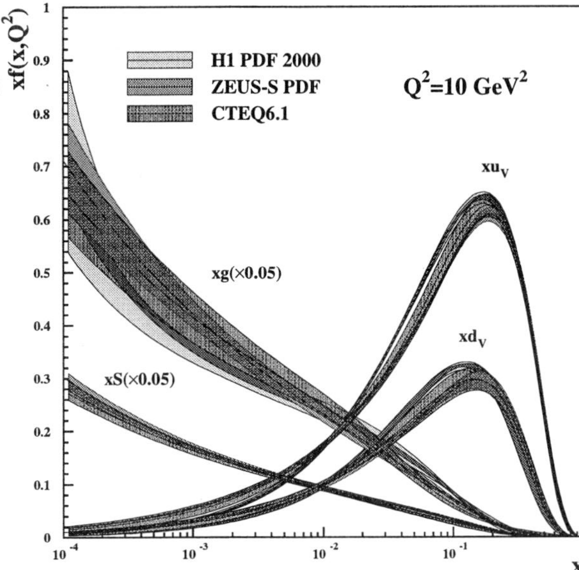

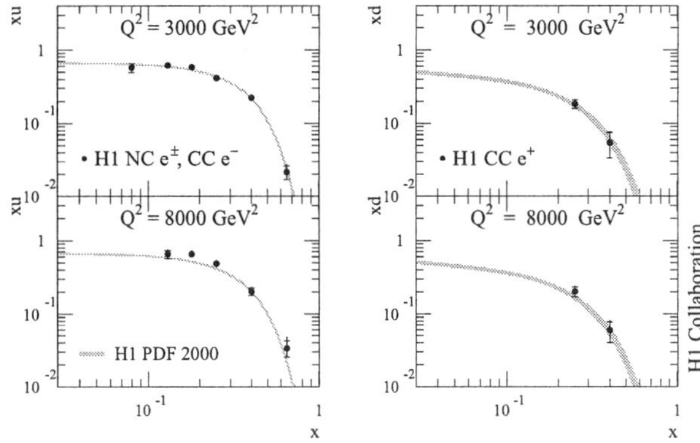

Figure 10. Results from the local extractions of parton densities by H1. The u density is obtained from NC $e^{\pm}$ and CC e^- data points for which it contributes in excess of 70% of the cross section. The d density is similarly extracted using CC e^+ data. The data are compared with the results of a QCD fit to all H1 NC and CC data.[6]

Figure 9. Parton densities as extracted by H1,[6] ZEUS[20] and CTEQ.[11] The sea quark and gluon densities are reduced by a factor of 20 for visibility.

different treatments of heavy quark evolution.[c] As $x \to 1$, the fixed-target data still give the best constraints. Deuteron running at HERA is desirable to test the assumption, used in all fits, that $\bar{d} - \bar{u} \to 0$ as $x \to 0$.[19]

Since the QCD fits rely on various assumptions, it is interesting to extract parton densities directly from the data at fixed values of x and Q^2, in a manner that is relatively insensitive to these assumptions. As shown in Fig. 10, the H1 collaboration has performed such an extraction of the u and d densities at high x using NC and CC data points for which the relevant parton contributes in excess of 70% to the measured cross section according to the H1 QCD fit.[6] The QCD fit is then used to correct for the remaining contributions. The resulting local extractions of the u and d densities are in good agreement with the predictions of the QCD fits using DGLAP evolution.

3.5. *Tests of the Gluon Density*

The fits to inclusive data rely on the DGLAP QCD evolution equations[25,26] to relate the scaling violations of $F_2^{\rm em}$ to the gluon density and require assumptions on the functional form of the gluon density at the starting scale for QCD evolution. It is important

also to constrain the gluon density from other complementary sources with different systematics, in order to test the overall consistency of the HERA data and the validity of the assumptions of DGLAP evolution and QCD hard scattering factorization. This has been done in several ways using hadronic final state data at HERA. Measurements of dijet and charm production cross sections are highly sensitive to the gluon density, since they proceed dominantly via the boson-gluon fusion process $\gamma^* g \to q\bar{q}$, the cross section for which is directly proportional to the gluon density at leading order of QCD. The gluon density can thus be extracted in a manner which is more sensitive to local variations.

Examples of jet data that are sensitive to the gluon density are given by Hirosky.[1] An example of recent charm data that constrain the gluon density can be found in Fig. 11, which shows the cross section for D^* production in DIS as a function of pseudorapidity, as measured by the ZEUS collaboration.[30] The data are compared with a theoretical prediction[31] based on NLO QCD matrix elements, interfaced to the gluon density from a fit to inclusive DIS measurements[20] and to fragmentation functions. The beautiful agreement of the data and theory confirms the gluon density from scaling violations and the validity of the NLO DGLAP theory at the 10% level. The theoretical errors, dominated by the choice of the charm quark mass m_c, are larger than the uncertainties on the data. Comparing the predictions using the ZEUS and CTEQ[13] parton densities shows that the data can be used to improve the

---

[c]The gluon densities are thus defined in different schemes and strictly speaking cannot be compared directly.

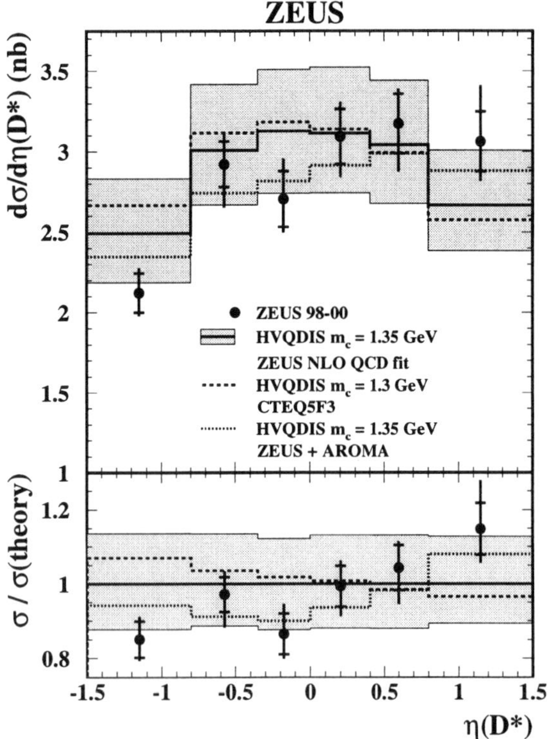

Figure 11. Cross section for D^* meson production in DIS, differential in pseudorapidity. The data are compared with the predictions of a NLO QCD calculation based on parton densities from a fit to DIS data,[20] together with an estimate of the theoretical uncertainties. The effects of switching to the CTEQ5F3 parton densities[13] or to Lund[28] rather than Petersen[29] fragmentation (ZEUS + AROMA) are also indicated.

constraints on the gluon density once the theoretical errors are better controlled.

By extrapolating the D^* cross sections in the measured range of pseudorapidity and transverse momentum to a charm production cross section integrated over the full phase space, it is possible to extract $F_2^{c\bar{c}}$, the charm contribution to the proton structure function F_2.[30,32] For $Q^2 \gg m_c^2$, such that the charm mass can be neglected, the ratio $F_2^{c\bar{c}}/F_2$ becomes close to 30% at low x, illustrating the importance of a proper treatment of the evolution of heavy quarks in any fit to HERA data.

4. The Low x Region

4.1. *Low x Physics*

The region of low x, newly accessed at HERA, has been the subject of much debate. The fast rise of the gluon density (Fig. 9) raises the question of whether unitarization effects may become important.[d] As the gluon density becomes large, the partons must ultimately begin to interact through processes such as gluon recombination ($gg \rightarrow g$).[34] This would lead to a taming of the low x rise of F_2 and a breakdown of the DGLAP approximation.

A full perturbative QCD expansion gives rise to evolution of parton densities with both $\ln Q^2$ and $\ln 1/x$. Standard DGLAP evolution is equivalent to a resummation of leading $\ln Q^2$ terms, such that the struck quark originates from a parton cascade ordered in virtuality. At sufficiently low x, evolution in $\ln 1/x$ must also become important, though it is not incorporated in the DGLAP approximation. Other approximations to QCD evolution may then become more appropriate. Examples are BFKL evolution,[35] which resums the $\ln 1/x$ terms to all orders, or CCFM evolution,[36] in which the partons are ordered in the angle at which they are emitted. CCFM evolution is equivalent to BFKL evolution for $x \rightarrow 0$, whilst limiting to the DGLAP equations at larger x.

It has been suggested that the inclusion of BFKL effects improves the description of low x inclusive measurements by QCD fits,[37] though no clear consensus exists on this question. There are also hints from various hadronic final state analyses that regions of phase space can be found at HERA for which standard DGLAP evolution is insufficient and CCFM evolution may provide a better description.[38]

Due to kinematic correlations (Fig. 1), low x values can only be accessed at low Q^2. The low Q^2 regime brings its own complications, such as possible higher twist contributions and the breakdown of convergence of perturbative QCD as the strong coupling increases. Around $Q^2 = 1$ GeV^2, the "confinement" transition takes place, such that the partons of asymptotic freedom are replaced by hadrons as the relevant degrees of freedom.

4.2. *F_2 at Low Q^2*

The data used in the measurements and QCD fits described in Sec. 2 cover the range $Q^2 \gtrsim 3$ GeV^2, where perturbative QCD can reliably be used. The ZEUS

---

[d]It has been argued that the conventional Froissart unitarity bound on hadronic total cross sections is not applicable to off-shell virtual photons.[33]

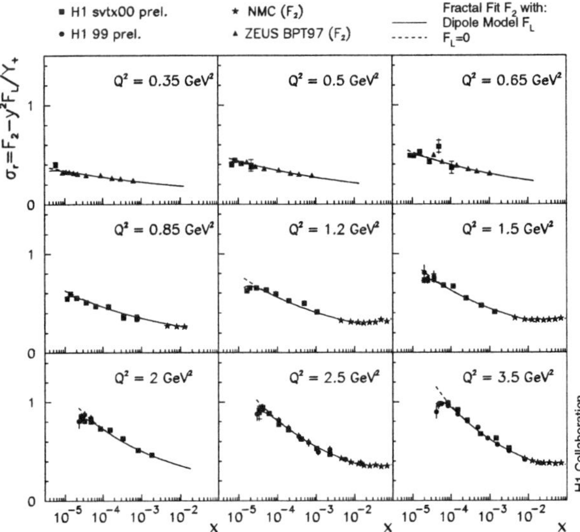

Figure 12. The reduced neutral current cross section $\tilde{\sigma}_{NC}$ at low Q^2. The data are compared with the predictions of a 'fractal' model of proton structure.[41]

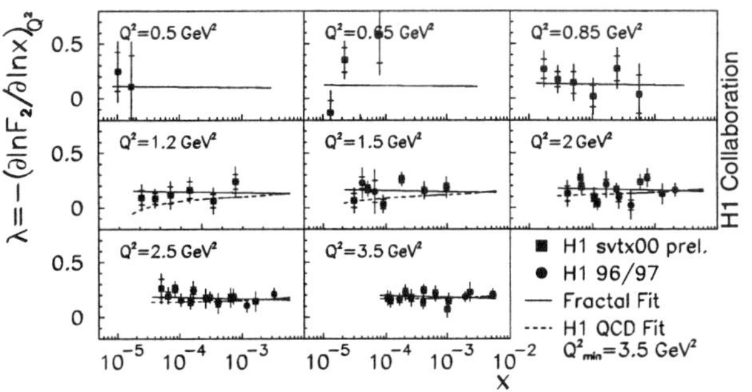

Figure 13. The logarithmic x derivative of F_2 in the low Q^2 region. The data are compared with the predictions of a 'fractal' model of proton structure.[41]

collaboration has obtained precise data (BPT97) in the range $0.0045 < Q^2 < 0.65$ GeV^2, using a silicon strip tracking detector and an electromagnetic calorimeter very close to the beampipe.[39] Previously, the intermediate region, $0.65 < Q^2 \lesssim 3$ GeV^2 has been only poorly explored at HERA, due to the acceptance limitations of the main detectors at small electron scattering angles. In order to improve this acceptance, a short run was taken in the year 2000 with the ep vertex shifted by 70 cm in the outgoing proton direction. The H1 collaboration has recently reported new inclusive NC measurements in the region $0.35 < Q^2 < 3.5$ GeV^2, using these data.[40] The resulting inclusive cross section measurements are shown in the form of the reduced cross section $\tilde{\sigma}_{NC}$ in Fig. 12. The new data span the transition from a fast rise of the cross section with decreasing x at $Q^2 = 3.5$ GeV^2 to a soft rise, similar to that observed in the energy dependence of hadron-hadron total cross sections,[42] at $Q^2 = 0.35$ GeV^2. At the lowest x values, a decrease in the cross section is observed due to the F_L term in Eq. (3) (see also Sec. 4.3).

In the double asymptotic limit,[43] the DGLAP equations can be solved with a solution whereby F_2 rises approximately as a power of x as x becomes small, such that $F_2 \sim x^{-\lambda}$. This feature is also predicted from the BFKL equations. Since unitarization effects would be expected to tame this growth, extracting λ has been suggested[44] as a means of searching for saturation effects. λ corresponds to the logarithmic x derivative of F_2 at fixed Q^2,

$$\lambda(x, Q^2) = (\partial \ln F_2 / \partial \ln x)_{Q^2} , \qquad (11)$$

which has been extracted locally from the differences between neighboring data points in x by the H1 collaboration.[40,45] The results in the kinematic region of the shifted vertex data are shown in Fig. 13. The data here and at larger Q^2 are consistent with no dependence of λ on x for fixed Q^2 and $x \lesssim 10^{-2}$, and thus with a monotonic rise of F_2 as x decreases with Q^2 fixed. There is thus no evidence for any taming of this rise in inclusive electroproduction data from HERA.

Since the logarithmic x derivative is compatible with independence of Q^2, the proton structure function at low x can indeed be parameterized as

$$F_2 = c(Q^2) \cdot x^{-\lambda(Q^2)} . \qquad (12)$$

H1 and ZEUS have both fitted their data to this form.[40,45,46] The results for $\lambda(Q^2)$ are shown in Fig. 14. Two distinct regions seem to be distinguished. For $Q^2 \gtrsim 3$ GeV^2, where partons are the relevant degrees of freedom, λ depends logarithmically on Q^2 and $c \sim 0.18$ is consistent with being constant. This behavior is well reproduced by DGLAP-based QCD fits. In contrast, for $Q^2 \lesssim 1$ GeV^2, there is evidence for a decrease in $c(Q^2)$ and for deviations of $\lambda(Q^2)$ from the logarithmic dependence on Q^2 as it tends to the value of 0.08 known to describe hadron-hadron[42] and photoproduction[47] total cross sections. In this region, the description by DGLAP breaks down as the confinement transition takes place on a distance scale of around 0.3 fm.

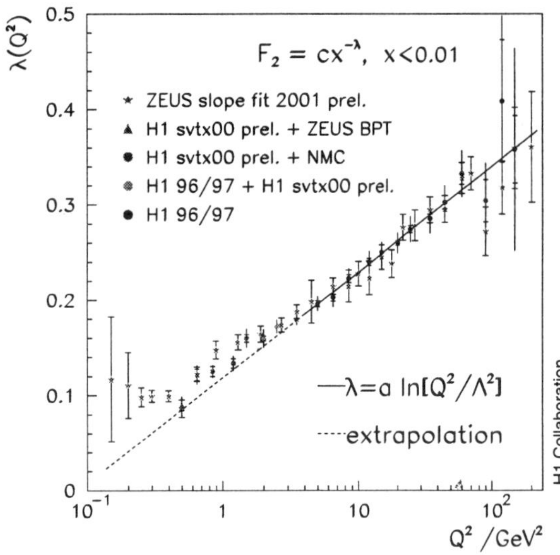

Figure 14. The results for λ from fits to low x data of the form $F_2 \sim x^{-\lambda}$. The data are compared with a parameterization in which λ grows logarithmically with Q^2.

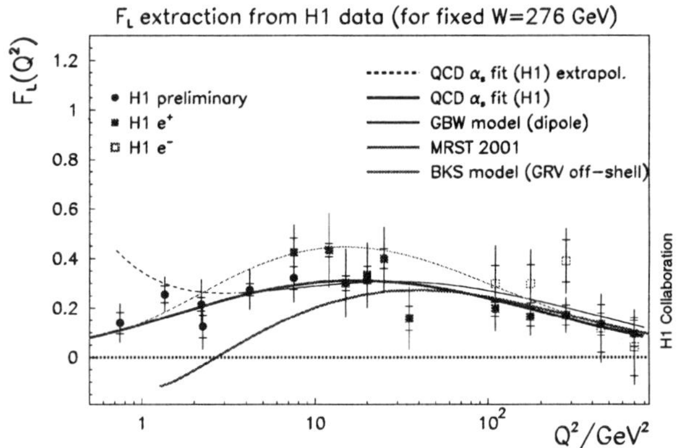

Figure 15. Summary of H1 F_L determinations shown at fixed photon-proton center-of-mass energy $W = 276$ GeV. The data are compared with the predictions of a QCD fit to H1 NC and CC data only,[6] a global QCD fit,[23] a phenomenological dipole model[50] and a model based on unintegrated parton densities and k_T factorization.[51].

4.3. F_L at Low Q^2

As can be seen from Fig. 12, the effects of F_L are visible in the inclusive reduced cross section at the lowest x, or highest y values. In this region, the scattered electron energy becomes very small and background from processes at $Q^2 \simeq 0$ in which a hadron is misidentified as the scattered electron becomes large. The H1 collaboration is able to make measurements for scattered electron energies as low as 3 GeV with the help of drift chambers and a silicon tracking detector accompanying the backward calorimeter. These detectors allow the event vertex to be reconstructed from the electron track at high y and enable the suppression of photoproduction background by ensuring that a track of the correct charge is linked to the electron candidate calorimeter cluster.

F_L is identically zero in lowest order QCD, but acquires a non-zero value at $\mathcal{O}(\alpha_s)$ due to gluon radiation. It is thus able to play a similar role to dijet and charm data (Sec. 3.5) in providing complementary information on the gluon density to that obtained from the scaling violations of F_2 assuming DGLAP evolution. This is particularly important at low x, where the Q^2 range of HERA F_2 measurements is rather small (see Fig. 1), dijet and charm measurements cannot be made due to kinematic restrictions and DGLAP evolution is most questionable. The

sensitivity to F_L at high y, visible in Fig. 12, has been exploited to determine F_L in the crucial region around $Q^2 = 1$ GeV^2.

The determination is made by fitting the reduced cross section to the form

$$\tilde{\sigma}_{NC} = F_2 - (y^2/Y_+) F_L \tag{13}$$
$$= c(Q^2) \cdot x^{-\lambda(Q^2)} - (y^2/Y_+) F_L , \tag{14}$$

where the second equality follows from Eq. (12) and at each Q^2 value, c, λ and F_L are free parameters. The results of the extraction are insensitive to the assumptions on the behavior of F_2 at the present level of accuracy. The shape of the low x turn-over of $\tilde{\sigma}_{NC}$ is driven by the y^2/Y_+ dependence, such that F_L can only be extracted at a single point in x. The results are shown as a function of Q^2 in Fig. 15, together with other F_L extractions using similar methods, spanning three orders of magnitude in Q^2.[5,6,48,49] The data are compared with a variety of predictions based on DGLAP QCD fits[22,23] and other phenomenological approaches.[50,51] The data show that F_L remains non-zero down to the lowest Q^2 values measured and already distinguish between the different models in the low x region.

Significant further progress in F_L measurements at HERA can only be made by reducing the proton beam energy, such that the F_2 and F_L terms in Eq. (13) can be separated through measurements at the same x and Q^2, but different y. This would remove the need for assumptions on the behavior of F_2

in the region where F_L effects are present and would allow measurements of the x dependence, providing further important discrimination between models.

5. Future Prospects

The HERA accelerator has recently restarted providing collisions, following a shutdown during which it was upgraded to provide a factor of around four increase in instantaneous luminosity. Spin rotators and polarimeters have also been placed around the electron ring, so that the effects of longitudinal polarisation of the electrons can be studied. In parallel, many upgrades have been made to the H1 and ZEUS detectors, including improved silicon tracking, forward tracking and track-based triggering. These improvements should improve the quality of data on charmed hadrons in particular and should extend the accessible phase space for many final state measurements towards higher x.

Over the next few years, the aim is to collect 1 fb^{-1} of data, representing a factor of 10 increase in statistics, equally shared between positron and electron running with positive and negative lepton helicities. A run with reduced proton beam energies in order to measure F_L and access the high x, intermediate Q^2 region is also planned. Further options for the running of HERA, for example replacing the proton beam with deuterons or heavier ions or polarising the proton beam in order to study nucleon spin at low x, do not currently form part of the future plans.

6. Summary

The data from the first phase of HERA running have now been fully analyzed from the point of view of high Q^2 inclusive charged and neutral current cross sections. The resulting data provide the best available constraints on the proton quark and gluon densities in the region $10^{-4} < x < 10^{-1}$, crucial for future experimentation at the Tevatron and LHC. Further improvements at larger x are possible in the future with higher luminosities and reduced proton energy running. The data on hadronic final states are highly sensitive to the QCD of hadronic interactions and complement the inclusive measurements, providing tests of the QCD evolution equations and competitive information on the gluon density. Here,

improvements in the precision of theoretical calculations are required in order to make significant further progress. There have been considerable recent developments in understanding the region of low x and Q^2 and testing the range of validity of DGLAP evolution. With the HERA-II run just beginning, the prospects are exciting for future measurements.

Acknowledgments

I would like to acknowledge the work of all members of the H1 and ZEUS collaborations, which has led to the many impressive results shown in this article. I would also like to thank J. Butterworth, V. Chekelian, E. Elsen, T. Greenshaw, B. Heinemann, V. Hudgson, M. Klein, T. Laštovička, K. McFarland, D. Naples, E. Rizvi, R. Thorne and R. Wallny for help with the preparation of this presentation and/or for proof-reading this manuscript.

References

1. R. Hirosky, these proceedings.
2. Y. Yamazaki, these proceedings.
3. E. Perez, these proceedings.
4. D0 Collaboration, *Phys. Rev. Lett.* **82**, 2451 (2001); CDF Collaboration, *Phys. Rev.* D **64**, 032001 (2001).
5. H1 Collaboration, *Eur. Phys. J.* C **19**, 269 (2001).
6. H1 Collaboration, *Eur. Phys. J.* C **30**, 1 (2003).
7. ZEUS Collaboration, *Phys. Lett.* B **539**, 197 (2002).
8. ZEUS Collaboration, *Eur. Phys. J.* C **C28**, 175 (2003).
9. ZEUS Collaboration, DESY 03-093, submitted to *Eur. Phys. J. C.*
10. ZEUS Collaboration, paper 630 submitted to International Europhysics Conference on High Energy Physics (EPS01), Budapest, Hungary, July 2001.
11. J. Pumplin *et al.*, *JHEP* **0207**, 012 (2002).
12. H1 Collaboration, *Phys. Lett.* B **568**, 35 (2003); ZEUS Collaboration, *Eur. Phys. J.* C **14**, 239 (2000).
13. H. Lai *et al.*, *Eur. Phys. J.* C **12**, 375 (2000).
14. A. Martin *et al.*, *Eur. Phys. J.* C **14**, 133 (2000).
15. R. Thorne, these proceedings.
16. BCDMS Collaboration, *Phys. Lett.* B **223**, 485 (1989).
17. NMC Collaboration, *Phys. Lett.* B **364**, 107 (1995).
18. K. Prytz, *Phys. Lett.* B **311**, 286 (1993).
19. T. Alexopoulos *et al.*, *"Electron-Deuteron Scattering at HERA, a Letter of Intent for an Experimental Programme with the H1 Detector"*, DESY-PRC 03/02 (available from http://www-h1.desy.de/h1/www/publications/H1eDLoI.pdf).
20. ZEUS Collaboration, *Phys. Rev.* D **67**, 012007 (2003).

21. CCFR/NuTeV Collaboration, *Phys. Rev. Lett.* **86**, 2742 (2001).

22. H1 Collaboration, *Eur. Phys. J.* C **21**, 33 (2001).

23. A. Martin *et al.*, *Eur. Phys. J.* C **23**, 73 (2002).

24. S. Alhekin, *Phys. Rev.* D **68**, 114002 (2003).

25. V. Gribov and L. Lipatov, *Sov. J. Nucl. Phys.* **15**, 438 & 675 (1972);
L. Lipatov, *Sov. J. Nucl. Phys.* **20**, 94 (1975);
G. Altarelli and G. Parisi, *Nucl. Phys.* B **126**, 298 (1977);
Y. Dokshitzer, *Sov. Phys. JETP* **46**, 641 (1977).

26. W. Furmanski and R. Petronzio, *Phys. Lett.* B **97**, 437 (1980).

27. CCFR Collaboration, *Phys. Rev. Lett.* **79**, 1213 (1997).

28. B. Andersson *et al.*, *Phys. Rep.* **97**, 31 (1983).

29. C. Petersen *et al.*, *Phys. Rev.* D **27**, 105 (1983).

30. ZEUS Collaboration, DESY-03-115, submitted to *Phys. Rev. D.*

31. B. Harris and J. Smith, *Phys. Rev.* D **57**, 2806 (1998).

32. H1 Collaboration, *Phys. Lett.* B **528**, 199 (2002).

33. S. Troshin and N. Tyurin, *Eur. Phys. J.* C **22**, 667 (2002).

34. L. Gribov, E. Levin and, M. Ryskin, *Nucl. Phys.* B **188**, 555 (1981);
L. Gribov, E. Levin and M. Ryskin, *Phys. Rep.* **100**, 1 (1983).

35. V. Fadin, E. Kuraev and L. Lipatov, *Sov. Phys. JETP* **44**, 443 (1976);
V. Fadin, E. Kuraev and L. Lipatov, *Sov. Phys. JETP* **45**, 199 (1977);
Y. Balistsky and L. Lipatov, *Sov. J. Nucl. Phys.* **28**, 822 (1978).

36. M. Ciafaloni, *Nucl. Phys.* B **296**, 49 (1988);
S. Catani, F. Fioriani and M. Marchesini, *Phys. Lett.* B **234**, 339 (1990);
S. Catani, F. Fioriani and M. Marchesini, *Nucl. Phys.* B **336**, 18 (1990);
M. Marchesini, *Nucl. Phys.* B **445**, 49 (1995).

37. R. Thorne, *Phys. Rev.* D **60**, 054031 (1999).

38. H1 Collaboration, *Nucl. Phys.* B **538**, 3 (1999);
ZEUS Collaboration, *Eur. Phys. J.* C **6**, 239 (1999);
ZEUS Collaboration, *Phys. Lett.* B **474**, 223 (2000);
H1 Collaboration, *Phys. Lett.* B **542**, 193 (2002).

39. ZEUS Collaboration, *Phys. Lett.* B **487**, 53 (2000).

40. H1 Collaboration, paper 975 submitted to International Conference on High Energy Physics (ICHEP02), Amsterdam, Holland, July 2002.

41. T. Laštovička, *Eur. Phys. J.* C **24**, 529 (2002).

42. A. Donnachie and P. Landshoff, *Phys. Lett.* B **296**, 227 (1992).

43. A. De Rujula *et al.*, *Phys. Rev.* D **10**, 1649 (1974);
R. Ball and S. Forte, *Phys. Lett.* B **335**, 77 (1994).

44. H. Navelet, R. Peschanski and S. Wallon, *Mod. Phys. Lett.* A **9**, 3393 (1994).

45. H1 Collaboration, *Phys. Lett.* B **520**, 183 (2001).

46. ZEUS Collaboration, *Eur. Phys. J.* C **7**, 609 (1999).

47. H1 Collaboration, *Zeit. Phys.* C **69**, 27 (1995);
ZEUS Collaboration, *Nucl. Phys.* B **627**, 3 (2002).

48. H1 Collaboration, paper 83 submitted to International Europhysics Conference on High Energy Physics (EPS03), Aachen, Germany, July 2003.

49. H1 Collaboration, paper 799 submitted to International Europhysics Conference on High Energy Physics (EPS01), Budapest, Hungary, July 2001.

50. K. Golec-Biernat and M. Wüsthoff, *Phys. Rev.* D **59**, 014017 (1999).

51. B. Badelek, J. Kwieciński and A. Staśto, *Zeit. Phys.* C **74**, 297 (1997).

DISCUSSION

Thomas Gehrmann (Zürich University): Concerning the measurement of $F_2^{c\bar{c}}$, you mention that you obtain $F_2^{c\bar{c}}$ from an extrapolation of the charmed hadron spectra. Could you please comment on how much $F_2^{c\bar{c}}$ is actually measurement, and how much is extrapolation? And: how big is the error due to the extrapolation?

Paul Newman: The extrapolations can be large. For the ZEUS measurement shown, they vary from a factor of 5 at low Q^2 to a factor of 1.5 at high Q^2. They are done using NLO QCD programs, the uncertainties are assessed and included in the quoted errors. They are not the dominant uncertainties, though assessing the full extrapolation errors is non-trivial.

Rik Yoshida (ZEUS spokesman, Argonne, paraphrased by PN): While the systematic errors for the extrapolation of the D^* cross section to $F_2^{c\bar{c}}$ are evaluated, the extraction of $F_2^{c\bar{c}}$ is necessarily a model dependent procedure. It is better to compare models of charm production with the data directly at the level of the measured differential cross sections. It is difficult to assess the correctness of models that only predict $F_2^{c\bar{c}}$ or the total charm cross section.

Paul Newman: That is a very good point. The usefulness of $F_2^{c\bar{c}}$ lies mainly in the illustration of the charm contribution to F_2 as a function of the more familiar variables x and Q^2.

Ikaros Bigi (University of Notre Dame du Lac): Do you see any evidence for intrinsic charm in the data, or what is the status of this ancient concept?

Paul Newman (paraphrased and extended): Our data are consistent with the charm component being entirely due to QCD evolution and thus related to the gluon density. With the present data, there is no need for an additional source. However, better data in the high x region where intrinsic charm contributions have previously been discussed are likely to become available at HERA-II, now that charm triggers and forward tracking have been improved.

PARTON DISTRIBUTIONS

R. S. THORNE

Cavendish Laboratory, University of Cambridge, Madingley Road, Cambridge, CB3 0HE, UK
E-mail: thorne@hep.phy.cam.ac.uk

I discuss our current understanding of parton distributions. I begin with the underlying theoretical framework, and the way in which different data sets constrain different partons, highlighting recent developments. The methods of examining the uncertainties on the distributions and those physical quantities dependent on them is analyzed. Finally I look at the evidence that additional theoretical corrections beyond NLO perturbative QCD may be necessary, what type of corrections are indicated and the impact these may have on the uncertainties.

1. Introduction

The proton is described by QCD – the theory of the strong interactions. This makes an understanding of its structure a difficult problem. However, it is also a very important problem – not only as a question in itself, but also in order to search for and understand new physics. Many important particle colliders use hadrons – HERA is an ep collider, the Tevatron is a $p\bar{p}$ collider, the LHC at CERN will be a pp collider, and an understanding of proton structure is essential in order to interpret the results. Fortunately, when one has a relatively large scale in the process, in practice only $> 1\ \mathrm{GeV}^2$, the proton is essentially made up of the more fundamental constituents – quarks and gluons (partons), which interact relatively weakly. Hence, the fundamental quantities one requires in the calculation of scattering processes involving hadronic particles are the parton distributions. These can be derived from, and then used within, the *factorization theorem* which separates processes into nonperturbative parts which can be determined from experiment, and perturbative parts which can be calculated as a power-series in the strong coupling constant α_S.

This is illustrated in the canonical example of deep-inelastic scattering. The cross section for the virtual photon-proton interaction can be written in the factorized form

$$\sigma(ep \to eX) = \sum_i C_i^{DIS}(x, \alpha_s(Q^2)) \otimes f_i(x, Q^2)$$

where Q^2 is the photon virtuality, $x = \frac{Q^2}{2m\nu}$, the momentum fraction of the parton (ν=energy transfer in the lab frame), and the $f_i(x, Q^2)$ are the parton distributions, i.e. the probability of finding a parton of type i carrying a fraction x of the momentum of the hadron. Corrections to the above formula are of $\mathcal{O}(\Lambda_{\mathrm{QCD}}^2/Q^2)$ and are known as higher twist. The parton distributions are not easily calculable from first principles. However, they do evolve with Q^2 in a perturbative manner, satisfying the evolution equation

$$\frac{df_i(x, Q^2)}{d\ln Q^2} = \sum_j P_{ij}(x, \alpha_s(Q^2)) \otimes f_j(x, Q^2)$$

where the splitting functions $P_{ij}(x, \alpha_s(Q^2))$ are calculable order by order in perturbation theory. The coefficient functions $C_i^P(x, \alpha_s(Q^2))$ describing a hard scattering process are process dependent but are calculable as a power-series, i.e. $C_i^P(x, \alpha_s(Q^2)) = \sum_k C_i^{P,k}(x)\alpha_s^k(Q^2)$. Since the $f_i(x, Q^2)$ are process-independent, i.e. *universal*, once they have been measured at one experiment, one can predict many other scattering processes.

Global fits[1-7] use all available data, largely structure functions, and the most up-to-date QCD calculations, currently NLO–in–$\alpha_s(Q^2)$, to best determine these parton distributions and their consequences. In the global fits input partons are parameterized as, e.g.

$$xf(x, Q_0^2) = (1 - x)^\eta (1 + \epsilon x^{0.5} + \gamma x)x^\delta$$

at some low scale $Q_0^2 \sim 1 - 5\ \mathrm{GeV}^2$, and evolved upwards using NLO evolution equations. Perturbation theory should be valid if $Q^2 > 2\ \mathrm{GeV}^2$, and hence one fits data for scales above $2 - 5\ \mathrm{GeV}^2$, and this cut should also remove the influence of higher twists, i.e. power-suppressed contributions.

In principle there are many different parton distributions – all quarks and antiquarks, and the gluons. However, $m_c, m_b \gg \Lambda_{\mathrm{QCD}}$ (and top does not usually contribute), so the heavy parton distributions are determined perturbatively. Also we usually assume $s = \bar{s}$, and that isospin symmetry holds, i.e. $p \to n$ leads to $d(x) \to u(x)$ and $u(x) \to d(x)$.

This leaves 6 independent combinations. Relating s to $1/2(\bar{u}+\bar{d})$ we have the independent distributions

$$u_V = u - \bar{u}, \ d_V = d - \bar{d}, \ \mathrm{sea} = 2(\bar{u}+\bar{d}+\bar{s}), \ \bar{d}-\bar{u}, \ g.$$

It is also convenient to define $\Sigma = u_V + d_V + \mathrm{sea} + (c+\bar{c})+(b+\bar{b})$. There are then various sum rules constraining parton inputs and which are conserved by evolution order by order in α_S, i.e. the number of up and down valence quarks and the momentum carried by partons (the latter being an important constraint on the gluon which is only probed indirectly),

$$\int_0^1 x\Sigma(x) + xg(x)\,dx = 1.$$

When extracting partons one needs to consider that not only are there 6 independent combinations, but there is also a wide distribution of x from 0.75 to 0.00003. One needs many different types of experiment for a full determination. The sets of data usually used are: H1 and ZEUS $F_2^p(x,Q^2)$ data[8,9] which covers small x and a wide range of Q^2; E665 $F_2^{p,d}(x,Q^2)$ data[10] at medium x; BCDMS and SLAC $F_2^{p,d}(x,Q^2)$ data[11,12] at large x; NMC $F_2^{p,d}(x,Q^2)$[13] at medium and large x; CCFR $F_2^{\nu(\bar{\nu})p}(x,Q^2)$ and $F_3^{\nu(\bar{\nu})p}(x,Q^2)$ data[14] at large x which probe the singlet and valence quarks independently; ZEUS and H1 $F_{2,charm}^p(x,Q^2)$ data;[15,16] E605 $pN \rightarrow \mu\bar{\mu} + X$[17] constraining the large x sea; E866 Drell-Yan asymmetry[18] which determines $\bar{d} - \bar{u}$; CDF W-asymmetry data[19] which constrains the u/d ratio at large x; CDF and D0 inclusive jet data[20,21] which tie down the high x gluon; and NuTev Dimuon data[22] which constrain the strange sea.

The determination of the different partons in given kinematic ranges can be split into a few different classes. We begin with **large x**. Here the quark distributions are determined mainly from structure functions, which are dominated by non-singlet valence distributions. Both the evolution of these non-singlet distributions and conversion to structure functions is quite simple involving no parton mixing

$$\frac{df^{NS}(x,Q^2)}{d\ln Q^2} = P^{NS}(x,\alpha_s(Q^2)) \otimes f^{NS}(x,Q^2)$$
$$F_2^{NS}(x,Q^2) = C^{NS}(x,\alpha_s(Q^2)) \otimes f^{NS}(x,Q^2).$$

Hence, the evolution of high x structure functions is a good test of the theory and of $\alpha_S(Q^2)$. The success is shown in Fig. 1. However - perturbation theory involves contributions to the coefficient functions

$\sim \alpha_S^n(Q^2)\ln^{2n-1}(1-x)$ and higher twist contributions are known to be enhanced as $x \rightarrow 1$. Hence, in order to avoid contamination of NLO theory one makes a cut $W^2 = Q^2(1/x-1)+m_p^2 \leq 10-15\,\mathrm{GeV}^2$.

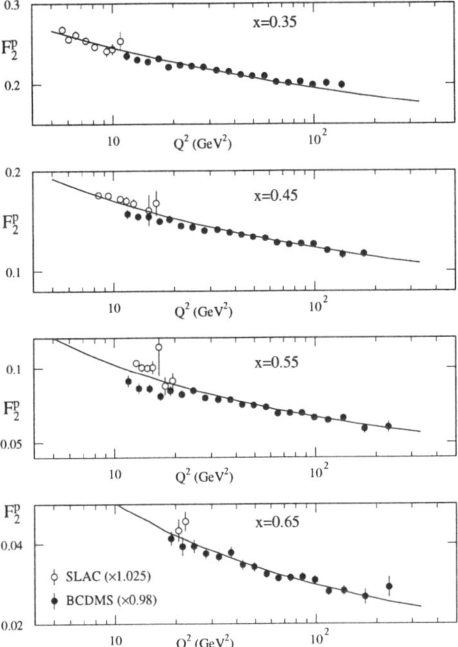

Figure 1. Description of large x BCDMS and SLAC measurements of F_2^p.

The extension to very **small x** has been made in the past decade by HERA. In this region there is very great scaling violation of the partons from the evolution equations and also interplay between the quarks and gluons. At each subsequent order in α_S each splitting function and coefficient function obtains an extra power of $\ln(1/x)$ (some accidental zeros in P_{gg}), i.e. $P_{ij}(x,\alpha_s(Q^2)), C_i^P(x,\alpha_s(Q^2)) \sim \alpha_s^m(Q^2)\ln^{m-1}(1/x)$, and hence the convergence at small x is questionable. The global fits usually assume that this turns out to be unimportant in practice, and proceed regardless. The fit is good, but could be improved. The large $\ln(1/x)$ terms mean that small x predictions are somewhat uncertain, as will be discussed later. Small x parton distributions are therefore an interesting field of study within QCD. They are also vital for understanding the standard production processes at the LHC, and perhaps some of the more exotic ones, as shown in Fig. 2, which demonstrates the range of x probed by the experiment.

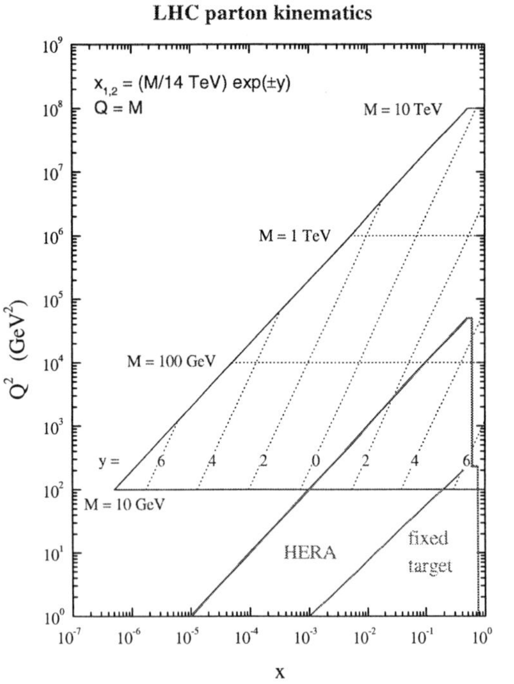

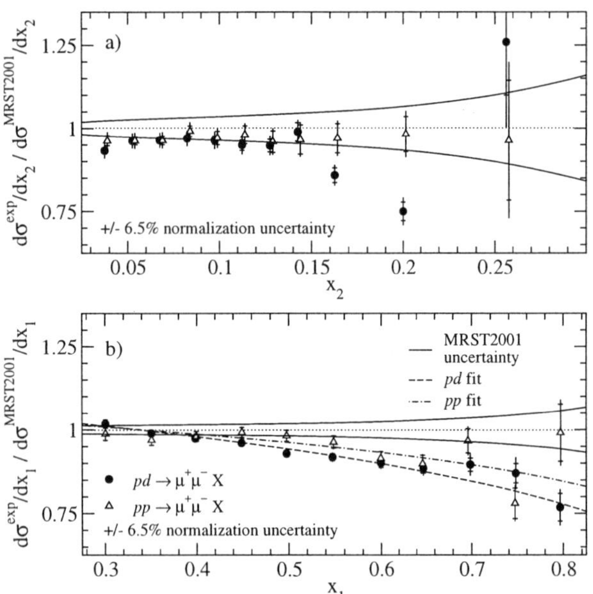

Figure 2. The range of x probed at HERA and the LHC as a function of the hard scale, e.g. particle mass and rapidity.

Figure 3. E866 fit to their Drell-Yan data as a function of x_1 (quark) and x_2 (antiquark).

The **high-x sea quarks** are determined by Drell-Yan data (assuming good knowledge of the valence quarks). There is new precise data from the E866/NuSea collaboration,[23] and their fit to these data shows a discrepancy with existing partons implying larger high-x valence quarks, as shown in Fig. 3. However, the fit performed by MRST (Fig. 4) and CTEQ displays no such discrepancy.

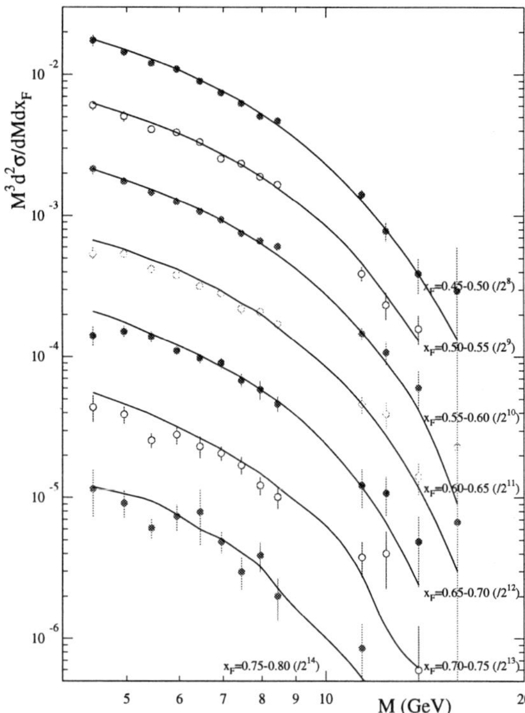

Figure 4. MRST fit to the E866 data for high values of x_F.

The **s(x) and $\bar{\text{s}}$(x) distributions** are probed using CCFR and NuTeV dimuon data, i.e. the processes

$$\nu + s \to \mu^- + c(\mu^+), \quad \bar{\nu} + \bar{s} \to \mu^+ + \bar{c}(\mu^-).$$

The quality of data is now such that one can examine the $s(x)$ and $\bar{s}(x)$ distributions separately. This has recently been performed in detail by CTEQ.[24] They find that $s(x) < \bar{s}(x)$ at quite small x, but since $\int (s(x) - \bar{s}(x))\, dx = 0$, (zero strangeness number) this leads to $\to \int x(s(x) - \bar{s}(x))\, dx = [S^-] > 0$, as demonstrated in Fig. 5. They obtain the rough constraint $0 < [S^-] < 0.004$. This is particularly significant because $NuTeV$ measure[25]

$$R^- = \frac{\sigma^\nu_{\text{NC}} - \sigma^{\bar{\nu}}_{\text{NC}}}{\sigma^\nu_{\text{CC}} - \sigma^{\bar{\nu}}_{\text{CC}}},$$

and in the standard model this satisfies $R^- = \frac{1}{2} - \sin^2\theta_W - (1 - \frac{7}{3}\sin^2\theta_W)\frac{[S^-]}{[V^-]}$. There is currently a 3σ discrepancy between this determination

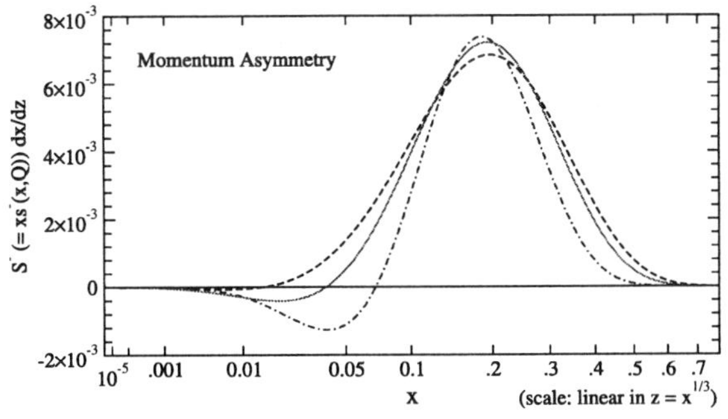

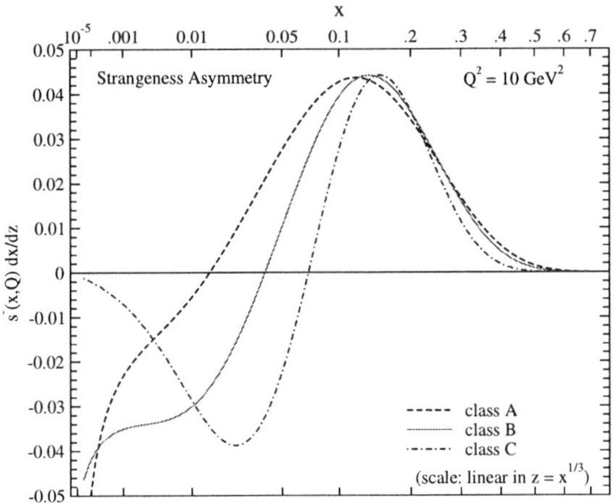

Figure 5. CTEQ strange momentum asymmetry (top) and number asymmetry (bottom).

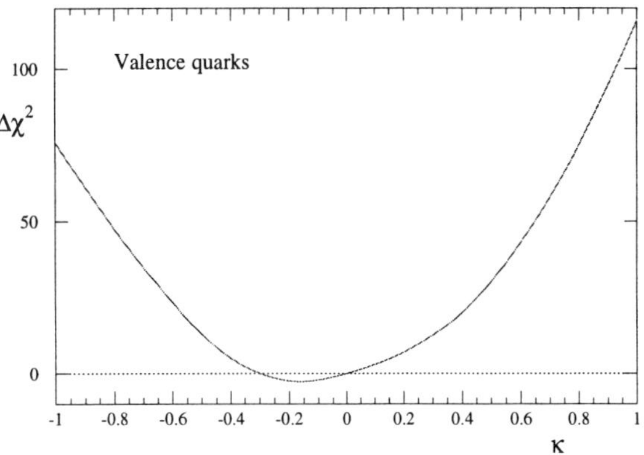

Figure 6. $\Delta\chi^2$ against the isospin violating parameter κ.

of $\sin^2\theta_W$ and others[26] but $[S^-] = 0.002$ reduces this anomaly from 3σ to 1.5σ. NuTeV themselves claim no such strange asymmetry when using partons obtained from fitting their own data,[22] so this is an issue which requires resolution.

MRST also look at the effect of **isospin violation**[27] since R^- also depends on this –

$$R^- = \frac{1}{2} - \sin^2\theta_W + \left(1 - \frac{7}{3}\sin^2\theta_W\right)\frac{[\delta U_v] - [\delta D_v]}{2[V^-]},$$

where $[\delta U_v] = [U_v^p] - [D_v^n]$, $[\delta D_v] = [D_v^p] - [U_v^n]$, and MRST use the simple parameterization

$$u_v^p(x) = d_v^n(x) + \kappa f(x), \qquad d_v^p(x) = u_v^n(x) - \kappa f(x),$$

where $f(x)$ is a simple function maintaining required conservation laws. The dependence on κ is shown in Fig. 6. The best fit value of $\kappa = -0.2$ leads to a similar reduction of the NuTeV anomaly, i.e. $\Delta sin^2\theta_W \sim -0.002$. But there is only a weak indication of this value and a fairly wide variation in κ is allowed.

The best determination of the **high-x gluon distribution** comes from inclusive jet measurements by D0 and CDF at the Tevatron. They measure $d\sigma/dE_T d\eta$ for central rapidity CDF or in bins of rapidity D0. At central rapidity the kinematic equality (at LO) is $x = 2E_T/\sqrt{s}$, and measurements extend up to $E_T \sim 400$ GeV ($x \sim 0.45$), and down to $E_T \sim 60$ GeV ($x \sim 0.06$). Gluon-gluon fusion dominates the hard cross section, but $g(x,\mu^2)$ falls off more quickly as $x \to 1$ than $q(x,\mu^2)$ so there is a transition from gluon-gluon fusion at small x, to gluon-quark to quark-quark at high x. However, as seen in Fig. 7 even at the highest x gluon-quark contributions are significant. Jet photoproduction at HERA will be another constraint in the future.

The above procedure completely determines the parton distributions at present. The total fit is reasonably good and for CTEQ6[2] is shown in Table 1 for the large data sets. The total $\chi^2 = 1954/1811$. For MRST the total $\chi^2 = 2328/2097$ – but the errors are treated differently, and different data sets and cuts are used. The same sort of conclusion is true for other *global fits*[3−7] (which use fewer data). However, there are some areas where the theory perhaps needs to be improved, as we will discuss later.

2. Parton Uncertainties

2.1. *Hessian (Error Matrix) Approach*

In this one defines the Hessian matrix H by

$$\chi^2 - \chi^2_{min} \equiv \Delta\chi^2 = \sum_{i,j} H_{ij}(a_i - a_i^{(0)})(a_j - a_j^{(0)}).$$

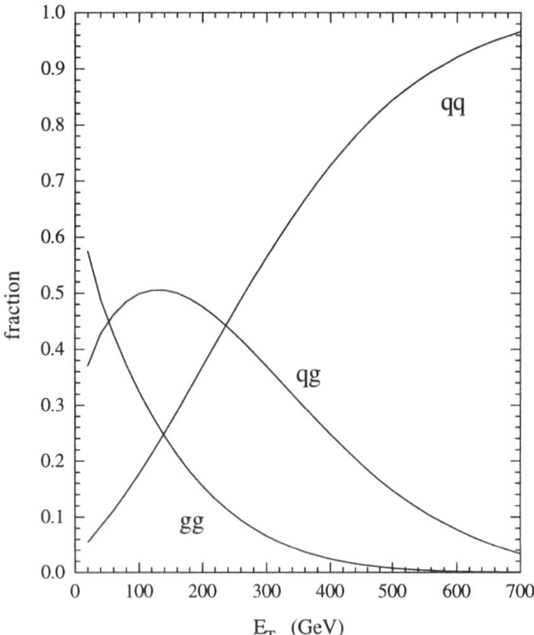

Figure 7. Fractional contributions to jet cross section from different parton-level contributions.

Table 1. Quality of fit to data for CTEQ6M.

Data Set	no. of data	χ^2
H1 ep	230	228
ZEUS ep	229	263
BCDMS μp	339	378
BCDMS μd	251	280
NMC μp	201	305
E605 (Drell-Yan)	119	95
D0 Jets	90	65
CDF Jets	33	49

H is related to the covariance matrix of the parameters by $C_{ij}(a) = \Delta\chi^2 (H^{-1})_{ij}$, and one can use the standard formula for linear error propagation,

$$(\Delta F)^2 = \Delta\chi^2 \sum_{i,j} \frac{\partial F}{\partial a_i} (H)^{-1}_{ij} \frac{\partial F}{\partial a_j}.$$

This has been employed to find partons with errors by Alekhin,[5] as seen in Fig. 8 and also by H1[6] (each with restricted data sets).

The simple method can be problematic with larger data sets and larger numbers of parameters due to extreme variations in $\Delta\chi^2$ in different directions in parameter space. This is solved by finding and rescaling the eigenvectors of H (CTEQ[28,29,2])

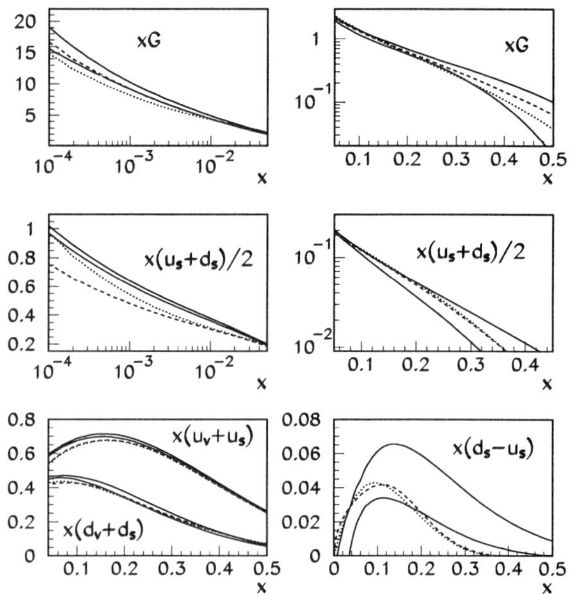

Figure 8. Results for Alekhin partons at $Q^2 = 9$ GeV^2 with uncertainties (solid lines) (dashed lines – CTEQ5M, dotted lines – MRST01).

leading to the diagonal form

$$\Delta\chi^2 = \sum_i z_i^2.$$

The uncertainty on a physical quantity is given by

$$(\Delta F)^2 = \sum_i \left(F(S_i^{(+)}) - F(S_i^{(-)}) \right)^2,$$

where $S_i^{(+)}$ and $S_i^{(-)}$ are PDF sets displaced along eigenvector directions by a given $\Delta\chi^2$. Similar eigenvector parton sets have also been introduced by MRST[31] and ZEUS. However, there is an art in choosing the "correct" $\Delta\chi^2$ given the complication of the errors in the full fit.[32] Ideally $\Delta\chi^2 = 1$, but this leads to unrealistic errors, e.g. values of $\alpha_S(M_Z^2)$ obtained by CTEQ using $\Delta\chi^2 = 1$ for each data set in the global fit are shown in Fig. 9, and are not consistent. CTEQ choose $\Delta\chi^2 \sim 100$, which is perhaps conservative. MRST choose $\Delta\chi^2 \sim 50$. An example of results is shown in Fig. 10.

2.2. Offset Method

In this method the best fit and parameters a_0 are obtained using only uncorrelated errors. The quality of the fit is then estimated by adding uncorrelated and correlated errors in quadrature. Roughly speaking systematic uncertainties are determined by letting each source of systematic error vary by 1σ and

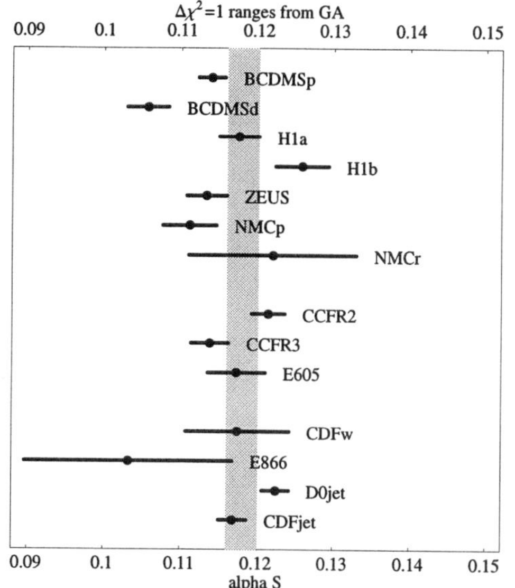

Figure 9. Values of $\alpha_S(M_Z^2)$ and their uncertainties using $\Delta\chi^2 = 1$ from CTEQ.

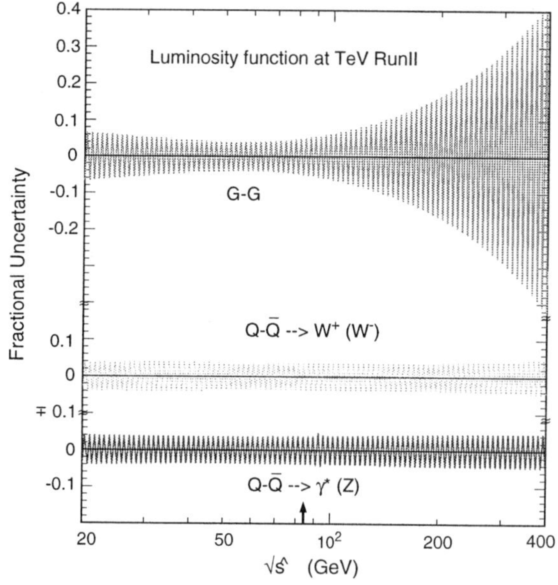

Figure 10. Luminosity uncertainties using the CTEQ Hessian approach.

adding the deviations in quadrature. This procedure is used by ZEUS,[7] and leads to an effective $\Delta\chi^2 > 1$. Some results are shown in Fig. 11.

2.3. *Statistical Approach*

In principle this involves the construction of an ensemble of distributions labelled by $\mathcal{F}$ each with prob-

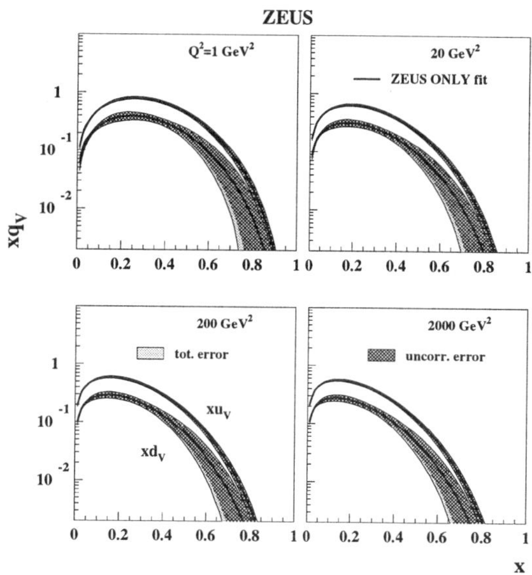

Figure 11. The valence partons extracted by ZEUS from a *global* fit and a fit to their own data alone (with some input assumptions). The latter illustrates a potential for a real constraint from HERA data alone in the future.

ability $P(\{\mathcal{F}\})$, where one can incorporate the full information about measurements and their error correlations into the calculation of $P(\{\mathcal{F}\})$. This is statistically correct, and does not rely on the approximation of linear propagation errors in calculating observables. However, it is inefficient, and in practice one generates N (N can be as low as 100) different distributions with unit weight but distributed according to $P(\{\mathcal{F}\})$.[4] Then the mean μ_O and deviation σ_O of an observable O are given by

$$\mu_O = \frac{1}{N} \sum_1^N O(\{\mathcal{F}\}), \quad \sigma_O^2 = \frac{1}{N} \sum_1^N (O(\{\mathcal{F}\}) - \mu_O)^2.$$

Currently this approach uses only proton DIS data sets in order to avoid complicated uncertainty issues, e.g. shadowing effects for nuclear targets, and also demands consistency between data sets. However, it is difficult to find many truly compatible DIS experiments, and consequently the Fermi2001 partons are determined by only H1, BCDMS, and E665 data sets. They result in good predictions for many Tevatron cross sections, e.g. inclusive jets and W and Z total cross sections. However, the restricted data sets mean there is restricted information – data sets are deemed either perfect or, in the case of most of them, useless – leading to unusual values for some parameters. e.g. $\alpha_S(M_Z^2) = 0.112 \pm 0.001$ and a very hard $d_V(x)$ at high x (together these two features facilitate a good fit to Tevatron jets independent of the high-x gluon). These partons would produce some extreme predictions, as seen later. Nevertheless, the approach does demonstrate that the Gaussian approximation is often not good, and therefore highlights shortcomings in the methods outlined in the previous sections. It is a very attractive, but ambitious large-scale project, still in need of some further development. In particular I feel it requires the inclusion of a wider variety of data in order to overcome the obstacle presented by the fact that most data sets in the global fit are not really as consistent as they should be in the strict statistical sense.

2.4. *Lagrange Multiplier Method*

This was first suggested by CTEQ[30] and has been concentrated on by MRST.[31] One performs the fit while constraining the value of some physical quantity, i.e. one minimizes

$$\Psi(\lambda, a) = \chi^2_{global}(a) + \lambda F(a)$$

for various values of λ. This gives a set of best fits for particular values of the quantity $F(a)$ without relying on the quadratic approximation for χ^2, as shown for σ_W in Fig. 12. The uncertainty is then determined by deciding an allowed range of $\Delta\chi^2$. One can also easily check the variation in χ^2 for each of the experiments in the global fit and ascertain if the total $\Delta\chi^2$ is coming specifically from one region, which might cause concern. In principle, this is superior to the Hessian approach, but it must be repeated for each physical process.

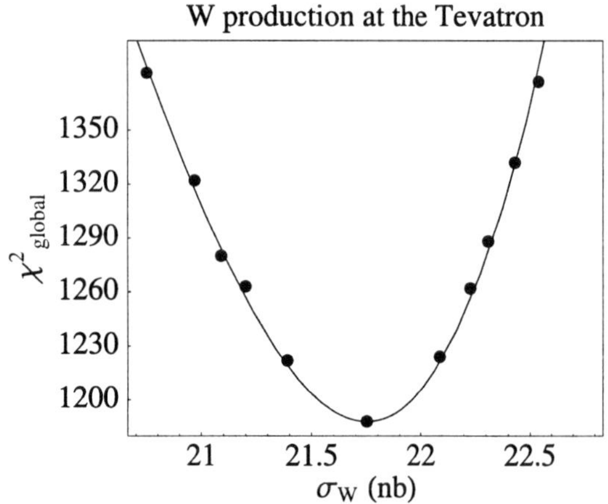

Figure 12. χ^2_{global} for CTEQ plotted against σ_W.

2.5. *Results*

I choose the cross section for W and Higgs production at the Tevatron and LHC (for $M_H = 115$ GeV) as examples. Using their fixed value of $\alpha_S(M_Z^2) = 0.118$ and $\Delta\chi^2 = 100$ CTEQ obtain

$$\Delta\sigma_W(\mathrm{LHC}) \approx \pm 4\% \quad \Delta\sigma_W(\mathrm{Tev}) \approx \pm 5\%$$

$$\Delta\sigma_H(\mathrm{LHC}) \approx \pm 5\%.$$

Using a slightly wider range of data, $\Delta\chi^2 \sim 50$ and $\alpha_S(M_Z^2) = 0.119$ MRST obtain

$$\Delta\sigma_W(\mathrm{Tev}) \approx \pm 1.2\% \quad \Delta\sigma_W(\mathrm{LHC}) \approx \pm 2\%$$

$$\Delta\sigma_H(\mathrm{Tev}) \approx \pm 4\% \quad \Delta\sigma_H(\mathrm{LHC}) \approx \pm 2\%.$$

MRST also allow $\alpha_S(M_Z^2)$ to be free. In this case $\Delta\sigma_W$ is quite stable but $\Delta\sigma_H$ almost doubles. Contours of variation in χ^2 for the predictions of these cross sections are shown in Fig. 13.

The same general procedure is also used by CTEQ[34] to look at the effect of new physics parameterized by the contact term

$$\pm(2\pi/\Lambda^2)(\bar{q}_L\gamma^\mu q_L)(\bar{q}_L\gamma_\mu q_L).$$

The curves in Fig. 14 show the fit to the $D0$ jet data, which is the most discriminating data set, for $\Lambda = 1.6, 2.0, 2.4, \infty$ TeV, and $A = -1$. For the highest values of Λ the fit even improves very slightly, but $\Lambda > 1.6$ TeV is clearly ruled out.

Hence, the estimation of uncertainties due to experimental errors has many different approaches and different types and amount of data actually fit. Overall the uncertainty from this source is rather small – only more than a few % for quantities determined by the high x gluon and very high x down quark. This is illustrated for the determinations of $\alpha_S(M_Z^2)$ in Table 2. There is generally good agreement, but their are some outlying values.

These outlying values of $\alpha_S(M_Z^2)$ show that different approaches can sometimes lead to rather different central values. This suggests that there are other matters to consider as well as the experimental errors on data. We also need to determine the effect of assumptions made about the fit, e.g. cuts made on the data, the data sets fit, the parameterization for input sets, the form of the strange sea, *etc.*. Many of these can be as important as the errors on the data used (or more so). This is demonstrated by the results from the LHC/LP Study Working Group[35] shown in Table 3, and by predictions for

Table 2. Values of $\alpha_s(M_Z^2)$ and its error from different NLO QCD fits.

Group	$\Delta\chi^2$	$\alpha_S(M_Z^2)$	
CTEQ6	$\Delta\chi^2 = 100$	$\alpha_s(M_Z^2) =$	$0.1165 \pm 0.0065(exp)$
ZEUS	$\Delta\chi^2_{eff} = 50$	$\alpha_s(M_Z^2) =$	$0.1166 \pm 0.0049(exp) \pm 0.0018(model) \pm 0.004(theory)$
MRST01	$\Delta\chi^2 = 20$	$\alpha_s(M_Z^2) =$	$0.1190 \pm 0.002(exp) \pm 0.003(theory)$
H1	$\Delta\chi^2 = 1$	$\alpha_s(M_Z^2) =$	$0.115 \pm 0.0017(exp) \, ^{+\,0.0009}_{-\,0.0005} \, (model) \pm 0.005(theory)$
Alekhin	$\Delta\chi^2 = 1$	$\alpha_s(M_Z^2) =$	$0.1171 \pm 0.0015(exp) \pm 0.0033(theory)$
GKK	CL	$\alpha_s(M_Z^2) =$	$0.112 \pm 0.001(exp)$

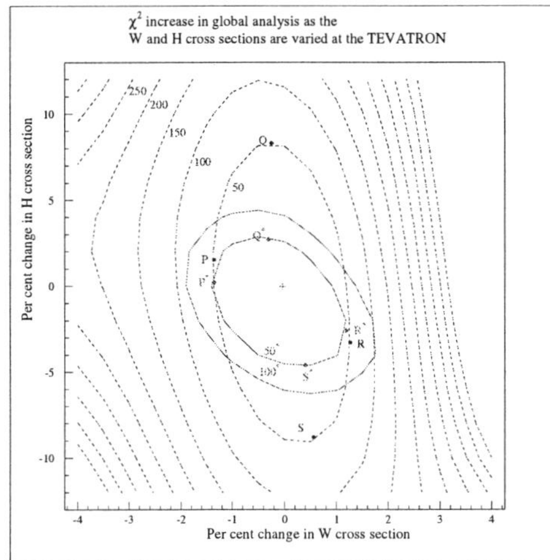

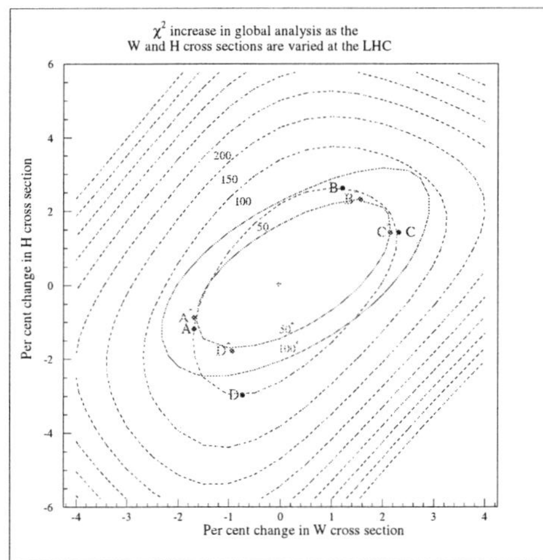

Figure 13. χ^2-plot for W and Higgs production at the Tevatron (top) and LHC (bottom) with α_S free (dashed) and fixed (solid) at $\alpha_S = 0.119$.

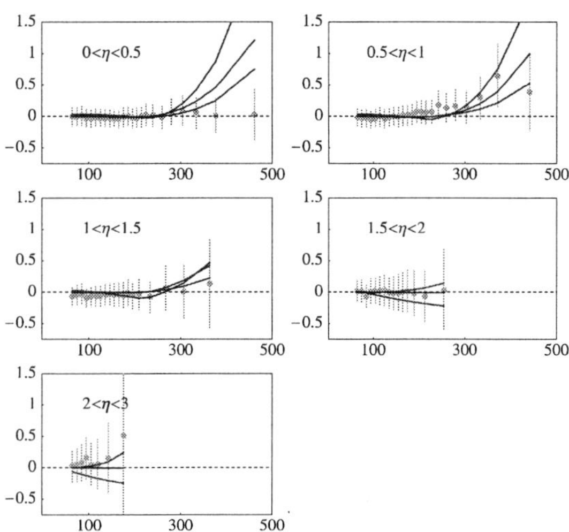

Figure 14. Fits to D0 jets for different values of A.

Table 3. Cross sections for Drell-Yan pairs (e^+e^-) with PYTHIA 6.206, rapidity < 2.5. The errors shown are the PDF uncertainties.

PDF set	Comment	xsec [pb]	PDF uncertainty
		$81 < M < 101$ GeV	
CTEQ6	LHAPDF	1065 ± 46	4.4%
MRST2001	LHAPDF	$1091 \pm ...$	3%
Fermi2002	LHAPDF	853 ± 18	2.2%

Higgs production at the Tevatron from MRST2001 and CTEQ6, and Fig. 16 – the comparison between the gluons for the two parton sets.

3. Theoretical Errors

3.1. *Problems in the Fit*

σ_W by MRST CTEQ and Alekhin[36] in Table 4. In both cases the discrepancies are mainly due to differences in detailed constraints (by data) on the quark decomposition. Differences between predictions are also shown by Fig. 15 – the predictions for W and

As well as the consequences of these assumptions we must consider the related problem of theoretical errors. Theoretical errors are indicated by some regions where the theory perhaps needs to be improved to fit the data better. There is a reasonably good

Table 4. Comparison of $\sigma_W \cdot B_{l\nu}$ for different partons.

PDF set	Comment	xsec [nb]	PDF uncertainty
Alekhin	Tevatron	2.73	$\pm$ 0.05 (tot)
MRST2002	Tevatron	2.59	$\pm$ 0.03 (expt)
CTEQ6	Tevatron	2.54	$\pm$ 0.10 (expt)
Alekhin	LHC	215	$\pm$ 6 (tot)
MRST2002	LHC	204	$\pm$ 4 (expt)
CTEQ6	LHC	205	$\pm$ 8 (expt)

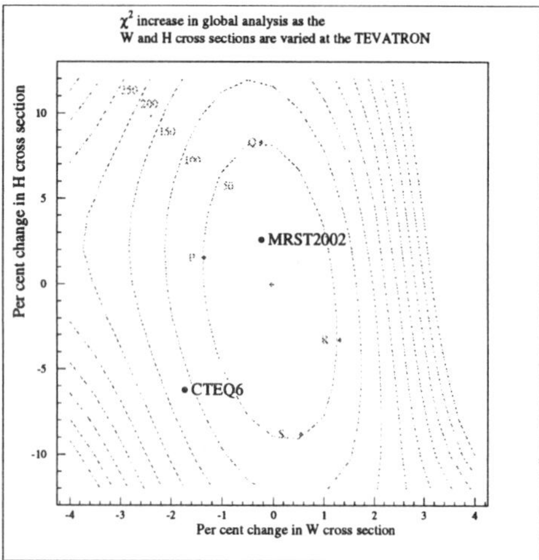

Figure 15. χ^2-plot for W and Higgs production at the Tevatron with α_S free. The predictions from CTEQ6 is marked.

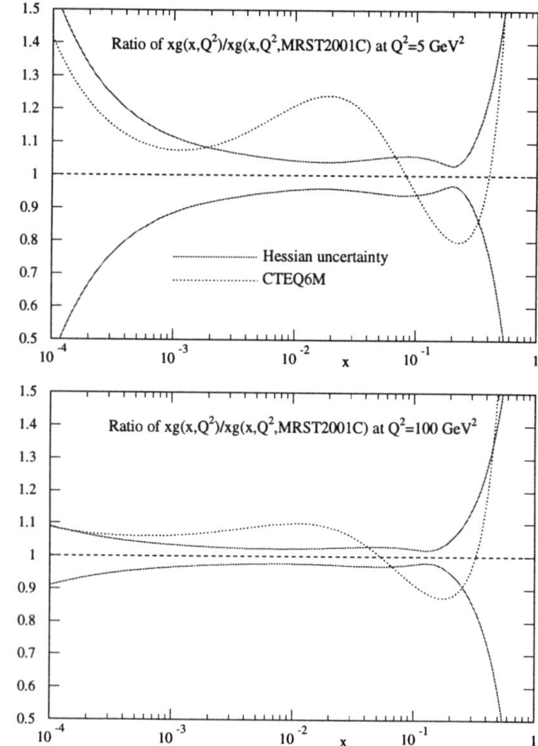

Figure 16. Fractional uncertainty in the MRST gluon compared with the difference in the central CTEQ6 gluon.

fit to HERA data, but there are some problems at the highest Q^2 at moderate x, i.e. in $dF_2/d\ln Q^2$, as seen for MRST and CTEQ in Fig. 17. Also the data require the gluon to be valence-like or negative at small x and low Q^2, e.g. the ZEUS gluon in Fig. 18, leading to $F_L(x, Q^2)$ being negative[1] at the smallest x, Q^2. However, it is not just the low x–low Q^2 data that require this negative gluon. The moderate x data need lots of gluon to get a reasonable $dF_2/d\ln Q^2$ and the Tevatron jets need a large high x gluon, and this must be compensated for elsewhere. In general MRST find that it is difficult to reconcile the fit to jets and to the rest of the data, Fig. 19, and that different data compete over the gluon and $\alpha_S(M_Z^2)$. The jet fit is better for CTEQ6 largely due to their different cuts on other data. Other fits do not include the Tevatron jets, but generally produce gluons largely incompatible with this data.

3.2. *Types of Theoretical Error, NNLO*

It is vital to consider theoretical errors. These include higher perturbative orders (NNLO), small

x $(\alpha_s^n \ln^{n-1}(1/x))$, large x $(\alpha_s^n \ln^{2n-1}(1-x))$ low Q^2 (higher twist), *etc.*. Note that renormalization/factorization scale variation is not a reliable method of estimating these theoretical errors because of increasing logs at higher orders.

In order to investigate the true theoretical error we must consider some way of performing correct large and small x resummations, and/or use what we already know about NNLO. The coefficient functions are known at NNLO. Singular limits $x \to 1$, $x \to 0$ are known for NNLO splitting functions as well as limited moments,[37] and this has allowed approximate NNLO splitting functions to be devised[38] which have been used in approximate global fits.[39] They improve the quality of the fit very slightly (mainly at high x) and $\alpha_S(M_Z^2)$ lowers from 0.119 to 0.1155. The gluon is smaller at NNLO at low x due to the positive NNLO quark-gluon splitting function. There is also a NNLO fit by Alekhin,[40] with some differences – the gluon is not smaller, probably due to the absence of Tevatron jet data in the fit and to a very different

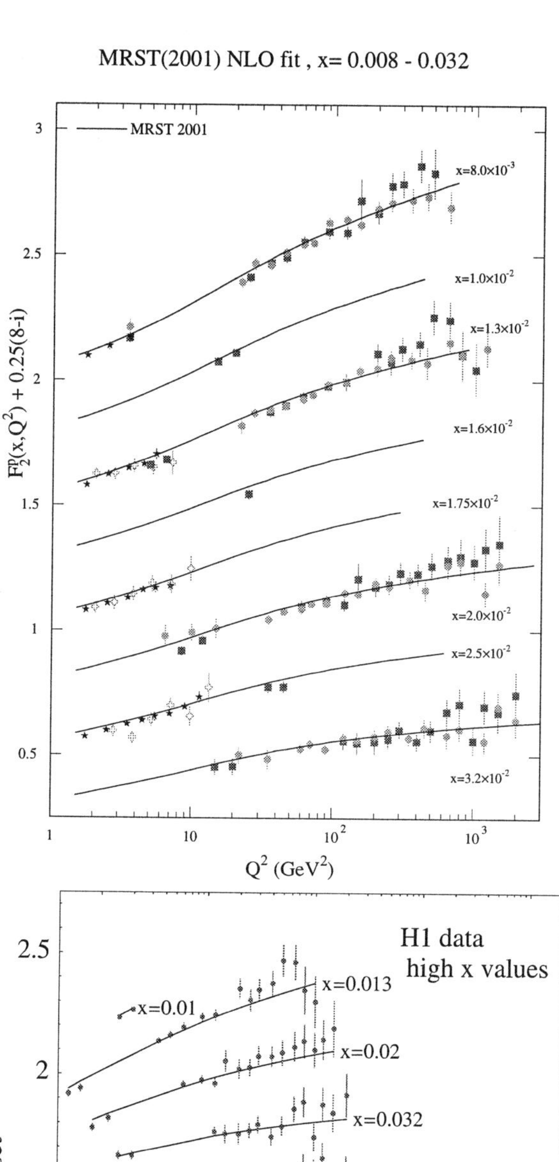

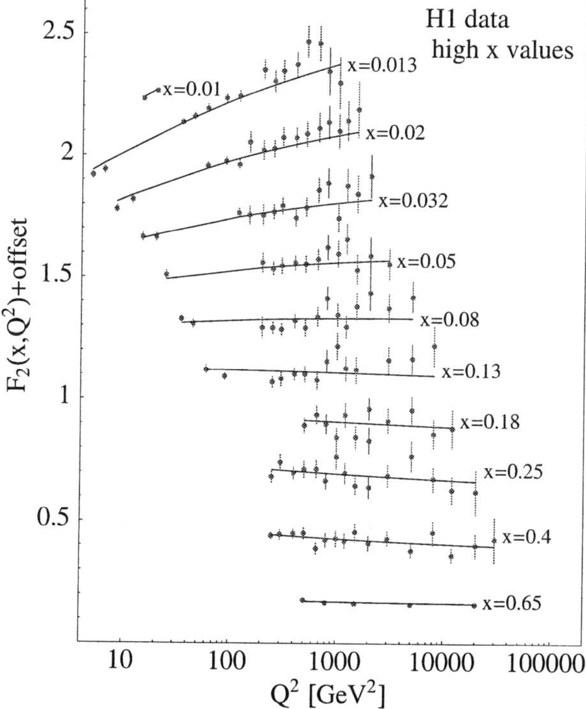

Figure 17. Comparison of MRST(2001) $F_2(x,Q^2)$ with HERA, NMC and E665 data (top) and CTEQ6 $F_2(x,Q^2)$ with H1 data (bottom).

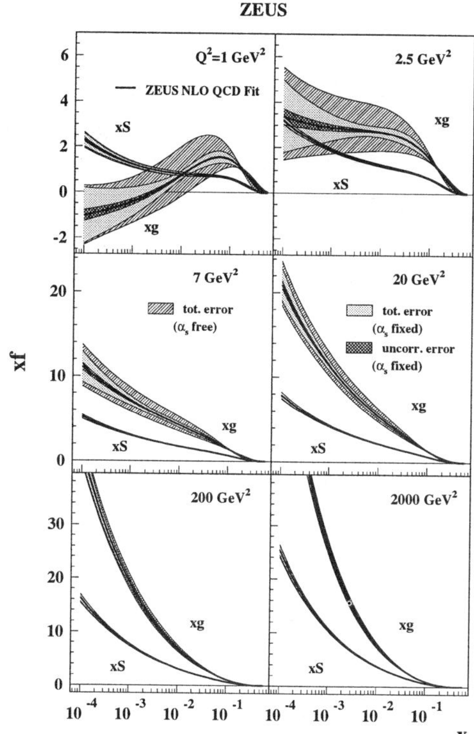

Figure 18. ZEUS gluon and sea quark distributions at various Q^2 values.

definition of the NNLO charm contribution. There is agreement in the reduction of $\alpha_S(M_Z^2)$ at NNLO, i.e. $0.1171 \to 0.1143$.

Using these NNLO partons there is reasonable stability order by order for the (quark-dominated) W and Z cross sections, as seen in Fig. 20. However, the change from NLO to NNLO is of order 4%, which is much bigger than the uncertainty at NLO due to experimental errors. Also, this fairly good convergence is largely guaranteed because the quarks are fit directly to data. There is greater danger in gluon dominated quantities, e.g. $F_L(x,Q^2)$, as can be seen in Fig. 21. Hence, the convergence from order to order is uncertain.

3.3. Empirical Approach

We can estimate where theoretical errors may be important by adopting the empirical approach of investigating in detail the effect of cuts on the fit quality, i.e. we try varying the kinematic cuts on data. The procedure is to change W_{cut}^2, Q_{cut}^2 and/or x_{cut}, re-fit and see if the quality of the fit to the remaining data

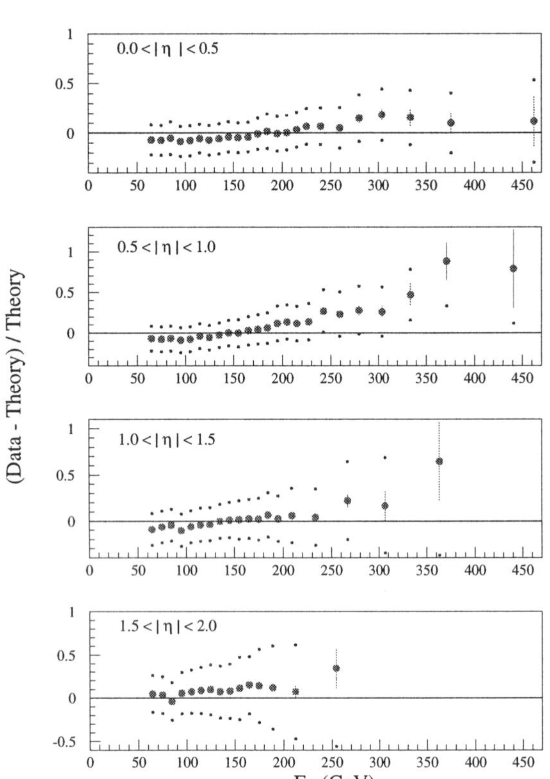

Figure 19. The MRST fit to D0 jet data. The points show the range of the systematic errors.

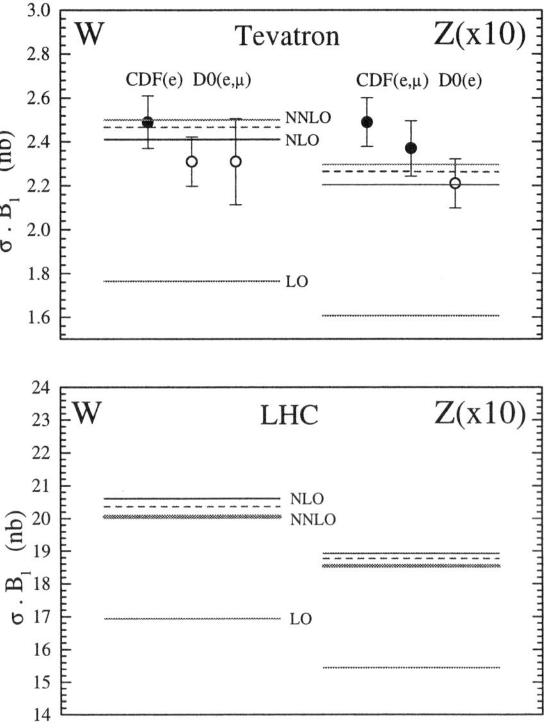

Figure 20. LO, NLO and NNLO predictions for W and Z cross sections.

improves and/or the input parameters change dramatically. (This is similar to a previous suggestion in terms of data sets.[41]) One then continues until the quality of the fit and the partons stabilize.[27]

For W^2_{cut} raising from 15 GeV^2 has no effect. When raising Q^2_{cut} from 2 GeV^2 in steps there is a slow, continuous and significant improvement for Q^2 up to > 10 GeV^2 (560 data points cut), suggesting that any corrections are probably higher orders not higher twist. The input gluon becomes slightly smaller at low x at each step (where one loses some of the lowest x data), and larger at high x. The value of $\alpha_S(M_Z^2)$ slowly decreases by about 0.0015. Raising x_{cut} leads to continuous improvement with stability reached at $x = 0.005$ (271 data points cut) with $\alpha_S(M_Z^2) \to 0.118$. There is an improvement in the fit to HERA, NMC and Tevatron jet data, and much reduced tension between the data sets. At each step the moderate x gluon becomes more positive, at the expense of the gluon below the cut be-

coming very negative and $dF_2(x, Q^2)/d\ln Q^2$ being incorrect. However, higher orders could cure this in a quite plausible manner. For example adding higher order terms to the splitting functions

$$P_{gg} \to .... + \frac{3.86\bar{\alpha}_S^4}{x}\left(\frac{\ln^3(1/x)}{6} - \frac{\ln^2(1/x)}{2}\right),$$

$$P_{qg} \to .... + \frac{5.12 N_f \bar{\alpha}_S^5}{6x}\left(\frac{\ln^3(1/x)}{6} - \frac{\ln^2(1/x)}{2}\right),$$

leaves the improved fit above $x = 0.005$ largely unchanged, but solves the problem below $x = 0.005$. Saturation corrections added to NLO and NNLO fits seem to make the situation worse. Hence, the cuts are suggestive of theoretical errors for small x and/or small Q^2. Predictions for W and Higgs cross sections at the Tevatron are still safe if $x_{cut} = 0.005$, since they do not sample partons at lower x. However, they change in a smooth manner as x_{cut} is lowered, due to the altered partons above x_{cut}.

There is a lot of work on explicit $\ln(1/x)$-resummations in structure functions and parton distributions for example,[42,43,44] but there is no complete consensus on the best approach. There is

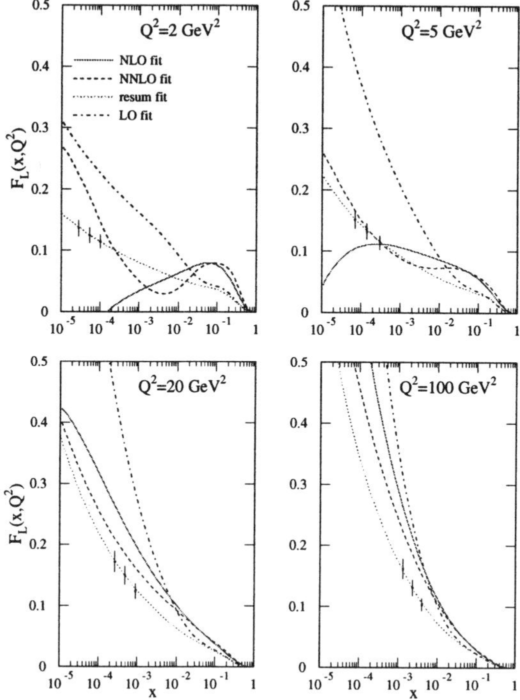

Figure 21. Comparison of the predictions for $F_L(x,Q^2)$ at LO, NLO and NNLO using MRST partons and also a $\ln(1/x)$-resummed prediction.[42]

also work on connecting the partons to alternative approaches at small x, e.g. dipole models[45] and pomerons.[46] These approaches can suggest improvements to the fits and changes in predictions, e.g. a resummed prediction[42] for $F_L(x,Q^2)$ is shown on Fig. 21. Accurate and direct measurements of $F_L(x,Q^2)$ and other quantities at low x and/or Q^2 (the predicted range and accuracy of $F_L(x,Q^2)$ measurements at HERA III is shown on Fig. 21) would be a great help in determining whether NNLO is sufficient or whether resummed (or other) corrections are necessary, or helpful for maximum precision.

4. Conclusions

One can perform global fits to all up-to-date data over a wide range of parameter space, and there are various ways of looking at uncertainties due to errors on data alone. There is no totally preferred approach. The errors from this source are rather small $- \sim 1 - 5\%$ except in a few regions of parameter space and are similar using various approaches.

The uncertainty from input assumptions e.g. cuts on data, parameterizations *etc.*, are comparable and sometimes larger, which means one cannot entirely believe one group's errors.

The quality of the fit is fairly good, but there are some slight problems. These imply that errors from higher orders/resummation are potentially large in some regions of parameter space, and due to correlations between partons these affect all regions (the small x gluon influences the large x gluon). Cutting out low x and/or Q^2 data allows a much-improved fit to the remaining data, and altered partons. Hence, for some processes theory is probably the dominant source of uncertainty at present and a systematic study is a priority, as is more data which would help determine our theoretical accuracy.

References

1. A.D. Martin *et al.*, *Eur. Phys. J.* **C23**, 73 (2002).
2. CTEQ Collaboration: J. Pumplin *et al.*, *JHEP* 0207:012 (2002).
3. M. Botje, *Eur. Phys. J.* **C14**, 285 (2000).
4. W.T. Giele and S. Keller, *Phys. Rev.* **D58**, 094023 (1998); W.T. Giele, S. Keller and D.A. Kosower, hep-ph/0104052.
5. S.I. Alekhin, *Phys. Rev.* **D68**, 014002 (2003).
6. H1 Collaboration: C. Adloff *et al.*, *Eur. Phys. J.* **C21**, 33 (2001).
7. A.M. Cooper-Sarkar, *J. Phys.* **G28** 2669 (2002); ZEUS Collaboration: S. Chekanov *et al.*, *Phys. Rev.* **D67**, 012007 (2003).
8. H1 Collaboration: C. Adloff *et al.*, *Eur. Phys. J.* **C13**, 609 (2000); H1 Collaboration: C. Adloff *et al.*, *Eur. Phys. J.* **C19**, 269 (2001).
9. ZEUS Collaboration: S. Chekanov *et al.*, *Eur. Phys. J.* **C21**, 443 (2001); ZEUS Collaboration: S. Chekanov *et al.*, *Eur. Phys. J.* **C28**, 175 (2003).
10. M.R. Adams *et al.*, *Phys. Rev.* **D54**, 3006 (1996).
11. BCDMS Collaboration: A.C. Benvenuti *et al.*, *Phys. Lett.* **B223**, 485 (1989); BCDMS Collaboration: A.C. Benvenuti *et al.*, *Phys. Lett.* **B236**, 592 (1989).
12. L.W. Whitlow *et al.*, *Phys. Lett.* **B282**, 475 (1992), L.W. Whitlow, preprint SLAC-357 (1990).
13. NMC Collaboration: M. Arneodo *et al.*, *Nucl. Phys.* **B483**, 3 (1997); *Nucl. Phys.* **B487** 3 (1997).
14. CCFR Collaboration: U.K. Yang *et al.*, *Phys. Rev. Lett.* **86**, 2742 (2001); CCFR Collaboration: W.G. Seligman *et al.*, *Phys. Rev. Lett.* **79** 1213 (1997).
15. ZEUS Collaboration: J. Breitweg *et al.*, *Eur. Phys. J.* **C12**, 35 (2000).
16. H1 Collaboration: C. Adloff *et al.*, *Phys. Lett.* **B528**, 1999 (2002).
17. E605 Collaboration: G. Moreno *et al.*, *Phys. Rev.* **D43**, 2815 (1991).

18. E866 Collaboration: R.S. Towell *et al.*, *Phys. Rev.* **D64**, 052002 (2001).

19. CDF Collaboration: F. Abe *et al.*, *Phys. Rev. Lett.* **81**, 5744 (1998).

20. D0 Collaboration: B. Abbott *et al.*, *Phys. Rev. Lett.* **86**, 1707 (2001).

21. CDF Collaboration: T. Affolder *et al.*, *Phys. Rev.* **D64**, 032001 (2001).

22. NuTeV Collaboration: M. Goncharov *et al.*, *Phys. Rev.* **D64**, 112006 (2001).

23. E866/NuSea Collaboration: J.C. Webb *et al.*, `hep-ex/0302019`.

24. S. Kretzer *et al.*, preprint BNL-NT-03/16.

25. NuTeV Collaboration: G.P. Zeller *et al.*, *Phys. Rev. Lett.* **88**, 091802 (2002).

26. P. Gambino, these proceedings.

27. A.D. Martin, R.G. Roberts, W.J. Stirling and R.S. Thorne, `hep-ph/0308087`, submitted to *Eur. Phys. J.*.

28. CTEQ Collaboration: J. Pumplin *et al.*, *Phys. Rev.* **D65**, 014011 (2002).

29. CTEQ Collaboration: J. Pumplin *et al.*, *Phys. Rev.* **D65**, 014013 (2002).

30. CTEQ Collaboration: D. Stump *et al.*, *Phys. Rev.* **D65**, 014012 (2002).

31. A.D.Martin *et al.*, *Eur. Phys. J.* **C28**, 455 (2003).

32. R.S. Thorne *et al.*, *J. Phys.* **G28**, 2717 (2002).

33. C. Pascaud and F. Zomer 1995 *Preprint* LAL-95-05.

34. D. Stump *et al.*, `hep-ph/0303013`.

35. D. Bourilkov, `hep-ph/0305126`.

36. S. Alekhin, `hep-ph0307219`.

37. S.A. Larin *et al.*, *Nucl. Phys.* **B492**, 338 (1997); A. Rétey and J.A.M. Vermaseren, *Nucl. Phys.* **B604**, 281 (2001).

38. W.L. van Neerven and A. Vogt, *Phys. Lett.* **B490**, 111 (2000).

39. A.D. Martin *et al.*, *Phys. Lett.* **B531**, 216 (2002).

40. S.I. Alekhin, *Phys. Lett.* **B519**, 57 (2001).

41. J. C. Collins and J. Pumplin, `hep-ph/0105207`.

42. R.S. Thorne, *Phys. Rev.* **D64**, 074005 (2001).

43. M. Ciafaloni, *et al.*, `hep-ph/0307188`.

44. G. Altarelli, *et al.*, `hep-ph/0306156`.

45. J. Bartels, *et al.*, *Phys. Rev.* **D66**, 014001 (2002).

46. A. Donnachie and P.V. Landshoff, *Phys. Lett.* **B550**, 160 (2002).

DISCUSSION

Bogdan Dobrescu (Fermilab): Very recently, CTEQ has done a global fit allowing $s \neq \bar{s}$, and MRST has done a global fit allowing an isospin violation. These two effects could in principle be correlated. Are there any plans for performing a global fit that allow both a strange asymmetry and isospin violation?

Robert Thorne: Yes, in principle those two effects could be correlated. In fact there are a lot of other things that you might have to worry about, for example uncertainties due to heavy target corrections. At the moment, for many of these we are just looking at one effect at a time. In the future it will be ideal, or even necessary, to do all of these in a global fit at the same time, but I think it will be a little while before that actually happens.

Regina Demina (University of Rochester): NLO calculations are used to predict many flavor PDF's. At the same time theory had only a marginal success in predicting heavy flavor production at Tevatron. New data from HERA starts probing b photoproduction. What are the plans for using this data in PDF fits in the future? What measurements can be done at the Tevatron to constrain these?

Robert Thorne: Well, in general heavy flavor should be generated highly perturbatively. There is a question about intrinsic heavy flavor but that would generally be important at very high x and relatively low Q^2, which isn't where either the Tevatron or HERA actually measure the heavy flavor sensitive quantities. In charm production the agreement generally seems to be good, whereas in beauty production the agreement somewhat conversely seems to be worse, where one would expect the theoretical prediction to be more reliable. For issues like the Tevatron beauty production there has been quite a lot of progress in things that might be going wrong such as the fragmentation functions. And there, one also has to get the best theoretical perturbative calculations which involves the resummation of any large logarithms. This is often not done in some of these comparisons. I think that this is something we have to look at very carefully, but some of the other speakers have talked about this so I didn't concentrate on it. As far as the measurements are concerned, as I said, at small x and relatively high scales it really should be perturbative. It is just a matter of what perturbative calculation would give the best result. If you want to look at something sensitive to intrinsic heavy flavor then it begins at high x, so if there is any way of probing bottom or charm at high x... but that is not really the natural region for the Tevatron or for HERA.

Sridhara Dasu (U. Wisconsin): You talked about F_2 predictions at various x, Q^2. However, in making global fits both MRST and CTEQ seem to ignore existing $F_L(x, Q^2)$ data from SLAC and CCFR. Although these data are at $1 < Q^2 < 10 \; GeV^2$, they do constrain the large x ($x > 0.1$) region. Why not use these data to get better gluon PDF's at $x > 0.1$?

Robert Thorne: We always have to make some decision on which data are put into the fit, and which data we check are consistent with the partons we obtain. In this case we check when we do have consistency. This breaks down at lower Q^2 presumably due to higher twist corrections, which can be larger at moderate x for F_L than F_2. Also, at high x even F_L is largely determined by valence quarks. Concerning the low x F_L extractions at HERA we essentially do include F_L in the fit since we actually fit to what is measured - the reduced cross section, which depends on both F_2 and F_L at high y.

Debbie Harris (Fermilab): You pointed out that NuSea data, when analyzed by NuSea, doesn't agree with MRST, yet MRST claims it does. At Tuesday's breakout session there was a discussion where NuTeV showed that their dimuon data also does not agree with the newest CTEQ fits allowing $s \neq \bar{s}$. The jury is still out, and more collaboration is needed.

Robert Thorne: I agree that more collaboration is needed particularly on the question of the strange asymmetry. MRST has yet to fit the dimuon data themselves in order to look for a s-$\bar{s}$ asymmetry. This topic should be a very interesting area of study.

MEASUREMENTS WITH POLARIZED HADRONS

T.-A. SHIBATA

Dept. Physics, Tokyo Institute of Technology, Oh-okayama, Meguro- ku, Tokyo, 152-8551, Japan
E-mail: shibata@nucl.phys.titech.ac.jp

Recent progress in physics with polarized hadrons is described with the emphasis on the spin structure of the nucleon. The nucleon spin problem, which was discovered by EMC in 1988, is now being studied in various experiments. Flavor separation of the quark helicity distributions has been made. Recent observations of asymmetries in deeply virtual Compton scattering (DVCS) and exclusive meson productions provide possibilities to access the total angluar momentum carried by quarks in the framework of generalized parton distributions. Single spin azimuthal asymmetries observed in semi-inclusive measurements provide a new handle to determine the transverse quark distributions which are basic but have never been measured so far.

1. Introduction

In the SU(6) framework of flavor and spin, the wave-function of the proton is described with three quarks uud, where the spins of the two quarks are parallel and the third quark spin is anti-parallel, which produces a total spin of 1/2. A sketch is given in Fig. 1. The expectation value of the u quark polarization is 2/3 and that of the d quark is -1/6. There is no component of the strange quark in the proton in this scheme.

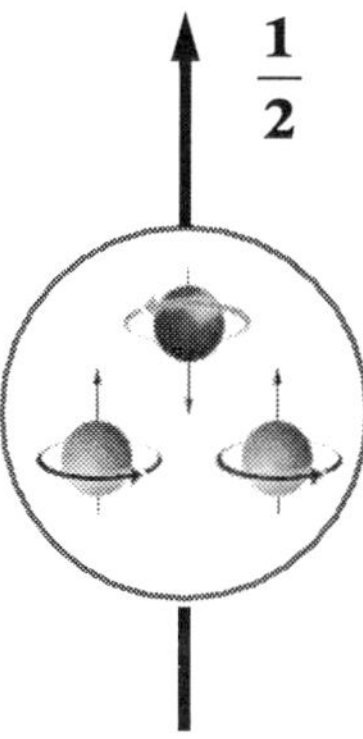

Figure 1. A sketch of the proton where the spins of the three quarks compose the proton spin 1/2.

This picture was however contradicted by the observation by EMC at CERN in 1988:[1] the quark spin contribution to the proton spin was found to be $(12 \pm 9 \pm 14)\%$ from a deep-inelastic scattering experiment with a longitudinally polarized muon beam on a longitudinally polarized proton target. This is often called the "proton spin problem". In later experiments on the proton and neutron, the quark spin contriution to the nucleon spin was measured to be 20-30%.

The nucleon is composed of quark spin, gluon spin and orbital angular momenta of quarks and gluons:

$$\frac{1}{2} = \frac{1}{2}\Sigma + \Delta G + L_q + L_g \qquad (1)$$

$$\Sigma = \Delta u + \Delta d + \Delta s + \Delta \bar{u} + \Delta \bar{d} + \Delta \bar{s}. \qquad (2)$$

Here, Δu for example denotes the helicity difference $u^{\rightarrow}(x) - u^{\leftarrow}(x)$ integrated over the full range of Bjorken x. First of all, the flavor separation of the quark spin contributions to the nucleon spin, namely the separation of $\Delta u, \Delta d, \Delta s, \Delta \bar{u}, \Delta \bar{d}, and \Delta \bar{s}$ is important. Deep-inelastic lepton scattering with a good identification capability of produced hadrons is the standard method for it. Combining with the knowledge on the fragmentation functions which describe the quark to hadron formation, one can tag the struck quark in the deep-inelastic scattering process. These measurements are called "semi-inclusive measurements". The diagram is shown in Fig. 2.

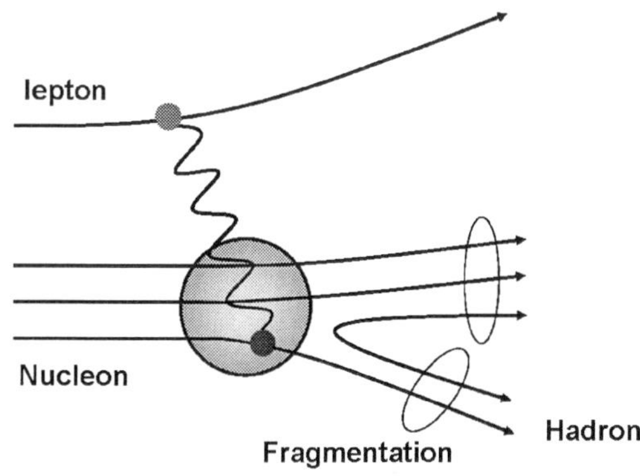

Figure 2. Semi-inclusive measurements of deep-inelastic lepton scattering. A part of produced the hadrons on the bottom are detected in experiments such as HERMES.

The cross sections for inclusive and semi-inclusive measurements are written respectively as

$$\sigma(x) \propto \sum_q e_q^2(x) \qquad (3)$$

$$\sigma_h(x,z) \propto \sum_q e_q^2\, q(x) D_q^h(z). \qquad (4)$$

Here, $q(x)$ is the quark distribution of each flavor, $D_q^h(z)$ the fragmentation function, $h = \pi^\pm, K^\pm, \dots$

The gluon spin contributions are being studied by COMPASS at CERN using muon deep-inelastic scattering. The typical beam energy is 160 GeV. Open charm production from photon-gluon fusion, shown in Fig. 3, are measured.

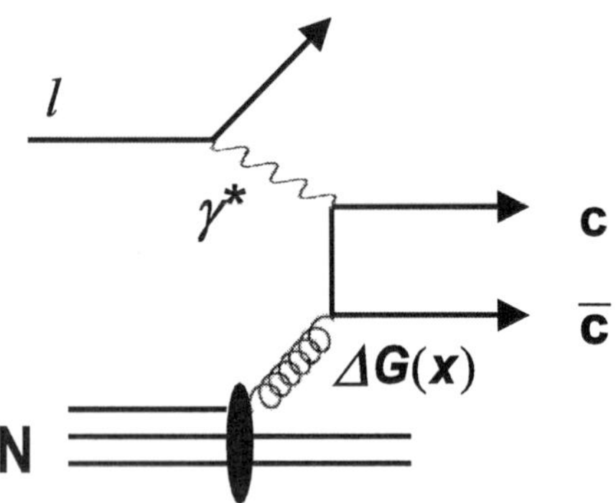

Figure 3. Photon-gluon fusion in deep-inelastic lepton scattering. The virtual photon couples to a quark or to an antiquark from the gluon dissociation.

Another approach to measure the gluon spin contribution is to use polarized proton-proton collisions at RHIC of BNL. Unlike lepton deep-inelastic scattering, major parts of the cross section are determined by strong interactions such as gluon-gluon, gluon-quark and quark-quark collisions. The details are shown in a later section.

2. Flavor Separation of Quark Helicity Distributions

Cross sections for deep-inelastic scattering are cross sections for absorption of a virtual photon.

A quark in the nucleon absorbs the photon by flipping its own spin direction. The left side in Fig. 4 is before the photon absorption, the right side after the photon absorption. The cross section is different for different spin states. The double spin asymmetry

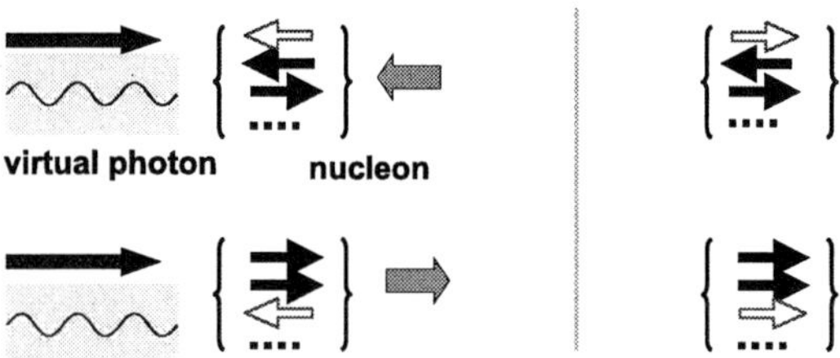

Figure 4. Absorption of a virtual photon by a quark in the longitudinally polarized nucleon.

A_1 arises from this cross section difference. A_1 is related to the cross section:

$$A_1(x) = \frac{\sigma_{\leftarrow}^{\rightarrow}(x) - \sigma_{\rightarrow}^{\rightarrow}(x)}{\sigma_{\leftarrow}^{\rightarrow}(x) + \sigma_{\rightarrow}^{\rightarrow}(x)}. \qquad (5)$$

The semi-inclusive asymmetry $A_1^h(x,z)$ can be written as

$$A_1^h(x,z) = \frac{\sum_q e_q^2\, \Delta q(x) D_q^h(z)}{\sum_q e_q^2\, q(x) D_q^h(z)}. \qquad (6)$$

Measurements of A_1 from semi-inclusive $\pi^\pm$ and $K^\pm$ measurements with a deuterium target at HERMES[2] are shown in Fig. 5. HERMES uses 27.6 GeV positron or electron beams at DESY-HERA. The data from SMC[3] with positive and negative hadron measurements are shown for comparison. The major parts of hadrons are pions. The data with the proton target[4] was also used in the analysis. A ring imaging Čerenkov detector plays an essential role in the particle identification of hadrons at HERMES.

The inclusive asymmetry increases with Bjorken x. The semi-inclusive asymmetries shown here also increase with Bjorken x reflecting the inclusive cross section. However, the K^- is made of sea quark only ($\bar{u}s$) and the asymmetry is consistent with zero.

The quark helicity distributions are extracted from the observed asymmetries: $A(x) = P(x)Q(x)$. $A(x) = (A_{1p}, A_{1p}^{\pi^+}, A_{1p}^{\pi^-}, A_{1d}, A_{1d}^{\pi^+}, A_{1d}^{\pi^-}, A_{1d}^{K^+}, A_{1d}^{K^-})$ is the asymmetry in inclusive and semi-inclusive measurements with the proton and deuterium targets while $Q(x) = (\frac{\Delta u}{u}, \frac{\Delta d}{d}, \frac{\Delta \bar{u}}{\bar{u}}, \frac{\Delta \bar{d}}{\bar{d}}, \frac{\Delta s}{s})$ is the quantity to be extracted. $\frac{\Delta \bar{s}(x)}{\bar{s}(x)}$ was allowed to move from minimum to maximum in the present analysis and its effect on the other five parameters were included in the systematic errors. $P(x)$ is a matrix. The matrix elements contain information on the fragmen-

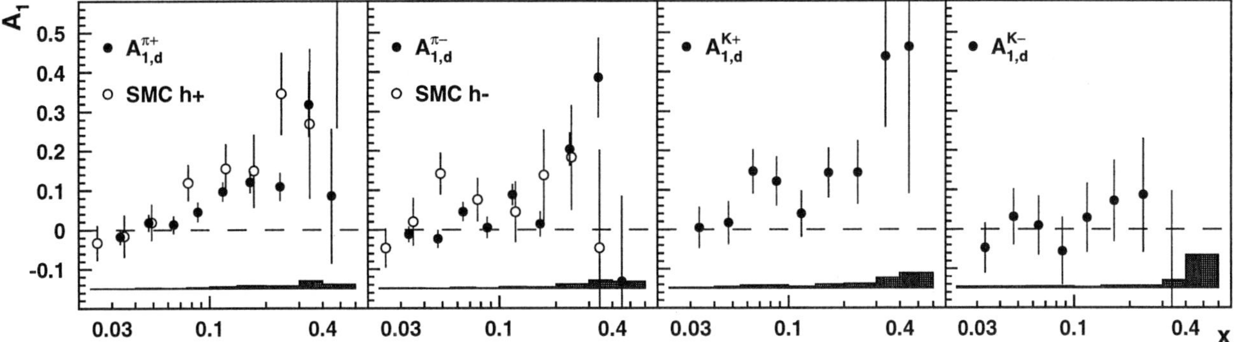

Figure 5. Double spin asymmetry A_1^h of semi-inclusive deep-inelastic scattering at HERMES with the deuterium target. The data of SMC for positive and negative hadrons are also shown for comparison.

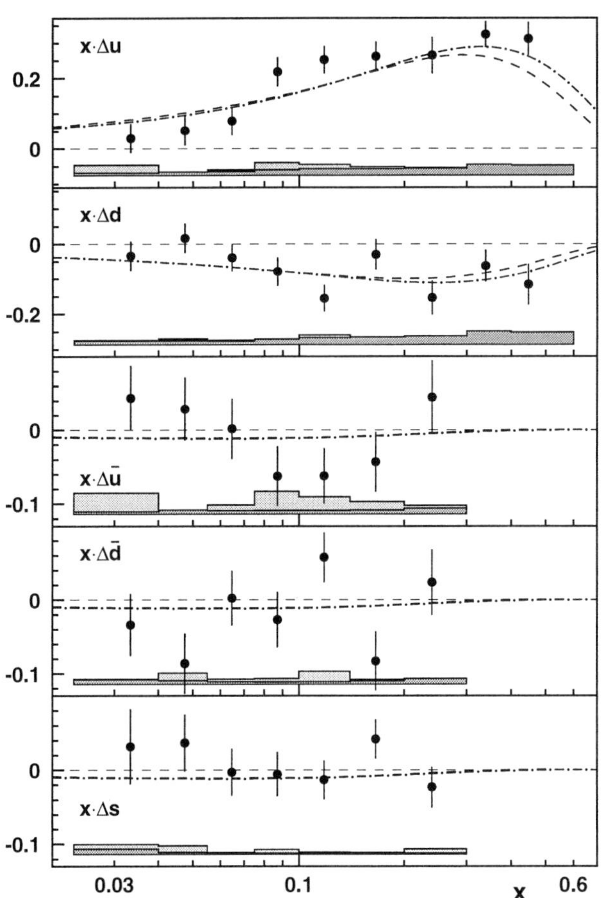

Figure 6. Flavor separation of the quark helicity distributions by HERMES.

tation. The elements are evaluated with a Monte Carlo which was tuned to approximate the hadron multiplicities measured at HERMES.

The results are shown in Fig. 6. The bands show systematic errors. The curves show QCD fits to inclusive measurements.[5] The helicity distribution of the u quark is positive while that of the d quark is negative. The helicity distributions of sea quarks are close to zero.

It should be noted that the flavor separation is done in each x bin independently except for small corrections for smearing due to the finite kinematical resolution of the spectrometer. It is not necessary to assume any functional forms in the measured and unmeasured regions of x. It is also not necessary to prefix the first moment of the quark helicity distribution, namely integration over Bjorken x, using information from other experiments such as neutron beta-decay and hyperon weak decays.

The measurements of the inclusive asymmetry A_1 with proton and neutron targets at large Bjorken x at JLab enables the flavor separation at large x since the sea quark contributions are small there.

3. Deeply Virtual Compton Scattering and Generalized Parton Distributions

The cross section for deep-inelastic scattering is given by the total absorption cross section of a virtual photon as mentioned before. It is therefore equivalent to the imaginary part of the forward elastic amplitude. "Elastic" in this case means Compton scattering.

Deeply virtual Compton scattering (DVCS) is similar, but is different in a sense that a real photon is produced in the final state as shown in Fig. 7.

A momentum fraction has to be taken out and put back as shown in Fig. 7. Therefore, the amplitude involved is not the forward one but is off-forward.

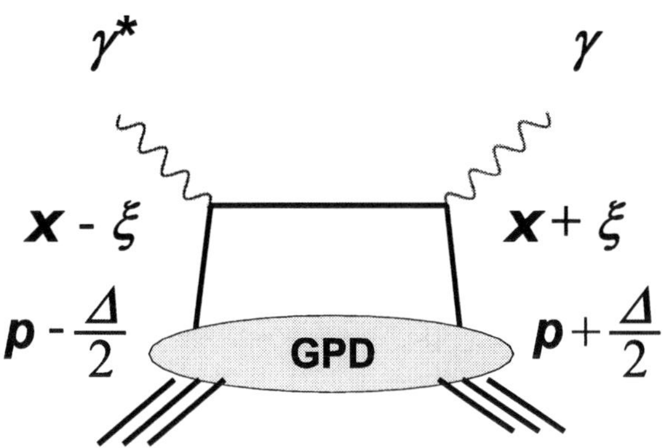

Figure 7. Diagram of deeply virtual Compton scattering (DVCS).

Recently DVCS received attention as it plays an important role in determining the generalized (sometimes called the off-forward or skewed) parton distributions. In the framework of the generalized parton distributions various observables such as DVCS, elastic form factors, quark distributions from deep-inelastic scattering etc. are coherently analyzed as shown in Fig. 8. There are four generalized parton distributions concerning quarks: $H^q(x,\xi,t), E^q(x,\xi,t), \tilde{H}^q(x,\xi,t), and \tilde{E}^q(x,\xi,t)$. In the forward limit where $t \to 0$ and $\xi \to 0$, $H^q(x,0,0)$ and $\tilde{H}^q(x,0,0)$ are reduced to $q(x)$ and $\Delta q(x)$, the ordinary quark distributions and helicity distributions, respectively. When integrated over x and summed over q, $H^q(x,\xi,t)$ and $E^q(x,\xi,t)$ are reduced to $F_1(t)$ and $F_2(t)$, the Dirac and Pauli nucleon form factors, respectively, while $\tilde{H}^q(x,\xi,t)$ and $\tilde{E}^q(x,\xi,t)$ are reduced to $g_A(t)$ and $h_A(t)$, the axial-vector and pseudo-scalar form factors, respectively.

Once the generalized parton distributions are determined, the total angular momentum contribution of the quarks J_q to the nuclon spin can be evaluated:[6]

$$\lim_{t \to 0} \frac{1}{2} \int_{-1}^{+1} dx\, x[H^q(x,\xi,t) + E^q(x,\xi,t)] = J_q. \quad (7)$$

Since the quark spin contribution $\frac{1}{2}\Sigma$ is already known, one can then determine the contributions of the orbital angluar mometum of the quarks L_q to the nucleon spin: $J_q = L_q + \frac{1}{2}\Sigma$. So far not many theoretical methods to access the orbital angluar momen-

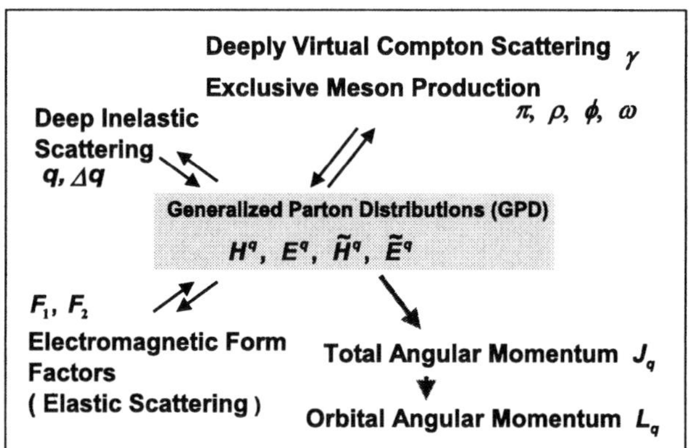

Figure 8. Relations of generalized parton distributions.

tum of quarks have been proposed, and the method mentioned above is an attractive one.

The processes to produce a direct photon are the DVCS and Bethe-Heitler processes. The latter involve the production of a photon from the initial or final lepton line. The cross section for direct photon production is therefore expressed as

$$\frac{d^4\sigma}{d\phi dt dQ^2 dx} \propto | A_{DVCS} + A_{BH} |^2 \quad (8)$$

$$= | A_{DVCS} |^2 + | A_{BH} |^2 + I \quad (9)$$

where I is the interference term between the DVCS and Bethe-Heitler amplitudes.

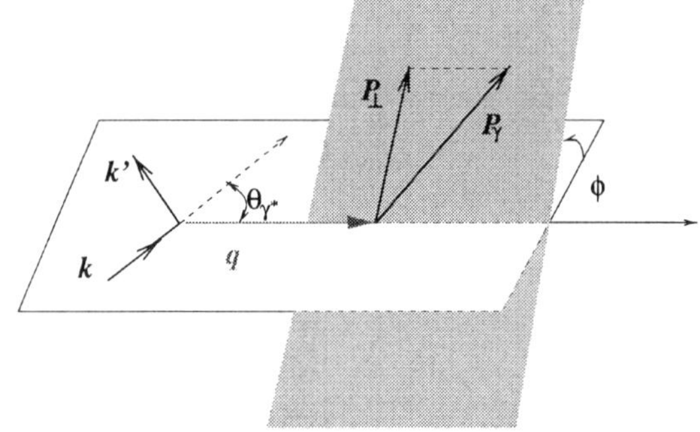

Figure 9. Definition of azimuthal angle ϕ.

The single spin asymmetry from a polarized positron beam with an unpolarized nucleon target is

$$\Delta\sigma_{LU} = \sigma(e^{\to}p) - \sigma(e^{\leftarrow}p) \propto -sin\phi \times Im\,I. \quad (10)$$

The beam charge asymmetry from positron and electron beams is

$$\Delta\sigma_{ch} = \sigma(e^+p) - \sigma(e^-p) \propto cos\phi \times Re\,I. \quad (11)$$

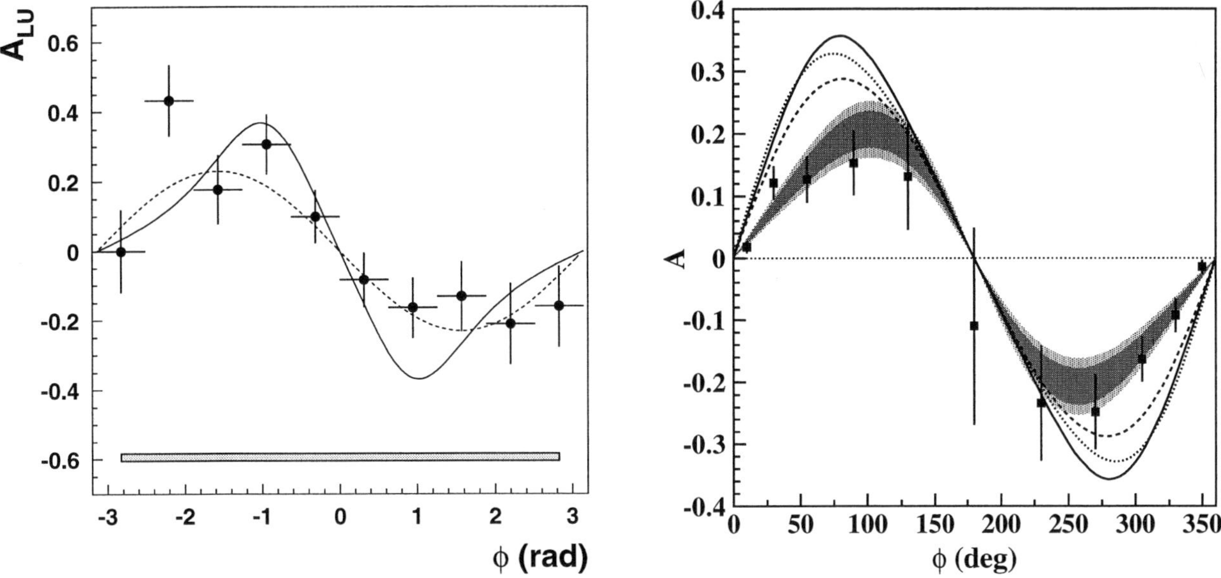

Figure 10. Beam spin asymmetry due to the interference between DVCS and Bethe-Heitler processes observed by HERMES (left) and CLAS (right)

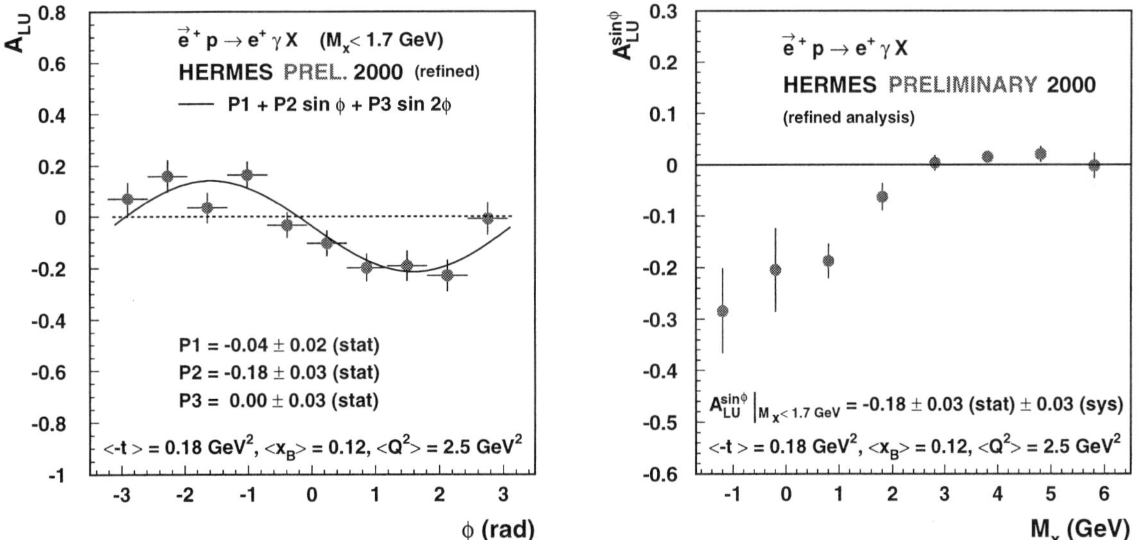

Figure 11. Beam spin asymmetry (left) and the sinφ moment (right) of DVCS as a function of the missing mass.

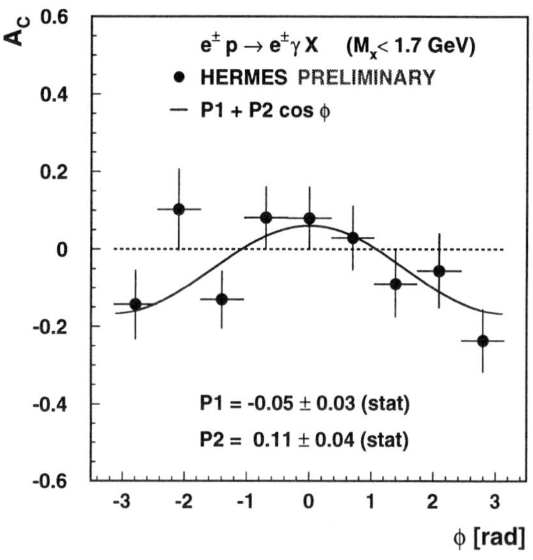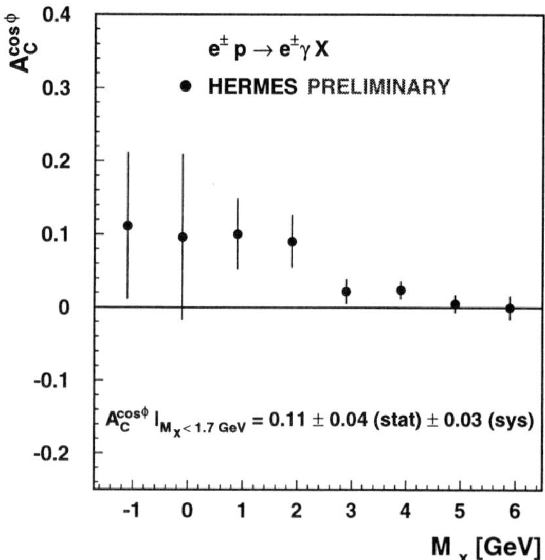

Figure 12. Beam charge asymmetry (left) and the $\cos\phi$ moment (right) of DVCS as a function of the missing mass.

The angle ϕ is the azimuthal angle defined in Fig. 9. The σ_{LU} was measured by HERMES[7] and CLAS.[8] The quantity σ_{ch} was measured by HERMES as both positron and electron beams are available at DESY. This way one can access the imaginary and real parts of the interference term. At higher energies the direct photon production is dominated by DVCS, so the DVCS cross section, not asymmetry, is measured by H1 and ZEUS.

The single spin asymmetries, A_{LU}, observed by HERMES and CLAS are shown in Fig. 10. The single spin asymmetry, A_{LU}, with updated statistics from HERMES and the missing mass distribution of its $\sin\phi$ moment are shown in Fig. 11. The beam charge asymmetry, A_c, from HERMES and the missing mass distribution of its $\cos\phi$ moment are shown in Fig. 12. The missing mass resolution is 0.8 GeV. In the region of the missing mass which corresponds to the nucleon mass, significant deviations of the moments from zero are observed.

To further extend the studies of the generalized parton distributions, exlusive meson productions[9] are useful. In these cases a meson is produced instead of a photon. When a pseudoscalar meson such as a π is produced, $\tilde{H}^q$ and $\tilde{E}^q$ are sensitive to these processes. When a vector meson such as a ρ, ϕ, or ω is produced, H^q and E^q are probed.

4. Quark Transversity Distributions

There are three leading twist (twist-two) quark distributions: ordinary quark distribution $q(x)$, helicity distribution $\Delta q(x)$, and transversity distrbution $\delta q(x)$. A sketch is given in Fig. 13.

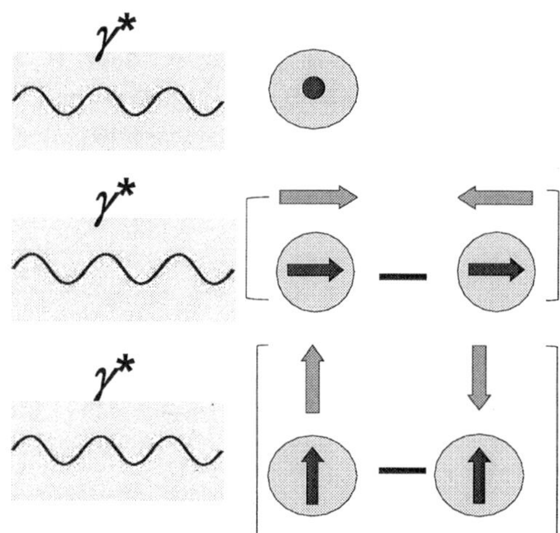

Figure 13. Three types of quark distributions: unpolarized distribution $q(x)$, helicity distribution $\Delta q(x)$, and transversity distribution $\delta q(x)$.

The spin averaged distribution is denoted by $q(x)$. It indicates a vector charge when integrated over x. $\Delta q(x)$ is the helicity difference. It represents the axial charge when integrated over x. $\delta q(x)$ is the helicity flip type, and represents the tensor charge. $\delta q(x)$ is chiral odd, so it is not accessed in inclusive deep-inelastic scattering. However, it is accessible with semi-inclusive measurements of deep inelastic scattering because $\delta q(x)$ can be accompanied with another chiral odd object, namely a chiral odd fragmentation function:

$$A_1^h \approx \sum_q e_q^2 \, \delta q(x) \cdot H(z)_{1,q}^{\perp h}. \qquad (12)$$

The quantity $q(x)$ has been very well measured, and $\Delta q(x)$ has been well measured, too. $\delta q(x)$ is the last unmeasured twist-two distribution, which attracted theoretical attention recently in parallel with the realization of the experimental feasibility in deep-inelastic scattering.

The three distributions are not independent. They are related by a Soffer bound:

$$\mid \delta q(x) \mid \leq \frac{1}{2} \mid q(x) + \Delta q(x) \mid . \qquad (13)$$

Since $\delta q(x)$ does not couple with the gluon, the Q^2 evolution should be different from that of $q(x)$ and $\Delta q(x)$. This is another interesting point.

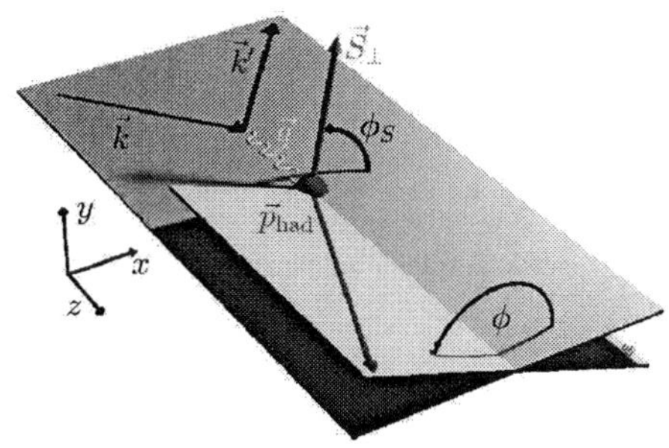

Figure 14. Definition of ϕ_s angle in the case of a transversely polarized target.

The data for which analyses were completed by HERMES are those using longitudinally polarized targets. The virtual photon emitted in each event from the incoming positron beam has a certain transverse component with respect to the positron beam axis. Therefore, the effect of transversity could show up already with the longitudinally polarized targets:

$$A_{UL}^{sin\phi} \sim S_L < sin\phi >_{UL} - S_T < sin\phi >_{UT} . \qquad (14)$$

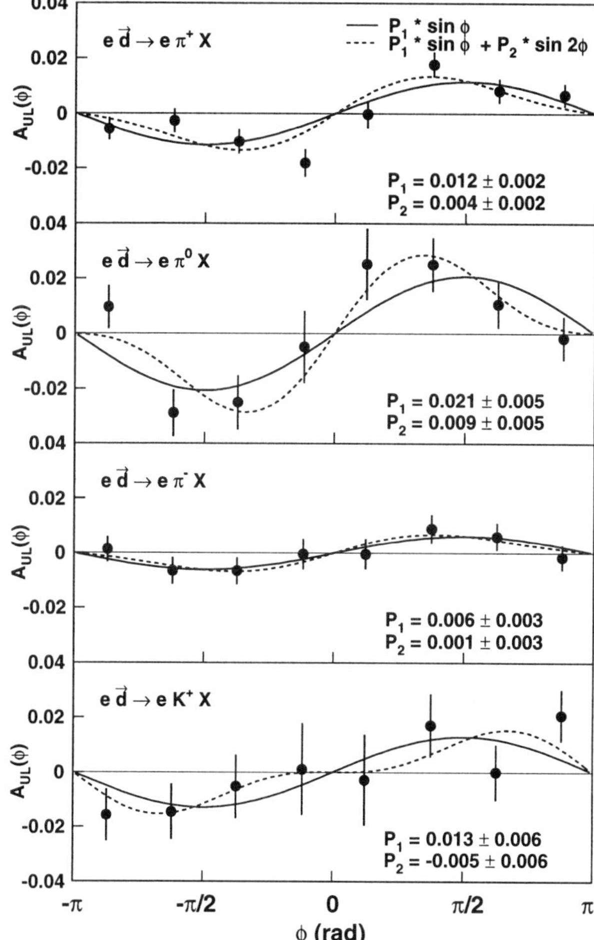

Figure 15. Single spin azimuthal asymmetries A_{UL} in semi-inclusive measurements of deep inelastic scattering

The transverse component of the virtual photon, S_T, is expressed as

$$S_T \approx \frac{2Mx}{Q}\sqrt{1-y} \qquad (15)$$

$$\leq sin\theta_\gamma \gtrsim 0.15. \qquad (16)$$

With the data being taken by HERMES since 2002 with a transversely polarized proton target, analyses can be further developed since there is an angle ϕ_s in addition to ϕ as shown in Fig. 14. The Collins effect is expected to show a $sin(\phi + \phi_s)$ moment while the Sivers effect is expected to show a $sin(\phi - \phi_s)$ moment. HERMES has already accumulated 700K events and COMPASS has also taken data. From the data taken so far using the longitudinally polarized target, the azimuthal angle dependence of the asymmetry A_{UL} was measured. The results on the proton target[10] and on the deuterium target[11], shown in Fig. 15, were obtained. In Fig. 15 the $sin\phi$ moment is positive for π^+, π^0 and K^+. It increases with Bjorken x, suggesting that the valence quark or u quark plays the dominant role. The $sin\phi$ moment is small for π^-.

With the new results from the transversely polarized target, it is expected that the physics of quark transversity distributions will be developed extensively.

JLab has also an experimental program to study this subject.

5. Polarized Proton-Proton Collisions

One of the aims of using polarized proton-proton collisions at RHIC of BNL is to study the gluon spin contribution to the proton spin. A longitudinally polarized proton beam is required to measure the helicity distributions of the partons in the proton. The physics goal or milestone is to achieve a beam polarization of 70% and an integrated luminosity of 320 pb^{-1} at $\sqrt{s} = 200$ GeV and 800 pb^{-1} at $\sqrt{s} = 500$ GeV. A beam polarization of about 30% and an integrated luminosity of about 0.4 pb^{-1} were achieved in 2003. The goals are expected to be reached several years from now.

The PHENIX collaboration plans to measure the gluon helicity distribution $\Delta G(x)$ from the asymmetry A_{LL} in direct photon production from quark-gluon collisions. The process is shown in Fig. 16 and is called "gluon Compton scattering". In the p_T region below 10 GeV the quark-gluon collision is dominant over gluon-gluon and quark-quark collisions. The STAR collaboration plans to measure the asymmetry in di-jet productions which also has a significant dependence on the gluon polarization in the proton.

The single spin asymmetry A_N in π^0 production at forward angles from transversely polarized proton collisions, $p^\uparrow + p \rightarrow \pi^0 + X$, has been measured by STAR as shown in Fig. 17. The asymmetry resembles the results from the earlier fixed target experiment E704 at Fermilab, in spite of the fact that $\sqrt{s}$ is now 200 GeV compared to 20 GeV at E704. The p_T is low also in the present measurement. Here again, the contributions of the Collins effect and the Sivers effect, as well as higher-twist contributions are being investigated.

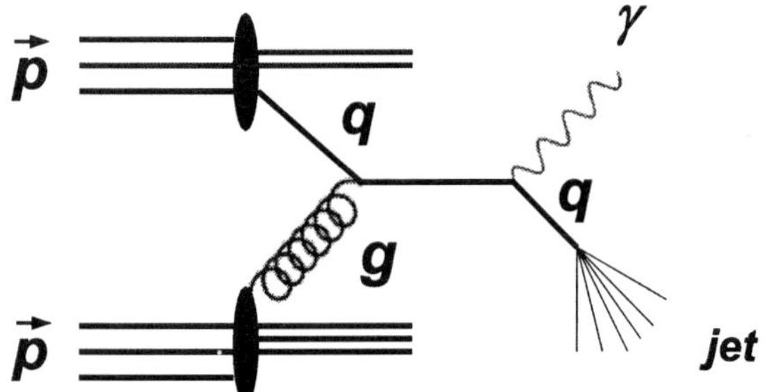

Figure 16. Diagram of $pp \rightarrow \gamma + X$. This process is called gluon Compton scattering.

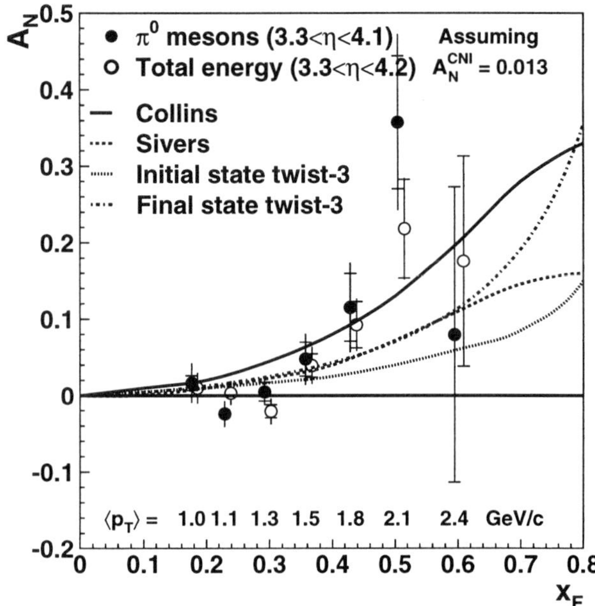

Figure 17. Single transverse beam spin asymmetry A_N in π^0 production at forward angles measured by STAR.

The asymmetry A_{LL} in π^0 production in the central rapidity region from longitudinally polarized proton collisions is being analysed by PHENIX.

6. Summary

To understand hadrons in terms of QCD, the spin structure of the nucleon is an important subject. Many experiments are currently running with e, μ, γ, p... beams. Flavor separation of the quark helicity distributions, $\Delta q(x)$, have been made using semi-inclusive measurements of produced hadrons in deep-inelastic lepton scattering. Asymmetries in DVCS and in exclusive meson production provide access to generalized parton distributions. It is expected that contributions of the orbital angular momentum of quarks to the nucleon spin can be evaluated in the framework of generalized parton distributions.

Single-spin azimuthal asymmetries have been observed in semi-inclusive measurements of deep-inelastic scattering. Quark transversity distributions, $\delta q(x)$, are basic in the sense that they are twist-two distributions but have never been determined so far. Results from transversely polarized targets will become available soon.

The polarized proton-proton collider now has longitudinally as well as transversely polarized beams. The single spin asymmetry A_N in π^0 productions at small angles from a transversely polarized beam have been obtained. The asymmetries at large p_T are being analyzed.

Spin physics with hadrons is a rapidly expanding field: many new ideas for measurements have been proposed and high precision data are being accumulated. We might have new surprises from these studies.

Acknowledgments

The author wishes to thank all the colleagues who made the large progress in spin physics in recent years possible.

References

1. J. Ashman *et al.*, *Phys. Lett.* B **206**, 364 (1988); J. Ashman *et al. Nucl. Phys.* B **328**, 1 (1989).
2. A. Airapetian *et al.*, hep-ex **0307064**, (2002).
3. B. Adeva *et al.*, *Phys. Lett.* B **420**, 180 (1998).
4. K. Ackerstaff *et al.*, *Phys. Lett.* B **464**, 123 (1999).
5. M. Glueck *et al.*, *Phys. Rev.* D **63**, 094005 (2001); J. Bluemlein and H. Boettcher, *Nucl. Phys.* B **636**, 225 (2002).
6. X. Ji, *Phys. Rev. Lett.* **78**, 610 (1997).
7. A. Airapetian *et al.*, *Phys. Rev. Lett.* **87**, 182001 (2001).
8. S. Stepanyan *et al.*, *Phys. Rev. Lett.* **87**, 182002 (2001).
9. A. Airapetian *et al.*, *Phys. Lett.* B **535**, 85 (2002).
10. A. Airapetian *et al.*, *Phys. Rev. Lett.* **84**, 4047 (2000); A. Airapetian *et al.*, *Phys. Rev.* D **64**, 097101 (2001).
11. A. Airapetian *et al.*, *Phys. Lett.* B **562**, 182 (2003).
12. A. Airapetian *et al.*, *Phys. Rev.* D **64**, 112005 (2001).

Ikaros Bigi (Notre Dame): Are there any recent data on the question of how the longitudinal polarization of the incoming lepton beam, either muon or neutrinos, are transferred into the final state by looking at polarizations of Λ, Ξ and even charm baryons?

Toshi-Aki Shibata: New data are being analyzed for longitudinal Λ polarization by HERMES in addition to the already published results.[12] COMPASS could also contribute to this subject.

DIFFRACTION AND VECTOR MESON PRODUCTION

Y. YAMAZAKI

High-energy accelerator research organisation (KEK), Oho 1-1, Tsukuba, 305-0801 Japan
E-mail: yuji.yamazaki@kek.jp

A review is given of the measurements of the diffractive process in recent years from two high-energy colliders, the HERA ep collider and the Tevatron $p\bar{p}$ collider. The energy dependence of the cross sections and the factorisation properties of diffractive processes are discussed.

1. Introduction

The diffractive process, defined as hadron-hadron scattering with an exchange of the vacuum quantum numbers and a small momentum transfer, has been understood in the framework of Regge theory in the past. The exchange was interpreted as the Pomeron trajectory, whose counterpart as a particle has not yet been observed. The concept of the particle-like Pomeron is therefore not satisfactory. The aim of studying diffractive processes in the era of high-energy colliders is to understand the partonic structure of the diffractive exchange in terms of perturbative QCD (pQCD).

The diffractive exchange has two distinguishing features. One is that it is a colour-singlet exchange. This means that the emission of hadrons from the exchange is suppressed. Another is that the diffractive events are observed with a small momentum transfer in both the transverse and longitudinal coordinates. The four-momentum of the exchange, t, and the longitudinal momentum fraction of the exchange, $x_{\mathbb{P}}$, are both small; $|t|$ is typically less than the square of the nucleon mass and $x_{\mathbb{P}}$ is smaller than 0.05. The two incoming hadrons are scattered through a small angle without losing any significant longitudinal momentum. As a consequence, the two diffractively dissociated systems are separated in rapidity space, forming a large rapidity gap (LRG).

In pQCD, the colour-singlet configuration of the exchange at leading order is described by either two gluons or two quarks with opposite colour charges. Assuming that the partons in the exchange originated from partons in the hadron emitting the exchange, the partonic content of the exchange at low-$x_{\mathbb{P}}$ is expected to be mainly gluons since they dominate in the low-x regime of the parent hadron, where x is the Bjorken variable, representing the longitudinal momentum fraction of the parton in the nucleon.

At lowest order, pQCD models describe the diffractive exchange as a gluon pair.

The partonic structure of the exchange can be studied using processes with a hard scale, like large Q^2, the negative of the four-momentum squared of the exchanged virtual photon in deep-inelastic scattering (DIS), large E_T in jet production or a heavy-quark mass. In this review, the data on such processes from the HERA ep collider and the highest energy $p\bar{p}$ collider, the Tevatron, are presented. This article concentrates on two topics, the factorisation properties of diffractive processes and the energy dependence of the diffractive cross sections, both of which give insight into the partonic structure of the diffractive exchange. Before reviewing the data, a short discussion on the parton densities of the nucleon and their relation to the diffractive cross section is given.

2. Diffractive Processes and Parton Densities in Hadronic Collisions

As already discussed, the partonic contents of the diffractive exchange in most of the pQCD models originate from the partons in the incoming hadrons, i.e. protons for HERA and the Tevatron. The partonic contents of the protons are primarily determined from the $F_2(x, Q^2)$ results measured using DIS interactions (see Fig. 1). A fast rise towards low x is observed in the F_2 data. The parton densities, extracted using NLO DGLAP fits of the F_2 data, show a corresponding increase towards lower x. In particular, the gluon density shows a faster increase than that of the quarks and dominates at low x. Such behaviour of F_2 is also expressed in the form of virtual photon-proton total cross section, $\sigma_{\text{tot}}^{\gamma^* p}$, as a function of Q^2 and W, the $\gamma^* p$ centre-of-mass energy through the relation $\sigma_{\text{tot}}^{\gamma^* p}(W, Q^2) \approx (4\pi\alpha^2/Q^2)F_2$. Since $W^2 \approx Q^2/x$ for small x, the increase of F_2 cor-

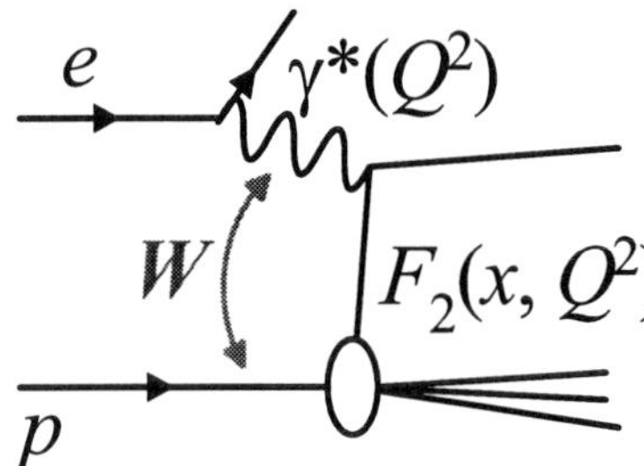

Figure 1. A diagram of DIS.

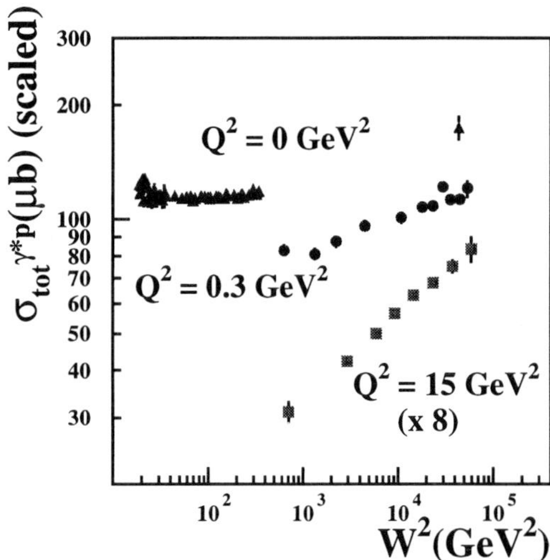

Figure 2. The total $\gamma^* p$ cross section measured at HERA.[1,2]

responds to the increase of the cross section towards high W. An example at $Q^2 = 15\,\mathrm{GeV}^2$ is shown in Fig. 2.

This behaviour of $\sigma_{tot}^{\gamma^* p}$ is in contrast to that of soft processes, e.g. the total photoproduction cross section, that is, the collision of a quasi-real photon with the proton at HERA. The lack of a hard scale in this process means that it is not possible to resolve partons in the proton. The W dependence of the cross section for this process shows only a slow increase. The size of the slope of the cross section in W is, therefore, considered to be evidence of a hard interaction. Such processes are expected to be described by pQCD. Similar argument can be applied for the centre-of-mass dependence of the cross sections in hadron-hadron collisions, where $\sqrt{s}$ is substituted in place of W.

Once the parton densities in the proton are known, a simple model where the probability to exchange a parton between two incoming particles is proportional to the parton density can be made. The energy dependence of the inelastic and diffractive processes from this model would be the following: the mechanism of the inelastic process for nucleon-nucleon collisions can be modelled by one-gluon exchange (Fig. 3(a)) in the low-x regime. This argument can also be applied for the low-x DIS where the virtual photon can be regarded as a quark-antiquark ($q\bar{q}$) dipole with a long lifetime when it collides with the nucleon. Under this assumption, the cross section behaviour is $\sigma(\sqrt{s}) \propto g(x)$ where $g(x)$ is the gluon density of the nucleon. Similarly, the diffractive process can be regarded as two-gluon exchange (Fig. 3(b)). The cross section behaviour could then be $\sigma(\sqrt{s}) \propto |g(x)|^2$, with a suppression factor of 1/9 to form a colour-singlet state. If the colour is not

cancelled between two gluons, which is naïvely expected to be the case in 8/9 of the interactions, the scatter results in a process with a multi-parton interaction (Fig. 3(c)). As the centre-of-mass energy of the scattering increases, the collisions will be increasingly dominated by low-x processes. Therefore, the probability to have two-gluon exchange should increase with $\sqrt{s}$ (or W). Indeed, assuming the inelastic cross section behaves like s^a, the diffractive cross section as well as the multi-parton interaction should increase approximately like s^{2a}. Therefore, the fraction of the cross section attributed to diffraction would increase with energy and may play an important role in understanding the total cross section behaviour of hadron-hadron collisions. In this article, the energy dependence of the diffractive cross sections are reviewed in detail.

It is conceivable that, with such a high density of partons in the nucleon at low x, the probability to exchange a gluon in a nucleon-nucleon crossing at a given impact parameter is close to unity. The probability to exchange three or more gluons could also be sizable. In this case, the colour combination of the multi-gluon state becomes complicated. For example, in the case of three-gluon exchange, an incoherent colour configuration would result in an inelastic multi-parton interaction as in the two-gluon case (Fig. 3(e)). In addition, two of the three glu-

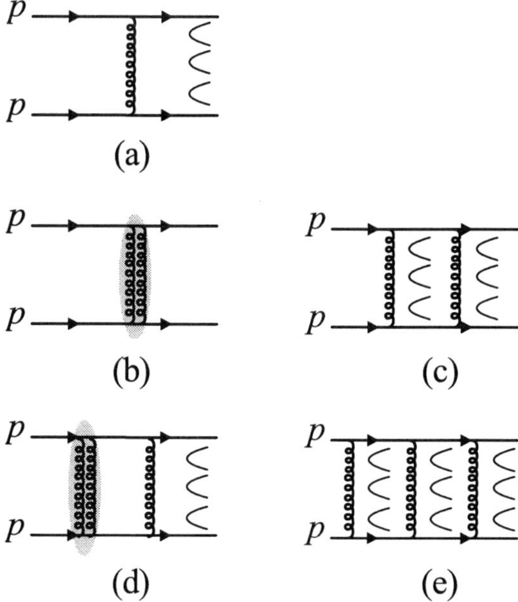

(a)

(b)   (c)

(d)   (e)

Figure 3. The diagram of the inelastic and diffractive events by multi-parton exchange: (a) inelastic scattering with one-gluon exchange; (b) diffractive exchange by two gluons; (c) inelastic multi-parton interaction by two gluons; (d) diffractive exchange with additional one gluon exchange; and (e) inelastic three-gluon exchange.

ons may form a colour-singlet state while the other is a colour-octet. In this case, the total colour is not neutral and is therefore considered to be an inelastic process (Fig. 3(d)) even though a diffractive state is formed. Thus, the fraction of the diffractive cross section to the total cross section may be suppressed at high energies where the gluon density in the proton is very large. In fact, measurements at the Tevatron shows the suppression of the cross section, in accord with this naïve expectation. This issue is also reviewed in detail below.

The above exercise shows that the two processes, the diffractive process and multi-parton interactions, should be discussed on the same theoretical basis. Both processes, copiously observed at HERA and Tevatron, are lacking a state-of-the-art understanding in terms of pQCD.

3. Inclusive Diffraction at HERA

The main diffractive process at HERA is in electroproduction, and is referred to as diffractive DIS

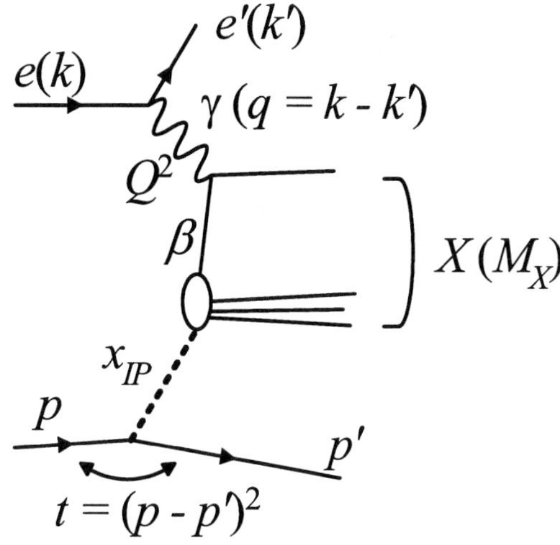

Figure 4. An example of the diagram of the diffractive process in deep-inelastic scattering.

(DDIS). The presence of the hard scale, Q^2, ensures that the virtual photon is point-like and that the photon probes the partonic structure of the diffractive exchange (Fig. 4), in analogy with the inclusive DIS processes. The primary measurement is the determination of the diffractive structure function, $F_2^{D(3)}$, which is derived from the diffractive cross section integrated over t,

$$\int dt \frac{d^4\sigma}{d\beta dQ^2 dx_{\mathbb{P}} dt} \simeq \frac{4\pi\alpha^2}{Q^2}(1 - y - y^2/2)$$
$$\times F_2^{D(3)}(\beta, Q^2, x_{\mathbb{P}})$$

where $\beta = x/x_{\mathbb{P}}$ is the longitudinal momentum fraction of the parton which couples to the photon in the diffractive exchange. $F_2^{D(3)}$ is sensitive to the quark content of the exchange. In order to determine the gluonic content, other processes such as jet or heavy-quark production are used. The gluonic content can also be obtained from the scaling violation of the F_2^D assuming DGLAP evolution.

In this section, three issues relating to F_2^D at HERA are reviewed: hard-scattering factorisation and the universality of the parton densities in the diffractive process; tests of the resolved Pomeron model and the W dependence of the diffractive cross sections.

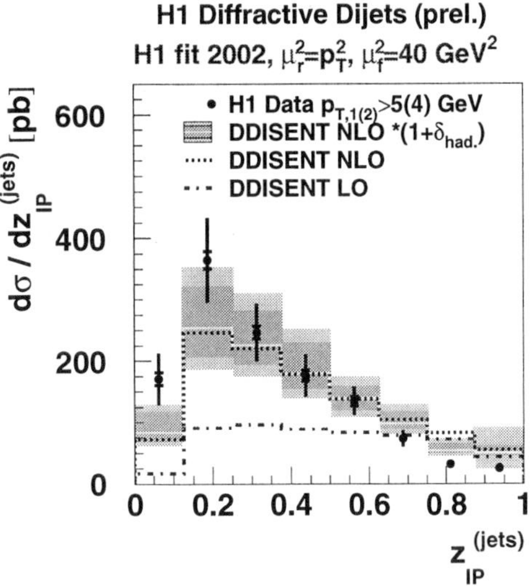

Figure 5. The measured dijet cross section in DDIS in comparison to the NLO calculation DISENT using NLO diffractive parton densities extracted from the H1 data.[5]

3.1. *Factorisation Properties in Diffractive DIS*

For hard QCD processes in general, like high-E_T jet production or DIS, the cross section can be factorised into two terms; the parton density and the hard parton-parton cross section. In the case of DIS, the cross section can be written as

$$\sigma = \sum_i f_i(\xi, \mu^2) \hat{\sigma}_{i\gamma}(\xi, \mu^2)$$

where i runs over all parton types, f_i is the parton density function for the i-th parton with longitudinal momentum fraction ξ, which is probed at the factorisation scale μ. $\hat{\sigma}_{i\gamma}$ denotes the cross section for the interaction of the i-th parton and the virtual photon. Such an expression, often referred to as the QCD factorisation theorem, is well supported by the data. If the theorem holds, only one parton per hadron is coupled to the hard scattering vertex.

The theorem is proven to be applicable also for DDIS,[3] namely

$$\frac{d\sigma(x, Q^2, x_{IP}, t)}{dx_{IP} dt} = \sum_i \int_x^{x_{IP}} dz$$
$$\times [\hat{\sigma}_{i\gamma}(z, Q^2, x_{IP}) f_i^D(z, Q^2, x_{IP}, t)]$$

where z is the longitudinal momentum fraction of the parton in the proton, $\sigma_{i\gamma}$ is again the hard scattering

parton-photon cross section for DDIS and f_i^D is the diffractive parton density for the i-th parton. f_i^D can be regarded as the parton density of the proton under the condition that the diffractive exchange occurs at a given (x_{IP}, t). If such a theorem holds, f_i^D should be universal for all hard processes, e.g. DIS, jet or heavy-quark production etc.. An experimental confirmation of this hypothesis was performed[4] in the following way: first, the diffractive parton densities are obtained from a DGLAP fit to the F_2^D data.[5] The extracted densities are then used in the cross section calculation for a process which is directly sensitive to the gluon density, like jet or heavy-quark production, to check the universality of the diffractive parton densities.

Figure 5 shows a comparison between diffractive dijet production in DIS and the NLO QCD calculations using the parameterisation of the extracted parton densities from $F_2^{D(3)}$. The cross section is shown as a function of $z_{IP}^{(jets)}$, the longitudinal momentum fraction of the partons in the diffractive exchange which participated in the hard scattering reconstructed from the dijet momenta. The prediction provides a good description of the data in both shape and normalisation. This shows that the extracted parton densities are universal, hence the factorisation theorem holds in diffractive DIS. Similar conclusions are obtained from the D^* production data.[4,6]

3.2. *Tests of the Resolved Pomeron Model*

The presence of a LRG in the diffractive processes and the success of Regge theory in describing the diffractive data implies that the exchange is a particle-like state with a long lifetime and has a partonic structure. This hypothesis, called the resolved Pomeron model, assumes factorisation of the diffractive exchange,

$$F_2^D \propto f_{IP/p}(x_{IP}, t) \cdot F_2^{IP}(\beta, Q^2) \tag{1}$$

where $f_{IP/p}$ is the flux of the "Pomeron" (i.e. the diffractive exchange) and $F_2^{IP}(\beta, Q^2)$ is the structure function of the Pomeron. The Pomeron flux depends only on the kinematic variables of the nucleon-Pomeron vertex, x_{IP} and t.

The validity of this model was examined as follows: Eq. (1) shows that $F_2^{D(3)}$ at a given (β, Q^2) can be written as $b \cdot f_{IP/p}(x_{IP})$ where $b = F_2^{IP}(\beta, Q^2)$ and after integrating over t. This shows that the cross

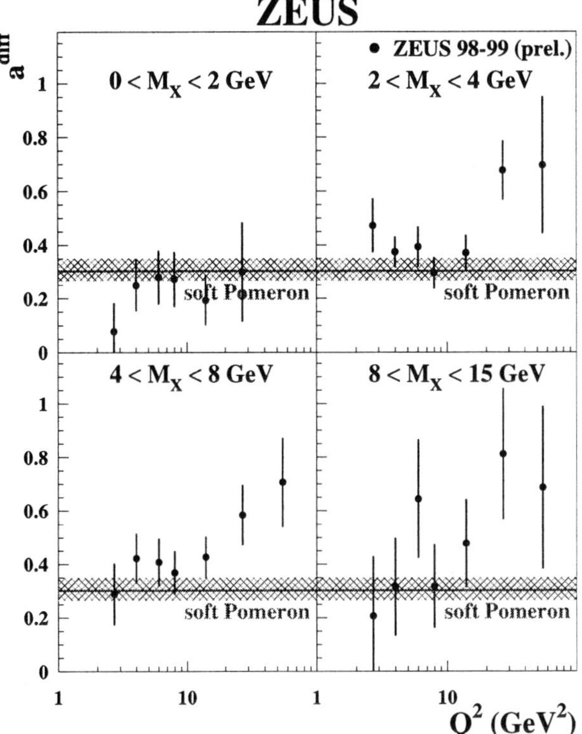

Figure 6. The exponent of the power-like fit $\sigma \propto W^{a_{\text{diff}}}$ on the diffractive cross sections $d\sigma/dM_X$ for four regions in M_X.[7] The horizontal line and the band show the value and error expected from the soft scattering.

section dependence of $f_{\mathbb{P}/p}$ on $x_{\mathbb{P}}$ is the same for any fixed value of (β, Q^2) apart from the normalisation which is given by b. In order to quantify the possible change of shape in $x_{\mathbb{P}}$, the $x_{\mathbb{P}}$ dependence or, equivalently, the W dependence of the cross section for different values of (β, Q^2) are compared by parameterising the flux as a function of $x_{\mathbb{P}}$ or W: $f_{\mathbb{P}/p} \propto x_{\mathbb{P}}^{-a/2} \propto W^a$. Figure 6 shows the parameter a or a_{diff} as a function of Q^2 for different ranges in the diffractive mass, M_X, of the photon dissociation system X. The values are mostly above the expectation from soft scattering, i.e. the dependence of the total hadron-hadron or photon-hadron cross sections. The steeper rise of the diffractive cross sections indicates the onset of pQCD behaviour. In addition, the rise of a_{diff} as a function of Q^2 is seen for $4 < M_X < 8$ GeV. This indicates that the flux of the diffractive exchange depends on Q^2, demonstrating the break down of the resolved Pomeron model. This Q^2 dependence is expected from pQCD where the cross section goes as $|g(x, Q^2)|^2$, thus the rise of the cross section with $W \sim 1/\sqrt{x}$ is a function of

Q^2, as in the case of inclusive DIS. Note, however, that the violation of the resolved Pomeron model is small.

3.3. *The Energy Behaviour of the Diffractive Cross Sections at HERA*

The rise of the inclusive diffractive cross sections with W at HERA is found to be steeper than that for soft scattering as described above. The rise in W can be further investigated by taking the ratio of the diffractive to the inclusive DIS cross section. The QCD models explain the diffraction process as the exchange of a diffractive state between the nucleon and the $q\bar{q}$ dipole, which originates from a virtual photon. The model with a pure two-gluon exchange, as discussed in Sec. 2, should show a rise, which is approximately twice as fast as that in the inclusive DIS: $\sigma_{\text{tot}} \sim g(x)$ while $\sigma_{\text{diff}} \sim |g(x)|^2$. This is under the assumption that both gluons obey the density probed at the factorisation scale μ^2.

Other types of models, like the semi-classical model[8] and the soft colour interaction model[9] assume that one gluon is probed with μ^2 and another soft gluon neutralises the colour-octet state of the hard gluon. Since the density of the soft gluon is expected to be approximately independent of $\sqrt{s}$ or W, the dependence of the cross section for the diffractive process is similar to the inelastic one: $\sigma_{\text{diff}} \propto g_{\text{hard}}(x) \cdot g_{\text{soft}}(\sqrt{s}) \simeq g(x)$. A similar prediction is made by the saturation model[10] where the soft behaviour of the $\gamma^* p$ cross section at very low Q^2, namely the slow rise with W, is expected to continue up to higher Q^2 than in the inclusive case. In both types of models, the ratio $\sigma_{\text{diff}}/\sigma_{\text{tot}}$ is predicted to be approximately constant in W.

Figure 7 shows the measured ratio $\sigma_{\text{diff}}(Q^2, W, M_X)/\sigma_{\text{tot}}(Q^2, W)$ as a function of W in regions of Q^2 and M_X. The ratio is flat as a function of W except for very low M_X where the ratio decreases with W. The energy dependence of the diffractive process is not steeper than that of the total cross section, in contradiction to the expectation from the two-gluon exchange models. This implies that the two exchanged partons do not have the same property as the hard partons exchanged in the DIS processes. This suggests the presence of soft phenomena in the colour-singlet parton pair.

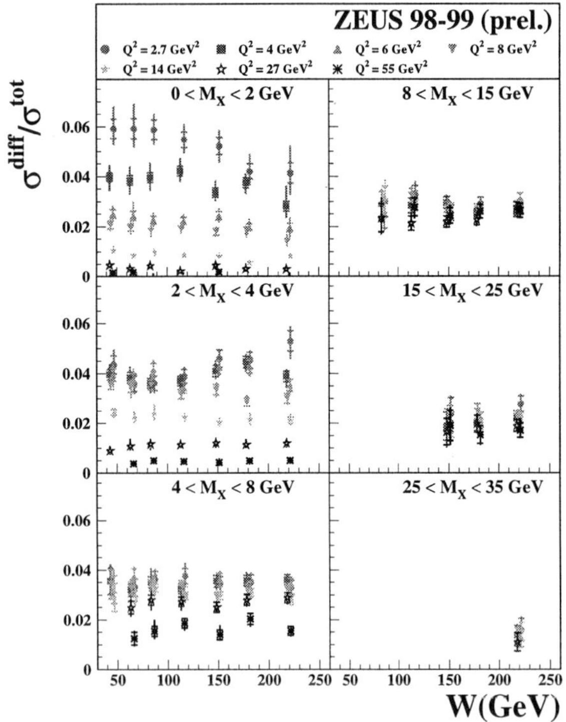

Figure 7. The ratio $\sigma_{\text{diff}}/\sigma_{\text{tot}}$ as a function of W, differentially in Q^2 and W.[7]

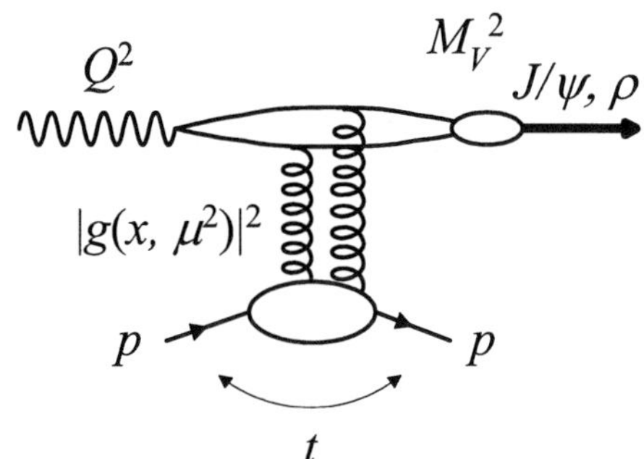

Figure 8. A typical diagram of the pQCD models for vector meson production at HERA.

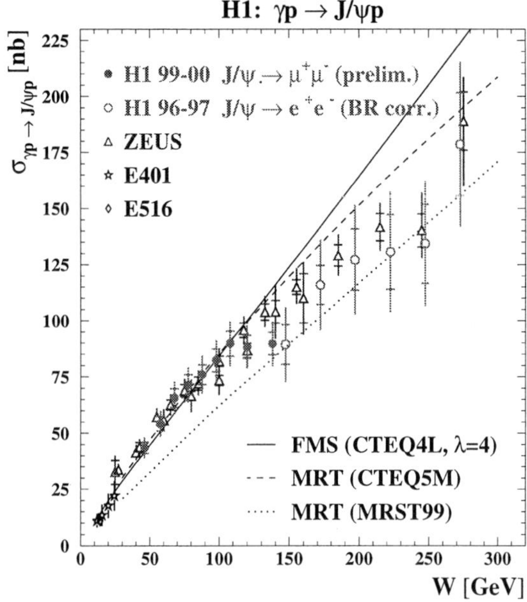

Figure 9. The W dependence of the total cross section of the J/ψ in photoproduction at HERA,[11] in comparison with two theoretical models: FKS[12] and MRT.[13]

4. Vector Meson Production at HERA

Vector meson production in diffractive scattering is part of the inclusive diffraction but with a particular final state. The pQCD models use the same framework: a virtual photon dissociating into a $q\bar{q}$ dipole, exchanging two gluons or a higher order state, then dissociating into a vector meson (Fig. 8).

4.1. *Energy Behaviour of Vector Meson Production*

The cross section is expected to be proportional to $|g(x)|^2$, similar to inclusive DDIS. The energy dependence of the cross section is expected to show a steep rise. Figure 9 shows the cross section for J/ψ in photoproduction at HERA[11] in comparison to pQCD. The different models provide a good description of the rise in W. The power-law parameterisation of the W dependence gives $\sigma \propto W^{0.7-0.8}$, which is close to the value expected from the gluon density at $\mu^2 \simeq 10\,\text{GeV}^2$ given by $M^2_{J/\psi}$. The figure

also shows the effect of using different parton densities, indicating that the cross section is sensitive to choice of the densities, in particular the gluon density.

Similar behaviour was also found in the W dependence of deeply-virtual Compton scattering (DVCS)[14,15] where the vector meson is actually a real massless photon. The hard scale of this process is given by Q^2. The measured cross sections at $\langle Q^2 \rangle \simeq 8\,\text{GeV}^2$ are well described by a $W^{0.8}$ dependence; similar to that observed in J/ψ photopro-

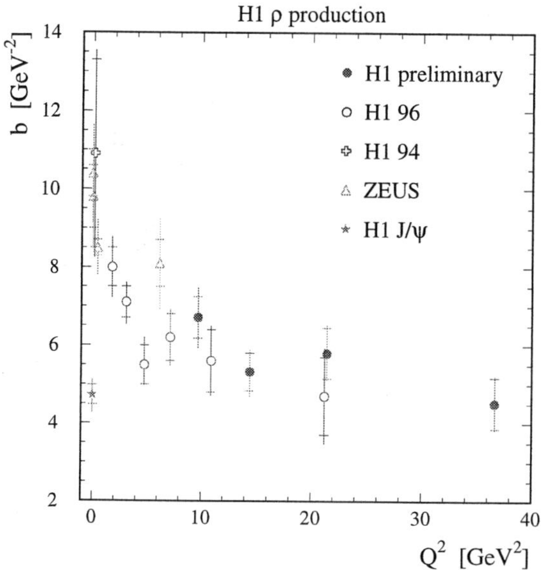

Figure 10. The B-parameter of the momentum distribution of the scattered proton, parameterised as $d\sigma/dt \propto e^{Bt}$, for ρ production at HERA.[16] The values are shown as a function of Q^2. The value from J/ψ photoproduction is also shown.

duction. These two measurements show that vector meson production with a sufficiently large hard scale can basically be understood by the exchange of two perturbative gluons. This is in contrast to inclusive DDIS where the energy dependence indicates the presence of soft physics.

4.2. *Forward Slope in the Vector Meson Production*

The observations described in the previous section imply that the interaction is point-like since it can be described by pQCD. In diffractive scattering, the size of the interaction can be determined from the scattering angle of the outgoing nucleon, measured using either the forward proton spectrometer, or from the final-state particles in case of an exclusive process like vector meson production. Typically the diffractive process shows an exponential fall-off with the transverse momentum of the scattered nucleon like e^{Bt} ($t \simeq -p_T^2$ for $x_{\mathbb{P}} \ll 1$). The value of B indicates the size of the interaction.

The slope was measured for ρ production at HERA[16] as a function of Q^2, as shown in Fig. 10. The slope B is found to be about 10 GeV^{-2} for photoproduction ($Q^2 \approx 0$), where the hard scale is given

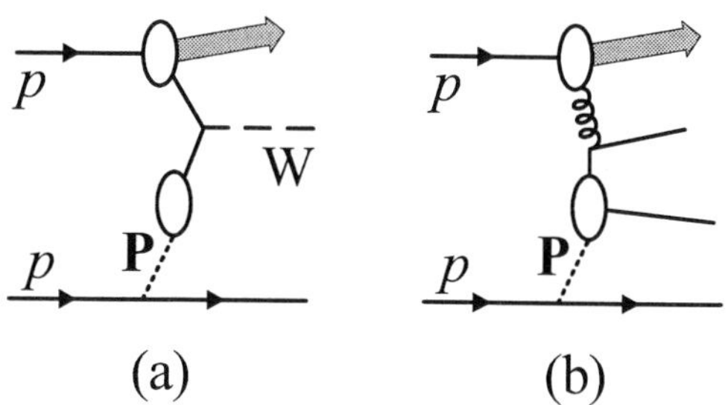

Figure 11. Example of the diagrams of single diffraction at the Tevatron: (a) W production (b) the jet and HQ production. **P** indicates the diffractive exchange.

by neither Q^2 nor the mass of the vector meson. The slope decreases towards higher Q^2 and asymptotically approaches a value of about 4.5 GeV^{-2}. This value of B is similar to the proton radius, meaning that the size of the dipole from the virtual photon is very small. In the case of photoproduction, the size of the dipole is large enough to establish an interaction even if the distance between the photon and the proton is more than twice the size of the proton radius. The B value for J/ψ photoproduction is also similar to the asymptotic value observed in ρ electroproduction. These observations indicate that the interaction is point-like provided a hard scale is present, supporting further the validity of the pQCD description of the data.

5. Hard Diffraction at the Tevatron

At the Tevatron, the diffractive process occurs in various combinations of the LRG configuration since the both incoming protons can emit the diffractive exchange. The partonic content of the exchange is mainly studied using single-diffractive processes, where the colour-octet parton emitted from one proton undergoes a hard scatter with the diffractive exchange emitted from the other proton; the partons in the exchange are probed by the colour-octet parton (see Fig. 11).

Various hard processes have been studied in Tevatron Run I data, such as W production (Fig. 11(a)),[17] which is sensitive to the quark content of the exchange, and dijet[18,19] and b-quark production[20] (Fig. 11 (b)), which are more sensitive to the gluonic content.

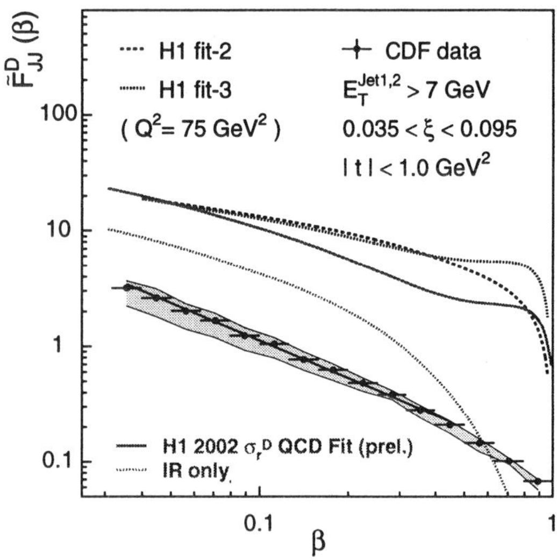

Figure 12. The measured dijet cross section in single-diffractive processes,[18] in comparison with the LO prediction of the cross section using the diffractive PDF's obtained with HERA F_2^D data.[4]

5.1. *Hard Diffraction at the Tevatron from Run I Data and the Factorisation Breaking*

One of the most striking features of the hard diffractive process measured at the Tevatron is a large suppression of the cross section with respect to the prediction based on the diffractive parton densities obtained from the HERA F_2^D data. Figure 12 shows the comparison of the dijet cross section in the single-diffractive process at the Tevatron[18] to the prediction using the parton density parametrisation discussed in Sec. 3.1. Although the prediction reproduces the shape of the data in the low-β region, the magnitude of the cross section is smaller by about a factor 5 to 10. A similar degree of suppression was observed in the other hard diffractive processes.[17−19] This indicates a strong factorisation breaking between HERA and the Tevatron: the diffractive parton densities are not universal between these two environments.

The reason for the breaking is not yet clearly known. It is currently attributed to re-scattering between the two beam remnants where one or more colour-octet partons are exchanged. This destroys the already formed colour-singlet state, as discussed in Sec. 2. The probability to preserve the colour-singlet state is often denoted as the "gap survival probability" S^2, which decreases as the re-scattering cross section increases.

There has been rapid progress in understanding the nature of the re-scattering on the theoretical side. Various attempts have been made in predicting the survival probability S^2: the renormalised Pomeron model,[21] where the Pomeron flux above 1 was renormalised to unity; the estimation of the re-scattering probability from the soft hadron-hadron cross sections;[22] estimating the re-scattering probability from the multi-parton interaction model tuned to Tevatron data;[23] and the soft-colour interaction model,[9,24] where the soft colour exchange between the remnants gives the probabilities of both re-scattering and the formulation of the colour-singlet state. Many of these models attempt to establish a unified understanding of the multi-parton phenomena in both diffractive and multi-parton scattering.

All these models are equally valid since they give a good description of the existing Tevatron data. The model should have, however, an ability to predict S^2 for higher energy reactions like those at the LHC. The predictive power of the models can be tested by comparing with the measurements of various processes that have not been explored so far. Providing such measurements is an important task for the experimentalists. Two examples of such measurements are discussed in the following sections.

5.2. *Factorisation Test with Hard Photoproduced Diffraction at HERA*

The first example is testing of QCD factorisation by looking at the diffractive dijet photoproduction at HERA, where the hard scale is provided by the E_T of the jets. Factorisation is expected not to hold in photoproduction events, where the resolved process has a photon remnant, allowing re-scattering. On the other hand, the direct process does not have a beam remnant and the suppression of diffractive events is expected to be much smaller than in the resolved process.

Figure 13 shows the dijet cross section in diffractive photoproduction as a function of x_γ^{jets}, the longitudinal momentum fraction of the parton that participated in the hard scatter, reconstructed from the dijet momenta. Resolved events dominate in the low-x_γ^{jets} region while the direct process is concentrated at x_γ^{jets} close to one. The cross sections were com-

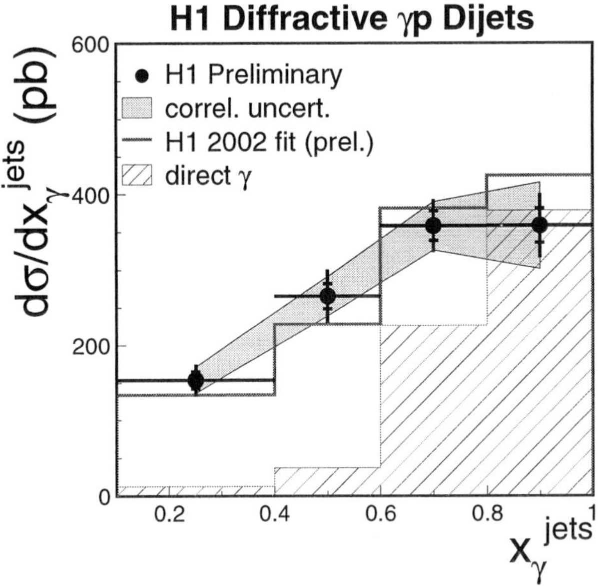

Figure 13. Dijet cross section in photoproduction at HERA[25] as a function of x_γ^{jets}. The solid line shows the prediction from the RAPGAP model.

pared to the leading-order RAPGAP Monte Carlo. No significant deviation was found from the model in the x_γ^{jets} dependences, implying that the resolved process is not suppressed. Also the magnitude of the cross section is reproduced by RAPGAP when using the diffractive parton densities obtained from the HERA DDIS data. This means that the overall cross sections are not suppressed with respect to DIS, implying that factorisation is not significantly broken between DDIS and photoproduction at HERA. This is in strong contrast to the large suppression observed in the Tevatron data. This measurement should constrain the models of the gap survival probability.

5.3. *Exclusive Dijet Production*

Another example of a newly explored process is the exclusive dijet production through the so-called double Pomeron exchange (DPE) at the Tevatron. This process produces the diffractive dissociative system X through a fusion of two diffractive exchanges (Fig. 14(a)). The system X can then decay into two partons resulting in exclusive dijet production.

This process is of particular interest since the same final state is expected for diffractive Standard Model Higgs production. The system X may produce a light scalar Higgs through a top quark loop,

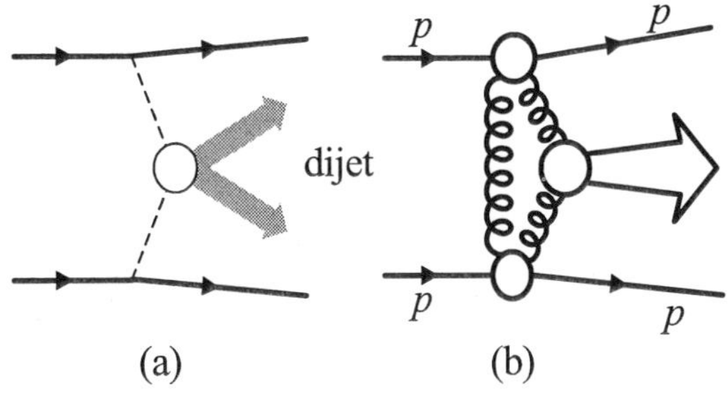

Figure 14. (a) Diagram of the double Pomeron exchange with a dijet in the final state. (b) An example of the pQCD diagram of the double Pomeron exchange with exclusive production.

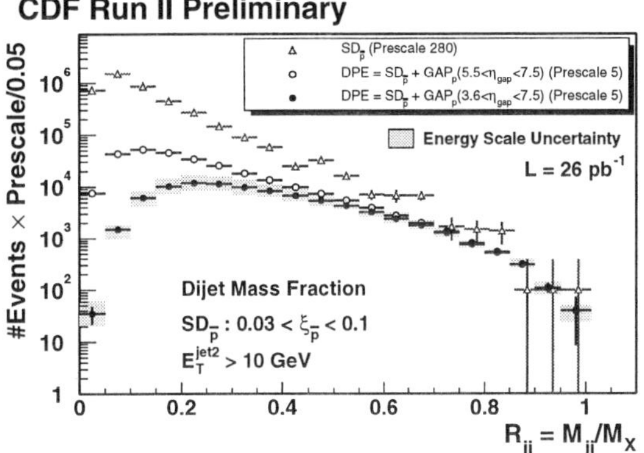

Figure 15. The dijet cross section with two rapidity gaps as a function of $R_{jj} = M_{jj}/M_X$, where M_{jj} is the dijet mass and M_X is the mass of the system X, produced by the fusion of two diffractive exchanges.[26]

which would then decay mainly into $b\bar{b}$. The cross section is estimated in QCD assuming that the dominant diagram is that shown in Fig. 14(b). The estimation of the cross section of such processes, however, is difficult for the following reasons: (1) the diagram in Fig. 14(b) has a gluon loop, which means that the diagram is not factorisable and hence the diffractive parton densities cannot be used; (2) the cross section may be suppressed by the survival probability S^2. The value of S^2 depends on the details of the process; and (3) initial- and final-state radiation suppress exclusive dijet production. The measurements should help to constrain the model of this process; more specifically, it gives a stringent test of the theoretical framework for estimating diffractive dijet production through the diagram shown in Fig. 14(b).

The exclusive dijet cross section has been measured by the CDF collaboration using the Tevatron Run II data. Prior to the start of Run II, the CDF collaboration has improved many of its forward detectors, which are required for measuring the rapidity gap near the outgoing protons. The cross sections were measured as a function of $R_{jj} = M_{jj}/M_X$, where M_{jj} is the dijet mass and M_X is the mass of the system X produced by the fusion of the two diffractive exchanges. The presence of rapidity gaps are requested at both ends of the system X.

Exclusive dijet production should be concentrated near $R_{jj} = 1$. There is, however, no clear peak found in the R_{jj} spectrum near unity. The region corresponding to the exclusive region was defined as $R_{jj} > 0.8$. The dijet cross section for $R_{jj} > 0.8$ and $E_T^{jet} > 25\,\text{GeV}$ is $34 \pm 5(\text{stat.}) \pm 10(\text{syst.})$ pb. The value gives an upper limit of exclusive dijet production. The smooth transition of the R_{jj} spectrum from the inclusive to the exclusive region suggests that the exclusive process is part of the cross section where the effect of initial- and final-state QCD radiation is small. This data should also play an important role in understanding the cross section for exclusive dijets, which is a background for diffractive Higgs production at the LHC.

6. Data Expected from the Near Future

In addition to the already mentioned detector upgrade in CDF, both the D0 and H1 collaborations are commissioning upgraded detectors for diffractive physics. D0 has installed a forward proton spectrometer system consisting of nine Roman pots. The benefits of the new detector are a very high acceptance also at low $x_{\mathbb{P}}$ (below 0.02) and towards high-t. The initial alignment of the spectrometer has been completed.

H1 has also installed a new system of Roman pots in the proton beamline downstream of the main detector at about 200 m. The pot locations have been optimised so that the acceptance is high for the diffractively produced protons. The spectrometer system indeed has a very high acceptance, typically above 80%, at low-t in the $x_{\mathbb{P}}$ range of the diffractive regime. The pots are installed and ready for commissioning after HERA resumes running after the shutdown, which is expected to start in September 2003.

7. Summary

In this contribution some important aspects of the diffractive process for interpreting diffraction in the framework of pQCD have been reviewed, namely the factorisation properties and the energy dependence of the cross sections. The experimental results can be summarised as follows.

- The resolved Pomeron model is mildly broken, implying that diffractive exchange is not a particle. A framework not based on the Regge theory is clearly necessary.

- The factorisation theorem holds for diffractive DIS. The extracted diffractive parton densities are universal within DIS.

- The energy dependence of vector meson production is as expected from two-gluon exchange models taking the gluon density from the F_2 data.

- The energy dependence of the cross section for inclusive diffractive DIS, however, shows a flat ratio to the total $\gamma^* p$ cross section. This means that the rise of the inclusive diffractive DIS cross section is slower than the expectation of the two-gluon model.

- A large suppression of the hard diffractive cross sections at the Tevatron was observed with respect to the predictions based on the diffractive parton densities obtained at HERA. This implies a strong breaking of factorisation between HERA and the Tevatron. It is attributed to the re-scattering of partons between the two beam remnants.

While the first three items show the success of the pQCD approach, the latter two suggest that calculations based on non-perturbative QCD are necessary for explaining the features observed in the data. Rapid developments have been made on the theoretical side addressing these questions, one of which is the survival probability of the rapidity gap which was reviewed in this article. In order to examine these new ideas, more precise data and new measurements should be provided by the experiments. The high-statistics run of Tevatron Run II and HERA II as well as the improved forward detectors should provide these much needed measurements.

8. Acknowledgments

The author would like to thank K. Goulianos, K. Terashi, C. Royon, P. Newman, M. Arneodo, R. Yoshida, A. Bruni, J. Cole, K. Borras and other members of the four collaborations, CDF, D0, H1 and ZEUS, for valuable discussions and suggestions. The author is also thankful to G. Ingelman, who gave important insight into the soft-gluon exchange models through discussion.

References

1. ZEUS collaboration, S. Chekanov *et al.*, *Phys. Lett.* B **549**, 32 (2002).
2. ZEUS collaboration, S. Chekanov *et al.*, *Eur. Phys. J.* C **21**, 443 (2001).
3. J.C. Collins, *Phys. Rev.* D **57**, 3051 (1998).
4. H1 collaboration, "Measurement and NLO DGLAP QCD Interpretation of Diffractive Deep-Inelastic Scattering at HERA", Contributed paper ♮172 to this Symposium.
5. H1 collaboration, "Measurement of the Inclusive Cross Section for Diffractive Deep-Inelastic Scattering at High Q^2", Contributed paper ♮173 to this Symposium.
6. ZEUS collaboration, S. Chekanov *et al.*, DESY-03-094.
7. ZEUS collaboration, "Deep Inelastic Diffractive Scattering with the ZEUS Forward Plug Calorimeter using 1998-1999 Data", Contributed paper ♮107 to this Symposium.
8. W. Buchmüller, M. McDermott and A. Hebecker, *Phys. Lett.* B **410**, 304 (1997).
9. A. Edin, G. Ingelman and J. Rathsman, *Phys. Lett.* B **366**, 371 (1996); J. Rathsman, *Phys. Lett.* B **452**, 364 (1999).
10. K. Golec-Biernat, M. Wüsthoff, *Phys. Rev.* D **59**, 014017 (1999); K. Golec-Biernat, M. Wüsthoff, *Phys. Rev.* D **60**, 114023 (1999).
11. H1 collaboration, "Elastic Photoproduction of Jpsi Mesons", Contributed paper ♮188 to this Symposium.
12. L. Frankfurt, W. Koepf and M. Strikman, *Phys. Rev.* D **54**, 3194 (1996).
13. A. Martin, M. Ryskin and T. Teubner, *Phys. Rev.* D **55**, 4329 (1997).
14. ZEUS collaboration, DESY-03-059.
15. H1 collaboration, "Deeply Virtual Compton Scattering at HERA", Contributed paper ♮193 to this Symposium.
16. H1 collaboration, "Elastic electroproduction of rho mesons at high Q^2", Contributed paper ♮175 to this Symposium.
17. CDF Collaboration, F. Abe *et al.*, *Phys. Rev. Lett.* **78**, 2698 (1997).
18. CDF Collaboration, T. Affolder *et al.*, *Phys. Rev. Lett.* **84**, 5043 (2000).
19. D0 collaboration, B. Abbott *et al.*, *Phys. Lett.* B **531**, 52 (2002).
20. CDF Collaboration, T. Affolder *et al.*, *Phys. Rev. Lett.* **84**, 232 (2000).
21. K. Goulianos, *Phys. Lett.* B **358**, 379 (1995); K. Goulianos and J. Montanha, *Phys. Rev.* D **59**, 114017 (1995).
22. A.B. Kaidalov *et al.*, *Eur. Phys. J.* C **21**, 521 (2001); E. Gotzman, E. Levin and U. Maor, *Phys. Rev.* D **60**, 094011 (1999); A. Bialas and R. Peshanski, hep-ph/0306133.
23. B. Cox, J. Forshaw and L. Lönnblad, hep-ph/0012310; B. Cox, J. Forshaw and B. Heinemann, *Phys. Lett.* B **540**, 263 (2002).
24. R. Enberg, G. Ingelman and N. Tîmneanu, *Phys. Rev.* D **64**, 114015 (2001).
25. H1 collaboration, "Dijets in Diffractive Photoproduction at HERA", Contributed paper ♮170 to this Symposium.
26. CDF collaboration, available on http://www-cdf.fnal.gov/ physics/new/qcd/dpe_run2/dpe_ex_run2.html.

DISCUSSION

Silvia Tentindo-Repond (Florida State University): In Run I of the Tevatron we had diffractive results from CDF and D0 concerning W and Z gauge boson production. What is the status of Run II?

Yuji Yamazaki: Of course there are already data. However, the cross sections are not yet available to the public. W and Z production turned out to be processes where the cross section measurement is very difficult, therefore I imagine it will take time to prepare new results. For example, the uncorrected number of events agrees between the two Tevatron experiments, but they differ at the cross section level. They are working on the cross section result and I believe we will have them soon.

High energy collisions of heavy nuclei at the Relativistic Heavy-Ion Collider permit the study of nuclear matter at extreme densities and temperatures. Selected experimental highlights from the early RHIC program are presented. Measurements of the total multiplicity in heavy-ion collisions show a surprising similarity to measurements in e^+e^- collisions after nuclear geometry is taken into account. RHIC has sufficient center-of-mass energy to use large transverse momentum particles and jets as a probe of the nuclear medium. Signatures of "jet quenching" due to radiative gluon energy loss of fast partons in a dense medium are observed for the first time at RHIC. In order to account for this energy loss, initial energy densities of 30-100 times normal nuclear matter density are required.

1. Introduction

Quantum Chromodynamics, the theory of the strong interaction, is predicted to have a rich phase diagram. At normal nuclear densities and at low temperature, quarks are confined in color-neutral mesons and baryons. At very high baryon densities and low temperatures, i.e. conditions that may be found in the interior of neutron stars, exotic states such as strange quark matter or color superconductors might be found. These high density, low temperature states are of astrophysical importance but are inaccessible in the laboratory.

At low baryon densities and high temperatures, QCD may have a phase transition to a quasi-deconfined state of quarks and gluons. This state is known as the Quark-Gluon plasma. Lattice QCD calculations predict that this phase transition occurs when the system reaches 150-170 MeV.[1]

RHIC was built to collide nuclei at ultra-relativistic energies. Such collisions may create an extended system with sufficient energy density and temperature for Quark-Gluon plasma formation. Nuclear collisions at lower energies (BNL/AGS and CERN/SPS) have shown novel collective phenomena associated with highly excited nuclear matter. Experimental results from Pb+Pb collisions at the SPS have shown that quark and gluon degrees of freedom are important during the early stages of the collision.[2] At SPS energies, however, the system probably spends insufficient time in the "Quark-Gluon Matter" phase to equilibrate. Hence these measurements do not allow for the investigation of the bulk properties of hot and dense QCD.

At RHIC, the system may live in a pre-hadronic phase for several fm/c, thereby raising the possibility that it can equilibrate in the Quark-Gluon plasma phase before hadronizing and disintegrating. I will review recent experimental results from RHIC, focusing on what they tell us about the existence of the Quark-Gluon plasma and the properties of such a state.

2. The Accelerator and the Experiments

The Relativistic Heavy-Ion Collider began taking data in the summer of 2000. The first collisions were of Gold nuclei at $\sqrt{s} = 56$ GeV, followed by the first scientific run with Au nuclei at $\sqrt{s} = 130$ GeV. The year 2001 saw full energy Au+Au collisions at $\sqrt{s} = 200$ GeV and the first proton-proton collisions at the same energy. The proton-proton program at RHIC is unique in that it is the first collider where protons can be longitudinally and transversely polarized in order to study the spin structure of the proton. The 2002-2003 RHIC run was devoted to understanding cold nuclear matter effects on particle production using d+Au collisions.

In a central (small impact parameter) Au+Au collision at RHIC, over 4000 charged particles are created.[3] No single experiment is able to fully measure and identify each of these particles on a event-by-event basis. Instead each of the four experimental collaborations focuses on a different subset of the possible measurements.

PHOBOS [4] is a table-top sized collection of silicon multiplicity detectors that specializes in counting the total multiplicity of particles produced over 4π on an event-by-event basis. A mid-rapidity spectrometer gives momentum information in a limited region of phase-space.

BRACHMS [5] uses two magnetic spectrometers to measure identified particle inclusive distributions over the largest possible rapidity range. In particular, the forward spectrometer of BRAHMS can be used to understand baryon stopping in these collisions.

STAR [6] is a large acceptance magnetic spectrometer that uses a large cylindrical Time-Projection Chamber to track particles over full azimuth for the near mid-rapidity region ($|\eta| < 1.5$). Particles are identified using topological decay or via smaller acceptance particle identification detectors (time-of-flight and Ring-Imaging Čerenkov).

PHENIX [7] focuses on rarer leptonic obervables as well as hadronic measurements over a moderately large kinematical range. Leptonic observables are identified at mid-rapidity using Ring-Imaging Čerenkovs and Electromagnetic Calorimeters, and muons are identified at forward rapidity using muon spectrometers.

The diversity of experimental approaches shows that the understanding of hot and dense nuclear matter requires a variety of different measurements, with no single measurement able to capture the complete picture.

3. Total Particle Multiplicities

In an ideal Quark-Gluon plasma, the effective degrees of freedom are quarks and gluons. Assuming a massless Stefan-Boltzmann ideal gas, the energy density ϵ of a system is given by,

$$\epsilon_{SB} = \frac{\pi^2}{30}(N_{bosons} + \frac{7}{8}N_{fermions})T^4. \qquad (1)$$

In an idealized Quark-Gluon plasma with massless quarks, there are 3 (flavor) $\times$ 2 (spin) $\times$ 3 (color) $\times$ 2 (quark-antiquark) = 36 fermionic quark states and 8 (color) $\times$ 2 (spin) = 16 bosonic gluon states. In a pion gas there are only 3 bosonic degrees-of-freedom (a resonance gas will have more). Hence, an early prediction is that the Quark-Gluon plasma should have a much higher energy (entropy) density than a corresponding hadronic system at the same temperature. If the system evolves adiabatically, the initial partonic entropy is transfered to pionic entropy and Quark-Gluon plasma formation was predicted to enhance total particle production relative a non-QGP reference (i.e. proton-proton collisions).

The PHOBOS collaboration has measured the total multiplicity produced in Au+Au collisions at RHIC and compared these data to e^+e^- and $\bar{p} + p$ reference data. When making such a comparison, nuclear geometry must be taken into account. There are two natural scaling expectations for particle production in nucleus-nucleus collisions:

N_{part} total number of participating projectile and target nucleons; and

N_{binary} total number of inelastic nucleon-nucleon collisions.

The quantities are estimated by the RHIC collaborations using measured data and Glauber calculations.[8]

Figure 1 shows the charged multiplicity as a function of pseudo-rapidity in Au+Au, e^+e^- and $\bar{p} + p$ collision at similar $\sqrt{s}$.[9] In e^+e^- the rapidity density is measured along the thrust axis. In Au+Au collisions, the rapidity density is divided by the number of participating nucleon-nucleon pairs. A remarkable similarity is observed between e^+e^- and Au+Au (except perhaps at large pseudo-rapidity where the participating baryons may have some effect). The pseudo-rapidity density is lower for $\bar{p} + p$ compared to the other systems at the same center-of-mass energy.

Integrating these distributions, one obtains the total multiplicity. This is shown for the three systems as a function of $\sqrt{s}$ in Fig. 2. Above $\sqrt{s} \approx 20$ GeV, e^+e^- is very similar to Au+Au, while the total multiplicity produced in $\bar{p} + p$ is consistently lower. The discrepancy between e^+e^- and $\bar{p}+p$ has already been understood as a consequence of the "leading particle effect".[10] The total multiplicity is proportional to the energy available for particle production, so the energy carried by the spectator fragments must be subtracted to calculate $\sqrt{s}_{\text{eff}}$ on and event-by-event basis. After accounting for the leading particle effect, there is a universal behavior in the the total multiplicity versus $\sqrt{s}_{\text{eff}}$ in Au+Au, e^+e^- and $\bar{p} + p$ collisions.

If the total multiplicity is indeed proportional to the entropy, these results would indicate that in central Au+Au collisions there is no excess entropy pro-

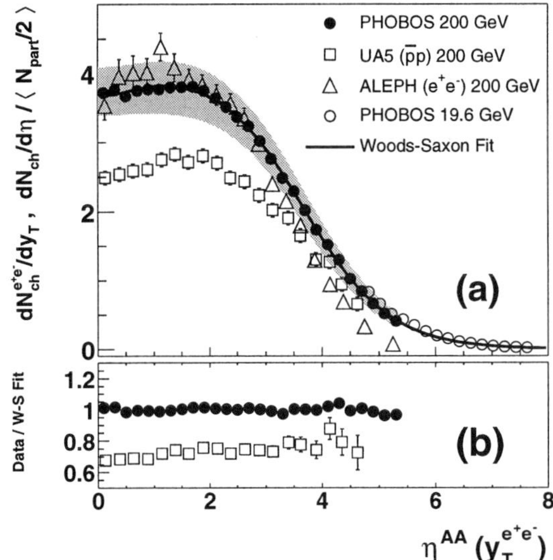

Figure 1. The charged particle pseudo-rapidity density as a function of η in central Au+Au, hadronic e^+e^-, and $\bar{p}+p$ collisions.[9] For the Au+Au collisions, the pseudo-rapidity density is divided by the number of participating nucleon pairs. In e^+e^- the rapidity density along the thrust axis is plotted. The bottom panel shows the ratio of the data to the Woods-Saxon fit.

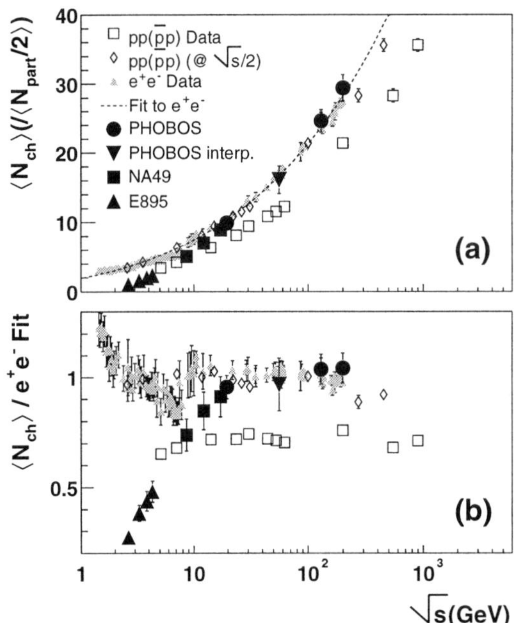

Figure 2. Top: Total charged multiplicity per participant pair in central Au+Au, e^+e^- and p+p($\bar{p}$) collisions as a function of the center-of-mass energy.[9] The p+p($\bar{p}$) is shown at both the beam center-of-mass energy (open squares) and at $\sqrt{s}/2 \approx \sqrt{s}_{\rm eff}$ (open diamonds). Bottom: Ratio of data to the e^+e^- fit.

duction per participating nucleon-nucleon pair compared to simpler systems. This places severe constraints on the dynamical evolution of the system, in particular on whether the system is able to reach an ideal Quark-Gluon plasma at some point during its evolution. Other observables from RHIC point to a large amount of particle interactions and collective behavior that develops after the initial nucleus-nucleus collision.[11] Apparently, the total multiplicity produced in these collisions is unaffected by the complex dynamical evolution of the system.

4. "High" p_T Physics and "Jet Quenching"

RHIC is the first heavy-ion research facility that has sufficient center-of-mass energy to produce high E_T jets and other hard probes in numbers sufficient for detection. Jets, in particular, are thought to be a sensitive probe of the color-charge density of the medium through which the fast parton travels. A colored object passing through a colored medium may interact coherently with the colored fields.[12] This is the QCD analogue of the Landau-Pomeranchuk-Migdal effect observed in QED. This coherent interaction leads to much greater energy losses than would be expected from stochastic Bethe-Bloch type processes. As a result of the coherence there may be a non-linear dependence of the total energy loss on the path length of nuclear matter traversed. For the color-charge densities predicted in central RHIC collisions, effective energy losses of order 10 GeV/fm are possible.

Quantitative measurements of parton energy loss would allow extraction of the color-charge density of the system at the earliest, hot and dense phase. These densities can then be compared to theoretical estimates of the densities expected in the case of Quark-Gluon plasma formation. The observation of parton energy does not indicate either deconfinement or thermalization and is thus not directly related to Quark-Gluon plasma formation. Instead, parton energy loss is a powerful probe for ensuring that the necessary gluon densities for Quark-Gluon plasma formation are achieved in these collisions, and may be combined with other measurements to map out the time evolution of the system and phase diagram of nuclear matter.

Several experimental observables have been used

at RHIC to study the propagation of fast partons in dense nuclear matter and look for signatures of parton energy loss. These observables include inclusive particle production at high p_T, two-particle correlations at high p_T, and the correlations between high p_T particle production and the impact parameter direction. Because of the large multiplicities in a central Au+Au collision at RHIC, full jet reconstruction is difficult if not impossible. Even if full jet reconstruction was possible, parton energy loss would redistribute the observed jet energy into lower momentum gluons and would thus not affect the total rate of jet production (i.e. energy loss is mostly an in-medium modification of the fragmentation function).

Measurement of inclusive particle production at moderate to high p_T (2-15 GeV/c) have been made by all four RHIC experiments. Inclusive high p_T particle production is sensitive to the leading particles produced in jet fragmentation. A softening of the in-medium fragmentation function (i.e. parton energy loss) will tend to reduce the number of produced high p_T particles. As mentioned in the previous section, particle production in nuclear collisions has two scaling expectations, N_{part} and N_{binary}. Since hard processes have very short associated time scales, each nucleon-nucleon collision occurs incoherently and the production of hard partons should scale with N_{binary}. Thus to measure the nuclear matter effects on high p_T particle production the nuclear modification factor R_{AA} is constructed,

$$ R_{AA} = \frac{d^2 N^{AA}/dp_T d\eta}{T_{AA} d^2 \sigma^{NN}/dp_T d\eta} \, , \qquad (2) $$

where $T_{AA}=\langle N_{binary}\rangle/\sigma_{inel}^{NN}$ from a Glauber calculation accounts for the nuclear collision geometry. In the absence of nuclear matter effects, R_{AA} should approach unity at moderate p_T (2–3 GeV/c).

Moderate to high p_T particle production in proton-nucleus collisions was studied extensively using fixed target experiments at Fermilab. It was found that there is an "anomalous nuclear enhancement" above p_T=2–3 GeV/c (also called the "Cronin effect").[13] This corresponds to $R_{pA} > 1$. This behavior has been understood qualitatively in terms of multiple soft parton interactions in addition to the hard parton scattering that leads to the production of high p_T particles. The recent deuteron-gold run at RHIC investigated whether the expected

Cronin enhancement is still observed at the much larger RHIC center-of-mass energy, or whether other effects such as saturation of gluon parton distribution functions at low Bjorken-x would tend to suppress particle production at moderate to high p_T.

Figure 3 shows PHENIX results on the nuclear modification factor plotted as function of p_T for d+Au and Au+Au collisions.[14] The top panel shows R_{dA} (R_{AA}) for charged hadrons, and the bottom panel shows a comparison of R_{dA} for charged hadrons and π^0. For both charged hadrons and neutral pions, there is an enhancement above unity at high p_T in d+Au collisions. This is consistent with earlier Fermilab observations. Comparing the charged hadrons to the neutral pions, the nuclear enhancement is larger for charged hadrons. A similar effect was observed at Fermilab, with heavier particles (kaons and protons) showing larger high p_T enhancements in proton-nucleus collisions. This flavor dependence of the Cronin effect is not understood theoretically.

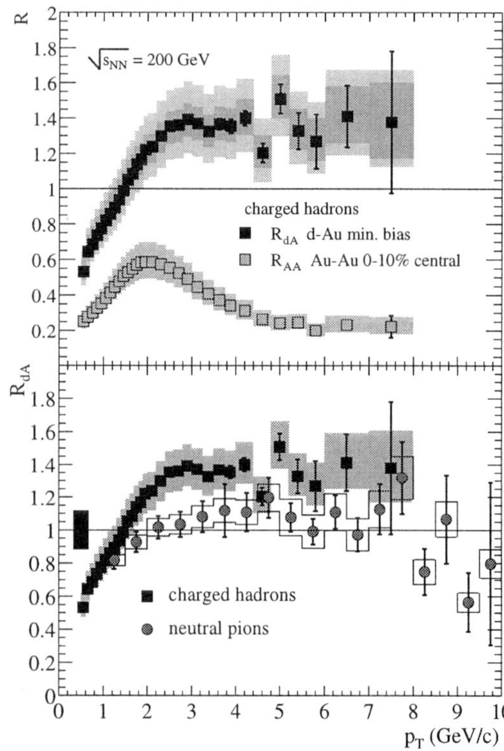

Figure 3. Nuclear modification factor as a function of p_T for charged hadrons and neutral pions as measured by the PHENIX collaboration.[14] The top panel compares d+Au and central Au+Au charged hadron measurements, while the bottom panel compares charged hadrons and neutral pions in d+Au collision.

The top panel compares the nuclear modification factor for d+Au and central Au+Au collisions. While the d+Au collisions show an enhancement, the Au+Au collisions show a dramatic deficit at high p_T compared to the binary collision scaled proton-proton reference. This suppression of high transverse momentum particle yields in central collisions of heavy nuclei was observed for the first time at RHIC.[15] Since this suppression is observed in central Au+Au collisions but not in d+Au collisions, it must be due to the interaction of partons or their fragmentation products with the matter produced in the collisions.

Both STAR and PHENIX have investigated the flavor dependence of the high p_T particle suppression.[16,17] Figure 4 shows the nuclear modification factor for Λ, K_s^0, charged kaons, and charged hadrons. In this case, the nuclear modification factor is the ratio of central (small impact parameter) and peripheral (large impact parameter) Au+Au collisions, with the number of binary collisions for each impact parameter calculated from the Glauber model. Here the baryons (PHENIX observes a similar effect for protons) show a rather different behavior compared to the lighter mesons. The Λ nuclear modification factor approaches unity at moderate p_T (2-4 GeV/c) but seems to have similar suppression as the kaons and charged hadrons at larger p_T.

Several scenarios have been proposed to describe this rather different behavior for baryons and mesons. In a quark coalescence picture, the moderate p_T baryons have a substantial contribution from processes where three quarks recombine to form a baryon. At larger p_T both the baryons and mesons are products of parton fragmentation. Another possibility is that the strong radial pressure gradients produced in a central RHIC collision accelerate soft baryons to moderate p_T ("radial flow"). This process is well established at lower transverse momentum, but the quantitative estimates of the contribution from this process vary. Finally, it has been noted that the anomalous nuclear enhancement observed in proton-nucleus collisions is larger for baryons compared to mesons. The same effect is being seen in nucleus-nucleus collisions and therefore may have the same hitherto unexplained origin. To disentangle the various scenarios for moderate p_T particle production in nucleus-nucleus collisions at RHIC, more experimental measurements are needed. These include

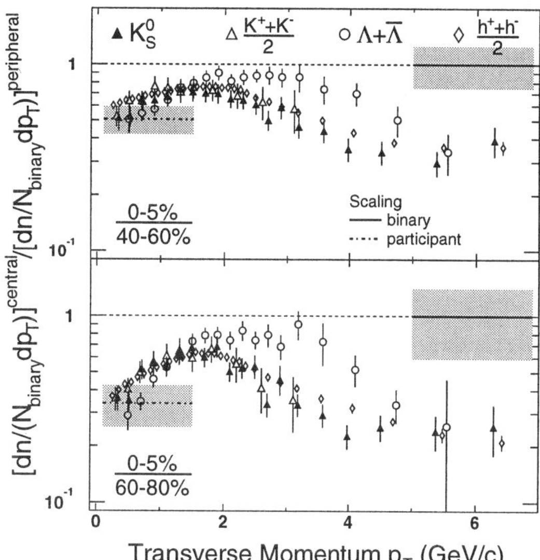

Figure 4. Nuclear modification factor for charged K, neutral K, and Λ as measured by the STAR collaboration.[17] The nuclear modification factor here is the ratio of central (0-5%) and peripheral (40-60% [top] 60-80 % [bottom]) Au+Au collisions.

a comparison of R_{AA} for heavy mesons and baryons to see if the lack of baryon suppression at moderate p_T is due to particle mass or quark content. Measurements of baryon and meson spectra in the d+Au collisions are also needed as the Fermilab measurements cannot be extrapolated to RHIC energies without a a theoretical model. The first such measurements show that the baryon excess at moderate p_T in central Au+Au collisions is not fully realized in d+Au collisions.[18]

The observed suppression of high p_T particle production is consistent with the predictions of jet quenching, but without direct evidence for jet production in nucleus-nucleus collisions it is difficult to draw any firm conclusions. Jets have been observed in e^+e^-, $p+p$, and proton-nucleus collisions, but had not been observed in previous experiments with nucleus-nucleus collisions due to insufficient center-of-mass energy. While full jet reconstruction is impossible in a central nucleus-nucleus collisions at RHIC energies, both STAR and PHENIX have presented evidence for jet production using two-particle correlations.[19,20]

Jets are characterized by a collimated spray of particles. In a typical jet finding algorithm, particle clusters are identified and the energy is summed to obtain the total jet (parton) energy. In rare cases, a parton fragments such that a single particle carries a large fraction of the energy. These are atypical jets, although they contribute a majority of the high p_T particles observed in a proton-proton collision. In other words, triggering on a leading particle selects a highly biased sample of jets. It is this highly biased jet sample that the RHIC experiments are able to study.

Figure 5 shows the relative azimuthal separation between moderate ($p_T > 2$ GeV/c) and high ($4 < p_T < 6$ GeV/c) transverse momentum charge hadrons in proton-proton, d+Au, and central Au+Au collisions.[21] The $p+p$ measurements show the azimuthal correlations expected from jet production, i.e. a narrow correlation at small relative azimuth and a wider correlation at $\Delta\phi = \pi$ due to back-to-back dijet events. The relative azimuthal distribution looks quite similar to measurements made by UA1.[22] In d+Au collisions, a similar correlation structure is observed indicating the presence of jets and dijets.

The small relative azimuth correlations in central Au+Au collisions are nearly identical to those observed in $p+p$ and Au+Au collisions. This is the first direct evidence of jet production in nucleus-nucleus collisions. Further tests, including the observation of charge-ordering between the trigger and next-to-leading charged hadron,[20] further support the hypothesis of jet production in nucleus-nucleus collisions.

The back-to-back correlations due to dijets, however, are absent in the most central nucleus-nucleus collisions. The back-to-back hadron pairs, however, are observed in peripheral Au+Au collisions.[20] Apparently, when a dijet pair is formed in a central Au+Au collision at RHIC it is highly unlikely that the fragmentation products of both jets escape. This is important additional evidence for the observation of jet quenching at RHIC.

Theoretical attempts to describe both the suppression of high p_T particle production and the absence of correlated back-to-back dihadron pairs require the existence of highly dense medium.[23,24] Early in the evolution of the system, the density must reach 30-100 times normal nuclear matter density ($\approx$ 5-15 GeV/fm^3). It is difficult to conceive of such a system in terms of hadrons, as the spacing between individual hadrons would be much smaller than the size of the hadrons. The existence of such

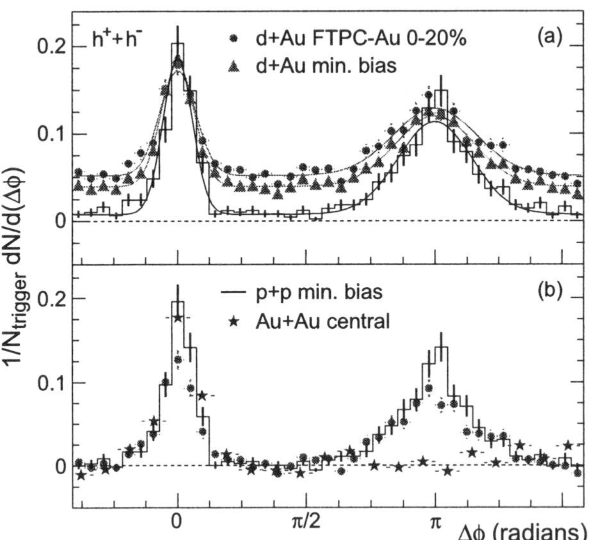

Figure 5. Per trigger particle relative azimuthal distributions in $p+p$, d+Au and Au+Au collisions measured by the STAR collaboration.[21] The top panel compares $p+p$ and d+Au collisions, while the bottom panel compares $p+p$, d+Au and central Au+Au collisions with appropriate background subtraction applied to each distribution. The azimuthal distributions for both minimum bias and small impact parameter (FTPC-Au 0-20%) d+Au collisions are shown. The trigger hadron has $4 < p_T < 6$ GeV/c, and the associated hadrons have $p_T > 2$ GeV/c. The correlations from within the same jet are at $\Delta\phi \approx 0$, while the dijet-like correlated pairs have $\Delta\phi \approx \pi$.

a dense system in itself, however, is not proof of a Quark-Gluon plasma. In order to properly be called a Quark-Gluon plasma, the system must reach thermal equilibrium in a high density phase, and show thermodynamic parameters consistent with a collection of deconfined, weakly-interacting quarks and gluons.

5. Conclusions and Outlook

This talk focused on only two of the many interesting results to come from RHIC so far. Studies of particle production in Au+Au collisions at RHIC show a surprising universality in the total multiplicity compared to simpler systems (e^+e^- and $p+p$) once nuclear geometry is taken into account. This similarity suggests that the number of particles produced in any high energy collision is governed only by the total energy available for particle production. In the case of heavy-ion collisions, particles (or partons) interact many times leading to strong dynamical correlations and collective motion. These interactions, however, do not seem to affect the total particle production.

The other novel observation from RHIC is the first observation of a strong nuclear suppression of high p_T particle production in central Au+Au collisions. The data show a factor of ≈ 5 suppression of single inclusive high p_T yields in central Au+Au collisions compared to the binary collision scaling expectation. In addition, there is an absence of back-to-back high p_T hadron pairs. To explain these observations, very high initial energy densities are required.

Whether or not these energy densities lead to the production of a thermalized Quark-Gluon plasma is an open question requiring other experimental signatures and theoretical investigations. The early experimental results have established that the system produced in Au+Au collisions at RHIC spends some time at densities above the predicted energy density for the phase transition and that there are early interactions among the particles. Whether these interactions can lead to an equilibrated system in the Quark-Gluon plasma phase is uncertain. Hydrodynamical models that assume early equilibration and a QGP equation of state have been quite successful in describing much of the RHIC data,[25] but the success of these models does not constitute proof of equilibration or allow unambiguous determination of the equation of state.

In addition, the current data are unable to constrain the temperature of the system during its early hottest phase, or to even answer when, in the evolution of the system, temperature becomes a well defined quantity. In order to constrain the temperature, leptonic obervables may be used. The dissociation of J/Ψ is thought to be highly temperature dependent and future high luminosity RHIC runs will measure J/Ψ yields with sufficient precision to constrain the temperature. There is also hope that photon or virtual photon (dilepton) radiation directly from the early hottest phase can be measured. By correlating measurements sensitive to the temperature, energy density, and pressure, it may be possible to map out the equation of state of hot nuclear matter and make quantitative measurements of the bulk thermodynamic properties of QCD. Future experimental programs at RHIC the LHC will accomplish this goal.

Acknowledgments

I would like to thank Peter Steinberg and Peter Jacobs for help preparing the talk and manuscript. This work was supported by the Office of Science, Nuclear Physics, U.S. Department of Energy under Contract No. DE-AC03-76SF00098.

References

1. For a recent review see F. Karsch and E. Laermann, hep-lat/0305025.
2. U. Heinz and M. Jacob, nucl-th/0002042.
3. I.G. Bearden *et al.*, *Phys. Rev. Lett.* **88**, 202301 (2002).
4. B.B. Back *et al.*, *Nucl. Instrum. Meth.* A **499**, 603 (2003).
5. M. Adamczyk *et al.*, *Nucl. Instrum. Meth.* A **499**, 437 (2003).
6. K.H. Ackermann *et al.*, *Nucl. Instrum. Meth.* A **499**, 624 (2003).
7. K. Adcox *et al.*, *Nucl. Instrum. Meth.* A **499**, 469 (2003).
8. B.B. Back *et al.*, *Phys. Rev.* C **65**, 031901(R) (2002); K. Adcox *et al.*, *Phys. Rev. Lett.* **86**, 3500 (2001); I.G. Bearden *et al.*, *Phys. Lett.* B **523**, 227 (2001); C. Adler *et al.*, *Phys. Rev. Lett.* **89**, 202301 (2002).
9. B.B. Back *et al.*, nucl-ex/0301017.
10. M. Basile *et al.*, *Phys. Lett.* B **92**, 367 (1980); M. Basile *et al.*, *Phys. Lett.* B **95**, 311 (1980).
11. C. Adler *et al.*, *Phys. Rev. Lett.* **87**, 182301 (2001); K. Adcox *et al.*, nucl-ex/0307010.
12. R. Baier, D. Schiff, and B.G. Zakharov, *Annu. Rev. Nucl. Part. Sci.* **50**, 37 (2000).
13. D. Antreasyan *et al.*, *Phys. Rev.* D **19**, 764 (1979).
14. S.S. Adler *et al.*, *Phys. Rev. Lett.* **91**, 072303 (2003).
15. K. Adcox *et al.*, *Phys. Rev. Lett.* **88**, 022301 (2002); C. Adler *et al.*, *Phys. Rev. Lett.* **89**, 202301 (2002).
16. S.S. Adler *et al.*, *Phys. Rev. Lett.* **91**, 172301 (2003).
17. J. Adams *et al.*, nucl-ex/0306007.
18. J. Adams *et al.*, nucl-ex/0309012.
19. M. Chiu *et al.*, *Nucl. Phys.* A **715**, 761c (2003).
20. C. Adler *et al.*, *Phys. Rev. Lett.* **90**, 082302 (2003).
21. J. Adams *et al.*, *Phys. Rev. Lett.* **91**, 072304 (2003).
22. G. Arnison *et al.*, *Phys. Lett.* B **118**, 173 (1982).
23. T. Hirano and Y. Nara, *Phys. Rev. Lett.* **91**, 082301 (2003).
24. X.N. Wang, nucl-th/0305010.
25. For a review see nucl-th/0305084.

DISCUSSION

Enrico Predazzi (University of Torino): Coming to the question you raised about the analogies between multiplicities in e^+e^-, in $p+p$ and in nucleus+nucleus collisions. As you mentioned, this is related to the leading particle effect, and we have proven long ago that this effect is a consequence of unitarity, or if you prefer of diffraction. So there is no surprise that the two results are similar, even so they come from so widely different initial states. The dynamics of Au+Au is washed away, and whatever remains is just basically diffraction. But we can talk later if you prefer.

David Hardtke: You know there is a great industry of applying saturation models to RHIC multiplicities, and saturation models are one way of enforcing unitarity. So this might not be a surprise, but based on initial predictions of massive entropy production it is a bit of a surprise to me. We need some theorists to sit down and figure out why there is this universality in particle production.

David Christian (Fermilab): One of the early indications from AGS and SPS was real enhanced strangeness production. Can you comment on measurements of strangeness production in the various experiments?

David Hardtke: The current understanding of strangeness production is the following. The idea was that if you have a Quark-Gluon plasma you lower the energy threshold for strange quark production to the same as that for light quarks and you produce much strangeness. At the end we came to the conclusion that in fact strangeness is not enhanced in heavy-ion collisions. It is actually suppressed in $p+p$ collisions, because there are a lot of statistical models of particle production. So people applied canonical and grand canonical formulations to parti-cle production, and what they found is that the $p+p$ collisions are described well by a canonical ensemble, whereas A+A collisions are described well by a grand canonical ensemble. So when you go from a canonical ensemble to a grand canonical ensemble you tend to increase the production of rare stuff like strangeness. So basically what you see in $p+p$ collisions is a canonical suppression of strangeness, and we think that in A+A collisions you just go into the true vacuum value of strangeness production. So it's no longer such a good direct signature of Quark-Gluon plasma, based on recent theoretical work.

Tord Ekdoef (Uppsala University): It appears to me as if, unlike a few years ago, you don't have a predicted evidence or signal of Quark-Gluon plasma. Is this correct?

David Hardtke: There are many predicted signals, but whenever we measure one some theorist comes a long and says: "Aha, but it is not a predicted signal!" The problem is that it is very difficult to distinguish between very high density hadronic matter and very high density quark and gluon matter. From the theoretical standpoint they basically look the same. So signatures such as J/ψ production, jet quenching etc. are all just sensitive to the very high density of quarks and gluons, but you don't know if they are combined in hadrons or free quarks and gluons. So there is no specific signature that can tell us which one we have, but the hope is that by correlating different signatures that are sensitive to energy, temperature, etc.. you can put enough constraints to finally eliminate the purely hadronic model. Most people believe that we have a Quark-Gluon plasma, but there is not a single definitive experiment that can prove it.

COSMOLOGY AND ASTROPARTICLE PHYSICS

COSMOLOGICAL IMPLICATIONS OF THE FIRST YEAR WILKINSON MICROWAVE ANISOTROPY PROBE RESULTS

L. VERDE

Dept. of Physics and Astronomy, University of Pennsylvania, Philadelphia, PA, USA
E-mail: lverde@physics.upenn.edu

The Wilkinson Microwave Anisotropy Probe (WMAP) team has recently analyzed and released the first-year data. We will review the implications for cosmology of these results. The highlight is that cosmology now has a standard cosmological model. With only 6 parameters the model fits not only WMAP data remarkably well, but also a host of other astronomical observations. We also present the results on neutrino mass limits and on dark energy properties from a joint likelihood analysis of WMAP data with small-scale CMB experiments and large-scale structure surveys. The data and supplementary information are publicly available and can be found on the experiment web site at http://lambda.gsfc.nasa.gov.

1. Introduction

The first year results from the Wilkinson Microwave Anisotropy probe (WMAP) were announced on February 11 2003. On the same day the telescope was renamed in honor of Prof. David Wilkinson, member of the science team and pioneer in the study of cosmic microwave background (CMB) radiation. WMAP was launched from Cape Canaveral on June 30 2001. It arrived at the L2 Lagrangian point in October 2001 (in practice the satellite was stable enough for CMB data-taking to commence at the beginning of August). The primary goal of the WMAP mission is to produce a high-fidelity all-sky polarization-sensitive map of the CMB radiation to determine the cosmology of our Universe. · After a year of observation it has produced a full sky map of the microwave sky in 5 frequencies, with a resolution a factor 30 higher than the previous full sky map as produced by the COBE satellite in 1992 (see Fig. 1). This is the cleanest picture of the early Universe; the structures on the CMB –the pattern of hot and cold spots– carry information about the composition, geometry, age, etc. of our Universe. The WMAP data release was accompanied by 13 papers[1−13] here we summarize the main results.

2. What does WMAP Measure that contains Cosmological Information?

WMAP data have yielded new results relative to several different epochs in the history of the Universe. With unprecedented precision we now know that CMB light comes to us from 380,000 years after the big bang, the first stars appeared 200 mil-

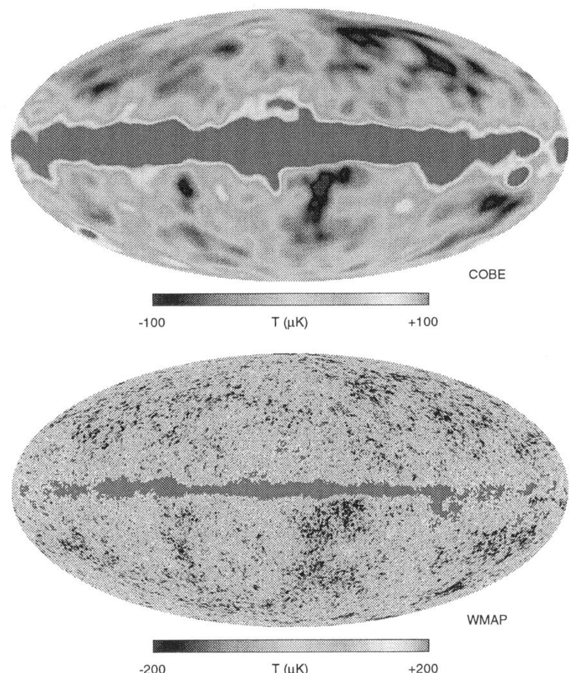

Figure 1. A comparison of the COBE 90 GHz map (upper panel[14]) and the W-band WMAP map (lower panel[1]).

lion years after the big bang, and the the Universe is 13.7 Gyr old. We have also learnt about the Universe composition and the properties of the seeds of cosmological structure formation.[1] What is WMAP measuring that contains all this cosmological information?

By looking at the CMB we see the leftover heat from the big bang; the hot and cold spots in a CMB temperature map correspond to fluctuations in the underlying density field.

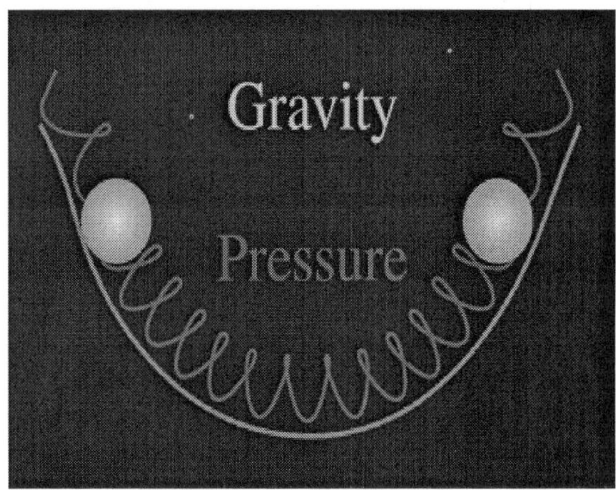

Figure 2. From http://background.uchicago.edu/~whu/ intermediate/intermediate.html: analogy for the acoustic oscillations in the baryon-photon fluid.

The radiation that we see from the CMB comes from the last scattering surface: when the Universe recombines, the CMB photons can travel to us virtually unperturbed thus giving us a snapshot of the photon-baryon fluid in the early Universe. On large scales, where no physical processes could have taken place, the anisotropies that we see correspond to "primordial ripples", the seeds from which the action of gravity grew cosmological structures such as galaxies and clusters of galaxies that we observe today.

On smaller scales two competing processes are at play in the photon-baryon plasma: photons exert radiation pressure which contrasts the compression from gravity resulting in acoustic oscillations. This can be visualized[15] imagining two masses connected by springs in a potential well (Fig. 2). The potential well represents an overdensity region, the depth of the potential well depend on the total mass (actually energy-density); the spring represents the radiation pressure while the masses represent the baryons. Of course, underdense regions can be thought of as potential "hills" in the same analogy.

The sound waves in the photon-baryon plasma stop oscillating at recombination, when support from radiation pressure stops. It is a snapshot of these oscillations that we see in the CMB. In other words by looking at the CMB we are "seeing sound".[16]

We can push the analogy further: as a violin string has a fundamental mode and overtones which depend on the length of the string, so the horizon size at the last scattering surface defines a fundamental mode (and its overtones) for the sound waves in the photon-baryon plasma. Thus the angular power spectrum[a] of the temperature anisotropies in the CMB carry the imprint of the fundamental mode and its overtones. The modes that the snapshot catch at extrema of the oscillations correspond to enhanced temperature fluctuations on scales given by the mode's wavelength. While the fundamental mode corresponds to a compression, the first overtone corresponds to a rarefaction and so on (this is easy to visualize by going back to the analogy of the springs and masses in a potential well).

These ideas are not new: they were worked out independently and almost simultaneously, on the two sides of the iron curtain in 1970.[17,18] However it was not until more than 20 years later that it was clear that one could learn about cosmology by looking at these acoustic fluctuations in details.[19−21]

How would one extract cosmological information from a CMB map in practice? First, we want to compress the CMB maps, for WMAP for example these are Mega-pixel maps, to study cosmology. The details of our method can be found in Hinshaw et $al.$ (2003).[3] We can express the temperature fluctuations in the CMB sky ($\delta T(\theta, \phi)$, where (θ, ϕ) denotes the position angle) by expanding it in spherical harmonics:

$$\delta T(\theta, \phi) = \sum_{\ell,m} a_{\ell m} Y_{\ell m}(\theta, \phi) \ . \tag{1}$$

If the anisotropies form a gaussian random field[b], that is if the real and imaginary parts of each $a_{\ell m}$ are independent normal deviates, all the statistical information is contained in the *angular power spectrum*:

$$C_\ell = \frac{1}{2\ell + 1} \sum_m |a_{\ell m}|^2 \ . \tag{2}$$

Figure 3 shows the CMB temperature anisotropy angular power spectrum before and after WMAP. In the left panel of Fig. 3 the points with error-bars correspond to different CMB experiments, in the right panel, WMAP band-powers data are shown with the error bar due to instrumental noise. In both panels the solid line is the best fit model to WMAP

---

[a] Harmonic transform of the two point correlation function.
[b] We find no evidence for deviation from gaussianity.[6]

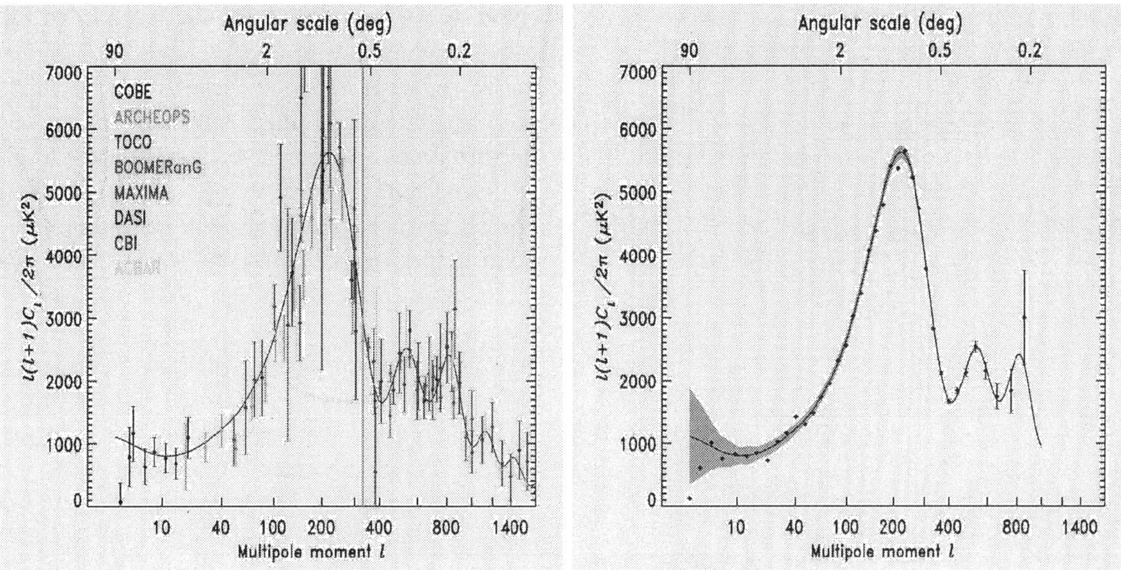

Figure 3. Left:[3] compilation of the CMB angular power spectrum before WMAP, the solid line is the best fit model for WMAP data. Right:[3] WMAP angular power spectrum: the gray band is the cosmic variance error-bar while the error-bars on the data points are the noise-error bars. Notice the first peak (corresponding to a compression) and the second peak (corresponding to a rarefaction).

data. The gray area in right panel of Fig. 3 shows the cosmic variance error. Cosmic variance error is due to the fact that we can observe only one Universe. Given a cosmological model with some specified cosmological parameters and thus a power spectrum, different realizations of the CMB sky drawn from that model will have slightly different power spectra. Since the CMB we can measure is just one realization of the true (unknown) underlying model, there is some cosmic variance error associated with the best fit C_ℓ.

Some features are evident from the right panel of Fig. 3: on large angular scales (small multipole ℓ), to good approximation we see the primordial ripples. On smaller scales (larger ℓ) we see a series of "acoustic peaks" corresponding to the acoustic oscillations in the photon-baryon fluid. The peak on scales of about a degree ($\ell \sim 200$) corresponds to the fundamental mode: these scales are so large that at recombination the mode had time to go through only one compression, the second peak corresponds to the first overtone and on these scales the mode went through a compression and a rarefaction and was "frozen" then; the third peak corresponds to a compression and so on. Since the horizon size at the last scattering surface is a well defined quantity, a

"standard rod", the angle it subtends must be related to the geometry of the space between the observer (WMAP satellite at L2, today) and the last scattering surface (at $z \sim 1100$). If a standard rod subtends an angle, say, α in a flat, euclidean space (you can visualize this in 2 dimensions as a flat sheet of paper), it will subtend an angle **larger** than α in a positively curved, closed space (which you can visualize in 2 dimensions as the surface of a sphere) and an angle **smaller** than α in a negatively curved, hyperbolic space. Since, in Einstein's general relativity theory, it is the matter-energy content that shapes the geometry of space-time, the position of the first acoustic peak tells us about the total matter-energy density content of the Universe. Cosmologists use the parameter Ω_{tot}: the ratio of the matter-energy density to the critical one necessary to make the geometry of space flat.

The height of the first peak must also be related to the depth of the potential wells as deeper wells will create stronger compression. Since the depth of the potential wells depend on the amount of mass, the height of the first peak tells us Ω_{matter}: the ratio of the matter density to the critical matter-energy density. Notice here that Ω_{tot} need not coincide with Ω_{matter}, as vacuum energy (or Einstein's

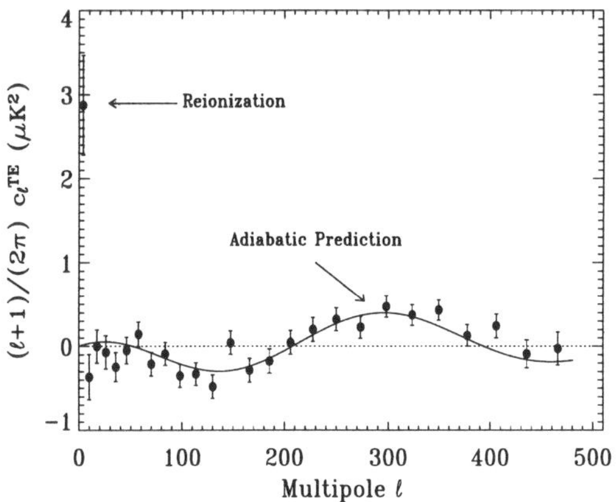

Figure 4. The temperature-polarization cross correlation power spectrum.[12]. The solid line is the prediction from the temperature data for adiabatic initial conditions. The excess power at $\ell < 10$ is due to early star formation.

biggest blunder, a cosmological constant) would contribute to Ω_{tot} by shaping space-time, but not to Ω_{matter}.

The analogy with the oscillations of the spring with a mass attached will help us understand that the relative height of the first and second peak tell us about the baryon content of the Universe. A larger baryon content correspond to a larger mass attached to the spring (while it is the total mass content, baryonic and non-baryonic that determines the depth of the potential wells): the larger the baryon content the bigger would be the ratio between compression and rarefaction, thus larger relative height of the first to second peak.[c]

2.1. *Information in the Polarization*

Cosmological information is not just enclosed in the temperature anisotropy but also in the polarization of the CMB radiation. The CMB polarization is produced by Thompson scattering of a quadrupolar radiation pattern. At decoupling the quadrupole is produced by velocity gradients. Since both the velocity field and the temperature anisotropies are created by density fluctuations, a component of the polarization should be correlated with the temperature anisotropy. In particular on scales ~ 1 degree, primordial adiabatic initial conditions, such as

---

[c]For more details see Page *et al.* (2003).[9]

those set up by inflation, predict an anti-correlation between temperature and polarization signal. This anti-correlation is not expected if initial conditions are of a different nature; for example isocurvature perturbations or perturbations originating from a casual seed model (such as e.g. topological defects). For primordial adiabatic initial conditions the temperature power spectrum gives precise predictions for the temperature-polarization cross correlation power spectrum. This prediction is exactly what we have seen in the data[12] (see Fig. 4): this is a triumph for the standard cosmological model.

I should note here that the WMAP team reported the temperature-polarization cross correlation. Last year, the first detection of the CMB polarization signal was announced;[22] now the measurement is so precise that the temperature-polarization cross correlation data can be used to learn about cosmology.

There is a second effect that can be seen in the polarization (TE) data: an excess signal on the largest scales, which is not predicted by the temperature data alone. If the first stars form at high redshift ($z \sim 20$), their light reionizes the Universe; free electrons scatter CMB photons introducing some optical depth (denoted by τ) and uniformly suppressing the power spectrum of the temperature fluctuations by $\sim 30\%$. Free electrons see the local $z \sim 20$ CMB quadrupole and polarize the CMB at large scales where no other mechanism of polarization operates. The excess polarization signal is a signature of the formation of the first stars in the Universe, such a large signal implies that stars started forming much earlier than most people previously thought.[12]

3. Interpretation of WMAP Data

Without going into technical details of the data analysis,[8] let us briefly outline the method of the analysis before going into the results. The analysis path follows these steps:

a) select a set of cosmological parameters;

b) compute for these parameter the "model" power spectrum C_ℓ^{theory}: for this step we use the publicly available software CMBFAST[23] but other packages are also avalilable;

c) compare to the measured power spectrum and compute the likelihood; and

Table 1. LCDM best fit parameters to WMAP data;[5] left: format for physicists, right: format for astrophysicists. All errors are 1σ.

Dark matter	$(2.25 \pm 0.38) \times 10^{-27} \text{kg/m}^3$	Ω_c	0.26 ± 0.07
Atomic density	$(2.7 \pm 0.1) \times 10^{-7}\ \text{cm}^{-3}$	$\Omega_b h^2$	0.024 ± 0.001
Age	13.4 ± 0.3 Gyr	h	0.72 ± 0.05
σ_8	0.9 ± 0.1	σ_8	0.9 ± 0.1
n_s	0.99 ± 0.04	n_s	0.99 ± 0.04
z_{reion}	17	τ	0.17 ± 0.04

d) repeat to find confidence regions.

With this procedure WMAP data can be analyzed alone and/or in combination with other, complementary, data sets.

The simplest (and most popular) cosmological model (the so called LCDM model) has 6 parameters, and corresponds to a flat, low density Universe composed of baryons, dark matter and dark energy. The 6 parameters are: the dark matter density (parameterized by Ω_c), the physical baryon density ($\Omega_b h^2$), the Hubble parameter ($H_0 = 100$ Km/s/Mpc$h$), the spectral slope of the primordial power spectrum (n_s), the amplitude of fluctuations (parameterized by the present-day r.m.s. fluctuations smoothed on 8 Mpc/h spheres σ_8), and the optical depth to the last scattering surface τ). For physicists these 6 parameters are equivalent to a different set of more familiar parameters (see Table 1). Following Occam's razor, we start analyzing this simple model and then add complications (i.e. extra parameters) one at the time. For this model WMAP data (temperature and temperature-polarization spectra) tell us that:

a) there is no evidence for deviations from gaussianity of the CMB maps:[6] the CMB looks gaussian (thus supporting our initial assumption that all statistical information about CMB anisotropies of the mega-pixel maps can be "compressed" into a power spectrum);

b) 15% of the CMB light was rescattered since the Universe reionized early: the estimated reionization redshift is $z \sim 20$ or 200 million years after the Big Bang;[12] and finally, probably the most important result

c) the simple, flat, LCDM model fits remarkably well this new set of observations of unprecedented precision: only 6 parameters fit 1346

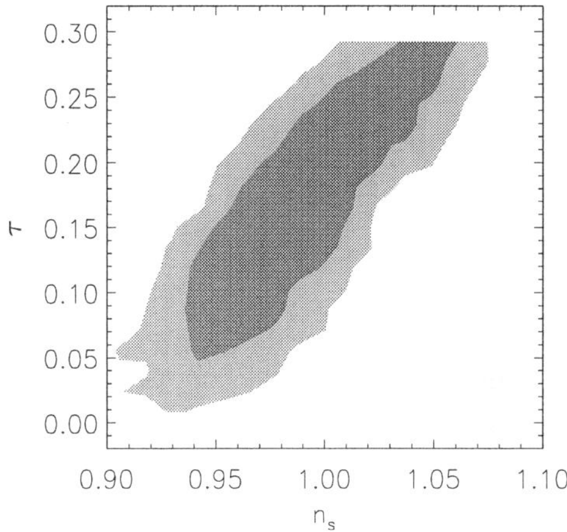

Figure 5. Main degeneracy among cosmological parameters for WMAP data.[5] Degeneracies increase the error-bars on marginalized quantities, but this will quickly improve with more years of operation.

data points (or, to be even more ambitious, one could say that 6 parameters fit the million-pixel maps).[5]

The best fit values for the cosmological parameters are reported in Table 1.

Another (derived) parameter that might be of interest is the baryon-photon ratio:[5] $\eta = 6.5^{+0.4}_{-0.3} \times 10^{-10}$.

Remarkably, these parameters fit not only WMAP data and other CMB experiments, but also a host of other cosmological observations. For example the Hubble parameter constraint (a measurement of the Universe's expansion rate), is in agreement with the Hubble space telescope measurement obtained from observations of stars of known intrinsic luminosity in nearby galaxies; and the age determination is in good agreement with other determinations based on the ages of the oldest stars.

When using WMAP data alone some degeneracies among cosmological parameters remain, the main degeneracy being between the primordial power spectrum spectral slope and the optical depth to the last scattering surface (see Fig. 5). Degeneracies among cosmological parameters increase the error-bars on marginalized quantities, but this will quickly improve with more years of operation.

4. Combining WMAP Data with other Data Sets

WMAP data can be combined with external, complementary data sets. In doing so we can achieve two goals. First, and more important, we can test the consistency of the cosmological model; second, if all data sets are consistent with the model, we can lift degeneracies among cosmological parameters.

We combine WMAP data with small-scale CMB experiments (CBI[25,26] and ACBAR[27]), with the power spectrum of large-scale structure data probed by the Anglo-Australian Two degree field galaxy redshift survey (2dFGRS[28,29]) and by the Lyman-alpha forest,[30,31] although the results presented here have been obtained without the Lyman-alpha data. The data set compilation is shown in Fig. 6. While the CMB probes the Universe on large scales and at redshift $z \sim 1100$, the 2dFGRS galaxy survey provides a three-dimensional map of the galaxy distribution in the local Universe, it probes smaller scales and $z \sim 0$. The Lyman-alpha forest is a series of absorption features in the spectra of distant quasars, caused by the cosmological structures intervening along the line of sight: from the correlation properties for the absorption lines it is possible to reconstruct the three-dimensional power spectrum of these intervening structures at $z \sim 3$ and probe scales much smaller than those accessible from CMB or galaxy surveys. As Fig. 6 illustrates, these observables are complementary in scale and in redshift.

5. Beyond the Simple LCDM Model

Armed with the statistical power of these external data sets, we can try to test models beyond the simple six-parameter LCDM model. For example we can drop the assumption that the universe is spatially flat, and introduce an extra parameter Ω_Λ, where

$\Omega_m + \Omega_\Lambda = \Omega_{tot}$, or we can drop the assumption that neutrinos are nearly massless.

5.1. *Flatness*

If we allow the geometry of the universe to deviate from flat but assume that the dark energy component is in the form of a cosmological constant, we obtain the constraints shown in Fig. 7.

As Fig. 7 shows CMB data alone (orange confidence region – from top left corner to bottom right corner – show the joint 2σ confidence level in the Ω_m-Ω_Λ plane) favors flat or nearly flat models. But dropping the flatness assumption creates new degeneracies among cosmological parameters: for example the Ω_m constraint is now weaker. The addition of external data sets lift this degeneracy: the (darker) red confidence contours are for CMB and Hubble space telescope constraint on the Hubble parameter. The green (from botton left corner to upper right corner) confidence contours are the 1, 2 and 3σ confidence levels obtained from supernovae data (see R. Kirshner's presentation). The blue (vertical) contours are the 1, 2 and 3σ constraints obtained from analysis of the 2dFGRS.[32] In total we obtain a constraint on the flatness of the Universe $\Omega_{tot} = 1.02 \pm 0.02$ (68% confidence level). It is quite remarkable that different measurements that rely on different physics, different observables and were carried out independently by different research groups, agree to better than 1σ level! One conclusion that we can draw from this is that we (and all chemistry) are a small minority (4%) of the Universe: dark energy makes up 73% and dark matter 23%.

5.2. *Dark Energy Properties*

Since we find no evidence for deviations from a flat geometry, we can assume $\Omega_{tot} = 1$ and proceed to constrain the properties of dark energy. Dark energy properties can be parameterized by its equation of state $w = -P/\rho$ where P denotes pressure and ρ the density. For a cosmological constant $w \equiv -1$ while for the alternative candidate, "quintessence", a dynamic, time evolving and spatially varying energy component, $w \neq -1$. Although in principle for a quintessence model w need not to be constant, we will assume, as a first approximation, that it is constant in time; the w measurement will thus refer to an "effective" value for w, some sort of weighted

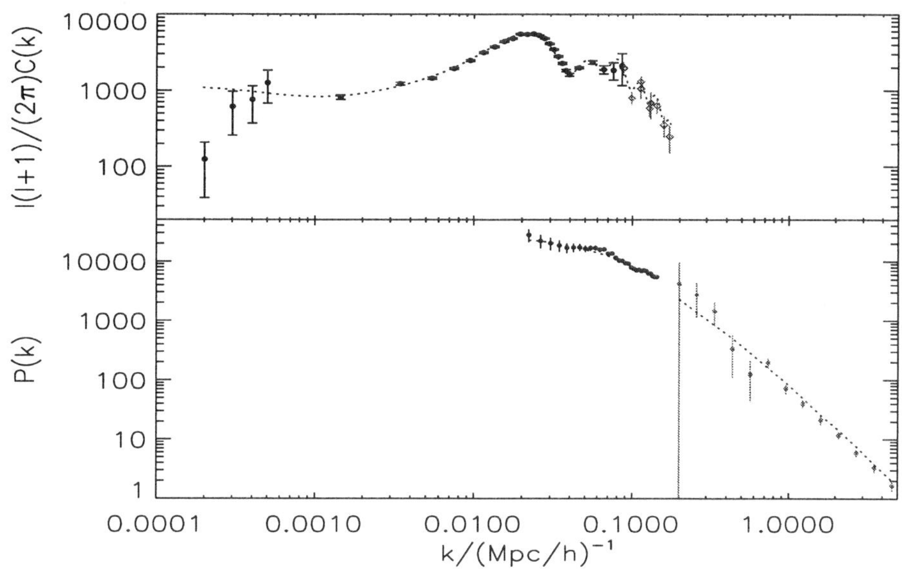

Figure 6. Top panel:[24] WMAP, small-scale CMB experiments (CBI and ACBAR), the multipole ℓ has been converted in wavenumber k. Bottom panel:[24] 2dFGRS data and Lyman-alpha. The data set considered are complementary in scale and in redshift. The dotted line is the best fit model to WMAP data. The extrapolation of this model to low redshift and smaller scales fit the data remarkably well.

average of the w values from $z = 0$ to $z = 1100$. In Fig. 8 we show the joint likelihood contours in the $w - \Omega_m$ plane for the supernovae data (green confidence region; from upper left to lower right), CMB, that is WMAP + CBI + ACBAR (orange; from bottom left to top right) and CMB+Hubble constant determination from the Hubble space telescope (darker region). As Fig. 8 shows, by combining different data sets we find no evidence for the dark energy being something different from a cosmological constant. Our constraint is:[5] $w = -0.98 \pm 0.12$.

5.3. *Neutrino Mass*

Up to now we have assumed that neutrinos are (nearly) massless, but we can try to constrain the neutrino mass in the context of a flat LCDM model. From current observations the CMB alone is not sensitive to the neutrino mass. The left hand side panel of Fig. 9 shows the – virtually indistinguishable – CMB power spectra for 2 models: one where neutrinos are massless the other one where there are three degenerate neutrinos species each of them with a mass of 0.6 eV. Neutrinos stream freely out of the potential wells thus if they have mass they tend to erase fluctuations on small scales, and thus to

suppress the growth of cosmic structures on those scales. The right-hand-side of Fig. 9 shows the matter power spectrum (at $z = 0$) on scales probed by large scale structure surveys, for the same two models; the error bars show the typical error for a data point on those scales. The shape and amplitude of the large-scale structure power spectra are different in the two models and it is clear that the two models can be easily distinguished if the matter power spectrum amplitude is known. All we can measure for example from a galaxy survey, however, is the power spectrum of the galaxy distribution and, in principle, it is not guaranteed that it will coincide with that of the underlying dark matter. Nevetheless the analysis of higher-order correlations[32] of the 2dFGRS shows that it is possible to constrain the relation between clustering of mass and that of galaxies. In other words, we have a way to measure the power spectrum of galaxies, infer that of matter and use this to place constraints on the neutrino mass. In doing so we obtain: $\Omega_\nu h^2 < 0.0067$ at the 95% confidence level when combining the CMB data (WMAP+CBI+ACBAR) with 2dFGRS.[5] For three degenerate neutrino species, this implies $m_\nu < 0.23$ eV. If we add the Lyman-alpha data this constraint remains virtually unchanged.

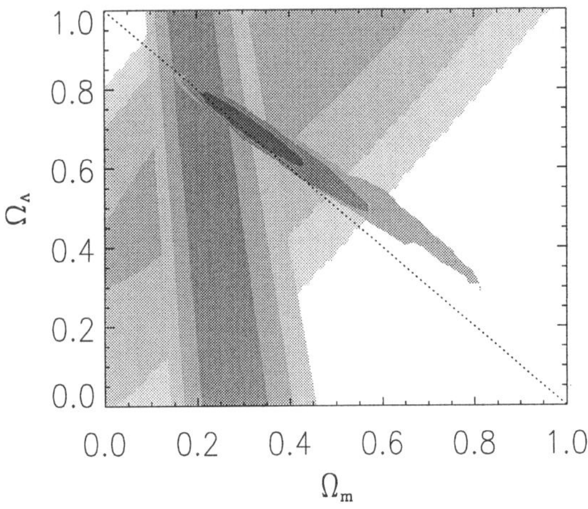

Figure 7. Joint likelihood contours in the $\Omega_m - \Omega_\Lambda$ plane:[24] Green (from botton left corner to upper right corner) Supernovae (see R. Kirshner's presentation); orange (from top left to bottom right) CMB that is WMAP+CBI+ACBAR, only the 2σ level is visible, the 1σ level coincide with the 2σ level (light red) for CMB+Hubble parameter determination from the Hubble space telescope; blue (vertical) 1, 2, and 3σ contours from the 2dFGRS. These different, independent data sets seem to agree to better than the 1σ level.

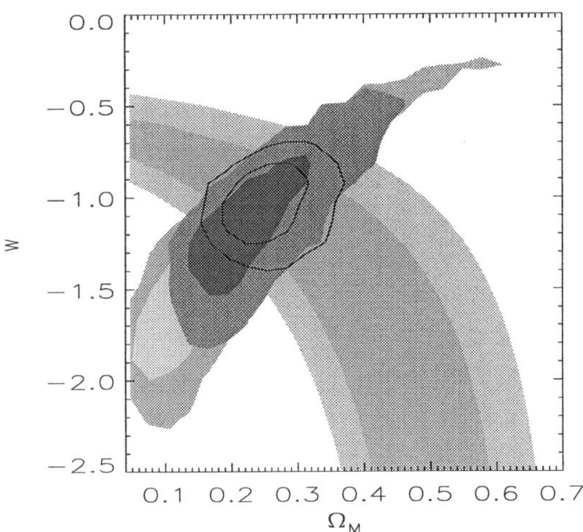

Figure 8. Joint likelihood contours in the $w - \Omega_m$ plane:[5] Green (from upper left to lower right) Supernovae (see R. Kirshner contribution); orange (from bottom left to top right) CMB (that is WMAP+CBI+ACBAR), dark red CMB+Hubble constant determination from the Hubble space telescope. Transparent contours are joint for all data sets: we see no evidence for deviation from $w = -1$.

6. Intriguing Features

We find that there are two intriguing discrepancies. First, given the best fit model, the reduced chi-square for the temperature power spectrum is 1.09: only in 3% of the cases, if the true underlying model was given by our best fit model, a realization of it in the sky would have a reduced chi-square as large or larger. While it is possible to explore the consequences for the inflationary paradigm of this finding,[7] it might be that the excess chi-square could be due to our underestimating the covariance matrix at the percent level. This effect will be accounted for in a forthcoming work, and will allow us to decide whether the "bad" reduced chi-square is due to inadequacy of our modeling or a sign of new physics. The other intriguing discrepancy is more evident from the 2 point correlation function of the temperature data (Fig. 10): there seems to be a lack of correlation at scales larger than about 60 degrees (black/darker line), a feature already evident from the COBE data. This lack of power on large scales manifests itself in the power spectrum as the multipoles at $\ell = 2$ and 3 being much lower than the

theory prediction (see Fig. 3). The statistical significance of the lack of large-scale power is difficult to interpret from Fig. 10 because of correlations between data points; Monte Carlo simulations are needed to assess the statistical significance of any deviation from the LCDM model[5,1] (green/lighter line). We find that in $\sim 0.2\%$ of the cases the LCDM model shows this lack of large-scale power. If the Universe was finite and smaller than the volume within the decoupling surface then one would expect to see a lack of correlation on very large scales, similar to what we see here. However, if the Universe was finite there should be several pairs of matched circles detectable in the sky.[33,34] In two independent searches, none has been found.[35,36]

7. Conclusions

We have presented the highlights of the implications for cosmology of the WMAP first year results. The standard LCDM model works remarkably well: only 6 parameters fit not only WMAP data but also small-scale CMB experiments (CBI and ACBAR) and large-scale structure data (2dFGRS, Lyα forest power spectrum) and derived parameters such as the age of the Universe, the Hubble constant value or

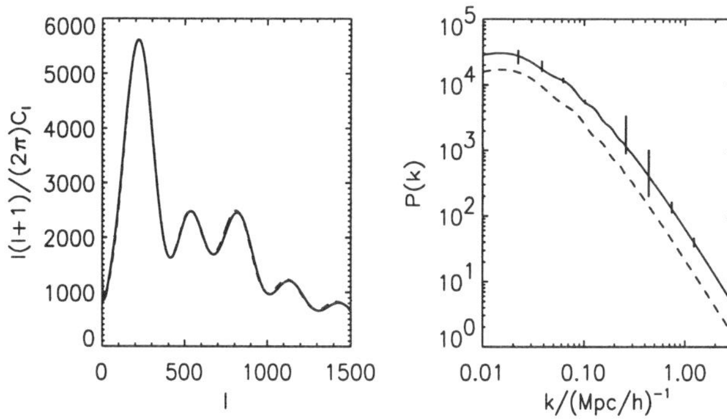

Figure 9. Two models:[8] one with massless neutrinos the other with three degenerate neutrino species each with a mass of 0.6 eV. These two models are virtually indistinguishable from the CMB power spectrum (left) but the matter power spectra are very different on scales probed by large-scale structure surveys (right). It is clear that the two models can be easily distinguished if the matter power spectrum shape and amplitude are known.

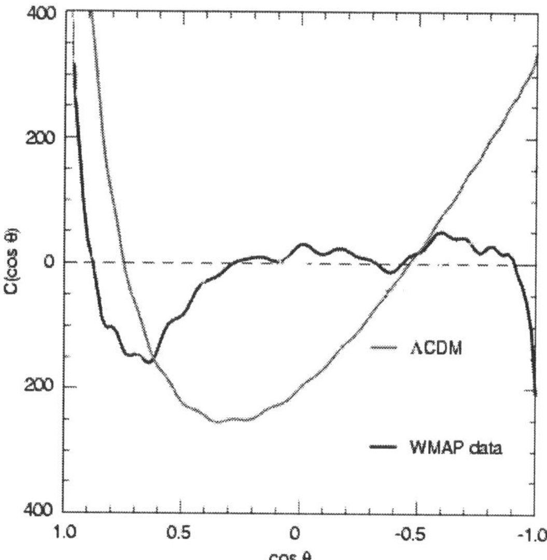

Figure 10. The correlation function for WMAP data (black/darker line), and the best fit LCDM model (green/lighter line).[5] In the WMAP data there seems to be a lack of correlation at scales larger than about 60 degrees compared to the prediction of the LCDM model.

the amplitude of fluctuations are in excellent agreement with other independent astrophysical determinations. We have extrapolated the WMAP observations at $z = 1088$ forwards in time to $z \sim 3$ (Lyα power spectrum) and $z \sim 0$ (e.g. 2dFGRS power spectrum, Supernovae data, HST key project data). This extrapolation describes the observations so well that we feel confident to attempt to extrapolate it backwards, to $z \gg 1088$, and try to shed some light on the dynamics of inflation.[7]

Cosmology is now at a similar stage to particle physics three decades ago, when we converged on the current standard model. The Standard Model of particle physics fits a wide range of data, but does not answer many fundamental questions e.g. "what's the origin of mass?" Cosmology has now a standard model, the LCDM model: a flat universe composed of non-baryonc matter, baryons and vacuum energy. The standard cosmological model has deep open questions e.g. "what's dark energy?", "what's dark matter?" Over the coming years improving CMB, large-scale structure, gravitational lensing and supernovae data etc. will provide ever more rigorous tests of the model.

Acknowledgments

The WMAP mission is made possible by the support of the Office of Space Science at NASA Headquarters.

I would like to thank my WMAP science team collaborators in this work: C. Barnes, C. Bennett (PI), M. Halpern, R. Hill, G. Hinshaw, N. Jarosik, A. Kogut, E. Komatsu, M. Limon, S. Meyer, N. Odegard, L. Page, H. Peiris, D. Spergel, G. Tucker, J. Weiland, E. Wollack, and E. Wright.

The data set, the WMAP temperature and temperature-polarization angular power spectra are available at http://lambda.gsfc.nasa.gov. If using these data set please refer to Hinshaw et al. (2003)[3] for the TT power spectrum and Kogut et al. (2003)[12] for the TE power spectrum. We also make available a subroutine that computes the likelihood for WMAP data given a set of C_ℓ, please refer to Verde et al. (2003)[8] if using this routine.

References

1. C. L. Bennett et al., *ApJS* **148**, 1 (2003).
2. C. L. Bennett et al., *ApJS* **148**, 97 (2003).
3. G. F. Hinshaw, *ApJS* **148**, 135 (2003).
4. G. F. Hinshaw, *ApJS* **148**, 63 (2003).
5. D. N. Spergel et al., *ApJS* **148**, 175 (2003).
6. E. Komatsu et al., *APJS* **148**, 119 (2003).
7. H. V. Peiris, *ApJSS* **148**, 213 (2003).
8. L. Verde et al., *ApJS* **148**, 195 (2003).
9. L. Page, *ApJS* **148**, 233 (2003).

10. L. Page, *ApJS* **148**, 39 (2003).

11. C. Barnes *et al.*, *APJS* **148**, 51 (2003).

12. A. Kogut *et al.*, *APJS* **148**, 161 (2003).

13. N. Jarosik *et al.*, *ApJS*148292003.

14. C. L. Bennett *et al.*, *ApJLett* **464**, L1 (1996).

15. W. Hu, http://background.uchicago.edu/~whu/ intermediate/intermediate.html.

16. W. Hu, N. Sugiyama, J. Silk, *Nature* **386**, 37 (1997).

17. R. A. Sunyaev, YA. B. Zeldovich, *AP&SS* **7**, 3 (1970).

18. J. P. E. Peebles, J. T. Yu, *ApJ* **162**, 815 (1970).

19. M. Kamionkowski, D. N. Spergel, N. Sugiyama, *ApJLett* **426**, 57 (1994).

20. G. Jungman, M. Kamionkowski, A. Kosowsky, D. N. Spergel, *Phys. Rev.* D **54**, 1332 (1996a).

21. G. Jungman, M. Kamionkowski, A. Kosowsky, D. N. Spergel, *Phys. Rev.* Lett. **76**, 1007 (1996b).

22. J. Kovak *et al.*, *ApJ* **astro-ph**, 0209478 (2003).

23. U. Seljak, M. Zaldarriaga, *ApJ* **469**, 437 (1996).

24. L. Verde *et al.*, *N. Astr. Rev.* , in press, (2003).

25. B. S. Mason *et al.*, *ApJ* **591**, 540 (2003).

26. T. J. Pearson *et al.*, *ApJ* **591**, 556 (2003).

27. C. L. Kuo *et al.*, *ApJ* **astro-ph**, 0212289 (2003).

28. M. Colless *et al.*, MNRAS **328**, 1039 (2001).

29. W. J. Percival *et al.*, *MNRAS* **327**, 1297 (2001).

30. R. Croft *et al.*, *ApJ* **581**, 20 (2002).

31. N. Gnedin, A. J. Hamilton, *MNRAS* **334**, 107 (2002).

32. L. Verde *et al.*, *MNRAS* **335**, 432 (2002).

33. N. J. Cornish, PRD **57**, 5982 (1998).

34. N. J. Cornish, D. N. Spergel, *PRD* **62**, 87304 (2000).

35. de Olivera Costa *et al.*, *preprint* **astro-ph**, 0307282 (2003).

36. N. J. Cornish, D. N. Spergel, G. D. Starkman, E. Komatsu, *preprint* **astro-ph**, 3010233 (2003).

DISCUSSION

Peter Rosen (DOE): What are the errors on the neutrino mass numbers that you gave?

Licia Verde: The number I quoted is the 95% upper limit on the neutrino mass, assuming there are 3 degenerate neutrino species. You may want to remember the value for $\Omega_\nu h^2$, which is probably more useful to you.

Chang Kee Jung (SUNY at Stony Brook): Do you see any more improvement in that number? Is there more data or any other future more precise measurement? There must be some kind of systematic limit on how well you can do with the neutrino mass.

Licia Verde: That limit can be improved in the very short term in 2 ways: first of all: better signal-to-noise. WMAP is still up there and the 2 years worth of data are "in the can" as of a few days ago. Better signal-to-noise can reduce the error bars on the other cosmological parameters this will shrink the degeneracies and help everything else.

Also, since the publication of the WMAP results, people have started analyzing gravitational lensing observations that directly probe the dark matter power spectrum at lower redshift. Basically what we are using is the fact that neutrinos stream freely so they tend to suppress or erase fluctuations on small scales. Then you compare the fluctuations on this side of the plot from the CMB (Fig. 6 on the left-hand-side) and the fluctuations on this side on the larger scale structures scales (Fig. 6 on the right-hand-side) and get a constraint from the growth of the fluctuation and the shape (see Fig. 9).

When you look at galaxies you have the uncertainties that you don't really know how to trace the mass. We have some handle on that but that grows the error bars. By looking at gravitational lensing you don't have that uncertainty. By using those data the constraint I think can be improved, but I didn't have the chance to work through the data yet.

As the data set improve and the statistical errorbars shrink, the big challenge will be to have good control of systematics. The CMB is a very "clean" data set because physics at the last scattering surface is simple and well understood. On the other hand, the physics that governs large-scale structure is connected to galaxy formation and evolution, is complicated and highly non-linear. Better understanding and modeling of the systematics induced by galaxy formation and evolution will be vital to further improve this measurement.

S. Sumowidagdo (Florida State University): Is there any chance for WMAP to observe tensor perturbation?

Licia Verde: We put some limits on tensors which we can discuss in more details in the afternoon discussion session. The polarization data will significantly improve: we will publish soon not just the power spectrum of the cross correlation between temperature and polarization but the polarization auto-correlation power spectrum and that will improve the current constraints. Having said this, a detection is in principle possible, but only for models with significant tensor modes.

BRIGHT STARS, DARK ENERGY

R. KIRSHNER

Harvard-Smithsonian Center for Astrophysics, 60 Garden St., Cambridge, MA 02138, USA
E-mail: rkirshner@cfa.harvard.edu

NO CONTRIBUTION RECEIVED

ASTROPARTICLE THEORY:
SOME NEW INSIGHTS INTO HIGH ENERGY COSMIC RAYS

E. ROULET

CONICET, Centro Atómico Bariloche, 8400, Argentina
E-mail: roulet@cab.cnea.gov.ar

Some new developments obtained in the last few years concerning the propagation of high energy cosmic rays are discussed. In particular, it is shown how the inclusion of drift effects in the transport diffusion equations leads naturally to an explanation for the knee, for the second knee and for the observed behavior of the composition and anisotropies between the knee and the ankle. It is shown that the trend towards a heavier composition above the knee has significant impact on the predicted neutrino fluxes above 10^{14} eV. The effects of magnetic lensing on the cosmic rays with energies above the ankle are also discussed, analyzing the main features of the different regimes that appear between the diffusive behavior that takes place at lower energies and the regime of small deflections present at the highest ones.

1. Introduction

Since their discovery in 1912, cosmic rays (CR's) were of great help for particle physics, providing a source of high energy particles for free, which only required the construction of detectors in order to observe different kinds of interesting phenomena. In this way, positrons, muons, pions, kaons and hyperons were discovered in the period 1930–1950. However, when after the '50s man-made accelerators reached energies beyond the GeV, particle physics moved back to the labs and cosmic ray research became focused on the study of the CR's themselves (rather than on the products of their interactions), trying to understand their origin, the mechanisms responsible for their acceleration and the way they propagate from the sources up to us.

There are only a few observable quantities associated with the CR fluxes. These are the energy spectrum, the CR composition and the anisotropies in arrival directions, and it is through their study that the CR mysteries have to be unraveled. For instance, the fact that the spectrum is essentially a power law and not a thermal one is what led Fermi to suggest that the CR acceleration was a stochastic process. Also, looking at low energy cosmic rays, the study of the abundances of spallation products (like Li, Be and B, which are produced mainly by spallation of C, N and O) gives information about the amount of matter traversed by the CR's, while the abundances of unstable isotopes (e.g. ^{10}Be) gives information on the time spent by CR's in the Galaxy. On the other hand, the anisotropies observed on the low energy CR's arriving from the East and from the West gave indications that they were caused by the deflections produced by the geomagnetic field on the positively charged CR's.

Many puzzles are also associated with the observed properties of the CR's with very high energies, those above 10^{15} eV and up to the highest ones exceeding 10^{20} eV. In particular, we would like to know what causes the spectral changes observed, what is the origin of the observed anisotropies and the changes in composition as a function of energy and how CR sources would look like at ultrahigh energies. As will be discussed below, a crucial issue in this respect is to understand in detail the propagation of CR's through the magnetic fields present in the Galaxy, since this is essential to determine their properties when we finally observe them.

2. The Cosmic Ray Spectrum

The differential flux of CR's changes by more than 30 orders of magnitude in the energy range from 10^9 eV to 10^{20} eV, following essentially a power law $E^{-\alpha}$ which shows only some small but noticeable breaks in the power index α. One has indeed $\alpha \simeq 2.6$–2.7 at low energies, with a steepening to $\alpha \simeq 3$ at the so-called *knee*, which takes place at $E_k \simeq 3 \times 10^{15}$ eV. A second steepening to $\alpha \simeq 3.3$, referred to as the *second knee*, has been reported at $E_{sk} \simeq 4 \times 10^{17}$ eV, while a hardening to $\alpha \simeq 2.7$ takes place at $E_a \simeq 5 \times 10^{18}$ eV, being known as the *ankle*. These general features are apparent in the data[1] shown in Fig. 1, which displays the quantity $E^3 \times \mathrm{d}\Phi/\mathrm{d}E$. The observed spectrum is just a reflection of the original source spectra which have been

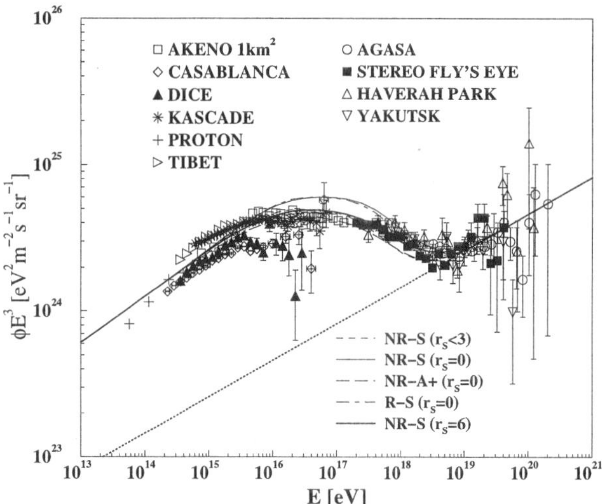

Figure 1. CR spectrum as measured by different experiments and predictions of the diffusion/drift scenario for different galactic models. The dotted line is the assumed extragalactic flux.

reshaped by the energy dependence of the confinement time of CR's in the Galaxy, since for increasing energies CR's escape more readily and hence their local density becomes less enhanced. To understand this last effect we will then discuss in some detail the transport of CR's through the magnetic fields permeating the Galaxy. Beyond the ankle the CR's are no longer confined inside the Galaxy (and most likely there are no galactic sources reaching such high energies) so that a transition to an extragalactic component should be taking place there.

2.1. *Cosmic Ray Diffusion and Drift*

Our Galaxy is believed to have a regular and a random magnetic field, both with local strength of a few μG. The regular component is aligned with the spiral arms, reversing its directions between consecutive arms and having a typical scale height of ~ 1 kpc. Also a more extended halo component is believed to exist, with a typical scale height of a few kiloparsecs but its symmetry properties are less well established. On the other hand, the random component has a spectrum not very different from a Kolmogorov one (as would result if its origin is associated with the turbulence in the interstellar medium (ISM)), i.e. with a magnetic energy density satisfying $dE_r/dk \propto k^{-5/3}$ in Fourier space, with a maximum scale of turbulence of order $L_{max} \simeq 100$ pc.

When particles of charge Ze propagate across

a regular field $\mathbf{B}_0$, they describe helical trajectories characterized by a pitch angle θ, so that the component of the velocity parallel to $\mathbf{B}_0$ is $v_\parallel = c \cos\theta$, and a Larmor radius given by

$$r_L = \frac{pc}{ZeB_0} \simeq \frac{E/Z}{10^{15}\,\text{eV}} \frac{\mu\text{G}}{B_0}\text{pc} \qquad (1)$$

(the radius of the helical trajectory is $r_L \sin\theta$). In the presence of the random component $\mathbf{B}_r$, the CR's will scatter off the magnetic field irregularities with associated scales of order r_L, changing their pitch angle but not their velocity. This will lead to a random walk and a diffusion along the magnetic field direction characterized by a diffusion coefficient

$$D_\parallel = \frac{\langle \Delta x_\parallel^2 \rangle}{2\Delta t} = \frac{c}{3}\lambda_\parallel, \qquad (2)$$

where $\lambda_\parallel$ is the pitch angle scattering length, which depends on how much power there is in the magnetic field modes at scales $\sim r_L$, i.e.

$$\lambda_\parallel \propto \left. \frac{r_L}{dE_r/d\,\ln k} \right|_{k=2\pi/r_L}. \qquad (3)$$

Hence, for a Kolmogorov spectrum one has $D_\parallel \propto E^{1/3}$, and in general if the spectrum of the random magnetic field energy satisfies $dE_r/dk \propto k^{m-2}$, one has $D_\parallel \propto E^m$.

The diffusion orthogonal to the regular magnetic field direction is typically much slower (unless the turbulence level is very high, in which case parallel and perpendicular motions become similar), and it is due to both pitch angle scattering and to the wandering of the magnetic field lines themselves, which drag with them the diffusing particles in the direction perpendicular to $\mathbf{B}_0$. Its evaluation has then to be done numerically, and recent results[2] show that for fixed levels of turbulence (i.e. for given values of $\sqrt{\langle \mathbf{B}_r^2 \rangle}/B_0$) the associated diffusion coefficient $D_\perp$ has a similar energy dependence as $D_\parallel$ as long as $r_L < L_{max}$.

The third ingredient is the antisymmetric (or Hall) diffusion,[3] which is associated to the drift of the CR's moving across the regular magnetic field. What is relevant here is the macroscopic drift associated to the gradient of the CR density, which leads to a current

$$\mathbf{J}_A = D_A \frac{\mathbf{B}_0}{B_0} \times \boldsymbol{\nabla} N. \qquad (4)$$

Notice that this macroscopic current is orthogonal to both $\mathbf{B}_0$ and $\boldsymbol{\nabla} N$, and its relation to the microscopic

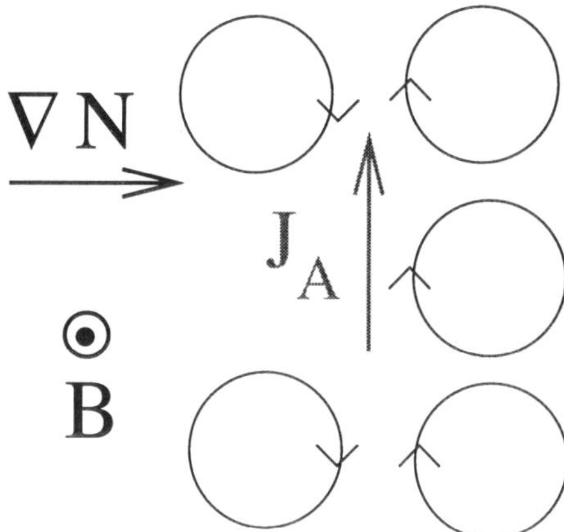

Figure 2. Illustration of the physical origin of the macroscopic drift.

drift associated to gradients and curvature in $\mathbf{B}_0$ is subtle.[4] In particular, the macroscopic drift is also present for a constant $\mathbf{B}_0$, and its origin is illustrated in Fig. 2. The antisymmetric diffusion coefficient is just given by

$$D_A \simeq \frac{r_L c}{3} \propto E. \qquad (5)$$

The CR density distribution in the Galaxy is then determined by the equation

$$\boldsymbol{\nabla} \cdot \mathbf{J}_D = Q, \text{ with } J_{Di} = -D_{ij}\nabla_j N, \qquad (6)$$

where Q describes the distribution of sources and the diffusion tensor is (adopting here the z-axis along the direction of $\mathbf{B}_0$)

$$D_{ij} = \begin{pmatrix} D_\perp & D_A & 0 \\ -D_A & D_\perp & 0 \\ 0 & 0 & D_\| \end{pmatrix}. \qquad (7)$$

To solve Eq. (6) it is convenient to assume for simplicity that the Galaxy has cylindrical symmetry and that the magnetic field is azimuthal, which is not far from being true, in which case $D_\|$ plays no role and one has

$$\mathbf{J}_D \simeq -D_\perp \boldsymbol{\nabla} N + D_A \frac{\mathbf{B}_0}{B_0} \times \boldsymbol{\nabla} N, \qquad (8)$$

so that the turbulent diffusive component is orthogonal to the isodensity contours while the drifts are parallel to them.

Hence, adopting a realistic configuration of magnetic fields, obtaining from them the diffusion tensor everywhere in the Galaxy (using in particular fits to the numerical results obtained for $D_\perp$ for different turbulence levels,[2]) assuming a distribution of CR sources and then numerically solving the diffusion

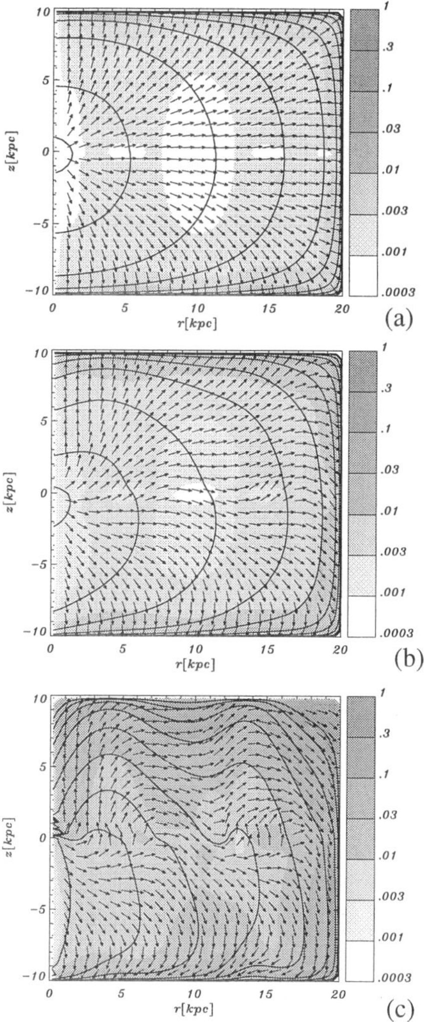

Figure 3. CR isodensity contours for $E/Z = 10^{14}$, 3×10^{15} and 5×10^{17} eV (from top to bottom). The arrows indicate the directions of the diffusion currents while the shading the level of anisotropy.

equations, one can obtain[5,6] the CR density distribution in the Galaxy and the resulting diffusion currents. This is exemplified in Fig. 3, which shows the isodensity contours (every two solid contours correspond to a change in density by an order of magnitude) for a source of constant strength in the inner 3 kpc of the galactic plane. The arrows indicate the direction of the diffusion currents, with the component parallel to the contours arising from the drifts. Notice that since the motion of a charged particle in a magnetic field is determined just by its rigidity, the relevant variable is the ratio E/Z. The three panels shown correspond to $E/Z = 10^{14}$, 3×10^{15} and 5×10^{17} eV, and it is clearly seen that for energies below ZE_k the dominant effect arises from the perpendicular diffusion, while for larger energies the drifts are responsible for the dominant escape mechanism. The transition between these two regimes is naturally understood from the different energy dependence of the two diffusion coefficients ($D_\perp \propto E^{1/3}$ while $D_A \propto E$). Actually, both coefficients become comparable, $D_\perp \simeq D_A$, at an energy of the order of ZE_k, and this crossover naturally generates the observed break in the spectrum.

2.2. *The Knee*

If the source spectrum is taken as a power law, $dQ/dE \propto E^{-\alpha_s}$, it is seen from the diffusion equation that the observed spectrum will be affected by the energy dependence of the diffusion coefficients. This just reflects the fact that the CR confinement time in the Galaxy is $\tau_e \propto D^{-1}$. Hence, below the knee, where transverse turbulent diffusion dominates, the observed spectral index will be $\alpha \simeq \alpha_s + 1/3$, while in the drift dominated regime one has instead $\alpha \simeq \alpha_s + 1$. Since this transition takes place at different energies for CR's with different charges, one expects that first the proton component will steepen its spectrum at an energy that will be identified with E_k, and then the heavier components will suffer a similar change in their spectrum but at energies ZE_k. This is illustrated in Fig. 4, which shows the local density of the different CR components obtained[5] in the scenario described in Fig. 3. The CR composition is normalized with existing satellite measurements of lower energies, typically below 10^{14} eV, and the individual spectra of H, He, O and Fe are displayed, together with the total flux.

It is seen that although each individual component changes its spectral index by $\Delta\alpha_i \simeq 2/3$, the total flux has a softer steepening, which is consistent with the measured change of $\Delta\alpha \simeq 0.3$. Here the source spectrum is assumed to have a constant spectral index $\alpha_s \simeq 2.3$ (which is consistent with expectations from first order Fermi acceleration in strong shocks, which typically predict $\alpha_s \simeq 2$–2.4) and the spectral change in the observed spectrum just results from not ignoring the physical effects of the CR drifts. Let us mention that the energy at which the break occurs depends on several things, like the assumed distribution of sources, the location

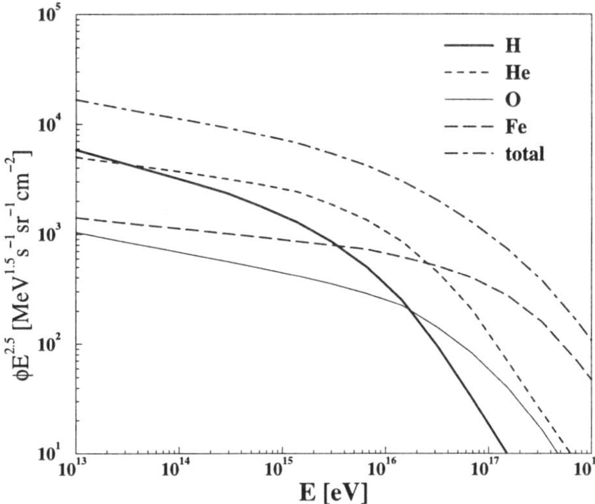

Figure 4. Predicted individual and total spectra in the diffusion/drift scenario.

in the Galaxy where the flux is measured, the symmetry of the regular magnetic field, the level of turbulence and the maximum scale of turbulence (scaling in particular as $L_{max}^{-1/2}$). It is very reassuring that for the realistic models adopted it just falls in the energy range where the knee is observed.

2.3. *The Second Knee and the Ankle*

Since the heaviest CR's having a significant abundance are the Fe nuclei, once one considers energies above $26 \times E_k \simeq 10^{17}$ eV the light components will be already very suppressed and hence the composition will be dominated by these heavy nuclei. When drifts then start to affect this component, the total spectrum will gradually steepen to $\alpha \simeq \alpha_s + 1 \simeq 3.3$, which is just the behavior observed at the so-called second-knee. This is illustrated by the different lines depicted in Fig. 1, which show the total spectrum obtained[7] with different magnetic field and source models. In addition, the straight dotted line represents the contribution from extragalactic CR's, with an assumed spectrum $\propto E^{-2.7}$ and normalized to fit observations above the ankle (at energies beyond 5×10^{19} eV, the effects of the GZK suppression by interactions with the CMB photons, not included here, should also be accounted for). It is seen that all the main features of the spectrum below the ankle find a natural explanation in terms of the CR diffusion process.

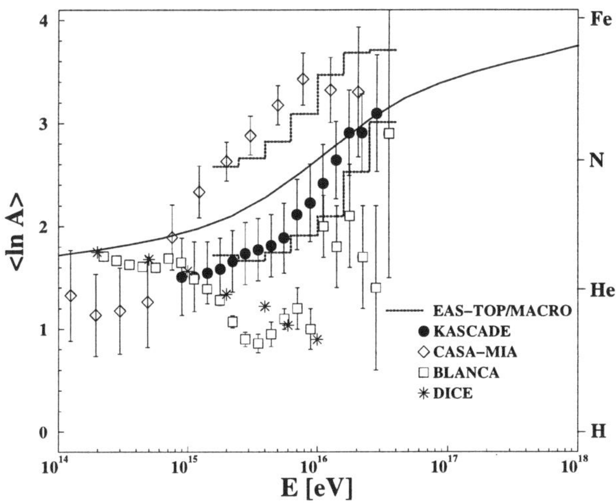

Figure 5. Measured composition and predictions (solid line) of the diffusion/drift scenario.

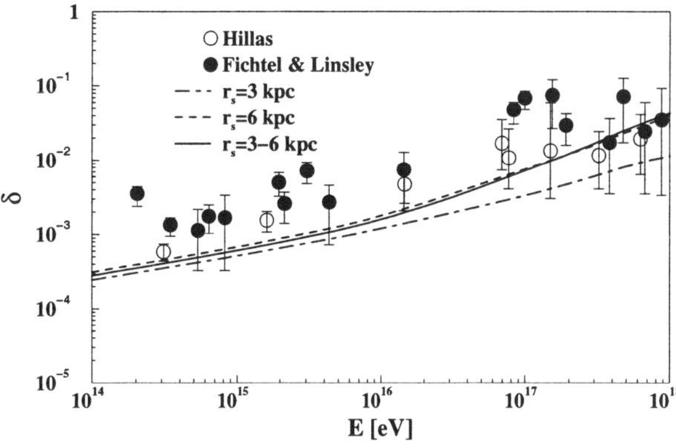

Figure 6. Measured anisotropies (amplitude of the first harmonic in right ascension) and predictions of the diffusion/drift scenario for different assumed source distributions.

2.4. *Composition and Anisotropies*

As regards the composition, the behavior just described implies that the CR's become increasingly heavy beyond the knee. The predictions[5] for the average mass number ($\langle \ln A \rangle$) are displayed in Fig. 5, together with the observational data on this quantity. Although there was a large spread on the measured values, the most recent results from the KASCADE[8] and the MACRO/EAS-TOP[9] collaborations are in quite good agreement between them and with the theoretical predictions of the diffusion/drift scenario. Moreover, KASCADE has reported a measurement of the spectra of four different mass groups (H, He, C, Fe) and they are remarkably consistent with the predictions depicted in Fig. 4. On the other hand, the MACRO/EAS-TOP experiment has found that the change in the spectral slope of the individual components is $\Delta\alpha_i = 0.7 \pm 0.4$, consistent with the value of 2/3 induced by the drifts.

Regarding the anisotropies, if the knee is actually due to a change in the escape mechanism from the Galaxy (rather than a change in the intrinsic source spectra), one expects to find a correlation between the change in the behavior of the energy spectrum and of the CR anisotropies. This is in fact what seems to be observed, and the predicted[6] anisotropies are in good agreement with the observational results, as is apparent from Fig. 6.

Let us mention that several possible mechanisms have been proposed in the past to explain the knee.

For instance, that it could be associated to the heavy CR components being photodisintegrated by interactions with optical or UV photons inside the sources.[10] This scenario would predict for instance that the composition should become lighter above the knee, at odds with the latest observations mentioned previously. It has also been suggested that the knee could be related to a change in the efficiency of CR acceleration in the sources. This is attractive since the traditional acceleration scenario based on supernovae exploding in the ISM is expected to lead to maximal energies of about $E/Z \sim 10^{14}$–10^{15} eV. However, these scenarios (and also the one with photodisintegrations) would have difficulties to account for the CR anisotropies observed, for the origin of the second knee and, in any case, since drift effects are anyhow present, they will have to be taken into account and would lead to further suppressions beyond those originating in the sources. Other acceleration mechanisms, such as supernovae exploding in the wind of their progenitors or one-shot acceleration in pulsars, may naturally reach larger values of E/Z, so that the assumption that α_s remains constant beyond E_k is quite plausible.

3. Cosmic Ray Induced Neutrino Fluxes

When CR's interact in the upper atmosphere producing particle cascades, a very important product which results are the fluxes of muon and electron neutrinos, which arise mainly from the decays of charged pions and kaons. The study of these atmospheric neutrinos has been of paramount importance in recent years, allowing in particular to establish the existence of neutrino flavor oscillations, and hence of neutrino masses and mixings. One may then say that in this respect CR's have returned to be a source of particle beams of great interest for particle physics.

Also the interactions of galactic CR's with the gas present in the ISM is expected to yield fluxes of high energy neutrinos from directions close to the galactic plane (ISM ν's). Actually, the associated fluxes of high energy photons from π^0 decays have already been observed by EGRET. In addition, several galactic (supernova remnants, microquasars, ...) and extragalactic (active galaxies, gamma ray bursts, ...) objects are also potential sources of very high energy neutrinos, being the target of the ongoing and projected neutrino observatories (Amanda, Baikal, Antares, Auger, Icecube, ...). Since the flux sensitivities achieved are expected to improve by more than two orders of magnitude in the next decade,[11] this will open a new window to explore the Universe by means of the high energy neutrinos. These messengers have the great advantage of arriving undeflected and unattenuated from their sources, containing then precious information about their production sites.

CR's with energies beyond E_k will produce atmospheric neutrinos (and ISM ν's) with very high energies, typically $E_\nu > 10^{14}$ eV. Atmospheric neutrinos with these energies are particularly interesting because they are expected to arise mainly from the decays of mesons containing charm quarks, since these decay "promptly" after being produced while, due to the relativistic time dilation, the pions and kaons have decay lengths much larger than 10 km at these energies, and are hence stopped by the atmosphere before they can decay, so that the neutrinos they produce have much lower energies.

To evaluate the fluxes of prompt neutrinos is quite delicate, since it requires[12] taking into account next-to-leading-order processes in the charm production cross section, which also turns out to depend on the behavior of the parton distribution functions (PDF's) at very low values ($x < 10^{-5}$) of the fraction of momentum carried by the partons, and these values are below those that can be measured with present accelerators. It has been actually suggested that the observations of very high energy atmospheric neutrinos could be used to study the PDF's in the range $10^{-9} < x < 10^{-4}$. Moreover, it is important to know these fluxes in detail because they constitute the main background for the search

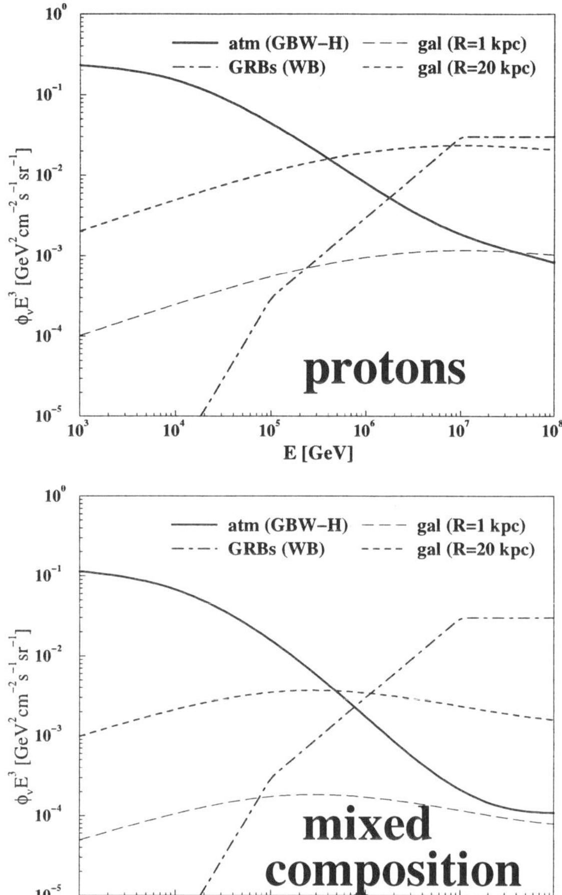

Figure 7. CR induced neutrino fluxes under the assumption that the CR are only protons (top) or have a mixed composition (bottom), as results in rigidity dependent explanations of the knee. The solid line is the flux of atmospheric neutrinos, the dashed lines are the ISM ν's from the direction of the galactic center (upper) or perpendicular to the galactic plane (lower line). Also indicated is the ν flux from GRB's predicted by Waxman and Bahcall.

of some astrophysical neutrino sources.

To predict these fluxes it actually turns out that one needs a detailed knowledge of the CR composition beyond the knee, since a heavy CR nucleus of mass A behaves just as a collection of A nucleons of energy E/A. Hence, if the CR's are heavy they will produce fluxes of neutrinos of much lower energies, which combined with the steepness of the CR spectrum implies a much reduced neutrino flux. This is illustrated in Fig. 7, which shows the neutrino fluxes predicted under the (standard) assumption that CR's consist mainly of protons as well as the predictions adopting the composition which re-

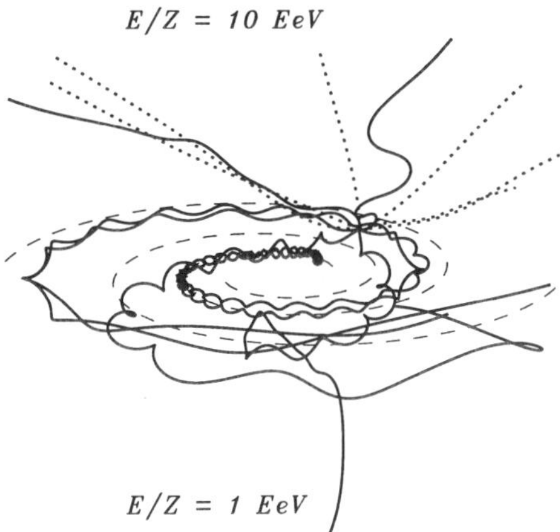

Figure 8. CR trajectories reaching the Earth for energies $E/Z = 10^{18}$ eV (1 EeV, solid lines) and $E/Z = 10^{19}$ eV (10 EeV, dotted lines). The dashed lines indicate the location of the galactic spiral arms.

sults in models where the knee is a rigidity dependent effect (such as the diffusion/drift scenario just discussed), and for which the composition becomes heavier above the knee.[13] Clearly this implies that the CR induced neutrino fluxes will be harder to detect, but also that the background for the search of other astrophysical neutrino sources (such as the GRB prediction also shown in the figure) will be reduced.

4. Magnetic Lensing

CR's have a diffusive propagation in the Galaxy as long as their Larmor radius stays smaller than the maximum scale of the turbulence, i.e. for E/Z below a few times 10^{17} eV (see Eq. (1)). It is only at higher energies that the trajectories gradually straighten up and that it may become possible to attempt to do "CR astronomy", i.e. to try to identify the images of individual CR sources. This is illustrated in Fig. 8, which shows[14] several trajectories of CR's reaching the Earth. At energies $E/Z = 10^{18}$ eV CR's are strongly deflected and there is no way in which the direction of arrival to the Earth can give information about the original source direction, while for $E/Z = 10^{19}$ eV the deflections start to become manageable.

The transition between the diffusive behavior to the regime of small deflections is actually very rich in new phenomena and several qualitatively different regimes appear.

4.1. *The Regime of Small Deflections*

Starting from the high energy end, we can expect that CR's arriving from distant sources will suffer only small deflections, so that an image with a slight (rigidity dependent) offset with respect to the original source position should be observed. If the intervening magnetic field and the CR charge were known, this effect could in principle be corrected for in order to infer the source coordinates.[15,16,17] To have an idea about the deflections involved, one can keep in mind that the deflection in the direction of the trajectory of a CR traversing a distance L across a constant magnetic field B is

$$\delta \simeq 3.2° \frac{10^{20} \text{ eV}}{E/Z} \frac{L}{3 \text{ kpc}} \frac{B}{2 \text{ }\mu\text{G}}. \tag{9}$$

Hence, the regular galactic magnetic field can produce sizeable deflections even for CR's with the highest energies observed, specially if their composition is heavy. It has to be said however that almost nothing is known about the CR composition above the ankle, and this is one of the issues that the future large statistics observatory AUGER is expected to clarify.

Notice that although the random magnetic field has an rms strength similar to the regular magnetic field strength, its effects on the overall CR deflections is smaller, due to its short coherence length[a] L_c, which results in a deflection growing only as $\sqrt{L/L_c}$, where L/L_c is just the number of magnetic "domains" traversed. Indeed, the rms deflection it induces is

$$\delta_{rms} = 0.6° \frac{10^{20} \text{ eV}}{E/Z} \frac{B_{rms}}{4 \text{ }\mu\text{G}} \sqrt{\frac{L}{3 \text{ kpc}}} \sqrt{\frac{L_c}{50 \text{ pc}}}. \tag{10}$$

The random component has however important effects on the lensing of CR's. Let us also mention that extragalactic magnetic fields can further affect the CR trajectories.[18,19,20] Their effects could be significant if they have sizeable strength and large coherence lengths (i.e. if $B \times \sqrt{L_c} > 10^{-9} \text{ G}\sqrt{\text{Mpc}}$).

4.2. *The Lensing Regime*

When the deflections produced by the magnetic fields are such that the CR's can arrive to the observer through paths which deviate from the straight one

---

[a]For a Kolmogorov spectrum, one has $L_c \simeq L_{max}/5$.

by more than the scale of coherence of the magnetic field (e.g. a few kpc for the regular field or the scale L_c for the random component), it becomes possible for the CR's to reach the observer through different, uncorrelated, paths. This just means that several different images of the same source will appear. This phenomenon is quite similar to the well known gravitational lensing phenomena, but here it is the magnetic deflection that replaces the gravitational one. Contrary to this last one, the magnetic deflections are energy dependent, so that the multiple images would look different at different energies. Anyhow, many magnetic lensing phenomena look as a function of energy very similar to some gravitational lensing phenomena as a function of time when the relative motion of the lens is relevant, such as in the microlensing events.

Besides producing multiple images, the magnetic field can also lead[14,21] to focusing effects which magnify or demagnify the original source fluxes. In particular, when new images appear at a certain energy E_0, they always appear in pairs (which moreover have opposite parities, i.e. one of the images will be inverted, something which would be however unobservable in practice), and they are magnified by a factor $A \propto (E_0 - E)^{-1/2}$, which diverges[b] for $E \to E_0$. This can lead to huge enhancements in the CR fluxes in narrow ranges of energies.

The multiple image formation is illustrated[22] in Fig. 9, which shows how a particular source will look like at different energies, and how for decreasing energies new images are produced and continuously displaced. The lensing effects for this same source are illustrated in the second panel, which shows the amplifications of the different images, clearly displaying the presence of lensing peaks when new images appear. These peaks can lead to interesting observational effects, since they will lead to an enhanced number of events in a narrow angular window, corresponding to the direction where the new pair of images appear in the sky, which will also be clustered in energy (or, actually, in rigidity). It is remarkable that this clustering in rigidity seems to be already present[22] in the clusters of pairs and triplets of events with small angular separations observed[23] at the highest energies ($E > 4 \times 10^{19}$ eV). The in-

---

^{b}The average magnification in a finite energy bin stays of course bounded.

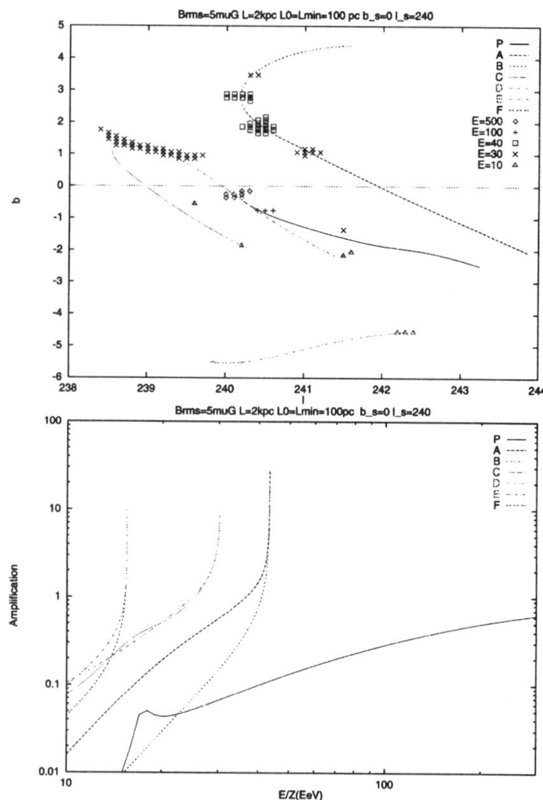

Figure 9. Images (top panel) of a CR source located near $(b, \ell) \simeq (0°, 240°)$. The different symbols correspond to the images obtained through a ray-shooting simulation (assuming an extended source) at energies $E/Z = 500, 100, 40, 30$ and 10×10^{18} eV. New secondary images appear in pairs below some critical energies, and for decreasing energies get displaced along the lines indicated. The bottom panel shows the magnification of the fluxes of the different images.

crease in statistics that will be achieved with Auger will allow these predictions to be tested with higher significance.

4.3. *The Scintillation Regime*

For typical galactic magnetic field models, multiple images of extragalactic CR sources appear at energies $E/Z \sim$ few $\times 10^{19}$ eV. As smaller energies are considered, one can show[22] that the number of secondary images grows exponentially, increasing to about 10^2 in a decade of energy and this trend continues for lower energies. The lensing peaks associated with the new images become increasingly narrow, and the result is that what is observed is just a blurred image, consisting of a very large number of secondary images, with an angular extent determined by δ_{rms} and with an average magnification which approaches unity. This behavior is illustrated

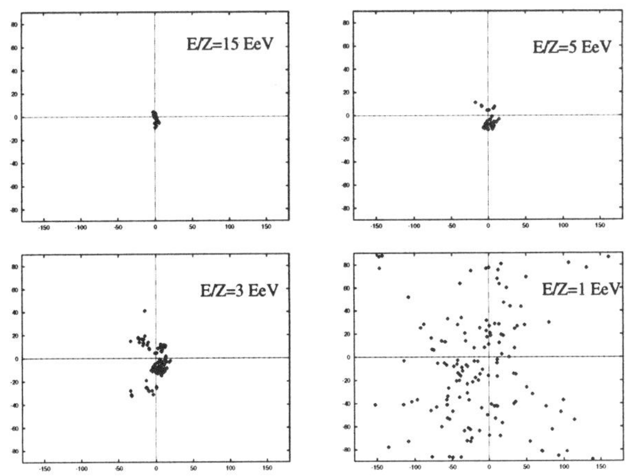

Figure 10. Images of a CR source in the scintillation regime.

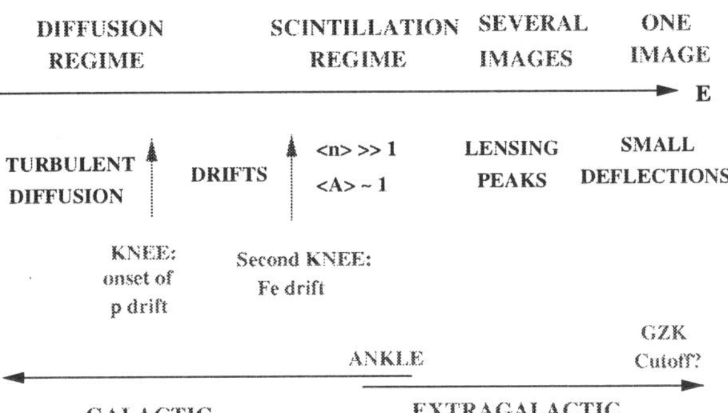

Figure 11. Sketch of the different CR propagation regimes.

in Fig. 10, which shows the results of a ray-shooting simulation displaying the way a given CR source will look like at different energies in the regime in which a large number of images are present.

This "scintillation" regime is reminiscent to the twinkling of stars caused by the refractive effects of the atmospheric turbulence, although in the present case it is the galactic magnetic turbulence that is responsible for the effects observed. This regime is different from the diffusion one, since the deflections δ_{rms} need not be large. Hence, only some small random deflections in the direction of the trajectories are being produced, but in order that the spatial diffusion regime sets in one has to still go down in energy until $r_L < L_{max}$, so that the resonant scattering leading to pitch angle scattering can take place.

A brief summary of the different propagation regimes discussed is shown in Fig. 11.

It is interesting that we are at a time when many of these ideas will be tested with more precise measurements of the CR's both in the region of the knee (with e.g. KASCADE-GRANDE) and at the highest energies (with AUGER), and this will certainly help to unveil some of the many mysteries that CR's still present to us.

Acknowledgments

This work is based on several collaborations done with J. Candia, L. Epele, D. Harari and S. Mollerach. It is partially supported by the Fundación Antorchas and the John Simon Guggenheim Foundation.

References

1. M Nagano and A Watson, *Rev. Mod. Phys.* **72**, 689 (2000).
2. F Casse, M Lemoine and G Pelletier, *Phys. Rev. D* **65**, 023002 (2002).
3. V S Ptuskin *et al.*, *Astron. Astrophys.* **268**, 726 (1993).
4. R A Burger, H Moraal and G M Webb, *Ap&SS* **116**, 107 (1985).
5. J Candia, E Roulet and L Epele, *JHEP* **0212**, 033 (2002).
6. J Candia, S Mollerach and E Roulet, *Journal of Cosmology and Astropart. Phys.*, *JCAP*, **0305**, 003 (2003).
7. J Candia, S Mollerach and E Roulet, *JHEP* **0212**, 032 (2002).
8. A Haungs, *J. Phys.* G **29**, 809 (2003).
9. M Aglietta *et al.* (EAS-TOP/MACRO Collaborations), `astro-ph/0305325`.
10. A M Hillas, *16 ICRC Kyoto*, vol. **8** (1979) 7; S Karakula and W Tkaczyk, *Astropart. Phys.* **1**, 229 (1993); J Candia, L Epele and E Roulet, *Astropart. Phys.* **17**, 23 (2002).
11. C Spiering, *J. Phys.* G **29**, 843 (2003).
12. A D Martin, M G Ryskin and A M Staśto, `hep-ph/0302140`; L Pasquali, M H Reno and I Sarcevic, *Phys. Rev. D* **59**, 034020.
13. J Candia and E Roulet, *Journal of Cosmology and Astropart. Phys.*, *JCAP*, **0309**, 005 (2003).
14. D Harari, S Mollerach and E Roulet, *JHEP* **08**, 022 (1999).
15. T Stanev, *Astrophys. J* **479**, 290 (1997).
16. G Medina Tanco, E De Gouveia dal Pino and J Horvath, *Astrophys. J.* **492**, 200 (1998).
17. D Harari, S Mollerach and E Roulet, *JHEP* **07**, 006 (2002).
18. G Sigl, M Lemoine and P Biermann, *Astropart. Phys.* **10**, 141 (1999).
19. G Medina Tanco, *Astrophys. J.* **505**, L79 (1998).
20. T Stanev, D Seckel and R Engel, `astro-ph/0108338`.
21. D Harari, S Mollerach and E Roulet, *JHEP* **02**, 035 (2000).
22. D Harari, S Mollerach, E Roulet and F Sánchez, *JHEP* **03**, 045 (2002).
23. Y Uchihori *et al.*, *Astropart. Phys.* **13**, 151 (2000).

WIMP/NEUTRALINO DIRECT DETECTION

M. DE JÉSUS

IPN Lyon-UCBL, IN2P3-CNRS, 4 rue Enrico Fermi, 69622 Villeurbanne Cedex, France
E-mail: dejesus@in2p3.fr

The most popular candidate for non-baryonic dark matter is the neutralino. More than twenty experiments are dedicated to its direct detection. This review describes the most competitive and promising experiments with different detection techniques. The most recent results are presented with some prospects for the near future.

1. Introduction

The existence of dark matter in the Universe is now well established in the astro-particle community reenforced by the recent astrophysical observations of the satellite experiment WMAP:[1] about 27% of the mass-energy of the Universe is composed of matter. Ordinary matter (baryons) contributes about 4% of this total mass density of the Universe and only $\simeq 1\%$ is visible according to the most recent measurements of the amount of deuterium in high red-shift clouds of gas and of the CMB.[2] Hence about 90% of this dark matter is not baryonic. We have to distinguish two categories, hot and cold dark matter particles referring to their velocity at the matter-radiation decoupling time in the early Universe. Hot dark matter implies moving relativistically and cold moving non-relativistically. Neutrinos with non-zero masses are hot dark matter candidates, however WMAP[1] results combined with other experiments and observations lead to a contribution $< 1.5\%$ for light neutrino species.

So the bulk of the non-baryonic dark matter is cold dark matter (CDM). Among the numerous solutions proposed by theorists, axions and neutralinos are favorites. Neutralinos are candidates of the generic class of Weakly Interacting Massive Particles (WIMP). Axions are particles proposed to solve the strong CP-violation problem in the Peccei-Quinn theory.[3] Astrophysical considerations combined with experimental constraints require an axion mass in the range 10^{-3} to 10^{-6} eV/c^2. More detailed discussions about axions can be found in papers listed in the References.[4−6] This paper will be dedicated to WIMP/neutralino detection. The neutralino is the lightest supersymmetric particle, a linear combination of the supersymmetric partners of the photon, Z and Higgs bosons, in the minimal supersymmetric extension of the Standard Model:

$$\chi^0 = a\tilde{\gamma} + b\tilde{Z} + c\tilde{H}_1^0 + d\tilde{H}_2^0. \tag{1}$$

Its mass is constrained to lie in the range 45 GeV $< m_\chi < 3$ TeV, where the lower bound comes from accelerator results from LEP and the upper bound is given by astrophysical constraints such as the age of the Universe or unitarity. Locally our galaxy is supposed to be imbedded in a WIMP halo.

Many experiments are dedicated to direct and indirect detection of WIMPs, two complementary techniques. Direct detection experiments measure the energy deposited by elastic scattering of a neutralino of our own galaxy off a target nucleus. For masses larger than $\simeq 200$ GeV, indirect detection of dark matter particles through their annihilation products may be more suitable. In this paper we will concentrate on the case of direct detection techniques. For a complete description of indirect approaches we refer to other papers listed in the References.[7−12]

In the direct detection approach the expected event rate depends on various parameters coming from astrophysics, particle physics and nuclear physics; it can range from 1 to 10^{-5} events/kg/day. The measured signal is very low (few keV) depending on the masses of the incident particle and of the scattered nucleus, but also on the nuclear recoil relative efficiency (quenching factor) in producing charge, light or heat. Hence WIMP direct searches put strong constraints on experimental background environments, and require detectors with very low energy thresholds. In this review we present the different possible signatures for disentangling a WIMP signal from the background. Different experimental approaches are described and illustrated by a few experiments. The current limits in the exclusion plot and near future prospects will be also presented.

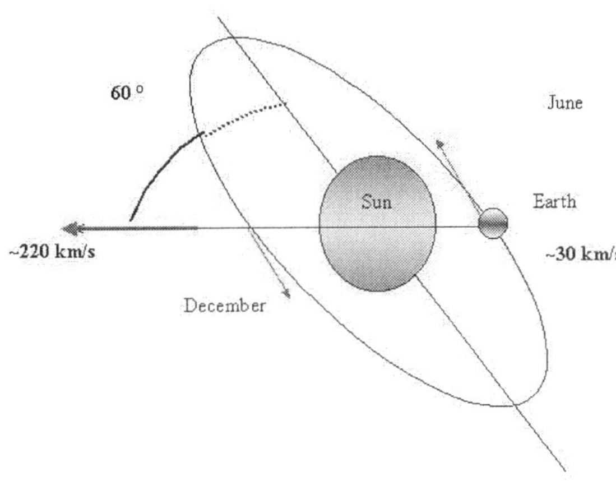

Figure 1. Annual modulation.

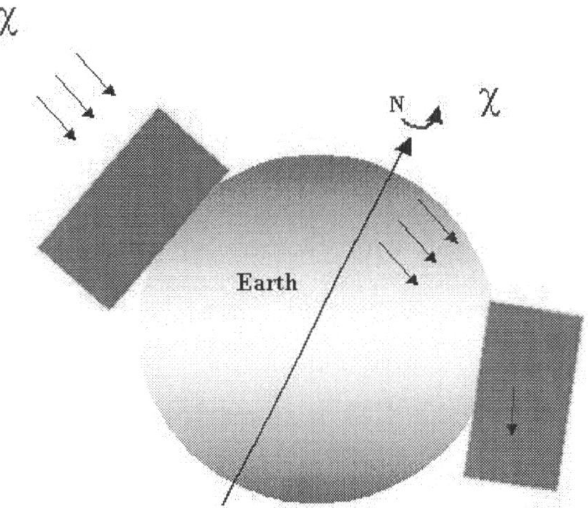

Figure 2. Diurnal modulation.

2. WIMP/Neutralino Direct Detection Physics Principles

As mentioned previously, the WIMP-nucleus interaction rate depends on various parameters. First we have to define a WIMP halo model. For simplicity the approximation of a maxwellian velocity distribution in the galactic frame is made (see Kamionkowski and Kinkhabwala (1998)[13] for a review on alternative halo models). Next, a supersymmetric model is chosen for predicting the WIMP interaction with quarks of the nucleons inside the target nucleus. Depending on the chosen model the WIMP-nucleus cross section has two components:[14,15] spin-dependent and spin-independent. The spin-independent term couples to the mass of the nucleus and the spin-dependent couples to its spin. The nuclear form factor depends on the nature of the interaction. The spin-dependent case is the most complicated one, requiring detailed nuclear models (for more details see dedicated papers[15,17]). In the following we will restrict this review to the simplest spin-independent case which is supposed to dominate in most models for massive target nuclei. Taking into account these previous considerations the interaction rate can be expressed as follows :

$$\frac{dR}{dQ} = \frac{\sigma_0 \rho_h}{2m_r^2 m_\chi} F^2(Q) \int_{v_{min}}^{\infty} \frac{f(v)}{v} dv \qquad (2)$$

where m_r is the WIMP-nucleus reduced mass $m_\chi m_N/(m_\chi + m_N)$, m_χ is the WIMP mass, and m_N

is the nucleus mass. $\rho_h = 0.3$ GeV/c^2/cm^3 is the assumed halo WIMP density at the position of the solar system, $f(v)$ is the dark matter velocity distribution, with an average r.m.s. velocity $v_0 = 220$ km/s, truncated above the escape velocity of the galaxy $v_{esc} \simeq 575$ km/s, and σ_0 is the total nucleus-WIMP interaction cross section and $F(Q)$ is the nuclear form factor.

2.1. Exclusion Plot $\sigma(m_\chi)$

In order to reliably compare supersymmetric models with results obtained by different experiments using different techniques a $\sigma_0(m_\chi)$ plot is built in the following way. The cross sections $\sigma_0(m_\chi)$ are normalized to a single nucleon $\sigma(m_\chi)$ to allow comparisons between different target nuclei. The measured nuclear recoil event rate is compared to a theoretical spectrum calculated for a given WIMP mass and cross section. If an experiment observes a signal then we build a $\sigma(m_\chi)$ contour plot. If the observed events cannot be unambigously associated with a WIMP signal an exclusion limit is calculated. WIMP signals have distinctive signatures that backgrounds are not supposed to be able to mimic. Three different signatures are proposed.

2.2. Annual Modulation

As a result of the Earth motion around the Sun the count rate in detectors should show an annual modulation (Fig. 1). The Earth's velocity relative to the galaxy varies, in June the Earth and the Sun velocities add up whereas in December they subtract.[18] The maximum amplitude of this effect in the signal is about 7%. We will report later that the DAMA collaboration using NaI scintillating crystals is the first experiment and at the moment the only one, claiming evidence of a WIMP annual modulation signal.

2.3. *Diurnal Modulation and Directionality*

Another possible modulation in the WIMP signal is the night and day variation, this effect is due to the shielding of the detector by the Earth from the incident flux. For masses close to 50 GeV and under certain assumptions the diurnal modulation can be larger than the annual one.[16] However the most interesting daily signature coupled with the annual one is the directionality of the WIMP wind as illustrated in Fig. 2. This effect is also larger than the annual one. The validation of the principle has been performed by the DRIFT-I experiment with a 1 m^3 low pressure TPC[19] prototype.

2.4. *Target Atomic Mass Effect*

Observed together annual and diurnal modulations are unambigous methods to distinguish WIMP and background signals, but they are very difficult to see. In the spin-independent case, the easiest method is to use different target materials as the event rate depends on the target atomic mass. To give an estimate of this effect we can use the Smith and Lewin[14] calculated integral rate R_0 with no form-factor correction and an average recoil energy E_R:

$$R_0 \simeq 5.87 \times v_0 \sigma_{m_\chi} \rho_h A^3 \frac{m_\chi}{(m_A + m_\chi)^2} \ /\text{kg/d}, \quad (3)$$

$$E_R \simeq 2 \times 10^{-6} m_\chi^2 (\frac{v_0}{c})^2 \frac{m_A}{(m_A + m_\chi)^2} \ \text{keV}. \quad (4)$$

Table 1 shows R_0 and E_R values for different targets, and for a given WIMP mass of $m_\chi \simeq 50$ GeV and $\sigma_{m_\chi} \simeq 7 \times 10^{-6}$ pb. Naïvely if we consider the event rate it seems to be more advantageous to use high mass nuclei, but looking at the recoil energy as the target atomic number (A) increases, the average deposited energy tends to decrease. So the choice of a target is a compromise between these two quantities. Moreover we can see, for example, that germanium is

Table 1. Integrated event rate (R_0) and average energy deposition ($< E_R >$) for different target atomic masses (A), with no form-factor correction and $m_\chi = 50$ GeV, $v_0 = 220$ km/s and $\sigma_{n\chi} = 7 \times 10^{-6}$ pb.

	A	R_0	$< E_R >$
H	1	5.10^{-5}	1
Na	23	0.3	11
Si	28	0.5	12
Ge	73	3	13
I	127	8	11
Xe	131	9	11
Pb	210	18	8

more efficient than silicon for WIMP detection while they have similar cross sections for neutrons.

Another important point is the possible neutron multiple scattering in the detector, which is impossible for a WIMP. We will see hereafter this method is used by the CDMS collaboration,[20] with germanium and silicon targets as illustrated in Fig. 6.

3. WIMP/Neutralino Direct Detection Techniques

WIMP detectors are constrained by three important requirements: low threshold, ultra low background and a high mass detector. When a WIMP interacts with a nucleus, the nuclear recoil can induce different signals (Fig. 3): heat, ionization and scintillation. During the last decade important technical developments were based on one or two of these different physics processes.

3.1. *Quenching Factor*

A relevant parameter in WIMP direct detection is the relative efficiency of nuclear recoil called quenching factor. It is the ratio of the number of charge carriers produced by a nuclear recoil due to the WIMP interaction over an electron recoil of the same kinetic energy (electron equivalent energy or "eee"). For scintillating materials the quenching factor is defined as the ratio between the light produced by a nuclear recoil and by an electron recoil. While in conventional detectors this factor is usually below 30%

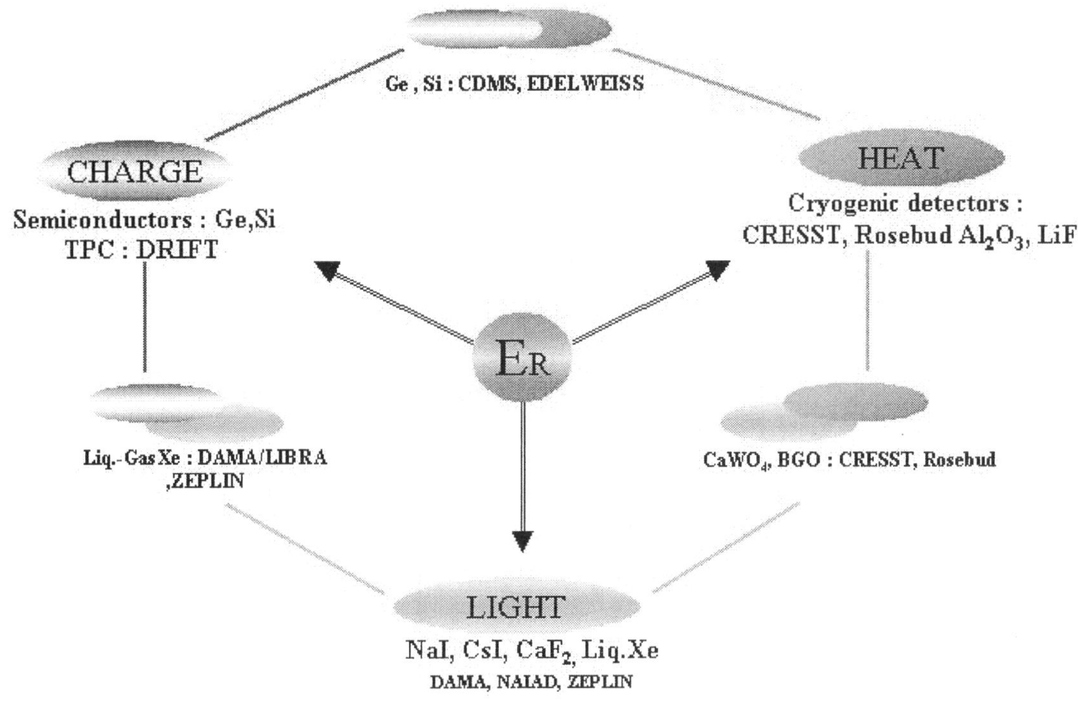

Figure 3. Illustration of the different techniques developed for the WIMP direct detection.

(e.g. measured to be $\simeq 0.3$ for germanium,[21] $\simeq 0.25$ for sodium and $\simeq 0.08$ for iodine[22]), for cryogenic detectors described hereafter it has been measured to be around one for recoiling nuclei independent of energy.[23–25]

3.2. *Classical Detectors: Semiconductors and Scintillators*

Germanium diodes initially used in double-beta-decay experiments were the first detectors used to search for WIMPs, since they have very low thresholds and very good resolutions. Experiments like IGEX[26,27] and HDMS,[28] with about 2 kg of enriched ^{76}Ge, achieved very low background count rates (<0.2 evt/kg/day in the interval 10-40 keV) and $E_{thr} \simeq 4-10$ keV-ee (equivalent to $\simeq 15-30$ keV recoil).

Large masses were easily achievable with scintillators like NaI or liquid xenon in a very pure environment. The DAMA experiment has operated more than 100 kg of NaI (each crystal weighting about 9.7 kg with energy threshold of $\simeq 2$ keV-ee, i.e. 22 keV recoil) for several years in the Gran Sasso un-derground laboratory. They accumulated data during the last 7 years and in 1997, after 4 years of data taking, they announced evidence for an annual modulated WIMP signal. The DAMA group claim their observation is compatible with a signal induced by a WIMP of $\simeq 52$ GeV mass and a WIMP-nucleon cross section of $\simeq 7.2$ pb. The DAMA collaboration has published[29], this last summer, the last 3 years campaign totaling 7 years and confirms their observation of an annual modulation signal as illustrated in Figs. 4 and 5. Right now none of the currently running dark matter experiments confirms this signal as we can see in the current exclusion plot in Fig. 9. Independent experiments with NaI detectors (NAIAD[30] in the Boulby mine, ANAIS[32] in Canfranc, ELEGANT[33] in Oto Cosmo Observatory) are currently running. The NAIAD[31] experiment's most recent results begin to exclude the DAMA $\sigma(m_\chi)$ region in the spin-independent exclusion plot.

As we have seen previously despite the very high purity level of classical detectors, they suffer ultimately from a lack of power discrimination between electron and nuclear recoils.

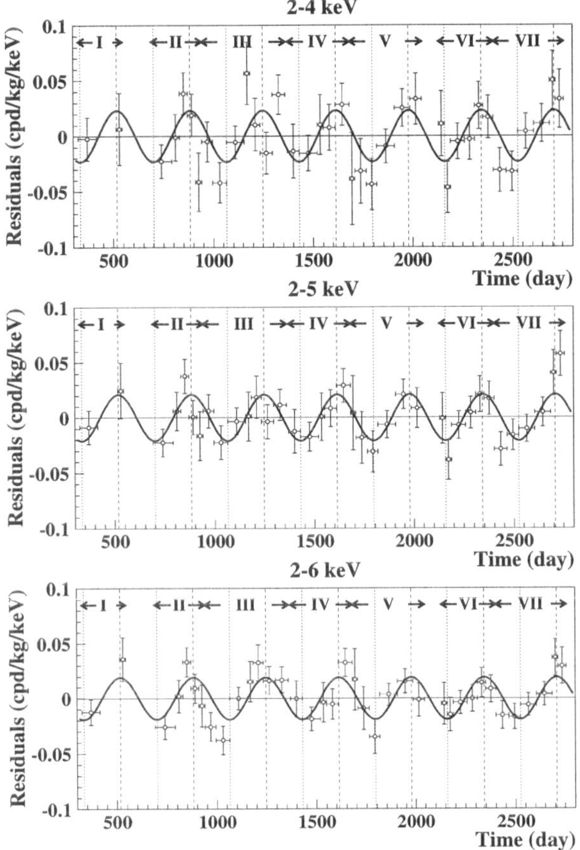

Figure 4. DAMA model independent residual count rates as a function of time for 7 years and three energy intervals (2-4), (2-5) and (2-6) keV-ee.

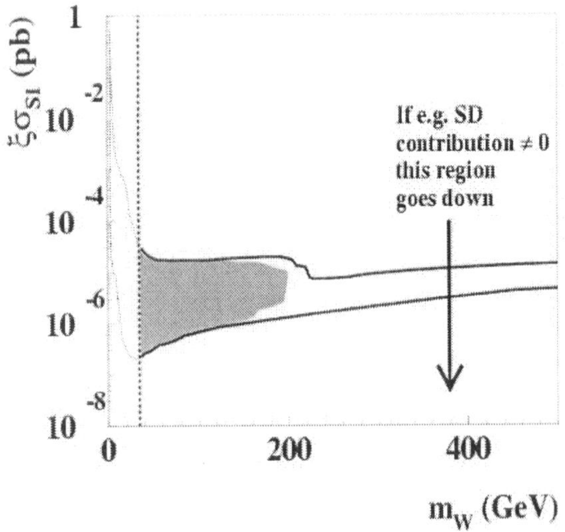

Figure 5. Dama limit for the dominant spin-independent case obtain with 7 years of data taking. This contour plot is obtained with different WIMP-halo models, see Bernabei *et al.*[29] for a detailed discussion.

The first discrimination method used is based on a pulse shape analysis. It is a statistical method where the measured quantity is the rise-time of the light signal which depends on the nature of the recoiling particle. This discrimination method is used with sodium iodide crystals (DAMA, NAIAD) but is also successfully used with liquid scintillators like liquid xenon.

With a 3.1 kg liquid-Xenon detector the ZEPLIN-I[34] collaboration has reached preliminary sensitivities which could exclude the DAMA zone. However some problems remain: a relatively high electronic background rate has to be understood, there is no nuclear recoil calibration for the low energy part of the spectrum (<50 keV-ee), and a poor energy resolution compared to bolometers. Some of these points should be answered in the next few months as the experiment in now currently running deep underground in the BOULBY mine.[14]

The DAMA/LIBRA collaboration is currently running a new NaI detector mith a larger mass ($\approx$ 250 kg) as well as a liquid-Xenon detector.

The future projects ZEPLIN-II and -III aim to develop a discrimination technique with a two phase liquid-gas Xenon detector with charge and light signals.

3.3. *Cryogenic Detectors*

Since the beginning of the 90's important developments were also made in new directions like cryogenic detectors. They are made of a crystal with a thermometer glued on it, operating at very low temperature (few tens of milli-Kelvin). Very low thresholds were reached by the CRESST-I experiment[36] with a 262 g sapphire calorimeter (resolutions of $\simeq$ 133 eV at 1.5 keV and thresholds $\simeq$ 500 eV).

The most impressive results were obtained with mixed techniques allowing the simultaneous measurement of two components heat-light or heat-charge. The two combined observables are a powerful tool to distinguish a nuclear recoil induced by a WIMP or a neutron interaction from electron recoils induced by a gamma or an electron interaction (quenching factor described previously). It is an event-by-event discrimination method. Again different approaches were explored by different worldwide collaborations. For cryogenic detectors the CDMS and EDELWEISS collabora-

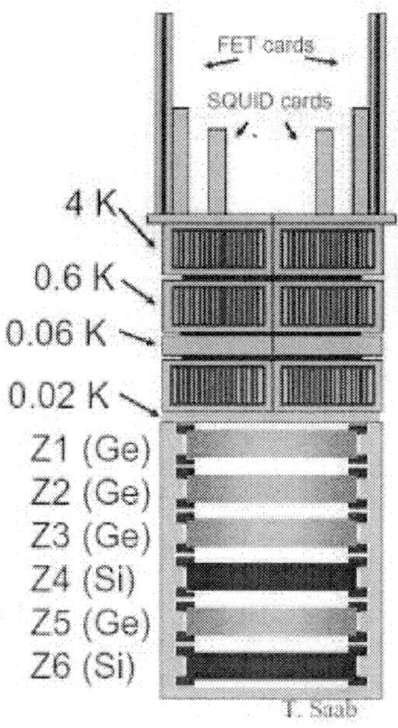

Figure 6. CDMS detector tower.

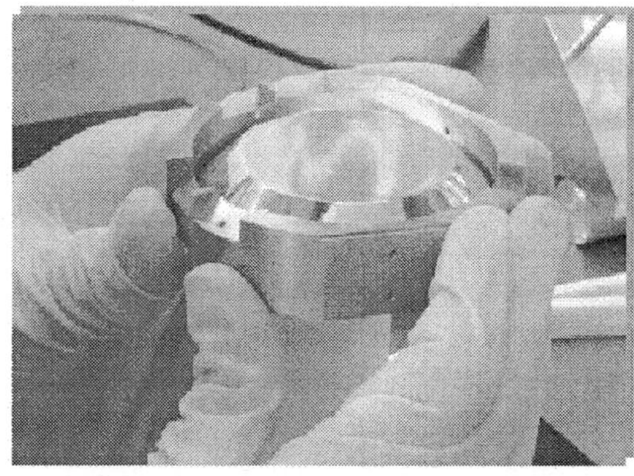

Figure 7. EDELWEISS 320 g Ge detector.

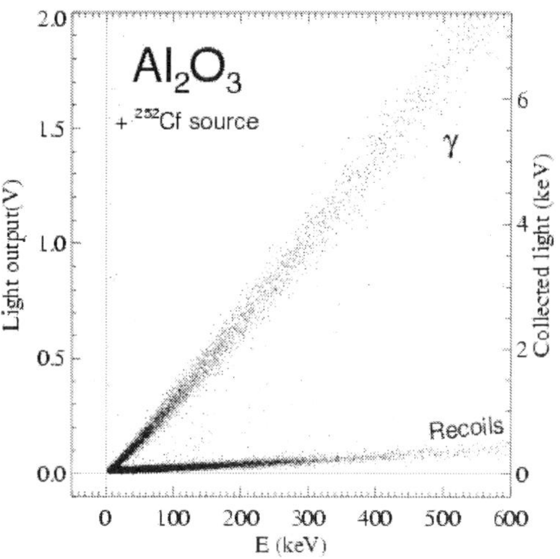

Figure 8. Discrimination between gammas and nuclear recoils in a 50 g sapphire bolometer at 20 mK by the ROSEBUD collaboration.

tions investigate the heat-ionization way, and the CRESST and ROSEBUD collaborations explore the heat-light channels.

The CDMS collaboration was the first[37,38] to operate a detector giving simultaneously ionization and heat signals with a germanium crystal. Until 2002 the experiment was running in the shallow site in Stanford with poor muon shielding inducing an important neutron background. Despite this limitation they derived competitive dark matter limits and were leaders for several years. They could subtract the neutron background using a Monte Carlo simulation and also take advantage of the fact that they run si-multaneously two different targets: germanium and silicon.[20,39] During the year 2003 the CDMS-II experiment is being installed in the deep underground Soudan mine where the muon flux is reduced by 5 orders of magnitude, thereby reducing the neutron background by a factor 400. They are currently operating 2 towers (Fig. 6) of 3×165 g Ge and 3×100 g Si detectors and 18 more detectors are under fabrication totaling 4 kg of germanium. The CDMS collaboration expects to improve its current sensitivity ($\simeq 1$ evt/kg/day) by two orders of magnitude.

The currently best spin-independent published limit was obtained by the EDELWEISS collaboration cumulating 32 kg-d. The EDELWEISS experiment is installed in the underground laboratory of Modane in the French-Italian Alps. They operate similar detectors to those of CDMS germanium crystals (Fig. 7) with different technologies for the electrodes[40] running at $\simeq 18$ mK. Three 320 g detectors are running simultaneously. During the last campaign in June 2003, 2 events were observed in the nuclear recoil zone whose origin is under investigation. More data is being analysed, but the EDELWEISS-I stage data taking will soon be finished. For the next stage a larger cryostat with a detection volume of 100 litres is built and is currently being tested. This cryostat benefits from an original technology developed at the CRTBT-Grenoble laboratory. The EDELWEISS-II installation will take place a year from now. The

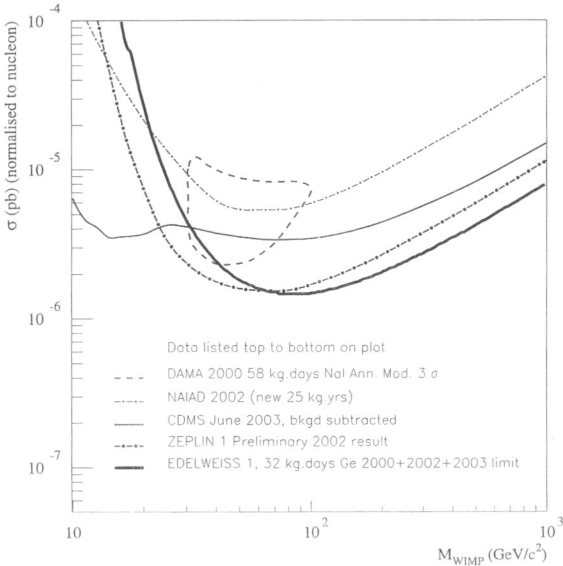

Figure 9. Current spin-independent limits for the most competitive experiments. The WIMP halo parameters used are $\rho_h = 0.3$ GeV/c^{-2}cm^{-3}, $v_0 = 220$ km/s. The closed contour corresponds to the 3σ allowed region of the DAMA first four years worth of data obtained with the same WIMP halo parameters.

first step will operate 21×320 g germanium detectors with NTD thermometers and 7×200 g NbSi thin film germanium detectors developed by the group of the CSNSM laboratory.[41] A muon veto made of 140 m^2 plastic scintillator will be added. It should reject the neutron background induced by cosmic muons in the inner lead shielding, which has been evaluated to be two orders of magnitude below the present EDELWEISS-I sensitivity $\simeq 0.2$ evt/kg/day. Such a background has to be clearly identified and rejected since the expected event rate for the EDELWEISS-II stage is about 10^{-2} evt/kg/day. In a second step, up to 120 detectors will operate simultaneously.

The CRESST-II[42] and ROSEBUD[43] experiments involve scintillating crystals as cryogenic detectors. They operate in the same way; the heat is measured with a thermometer glued on the scintillator and the light is collected with a second thin but large surface crystal. The main advantage of such a method is the large possibility for scintillating target materials: CaWO$_4$, PbWO$_4$, Al$_2$O$_3$, BaF, BGO, ... and for large volumes. A few years ago S. Pécourt et al.[44] characterized the phonon channel of a 1 kg Al$_2$O$_3$ bolometer and recently the same team[43] has succeeded in measuring the light output of a 50 g

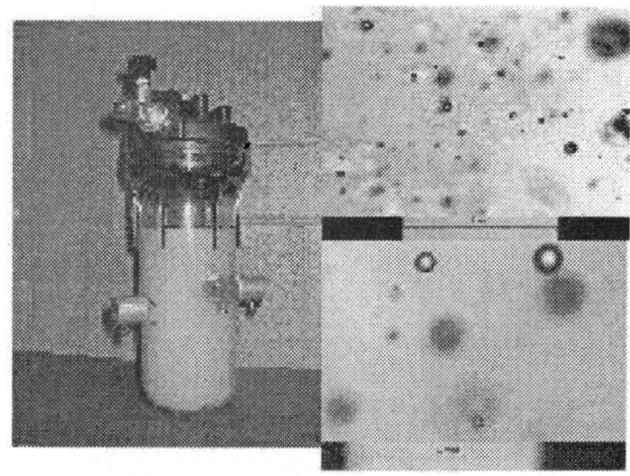

Figure 10. PICASSO new 1 liter module.

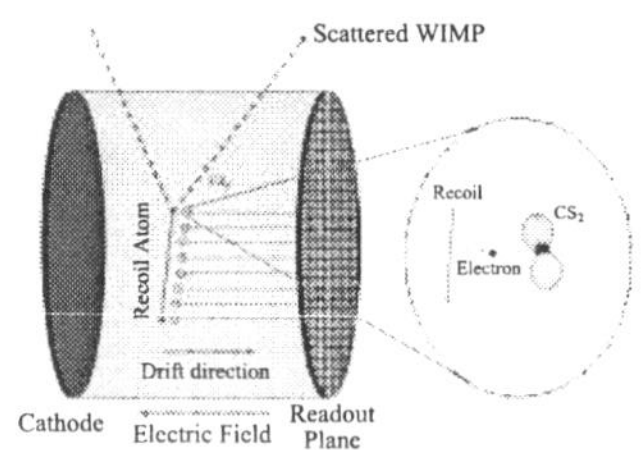

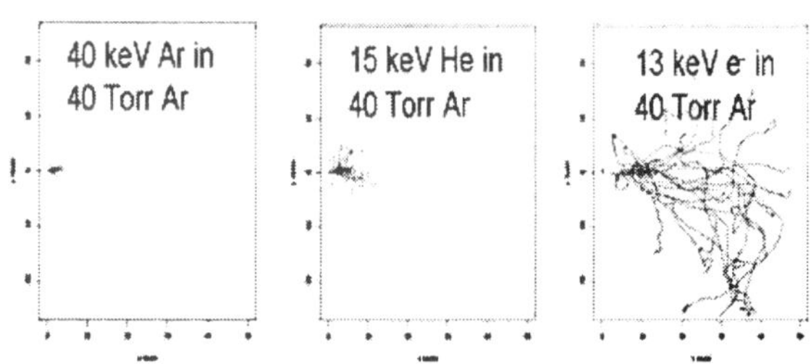

Figure 11. DRIFT-1 ionization tracks for three different types of recoiling particles : argon, helium and electrons.

Al$_2$O$_3$ bolometer (Fig. 8).

The CRESST-II[42] experiment should operate 33×300 g modules of CaWO$_4$ totaling about 10 kg.

3.4. *New Promising Techniques*

In addition to the techniques described above, illustrated by currently running and near future experiments, other promising techniques are under investigation.

The PICASSO[45,46] and SIMPLE[47] experiments have choosen to adapt a well known technology used in neutron dosimetry, to develop a counter for WIMP induced nuclear recoils. The method is based on small superheated Freon droplets imbedded in a gel matrix at room temperature. The nuclear recoil of ^{19}F induces the explosion of a droplet, creating an acoustic shock wave measured with piezoelectric transducers. By varying the temperature of the gel the energy threshold can be triggered in such a way that the electron recoil induced by gamma background can be supressed. Calibration is made at different pressures and temperatures with mono-energetic neutrons produced by a Van de Graff Tandem. The use of ^{19}F (spin-1/2 isotope) makes the search for spin-dependent neutralinos particularly interesting. A first generation of detectors, 16 modules of 8 ml, lead to the published limit of the PICASSO collaboration.[45,46] They are currently running the second generation of modules with a larger volume (Fig. 10) in an improved low background environnement in the SNO underground laboratory: PIC@SNO. New purification techniques were developed especially for the PICASSO experiment.[48] Despite a very good background discrimination the main disadvantage of such an integrating detector is the necessity to run the experiment at different threshold energies in order to measure the deposited energy spectrum.

To take advantage of the directionality which appears as the clearest signature of WIMPs, the UKDMC collaboration has developed and is currently running successfully, the DRIFT-I detector. It consists of a 1 m^3 low pressure TPC filled with a $Xe - CS_2$ gas mixture. The principle of the TPC is well known, the innovation is the use of CS_2 negative ions instead of e^- as charge carriers reducing the diffusion in order to achieve millimetric track resolution (Fig. 11). Important improvements on the read-out techniques such as MICROMEGAS,[35] in order to increase the pressure and hence the target mass, are underway. Other possible target gases are also being studied to prepare for the next generation DRIFT-II and -III detectors with a larger gas mass for the TPC of the order of 100 kg.

4. Conclusions

The current experimental spin-independent limit turns around 10^{-6} pb which corresponds to a count rate of about 0.2 to 1 evt/kg/day. To achieve this limit it took about 10 years for most of the currently running first generation experiments to develop these detectors. The next generation under construction, most of which are in the final stages, aim to improve

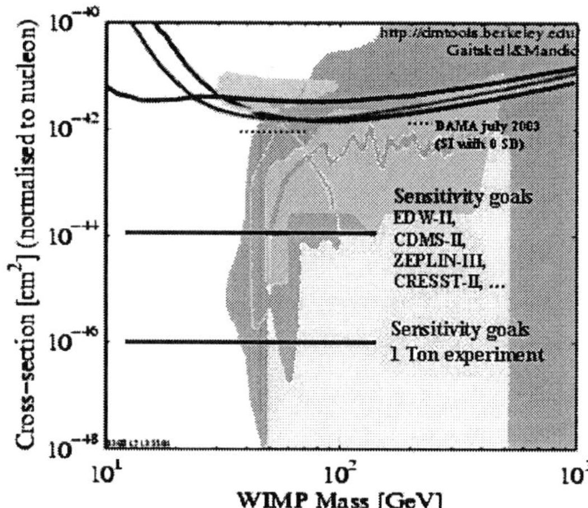

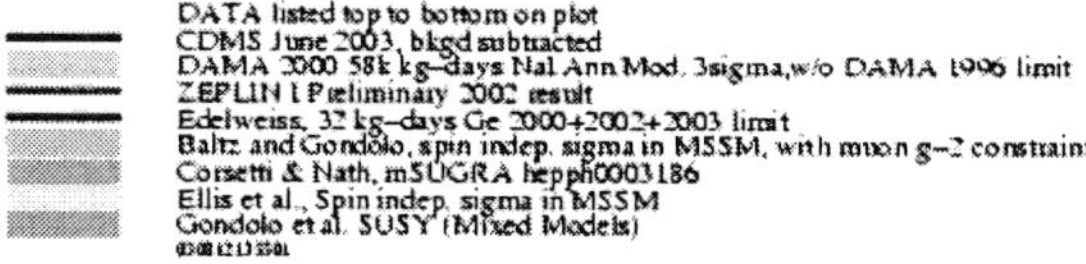

Figure 12. Projected limits for some of the next generation experiments. The colored regions represent different SUSY model calculations.

this limit by two orders of magnitude, that means a count rate around 10^{-2} evt/kg/day. This has a price: lowering the sensitivity by about two orders of magnitude implies increasing the target mass by about the same factor (for example EDELWEISS-I worked with 3×320 g Ge and EDELWEISS-II should run at the end with 120×320 g Ge detectors).

With this scaling the ultimate neutron background induced by muons can no longer be neglected. It is the reason why experiments like EDELWEISS-II, CDMS-II and CRESST-II will use a muon veto.

The next five years are very promising: a clarification of the DAMA annual modulation signal is essential. Indirect Earth-based and Space experiments like Antares, HESS, AMS and GLAST should give independent cross checks. Meanwhile accelerator physics will explore an important part of SUSY space parameters on the exclusion plot (Fig. 12).

Nevertheless the one ton scale experiment will probably involve larger international collaborations. The technical challenge will be to build an experiment able to achieve the extremely low background necessary to cover most of the prediction of mSUGRA models.

Acknowledgments

I would like to thank the in2p3 for its financial support.

References

1. A.B. Lahanas, N.E. Mavromatos and D.V. Nanopoulos, *hep-ph/0308251* (2003).
2. G. Huey *et al.*, *astr-ph/03007080* (2003).
3. R.D. Peccei and H.R. Quinn, *Phys. Rev.* D **16**, 1791 (1977).
4. M. Dine *et al.*, *Phys. Lett.* B **104**, 199 (1981).
5. G.G. Raffelt, *Phys. Rev.* D **66**, 010001 (2002).
6. S.J. Asztalos *et al.*, *Astrophys. J.* **571**, L27 (2002).
7. V. Bertin, E. Nezri and J. Orloff, *JHEP* **0302**, 046 (2003).
8. G. Dudas *et al.*, *Phys. Rev.* D **67**, 023505 (2003).
9. I.F.M. Albukerque *et al.*, *Phys. Rev.* D **37**, 1353 (1988).
10. L. Pieri and E. Branchini, to appear in *proceedings of the 28th International Cosmic Ray Conference* (2003).
11. E.A. Baltz *et al.*, *Phys. Rev.* D **65**, 063511 (2002).
12. M. Fujii and M. Ibe, *hep-ph/0308118* (2003).
13. M. Kamionkowski and A. Kinkhabwala, *Phys. Rev.* D **57**, 3256 (1998).
14. P.F. Smith and J.D. Lewin, *Phys. Rep.* **187**, 203 (1990).
15. G. Jungman, M. Kamionkowski and K. Griest, *Phys. Rep.* **267**, 195 (1996).
16. D.N. Spergel, *Phys. Rev.* D **37**, 1353 (1988).
17. M.T. Ressel and D.J. Dean, *Phys. Rev.* C **56**, 535 (1997).
18. F. Hasenbalg *et al.*, *Astropart. Phys.* **9**, 339 (1998).
19. C.J. Martoff *et al.*, *Nucl. Instr. Meth.* A **440**, 355 (2000).
20. D. Akerib *et al.*, *Phys. Rev.* D **66**, 122003 (2002).
21. Y. Messous *et al.*, *Astropart. Phys.* **3**, 361 (1995).
22. G. Gerbier *et al.*, *Astropart. Phys.* **11**, 287 (1999).
23. J.W. Zhou *et al.*, *Nucl. Instr. Meth.* A **349**, 225 (1994).
24. A. Alessandrello *et al.*, *Phys. Lett.* B **408**, 465 (1997).
25. E. Simon *et al.*, *Nucl. Instr. Meth.* A **507**, 643 (2003).
26. C. E. Aalseth *et al.*, *Phys.Rev.* D **65**, 092007 (2002).
27. A. Morales *et al.*, *Phys. Lett.* B **532**, 8 (2002).
28. H.V. Klapdor-Kleingrothaus *et al.*, *Astropart. Phys.* **18**, 525 (2003).
29. R. Bernabei *et al.*, *Riv. N. Cim.* **26**, 1 (2003).
30. B. Ahmed *et al.*, *Astropart. Phys.* **19**, 691 (2003).
31. N.J.T. Smith, *Private communication*, July 2003.
32. S. Cebrian *et al.*, *Nucl. Phys. Proc. Suppl.* **114**, 111 (2003).
33. H. Ejiri *et al.*, in *Proceedings of the Second International Workshop on the Identification of Dark Matter*, pp. 323, eds. N. J. C. Spooner and V. Kudryavtsev, (World Scientific, Singapore, 1999).
34. T. J. Sumner *et al.*, to appear in *5th Int. Symp. Sources and Detection of Dark Matter and Dark Energy in the Universe*, (Marina del Rey, 2002).
35. R. Luscher, *astro-ph/0305310* (2003).
36. G. Anglober *et al.*, *Astropart. Phys.* **18**, 43 (2002).
37. T. A. Shutt *et al.*, *Phys. Rev. Lett.* **69**, 3425 (1992).
38. T. A. Shutt, *Ph.D. Thesis*, Department of Physics, University of California at Berkeley, 1993.
39. D. Akerib *et al.*, *hep-ex/0306001* (2003).
40. A. Benoit *et al.*, *Phys. Lett.* B **545**, 43 (2002).
41. A. Juillard, in *Proceedings of the 10th International Workshop on Low Temperature Detectors*, (Genoa, Italy, 7-11 July 2003).
42. G. Angloher, to appear in *Eighth International Workshop on Topics in Astroparticle and Underground Physics*, (University of Washington, Seattle, Washington, September 5-9 2003).
43. P. de Marcillac, in *Proceedings of the 10th International Workshop on Low Temperature Detectors*, (Genoa, Italy, 7-11 July 2003).
44. S. Pécourt *et al.*, *Nucl. Instr. Meth.* A **438**, 333 (1999).
45. R. Gornea *et al.*, *IEEE Transactions on Nuclear Science*, Nov. 2001.
46. M. Barnabé-Heider *et al*, in *7th Int. Conf. on Advanced Technology and Particle Physics*, (Villa Olmo, Como, October 15-19 2001, World Scientific, Singapore, 2003).
47. J.I. Collar *et al.*, *New J. Phys.* **2**, 14 (2000).
48. PICASSO collaboration, *Private communication*, Oct. 2003.

DISCUSSION

Stan Wojcicki (Stanford University): Why has the DAMA allowed region on the WIMP mass versus cross section plane increased, and now allows for significantly smaller values of the cross section?

Maryvonne De Jésus: The DAMA allowed region on the WIMP mass versus cross section plane is increased because it is calculated with different values for some of the halo parameters, for example, a different WIMP velocity and a different local WIMP density.

Chang Kee Jung (SUNY at Stony Brook): Why is the dark matter halo in our galaxy assumed to be stationary against the motion of the rotating solar system? Namely, wouldn't a rotating dark matter halo be more natural? If so, how does this affect the experiments?

Maryvonne De Jésus: The dark matter halo in our galaxy is assumed to be stationary against the motion of the rotating solar system for convenience. This assumption is a simple model commonly used by experimentalist to compare data in the same WIMP mass versus cross section plane. But some authors, like A. Green, investigated other non-co-rotating halos and deduced that WIMP interaction rates strongly depend on the halo model.

Mike Albrow (Fermilab): The DAMA time variations are clearly not statistical fluctuations. Are there conventional explanations (excluding that it has seen dark matter) for it?

Maryvonne De Jésus: The Dama collaboration check all possible background origins which could give rise to an annual modulated signal similar to that observed. They declared everything is understood and the observed signal can't be mimicked by a background signal and hence the signal is produced by galactic WIMPS.

Bennie Ward (Baylor University & the University of Tennesse): The two events which you said could be interpreted as neutron-induced were, nonetheless, not so treated. Could you explain the corresponding logic of their treatment?

Maryvonne De Jésus: The two events observed by the EDELWEISS collaboration during the last campaign in June 2003 could be explained by some holes in the neutron shielding which appeared recently. This shielding is made of paraffin which is very sensitive to temperature variations. Other possible background origins are still under investigation.

NEUTRINO PHYSICS

NEUTRINO MASSES FROM DOUBLE-β DECAY AND KINEMATICS EXPERIMENTS

G. GRATTA

Stanford University, Dept. of Physics, Stanford, CA 94305, USA
E-mail: gratta@hep.stanford.edu

NO CONTRIBUTION RECEIVED

RESULTS AND STATUS OF CURRENT ACCELERATOR NEUTRINO EXPERIMENTS

K. NISHIKAWA

Kyoto University, Dept. of Physics, Kitashirakawa-Oiwakecho, Sakyo-ku, Kyoto 606-8502, Japan
E-mail: nishikaw@scphys.kyoto-u.ac.jp

NO CONTRIBUTION RECEIVED

REACTOR NEUTRINO EXPERIMENTS

K. INOUE

Research Center for Neutrino Science, Tohoku University
Aramaki Aoba, Aoba, Sendai, Miyagi 980-8578, Japan
E-mail: inoue@awa.tohoku.ac.jp

Previous searches for neutrino oscillations with reactor neutrinos have been done only with baselines less than 1 km. The observed neutrino flux was consistent with the expectation and only excluded regions were drawn on the neutrino-oscillation-parameter space. Thus, those experiments played important roles in understanding neutrinos from fission reactors. Based on the knowledge from those experiments, an experiment with about a 180 km baseline became possible. Results obtained from this baseline experiment showed evidence for reactor neutrino disappearance and finally provide a resolution for the long standing solar neutrino problem when combined with results from the solar neutrino experiments. Several possibilities to explore the last unmeasured mixing angle θ_{13} with reactor neutrinos have recently been proposed. They will provide complementary information to long baseline accelerator experiments when one tries to solve the degeneracy of oscillation parameters. Reactor neutrinos are also useful to study the neutrino magnetic moment and the most stringent limits from terrestrial experiments are obtained by measuring the elastic scattering cross section of reactor neutrinos.

1. Reactor Neutrinos

The first observation of neutrino existence was carried out by using a nuclear power reactor in 1956. The experiment led by Reines was called "Poltergeist" and located 11 m distance from a powerful reactor (700 MW) at Savannah River, 12 m underground. It used a large amount of proton target (200 liters of water), liquid scintillator (1400 liters of LS) and 55 photo-multiplier tubes for signal detection. The underground site was important to reduce cosmic ray induced backgrounds. A modern experiment in 2003 is still a simple extension of Poltergeist but uses a 70 GW (effective) power reactor, a proton target (1000 ton LS) and 1879 PMT's at a distance of about 180 km and 1000 m underground. This extension of many orders of magnitude became possible because of many important improvements in the knowledge of reactor neutrinos.

In the observation of reactor neutrinos, 4 fissile nuclei (^{235}U, ^{239}Pu, ^{238}U and ^{241}Pu) are important and the others contribute only at the 0.1% level. Fission fragments from these nuclei sequentially beta-decay and emit antielectron-neutrinos. The purity of the "anti"neutrinos is very high and an electron-neutrino contamination is only at the 10 ppm level above 1.8 MeV. These four nuclei release similar energy when they undergo fission[1] (^{235}U 201.7$\pm$0.6, ^{239}Pu 210.0$\pm$0.9, ^{238}U 205.0$\pm$0.9 and ^{241}Pu 212.4$\pm$1.0 MeV). Thus the fission rate is strongly correlated with the thermal power output that is measurable at much better than 2% even without any special care. Then, one fission causes about 6 neutrino emissions on average and therefore, the neutrino intensity can be roughly estimated to be $\sim 2 \times 10^{20}\, \bar{\nu}_e/\mathrm{GW}_{th}/\mathrm{sec}$. Fission spectra reach equilibrium within a day above $\sim$2 MeV. However, attention to the long-lived nuclei such as:

$$^{106}Ru \xrightarrow{T_{1/2}=372d} Rh \xrightarrow[E_{max}=3.541MeV]{} Pd$$

$$^{144}Ce \xrightarrow{T_{1/2}=285d} Pr \xrightarrow[E_{max}=2.996MeV]{} Nd$$

is necessary.[2] They affect the correlation between thermal power and neutrino flux.

The beta spectra from ^{235}U, ^{239}Pu and ^{241}Pu have been measured with a spectrometer irradiating thermal neutrons at ILL.[3] They fitted the observed beta spectra from 30 hypothetical beta branches and converted each branch to a neutrino spectrum.[4] In the case of ^{238}U, it doesn't undergo to fission with thermal neutrons and only a theoretical calculation[5] is available. This calculation traces 744 unstable fission products and obtains the corresponding neutrino spectrum. The error on the calculated spectrum is larger than the measurement, but it contributes only $\sim$8% on average for ordinary reactor cores. And knowing the time evolution of the fuel composition, the uncertainty of the neutrino event rate coming from the calculation of these spectra is only $\sim$2.3%.

The neutrino reaction used in "Poltergeist" and following experiments is the inverse beta decay:

$$\bar{\nu}_e + p \rightarrow e^+ + n.$$

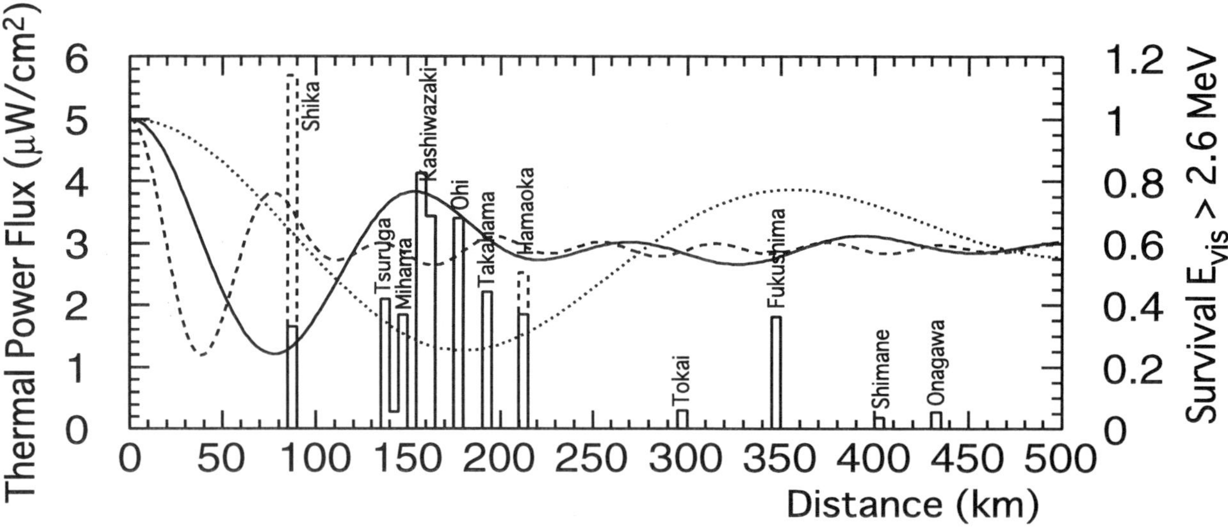

Figure 1. Distribution of nuclear power reactors as a function of distance from the KamLAND site. The solid histogram is the current operation and dashed histogram is an expected operation in 2006 (Shika at 88 km increases by a factor 3). Height of the histogram shows the thermal power flux contribution at Kamioka. Also shown as solid ($\Delta m^2 = 7 \times 10^{-5}$ eV^2), dashed (3×10^{-5}) and dotted (1.4×10^{-4}) lines are the survival probability of $\bar{\nu}_e$ as a function of distance (all for $\sin^2 2\theta = 0.84$). The probability is calculated for events above 2.6 MeV in visible energy.

Advantages of this reaction are the rich target number, low reaction threshold

$$E_{lab} = \frac{(M_n + m_e)^2 - M_p^2}{2M_p} = 1.806 \text{ MeV},$$

very precisely known cross section and the delayed two-fold coincidence signal. The cross section is well related to the neutron life time ($n \to p + e^- + \bar{\nu}_e$) as follows.

$$\sigma_{tot}^{(0)} = \frac{2\pi^2/m_e^5}{f_{p.s.}^R \tau_n} E_e^{(0)} p_e^{(0)}$$

Recent precise studies on neutrons provided the very accurate lifetime of $\tau_n = 885.7 \pm 0.8$ sec and as a result the precision of the neutrino cross section on proton is 0.2% with order 1/M corrections (Coulomb, weak magnetism, recoil, inner and outer radiative corrections).[6] The prompt signal is the positron and its annihilation gammas with a material electron, thus the prompt signal has energies always larger than 1 MeV (two electron masses). The delayed signal is the capture of the neutron on an environmental atom such as hydrogen, gadolinium, cadmium and ^{3}He. It provides clear tagging and very good background discrimination in timing, position and energy.

The validity of the neutrino spectra and cross section calculations have been experimentally tested. The most accurate measurement has been performed by the Bugey experiment.[7] It measured an overall reaction rate with 1.4% accuracy and the result ($\sigma_f = 5.750 \times 10^{-43}$ cm^2/fission $\pm$ 1.4%) was in very good agreement with the calculation ($\sigma_{V-A} = 5.824 \times 10^{-43}$ cm^2/fission $\pm$ 2.7%). The ratio of them is $\sigma_f/\sigma_{V-A} = 0.987 \pm 1.4\% \pm 2.7\%$. Also, Bugey-3 tested models of neutrino spectra and the ILL spectra showed an excellent agreement.[8] Finally, a few % precision became possible without a near detector for flux normalization with the above calculation models and procedures.

2. KamLAND

Reactor neutrino experiments before KamLAND had baselines only up to 1 km. All results from such short baselines were consistent with the calculation and oscillation parameters above 8×10^{-4} eV^2, and full mixing has been excluded with them. On the other hand, all solar neutrino experiments showed solar neutrino deficits and a combined neutrino oscillation analysis favored the large mixing angle solution below several $\times 10^{-4}$ eV^2. The only exception was the neutral current measurement at SNO and it was perfectly consistent with the standard solar model prediction. It resulted in a more than 5σ positive neutrino flux other than electron-neutrinos and is evidence for neutrino flavor transformation of solar electron-neutrinos to the other neutrino types. These results are well explained by the neutrino oscillation hypothesis but other models such as resonant spin flavor precession, neutrino decay, flavor-changing non-standard interactions and CPT violation models were still valid. Even assuming the neutrino oscillation, mass squared difference could vary from 10^{-10} to several $\times 10^{-4}$ eV^2. In order to explore these small mass difference region, baselines more than 100 km were required. For an experiment with such a baseline, much more powerful reactors, a bigger detector and a deeper underground site were necessary.

The total power generation with nuclear fission reactors in the world amounts to ~ 1.1 TW (thermal). It corresponds to $\sim 2 \times 10^{23}$ $\bar{\nu}_e$ creations/sec. Surprisingly, 70 GW (7% of world total) is generated at 130-240 km distance from the Kamioka site where former Kamiokande and Super-Kamiokande exist. Reactor neutrino flux at the Kamioka site is $\sim 5 \times 10^6$ cm^{-2}s^{-1} and requires order kiloton underground detector for a practical experiment. KamLAND took over the cavity for the former Kamiokande experiment located 1000 m underground from the top of the mountain (2700 mwe (meters-of-water-equivalent)). Figure 1 shows the expected oscillatory pattern of neutrino survival probability for the LMA parameters. The effective distance of the reactors is ~ 180 km and it is very sensitive to the difference of Δm^2. A new Shika reactor is also planned to start at a distance of 88 km.

In order to estimate the expected neutrino spectrum at the Kamioka site, histories of thermal power output, new fuel volume ratios, fuel enrichments and burn-up information are obtained from 52 commercial reactor cores ($\sim 97\%$ contribution) and histories of electric power output for 18 Korean reactors ($\sim 2.5\%$ contribution) are also taken into account.

KamLAND is a monolithic liquid scintillator detector as shown in Fig. 2. It uses 1200 m^3 LS suspended with a 135 μm-thick balloon in 1800 m^3 of buffer oil. They are contained in an 18 m diameter stainless steel tank. The LS is a mixture of 80% do-

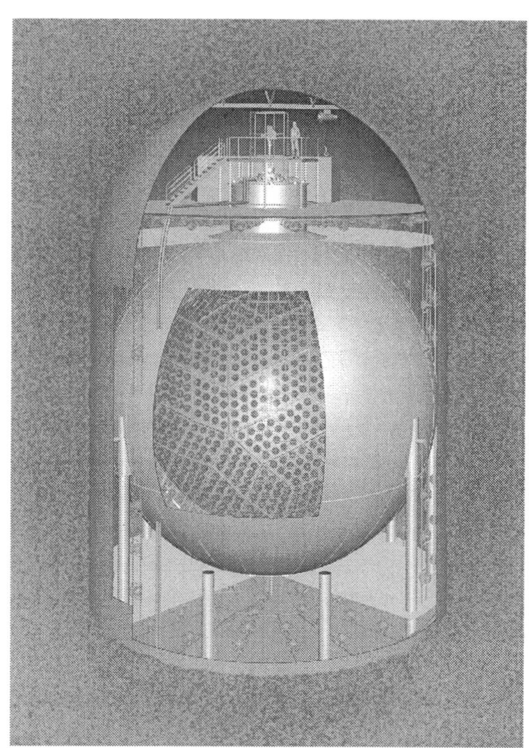

Figure 2. The KamLAND detector.

decane, 20% 1-2-4-trimethylbenzene and 1.5 g/liter PPO. The density of the LS is 0.78 g/cm^3. The buffer oil is a mixture of 50% dodecane and 50% isoparaffin and the density is controlled to be 0.04% lighter than the LS. The light output of the LS is ~ 8000 photons/MeV and its attenuation length is effectively ~ 10 m. These photons are monitored by the newly developed 1325 17''-tubes and 554 old 20''-tubes from the Kamiokande experiment. Total photo-coverage is 34% but only the 17''-tubes (22% coverage) are used for the first results.

The event trigger is set at 0.7 MeV (visible) providing a sufficiently low threshold for anti-neutrino reactions. Also a delayed trigger (0.4 MeV) is enabled for 1 msec after each prompt triggers to perform real-time impurity measurements of

$$^{214}Bi \xrightarrow{\beta,\gamma} {}^{214}Po(\tau = 237\mu sec) \xrightarrow[7.687MeV]{\alpha} {}^{210}Pb.$$

The latter alpha decay is observed with a quench of about a factor 14 and the delayed threshold is necessary to tag this delayed coincidence event. The measured impurity level at KamLAND is listed in Table 1 together with the requirements.

The achieved impurity level is much lower than

Table 1. Requirements and achievements of radioactive impurities.

Impurities	Achievements	Req.(reactor)	Req.(solar)
^{222}Rn	$0.03\ \mu$Bq/m^3		
^{238}U	$3.5 \pm 0.5 \times 10^{-18}$ g/g	10^{-13} g/g	10^{-16} g/g
^{232}Th	$5.2 \pm 0.8 \times 10^{-17}$ g/g	10^{-13} g/g	10^{-16} g/g
^{40}K	$< 2.7 \times 10^{-16}$ g/g	10^{-14} g/g	10^{-18} g/g
^{85}Kr	~ 1 Bq/m^3		$1\ \mu$Bq/m^3
^{210}Pb	~ 100 mBq/m^3		$1\ \mu$Bq/m^3
On the balloon		Equiv. mine dust	
^{222}Rn	4.0×10^{-4} Bq		
^{238}U	3.1×10^{-8} g	0.9 g	
^{232}Th	9.7×10^{-4} g	0.1 g	

the requirements for the reactor neutrino measurement and it is even cleaner than that for a solar neutrino observation. Thanks to this being the world's cleanest environment, accidental coincidence backgrounds are very low, 0.0086 ± 0.0006 events for a 2.6 MeV analysis threshold in a 162 ton-yr data sample and 1.81 ± 0.08 events for a 0.9 MeV threshold. The low energy accidental backgrounds are dominated by a combination of ^{210}Bi and ^{208}Tl. Cosmic ray muons are reduced by a factor 10^5 with 2700 mwe rock overburden, but this can cause correlated backgrounds and they are still the main source of backgrounds. One type of correlated backgrounds are fast neutrons from outside the detector. By tagging outer-detector-penetrating (OD) muons, the event rate of fast neutrons are measured. By considering the OD inefficiency and extrapolating contributions to rock-through-muons, total fast neutron backgrounds are estimated to be less than 0.5 events. Another type of correlated events is spallation products such as ^{8}He and ^{9}Li. They are beta-decay nuclei but they also emit neutrons at the same time and discrimination of them against the actual neutrino signal can be done only by looking at a space-time correlation of the parent muons. Considering a rejection efficiency of a spallation cut, the amounts of background from such spallation is estimated to be 0.94 ± 0.85 and 1.1 ± 1.0 events for the 2.6 and 0.9 MeV threshold, respectively. Finally, the amounts of total background are estimated to be 2.9 ± 1.1 and 1 ± 1 for the two thresholds.

There is also believed to be a neutrino background called "geo-neutrinos." Radioactivity con-

Table 2. Event selections.

(1)	fiducial cut	$R < 5m$, 3.46×10^{31} protons
(2)	timing correlation	$0.5 < \Delta T < 660\mu s$, $\tau = 212\ \mu s$
(3)	vertex correlation	$\|\vec{r}_{prompt} - \vec{r}_{delayed}\| < 1.6$ m
(4)	delayed energy	$1.8 < E < 2.6$ MeV
(5)	thermometer cut	$\sqrt{x^2 + y^2} > 1.2$ m
(6)	spallation cut	2 sec, all volume ($dQ > 10^6$ p.e.)
		2 sec, $L < 3$ m ($dQ < 10^6$ p.e.)
(7)	energy threshold	$E_{vis} > 2.6$ MeV

tained in the earth is thought to be the major source (20 TW) of observed heat flow (44 TW) on the surface of the earth, and 16 TW out of it is coming from the ^{238}U and ^{232}Th decay series. In the series of their decay, anti-electron-neutrinos are also emitted. "A model guess" predicts about 9 and 0.04 events for 0.9 and 2.6 MeV analysis thresholds. If these neutrinos are eventually observed, it would be the important start of "Neutrino geophysics", but for the moment, the 2.6 MeV analysis threshold is employed to eliminate an uncertainty of geo-neutrino flux.

Event selections used for the first results are listed in Table 2. The detection efficiency up to number (5) is 78.3% and the dead time from the spallation cut is 11.4%.

Energy calibrations have been performed by suspending radioactive sources along the z-axis. Also uniformly distributing spallation products such as neutron capture and ^{12}B were used to monitor uniformity and time variation of the energy scale. The

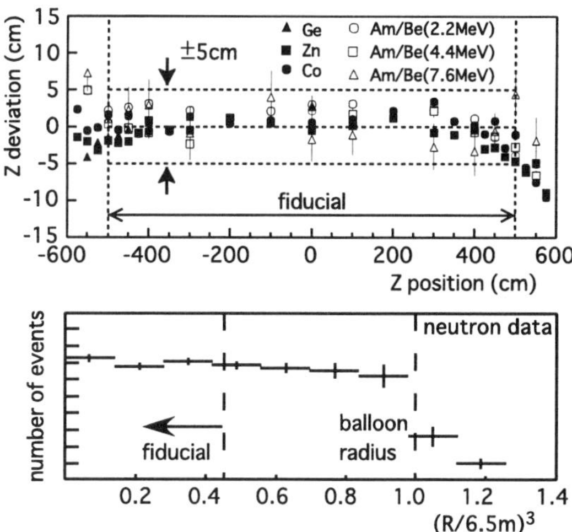

Figure 3. Vertex calibrations. The upper panel shows systematic biases of reconstructed positions along the z-axis. Biases are less than 5 cm within the fiducial region -500 cm to 500 cm for all energies. The lower panel is the R^3 distribution of uniformly distributed spallation neutron events. The uniform distribution appears as a flat distribution.

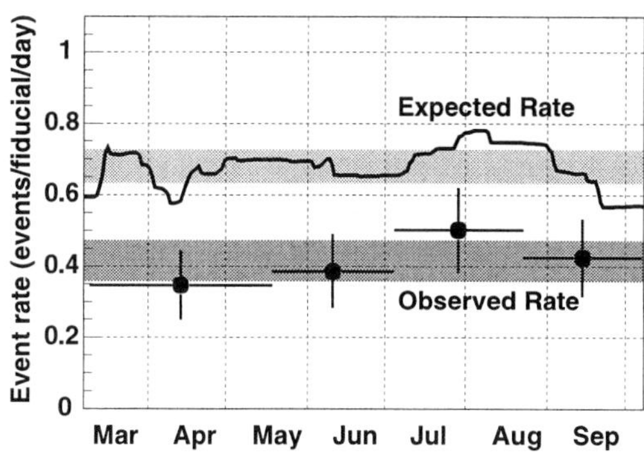

Figure 4. Reactor neutrino event rate. The plots and line are the observed and expected event rate, respectively, and the gray hatches are their averages. The structure in the expected rate reflects the change of reactor operations.

position dependence was smaller than 1.4% and the time variation was smaller than 0.6%. A quenching effect of the LS was also studied with alpha decays from ^{222}Rn, a wide range of gamma ray sources and beta decay of spallation products. The total uncertainty of the energy scale at the threshold energy 2.6 MeV is 1.91% which translates to a 2.1% systematic error in the neutrino event rate.

The largest systematic error comes from the determination of the fiducial volume. The vertex reconstruction was tuned with various radioactive sources along the z-axis and the systematic biases of vertices are smaller than 5 cm for the calibrated energy ranges from 1 MeV to 7.6 MeV as seen in the Fig. 3. A five cm bias corresponds to a 3% fiducial volume error. However, it is also calibrated by the vertex distribution of uniformly distributed spallation events. The ratio of number of events in the fiducial volume to the total volume was compared with the volume ratio. Distributions of neutron capture and ^{12}B events were consistent with the volume ratio at ±4.1% and ±3.5% accuracies, respectively. Those accuracies are dominated by the statistical uncertainty of the small number of spallation events. Currently, the fiducial volume error is conservatively estimated with the worst precision data. A breakdown of the systematic error is listed in Table 3.

From March 4 to October 6, 2002, KamLAND acquired 145.1 live days of data, equivalent to a 162 ton-yr exposure. Figure 4 shows the observed and the expected neutrino event rate above 2.6 MeV. The observed rate was always smaller than the expected. Combining the data from the entire period, the expected reactor neutrino signal is 86.8 ± 5.6 events and the estimated background rate is 1 ± 1 events. Only 54 events were observed. This shows clear evidence for reactor neutrino disappearance at the 99.95% confidence level. The ratio of observed to expected rate is:[9]

$$R = 0.611 \pm 0.085(stat) \pm 0.041(syst).$$

The expected signal for the 0.9 MeV threshold is 124.8±7.5 events and the background rate is 2.9±1.1 events without the uncertain geo-neutrino contribution of ~9 events. The observed number of signal events, 86, is consistent with neutrino disappearance. Figure 5 shows the observed ratios of reactor neutrino fluxes from KamLAND and previous experiments. The band shows the allowed range of the oscillatory pattern from various LMA parameters and the deficit at KamLAND is in good agreement with the LMA solution. The other oscillation solutions such as the SMA, LOW and VAC are shown as a dashed line and they are excluded at the 99.95% confidence level with the KamLAND result alone under the assumption of the CPT invariance. Similarly the other possibilities such as RSFP are also eliminated under the CPT hypothesis and the solar neutrino problem has been finally resolved as the LMA solution by this process of elimination.

Table 3. A breakdown of the KamLAND systematic error.

	0.9 MeV	2.6 MeV	
Thermal power	2.0	2.0	Japanese reactors contribute $\sim$90% of the neutrino flux.
Korean reactors	0.25	0.25	Only electric power is known but contribution is $\sim$2.5%
Other reactors	0.35	0.35	Contribution is only 0.7%
Burn-up effect	1.0	1.0	Fraction of ^{235}U/^{238}U/^{239}Pu/^{241}Pu
Long-life nuclei	0.5	0.002	Contribution of ^{106}Ru and ^{144}Ce
Time-lag of beta decay	0.3	0.3	< 1 day time lag for an equilibrium
Neutrino spectra	2.3	2.5	See the References[4,5]
Cross section	0.2	0.2	See the References[6]
Total LS mass	2.1	2.1	1171 ± 25 m^3
Fiducial volume ratio	4.1	4.1	Vertex distribution of spallation neutron
Energy threshold	-	2.1	Position 1.4%, time 0.6%, quench 1.02%, dark 0.4%
Efficiency of cuts	2.1	2.1	Capture time, space correlation, energy window
Live time	0.07	0.07	
Total	6.0	6.4	

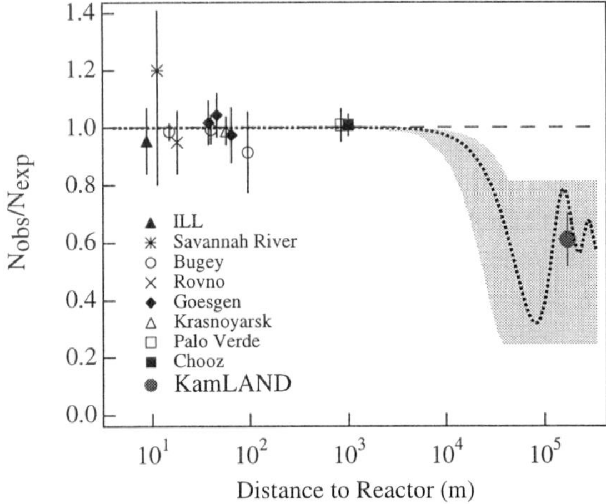

Figure 5. Observed neutrino rate at various baselines. The bands are for the various LMA parameters and the dotted line is for $(\sin^2 2\theta, \Delta m^2) = (0.833, 5.5 \times 10^{-5}$ eV$^2)$. The dashed lines are for no-oscillation, the SMA, LOW and VAC solutions.

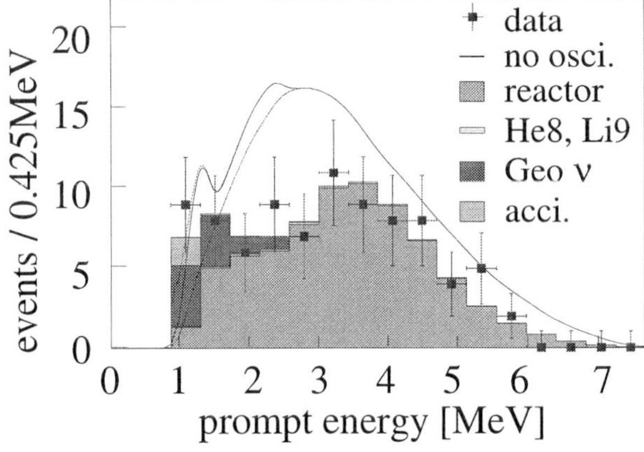

Figure 6. Energy spectrum at KamLAND. The curves are for no oscillation and the histograms are for the best-fit oscillation spectrum and the various background estimations.

The observed energy spectrum and the background estimations are shown in Fig. 6 with the best-fit oscillation spectrum at $(\sin^2 2\theta, \Delta m^2) = (1.0, 6.9 \times 10^{-5}$ eV$^2)$. If a 0.9 MeV threshold is used and treating the number of geo-neutrinos as a free parameter, the best fit for $\sin^2 2\theta$ becomes 0.91. The best-fit parameters for the number of geo-neutrinos are 4 for ^{238}U and 5 for ^{232}Th. However, they are consistent with 0 at 95% confidence level and it is still just a hint of geo-neutrinos. Using the spectrum shape information with the rate, the allowed oscillation parameter region has been obtained as shown in Fig. 7. The two bands overlap with the LMA solution from the solar neutrino experiments.

In order to claim neutrino oscillation with KamLAND alone, the oscillatory pattern should be found in the energy spectrum or distance dependence. The current spectrum is still consistent with constant suppression and thus the result is not evidence for neutrino oscillation, yet. In the future with more statistics, KamLAND can pin-point the Δm^2 with a greater precision (better than 5%). But, a precision measurement of $\sin^2 2\theta$ requires another measurement such as low energy solar neutrino observations or a precise measurement of NC/CC ratio at the SNO experiment. The new Shika reactor planned to start in 2006 may provide the first observation of an oscillation dip if LMA1 ($\Delta m^2 \sim 7 \times 10^{-5}$ eV^2) is the true solution. Three years of the new Shika ON data minus three years of OFF data (see Fig. 8)

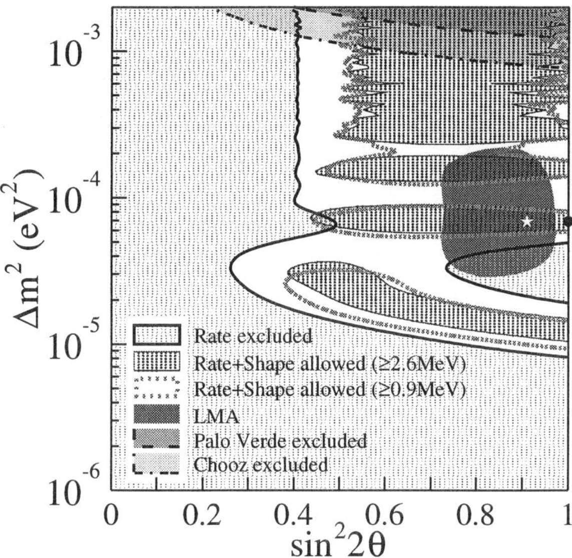

Figure 7. Allowed region of oscillation parameters. Region obtained from both 2.6 and 0.9 MeV analysis are shown.

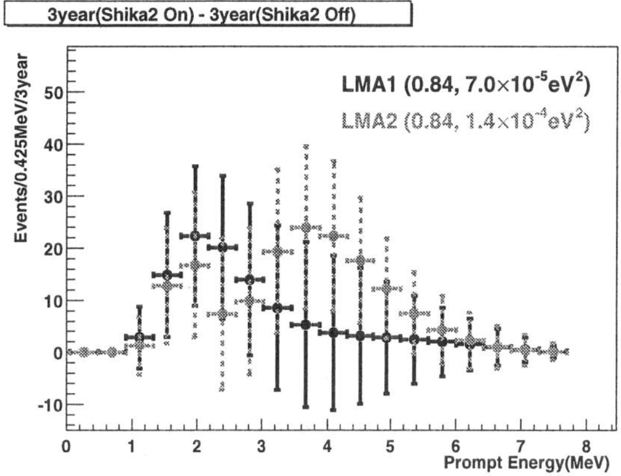

Figure 8. Expected spectrum from three years new Shika ON minus three years OFF with 5.5 m radius fiducial volume. Solids are for LMA1 $(0.84, 7 \times 10^{-5}\text{eV}^2)$ and dots are for the LMA2 $(0.84, 1.4 \times 10^{-4}\text{eV}^2)$ parameters.

will provide a new baseline measurement at 88 km. And the expected ratios from the LMA1 and LMA2 $(\Delta m^2 \sim 1.4 \times 10^{-4} \text{ eV}^2)$ are $\frac{121\pm36}{173} \sim 0.70$ and $\frac{45\pm37}{173} \sim 0.26$, respectively.

3. θ_{13} Experiments

Neutrino mixing is commonly parameterized by the Maki-Nakagawa-Sakata matrix. The matrix contains 6 parameters (3 mixing, one Dirac CP phase and two Majorana CP phases) and 4 out of 6 (3 mixing and Dirac CP) are accessible with neutrino oscillation studies. The angle θ_{23} is measured by atmospheric neutrinos and a long baseline accelerator experiment to be almost maximal ($\theta_{23} \sim \pi/4$). The angle θ_{12} is measured by solar neutrino experiments and the KamLAND reactor neutrino experiment and it is known to be large ($\theta_{12} \sim \pi/6$). Only the third angle θ_{13} is unmeasured yet. The Dirac CP phase appears with θ_{13} and the size of the angle determines an accessibility to the CP phase in future experiments.

Accelerator experiments are aiming to observe a finite value of θ_{13} via an appearance reaction of $\nu_\mu \to \nu_e$. Its probability is related to the CP phase, the sign of $\theta_{23} - \pi/4$ and the mass hierarchy (the sign of Δm_{31}^2). This means there is a potential to measure those variables, but on the other hand, the θ_{13} measurement is affected by them. This is known as the "parameter degeneracy".

The reactor neutrino experiments observe disappearance of anti-electron-neutrinos and the probability doesn't depend on the CP phase nor the sign of $\theta_{23} - \pi/4$ and Δm_{31}^2. The neutrino survival probability for reactor neutrino experiments, for the normal hierarchy case, is given by the following expression.

$$\begin{aligned}
P = 1 &- 4s_{13}^2 c_{13}^2 \sin^2 \Delta_{31} \\
&- 4c_{13}^4 s_{12}^2 c_{12}^2 \sin^2 \Delta_{21} \\
&+ 2s_{13}^2 c_{13}^2 s_{12}^2 [\cos(2\Delta_{31} - 2\Delta_{21}) - \cos 2\Delta_{31}] \quad (1)
\end{aligned}$$

For the inverted hierarchy case, s_{12}^2 in the last term is replaced with c_{12}^2. The first and second terms are relevant to θ_{13} and θ_{12} measurements and valid at distances of order 1 km and 100 km, respectively. The third term is small but related to the mass hierarchy and is relevant for distances of the order of a few tens of km. Thus, the reactor experiment can be a pure θ_{13} measurement when a proper baseline (order 1 km) is chosen. Experiments with this baseline have already been performed by CHOOZ and Palo Verde, and the current best limit is obtained from the CHOOZ experiment.[10] Looking at the CHOOZ experiment, the systematic error is dominated by the understanding of reactor neutrinos (1.9% from reactor cross section, 0.7% from reactor power and 0.6% from energy release/fission out of a 2.9% total systematic error). In order to improve the situation, a Near/Far detector system is necessary to cancel these uncertainties. On the other hand, the Palo Verde experiment[11] is dominated by a background estimation error (3.3% out of a 5.3% total error). This is because the Palo Verde site is shallow (32 mwe) and the S/N ratio is only 0.5 ~ 1. This implies that the site should be as deep as or deeper than the CHOOZ (300 mwe).

Several experiments are proposed to focus on the cancellation of the systematic error with a Near/Far system and reduction of backgrounds by going deeper. The first proposal was from Krasnoyarsk.[12] The site is 600 mwe underground and two 46-ton detectors are located at distances of 115 m and 1000 m from a 1.6 GW reactor. The design value of their systematic error is 0.8% for a rate analysis and 0.5% for a shape analysis. Using a shape analysis, the expected sensitivity is one order of magnitude better than the CHOOZ result.

Another proposal is from Kashiwazaki.[13] Kashiwazaki is the most powerful reactor complex (24.3 GW) in the world. It consists of two clusters of reactor cores. One has 4 cores and another has 3 cores. It plans to have two near detectors for two clusters at 300 ~ 350 m and one far detector at a distance of about 1300 m. The overburdens are 200 mwe for the near detectors and 400 mwe for the far detector, requiring digging 6 m diameter shafts for each detector. The size of the detector is 8.5 tons. The design value of the systematic error is 0.5 ~ 1% and the expected sensitivity to $\sin^2 2\theta$ is about a factor 7-10 better than the CHOOZ result ($0.15 \to 0.016 \sim 0.025\ @\Delta m^2 = 2.6 \times 10^{-3} \text{eV}^2$) with 2 years of data taking as shown in Fig. 9. The site use is already permitted and a possible fastest schedule is for data-taking to start in 2007.

The US activities can be found in the References.[14] Site selection is under way and one candidate is "Diablo Canyon Nuclear Power Plant." The key point of their study is a movable far detector to calibrate near/far detectors head to head. The planned start time is the year 2008.

Activities in Europe can be found in the References.[15] Site selection is under way. One different type of experiment is also proposed.[16] It uses a 20-30 km baseline with a big detector (112 tons). An outstanding feature of this experiment appears if the LMA2 is the right solution, $\Delta m^2_{31} < 2.5 \times 10^{-3} eV^2$ and $\sin^2\theta_{13} > 0.03$. In such a case, the experiment can determine the mass hierarchy with a 125 GW-kt-yr data set.

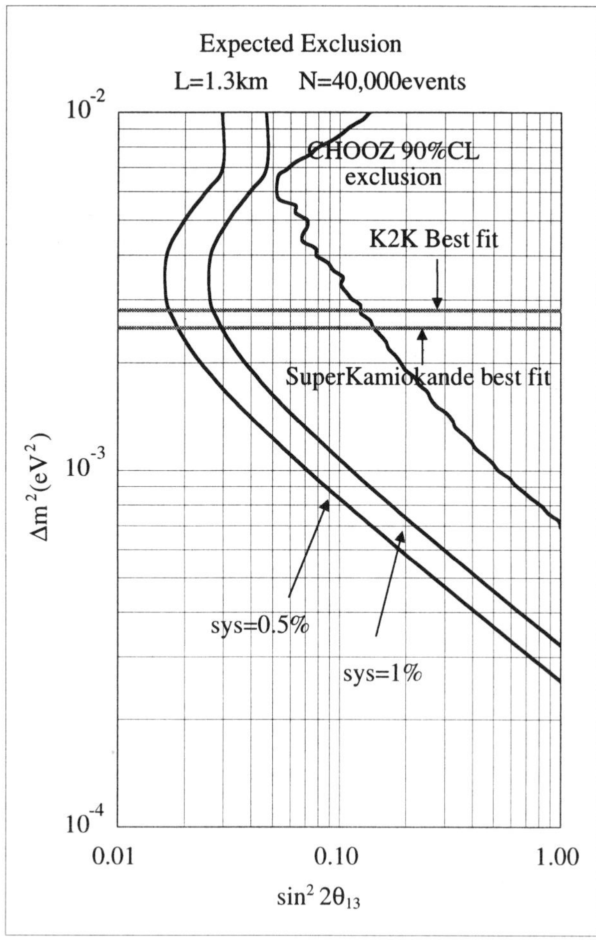

Figure 9. Expected sensitivity of Kashiwazaki experiment with 2 years of data.

4. Search for Neutrino Magnetic Moment

If neutrinos have magnetic moments, their elastic scattering cross section with electrons can be written as follows:

$$\frac{d\sigma}{dT} = \frac{G_F^2 m_e}{2\pi}[(g_V + g_A)^2$$
$$+ (g_V - g_A)^2(1 - \frac{T}{E_\nu})^2 + (g_A^2 - g_V^2)\frac{m_e T}{E_\nu^2}]$$
$$+ \frac{\pi\alpha^2\mu_\nu^2}{m_e^2}\frac{1 - T/E_\nu}{T}. \qquad (2)$$

The last term comes from the magnetic moment interaction. The effect of the magnetic moment becomes larger as the recoil energy goes lower. In order to investigate the neutrino magnetic moment, experiments looked for an excess in the low energy recoil. Figure 10 shows the expected differential cross section for the Standard Model weak interaction and for a magnetic moment interaction in the case of

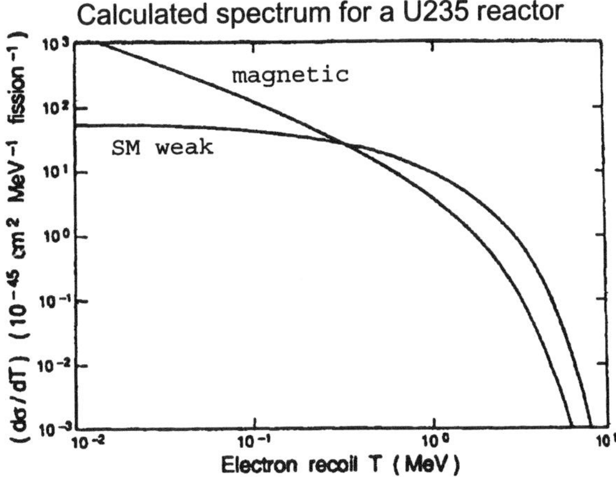

Figure 10. Differential cross section for the Standard Model weak interaction and a magnetic moment interaction with $\mu_\nu = 10^{-10}\mu_B$.

$\mu_\nu = 10^{-10}\mu_B$.

The TEXONO experiment in Taiwan measured the recoil energy spectrum above 5 keV at 28 m from one of the Kuo-Sheng reactors (2 × 2.9 GW). They used an ultra-low-background high-purity Germanium detector (1.06 kg) with NaI(Tl) and CsI(Tl) anti-Compton detectors. By fitting reactor ON spectra to $\phi_{OFF} + \phi_{SM} + \kappa^2 \phi_{MM}[10^{-10}\mu_B]$, they obtained $\kappa^2 = -0.4 \pm 1.3(stat) \pm 0.4(sys)$ and translated the result to a limit on the neutrino magnetic moment of less than $1.3 \times 10^{-10}\mu_B$ at the 90% confidence level.[17] The sensitivity of the experiment is limited by an uncertainty of the fission neutrino spectra below 1.8 MeV where various long-lived nuclei become significant and no precise estimation is available now.

The MuNu experiment[18] set their analysis threshold at 900 keV which corresponds to a well-understood neutrino energy of $E_\nu > 1.8$ MeV. The effect of the magnetic moment is smaller at this energy and the experiment must have high statistics and be very precise to see the effect. The MuNu experiment uses a 2.75 GW reactor in Bugey and a CF$_4$ time projection chamber with target mass of 11.4 kg. The experiment can measure the recoil direction and a subtraction of uniform background from the forward direction signal (uniform BG + signal) can be done. They obtained a limit of $\mu_\nu < 1.0 \times 10^{-10}\mu_B$ at the 90% confidence level from 66.6 days of reactor ON data.

5. Summary

KamLAND has observed evidence for reactor neutrino disappearance at a distance of ∼180 km at the 99.95% confidence level. Assuming CPT invariance, the result is only compatible with the LMA solution of the neutrino oscillation hypothesis. In a process of elimination, the long-standing solar neutrino problem has been finally solved.

The last unmeasured mixing angle can be explored with reactor experiments down to $\sin^2 2\theta_{13} \sim 0.02$ which is comparable to accelerator long baseline experiments. Reactor experiments are relatively low cost and quicker and are complementary to accelerator experiments when solving the parameter degeneracy. Various possibilities are being discussed and international collaborations are being formed.

Direct searches for the neutrino magnetic moment are extensively performed with reactor neutrinos and the best limit, so far, is $\mu_\nu < 1.0 \times \mu_B$ from MuNu experiment at the 90% confidence level.

References

1. M. F. James, *J. Nucl. Energy* **23**, 517 (1969).
2. V. I .Kopeikin *et al.*, Physics of Atomic Nuclei, **64-5**, 849 (2001).
3. M. Mampe *et al.*, *Nucl. Instrum. Methods* **154**, 127 (1978).
4. K. Schreckenbach *et al.*, *Phys. Lett.* B **160**, 325 (1985); A. A. Hahn *et al.*, *Phys. Lett.* B **218**, 365 (1989).
5. P. Vogel et al., *Phys. Rev.* C **24**, 1543 (1981).
6. P. Vogel and J. F. Beacom, *Phys. Rev.* D **60**, 053003 (1999); A. Kurylov *et al.*, *Phys. Rev.* C **67**, 035502 (2003).
7. Y. Declais *et al.*, *Phys. Lett.* B **338**, 383 (1994).
8. B. Achkar *et al.*, *Phys. Lett.* B **374**, 243 (1996).
9. The KamLAND Collaboration, *Phys. Rev. Lett.* **90**, 021802 (2003).
10. M. Apollonio *et al.*, *Eur. Phys. J.* C **27**, 331 (2003).
11. F. Boehm *et al.*, *Phys. Rev.* D **61**, 112001 (2001).
12. V. Martemyanov *et al.*, hep-ex/02111070.
13. H. Minakata *et al.*, hep-ph/0211111; F. Suekane *et al.*, hep-ex/0306029.
14. For example: http://home.fnal.gov/~link/theta_13/ and http://kmheeger.lbl.gov/theta13/.
15. For example, http://bama.ua.edu/~busenitz/ rnu2003_talks/lasserre3.ppt.
16. S. Schonert *et al.*, hep-ex/0203013; S.Choubey *et al.*, hep-ph/0306017.
17. H. B. Li *et al.*, *Phys. Rev. Lett.* **90**, 131802 (2003).
18. The MuNu collaboration, *Phys. Lett.* B **564**, 190 (2003).

DISCUSSION

Carlos Wagner (Argonne): You mentioned the bound on θ_{13} of 10 degrees. Does the bound include the new Super-Kamiokande result?

Kunio Inoue: No, it does not include the new result.

Hugh Montgomery (Fermilab): This is about the third talk in the last 6 months where the picture is being painted that the next reactor experiment can be built rather quickly. Would you like to comment on the associated running time to get to results.

Kunio Inoue: For Kashiwazaki it is about 2 years. In 2 years, they can accumulate about 40,000 neutrinos in the far detector that corresponds to a 0.5% statistical error and that is good enough to improve the θ_{13} region down to 0.02.

Hugh Montgomery (Fermilab): So that's down to $\sin^2 2\theta$ of 0.02?

Kunio Inoue: Yes.

Luc Declais (IPNL, France): When you compare in an experiment with two or three detector positions like in Japan, you still have to compare detectors that need to be identical but cannot be completely identical. So, it is not as easy to reduce the systematics as it has been quoted in some papers. For example, what is the energy calibration difference needed between two detectors in order to avoid wriggles when you compare spectra?

Kunio Inoue: For the energy calibration you don't set an energy threshold, so it's just the reaction rate and there's no error from the energy calibration. We set the threshold at 1 MeV so we take all data. And for the fiducial volume errors, we don't set a fiducial volume, actually, it is defined by the volume of the scintillating detector. So gadolinium is loaded only in the inner detector and all events that come from that gadolinium are measured, but of course we can't measure like in a far- and near-detector head-to-head comparison, so the largest error comes from their difference and it is about 0.5% as quoted in the proposal.

Luc Declais (IPNL, France): But this is only related to the total number of events. But in order to achieve such low sensitivity for θ_{13}, you need to compare the shape of the energy spectrum you measure, to do so you need to have a very good comparison between the energy calibrations of the detector to another detector.

Kunio Inoue: That's right. In the Krasnoyarsk case, they are going to use the shape but in the Kashiwazaki case, it assumes only systematic errors and if we can use the energy spectrum it will improve. Currently it's probably better to consider this to be a 1% systematic error for a conservative case.

Louis William (LANL): What is the possibility that KamLAND will be able to make a solar neutrino measurement?

Kunio Inoue: Currently we are putting the biggest effort in reducing Krypton-85 and Lead-210. We can reduce Krypton by bubbling it, it is very easy. We know Lead-210 can be reduced by distillation, but it costs a lot. We're still looking at water extraction methods to remove Lead-210. And we think we can eventually detect solar neutrinos.

REVIEW OF SOLAR NEUTRINO EXPERIMENTS

A. BELLERIVE

Ottawa-Carleton Institute for Physics, Department of Physics,
Carleton University, 1125 Colonel By Drive, Ottawa, K1S 5B6, Canada
E-mail: alain_bellerive@carleton.ca

This paper reviews the constraints on the solar neutrino mixing parameters with data collected by the Homestake, SAGE, GALLEX, Kamiokande, SuperKamiokande, and SNO experiments. An emphasis will be given to the global solar neutrino analyses in terms of matter-enhanced oscillation of two active flavors. The results to-date, including both solar model dependent and independent measurements, indicate that electron neutrinos are changing to other active types on route to the Earth from the Sun. The total flux of solar neutrinos is found to be in very good agreement with solar model calculations. Future measurements will focus on greater accuracy for mixing parameters and on better sensitivity to low neutrino energies.

1. Introduction

The deficit of neutrinos detected coming from the Sun compared with our expectations based on laboratory measurements, known as the Solar Neutrino Problem, has remained one of the outstanding problems in basic physics for over thirty years. It appeared inescapable that either our understanding of the energy producing processes in the Sun is seriously defective, or neutrinos, some of the fundamental particles in the Standard Model, have important properties which have yet to be identified. It was indeed argued by some that we needed to change our ideas on how energy was produced in fusion reactions inside the Sun. Others suggested that the problem arose due to peculiar characteristics of neutrinos such as oscillations and matter effects. It is then useful to review the evolution of our understanding from the data collected by various solar neutrino experiments. For completeness, new results from the Sudbury Neutrino Observatory not presented at the conference are included in the discussion presented here.

2. Solar Neutrinos

Hans Bethe suggested that the energy in the Sun was produced by nuclear reactions that allow hydrogen to be transformed into helium.[1] This nuclear time scale is about 10^{10} years and corresponds to the epoch over which a star evolves in the main sequence. The Sun evolves slowly by adjusting its temperature so that the average thermal energy of a nucleus is small compared to the Coulomb repulsion an ion feels from potential fusion partners. The large Coulomb repulsion slows the nuclear rate to an astronomically long time scale. Hence the rate for nuclear reactions in the solar interior is dominated by Coulomb barriers. Through the reactions listed in Table 1, four protons combine to form the helium nucleus containing two protons and two neutrons. As the protons only fuse to make helium under the very high density, high temperature conditions present at the centre of the Sun, it is virtually impossible to reproduce these conditions in the laboratory and hence study directly the nuclear solar fusion hypothesis. Considering also that it takes a photon about ten thousand years to reach the surface of the Sun from the core, investigation of the electromagnetic spectrum considering the latter time scale dilutes all clues about the origin and production mechanism of the photons.

However, somehow in the fusion process, two of the protons have to become neutrons. The only reactions that allow this to happen are caused by weak interactions responsible for nuclear beta decay and, each time a neutron is formed, there must be an associated electron neutrino produced. Neutrinos can travel directly from the core of the Sun to the Earth

Table 1. Neutrino production from fusion reactions in the Sun.[2] The total solar flux at the Earth is 6.5×10^{10} neutrinos per cm^2 and per second. The majority of the solar neutrinos come from the pp chain (more than 91%); while the ^7Be, pep, and ^8B chains correspond to about 7%, 0.2%, and 0.008% of the total flux, respectively. The hep contribution is minuscule and mostly neglected.

Reaction	Label	Flux (cm^{-2} s^{-1})
$p + p \rightarrow {}^2H + e^+ + \nu_e$	pp	5.95×10^{10}
$p + e^- + p \rightarrow {}^2H + \nu_e$	pep	1.40×10^8
$^3He + p \rightarrow {}^4He + e^+ + \nu_e$	hep	9.3×10^3
$^7Be + e^- \rightarrow {}^7Li + \nu_e$	^7Be	4.77×10^9
$^8B \rightarrow {}^8Be^* + e^+ + \nu_e$	^8B	5.05×10^6

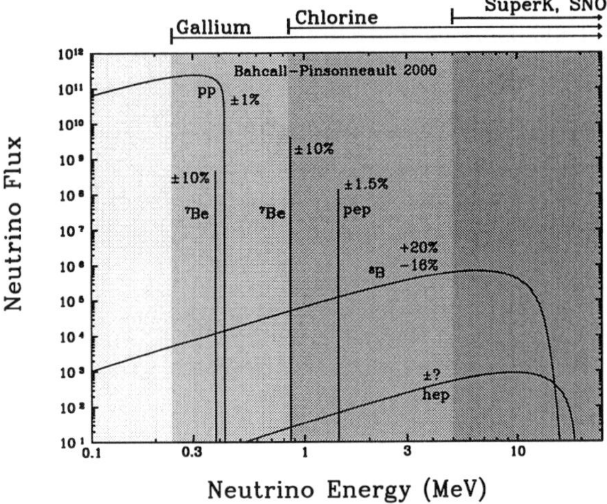

Figure 1. The solar neutrino spectra predicted by the SSM. The neutrino fluxes at one astronomical unit from continuum sources are given in units of $\mathrm{cm}^{-2}\,\mathrm{s}^{-1}\,\mathrm{MeV}^{-1}$, and the line fluxes are given in $\mathrm{cm}^{-2}\,\mathrm{s}^{-1}$. Courtesy of J.N. Bahcall from http://www.sns.ias.edu/~jnb/.

in a few minutes and hence provide a direct way to study processes by which protons form helium in the Sun. As early as 1949 Luis Alvarez proposed that the hypothesis of nuclear reactions powering the Sun could be tested by measuring the solar neutrino flux.

The number of neutrinos that we should expect is equal to twice the ratio of the energy received at the Earth in the form of sunshine, to the energy released when the four protons produce helium. The resulting number is huge: almost 10^{11} neutrinos pass through each square centimeter on Earth every second. Even with such numbers the detection proved a formidable challenge because of the very small scattering cross section of neutrinos on ordinary matter. The attraction of the measurement is nevertheless clear!

The detailed prediction of the electron neutrino flux created by the thermonuclear reactions in the interior of the Sun was performed by John Bahcall and his collaborators from the 1960's until now. Their calculations are referred to as the Standard Solar Model (SSM). In this paper, the Bahcall-Pinsonneault calculations[2] are used to compare experimental results and theoretical predictions. The solar neutrino spectra predicted by the SSM are shown in Fig. 1.

3. Neutrino Oscillations

It is known that neutrinos exist in different flavors corresponding to the three charged leptons: the electron, muon, and tau particles. If neutrinos have masses, flavor can mix in a charged-current interaction mediated by the W boson. The neutrino emitted in a weak interaction is then a superposition of mass eigenstates

$$\nu_\ell = \sum_{i=1}^{n} U_{\ell i} |\nu_i\rangle . \tag{1}$$

The charged-current interactions in the leptonic sector are then described by the mixing matrix U

$$U = \begin{pmatrix} U_{e1} & U_{e2} & \cdots & U_{en} \\ U_{\mu 1} & U_{\mu 2} & \cdots & U_{\mu n} \\ U_{\tau 1} & U_{\tau 2} & \cdots & U_{\tau n} \end{pmatrix} . \tag{2}$$

Here the neutrino mass eigenstates are denoted by ν_i with $i = 1, 2, \cdots, n$, while the charged lepton flavor eigenstates are labeled (e, μ, τ). In the case of three generations of neutrino, the matrix U is called the Maki-Nakagawa-Sakata-Pontecorvo (MNSP) matrix[3] and appears analogous to the Cabibbo-Kobayashi-Maskawa (CKM) mixing matrix[4] in the quark sector. The MNSP can be factorized as

$$U = U_{12} \times U_{23} \times U_{13} , \tag{3}$$

with

$$U_{12} = \begin{pmatrix} c_{12} & s_{12} & 0 \\ -s_{12} & c_{12} & 0 \\ 0 & 0 & 1 \end{pmatrix} , \tag{4}$$

$$U_{23} = \begin{pmatrix} 1 & 0 & 0 \\ 0 & c_{23} & s_{23} \\ 0 & -s_{23} & c_{23} \end{pmatrix} , \tag{5}$$

$$U_{13} = \begin{pmatrix} c_{13} & 0 & s_{13}e^{i\delta} \\ 0 & 1 & 0 \\ -s_{13}e^{-i\delta} & 0 & c_{13} \end{pmatrix} , \tag{6}$$

where $c_{ij} = \cos\theta_{ij}$, $s_{ij} = \sin\theta_{ij}$, and i, j denote the lepton generations. Possible CP-violation is naturally embedded in the phase δ.

The leptonic mixing matrix naturally allows for flavor oscillations of the neutrinos. The most general form for solar neutrino oscillations can be simplified where only two neutrinos participate in the oscillation. The large neutrino flavor mixing between the second and third generation inferred from atmospheric neutrino data[5] in conjunction with the absence of an oscillation signal in the CHOOZ reactor neutrino experiment[6] requires a small component of one of the three mass eigenstates to the electron flavor eigenstate. Hence the survival probability for an electron neutrino to propagate in time can take the approximate form

$$P_{e\alpha} = \delta_{e\alpha} - (2\delta_{e\alpha} - 1) \sin^2 2\theta \sin^2 (1.27 \frac{\Delta m^2 L}{E}). \quad (7)$$

The mixing angle is represented by θ, L is the distance between the production point of ν_e and the point of detection of ν_α, E is the energy of the neutrino, and $\Delta m^2 \equiv m_2^2 - m_1^2$ is the difference in the squares of the masses of the two states ν_2 and ν_1 which are mixing. The function $\delta_{e\alpha}$ is the usual Kronecker delta. The numerical constant 1.27 is valid for L in meters, E in MeV, and Δm^2 in eV^2. Consequently, the electron neutrino of energy E produced in a weak interaction inside the Sun can then be described in vacuum by the Hamiltonian

$$H = \begin{pmatrix} -\frac{\Delta m^2}{4E} \cos 2\theta & \frac{\Delta m^2}{4E} \sin 2\theta \\ \frac{\Delta m^2}{4E} \sin 2\theta & \frac{\Delta m^2}{4E} \cos 2\theta \end{pmatrix}, \quad (8)$$

such that

$$i \frac{d}{dt} \begin{pmatrix} \nu_e \\ \nu_\beta \end{pmatrix} = \frac{1}{2} H \begin{pmatrix} \nu_e \\ \nu_\beta \end{pmatrix}. \quad (9)$$

The notation $\beta = \mu\tau$ is often used to represent the other type of flavor present in the solar flux. In the context of the Solar Neutrino Problem, this suggests the investigation of (i) the disappearance of pure electron neutrinos produced in nuclear reactions when they reach the Earth, or (ii) the appearance of neutrinos of another flavor in the solar beam. The Earth-Sun distance is set by the planetary equations of Kepler. The energy of the neutrino depends on the type of nuclear reaction (c.f. Table 1) which produced the electron neutrino. By studying the time evolution of the solar neutrinos, all the physics is then embedded in the parameters θ and Δm^2. The full parameter space is covered with $\Delta m^2 \geq 0$ and $0 \leq \theta \leq \frac{\pi}{2}$, or $0 \leq \theta \leq \frac{\pi}{4}$ and either sign for Δm^2. Note that the survival probability of Eq. (7) is invariant under the transformations $\Delta m^2 \to -\Delta m^2$ and $\theta \to \frac{\pi}{2} - \theta$. These transformations redefine the mass eigenstates by $\nu_1 \leftrightarrow \nu_2$. This situation implies that there is a two-fold discrete ambiguity in the interpretation of $P_{e\alpha}$ in the two-neutrino oscillation scheme: the two different sets of physical parameters $(\Delta m^2, \theta)$ and $(\Delta m^2, \frac{\pi}{2} - \theta)$ give the same transition probability in vacuum.

The observation of vacuum oscillations does not determine whether ν_1 or ν_2 is heavier. Hence, the measurement of $P_{e\alpha}$ in vacuum cannot differentiate whether the larger component of ν_e resides in the heavier or in the lighter neutrino mass eigenstate. Thus, a solution to the Solar Neutrino Problem implies the determination of one angle $\theta \equiv \theta_{12}$, one mass difference $\Delta m^2 \equiv \Delta m_{12}^2$, and the sign of Δm^2. This corresponds to the extraction of the three MNSP elements: U_{e1}, U_{e2}, and U_{e3}.

4. Matter Effects

As was realized by Mikheev and Smirnov,[7] based on the formalism of Wolfenstein,[8] the two-fold symmetry is lost when mixed neutrinos travel through regions of dense matter. In the Sun (or the Earth), neutrinos can undergo forward scattering with the particles in the medium. These interactions are, in general, flavor dependent. This possible modification of the oscillation pattern in matter is referred to as the Mikheev-Smirnov-Wolfenstein (MSW) effect. When neutrinos propagate through matter the Hamiltonian must be modified to include scattering of the electrons in the matter, both through neutral-current interactions (which affect all types of neutrinos equally and lead to no observable changes in the oscillation pattern) and through charged-current interactions, which at solar neutrino energies affect only the electron neutrinos.

The conditions that have to be satisfied for the MSW effect to occur involve the neutrino energy, E, and the local electron density, N_e. As the process is a resonant one, the effect can be strongly energy dependent. The evolution equation is still given by Eq. (9), but

$$H = \begin{pmatrix} -\frac{\Delta m^2}{4E} \cos 2\theta + \sqrt{2} G_F N_e & \frac{\Delta m^2}{4E} \sin 2\theta \\ \frac{\Delta m^2}{4E} \sin 2\theta & \frac{\Delta m^2}{4E} \cos 2\theta \end{pmatrix}, \quad (10)$$

where G_F is the Fermi coupling constant, while θ and Δm^2 are the usual mixing parameters. In the MSW oscillation scheme, the matter mixing angle term, $\sin^2 2\theta_m$, is related to the vacuum mixing angle, $\sin^2 2\theta$, by:

$$\sin^2 2\theta_m = \frac{\sin^2 2\theta}{(w - \sin^2 2\theta)^2 + \sin^2 2\theta}, \qquad (11)$$

with

$$w = -\sqrt{2}\, G_F N_e E / \Delta m^2 . \qquad (12)$$

Thus, neutrinos created as electron-type in the centre of the Sun could emerge from the solar surface as mixed electron, muon, and tau neutrinos. The model also implies that, under certain conditions, muon or tau neutrinos striking the Earth could turn back into electron neutrinos leading to the intriguing possibility that the Sun might appear brighter to neutrino detectors at night than during the day.

5. Chlorine Experiment

The exploration of solar neutrinos started in the mid-1960's with Ray Davis.[9] It led to the first experiment that successfully detected neutrinos coming from the Sun. The experiment of Davis and his team was carried out deep underground in the Homestake mine in the US. The detector was based on a concept first proposed by Bruno Pontecorvo at Chalk River in 1946, in which neutrino reactions on chlorine are measured. Neutrinos striking chlorine can make an isotope of argon through the reaction

$$\nu_e + {}^{37}Cl \rightarrow e^- + {}^{37}Ar ,$$

with an energy threshold of 0.814 MeV. This reaction is rare and does not happen very often. In fact, about one atom of argon is produced each week in a tank containing 100,000 gallons of the dry-cleaning fluid, perchlorethylene. The challenge of this radiochemical experiment is to extract the few atoms of argon and count them by noting their decay back to chlorine which occurs with a half-life of 35 days. To carry out the low atom counting, the tank of dry-cleaning fluid is left for about a month and then purged with helium gas to sweep out the two or three atoms of argon. These atoms must then be separated from the helium by freezing them in a cold trap. They are then transferred to a low background counter where any decays are recorded over a period of several months. The first results were announced in 1968. The measurement clearly showed argon atoms produced by neutrinos, but the number was only one quarter of

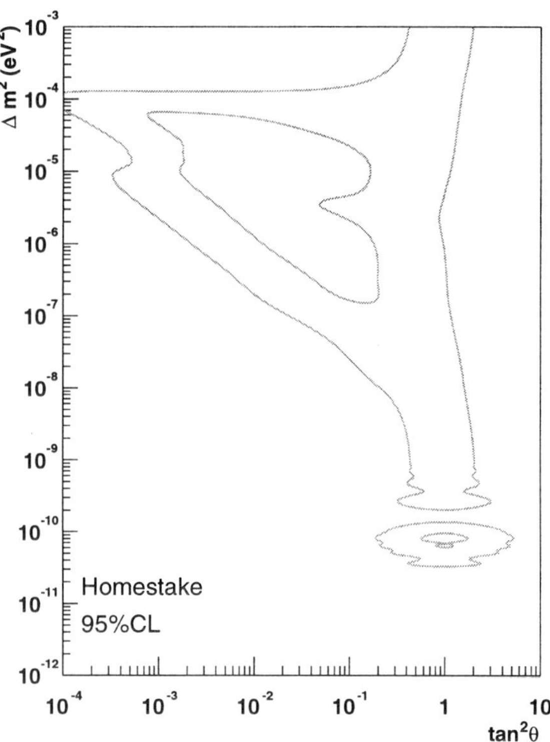

Figure 2. Mixing plane constraint provided by the chlorine experiment. The inside of the covariance regions is allowed at the 95% CL.

the predicted production rate. The chlorine experiment took data until 1995 and yielded[10]

$$\Phi_{Cl} = 2.56 \pm 0.16 \pm 0.16 \text{ SNU} , \qquad (13)$$

while the SSM predicted rate was

$$\Phi_{Cl}(SSM) = 7.6 \,^{+1.3}_{-1.1} \text{ SNU} . \qquad (14)$$

A SNU (Solar Neutrino Unit) is the product of the solar neutrino fluxes (measured or calculated) and the calculated cross sections. Hence one SNU equals one capture per second and per 10^{36} target atoms. The constraint on the oscillation parameters $\Delta m^2 - \tan^2 \theta$ from the Homestake experiment is shown in Fig. 2. One uses $\tan^2 \theta$ instead of $\sin^2 2\theta$ so that $\tan^2 \theta < 1$ corresponds to $m_2 > m_1$ (normal hierarchy) and $\tan^2 \theta > 1$ to $m_2 < m_1$ (inverted hierarchy). The allowed region is obtained by comparing the measured and the calculated SSM solar neutrino fluxes (see Table 1). Ray Davis was awarded the 2002 Nobel Prize in physics for his pioneering work which provided the first evidence that the electron neutrino flux at the Earth, created by the thermonuclear reactions that power the Sun, is substantially less than would be predicted by the SSM.

6. Gallium Experiments

While the chlorine detector was mainly sensitive to the highest energy neutrinos (c.f. Fig. 1), two gallium experiments, one at the Baksan laboratory[11] in Russia and one at the Gran Sasso laboratory[12] in Italy, were set up to test the oscillation hypothesis at lower energy. The highest energy neutrinos and the production rate for these depends strongly on the solar central temperature; so a small change in the conditions was argued to give large changes in the predicted rates. On the other hand, the lower energy neutrino flux is expected to follow directly from the solar luminosity as discussed above. The motivation of the gallium experiments was then to disentangle which MSW neutrino oscillation scenario causes the Solar Neutrino Problem.

Like the ^{37}Cl detector, the gallium detectors could only detect electron type neutrinos because they looked for the reaction

$$\nu_e + {}^{71}Ga \rightarrow e^- + {}^{71}Ge .$$

The energy threshold of the ^{71}Ga detectors is 0.233 MeV and hence allows the interaction of pp, ^7Be, ^8B, and pep neutrinos. The Russian-American group (SAGE) used a liquid metal target which contained 50 tons of gallium; while the European group (GALLEX/GNO) used 30 tons of natural gallium in an aqueous acid solution. Small proportional counters are used to count the germanium from the radiochemical target. The ^{71}Ge electron capture decay occurs with a half-life of 16.5 days. The Auger electrons and X-rays produce the typical L-peak and K-peak energy distribution. As a cross-check, both peaks are counted separately. A calibration with strong ^{51}Cr neutrino sources provides a nice verification for low atom chemical extraction and counting techniques. Improvements like new electronics, better radiation shielding, and improved calibration of counters are being pursued to continue regular data runs.

Both experiments found about half of the expected rate. The most recent results of SAGE[13] were presented at LowNu 2003 and yield for the period 1990-2003:

$$\Phi_{Ga} = 64.5\,^{+6.8}_{-6.5}\,^{+3.7}_{-3.2}\ \text{SNU} \quad \text{[L-peak]},$$

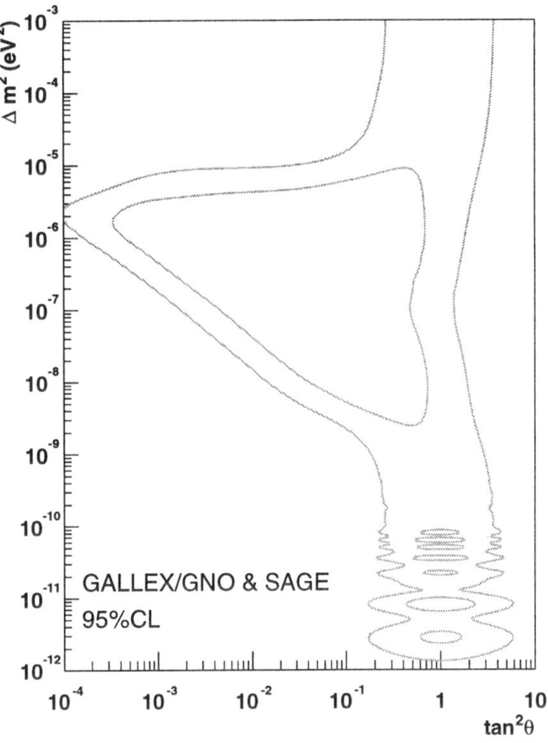

Figure 3. Constraint on the solar oscillation parameters from the gallium experiments. The inside of the covariance regions is allowed at the 95% CL. This exclusion plot uses the most recent results from SAGE and GALLEX/GNO.

$$\Phi_{Ga} = 73.3\,^{+5.9}_{-5.7}\,^{+3.7}_{-3.2}\ \text{SNU} \quad \text{[K-peak]}, \qquad (15)$$
$$\Phi_{Ga} = 69.6\,^{+4.4}_{-4.3}\,^{+3.7}_{-3.2}\ \text{SNU} \quad \text{[Overall]} .$$

The GALLEX/GNO results for 1991-2002 were summarized at the Neutrino 2002 conference[14] and they were not updated for this symposium:

$$\Phi_{Ga} = 77.5 \pm 6.2 \pm 4.5\ \text{SNU} \quad \text{[GALLEX]},$$
$$\Phi_{Ga} = 65.2 \pm 6.4 \pm 3.0\ \text{SNU} \quad \text{[GNO]}, \qquad (16)$$
$$\Phi_{Ga} = 70.8 \pm 4.5 \pm 3.8\ \text{SNU} \quad \text{[GALLEX/GNO]} .$$

Yet again the data are incompatible with the SSM since the expected rate is

$$\Phi_{Ga}(\text{SSM}) = 129\,^{+9}_{-7}\ \text{SNU} . \qquad (17)$$

The constraint of the SAGE and GALLEX/GNO experiments on the MSW plane is shown in Fig. 3. The allowed region from the combination of the gallium results is obtained by comparing the measurements with the calculations of the SSM (see Table 1).

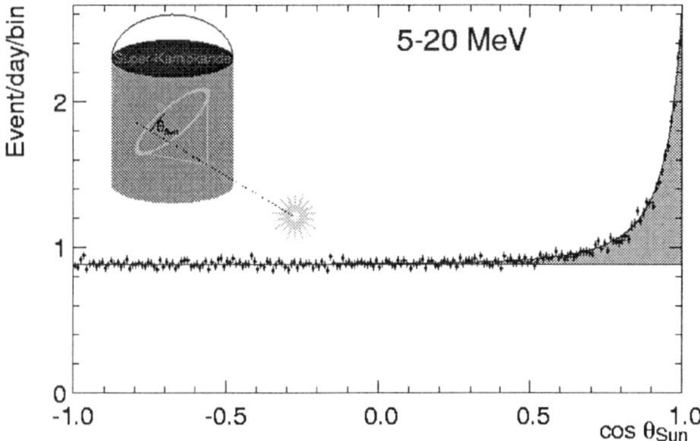

Figure 4. Angular distribution of solar neutrino candidates. SK data as of December 2002.

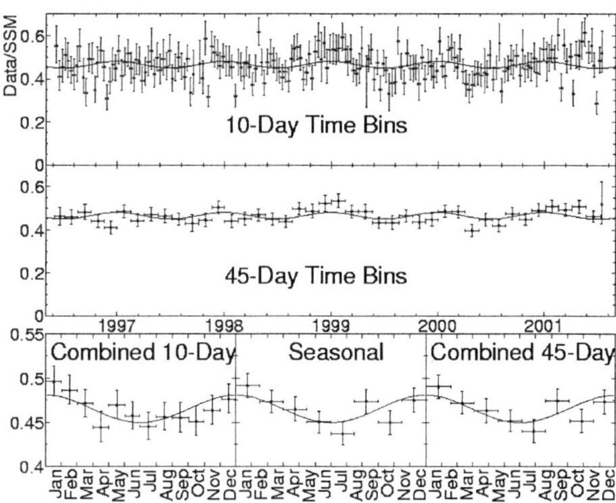

Figure 5. Time variation of the ES flux scaled by the SSM prediction. The curves represent the expected flux modulation due to the eccentricity of the Earth's orbit. SK data as of December 2002.

7. Kamiokande and SuperKamiokande

Following the first observations from the chlorine experiment the first priority was obviously an experimental confirmation of the solar-neutrino deficit. This was provided in 1987 by the Kamiokande water Čerenkov detector[15] in Japan, which also saw a significant (but, interestingly enough, not an identical) suppression of the measured rate of neutrinos from the Sun. This great achievement was rewarded by the 2002 Nobel Prize in physics to Masatoshi Koshiba. The main advantage of the Kamiokande detector is the real-time nature of the neutrino interactions viewed in the active fiducial volume (2,140 tons of ultra-pure light water) by 948 photomultiplier tubes (PMT). The Kamiokande Collaboration demonstrated that the neutrinos are actually coming from the direction of the Sun by reconstructing the direction of flight of the incident neutrinos from the neutrino-electron scattering (ES) reaction $\nu_x + e^- \to \nu_x + e^-$. Light water detectors are mainly sensitive to ν_e, but also to ν_μ and ν_τ, with a reduced cross section $\sigma(\nu_{\mu\tau} e^- \to \nu_{\mu\tau} e^-) \simeq 0.15 \times \sigma(\nu_e e^- \to \nu_e e^-)$.

The follow-up of the Kamiokande project is called the SuperKamiokande (SK) experiment.[16] It was built to investigate in more detail the nature of atmospheric and solar neutrino oscillations. The SK detector is a huge, 40 m in diameter and 40 m high, circular cylinder filled with 50,000 tons of ultra-pure light water. The SK detector operated at an energy threshold of 5 MeV and hence permitted the study of the ^8B neutrinos. It is divided into an outer detector to veto incoming cosmic ray muons and to shield external low energy background; and an inner detector (32,000 tons, of which 22,500 tons is the active fiducial volume) viewed by 11,146 PMT. As in Kamiokande, solar neutrinos are observed by detecting Čerenkov photons emitted by the electrons resulting from ES events. The event rate was about 15 events per day (substantially larger than the rate in the radiochemical experiments).

The number of Čerenkov photons collected by the PMT can be calibrated as a measurement of the electron energy, while the position and times of the hit phototubes can be used to reconstruct the $\nu_x - e^-$ interaction vertex and the electron direction. As mentioned before and as depicted in Fig. 4, the electron direction shows a strong peak pointing directly away from the Sun and enables discrimination against low-energy backgrounds in the detector arising from radioactive contaminants and spallation products produced by penetrating cosmic ray muons.[17,18]

The SK data allows measurements of the time dependence of the ES flux. It led to the measurement of the day/night rate asymmetry[18]

$$A_{\rm DN} = 2\frac{\Phi_{\rm D} - \Phi_{\rm N}}{\Phi_{\rm D} + \Phi_{\rm N}} = -0.021 \pm 0.020\,^{+0.013}_{-0.012}, \quad (18)$$

and the precise determination of the ES neutrino flux[18]

$$\Phi_{\rm ES} = (2.35 \pm 0.02 \pm 0.08) \times 10^6 \text{ cm}^{-2}\text{s}^{-1}. \quad (19)$$

The energy shape of the recoil electron agrees well, within experimental errors, with that predicted from the neutrino spectrum from the beta decay of ^8B. The measurement of the absolute flux, however, is about 46.5% of that predicted by the SSM.

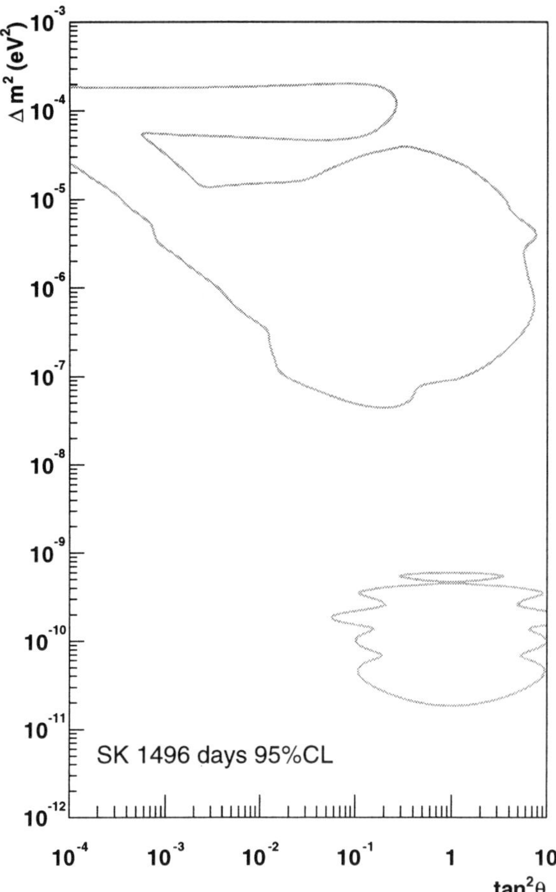

Figure 6. SSM flux independent excluded area using the SuperKamiokande zenith spectrum shape alone. The inside of the covariance regions is excluded at the 95% CL.

The real-time data of SK also provides the framework to study in detailed the shape of the zenith angle spectrum (i.e. the time and shape variation of the ES energy spectrum). The SK Collaboration looked at seasonal effects using 10 and 45 day bins.[17,18] This is shown in Fig. 5. The time modulation of the measured ES flux is unfortunately consistent with the eccentricity of the Earth's orbit around the Sun. Therefore, beyond the rate difference between Φ_{ES} of Eq. (19) and the SSM prediction, SK does not see any signature for oscillation from solar neutrinos. On the other hand, it is remarkable to visualize the complementarity of the radiochemical and SK experiments. Figure 6 shows the excluded regions of the mixing parameters from the SK Collaboration using the zenith spectrum shape alone. The lack of spectral distortion and daily variation breaks some degeneracy of the allowed regions of the chlorine and the gallium experiments.

Recently, SK looked for anti-neutrinos via the reaction $\bar{\nu}_e + p \rightarrow n + e^+$ after removing solar neutrino candidates with the effective cuts $E > 8$ MeV and $\cos\theta_{sun} < 0.5$. From an accurate estimation of the intrinsic spallation background, which remains after the $\bar{\nu}_e$ selection criteria, they performed the first search for low energy $\bar{\nu}_e$ from the Sun (which is very relevant if neutrinos have a magnetic moment). In the absence of a signal, they reported an upper limit for the conversion probability to $\bar{\nu}_e$ of the ^8B solar neutrinos. This conversion limit is 0.8% (90% CL) of the SSM neutrino flux in the range of 8-20 MeV.[19]

8. Sudbury Neutrino Observatory

The Sudbury Neutrino Observatory (SNO) is a 1,000 ton heavy-water Čerenkov detector[20] situated 2 km underground in INCO's Creighton mine in Canada. Another 7,000 tons of ultra-pure light water is used for support and shielding. The heavy water is in an acrylic vessel (12 m diameter and 5 cm thick) viewed by 9,456 PMT mounted on a geodesic structure 18 m in diameter; all contained within a polyurethane-coated barrel-shaped cavity (22 m diameter by 34 m high). The SNO detector has been filled with water since May 1999 and is moving toward the Neutral-Current Detector (NCD) phase of its scientific program. The solar-neutrino detectors in operation prior to SNO were mainly sensitive to the electron neutrino type; while the use of heavy water by SNO allows the flux of all three neutrino types to be measured. Electron neutrinos can interact through a charged-current interaction while all neutrinos can interact through a neutral-current reaction. The determination of these reaction rates is a critical measurement in determining if neutrinos oscillate in transit between the core of the Sun and their observation on Earth.

Neutrinos from ^8B decay in the Sun are observed in SNO from Čerenkov processes following these reactions:

- Charged-current (CC) reaction, specific to electron neutrinos:

$$d + \nu_e \to p + p + e^- .$$

This reaction has a Q value of 1.4 MeV and the electron energy is strongly correlated with the neutrino energy, providing potential sensitivity to spectral distortions.

- Neutral-current (NC) reaction, equally sensitive to all non-sterile neutrino types ($x = e, \mu, \tau$):

$$\nu_x + d \to n + p + \nu_x .$$

This reaction has a threshold of 2.2 MeV and is observed through the detection of neutrons by three different techniques in separate phases of the experiment.

- Elastic-scattering (ES) reaction:

$$\nu_x + e^- \to \nu_x + e^- .$$

This reaction has a substantially lower cross section than the other two and as mentioned before is predominantly sensitive to electron neutrinos.

The relations

$$\Phi_{\rm CC} = \phi_e ,$$

$$\Phi_{\rm ES} = \phi_e + 0.15\phi_{\mu\tau} , \qquad (20)$$

$$\Phi_{\rm NC} = \phi_e + \phi_{\mu\tau} ,$$

give SNO the status of an appearance experiment. The SNO experimental plan calls for three phases of about two years each wherein different techniques will be employed for the detection of neutrons from the NC reaction. During the first phase, with pure heavy water, neutrons were observed through the Čerenkov light produced when neutrons were captured on deuterium, producing 6.25 MeV gammas. In this phase, the capture probability for such neutrons was about 25% and the Čerenkov light is relatively close to the threshold of about 5 MeV for the electron energy, imposed by radioactivity in the detector. For the second phase, about 2 tons of $NaCl$ was added to the heavy water and neutron detection was enhanced through capture on Cl, with about 8.6 MeV gamma energy release and about 83% capture efficiency. Here, results from the pure D_2O and salt phases will be reported. For the third phase, the salt will be removed and an array of ^3He-filled

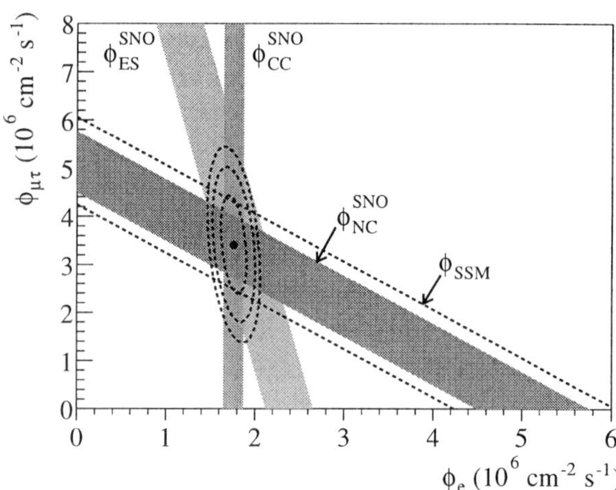

Figure 7. Flux of ^8B solar neutrinos which are of μ or τ flavor versus the flux of electron neutrinos deduced from the three neutrino reactions in SNO. The diagonal bands show the total ^8B flux as predicted by the SSM (dashed lines) and that measured with the NC reaction in SNO (solid band). The intercepts of these bands with the axes represent the $\pm 1\sigma$ errors. The bands intersect at the fit values for ϕ_e and $\phi_{\mu\tau}$, indicating that the combined flux results are consistent with neutrino flavor transformation at the 5.3σ level.

proportional counters will be installed to provide direct detection of neutrons with a capture efficiency of about 45%.

During the pure D_2O phase of the experiment, the signal was extracted with a statistical analysis technique based on the direction, $\cos\theta_{\rm sun}$, the position, R, and the kinetic energy, T_e, of the event assuming the SSM energy spectrum shape.[21] The final selection criteria were $T_e \geq 5$ MeV and $R \leq 550$ cm. An extended maximum-likelihood fit yields[22]

$$\Phi_{\rm CC} = 1.76\,^{+0.06\ +0.09}_{-0.05\ -0.09} \times 10^6 \ {\rm cm}^{-2}{\rm s}^{-1} ,$$
$$\Phi_{\rm ES} = 2.39\,^{+0.24\ +0.12}_{-0.23\ -0.12} \times 10^6 \ {\rm cm}^{-2}{\rm s}^{-1} , \qquad (21)$$
$$\Phi_{\rm NC} = 5.09\,^{+0.44\ +0.46}_{-0.43\ -0.43} \times 10^6 \ {\rm cm}^{-2}{\rm s}^{-1} .$$

The excess of the NC flux over the CC and ES fluxes implies neutrino flavor transformation. There is also a very nice agreement between the SNO NC flux and the total ^8B flux of $5.05^{+1.01}_{-0.81} \times 10^6$ cm^{-2}s^{-1} predicted by the SSM. The simple change of variables in Eq. (20) resolves the data directly into electron and non-electron components[22]

$$\phi_e = 1.76\,^{+0.06\ +0.09}_{-0.05\ -0.09} \times 10^6 \ {\rm cm}^{-2}{\rm s}^{-1} , \qquad (22)$$
$$\phi_{\mu\tau} = 3.41\,^{+0.45\ +0.48}_{-0.45\ -0.45} \times 10^6 \ {\rm cm}^{-2}{\rm s}^{-1} . \qquad (23)$$

As depicted in Fig. 7, where the error ellipses represent the 68%, 95%, and 99% joint probability con-

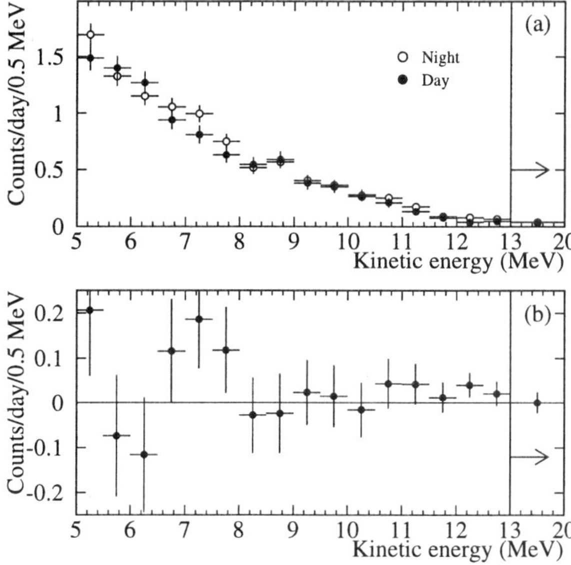

Figure 8. SNO (a) energy spectra for day and night. All signals and backgrounds contribute. The final bin extends from 13.0 to 20.0 MeV. (b) Difference, night - day, between the spectra. The day rate was 9.23 ± 0.27 events/day, and the night rate was 9.79 ± 0.24 events/day.

tours for ϕ_e and $\phi_{\mu\tau}$, there is clear evidence of solar neutrino flavor transformation at 5.3 standard deviations.

Allowing a time variation of the total flux of solar neutrino leads in SNO to day/night measurements which are sensitive to neutrino type[23]

$$A_{\mathrm{DN}}(\text{total}) = \left(-24.2 \pm 16.1 \, {}^{+2.4}_{-2.5}\right)\%, \qquad (24)$$

$$A_{\mathrm{DN}}(e) = \left(12.8 \pm 6.2 \, {}^{+1.5}_{-1.4}\right)\%. \qquad (25)$$

By enforcing no asymmetry in the total rate, i.e. $A_{\mathrm{DN}}(\text{total}) = 0$, the day/night asymmetry for the electron neutrino is[23]

$$A_{\mathrm{DN}}(e) = \left(7.0 \pm 4.9 \, {}^{+1.3}_{-1.2}\right)\%. \qquad (26)$$

The day and night energy spectra for all accepted events are shown in Fig. 8. Backgrounds were subtracted separately for day and night as part of the signal extraction. No systematic has been identified, in either signal or background regions, that would suggest that the small difference between day and night is other than a statistical fluctuation.

Even if they were not ready for this conference, SNO published their first results of the salt phase[24] shortly after. The measurements were made with dissolved $NaCl$ in the heavy water to enhance the

sensitivity and signature for neutral-current interactions. Neutron capture on ^{35}Cl typically produces multiple γ rays while the CC and ES reactions produce single electrons. The greater isotropy of the Čerenkov light from neutron capture events relative to CC and ES events allows better statistical separation of the event types. More importantly, this separation allows a precise measurement of the NC flux to be made independently of assumptions about the CC and ES energy spectra. The degree of the Čerenkov light isotropy is determined from the pattern of PMT hits. To minimize the possibility of introducing biases, SNO performed a blind analysis procedure for the more model independent determination of the total active (ν_x) ^8B solar neutrino flux. The salt analysis was performed on the new data set, statistically separating events into CC, NC, ES, and external-source neutrons using an extended maximum-likelihood fit based on the distributions of isotropy, cosine of the event direction relative to the vector from the Sun, $\cos\theta_{\mathrm{sun}}$, and radius, R, within the detector. This analysis differs from the previous analyses of the pure D_2O data[22,23] since the spectral distributions of the ES and CC events are not constrained to the ^8B shape, but are extracted from the data. Čerenkov event backgrounds from $\beta - \gamma$ decays were reduced with an effective electron kinetic energy threshold $T_e \geq 5.5$ MeV and a fiducial volume with radius $R \leq 550$ cm. The extended maximum-likelihood analysis gives the following ^8B fluxes[24]

$$\Phi_{\mathrm{CC}} = 1.59 \, {}^{+0.08}_{-0.07} \, {}^{+0.06}_{-0.08} \times 10^6 \ \mathrm{cm}^{-2}\mathrm{s}^{-1},$$

$$\Phi_{\mathrm{ES}} = 2.21 \, {}^{+0.31}_{-0.26} \pm 0.10 \times 10^6 \ \mathrm{cm}^{-2}\mathrm{s}^{-1}, \qquad (27)$$

$$\Phi_{\mathrm{NC}} = 5.21 \pm 0.27 \pm 0.38 \times 10^6 \ \mathrm{cm}^{-2}\mathrm{s}^{-1}.$$

These fluxes are in agreement with previous SNO measurements and the SSM. The ratio of the ^8B flux measured with the CC and NC reactions then provides a strong signature of solar neutrino oscillations

$$\frac{\Phi_{\mathrm{CC}}}{\Phi_{\mathrm{NC}}} = 0.306 \pm 0.026 \pm 0.024. \qquad (28)$$

The salt shape-unconstrained fluxes presented here combined with shape-constrained fluxes and day/night energy spectra from the pure D_2O phase[22,23] place impressive constraints on the allowed neutrino flavor mixing parameters. Two-flavor active neutrino oscillation models predict the CC, NC, and ES rates in SNO. In the fit, the ratio f_B of the total ^8B flux to the SSM value is a free parameter

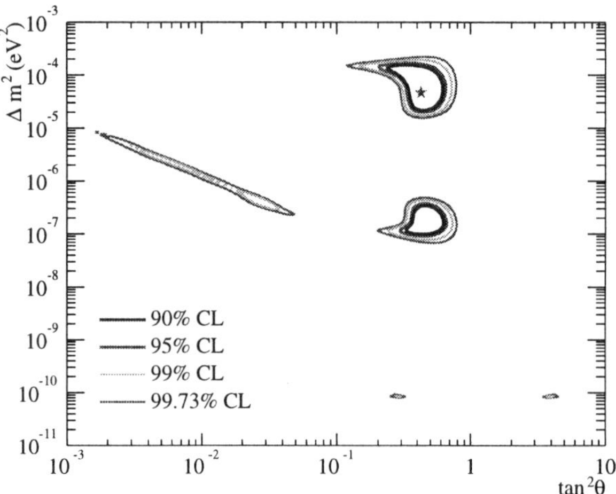

Figure 9. SNO-only neutrino oscillation contours, including pure D_2O day/night spectra, salt CC, NC, ES fluxes, with ^8B flux free and hep flux fixed. The best-fit point is $\Delta m^2 = 4.7 \times 10^{-5}$, $\tan^2\theta = 0.43$, $f_B = 1.03$, with χ^2/d.o.f.=26.2/34. The inside of the covariance regions is allowed.

Table 2. Overview of all the solar neutrino data. Only experimental errors are considered in the ratio of the data to the SSM predictions.

Experiment	Reaction	Ratio data/SSM
Chlorine	CC	0.34 ± 0.03
SAGE+GALLEX/GNO	CC	0.55 ± 0.03
SuperKamiokande	ES	0.47 ± 0.02
SNO	CC	0.35 ± 0.02
SNO	ES	0.47 ± 0.05
SNO	NC	1.01 ± 0.13

together with the mixing parameters. A combined χ^2 fit to SNO D_2O and salt data alone yields the allowed regions in Δm^2 and $\tan^2\theta$ shown in Fig. 9.

9. Global Fits

This section summarizes the solar neutrino data in a global analysis of all experiments. Table 2 shows the experimental results compared with the SSM prediction. The data indeed suggest an energy dependence due to the different energy thresholds of the different detection techniques.

The global analysis presented here includes the chlorine results,[10] the updated gallium flux measurements,[13,14] the SK zenith spectra,[17,18] and the D_2O and salt results from SNO.[22,23,24] The free parameters in the global fit are the total ^8B flux, the difference of the squared mass Δm^2, and the mixing angle θ. The higher energy hep flux is fixed at

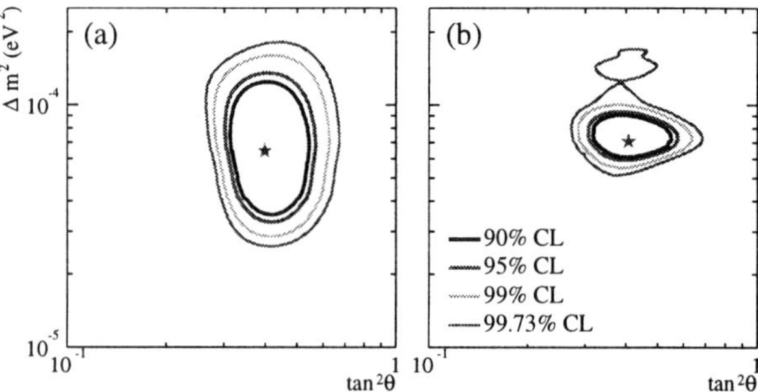

Figure 10. Allowed region of the MSW plane determined by a χ^2 fit to (a) the chlorine, gallium, SK, and SNO experiments. The best-fit point is $\Delta m^2 = 6.5 \times 10^{-5}$, $\tan^2\theta = 0.40$, $f_B = 1.04$, with χ^2/d.o.f.=70.2/81. (b) Solar global + KamLAND. The best-fit point is $\Delta m^2 = 7.1 \times 10^{-5}$, $\tan^2\theta = 0.41$, $f_B = 1.02$. In the MSW analyses, the ratio (f_B) of the total ^8B flux to the SSM value is a free parameter, while the total hep flux is fixed to the SSM prediction. The inside of the covariance contours is the allowed region.

9.3×10^3 cm^{-2} s^{-1}. Contours are generated in Δm^2 and $\tan^2\theta$ for $\Delta\chi^2 = 4.61$ (90% CL), 5.99 (95% CL), 9.21 (99% CL), and 11.83 (99.73% CL). As presented in Fig. 10(a), the combined results of all solar neutrino experiments can be used to determine a unique region of the MSW plane. In this global fit of all solar neutrino data, the allowed regions in parameter space shrink considerably and the Large Mixing Angle (LMA) region is selected.

A global analysis including the KamLAND reactor anti-neutrino results[25] shrinks the allowed region further, with a best-fit point of $\Delta m^2 = 7.1^{+1.2}_{-0.6} \times 10^{-5}$ eV^2 and $\theta = 32.5^{+2.4}_{-2.3}$ degrees, where the errors reflect 1σ constraints on the 2-dimensional region. This is summarized in Fig. 10(b). With the new SNO measurements the allowed region is constrained to only the lower band of LMA at $> 99\%$ CL. The best-fit point with one dimensional projection of the uncertainties in the individual parameters (marginalized uncertainties) has $\Delta m^2 = 7.1^{+1.0}_{-0.3} \times 10^{-5}$ eV^2 and $\theta = 32.5^{+1.7}_{-1.6}$ degrees. This disfavors maximal mixing ($\tan^2\theta = 1$) at a level equivalent to 5.4 standard deviations.

10. Future

The upcoming generation of real-time low energy solar neutrino experiments will tackle new frontiers and allow the direct study of the mono-energetic ^7Be solar neutrinos. The goal of the Borexino and Kam-

LAND experiments is in fact to measure the time-dependence of the neutrino-electron scattering rate from neutrinos produced by electron capture on ^7Be in the Sun. The produced ES electrons are below threshold for production of Čerenkov light and thus liquid scintillator is used as detector target for enhancing the detection capability. The challenge at such a low threshold is the unprecedented low level of radioactivity required; namely less then 10^{-16} g/g of ^{238}U and ^{232}Th and less than one part in 10^{18} of ^{14}C. One of the dominant backgrounds to be removed is radioactive Krypton (^{85}Kr). The major interest in the results for the next generation of low energy neutrino experiments arises when one compares the total ^7Be rate to the day/night rate asymmetry, $A_{\rm DN}$. This will give an excellent independent discrimination between different MSW solutions. It will therefore over-constrain the $\Delta m^2 - \tan^2 \theta$ plane in a global analysis of all the solar neutrinos experiments.

The Borexino detector[26] is located in Hall C of the Gran Sasso Laboratory. The time for the start of data taking is unclear due to an environmental problem encountered in 2002. Borexino is a 300 ton liquid-scintillator based detector with 100 tons of active fiducial mass in a 8.3 m diameter spherical nylon bag surrounded by a 2.6 meter thick spherical shell filled with buffer oil. The liquid scintillator and buffer liquid are viewed by 2,240 PMT which are mounted inside a 13.5 m diameter stainless steel tank; which is in turn surrounded by a 18 m spherical tank filled with ultra-pure light water to act as a radiation shield. The expected energy threshold is about 800 keV.

The KamLAND detector is located in the cavity used for the original Kamiokande experiment. The primary goal of KamLAND[25] is to investigate the oscillation of $\bar{\nu}_e$ emitted from distant nuclear reactors.[27] The investigation of ^7Be neutrinos comes for free in a later stage when their scientific program will shift from a coincidence experiment to an ES low energy experiment. The radioactive background remains of course the main concern for solar neutrino physics. KamLAND hosts 1,000 tons of liquid scintillator contained in a 13 m diameter spherical balloon, which is in turn surrounded by a 18 m spherical tank with 1,879 inward facing phototubes. The space between the balloon and the tank is filled with buffer oil and the containment tank is immersed in a 3,200 ton water Čerenkov detector instrumented with 225 PMT.

The other important next generation of real-time solar neutrino experiments should detect the fundamental pp neutrinos, which constitute about 91% of the total neutrino flux predicted by the SSM. Unfortunately, there are no approved pp solar neutrino experiments at the present time, although there are a number of promising proposals under development and R&D is ongoing.

11. Summary

Solar neutrino oscillation is clearly established by the combination of the results from the chlorine, gallium, SK, and SNO experiments. The real-time data of SK and SNO do not show large energy distortion nor time-like asymmetry. SNO provided the first direct evidence of flavor conversion of solar electron neutrinos by comparing the CC and NC rates. Matter effects explain the energy dependence of solar oscillation, and Large Mixing Angle (LMA) solutions are favored.

The global analysis of the solar neutrino detectors and reactor neutrino results yields $\Delta m^2 = 7.1^{+1.0}_{-0.3} \times 10^{-5}$ eV^2 and $\theta = 32.5^{+1.7}_{-1.6}$ degrees. Maximal mixing is rejected at the equivalent of 5.4 standard deviations and confirms the region $\tan^2 \theta < 1$, which corresponds to the normal mass hierarchy of $m_2 > m_1$ (i.e. $\Delta m^2 > 0$).

Solar neutrino data demonstrates that neutrinos have mass and that the minimal Standard Model is incomplete. Unlike the quark sector where the CKM mixing angles are small, the lepton sector exhibits large mixing. The neutrino masses and mixing may play significant roles in determining structure formation in the early universe as well as supernovae dynamics and the creation of matter. The coming decade will be exciting for neutrino physics helping to delineate the *New* Standard Model that will include neutrino masses and oscillations. This will lead to precision measurements of the leptonic mixing matrix, determination of neutrino masses, and investigation of CP and CPT properties in the lepton sector. After 30 years of hard labor from the nuclear and particle physics community, the Solar Neutrino Problem is now becoming an industry for precise measurements of neutrino oscillation parameters with the next generation of solar neutrino and long baseline neutrino experiments.

Acknowledgments

This article builds upon the careful and detailed work of many people. Special thanks to E. Bellotti, M. Boulay, J. Formaggio, V. Gavrin, K. Graham, K. Heeger, R. Hemingway, A. Ianni, A. Marino, M. Nakahata, A. Poon, Y. Takeuchi, and J. Wilkerson. This research has been financially supported in Canada by the Natural Sciences and Engineering Research Council (NSERC), the Canada Research Chair (CRC) Program, and the Canadian Foundation for Innovation (CFI). I am grateful to the SNO Collaboration for giving me the opportunity to contribute to the 2003 Lepton Photon Symposium.

References

1. H.A. Bethe, *Phys. Rev.* **55**, 436 (1939).
2. J.N. Bahcall, H.M. Pinsonneault, and S. Basu, *Astrophys. J.* **555**, 990 (2001).
3. Z. Maki, M. Nakagawa, S. Sakata, *Prog. Theor. Phys.* **28**, 870 (1962); B. Pontecorvo, *Sov. Phys. JETP* **26**, 984 (1968).
4. N. Cabibbo, *Phys. Rev. Lett.* **10**, 531 (1963); M. Kobayashi, and T. Maskawa, *Prog. Theor. Phys.* **49**, 652 (1973).
5. Y. Fukuda *et al.*, *Phys. Rev. Lett.* **81**, 1562 (1998).
6. M. Apollonio, *Phys. Lett.* B **466**, 415 (1999).
7. S.P. Mikheyev and A.Yu. Smirnov, *Sov. J. Nucl. Phys.* **42**, 913 (1985).
8. L. Wolfenstein, *Phys. Rev.* D **17**, 2369 (1978).
9. R. Davis, Jr., *Phys. Rev. Lett.*, 302 (1964).
10. B.T. Cleveland *et al.*, *Ap. J.* **496**, 505 (1998).
11. J.N. Abdurashitov *et al.*, *Phys. Rev.* C **60** (1999).
12. W. Hampel *et al.*, *Phys. Lett.* B **447**, 127 (1999); M. Altmann *et al.*, *Phys. Lett.* B **490**, 16 (2000).
13. V. Gavrin, 4th International Workshop on Low Energy and Solar Neutrinos, Paris, May 19–21, 2003.
14. T. Kirsten, XXth Int. Conf. on Neutrino Physics and Astrophysics, Munich, May 25–30, 2002; *Nucl. Phys.* B (Proc. Suppl.) **118** (2003).
15. Y. Suzuki, *Nucl. Phys.* (Proc. Suppl.) **34**, 54 (1995).
16. M. Nakahata *et al.*, *Nucl. Inst. Meth.* A **421**, (1999).
17. S. Fukuda *et al.*, *Phys. Rev. Lett.* **86**, 5651 (2001).
18. S. Fukuda *et al.*, *Phys. Lett.* B **539**, 179 (2002).
19. S. Fukuda *et al.*, *Phys. Rev. Lett.* **90**, 171302 (2003).
20. J. Boger *et al.*, *Nucl. Inst. Meth.* A **449**, 172 (2000).
21. C.E. Ortiz *et al.*, *Phys. Rev. Lett.* **85**, 2909 (2000).
22. Q.R. Ahmad *et al.*, *Phys. Rev. Lett.* **89**, 011301 (2002).
23. Q.R. Ahmad *et al.*, *Phys. Rev. Lett.* **89**, 011302 (2002).
24. Q.R. Ahmad *et al.*, submitted to *Phys. Rev. Lett.*, Sept. 2003, nucl-ex/0309004.
25. K. Eguchi *et al.*, *Phys. Rev. Lett.* **90**, 021802 (2003).
26. G. Bellini, XXth Int. Conf. on Neutrino Physics and Astrophysics, Munich, May 25–30, 2002; *Nucl. Phys.* B (Proc. Suppl.) **118** (2003); G. Alimonti *et al.*, *Astropart. Phys.* **16**, 205 (2002).
27. Kunio Inoue in these proceedings.

DISCUSSION

Luc Declais (IPNL, France): Can you say more about the interest of real-time low energy pp neutrino experiments for solar neutrinos, because GNO will measure the total amount of neutrinos above a given threshold with an accuracy of 7%? So is more or less everything is known?

Alain Bellerive: When I first wrote the summary for future experiments, the last point was "The importance of pp real-time neutrino experiments" and I changed it to "Real-time low energy neutrinos are the ultimate probe of the Sun and test of the Standard Solar Model". So you're right, at the moment if you can pin down the ^7Be to really high accuracy, I think that it will provide a really good proof of matter effects in the context of the SSM. The pp neutrinos are nevertheless an important piece of the puzzle since the majority of the neutrinos from the Sun are produced via the pp reaction.

NEUTRINO PHYSICS: OPEN THEORETICAL QUESTIONS

A. Y. SMIRNOV

International Centre for Theoretical Physics, Strada Costiera 11, 31014 Trieste, Italy
and
Institute for Nuclear Research, Russian Academy of Sciences,Moscow, Russia
E-mail: smirnov@ictp.trieste.it

We know that neutrino mass and mixing provide a window to physics beyond the Standard Model. Now this window is open, at least partly. And the questions are: what do we see, which kind of new physics, and how far "beyond"? I summarize the present knowledge of neutrino mass and mixing, and then formulate the main open questions. Following the bottom-up approach, properties of the neutrino mass matrix are considered. Then different possible ways to uncover the underlying physics are discussed. Some results along the line of: seesaw, GUT and SUSY GUT are reviewed.

1. Introduction

This review is devoted to neutrino masses and mixing. It covers experimental results, their interpretation and implications. It is in this area that enormous progress has been achieved during the last few years.

The field develops fast, and already after the Symposium a number of important results have been published including the SNO salt phase data, new analysis of the Heidelberg-Moscow experimental results, *etc.*.

In Sec. 2 the main achievements in reconstruction of the neutrino mass and mixing spectrum are summarized. The open theoretical questions are formulated in Sec. 3. In Sec. 4, following the bottom-up approach, the neutrino mass matrix is reconstructed and its properties are studied. In Sec. 5 the ways we may go in answering the open questions are outlined.

2. What Have We Learned?

2.1. *Solar Neutrinos*

The latest SNO salt phase results[1] have further confirmed the correctness of the Standard Solar Model (SSM) neutrino fluxes[2] and the realization of the MSW large mixing (LMA) conversion mechanism[3] inside the Sun.[1,4–10] In Fig. 1 we show the allowed region of the oscillation parameters $\tan^2\theta_{12}$ and Δm_{12}^2 from the 2ν combined analysis of the solar neutrino and KamLAND[11] results. The best-fit values of the parameters are

$$\Delta m_{12}^2 = 7.1 \times 10^{-5}\text{eV}^2, \quad \tan^2\theta_{12} = 0.4. \tag{1}$$

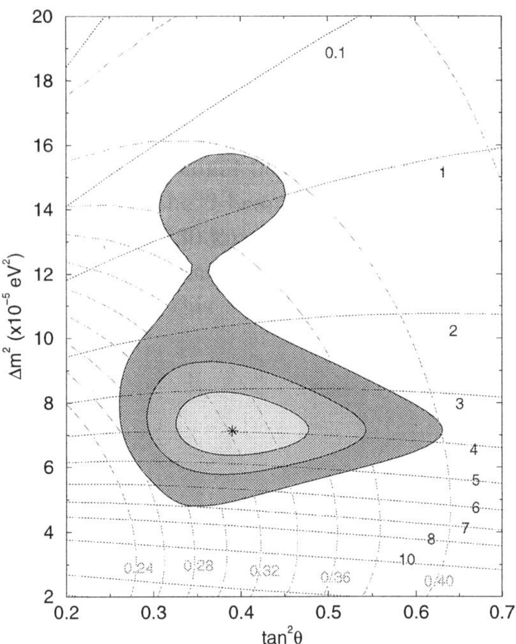

Figure 1. The allowed regions of oscillation parameters from the combined fit of the solar neutrino data and the KamLAND spectrum at 1σ, 2σ, 3σ CL.[10] Shown are also the contours of constant CC/NC ratio (dotted lines) and the Day-Night asymmetry (dashed lines) at SNO (numbers on the curves in %).

Combined fit of the solar, KamLAND[11] and CHOOZ[12] results favors nearly zero 1-3 mixing: $\sin^2\theta_{13} \sim 0$.[9,10] Basically the data have selected the l-LMA region with $\Delta m_{12}^2 < 10^{-4}$ eV^2 (the h-LMA region is accepted now at 3σ only), and strongly disfavored maximal 1-2 mixing. The upper bound is

$$\tan^2\theta_{12} < 0.64 \quad (3\sigma). \tag{2}$$

As a result of these improvements, the physics of the conversion is now even determined quantitatively.[5,10] In particular, recent results show relevance of the notion of resonance, they fix the relative strength of the effects of the adiabatic conversion and the oscillations as function of the neutrino energy.[10]

In Fig. 1 we show also the contours of constant CC/NC ratio and Day-Night asymmetry of the CC-events at SNO. They allow one to evaluate an impact of future SNO measurements. The KamLAND operation will allow one to eventually determine Δm^2_{12} with about 10% accuracy.

Are there any data which indicate deviation from the LMA picture? In this connection we consider two generic features of the LMA-MSW solution:

- the predicted Ar-production rate, $Q_{Ar} = 2.96 \pm 0.25$ SNU, is about 2σ higher than the Homestake[13] result; and

- the upturn of the spectrum at low energies, that is, the increase of the ratio N^{obs}/N^{SSM} with decrease of energy, is expected which can be as large as $10-15\%$. However, the latest SNO as well as the previous SNO and SuperKamiokande[14] spectral data do not show the upturn, being in agreement with the absence of distortion.

Both problems can be resolved simultaneously, if a light sterile neutrino exists with very small active-sterile mixing:[15]

$$\frac{\Delta m^2_{01}}{\Delta m^2_{21}} = 0.05 - 0.2, \quad \sin^2 2\alpha = 10^{-5} - 10^{-3}. \quad (3)$$

Such a mixing produces a dip in the survival probability (Fig. 2) which suppresses both the Ar-production rate and the upturn of spectrum. The best description of the data would correspond to the dip at relatively high energies (panel for $R_\Delta = 0.10$) when the CNO- and pep-neutrino fluxes and the low energy part of the boron neutrino spectrum are suppressed.

Such a possibility can be tested in the future low energy neutrino experiments: BOREXINO,[16] KamLAND, MOON, etc.,[17] as well as in further measurements of the spectrum by SNO and SK.

2.2. Atmospheric Neutrinos

A recent refined analysis of the SuperKamiokande data in terms of $\nu_\mu - \nu_\tau$ oscillations gives[18] at 90% CL

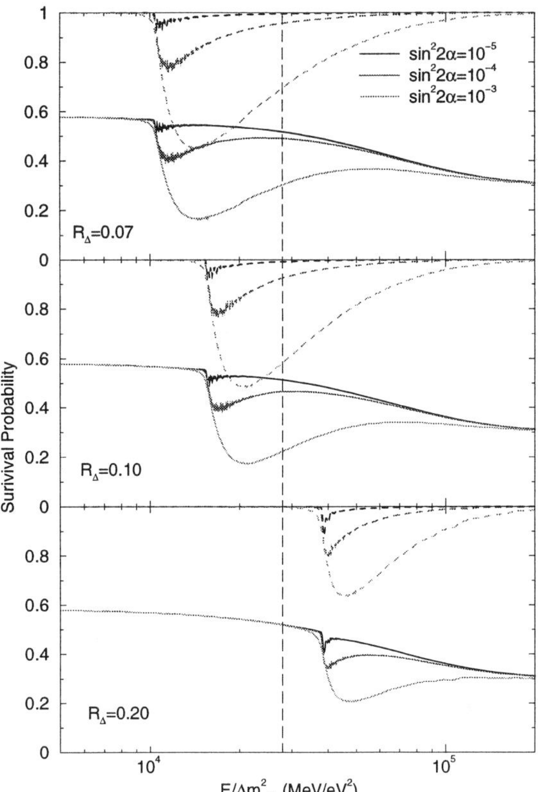

Figure 2. The survival probabilities of the electron neutrino (solid line) and the active neutrinos (dashed line) for different values of the sterile-active mixing.[15] The panels correspond to three different values of $R_\Delta \equiv \Delta m^2_{01}/\Delta m^2_{21}$. Vertical dashed line indicates the position of the 1-2 resonance.

$$\Delta m^2_{13} = (1.3 - 3.0) \times 10^{-3} \text{eV}^2, \quad \sin^2 2\theta_{23} > 0.9 \quad (4)$$

with the best fit at $\Delta m^2_{12} = 2.0 \times 10^{-3}$ eV^2 and $\sin^2 2\theta_{12} = 1.0$. Combined analysis of the CHOOZ and the atmospheric neutrino data puts the upper bound on the 1-3 mixing[19]

$$\sin^2 \theta_{13} < 0.067 \quad (3\sigma). \quad (5)$$

The open question is whether oscillations of the atmospheric ν_e exist? There are two possible sources of these oscillations: (i) non-zero 1-3 mixing and "atmospheric" Δm^2_{13}, and (ii) solar oscillation parameters in Eq. (1). Also their interference should exist.[20] After confirmation of the LMA-MSW solution we can definitely say that oscillations driven by the LMA parameters (the LMA oscillations) should show up at some level. Relative modification of the ν_e flux due to the LMA oscillations can be written as[20]

$$\frac{F_e}{F_e^0} - 1 = P_2(r \cos^2 \theta_{23} - 1), \quad (6)$$

where $P_2(\Delta m_{12}^2, \theta_{12})$ is the 2ν transition probability and $r \equiv F_\mu^0/F_e^0$ is the ratio of the original ν_μ and ν_e fluxes. In the sub-GeV region, where P_2 can be of the order 1, the ratio equals $r \approx 2$, so that the oscillation effect is proportional to the deviation of the 2-3 mixing from the maximal value: $D_{23} \equiv 1/2 - \sin^2\theta_{23}$. In Fig. 3 we show the ratio of numbers of the e-like events with and without oscillations as function of the zenith angle of the electron. For the allowed range of $\sin^2\theta_{23}$ and the present best-fit value of Δm_{12}^2 the excess can be as large as 5 - 6%. The excess increases with decreasing energy.

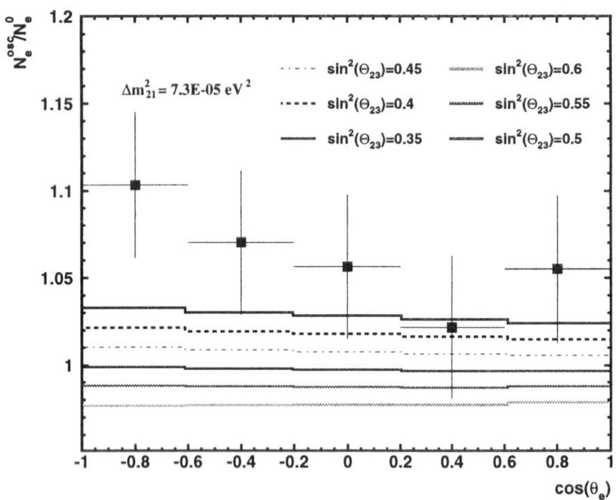

Figure 3. The ratio of numbers of the e-like events with and without oscillations as function of the zenith angle of the electron for different values of $\sin^2\theta_{23}$.[20] Other parameters are $\sin^2 2\theta_{12} = 0.82$, $\sin\theta_{13} = 0$ and $\Delta m_{12}^2 = 7.3 \times 10^{-5}$ eV^2. Also shown are the SuperKamiokande experimental points.

Future searches for the excess can be used to restrict or measure D_{23}. In fact, the latest analysis, (without renormalization of the original fluxes) shows some excess of the e-like events at low energies and the absence of excess in the multi-GeV sample, thus giving a hint of non-zero D_{23}. Establishing this deviation has important consequences for understanding the origins of neutrino masses and mixing.

Non-zero 1-3 mixing generates the interference effect between the LMA oscillations amplitudes.[20] The interference contribution does not contain the "screening" factor, in Eq. (6), and can reach $2-4\%$ for the allowed values of $\sin\theta_{13}$. This produces an uncertainty in the determination of D_{23}. So, D_{23} can be measured if either a large excess is found or/and a stronger bound on the 1-3 mixing is established.

2.3. *Mass Spectrum and Mixing*

Information obtained from the oscillation experiments allows us to make significant progress in the reconstruction of the neutrino mass and flavor spectrum (Fig. 4).

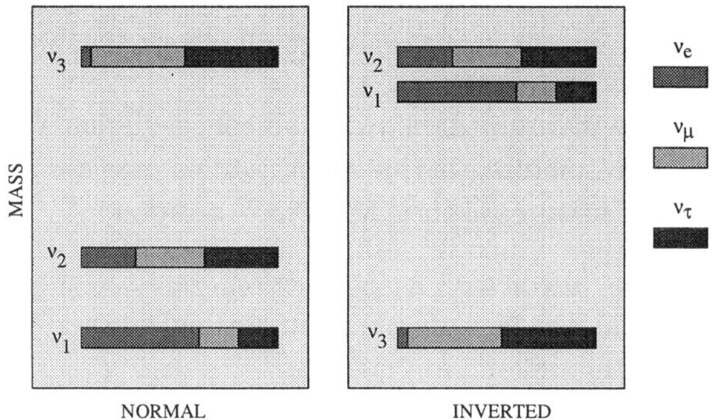

Figure 4. Neutrino mass and flavor spectra for the normal (left) and inverted (right) mass hierarchies. The distribution of flavors (colored parts of boxes) in the mass eigenstates corresponds to the best-fit values of mixing parameters and $\sin^2\theta_{13} = 0.05$.

The unknowns are:

(i) admixture of ν_e in ν_3: U_{e3};

(ii) type of mass spectrum: hierarchical; non-hierarchical with certain ordering; degenerate, which is related to the value of the absolute mass scale, m_1; and

(iii) type of mass hierarchy (ordering): normal, inverted.

Using a global fit of the oscillation data one can find intervals for the elements of the PMNS mixing matrix $\|U_{\alpha i}\|$:

$$\begin{pmatrix} 0.79 - 0.86 & 0.50 - 0.61 & 0.0 - 0.16 \\ 0.24 - 0.52 & 0.44 - 0.69 & 0.63 - 0.79 \\ 0.26 - 0.52 & 0.47 - 0.71 & 0.60 - 0.77 \end{pmatrix}, \quad (7)$$

where columns correspond to the flavor index and rows to the mass index.[21]

Now we are in a position to construct the leptonic unitarity triangle, although the finite size of one angle is still unknown. For practical reason (no intensive ν_τ beams) we consider the triangle which employs the e- and μ- rows of the mixing matrix (Fig. 5). The triangle is not degenerate in spite of the strong bound on the 1-3 mixing.

Is it possible to reconstruct the triangle using results from future experiments? Can we use the triangle to determine the CP-violation phase, δ? The

400

area of the triangle is related to the Jarlskog invariant $J_{CP} \equiv Im[U_{e1}U_{\mu2}U_{e2}^*U_{\mu1}^*]$: $S = J_{CP}/2$. Reconstruction of the triangle is complementary to measurements of the neutrino-antineutrino asymmetries in oscillations. Interestingly, the main problem here is the coherence: the same coherence which leads to the oscillations. For the triangle method we need to study interactions of the mass eigenstates, whereas in practice we deal with flavor (coherent) states. So, breaking of the coherence, averaging of oscillations, experiments with the beams of mass eigenstates and measurements of the survival (rather than transition) probabilities are the key elements of the method.[22]

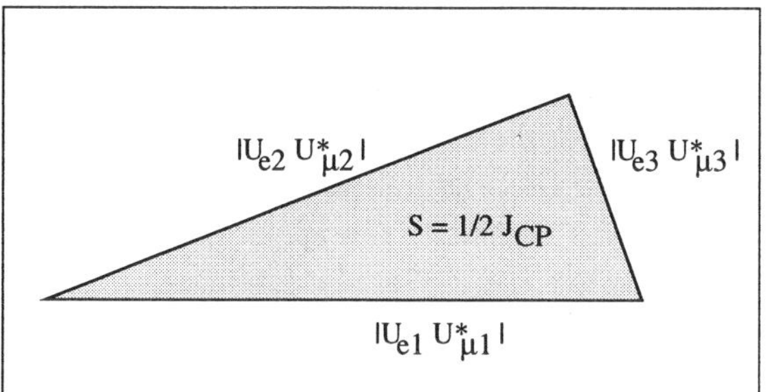

Figure 5. Possible leptonic unitarity triangle. We take the best-fit values of θ_{12}, and θ_{23} and $\sin\theta_{13} = 0.16$.

2.4. Neutrinos from SN1987A

After confirmation of the LMA-MSW solution we can definitely say that the effect of flavor conversion has already been observed in 1987. One must take into account the conversion effects in analysis of SN1987A[23] and future supernova neutrino data.

In terms of the original fluxes of the electron and muon antineutrinos, $F^0(\bar{\nu}_e)$ and $F^0(\bar{\nu}_\mu)$, the electron antineutrino flux at the detector can be written as

$$F(\bar{\nu}_e) = F^0(\bar{\nu}_e) + \bar{p}\Delta F^0, \qquad (8)$$

where $\Delta F^0 \equiv F(\bar{\nu}_\mu) - F(\bar{\nu}_e)$, and $\bar{p}$ is the permutation factor. In assumptions of the normal mass hierarchy (ordering) and the absence of new neutrino states, $\bar{p}$ can be calculated precisely: $\bar{p} = 1 - P_{1e}$, where P_{1e} is the probability of $\bar{\nu}_1 \rightarrow \bar{\nu}_e$ transition inside the Earth.[24,25] It can be written as $\bar{p} = \sin^2\theta_{12} + f_{reg}$, where f_{reg} describes the effect of oscillations (regeneration of the $\bar{\nu}_e$ flux) inside the Earth. Due to the difference in distances traveled by neutrinos to Kamiokande, IMB and Baksan detectors inside the Earth: 4363 km, 8535 km and 10449 km cor-

respondingly, the permutation factors differ for these detectors (Fig. 6). The Earth matter effect can partially explain the difference between the Kamiokande and the IMB spectra of events.[25]

In contrast to $\bar{p}$, the original fluxes, and consequently ΔF^0, are not well known, and one can not make precise predictions of the flux modification in Eq. (8).

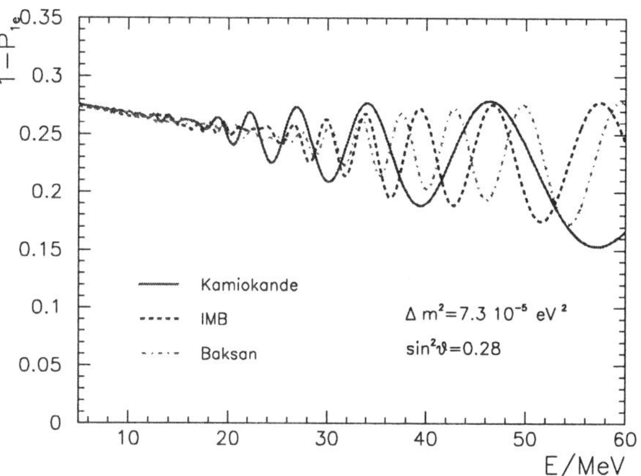

Figure 6. The permutation factor $\bar{p} = 1 - P_{1e}$ as a function of neutrino energy for Kamiokande II, IMB and Baksan detectors.[26]

For the inverted mass hierarchy and $\sin^2\theta_{13} > 10^{-5}$ one would get a stronger permutation, $\bar{p} = 1$, and therefore a harder $\bar{\nu}_e$ spectrum, as well as the absence of the Earth matter effect. This is disfavored by the data,[27] though in view of small statistics and uncertainties in the original fluxes it is not possible to make a firm statement.

2.5. Absolute Scale of Mass

From the oscillation results we can put only a lower limit on the heaviest neutrino mass:

$$m_h \geq \sqrt{\Delta m_{13}^2} > 0.04 \text{ eV}, \qquad (9)$$

where $m_h = m_3$ for the normal mass hierarchy, and $m_h = m_1 \approx m_2$ for the inverted hierarchy. The neutrinoless double beta decay is determined by the combination

$$m_{ee} = |\sum_k U_{ek}^2 m_k e^{i\phi(k)}|, \qquad (10)$$

where $\phi(k)$ is the phase of the k eigenvalue. Figure 7 summarizes the present knowledge of the absolute mass scale. Shown are the allowed regions in the plane of m_{ee} probed by $\beta\beta_{0\nu}$ decay and the mass of lightest neutrino probed by the direct kinematical methods and cosmology. The best present bound on m_{ee} is given by the Heidelberg-Moscow experiment: $m_{ee} < 0.35 - 0.50$ eV,[29] part of collaboration claims evidence of a positive signal.[30] Interestingly,

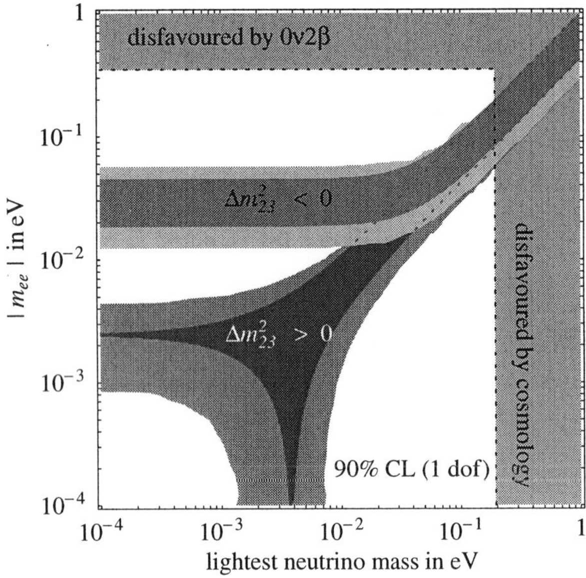

Figure 7. The 90% CL range for m_{ee} as a function of the lightest neutrino mass for the normal ($\Delta m_{23}^2 > 0$) and inverted ($\Delta m_{23}^2 < 0$) mass hierarchies.[28] The darker regions show how the allowed range for the present best-fit values of the parameters with negligible errors.

the present double beta decay measurements and cosmology have similar sensitivities $m_{ee} \sim m_1 \sim (0.2-0.5)$ eV. The latter corresponds to the degenerate mass spectrum: $m_1 \approx m_2 \approx m_3 \equiv m_0$. Analyses of cosmological data (with WMAP) result in the 95% CL upper bounds $m_0 < 0.23$ eV,[31] $m_0 < 0.6$ eV[32] and $m_0 < 0.34$ eV.[33] Independent analysis which includes the X-ray galaxy cluster data gives non-zero value $m_0 = 0.20 \pm 0.10$ eV.[34]

Future improvements of the upper bound on m_{ee} have the potential to distinguish between the hierarchies: According to Fig. 7, if the bound $m_{ee} < 0.012$ eV is established, the inverted hierarchy will be excluded at 90 % CL.

2.6. *LSND*

The situation with this ultimate neutrino anomaly[35] is really dramatic: all suggested physical (not related to the LSND methods) solutions are strongly or very strongly disfavored now. At the same time, being confirmed, the oscillation interpretation of the LSND result may change our understanding the neutrino (and in general fermion) masses.

A recent analysis performed by the KARMEN collaboration[36] has further disfavored a scenario[37] in which the $\bar{\nu}_e$ appearance is explained by the anomalous muon decay $\mu^+ \to \bar{\nu}_e \bar{\nu}_i e^+$ ($i = e, \mu, \tau$).

The CPT-violation scheme[38] with different mass spectra of neutrinos and antineutrinos is disfavored by the atmospheric neutrino data.[39] No compatibility of LSND and "all but LSND" data have been found below 3σ.[40]

The main problem of the (3 + 1) scheme with $\Delta m^2 \sim 1$ eV^2 is that the predicted LSND signal, which is consistent with the results of other short base-line experiments (BUGEY, CHOOZ, CDHS, CCFR, KARMEN) as well as the atmospheric neutrino data, is too small: the $\bar{\nu}_\mu \to \bar{\nu}_e$ probability is about 3σ below the LSND measurement.

Introduction of the second sterile neutrino with $\Delta m^2 > 8$ eV^2 may help.[41] It was shown[42] that a new neutrino with $\Delta m^2 \sim 22$ eV^2 and mixings $U_{e5} = 0.06$, $U_{\mu5} = 0.24$ can enhance the predicted LSND signal by $(60-70)\%$. The (3 + 2) scheme has, however, problems with cosmology and astrophysics.

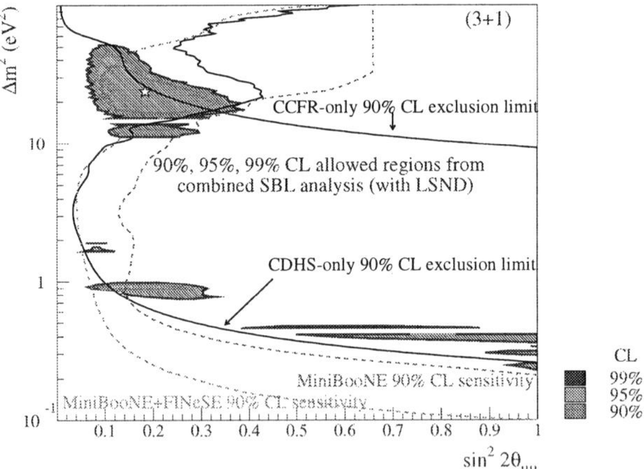

Figure 8. The allowed regions of parameters of the (3 + 1) scheme, Δm_{14}^2 and $\sin^2 2\theta_{\mu\mu} \approx 4|U_{\mu4}|^2$, at different confidence levels. Shown are the 90% sensitivity limits of the Mini-BooNE and the MiniBooNE+FINeSE experiments.

The generic prediction of (3 + n) schemes is the ν_μ oscillation disappearance at the level of existing upper bouds from CDHS,[43] CCFR,[44] and NOMAD[45] experiments. New searches of ν_μ disappearance are being performed by the MiniBooNE experiment[46] and planned by the proposed experiment FINeSE[47] (see Fig. 8, where the sensitivity region of these searches is shown[47]).

The combination of the two described solutions, namely the $3 + 1$ scheme with CPT-violation, has been considered.[48]

2.7. *Known and Unknown*

Information described in the previous sections can be summarized in the following way.

1. The observed ratio of the mass squared differences, $\Delta m_{12}^2/\Delta m_{23}^2 = 0.01 - 0.15$, implies that there is no strong hierarchy of masses:

$$\frac{m_2}{m_3} > \sqrt{\frac{\Delta m_{12}^2}{\Delta m_{23}^2}} = 0.18^{+0.22}_{-0.08}. \tag{11}$$

For charge leptons the corresponding ratio is 0.06.

2. There is the bi-large or large-maximal mixing between the neighboring families (1 - 2) and (2 - 3). Still rather significant deviation of the 2-3 mixing from the maximal one is possible.

3. Mixing between remote (1-3) families is weak.

Several key elements are unknown yet leading to a variety of possible interpretations.

Knowledge of the absolute mass scale, type of mass spectrum, and type of mass hierarchy is of the highest priority. The 1-3 mixing has important phenomenological consequences; its value is a test of the mechanisms of the lepton mixing enhancement. The CP-violating Majorana phases are extremely important for the structure of neutrino mass matrices. Deviations of the 2-3 and 1-2 mixings from maximal values play a crucial role in understanding the origins of neutrino masses. The existence of new neutrino states (their search should be a permanent item in the scientific agenda) may change completely our approaches to the underlying theory.

These are phenomenological and experimental questions we will deal with during the next $20 - 30$ years.

3. Open Theoretical Questions

What does all this (results on neutrino masses and mixing) mean?

Among old, still open, questions are the following: Why are neutrino masses so small in comparison with the charged lepton and quark masses? What is the origin of neutrino mass? Is it the same as the one for quarks and charged leptons?

What are the relations between neutrino masses and other mass/energy scales in nature, e.g. the scale of cosmological constant or dark energy?

Why is the lepton mixing large? Why is it so different from quark mixing? Is the 2-3 mixing exactly maximal? What are the relations between different mixing angles (if any)? How is the observed pattern of lepton mixing is generated?

In the quark sector the smallness of mixing is related to the strong mass hierarchy. What are the relations between the lepton masses and lepton mixing?

Do neutrinos show certain flavor or horizontal symmetry? If so, is this symmetry consistent with the pattern of quark masses and mixing?

Are the results of neutrino masses and lepton mixing consistent with the quark-lepton symmetry and Grand Unification?

If new light sterile neutrinos exist what is their nature and why are they light?

What are the implications of the neutrino results for GUT, SUSY, models with extra dimensions, and strings? *Vice versa:* what can these beyond the SM theories tell us about neutrinos?

4. Bottom-Up

One can try the "top-down" approach confronting immediately a proposed model with experimental results. Inversely, to get some hints in answering the above questions, it may be worthwhile to try to move bottom-up.

4.1. *Neutrino mass matrix*

There are several steps in the bottom-up approach.

1). Take the results on Δm_{ij}^2, θ_{ij}, m_{ee}, etc..

2). Reconstruct the neutrino mass matrix in the flavor basis (where the charge lepton mass matrix is diagonal) assuming also that neutrinos are Majorana particles. Notice that the mass matrix unifies information contained in masses and mixing angles and this may provide some more hints toward the underlying theory.

3.) Identify the symmetry basis (which may differ from the flavor basis) and the symmetry scale.

Take into account the renormalization group effects.

4.) Identify the symmetry (as well as mechanism of symmetry violation, if needed) and underlying dynamics.

Let us make the first step in the bottom-up approach. The mass matrix in the flavor basis can be written as

$$m = U^* m^{diag} U^+, \qquad (12)$$

where $U = U(\theta_{ij}, \delta)$ is the mixing matrix, δ is the Dirac CP-violating phase, and

$$m^{diag} = diag(m_1 e^{-2i\rho}, \ m_2, \ m_3 e^{-2i\sigma}). \qquad (13)$$

Here ρ and σ are the Majorana phases. The mass eigenvalues equal $m_2 = \sqrt{m_1^2 + \Delta m_{12}^2}$, and $m_3 = \sqrt{m_1^2 + \Delta m_{13}^2}$.

The results of reconstruction of the mass matrix[49,50] are shown in Figs. 9, 10, and 11 as the $\rho - \sigma$ plots for the absolute values of the 6 independent matrix elements. They correspond to three extreme cases: normal mass hierarchy, quasi-degenerate spectrum and inverted mass hierarchy. The figures illustrate a variety of possible structures. In particular, for the normal mass hierarchy (Fig. 9) there is clear structure with the dominant $\mu-\tau$ block. Interesting parameterizations of the mass matrix (up to an overall mass factor) are

$$\begin{pmatrix} 0 & 0 & \lambda \\ 0 & 1 & 1 \\ \lambda & 1 & 1 \end{pmatrix}, \qquad \begin{pmatrix} \lambda^2 & \lambda & \lambda \\ \lambda & 1 & 1 \\ \lambda & 1 & 1 \end{pmatrix}, \qquad (14)$$

where $\lambda \sim 0.2$. Also the matrix similar to the first one in Eq. (14) with $m_{12} \sim \lambda$ and $m_{13} \approx 0$ is possible.

In the case of a quasi-degenerate spectrum, the interesting dominant structures are

$$\begin{pmatrix} 1 & 0 & 0 \\ 0 & 1 & 0 \\ 0 & 0 & 1 \end{pmatrix}, \qquad \begin{pmatrix} 1 & 0 & 0 \\ 0 & 0 & 1 \\ 0 & 1 & 0 \end{pmatrix}. \qquad (15)$$

These matrices are realized for values of phases in the corners of the plots: $\rho, \sigma = 0, \pi$ (the first matrix) or at $\rho = 0, \pi$, $\sigma = \pi/2$ (the second one) which corresponds to definite CP-parities of the mass eigenstates. Also the "democratic" structure with equal moduli of elements is possible for the non-trivial values of phases.[51] Changing the phases one can get any intermediate structure between those in Eqs. (14) and (15).

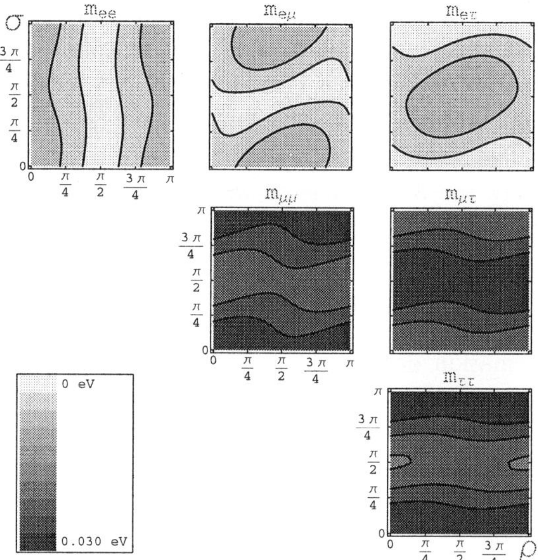

Figure 9. The Majorana mass matrix for the normal mass hierarchy: $m_3/m_2 = 5$, $m_1 \approx 0.006$ eV. We show contours of constant mass in the $\rho - \sigma$ plots for the moduli of mass matrix elements. We take for other parameters $\Delta m_{12}^2 = 7 \times 10^{-5}$eV^2, $\Delta m_{13}^2 = 2.5 \times 10^{-3}$eV^2, $\tan^2 \theta_{12} = 0.42$, $\tan \theta_{23} = 1$, $\sin \theta_{13} = 0.1$, and $\delta = 0$.

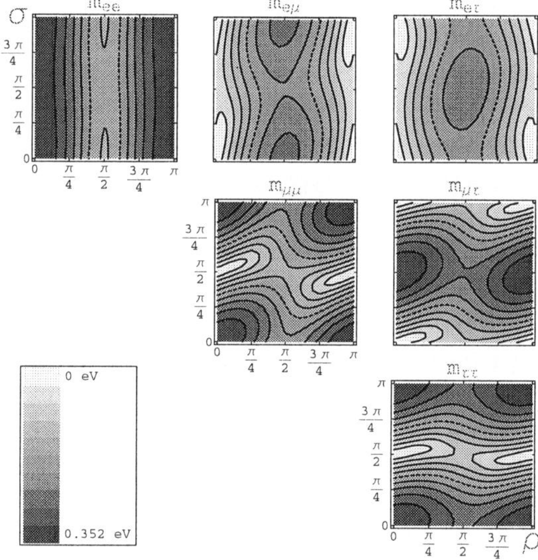

Figure 10. The same as in Fig. 9 for the quasi-degenerate spectrum: $m_3/m_2 = 1.01$, $m_1 \approx 0.35$ eV.

In the case of the inverted hierarchy, generically the ee- element is not small. Among interesting examples are

$$\begin{pmatrix} 0.7 & 1 & 1 \\ 1 & 0.1 & 0.1 \\ 1 & 0.1 & 0.1 \end{pmatrix}, \qquad \begin{pmatrix} 1 & 0.1 & 0.1 \\ 0.1 & 0.5 & 0.5 \\ 0.1 & 0.5 & 0.5 \end{pmatrix}. \qquad (16)$$

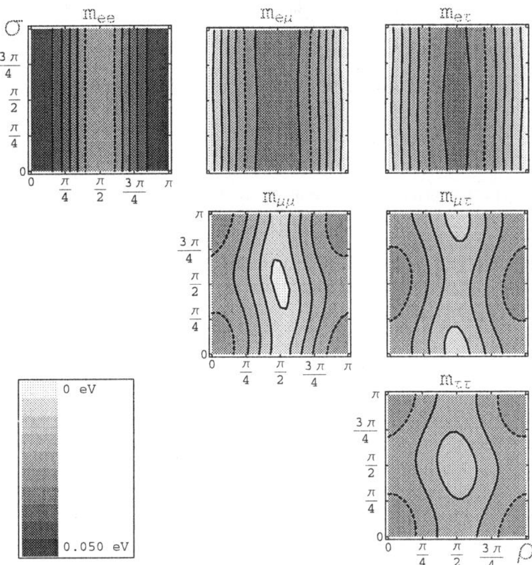

Figure 11. The same as in Fig. 9 for the inverted mass hierarchy: $m_3/m_2 = 0.1$, $m_3 \approx 0.005$ eV.

In the SM and MSSM the renormalization group effects do not change the structure of the mass matrix: the corrections to a given element are proportional to the element itself. Furthermore, the corrections are small even in the SUSY case (below 0.1%). So, unless some new interactions exist, the mass matrix determined at low energies does not change structure when running up to the scale where the corresponding mass operators are formed or up to the symmetry scale.

In contrast to the matrix structure, the radiative corrections are important for the oscillation observables in the case of partially or quasi degenerate mass spectra.

Scanning the $\rho - \sigma$ plots shown in Figs. 9, 10, and 11, one can make the following observations.

1). A large variety of different structures is still possible, depending strongly on the unknown m_1, type of mass hierarchy and Majorana phases. The dependence on $\sin\theta_{13}$ and δ is weak.

2). Generically the hierarchy of elements is not strong: within 1 order of magnitude. At the same time matrices with one or two exact zeros are not excluded.[52]

3). Matrices are possible with:

- dominant (i) diagonal elements ($\sim I$), (ii) $\mu\tau$-block, (iii) e-row elements, (iv) $ee-, \mu\tau-, \tau\mu-$ elements (triangle structure),

- democratic structure,

- flavor alignment,

- non-hierarchical structures with all elements of the same order,

- flavor disordering,

- zeros at different places, and

- equalities of various element.

4). Typically, the hierarchical structures appear for the Majorana phases near 0, $\pi/2$, or π.

5). Matrices can be parameterized in terms of powers of small parameter $\lambda = 0.2 - 0.3$ consistent with the Cabibbo mixing.

In a significant part of the parameter space the matrix does not show any regularities, and relative values of its elements appear as random numbers which spread within one order of magnitude. This supports the idea of "Anarchy".[53,54] Consideration of the anarchy of elements is a test of the possible complexity of the neutrino mass matrix. The case of anarchy can be imitated if neutrino masses have several different contributions, and even if each of them has well defined structure or symmetry, the sum may show up as a matrix with disorder. In this connection one can consider representations of the mass matrix as the sum of matrices, given in Eq. (15), as well as the democratic matrix with certain coefficients.[55]

What is more fundamental: oscillation observables or neutrino mass matrix in some basis? The answer may depend on the type of mass spectrum. In the case of hierarchical spectrum the observables are visibly imprinted into the structure of the mass matrix. In contrast, for the quasi-degenerate spectrum they are just very small perturbations of the dominant structure which is determined by the non-oscillatory parameters: the absolute mass scale and the Majorana CP-violating phases. Then the oscillation parameters can be a result of interplay of some small, in particular, radiative corrections.

In the case of Majorana neutrinos, the elements of the mass matrix are physical parameters: they can be immediately measured in the neutrinoless beta decay and, in principle, in other similar processes. In practice, it is not possible to reconstruct the mass matrix from experiment completely. Even in the most optimistic case the phase σ will be undetermined and Figs. 9, 10, and 11, give an idea of the remaining uncertainty. Only in the case of the g mass hierarchy does the dependence on σ disappear. The hope is that even without complete reconstruction of the mass matrix we will be able to uncover the underlying physics.

4.2. *Neutrino Mass and Horizontal Symmetry*

Do the results on neutrino masses and mixing indicate certain regularities or symmetry? Can the dominant structures of the mass matrix be explained by a symmetry with the sub-dominant elements appearing as a result of violations of the symmetry? Is the neutrino mass matrix consistent with symmetries suggested for quarks?

The following symmetries have been considered.

1). $L_e - L_\mu - L_\tau$.[56] This symmetry supports, in particular, the structure with an inverted mass hierarchy. However, the rather large element m_{ee} (Fig. 11) shows strong violation of this symmetry.

2). Discrete symmetries: A_4,[57] S_3,[58] Z_4,[59], and D_4.[60] They reproduce successfully the dominant structures in Eq. (15) as well as the "democratic" matrix.

However, both classes of symmetries 1) and 2) typically treat quarks and leptons differently.

3). $U(1)$.[61] In the Froggatt-Nielsen context[62] this symmetry can describe mass matrices of both quarks and leptons. However, the claimed predictability of this approach can be questioned: the $U(1)$ charges should be considered as discrete free parameters. Furthermore, precise description of data usually requires coefficients (prefactors) of the order 1 (1/2 - 2) in front of powers of the expansion parameter. The outcome is that the mixing pattern depends substantially on values of these unknown prefactors.

4). $SU(2)$,[63] $SO(3)$,[64] and $SU(3)$[65] require a complicated Higgs sector to break the symmetry. Often models are too restrictive and predictions are on the borders of allowed regions.

The question is still open. Different symmetries are consistent with the neutrino data. But realizations of these symmetries in specific models are not simple. The hope is that future neutrino data (better knowledge of the mass matrix) can discriminate among possibilities.

5. How We May Go...

5.1. *Neutrality and Mass*

In answering the questions of Sec. 3 one can implement the "minimalistic" approach, that is, to try to relate features of the neutrino masses and mixings with already known differences of characteristics of neutrinos and other fermions.

The main feature of neutrinos is neutrality:

$$Q_\gamma = Q_c = 0. \qquad (17)$$

It leads to the following possibilities:

- neutrinos can be Majorana particles;

- they can mix with singlets of the SM symmetry group; and

- the right-handed components (RH), if they exist, are singlets of $SU(3) \times SU(2) \times U(1)$. So, their masses are unprotected by the symmetry and therefore can be large.

In turn, properties of the RH components open two other possibilities. The RH neutrinos can:

- have large Majorana masses: $M_R \gg V_{EW}$ (which leads to the see-saw); and

- propagate in (large, or warped, or infinite) extra dimensions, or be located on the "hidden" (not ours) brane in contrast to other fermions.

Introduction of the RH neutrino has a number of attractive features,[66] in particular, it allows one to extend the electroweak symmetry to the gauged $SU(2)_L \times SU(2)_R \times U(1)_{B-L}$.

Is this enough to explain the properties of the mass spectrum and mixings?

5.2. *Effective Operator*

Suppose the SM particles are the only light degrees of freedom. Then at low energies (after integrating out the heavy degrees of freedom) one can get the operator:[67]

$$\frac{\lambda_{ij}}{M}(L_i H)^T (L_j H), \quad i,j = e, \mu, \tau, \qquad (18)$$

where L_i is the lepton doublet, λ_{ij} are the dimensionless couplings and M is the cut-off scale. After EW symmetry breaking it generates the neutrino masses

$$m_{ij} = \frac{\lambda_{ij} \langle H \rangle^2}{M}. \qquad (19)$$

For $\lambda_{ij} \sim 1$ and $M = M_{Pl}$ we find $m_{ij} \sim 10^{-5}$ eV.[68] Three important conclusions immediately follow from this consideration.

1). The Planck scale (gravitational) interactions are not enough to generate the observed values of the masses. So, new scales of physics below M_{Pl} should exist.

2). Contributions to the neutrino masses of the order $\sim 10^{-5}$ eV are still relevant for phenomenology. Sub-dominant structures of the mass matrix can be generated by the Planck scale interactions.[69]

3). The neutrino mass matrix can get observable contributions from all possible energy/mass scales from the EW scale to the Planck scale. As a consequence, the structure of the mass matrix can be rather complicated.

5.3. See-saw

The see-saw (type I) mechanism[70] implements the neutrality in full strength (Majorana nature, heavy RH components). Introducing the Dirac mass matrix, $m_D = Y v_{EW}$, where Y is the matrix of Yukawa couplings and v_{EM} is the electroweak VEV, we have

$$m = -m_D^T M_R^{-1} m_D \quad (type\,I). \qquad (20)$$

If the $SU(2)$ triplet, Δ_L, exists which develops a VEV $\langle \Delta_L \rangle$, the left-handed neutrinos can get a direct mass m_L via the interaction $f_\Delta L^T L \Delta_L$. If Δ_L is very heavy, it can develop the induced VEV from interactions with a doublet: $\langle \Delta_L \rangle = v_{EW}^2/M$. So that

$$m_L = f_\Delta \frac{v_{EW}^2}{M} \quad (type\,II), \qquad (21)$$

and here we deal with the see-saw of VEV's.[71]

In $SO(10)$ with 126_H-plet of Higgses we have $M_R = f v_R$, where f is the Yukawa coupling of the matter 16-plet with 126_H and v_R is the VEV of the $SU(5)$ singlet component of 126_H. Now $f_\Delta = f$, and the general mass term which contains both types of contributions can be written as

$$m = \frac{v_{EM}^2}{v_R}(f\lambda - Y^T f^{-1} Y). \qquad (22)$$

Here λ is the coupling of 10- and 126-plets. According to this expression the flavor structure of the two contributions may partially correlate.

The number of RH neutrinos can differ from 3. Two possibilities have been explored:

"*... less than 3*": which corresponds to the 3×2 see-saw in the case of two RH neutrinos.[72] Such a possibility can be realized in the limit when one of the RH neutrinos is very heavy: $M \sim M_{Pl}$, being,

e.g. unprotected by the $SU(2)_H$ horizontal symmetry. It leads to one exactly massless LH neutrino and smaller number of free parameters.

One can further reduce the number of unknown parameters postulating zeros in the Dirac matrix m_D.[52] This can lead to the predictions for $\sin\theta_{13}$, m_{ee}, as well as for relations between δ and the phase responsible for leptogenesis.

"*... more than 3*": additional singlets of the SM may not be related to the family structure. Alternatively, three additional singlets, S, which belong to families, can couple to the RH neutrinos. In the latter case the double see-saw can be realized.[73] In the basis (ν, ν^c, S), the mass matrix may have the form

$$\begin{pmatrix} 0 & m_D & 0 \\ m_D^T & 0 & M \\ 0 & M^T & \mu \end{pmatrix} \qquad (23)$$

which leads to the light neutrino masses:

$$m = -m_D^T (M^{-1})^T \mu M^{-1} m_D. \qquad (24)$$

Two interesting limits are: (i) $\mu \ll M$, it allows one to reduce all high mass scales for the same values of the light neutrino masses, (ii) $\mu \gg M$, e.g. $\mu = M_{Pl}$, and $M = M_{GU}$: in this case the intermediate mass scale, $M_{GU}^2/M_{Pl} = 10^{12} - 10^{14}$ GeV for the masses of RH neutrinos can be obtained.

5.4. Grand Unification and Neutrino Mixing

GU theories provide a large mass scale comparable to the scale of RH neutrino masses.[74] Furthermore, one can argue that GUT + see-saw can naturally lead to the large lepton mixing in contrast to the quark mixing. The arguments go like this:

1. Suppose that all quarks and leptons of a given family are in a single multiplet F_i (as 16 of SO(10)).

2. Suppose that all Yukawa couplings are of the same order thus producing matrices with generically large mixing.

3. If the Dirac masses are generated by an unique Higgs multiplet, say 10_H of SO(10), the mass matrices of the up and down components of the weak doublets have identical structures, and so, will be diagonalized by the same rotations. As a result:

- no mixing appears for quarks, and

- masses of up and down components will be equal to each other (this needs to be corrected).

4. In contrast to other fermions, the RH neutrinos acquire Majorana masses via the additional Yukawa couplings (with 126_H of SO(10)).

5. If those (Majorana type) Yukawa couplings are also of the generic form, they produce M_R with large mixing which leads then to large lepton mixing.

The problem of this scenario is the strong hierarchy of the quark and lepton masses. Indeed, taking the neutrino Dirac masses as $m_D = diag(m_u, m_c, m_t)$ in a spirit of GU, we find that for generic M_R the see-saw type I formula (20) produces strongly hierarchical mass matrix with small mixings. Possible solutions are:

1. a special structure of M_R which compensates the strong hierarchy in m_D;

2. a substantial difference in the Dirac matrices of quarks and leptons: $m_D(q) \neq m_D(l)$; or

3. a type II see-saw for which there is no relation to m_D.

In what follows we will comment on these three possibilities.

See-Saw enhancement of mixing.[75] Can the same mechanism (see-saw) which explains the smallness of the neutrino mass also explain the large lepton mixing? So, that the large mixing appears as an artifact of the see-saw?

The idea is that due to the (approximate) quark-lepton symmetry, the Dirac mass matrices of the quarks and leptons have the same (similar) structure $m_D \sim m_{up}$, $m_l \sim m_{down}$ leading to small mixing in the Dirac sector. However, the special structure of M_R (which has no analogue in the quark sector) leads to an enhancement of lepton mixing. Two different possibilities have been found:

- strong (nearly quadratic) hierarchy of the RH neutrino masses: $M_{iR} \sim (m_{iup})^2$; and

- strong interfamily connection (pseudo Dirac structures) like

$$M_R = \begin{pmatrix} A & 0 & 0 \\ 0 & 0 & B \\ 0 & B & 0 \end{pmatrix}. \qquad (25)$$

In the three neutrino context both possibilities can be realized simultaneously, so that the pseudo

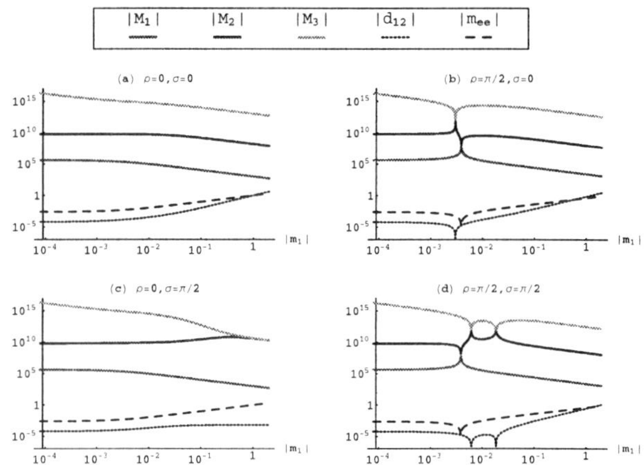

Figure 12. The masses of the RH neutrinos in GeV as functions of the lightest neutrino mass $|m_1|$ in eV (solid lines) for different values of the Majorana phases of light neutrinos.[75] We take $\sin\theta_{13} = 0$ and the best-fit values of other oscillation parameters. Shown is also the dependence of m_{ee} (in eV) on $|m_1|$ (thin dashed line).

Dirac structure leads to maximal 2-3 mixing, whereas the strong hierarchy $A \ll B$ enhances the 1-2 mixing.

In Fig. 12 we show dependences of the RH neutrino masses reconstructed from the low energy data on the lightest neutrino mass for different values of the Majorana phases. According to this figure

1). In the largest part of the parameter space (m_1, ρ, σ) there is a very strong (4 - 5 orders of magnitude) mass hierarchy of the RH neutrinos.

2). The lightest mass is typically below 10^5 GeV, thus strongly violating the lower bound on the mass from the condition of successful leptogenesis: $M_1 > 4 \times 10^8$ GeV.[77]

3). At certain points the level crossings occur. At these points (i) there is a strong degeneracy of mass eigenstates: in particular, $M_1 = M_2$, (ii) the lightest mass can be as large as 10^8 GeV and (iii) the lepton asymmetry can be resonantly enhanced[78] up to the required value.

Large mixing and type II see-saw. In general, the structure of neutrino mass matrix generated by the type II (triplet) see-saw is not related to structures of matrices of other fermions.

In some particular cases, however, the relations can appear leading to interesting consequences. In the SO(10) model the 126_H Higgs multiplet can play a double role: (i) generate neutrino masses $m_L = Y_{126}v_\Delta$, where v_Δ is the VEV of the SU(2) triplet in 126_H; and (ii) give contributions to the quark and charged lepton masses (if doublets contained in 126 get VEV's) reproducing the Georgi-Jarlskog mass relation for the first and second generations.

Since $m_b - m_\tau \propto (Y_{126})_{33}$, the contribution of 126_H destroys the $b-\tau$ unification unless $(Y_{126})_{33} \lesssim (Y_{126})_{23}$. The latter leads to large (but not necessarily maximal) 2-3 lepton mixing. In this context, the $b-\tau$ unification implies large 2-3 mixing.[79]

In models of this type a successful leptogenesis is possible with participation of the scalar triplet.[80] The model has been generalized also to 3 generations, leading typically to 1-2 mixing at the larger side of the allowed region.[81]

Single RH neutrino dominance.[82] The large neutrino mixing and relatively strong mass hierarchy implied by the solar and atmospheric neutrino data can be reconciled if only one RH neutrino gives the dominant contribution to the see-saw. (This leads to the submatrix of m_L with nearly zero determinant.) There are two different realizations of this possibility. In one case the large mixing originates from the large mixing in the Dirac neutrino mass matrix m_D: two LH neutrinos have nearly equal couplings to the dominating RH component. Suppose that $(m_D)_{23} \approx (m_D)_{33} = m$, $(m_D)_{13} = \lambda m$ ($\lambda \approx 0.2$) and all other elements of m_D are much smaller. Then if only $(M^{-1})_{33}$ is large in the inverted matrix, the see-saw leads to the mass matrix which reproduces the second structure in Eq. (14).

In another version, the dominance is realized when two RH neutrinos are much heavier than the third (dominating) one and no large mixing in m_D appears. This is equivalent to the strong mass hierarchy case of the see-saw enhancement mechanism. A realization requires $(m_D)_{22} \approx (m_D)_{23} \ll (m_D)_{33}$, and $(M^{-1})_{22}$ being the dominant element.

Lopsided models.[83] Large lepton mixing follows from the charge lepton mass matrix which should be non-symmetric (no left-right symmetry). This does not contradict the Grand Unification: in SU(5) the LH components of leptons are unified with the RH components of quarks: $5 = (d^c, d^c, d^c, l, \nu)$. Therefore large mixing of the LH leptonic components is accompanied by large mixing of the RH d-quarks which is unobservable. Introducing the Dirac mass matrix of the charged leptons with the only large elements $(m_l)_{33} \sim (m_l)_{23}$, one can obtain the large 2-3 lepton mixing. This scenario can also be realized in $SO(10)$, if the symmetry is broken via $SU(5)$. A double lopsided matrix for both large mixing is also possible.

Radiative enhancement of mixing.[84] The idea is that the difference between the quark and lepton mixings is a result of different renormalization group effects. The lepton mixing is small (similar to quark mixing) at the GU scale but running to low energies leads to its enhancement.

The main requirement of such an enhancement is that the neutrino mass spectrum is quasi-degenerate (and this is the key point which distinguishes quarks and leptons). The enhancement occurs when neutrinos become even more degenerate at low energies. For instance, running of the 2-3 mixing is described by

$$\frac{d\sin\theta_{23}}{dt} \sim (\sin\theta_{12}U_{\tau 1}D_{31} - \cos\theta_{12}U_{\tau 2}D_{32}), \quad (26)$$

where $t \equiv 1/8\pi^2 log(q/M)$, $D_{ij} \equiv (m_i + m_j)/(m_i - m_j)$, and m_i are the mass eigenvalues. The minus sign in the denominators of D_{ij} plays the key role.

The mechanism requires fine tuning of the initial mass splittings and radiative corrections. In principle, in the MSSM both the 1-2 and 2-3 mixings can be enhanced in this way. In the SM only 1-2 mixing can be enhanced.

The fine-tuning problem can be avoided if the masses are generated from the Kähler potential. In this case large mixing appears as the infrared fixed point.[85]

Another possible application of the radiative effects is the generation of small oscillation parameters like Δm_{12}^2, and $\sin\theta_{13}$.[86] Again this can be realized only in the case of a mass spectrum with degeneracy.

5.5. *How To Test the See-Saw Mechanism?*

This is the key question which implies essentially the test of existence of the heavy Majorana RH neutrinos. There are two (known) possibilities.

1). Leptogenesis.[87] For the hierarchical RH neutrino spectrum and in assumption of the type I see-saw, it gives bounds on (i) the mass of the lightest RH neutrino M_{R1}, (ii) the effective parameter $\tilde{m}_1$ which determines the washout effect. Notice that the leptogenesis probes the combination of the Yukawa couplings $(YY^\dagger)_{ii}$.

2). The RH neutrinos can produce renormalization effects above the scale of their masses: between M_R and, say, the GUT scale. In particular, they can renormalize the $m_b - m_\tau$ mass relation[88] which leads to the observable effect in the assumption of $m_b - m_\tau$ unification at the GUT scale.

Another possibility is that the renormalization due to RH neutrinos modifies masses and mixing of the light neutrinos, e.g. enhances the lepton mixing.[89]

5.6. *SUSY See-Saw*

Additional possibilities to test the see-saw mechanism appear if SUSY is realized. The part of superpotential relevant for the see-saw can be written as

$$W_{lep} = l^{cT} Y_l L H_1 + \nu^{cT} Y L H_2 + \frac{1}{2}\nu^{cT} M_R \nu^c. \quad (27)$$

Structures relevant for the see-saw are imprinted into properties of the slepton sector. So, studying the properties of sleptons (masses, decay rates, etc.) one can get information about the neutrino mass generation.

Certain predictions can be made in the assumptions of universal soft SUSY breaking masses (m_0, A_0) at high (GUT ?) scale M_X, and the absence of new particles/interactions up to M_X.

Due to renormalization group effects the Yukawa couplings (27) give contributions to the masses of left sleptons at low energies:[90]

$$(m_S^2)_{ab} = m_a^2 \delta_{ab} - \frac{3m_0^2 + A_0^2}{8\pi^2}(Y^\dagger)_{ai}(Y)_{ib} log\left(\frac{M_X}{M_{iR}}\right). \quad (28)$$

The contribution splits masses of sleptons of different flavors and sleptons-antisleptons as well as leading to mixing of sleptons (the off-diagonal terms in Eq. (28)) which is related to the mixing of neutrinos.

In turn, these contributions to the slepton masses produce a number of observable effects:

1. rare leptonic decays: the one-loop mixing of sleptons of different flavors induces the flavor violating decays: $\mu \to e\gamma$, $\tau \to \mu\gamma$, $\tau \to e\gamma$;[90]

2. sneutrino flavor oscillations;[91]

3. slepton decays;[92]

4. sneutrino-antisneutrino oscillations;[93] and

5. contribution to the electric dipole moments of charged leptons.[94]

Up to a log factor these effects are determined by the combination $\sim (Y^\dagger Y)$. Notice that another combination: $(Y^T M^{-1} Y)$ enters the see-saw (type I) formula. It was shown[95] that knowledge of these combinations allows, in principle, one to reconstruct parameters of the RH neutrino sector (masses, phases). For this, in turn, one needs to reconstruct completely the mass matrix of light neutrinos, discover SUSY and measure rare processes with high enough accuracy. This looks practically impossible, at least now.

Partial tests of the see-saw can be done by studying the rare decays induced by the slepton mixing. The branching ratio equals

$$B(\mu \to e\gamma) = \frac{\alpha^3}{G_F^2 m_{SUSY}^8}|(m_S^2)_{\mu e}|^2 \tan^2\beta, \quad (29)$$

where $m_{SUSY} = m_{SUSY}(m_0, m_{1/2})$ is the effective SUSY mass parameter, $m_{1/2}$ is the gaugino mass, $\tan\beta$ is the ratio of MSSM Higgs doublet VEV's, and $(m_S^2)_{\mu e}$ is given in Eq. (28).

If the large lepton mixing originates from the Dirac mass matrix (lopsided models, versions of the single RH neutrino dominance), the Yukawa couplings $Y_{\mu i}$, Y_{ei} are large and for $m_{SUSY} \sim 200$ GeV the branching ratio in Eq. (29) turns out to be $10^{-12} - 10^{-11}$ - at the level of the present experimental bound. Determination of m_{SUSY} in terms of m_0 and $m_{1/2}$ beyond the leading log approximation has further enhanced the branching ratio.[96]

5.7. *Other Mechanisms*

What are other possibilities apart from the see-saw? The incomplete list includes.

1. Various radiative mechanisms: The Zee (one loop) mechanism[97] is essentially excluded in its minimal version by data.[98] One loop generation also occurs in the SUSY models with trilinear R-parity violating couplings. Neutrino masses can be generated in two loops as suggested by Zee[97] and Babu.[99]

2. Neutrino mass generation by the bi-linear R-parity violation terms.[100] This mechanism is a combination of the see-saw and radiative effects: neutrino mass appears as the see-saw due to mixing of neutrinos with neutralinos (Higgsino) and the latter is generated by running from the high mass scales.

3. Mechanisms related to the existence of extra dimensions. There are different scenarios: (i) large extra dimensions (ADD)[101] where the Dirac neutrino mass is suppressed by the large volume of

extra dimensions, (ii) warped extra dimensions (RS), where the RH neutrinos can be zero modes localized on the hidden brane, thus leading again to the small Dirac neutrino mass,[102] and (iii) infinite extra dimensions.[103]

4). There are several new proposals which implement various realizations of the see-saw mechanism.

Models with dynamical symmetry breaking[104] reproduce the low scale see-saw mechanism with the RH neutrino masses below the symmetry breaking scale, and correspondingly, with small Dirac masses (much smaller than the masses of quarks and charged leptons).

Also in models with "Little Higgs" [105] the neutrinos get masses via the low scale see-saw.

In models with dimensional deconstruction[106] the see-saw scale (masses of the RH neutrinos) is determined by the inverse lattice spacing. The Dirac mass matrix is nearly diagonal and the large (maximal) mixing follows from the pseudo-Dirac structures in the RH mass matrix which correspond to $L_e - L_\mu - L_\tau$ symmetry. Essentially, the maximal mixing appears because different lepton families belong to different sites of the lattice and the link scalar fields (singlets of SM) couple with the RH components of neutrinos, thus producing the off-diagonal (in flavor space) Majorana mass terms.

These alternative mechanisms imply deviation from minimality. They can accommodate the neutrino masses and produce some interesting features. However, they do not really lead to a better understanding of the experimental results and require introduction of additional elements (physics) beyond the Standard Model with RH neutrinos.

Turning the arguments: the neutrino data can be used to put limits on the suggested alternative mechanisms, and consequently, on physics beyond the SM. In such a way the neutrinos can probe extra dimensions, dynamical symmetry breaking, etc..

6. Conclusions

During last several years enormous progress has been achieved in the determination of the neutrino masses and mixings and in studies of the neutrino mass matrix. Still, large freedom exists in the possible structures of the mass matrix which leads to very different interpretations of the results. There are no definite hints from the bottom-up approach yet, and more information is needed, in particular, on the type of mass spectrum.

The main question (still open) is: what is behind the obtained results? What is the underlying physics? Preference? Probably, the see-saw associated to the Grand Unification. The context of $SO(10)$ looks rather appealing in spite of known problems. Other mechanisms (being in a less advanced stage of development) are not excluded and can give leading or sub-leading contributions to neutrino mass.

How can ideas about neutrinos be checked? Future experiments will perform precision measurements of neutrino parameters. Apart from this to understand the underlying physics we will certainly need results from the non-neutrino experiments:

- astrophysics and cosmology;

- searches for rare processes like flavor violating lepton decays, proton decay, etc.; and

- future high energy colliders.

Acknowledgments

I am grateful to E. Kh. Akhmedov, P. de Holanda, M. Frigerio, C. Lunardini and O. Peres for fruitful discussions and help in preparation of this talk.

References

1. SNO collaboration (Q. R. Ahmad *et al.*), nucl-ex/0309004.

2. J. N. Bahcall, M.H. Pinsonneault and S. Basu, *Astrophys. J.* **555**, 990 (2001).

3. L. Wolfenstein, *Phys. Rev.* D **17**, 2369 (1978); in "*Neutrino-78*", Purdue Univ., C3, (1978). S. P. Mikheyev and A. Yu. Smirnov, *Yad. Fiz.* **42**, 1441 (1985); *Nuovo Cim.* C **9**, 17 (1986); *Sov. Phys. JETP*, **64**, 4 (1986).

4. A. B. Balantekin and H. Yüksel, hep-ph/0309079.

5. G.L. Fogli, E. Lisi, A. Marrone, A. Palazzo, hep-ph/0309100.

6. M. Maltoni, T. Schwetz, M. A. Tortola, J.W.F. Valle, hep-ph/0309130 (v.2).

7. P. Aliani, V. Antonelli, M. Picariello, E. Torrente-Lujan, hep-ph/0309156.

8. P. Creminelli, G. Signorelli, A. Strumia, hep-ph/0102234, v.5, Sept. 15 (2003).

9. A. Bandyopadhyay, S. Choubey, S. Goswami, S. T. Petcov, D.P. Roy, hep-ph/0309174.

10. P. C. de Holanda, A.Yu. Smirnov, hep-ph/0309299.

11. K. Eguchi *et al.*, (KamLAND), *Phys. Rev. Lett.* **90**, 021802 (2003).

12. CHOOZ Collaboration, M. Apollonio *et al.*, *Phys.*

Lett. B **466**, 415 (1999); *Eur. Phys. J.* , C **27**, 331 (2003).

13. B. T. Cleveland *et al.*, *Astroph. J.* **496**, 505 (1998).

14. Y. Fukuda *et al.*, (Super-Kamiokande) *Phys. Rev. Lett.* **86**, 5651 (2001), *ibidem*, **82**, 5656 (2001), *Phys. Lett.* B **539**, 179 (2002); hep-ex/0309011.

15. P. C. de Holanda, A.Yu. Smirnov, hep-ph/0307266.

16. BOREXINO Collaboration, G. Alimonti *et. al.*, *Astropart. Phys.* **16**, 073022 (2002).

17. See talks at Low 4Nu *4th International Workshop on Low Energy and Solar Neutrinos*, Paris, France, May 19-21 (2003), http://cdfpc53.in2p3.fr/LowNu2003/.

18. Super-Kamiokande Collaboration, Y. Hayato, talk given at *the HEP2003 International Europhysics Conference* (Aachen, Germany, 2003), website: eps2003.physik.rwth-aachen.de .

19. G.L. Fogli, E. Lisi, A. Marrone, D. Montanino, A. Palazzo, A.M. Rotunno, hep-ph/0308055.

20. O. L. G. Peres, A.Yu. Smirnov, *Phys. Lett.* B **456**, 204 (1999); hep-ph/0309312.

21. M. C. Gonzalez-Garcia, C. Pena-Garay, hep-ph/0306001.

22. Y. Farzan, A.Yu. Smirnov, *Phys. Rev.* D **65**, 113001 (2002).

23. K. Hirata *et al.*, *Phys. Rev. Lett.* **58**, 1490 (1987); R. M. Bionta *et al.*, *Phys. Rev. Lett.* **58**, 1494 (1987); E. N. Alekseev, *et al.*, *JETP Lett.* **45**, 589 (1987).

24. A. S. Dighe, A. Yu. Smirnov, *Phys. Rev.* D **62**, 033007 (2000).

25. C. Lunardini, A.Yu. Smirnov, *Phys. Rev.* D **63**, 073009, (2001); M. Kachelriess *et al.*, *Phys. Rev.* D **65**, 073016 (2002).

26. C. Lunardini, A.Yu. Smirnov, (in preparation).

27. A. Yu. Smirnov, D. N. Spergel, J. N. Bahcall, *Phys. Rev.* D **49**, 1389 (1994); H. Minakata, H. Nunokawa, *Phys. Lett.* B **504**, 301 (2001).

28. F. Feruglio, A. Strumia, F. Vissani, *Nucl. Phys.* B **637**, 345 (2002), *Addendum-ibid.*, B **659**, 359 (2003).

29. H.V. Klapdor-Kleingrothaus *et al.*, *Eur. Phys. J.* A **12**, 147 (2001), A. M. Bakalyarov *et al.*, talk given at *the 4th International Conference on Non-accelerator New Physics* (NANP 03), Dubna, Russia, 23-28 Jun. 2003, hep-ex/0309016.

30. H.V. Klapdor-Kleingrothaus *et al.*, *Mod. Phys. Lett.* A **16**, 2409 (2001).

31. D. N. Spergel *et al.*, *Astrophys. J. Suppl.*, **148**, 175 (2003), [astro-ph/0302209].

32. O. Elgaroy, O. Lahav, *JCAP* **0304**, 004 (2003).

33. S. Hannestad, *JCAP* **0305**, 004 (2003).

34. S. W. Allen, R. W. Schmidt and S. L. Bridle, astro-ph/0306368.

35. A. Aguilar *et al.*, (LSND Collaboration) *Phys. Rev.* D **64**, 112007 (2001).

36. B. Armbruster, *et al.*, (KARMEN), *Phys. Rev. Lett.* **90**, 181804 (2003).

37. K. S. Babu and S. Pakwasa, hep-ph/0204226.

38. G. Barenboim, L. Borissov, J. Lykken, hep-ph/0212116.

39. A. Strumia, *Phys. Lett.* B **539**, 91 (2002).

40. M.C. Gonzalez-Garcia, M. Maltoni T. Schwetz, *Phys. Rev.* D **68**, 053007 (2003).

41. O. L. G. Peres, A.Yu. Smirnov, *Nucl. Phys.* B **599**, 3 (2001).

42. M. Sorel, J. Conrad, M. Shaevitz, hep-ph/0305255.

43. F. Dydak *et al.*, *Phys. Lett.* B **134**, 218 (1984).

44. A. Romosan *et al.*, *Phys. Rev. Lett.* **78**, 2912 (1997).

45. P. Astier *et al.*, [NOMAD Collaboration] *Nucl. Phys.* B **611**, 3 (2001).

46. A. Bazarko, hep-ex/0210020.

47. J. Conrad, private communication.

48. V. Barger, D. Marfatia, K. Whisnant, hep-ph/0308299.

49. See for review: G. Altarelli, F. Feruglio, *Phys. Rept.* **320**, 295 (1999), *Phys. Lett.* B **439**, 112 (1998).

50. M. Frigerio, A. Yu. Smirnov, *Nucl. Phys.* B **640**, 233 (2002), *Phys. Rev.* D **67**, 013007 (2003).

51. H. Fritzsch and Z. Z. Xing, *Phys. Lett.* B **372**, 265 (1996), *ibidem* **440**, 313 (1998); G.C. Branco, J.I. Silva-Marcos, *Phys. Lett.* B **526**, 104 (2002).

52. P. H. Frampton, S. L. Glashow and D. Marfatia, *Phys. Lett.* B **536**, 79 (2002).

53. L. J. Hall, H. Murayama and N. Weiner, *Phys. Rev. Lett.* **84**, 2572 (2000); N. Haba and H. Murayama, *Phys. Rev.* D **63**, 053010 (2001); A. de Gouvea and H. Murayama, hep-ph/0301050; J. R. Espinosa, hep-ph/0306019.

54. F. Vissani, *Phys. Lett.* B **508**, 79 (2001); M. Hirsch and S. F. King, *Phys. Lett.* B **516**, 103 (2001); G. Altarelli, F. Feruglio and I. Masina, *JHEP* **0301**, 035 (2003).

55. P.F. Harrison, W.G. Scott, *Phys. Lett.* B **557**, 76 (2003); E. Ma, hep-ph/0308282.

56. S. T. Petcov, *Phys. Lett.* B **110**, 245 (1982), R. Barbieri *et al.*, *JHEP* **9812**, 017 (1998).

57. E. Ma, G. Rajasekaran, *Phys. Rev.* D *64* 113012, (2001); K.S. Babu, E. Ma, J.W.F. Valle, *Phys. Lett.* B **552**, 207 (2003).

58. J. Kubo *et al.*, *Prog. Theor. Phys.* **109**, 795 (2003).

59. E. Ma, G. Rajasekaran, hep-ph/0306264.

60. W. Grimus, L. Lavoura, hep-ph/0305046.

61. J. Bijnens C. Wetterich, *Nucl. Phys.* B **292**, 443 (1987); M. Leurer, Y. Nir and N. Seiberg, *Nucl. Phys.* B **398**, 319 (1993), *ibidem*, **420**, 468 (1994); L. E. Ibanez and G.G. Ross, *Phys. Lett.* B **332**, 100 (1994); P. Binetruy and P. Ramond, *Phys. Lett.* B **350**, 49 (1995), for references and recent discussion see G. Altarelli and F. Feruglio, hep-ph/0306265.

62. C. D. Froggatt and H. B. Nielsen, *Nucl. Phys.* B **147**, 277 (1979).

63. R. Kuchimanchi and R. N. Mohapatra, *Phys. Rev.* D **66**, 051301 (2002).

64. R. Barbieri, L. J. Hall, G. L. Kane and G. G. Ross, hep-ph/9901228.

65. G. Kribs, hep-ph/0304256; S. F. King, *Phys. Lett. B* **520**, 243 (2001); S.F. King, G.G. Ross, *Phys. Lett. B* **574**, 239 (2003).

66. R. E. Marshak and R. N. Mohapatra, *Phys. Lett. B* **91**, 222 (1980).

67. S. Weinberg, *Phys. Rev. Lett.* **43**, 1566 (1979).

68. R. Barbieri, J. Ellis and M. K. Gaillard, *Phys. Lett. B* **90** 249 (1980); E. Kh. Akhmedov, Z. G. Berezhiani, G. Senjanović, *Phys. Rev. Lett.* **69**, 3013 (1992).

69. F. Vissani, M. Narayan, V. Berezinsky, *Phys. Lett. B* **571**, 209 (2003).

70. M. Gell-Mann, P. Ramond and R. Slansky, in *Supergravity*, eds P. van Niewenhuizen and D. Z. Freedman (North Holland, Amsterdam 1980); P. Ramond, *Sanibel talk*, retroprinted as hep-ph/9809459; T. Yanagida, in *Proc. of Workshop on Unified Theory and Baryon number in the Universe*, eds. O. Sawada and A. Sugamoto, KEK, Tsukuba, (1979); S. L. Glashow, in *Quarks and Leptons*, Cargèse lectures, eds M. Lévy, (Plenum, 1980, New York) p. 707; R. N. Mohapatra and G. Senjanović, *Phys. Rev. Lett.* **44**, 912 (1980).

71. R. N. Mohapatra and G. Senjanović, *Phys. Rev. D* **23**, 165 (1981), C. Wetterich, *Nucl. Phys. B* **187**, 343 (1981).

72. P. H. Frampton, S. L. Glashow, T. Yanagida, *Phys. Lett. B* **548**, 119 (2002).

73. R. N. Mohapatra and J. W. F. Valle, *Phys. Rev. D* **34**, 1642 (1986).

74. For recent review of neutrino masses in GUT see M-C. Chen and K. T. Mahanthappa, hep-ph/0305088.

75. A. Yu. Smirnov, *Phys. Rev. D* **48** 3264 (1993); M. Tanimoto, *Phys. Lett. B* **345**, 477 (1995); T.K. Kuo, Guo-Hong Wu, Sadek W. Mansour, *Phys. Rev. D* **61**, 111301 (2000); G. Altarelli F. Feruglio and I. Masina, *Phys. Lett. B* **472**, 382 (2000); S. Lavignac, I. Masina, C. A. Savoy, *Nucl. Phys. B* **633**, 139 (2002). A. Datta, F. S. Ling and P. Ramond, hep-ph/0306002; M. Bando, *et al.*, hep-ph/0309310.

76. E. K. Akhmedov, M. Frigerio, A. Yu. Smirnov, *JHEP* **0309**, 021 (2003).

77. W. Buchmuller, P. Di Bari and M. Plümacher, *Nucl. Phys. B* **643**, 367 (2002).

78. A. Pilaftsis, T. E. J. Underwood, hep-ph/0309342.

79. B. Bajc, G. Senjanović, F. Vissani, *Phys. Rev. Lett.* **90**, 051802 (2003).

80. T. Hambye, G. Senjanović, hep-ph/0307237.

81. H. S. Goh, R. N. Mohapatra and S-P. Ng, hep-ph/0303055.

82. S. F. King, *Phys. Lett. B* **439**, 350 (1998), *Nucl. Phys. B* **562**, 57 (1999); S. Davidson and S. F. King, *Phys. Lett. B* **445**, 191 (1998); for recent discussion see S. F. King, hep-ph/0310204.

83. K. S. Babu and S. M. Barr, *Phys. Lett. B* **381**, 202 (1996); S. M. Barr, *Phys. Rev. D* **55**, 1650 (1997). C. H. Albright and S. M. Barr, *Phys. Rev. D* **58**, 013002 (1998); G. Altarelli and F. Feruglio, *Phys. Lett. B* **439**, 112 (1998), *JHEP* **9811**, 021 (1998).

84. K. S. Babu, C. N. Leung and J. Pantaleone, *Phys. Lett. B* **319**, 191 (1993); J. R. Ellis and S. Lola, *Phys. Lett. B* **458**, 310 (1999); N. Haba, *et al., Eur. Phys. J.* **C10**, 677 (1999); N. Haba, N. Okamura, M. Sugiura, Prog. Theor. Phys. **103**, 367 (2000); J. A. Casas *et al., Nucl. Phys. B* **556**, 3 (1999), *ibidem*, **569**, 82 (2000), *ibidem* **573**, 659 (2000); P.H. Chankowski, W. Krolikowski, S. Pokorski, *Phys. Lett. B* **472**, 109 (2000); K. R. Balaji *et al., Phys. Rev. Lett.* **84**, 5034 (2000); T. K. Kuo J. Pantaleone and G. H. Wu, *Phys. Lett. B* **518**, 101 (2001); S. Antusch, M. Drees, J. Kersten, M. Lindner, M. Ratz, *Phys. Lett. B* **519**, 238 (2001).

85. J. A. Casas, J. R. Espinosa and I. Navarro, hep-ph/0306243.

86. S.T. Petcov, A.Yu. Smirnov, *Phys. Lett. B* **322**, 109 (1994); A. S. Joshipura, *Phys. Lett. B* **543**, 276 (2002); A. S. Joshipura, S. D. Rindani *Phys. Rev. D* **67**, 073009 (2003), *ibidem, Phys. Rev. D* **67**, 091302 (2003); S. Antusch, J. Kersten, M. Lindner, M. Ratz, hep-ph/0305273.

87. M. Fukugita and T. Yanagida, *Phys. Lett. B* **147**, 45 (1986).

88. F. Vissani, A. Yu. Smirnov, *Phys. Lett. B* **341**, 173 (1994). A. Brignole, H. Murayama, R. Rattazzi, *Phys. Lett. B* **335**, 345 (1994).

89. M. Lindner, S. Antusch, J. Kersten, M. Lindner, M. Ratz, *Phys. Lett. B* **544**, 1 (2002).

90. F. Borzumati and A. Masiero, *Phys. Rev. Lett.* **57**, 961 (1986).

91. N. Arkani-Hamed, H-C. Cheng, J. L. Feng, L. J. Hall, *Phys. Rev. Lett.* **77**, 1937 (1996).

92. I. Hinchliffe, F.E. Paige, *Phys. Rev. D* **63**, 115006 (2001).

93. Y. Grossman, H. E. Haber, *Phys. Rev. Lett.* **78**, 3438 (1997).

94. A. Romanino, A. Strumia, *Nucl. Phys. B* **622**, 73 (2002); J. R. Ellis, J. Hisano, S. Lola (CERN), M. Raidal, *Nucl. Phys. B* **621**, 208 (2002); for recent discussion see I. Masina, hep-ph/0304299.

95. S. Davidson, A. Ibarra, *JHEP* **0109**, 013 (2001).

96. S. T. Petcov, *et al.*, hep-ph/0306195.

97. A. Zee, *Phys. Lett. B* **93**, 389 (1980), *ibidem* **161**, 141 (1985);

98. For the latest discussion see, *e.g.* X. G. He hep-ph/0307172.

99. K. S. Babu, *Phys. Lett. B* **203**, 132 (1988).

100. L. J. Hall and M. Suzuki, *Nucl. Phys. B* **231**, 419 (1984), A. Joshipura, M. Nowakowski, *Phys. Rev. D* **51**, 2421 (1995); A. Yu. Smirnov, F. Vissani, *Nucl. Phys. B* **460**, 37 (1996); R. Hempfling, *Nucl. Phys. B* **478**, 3 (1996); for the latest discussion see M. A. Diaz, *et al.*, hep-ph/0302021.

101. N. Arkani-Hamed, S. Dimopoulos, G. R. Dvali and J. March-Russell, *Phys. Rev. D* **65**, 02432 (2002); K. R. Dienes, E. Dudas and T. Ghergetta, *Nucl.*

Phys. B **557**, 25 (1999).

102. Y. Grossman and M. Neubert, *Phys. Lett.* B **474**, 361 (2000).

103. G. R. Dvali, G. Gabadadze, M. Porrati, *Phys. Lett.* B **485**, 208 (2000).

104. T. Appelquist and R. Shrock, *Phys. Lett.* B **548**, 204 (2002); T. Appelquist, M. Piai, R. Shrock, hep-ph/0308061.

105. F. Bazzocchi *et al.*, hep-ph/0306184.

106. G. Seidl, hep-ph/0301044; K. S. Balaji, M. Lindner and G. Seidl, *Phys. Rev. Lett.* **91**, 161803 (2003).

DISCUSSION

William Lois (LANL): Can some type of SEE-SAW model explain a light sterile neutrino that would influence the solar neutrino data?

Alexei Smirnov: We have not studied this question. However, I think there is no problem in constructing a kind of singular see-saw model which accommodates a sterile neutrino with the proposed properties.

FUTURE FACILITIES

FUTURE EXPERIMENTS WITH NEUTRINO SUPERBEAMS, BETA-BEAMS, AND NEUTRINO FACTORIES

D. A. HARRIS

Fermi National Accelerator Laboratory P.O. Box 500 Batavia, Illinois, USA
E-mail: dharris@fnal.gov

This report describes the goals of the next generations of accelerator-based neutrino experiments, and the various strategies that are being considered to achieve those goals. Because these next steps in the field are significantly different from the current or previous steps, novel techniques must be considered for both the detectors and the neutrino beams themselves. We consider not only conventional neutrino beams created by decays of pions, but also those which could be made by decays of beams of relativistic isotopes (so-called "β-beams") and also by decays of beams of muons (neutrino factories).

1. Introduction

The current generation of neutrino experiments (K2K, MINOS, MiniBooNE, OPERA, and ICARUS) is focusing on verifying the oscillation framework and making the first precision measurements of the neutrino mass splittings. Specifically, they are each optimized to search for a particular signal which is predicted to be several statistical standard deviations away from zero, and which is predicted based on previous experimental evidence. The next step in this field, however, is to search for the last unmeasured element in the leptonic mixing matrix, and here there is essentially no theoretical guidance for how big a signal to expect.

Assuming that there are only three generations of neutrinos contributing to oscillations, the leptonic mixing matrix U, which converts from the flavor to the mass eigenstates, can be parameterized with three angles θ_{12}, θ_{23}, and θ_{13} and a CP-violating phase δ.

$$U = \begin{pmatrix} C_{12}C_{13} & S_{12}C_{13} & S_{13}e^{-i\delta} \\ -S_{12}C_{23} + ... & C_{12}C_{23} + ... & S_{23}C_{13} \\ S_{12}S_{23} + ... & -C_{12}S_{23} + .. & C_{23}C_{13} \end{pmatrix} \quad (1)$$

where $S_{ij} = \sin\theta_{ij}$ and $C_{ij} = \cos\theta_{ij}$. The first two mixing angles (θ_{12}, θ_{23}) are known to be large from the solar and atmospheric disappearance measurements, respectively. The third mixing angle, θ_{13}, can be measured by either looking for electron neutrino disappearance (for example, at a reactor experiment at a few kilometers distance), or by measuring $\nu_\mu \rightarrow \nu_e$ at the "atmospheric mass splitting" (Δm_{23}^2). The current most sensitive limit for $\sin^2 2\theta_{13}$ is roughly 0.1, and comes from the CHOOZ reactor experiment.[1]

In the limit where the solar mass splitting (Δm_{12}^2) is zero, the probability for muon neutrinos to oscillate to electron neutrinos in vacuum is

$$P(\nu_\mu \rightarrow \nu_e) = \sin^2\theta_{23}\sin^2 2\theta_{13}\sin^2\left(\frac{\Delta m_{23}^2 L}{4E}\right) \quad (2)$$

where L and E are the distance the neutrino has travelled and its energy, respectively.

So by looking for electron appearance in a muon neutrino beam (or muon appearance in an electron neutrino beam), one can get an indication for $\sin^2 2\theta$ to be non-zero. However, once a signal for $\nu_\mu \leftrightarrow \nu_e$ is seen, there is much more to be measured in this sector. At that point all the additional terms not given in Eq. 2 become important. By looking at the oscillation probability for both neutrinos and antineutrinos one can determine the neutrino mass hierarchy, and also start to search for CP-violation.

For neutrinos propagating through the vacuum, the asymmetry in the neutrino and antineutrino probabilities can be expressed as follows:

$$A(\mu e) = \frac{P(\nu_\mu \rightarrow \nu_e) - P(\bar{\nu}_\mu \rightarrow \bar{\nu}_e)}{P(\nu_\mu \rightarrow \nu_e) + P(\bar{\nu}_\mu \rightarrow \bar{\nu}_e)} \quad (3)$$

$$= \frac{\Delta m_{12}^2 L}{E}\frac{\sin\delta}{\sin\theta_{13}} < 1 \quad (4)$$

where L and E are the experiment baseline and neutrino energy, respectively, Δm_{12}^2 is the mass squared difference that has been seen in the solar neutrino sector, and δ is the CP-violating phase in the leptonic mixing matrix. Note that although the asymmetry must by definition be less than one, as θ_{13} gets smaller the asymmetry actually gets larger. Also, if an experiment is planned to run with longer baseline, say the second oscillation maximum ($\Delta m_{23}^2 L/4E = 3\pi/2$) instead of the first ($\Delta m_{23}^2 L/4E = \pi/2$), then the CP-violating effects are three times as large as at the first.[2]

When electron neutrinos and antineutrinos travel through the earth, they scatter off electrons in the earth, which itself causes an asymmetry, even without any CP-violating phase in the mixing matrix. In the limit of $\Delta m_{12}^2 = 0$ (which would also imply no CP-violation), the asymmetry for an experiment running near the oscillation maximum (i.e. where $\Delta m_{23}^2 L/4E$ is near $\pi/2$) is:[3]

$$A(\mu e) = \frac{2E}{E_R}\left(1 - \left[\frac{\pi}{2}\right]^2 \frac{E - E_{om}}{E}\right) \quad (5)$$

$$E_R = \frac{\Delta m_{23}^2}{2\sqrt{2}G_F\rho_e} \approx 11\ GeV$$

$$E_{om} = \frac{\Delta m^2 L}{2\pi}$$

In the experiments to be described in this document, the matter effects alone will cause this asymmetry to be anywhere from a few per cent to close to unity. By measuring the sign of these matter effects, one can determine whether or not Δm_{23}^2 is positive or negative, which will determine whether or not neutrinos follow the same mass pattern as the charged fermions. Most theories predict that the mass hierarchy is "normal" (or Δm_{23}^2 is positive), but that fact makes this hierarchy even more important to determine experimentally.

There is also information in the precise value of the atmospheric mixing angle, otherwise known as θ_{23}, which is currently within experimental errors of $\pi/4$. How much this angle differs from $\pi/4$ is important for two reasons. The fact that it is so near to $\pi/4$ in the first place implies some underlying symmetry, whose origin we do not yet know, and yet how different θ_{23} is from $\pi/4$ implies how strongly that symmetry could be broken. Imagine how different the field of particle physics would be today if we did not know how far the long-lived neutral kaon was from not being simply an exactly equal admixture of K^0 and $\bar{K}^0$! Also, from a purely experimental point of view, the farther θ_{23} is from $\pi/4$ the harder it becomes to extract the other matrix elements from measurements of probabilities. This is because while the ν_μ disappearance probability is a function of $\sin^2 2\theta_{23}$, the appearance probabilities are functions only of $\sin^2 \theta_{23}$, which by definition is more poorly constrained than θ_{23}, for values near $\pi/4$. So, although the primary motivation of the experiments

described in this report is a search for θ_{13} and not an improved measurement of $\sin^2 2\theta_{23}$, one would want and expect a new level of precision here as well.

Before describing each of the experiments individually, however, this report will discuss general techniques for making neutrino beams in the first place, and then describe a few detectors which are being considered in either the near or far term future.

2. Techniques for Making Neutrino Beams

There are currently three vastly different strategies for making neutrino beams, and can be classified by what particle is decaying to produce the neutrinos. Conventional beams, made from pion decays, have been used in particle physics for decades. However, this technique inherently makes a beam which contains both muon and electron neutrinos, which ultimately limits how small a $\nu_\mu \to \nu_e$ oscillation probability can be measured. Beta-beams, made from decays of beams of radioactive isotopes, are purely ν_e or $\bar{\nu}_e$, depending on what isotope is decaying.[4] Finally, neutrino factory beams come from positively or negatively charged muon decays, and contain roughly equal amounts of ν_μ and $\bar{\nu}_e$, or of $\bar{\nu}_\mu$ and ν_e, depending on the charge of the muons.[5] Although these last beams seem to be the most mixed, in fact they are the easiest to use from the detector standpoint: by measuring the presence of a muon in the far detector one has measured the flavor of the final-state neutrino, and by measuring that muon's charge one has determined the flavor of the initial state neutrino in the beam, since the electron-flavor neutrinos always have the opposite helicity to the muon-flavor neutrinos.

2.1. *Conventional Neutrino Beams*

Conventional neutrino beams are created by bombarding a target with as many protons as can be provided, and then focusing the produced mesons (mostly pions and some kaons) into an evacuated decay region where they are allowed to decay. The mesons will decay primarily to muons and muon neutrinos, and because the decay is a two-body decay the energy of the neutrino is directly related to the parent meson energy and the angle between the parent meson direction and the neutrino. In order to produce a beam of antineutrinos, the polarity of the focusing devices must be reversed, and one arrives at a $\bar{\nu}_\mu$ beam. In either case there is always some contamination of ν_e or $\bar{\nu}_e$, since both the kaons and daughter muons can undergo three-body decays which produce electron neutrinos.

In the near term future, two experiments are being proposed which use detectors which are not located on the beamline axis. Because of the two-body nature of the pion decay, there is an angle with respect to the focused pion direction where a broad band of pion energies will produce a narrow band of neutrino energies. This is in contrast to an "on-axis" neutrino experiment, where the neutrino energy at the far detector is directly proportional to the pion energy in the beamline. So although the total event rates at an off-axis location are considerably reduced, the ν_μ event rate in a narrow energy band is in fact much larger than in the same energy band for an on-axis detector. Because the electron neutrinos are produced in three-body decays, this event peaking is not nearly as strong, and as a result not only the signal in a narrow energy band improves, but the signal-to-background ratio is much better than in the on-axis case.

For the farther term future, "on-axis" beams are again being considered, where the primary motivation is not to extend the search for ever smaller values of θ_{13}, but to measure, once $\nu_\mu \to \nu_e$ is found, the neutrino mass hierarchy through matter effects and to begin the search for CP-violation in the lepton sector.

2.2. Beta-Beams

Beta-beams are a relatively new idea for making extremely pure electron neutrino or antineutrino beams using accelerated beams of radioactive ions.[4] For example, if one were to produce and accelerate an intense beam of ^6He (^{18}Ne) ions to 139 (55) GeV/u, then one would have a wide-band energy neutrino (antineutrino) beam with an energy up to about 1 GeV. The average neutrino beam energies are on the order of a few hundred MeV, and therefore a far detector would have to be located about one hundred km away to address the atmospheric oscillation frequency. This proposal is being studied in depth at CERN, where the ISOL system would be used to produce roughly 10^{17} ions per year. These ions would

then be accelerated in two cyclotrons, the Proton Synchrotron (PS), and finally the Super Proton Synchrotron (SPS), after which they would be injected into a decay ring. To maximize the number of ions decaying while pointing towards a far detector, the ring could consist of two straight sections that are 2,500 m long, connected by short arcs.[6]

Because the β-beam neutrino energies and event rates at a far detector are comparable with those of atmospheric neutrinos, the relativistic ions must be bunched once they arrive at the decay ring. Although this complicates the acceleration, it allows for twice as many measurements to be made. By storing both the ^6He and ^{18}Ne ions in the ring at the same time (in which case they would have to be at the same momentum), one could produce bunched beams of either neutrinos or antineutrinos. By using relative timing between the beamline and the far detector events one could then measure neutrino and antineutrino oscillation probabilities in the same run.[7]

2.3. Neutrino Factories

Finally, the term "neutrino factory" has been used to describe neutrino beams created by the decays of high energy muons which are again stored in elongated rings. A neutrino factory would produce beams of equal numbers of ν_μ and $\bar{\nu}_e$ ($\bar{\nu}_\mu$ and ν_e) when negative (positive) muons are stored in the ring. With positively charged muons circulating in the ring, one could measure $\nu_e \to \nu_\mu$ simply by looking for a neutrino interaction in the far detector with a negatively charged muon emerging. The $\bar{\nu}_\mu$ neutrinos which are also in this same beam would produce only positively charged muons. Although neutrino interactions frequently produce pions of both charges, these pions can very easily be distinguished from muons once the muons are above a few GeV. Neutrino factories are therefore being considered with muon energies from 20 GeV up to 50 GeV, which produce wide band neutrino energies with an average of about a third of the muon energy. For most studies, the number of muons decaying in the ring per year is on the order of 10^{20}. The baselines considered for these beams range from about 700 km to 8000 km.

A neutrino factory begins with a high intensity beam of protons (like the one needed for a neutrino superbeam, for example) which is sent to a target which has been optimized to make low energy pions. The produced pions are then collected through solenoid focusing while they decay, at which point they are a dense cloud of muons (with a broad range of energies). By phase rotation the muon energy spread is reduced, at which point the muons can be accelerated. Because of the short muon lifetime the muons must be accelerated very quickly, and herein lies the biggest technological challenge in building a neutrino factory.

However, the excitement that has been generated in this field over the possibility of building a neutrino factory is due to the fact that for a given proton power, a 30 GeV neutrino factory would produce about a factor of three more neutrinos than a superbeam at a comparable energy.[8] Also, the oscillation searches done with that beam would have more than an order of magnitude lower backgrounds, simply by using a detector that could distinguish the charge of an outgoing muon.

2.4. *Beamline Summary*

Figure 1 shows the neutrino fluxes per Megawatt (with the exception of the β-beam) for several of the proposals to be described in this report. Because the fluxes that are relevant are those at the location of the experiment, the fluxes given are for each of the different baselines being considered.

3. Techniques for Detecting Neutrinos

Clearly for an oscillation experiment, a detector must be able to identify both the flavor and the energy of the incoming neutrino. The separation between an electron and a muon is trivial in most detectors, but the challenge is in both reconstructing the neutrino energy, and separating muons or electrons from other final-state particles which are often produced in neutrino interactions (i.e. charged or neutral pions).

Before listing the detectors being considered for the next generation of neutrino experiments, it is useful to list the different interactions that accelerator-produced neutrinos (i.e. at or above a few hundred MeV in energy) can have with matter. First of all, there is the quasielastic interaction, which dominates the total cross section below about 700 MeV: this occurs when a neutrino simply exchanges a W with a target neutron to create a proton, and the nucleus remains "intact" throughout the interaction. The neutrino energy in this case can be simply computed by measuring the final-state lepton and its outgoing angle with respect to the incoming neutrino direction, which is known. The next process which becomes important as the neutrino energy increases is resonance production, where instead of producing a neutron from a proton, a resonance (Δ) is produced, which can than decay strongly to a proton and a pion. This process becomes important around about 1 GeV. Finally, there is deep-inelastic scattering, where the target nucleus is completely broken up, and there are many final-state particles, whose energies when summed equal the incoming neutrino energy. One rule of thumb is that all three process contribute about equally to the total cross section at about 1 GeV, but above that the DIS cross section rises linearly with neutrino energy while the resonance and quasielastic cross sections remain constant.

Although the previous paragraph discusses charged current interactions where there is at least a final-state charged lepton, there are analogous neutral current (NC) interactions, and these can provide significant background when one is searching for $\nu_\mu \to \nu_e$. In neutral current interactions the outgoing lepton is a neutrino, and there is often a neutral pion produced, which, depending on the detector, could be mistaken for an electron. Luckily, however, these pions have a steeply falling energy distribution, and because of the missing neutrino energy the total visible energy in the event is less than that of the incoming neutrino energy. Nevertheless, because the probability one is trying to measure may be extremely small, large rejection factors are needed. In order to reduce the neutral current background to a level near that of the intrinsic electron neutrino contamination in the beam, a detector would have to have a rejection factor of about one in 500.

It is important to point out, however, that although it is straightforward to predict which processes will give a background to these neutrino oscillation experiments, the cross sections for many of those processes are currently far from well-measured. An example of one of the best-measured processes that will give backgrounds to conventional oscillation searches is the process $\nu_\mu n \to \mu^- p\pi^0$, shown in Fig. 2.[9] This process is extremely important in

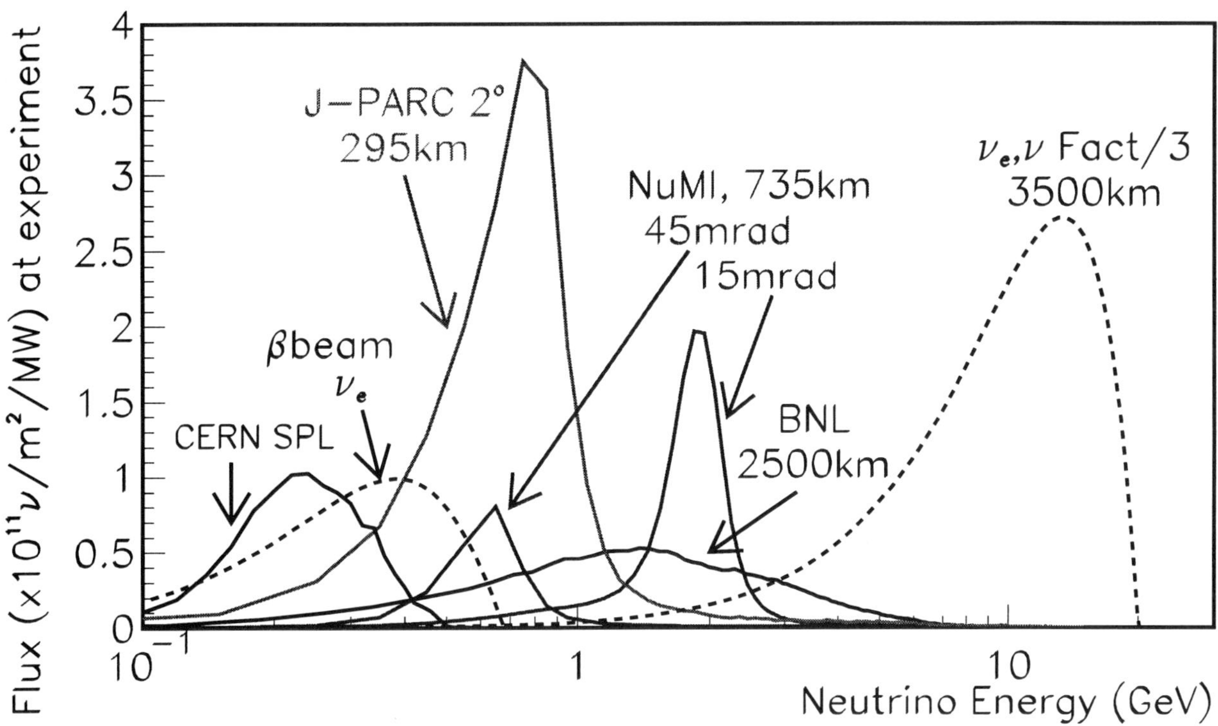

Figure 1. Neutrino fluxes for a variety of beamlines: superbeams, β-beams, and neutrino factories are shown, where the ν_e fluxes are given in dotted lines, and ν_μ fluxes in solid lines. The fluxes are normalized to 1 MW proton power, except for the β-beam, which uses considerably less proton power but is limited by different factors.

$\nu_\mu \to \nu_e$ searches because detectors can potentially mistake π^0's for electrons. It turns out that nuclear effects are important here, and the neutral current analog cross section ($\nu_\mu n \to \nu_\mu n \pi^0$) is even more poorly constrained, since one cannot measure in that case the incoming neutrino energy.

3.1. *Water Čerenkov*

It is largely due to the success of water Čerenkov devices as neutrino detectors that we know as much as we do about neutrino masses and mixing. Because the neutrino target material (water) is extremely cheap and only the surface of the detector must be instrumented, this technique allows one to afford an extremely massive detector. In fact, the most massive water Čerenkov detector (Super-Kamiokande) of today is a factor of ten heavier than the next most massive detector made of steel-scintillator (MINOS). Water Čerenkov detectors can identify neutrino interactions through the detection of rings of Čerenkov light coming from the outgoing charged lepton. The total light measured determines the lepton energy,

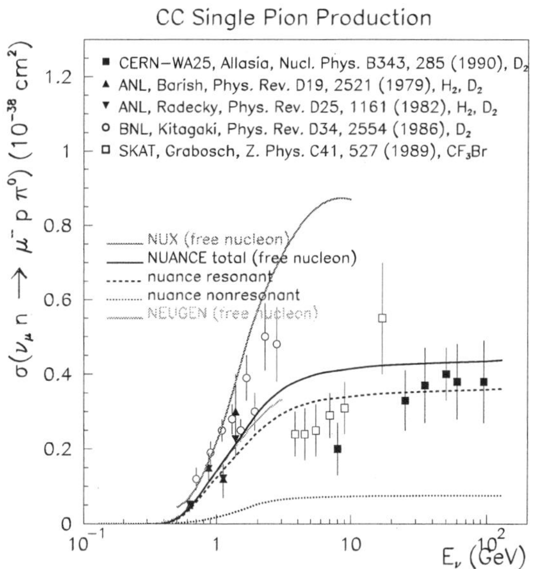

Figure 2. Compilation of cross sections for charged current single pion production, and predictions for different neutrino event generators.[9]

and the position of the ring determines the outgoing lepton direction, which combined with the lepton energy gives an accurate measurement of the incoming neutrino energy for quasielastic events. Finally, the character of the ring itself can be used for particle identification: a muon scatters very little and therefore leaves a ring with very sharp edges, while an electron scatters much more and makes a ring with "fuzzy" edges, as shown in the upper left event display in Fig. 3.[10]

As the neutrino energy increases, however, and other process begin to dominate the cross section, water Čerenkov detectors do not measure the total neutrino energy nearly as well. This is because there are other final-state particles in the event, for example, charged pions or protons, which are below the Čerenkov threshold for water, yet which still carry off a substantial amount of energy. Also, at higher energies, the background rejection is considerably worse. The production of π^0's from neutral current events is a larger problem. At low energies (about 1 GeV or so), the two rings which come from converted photons from π^0 decays are easily distinguished, and only a small fraction of them overlap.[11] However, as the π^0 energy increases, the two photons are emitted close together, and eventually the resulting rings have a very high probability of overlapping.

3.2. *Fine-Grained Calorimetry*

Another technique which has long been in use to detect neutrinos is calorimetry: interleaving planes of absorber with planes of active material. The flavor of the neutrino interacting can be determined by the longitudinal energy deposition of tracks in the event: a muon tends to make a long minimum ionizing track, while an electron makes a much shorter highly ionizing track (see the upper right event display in Fig. 3[12]). Since the active material is sensitive to the charge crossing the plane, the thresholds for particle detection can be significantly lower than that of a Čerenkov detector, and depend largely on the segmentation of the passive material. Therefore, fine-grained calorimeters do a good job of measuring the neutrino energy, and in fact can do even better as the final-state particles become more energetic and are themselves better measured. As an example, a detector consisting of particle board interleaved every third of a radiation length with readout has been

shown to measure neutrino energies to about 15% at 2 GeV.[12]

The extent to which fine-grained calorimeters can reject neutral current backgrounds depends on the granularity (both transverse and longitudinal) of the detector. A neutral pion will again have two electromagnetic clusters rather than one, and it is unlikely that the two photons would start showering at the same location, which would be the case for a single electron shower. The same fine-grained calorimeter described has been shown in simulations to achieve a NC background rejection of better than 1 in 500, in a narrow band 2 GeV neutrino beam.

3.3. *Liquid Argon TPC*

Finally, a detector which could have superior energy resolution and background rejection compared to either of the technologies discussed above is a Liquid Argon Time Projection Chamber, such as is being used by the ICARUS experiment.[13] By instrumenting ultra-pure liquid argon with planes of wires in two dimensions, and reading out the signals induced by ionization as a function of time, one can reconstruct a three-dimensional record of a neutrino interaction. With fine enough wire spacing, this kind of detector produces bubble-chamber like images (see lower right event display in Fig. 3[13]), with particle identification through dE/dx measurements and precision tracking. In particular, a π^0 could be rejected from an electron simply by seeing if the first few radiation lengths of the track were consistent with one or two highly-ionizing particles.

However, we are just beginning as a field to see how a large scale detector of this technology will perform. A small prototype was placed in the WANF beam during the CHORUS/NOMAD run, and high energy neutrino interactions have been recorded where neutral pions can cleanly be identified. Also, a 600 ton module was tested on cosmic rays for several months in 2001, and the tracks seen there also agree well with the Monte Carlo predictions. In order to get enough fiducial mass for a next generation experiment, however, we must first see how large a single volume can be made which still functions effectively, and also it is important to study a small prototype in a low energy neutrino beam to again confirm the simulations at those energies.

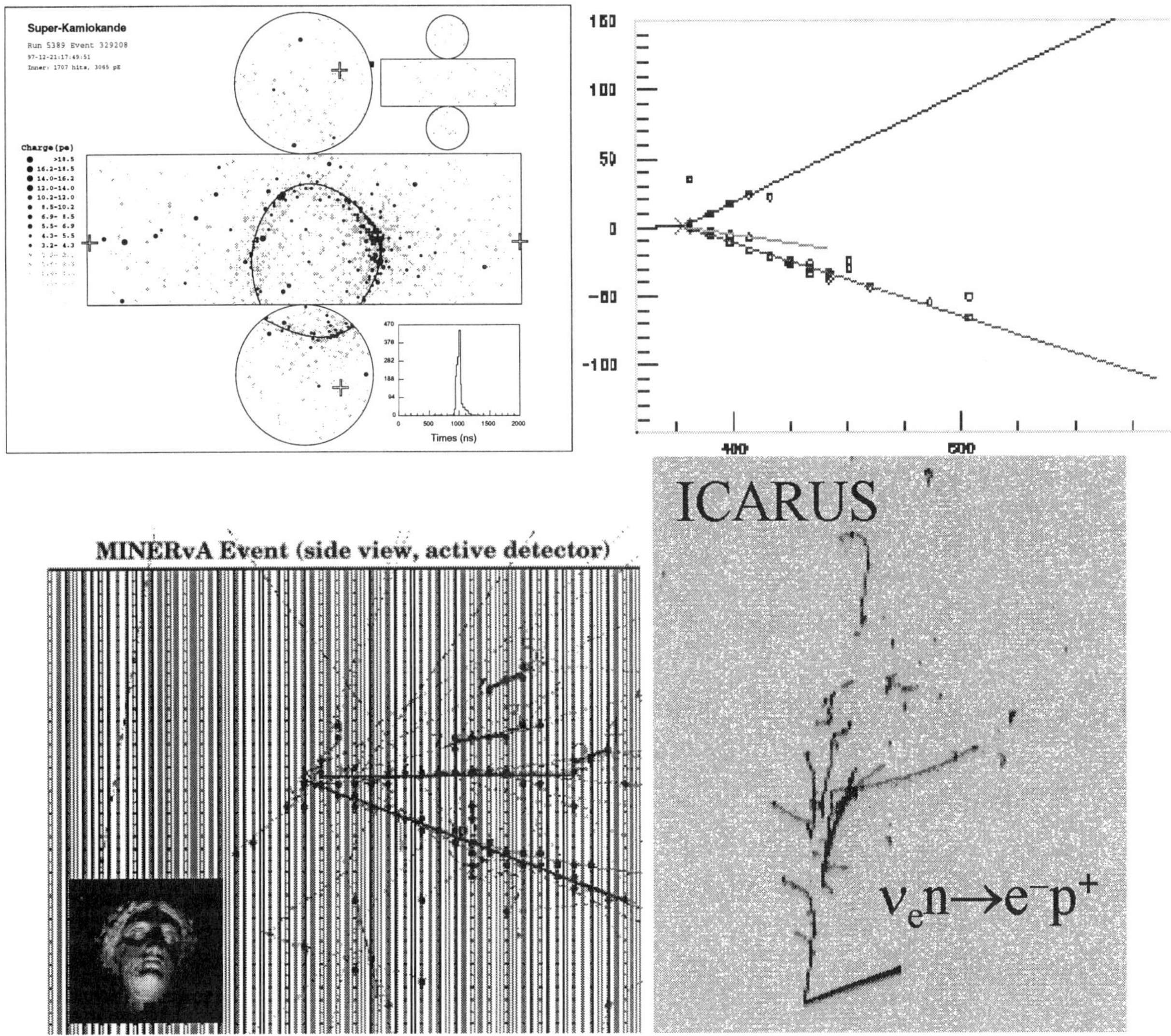

Figure 3. Neutrino event displays for a variety of different detectors: from upper left, clockwise: a ν_e quasielastic event in water Čerenkov,[10] low-Z calorimeter,[12] liquid argon TPC,[13] and finally a potential background event with a π^0 in an all scintillator detector.[14]

3.4. *Near Detectors*

Figure 3 shows what signal events would look like for a water Čerenkov detector,[10] a fine grained calorimeter,[12] and a liquid argon TPC.[13] The neutral current background is expected to be negligible for the liquid argon TPC, but for the other detector technologies this background can be as large as the intrinsic ν_e's in the beamline. For an accurate far detector prediction one needs not only a near detector measurement of the number of total background events surviving all cuts, but also an understanding of how that number changes when extrapolating to the far detector. To do this, we as a field need accurate dedicated cross section experiments, using detectors which can do more than simply identify an event as a signal candidate. The K2K experiment has initiated this program by their suite of near detectors at KEK, and at this conference K2K has presented an updated measurement of the ratio of π^0/μ production.[11] Another example being proposed, which would operate at much higher neutrino energies is the Minerva experiment,[14] which would use the NuMI Beamline at Fermilab with a fully active scintillator target. A sample event display from a potential background source (π^0 production) is also shown in Fig. 3. Clearly by measuring much more about the various processes we can vastly improve our models of neutrino interactions, and then the predictions for a far detector can be made accurately.

Although it will not be stated explicitly, each of the following experiments being described will plan to have at least one near detector in their beamline, and in some cases more than one.

4. Determining the Number of Neutrinos

The first question that must be answered before we really understand what future experiments can do, is whether or not the appearance of $\bar{\nu}_e$'s in a $\bar{\nu}_\mu$ beam at Los Alamos is due to oscillations or not. If MiniBooNE confirms the LSND neutrino signal, then there is much more to measure and many more basic questions that must be asked than what is usually described in future experiment proposals.

One example of a new signature which could be implied if MiniBooNE sees a signal, is a sizable disappearance probability at the "LSND Δm^2". However, in order to measure the disappearance to the required precision, MiniBooNE would have to add a second detector to the beamline. The optimal location for that second detector depends on what Δm^2 is actually measured by MiniBooNE.

Another implication if there are sterile neutrinos is that there could be sizable CP-violating effects in the $\nu_\mu \rightarrow \nu_\tau$ channel and/or the $\nu_e \rightarrow \nu_\mu$ channel, again at the "LSND Δm^2".[15] There are also models where CPT is violated, and the disappearance signature at the atmospheric Δm^2 would be different between neutrinos and antineutrinos.[16]

It is clear that independent of the theories that are being considered today, if MiniBooNE confirms the LSND signature to be due to neutrino oscillations then there is much more to do than is even outlined here. In the interest of brevity, the remainder of this document will make the assumption that this does not happen.

5. Near Term θ_{13} Experiments: Phase I

The two experiments described next appear at first glimpse to be rather similar: they are both optimized to search for $\nu_\mu \rightarrow \nu_e$ in a conventional beam, and expect to run near the first oscillation maximum, or where $\Delta m_{23}^2 L/4E$ is about $\pi/2$. To do this they both use off-axis neutrino beams, and both take advantage of significant investments in neutrino physics that have already been made. The experiments themselves are first described, but then we consider what could be learned by having both measurements, in the case where a signal is seen.

5.1. *J-PARC to Super-Kamiokande*

A new hadron facility is being constructed at Jaeri, called J-PARC.[17] It will have a 0.8 MW proton source, with an initial proton energy of 40 GeV which could eventually be raised to 50 GeV. Connected with that facility is a neutrino beamline, which would be used in conjunction with the Super-Kamiokande water Čerenkov detector.[18] By using the off-axis technique and also building a decay pipe that is trapezoidal in shape, this experiment has the flexibility to run at several different off-axis angles, which correspond to different neutrino energies, ranging from a 550 MeV beam to a 700 MeV beam. The resulting electron neutrino background is roughly 0.2%, at the muon neutrino peak, and slighly higher when integrated over the entire width of the peak.

The first stage of this experiment expects to be sensitive to values of θ_{13} up to about a factor of 10 past the CHOOZ limit, after 5 years of running. The experiment has recently been included in Phase I of the J-PARC running, and hopes to start taking high intensity neutrino data in 2008. The construction of the neutrino beamline is expected to begin in 2004, and of course, the Super-Kamiokande detector is already in place and fully functional.

A later stage, which involves a more powerful proton source and a larger detector, is also under consideration. This stage will focus on CP-violation and therefore has to run for significant periods in "antineutrino" mode.

5.2. *NuMI Off-Axis Experiment*

While the NuMI beamline is busy making an on-axis neutrino experiment to measure Δm_{23}^2 (MINOS), it will also be producing "off-axis" neutrino beams at sites all over northern Minnesota and southern Ontario.[12] If one were to place a detector at roughly 20 mrad from the NuMI beamline, it would see a very narrow-band beam at 2 GeV. At 40 mrad from the beamline, the narrow-band beam is closer to 1 GeV. Because the NuMI beamline will only be getting protons (and hence making neutrinos) for 10 microseconds out of every 2 seconds, it is possible to consider placing a detector at the surface of the earth, and using timing to reject cosmic ray backgrounds. Because one is not constrained by needing a large underground cavern, there is a huge flexibility of experiments that can be proposed: one can vary both the energy and the baseline, somewhat independently, by simply changing the off-axis angle.

A first stage of a NuMI off-axis experiment is being proposed, and involves placing a fine-grained calorimeter at a distance of about 800 km from the neutrino source at Fermilab, and at an off-axis distance of 10 km, corresponding to about 12.5 mrad.[12] At this location the electron neutrino background is 0.5% when integrated over the peak, which is itself about 20% FWHM. For a neutrino beam of this energy and baseline, the matter enhancement (or suppression) is about 20%.

A later stage of this experiment would also involve running in antineutrino mode with an upgraded proton source, but also there is the flexibility to run at different off-axis angles and baselines, depending on what the first generation sees.

6. When the Whole is Greater than the Sum of its Parts

In order to understand the motivation for building two long baseline neutrino experiments rather than simply running one of them for twice as long (or building it to be twice as powerful), it is helpful to look at the oscillation probability to all orders.[19] One useful expansion of the full probability in the earth is as follows:[20]

$$P = P_1 + P_2 + P_3 + P_4 \tag{6}$$

$$P_1 = \sin^2\theta_{23}\sin^2 2\theta_{13}\left(\frac{\Delta_{13}}{B_\pm}\right)^2\sin^2\frac{B_\pm L}{2}$$

$$P_2 = \cos^2\theta_{23}\sin^2 2\theta_{12}\left(\frac{\Delta_{12}}{A}\right)^2\sin^2\frac{AL}{2}$$

$$P_3 = J\cos\delta\left(\frac{\Delta_{12}}{A}\right)\left(\frac{\Delta_{13}}{B_\pm}\right)$$

$$\times\cos\frac{\Delta_{13}L}{2}\sin\frac{AL}{2}\sin\frac{B_\pm L}{2}$$

$$P_4 = \mp J\sin\delta\left(\frac{\Delta_{12}}{A}\right)\left(\frac{\Delta_{13}}{B_\pm}\right)$$

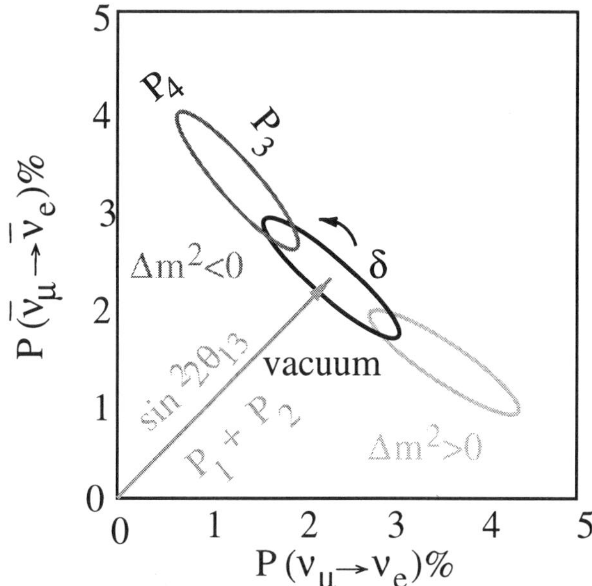

Figure 4. Neutrino versus antineutrino probabilities for $\nu_\mu \to \nu_e$, at a given baseline and neutrino energy, for a single value of $\sin^2 2\theta_{13}$.

$$\times\sin\frac{\Delta_{13}L}{2}\sin\frac{AL}{2}\sin\frac{B_\pm L}{2}$$

where

$$\Delta_{ij} = \frac{\Delta m_{ij}^2}{2E_\nu}$$

$$A = \sqrt{2}G_F n_e$$

$$B_\pm = |A \pm \Delta_{13}|$$

$$J = \cos\theta_{13}\sin 2\theta_{12}\sin 2\theta_{13}\sin 2\theta_{23}$$

and the $\pm$ signifies neutrinos or antineutrinos.

As an example of the ambiguities associated with one neutrino and antineutrino measurement, all the possible values of the neutrino and antineutrino probabilities for one single value of θ_{13} at one experiment are shown in Fig. 4. So although one measurement of neutrino oscillation probablity can set a lower limit on how large θ_{13} is, it is far from a determination of the mixing angle. It has been shown that even with a neutrino and antineutrino measurement at one energy and baseline, there can be several solutions of θ_{13} and δ, because of imperfect knowlege of the mass hierarchy, or how far $\sin^2 2\theta_{23}$ is from $\pi/4$.[21]

So clearly, to get to the remaining underlying physics, namely the mass hierarchy and CP-violation in the lepton sector, we need more than one neutrino and one antineutrino measurement of $\nu_\mu \to \nu_e$. We must measure both these transitions at least once but preferably at more than one energy, and one of the sets of measurements must be over a large enough baseline that the matter effects in the earth are significant (at least 700 km).

7. Phase II Experiments: The Next Decade in $\sin^2 2\theta_{13}$

What kinds of experiments are done in the longer term future depend very much on what the first generation of experiments sees. One can imagine several scenarios, depending on which experiment sees evidence for θ_{13} first. It is helpful to group them in ranges of $\sin^2 2\theta_{13}$, where for each range a different set of experiments may be able to make the relevant measurements. Table 1 shows for different ranges of $\sin^2 2\theta_{13}$, which experiments (or combinations of experiments) can do which physics.

Consider the discovery of $\nu_\mu \to \nu_e$: if $\sin^2 2\theta_{13}$ is within a factor of 2 of the CHOOZ limit, the MINOS or CNGS experiments should be able to get the first evidence. If $\sin^2 2\theta_{13}$ is not seen in MINOS or CNGS, but is within a factor of roughly 10 of the CHOOZ limit, the NuMI off-axis and/or J-PARC would be able to see the first evidence. Then, if it is not seen at these next generation experiments, then upgrades to either NuMI off-axis or J-PARC to Super-Kamiokande would be recommended (Phase II). Recall that to reach another factor of 10 improvement in the limit, however, one must increase the product of the proton power and the detector mass by a factor of 100! Finally, if they are not seen at any of those next generation experiments, one would turn to a neutrino factory where still another factor of 10 or 100 in $\sin^2 2\theta_{13}$ could be accessed. The first row of Table 1 shows this progression.

But of course there is much more to be learned than simply determining that θ_{13} is non-zero. If it is discovered in MINOS or CNGS, then NuMI off-axis and J-PARC experiments combined have a chance of seeing either the mass hierarchy, or a hint of CP-violation, if it is maximal, as shown in Fig. 5, and described in references[22,23] and more generally in the collection of references in note.[19] This is represented by the first column of Table 1.

If it is discovered in J-PARC to Super-Kamiokande or NuMI off-axis, then there are a host of Phase II experiments being proposed which would

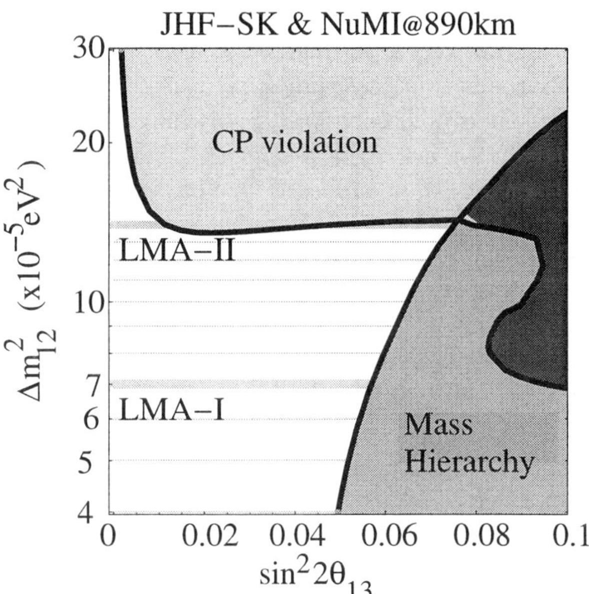

Figure 5. Ranges in Δm^2_{12} and $\sin^2 2\theta_{13}$ parameter space where maximal CP-violation or the mass hierarchy could be determined.[22,23]

have a chance of seeing CP-violation and determining the mass hierarchy in a broader range of parameter space than what is shown in Fig 5. Those experiments are described in the following section, and would appear in the second column of Table 1.

7.1. *Upgrades to J-PARC or NuMI Off-Axis*

In order to get a substantial improvement in the reach of any of the experiments described so far, one must improve not only the proton source, but also the detector mass. Of course it is only the product of the two which determines the statistics (and hence sensitivity) of the experiment, but the costs of improving either by itself are far from linear. For example, beamlines are currently in use which could withstand about 1 MW of power, but much work must still be done to understand what kinds of targets could withstand 4 MW of proton power. Some of this work is already being carried out by the muon storage ring collaboration, because this is the same power that is needed for a neutrino factory. Similarly, although one could "simply" build 10 Super-Kamiokande detectors, and the cost there would scale roughly linearly, there is some gain to be made by making the detector volume larger. Since

Table 1. Summary of which experiment can contribute to which decade of $\sin^2 2\theta_{13}$.

Physics	Value of $\sin^2 2\theta_{13}$			
	$> 4 \times 10^{-2}$	$> 1 \times 10^{-2}$	$> 10^{-3}$	$> 10^{-4}$
Seeing $\theta_{13} \neq 0$	MINOS CNGS	Conventional Superbeams Phase I	Conventional Superbeams Phase II	ν Factory $L \geq 3500\ km$
Mass Hierarchy	Combinations of Phase I Superbeams	Combinations of Phase II Super/β-beams	Combinations of ν Factory and Super/β-beams	ν Factory $L \sim 7700\ km$
Evidence for CP-violation	Combinations of Phase I Superbeams	Combinations of Phase II Super/β-beams	Combinations of ν Factory and Super/β-beams	Combinations of ν Factory 2 baselines

one is only instrumenting the surface of the detector, and the dominant cost of the detectors is in the phototubes, the cost would (ideally) go only as the 2/3 power of the mass. Also, for larger detectors, the ratio of fiducial volume to total volume is closer to 1: it is worth noting that although the Super-Kamiokande detector is 50 kton, the fiducial volume is only 22.5 kton, less than half.

Although the target design is not understood, the remainder of the J-PARC to Super-Kamiokande beamline is being constructed to be able to withstand a factor of 8 increase of proton power above what is assumed for Phase I. The proton source itself would have to be increased by that factor of 8, however. An upgrade of the detector, called HyperKamiokande, is now being studied. This detector would have a factor of 20 more mass than the Super-Kamiokande detector, or 1 Mton of water. The HyperKamiokande detector would be located in the same mountain range as the Super-Kamiokande detector, and in order for both detectors to see the same range of neutrino beam energies (off-axis), the beamline itself points into the earth at a higher angle than would be required for an on-axis beam at either of those two locations. Currently there are new technologies for photo-sensors which are being studied, since they not only dominate the cost, but also would need to be more robust, since the detector volume (and hence water pressure!) is increasing in future proposals.[24]

Similarly, in order to produce more neutrinos in the NuMI Beamline, the Fermilab proton source would need to be upgraded. There are currently two designs for an upgraded Booster, which would provide 8 GeV protons that then get accelerated in the Main Injector.[25] One is an upgraded proton synchrotron which is similar in size to the current Booster, and the other is a proton linac which would use very novel acceleration techniques which are of interest to the linear collider community. Either of those designs, combined with a modest upgrade of the Main Injector and the NuMI Beamline itself, could bring the NuMI facility up to 2 MW, or a factor of 5 above what the first generation experiment hopes to see. The way in which the NuMI off-axis detector would be upgraded depends slightly on what the first generation of experiments finds. If CP- violation and matter effects both act in the same direction to make the neutrino and antineutrino probabilities very different from one another, then the best bet would be to add more detector at the same location – either by copying the existing design, or upgrading with a more sensitive detector with better background rejection (for example, liquid Argon).

However, if the first generation of the NuMI off-axis experiment sees virtually the same neutrino and antineutrino probabilities, this means the CP-violating effects and matter effects are cancelling one another. In order to make progress one would use a different location – for example, by going to the second appearance maximum (at for example, the same baseline but one third the neutrino energy as the first experiment) the CP-violating effects would be a factor of 3 larger, but the matter effects would be a factor of 3 smaller. And in this case, by combining both first and second generation results not only could the mass hierarchy be determined, but also there is a chance of seeing maximal CP-violation.[26]

7.2. Brookhaven Proposal

In contrast to the proposals described thus far, there is a proposal from a Brookhaven-based study group that aims to use one broad-band neutrino beam at an extremely long distance.[27] This experiment proposes that the Brookhaven AGS, which produces 28 GeV protons, be upgraded to 1 MW, where targeting is still manageable with today's technology. The detector for this proposal would be a very large underground water Čerenkov device, which would presumably be part of a National Underground Science Laboratory. One design for this detector, the UNO detector[28] would have a fiducial mass of 450 kton, a factor of 20 over the Super-Kamiokande fiducial mass.

Because the signal distribution is now spread over a broad energy region, the experiment must take particular care in improving the signal-to-background ratio. The first step is to reduce the background by using only single-ring (quasi-elastic-like) events for all neutrino energies, since π^0's tend to make two rings. The second step is to increase the signal probability – this is done in two ways: at high energies, the matter effects will enhance the probability of either neutrinos or antineutrinos many times above what it would be at low energies. At lower energies in this experiment, the baseline is so long that these neutrinos are at the second and third oscillation maxima, where the CP-violating effects are 3 or 5 times as large. Although the experiment would start in neutrino running, if at first the evidence for $\nu_\mu \to \nu_e$ is not seen at high energies, then they would switch to antineutrino running where the matter enhancement would then be important.

Because of the broad-band neutrino beam, the experiment would have a sensitivity to more than just maximal CP-violation. If δ is close to 0, then the $\cos\delta$ terms in the probability become important, and those can contribute at the oscillation minima, or at energies where $\Delta m^2 L/4E$ is equal to π or 2π.

7.3. *CERN SPL and Beta-Beams*

There is a proposal based at CERN which is focused not only on CP-violation measurements with a very low energy conventional beamline, but also on T-violation by comparing muon appearance from an electron neutrino beam with the conventional neutrino beam results.[29] The intent is to build two beamlines, one conventional, which uses the 2.2 GeV protons from the CERN SPS, and one β-beam, which uses the ISOL technology, and pointing them to the same far detector. This plan would take advantage of the extra mountain tunneling that is currently planned in Frejus, and again a very large water Čerenkov detector is being used for the proposal. Because the neutrino cross section at these energies is extremely small, the plan is to start with a 4 MW proton source, and again a detector with 450 kton of fiducial mass.

At extremely low neutrino energies (some few hundred MeV) the neutral current background is low not only because the relevant cross section ratio (π^0/μ) is lower than at higher energies, but also because the π^0's that are produced are so low in energy that when they decay to two photons, they virtually always make two distinct rings.

One downside of these very low energies, however, is that the antineutrino to neutrino cross section ratio is even lower than a third, which is the ratio at 1 GeV. Therefore, the sensitivities quoted for this experiment are gained by running for 2 years in neutrino mode and an additional 8 years in antineutrino mode for a conventional beam, with a similar total running time for the β-beam.

Finally, for a 130 km baseline the matter effects are extremely tiny, and as such do not affect the precision with which this experiment might be able to see CP-violation. However, by the same token this experiment by itself would not be able to determine the mass hierarchy.

8. The Last Decade in $\sin^2 2\theta_{13}$: Neutrino Factories

Once again, the need for the next generation of neutrino experiments changes as a function of what the previous generation has seen. If θ_{13} is so small that it has just barely been seen by the second generation J-PARC or NuMI off-axis experiments, then it will take a neutrino factory with a baseline of (very roughly) 3500 km to address the question of the neutrino mass hierarchy and CP-violation (as represented by the third column in Table 1). If θ_{13} is too small to be seen even with conventional beams, then a neutrino factory, running with a 4 MW proton source and a 50 kton magnetized detector (such as the design used by MINOS), would still be able to push the sensitivity of $\sin^2 2\theta_{13}$ another two orders of magnitude. There have been many studies over the past 5 years addressing the capabilities of neutrino factories: reference[30] describes many of the early ones, and references[19] are more recent articles describing the tricky business of getting from measured probabilities to the mixing angles themselves.

One strategy to get the farthest reach on θ_{13} was obtained by looking carefully at the probability shown in Eq. 6. If one were to design an experiment such that $\sin(AL/2)$ is equal to 0, in other words, $AL/2 = 2\pi$, then only the first term P_1 would contribute to the oscillation probability, and the matter enhancement would be enormous.[31] This condition is satisfied for all energies and Δm^2 values, for a baseline of about 7700 km, and depends only on the density of electrons in the earth. Of course, building a storage ring with such a steep slope will be challenging from a civil engineering standpoint, but the detector technology is extremely straightforward. For this experiment, it is enough to measure one probability – if oscillations are seen at this baseline in ν_e (μ^+) running then one has determined the mass hierarchy *and* measured $\sin^2 2\theta_{13}$ with one measurement. If it is not seen then one would switch to antineutrinos (with μ^-'s in the ring), where again a large probability would be predicted, even for small values of θ_{13}.

Although we will not know the primary motivation for a neutrino factory until the next generation of experiments run, we know now that it has the potential to provide the most precise measurements of θ_{13} and the CP-violating phase, or possibly even the only measurement of θ_{13} and the mass hierarchy. For this reason alone it is important to continue R&D to understand how to build a neutrino factory. Since the last time this conference was held there has been much progress in designing a neutrino factory that the field can afford. There have been advances in design of phase rotation, cooling, and acceleration, such that the current design can boast of substantial savings in all of the cost drivers in the facility.[32]

9. Conclusions

Table 2 shows the various experiments that have been described in this report. They vary widely in terms of physics reach, time scales for construction and completion, and technical feasibility. Although it is not clear today which path this field will ultimately take, there are two extremely good near term opportunities for the next generation of measurements: both the J-PARC to Super-Kamiokande

and the NuMI off-axis proposals should see evidence for $\sin^2 2\theta_{13}$ if it is within a factor of ten of the current CHOOZ limit. Once that first evidence is seen then the debate can truly begin on how to best design a program to get to the neutrino mass hierarchy and CP-violation in the lepton sector. The experimental challenges for both accelerator and particle physicists are high but the rewards for these measurements will be higher still.

Acknowledgments

This report describes the work of hundreds of people working all over the world to figure out how we can get to this exciting physics. I am indebted to them, and in particular to the following people for their help in putting together this presentation: Milind Diwan, Gary Feldman, Steve Geer, Atsuko Ichikawa, Yoshi Kuno, Ken Long, Kevin McFarland, Mauro Mezzetto, Ko Nishikawa, Bob Palmer, Brett Viren, and Walter Winter.

References

1. M. Apollonio *et al.*, *Phys. Lett.* B **466**, 415 (1999).
2. W.J. Marciano, hep-ph/0108181, August (2001).
3. S.Parke first derived this for a talk at WIN2003, October (2003).
4. P. Zucchelli, *Phys. Lett.* B **532**, 166 (2002).
5. S.Geer, *Phys. Rev.* D **57**, 6989 (1998).
6. M. Lindroos *et al.*, CERN/PS-2002-078 (2002).
7. J. Bouchez, presented at NuFact03 (2003).
8. C. Albright *et al.* (Eds. S. Geer and H. Schellman), Report to the Fermilab Directorate FERMILAB-FN-692, April (2000), hep-ex/0008064.
9. See G.P.Zeller, talk given at NuINT02 for a summary of the current status of low energy neutrino cross section data.
10. "Evidence for Neutrino Mass from Observations of Atmospheric Neutrinos with Super–Kamiokande", dissertation of Mark Messier, Boston University (1999).
11. K. Nishikawa, Lepton Photon transparencies and proceedings.
12. D. Ayres *et al.*, hep-ex/0210005 and http://www-off-axis.fnal.gov.
13. Updated Icarus Technical Design Report, CERN/SPSC 2002-027, see also http://pcnometh4.cern.ch/ http://budoe.bu.edu/~messier/thesis/
14. See http://www.pas.rochester.edu/minerva
15. V.Barger, S.Geer, R.Raja, K.Whisnant, *et al.*, *Phys. Rev.* D **63**, 033002 (2001).

Table 2. Summary of various experiments described in this report.

Beam Name	Peak Energy (GeV)	Baseline (km)	Mass (kton)	Power (MW)	$\sin^2 2\theta_{13}$ Sens.[a]	δ[b]	Matter Effect
OPERA[33]	17	732	1.8	0.15	0.04	-	
ICARUS[33]	17	732	2.4	0.15	0.03	-	
MINOS[34]	3.5	735	5	0.4	0.05	-	
CNGS Modified[35]	2	732	2.35	.15	~ 0.02		$\geq$CP
JHF2SK	0.8	295	22.5	0.8	0.006	-	-
NuMI OA	2	850	50	0.4	0.004	-	$\geq$CP
SJHF2HK	0.8	295	450	4	$\sim 0.001^c$	$|\delta| > 20°$	<CP
SNUMI-OA	1 or 2	> 800	100	2	$\sim 0.001^c$	135 ± 20	$\geq$CP
BNL2NUSL	1.5	> 2500	500	1	0.004 CP	45 ± 20	> & <
CERN SPL	0.2	130	400	4	0.0016	90 ± 30	$\ll$CP
β-Beam	0.2	130	400	.04		T-viol.	$\ll$CP
ν Factory	15	~ 3000	50	4	$few \times 10^{-4}$	90 ± 20	$\gg$CP
ν Factory	15	~ 7700	50	4	$< 10^{-4}$	-	Huge

[a] At $\Delta m_{32}^2 = 3 \times 10^{-3}$ eV^2, at 90% CL

[b] All evaluated at different regions of parameter space!

[c] Assume 5% systematic uncertainty!

16. G.Barenboim and J.Lykken, *Phys. Lett.* B **554**, 73 (2003).

17. J-PARC web page: `http://neutrino.kek.jp/jhfnu/`

18. Itow *et al.*, J-PARC to Super-Kamiokande Letter of Intent, `hep-ex/0106019`.

19. V. Barger, D.Marfatia, K.Whisnant, *Phys. Lett.* B **560**, 75 (2003) and K. Dick, M. Freund, P. Huber, M. Lindner, *Nucl. Phys.* B **598**, 543 (2001), and J. Burguet-Castell, M.B.Gavela, J.J. Gomez-Cadenas, P. Hernandez, O. Mena, *Nucl. Phys.* B **646**, 301 (2002).

20. H. Minakata, H. Nunokawa, *JHEP* **0110**, 001 (2001).

21. V. Barger, D.Marfatia, K.Whisnant, *Phys. Rev.* D **65**, 073023 (2002).

22. P. Huber, M. Lindner, W. Winter, *Nucl. Phys.* B **654**, 3 (2003).

23. H. Minakata, H. Nunokawa, S. Parke, *Phys. Rev.* D **68**, 013010 (2003).

24. K. Nakamura, talk at Neutrinos and Implications for Physics Beyond the Standard Model, Stony Brook, October 2002, `http://insti.physics.sunysb.edu/itp/conf/neutrino.html`

25. Proton Driver Design Study `http://www-bd.fnal.gov/pdriver/8GEV`

26. G. Feldman, July 3 HEPAP Meeting, `http://doe-hep.hep.net/HEPAP/Agendajuly03.html`

27. M.V. Diwan *et al.*, `hep-ph/0303081`, March 2003; M.V.Diwan *et al.*, `hep-ex/0211001` October 2002; and D.Beavis *et al.*, `hep-ex/0205040`, April 2002.

28. UNO whitepaper,"Physics Potential and Feasibility of UNO", June 2001 SBHEP01-3 available at `http://superk.physics.sunysb.edu/uno/`

29. The physics potential of both the CERN SPL beam and β-beams is described in the CERN yellow report M. Apollonio *et al.*, `hep-ph/0210192`.

30. There are many articles discussing neutrino factory capabilities, see for example J.J. Gomez-Cadenas and D.A. Harris, *Ann. Rev. Nucl.* and *Part. Sci.***52**, 253 (2002) and references therein.

31. See for example, P. Huber and W. Winter, *Phys. Rev.* D **68**, 037301 (2003); A. Asratyan *et al.*, `hep-ex/0303023`.

32. MUCOOL R&D web page: `http://www.fnal.gov/projects/muon_collider/cool/cool.html`, and papers submitted to Accelerator R&D session at EPS 2003: `http://eps2003.physik.rwth-aachen.de/`

33. M. Komatsu, P. Migliozzi, F. Terranova, *J. Phys.* G **29**, 443 (2003).

34. M.Diwan, M.Messier, B.Viren, L.Wai, NUMI-L-714

35. A. Rubbia, P. Sala, *JHEP* **209**, 4 (2002).

36. Mauro Mezzetto, private communication.

37. "Parameters of Radiological Interest for a β-beam decay ring" M. Magistris and M. Silari, Technical Note in preparation.

DISCUSSION

Antonio Ereditato (INFN-Napoli): A remark about your concerns on the LAr TPC technique: a 600 ton ICARUS module has been successfully tested at the surface with cosmic rays. At the same web site from which you picked up MC events, you may find a "real event" gallery and a list of published papers using these events.

Deborah Harris: I agree that much progress has been made recently with event reconstruction. It will still be very important however to see how well this detector does in discriminating between π^0's and single electrons. I would certainly be very happy if this technology were the one that gets used in the long term future!

Peter Rosen (DOE): Have you thought much about the measurement of neutral currents? One of the beautiful features of solar neutrinos is SNO's ability to do that.

Deborah Harris: The superbeam experiments described here have really been optimized for ν_μ to ν_e searches. At the limit they have considered how well the ν_μ disappearance probability could be measured. However, as far as I know the only analysis done on using neutral current events, either to search for ν_τ "appearance" or to search for the sterile neutrino sector, is the MINOS experiment.

Joe Formaggio (Univ. of Washington): Doesn't the atmospheric neutrino background [for a β-beam experiment] get reduced by beam timing?

Deborah Harris: If you were nominally trying to build a β-beam, you would simply fill the decay ring with relativistic ions, and then you would have basically no timing. However, by working out how to impose a bunch structure in a β-beam you not only reduce the atmospheric background, but you also allow the possibility of putting He and Ne (and thereby making ν_e and $\bar\nu_e$) in different bunches.

Bennie Ward (Univ. of Tennessee): In the study of CP-violation in the quark sector, we're used to discovery in terms of many, many sigma. Could you comment on the use of a 3 sigma discovery criterion in the lepton sector?

Deborah Harris: I think it's a measure of how difficult these experiments are to do in the first place. With superbeams and what we know of today's technology, there is only a small region of parameter space where even maximal CP-violation can be seen, because of the inherent backgrounds and how well one can reasonably expect to ever know them. In contrast, neutrino factory beams provide measurements with two orders of magnitude lower backgrounds and high rates, and is certainly the only place we know of now where CP-violating effects could possibly be seen at "many, many sigma".

Ed Witten (Princeton University): I would think that the radiation levels at a β-beam facility would be prohibitively large – is this something that has been addressed?

Deborah Harris: This is, as you suspect, a serious concern to people designing β-beams. The estimated radioactivity level for a facility with the fluxes described in this report is 8.5 Watt/m, while 1 Watt/m is the standard security limit for an accelerator.[36] What is even more worrisome is that this radioactivity tends to accumulate at the end of the straight sections. Although special care needs to be taken for the proper disposal of the materials in the ring after the experiment is over, with proper design the doses to the magnets themselves can be reduced to an acceptable level. See reference[37] for a recent calculation of the dose levels.

DETECTOR R&D

T. BEHNKE

DESY, Notkestrasse 85 D-22607 Hamburg, Germany
E-mail: Ties.Behnke@desy.de

The next big project in high energy physics should be a high energy e^+e^- linear collider, operating at energies up to around 1 TeV. A vigorous R&D program has started to prepare the grounds for a detector at such a machine. The amounts of precision data expected at this machine make a novel approach to the reconstruction of events necessary; the particle flow ansatz. This in turn influences significantly the design of a detector for such an experiment. Apart from work ongoing for the linear collider detector, preparations are under way for an update of the LHC. This requires extremely radiation hard detectors. In this paper the state of the different detector development projects is reviewed.

1. Introduction

Over the last few years a consensus has emerged within the particle physics communities worldwide that the next big machine to be built should be an electron-positron linear collider,[1] operating at en ergies between a few 100 GeV and approximately one TeV. In a series of international workshops the physics community has evaluated both the physics programme to be performed at such a machine, and the resulting requirements for both the machine and the detector at this machine. They have been documented among others in the TESLA technical design report, the SNOWMASS report and the JLC roadmap book.[2-4]

The consensus achieved within the community builds upon a long and very successful history of first developing, then of testing the Standard Model of particle physics at ever more powerful accelerators. Machines like LEP, HERA, the Tevatron, and the B factories have collected impressive data samples and performed precision tests of the Standard Model at energy scales approximately up to the weak scale. Starting in 2007, the Large Hadron Collider, located at CERN, will continue the detailed investigation of the Standard Model and the search for physics beyond the Standard Model. It is hoped that at the LHC the last missing piece in the Standard Model puzzle, the Higgs Boson, will be discovered as well as signs of possible new physics beyond the Standard Model. The LHC will be the first collider which will push the energy scale well beyond the electroweak scale, close to the TeV energy scale.

The LHC by itself however will not be able to fully reveal the physics at these high energy scales. It should be complemented by an electron-positron collider reaching energies of about one TeV. Such a machine will be able to investigate in detail the structure of the electroweak symmetry breaking and study the Higgs Boson, if it exists. In addition nearly all scenarios for physics beyond the Standard Model will have observable signatures in the energy range covered by such a linear collider. Thus such a machine will complement the studies which can be done at the LHC.

Over the last decade tremendous work has been done to develop detectors for the LHC. The LHC exposes detectors to previously unheard-of radiation doses, thus forcing the development of very radiation hard technologies. Upgrades already now planned for the LHC will push the expected doses higher by another order of magnitude, thus requiring another big step in the development of radiation hard detector technologies.

In contrast the experimental conditions at the electron-positron linear collider are relatively benign. The full realization of the physics potential of the machine however implies that a new level of precision in the reconstruction of the complete event should be obtained. Over the last years a fully integrated approach to event reconstruction has been studied in detail, where the individual parts of the detector are optimally combined in the reconstruction to achieve the best possible reconstruction of the underlying event.

In this article the state of the developments of detector technologies and reconstruction strategies for the next generation of colliders is discussed. Particular emphasis is given to the detector at a linear electron-positron collider, but work ongoing in the preparation of the upgrade of the LHC is also discussed. While an impressive amount of detector

R&D has been done and is still being done for the first generation of detectors at the LHC this is not the subject of this review.

2. Event Reconstruction at a Lepton Collider

The physics programme anticipated for the planned linear electron-positron collider demands large amounts of data collected over a wide range of center-of-mass energies reaching from 90 GeV at the Z^0 resonance, to close to 1 TeV, at the maximum energy such a machine can reach. The type of measurements which the detector has to be capable of performing ranges from the reconstruction of individual particles in the final state, over the reconstruction of event properties as e.g. the jet structure in multi-jet events, to the precision measurements of vertices in heavy flavor decays.

A typical process to be studied at this machine is the decay of a light Higgs boson. If a light Higgs particle exists, the reconstruction of its properties, including the coupling to different types of final-state particles, requires a data sample approaching several ab^{-1}. For Higgs boson masses below approx. 140 GeV bottom quarks in the final state play an important role, at higher masses, complicated multi-jet final state dominate. The same is true for Standard Model processes, as e.g. top decays, which again result in multi-jet final states. The reconstruction of the Higgs mass or the properties of these particles requires, that these jets are measured superbly, thus stressing the capability of the detector to reconstruct the jet parameters.

A typical list of requirements for a detector at a linear electron-positron collider can be summarized in the following list.

- The detector should have an excellent capability to reconstruct the overall event. The method currently favoured is the so-called Particle Flow Technique, which will be discussed in more detail below.[5,6] Since particle flow is based on the optimal combination of tracking and calorimeter information this implies:

  - excellent track reconstruction capability. The detector should be able to find and reconstruct tracks from charged particles with close to 100% efficiency over a large solid angle. A benchmark condition is that the momentum resolution should be such that the mass resolution achieved for the recoiling muons in the reaction $HZ \to H\mu^+\mu^-$ is not dominated by detector effects.[7]

  - excellent reconstruction of charged and neutral particles in the calorimeter. The separation of neutral and charged particles is particularly important, thus stressing the spatial resolution of the calorimeter. To maximize the separation between charged and neutral particles a strong central magnetic field is necessary.

  - excellent hermeticity of the detector, to be able to fully reconstruct the event. In addition the sensitivity to many new physics signals is greatly enhanced in a hermetic detector.

- Many properties of the event can be reconstructed if the flavor of the final state is known. For heavy flavors this can be done with excellent efficiency and purity if lifetime information is available for the decaying particles. A supreme vertex detector therefore is needed, to reconstruct long-lived particles with excellent resolution.[8]

Over the years a number of different detector concepts have been studied in some detail. Two different approaches have emerged, a so-called large and a small detector type. Both have in common an excellent electromagnetic and hadronic calorimeter, arranged inside a central solenoidal magnetic field. They differ in the strategy of the tracking system. The large detector relies on a combination of a high resolution vertex detector, typically a pixel device, with a large volume gas-based drift detector, typically a TPC. The small detector on the other hand uses only Si based detectors for the tracking. The high resolution vertex detector is the core detector for both vertexing and for track finding, and is supplemented by a small number of Si-detector layers on the outside, to add momentum measurement capabilities and to bridge the gap between the vertex and the calorimetric detector.

2.1. *The Particle Flow Concept*

The concept of particle flow, sometimes also called energy flow, has been used in a number of detectors throughout the last decade. Particle flow works best at moderate energies of the individual particles, below about 100 GeV. In this regime the tracker is much superior in momentum resolution to a calorimeter for charged particles, and thus charged particles are primarily measured with the tracking system. Neutral particles on the other hand are reconstructed in the calorimeter. The particle flow algorithm then is the optimal combination of the information from the two detector systems, to achieve the best overall reconstruction possible.

Although this concept found wide spread acceptance already during the running of LEP, none of the detectors at LEP or at any other large collider were designed from the beginning with the concept of particle flow as one of the main requirements.

The goal of a particle flow reconstruction can be summarized very simply: reconstruct the four-vectors of all particles which participate in the event. This goal is typically achievable for charged particles, but much more difficult for neutral particles. In the past the reconstruction of neutral particles was often limited to the reconstruction of the energy carried by all - or a subset of - neutral particles, rather than the individual reconstruction of neutral particles. The reconstruction of photons buried in jets e.g. is nearly impossible with a typical LEP detector. This is one of the explanations why a detector at e.g. LEP[9] can reconstruct the energy of jets only with something like $60\%/\sqrt{E}$.

Monte Carlo studies show that an ideal reconstruction algorithm,[6] which finds each particle and measures its energy, its direction and its momentum correctly, could reach a jet energy resolution of $14\%/\sqrt{E}$.

The importance of excellent jet energy resolution is demonstrated quite clearly in the reconstruction of WW or ZZ final states. These are important signatures both for Standard Model and for new physics signals, and a clear separation between WW and ZZ final states is very important. With a jet energy resolution of $60\%/\sqrt{E}$ a separation of both final states is nearly impossible with good purity and efficiency. At about $30\%/\sqrt{E}$ the two signals are much more clearly separated. The different behaviour is shown

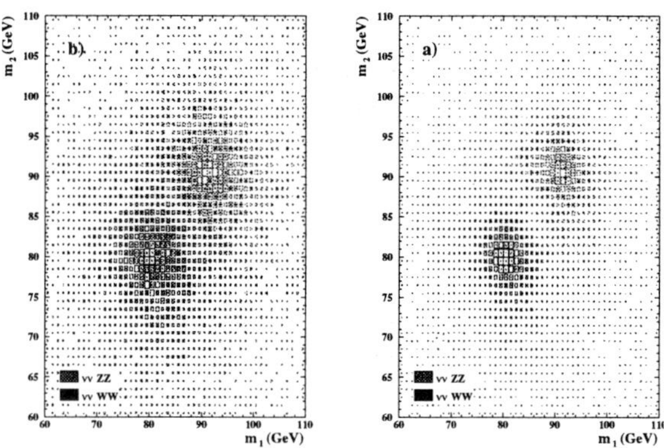

Figure 1. Reconstructed mass for hadronic WW and ZZ final states, as reconstructed[2] for an energy resolution of (left) $60\%/\sqrt{E}$, (right) $30\%/\sqrt{E}$.

in Fig. 1. Over the years a resolution of $30\%/\sqrt{E}$ has come to be accepted as a good compromise between the theoretically possible and the practically achievable resolution.

3. Detector Developments for the Linear Collider Detector

The requirements listed in the previous detector have been realised in a number of different detector concepts. A side view of a particular detector, the proposal for the TESLA collider published in the TESLA TDR[2] is shown in Fig. 2. In this section the different subsections of the proposed detectors, and the technological challenges are discussed.

3.1. *Vertex Detector*

The detector closest to the interaction point is the vertex detector, constructed from Si pixel detectors. It has to meet very stringent requirements on the resolution, achieve excellent stand-alone pattern recognition, and ensure a robust and reliable operation in this environment prone to significant background induced from the beams.

A number of different technologies are under discussion for this detector. All have in common that they are based on pixel detectors, with pixels not larger than around $50 \times 50\ \mu m^2$. This is to ensure that the occupancy of these devices during operation stays below a few percent. Most hits seen in this detector will be background induced, primarily from beam-beam generated backgrounds. In addition the devices are exposed to a fairly large -

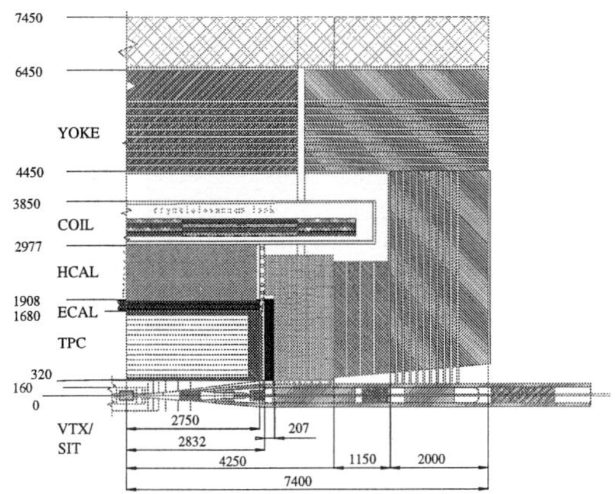

Figure 2. Sideview of one quadrant of the proposed TESLA detector.

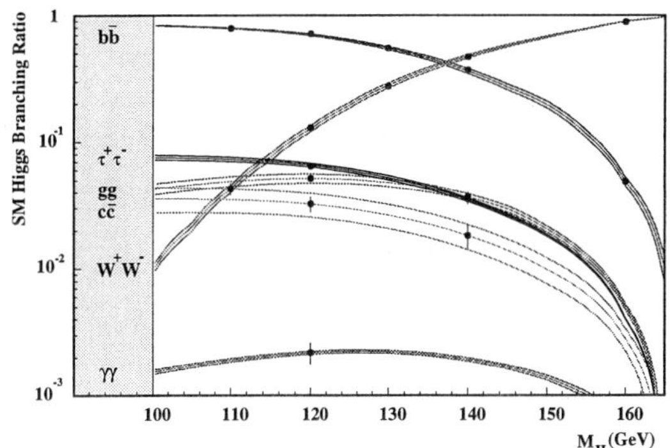

Figure 3. Predicted and reconstructed[10] branching ratio, using the simulation, of the Higgs boson into different fermion species.

though small by LHC standards - flux of neutrons of up to 10^{10} $n\,\mathrm{cm}^{-2}\mathrm{s}^{-1}$, coming from the interaction of the beam and its halo with the beam elements close to the interaction point. Technologies which are under investigation are based on CCD sensors, CMOS sensors, DEPFET sensors, and some other newer technologies. Although initial studies were also performed for the use of LHC-type pixel detectors at such a machine, they are not vigorously pursued since these detectors are much thicker than the other technologies under discussion.

The requirements on the vertex detector are defined primarily through the need for excellent tagging of secondary vertices. At SLD the power of a CCD type pixel detector has been convincingly demonstrated, resulting in outstanding efficiency and purity for the tagging of bottom hadrons. At the linear collider this capability is extended, by reducing the size of the pixels, and reducing the inner radius of the first layer, to improve significantly on the tagging of charmed hadrons. The possible performance of such a device is demonstrated in Fig. 3. Here the predicted and reconstructed branching ratios of a light Higgs boson are shown, for different final states. Even the decays into charm or tau final states can be clearly reconstructed.[10]

3.1.1. *CCD vertex detectors*

The CCD technology is a mature technology. Its applicability in a collider experiment has been demonstrated at the SLD experiment, where a detector with some 400 million channels was installed. Compared to that installation a detector at a linear col-

lider has to be read out much faster. At the same time the material present in the detector should be reduced as much as possible, to minimize the multiple scattering of long lived particles in the detector and thus increasing the precision which which its lifetime can be reconstructed. For a recent review see the References.[11]

Since the colliding electron and positron bunches in a linear collider are arranged in long trains, with bunch distances of at most a few hundred ns, a CCD detector can not be read out in between bunches. The signals will be integrated over many bunch crossings. To minimize this number the readout speed has to be increased as much as possible. At the moment a 50 MHz clock is foreseen, an increase over the SLD system of nearly a factor of 10. To further speed up the readout a column-parallel readout is planned, where rather than reading out the complete chip serially into one line, each column of pixels is read out into one independent readout chip. Recent results indicate that such performance can indeed be achieved in prototype chips.

Another area of intense R&D for CCD detectors is the thickness of the devices. Since the active part of the CCD is concentrated in a rather thin layer of Si on one side of the chip, it naturally lends itself to thinning of the detector. The influence of the thickness of the detector on the determination of the impact parameter is shown in Fig. 4.

3.1.2. *CMOS detector technology*

Over the last few years the CMOS technology has been intensely studied in view of their applicability to vertex detectors. In a CMOS based vertex detector

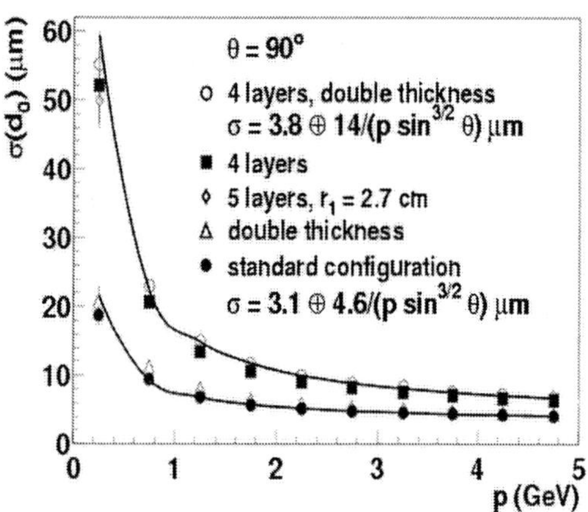

Figure 4. Impact parameter distribution for a four and a three layer vertex detector, and for different material distributions.[8]

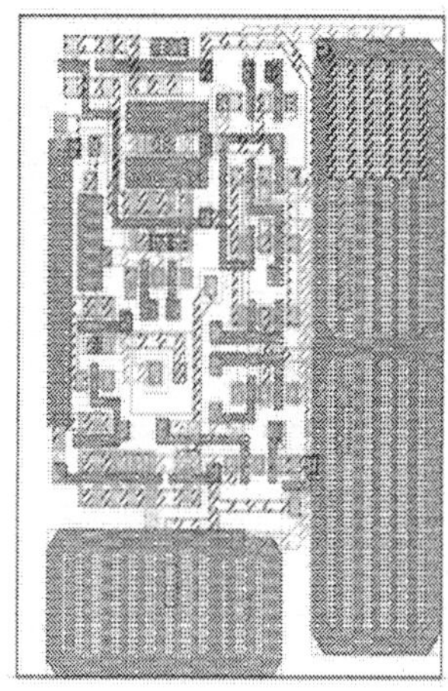

Figure 5. Enlarged view of one pixel of the MAPS silicon detector.

the charge produced under a pixel is not transported to the edge of the chip, as it is done in a CCD, but is processed by electronics integrated into the pixel. These types of detectors are known as MAPS (Monolithic Active Pixel Sensor) devices.[12] A view of the masks of a single pixel is shown in Fig. 5. After processing the signal is stored in the pixel until it is read out by some electronics located on the edges of the device.

Compared to a CCD detector a MAPS is operationally simpler. Since no charge needs to be transported over long distances it is expected to be more radiation hard. However very little experience exists about the use of such devices as particle detectors. Prototype chips have been operated very successfully during the last years. However there are still a number of open technical questions, which will be addressed over the next few years. As for the CCD the question of the material present in the sensors needs to be settled. While the potential for thinning of the devices is similar to the one for the CCD, no actual tests have been performed so far. For the use of a detector in the confined space close to the beampipe, the power consumption of the devices is going to be very important. As each pixel is an active device, potentially more power will be dissipated along the active area of the detector than is the case for the CCD.

A number of groups have formed over the last few years which design, produce and test prototype structures. Recently the sixth generation of test chips were produced and have been tested with

beam. The planned vigorous R&D program over the next few years should result in enough information to decide whether these devices are an alternative to a CCD-based detector.

3.1.3. *Other technologies*

A number of other technologies are under investigation at present. The DEPFET pixel detector again places active elements into the pixel. The charge created in a depleted region of the Si wafer is drifted into a potential well, where it is stored. It is then read out with a transistor integrated into the chip. Multiple readout is possible. Since the amount of electronics integrated into one pixel is very small, power consumption per pixel is very low. As for the other options, readout speed and thinning of the sensors are areas where development work is needed.[13]

A number of variants of the CMOS technology exist. In the FAPS scheme the limited readout speed of the MAPS chip is addressed by implanting more than one storage capacitor into one pixel, thus allowing the storage of more than one signal on one chip, which are then read out serially.

3.1.4. *Comparison of the different technologies*

At the moment no conclusion can be reached as to which of the different technologies will in the end be chosen. All are under active development, and all face challenges to meet the requirements for use at a linear collider detector. Of particular importance is the readout speed, which directly determines how many bunch crossings are integrated into one readout cycle. Thinning of the detectors is another active area of research, where no final answers exist for any of the technologies. For use in a real detector the real-estate needed for the final detector system - that is the active area and the inactive area needed for readout - is another important criterion. A vigorous research program has started at different laboratories around the world to try and answer these questions.

3.2. *Tracking Detectors*

The two different detector options being studied differ primarily in their choice of technology for a main tracker: gaseous detector versus Si detectors. Depending on the choice the role the tracking detector will play is slightly different.

In the case of a large volume gaseous detector, as e.g. proposed for the TESLA detector, the tracker will be one of the main devices used in track finding and track reconstruction. This will ease the expectations for the vertex detector for track finding together with the hope of making the whole system more robust against backgrounds and tracking inefficiencies.

In the model of a "small" detector the tracking is done by a combination of the vertex detector and a few layers of Si strip detectors. The main part of the tracking is left to the vertex detector. The Si strip detectors are primarily used to measure the momentum of tracks in the strong magnetic field, by adding enough lever-arm to the small vertex detector. This detector concept is based on the excellent experience at the SLD detector, where the high granularity CCD vertex detector has proven itself to be a very robust and powerful tracking device.

3.2.1. *The TPC tracker*

A TPC has been used in a number of large scale experiments as the central tracking device. For the linear collider a TPC has been proposed because it offers a large number of space points with good single hit spatial resolution (around 100 μm seems achievable), and reasonable double track resolution.[2] This concept has been shown to result in a tracking system with high track finding efficiency. The large number of points reconstructed in the TPC makes for a very robust system, with a lot of build-in redundancy. In addition a gas detector offers the capability to identify different particle types via dE/dx, the specific

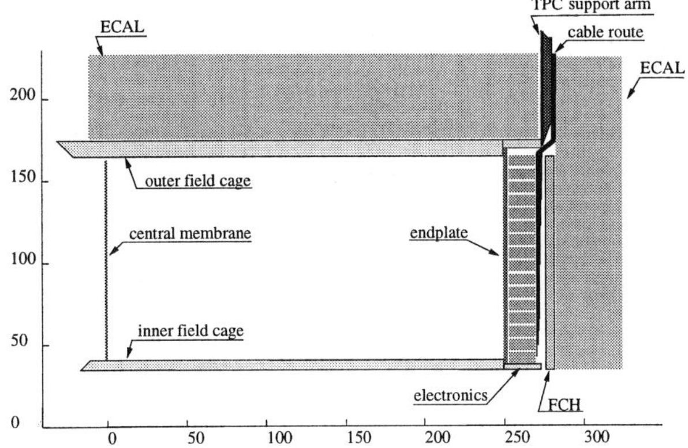

Figure 6. One quarter view of a TPC as proposed for a detector at a linear collider.

energy loss. In Fig. 6 a side view of a proposed TPC is shown.

A major area of development is the readout technology. Traditionally wire chambers have been used to detect the charge produced in the TPC volume. Wire chambers are a proven technology, resulting in a robust system. There are however a number of disadvantages connected to a wire chamber readout. First of all, the wires require a sophisticated mechanical structure to support their tension. This introduces material in an area in front of the endcap calorimeter, where the amount of material should be minimized. Wire chambers are in addition limited to a minimum spacing between the readout elements, the wires, which is in the region of mm. This implies intrinsically different granularities in the directions along or perpendicular to the wire. This results in a reduced resolution and increased systematic errors.

Recently micro-gas detectors have received a lot of attention. Different types of such detectors are GEM (Gas Electron Multiplier) or micromegas.[14,15] A GEM consists of a insulating foil, typically Kapton, clad on both sides with a thin layer of copper. Holes are present in the foil and the copper connecting both sides. If a potential difference is applied to the two copper surfaces, a strong electric field develops in the holes. If such a foil is placed perpendicular to the drift direction of the electrons, and if proper potentials are applied so that the electrons are forced into the holes, gas amplification takes place inside the holes. The charge after amplification is extracted from the hole on the opposite side, and detected by pads connected to charge sensitive amplifiers. Micromegas operate on a similar principle, except that the GEM foil is replaced by a metallic grid, placed on top of the pad plane at a short distance. The gas amplification takes place in the area between the grid and the pad plane.

Micro-gas detectors are investigated as possible replacements of the wire chambers in a TPC. In this case a system of MPGDs is placed at the end of the drift volume, charge amplification takes place in these devices, and the charges are detected on pad planes as they are in more conventional wire chamber TPC. The main advantages of a MPGD TPC is that the readout is mechanically more simple, and might result in a device with less material than the conventional chamber. In addition these devices are intrinsic two-dimensional, and should introduce smaller systematic errors than the wire TPC.

During the last few years an international group has formed with the goal of investigating and developing the technology of a TPC equipped with micro pattern gas detectors.[16] The group includes members mostly from Europe and the Americas, with links to Asian groups forming. Several prototype TPCs have been built, equipped with GEMs or with Micromegas. The principle of the operation has been demonstrated, and resolutions have been studied. Recently for the first time small prototype chambers were exposed to large magnetic fields as they are expected in a real detector. While results are still very preliminary, no adverse effects of these magnetic fields on the operation of the MPGD or the TPC were found. As an example the measured ion feedback in a prototype TPC is shown in Fig. 7, as a function of the magnetic field. The behavior of the feedback on the field is not very strong, and very smooth. Although no quantitative model exists at the moment to explain the observed behavior, qualitatively is does not present any surprises.

The developments of TPCs are not limited to the area of detector development for the linear collider. In fact TPCs are used in a wide area of different applications. An interesting new device under construction is the TPC to be used in the ICARUS experiment in Gran Sasso.[18] Here the drift volume is filled with liquid Argon, resulting in extremely long drift times of up to 1 ms for the 1.5 m drift distance.

3.2.2. *The all-Si tracker*

The alternative tracker option is based on a few layers of Si strip detectors arranged in cylinders and

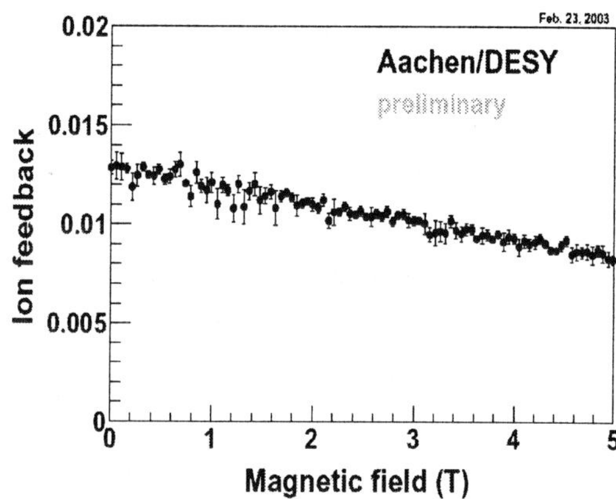

Figure 7. Ion feedback measured in a TPC prototype chamber, as a function of the applied magnetic field.[17]

disks to fill the space between the vertex tracker and the calorimeter.[19] Based on studies done for the LHC experiments it is expected that the excellent resolution obtainable with these devices compensates for the fact that there are only a few points, thus maintaining a good tracking efficiency. In the existing designs particular emphasis is placed on good reconstruction in the plane perpendicular to the beam, measuring primarily the curvature of the tracks in the magnetic field. For cost reasons only one or two of the layers are instrumented to measure both the $r - \phi$ and the z coordinate.

The main challenges of this design are the mechanical design of a large area Si strip detector system, minimizing the material present in the support structures, and at the same time optimizing the stability of the device.

While the tracking and the measurement of the parameters of isolated tracks can be done in such a detector as well as in a TPC based tracker, it is not yet known, whether such a tracker influences the particle flow reconstruction in any way. The detection of long lived particles and the tagging of photon conversions are two topics, which will be studied in great detail in the near future, and which will help to understand the relative merits of the different detector concepts.

3.3. *The Electromagnetic Calorimeter*

In the framework of a particle flow reconstruction algorithm the ideal calorimeter would provide a three-dimensional picture of the shower developing inside the detector. This ideal detector can be approxi-

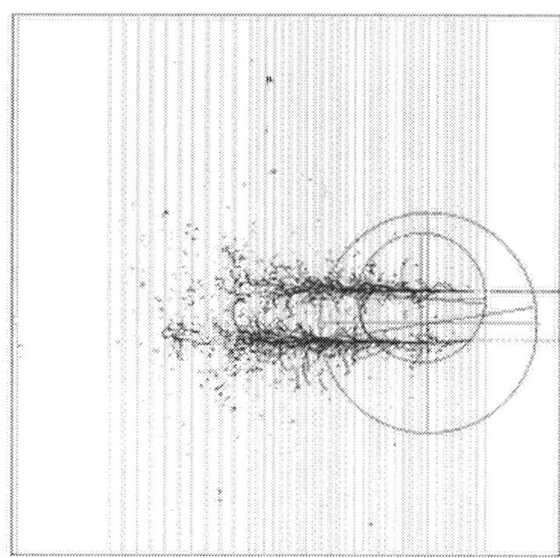

Figure 8. Signals deposited by two closeby photons in the Si-W sampling calorimeter.[6] The distance between the two photons is approximately 1 cm.

mated by a sampling calorimeter with the typical size of a cell given by approximately the Molière radius in the material. If this is supplemented by many samples along the direction in which the shower develops, a detailed reconstruction of individual showers becomes possible. The reconstruction of close-by charged and neutral particles is helped by the immersion of the device in a strong magnetic field, separating charged and neutral particles. The size of the device and the resolution obtainable are given primarily by the size of the Molière radius. A tungsten sampling calorimeter therefore is an attractive option.[2] In Fig. 8, the resolution with which two photons from a conversion would be seen, is illustrated.

The Molière radius in a W-based calorimeter is around 1 cm, depending on the thickness of the sensitive detector. Most studies at the moment consider a Si detector as the most promising option for such a device. Si detectors are thin, can be easily subdivided into small cells, and can be produced to cover large areas. The main problem is the cost - a Si based electromagnetic calorimeter will be rather expensive, and becomes a major cost factor for a linear collider detector.

The current R&D activities concentrate on finding a technological solution for building a large area readout plane from Si, with low enough noise and fast enough readout to be a viable option. At the moment it is thought that a 2.5 mm thick readout gap is necessary between Tungsten plates. However developments are under way to reduce this to close to

1.5 mm. The typical cell sizes which are considered are between 5×5 mm^2 and 10×10 mm^2. In Europe an option is being investigated where a second readout board on top of the Si detector will pick up the signals from the individual cell, and route them to the edge of each calorimeter module. In the US the modules are divided into "wafers" of approximately $8''$ diameter, with a readout chip located centrally on the chip. In either case the transfer of the charge to the readout over a fairly long distance presents a significant challenge.

The CALICE collaboration, which has members from Europe, the Americas and Asia, plans to build a small scale prototype of such a detector by the end of 2004. This detector will then be extensively studied in a series of test beam experiments, together with a HCAL prototype, described below.[20]

3.4. *The Hadronic Calorimeter*

In a calorimeter optimized for the particle flow reconstruction the hadronic calorimeter plays an important role as well. Ideally this device is built as a tracking calorimeter as well, which allows the detailed reconstruction of the development of the hadronic shower. Compared to the electromagnetic calorimeter, the reconstruction of a hadronic shower is in some respect more challenging, since the presence of neutrons in the shower result in a much less localized nature of the shower, and thus in more problems identifying the components of a shower. In addition a good hadronic calorimeter not only should be able to reconstruct the hadronic part of the shower, but should also be sensitive to the always present electromagnetic component in each hadronic shower.

At the moment two different concepts for such a calorimeter are being discussed. The more conventional approach uses a sampling calorimeter with e.g. a scintillator based readout to record the position and the size of the energy deposits.[20] A different approach concentrates fully on the spatial information by reading out the calorimeter in an extremely finely grained way, but only recording whether or not a cell has been hit. This so called digital approach can work, because a strong correlation exists between the total energy deposited in the calorimeter, and the number of cells hit.[20] This is demonstrated in Fig. 9. Shown is the result of a simulation study correlat-

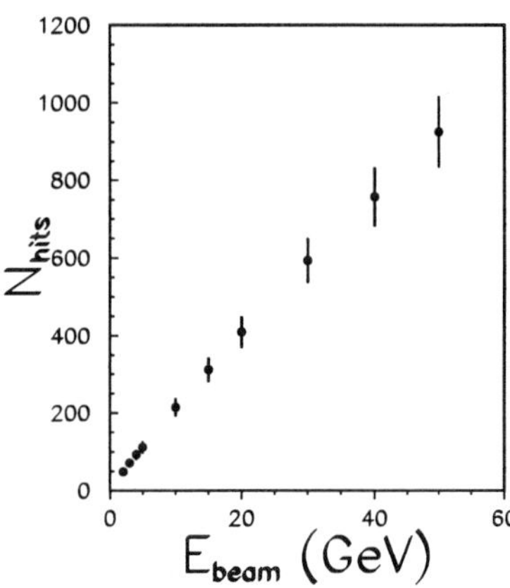

Figure 9. Correlation between the number of cells with hits, and the total energy in the calorimeter, for a hadronic calorimeter.

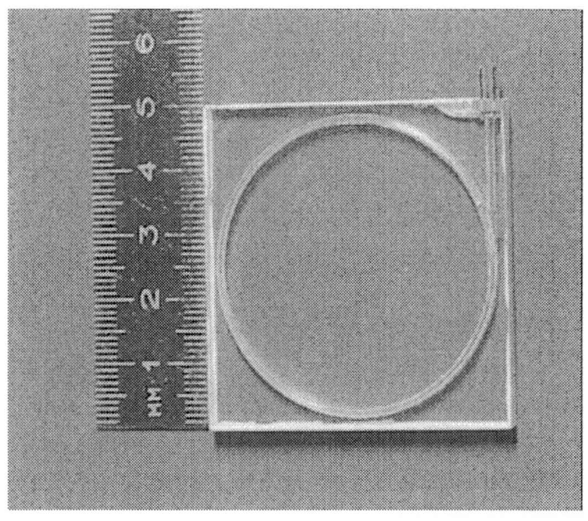

Figure 10. Photograph of a single 5×5 cm^2 scintillator tile equipped with a SiPM. The SiPM is visible at the top right corner of the tile.

ing the number of hits seen with the total energy deposited.

The first conventional calorimeter option will most likely be realised as a sampling calorimeter with scintillator readout. Even here the particle flow concept requires that the granularity is very good, so that the hadronic components of the shower can be resolved, and that the electromagnetic components can be correctly attributed to the correct shower. Typical cell sizes considered are around 5×5 cm^2. The readout of a correspondingly large number of tiles presents a significant challenge. Conventional photo-tubes are disfavored because of cost, and because they do not operate in strong magnetic fields. Long clear fibers would need to be routed outside the central coil, into a low field region, with the corresponding significant complication and cost. A novel approach is being investigated whereby Si-based miniaturized photo-multipliers are integrated directly into the tile. Such devices, which became recently available from a Russian group, offer something like 1000 pixels in a 1 mm^2 area. Each pixel by itself can not record analog information. However the number of pixels which register a photon is a measure of the total amount of light, and thus these devices can be used in an analog readout scheme. Recent tests have confirmed that the integration into the tile is possible, and that a reliable measurement of the amplitude is possible. These devices would significantly simplify the problem of reading the large number of cells. A sample tile with the SiPM inte-

grated is shown in Fig. 10.

The digital HCAL option requires a readout system which can record energy deposits in cells of typically 1×1 cm^2 over large areas. Si detectors are much too expensive for such things. Alternative solutions based on Resistive Plate Chambers (RPC's) or on GEM based systems are under investigation. Given that the rate requirements for a detector at a linear collider are not very high, it is expected that resistive plate chambers should be adequate for this device.

An equally big challenge to the development of the necessary hardware for such a tracking calorimeter is the development of reconstruction algorithms to exploit the potential of these devices. Conventional calorimeter reconstruction methods are no longer adequate for the large granularity in a particle flow detector. Novel approaches are needed, which in many way resemble pattern reconstruction algorithms developed for tracking systems. This work is still in its infancy, but its fast development is essential should the particle flow calorimeter become a realistic option. They are needed to both gauge the performance of these devices, and understand beam test data, and are essential in designing and in optimizing the detector in terms of cell size, absorber thickness etc.. An important goal of the first generation of test beam experiments planned for 2004/2005 therefore will be the verification of simulation tools, so that reliable predictions for the performance of the complete system can be made.

4. Detector Developments for the SLHC

For a possible upgrade of the LHC collider, radiation hardness of the detector components will be an even greater challenge than it is already for the LHC. It is expected that at the SLHC (Super-LHC) the 1 MeV equivalence dose of neutrons at a distance of 4 cm from the beam line will increase from an expected $\Phi(R = 4$ cm$) = 3 \times 10^{15}$ cm^{-2} to $\Phi(R = 4$ cm$) = 1.6 \times 10^{16}$ cm^{-2} - a rate which none of the known detector technologies will be able to tolerate.

A number of groups have started a R&D activity with the goal of significantly improving the radiation hardness of Si based detectors. A systematic program has started, based upon the work done for the first stage of the LHC experiments, to explore the mechanism responsible for radiation damage, and to find ways to increase the radiation hardness. Recently significant progress has been reported when using epitaxial Si detectors grown on Czochralski substrates. Oxygenated silicon has for some time been known to improve the radiation hardness of Si - in particular with respect to photon irradiation. The new epitaxial detectors seem, in addition, to show a significantly improved radiation hardness under hadronic irradiation. From preliminary results shown by the group it appears that for a hadronic fluence of a few times 10^{16} cm^{-2} the charge collection efficiency drops by approximately 40%. For the first time a material has been demonstrated which is still functional after exposure to the full SLHC dose.[21]

5. Summary

The next generation of experiments in High Energy Physics, most notably at the electron-positron linear collider, pose many challenges to the detector developer. In particular the concept of a detector optimized for particle flow requires a significantly more powerful calorimeter, with up-to-now unheard of segmentation. Such a device together with high resolution tracking devices are the subject of an intense and international R&D effort.

Significant progress has been made in the past years in the radiation hardness of Si detectors. This is of particular importance for a possible LHC upgrade project.

In summary detector R&D is a very active and interesting field, where many new results are to be expected over the next few years. When in a few years a decision on the linear collider is expected enough detector R&D should have been done to make a rational decision for a detector at such a machine.

6. Acknowledgments

I wish to thank many colleagues from the different linear collider physics and detector R&D groups around the world, who contributed material for this review, and were available for questions and discussions. I wish to thank the organizers of this conference for a productive and pleasant conference.

References

1. P. Grannis, S. Komamiya, T. Matsui and F.Richard (eds.), *Understanding Matter, Energy, Space and Time: The case for the e^+e^- linear collider*, see http://sbhep1.physics.sunysb.edu/~grannis/wwlc_report.html.
2. F. Richard, J.R. Schneider, D. Trines and A. Wagner (eds.), *TESLA: Technical Design report, DESY 2001-01, ECFA 2001-209, TESLA Report 2001-023, TESLA-FEL 2001-05* (2001).
3. American Linear Collider Working Group, *Linear Collider Physics, BNL-52627, CLNS 01/1729, FERMILAB-Pub-01/058-E, LBNL-47813, SLAC-R-570, UCRL-ID-143810-DR, LC-REV-2001-074-US* (2001).
4. Asian Comittee for Future Accelerators, Japan High Energy Physics Committee, High Energy Accelerator Research Organisation, *GLC Project* (2003).
5. H. Videau and J.C. Brient, *A Si-W calorimeter for linear collider physics*, prepared for 10th International Conference on Calorimetry in High Energy Physics (CALOR 2002), Pasadena, California, 25-30 Mar 2002.
6. V. Morgunov, *Calorimetry design with energy-flow concept (imaging detector for high-energy physics)*, prepared for 10th International Conference on Calorimetry in High Energy Physics (CALOR 2002), Pasadena, California, 25- 30 Mar 2002.
7. J. Hauschildt, *LC note LC-DET-2001-036* (2001), see http://www-flc.desy.de/lcnotes.
8. S.M. Xella-Hansen *et al.*, *LC-DET-2003-061* (2003), see http://www-flc.desy.de/lcnotes.
9. M. N. Minard, *Jet energy measurement with the ALEPH detector at LEP2*, prepared for 10th International Conference on Calorimetry in High Energy Physics (CALOR 2002), Pasadena, California, 25-30 Mar 2002.
10. M. Battaglia and K. Desch, *hep-ph/0101165* (2001).

11. C. Damerell and K. Stefanov, *LC-DET-2003-060* (2003), see http://www-flc.desy.de/lcnotes.

12. M. Winter, *New developments for silicon detectors* prepared for 31st International Conference on High Energy Physics (ICHEP 2002), Amsterdam, The Netherlands, 24-31 Jul 2002.

13. P.Fisher *et al.*, *LC-DET-2001-004* (2001), see http://www-flc.desy.de/lcnotes.

14. R. Bouclier *et al.*, *The gas electron multiplier (GEM), IEEE Trans. Nucl. Sci.*} **44**, *646 (1997)*.

15. Y. Giomataris, P. Rebourgeard, J. P. Robert and G. Charpak, *Nucl. Instrum. Meth. A* **376**, 29 (1996).

16. The LC TPC group, LC-DET-2002-008, see http://www-flc.desy.de/lcnotes.

17. P. Wienemann, DESY, private communication.

18. F. Arneodo *et al.*, *Nucl. Phys. Proc. Suppl.* **54B**, 95 (1997).

19. C. J. Damerell *et al.*, *Ideas for the NLC detector, SLAC-REPRINT-1996-009* (1996).

20. S. Valkar *et al.*, *A Calorimeter for the e+ e- Tesla Detector: Proposal for R&D, DESY-PRC-RD-01-02* (2001).

21. G. Kramberger *et al.*, report to be published.

DISCUSSION

David Miller (University College London): It's a comment on an early graph where you showed the WW and ZZ separation, and it is just a plea that people must not think that WW and ZZ are the background at the linear collider. WW and ZZ are where we find out what is going on in electroweak symmetry breaking. It's a signal.

Ulrich Goerlach (Univ. of Strasbourg): You mentioned very briefly the MAPS detectors. Although you didn't explain in detail their functioning, I was a little bit surprised about your statement that they may lead to a larger material budget. I do not agree with this. Since MAPS use only a very small part of the silicon as an active medium, they have the potential to be thinned down to a few tens of a micron, leading to very thin detectors, and thus to a very low material budget.

Ties Behnke: The point is, that as in most of these detector technologies, it is not known how far we can push the thinning of these different detectors. In fact, even for the CCD technology, where people thought for a while, they understood how to do the thinning, problems turned up when they actually tried to build a test device. I think it is just one of the very important points which needs to be addressed by a vigorous R&D program. We just have to see at the end what comes out, and how well these different technologies perform.

Paula Collins (CERN): I just have a comment and a question. The comment is that for Czochralski silicon we have recently tested irradiated Czochralski silicon with fast electronics in a test beam and the results will be coming out very shortly. The question is then: a year ago in the linear collider community there was the concept floating around of using silicon drift detectors for the TPC. I'd like to know if this is still something which is favored? As I understood the time stamp is very useful for sorting out the integration of many events in the CCD detector, and that was the advantage of it.

Ties Behnke: To be honest, I am not aware of any vigorous program in that direction. The standard TPC does give you enough timing information to separate bunches at least at TESLA, where you have some 300 ns or so time. From this point of view a potentially better timing resolution is not needed. For the warm machines, where the timing between bunches is much shorter, a separation of individual events from each pulse will be very hard in any technology under discussion, and probably is not needed.

Alan Sill (Texas Tech University): Now I have built detectors, tracking detectors for many years. I love tracking detectors, so don't get the sense of this question wrong. But I think you have really repeated a lot of the statements that have been made for many years about linear collider detectors that really need to be challenged. For example the idea that you get resolution by counting all the little particles. And you see where this leads you. You are driven toward very complex detectors, very expensive ones, especially calorimeters. A sweep the table argument, that's much simpler, is to say let's get the energy resolution in the calorimeter. Let's build a high-precision calorimeter that gets a lot of energy resolution. In that 60% of the charged particles are of course a lot of hadrons. Is there any attempt in the linear collider community to challenge some of these statements that have been made for many years, but maybe need to be challenged a little bit?

Ties Behnke: There is a lot of activity trying to really find out whether the statements that I made are stringent or not. It has been established using simulations that the particle flow algorithm applied to a detector with granularities as I described, can give you the resolution we want for the jet energy - I showed the example of WW and ZZ separation. The resolution which has been achieved in existing detectors is about a factor of two worse. Exactly how far one can push the conventional approach, or how much one can simplify the detectors I discussed, and still achieve decent resolution, is not really known. The trouble is that these questions are not easily answered. You really have to go through the complete reconstruction procedure if you want to study whether such a system does or does not work. You have to develop reconstruction programs both for the tracker and for the calorimeter. You have to really do a proper job in simulating and reconstructing the events, which just takes time. These efforts are ongoing. There aren't complete results yet. But, yes, people do look quite critically whether you really need this tremendous granularity in the calorimeter, with the correspondingly large cost, or if you can get away with a much simpler device.

THE LARGE HADRON COLLIDER: A STATUS REPORT

L. MAIANI

CERN, CH-1211, Geneva, Switzerland
E-mail: Luciano.Maiani@cern.ch

The status of LHC construction, machine and detectors, is reviewed, with particular emphasis on the expected physics and on the industrial production of machine components.

Summary

1. Introduction

2. New Physics at High Energy

3. LHC Construction

4. LHC Detectors

5. Computing

6. Conclusions

1. Introduction

The installation of a Large Hadron Collider in the LEP tunnel, was first considered at the beginning of the 80's and it was approved by the CERN Council in December 1996. The tunnel has a circumference of 27 km. The nominal beam energy was determined to be 7 TeV, corresponding to a nominal magnetic field in the cryogenic dipoles of 8.3 Tesla.[1]

From the outset, it was decided that the LHC should accelerate both protons and heavy ions. The LHC energy range and luminosity will allow it to aim at several physics goals:

i search for the Higgs boson in the whole mass region where it is predicted by the Standard Theory, from the LEP reach up to almost 1 TeV;

ii search for new particles, such as those associated with Supersymmetry, indicated as the remedy to the "unnaturalness" of the Standard Theory itself;

iii high precision study of CP-violation in B meson decays; and

iv study of the physical properties of the new phase of nuclear matter at high temperature (quark-gluon plasma) that supposedly is produced in relativistic heavy ion collisions.

In 1996, a cost was stated for the LHC proper (the hardware that goes in the LEP tunnel and the civil engineering works to adapt the tunnel and to build the experimental areas) assuming that all other expenses would be covered by the normal operational budget of the Laboratory.

During the years 2000 and 2001 the LHC programme underwent a full cost and schedule review, including the costs associated with the refurbishing of the injection line (which uses the existing PS and the SPS machines), the cost-to-completion of the four detectors, the cost of the computing infrastructure, as well as overall personnel costs and contingency. The result was approved by the Council in December 2002 to form the basis of the present CERN planning and is as follows.

- **Machine**: 4.8 BCHF. This includes:
  - materials and CERN personnel expenses for the whole LHC; and
  - 330 MCHF for residual contingency and for a reserve for the cost escalation of LHC contracts.

- **Detectors**: 1.2 BCHF, of which 20% is supported by the CERN budget.

The figures given above include special contributions that CERN has received from the Host States (Switzerland and France) and from several CERN non-Member States (US, Japan, Russia, Canada, India), the latter adding to about 0.75 BCHF. These contributions are mostly in-kind, with items provided by industry and by the main particle physics laboratories in these countries. The world-wide participation in the machine and detector construction makes the LHC the first really global particle physics project.

The construction of such a large machine is requiring an unprecedented effort from the Laboratory. The last approved Long Term Plan, which covers the period 2003-2010, indicates that around 80% of

CERN resources (personnel plus materials) are allocated to the LHC programme (directly $\approx 50\%$, indirectly $\approx 30\%$).

In this talk, I'd like to first review briefly the physics that can be expected from the LHC. For time reasons, I will restrict this to the High Energy frontier, namely the physics with ATLAS and CMS.

Then I will give you a tour of the civil engineering of the detector caverns, the industrial production and the installation of the LHC, machine and detectors. With the main contracts in place since last year, we are now assisting the very exciting ramping up of the industrial production, due to reach cruising speed during the next year.

After the difficult discussions we had with the CERN Council over the last two years, and a drastic reformulation of CERN internal strategies, the LHC financial set up is now considered to be on solid, if very tight, ground.

Given all these results, Management is in the position to confirm the LHC schedule already announced in March 2002, namely:

- completion of the LHC machine in the last quarter of 2006;

- first beams injected during the spring of 2007; and

- first collisions in mid 2007.

2. New Physics at High Energy

At the LHC scale of exchanged momenta, valence quarks in the proton have lost much of their importance. Gluons and the sea of quark-antiquark pairs dominate. Unlike at lower energy, proton-proton and proton-antiproton collisions are much alike at these energies. These features and the very high luminosity (target luminosity $10^{34}\mathrm{cm}^{-2}\mathrm{s}^{-1}$) will make the LHC a real factory for any conceivable particle, from Higgs bosons to B mesons ... to mini black holes, should they exist. Table 1 shows the yearly production of several kinds of particles, compared to the total statistics that will be collected at previous machines by 2007.

The LHC will be a very versatile machine indeed, with an expected lifetime spanning over two decades, possibly extended by a luminosity upgrade, which is just being considered these days.

2.1. *Searching for the Higgs Boson*

There is little doubt that the LHC and its planned detectors will be able to detect the Standard Theory Higgs boson over the whole mass range where it is predicted,[2] from around 114 GeV (the LEP limit) up to about 1 TeV, with a confidence limit above 5σ in one year at low luminosity ($3 \times 10^{33}\mathrm{cm}^{-2}\mathrm{s}^{-1}$), and well above in one year at the nominal luminosity (30 fb^{-1}). The lowest mass region is particularly difficult, which shows the usefulness of the magnificent job done by the accelerator physicists and by the LEP collaborations to extend the lower bounds as much as possible. Of particular importance in this region is the recently studied W or Z fusion channel:

$$qq \to qqH \to qq\tau^+\tau^- \tag{1}$$

with one of the quarks in the final state tagged as a forward jet.

The mass of the Higgs boson will be measured with very good precision, much the same as the W mass was measured at the Tevatron. At the end of the day, after ten years at nominal luminosity (i.e. an integrated luminosity of 300 fb^{-1}) one may be able to attain errors of the order of 10^{-3}, in most of the low mass range, to order of per cent, in the higher mass range.

2.2. *Higgs Boson Couplings*

The real signature of spontaneous symmetry breaking is, of course, that the values of the Higgs boson couplings to the various particles should be in the ratio of the corresponding masses. The absolute values of the couplings are hard to measure in a hadron collider, due to not well known production cross sections and total Higgs boson width. However, a lot of uncertainties disappear in the ratios and coupling constant ratios would give anyway decisive evidence that the observed particle is indeed the long sought Higgs boson. Precisions of 10-20% can be obtained for a number of important channels. Increasing the LHC luminosity by a factor 10 (as considered at present for the LHC upgrade) could reduce errors by a factor of 2 and give some access to the triple Higgs coupling.[3] It is here that a low energy electron-positron linear collider (LC) could make significant progress, measuring absolute couplings to a precision of a few percent.

Table 1. Expected event production rates in ATLAS or CMS for some (known and new) physics processes at the initial "low" luminosity of $\mathcal{L} = 10^{33}$ cm^{-2} s^{-1}.

Process	Events/s	Events per year	Total statistics collected at previuos machines by 2007
$W \to e\nu$	15	10^8	10^4 LEP/10^7 Tevatron
$Z \to e^+e^-$	1.5	10^7	10^7 LEP
$t\bar{t}$	1	10^7	10^4 Tevatron
$b\bar{b}$	10^6	10^{12} - 10^{13}	10^9 Belle/Babar ?
$H, m = 130$ GeV	0.02	10^5	?
Gluino pairs, $m = 1$ TeV	0.001	10^4	$-\ -\ -$
Black holes, $m > 3$ TeV ($M_D = 3$ TeV, $n = 4$)	10^{-4}	10^3	$-\ -\ -$

2.3. *The Standard Theory Becomes "Unnatural" above the TeV Region*

Scalar particle masses are not protected against becoming of the order of the largest mass of the theory, the cut-off, because of quantum corrections.[4] Only an unreasonable fine-tuning between the bare mass and its quantum corrections could make $M_H \ll M_{\text{Planck}}$ as required by observation. Presently three different solutions have been developed:

- the Higgs boson is not elementary but rather a fermion-antifermion state, bound by new strong forces at the TeV scale[5] (dubbed by the generic name Technicolor); this solution is strongly disfavoured by the LEP/Tevatron data;

- Supersymmetry in the TeV region.[6] In softly broken SUSY,[7] the Higgs mass is determined by the scale of SUSY breaking. Minimal SUSY models are compatible with LEP and Tevatron data; a (quasi) stable neutral lightest supersymmetric particle seems today the best candidate for the cosmological cold dark matter indicated by astronomical observations; and

- additional space dimensions with a very large radius,[8] where gravity is characterized by a mass which is itself of the order of TeV: in this case the Planck mass has no fundamental meaning and even unprotected masses are of the order of TeV.

The LHC will be able to shed decisive light on these alternatives.

2.4. *Searching for SUSY*

As for particle content, MSSM features:

- two Higgs doublets, with five physical scalar bosons, three neutral: h, H, A and two charged ones: $H^{\pm}$; and

- the SUSY partners of known particles:
  - spin 0: scalar quarks, scalar leptons; and
  - spin 1/2: gluinos, gauginos, Higgsino.

The decay chains of the SUSY partners are likely to contain several neutral, invisible particles, giving rise to spectacular missing energy signal.

The simplest case is that of the Minimal Supersymmetric Standard Models (MSSM) of the restricted mSUGRA type, characterized by two mass parameters: m_0, $m_{1/2}$, and one sign, μ. The allowed region in parameter space is restricted by:[9]

- the non-observation of SUSY signals at LEP and Tevatron; and

- the values of the cosmological observables reflecting cold dark mass distribution.

If the lightest supersymmetric particle provides the cosmological dark mass and its parameters are in the typical region (i.e. no large conspiracies between different contributions) SUSY can be discovered very quickly at the LHC. At full luminosity, LHC can discover squarks and gluinos up to about 3 TeV.[10] Masses can also be determined, in several instances with good precision (few to several percent) by the observation of end points in the mass distribution of different particle combinations (dileptons, jets, etc.).

2.5. *MSSM Benchmarking of Colliders*

The masses of the SUSY partners can take large values when special cancellations between different parameters take place. In some cases, they could even escape detection at the LHC. This has been studied in Ref. 9. A number of points in parameter space have been chosen and for each of them the authors have computed the number of SUSY particles that will be observed at the LHC and other colliders. In line with the previous considerations, the LHC will see squarks and gluinos in most cases, in some cases also sleptons. However there are some "nasty" cases where SUSY particles will be out of reach and only a Standard Theory Higgs boson will be seen.

The situation with a "low energy" electron-positron collider of 0.5 TeV has also been considered. In generic cases, sleptons will be seen much more efficiently than with the LHC, as had to be expected, but the bad cases remain. Indeed a real resolution of the issue requires collider energies as high as 3 or 5 TeV, such as may be attainable with the double beam acceleration method (CLIC).

2.6. *Extra Dimensions at mm Scale?*

The quest for a unified theory of General Relativity and Quantum Mechanics has led to a revival of the old idea, introduced by Kaluza and Klein in the thirties, that space time must have extra space dimensions, besides the familiar 3 space + 1 time dimensions. Kaluza and Klein assumed that the extra dimensions are compact, with radius R, so that particles could not feel the extra space dimensions until their quantum wavelength was smaller than R. Recent ideas,[8] combine Kaluza Klein extra dimensions with a new notion, namely that matter and gauge fields can be confined to a lower dimensional manifold (a D-brane), while gravity extends instead to all space. This opens up the revolutionary idea that R may be very large, on the particle physics scale, and we would not detect the extra dimensions simply because the particles we are made of are dynamically confined to a D-brane with 3+1 dimensions. The gravitational potential in the full $4+\delta$ space-time (δ is the number of extra dimensions) is characterized by a constant M_D, with dimension of a mass. M_D is related to M_{Planck} according to:

$$R = \frac{1}{M_D} \left(\frac{M_{Planck}}{M_D} \right)^{\frac{2}{\delta}}. \tag{2}$$

The upshot is that indeed M_D could be of the order of, say, 1 TeV (therefore no naturalness or hierarchy problem).

There are two ways we can put this idea to a test. First, at distances of order of R, gravitational forces should show deviations from the Newton $1/r^2$ law. Present results do exclude this down perhaps to a micron, which is still roughly compatible with the idea. Second, in high energy collisions, at exchanged momenta $\gg 1/R$, gravitons and excited gravitons would be radiated with large cross sections (and leave the apparatus undetected), causing deviations from the expected, QCD or QED cross sections. The idea has been put to test already at LEP. One can obtain larger limits on M_D at the LHC,[11] from the observation of the reaction:

$$q + q \to \text{gluon} + (\text{unobserved particles}). \tag{3}$$

Related to the same idea of a strong gravity in $4+\delta$ dimensions is the possibility of producing mini black holes in LHC collisions. Tiny black holes of $M_{BH} \approx 1$ TeV can be produced if partons pass at a distance smaller than their Schwartschild radius, which would also be of the order of 1 TeV^{-1}. One expects large partonic cross sections. For $M_D = 3$ TeV and $\delta = 4$, one finds $\sigma(pp \to BH) \approx 100$ fb, leading to 1000 (very spectacular) events in 1 year at low luminosity.

In conclusion:

- the LHC will give a definite answer to the Higgs boson problem and will explore the TeV region with very good efficiency;

- a sub TeV Linear Collider is needed for precision Higgs boson physics particularly in the case of a light Higgs boson;

- the LC would be also useful to distinguish SM from Minimal Supersymmetric SM; and

- multi-TeV capability is needed to really understand Supersymmetry ... or to identify other alternatives.

Pursuing accelerator R&D on Multi-TeV accelerators, CLIC and the Very Large Hadron Collider/Eloisatron is vital for the future of particle physics.

3. LHC Construction

Decisive progress has been made during the last 12 months, in all sectors of the LHC project. Old concerns have been overcome, for example those due to the insolvencies of some crucial companies, new concerns are appearing but no potential show stopper is in sight. With the help of some pictures, I will give you a guided tour through the LHC work: civil engineering, production of machine hardware components and the new activity, started this year, namely LHC installation in the tunnel.

3.1. *Civil Engineering*

The preparation of the experimental caverns and related structures is reaching its final phase, with more than 80% of the work completed.

The ATLAS cavern was inaugurated on June 4, 2003 and handed over to the experimental collaboration. The situation in July is shown in Fig. 1, with the 13 floor metallic structure by now installed on both sides.

The CMS cavern is completely excavated. Both the main and service shafts have revealed some cracks and water leaks, due presumably to movements of the rock produced by the excavation of the main cavern below. The repair may entail some extra cost and some delay in CMS installation, which are at present under evaluation.

3.2. *Machine Components*

With all main contracts assigned, the real problem is the follow-up of the industrial production, to make sure that companies respect the promised schedule. The situation is made more delicate by the fact that CERN, in many cases, has the double role of supplier and end customer (e.g. supplying the superconducting cables to the dipole cold mass assemblers). Adequate stocks must be built up at CERN, any delay

Figure 1. ATLAS cavern: inside.

in the intermediate steps will entail additional costs. The crux of the matter is: vendors lie!

To monitor the progress in production, we have introduced the *LHC Dashboard*. For the main LHC components (superconducting cables, dipole assembly, etc.) the Dashboard shows:

- the contractual evolution, as defined in February 2002 when the present LHC schedule was agreed;

- the number of delivered components vs. time; and

- the "just in time delivery curve".

The LHC Dashboard can be accessed from the LHC project home page[12] and is updated every month. The site has been very frequently visited in the last year and we hope it contributes to increase the transparency of the process.

3.2.1. *Superconducting cables*

Cable production was quite a concern in 2002, due to the difficult start up of the production in some companies and to the breaking of the cabling machine in Brugg. The situation is considerably better now, with a reasonable production rate. As of August 31, about 42% of the inner cable billets and 51% of the outer cable billets have been manufactured and approved. This is enough to build magnets for slightly more than 3 octants (462 dipoles).

Figure 2. Cryostatted dipoles stored at CERN.

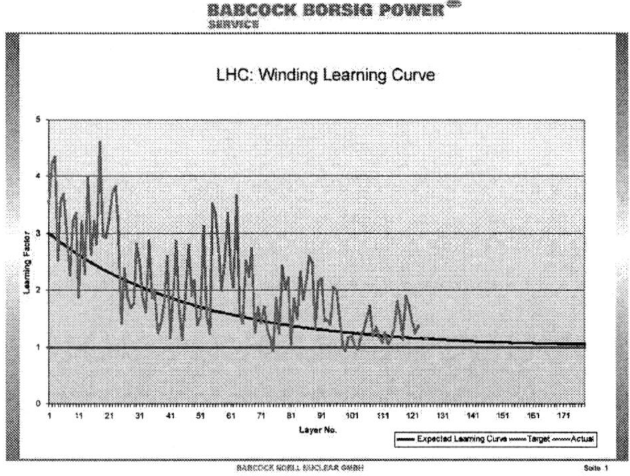

Figure 3. Nöll Nuclear: the learning curve for the winding of the LHC main dipoles coils.

3.2.2. *Cryogenic dipole assembly I*

The next crucial item is the assembly of the cryogenic dipoles by three European companies. The ramping up of the slope in the production of collared coils is impressive: 23 collared coils were received and approved in July, i.e. one per working day. This is not far from the expected cruising rate of 35 collared coils/month. The present bottleneck is in the cold testing of the full (cryostatted) magnet. Delays in the delivery to CERN of the cold feed boxes made it so that until summer we had available only 4 test stations, of the 12 foreseen. In July two more cold feed boxes had been commissioned and two new ones had arrived. We plan to reach the 12 units by the end of the year, which should enable us to eliminate the backlog. In the meanwhile, cryostatted dipoles are being stored at CERN, Fig. 2.

Responding to various rumours for the contrary, I want to state explicitly that all dipoles will be cold tested before installation in the tunnel, in particular to make sure that short circuits are not created by spurious metallic fragments in the coils, under the enormous electromagnetic forces that develop when cold dipoles are fully powered at 12000 A.

3.2.3. *Cryogenic dipole assembly II*

The companies that make the cold mass assembly had to ramp up in available space, tooling and personnel over the last year. They did it, remarkably well. Recruitment of young, enthusiastic people has taken place in all three companies and technology

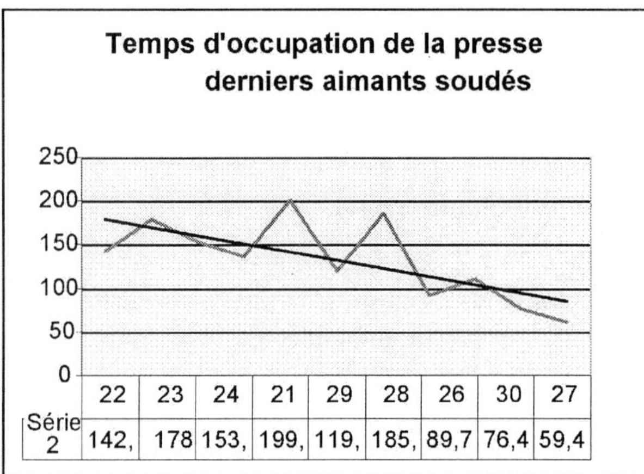

Figure 4. ALSTOM: the learning curve for the welding of the LHC main dipole cold mass.

transfer from CERN can be considered complete. Figure 3 shows the learning curve for coil winding at Nöll Nuclear. Upper (less efficient) peaks correspond to new recruits, lower (more efficient) peaks to senior people. The learning curve is approaching the expected asymptotic value, on which unit prices have been based in the contract! A similar learning curve applies to the very delicate soldering process of the two half shells that close the cold mass, as shown in Fig. 4 for Alstom. The theoretical 50 hours time is now being approached.

It is reassuring that the quality of the magnets we are receiving is quite good. Figure 5 gives the histogram of the quenches to reach the nominal 8.3 Tesla field, for the 50 cold tested dipoles: 80% get

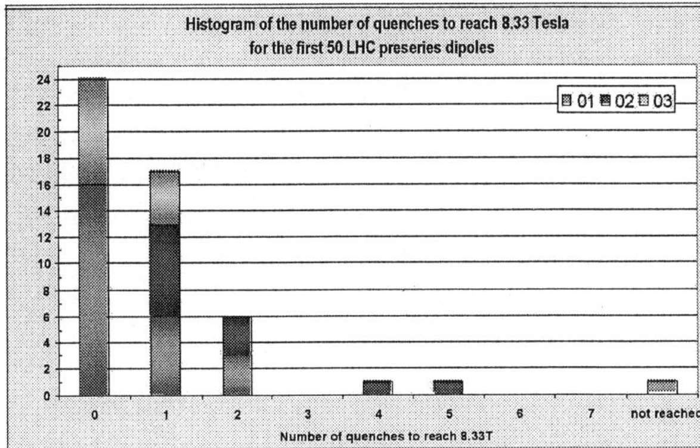

Figure 5. Histogram of the number of quenches to reach the nominal LHC field (8.3 Tesla) for the first 50 dipoles.

to the nominal field with not more than one quench, 50% of them showing no quench at all. One dipole out of 50 did not reach the nominal field because of a construction fault. We seem to have a robust and reliable design.

3.2.4. *Quadrupole subcomponents*

The corrector magnet supply for the Short Straight Section quadrupoles is still a problem, due to a failure of a subcontractor to comply with the agreed schedule. To compensate for the delay, the contract for the assembly of SSS quadrupoles has been extended for (up to) one more year, with no effect on the overall planning but some extra cost.

3.3. *Installation*

The LHC tunnel is being prepared to receive the LHC components. Sector 7-8, where the LHC-b cavern (previously DELPHI) is located, was ready on schedule. Installation of the cryogenic line started on June 21, with a six week delay, which we hope to recuperate. Information on the general installation progress can be found by clicking on *General Coordination* (also updated monthly) in the LHC Dashboard. To conclude:

- the insolvency cases occurred in 2002 have been dealt with without impacting the overall project schedule (but with some increase in the cost to completion);

- superconducting cable production has about reached its nominal rate;

- cryogenic-dipole production is ramping up in all 3 firms;

- installation is proceeding with some initial delay; and

- the LHC Dashboard web address, with component production and master schedule, is given in Ref. 12.

The present concerns are:

- extra costs in civil engineering;

- cracks at CMS shafts;

- production of corrector magnets for the SSS quadrupoles; and

- late production of cold feed boxes, delaying dipole cold testing at CERN.

4. LHC Detectors

Unlike the LHC machine, the construction of detector parts has been going on in a diffused way, in the large network of the collaborating institutions. The advancements made in civil engineering over the last two years have made it possible to start the assembly of the detectors at CERN and, in some case, their integration with the machine. This is, of course, a formidable challenge that is putting extreme strain on the Laboratory infrastructure and manpower and is requiring special coordination and monitoring tools. The result, however, is that CERN is now starting to be populated by gigantic, sophisticated, "toys" being assembled in different places.

Naturally, concerns abound, but it is encouraging to see, also here, that old concerns have mostly been overcome and progress is taking place.

The schedule remains very tight. The extra costs-to-completion of ATLAS, CMS and, to a lesser extent, ALICE, which have been declared in 2001, are at present covered by the Funding Agencies only up to some 70-80%. This will make it necessary to stage and de-scope the detectors, at least in the first go, leading to painful and risky choices. The more technical concerns have been summarized by R. Cashmore at the June CERN Council:

- ATLAS
  -Barrel Toroid Schedule;

-TRT schedule; and

-production of DMILL electronics by ATMEL.

- **CMS**

  -ECAL production; and

  -silicon tracker module mass production.

- **ALICE**

  -TPC production.

All in all, the collaborations are confident that they will provide properly working detectors at the start up of the machine.

5. LHC Computing

The enormous flux of information coming out of LHC collisions and the need to transfer all these data to a world-wide network of institutions make the LHC the ideal testing ground of a new idea, developed at the end on the 90's, the GRID.[13]

By this, we mean an infrastructure, based on the existing high band telecommunication network, with imbedded nodes (Tiers) which can store the data and the applications and send them to a diffused population of users. The name GRID, in fact, is coined after the Power GRID, whereby a user draws energy just by plugging in his electrical appliance, but he neither owns the source of the energy, nor even knows where the energy is coming from.

Over the last years, a wealth of projects to develop the GRID has started, in Europe and in the US. CERN has promoted the EU funded Data GRID project.

In 2001, CERN launched a specific project for LHC computing, the LHC Computing GRID (LCG), which is now deploying a prototype service.

Data GRID has been followed by a wider scope project, with large European participation, called Enabling GRID for E-science in Europe (EGEE) and aimed at expanding GRID applications to other sciences and industry. EGEE is funded within the 6th Framework Programme by 32 MEuro over two years, an important sign of interest from the side of the European Community. LCG will use the middleware produced by the EGEE project and by the analogous US projects (Globus, Condor, PPDG and GriPhyN).

6. Conclusions: Where Are We?

In September 2001, a mid-project analysis of the whole LHC programme showed a sizable extra cost to completion, above the cost stipulated in 1996. I can summarize the subsequent steps taken to remedy the situation as follows.

- **Sept. 2001**

  - An extra cost-to-completion of the LHC programme was declared, of about 800 MCHF (machine, detectors, computing, missing extra contributions with respect to the 1996 plan).

- **Dec. 2001**

  - The main remaining contracts were adjudicated (cold mass assembly, cryogenic line).

- **March 2002**

  - LHC commissioning was rescheduled to April 2007, to comply with the industrial production rate of the main components (e.g. cables).

- **June 2002**

  - Following internal reviews and the recommendations by an External Review Committee, Management proposed a "balanced package" of measures, *to absorb the extra cost in a constant CERN budget.* In round figures, the package was as follows:
  - 400 MCHF from programme reduction, focusing of personnel on LHC, savings, extra external resources (e.g. for computing);
  - 100 MCHF from rescheduling the LHC to 2007; and
  - 300 MCHF, full repayment of the LHC reported, from 2008 to 2010.

- **Dec. 2002**

  - The Management's Long Term Plan for the years 2003-2010 was approved by the Council.
  - A long term loan (300 MEuros) was obtained by the European Investment Bank, to cover the LHC cash-flow peak.

Thus, in December 2002 we could conclude that the LHC was back on track, with a sound, if very tight, financial plan, including the means to overcome the cash-flow we shall face in the coming years.

Today the progress of the LHC is gauged by *new, specific control tools* in addition to the classical peer committee reviews:

- **Machine**:
  - Earned Value Management for the materials budget;
  - Cost & Schedule Annual Review; and
  - LHC Dashboard to report monthly on production and installation progress.

- **Detectors**:
  - regular integration reviews; and
  - periodic machine-detector meetings.

In addition:

- the LHC cost has remained stable over the last year;

- the production of machine and detector components, installation and integration are approaching the cruising speed; and

- several old concerns have been overcome; new concerns appear, but no show stopper.

In conclusion, I can say that CERN has profited from the cost-to-completion crisis in 2001 to enforce real changes: we have a much a leaner programme, but have also succeeded in making CERN into a well-focused Laboratory, an indispensable condition to carry forward such a large project under tight constraints.

Drawing from the results I have just presented, and with fewer reservations than last year, CERN Management can confirm the following LHC schedule:

- completion of the LHC machine in the last quarter of 2006;

- first beams injected during spring 2007; and

- first collisions in mid 2007.

Thinking about LHC upgrades has started!

Acknowledgements

I am greatly indebted to Fabiola Gianotti for providing much of the physics material presented here and to Lyn Evans, Roger Cashmore and Lucio Rossi for sharing with me their deep knowledge of the LHC. Of course any imprecision on both physics and LHC construction is exclusively my responsibility. I would like to thank Verónica Riquer for assistance in the preparation of material for this conference and for the careful review of the text. The computing support of Tony Shave is also acknowledged.

References

1. L.R. Evans *Beam Physics at LHC*, Particle Accelerator Conference, PAC 2003, Portland USA, 2003, CERN-LHC-PROJECT-REPORT-635;
   L.R. Evans *The Large Hadron Collider: Present Status and Prospects*, Presented at 18th International Conference on High Energy Accelerators HEACC 2001, Tsukuba, Japan 2001, CERN-OPEN-2001-027 and references therein;
   L.R. Evans, *IEEE Trans. on Applied Superconductivity* **10**, 44 (2002).

2. See e.g. ATLAS Collaboration, *Detector and Physics Performance*, Technical Design Report, CERN/LHCC/99-15;
   S. Asai *et al.*, *Prospects for the Search of a Standard Model Higgs Boson in ATLAS Using Vector Boson Fusion*, ATLAS Scientific Note, SN-ATLAS-2003-024;
   D. Denegri *et al.*, *Summary of the CMS discovery potential for the MSSM SUSY Higgs bosons*, CMS NOTE 2001/032, hep-ph/0112045;
   R. Kinnunen, for CMS Collaboration, *Higgs Physics at the LHC*, Presented at 10th International Conference on Supersymmetry and Unification of Fundamental Interactions (SUSY02), Hamburg, Germany 2002, Published in Hamburg 2002, Supersymmetry and Unification of Fundamental Interactions, vol.1 p. 26-42.

3. F. Gianotti, M. Mangano, T. Virdee *et al.*, *Physics Potential and Experimental Challenges of the LHC Luminosity Upgrade*, CERN-TH/2002-078, hep-ph/0204087.

4. G. 't Hooft, Recent Developments in Gauge Theories, Proceedings of the NATO Advanced Study Institute, Cargese, 1979, eds. G. 't Hooft *et al.*, (Plenum Press, NY, 1980).

5. E. Farhi, L. Susskind, *Phys. Rept.* **74**, 277 (1981);
   For recent reviews see e.g. K. Lane, *Two Lectures in Technicolor*, FERMILAB-PUB-02-040-T (February 2002), hep-ph/0202255;
   R.S. Chivukula, *Lectures on Technicolor and Compositeness*, Lectures given at Theoretical Advanced Study Institute in Elementary Particle Physics (TASI 2000): Flavour Physics for the Millennium, Published in Boulder 2000, Flavour Physics for the Millennium, p. 731-772.

6. L. Maiani, *All You Need to Know About the Higgs Boson*, Proceedings of the 1979 Gif-sur-Yvette Summer School on Particle Physics, eds. M. Davier *et al.*;
   E. Witten, *Phys. Lett.* B **105**, 267 (1981).

7. S. Dimopoulos, H. Georgi, *Nucl. Phys.* B **193**, 150 (1981);
For a recent discussion, see e.g. M.E. Peskin, *Supersymmetry: The Next Spectroscopy*, Presented at Werner Heisenberg Centennial Symposium Developments in Modern Physics, Munich, Germany 2001, SLAC-PUB-9613, hep-ph/0212204.

8. N. Arkani-Hamed, S. Dimopoulos, *Phys. Rev.* D **59**, 086004 (1999);
G.F. Giudice, R. Rattazzi, J.D. Wells, *Nucl. Phys.* B **544**, 3 (1999);
E.A. Mirabelli, M. Perelstein, M.E. Peskin, *Phys. Rev. Lett.* **82**, 2236 (1999);
See also L.J. Hall, *Beyond the Standard Model*, Proceedings of the 30th International Conference on High Energy Physics (ICHEP2000), Osaka, Japan, 2000, eds. C.S. Lim, Taku Yamanaka, (World Scientific, Singapore, 2001).

9. M. Battaglia, A. De Roeck, J. Ellis, F. Gianotti, K.A. Olive, L. Pape , *Updated Post-WMAP Benchmarks for Supersymmetry*, CERN-TH-2003-138, UMN-TH-2005-03, FTPI-MINN-03-16, hep-ph/0306219.

10. S. Abdullin *et al.*, *J.Phys.G:Nucl. Part. Phys.* **28**, 469 (2002);
S. Abdullin *et al.*, *Search for SUSY at large* $\tan\beta$ *in CMS, the low luminosity case*, CMS NOTE 1999/018.
For ATLAS, see the first paper under Ref. 2.

11. L. Vacavant and I. Hinchliffe, *Model Independent Extra-dimension Signatures with ATLAS*, ATLAS Internal Note ATL-PHYS-2000-016, hep-ex/0005033.

12. http://lhc-new-homepage.web.cern.ch/lhc-new-homepage/.

13. C. Kesselman and I. Foster (eds.), *The Grid: Blueprint for a New Computing Infrastructure*, (Morgan Kauffman, San Francisco, CA, 1999).

DISCUSSION

Robert Zwaska (University of Texas, Austin): What are the status and prospects for LHCb?

Luciano Maiani: No particular comments, LHCb is proceeding well and all indications are that it will be ready by April 2007.

PHYSICS OF THE LINEAR COLLIDER

F. RICHARD

Laboratoire de l'Accélérateur Linéaire, IN2P3-CNRS et Université de Paris-Sud,
F-91405 Orsay Cedex, France
Email: richard@lal.in2p3.fr

This presentation intends to illustrate the specific capabilities of an e^+e^- sub-TeV collider to provide answers on the basic issues in physics: the origin of mass, hierarchy of masses, and cosmological problems. Some foreseeable scenarios are discussed with a possible synergy with the LHC.

1. Introduction

In this presentation I intend to summarize the main physics prospects of the Linear Collider (LC) presently under consideration in North America, Asia and Europe. These prospects have been studied by 3 communities and there exist various documents describing them in detail.[1]

After recalling briefly the baseline for the future LC, I will mention some important features of the detectors under consideration and the requirements needed for the machine parameters.

Concerning the physics, I will focus on 3 aspects.

1. The mechanism of electro-weak symmetry breaking (EWSB), in other words what is the origin of mass in particle physics. This aspect will be my main emphasis.

2. The problem of mass hierarchy, in particular in the Higgs sector, and the need for new mechanisms beyond the Standard Model (SM) like Supersymmetry (SUSY).

3. The input on cosmology of this model which can explain the origin of Dark Matter in the universe.

In a short presentation one can only give a very partial view of ongoing studies performed in the 3 regions. Examples of uncovered or very partially covered topics are: e^+e^- physics at MultiTeV (CLIC scheme[2]), the various SUSY breaking scenarios, SUSY and CP-violation, SUSY and the neutrino sector, extra dimensions with different schemes either alternate or combined with SUSY, e^-e^-, γe and $\gamma\gamma$ physics, precise tests of QCD with a LC.

2. The TeV LC

The present goal is to construct an e^+e^- LC covering an energy between the Z boson mass and 500 GeV, with polarized electrons (at least 80%) and collecting 500 fb^{-1} in the first four years of running. This LC should be able to reach, in a second stage, an energy ~ 1 TeV, collecting about 500 fb^{-1}/year.

Various options are considered, which would require additional equipment and whose priorities will depend on the physics priorities emerging after LHC and LC operation:

1. positron polarization, at the 60% level, is not easy to implement but is needed to fully exploit the precision of a Z factory (the GigaZ scheme). This polarization is also needed for transverse polarization measurements which have recently been emphasized;[3]

2. an e^-e^- collider, is possible with reduced hardware changes; and

3. a $\gamma\gamma$ (or a γe collider), which could operate with a maximum energy $\sim 80\%$ of the nominal energy. This scheme would require major changes in the interaction region. As for the e^-e^- case, there will be a reduction in luminosity at the same energy but a high degree of polarization is feasible. Contrary to e^+e^-, a zero crossing angle is not possible even with the supraconductive technology.

3. The Detector

Given the LC luminosity, two orders of magnitude above LEP, in several analyses the precision will be limited by systematic uncertainties. Part of these uncertainties come from detector limitations but this will improve using better technologies.

1. Improved vertexing which will allow a clean and efficient separation of charm quarks and also tagging of tau leptons. The improvement with respect to LEP is due to a small radius beam pipe and to the resolving power of thin pixel detectors.

2. Improved energy flow: the aim is to improve the jet energy resolution by about a factor of 2 with respect to LEP/SLD with fine segmentation of the calorimeters which are inside the magnetic coil. This improvement should allow one to cope efficiently with 6/8 jet topologies from ZHH and ttH channels. This resolution will also allow a clean separation between ZZ and WW hadronic final states to isolate the $WW\nu\bar{\nu}$ channel.

3. Momentum resolution will be improved by a factor 10 with respect to LEP/SLD, with polar angle coverage down to 100 mrad.

4. Hermeticity on energetic γ's and electrons should go down to about 5 mrad with instrumented masks and good segmentation of the very forward calorimeter.

A detailed discussion of the new technologies implied can be found in the presentation given by T. Behnke at this conference.

A limiting factor will also come from our knowledge of the differential luminosity, of the polarization and of the energy calibration. Physics groups and machine experts are actively investigating these issues.

4. Which Physics Scenario for EWSB ?

This is clearly the central issue for the LHC and the LC. While there is no doubt that the LHC should discover a SM/MSSM type Higgs boson, one should be prepared for unconventional answers from Nature and I will illustrate this schematically in the following sections.

From LEP/SLD/Tevatron precision measurements (PM) one can derive 2 important consequences.

1. The Standard Model (SM) or its supersymmetric (SUSY) minimal extension (MSSM) are compatible with PM.

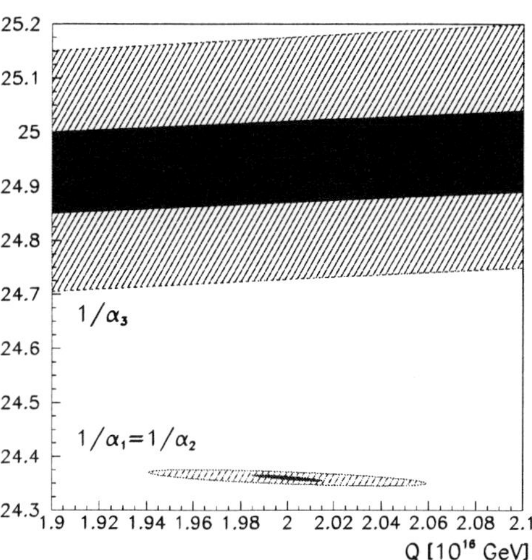

Figure 1. This figure shows that while LEP/SLD data (hatched) are compatible with SUSY-GUT, there is an indication of a discrepancy which, if real, should become clear with the precision of GigaZ (full).

2. Unification of the 3 interactions occurs at $\sim 2\times10^{16}$ GeV provided that the MSSM mass spectrum is at ~ 1 TeV (see Fig. 1[4]).

A light Higgs is therefore expected, below 130 GeV as a consequence of MSSM, below 250 GeV as a consequence of PM assuming that there is no contribution other than SM.

With a critical view[5] of PM, one can notice that $\sin^2\theta_W$ from the charge asymmetry on b quarks measured at LEP1 is hardly consistent with the value obtained at SLD from polarization asymmetry.[6] So far no compelling explanation has emerged from theory nor experiment. New physics interpretations, although not impossible,[7] are severely limited by the absence of a significant deviation on the b quark cross section.

Another discrepancy, found in NuTeV, is still under investigation, but it should be noticed that this determination of $\sin^2\theta_W$ has no sensitivity to the value of the Higgs boson mass.[6]

When removing from the fit the b quark asymmetry result one finds that the Higgs mass value indicated by the data is too low given the direct search limit. At this stage one can remark that the effect is at the 2 sigma level and that the recent update on the top mass found by D0 with Run 1 data and the shift in the W mass after a reanalysis of ALEPH data should even reduce it further. In our opinion one should therefore wait for the new top mass measurements at FNAL before drawing any definite conclusion on this discrepancy.

Another possibility could be that there are non SM/MSSM contributions coming from an alternate scheme (see elsewhere[8] for a general discussion). I will therefore consider, for simplicity, that a LC could be dealing with three EWSB scenarios.

1. There is a light Higgs boson consistent with MSSM. There will then be emphasis on the precise measurement of Higgs couplings based on the collection of $\sim 10^5$ HZ events.

2. There is a Higgs boson but with a mass incompatible with SM/MSSM. A LC would then focus its effort in understanding the underlying mechanism and search for direct or indirect signals of new physics.

3. There is no Higgs boson, implying that the longitudinally polarized gauge bosons will strongly interact. A LC would then run at its maximal energy and focus on WW final states.

I will illustrate with some examples how a sub-TeV LC can provide the appropriate answers in these three scenarios.

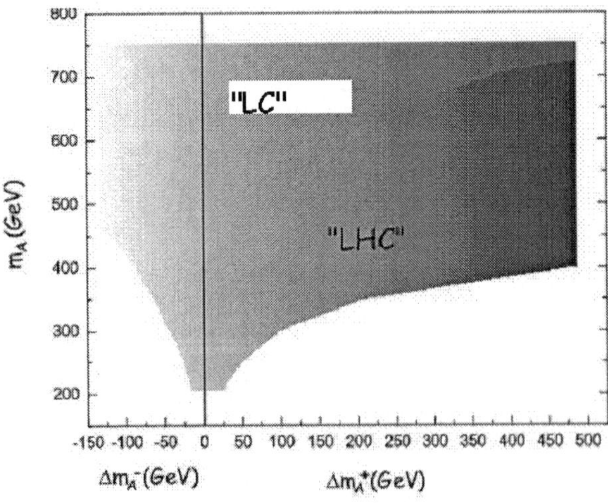

Figure 2. Sensitivity to the pseudoscalar Higgs boson mass at LC and LHC. This figure assumes an integrated luminosity of 300 fb^{-1} and the combination of 2 experiments for LHC.

4.1. MSSM Scenario

The first questions are obviously: are we dealing with a CP-even scalar, are the coupling to fermions and to bosons consistent with MSSM?

1. A LC will identify unambiguously the spin-parity of a Higgs boson by measuring the shape of the HZ cross section near threshold and the angular distributions provided by the ZH channel.

2. All fermionic (except for the top) and bosonic couplings will be measured at the % level for tree level couplings, at 5% for the gluonic width Γ_{gg} and 20% for $\Gamma_{\gamma\gamma}$. For the latter one can reach the % level using the photon collider scheme.

3. The Higgs coupling to the top quark can be measured with a $7-15\%$ accuracy in the $120-200$ GeV mass range. For the Higgs self-coupling λ_{HHH} the accuracy would be 10% (20%) at 800 GeV (500 GeV) center-of-mass energy.

With such accuracies one can detect with high sensitivity the presence of a non-standard component.

1. In the MSSM itself, it is possible to detect the influence of the heavier Higgs bosons. As shown in Fig. 2, one can estimate[9] the mass of these bosons up to about 700 GeV. This goes beyond the mass reach of an e^+e^- LC but, given this information, could motivate a photon collider scheme in which these Higgs bosons can be singly produced.

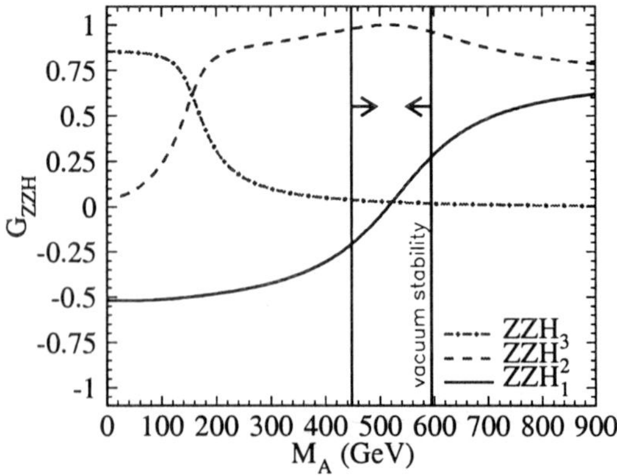

Figure 3. Variation of the ZZH coupling with the mass of the pseudoscalar A, in an NMSSM model, for the 3 scalar states.

2. Beyond MSSM one could have CP-violation in the Higgs sector or NMSSM with an additional Higgs isosinglet which can mix with the isodoublets. This could result in a significant drop in the ZZH coupling as shown[10] in Fig. 3. In this respect a LC is very robust and can stand a reduction of a factor 100 in the ZH cross section.

3. Similarly it is worth recalling that the Higgs detection does not depend on the final-state branching ratios, in particular the LC can very well detect an "invisible" Higgs, which may occur in various schemes.[11] Moreover if Γ_{inv} is at the 2% level, the LC could still give a 5 standard deviation evidence on the presence of an invisible channel. An example[12] is given within SUSY where one assumes an unusual hierarchy between gaugino masses M1 and M2 (M1~M2/10), such that the limit from LEP2 on the chargino does not eliminate the possibility of a very light LSP. Figure 4 shows the large effects possible in this scheme.

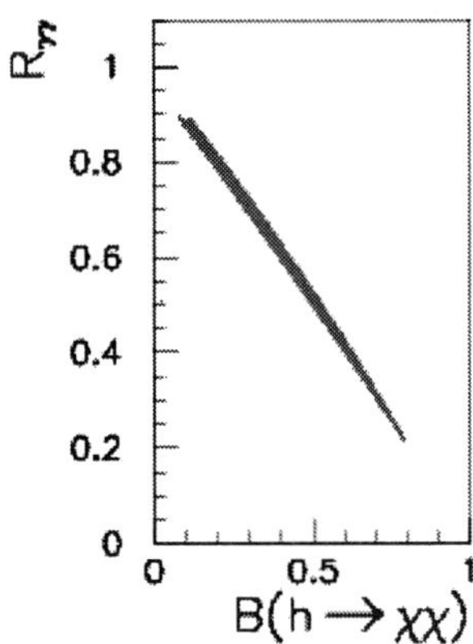

Figure 4. Variation of $R_{\gamma\gamma}$, the rate at LHC of Higgs into a pair of photons normalized to the SM rate, versus the branching ratio of the Higgs into a pair of LSP in a model with non universal gaugino masses.

4. Theories with large extra dimensions also provide valid schemes for EWSB. Within these theories the radion is a scalar field introduced to stabilize the small dimensions. It can therefore mix with the Higgs boson and, accordingly, modify the couplings of the Higgs to bosons and fermions as seen in Fig. 5. This figure[13] shows that the modification is similar for W and fermions and therefore cancels in the ra-

tio. A LC however has access to the couplings and would measure this variation. The gluon pair BR should also provide a valid input in the interpretation of this effect.

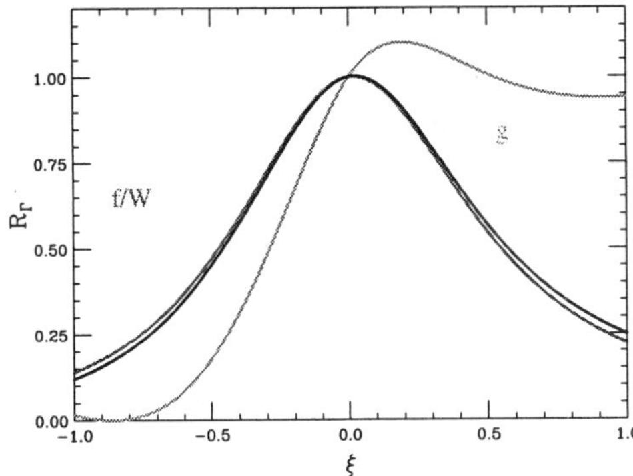

Figure 5. Ratio of the Higgs widths to their SM value for fermions, W bosons and gluons as a function of the mixing parameter in a radion model, assuming a Higgs mass of 125 GeV.

4.2. Quantum Level Consistency

The accuracy on the indirect determination of the Higgs mass can be improved by an order of magnitude with respect to LEP1/SLD. GigaZ should measure $\sin^2\theta_W$ with an error $\sim 10^{-5}$ provided one has polarized positrons. Figure 6 shows the other limiting accuracies and how they should evolve.

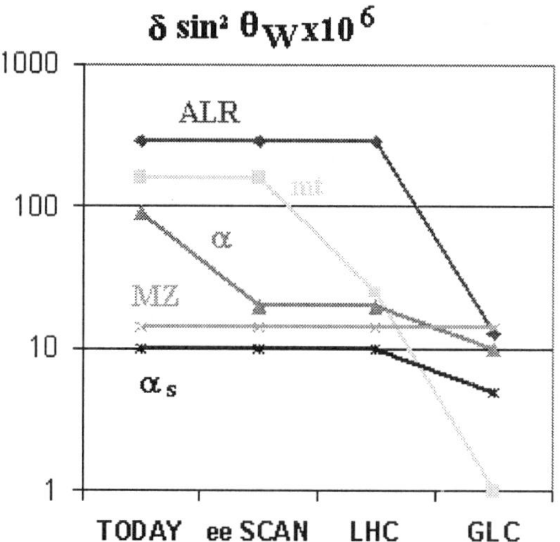

Figure 6. Predicted evolution of the experimental and theoretical accuracy on $\sin^2\theta_W$ with the expected progress on the polarization asymmetry, the top mass measurement, the decrease in uncertainty on $\alpha(M_Z)$ provided by e^+e^- scans and the improved precision on α_s at GigaZ.

The B- and K-factories should provide the input needed to settle the issue on $\alpha(M_Z)$ while the top mass accuracy with a LC should eliminate this source of error.

In parallel, there is a continuous and fruitful effort in reducing the theoretical uncertainty which is now coordinated in the "LoopVerein" working group.[14] As an example, Fig. 7 indicates the continuous progress[15] achieved on theoretical errors for the luminosity. For measurements which concern $\sin^2\theta_W$, the present theoretical uncertainty[16] of $\sim 6\times10^{-5}$ has to be reduced to match the expectations given in Fig. 6.

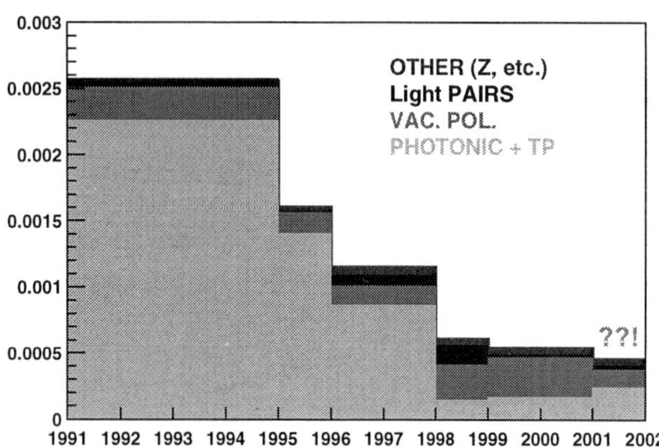

Figure 7. Components of the luminosity theoretical error at LEP1.

With GigaZ one could therefore test at the 5% level the equality:

$$M_H^{Direct} = M_H^{Indirect}$$

and draw valid conclusions on the global consistency of SM or MSSM.

An alternative to GigaZ, if positron polarization is not available, would be an improved measurement of the W mass using a threshold scan. The critical issue, there, is energy calibration: one has to extrapolate the beam measurement from the Z mass to the W threshold with a precision of $\sim 5\times10^{-5}$. One expects $\delta M_W \sim 6$ MeV, which gives a 10% accuracy on the indirect Higgs boson mass.

4.3. Non-MSSM Higgs Scenario

Let us assume that a Higgs boson has been found with a mass inconsistent with PM, meaning above 200 GeV. This mass is therefore also inconsistent with MSSM. While PM are still constraining the gauge sector since, for instance, one could measure the BR of the Higgs into WW at 5% with M_h=250 GeV, the main mission of a LC would then be to find the "guilty part".

If there is direct evidence for new physics at the LHC, e.g. if one observes a candidate Z', the LC can decipher the message with precision measurements in the channel $e^+e^- \to f\bar{f}$.

With measurements in the TeV range, the LC provides high discrimination against the various Z' predicted by several symmetry groups[17] as shown in Fig. 8. Knowing the Z' mass from LHC, LC PM at high energy will extract the vector and axial couplings of this Z'.

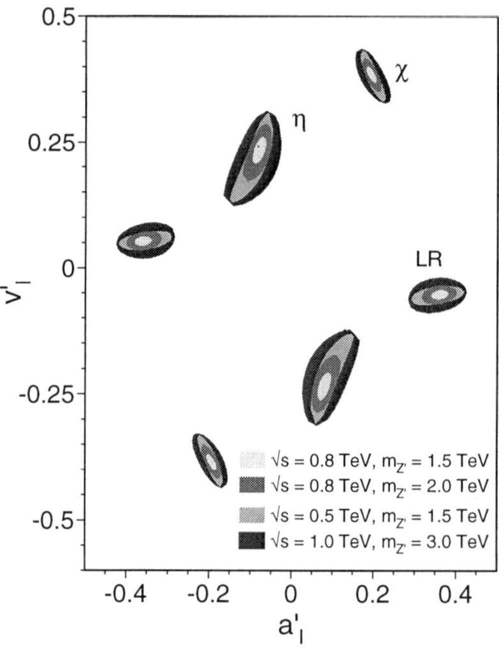

Figure 8. Accuracies expected on Z' vectorial and axial couplings from a LC operating up to 800 GeV for various extensions of the SM.

Similarly one can use the GigaZ to measure Z-Z' mixing, providing extra information.[18] These measurements should also allow one to restore the consistency of PM with the observed Higgs mass.

As shown in Fig. 9 the mass domain of a LC covers[18] and, in some instances, surpasses, the LHC domain. It may therefore turn out that only the LC information will be left to solve the puzzle of a Higgs mass inconsistent with PM.

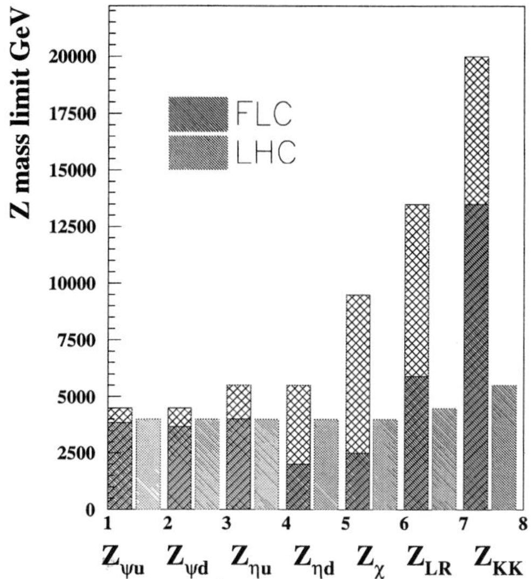

Figure 9. Mass domain covered by LHC (shaded/pink) and LC (hatched/blue) for various scenarios predicting a Z'. In the case of the LC the dark hatched part corresponds to the limit of sensitivity through mixing, while the light hatched part corresponds to the sensitivity through interference for a LC operating at 800 GeV.

In some cases the Z' can decouple from ordinary fermions as in the UED model (Universal Extra Dimensions[19]) where the new particles are mass degenerate Kaluza-Klein excitations of ordinary particles which carry a conserved quantum number and therefore need to be pair-produced. Remarkably, given the absence of direct coupling to fermions, this model has much weaker mass bounds than usual extensions with new Z' or extra dimensions and therefore it is not yet excluded that there could be pair-production at a TeV LC. If not, there still remains the effect on weak isospin violation in the top sector which contributes to the ρ parameters and therefore is measurable at the GigaZ.

4.4. The Little Higgs Scenario

Among the possible non-MSSM extensions of the SM, the "Little Higgs" model[20] has received special attention since it offers a viable solution to the hierarchy problem in the Higgs sector. In this scheme, the Higgs boson is a pseudo-goldstone boson originating from a symmetry broken at $10-100$ TeV.

There is a perturbative theory below this scale and quadratic divergences of the Higgs mass are cancelled at first order by the contribution of new particles originating from the new symmetries. In particular there could be a light U(1) gauge boson, the B', which contributes to the ρ parameter in the minimal model called "Littlest Higgs".

If the LHC finds this B' then, as stated previously, the LC will allow one to identify its origin. If not, the LC can predict[18] the mass of this object from PM and indicate which improvements in luminosity/energy are needed at the LHC (or at future colliders) to discover it.

To conclude on this non-MSSM Higgs scenario, one can say that it shows in a quantitative form how there could be a strong synergy between the LC and the LHC and that it suggests that we may need the full information of a LC from the Z-pole to the maximum energy.

4.5. No Higgs Scenario

In the case of no Higgs or, equivalently of a very heavy Higgs boson, with a mass above 1 TeV, one needs, even more than for the previous scenario, some kind of conspiracy cancelling the contribution of the Higgs term in PM.

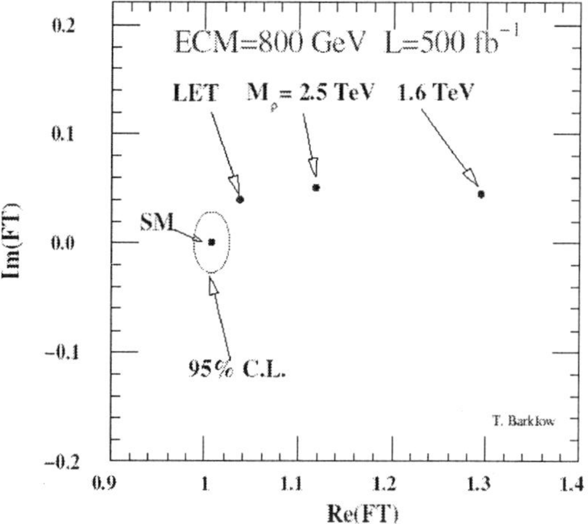

Figure 10. Sensitivity on the W_L form factor FT with a LC operating at 800 GeV and an integrated luminosity of 500 fb^{-1} when there is a ρ resonance at 1.6 TeV or 2.5 TeV or, with no resonance, the effect as predicted by the Low Energy Theorem.

463

The most direct manifestation of the heavy Higgs should occur in the gauge sector, where longitudinally polarized W bosons should strongly interact as a result of the absence of a Higgs exchange term cancelling the divergence in the process $W_L W_L \rightarrow W_L W_L$.

A likely scenario would be that the strong interaction, as in QCD, results into a ρ-type resonance with a mass below $\Lambda_{EWSB} = 4\pi v \sim 3$ TeV. Using the channel $e^+ e^- \rightarrow W^+ W^-$ and extracting the longitudinal part, one can observe a clear effect[21] with a LC operating at 800 GeV as shown in Fig. 10. Figure 11 displays the energy dependence of the signal if there is a resonance. Even if there is no resonance (Fig. 10), one should still observe a minimal effect, the so-called LET term predicted on a model-independent basis (Low Energy Theorem).

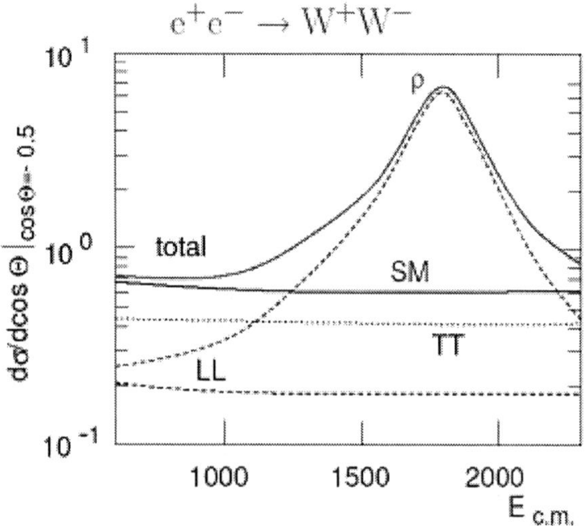

Figure 11. Variation with the LC energy of the backward differential cross section for W^+W^- with a ρ resonance. The upper dashed (red) curve corresponds to the LL component part to be compared to the lower dashed curve (blue) curve corresponding to the SM.

Table 1 summarizes the evidence foreseen at the LC and the LHC. One can clearly see an advantage for the LC if there is a resonance, while the 2 machines are roughly equivalent otherwise.

In case there is no resonance in the J=1 and I=1 channel, one could alternatively use the channels $e^+ e^- \rightarrow \nu\bar{\nu}W^+W^-$ or $\gamma\gamma \rightarrow W^+W^-$ to access other quantum numbers.

One can draw similar conclusions in the language of triple gauge couplings, TGC, which should also

Table 1. Sensitivity of the LC and LHC to the presence of a strong interaction component in W^+W^- for different luminosities, energies and assuming the presence of a ρ resonance or without it (LET).

	$\sqrt{s}$ GeV	$\mathcal{L}$ fb^{-1}	M_ρ 1.6 TeV	LET
LC	0.5	300	16 σ	3 σ
LC	0.8	500	38 σ	6 σ
LC	1.5	200	204 σ	5 σ
LHC	14	100	6 σ	5 σ

manifest deviations in this scenario. One has 5 TGC preserving parity, custodial symmetry (to avoid effects on PM). Three are measured at GigaZ and in $e^+ e^- \rightarrow W^+W^-$, 2 with $e^+ e^- \rightarrow \nu\bar{\nu} W^+W^-$. These couplings can be expressed as:

$$\alpha = \frac{\Lambda_{EWSB}^2}{\Lambda^2}.$$

To be significant, a limit on Λ should be above Λ_{EWSB}. This condition is satisfied for the first 3 couplings (also true for the other 2) with the LC as shown in Fig. 12. Note that there is a significant improvement on the determination of these parameters with the input of the GigaZ.

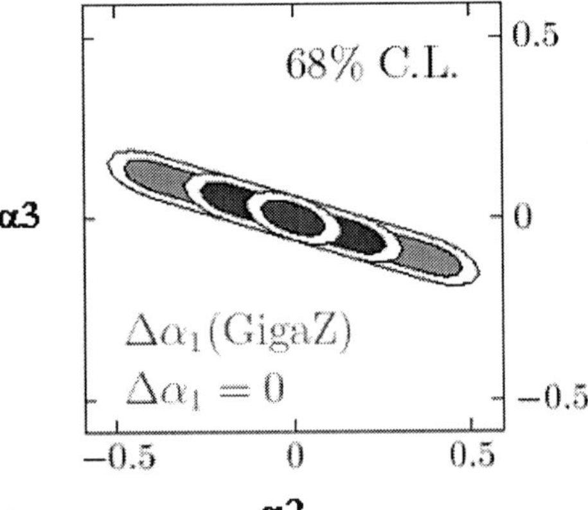

Figure 12. Sensitivity to the triple gauge boson couplings α_3 and α_2 with LC. The central ellipse assumes no error on α_1, while the middle (dark red) one assumes a precision given by GigaZ.

5. The SUSY Scenario

Although there is very little experimental support for it, SUSY is considered as the leading candidate theory beyond the SM. On the theoretical side it provides a consistent way to understand EWSB through the Higgs mechanism (no hierarchy problem and

EWSB naturally driven by the large Yukawa coupling of the top quark). On the experimental side it allows unification of the 3 forces and it provides natural links to cosmology, in particular a mechanism for generating Dark Matter (DM) in the universe. The g-2 deviation with respect to the SM, which would constitute a precious indication of a SUSY effect, is still uncertain given the 2 contradictory results[22] obtained for the hadronic correction. If confirmed it would favor a SUSY solution with light sleptons and light gauginos.

The basic issues for this theory, after it is revealed at the LHC or Tevatron, will be:

1. to fully confirm the SUSY hypothesis by measuring the spin and couplings (precisely predicted in this theory) of these particles. Recall that, e.g. in the UED framework,[23] one is able to fake the presence of SUSY but unable to pass the above criteria; and

2. to understand the SUSY breaking (SSB) mechanism, for which there is plethora of proposed schemes, none of them clearly emerging as preferable to the others.[20]

For pedagogical reasons only, I will use the simplest of SSB schemes, that is mSUGRA in which one can represent the parameter space in two dimensions, in terms of the common scalar and gaugino masses at GUT m0 and M1/2. $\tan\beta$ is a free parameter and μ is derived in absolute value by imposing EWSB. Figure 13 shows[24] that various experimental conditions, in particular the need to provide the adequate amount of neutral DM and the need to generate EWSB, put severe restrictions on these parameters. One can distinguish, in Fig. 13, four allowed domains.

1. The *blob* domain, studied so far in LC and LHC studies. This region in which SUSY parameters are moderate is favored by the advocates of *Fine-Tuning criteria* (FT), meaning that one expects that the SUSY-EWSB generation of the *W* and *Z* mass should not result from fine-tuned cancellations between large SUSY masses. In this region the DM candidate is a Bino and the presence of light sleptons allows one to get sufficient t-channel annihilation to keep DM under control.

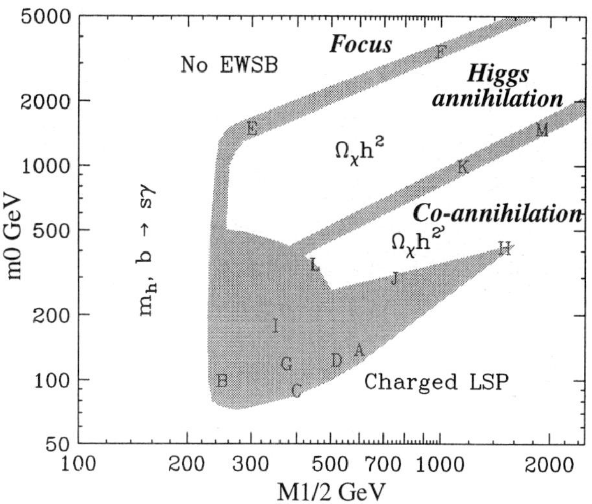

Figure 13. Allowed regions for the m0 (scalar mass) and M1/2 (gaugino mass) parameters within mSUGRA taking into account EWSB, radiative decay of the *b* quark and the DM constraints. The letters indicate the points which are proposed as benchmarks for the LHC and the LC. There are 3 regions called "focus" (large m0 solutions), "co-annihilation" (sleptons almost degenerate with LSP) and "Higgs annihilation" (LSP about 1/2 the heavy Higgs mass) which allow large values of these parameters.

2. In the *co-annihilation* domain, the Bino LSP gets heavier and, for DM, one needs to assume that the LSP is almost degenerate in mass with the sleptons, in particular with the stau particle which is usually the lightest slepton. The recent results from WMAP, have further restricted this domain as will be discussed later (see Fig. 15).

3. The *Higgs annihilation* domain corresponds to a region of parameters for which the heavier Higgs particle have a mass close to twice the LSP mass, in such a way that efficient s-channel annihilation can occur to reduce the amount of DM. This solution allows one to reach high SUSY masses, not only beyond LC reach but even beyond the LHC.

4. The *focus* domain is remarkable in the sense that it can still claim absence of FT with large values of m0. This is so since the coefficient of m0 in the EWSB equation is minute.[25] In this scenario, DM can be controlled by noting that the LSP can be a Higgsino if μ is small, favoring s-channel annihilation. Such a scenario could be quite peculiar since one might only see the 2 first lightest neutralinos and the first chargino, all with masses $\sim \mu$.

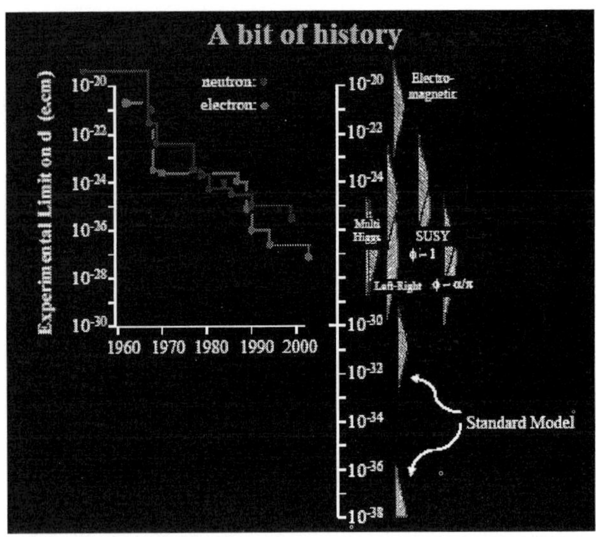

Figure 14. Experimental limit on EDM for neutrons and electrons versus time. On the right-hand-side are indicated regions predicted by various theories. The SUSY prediction assumes moderate scalar masses.

Before leaving this part of the discussion, let me point out an important experimental aspect of these arguments on the LSP sector. In scenario 2 and 4 there could be a degeneracy in mass between the sleptons and the LSP and between the gaugino themselves. This conclusion, simply expressed within mSUGRA is certainly quite general[26] and deserves attention. As shown at LEP2, techniques can be used to detect the charginos in such cases given the large production cross section involved. This might not be the case for the slepton with a much reduced cross section.

5.1. *The Flavor Sector*

Previous discussion could lead to the conclusion that nature will choose the less FT solution, that is the "blob" sector. This conclusion does not take into account the so-called *flavor problem* of SUSY.

This problem extends to the various observables where experimental information puts very severe limitations on SUSY parameters: FCNC constraints, CP-violation in the K sector, EDM limits for electrons and neutrons (in rapid progress[27] as shown in Fig. 14) and proton lifetime limits. Several authors propose to avoid these constraints by setting the scalar masses, at least for the first two families, to very high value. This could happen in the framework of the "focus" scenario without drastic FT.

There are other ways to avoid some of the problems and restore the "blob" scenario. One can for example postulate a hidden symmetry (of the Left-Right type[28]) which forces the SUSY phases to 0, or assume a cancellation mechanism[29] which reduces the importance of CP-violation in the EDM process. The later seems to also imply some kind of fine-tuning given that one already needs to assume a very small phase for the μ term.

5.2. *The Three Scenarios*

From previous discussion and for the sake of simplicity, one may foresee 3 types of scenarios representing different challenges for a LC.

1. All SUSY particles are very heavy except the lightest Higgs boson h and, possibly, the two lightest neutralinos and the lightest chargino accessible at the LC and the gluino at the LHC. This would correspond to a "focus" type solution. The LSP could be a Higgsino with $|\mu| < M_1$ or a Wino if the usual inequality $M_1 < M_2$ is violated. In both cases there would be a substantial s-channel annihilation. The three lightest gauginos are almost degenerate in mass but, as shown at LEP2, observable at the LC.

2. Same as 1 but, in addition, the third generation of scalar particles is light. Then co-annihilation can take place between the LSP and the stau particle. As has been pointed out,[30] the new results from WMAP would require a very close degeneracy and therefore the stau would be extremely hard to observe. The good news however is that the WMAP results imply that the LSP cannot be heavier than ~ 500 GeV and therefore would fall within the range of observation of a LC.

3. The "blob" scenario which allows a wide range of possibilities subject however to the Higgs mass limit of LEP2 and, possibly, to the g-2 measurement from BNL. Figure 15 indicates a recent update[30] illustrating these features.

5.3. *DM at the LC*

If within reach, the LC will accurately measure the mass of the LSP and its couplings, in other words its Higgsino/Wino/Bino content.

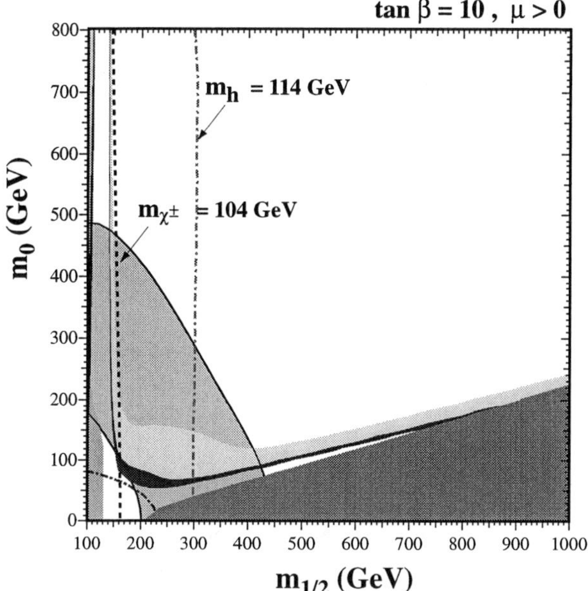

Figure 15. Allowed regions within mSUGRA assuming $\mu>0$ and $\tan\beta=10$. Indicated also is the mass limit from LEP2 on charginos and on the Higgs boson mass. The small bands correspond to DM solutions with (darkest shaded region/dark blue) and without (lightest shaded region/pale green) WMAP derived results. The ~ 300 GeV wide band (decreasing from the mid-left side to the bottom/pink) is favoured by an interpretation of the g-2 results, still under investigation.

There are various tools to perform such an analysis, threshold scans, polarization asymmetries which I will not describe here but can be found elsewhere.[17]

These measurements will constitute an input for cosmology as will be illustrated shortly. They will also allow one to precisely interpret the non-accelerator searches on primordial DM performed with various techniques.[31]

Taking point B within mSUGRA (see Fig. 13), a case treated with LHC inputs,[32] I have estimated the accuracy on $\Omega_{DM}h^2$ from SUSY using the code MicroMEGAs,[33] given that a LC can measure the LSP and τ slepton masses with accuracies of 0.1 GeV and 0.6 GeV respectively.[17] Any significant discrepancy with cosmology may reveal extra sources of DM (e.g. axions or very heavy objects produced in the early phases of the universe).

It is also worth recalling[17] that the chargino and neutralino measurements, if there is a Bino component, provide an indirect sensitivity on the masses of selectrons and sneutrinos masses which are precise up to masses ~ 1 TeV, that is well above the reach of

Table 2. Relative precision on DM obtained by cosmology in the present phase (WMAP) and foreseen (Planck), to be compared with the accuracy foreseen on its determination at the LHC and at the LC assuming point B as defined in Fig. 13.

WMAP	7%
LHC	$\sim 15\%$
Planck	$\sim 2\%$
LC	$\sim 3\%$

a TeV linear collider.

6. LC and the GUT Scale

As already stated, understanding the origin of SSB will be the major goal of our field once SUSY is discovered at the Tevatron or LHC. To reach this goal, it will be essential to measure SUSY masses and couplings precisely and in a model independent framework. As an example, in the gaugino sector, the LC should allow one to measure precisely M1 and M2 while M3 is provided by the LHC from the gluino mass. These three quantities would be extrapolated to the GUT scale and provide a crucial test of SSB. Figure 16[4] gives an example of such a confrontation in the case of a string inspired model.

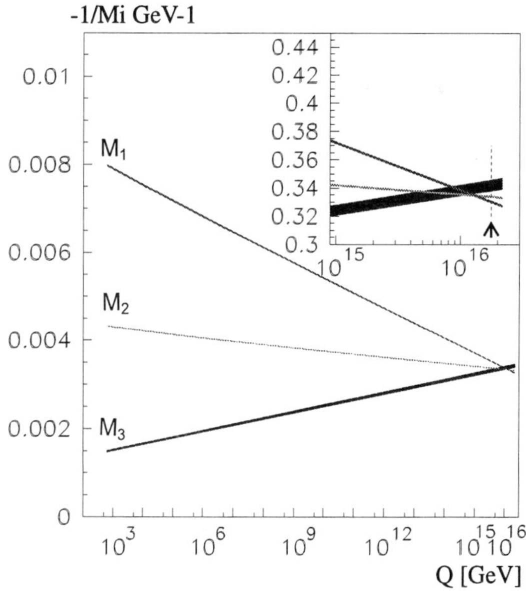

Figure 16. Evolution of the gaugino masses up to the GUT scale with a window showing the breaking of universality due to loop corrections in a string inspired model.

One can observe that while the convergence is satisfied, it occurs at a mass below the GUT scale given by the couplings. This shift is due to loop corrections present in such a scheme and the LC accuracy is needed to observe this subtle effect present in string theories.[34] This example shows how a LC may turn out to be crucial to understand the underlying fundamental theory.

Moreover one can observe in Fig. 16 the unmatched accuracy of M3. Recently[35] it has been shown that this accuracy can be greatly improved if the LHC can use LC information, in particular the mass determination of the LSP. One could therefore reduce the error given in Fig. 16 by a factor 3.

This example illustrates the potential of a LC+LHC combination to reach a very ambitious goal, pioneered by LEP/SLD: explore the GUT/string scale to reconstruct the fundamental parameters of an underlying theory which no accelerator will ever be able to test directly. Studies of this type are systematically being done in the framework of LHC/LC collaborative studies.[35]

7. Summary: Why Do We Need a LC ?

From this very incomplete description of the prominent topics which will be addressed by a LC, one can derive the following list of goals.

1. To provide the full picture on an SM/MSSM Higgs.

2. To provide an answer on the issue of EWSB even in more difficult or unexpected situations, e.g. a reduced cross section, a heavy Higgs or no Higgs scenario.

3. To access to the SSB mechanism by combining LC and LHC measurements and, even more ambitiously, to access the GUT scale and to the parameters of the underlying string theory.

4. To predict very precisely, within SUSY, the amount of DM and confront it to the observation by cosmology.

5. To interpret unambiguously an unexpected discovery at LHC, e.g. a Z' or a Kaluza Klein excitation.

6. To estimate mass scales beyond LC/LHC reach:

- observing deviations on PM translated, e.g. into a heavy Higgs, a heavy sfermion or a Z' mass; and

- testing the theory at the quantum level and eventually predicting new mass scales as has already occurred with LEP/SLD/Tevatron for the Higgs mass.

These various examples indicate how PM at a LC may reveal new frontiers in energy which would call for an upgraded LHC or for a new generation of colliders like CLIC or VLHC.

Acknowledgments

It is a pleasure to thank and congratulate the team of this Symposium for the excellent organization of the meeting.

References

1. See the following web sites and documents referred therein:
   - For America `http://blueox.uoregon.edu/~lc/alcpg/`
   - For Asia `http://acfahep.kek.jp`
   - For Europe `http://www.desy.de/conferences/ecfa-desy-lcext.html`
2. CLIC Physics Study Group: `http://deroeck.home.cern.ch/deroeck/clic/spin2.html`
3. T.G. Rizzo, JHEP 0308:051,2003, hep-ph/0306283.
4. G.A. Blair *et al.*, *Eur. Phys. J.* C 27, 263-281 (2003).
5. M.S. Chanowitz, *Phys. Rev.* D **66** 073002,2002, hep-ph/0207123.
6. Talk of P. Gambino at this conference.
7. D. Choudhury *et al.*, *Phys. Rev.* D **65** 053002 (2002), hep-ph/0109097.
8. M.E. Peskin, J.D. Wells, *Phys. Rev.* D **64** 093003 (2001), hep-ph/0101342.
9. E. Gross *et al.*, to be published.
10. D.J. Miller *et al.*, hep-ph/0304049.
11. Examples of scenarios with a Higgs decaying invisibly can be found in:
    -A. Djouadi, M. Drees hep-ph/9703452 (gravitino)
    -A.S. Joshipura, S.D. Rindani hep-ph/9205220 (Majorons)
    -G.F. Giudice *et al.*, hep-ph/0002178 (graviscalar)
    -N. Arkani-Hamed *et al.*, hep-ph/9811448 (KK of ν_R)
    -A. Widom, Y.N. Srivastava hep-ph/0004231 (gravitons).
12. F. Boudjema *et al.*, hep-ph/0206311.
13. J.L. Hewett and T.G. Rizzo, SNOWMASS-2001-P338, hep-ph/0112343.

14. `http://wwwth.mppmu.mpg.de/ members/loopverein/TheLoopVerein.html` and references therein.

15. S. Jadach, hep-ph/0306083.

16. See for instance W. Hollik, talk given at the Mini-Workshop on Electroweak Precision Data and the Higgs Mass, Zeuthen, Germany, 28 Feb - 1 Mar 2003, DESY-PROC-2003-01.

17. TESLA TDR part 3. Physics at an $e^+ e^-$ Linear Collider, hep-ph/0106315.

18. F. Richard, hep-ph/0303107.

19. T. Appelquist and Ho-Ung Yee, *Phys. Rev. D* **67** 055002 (2003), hep-ph/0211023.

20. Talk of G.F. Giudice at this conference.

21. Proceedings of the 5th International Linear Collider Workshop (LCWS2000) and Reference.[17]

22. M. Davier *et al.*, LAL-03-50, hep-ph/0308213.

23. Hsin-Chia Cheng *et al.*, *Phys. Rev. D* **66** 056006 (2002), hep-ph/0205314.

24. J.R. Ellis, hep-ex/0210052.

25. J.L. Feng, K. T. Matchev, and T. Moroi, *Phys. Rev.* D **61** 075005 (2000).

26. D. Hooper and T. Plehn, *Phys. Lett. B* **562** 18-27 (2003), hep-ph/0212226.

27. See B. E. Sauer, status of EDM measurements in `http://frontierscience.lnf.infn.it/ 2002/talks/sauer.pdf`

28. K.S. Babu *et al.*, *Phys. Rev. D* **65** 016005 (2002), hep-ph/0107100.

29. T. Ibrahim and P. Nath, *Phys. Rev. D* **58** 111301 (1998) [Erratum-ibid. D 60, 099902 (1999)], hep-ph/9807501.

30. J.R. Ellis, *Phys. Lett. B* **565** 176-182.

31. Talk given by M. de Jesus at this Conference.

32. M. Battaglia *et al.*, hep-ph/0306219.

33. G. Belanger, *et al.*, hep-ph/0112278.

34. P. Binetruy, M.K. Gaillard and B.D. Nelson, *Nucl. Phys. B* **604** 32 (2001).

35. LHC/LC Study Group `http://www.ippp.dur.ac. uk/~georg/lhclc/` and references therein.

DISCUSSION

Chang Kee Jung (Stony Brook): What is the most current MGUT apparent value with an error? What is the MGUT value newly derived by this particular string model shown in your talk?

Francois Richard: In the plot shown, the gaugino masses unify at about 10^{16} GeV in contrast with the coupling constants which unify at twice this value. The claim is that such an effect can be seen at a LC. In this particular case the effect originates from loop corrections within a string inspired theory.

Thomas Germann (Zurich University): You mention the option of using transverse polarization. What are the physics observables to be probed with this?

Francois Richard: I had in mind the possibility of measuring polarization asymmetries which are insensitive to vector contributions but can reveal tensor terms originating from low scale gravity contributions. This topic has been recently advocated by T. Rizzo *et al.*, and they find a sensitivity of this method up to very high mass scales.

Andreas Kronfeld (FNAL): If I understood your remarks on dark matter correctly, cosmology could determine the total component of dark matter at the 2% level. Meanwhile, LC could determine the contribution of a single identified component (e.g. the LSP) with similar precision. Then you could imagine $DM_{Total}= (30.0\pm0.5)\%$ and $DM_{LSP}= (20.0\pm0.5)\%$. That would be fascinating. Did I understand this correctly?

Francois Richard: Yes, there is no reason that the two numbers coincide but it matters that the precisions are similar for an optimal comparison.

LINEAR COLLIDER OPTIONS:
STATUS OF THE R&D AND PLANS FOR TECHNOLOGY SELECTION

M. TIGNER

Laboratory for Elementary-Particle Physics, Cornell University, Ithaca, NY 14853, USA
E-mail: mt52@cornell.edu

R&D for the linear collider, LC, is well along towards the point where a technology selection can be made. The status of the R&D is described briefly and the evolving procedure for making a selection is presented together with the current ideas on how to proceed.

1. Background

1.1. *International Framework*

While it is true that much of the initiative behind the LC is from the grass roots, there is a considerable, international organizational framework that has been built up over the years and which is now serving us well. Much remains to be done in getting the government support agencies involved in a fundamental way but we have a good working foundation in place. One way of understanding it is to look at Fig. 1 which displays the framework graphically and gives the times of formation of essential elements. The parent international organization, IUPAP, was founded in 1922. Details of its purpose and activities can be found at www.IUPAP.org. Seeing future trends in HEP already in 1975, ICFA, the International Committee on Future Accelerators was formed and comprises of members from countries active in HEP. At that time it was already apparent that not only was international participation in construction and operation of detectors going to remain central to the field but that international collaboration in the conception and provision of new accelerator facilities would someday be necessary. That time seems to have arrived. In 2002, Hirotaka Sugawara, then Director of KEK and the ICFA Chair, led us in forming the International Linear Collider Steering Committee, ILCSC. The aim of which is to aid in the first instance with allowing all competent and interested participants in all regions to participate in the conception and creation of the accelerator as well as the detector complement. It relies on the regional steering groups, ALCSG, ELCSG and USLCSG for major inputs and coordination of the work in the regions, now generally defined as Asia, Europe and North America but certainly expandable as this global project evolves. The primary work of the ILCSC needs to be carried out by subcommittees. As seen in Fig. 1, there are currently three such subcommittees now active. In future this number will change.

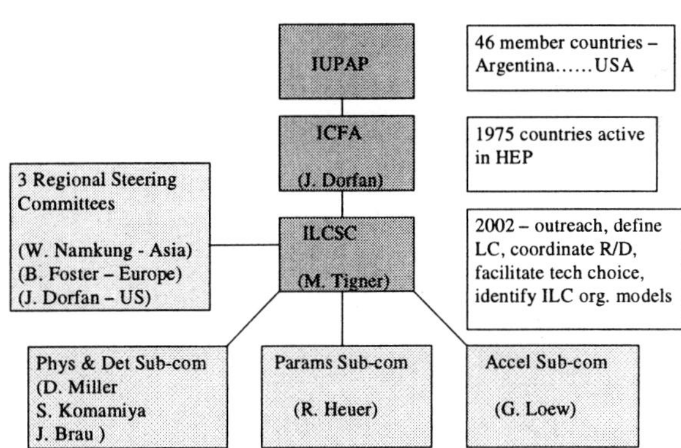

Figure 1. International framework for current LC R&D.

1.2. *The International Technical Review Committee*

In 1994 the Inter Laboratory Collaboration for R&D Towards a Linear Collider created the International Linear Collider Technical Review Committee, popularly known as the TRC, under Greg Loew of SLAC. The purpose was to document the status of R&D on the then 8 e^+e^- collider concepts. They issued their landmark report in 1995. In 2001 ICFA itself reconvened the TRC, again under Greg Loew. This TRC had a Steering Committee comprising R. Brinkmann, DESY, K. Yokoya, KEK, T. Raubenheimer, SLAC, G. Guignard, CERN and Working Groups comprising 37 members who undertook the enormous task of reviewing the status of the now 4 possible options: superconducting (e.g. TESLA), normal conducting

X-band (e.g. NLC/JLC), combined normal conducting X- and C-bands (e.g. JLC-C) and the CLIC concept at 30 GHz.

2. R&D Status

2.1. *Scorecard and Calendar*

The TRC report was delivered earlier this year and defines and ranks the R&D needed for choosing the technology to go forward with. These R&D items range from those needed for basic feasibility assessment to those needed for design and cost optimization and rating them with ranks R1–R4, R1 being the category needed for feasibility assessment. It is important to note that most of the "press" have focused on achievable accelerating gradients but many other things are of prime importance. While it is true that the energy advance needed beyond the SLC is a factor of 5 to 10, the increase in luminosity needed is 10^4 and that has not much to do with gradient.

The report gave a scorecard for each of the 4 options at each of the R levels. At the R1 level, i.e. feasibility demonstration still needed, four categories were defined: modulators, klystrons, RF distribution system and accelerator structures. Of these: TESLA needs to demonstrate accelerator structure gradient for its 800 GeV version; X- and C-band versions need to demonstrate RF distribution as well as accelerator structures; whereas the 30 GHz version needs feasibility demonstration in all four categories.

Each of the proponents for superconducting and normal conducting X and C-band has scheduled demonstrations addressing the R1 ratings for their offerings. They are committed to show results this fall and early next year so that the technology selection can be completed as early as possible in 2004.

3. Technology Selection and Beyond

3.1. *Committee of Wise Persons*

The ILCSC has recommended that a technology recommendation be made by a committee of wise persons selected from the world scientific community. ICFA has accepted this recommendation. The regional steering groups are now hard at work with their communities to bring their nominations before ILCSC and ICFA in time that the wise persons can begin work in early 2004. While details of their procedure must await their appointment, it is inevitable that they will hold meetings at the primary laboratories of the proponents and have tutorial sessions to familiarize themselves with the subject of the linear collider and what is needed to do the science envisioned. The resultant make-up of the wise person committee, WPC, and its proposed procedure will be widely communicated as soon as it is available.

3.2. *PreGlobal Design Group*

It is generally agreed among those responsible for the current LC R&D programs that the community needs to join in turning the technology choice into a concept design based on that choice and on the extensive work that has gone on to date. Further, as no additional resources are likely to be made available immediately, we will need to do that with the resources already in place, coordinated by a central organization, now referred to as the PreGlobal Design Group working through regional managers of some sort. To draft the mandate of such a group and organization for eventual action by ICFA, the ILCSC has appointed a task force consisting of the Chairs of the regional steering groups and a lab director from each of the regions. It now seems likely that the PreGlobal Design Group would be put into action immediately following the technology choice. However this is still under active discussion.

DISCUSSION

Bennie Ward (Baylor University & University of Tennessee): From your transparencies, I could not see at what point the funding agencies enter the process that you have presented. Could you please comment?

Maury Tigner: Connections with funding agencies differ greatly in the three regions: Asia, Europe and the US. In Europe they differ from country to country. In general, the agencies are already involved through the need for them to approve expenditures. In July many of the agency representatives met in London to begin discussions on how to govern and finance the LC project internationally.

Hugh Montgomery (Fermilab): I've not understood from today's talks when, with respect to the technology discussion, the global study group would be populated. What is your view of how that will happen?

Maury Tigner: This is a matter very much under discussion at the moment. There is now a Task Force in place to recommend the mandate and organization of this global study group. At a minimum it will be activated immediately upon achieving a technology recommendation.

Tony Liss (University of Illinois): There is also a university-based R&D group, and I was wondering if you could comment on how they fit into this work, if at all?

Maury Tigner: University groups are very much a part of this activity and are already performing important parts of the R&D. That will only grow in the future.

Maria Spiropulu (University of Chicago): You gave us a list of committees and a list of times by which there will be a technology decision. Who will force that this timing is kept?

Maury Tigner: There is no external authority that has the will or competence to do this. It must be generated within our community. Because of the importance of getting this job done I firmly believe that our self discipline will suffice.

Bruce Yabsley (Virginia Tech): This is perhaps an ICFA question. I've heard concern expressed that in this security environment, the U. S. might be less willing to fund a facility outside the U. S., or if it was on U. S. soil, might be unwilling to relinquish control to the extent that would be acceptable to the rest of the community. Is that a concern that is shared by the committee, or is the perspective different from where you are standing?

Maury Tigner: The current Administration's science officials are very keen on this being a truly international enterprise wherever it is located and have repeated that to us many times.

FUTURE DIRECTIONS

SUPERSYMMETRY AND OTHER SCENARIOS

E. WITTEN

School of Natural Sciences, Institute for Advanced Study, Princeton, NJ 08540, USA
E-mail: witten@ias.edu

In this talk, I survey the merits and demerits of Supersymmetry, as well as other approaches to the gauge hierarchy problem.

1. Introduction

It has come to be a familiar story – most aspects of the Standard Model are increasingly well tested. This even includes CP-violation and the properties of the top quark, as has been reported earlier this week.

The one part of the Standard Model about which we really have no clear experimental information is the mechanism of electroweak symmetry breaking. Why are the weak interactions not obvious in everyday life, in the same sense that electromagnetism is obvious?

For a long time, this question has been the key indication that something really new is in store. Experiment in the next few years is surely approaching the decisive stage – either here at Fermilab, or at the LHC.

The most simple possibility is the original electroweak theory, which assumed a single elementary Higgs boson and nothing else. On the whole, although there are some very small discrepancies, this is in very good agreement with experimental data, which moreover suggest (in the context of the pure Standard Model) that the Higgs mass is no more than about 200 GeV/c^2. Some alternative theories predict a Higgs particle plus many additional things; some predict no Higgs particle but many other things instead. The pure Standard Model with only the Higgs is really the only picture that *doesn't* predict a host of new particles for current and planned accelerators.

The pure Standard Model with only the Higgs has numerous virtues:

- it is simple;

- it agrees quite well with a mass of experimental data; and

- it explains a lot of things that would otherwise

be puzzles, like why Flavor-Changing-Neutral-Currents and baryon-, lepton-, and CP-violating interactions are so suppressed.

Basically, and despite a lot of ingenuity that has gone into this, we don't know a completely satisfactory extension of the Standard Model.

Despite this, most physicists (including myself) remain convinced that the minimal Standard Model with only the Higgs is unlikely to be the full story. The main reason for this is the "hierarchy problem." A scalar field ϕ can have a bare mass term m^2. Moreover, the quantity m^2 is not stable against quantum corrections; in the Standard Model, the renormalization of m^2 is quadratically divergent, so that if the Standard Model is somehow cut off at a mass scale M, the one-loop renormalization is of order αM^2 (where α is the fine structure constant). This is unnatural for $m^2 \ll \alpha M^2$.

In a model with spontaneous electroweak symmetry breaking, the problem really affects not only the Higgs mass, but also its expectation value, and hence it affects the masses of other particles that get their masses from gauge symmetry breaking – the W and Z, and the quarks and charged leptons. So it is unnatural to have the W and Z at 80 or 90 GeV/c^2, and the Higgs below 200 GeV/c^2, unless the Standard Model is somehow "cut off" and embedded in a richer structure that tames the ultraviolet divergence in the Higgs boson mass – at an energy no bigger than about 1 TeV.

Extensions of the Standard Model differ largely in how this is done. The question has central importance for the future of physics, because different outcomes in the exploration of electroweak symmetry breaking will tend to lead us in very different directions.

We have grappled with these issues for many years. Meanwhile, observation appears to have presented us with another phenomenon that we should discuss here as it seems fine-tuned in a similar way. This is the acceleration of the cosmic expansion, which points to a tiny but non-zero cosmological constant, or maybe a more complicated form of "dark energy." This actually poses a fine-tuning problem similar to the problem of the Higgs boson. In the Standard Model, the energy of the vacuum is quartically divergent. The simplest approximation is simply to add up zero point energies

$$\pm \frac{1}{2}\hbar\omega \tag{1}$$

for every Bose mode or Fermi mode of momentum k and energy $\hbar\omega = \sqrt{(\hbar ck)^2 + (mc^2)^2}$. The integral over k

$$\pm \int d^3k \sqrt{(\hbar ck)^2 + (mc^2)^2} \tag{2}$$

is quartically divergent, so the best we can say is that the energy of the vacuum is expected to be of order M^4, where M is the cut-off energy at which "something else" happens and the contributions to the vacuum energy are cut off. Experiment appears to point to a vacuum energy Λ of order $(10^{-3}\ \text{eV})^4$, where the mass scale 10^{-3} eV is way below any possible Standard Model cut-off. It actually is relatively close to what appears to be the neutrino mass scale, but so far no one has had much success in explaining this. A nice explanation of the fine-tuning of the vacuum energy has not yet emerged.

This puzzle has lent comfort to one line of thought *which I personally hope is wrong* but which I have to mention in any discussion of fine-tuning. This is the "anthropic theory," according to which the smallness of the cosmological constant is not a consequence of the laws of nature in the usual sense. According to this picture, the laws of nature allow for a plethora of physical states, which are realized in different parts of the universe, and have widely differing values of the cosmological constant. But we live in a region in which the cosmological constant is small, simply because elsewhere the Universe expands and cools too rapidly for life to emerge.

Once one starts to admit anthropic interpretations of fine-tuning problems like the cosmological constant, is is clear that such a proposal might be made for other fine-tuning problems, such as the problem of the Higgs boson mass. Certainly, we would not be here if the Higgs boson mass, and hence also the W and Z and quark and lepton masses, were

greatly bigger. If they were near the Planck scale, for example, any collection of more than a few elementary particles would collapse to a Black Hole. More generally, if the elementary particle masses were scaled up by a factor N, the number of elementary particles in a star or planet would scale down like N^{-3}, and for very modest N the stars would stop shining. If experiment will uncover a Higgs boson at the range of masses suggested by the Standard Model, but (even at LHC energies) no further structure emerges that will explain how Nature solved the fine-tuning problem, this will certainly be viewed by some as support for anthropic explanations of fine-tuning.

On the other hand, at the moment, (virtually) no one seems to be predicting this. Physicists who do favor anthropic explanations tend rather to argue that if a fine-tuning problem, like the Higgs mass, can have a rational explanation, then regions of the Universe in which such a mechanism is manifested are far more abundant than regions in which the Higgs mass is small "accidentally."

So everyone seems to agree on one thing: we want from accelerators not just a Higgs boson, but a mechanism that will "stabilize" the scale of electroweak symmetry breaking and explain why the Higgs boson, and the rest of the particles, are not much heavier. But what? Numerous suggestions have been made:

- "Higgsless" models – based on dynamical symmetry breaking;

- models with branes and large extra dimensions, plus possible strong dynamics;

- "Little Higgs" – Higgs as a pseudo-Goldsone boson; and

- Supersymmetry.

One thing they all have in common is that there is no perfect model – all known approaches are at risk of spoiling some Standard Model successes. It seems impractical to review all the options. The range of models considered has grown too widely in the last few years. Many of the new proposals at the moment are scenarios more than models. Instead, I will concentrate in the last part of this talk in explaining the virtues, but also the drawbacks, of one approach that I think is especially interesting. This is Supersymmetry. First the virtues:

- SUSY can make a "small" Higgs mass natural;

- SUSY is part of a larger vision of physics, not just a technical solution;

- the measured value of $\sin^2 \theta_W$ favors SUSY GUT's;

- SUSY survives electroweak tests; and

- the top quark mass has turned out to be heavy, as needed for electroweak symmetry breaking in the context of SUSY.

SUSY is a unique new symmetry that relates bosons to fermions, in a sense explaining why fermions exist. Relating bosons to fermions also makes it possible to explain the smallness of the Higgs mass, since we do know why the smallness of fermion masses can be natural. So that is at least the germ of how SUSY solves the fine-tuning problem.

SUSY inherits the successes of Grand Unification, because given modern measurements of $\sin^2 \theta_W$, as well as bounds on the proton lifetime, the supersymmetric version of Grand Unification is the one that works.

So here we really must remember the merits of Grand Unification, which are substantial in their own right.

- It makes sense of the quark and lepton quantum numbers, which look like quite a mess in the Standard Model. A generation of Standard Model quarks and leptons

$$\begin{pmatrix} u \\ d \end{pmatrix}_{1/3} \oplus \bar{u}_{-4/3} \oplus \bar{d}_{2/3} \oplus \begin{pmatrix} \nu \\ e^- \end{pmatrix}_{-1} \oplus e_2^+ \quad (3)$$

turns into a simple $\bar{5} + 10$ of SU(5), or 16 of SO(10).

- The unification scale M_{GUT} inferred from low energy data is relatively close to the Planck scale, but high enough to avoid disaster with the proton lifetime.

- The neutrino mass scale suggested in the late 1970's based on GUT's, $m_\nu \sim M_W^2/M_{GUT} \sim 10^{-2}\,\mathrm{eV}$, has apparently turned out to be about right.

- Grand Unification fits neatly with strings and Quantum Gravity.

- The observed fluctuations in the cosmic microwave radiation are naturally (but speculatively) interpreted in terms of an inflationary epoch close to the GUT scale.

In short, Grand Unification is a really nice story. But it really only makes sense with Supersymmetry, for two reasons:

- the measured value of $\sin^2 \theta_W$ agrees with Grand Unification only if Supersymmetry is included; and

- the unification scale and proton lifetime come out to be too small without SUSY.

So the successes of GUT's encourage the search for Supersymmetry, and discovery of Supersymmetry would enhance the attractiveness of GUT's.

As I have tried to argue, SUSY is not just a technical solution to problems like the hierarchy problem. It is:

- a unique new symmetry principle;

- part of an attractive larger picture in GUT's;

- and actually, an essential part of an even more ambitious picture in string theory.

In fact, the concept of Supersymmetry emerged historically at least in part because of its role in string theory. Experimental discovery of Supersymmetry would certainly give string theory a big boost, and learning how Supersymmetry is broken might very well give string theorists crucial clues about how to proceed. Moreover, while some alternative theories of the smallness of the electroweak scale – like models of composite Higgs bosons – have repeatedly run into trouble, Supersymmetry is comfortably consistent with the precision electroweak tests.

2. Problems

For good or ill, the SUSY models considered today are the same ones that were considered viable twenty years ago. In fact the old models remained viable because the top quark turned out to be sufficiently heavy, as was required for electroweak symmetry breaking. It really is not entirely good that the models of today are the same ones as twenty years ago. It means that the models have held up, but also that certain problems that troubled theorists twenty years ago have still not been solved!

For today, we are not going to dwell on the good news. We also want to look at the drawbacks of Supersymmetry. The most obvious drawback is simply that Supersymmetry hasn't been found yet, though we have been hoping for a long time. Looking back, for example, to the summary talk (by David Gross) at Lepton-Photon 1993, I see that ten years ago SUSY was already described as the "standard non-standard theory." He also gave a list of its successes and drawback rather similar to what I am explaining today. That is beginning to be a long time.

It is disappointing that we have not found SUSY yet, but for the most part it is perhaps not too surprising. If charged superpartners are just a little bit above M_Z, we would not have seen them yet. Superpartners get masses from electroweak breaking *and* SUSY breaking so it is natural for them to be a bit above the Z, which gets mass only from electroweak breaking. But there is perhaps one particle whose absence until now is a little embarrassing, namely the Higgs boson.

Assuming the minimal supersymmetric spectrum, one has at tree level

$$M_{\text{Higgs}} < M_Z \sim 91 \text{ GeV}/c^2. \qquad (4)$$

We may compare this to experiment,

$$M_{\text{Higgs}} > 114 \text{ GeV}/c^2. \qquad (5)$$

Actually, there is a large radiative correction due to the heavy top quark, and the theoretical bound on the Higgs mass is usually quoted as

$$M_{\text{Higgs}} < 130 \text{ GeV}/c^2. \qquad (6)$$

So there is not quite a contradiction. But rather optimistic assumptions go into getting the radiative correction so large. One needs couplings not favored by many of the models, and/or superpartner masses so large as to make the smallness of M_Z look a little unnatural. Though there is no contradiction yet, it would certainly clarify things a lot to know what M_{Higgs} is. And it would really be nice if it turned out to be 115 GeV/c^2, the value hinted at by LEP.

At a different level, Supersymmetry would have been more convincing if it had achieved some simplification in the Standard Model. For example, could the Higgs boson be a superpartner of the electron? Unfortunately, no: models that tried things like that did not work. So the Minimal SUSY Standard Model essentially doubles the spectrum.

SUSY (like many attempts to resolve the fine-tuning problem) actually complicates some successes of the Standard Model. For instance, one triumph of the Standard Model is to naturally conserve baryon and lepton number, because there are no renormalizable (perturbative) couplings of Standard Model fields that violate those symmetries. This is lost with Supersymmetry, where renormalizable interactions causing catastrophic proton decay are possible. The most commonly adopted solution to this problem is to assume a new symmetry called R-parity; this is possible but not obviously compelling. Supersymmetry also potentially undoes some of the successes of the Standard Model in suppressing Flavor-Changing-Neutral-Currents and CP-violation, by introducing troublesome new loop diagrams involving superpartners. And Supersymmetry introduces at the GUT scale a new scenario for proton decay via dimension-five operators. This is troublesome for many models given modern experimental limits on the proton lifetime.

And how is SUSY broken? There are two major approaches:

- Gravity Mediation – Supersymmetry is broken at a very high scale and SUSY breaking is mediated to the Standard Model via Supergravity interactions; and

- Gauge Mediation – Supersymmetry is broken at 100 TeV or so and its breaking is communicated to the known world via gauge forces.

Each type of model has its virtues, and neither has yet given a clear path to solving all the problems. For example, thinking about the cosmological constant might lead us to favor gravity mediation. The reason can be seen by considering the potential energy in Supergravity:

$$V = \mid DW/D\varphi \mid^2 - G_N \mid W \mid^2 \qquad (7)$$

Here G_N is Newton's constant. To make V small, a cancellation between the two terms is needed. This cancellation certainly involves gravity, as the second term is explicitly proportional to G_N. The inevitable role for gravity might make gravity mediation seem more natural.

If we instead consider excessive new sources of Flavor-Changing-Neutral-Currents and CP-violation, we find that gauge mediation gives much more obvious ways to eliminate them. In short, we don't have a fully convincing picture of Supersymmetry breaking. That is actually one of the things that makes Supersymmetry an exciting target for experiments. If we had a convincing, workable picture of what the weak scale Superworld would really look like, we'd be more convinced that it is there, but we'd have less to learn by finding it. As it is, it would be quite dramatic to learn how nature did solve all the problems. And if Supersymmetry is discovered, each perplexing question about how Supersymmetry might work in the real world will turn into an opportunity to learn a fundamental new lesson about nature.

In short, discovering Supersymmetry (or any other solution of the hierarchy problem, since all known options raise vexing problems) would put experiment ahead, as was the normal state of affairs in the days before the emergence of the Standard Model. Unraveling the details of the weak scale Superworld – with its host of new particles and new interactions – will be quite a long and complex project. It will provide an excellent target for the precision of lepton colliders, as well as the higher energy of proton colliders.

Ikaros Bigi (Notre Dame): You mentioned the prediction that the Higgs should be 130 GeV/c^2. If it were not found, would you with your long-established love of Supersymmetry give up, or will you find another option?

Edward Witten: It really depends on what is found by the experiments that hypothetically show that weak-scale Supersymmetry is not present. Hopefully, in the scenario you suggest, something else will turn up that will give us a good hint.

Maria Spiropulu (Chicago): I was surprised to see in your talk a discussion about the anthropic principle. Why did you think about that? Why did you put that in this talk?

Edward Witten: I hope that the anthropic point of view turns out to be wrong because I hope we will be able to understand the Universe better than this point of view would probably allow. I felt I should discuss it in the talk because of the conceivable analogy between the fine-tuning problem involved in the vacuum energy and the fine-tuning problem associated with the Higgs mass.

Bogdan Dobrescu (FNAL): You have mentioned that the discovery of SUSY would be nice among other things because it would support the case for string theory. Could you comment on the possibility of experimental tests of string theory, whether you are suggesting that if SUSY were not discovered it would make an important problem for string theory?

Edward Witten: Personally, I think that string theory is on the right track. Its elegance in incorporating the physics we know, reconciling quantum mechanics with gravity, and giving new insights about familiar theories convinces me of that. But there is a big gap between what we know and what we'd need to know to understand nature, and it isn't clear to what extent we can make rapid progress. Certainly, experimental discovery of Supersymmetry would improve the odds a lot. Other experiments – ranging from proton decay experiments to the search for a gravitational wave signature from inflation – can also make important contributions.

OUTLOOK: THE NEXT TWENTY YEARS

H. MURAYAMA

Department of Physics, University of California, Berkeley, CA 94720

and

Theoretical Physics Groups, Lawrence Berkeley Laboratory, University of California, Berkeley, CA 94720

and

School of Natural Sciences, Institute for Advanced Study, Princeton, NJ 08540

E-mail: murayama@ias.edu

I present an outlook for the next twenty years in particle physics. I start with the big questions in our field, broken down into four categories: horizontal, vertical, heaven, and hell. Then I discuss how we attack the big questions in each category during the next twenty years. I argue for a synergy between many different approaches taken in our field.

1. Introduction

I was asked specifically not to give a summary talk of the Symposium, but rather an "outlook" talk. Obviously I will refer to many excellent talks given during this Symposium, but I will not try to cover everything that had been said. If you find that your favorite subject is not covered in my talk, you are welcome to come up here and complain to Keith Ellis.

What I will try to do in this talk is present my own perspective on how our field may develop in the next twenty years. Whenever we talk about the future, a very natural question is whether it is bright, as bright as the illumination of the Chicago skyscrapers we admired at the time of the banquet (Fig. 1), or dark, as dark as parts of the East Coast were this week (Fig. 2). You will see my verdict at the end of the talk. But before getting to the verdict, I'd like to talk about the current situation of our field.

One way to look at our field of particle physics is that it has specialized into so many different subfields. During this Symposium alone, we heard about many exciting subjects and experiments. Here is an *incomplete* list in a random order: $0\nu\beta\beta$, B-physics, proton decay, LHC, Higgs, reactor antineutrinos, K-physics, Lepton Flavor Violation, Cosmic Microwave Background, string theory, e^+e^-, top quark, accelerator-based neutrinos, lattice QCD, charm physics, Dark Energy, hidden dimensions, neutrino factory, hadron physics, Supersymmetry, Dark Matter, atmospheric neutrinos, Linear Collider, exotics, solar neutrinos, and the increasingly important topics of outreach and politics. People

Figure 1. The bright skyline of Chicago.

Figure 2. The dark skyline of New York City during the blackout.

may say that the field is completely fragmented. On the other hand, I have a somewhat different view. I think that most of them are heading to a synergy at an important energy scale: TeV. I don't mean that TeV-scale physics will solve all the puzzles we are facing. What I mean is that any of these interest-

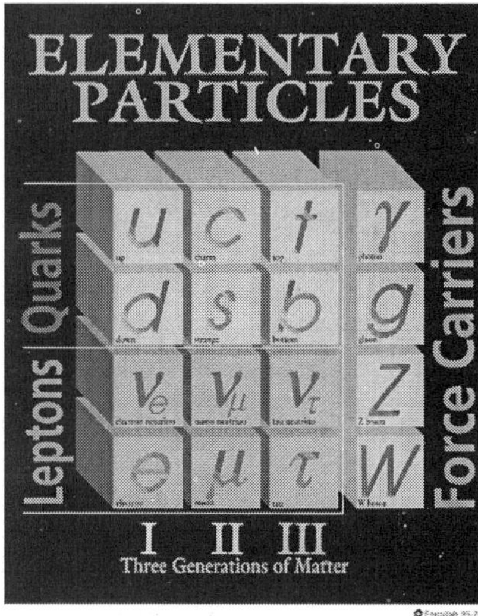

Figure 3. The table of particles in the Standard Model.[1] Vertical Questions are concerned with particles in a single generation, while Horizontal Questions refer to the relationship sideways in the table.

fermion masses

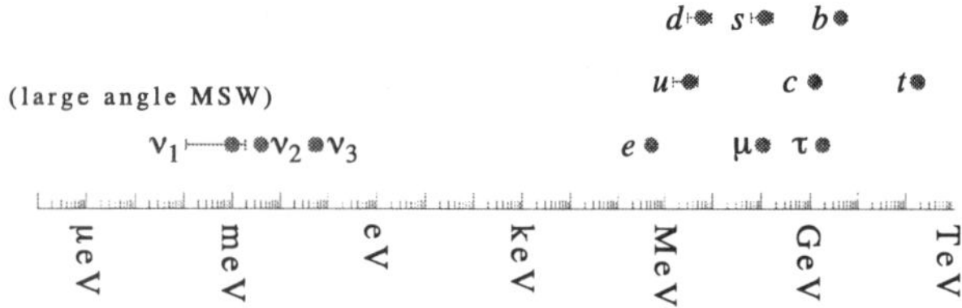

Figure 4. The mass spectrum of quarks and leptons we don't understand.

Table 1. The quantum numbers of quarks and leptons we don't understand. Here, all particles are shown in their left-handed chirality states.

$$Q(3, 2, +\tfrac{1}{6}), \qquad u(\bar{3}, 1, -\tfrac{2}{3}), \qquad d(\bar{3}, 1, +\tfrac{1}{3}),$$
$$L(1, 2, -\tfrac{1}{2}), \qquad e(1, 1, +1)$$

They concern properties within each family of particles (Fig. 3). Why are there three unrelated gauge forces? Why is the strong interaction strong? Why are all electric charges quantized in the same unit? What physics guarantees the seemingly miraculous anomaly cancellation? What physics explains the quantum numbers of quarks and leptons we see (Table 1)? Is there a unified description of all forces? Why is $m_W \ll M_{\mathrm{Planck}}$? (Hierarchy Problem.)

Recently, we added many *questions from the heaven* (Fig. 5). What is Dark Matter? What is Dark Energy? Why are we at the special moment when the energy densities of Dark Matter and Dark Energy are the same within a factor of two? ("Why now?" problem.) What exactly was the Big Bang? Why is the Universe so big? (Flatness problem, horizon problem.) How were galaxies and stars (and eventually us) created?

If there are questions from the heaven, there are also *questions from the hell*. To the best of the collective knowledge of Homo sapiens, we live at the bottom of a strange potential with a wine bottle shape: that's the hell we are in (Fig. 6). Because of this potential, the Bose–Einstein condensate (BEC) of the Higgs boson is supposed to be present in our Universe, and we are swimming in this BEC. What is this Higgs boson thing? Why does it have this strange potential with a negative mass-squared? Why is there only one scalar particle in the Standard Model, designed to do its most mysterious part? Is it elementary or composite? Is it *really* condensed in our Universe?

We do not have the right to expect that *any* of these questions can be answered within our lifetime (or ever). Nonetheless there is a good potential for us to answer some or many of them. How exactly do we do it? I will refer to Supersymmetry as an example many times in my talk, but I expect similar stories with any scenario of TeV-scale physics. In any case, TeV is the key.

ing physics topics must once go through the study of the TeV scale before they reach their own destinations. It is a hub where everybody has to transfer to another flight.

Why do I think so? To see this, let me start enumerating the big questions in our field. I broke them down to four categories.

The first category is what I call the *horizontal questions*. They are about relationships between the three families of elementary particles (Fig. 3). Why are there three generations and no more? What physics determines their masses and mixings (Fig. 4)? What is the energy scale of that physics? Why do neutrinos have mass and yet they are so light? What is the origin of CP-violation? What is the origin of the matter-antimatter asymmetry in our Universe?

The second category is the *vertical questions*.

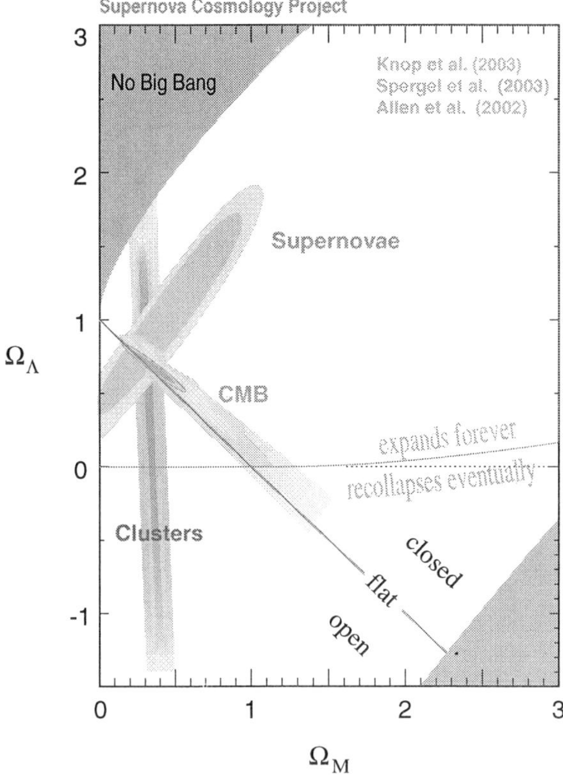

Figure 5. The unknown constituents of the universe.

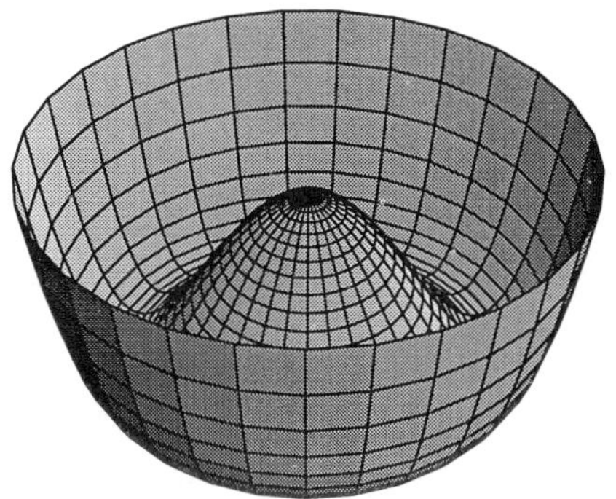

Figure 6. The hell of the Universe we live in and don't know why.

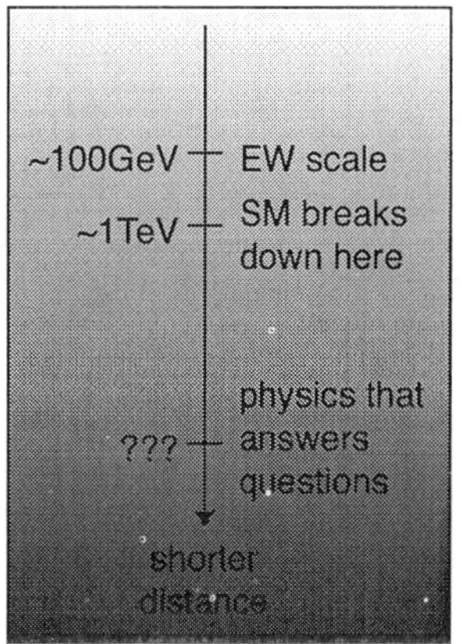

Figure 7. We would like to access physics at a short distance that answers some of the big questions. But before getting there, the Standard Model breaks down around a TeV and everything at shorter distances is grayed out.

I have just finished the introduction. Now I move on to discuss each category of questions: hell, heaven, vertical and horizontal. My verdict on the future of our field follows after that.

2. Hell

What we know is that the Standard Model of particle physics is completely incapable of answering the big questions I've listed. What we want to do is to look for physics beyond the Standard Model that answers these big questions. By definition, that is physics at shorter distances. In order to talk about the new physics that appears at a some small distance scale, the Standard Model must survive down to whatever that short distance scale is. The problem is that it doesn't. This is the hierarchy problem. It is the main obstacle for us to address the big questions. We can't even get started! (Fig. 7)

To illustrate the reason why we can't even get started, let us rewind the video back to the end of the 19th century. *Once upon a time*, there was a hierarchy problem.[2] It was a crisis about the mass

of the electron. We know like charges repel. It is hard to keep electric charge in a small pack because it repels itself. On the other hand, we know the electron is basically point-like. Our best limit is that the "size" of the electron is less than something like 10^{-17} cm. The problem is that, if you want to keep the charge in such a small pack, you need a lot of energy. A naïve guess is that you need at least

$$\Delta E \sim \frac{\alpha}{r_e} \sim 1 \text{ GeV} \frac{10^{-17} \text{ cm}}{r_e}. \tag{1}$$

But we know we can't afford it. The energy carried by an electron is just $E = mc^2 = 0.511$ MeV, nowhere close to what we need. In fact, the best we can do is to pack the charge down to about 10^{-13} cm, which is the so-called classical radius of the electron. In other words, the classical theory of electromagnetism breaks down around this distance scale, and we cannot discuss physics below 10^{-13} cm. We can't get started!

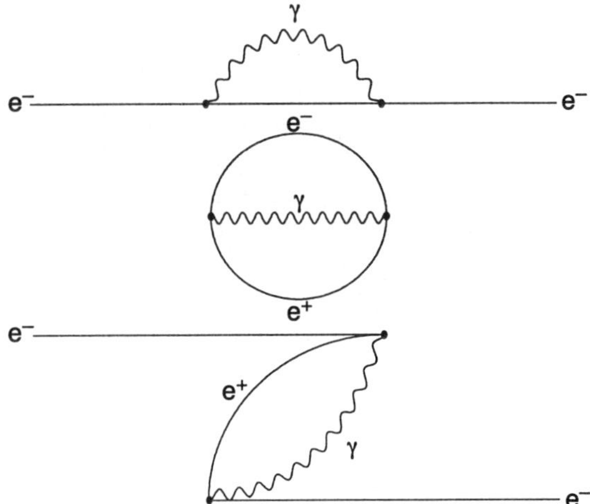

Figure 8. Top: The electron sees the Coulomb field it created itself. Center: The vacuum is full of "bubbles" in which an electron-position pair is created spontaneously and annihilate back to the vacuum within the time allowed by the uncertainty principle. Bottom: An electron may decide to annihilate the positron in the "bubble" while the electron originally in the "bubble" remains as a real particle.

But we don't talk about this problem anymore, because there was a resolution. Antimatter came to the rescue. We solved the crisis by doubling the number of particles. Here is how it works.

The electron creates a Coulomb field around itself, and it feels its own field. Namely, it repels itself (Fig. 8, top). But we discovered antimatter. Moreover, we discovered that the world is quantum mechanical. Once you have these two ingredients, there is an inevitable consequence. The "vacuum" we see isn't empty at all. It constantly creates pairs of electrons and positrons, together with a photon. Of course, energy conservation forbids it, but quantum mechanics allows us to borrow energy as long as nobody notices it. The created pair must annihilate back to the vacuum within the time allowed by the uncertainty principle (Fig. 8, center). Such pairs are called "vacuum bubbles."

When you place an electron in this fluctuating vacuum, it "sees" a positron nearby. Sometimes, it decides to annihilate the positron in the bubble. Then the electron that was originally a part of the bubble now remains as a "real" particle (Fig. 8, bottom). It turns out that this process also contributes to the energy of the electron with a *negative* sign, that nearly exactly cancels the self-repelling energy we were worried about. The grand total is roughly

$$\Delta m_e c^2 \sim m_e c^2 \times \frac{\alpha}{4\pi} \log(m_e r_e). \qquad (2)$$

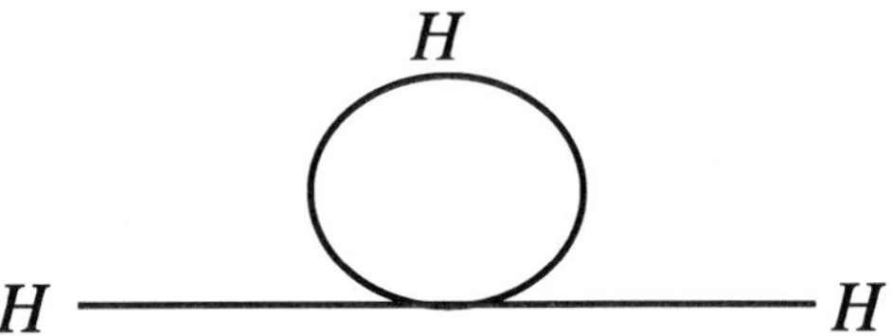

Figure 9. The self-repulsion of the Higgs boson makes it hard to be contained in a small size.

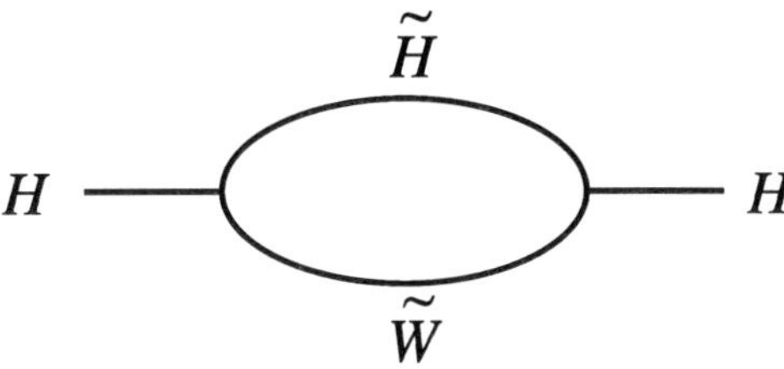

Figure 10. The Supersymmetric attraction diagram cancels the Higgs self-repulsion diagram.

This is nice. First of all, the additional energy you need is proportional to the original energy (the rest energy $m_e c^2$), and we are talking about a percentage correction. Second, even if you take the smallest size imaginable, namely the Planck size $r_e \sim 10^{-33}$ cm, the size of the correction is only about 10%. Now we can get started to think about physics below 10^{-13} cm.

The problem we are facing now is very similar. The minute you think that we are swimming in the Higgs BEC, you should ask if Higgs can be contained in a small package. It turns out that the Higgs also repels itself because of its self-interaction (Fig. 9). It requires a lot of energy to contain itself. The theory breaks down again, this time around 10^{-17} cm. We are stuck. We can't get started to address the big questions. We can't "see" the interesting physics at shorter distances that answers the big questions.

One way to solve this problem is to assume that history repeats itself. We double the number of particles again. The new particles cancel the contribution from the Higgs self-repelling energy (Fig. 10). This is the idea of Supersymmetry, which makes the Standard Model consistent with whatever physics there is at shorter distances. Indeed, the correction to the Higgs energy is

$$\Delta m_H^2 \sim \frac{\alpha}{4\pi} m_{SUSY}^2 \log(m_H r_H), \qquad (3)$$

where r_H is the "size" of the Higgs boson. Of course Supersymmetry is not the only solution, but it is

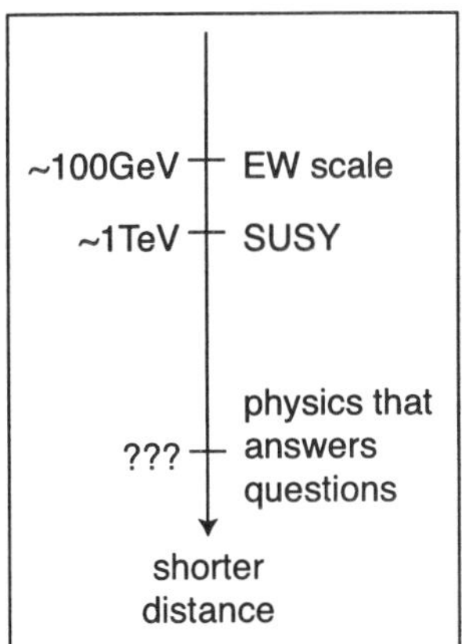

Figure 11. Once the hierarchy problem is solved, we will be allowed to talk about physics at shorter distances to address the big questions.

Figure 12. A nice picture of Fermi, except the definition of the fine structure constant is wrong. What was he thinking?

true that any solution of this kind appears at the TeV scale.

Once the hierarchy problem is solved, we can finally get started. It opens the door to the answers to the big questions (Fig. 11). The sky clears up and we can start "seeing" physics at shorter distances. An even more interesting possibility is that the solution itself provides additional probes to physics at shorter distances. We will talk about some examples soon.

In fact, the importance of the TeV-scale has been known since 1933. When Fermi (Fig. 12) wrote down his theory of nuclear beta decay, he knew the relevant energy scale: $G_F^{-1/2} \simeq 300$ GeV. It is truly exciting that we are finally getting to this energy scale!

For a long time, theorists, including myself, had been talking about three major directions to solve the hierarchy problem. One is Supersymmetry, which I already talked about. It is the idea that the history repeats itself. Just like antimatter solved the crisis of the electron mass, we double the number of particles. The second direction is to learn from Cooper pairs. The Higgs condensate is a composite made of two fermions. This idea is often called technicolor. These two ideas are two decades old, and many of you have witnessed a nearly religious war between the two camps. The third direction is relatively recent: physics *ends* at a TeV. The TeV-scale is the ultimate scale of physics where quantum gravity manifests itself. It may be superstrings. We may produce blackholes at accelerators or by cosmic rays. This is possible if there are hidden dimensions curled up in small sizes, somewhere between 10 μm

to 10^{-17} cm.

But the fact that the third direction was proposed relatively recently suggests that there are many more possibilities we theorists haven't thought of. Indeed, just the last two years have seen an outbreak of new ideas. The Higgs boson may be like pions in QCD, a pseudo-Nambu-Goldstone boson. Models based on this idea are called the "little Higgs" theories.[3] Or maybe the Higgs boson is actually a gauge boson. Extra-dimensional components of a gauge field do not appear to have spins to a four-dimensional observer because they are spinning in extra dimensions. This idea is sometimes called Gauge-Higgs Unification.[4] Or maybe there is no Higgs after all, and the reason why W and Z are massive is because they are Kaluza–Klein bosons, running along the extra dimension to acquire their "rest" (for a 4D observer) energies. This idea came to be known as "Higgsless" theories.[5] Most recently, I'm pushing the idea of "technicolorful supersymmetry."[6] You see, I'm pretty ecumenical.

Clearly the landscape of theories is getting more and more complicated (Fig. 13). As the solution to the problem draws near experimentally, theorists are proposing more possibilities. It is increasingly clear that all the theoretical possibilities we have talked about, have designed the experiments around, and

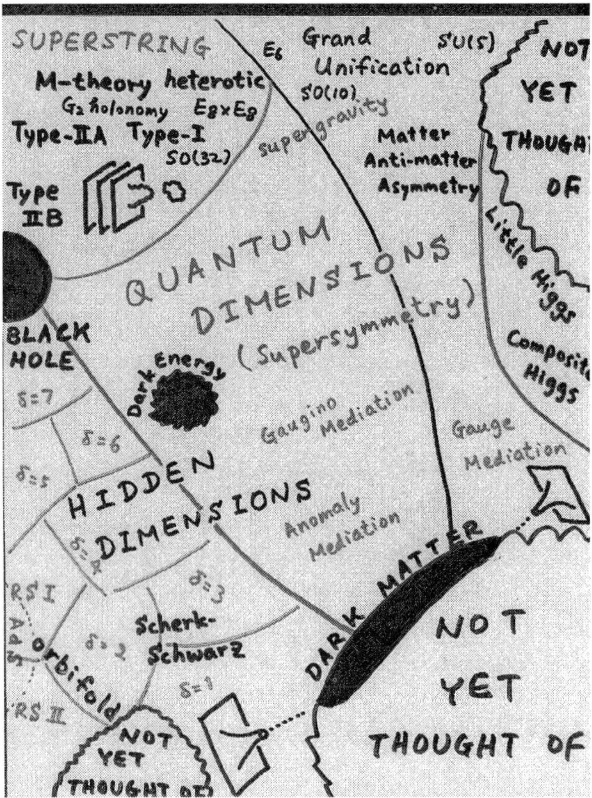

Figure 13. The landscape of theories that has many uncharted territories. The task for experiments is to zoom down to a point on this map.

ran Monte Carlo on, are only a small portion of the land of all theories. There are many islands and continents already labeled on the map, but much of the land is uncharted.

The task for the future experiments is enormous: to zoom in to a point on the map most of which is still uncharted. We need to identify the physics responsible for the Higgs BEC, and it is quite likely that we haven't thought of the right solution yet. We are all excited about the LHC, where we will discover particles and new phenomena that address this issue. Many possibilities will be ruled out. However, new interpretations will necessarily emerge. Then the race will be on. Theorists come up with new interpretations. Experimentalists exclude new interpretations. It will be a *long* period of elimination. As is always the case, the crucial information is in the *details*. We would like to elucidate the physics by reconstructing the Lagrangian of the "true theory" term by term from measurements.

In this process, the absolute confidence behind our understanding is crucial, especially when we wit-

ness a major discovery. Just for the sake of discussion, let us say that Supersymmetry happens to be the "true theory." It is relatively easy to reach, what I'd like to call, "New York Times-level confidence." We will see a headline like "*The Other Half of the World Discovered.*" But everybody in this auditorium knows that there is a long way to go from this level of confidence to the other level of confidence, which I'd like to call "Halliday–Resnik-level confidence." It will take an incredible level of confidence to put a paragraph like this one in the freshman physics textbook:

> *We have learned that all particles we observe have unique partners of different spin and statistics, called superpartners, that make our theory of elementary particles valid to small distances.*

Upon seeing this slide, one of my colleagues in Berkeley was impressed by the fact that Halliday and Resnik can turn something as exciting as the discovery of Supersymmetry into something this dry and dull. Well, *that* wasn't my point. My point is that we need to go through many detailed, precise, unambiguous measurements for us to reach this level of confidence.

Again for the sake of discussion, let us say that hidden dimensions happen to be the "true theory." We will see events where the high-energy collisions on our three-dimensional sheet will produce some particles we can see and other particles that disappear into the extra dimensions, such as the graviton (Fig. 14, top). Then we find that the energy and momentum are not balanced apparently. There is clearly something exciting going on. However, such a discovery wouldn't establish the theory. We'd like to know how many of such extra dimensions there are, for instance. One way to address this question is to measure the rates of this kind of events at two different energies at an e^+e^- Linear Collider. The energy dependence of the rates can tell us the number of extra dimensions (Fig. 14, bottom).

Let us pick Supersymmetry again. In this case, the Tevatron and/or LHC will expand our sensitivity in the parameter space greatly beyond where we are, and many precise measurements will be performed at the LHC (Fig. 15). However we will still like to know if the new particles truly have the same quantum numbers as the particle we already know, with their

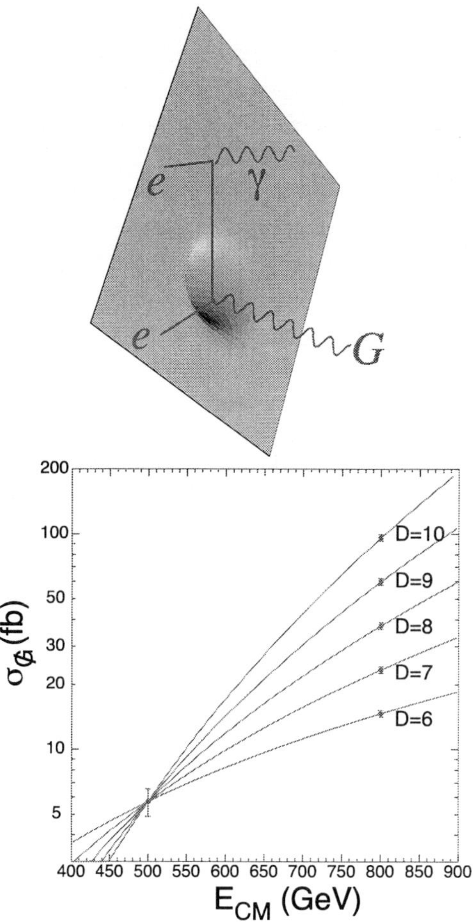

Figure 14. Top: Emission of a graviton into the hidden dimensions. Bottom: The energy dependence of the rates for various number of dimensions.[7]

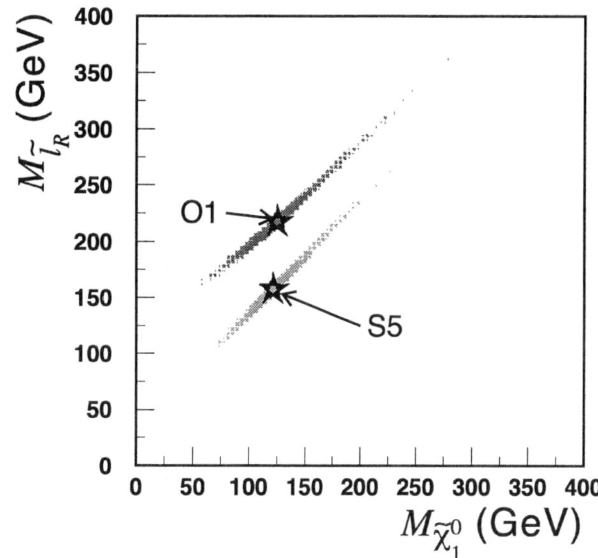

Figure 15. Precision measurement of superparticle masses at the LHC.[8]

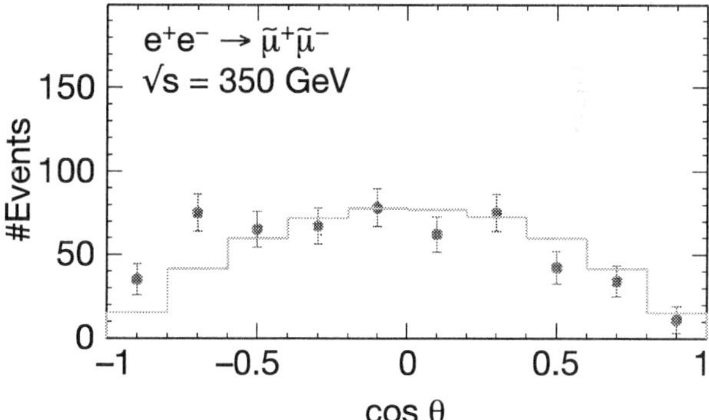

Figure 16. Test if the smuon has spin zero at the LC.[9] The spin one case would show an upside down angular distribution.

spins differing by 1/2. Again the Linear Collider can determine the quantum numbers and spins, and if they have the correct couplings, etc. (Fig. 16). This way, we will establish Supersymmetry with absolute confidence.

3. Heaven

I now turn to the questions from the Heaven. One of the major results this year reported at this Symposium is the study of cosmic microwave background anisotropies by the WMAP satellite. From a global fit, they have reported precise measurements of important cosmological parameters that include[12]

$$h = 0.71 \pm 0.04,$$
$$\Omega_M h^2 = 0.135 \pm 0.009,$$
$$\Omega_b h^2 = 0.0224 \pm 0.0009,$$

$$\Omega_{total} = 1.02 \pm 0.02.$$

This is yet another big step in precision cosmology. To me, the most important information is that the case for non-baryonic dark matter is now as strong as 12σ: $|\Omega_M - \Omega_b|h^2 = 0.113 \pm 0.009$.

People have looked for dim stars or big planets that make up Dark Matter in the halo of our galaxy, dubbed MACHOs (Massive Compact Halo Objects). The search resulted in a strong upper limit on the halo fraction of such astronomical objects.[13] Instead, we are led to WIMPs (Weakly Interacting Massive Particles). They are stable heavy particles produced in the early Universe when the temperature was as high as their mass. As the universe cooled, the temperature was so low that they were no longer created. They started to annihilated with each other, but as

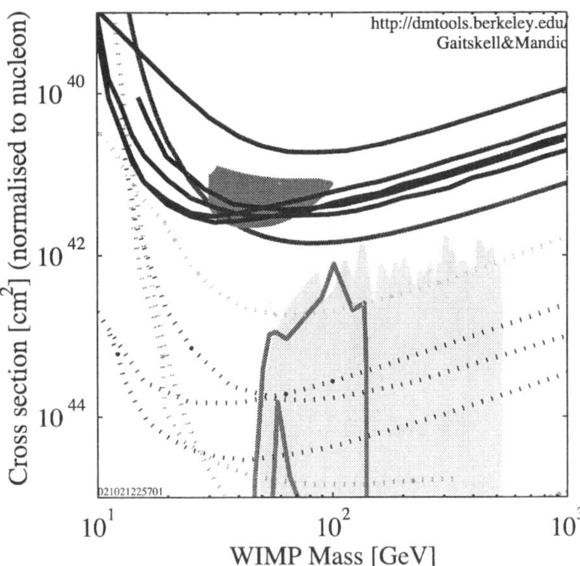
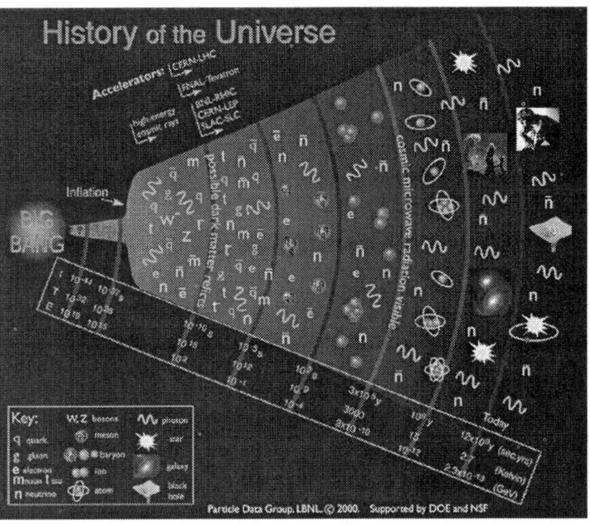

Figure 17. The next-generation Dark Matter search experiments will get into the interesting portion of the WIMP parameter space.[10]

Figure 18. Schematic history of Universe. We have a pretty good grip on physics back to about a second after the Big Bang thanks to CMB and nucleosynthesis. The quark-gluon plasma is physics back at 10^{-5} sec. An agreement between accelerator-based data on Dark Matter and cosmological data would provide understanding much closer to The Beginning, back to 10^{-12} sec after the Big Bang.

the universe expands, they saw fewer and fewer of each other and beyond some point they could no longer find each other to annihilate. In this way, they are left-overs from the near complete annihilation. The amount of energy density left over is[11]

$$\Omega_M = \frac{0.756(n+1)x_f^{n+1}}{g^{1/2}\sigma_{ann}M_{Pl}^3}\frac{3s_0}{8\pi H_0^2} \approx \frac{\alpha^2/(\mathrm{TeV})^2}{\sigma_{ann}}. \quad (4)$$

It is very interesting that weakly coupled (as weak as α) particles at the TeV scale can provide the correct energy density to explain the Dark Matter.

A stable, weakly-coupled particle would be an excellent candidate for Dark Matter. Actually, it should be very weakly coupled because ordinary neutrinos would be too strongly coupled and are excluded by the negative search results. There are no such candidate particles in the Standard Model. The candidate most talked about is the Lightest Supersymmetric Particle (LSP), which is the superpartner of the photon or Z in most models. Indeed, the direct search experiments so far have made only a small foray, but the next generation experiments will take a significant bite out of the interesting part of the parameter space (Fig. 17). This way, we will know that Dark Matter is indeed there floating in the halo of our galaxy. On the other hand, we would also like to know what it is. For this purpose, we'd like to produce ample quantities of Dark Matter in the laboratory to study its properties in detail.

I have argued that the Dark Matter is likely be a TeV-scale electrically neutral weakly interacting particle. There are many such candidates: Lightest Supersymmetric Particle, Lightest Kaluza–Klein particle in universal extra dimension, etc. Given that I expect new particles at the TeV-scale to address the "Hell" problems, it is quite conceivable that one of those particles is stable (or long-lived enough) to be the Dark Matter. If so, it will be accessible at accelerators, such as the LHC and LC. Precision measurements of its mass and couplings to other particles at LHC and LC will allow us to calculate its cosmic abundance. If that calculation based on accelerator experiments turns out to agree with the cosmological observations, it would be a major triumph of modern physics. We will understand the universe all the way back to when it was only about 10^{-12} sec old after Big Bang (Fig. 18)!

The Dark Energy is even more mysterious and we should be ready for more surprises. One big question is why we seem to see nearly equal amounts of Dark Energy and Dark Matter now. This is the notorious "Why Now?" problem. We seem to live at a very special moment in the evolution of universe. It almost feels like we are stepping back from the heliocentric view of Copernicus to the geocentric world

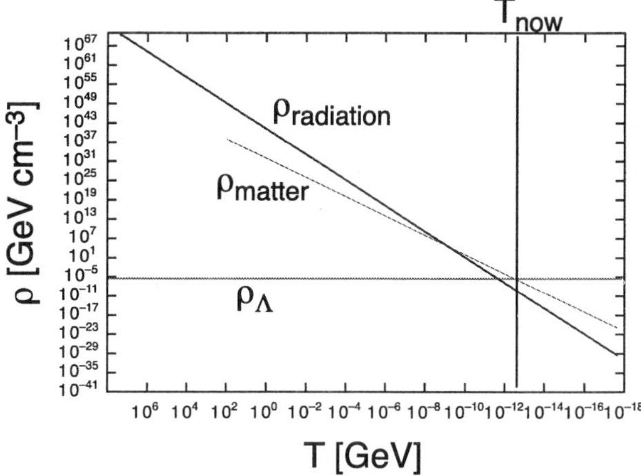

Figure 19. The triple coincidence of three energy densities.[14]

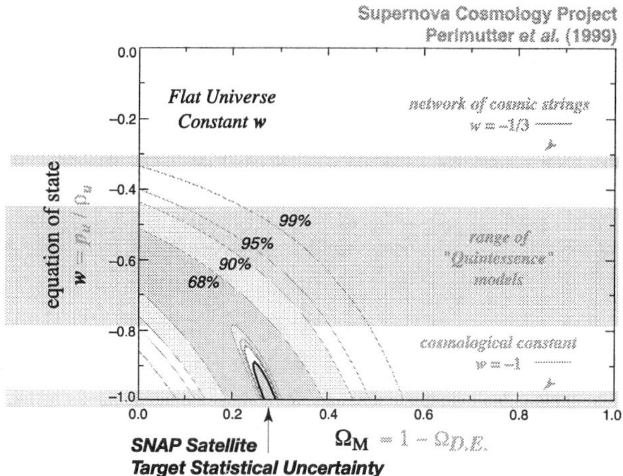

Figure 20. The projective accuracy of the equation of state of the Dark Energy by SNAP.[16]

of Ptolemy. We physicists all hate the idea that we are special.

Given that the problem is so big, it is useful to step back a little bit and look at the situation globally. Then we find that it is not just Dark Matter and Dark Energy; the "radiation," which basically refers to CMB photons and neutrinos, has a similar energy density as well (Fig. 19). It is actually a triple coincidence problem. We have three lines with different slopes that meet at a single point. Leaving $O(1)$ numerical constants aside, the radiation energy density is $\rho_{\rm rad} \sim T^4$, while the Dark Matter energy density is $\rho_M \sim m^2 T^3 / M_{Pl}$, where $m \sim 1$ TeV gives the correct amount as we have seen earlier. In order for the Dark Energy to meet with both of them, we need the Dark Energy density to be $\rho_\Lambda \sim ({\rm TeV}^2/M_{Pl})^4$. Indeed, the observation suggests $\rho_\Lambda \approx (2\ {\rm meV})^4$, while ${\rm TeV}^2/M_{Pl} \approx 0.5$ meV, tantalizingly close. It looks like figuring out TeV-scale physics is crucial for the Dark Energy problem, too.

The parameter we would most like to measure about the Dark Energy is its equation of state. Marc Kamionkowski once told me that the "equation of state" is a misnomer. It is not an equation, but rather a ratio of the pressure to the energy density $w = p/\rho$. Due to some reason, it is called the *equation* of state, but it is just a number. In any case, the cosmological constant corresponds to $w = -1$, while an evolving dynamical system typically has $w > -1$. A dedicated high-statistics study of high-redshift supernovae, complemented by the study of nearby ones to pin down the systematic issues would be extremely useful: such as SuperNova Acceleration

Probe (SNAP) using a dedicated satellite. It will determine the "equation of state" at a high accuracy. My favorite candidate for Dark Energy, a frustrated network of domain walls[15] that leads to $w = -2/3$, will be cleanly distinguished from the cosmological constant once SNAP happens (Fig. 20). Once we know the equation of state, we will get the first glimpse of the nature of the Dark Energy. Where to go from there will depend on what we find.

4. Vertical

Now on to the Vertical Questions.

Einstein once asked a very simple question. Is there an underlying simplicity behind the vast phenomena in Nature? He dreamed of finding a *unified* description of all phenomena we see. But he failed to find a unified theory of electromagnetism and his theory of gravity, general relativity.

Indeed, trying to come up with a more universal, more fundamental, more unified theory is in the blood of all of us physicists. An early example of unification is Sir Newton: he unified apples and planets. It was a revolutionary thought: the same law of physics applied to both terrestrial bodies, like an apple, and celestial bodies, like planets. Out of this unification came two important theories, Newton's law of mechanics, and the inverse-square law of gravity. A more familiar example is Maxwell, who unified electric and magnetic forces. At the time of Einstein, there were also strange phenomena in atomic physics that led to quantum mechanics. In addition,

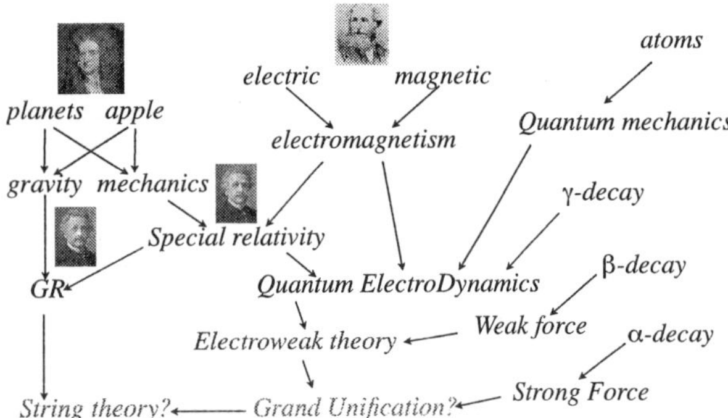

Figure 21. A brief history of unification in physics.

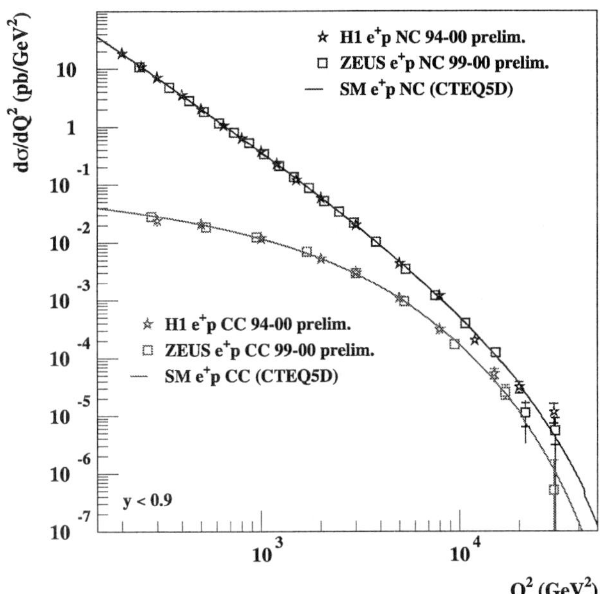

Figure 22. We are heading towards unification of the electromagnetism and the weak force.[17]

there were even more mysterious phenomena in nuclear physics, such as α-decay, β-decay, and γ-decay.

Later, there was an important unification in physics which somehow people don't talk about much. It is Quantum ElectroDynamics (QED), that unifies special relativity, electromagnetism, quantum mechanics, and some other phenomena such as nuclear γ-decay. It is an incredibly successful theory that predicts the magnetic moment of the electron down to its twelfth digit. It is equally incredible that experiments can measure it down to its twelfth digit, and all twelve digits agree with each other. This is a great triumph for the general idea of unification in physics.

The other phenomena led to discoveries of new forces. The nuclear β-decay was the first manifestation of the weak force, while the α-decay was that of the strong force. We are now just about to achieve the next layer of unification, between QED and the weak force. Beyond that, we are still at the stage of speculation. The strong force may be further unified with the electroweak forces into a single force; it is called the grand unification. We also would like to see gravity unified with the other forces. Currently the best bet is string theory.

We are indeed just about to achieve the next layer of unification. Figure 22 shows the strengths of electromagnetic and weak forces as a function of the energy scale. The first manifestation of the weak force, nuclear β-decay, was measured at much lower energies, off the scale in Fig. 22, where the strengths of the two forces were many orders of magnitude different. However our predecessors figured out that they are supposed to be of the same kind; an amazing insight. After many decades we inched up in

energy, and are finally approaching the energy scale where they indeed become the same. It has been a long-term goal since the 1960's and we are getting there! However the important missing link is the Higgs boson as we talked about already.

Beyond the unification of electromagnetism and the weak force at the TeV-scale, how do we gain any information about the next layer, the grand unification? We have all seen during the Symposium that the strengths (gauge couplings) of the three forces, $SU(3)$, $SU(2)$, and $U(1)$, appear to become equal at a very high-energy scale 10^{16} GeV, if the Standard Model is Supersymmetric. The energy scale appears so remote that we may not gain any further information. However, if Supersymmetry is discovered, and the masses of the new particles are measured to a high precision by combining the data from the LHC and the LC, we will have a quantitative test of grand unification. The superpartners of the gauge bosons, gauginos, should have masses that unify at the same energy scale where the gauge couplings unify. This is a highly non-trivial test of whether forces unify. If this happens, we would definitely want to see proton decay! This is a wonderful example of how, once the hierarchy problem is solved, the solution itself will provide new probes to physics that directly address the big questions.

Once the hierarchy problem is solved, we can

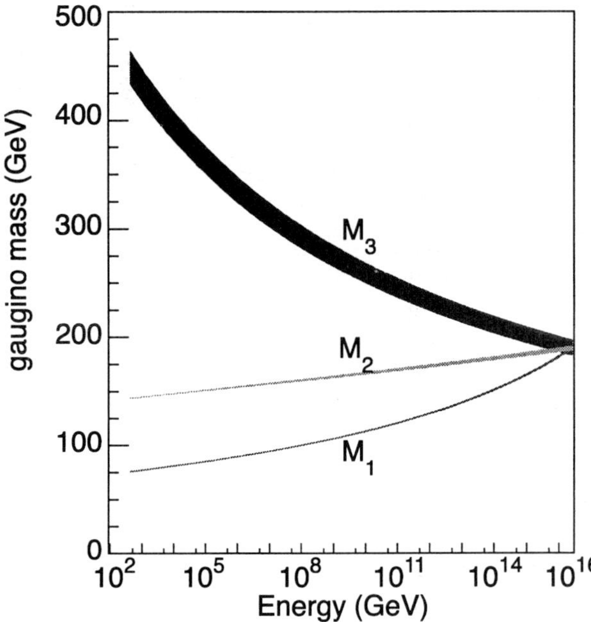

Figure 23. If Supersymmetry is found, it will provide a further probe to shorter distance scale physics, such as testing grand unification using the gaugino masses.[18]

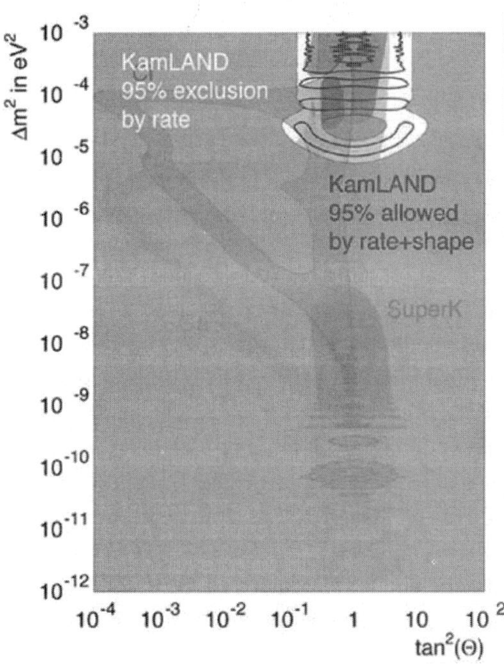

Figure 24. Solar neutrino data and reactor data converged on the Large Mixing Angle solution.[24]

systematically look for effects from physics at high energies. They can be parameterized as effective operators added to the Standard Model,

$$\mathcal{L} = \mathcal{L}_{\mathrm{SM}} + \frac{1}{\Lambda}\mathcal{L}_5 + \frac{1}{\Lambda^2}\mathcal{L}_6 + \cdots \quad (5)$$

where Λ is the high energy scale of new physics. The effects in $\mathcal{L}_5$ are suppressed by a single power of the high energy scale, $\mathcal{L}_6$ by two powers, etc.. What terms there can be have been classified systematically by Weinberg, and there are many terms suppressed by two powers:

$$\mathcal{L}_6 \supset QQQL,\ \bar{L}\sigma^{\mu\nu}W_{\mu\nu}He,\ W_\nu^\mu W_\lambda^\nu B_\mu^\lambda,$$
$$\bar{s}d\bar{s}d,\ (H^\dagger D_\mu H)(H^\dagger D^\mu H),\cdots. \quad (6)$$

The examples here contribute to proton decay, $g-2$, the anomalous triple gauge boson vertex, $K^0\text{–}\overline{K}^0$ mixing, and the ρ-parameter, respectively. It is interesting that there is only one operator suppressed by a single power:

$$\mathcal{L}_5 = (LH)(LH). \quad (7)$$

After substituting the expectation value of the Higgs, the Lagrangian becomes

$$\mathcal{L} = \frac{1}{\Lambda}(LH)(LH) \to \frac{1}{\Lambda}(L\langle H\rangle)(L\langle H\rangle) = m_\nu\nu\nu, \quad (8)$$

nothing but the neutrino mass.

Therefore the neutrino mass plays a very unique role. It is the lowest-order effect of physics at short distances. This is a very tiny effect. Any kinematical effects of the neutrino mass are suppressed by $(m_\nu/E_\nu)^2$, and for $m_\nu \sim 1$ eV which we now know is already too large and $E_\nu \sim 1$ GeV for typical accelerator-based neutrino experiments, it is as small as $(m_\nu/E_\nu)^2 \sim 10^{-18}$. At first sight, there is no hope to probe such a small number. However, any physicist knows that interferometry is a sensitive method to probe extremely tiny effects. For interferometry to work, we need a coherent source. Fortunately there are many coherent sources of neutrinos in Nature: the Sun, cosmic rays, reactors (not quite Nature), etc.. We also need interference for an interferometer to work. Fortunately, there are large mixing angles that make the interference possible. We also need long baselines to enhance the tiny effects. Again fortunately there are many long baselines available, such as the size of the Sun, the size of the Earth, etc.. Nature was very kind to provide all the necessary conditions for interferometry to us! Neutrino interferometry, a.k.a. neutrino oscillation, is a unique tool to study physics at very high energy scales. Indeed, the naïve interpretation of the neutrino oscillation results we heard about during this conference suggests $\Lambda \sim 10^{15}$ GeV! This gives an important look at the physics of grand unification.

5. Horizontal

The Horizontal Questions are about the flavor. As we witnessed during this conference, this is a historic era in flavor physics. In the lepton sector, Cowan and Reines detected neutrinos from a nuclear power reactor back in 1956, but we hadn't learned much about the nature of neutrinos for decades. In 1998, SuperKamiokande announced the discovery of oscillation in atmospheric neutrinos.[19] In 2002, SNO established the flavor conversion in solar neutrinos.[20] Later the same year, KamLAND decided the solution to the solar neutrino problem (Fig. 24).[21]

In the quark sector, the progress is similarly spectacular. Back in 1964, Fitch and Cronin discovered indirect CP-violation in K^0–$\overline{K}^0$ mixing. Again we didn't learn much beyond it for decades. Then in 1998, CPLEAR established T-violation in the same system.[22] In 1999, KTeV and NA48 established the direct CP-violation, ε'/ε.[23] In 2001, BaBar and Belle established indirect CP-violation in the B_d system, the first CP-violation in a system other than kaons.[25] These results combined to confirm the Kobayashi–Maskawa theory of CP-violation (Fig. 25).

But there are even more questions to be answered. The main question is this: *What distinguishes different generations?* Three generation of particles have the same quantum numbers, but they look very different. They have masses hierarchically different by many orders of magnitude. They mix rather little. Both hierarchical masses and small mixings go against our "common sense" in quantum mechanics. If two states share exactly the same set of quantum numbers, you expect them to have similar energy levels, and you expect them to mix a lot. The lack of both suggests that there is an ordered structure behind the flavor.

Many theorists including myself think that there are probably hidden flavor quantum numbers that distinguish different generations. A quantum number means a new symmetry according to Noether's theorem: a flavor symmetry. This new symmetry must allow the top quark Yukawa coupling because

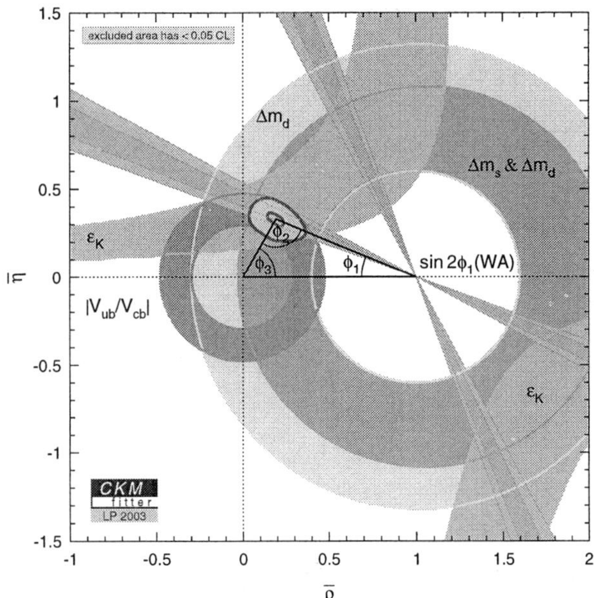

Figure 25. Consistency of various CKM parameter measurements shows the success of Kobayashi–Maskawa theory.[26]

it is $O(1)$. On the other hand, it forbids all the other Yukawa couplings so that all other quarks start out massless. But this symmetry must be only approximate, and is broken a little. This small symmetry breaking allows other Yukawa couplings, generating small and hierarchical Yukawas. Because different generations have different quantum numbers, they are not allowed to mix. Again the small symmetry breaking allows them to mix by small amounts.

Here is a toy model of a simple $U(1)$ flavor symmetry.[27] I basically introduce a new charge to all particles. This symmetry is broken by a small parameter $\langle \epsilon \rangle \sim 0.02$ of charge -1. Let me assign charges to quarks and leptons in the following way,

$$\mathbf{10}(Q, u_R, e_R)(+2, +1, 0) \tag{9}$$

$$\mathbf{5}^*(L, d_R)(+1, +1, +1). \tag{10}$$

Here, I used grand-unified terminology of decuplet and quintet, and the three charges refer to the first, second, and third generation, respectively. This charge assignment keeps the top quarks, both left- and right-handed, neutral, and the top quark Yukawa coupling is allowed, while all other entries of the Yukawa matrices are forbidden. However, using

ϵ, we can fill in other entries as well. We find

$$M_u \sim \begin{pmatrix} \epsilon^4 & \epsilon^3 & \epsilon^2 \\ \epsilon^3 & \epsilon^2 & \epsilon \\ \epsilon^2 & \epsilon & 1 \end{pmatrix}, \qquad (11)$$

$$M_d \sim \begin{pmatrix} \epsilon^3 & \epsilon^3 & \epsilon^3 \\ \epsilon^2 & \epsilon^2 & \epsilon^2 \\ \epsilon & \epsilon & \epsilon \end{pmatrix}, \qquad (12)$$

$$M_l \sim \begin{pmatrix} \epsilon^3 & \epsilon^2 & \epsilon \\ \epsilon^3 & \epsilon^2 & \epsilon \\ \epsilon^3 & \epsilon^2 & \epsilon \end{pmatrix}. \qquad (13)$$

It is easy to find the hierarchical mass eigenvalues,

$$m_u : m_c : m_t \sim m_d^2 : m_s^2 : m_b^2$$
$$\sim m_e^2 : m_\mu^2 : m_\tau^2 \sim \epsilon^4 : \epsilon^2 : 1, \qquad (14)$$

which works pretty well especially given how simple the toy model is. The mixing angles are also suppressed by powers in ϵ, and they all work out within a factor of 5 or so.

It is exciting that new data from neutrinos are already providing significant new information about flavor symmetries. As you know, neutrino data has been full of surprises. All mixing angles are large, except for U_{e3} which has not been measured. In particular, the atmospheric neutrino mixing appears maximal. Two mass-squared splittings are not very different, $\Delta m_{solar}^2 / \Delta m_{atmospheric}^2 \sim 1/30$. The hierarchy in masses rather than (masses)^2 is the square root of this, $\sqrt{1/30} = 0.2$. This isn't much of a hierarchy. Now we can ask the question of whether there is a symmetry or structure behind the neutrinos.

As far as we can tell, we don't need any symmetry or structure behind neutrinos, unlike the quark and charged leptons. If you run a Monte Carlo of random complex three-by-three matrices with the seesaw mechanism, you find that the maximal mixing is the most likely outcome if plotted against $\sin^2 2\theta$, and the peak in $\Delta m_{solar}^2 / \Delta m_{atmospheric}^2$ is about 1/10. Apparently no particular structure in the neutrino mass matrix is needed. I called this observation "anarchy" in neutrinos. Actually, the charge assignments I discussed earlier did not distinguish the three generation of neutrinos at all; and we do expect anarchy that is consistent with the current data.

Of course there are other proposals to understand the neutrino masses and mixings together with quarks and charged leptons. Table 2 shows a list of proposed models of flavor symmetries as of October

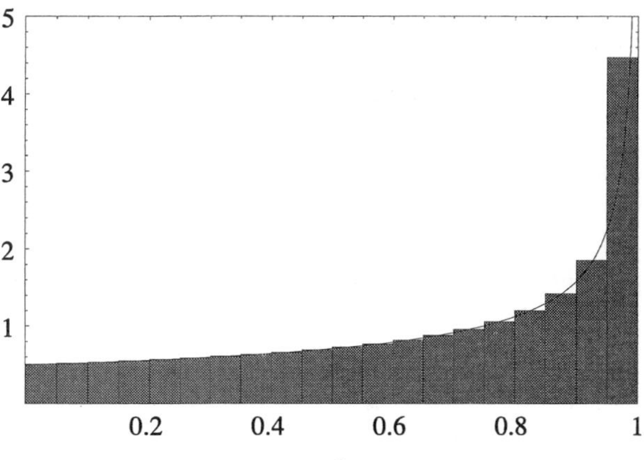

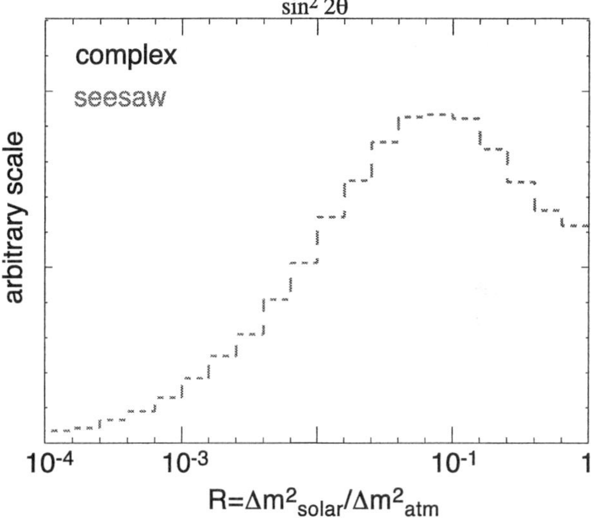

Figure 26. Random three-by-three matrices show distributions quite consistent with the observed patterns of neutrino masses and mixings.[27] Top: $\sin^2 2\theta_{23}$. Bottom: $\Delta m_{solar}^2 / \Delta m_{atmospheric}^2$.

2002. By December, KamLAND excluded the third and fourth flavor symmetries because they predicted other solutions to the solar neutrino problem. The survivors will be put to further tests soon. They predict different orders of magnitude for θ_{13}, $O(1)$, $O(\lambda)$, or $O(\lambda^2)$. If a more precise measurement of θ_{13} turns out to give $\sin^2 2\theta_{23} = 1.00 \pm 0.01$, we would probably want a reason why it is *so* maximal, implying a new symmetry in the neutrino sector.

We'd like to push this program further to narrow down the choice of flavor symmetries. We basically need more and more flavor parameters. In fact, any TeV-scale physics would have a new flavor structure that affects flavor physics significantly. Let me take Supersymmetry as an example. Squarks and sleptons come with their own mass matrices, in addition

Table 2. Compilation of different flavor symmetry models and their predictions for neutrino masses and mixings as of Oct. 2002.[28] The third and fourth rows were excluded by Dec. 2002. The next benchmark is U_{e3}.

Model	parameters	d_{23}	$\Delta m_{12}^2/|\Delta m_{23}^2|$	U_{e3}	$\tan^2\theta_{12}$	$\tan^2\theta_{23}$
A	$b=0$	$O(1)$	$O(1)$	$O(1)$	$O(1)$	$O(1)$
SA	$b=1$	$O(1)$	$O(d_{23}^2)$	$O(\lambda)$	$O(\lambda^2/d_{23}^2)$	$O(1)$
H_{II}	$a=1, b=2$	$O(\lambda^2)$	$O(\lambda^4)$	$O(\lambda^2)$	$O(1)$	$O(1)$
H_{I}	$a=1, b=2$	0	$O(\lambda^6)$	$O(\lambda^2)$	$O(1)$	$O(1)$
IH		$O(\lambda^4)$	$O(\lambda^2)$	$O(\lambda^2)$	$1+O(\lambda^2)$	$O(1)$

$$\begin{pmatrix} \tilde{s}_R \\ \tilde{s}_R \\ \tilde{s}_R \\ \tilde{\nu}_\mu \\ \tilde{\mu} \end{pmatrix} \leftrightarrow \begin{pmatrix} \tilde{b}_R \\ \tilde{b}_R \\ \tilde{b}_R \\ \tilde{\nu}_\tau \\ \tilde{\tau} \end{pmatrix}$$

Figure 27. The large $\nu_\mu \to \nu_\tau$ mixing suggests a large mixing of the whole $SU(5)$ multiplets and also of their superpartners.

to quark and lepton mass matrices. Off-diagonal elements in squark/slepton mass matrices violate flavor. Therefore, a flavor symmetry that distinguishes different generations will automatically suppress the off-diagonal elements. If we can probe such small flavor violations in Supersymmetric loop diagrams, we would like to identify patterns in them, and eventually deduce the required symmetry behind them. Basically, we try to repeat what Gell-Mann and Okubo did in baryon and meson masses to identify the symmetry behind masses and mixings.

Different models differ in flavor quantum number assignments. Different quantum number assignments lead to different consequences for θ_{13}, the matter effect, CP-violation, B-physics, K-physics, Lepton Flavor Violation, proton decay, and practically anything we can imagine that involves flavor. This way, we hope to identify the underlying flavor symmetry. I admit this is a long shot. We even don't know the energy scale of the physics of flavor. It may turn out to be too remote to access directly in experiments. But I'd like to argue that this is

not necessarily bad. In archaeology, you don't reproduce the events in the laboratory. But once you have enough circumstantial evidence of fossils, artifacts, geological records, etc., that are consistent with a reasonable hypothesis, you eventually believe it. It may not be a formal proof at the level particle physicists are accustomed to, but it is nonetheless the next best thing. A good example is the cosmic microwave background. It is a wonderfully colorful, sexy fossil, and we can extract so much information out of it. We don't recreate the Big Bang, but we have already learned so much and we will learn even more from the CMB.

I'd like to emphasize that this program will be a collaboration of energy-frontier experiments and low-energy flavor experiments. We need to know the TeV-scale physics so that we know what runs inside the loops. We need to know their masses. Then the flavor data will allow us to extract flavor violations among the particles in the loops.

Here is one specific example we should pursue.[29] We'd like to know if quarks and leptons have a common origin of flavor. As already mentioned, one striking discovery was that the ν_μ and ν_τ mix a *lot*, maybe even maximally. Suppose you make it grand-unified. s_R lives in the same multiplet as ν_μ, and b_R with ν_τ. You'd expect a large mixing between s_R and b_R, too (Fig. 27). But mixing among right-handed quarks completely drops out from the CKM phenomenology because there is no right-handed charged current (as far as we know). It looks like we can't probe this question. On the other hand, if there is Supersymmetry, a large mixing between $\tilde{s}_R$ and $\tilde{b}_R$ is physical, and can induce $O(1)$ effects in $b \to s$ transitions through loop diagrams (Fig. 28, top and center). Especially in leptogenesis that relies on CP-violation in the neutrino sector, we expect CP-violation in $\tilde{s}_R$–$\tilde{b}_R$ mixing that may show up in B-physics.

For example, we may see CP-violation in B_s mixing that can be studied in $B_s \to J/\psi + \phi$. The rates in $B_d \to X_s \ell^+ \ell^-$ may differ from the Standard Model and CP-violation may be seen. CP-violation in $B_d \to \phi + K_S$ may be different from that in $J/\psi + K_S$ within all the other constraints, such as $b \to s\gamma$ (Fig. 28, bottom). The current situation for ϕK_S is somewhat confusing, with BaBar and Belle not consistent with each other.[26] If they will settle in the middle, however, that may be the first indica-

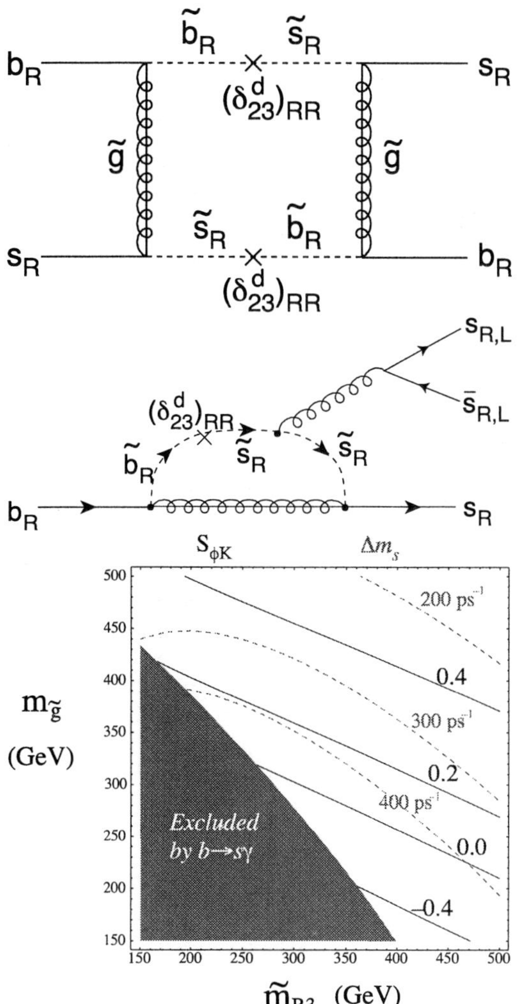

Figure 28. The impact of large $\tilde{s}_R$–$\tilde{b}_R$ mixing on B-physics. Top: possible contribution to the B_s mixing. Center: possible contribution to the $B_d \to \phi K_S$ decay. Bottom: $S_{\phi K}$ in solid lines, Δm_s in dotted lines, and the constraint from $b \to s\gamma$ in shaded region.[30]

tion of the common origin of flavor between quarks and leptons! I'm very much looking forward to more data.

After going through this program, suppose we identify a reasonable flavor symmetry that explains all data. Then we will be greedy enough to want to know what physics is behind the flavor symmetry. In the case of Gell-Mann–Okubo, once the $SU(3)$ flavor symmetry was identified, the next step was QCD. Clearly, we have to be very very lucky to get to that level. Nonetheless, it is useful to remember that the next level will crucially depend on what we find at the TeV-scale. For example, if the TeV-scale turns

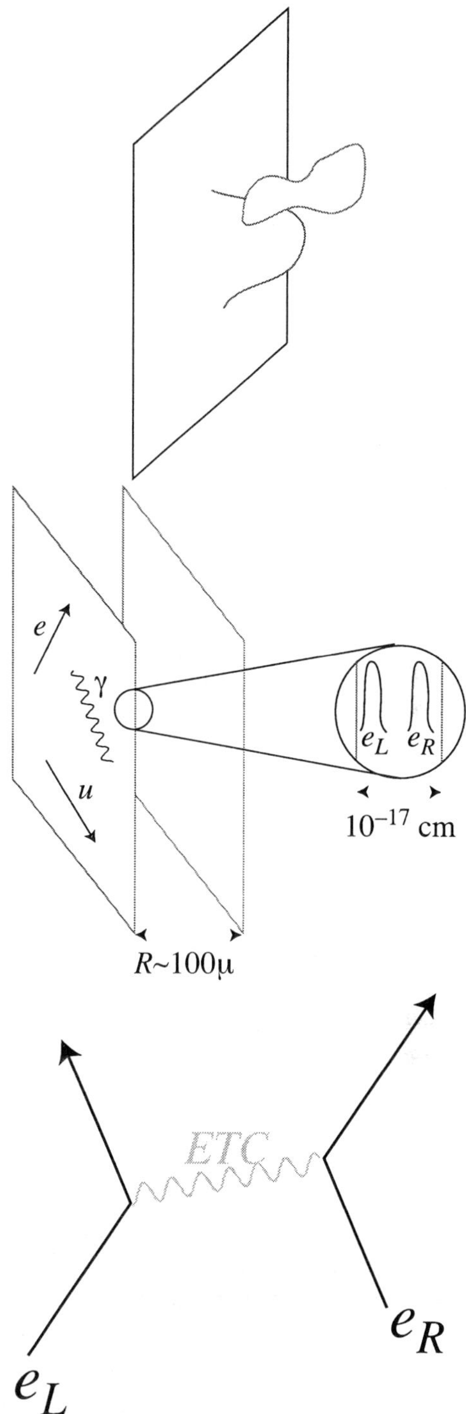

Figure 29. Different views on the origin of flavor symmetry depending on the outcome of the TeV-scale physics. Top: string origin in Supersymmetric models. Center: physical dislocation of different generations within a fat brane in models with hidden dimensions. Bottom: exchange of new massive gauge bosons at 100 TeV scale in technicolor models.

out to be Supersymmetry, the flavor symmetry may be a consequence of the anomalous $U(1)$ gauge group with the Green–Schwarz mechanism from string theory (Fig. 29, top).[31] If it is hidden dimensions, it may be that the three-dimensional brane we live on is fat enough so that different generations are physically dislocated within the brane, providing an effective flavor symmetry (Fig. 29, center).[32] If it is technicolor, it may be due to a new broken gauge interaction at the 100 TeV scale (Fig. 29, bottom).[33] I certainly can't see far enough to know how things will play out.

6. Conclusion

What I'm looking forward to seeing in the next twenty years is a synergy of many different approaches in particle physics. The big questions I've listed at the beginning of my talk are all very ambitious questions. They are elusive. There is no guarantee that we can answer them.

But we know what the main obstacle is. It is the cloud of the TeV-scale that is preventing us from obtaining clear views. And we are getting there. We have to make sure that there are many different approaches diverse enough to determine what is going on at the TeV-scale. They will converge to reveal the big picture. Even though what I'm saying is ambitious, it is *conceivable*. And this idea of synergy applies to any scenario of TeV-scale physics, as far as I can tell.

Given all this discussion, the outlook for the next twenty years is:

Bright!

References

1. http://www-visualmedia.fnal.gov/VMS_Site/gallery/stillphotos/1995/95-759D.jpg.
2. H. Murayama, "Supersymmetry," Talk given at 22nd INS International Symposium on Physics with High Energy Colliders, Tokyo, Japan, 8-10 Mar 1994. Published in Proceedings of INS Symposium, World Scientific, 1994. arXiv:hep-ph/9410285.
3. N. Arkani-Hamed, A. G. Cohen and H. Georgi, *Phys. Lett.* B **513**, 232 (2001) [arXiv:hep-ph/0105239]; N. Arkani-Hamed, A. G. Cohen, T. Gregoire and J. G. Wacker, *JHEP* **0208**, 020 (2002) [arXiv:hep-ph/0202089]; N. Arkani-Hamed, A. G. Cohen, E. Katz and A. E. Nelson, *JHEP* **0207**, 034 (2002) [arXiv:hep-ph/0206021].
4. N. S. Manton, *Nucl. Phys.* B **158**, 141 (1979). C. Csaki, C. Grojean and H. Murayama, *Phys. Rev.* D **67**, 085012 (2003) [arXiv:hep-ph/0210133].
5. C. Csaki, C. Grojean, H. Murayama, L. Pilo and J. Terning, arXiv:hep-ph/0305237. C. Csaki, C. Grojean, L. Pilo and J. Terning, arXiv:hep-ph/0308038. Y. Nomura, arXiv:hep-ph/0309189.
6. H. Murayama, arXiv:hep-ph/0307293.
7. J. A. Aguilar-Saavedra *et al.* [ECFA/DESY LC Physics Working Group Collaboration], "TESLA Technical Design Report Part III: Physics at an e+e- Linear Collider," arXiv:hep-ph/0106315.
8. H. Bachacou, I. Hinchliffe and F. E. Paige, *Phys. Rev.* D **62**, 015009 (2000) [arXiv:hep-ph/9907518].
9. T. Tsukamoto, K. Fujii, H. Murayama, M. Yamaguchi and Y. Okada, *Phys. Rev.* D **51**, 3153 (1995).
10. http://dmtools.berkeley.edu.
11. E.W. Kolb, Michael S. Turner, The Early Universe. Redwood City, USA: Addison-Wesley (1990) 547 p. (Frontiers in physics, 69).
12. D. N. Spergel *et al.*, *Astrophys. J. Suppl.* **148**, 175 (2003) [arXiv:astro-ph/0302209]. See also Licia Verde in this proceedings.
13. C. Alcock *et al.* [MACHO Collaboration], *THE ASTROPHYSICAL JOURNAL*, 499:L9-L12 (1998), arXiv:astro-ph/9803082. *ibid.*, *The Astrophysical Journal*, 550:L169-L172 (2001).
14. N. Arkani-Hamed, L. J. Hall, C. F. Kolda and H. Murayama, *Phys. Rev. Lett.* **85**, 4434 (2000) [arXiv:astro-ph/0005111].
15. A. Friedland, H. Murayama and M. Perelstein, *Phys. Rev.* D **67**, 043519 (2003) [arXiv:astro-ph/0205520].
16. http://snap.lbl.gov/target2.jpg.
17. M. Erdmann, talk presented at XX International Symposium on Lepton and Photon Interactions at High Energies, 23rd-28th July 2001, Rome Italy. Published in the proceedings, J. Lee-Franzini, (ed.), F. Bossi, (ed.) (Frascati), P. Franzini, (ed.) *Int. J. Mod. Phys.* A **17**, 2925-3549 (2002).
18. H. U. Martyn and G. A. Blair, arXiv:hep-ph/9910416.

19. Y. Fukuda *et al.* [Super-Kamiokande Collaboration], *Phys. Rev. Lett.* **81**, 1562 (1998) [arXiv:hep-ex/9807003]. See also Koichiro Nishikawa in this proceedings.

20. Q. R. Ahmad *et al.* [SNO Collaboration], *Phys. Rev. Lett.* **89**, 011301 (2002) [arXiv:nucl-ex/0204008]. See also Alain Bellerive in this proceedings.

21. K. Eguchi *et al.* [KamLAND Collaboration], *Phys. Rev. Lett.* **90**, 021802 (2003) [arXiv:hep-ex/0212021]. See also Kunio Inoue in this proceedings.

22. A. Angelopoulos *et al.* [CPLEAR Collaboration], *Phys. Lett.* B **444**, 43 (1998).

23. A. Alavi-Harati *et al.* [KTeV Collaboration], *Phys. Rev. Lett.* **83**, 22 (1999) [arXiv:hep-ex/9905060]. V. Fanti *et al.* [NA48 Collaboration], *Phys. Lett.* B **465**, 335 (1999) [arXiv:hep-ex/9909022].

24. http://hitoshi.berkeley.edu/neutrino.

25. B. Aubert *et al.* [BABAR Collaboration], *Phys. Rev. Lett.* **87**, 091801 (2001) [arXiv:hep-ex/0107013].

K. Abe *et al.* [Belle Collaboration], *Phys. Rev. Lett.* **87**, 091802 (2001) [arXiv:hep-ex/0107061].

26. Tom Browder in this proceedings.

27. N. Haba and H. Murayama, *Phys. Rev.* D **63**, 053010 (2001) [arXiv:hep-ph/0009174].

28. G. Altarelli, F. Feruglio and I. Masina, *JHEP* **0301**, 035 (2003) [arXiv:hep-ph/0210342].

29. D. Chang, A. Masiero and H. Murayama, *Phys. Rev.* D **67**, 075013 (2003) [arXiv:hep-ph/0205111].

30. R. Harnik, D. T. Larson, H. Murayama and A. Pierce, arXiv:hep-ph/0212180.

31. L. E. Ibanez and G. G. Ross, *Phys. Lett.* B **332**, 100 (1994) [arXiv:hep-ph/9403338]. P. Binetruy and P. Ramond, *Phys. Lett.* B **350**, 49 (1995) [arXiv:hep-ph/9412385].

32. N. Arkani-Hamed and M. Schmaltz, *Phys. Rev.* D **61**, 033005 (2000) [arXiv:hep-ph/9903417].

33. E. Eichten and K. D. Lane, *Phys. Lett.* B **90**, 125 (1980).

DISCUSSION

Bennie Ward (Baylor University & University of Tennessee): In your discussion of the hierarchy problem you did not mention the anthropic principle. Could you please comment?

Hitoshi Murayama: As Ed said in the previous talk, I don't see the anthropic principle as the solution to a physical question. I suppose you can't exclude it, however.

John Collins (Penn State): You said that the Standard Model breaks down at a scale of around a TeV. How do you reconcile this with the fact the renormalized Standard Model is consistent to much higher energies?

Hitoshi Murayama: It is a matter of definition what you mean by "breaks down." The Standard Model is certainly consistent as a renormalizable field theory, and can be applied to arbitrary high energies in that sense. However, we view it as a low-energy effective field theory rather than the ultimate theory of everything, and therefore it has an ultraviolet cut-off. My definition of the Standard Model breakdown is the fact that the perturbative corrections exceed the bare Higgs mass-squared parameter as the cut-off is raised beyond TeV. It is the same sense as when Landau and Lifshitz discussed the breakdown of classical electrodynamics at the classical radius of the electron.

Symposium Program

Monday 11th August 2003

OPENING REMARKS

8:30am   Welcome
         Michael Witherell, Director of Fermi National Accelerator Laboratory
8:35am   Welcome
         S. Peter Rosen, Office of Science, U. S. Department of Energy

Session 1 - COLLIDER AND ELECTROWEAK

Session Chair: Richard Keeler, Victoria

8:45am   Top Quark Measurements at the Tevatron
         Patrizia Azzi, Padova
9:20am   Electroweak Measurements at the Tevatron
         Terry Wyatt, University of Manchester
9:55am   Precision Electroweak Physics from Low to High Energies
         Paolo Gambino, CERN

Session 2 - COLLIDER SEARCHES

Session Chair: Rohini Godbole, Bangalore

11:00am   Higgs and Supersymmetry
         Michael Schmitt, Northwestern University
11:35am   Exotic Searches at Colliders
         Emmanuelle Perez, Saclay
12:10pm   Theoretical Predictions for Collider Searches
         Gian Giudice, CERN

Session 3 - QCD AND LATTICE

Session Chair: Dong-Sheng Du, Beijing, IHEP

2:15pm   QCD Theoretical Developments
         Thomas Gehrmann, Zurich
2:50pm   Hard QCD at Colliders
         Robert Hirosky, Virginia
3:25pm   Lattice Gauge Theory
         Peter Lepage, Cornell

Session 4 - RARE B AND K DECAYS

Session Chair: Robert Tschirhart, FNAL

4:30pm   Rare Kaon Decay Physics
         Augusto Ceccucci, CERN
5:15pm   Beyond the Standard Model Sensitivity in K and B Physics
         Yuval Grossman, Technion

Monday 11th August 2003

6:00pm   Reception, Physics Poster Session, Jazz by Chicago Hot Six

8:00pm   Special Session: Young Particle Physicists Town Meeting

Tuesday 12th August 2003

Session 5 - WEAK MIXING PHASES

Session Chair: Ikaros Bigi, Notre Dame

8:45am   CKM Phases (β/ϕ_1)
        Tom Browder, Hawaii

9:20am   CKM Phases ($\alpha, \gamma/\phi_2, \phi_3$)
        Hassan Jawahery, Maryland

9:55am   Mixings, Lifetimes, Spectroscopy and Production of B-quarks
        Kevin Pitts, Illinois

Session 6 - CKM MAGNITUDES AND QCD OF HEAVY HADRON DECAYS

Session Chair: Michael Danilov, ITEP

11:00am   CKM Matrix Element Magnitudes
        Klaus Schubert, Dresden

11:45am   QCD and Heavy Hadron Decays
        Gerhard Buchalla, Munich

1:00–2:00pm **Informal Breakout Session I**   Session 1 and 2 Moderator: Rohini Godbole, Bangalore

Session 3 Moderator: Kirill Melnikov, Hawaii

Session 7 - RARE DECAYS AND PENGUINS

Session Chair: Marina Artuso, Syracuse

2:15pm   Rare Hadronic B Decays
        John Fry, University of Liverpool

3:05pm   Radiative and Electroweak Rare B Decays
        Mikihiko Nakao, KEK

Session 8 - CHARM AND QUARKONIUM PHYSICS

Session Chair: Steve Olsen, Hawaii

4:30pm   Experimental Limits on New Physics from Charm Decay
        Bruce Yabsley, Virginia Tech

5:05pm   Heavy Quarkonium, Production and Spectroscopy
        Tomasz Skwarnicki, Syracuse

5:40pm   Lessons from Standard Charm Decays
        Jussara de Miranda, CBPF

Wednesday 13th August 2003   Free Day

6:30pm   Reception, Crystal Garden, Navy Pier

7:30pm   Banquet, Crystal Garden, Navy Pier

Thursday 14th August 2003

Session 9 - ASTROPARTICLE PHYSICS

Session Chair: Roberto Peccei, UCLA

8:45am   Cosmic Microwave Background Experiments
         Licia Verde, Princeton
9:20am   Dark Matter and Energy
         Robert Kirshner, Harvard
9:55am   Astroparticle Theory
         Esteban Roulet, Bariloche

Session 10 - NEUTRINO PHYSICS I

Session Chair: Stanley Wojcicki, Stanford

11:00am   Dark Matter Experiments
         Maryvonne de Jesus, Lyon
11:35am   Double-Beta Decay and Tritium Decay Experiments
         Giorgio Gratta, Stanford
12:10pm   Results and Status of Current Accelerator Neutrino Experiments
         Koichiro Nishikawa, Kyoto

1:00–2:00pm **Informal Breakout Session II**   Session 4, 5, 6, 7 and 8 Moderator: Fred Gilman, CMU
Session 9 Moderator: Roberto Peccei, UCLA

Session 11 - NEUTRINO PHYSICS II

Session Chair: William Louis, LANL

2:15pm   Reactor Neutrino Experiments
         Kunio Inoue, Tohoku
3:00pm   Solar Neutrino Experiments
         Alain Bellerive, Carleton

Session 12 - NEUTRINO PHYSICS III

Session Chair: Enrique Fernandez, Barcelona

4:30pm   Neutrino Physics: Open Theoretical Questions
         Alexei Smirnov, Moscow and ICTP
5:15pm   Future Experiments with Neutrino Superbeams, Betabeams and Factories
         Deborah Harris, FNAL

Special Session - EXTREME COMPUTING: THE DATA GRID AND THE FUTURE OF DISTRIBUTED COMPUTING

Session Chairs: Vicky White and Ruth Pordes, FNAL

7:30pm   Introduction and Overview of the Grid: Current Status of U. S. Grid R&D and Deployment
         Ian Foster, Argonne and Chicago
8:00pm   Grids in Europe and the LCG Project
         Ian Bird, CERN, LCG
8:30pm   Network Research Infrastructures: Back to the Future
         Bob Aiken, Cisco

Thursday 14th August 2003

8:50pm	Break
9:05pm	All Grid, All the Time
	Stephen Perrenod, SUN
9:25pm	Grid Services and Web Services: Harmonic Convergence
	David Martin, IBM
9:45pm	TeraGrid and High-End Computing: Lessons and Futures
	Daniel Reed, UIUC, NCSA, TeraGrid

Friday 15th August 2003

Session 13 - HADRON STRUCTURE I

Session Chair: Albrecht Wagner, DESY

8:45am	Deep Inelastic Scattering
	Paul Newman, University of Birmingham
9:20am	Parton Distribution Function
	Robert Thorne, Cambridge University
9:55am	Measurements with Polarized Hadrons
	Toshi-Aki Shibata, Tokyp Inst. Tech.

Session 14 - HADRON STRUCTURE II AND DETECTOR R&D

Session Chair: Halina Abramowicz, Tel Aviv

11:00am	Diffraction and Vector Meson Production
	Yuji Yamazaki, KEK
11:35am	Heavy-Ion Collisions
	David Hardtke, LBNL
12:10pm	Detector R&D
	Ties Behnke, DESY

Laboratory Poster Session/Fermilab Tours

1:30–4:30pm	Fermilab Tours
2:15–6:00pm	Lab Poster session

2:00–3:00pm **Informal Breakout Session III**   Session 10, 11 and 12 Moderator: William Louis, LANL

Session 13 and 14 Moderator: Halina Abramowicz, Tel Aviv

3:00–4:00pm   Detector R&D Moderator: Robert Kephart, FNAL

8:00pm   Public lecture, "Windows on the Universe: New Questions about Matter, Space and Time"
      Michael Witherell, FNAL

Saturday 16th August 2003

Session 15 - NEW ACCELERATORS

Session Chair: David Miller, UC, London

8:45am   Report from the Young Physicists
        Veronique Boisvert, CERN
9:00am   LHC
        Luciano Maiani, CERN
9:35am   Physics of the Linear Collider
        Francois Richard, Orsay
10:10am   Report from the Chair of ICFA
        Jonathan Dorfan, ICFA and SLAC
10:25am   Report from the Chair of C11
        Vera Luth, IUPAP-C11

Session 16 - FUTURE DIRECTIONS

Session Chair: Hirotaka Sugawara, KEK

11:00am   Linear Collider Options: Status of the R&D and Plans for Technology Selection
        Maury Tigner, ILCSC and Cornell
11:45am   Supersymmetry and Other Scenarios
        Edward Witten, IAS-Princeton

Session 17 - OUTLOOK

Session Chair: Cathy Newman-Holmes, FNAL

2:00pm   Outlook: The Next Twenty Years
        Hitoshi Murayama, UC, Berkeley

Physics Posters

Laboratory Posters

Contributed Papers

Electroweak Physics and Beyond

H1 Collaboration . . . . Search for New Physics in $e^{\pm}q$ Contact Interactions at HERA

Jadach, S.   hep-ph/0209268   Electric Charge Screening Effect in Single W Production with the Koralw Monte Carlo

Jadach, S.   hep-ph/0109279   Exact Differential $O(\alpha^2)$ Results for Hard Bremsstrahlung in e^+e^- Annihilation to 2F at and Beyond LEP-2 Energies

Jadach, S.   hep-ph/0109072   On Theoretical Uncertainties of the W Boson Mass Measurement at LEP2

KLOE Collaboration   hep-ex/0307051   Determination of $\sigma(e^+e^- \to \pi^+\pi^-)$ from Radiative Processes at DAΦNE

Landry, F.   hep-ph/0212159   Tevatron Run-1 Z Boson Data and Collins-Soper-Sterman Resummation Formalism

Nelson, C. A.   hep-ph/0304198   On the Top-Quark's Chiral Weak Moment

OPAL collaboration . . . . Measurement of the b Quark Forward-Backward Asymmetry around the Z^0 Peak using an Inclusive Tag

OPAL collaboration . . . . Measurement of Neutral-Current Four-Fermion Production at LEP2

OPAL collaboration . . . . Measurement of the b and c Quark Forward-Backward Asymmetries and Average B Mixing in Z^0 Decays using Leptons in Hadronic Events

OPAL collaboration . . . . Multi-Photon Production in e^+e^- Collisions at 181 - 209 GeV

OPAL collaboration . . . . Test of Non-commutative QED in the Process e^+e^- to $\gamma\gamma$ at LEP

OPAL collaboration . . . . Tests of the Standard Model and Constraints on New Physics from Measurements of Fermion-pair Production at 189-209 GeV at LEP

OPAL collaboration . . . . Study of Z Pair Production and Anomalous Couplings in e^+e^- Collisions at $\sqrt{s}$ between 190 and 209 GeV

OPAL collaboration . . . . Determination of the LEP Beam Energy using Radiative Fermion-pair Events

OPAL collaboration . . . . Measurement of the Mass of the W Boson in e^+e^- Collisions using the Fully Leptonic Channel

OPAL collaboration . . . . Measurement of Charged Current Triple Gauge Boson Couplings at LEP

OPAL collaboration . . . . Measurement of Triple Gauge Boson Couplings using Photonic Events with Missing Energy at $\sqrt{s} = 189$-207 GeV

OPAL collaboration . . . . W Boson Polarisation at LEP2

OPAL collaboration . . . . Constraints on Anomalous Quartic Gauge Boson Couplings using Acoplanar Photon Pairs at LEP-2

OPAL collaboration . . . . A study of $W^+W^-\gamma$ Events at LEP

OPAL collaboration . . . . Bose-Einstein Correlations in W-pair Hadronic Decays at LEP

OPAL collaboration . . . . Colour Reconnection Studies in $e^+e^- \to W^+W^-$ Events at $\sqrt{s}$=190-208 GeV using Particle Flow

QCD Phenomenology and Theory

Flavour Physics: Rare Physics and Beyond the Standard Model

Flavour Physics: CKM and QCD

Future Facilities

List of Participants

1	Abe, Kazuo	KEK	Japan
2	Abramowicz, Halina	Tel Aviv University	Israel
3	Adams, Mark	University of Illinois, Chicago	USA
4	Aguilar-Benitez, Manuel	CIEMAT	Spain
5	Aihara, Hiroaki	University of Tokyo	Japan
6	Akchurin, Nural	Texas Tech University	USA
7	Albert, Justin	California Institute of Technology	USA
8	Albright, Carl	Northern Illinois University	USA
9	Albrow, Michael	Fermi National Accelerator Laboratory	USA
10	Alekhin, S. I.	IHEP, Serpukhov	Russia
11	Anderson, Kelby	University of Chicago	USA
12	Andrieu, Michel	The Ultimate Particle	France
13	Apollinari, Giorgio	Fermi National Accelerator Laboratory	USA
14	Appel, Jeff	Fermi National Accelerator Laboratory	USA
15	Armstrong, Stephen	CERN	CERN
16	Artuso, Marina	Syracuse University	USA
17	Aston, David	Stanford Linear Accelerator Center	USA
18	Atramentov, Oleksiy	Iowa State University	USA
19	Avery, Paul	University of Florida	USA
20	Avvakumov, Sergey	Stanford University	USA
21	Aymar, Robert	CERN	CERN
22	Ayres, David	Argonne National Laboratory	USA
23	Azzi, Patrizia	INFN, Padova	Italy
24	Bacci, Cesare	INFN, Roma II	Italy
25	Bachacou, Henri	Lawrence Berkeley National Laboratory	USA
26	Banerjee, Sudeshna	Tata Institute of Fundamental Research	India
27	Bargassa, Pedrame	Rice University	USA
28	Bauer, Dan	Fermi National Accelerator Laboratory	USA
29	Bedeschi, Franco	INFN, Pisa	Italy
30	Behnke, Ties	DESY, Hamburg	Germany
31	Bellagamba, Lorenzo	INFN, Bologna	Italy
32	Bellerive, Alain	Carleton University	Canada
33	Bellettini, Giorgio	INFN, Pisa	Italy
34	Benvenuti, Alberto	CERN	Italy
35	Beretvas, Andy	Fermi National Accelerator Laboratory	USA
36	Berger, Edmond	Argonne National Laboratory	USA
37	Bernardi, Gregorio	LPNHE-Paris	France
38	Bernstein, Robert	Fermi National Accelerator Laboratory	USA
39	Bertani, Monica	Laboratori Nazionali di Frascati	Italy
40	Bertolucci, Sergio	Laboratori Nazionali di Frascati	Italy
41	Bhat, Pushpa	Fermi National Accelerator Laboratory	USA
42	Bianco, Stefano	Laboratori Nazionali di Frascati	Italy
43	Bigi, Ikaros	University of Notre Dame	USA
44	Binkley, Morris	Fermi National Accelerator Laboratory	USA
45	Black, Kevin	Boston University	USA
46	Blaylock, Guy	University of Massachusetts	USA

47	Blazey, Jerry	Northern Illinois University	USA
48	Bloom, Elliott	Stanford Linear Accelerator Center	USA
49	Bloom, Ken	University of Michigan	USA
50	Bocian, Dariusz	INP Cracow	Poland
51	Bock, Greg	Fermi National Accelerator Laboratory	USA
52	Boehnlein, Amber	Fermi National Accelerator Laboratory	USA
53	Boisvert, Veronique	CERN	CERN
54	Bortoletto, Daniela	Purdue University	USA
55	Branson, James	University of California, San Diego	USA
56	Brau, James E.	University of Oregon	USA
57	Bromberg, Carl	Michigan State University	USA
58	Browder, Tom	University of Hawaii	USA
59	Brown, Chuck	Fermi National Accelerator Laboratory	USA
60	Brubaker, Erik	University of California, Berkeley	USA
61	Buchalla, Gerhard	LMU	Germany
62	Buchanan, Norm	Florida State University	USA
63	Buckley-Geer, Elizabeth	Fermi National Accelerator Laboratory	USA
64	Buehler, Marc	University of Illinois, Chicago	USA
65	Buescher, Volker	University of Freiburg	Germany
66	Buesser, Karsten	DESY, Hamburg	Germany
67	Bussey, Peter J.	Glasgow University	UK
68	Butler, Joel	Fermi National Accelerator Laboratory	USA
69	Calder, Neil	Stanford Linear Accelerator Center	USA
70	Canelli, Florencia	University of Rochester	USA
71	Casey, Brendan	Brown University	USA
72	Cashmore, Roger	CERN	CERN
73	Ceccucci, Augusto	CERN	CERN
74	Chakraborty, Dhiman	Northern Illinois University	USA
75	Chang, Lay Nam	Virginia Tech	USA
76	Chang, Ngee-Pong	City College of New York	USA
77	Chang, Paoti	National Taiwan University	Taiwan
78	Chau, Nguyen Ham	Vietnam Cultural Window Magazine	Vietnam
79	Chen, Chunhui	University of Pennsylvania	USA
80	Cheng, Hai-Yang	Academia Sinica	Taiwan
81	Chertok, Max	University of California, Davis	USA
82	Cheung, Harry W. K.	Fermi National Accelerator Laboratory	USA
83	Chiang, Cheng-Wei	Argonne National Laboratory	USA
84	Childress, Sam	Fermi National Accelerator Laboratory	USA
85	Chiodini, Gabriele	INFN, Lecce	Italy
86	Chlebana, Frank	Fermi National Accelerator Laboratory	USA
87	Choi, Seong Youl	Chonbuk National University	Korea
88	Choi, Soo Kyung	Gyeongsang National University	Korea
89	Chomeiko, Nikolai	National Center for Particle Physics	Belarus
90	Chou, Aaron	Fermi National Accelerator Laboratory	USA
91	Chou, Chung-Hsien	Academia Sinica	Taiwan
92	Choudhary, Brajesh	Fermi National Accelerator Laboratory	USA
93	Christian, Dave	Fermi National Accelerator Laboratory	USA
94	Christova, Ekaterina	Inst. for Nucl. Research & Nucl. Energy	Bulgaria
95	Clare, Robert	University of California, Riverside	USA

96	Coca, Mircea N.	University of Rochester	USA
97	Cochran, James	Iowa State University	USA
98	Collins, John	Penn State University	USA
99	Conta, Claudio	INFN, Pavia	Italy
100	Conway, John	Rutgers University	USA
101	Cooper, John	Fermi National Accelerator Laboratory	USA
102	Copic, Katherine	University of Michigan	USA
103	Cortes, Jose Luis	Universidad de Zaragoza	Spain
104	Cox, Brad	Univeristy of Virginia	USA
105	Crawford, Glen	U. S. Department of Energy	USA
106	Culbertson, Ray	Fermi National Accelerator Laboratory	USA
107	Curatolo, Maria	Laboratori Nazionali di Frascati	Italy
108	Cutts, David	Brown University	USA
109	Dainton, J. B.	Liverpool University	UK
110	Danilov, Mikhail V.	ITEP, Moscow	Russia
111	Dasu, Sridhara	University of Wisconsin, Madison	USA
112	Davies, Gavin	Imperial College London	UK
113	de Gouvea, Andre	Fermi National Accelerator Laboratory	USA
114	de Jesus, Maryvonne	Universite Claude Bernard Lyon-1	France
115	de Jong, Sijbrand	NIKHEF/University of Nijmegen	Netherlands
116	de Miranda, Jussara	Centro Brasileiro De Pesquisas Fisicas	Brazil
117	Declais, Yves	Universite Claude Bernard Lyon-1	France
118	Del Re, Daniele	University of California, San Diego	USA
119	Dell'Agnello, Simone	Laboratori Nazionali di Frascati	Italy
120	Demina, Regina	University of Rochester	USA
121	Denisov, Dmitri	Fermi National Accelerator Laboratory	USA
122	Deryuzhkova, Oksana	Gomel State University	Belarus
123	Di Giacomo, Adriano	INFN, Pisa	Italy
124	Diehl, Markus	DESY	Germany
125	Dionisi, Carlo	INFN, Roma I	Italy
126	Dittmann, Jay	Fermi National Accelerator Laboratory	USA
127	Dixon, Lance	Stanford Linear Accelerator Center	USA
128	Dobrescu, Bogdan	Fermi National Accelerator Laboratory	USA
129	Dominguez, Aaron	Lawrence Berkeley National Laboratory	USA
130	Dominguez, Cesareo	University of Cape Town, ICTP	South Africa
131	Dorfan, Jonathan M.	Stanford Linear Accelerator Center	USA
132	Dorjkhaidav, Orlokh	Syracuse University	USA
133	Dosselli, Umberto	INFN, Padova	Italy
134	Drell, Persis	Stanford Linear Accelerator Center	USA
135	Du, Dong-sheng	Chinese Academy of Sciences	China
136	Dubois-Felsmann, Gregory	California Institute of Technology	USA
137	Duboscq, Jean	Cornell University	USA
138	Duflot, Laurent	Universite de Paris-Sud	France
139	Durante, Elisabetta	Il Sole 24 Ore	Italy
140	Eads, Michael	Northern Illinois University	USA
141	Eichten, Estia	Fermi National Accelerator Laboratory	USA
142	Eisele, Franz	University of Heidelberg	Germany
143	Ekelof, Tord	Uppsala University	Sweden
144	Ellis, Keith	Fermi National Accelerator Laboratory	USA

145	Erbacher, Robin D.	Fermi National Accelerator Laboratory	USA
146	Ereditato, Antonio	INFN, Napoli	Italy
147	Erwin, Albert	University of Wisconsin, Madison	USA
148	Estrada, Juan	Fermi National Accelerator Laboratory	USA
149	Evans, Harold	Columbia University	USA
150	Fabbri, Franco	Laboratori Nazionali di Frascati	Italy
151	Fajfer, Svjetlana	University of Ljubljana	Slovenia
152	Fang, Hung-Chung	University of California, Berkeley	USA
153	Farbin, Amir	University of Maryland	USA
154	Farrington, Sinéad M.	University of Glasgow	UK
155	Feldman, Gary	Harvard University	USA
156	Feligioni, Lorenzo	Boston University	USA
157	Fernandez, Enrique	Univ. Autonoma de Barcelona	Spain
158	Field, Richard	University of Florida	USA
159	Filthaut, Frank	University of Nijmegen	Netherlands
160	Firestone, Alexander	National Science Foundation	USA
161	Fleischer, Manfred	DESY	Germany
162	Fogli, Gianluigi	INFN, Bari	Italy
163	Folkerts, Petra	DESY	Germany
164	Formaggio, Joseph A.	University of Washington	USA
165	Foster, Bill	Fermi National Accelerator Laboratory	USA
166	Foster, Brian	University of Bristol	UK
167	Fox, Harald	Northwestern University	USA
168	Franzini, Paolo	INFN, Roma I	Italy
169	Fry, John	University of Liverpool	UK
170	Fu, Shaohua	Columbia University	USA
171	Fuess, Stuart	Fermi National Accelerator Laboratory	USA
172	Fusco, Piergiorgio	INFN, Bari	Italy
173	Galea, Cristina	NIKHEF	Netherlands
174	Gambino, Paolo	CERN	CERN
175	Gao, Mingcheng	Columbia University	USA
176	Gao, Yongsheng	Southern Methodist University	USA
177	Garbincius, Peter	Fermi National Accelerator Laboratory	USA
178	Garcia-Bellido, Aran	University of Washington	USA
179	Garfinkel, Arthur	Purdue University	USA
180	Garisto, Robert	Physical Review	USA
181	Gehrmann, Thomas	RWTH Aachen	Germany
182	Gerberich, Heather K.	Duke University	USA
183	Giao, Nguyen Mong	Institute of Physics	Vietnam
184	Gibson, Adam	University of California, Berkeley	USA
185	Giele, Walter	Fermi National Accelerator Laboratory	USA
186	Gillies, James	CERN	CERN
187	Gilman, Fred	Carnegie Mellon University	USA
188	Ginsburg, Camille	University of Wisconsin, Madison	USA
189	Giorgi, Marcello	INFN, Pisa	Italy
190	Giudice, Gian-Francesco	CERN	CERN
191	Gladney, Larry	University of Pennsylvania	USA
192	Glenzinski, Doug	Fermi National Accelerator Laboratory	USA
193	Godbole, Rohini M.	Indian Institute of Science	India

194	Goerlach, Ulrich	Inst. de Recherches Subatomiques (IReS)	France
195	Goers, Stefan	Universitat Bonn	Germany
196	Goldman, Terrance J.	Los Alamos National Laboratory	USA
197	Goldschmidt, Nathan J.	University of Michigan	USA
198	Golling, Tobias	University of Bonn	Germany
199	Golowich, Eugene	University of Massachusetts	USA
200	Gomez-Ceballos, Guillelmo	University of Cantabria	Spain
201	Gonzalez Sprinberg, Gabriel	Instiuto de Fisica, ICTP	Uruguay
202	Gorelov, Igor	University of New Mexico	USA
203	Goshaw, Alfred	Duke University	USA
204	Goussiou, Anna	Imperial College of Science & Technology	UK
205	Grab, Christoph	ETH, Zurich	Switzerland
206	Grady, William	Chicago Tribune	USA
207	Gratta, Giorgio	Stanford University	USA
208	Gray, Stephen	Cornell University	USA
209	Greenlee, Herb	Fermi National Accelerator Laboratory	USA
210	Gritsan, Andrei	Lawrence Berkeley National Laboratory	USA
211	Grivaz, Jean-Francois	IN2P3/CNRS et Universite de Paris-Sud	France
212	Gronberg, Jeff	Lawrence Livermore National Laboratory	USA
213	Grossman, Yuval	Stanford Linear Accelerator Center	USA
214	Grosso-Pilcher, Carla	University of Chicago	USA
215	Groysman, Yuriy	Lawrence Berkeley National Laboratory	USA
216	Gutierrez, Gaston	Fermi National Accelerator Laboratory	USA
217	Haas, Andrew	University of Washington	USA
218	Hadley, Nick	University of Maryland	USA
219	Haisch, Ulrich	Fermi National Accelerator Laboratory	USA
220	Hanagaki, Kazu	Fermi National Accelerator Laboratory	USA
221	Hara, Kohji	Osaka University	Japan
222	Hardtke, David	Lawrence Berkeley National Laboratory	USA
223	Harr, Robert	Wayne State University	USA
224	Harris, David	American Physical Society	USA
225	Harris, Deborah	Fermi National Accelerator Laboratory	USA
226	Harris, Robert	Fermi National Accelerator Laboratory	USA
227	Hauschild, Michael	CERN	CERN
228	Hauser, Jay	University of California, Los Angeles	USA
229	Hays, Christopher P.	Duke University	USA
230	Hayward, Helen	Liverpool University	UK
231	Hazumi, Masashi	KEK	Japan
232	Heeger, Karsten	Lawrence Berkeley National Laboratory	USA
233	Heinemann, Beate	University of Liverpool	UK
234	Heinemeyer, Sven	Technischen Universitat Munchen	Germany
235	Heintz, Ulrich	Boston University	USA
236	Heltsley, Brian	Cornell University	USA
237	Hemingway, Richard	Carleton University	Canada
238	Herten, Gregor	University of Freiburg	Germany
239	Heuer, Rolf-Dieter	University of Hamburg	Germany
240	Hidaka, Keisho	Tokyo Gakugei University	Japan
241	HIll, Christopher	University of California, Santa Barbara	USA
242	Hill, Richard	Stanford Linear Accelerator Center	USA

243	Hirosky, Robert	University of Virginia	USA
244	Hobbs, John	State Univ. of NY, Stony Brook	USA
245	Hosotani, Yutaka	Osaka University	Japan
246	Hou, George W. S.	Academia Sinica	Taiwan
247	Howcroft, Caius	Cambridge University	UK
248	Hsiung, Yee Bob	National Taiwan University	Taiwan
249	Hughes, Gareth	Lancaster University	UK
250	Huss, Daniel	Inst. de Recherches Subatomiques (IReS)	France
251	Huston, Joey	Michigan State University	USA
252	Hylen, James	Fermi National Accelerator Laboratory	USA
253	Ianni, Aldo	Laboratori Nazionali del Gran Sasso	Italy
254	Igi, Keiji	Kanagawa University	Japan
255	Iijima, Toru	Nagoya University	Japan
256	Ikado, Koji	Waseda University	Japan
257	Inoue, Kunio	Tohoku University	Japan
258	Jackson, Judy	Fermi National Accelerator Laboratory	USA
259	James, Catherine	Fermi National Accelerator Laboratory	USA
260	Jaros, John	Stanford Linear Accelerator Center	USA
261	Jawahery, Hassan	University of Maryland	USA
262	Jenkins, Elizabeth	University of California, San Diego	USA
263	Jensen, Doug	Fermi National Accelerator Laboratory	USA
264	Jensen, Hans	Fermi National Accelerator Laboratory	USA
265	Johnson, Denis	Vrije Universiteit Brussel	Belgium
266	Johnson, Marvin	Fermi National Accelerator Laboratory	USA
267	Jones, Roger	University of Lancaster	UK
268	Jostlein, Hans	Fermi National Accelerator Laboratory	USA
269	Jung, Chang Kee	State Univ. of NY, Stony Brook	USA
270	Juste, Aurelio	Fermi National Accelerator Laboratory	USA
271	Kado, Marumi	Lawrence Berkeley National Laboratory	USA
272	Kaefer, Daniela	RWTH Aachen	Germany
273	Kagan, Alex	University of Cincinnati	USA
274	Kang, Hyejoo	Stanford University	USA
275	Kang, Kyungsik	Brown University	USA
276	Karagoz Unel, Muge	Northwestern University	USA
277	Karshon, Uri	Weizmann Institute of Science	Israel
278	Kayser, Boris	Fermi National Accelerator Laboratory	USA
279	Keeler, Richard K.	University of Victoria	Canada
280	Kehoe, Robert	Michigan State University	USA
281	Kent, Steve	Fermi National Accelerator Laboratory	USA
282	Kephart, Robert	Fermi National Accelerator Laboratory	USA
283	Kephart, Thomas	Vanderbilt University	USA
284	Kilminster, Ben	University of Rochester	USA
285	Kim, Doris Yangsoo	Univ. of Illinois, Urbana-Champaign	USA
286	Kim, Minsuk	Kyungpook National University	Korea
287	Kim, Shinhong	University of Tsukuba	Japan
288	Kim, Young-Kee	University of Chicago	USA
289	Kirshner, Robert	Harvard-Smithsonian Cen. for Astro.	USA
290	Klanner, Robert	DESY	Germany
291	Klima, Boaz	Fermi National Accelerator Laboratory	USA

292	Klingenberg, Reiner	University of Dortmund	Germany
293	Koerner, Juergen G.	Universitat Mainz	Germany
294	Koetz, Ulrich	DESY	Germany
295	Komamiya, Sachio	University of Tokyo	Japan
296	Krasny, Mieczyslaw	CNRS, Universite Paris VI	France
297	Kretzer, Stefan	Brookhaven National Laboratory	USA
298	Kroll, Joseph	University of Pennsylvania	USA
299	Kronfeld, Andreas	Fermi National Accelerator Laboratory	USA
300	Kruecker, Dirk	DESY	Germany
301	Kulasiri, Ratnappuli Luminda	University of Cincinnati	USA
302	Kuno, Yoshitaka	Osaka University	Japan
303	Kutschke, Rob	Fermi National Accelerator Laboratory	USA
304	Lach, Theodore	Lucent Technologies	USA
305	Lam, Harry C. S.	McGill University	Canada
306	Lammel, Stephan	Fermi National Accelerator Laboratory	USA
307	Lanceri, Livio	INFN, Trieste	Italy
308	Landsberg, Greg	Brown University	USA
309	Lane, Kenneth	Boston University	USA
310	Larios, Francisco	CINVESTAV	Mexico
311	Le Diberder, Francois	Universite de Paris-Sud	France
312	Leder, Gerhard	Austrian Academy of Science	Austria
313	Lee, Sungwon	Texas A&M	USA
314	Lee, William	Florida State University	USA
315	Lee-Franzini, Juliet	Laboratori Nazionali di Frascati	Italy
316	Legendre, Marie	Saclay	France
317	Lepage, Peter	Cornell University	USA
318	Levy, Aharon	Tel Aviv University	Israel
319	Levy, Stephen	University of Chicago	USA
320	Lewis, Jonathan	Fermi National Accelerator Laboratory	USA
321	Li, Bing An	University of Kentucky	USA
322	Li, Hsiang-nan	Academia Sinica	Taiwan
323	Li, Qizong	Fermi National Accelerator Laboratory	USA
324	Li, Selina	Cornell University	USA
325	Libby, James	Stanford Linear Accelerator Center	USA
326	Limentani, Silvia	INFN, Padova	Italy
327	Limon, Peter	Fermi National Accelerator Laboratory	USA
328	Lin, C-J Stephen	Fermi National Accelerator Laboratory	USA
329	Liparteliani, Akaki	Tbilisi State University	Georgia
330	Lipkin, Harry J.	Weizmann Institute of Science	Israel
331	Lipton, Ron	Fermi National Accelerator Laboratory	USA
332	Liss, Tony	Univ. of Illinois, Urbana-Champaign	USA
333	Liu, Stanley	World Scientific	USA
334	Lockyer, Nigel	University of Pennsylvania	USA
335	Lopez, Angel	University of Puerto Rico, Mayaguez	USA
336	Lopez Castro, Gabriel	CINVESTAV	Mexico
337	Louis, William C.	Los Alamos National Laboratory	USA
338	Luci, Claudio	INFN, Roma I	Italy
339	Ludwig, Jens	Albert-Ludwigs Universitat Freiburg	Germany
340	Lusin, Sergei	Fermi National Accelerator Laboratory	USA

341	Luth, Vera	Stanford Linear Accelerator Center	USA
342	Lutz, Pierre	CNRS, Universite Paris VI	France
343	MacFarlane, David	University of California, San Diego	USA
344	Macri, Mario	INFN, Genova	Italy
345	Maiani, Luciano	CERN	CERN
346	Mallik, Usha	University of Iowa	USA
347	Mannel, Thomas	Universitat Karlsruhe	Germany
348	Mansoulie, Bruno	CEA Saclay	France
349	Mantsch, Paul	Fermi National Accelerator Laboratory	USA
350	Maravin, Yurii	Fermi National Accelerator Laboratory	USA
351	Marcellini, Stefano	INFN, Bologna	Italy
352	Martin, Victoria	Northwestern University	USA
353	Martinez, Roberto	National University of Colombia, ICTP	Colombia
354	Maruyama, Takasumi	University of Chicago	USA
355	Matteuzzi, Clara	Universita di Milano-Bicocca	Italy
356	Mazzucato, Edoardo	CEA Saclay	France
357	McBride, Patricia	Fermi National Accelerator Laboratory	USA
358	Meder, David	Universitat Mainz	Germany
359	Melanson, Harry	Fermi National Accelerator Laboratory	USA
360	Mellado, Bruce	University of Wisconsin, Madison	USA
361	Melnikov, Kirill	University of Hawaii	USA
362	Mendez, Hector	University of Puerto Rico, Mayaguez	USA
363	Merkel, Petra	Fermi National Accelerator Laboratory	USA
364	Messina, Andrea	University of Rome	Italy
365	Meyer, Holger	Fermi National Accelerator Laboratory	USA
366	Miao, Ting	Fermi National Accelerator Laboratory	USA
367	Mikuz, Marko	Jozef Stefan Institute	Slovenia
368	Miller, David J.	University College London	UK
369	Miller, Steven	Superconductor Week	USA
370	Mills, Corrinne	University of California, Santa Barbara	USA
371	Mitchell, Ryan	University of Tennessee	USA
372	Mnich, Joachim	RWTH Aachen	Germany
373	Mohr, Brian	University of California, Los Angeles	USA
374	Mommsen, Remigius	University of California, Irvine	USA
375	Montgomery, Hugh	Fermi National Accelerator Laboratory	USA
376	Morelos, Antonio	Univ. Autonoma de San Luis Potosi	Mexico
377	Morita, Youhei	KEK	Japan
378	Morozumi, Takuya	Hiroshima University	Japan
379	Moshe, Moshe	Technion-Israel Institute of Technology	Israel
380	Mrenna, Stephen	Fermi National Accelerator Laboratory	USA
381	Muciaccia, Mariateresa	INFN, Bari	Italy
382	Mukherjee, Asmita	Universitat Dortmund	Germany
383	Mulders, Martijn	Fermi National Accelerator Laboratory	USA
384	Mulhearn, Michael	Massachusetts Institute of Technology	USA
385	Murat, Pavel	Fermi National Accelerator Laboratory	USA
386	Murayama, Hitoshi	University of California, Berkeley	USA
387	Muresan, Raluca	Niels Bohr Institute	Denmark
388	Nachtman, Jane	Fermi National Accelerator Laboratory	USA
389	Nadolsky, Pavel	Southern Methodist University	USA

390	Nakamura, Kenzo	KEK	Japan
391	Nakano, Itsuo	Okayama University	Japan
392	Nakao, Mikihiko	KEK	Japan
393	Nambu, Yoichiro	University of Chicago	USA
394	Napier, Austin	Tufts University	USA
395	Nappi, Aniello	INFN, Perugia	Italy
396	Narain, Meenakshi	Boston University	USA
397	Nasretdinov, Kamil	Institute of Nuclear Physics	Uzbekistan
398	Nauenberg, Uriel	University of Colorado	USA
399	Nelson, Charles	State Univ. of NY, Binghamton	USA
400	Neuffer, David	Fermi National Accelerator Laboratory	USA
401	Nevzorov, Roman	ITEP, Moscow	Russia
402	Newman, Harvey	California Institute of Technology	USA
403	Newman, Paul R.	DESY	UK
404	Newman-Holmes, Catherine	Fermi National Accelerator Laboratory	USA
405	Newsom, Charles	University of Iowa	USA
406	Niczyporuk, Bogdan	Jefferson Lab	USA
407	Niebuhr, Carsten	DESY	Germany
408	Nielsen, Jason	Lawrence Berkeley National Laboratory	USA
409	Nishikawa, Koichiro	Kyoto University	Japan
410	Oakes, Robert	Northwestern University	USA
411	O'Fallon, John R.	U. S. Department of Energy	USA
412	Oh, Sunkun	Kunkuk University	Korea
413	Ohska, Tokio Kenneth	KEK	Japan
414	Olness, Fred	Southern Methodist University	USA
415	Olsen, James	Princeton University	USA
416	Olsen, Stephen	University of Hawaii	USA
417	Ootani, Wataru	University of Tokyo	Japan
418	Orava, Risto	University of Helsinki	Finland
419	Orejudos, William Gilbert	Lawrence Berkeley National Laboratory	USA
420	Oshima, Nobu	Fermi National Accelerator Laboratory	USA
421	Owens, Jeff	Florida State University	USA
422	Padilla, Cristobal	CERN	CERN
423	Pakvasa, Sandip	University of Hawaii	USA
424	Pan, Yibin	University of Wisconsin, Madison	USA
425	Papadimitriou, Vaia	Texas Tech University	USA
426	Parashar, Neeti	Louisiana Tech	USA
427	Parke, Stephen	Fermi National Accelerator Laboratory	USA
428	Parua, Nirmalya	State Univ. of NY, Stony Brook	USA
429	Pascoli, Silvia	University of California, Los Angeles	USA
430	Paul, Stephan	Technischen Universitat Munchen	Germany
431	Pauletta, Giovanni	University of Udine	Italy
432	Peach, K. J.	Rutherford Appleton Laboratory	UK
433	Peccei, Roberto	University of California, Los Angeles	USA
434	Peoples, John	Fermi National Accelerator Laboratory	USA
435	Perez, Emmanuelle	CEA Saclay	France
436	Perez, Miguel Angel	CINVESTAV	Mexico
437	Perez Martinez, Aurora	Instituto de Cibernetica, ICTP	Cuba
438	Peroni, Cristiana	INFN, Torino	Italy

439	Peruzzi Piccolo, Ida	Laboratori Nazionali di Frascati	Italy
440	Petrarca, Silvano	INFN, Roma I	Italy
441	Petyt, David	University of Minnesota	USA
442	Pietschmann, Herbert	University of Vienna	Austria
443	Piilonen, Leo	Virginia Tech	USA
444	Pilcher, James	University of Chicago	USA
445	Pitts, Kevin	Univ. of Illinois, Urbana-Champaign	USA
446	Pitzl, Daniel	DESY	Germany
447	Podobnik, Tomaz	University of Ljubljana	Slovenia
448	Ponce Gutierrez, William	Universidad de Antioquia, ICTP	Colombia
449	Pooth, Oliver	RWTH Aachen	Germany
450	Pordes, Stephen	Fermi National Accelerator Laboratory	USA
451	Pratt, T. S.	Liverpool University	UK
452	Predazzi, Enrico	INFN, Torino	Italy
453	Prell, Soeren	Iowa State University	USA
454	Prepost, Richard	University of Wisconsin, Madison	USA
455	Qi, Xiangrong	Fermi National Accelerator Laboratory	USA
456	Qian, Jianming	University of Michigan	USA
457	Quigg, Chris	Fermi National Accelerator Laboratory	USA
458	Quinn, Breese	Fermi National Accelerator Laboratory	USA
459	Ragazzi, Stefano	INFN, Milano	Italy
460	Raja, Rajendran	Fermi National Accelerator Laboratory	USA
461	Ramberg, Erik	Fermi National Accelerator Laboratory	USA
462	Ramirez, Eduardo	University of Puerto Rico, Mayaguez	USA
463	Rani, K. Jyotshna	Tata Institute of Fundamental Research	India
464	Rapidis, Petros	Fermi National Accelerator Laboratory	USA
465	Rapin, Divic	Ecole de Physique	Switzerland
466	Ratoff, Peter	University of Lancaster	UK
467	Raven, Gerhard	NIKHEF	Netherlands
468	Ray, Heather Lynn	University of Michigan	USA
469	Recksiegel, Stefan	Technische Universitaet Muenchen	Germany
470	Reed, Daniel	Univ. of Illinois, Urbana-Champaign	USA
471	Reidy, James	U. S. Department of Energy	USA
472	Repko, Wayne	Michigan State University	USA
473	Repond, Jose	Argonne National Laboratory	USA
474	Reyna, David	Argonne National Laboratory	USA
475	Riazuddin, Riaz	Quaid-i-Azam University, ICTP	Pakistan
476	Richard, Francois	Universite de Paris-Sud	France
477	Riesselmann, Kurt	Fermi National Accelerator Laboratory	USA
478	Ritchie, Jack	University of Texas	USA
479	Riu, Imma	University of Geneva	Switzerland
480	Roberts, B. Lee	Boston University	USA
481	Rodriguez Lopez, Jairo Alexis	National Univ. of Colombia, ICTP	Colombia
482	Rolli, Simona	Tufts University	USA
483	Romano, Francesco	INFN, Bari	Italy
484	Roodman, Aaron	Stanford Linear Accelerator Center	USA
485	Rosen, Jerome	Northwestern University	USA
486	Rosen, Peter	U. S. Department of Energy	USA
487	Rosenfeld, Rogerio	Universidade Estadual Paulista	Brazil

488	Rosner, Jonathon	University of Chicago	USA
489	Rott, Carsten	Purdue University	USA
490	Roulet, Esteban	Centro Atomico Bariloche	Argentina
491	Rovelli, Tiziano	INFN, Bologna	Italy
492	Roy, Probir	Tata Institute of Fundamental Research	India
493	Rubinstein, Roy	Fermi National Accelerator Laboratory	USA
494	Ruchti, Randy	University of Notre Dame	USA
495	Rumerio, Paolo	Northwestern University	USA
496	Saarikko, Heimo	University of Helsinki	Finland
497	Sacquin, Yves	Palais-Decouverte	France
498	Sakai, Yoshihide	KEK	Japan
499	Sala, Silvano	INFN, Milano	Italy
500	Sanders, Michiel	University of Manchester	UK
501	Savard, Pierre	University of Toronto	Canada
502	Schellman, Heidi	Northwestern University	USA
503	Schietinger, Thomas	Universite de Lausanne	Switzerland
504	Schlenstedt, Stefan	DESY	Germany
505	Schmitt, Michael	Northwestern University	USA
506	Schneekloth, Uwe	DESY	Germany
507	Schroeder, Henning	Universitat Rostock	Germany
508	Schubert, Klaus	Technical University Dresden	Germany
509	Schuler, Peter	DESY	Germany
510	Schwartz, Alan	University of Cincinnati	USA
511	Sciutto, Sergio	Universidad Nacional de La Plata	Argentina
512	Scott, Adam	University of California, Santa Barbara	USA
513	Scuri, Fabrizio	INFN, Pisa	Italy
514	Sefkow, Feliz	DESY	Germany
515	Seife, Charles	Science Magazine	USA
516	Seth, Kamal	Northwestern University	USA
517	Sevior, Martin	University of Melbourne	Australia
518	Shamim, Mansoora	Kansas State University	USA
519	Sharma, Vivek	University of California, San Diego	USA
520	Sheldon, Paul	Vanderbilt University	USA
521	Shepard, Paul F.	University of Pittsburgh	USA
522	Sher, Marc	College of William and Mary	USA
523	Shibata, Toshi-Aki	Tokyo Institute of Technology	Japan
524	Shipsey, Ian	Purdue University	USA
525	Shivpuri, R. K.	Delhi University	India
526	Siegrist, James	Lawrence Berkeley National Laboratory	USA
527	Simone, James	Fermi National Accelerator Laboratory	USA
528	Sinervo, Pekka	University of Toronto	Canada
529	Sirlin, Alberto	New York University	USA
530	Skuja, Andris	University of Maryland	USA
531	Skwarnicki, Tomasz	Syracuse University	USA
532	Slaughter, Jean	Fermi National Accelerator Laboratory	USA
533	Smirnov, Alexei	Abdus Salam Int'l. Cen. for Theor. Physics	Russia
534	Snider, Rick	Fermi National Accelerator Laboratory	USA
535	Snow, Gregory	University of Nebraska	USA
536	Solomey, Nickolas	Illinois Institute of Technology	USA

537	Son, Dongchul	Kyungpook National University	Korea
538	South, David	DESY	Germany
539	Spanier, Stefan	University of Tennessee	USA
540	Stamenov, Dimiter Borisov	Inst. for Nucl. Research & Nucl. Energy	Bulgaria
541	Stasto, Anna	DESY	Germany
542	Stewart, Iain	Massachusetts Institute of Technology	USA
543	Stone, Sheldon	Syracuse University	USA
544	Strang, Michael A.	University of Texas, Arlington	USA
545	Strohmer, Raimund	LMU Munich	Germany
546	Stuart, David	University of California, Santa Barbara	USA
547	Stutte, Linda	Fermi National Accelerator Laboratory	USA
548	Sugawara, Hirotaka	University of Hawaii	USA
549	Sugimoto, Shojiro	KEK	Japan
550	Szleper, Michal	Northwestern University	USA
551	Takayama, Fumihiro	University of California, Irvine	USA
552	Tanaka, Katsumi	Ohio State University	USA
553	Tanaka, Masashi	Argonne National Laboratory	USA
554	Tanaka, Reisaburo	Okayama University	Japan
555	Tanimoto, Naho	Okayama University	Japan
556	Thorne, Robert S.	University of Cambridge	UK
557	Tigner, Maury	Cornell University	USA
558	Timmermans, Jan	CERN	CERN
559	Toki, Walter	Colorado State University	USA
560	Tomalin, Ian R.	Rutherford Appleton Laboratory	UK
561	Tomoto, Makoto	Fermi National Accelerator Laboratory	USA
562	Tonelli, Diego	INFN, Pisa	Italy
563	Totsuka, Yoji	KEK	Japan
564	Tran, Minh-Tam	University of Lausanne	Switzerland
565	Treccani, Michele	University of Pavia	Italy
566	Trischuk, William	University of Toronto	Canada
567	Tschirhart, Robert	Fermi National Accelerator Laboratory	USA
568	Tung, Wu-Ki	Michigan State University	USA
569	Turala, Michal	Institute of Nuclear Physics, Krakow	Poland
570	Tzanakos, George	University of Athens	Greece
571	Vachon, Brigitte	Fermi National Accelerator Laboratory	USA
572	van den Bergen, Mieke	Interactions	Netherlands
573	van Middelkoop, Ger	NIKHEF	Netherlands
574	Varelas, Niko	University of Illinois, Chicago	USA
575	Velasco, Mayda	Northwestern University	USA
576	Verde, Licia	Princeton University	USA
577	Vidal, Rick	Fermi National Accelerator Laboratory	USA
578	von Toerne, Eckhard	Kansas State University	USA
579	Wagner, Albrecht	DESY	Germany
580	Wagner, Carlos	Argonne National Laboratory	USA
581	Wahl, Horst	Florida State University	USA
582	Walkowiak, Wolfgang	Universitat Siegen	Germany
583	Wallny, Rainer	University of California, Los Angeles	USA
584	Wang, Gensheng	Case Western Reserve University	USA
585	Wang, Jianchun	Syracuse University	USA

586	Ward, B. F. L.	University of Tennessee	USA
587	Wasserbaech, Steven	Utah Valley State College	USA
588	Waters, David	University College London	UK
589	Watts, Gordon	University of Washington	USA
590	Weber, Michael	Fermi National Accelerator Laboratory	USA
591	Weerts, Harry	Michigan State University	USA
592	Wegner, Martin	RWTH Aachen	Germany
593	Wester, William	Fermi National Accelerator Laboratory	USA
594	White, Andy	University of Texas, Arlington	USA
595	White, Herman	Fermi National Accelerator Laboratory	USA
596	Whitmore, James	National Science Foundation	USA
597	Whitmore, Julie	Fermi National Accelerator Laboratory	USA
598	Wicklund, Barry	Argonne National Laboratory	USA
599	Wilkinson, Guy	University of Oxford	UK
600	Wilson, Michael	University of California, Santa Cruz	USA
601	Wilson, Peter	Fermi National Accelerator Laboratory	USA
602	Wimpenny, Stephen	University of California, Riverside	USA
603	Witherell, Michael S.	Fermi National Accelerator Laboratory	USA
604	Witten, Edward	Institute for Advanced Study	USA
605	Wittlin, Jodi	Boston University	USA
606	Wojcicki, Stanley	Stanford University	USA
607	Wolbers, Stephen	Fermi National Accelerator Laboratory	USA
608	Womersley, John	Fermi National Accelerator Laboratory	USA
609	Won, Eunil	Harvard University	USA
610	Worcester, Matthew Peter	University of California, Los Angeles	USA
611	Worm, Steven	Rutgers University	USA
612	Wormser, Guy	IN2P3	France
613	Wright, Allison	Nature Publishing Group	UK
614	Wright, Douglas	Lawrence Livermore National Laboratory	USA
615	Wu, San Lan	University of Wisconsin, Madison	USA
616	Wyatt, Terry	University of Manchester	UK
617	Yabsley, Bruce	Virginia Tech	USA
618	Yagil, Avi	Fermi National Accelerator Laboratory	USA
619	Yamada, Ryuji	Fermi National Accelerator Laboratory	USA
620	Yamamoto, Hitoshi	Tohoku University	Japan
621	Yamanaka, Taku	Osaka University	Japan
622	Yamashita, Satoru	University of Tokyo	Japan
623	Yamazaki, Yuji	KEK	Japan
624	Yang, Haijun	Universitiy of Michigan	USA
625	Yang, Un-Ki	University of Chicago	USA
626	Yarba, Victor	Fermi National Accelerator Laboratory	USA
627	Yasuda, Taka	Fermi National Accelerator Laboratory	USA
628	Ye, Shuwei	University of Texas, Dallas	USA
629	Yeh, Gong Ping	Fermi National Accelerator Laboratory	USA
630	Yen, Yevgenij	Bergische Universitat Wuppertal	Germany
631	Yoh, John	Fermi National Accelerator Laboratory	USA
632	Yoshida, Rikutaro	Argonne National Laboratory	USA
633	Yu, Hoi-Lai	Academia Sinica	Taiwan
634	Yu, Shin-Shan	University of Pennsylvania	USA

635	Yuta, Haruo	Aomori University	Japan
636	Zeitnitz, Christian	Universitat Mainz	Germany
637	Zeller, Michael	Yale University	USA
638	Zeng, Qinglin	Colorado State University	USA
639	Zeyrek, Mehmet	Middle East Technical University	Turkey
640	Zhang, Lin	Texas Tech University	USA
641	Zheng, Hai	California Institute of Technology	USA
642	Zhu, Jun-Jie	University of Maryland	USA
643	Zucchelli, Stefano	INFN, Bologna	Italy
644	Zutshi, Vishnu	Northern Illinois University	USA
645	Zwaska, Bob	University of Texas	USA

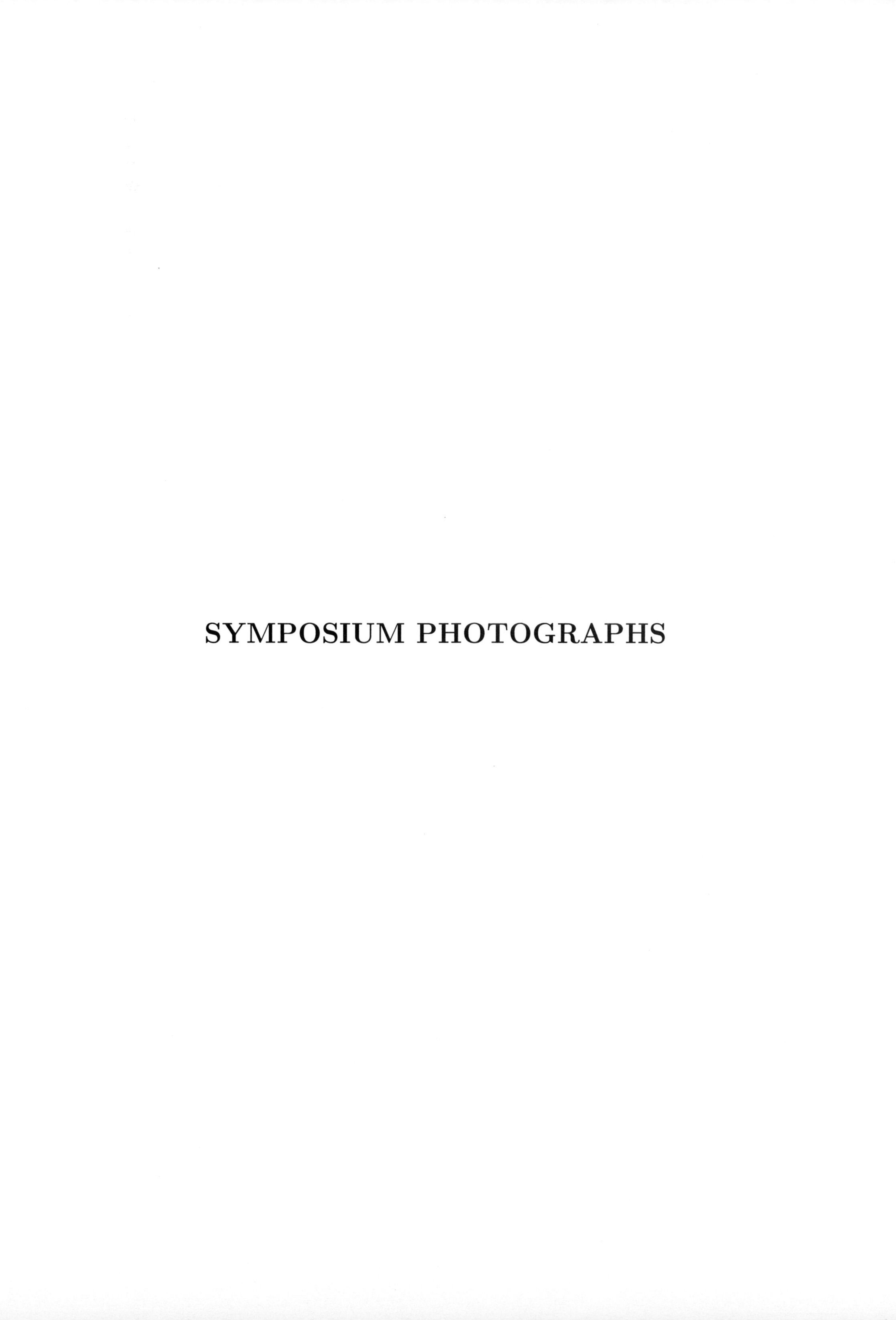

SYMPOSIUM PHOTOGRAPHS

PHYSICS POSTER SESSION

PHYSICS POSTER SESSION

GRID POSTER SESSION

DIRECTORS' PRESS CONFERENCE

TOURS OF FERMILAB

INTERACTIONS.ORG LAUNCH

SYMPOSIUM BANQUET

BEHIND THE SCENES

BEHIND THE SCENES

Author Index